The Princeton Guide to Ecology

The Princeton Guide to Ecology

ASSOCIATE EDITORS Stephen R. Carpenter
University of Wisconsin–Madison

H. Charles J. Godfray
University of Oxford

Ann P. Kinzig
Arizona State University

Michel Loreau
McGill University

Jonathan B. Losos
Harvard University

Brian Walker
CSIRO Sustainable Ecosystems

David S. Wilcove
Princeton University

MANAGING EDITOR Christopher G. Morris

Simon A. Levin EDITOR
Princeton University

PRINCETON UNIVERSITY PRESS
Princeton & Oxford

Published by Princeton University Press, 41 William Street, Princeton, New Jersey 08540
In the United Kingdom: Princeton University Press, 6 Oxford Street, Woodstock, Oxfordshire OX20 1TW

LIBRARY OF CONGRESS CATALOGING-IN-PUBLICATION DATA
The Princeton guide to ecology / Simon A. Levin, editor.
p. cm.
Includes bibliographical references and index.
ISBN 978-0-691-12839-9 (hardcover : alk. paper)
1. Ecology. 2. Ecology—Economic aspects.
I. Levin, Simon A.
QH541.p742 2009
577—dc22 2008049649

British Library Cataloging-in-Publication Data are available

This book has been composed in Sabon and Din

Printed on acid-free paper. ∞

press.princeton.edu

Printed in the United States of America

10 9 8 7 6 5 4 3 2 1

Contents

Preface

One can argue about when ecology was born as a science, although surely the writings of Charles Darwin and Alfred Russell Wallace created the essential context for the emergence of a new study of the interrelationships of species with each other and with their environments. The term "oekologie," combining the Greek words for "household" and "knowledge," was coined in 1866 by the remarkable German scientist, philosopher, and physician Ernst Haeckel and first was developed in scientific depth in the 1895 textbook by the Danish botanist Johannes Eugenius Buelow Warming, *Plantesamfund—Grundtræk af den økologiske Plantegeografi* [*Plant Communities: An Introduction to Ecological Plant Geography*]. Ecology has come a long way as a subject, from Eugen Warming to global warming.

Ecology has its roots in natural history and, indeed, in evolutionary thinking. But ecology itself has evolved considerably since its birth, building bridges to mathematics, to the physical sciences and engineering, to molecular biology, and, increasingly, to the social sciences. Just as we are beginning to appreciate not only the beauty of natural systems but also their essential role in providing an infinite range of goods and services on which humanity depends, we are reluctantly also learning that we are destroying those life-support systems and threatening the sustainability of the biosphere as we know it. Ecology, the unifying science in integrating knowledge of life on our planet, has become the essential science in learning how to preserve it.

This volume is an effort to present, in one readable collection, the diversity of ecology, from the basic to the applied. It is meant to serve both as a reader for anyone interested in learning more about the subject and as an essential reference for college and university courses on ecology and sustainability as well as for advanced high school students and the interested lay public. As such, it builds on the basic principles of autecology, population biology, and community and ecosystems science, which form the foundation for discussions regarding current threats to sustainability and how we can manage the biosphere responsibly. *The Princeton Guide to Ecology* is organized into seven sections tightly integrated with one another. The core textual material is supplemented by suggestions for further reading at the end of each article, by a glossary of key terms, and by a chronology that traces landmark events in ecology.

Ecology views biological systems as wholes, not as independent parts, while seeking to elucidate how these wholes emerge from and affect the parts. Increasingly, this holistic perspective, rechristened as the theory of *complex adaptive systems*, has informed understanding and improved management of economic and financial systems, social systems, complex materials, and even physiology and medicine—but essentially this means little more than taking an ecological approach to such systems, investigating the interplay among processes at diverse scales and the interaction between systems and their environments.

In many colleges and universities where ecology has flourished, botany and zoology have vanished as separate departments and been replaced by more integrative ones. Ecologists tend to organize their thinking across scales, from cells to organisms, from organisms to populations, from populations to communities, ecosystems, landscapes, and the biosphere. This view also dictates the organization of this volume, which begins with autecology, the study of the physiology, behavior, and life history of the primary integrative unit of ecology, the organism. From the organismal level, the next natural levels of organization are the population, then the community and ecosystem, and then finally landscapes and the biosphere.

With this basic foundation, the *Guide* then turns to more applied issues: understanding what biodiversity and the ecological systems in which they reside mean to us, as captured in the concept of "ecosystem services"; exploring the scientific basis for managing our natural systems and the resources we extract from them; and developing the theoretical principles underlying the conservation of natural resources. These chapters naturally reach out to other disciplines, including economics and the social sciences, for the partnerships that are essential in achieving a sustainable future for humanity.

If this ambitious effort has been successful, it is because of the exceptional quality of the authors and their contributions, and especially the remarkable set of associate editors who have cheerfully integrated their sections and worked closely with one another to assure transitions as seamless as could be imagined. Anne Savarese at Princeton University Press and managing editor Chris Morris have assured smooth logistics throughout and added their own keen insights at appropriate times. I also am delighted to acknowledge the inspiration of Sam Elworthy, former editor-in-chief of Princeton University Press, who conceived the idea of the *Guide* and convinced me to take on the project. As always, I am grateful for the unwavering support of Carole Levin, my wife and friend.

As we go to press, our happiness at the completion of this effort is mixed with sadness because of the untimely death on March 22, 2008, of our distinguished contributor, Robert Denno.

Simon A. Levin
Princeton, New Jersey
May 17, 2008

Contributors

David D. Ackerly, *Department of Integrative Biology, University of California, Berkeley*
I.3 PHYSIOLOGICAL ECOLOGY: PLANTS;
I.16 PHYLOGENETICS AND COMPARATIVE METHODS

Eldridge S. Adams, *Department of Ecology and Evolutionary Biology, University of Connecticut*
I.8 SOCIAL BEHAVIOR

Joseph Alcamo, *Center for Environmental Systems Research, University of Kassel*
VII.4 MANAGING THE GLOBAL WATER SYSTEM

Priyanga Amarasekare, *Department of Ecology and Evolution, University of Chicago*
II.6 COMPETITION AND COEXISTENCE IN ANIMAL COMMUNITIES
This research was funded by a grant from NSF (DEB-0717350).

Darren Bade, *Department of Biological Sciences, Kent State University*
III.12 FRESHWATER CARBON AND BIOGEOCHEMICAL CYCLES

Victoria J. Bakker, *Division of Physical and Biological Sciences, University of California, Santa Cruz*
V.2 POPULATION VIABILITY ANALYSIS

Marissa L. Baskett, *Department of Environmental Science and Policy, University of California, Davis*
VI.7 MARINE ECOSYSTEM SERVICES

Michael Begon, *School of Biological Sciences, University of Liverpool*
II.9 ECOLOGICAL EPIDEMIOLOGY

Michael A. Bell, *Department of Ecology and Evolution, Stony Brook University*
I.17 MICROEVOLUTION

Thomas Bell, *Department of Zoology, University of Oxford*
II.12 ECOLOGY OF MICROBIAL POPULATIONS

E. T. Borer, *Department of Zoology, Oregon State University*
III.6 TOP-DOWN AND BOTTOM-UP REGULATION OF COMMUNITIES
This work was completed as part of the Trophic Structure Comparisons Working Group supported by the National Center for Ecological Analysis and Synthesis, a Center funded by NSF (Grant #DEB-0072909), the University of California at Santa Barbara, and the state of California.

Mark S. Boyce, *Department of Biological Sciences, University of Alberta*
VII.3 WILDLIFE MANAGEMENT

Corey J. A. Bradshaw, *Research Institute for Climate Change and Sustainability, University of Adelaide*
V.1 CAUSES AND CONSEQUENCES OF SPECIES EXTINCTIONS

Cheryl J. Briggs, *Department of Ecology, Evolution and Marine Biology, University of California, Santa Barbara*
II.8 HOST–PARASITOID INTERACTIONS

Judith L. Bronstein, *Department of Ecology and Evolutionary Biology, University of Arizona*
II.11 MUTUALISM AND SYMBIOSIS

Barry W. Brook, *Research Institute for Climate Change and Sustainability, University of Adelaide*
V.1 CAUSES AND CONSEQUENCES OF SPECIES EXTINCTIONS

Joel S. Brown, *Department of Biological Sciences, University of Illinois at Chicago*
I.7 FORAGING BEHAVIOR

Ragan M. Callaway, *Division of Biological Sciences, University of Montana*
III.4 FACILITATION AND THE ORGANIZATION OF PLANT COMMUNITIES

Stephen R. Carpenter, *Department of Zoology, University of Wisconsin–Madison*
VII MANAGING THE BIOSPHERE

Just Cebrian, *Dauphin Island Sea Lab and Department of Marine Sciences, University of South Alabama*
III.9 ECOSYSTEM PRODUCTIVITY AND CARBON FLOWS: PATTERNS ACROSS ECOSYSTEMS

Jérôme Chave, *CNRS (Centre National de la Recherche Scientifique), Laboratoire Evolution et Diversité Biologique*
III.2 COMPETITION, NEUTRALITY, AND COMMUNITY ORGANIZATION

Ryan Chisholm, *Department of Ecology and Evolutionary Biology, Princeton University*
VII.8 THE ECOLOGY, ECONOMICS, AND MANAGEMENT OF ALIEN INVASIVE SPECIES

Scott L. Collins, *Department of Biology, University of New Mexico*
IV.5 BOUNDARY DYNAMICS IN LANDSCAPES
This research was supported by National Science Foundation support to the Sevilleta Long-term Ecological Research Program at the University of New Mexico (DEB 0620482).

Robert K. Colwell, *Department of Ecology and Evolutionary Biology, University of Connecticut*
III.1 BIODIVERSITY: CONCEPTS, PATTERNS, AND MEASUREMENT

Molly S. Cross, *Wildlife Conservation Society North America Program*
V.6 CONSERVATION AND GLOBAL CLIMATE CHANGE

Peter Daszak, *Consortium for Conservation Medicine*
VI.9 REGULATING SERVICES: A FOCUS ON DISEASE REGULATION

Diane M. Debinski, *Department of Ecology, Evolution and Organismal Biology, Iowa State University*
V.6 CONSERVATION AND GLOBAL CLIMATE CHANGE

Robert F. Denno, *late Professor of Entomology, University of Maryland*
II.7 PREDATOR–PREY INTERACTIONS

Daniel F. Doak, *Department of Zoology and Physiology, University of Wyoming*
V.2 POPULATION VIABILITY ANALYSIS

Martha Downs, *Environmental Change Initiative, Brown University*
VI.6 GRASSLANDS

Laurie E. Drinkwater, *Department of Horticulture, Cornell University*
VI.4 HUMAN-DOMINATED SYSTEMS: AGROECOSYSTEMS

Ray Dybzinski, *Department of Ecology, Evolution, and Behavior, University of Minnesota*
II.5 COMPETITION AND COEXISTENCE IN PLANT COMMUNITIES

Stephen P. Ellner, *Department of Ecology and Evolutionary Biology, Cornell University*
II.1 AGE-STRUCTURED AND STAGE-STRUCTURED POPULATION DYNAMICS

J. J. Elser, *School of Life Sciences, Arizona State University*
III.15 ECOLOGICAL STOICHIOMETRY

Paul Falkowski, *Department of Geological Sciences and Institute of Marine and Coastal Sciences, Rutgers University*
III.13 THE MARINE CARBON CYCLE

Myra E. Finkelstein, *Division of Physical and Biological Sciences, University of California, Santa Cruz*
V.2 POPULATION VIABILITY ANALYSIS

Joern Fischer, *Fenner School of Environment and Society, Australian National University*
IV.2 LANDSCAPE PATTERN AND BIODIVERSITY

Jonathan A. Foley, *Center for Sustainability and the Global Environment and Department of Environmental Studies and Atmospheric and Oceanic Sciences, University of Wisconsin–Madison*
VII.7 AGRICULTURE, LAND USE, AND THE TRANSFORMATION OF PLANET EARTH

Kevin J. Gaston, *Department of Animal and Plant Sciences, University of Sheffield*
I.12 GEOGRAPHIC RANGE

Rosemary Gillespie, *Department of Environmental Science, Policy and Management, University of California, Berkeley*
I.19 ADAPTIVE RADIATION

H. Charles J. Godfray, *Department of Zoology, University of Oxford*
II POPULATION ECOLOGY

Scott J. Goetz, *Woods Hole Research Center*
I.11 REMOTE SENSING AND GEOGRAPHIC INFORMATION SYSTEMS

Indur M. Goklany, *Office of Policy Analysis, U.S. Department of the Interior*
VI.12 TECHNOLOGICAL SUBSTITUTION AND AUGMENTATION OF ECOSYSTEM SERVICES

James R. Gosz, *Biology Department, University of New Mexico*
IV.5 BOUNDARY DYNAMICS IN LANDSCAPES
This research was supported by National Science Foundation support to the Sevilleta Long-term Ecological Research Program at the University of New Mexico (DEB 0620482).

Catherine Graham, *Department of Ecology and Evolution, Stony Brook University*
I.11 REMOTE SENSING AND GEOGRAPHIC INFORMATION SYSTEMS

D. S. Gruner, *Department of Entomology, University of Maryland*
III.6 TOP-DOWN AND BOTTOM-UP REGULATION OF COMMUNITIES
This work was completed as part of the Trophic Structure Comparisons Working Group supported by the National Center for Ecological Analysis and Synthesis, a Center funded by NSF (Grant #DEB-0072909), the University of California at Santa Barbara, and the state of California.

Nick Haddad, *Department of Zoology, North Carolina State University*
V.3 PRINCIPLES OF RESERVE DESIGN

Benjamin S. Halpern, *National Center for Ecological Analysis and Synthesis, University of California, Santa Barbara*
VI.7 MARINE ECOSYSTEM SERVICES

Ilkka Hanski, *Department of Ecology and Systematics, University of Helsinki*
II.4 METAPOPULATIONS AND SPATIAL POPULATION PROCESSES

Alan Hastings, *Department of Environmental Science and Policy, University of California, Davis*
II.3 BIOLOGICAL CHAOS AND COMPLEX DYNAMICS

Andrew Hector, *Institute of Environmental Sciences, University of Zurich*
III.14 BIODIVERSITY AND ECOSYSTEM FUNCTIONING

Philip Hedrick, *School of Life Sciences, Arizona State University*
I.15 POPULATION GENETICS AND ECOLOGY

Nicole Heller, *Department of Biology, Franklin & Marshall College*
III.18 RESPONSES OF COMMUNITIES AND ECOSYSTEMS TO GLOBAL CHANGES

Justin P. Henningsen, *Biology Department, University of Massachusetts*
I.4 FUNCTIONAL MORPHOLOGY: MUSCLES, ELASTIC MECHANISMS, AND ANIMAL PERFORMANCE

Ray Hilborn, *School of Aquatic and Fishery Sciences, University of Washington*
VII.2 FISHERIES MANAGEMENT

Richard J. Hobbs, *School of Plant Biology, University of Western Australia*
IV.2 LANDSCAPE PATTERN AND BIODIVERSITY
V.7 RESTORATION ECOLOGY

Robert D. Holt, *Department of Zoology, University of Florida*
III.3 PREDATION AND COMMUNITY ORGANIZATION

R. A. Houghton, *Woods Hole Research Center*
III.11 TERRESTRIAL CARBON AND BIOGEOCHEMICAL CYCLES

Terry P. Hughes, *Centre for Coral Reef Biodiversity, James Cook University*
IV.8 SEASCAPE PATTERNS AND DYNAMICS OF CORAL REEFS

Duncan J. Irschick, *Biology Department, University of Massachusetts*
I.4 FUNCTIONAL MORPHOLOGY: MUSCLES, ELASTIC MECHANISMS, AND ANIMAL PERFORMANCE

Anthony R. Ives, *Department of Zoology, University of Wisconsin–Madison*
II.2 DENSITY DEPENDENCE AND SINGLE-SPECIES POPULATION DYNAMICS

Jeremy B. C. Jackson, *Smithsonian Tropical Research Institute*
V.5 MARINE CONSERVATION

Peter Kareiva, *The Nature Conservancy*
VI.10 SUPPORT SERVICES: A FOCUS ON GENETIC DIVERSITY

David M. Karl, *Center for Microbial Oceanography: Research and Education, and Department of Oceanography, University of Hawai'i at Manoa*
IV.9 SEASCAPE MICROBIAL ECOLOGY: HABITAT STRUCTURE, BIODIVERSITY, AND ECOSYSTEM FUNCTION

A. Marm Kilpatrick, *Consortium for Conservation Medicine*
VI.9 REGULATING SERVICES: A FOCUS ON DISEASE REGULATION

Joel G. Kingsolver, *Department of Biology, University of North Carolina at Chapel Hill*
I.14 PHENOTYPIC SELECTION

Ann P. Kinzig, *School of Life Sciences, Arizona State University*
VI ECOSYSTEM SERVICES

Allan Larson, *Biology Department, Washington University*
I.13 ADAPTATION

Julien Lartigue, *Office of Oceanic and Atmospheric Research, National Oceanic and Atmospheric Administration*
III.9 ECOSYSTEM PRODUCTIVITY AND CARBON FLOWS: PATTERNS ACROSS ECOSYSTEMS

M. A. Leibold, *Section of Integrative Biology, University of Texas*
III.8 SPATIAL AND METACOMMUNITY DYNAMICS IN BIODIVERSITY

Ricardo M. Letelier, *College of Oceanic and Atmospheric Sciences, Oregon State University*
IV.9 SEASCAPE MICROBIAL ECOLOGY: HABITAT STRUCTURE, BIODIVERSITY, AND ECOSYSTEM FUNCTION

Danny Lewis, *Department of Entomology, University of Maryland*
II.7 PREDATOR–PREY INTERACTIONS
IV.9 SEASCAPE MICROBIAL ECOLOGY: HABITAT STRUCTURE, BIODIVERSITY, AND ECOSYSTEM FUNCTION

David B. Lindenmayer, *Center for Resource and Environmental Studies, Australian National University*
IV.2 LANDSCAPE PATTERN AND BIODIVERSITY

Nicolas Loeuille, *Université Paris 6 (Laboratory of Ecology)*
III.19 EVOLUTION OF COMMUNITIES AND ECOSYSTEMS

Michel Loreau, *Department of Biology, McGill University*
III COMMUNITIES AND ECOSYSTEMS

Jonathan B. Losos, *Department of Organismic and Evolutionary Biology, Harvard University*
I AUTECOLOGY

John A. Ludwig, *CSIRO Tropical Research Center*
IV.1 LANDSCAPE DYNAMICS

Pablo A. Marquet, *Center for Advanced Studies in Ecology and Biodiversity and Ecology Department, Catholic University of Chile*
III.16 MACROECOLOGICAL PERSPECTIVES ON COMMUNITIES AND ECOSYSTEMS

Pamela A. Matson, *School of Earth Sciences, Stanford University*
III.10 NUTRIENT CYCLING AND BIOGEOCHEMISTRY

Brian A. Maurer, *Department of Fisheries and Wildlife and Department of Geography, Michigan State University*
IV.6 SPATIAL PATTERNS OF SPECIES DIVERSITY IN TERRESTRIAL ENVIRONMENTS

Kevin McCann, *Department of Zoology, University of Guelph*
III.7 THE STRUCTURE AND STABILITY OF FOOD WEBS

Evelyn H. Merrill, *Department of Biological Sciences, University of Alberta*
VII.3 WILDLIFE MANAGEMENT

Clark A. Miller, *Consortium for Science, Policy, and Outcomes and Department of Political Science, Arizona State University*
VII.11 ASSESSMENTS: LINKING ECOLOGY TO POLICY

Chad Monfreda, *Center for Sustainability and the Global Environment, University of Wisconsin–Madison*
VII.7 AGRICULTURE, LAND USE, AND THE TRANSFORMATION OF PLANET EARTH

Paul R. Moorcroft, *Department of Organismic and Evolutionary Biology, Harvard University*
IV.4 BIODIVERSITY PATTERNS IN MANAGED AND NATURAL LANDSCAPES

Rebecca J. Morris, *Department of Zoology, University of Oxford*
II.10 INTERACTIONS BETWEEN PLANTS AND HERBIVORES

William F. Morris, *Department of Biology, Duke University*
I.10 LIFE HISTORY

William Murdoch, *Department of Ecology, Evolution, and Marine Biology, University of California, Santa Barbara*
VII.1 BIOLOGICAL CONTROL: THEORY AND PRACTICE

Shahid Naeem, *Department of Ecology, Evolution, and Environmental Biology, Columbia University*
VI.2 BIODIVERSITY, ECOSYSTEM FUNCTIONING, AND ECOSYSTEM SERVICES

Jon Norberg, *Department of Systems Ecology, Stockholm University*
VI.3 BEYOND BIODIVERSITY: OTHER ASPECTS OF ECOLOGICAL ORGANIZATION

Patrik Nosil, *Department of Zoology, University of British Columbia*
I.18 ECOLOGICAL SPECIATION: NATURAL SELECTION AND THE FORMATION OF NEW SPECIES

Sarah H. Olsen, *Center for Sustainability and the Global Environment, University of Wisconsin–Madison*
VII.6 MANAGING INFECTIOUS DISEASES

Megan O'Rourke, *Department of Ecology and Evolutionary Biology, Cornell University*
VI.4 HUMAN-DOMINATED SYSTEMS: AGROECOSYSTEMS

Elinor Ostrom, *Department of Political Science, Indiana University*
VII.10 GOVERNANCE AND INSTITUTIONS

Guayana I. Páez-Acosta, *Department of Global Ecology, Carnegie Institution for Science*
VI.5 FORESTS

Margaret A. Palmer, *Department of Entomology, University of Maryland*
VI.8 PROVISIONING SERVICES: A FOCUS ON FRESH WATER

Jonathan A. Patz, *Center for Sustainability and the Global Environment and Department of Population Health Sciences, University of Wisconsin–Madison*
VII.6 MANAGING INFECTIOUS DISEASES
VII.7 AGRICULTURE, LAND USE, AND THE TRANSFORMATION OF PLANET EARTH

Daniel Pauly, *Fisheries Center, Aquatic Ecosystems Research Laboratory, University of British Columbia*
IV.10 SPATIAL DYNAMICS OF MARINE FISHERIES

Oliver R. W. Pergams, *Department of Biological Sciences, University of Illinois at Chicago*
VI.10 SUPPORT SERVICES: A FOCUS ON GENETIC DIVERSITY

Nicolas Perrin, *Department of Ecology and Evolution, University of Lausanne*
I.6 DISPERSAL

Charles Perrings, *School of Life Sciences, Arizona State University*
VI.11 THE ECONOMICS OF ECOSYSTEM SERVICES

Debra P. C. Peters, *USDA–Agricultural Research Service, Jornada Experimental Range, New Mexico State University*
IV.5 BOUNDARY DYNAMICS IN LANDSCAPES
This research was supported by National Science Foundation support to the Sevilleta Long-term Ecological Research Program at the University of New Mexico (DEB 0620482).

David W. Pfennig, *Department of Biology, University of North Carolina at Chapel Hill*
I.14 PHENOTYPIC SELECTION

Alison G. Power, *Department of Ecology and Evolutionary Biology, Cornell University*
VI.4 HUMAN-DOMINATED SYSTEMS: AGROECOSYSTEMS

Robert L. Pressey, *School of Biological Sciences, University of Queensland*
V.4 BUILDING AND IMPLEMENTING SYSTEMS OF CONSERVATION AREAS

Navin Ramankutty, *Department of Geography and Earth System Science Program, McGill University*
VII.7 AGRICULTURE, LAND USE, AND THE TRANSFORMATION OF PLANET EARTH

Mark Rees, *Department of Animal and Plant Sciences, University of Sheffield*
II.1 AGE-STRUCTURED AND STAGE-STRUCTURED POPULATION DYNAMICS

David C. Richardson, *Marine-Estuarine-Environmental Sciences Program, University of Maryland*
VI.8 PROVISIONING SERVICES: A FOCUS ON FRESH WATER

Jon Paul Rodríguez, *Center for Ecology, Venezuelan Institute for Scientific Investigations (Instituto Venezolano de Investigaciones Cientificas - IVIC)*
VI.13 CONSERVATION OF ECOSYSTEM SERVICES

Howard Rundle, *Department of Biology, University of Ottawa*
I.18 ECOLOGICAL SPECIATION: NATURAL SELECTION AND THE FORMATION OF NEW SPECIES

Osvaldo E. Sala, *Department of Ecology and Evolutionary Biology, Center for Environmental Studies, and Environmental Change Initiative, Brown University*
VI.6 GRASSLANDS

Marten Scheffer, *Aquatic Ecology and Water Management Group, Wageningen University*
III.17 ALTERNATIVE STABLE STATES AND REGIME SHIFTS IN ECOSYSTEMS

D. W. Schindler, *Department of Ecology, University of Alberta*
VII.5 MANAGING NUTRIENT MOBILIZATION AND EUTROPHICATION

Oswald J. Schmitz, *Yale School of Forestry and Environmental Studies, Yale University*
III.5 INDIRECT EFFECTS IN COMMUNITIES AND ECOSYSTEMS: THE ROLE OF TROPHIC AND NONTROPHIC INTERACTIONS

Thomas W. Schoener, *Section of Evolution and Ecology, College of Biological Sciences, University of California, Davis*
I.1 ECOLOGICAL NICHE

R. J. Scholes, *CSIR Division of Water, Environment, and Forest Technology, South Africa*
VI.1 ECOSYSTEM SERVICES: ISSUES OF SCALE AND TRADE-OFFS

Anthony R. E. Sinclair, *Department of Zoology, University of British Columbia*
VII.3 WILDLIFE MANAGEMENT

Navjot S. Sodhi, *Department of Biological Sciences, National University of Singapore*
V.1 CAUSES AND CONSEQUENCES OF SPECIES EXTINCTIONS

Luis A. Solórzano, *Andes-Amazon Initiative, Gordon and Betty Moore Foundation*
VI.5 FORESTS

Judy Stamps, *Section of Evolution and Ecology, College of Biological Sciences, University of California, Davis*
I.5 HABITAT SELECTION

R. W. Sterner, *Department of Ecology, Evolution, and Behavior, University of Minnesota*
III.15 ECOLOGICAL STOICHIOMETRY

Stephanie A. Stuart, *Department of Integrative Biology, University of California, Berkeley*
I.3 PHYSIOLOGICAL ECOLOGY: PLANTS

John N. Thompson, *Department of Ecology and Evolutionary Biology, University of California, Santa Cruz*
II.13 COEVOLUTION

David Tilman, *Department of Ecology, Evolution, and Behavior, University of Minnesota*
II.5 COMPETITION AND COEXISTENCE IN PLANT COMMUNITIES

David J. Tongway, *Fenner School of Environment and Society, Australian National University*
IV.1 LANDSCAPE DYNAMICS

Joseph Travis, *Department of Biological Science, Florida State University*
I.9 PHENOTYPIC PLASTICITY

Will R. Turner, *Center for Applied Diversity Science, Conservation International*
V.4 BUILDING AND IMPLEMENTING SYSTEMS OF CONSERVATION AREAS

Peter M. Vitousek, *Department of Biological Sciences, Stanford University*
III.10 NUTRIENT CYCLING AND BIOGEOCHEMISTRY

Brian Walker, *CSIRO Sustainable Ecosystems*
IV LANDSCAPES AND THE BIOSPHERE

Reg Watson, *Fisheries Centre, University of British Columbia*
IV.10 SPATIAL DYNAMICS OF MARINE FISHERIES

Martin Wikelski, *Department of Ecology and Evolutionary Biology, Princeton University*
I.2 PHYSIOLOGICAL ECOLOGY: ANIMALS

Andy Wilby, *Department of Biological Sciences, Lancaster University*
III.14 BIODIVERSITY AND ECOSYSTEM FUNCTIONING

David S. Wilcove, *Department of Ecology and Evolutionary Biology, Princeton University*
V CONSERVATION BIOLOGY

F. I. Woodward, *Department of Animal and Plant Sciences, University of Sheffield*
IV.7 BIOSPHERE–ATMOSPHERE INTERACTIONS IN LANDSCAPES

Jianguo Wu, *School of Life Sciences, Arizona State University*
IV.3 ECOLOGICAL DYNAMICS IN FRAGMENTED LANDSCAPES

Anastasios Xepapadeas, *Economics Department, University of Crete*
VII.9 ECOLOGICAL ECONOMICS: PRINCIPLES OF ECONOMIC POLICY DESIGN FOR ECOSYSTEM MANAGEMENT

Erika Zavaleta, *Environmental Studies Department, University of California, Santa Cruz*
III.18 RESPONSES OF COMMUNITIES AND ECOSYSTEMS TO GLOBAL CHANGES

The Princeton Guide to Ecology

I

Autecology

Jonathan B. Losos

Autecology refers to how a single species interacts with the environment; its counterpart is synecology, which refers to how multiple species interact with each other. This latter term is mostly congruent with the field of community ecology, the subject of part III of this volume.

Integral to any discussion of autecology is the concept of the niche. This concept has a long and checkered history in the field of ecology, and the term itself has taken on different meanings through time (chapter I.1). In the most general sense, however, we may think of the niche of a population as the way members of that population interact with their environment, both biotic and abiotic. In other words, the term "niche" refers to where organisms live and what they do there.

The first step in considering how organisms interact with their environment is investigating how the specific phenotypic characteristics of members of a population allow them to exist in a particular environment. The environment poses a wide variety of challenges to organisms: for example, they must be able to obtain and retain enough water, withstand high or low temperatures, and obtain enough nutrients to survive. More than a century of research has revealed that species, and even populations of species, are often finely tuned to the specific conditions in the environment in which they live. In recent years, increasingly sophisticated approaches and instrumentation have allowed an exquisitely detailed understanding of the physiological basis of organismal function (chapters I.2–I.4).

Animals—and, in some sense, fast-growing plants—also can influence the way they interact with their environment through behavioral means. For example, animals can choose the habitat in which they occur and thus can determine, to some extent, the environment they experience throughout their lives (chapter I.5). Many organisms move from their birth site at a particular stage in life; although for plants and some animals, dispersal is passive, other species actively choose where to settle (chapter I.6).

Behavior, of course, is a key component of how most animals interact with their environment. Almost all aspects of the natural history of animals have a behavior component. In part I, we consider foraging (chapter I.7) and social behavior (chapter I.8). Other topics are included in parts II and VI of this volume.

Most plants have relatively little ability to determine the environmental conditions they experience. But plants often have another option available—they frequently exhibit substantial phenotypic plasticity, which allows a plant to alter its phenotype in an advantageous way to be better suited to its environment. Scientists have long appreciated this ability in plants, and zoologists have come to realize relatively recently that many animal species exhibit adaptive phenotypic plasticity as well (chapter I.9).

Organisms adapt in yet another way, by molding their life cycle—what is termed "life history"—to the particular environment in which they live (chapter I.10). Thus, species in environments in which resources are abundant and threats are common may have short generation times and early reproduction. Conversely, in environments in which resources are more scarce but threats are not as severe, a more successful strategy may be to defer reproduction and to invest in becoming better competitors for resources, delaying reproduction and ultimately producing fewer, but better provisioned, offspring.

No species occurs everywhere in the world. The behavior and physiological capabilities of a species determine where a species can and cannot occur. In the last few years, advances in remote sensing technology have provided the capability to visualize the distribution of environmental conditions with great precision over large spatial scales (chapter I.11). Combined with records of species occurrences and, ideally, an understanding of species' physiological capabilities, these geographic information systems approaches have opened new vistas for understanding how and why species occur where they do; these approaches are also of great importance in predicting how species will respond to rapidly changing environmental conditions (see parts IV and V). Of course, the distribution of a species is not only a function of its physiological capabilities and other aspects of its ecology. Rather, Earth geography and history also are important—a species cannot occupy an area that it has never had the opportunity to colonize. Consequently, biological and historical factors combine to determine the geographic range of any species (chapter I.12).

Integral to an understanding of how organisms interact with their environment is the concept of adaptation, the idea that natural selection has molded the characteristics of populations so that they are well suited to the particular circumstances in their environment (chapter I.13). Of course, this is not to say that organisms are optimally adapted to their current conditions, nor that every feature exhibited by a population represents an adaptation for some aspect of the environment. Quite the contrary, natural selection is only one of many processes that affect how populations evolve (chapters I.14 and I.15); in some circumstances, processes other than natural selection will predominate, leading populations to be less well adapted to their current circumstances.

Ecologists are increasingly interested in the evolutionary time scale. On one hand, it has become clear that, in many cases, we can understand the current state of species and of entire communities only by considering their history. Species are not blank slates, to be molded by selection to the optimum configuration for their environment; rather, they have a historical starting point, and selection can work to modify species only from this point (chapter I.13). Similarly, communities, too, have histories—the current state of a community is a result of which species have managed to get to a given locality and how those species interact once there. Methods to incorporate evolutionary information, in the form of phylogenies (or evolutionary trees), are now widely utilized and becoming increasingly sophisticated (chapter I.16). Conversely, evolutionary biologists have clearly demonstrated over the last several decades that evolutionary change can occur very rapidly (chapter I.17). Consequently, ecologists ignore evolution at their own peril—populations can adapt quickly enough that evolution can have effects even on ecological time scales.

Evolution is important in another respect. The components of ecological interactions are species. The study of speciation—how new species arise—has long been the province of evolutionary biologists, but in recent years it has become clear that ecology may play an important role in affecting rates of speciation. In particular, the concept of ecological speciation—the idea that speciation is intimately tied to ecological divergence—has gathered great support (chapter I.18). Hence, in this respect as well, ecological and evolutionary perspectives are strongly intertwined. Finally, over larger time scales, certain groups of organisms diversify greatly, producing not only a large number of species but also occupying a great variety of ecological niches. Some scientists consider this phenomenon, known as adaptive radiation, to be responsible for the majority of life's diversity (chapter I.19).

I.1

Ecological Niche

Thomas W. Schoener

OUTLINE

1. Three concepts of the ecological niche
2. The recess/role niche and seeking ecological equivalents
3. The population-persistence niche and mechanistically representing competition
4. The resource-utilization niche and understanding the evolution of species differences
5. Environmental niche modeling and analyzing niches on a macroscale
6. Conclusion

It may come as something of a surprise that *ecological niche*, a term so common in the popular media, has three distinct meanings among scientists, each with an associated conceptual basis: these are the recess/role niche, the population-persistence niche, and the resource-utilization niche.

GLOSSARY

character displacement. The situation in which two species are more different in geographic locations where they overlap than between locations where they occur alone

community. Those species populations occurring at some location

competition. Ecological interaction in which two or more species negatively affect one another by consuming common resources or by other harmful means

convergence. Development of increasing similarity over time, usually applied to species somewhat unrelated evolutionarily

niche dimension. Environmental variable along which a species' niche is characterized, e.g., food size, and typically represented as the axis of a graph

polymorphism. The existence of two or more forms, differing in morphology or some other way, in the same population

population. Those individuals of a species occurring at some location

population growth rate r. The per capita rate at which a population changes size, typically computed as the birthrate minus the death rate

1. THREE CONCEPTS OF THE ECOLOGICAL NICHE

The Recess/Role Niche

The first use of "ecological niche" appeared in a report on ladybugs written by R. H. Johnson nearly a century ago, although the term was used shortly thereafter by the zoologist Joseph Grinnell, who is generally given credit for its original development. The meaning was very close to figurative usage: the ecological niche of a species is its "role," "place," or more literally "recess" (in the sense of a "nook" or "cubbyhole") in an ecological community. Thus, the California thrasher, one of Grinnell's major examples, is a bird of the chaparral community that feeds mostly on the ground by working over the surface litter and eating both animal and plant items of a suitable size. Escape from predators is similarly terrestrial, with the well-camouflaged bird shuffling off through the underbrush on the rare occasions when it is threatened.

The idea that there exists a set of characteristic habitat and food types with accompanying behavioral, morphological, and physiological adaptations leads to the notion of ecological equivalents. These are defined as two or more species with very similar niche characteristics that occur in completely different localities. An example from Grinnell's writings is the kangaroo rat of North America, which "corresponds exactly" to the jerboa (another desert rodent species) of the Sahara. The existence of ecological equivalents would imply that rather invariant rules determine the niches available for occupancy in a particular kind of environment, e.g., a desert. Moreover, niches can be empty in the sense that a suitable species does not occur within a locality, perhaps because it never got there or was unable to evolve in situ.

But to what extent do ecological equivalents really exist? Decades after Grinnell's work, we now know (section 2, below) that although some examples can certainly be found, perhaps more commonly, species of similar environments (e.g., deserts) among distant localities are neither identical nor often even similar. Perhaps such considerations helped to engender the two other meanings of ecological niche, each with its accompanying set of ideas about how the ecological world works.

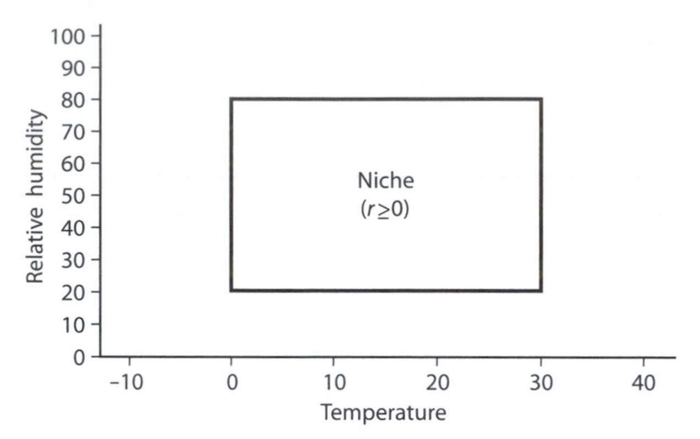

Figure 1. Example of Hutchinson's population-persistence niche. Rectangle encloses the ranges of temperature and humidity in which the species' population can persist (where $r \geq 0$).

The Population-Persistence Niche

The *population-persistence niche* has its roots in papers written in the mid-twentieth century by the ecologist and limnologist G. E. Hutchinson. This concept focuses on the species, in this case its population, rather than on the environment. Hutchinson formulates the ecological niche as a quantitative description of the range of environmental conditions that allow a population to persist in some location; the term *persist* means having a positive or at least zero (break-even) population growth rate, r (if r is negative, the population dwindles away to extinction). An example of an environmental condition is temperature; a second example would be humidity (for organisms on land) or salinity (for organisms in water). If we represent an environmental condition by the axis of a graph, a *range* is an interval along that axis, e.g., temperature from 0°C to 30°C (figure 1). A second interval, say for relative humidity, might range from 20% to 80% along the humidity axis. We can have as many different environmental axes as necessary to characterize the population growth rate. If r for a given axis is uncorrelated with the values of variables of the other axis (e.g., if the range of temperatures allowing $r \geq 0$ is the same for any value of humidity), then the niche is rectangular (as in figure 1); otherwise it will have other shapes. Hutchinson labeled his concept the cumbersome "n-dimensional hypervolume" (imagine three or more environmental axes). The more succinctly labeled *fundamental niche* is that portion of niche space where the species population can persist. The fundamental niche is visualized as being in the absence of other species that compete with the given species for resources and thereby affect its persistence. To account for this latter circumstance, Hutchinson defined the "realized niche" as that portion of the fundamental niche not overlapping the fundamental niches of competing species, plus that portion overlapping the competing species' niches where the given species can still persist (have $r \geq 0$).

Hutchinson's concept is important for several reasons. First, it provides a precise, quantitative way to characterize the ecological niche. Second, it focuses on what the species itself does rather than on the opportunity for a species to exist or not in a community (the latter being the "recess" concept of Grinnell). Thus, ecological equivalents are not necessarily expected and, if they do not occur, are not troubling to the concept: for Hutchinson, there are no "empty niches."

Such a precise formulation of the niche is not without its drawbacks, however. Chief among them perhaps is the difficulty of finding out what the population-persistence niche of a species actually is in nature. Presumably, for each point of the n-dimensional hypervolume—say for each value of temperature and humidity—one needs to culture populations or otherwise determine their population growth rate r; and one repeats this for different points until one has all combinations of temperature and humidity for which the population can persist. The difficulty of so doing for all but microorganisms (at best) is easy to imagine. A second problem is that certain niche characteristics as conceptualized by Grinnell are not easily ordered along an environmental axis. An example is food size: at any given real location, food comes in a variety of sizes (rather than there being one food size for each location). Of course, one can use average food size, but such a concept is not as plausible as using average temperature because animals come across a variety of food sizes on a daily basis. Animals of a particular body size (and therefore a particular size of feeding apparatus, e.g., mouth) have limitations on the extreme values of food size that can be consumed: items too large cannot be swallowed, and those too small cannot be handled deftly (or eaten in an energetically profitable way). Hence, a more detailed

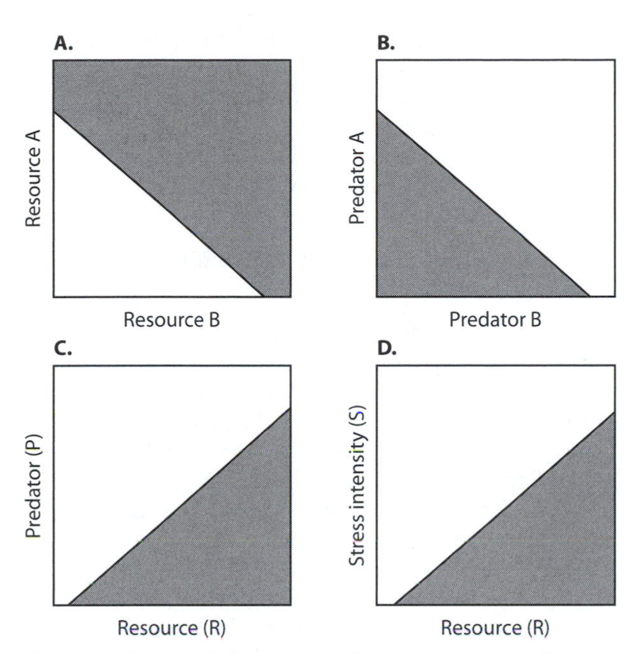

Figure 2. Chase-Leibold concept of the population-persistence niche. Each panel has two regions, a shaded region where $r \geq 0$ and an unshaded region where $r < 0$. The niche is the shaded region. (A) A species with two substitutable resources (the axes measure resource density); (B) a species with two predators; (C) a species with a predator and a resource; (D) a species with a stress and a resource. (Figure courtesy of J. M. Chase.)

description than the average food size available at a location is desirable. Third, Hutchinson's niche is one-sided in the sense that it assumes a rather passive species that does not affect other species in the community in a way that eventually feeds back onto the given species. Fourth, Hutchinson focuses almost exclusively on one type of ecological interaction, competition between species; for example, his distinction between the fundamental and the realized niche. In this way, his concept was not as inclusive as that of Grinnell.

In part as a reaction to the latter two drawbacks, Jonathan Chase and Mathew Leibold have substantially extended the population-persistence niche. In a recent but already very influential book, they define the niche as a joint specification of environmental conditions or variables that allow a species to have $r \geq 0$ along with the effects of that species on those environmental variables. Niche axes are quite broadly construed and can include a variety of factors that impact populations (and vice versa); examples include amount of a given resource, abundance of a given predator, and degree of a physical stress such as wind speed (figure 2). Thus, one can incorporate effects of species on environmental conditions, and one can specify a given region of niche space where a species has $r \geq 0$ (figure 2). Although this model represents a vast improvement in the concept of population-persistence niche, the operational difficulty of measurement still exists: determining the niche for figure 2 (Chase and Leibold) is not much easier than for figure 1 (Hutchinson).

The Resource-Utilization Niche

An eminently operational concept of the ecological niche, formulated by two evolutionary ecologists, Robert MacArthur and Richard Levins, is the *resource-utilization niche*, our third meaning. Like the population-persistence niche, the resource-utilization niche is quantitative and multidimensional, but it focuses entirely on what members of a species population in some locality actually do—in particular, how they use resources. The relative use (= utilization) of resources along a given niche axis can be described as a frequency distribution or histogram. Take, for example, the axis food size. We can (figure 3, top) draw a histogram showing the fraction of food of different sizes consumed by all members combined of a given population; e.g., the fraction of the total population's foods between 5 and 6 mm. If we have a second dimension, say feeding height, we can graph the fraction of food items eaten at different heights in the vegetation. The two can be combined as a joint distribution or three-dimensional histogram (figure 3, bottom), and this can be further generalized (although not easily graphed) for as many dimensions as ecologists find important to describe the population's resource use. A broad classification of the kinds of niche axes used for utilizations consists of habitat, food type, and time. Within habitat, microhabitat and macrohabitat are distinguished, whereby microhabitat has a smaller spatial scale (e.g., height in vegetation) than does macrohabitat (e.g., vegetation zone such as tropical rainforest or desert). Within food type, food size and hardness can be distinguished. Within time, daily and seasonal activity can be distinguished.

The resource-utilization niche immediately frees us from the problem with Hutchinson's formulation that certain environmental variables cannot be meaningfully described using only the average. Indeed, the resource-utilization niche is nothing more than a precisely formulated description of the natural history of a species: its habitat, food types, and activity times, among other things. Such natural history can include nonfeeding habitats and activity times for behaviors such as predator escape and mating, all characterizable on its niche axes. Thus, we have a niche concept that precisely encapsulates what ecologists measure anyway. Indeed, Grinnell, the originator of the *recess/role niche* concept, measured such things in his study

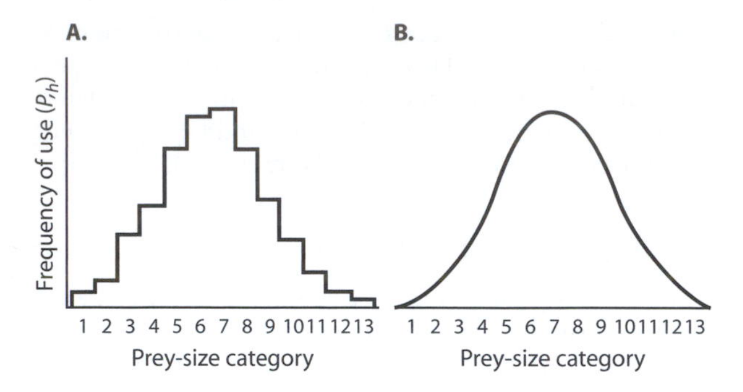

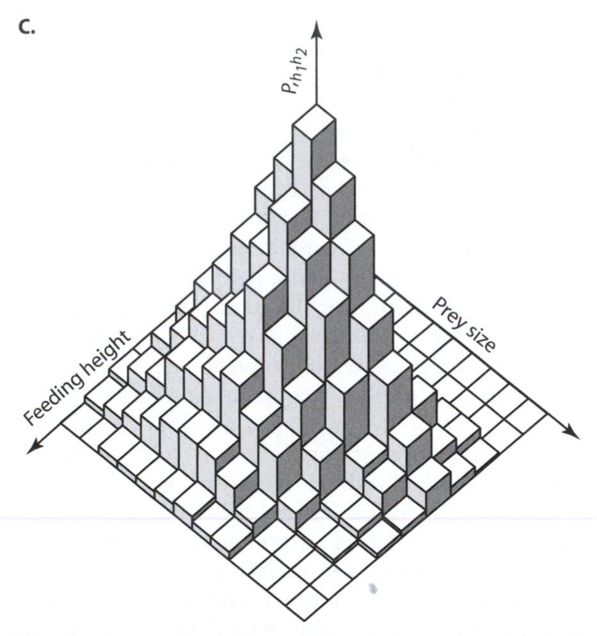

Figure 3. An example of the resource-utilization niche. (A) A one-dimensional niche, where the dimension is prey size. Numbers give prey-size categories, indexed by *h*; (B) the same utilization smoothed; (C) utilization of two resource dimensions, prey size and feeding height. (Redrawn from Schoener, 1986.)

organisms but with the assumption that, in so doing, he was discovering something about the availability of niches in the community—an availability or opportunity to which the species more or less had to conform. The resource-utilization niche, in contrast, assumes nothing about rigidly determined niche recesses in a community, nor about the necessity of ecological equivalents, nor about the existence of empty niches. The resource-utilization niche was formulated a decade or so after the population-persistence niche but, unlike the latter, has remained rather unchanged up to the present. This is despite the fact that, by emphasizing resources, it is seldom extended beyond resource use, unlike both the recess/role niche of Grinnell and the population-persistence niche of Chase and Leibold.

We now review seriatim the kinds of research engendered by the three concepts of the niche as well as a very recent research trend called ecological niche modeling that includes elements of all three.

2. THE RECESS/ROLE NICHE AND SEEKING ECOLOGICAL EQUIVALENTS

In an early study of grassland birds inhabiting far-flung locations—Kansas, Chile, and California—Cody found that each community contained about the same number of species and the same ecological types: three or four passerines (small "perching" birds), a larger vegetarian "grouse-like" species, both a long- and a short-billed wader, and two or three raptors. Twenty pairs of ecological equivalents were identified between the two Mediterranean systems: Chile and California. However, later studies by Cody in other Mediterranean systems including Sardinia and South Africa showed a weaker pattern, especially for the latter, whose floras were very different.

In contrast to birds, plants in Chilean and Californian systems showed little convergence at the community level; for example, woody vegetation in Chile comprises less of the total cover but more total species and has a greater diversity of height layers than in California. Nonetheless, the major growth forms (e.g., broad-leaved evergreen, broad-leaved deciduous) are similar, even with regard to number of species, although several forms present in Chile (e.g., spinose-stemmed shrubs) are absent from California—an apparent empty niche. Major resemblances between plant growth forms among plants with very different evolutionary lineages occur rather commonly among plants; a striking example is given by American cacti and African euphorbs.

Perhaps the least evidence for ecological equivalents after systematic search is among colubrid snakes of North, Central, and South America. Cadle and Greene find few ecological equivalents (and little evidence for community similarity); instead, a number of types (fossorial earthworm eaters, nocturnal arboreal lizard/frog eaters) in some communities are conspicuously absent in others.

Probably the most extensive work on convergence and ecological equivalents has been done on lizards. An initial study by Fuentes, again comparing Chile and California, found convergences in community characteristics as well as in individual niche traits—microhabitat, daily activity time, and food type. In a second major study, Pianka found less evidence for

Figure 4. An example of ecological equivalents: the horned toad (*Phrynosoma platyrhinos*) of North American deserts and the thorny devil (*Moloch horridus*) of Australian deserts. (From Pianka, E. R. 2000. Evolutionary Ecology. San Francisco: Harper & Row. Used by permission of Pearson Education, Inc.)

similarity in community characteristics than difference among lizards of the three warm-desert systems of North America, Australia, and Africa. Nonetheless, striking ecological equivalents sometimes exist, such as the amazing resemblance between the horned toad of North America and the thorny devil of Australia (figure 4).

Examples of ecological equivalents are most impressive when the species from widely different localities are relatively unrelated in terms of evolutionary descent: convergent evolution toward the same morphology and behavior would seem to support the idea of the niche as a functional optimum characteristic of particular types of communities (e.g., those in deserts) into which species repeatedly evolve. Nonetheless, a plausible hypothesis for lack of convergence is that major evolutionary stocks are so different that evolution is too constrained to produce much convergence. Melville, Harmon, and Losos recently examined two lizard families, the Iguanidae and Agamidae, of North America and Australia, respectively, which are closely enough related to belong to the same clade (Iguania) even though they have been geographically separate for as long as 150 million years. Using an approach that takes into account evolutionary relatedness, they found convergence in habitat use and locomotor morphology, including pairs of ecological equivalents, between the two deserts.

Another example of convergence among relatively closely related species is provided by the *Anolis* lizards of large West Indian islands: Cuba, Hispaniola, Jamaica, and Puerto Rico. Here, various ecomorphs—species occupying the same microhabitat—have independently evolved on the separate islands. Harmon, Kolbe, Cheverud, and Losos found that five functionally distinct morphological characters—body size, body shape, head shape, lamella (ridges on toes) number, and sexual size dimorphism—converge among the different islands as a function of habitat similarity. For example, lizards living on the ground and low trunks are more similar between Cuba and Hispaniola than either is to other ecomorphs (e.g., those living in tree crowns) co-occurring on the same island and to which they are more closely related.

A final recently discovered example of convergence occurs in a completely different group: orb-weaving spiders of the genus *Tetragnatha* of the Hawaiian islands. Blackledge and Gillespie found that spiders inhabiting different islands constructed remarkably similar webs. These convergences toward ecological equivalency, which they called "ethotypes" (ethology is the study of behavior, and this emphasizes the behavioral similarity), occurred independently in evolution. Like the Australian Iguania discussed above, the group as a whole consists of relatively closely related species.

In conclusion, although the evidence for ecological equivalents is certainly mixed, more and more examples are coming to light that make Grinnell's rather old concept seem alive if not completely well. As Schluter has suggested, to the extent that ecological equivalents exist and are independently evolved, morphology, physiology, and behavior must constrain the efficiencies with which resources and other factors characteristic of particular kinds of ecosystems (e.g., deserts) can be dealt with—ecological equivalents mark peaks in the adaptive landscape.

3. THE POPULATION-PERSISTENCE NICHE AND MECHANISTICALLY REPRESENTING COMPETITION

Maguire in 1973 may have been first to plot population growth rate r for real species as a function of niche dimensions and to make predictions about the competitive outcomes among them. In the 1950s, Birch had studied several species of beetle infesting stored grain in Australia; figure 5 shows Maguire's plot of Birch's data with respect to temperature and moisture. Isoclines of positive values of r down to zero (no population growth) show different patterns for the two species, such that *Calandra oryzae* has a higher r for lower temperatures and somewhat greater moistures than *Rhizopertha dominica*. The dashed line in figure 5 separates regions of niche space where one versus the other species has the higher r. Assuming no complications, an environment on one or the other side of the line will favor one or the other species of beetle in competition.

To illustrate their ideas about the population-persistence niche, Chase and Leibold replot data of Tilman for two species of diatoms, *Asterionella* and *Cyclotella* (figure 6). The situation is somewhat more complex than that shown in figure 2 because resources are not substitutable (which would mean that the populations can survive on either resource alone or on some combination) but rather are essential: figure 6A shows the general case, where a species must have a

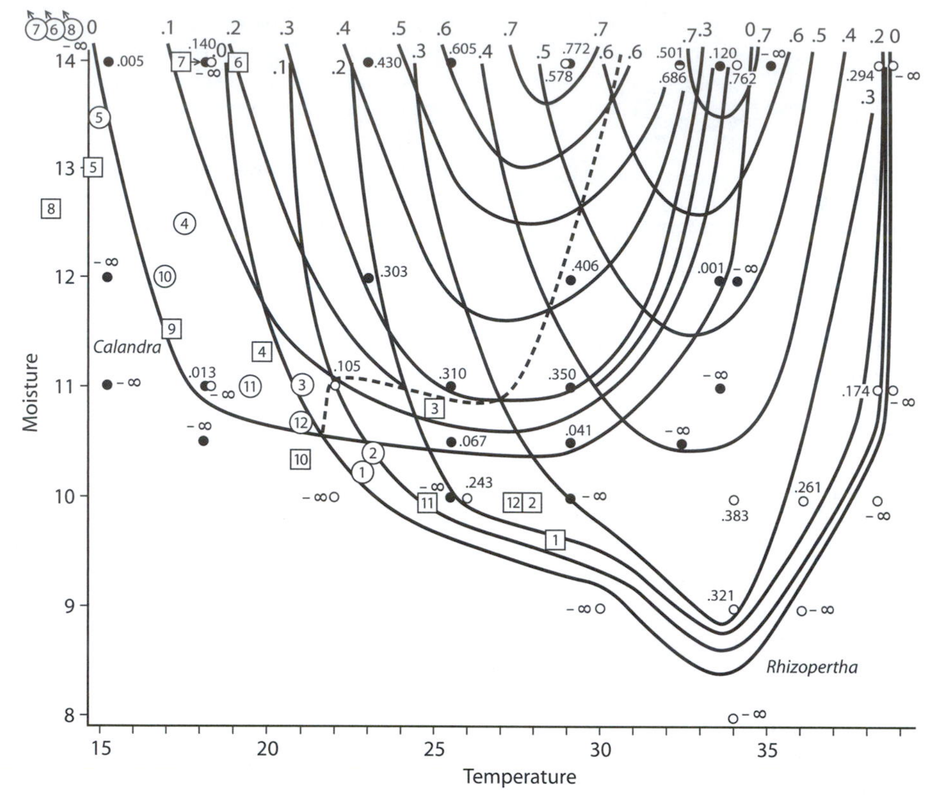

Figure 5. Maguire's illustration of the population-persistence niche for the dimensions temperature (C) and percentage moisture of wheat, using the beetles *Calandra oryzae* (small strain) and *Rhizopertha dominica*. Numbers in squares indicate average monthly conditions in Bourke, New South Wales, Australia; numbers in circles give same for Adelaide, South Australia. (Redrawn from Maguire, B., Jr. 1973. Niche response structure and the analytical potentials of its relationship to the habitat. American Naturalist 107: 213–246)

minimal amount of each resource in order that $r \geq 0$. For two such species, coexistence is possible if each species can just survive ($r = 0$) for a different one of the two resources. In Tilman's experiment, the resources are the nutrients silicate (SiO_2) and phosphate (PO_4), and the levels of each can be controlled in the laboratory. *Asterionella* is a specialist on SiO_2, and *Cyclotella* on PO_4. From the individual species growth curves on the separate resources, one can predict regions of niche space (plots of SiO_2 versus PO_4 concentration) where each species has a lower $r = 0$ and so is limited by a different resource. In that region (figure 6B), the species can coexist. Outside that region, one or the other species wins, depending on which resource is more abundant.

Such empirical studies are impressively successful in the highly controlled setting of the laboratory, but they are very difficult indeed to perform in the field. Chase and Liebold could find only one such field study, again by Tilman (and Wedin), in which several plant species vary in their ability to utilize nitrogen from the soil. These relative abilities were used rather successfully to predict competitive outcomes along a natural nitrogen gradient. Probably, practical difficulties largely explain why the population-persistence niche is a concept with mostly theoretical development. It seems most likely that it will be easiest to apply to organisms with the size and behavior that enable their populations to persist in small spatial units (sometimes called microcosms).

4. THE RESOURCE-UTILIZATION NICHE AND UNDERSTANDING THE EVOLUTION OF SPECIES DIFFERENCES

How similar can species be and still coexist? An answer was obtained in the last section for species having a small number of ecological requirements or resource types. What if species fed on a wide variety of resources, such as foods of different sizes found at different

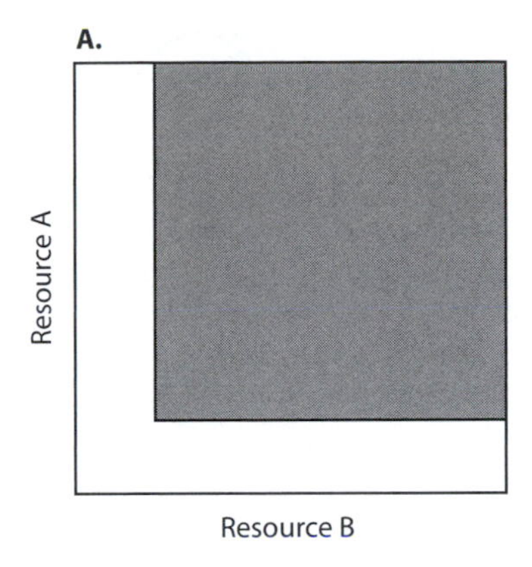

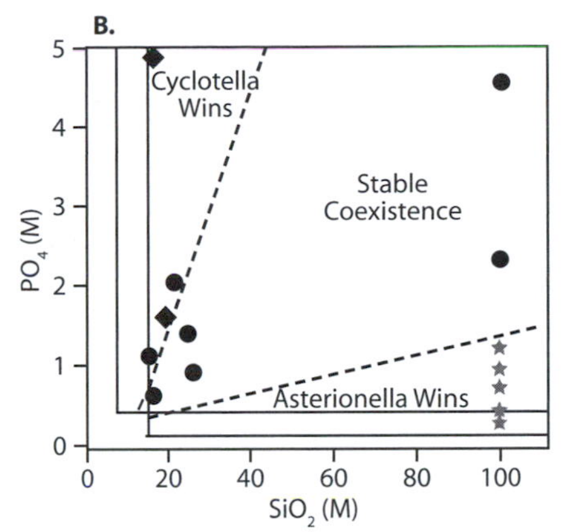

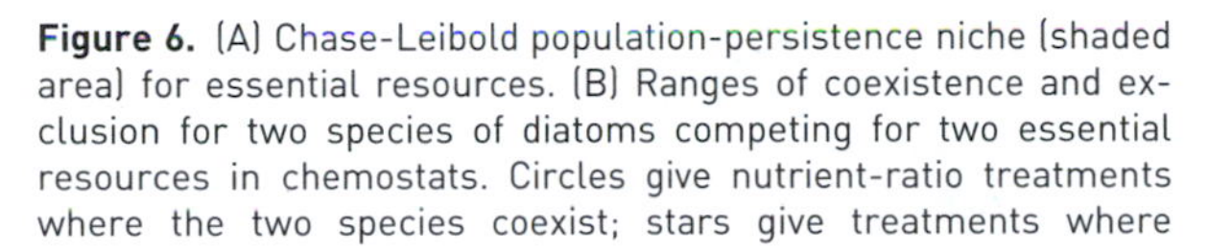

Figure 6. (A) Chase-Leibold population-persistence niche (shaded area) for essential resources. (B) Ranges of coexistence and exclusion for two species of diatoms competing for two essential resources in chemostats. Circles give nutrient-ratio treatments where the two species coexist; stars give treatments where *Asterionella* excludes *Cyclotella;* diamonds give treatments where *Cyclotella* excludes *Asterionella.* (Redrawn from Chase and Leibold, 2003; their figure 4.1 in turn redrawn from Tilman, D. 1977. Resource competition between planktonic algae: An experimental and theoretical approach. Ecology 58: 338–348)

vegetation heights and preferring different temperatures? This situation applies to predators, such as Grinnell's California thrasher, that eat a great variety of insects and other arthropods that in turn have their own populations with their own niche characteristics.

The 1967 paper in which MacArthur and Levins promoted the resource-utilization niche has as its main objective the understanding of how similar competing species can be and yet still coexist. It is sometimes said that species cannot coexist if they occupy the same niche, but the theory of MacArthur and Levins also posits that if the niches of the species are too similar (too much niche overlap), they still cannot coexist. To illustrate, imagine two species with the one-dimensional niche in figure 7; this dimension might be food size, and one species tends to eat larger food on average than the other. If the niches are too close (figure 7A, left), they are too similar (the niche overlap [shaded area] is too great), and the better competitor will eliminate the other from the community. That degree of closeness at which the species can just coexist (any closer and one is eliminated) is called the *limiting similarity* (figure 7A, middle); the niches can, of course, be farther apart and still allow coexistence (figure 7A, right).

Limiting similarity is measured in units of d/w, where d is the distance between peaks and w is the width of the niche (usually computed as the standard deviation of the utilization distribution; figure 7B). The larger the w, the more generalized the species; a specialist has a thin niche (small w; figure 7C).

In MacArthur and Levins's theory, a d/w slightly larger than 1.0 is the limiting similarity; much subsequent work has shown limiting similarity to vary greatly yet be about 1 (certainly to an order of magnitude). Indeed, sometimes real species differ by almost exactly this theoretical value. A sensational example is provided by two mud snails (*Hydrobia*) studied by Fenchel in Denmark. The snails ingest particles: diatoms and inorganic pebbles covered with minute sessile organisms. About 150 years before the study, a fjord collapsed, and one species invaded the other's range. The resource-utilization niches of the species displaced away from one another, apparently independently, numerous times, to $d/w \approx 1$ (figure 8, top left). Corresponding to this niche difference is a difference in body (shell) size such that larger species ingest larger particles (figure 8, top right), and the body sizes of the species had diverged (in a process called character displacement) to a ratio of 1.3–1.5 (figure 8, bottom). Consistent with the theory of Taper and Case (see below), this ratio is higher than the ratio of d's for the two resource utilizations of 1.2.

So far we have represented the resource-utilization niche as a distribution summing together the food-size or other niche characteristics of all individuals in a population. However, individuals may differ in their niche characteristics, sometimes just by chance opportunity (e.g., what they happen to come across to eat), but sometimes because they have different morphologies and behaviors that make them specialized for a certain portion of a niche axis (just as species can be specialized). Figure 9 shows the two extreme possibilities for such component individuals; note that each individual can be a generalist (figure 9, left) or a

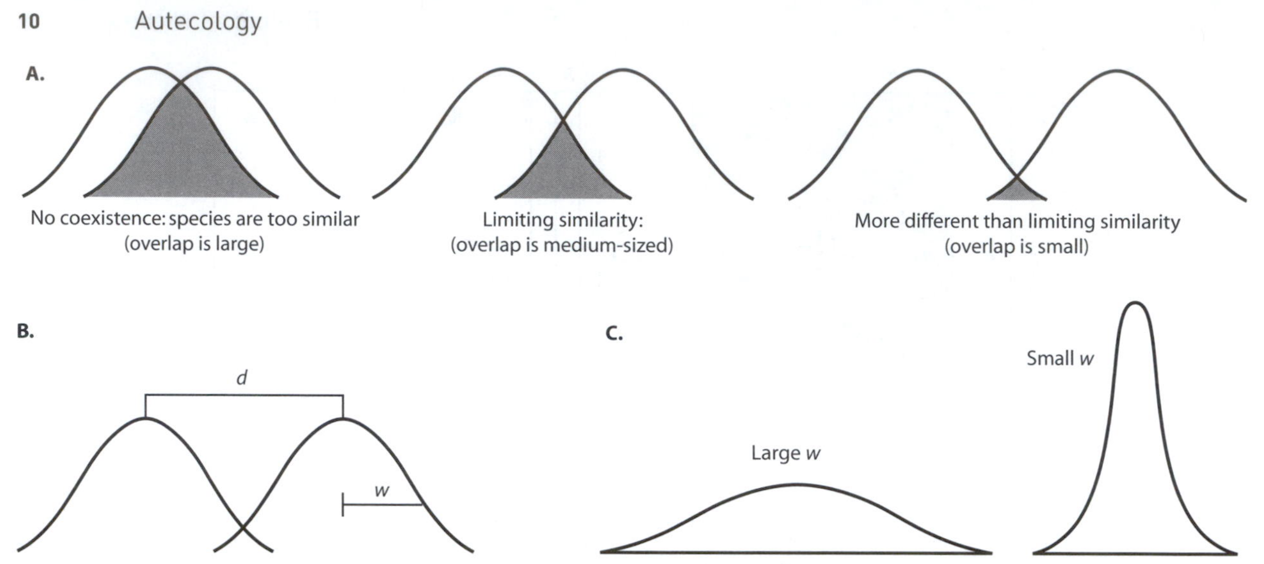

Figure 7. (A) One-dimensional resource-utilization niches of two species showing no coexistence because niches are too similar (left), limiting similarity—species can just coexist (middle), and greater than limiting similarity also allowing coexistence (right). (B) Niche distance *d* and niche width *w* for two species. (C) Niches of generalist (large *w*) and specialist (small *w*) species.

specialist (figure 9, middle), in either case producing the same utilization for all individuals combined (figure 9, right). What difference does it make which of the two situations one has? A series of specialist individuals may eventually allow the population as a whole to be more generalized in the absence of competing species, and this "polymorphism" might even lead to speciation (see chapter I.18). Such polymorphism, when measured in terms of those morphological characters corresponding to position on the niche axis (e.g., shell size corresponding to mean food-particle size), was uncommon in the literature at the time of Taper and Case's paper, and this was consistent with their theoretical model in which the proportion of different kinds of individual niches evolves once the competing species meet geographically. Recently, however, Bolnick, Svänback, Arágo, and Persson looked at the resource-utilization niches themselves rather than the morphological characters that reflect them. They found that the bigger the *w* for the total population, the bigger the between-phenotype niche width, measured as the standard deviation of the *d*'s of the niches of the component individuals. It remains to be seen exactly how these apparently somewhat contradictory trends will be reconciled.

5. ENVIRONMENTAL NICHE MODELING AND ANALYZING NICHES ON A MACROSCALE

A recent set of techniques, called environmental niche modeling (ENM), combines elements of all three niche concepts. The method characterizes the macrohabitat niche of a species by quantitatively summarizing geographic-information-system (GIS) information on climatic and similar variables at stations throughout the species' geographic range. Such macrohabitat niche information is then used to predict the potential geographic range of the subject species. Because of its focus on macrohabitat, the scale is similar to Hutchinson's version of the population-persistence niche. However, the method specifies the "empty niches" of Grinnell's recess/role niche as those localities having the niche characteristics of the subject species but where that species does not, in fact, occur. Finally, it allows quantification of niche similarity between species via measures of niche overlap used for the typically finer scale of resource-utilization niches of MacArthur and Levins.

One of the most successful applications of ENM so far examines the question of whether the more closely related species are, the more similar are their niches. The question is important because if the answer is yes, evolutionary history must have a major influence in determining niche characteristics relative to the influence of the community in which the species now occurs. A study by Knouft, Losos, Glor, and Kolbe on the 11 species of the *Anolis sagrei* group in Cuba found no evidence that niches were more similar, the more closely related the species (evolutionary relatedness is assessed using molecular genetics). A second study, by Warren, Glor, and Turelli showed along with the previous study that the most recently diverged species

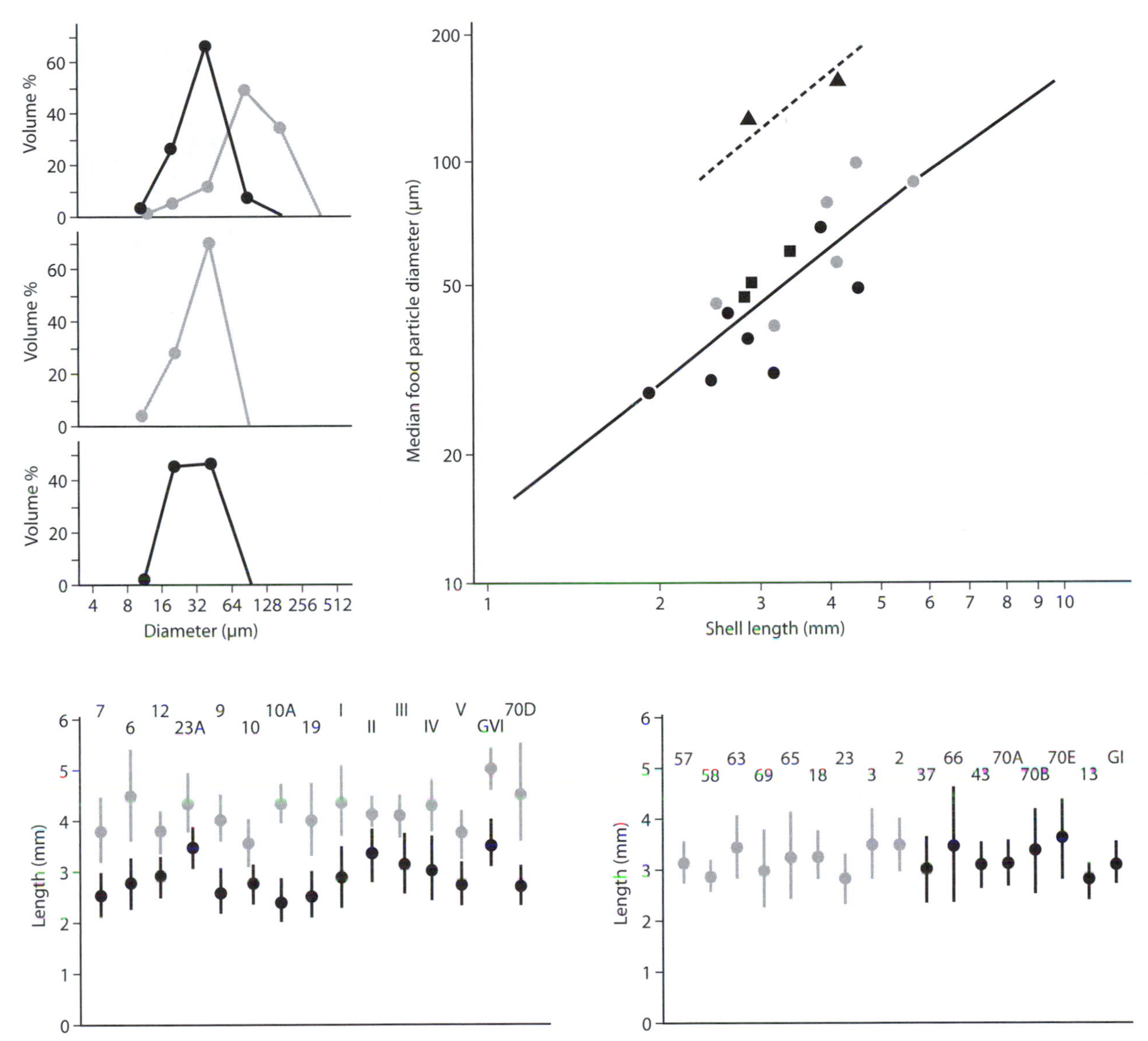

Figure 8. (top left) Resource-utilization niches for prey size among two species of gastropods (*Hydrobia ulvae,* gray circles; *H. ventrosa,* black circles) where the species overlap (top), where *H. ulvae* is alone (middle), and where *H. ventrosa* is alone (bottom). (top right) Median diameters of ingested food particles of four species of *Hydrobia* plotted against shell length. (bottom) Average lengths of *Hydrobia ulvae* (gray circles) and *H. ventrosa* (black circles) from 15 localities where the species co-occur (left) and 17 localities where one of the two species occurs alone. All samples from the Limfjord during summer 1974. [Redrawn from Fenchel, T. 1975. Character displacement and coexistence in mud snails (Hydrobiidae). Oecologia 20: 19–32]

had the greatest climatic-niche differences. The second study, however, gave somewhat more support for the hypothesis in general, in that niche similarity between closely related species of birds, butterflies, and mammals separated by the Isthmus of Panama was greater than expected by chance. However, somewhat contrary to the founding ENM study by Peterson, Soberon, and Sachez, niches were rarely identical, so the overall answer is in fact mixed, as is so often the case in ecology.

6. CONCLUSION

The research trends discussed in relation to the three niche concepts are summarized as an evolutionary tree in figure 10. In this diagram, the thicker arrows

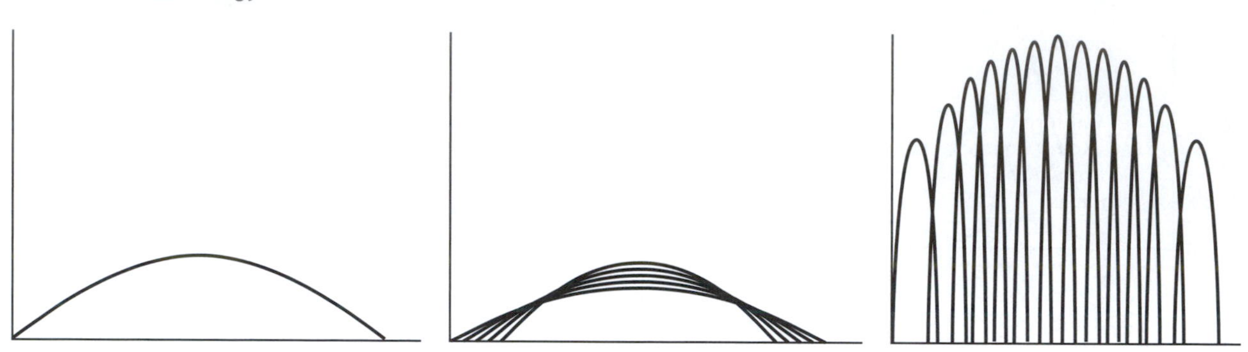

Figure 9. (left) Resource-utilization niche of a species population. (middle) Population of generalist individuals whose niches sum to the curve at top. (right) Population of specialist individuals whose niches sum to the curve at top. (After Klopfer, Peter H. 1962. Behavioral Aspects of Ecology. Princeton, NJ: Princeton University Press)

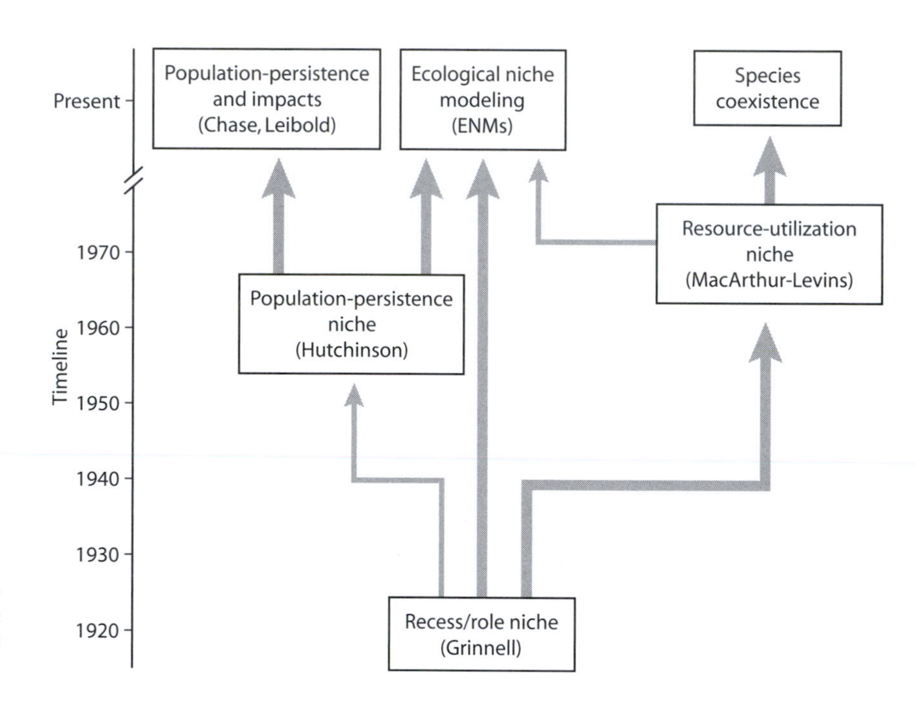

Figure 10. Timeline showing the development of niche concepts and the research programs stemming from them.

indicate a greater influence of one concept or research program on the next. Note that all three niche concepts, despite sometimes rather early beginnings, have stimulated research that is being actively pursued at the present time.

FURTHER READING

Chase, Jonathan M., and Mathew A. Leibold. 2003. Ecological Niches. Chicago and London: University of Chicago Press. *A recent major revision of the population-persistence niche concept.*

Grinnell, Joseph. 1917. The niche-relationships of the California thrasher. Auk 34: 427–433. *One of the founding papers of the recess/role niche concept.*

Harmon, Luke J., Jason J. Kolbe, James M. Cheverud, and Jonathan B. Losos. 2005. Convergence and the multidimensional niche. Evolution 59: 409–421. *A very recent study of convergence and ecological equivalents that emphasizes the niche.*

Hutchinson, G. Evelyn. 1957. Concluding remarks. Cold Spring Harbor Symposia on Quantitative Biology 22: 415–427. *The founding paper of the population-persistence niche concept.*

Knouft, Jason H., Jonathan B. Losos, Richard E. Glor, and Jason J. Kolbe. 2006. Phylogenetic analysis of the evolution of the niche in lizards of the *Anolis sagrei* group. Ecology 87: S29–S38. *A recent exemplar of the environmental-niche-modeling technique.*

MacArthur, Robert H., and Richard Levins. 1967. The limiting similarity, convergence and divergence of coexisting

species. American Naturalist 101: 377–385. *The founding paper of the resource-utilization niche concept.*
Mooney, Harold A., ed. 1977. Convergent Evolution and Chile and California. Stroudsburg, PA: Dowden, Hutchinson and Ross. *Contains major, detailed papers comparing two climatically similar regions.*
Schoener, Thomas W. 1986. Resource partitioning. In J. Kikkawa and D. J. Anderson, eds. Community Ecology—Pattern and Process. Oxford: Blackwell Scientific Publications, 91–126. *A review of how the resource-utilization niche is used in ecological research.*
Schoener, Thomas W. 1989. The ecological niche. In J. M. Cherrett, ed. Ecological Concepts: The Contribution of Ecology to an Understanding of the Natural World. London: Blackwell Scientific Publications, 79–113. *A detailed history of the development of niche concepts.*

I.2

Physiological Ecology: Animals

Martin Wikelski

OUTLINE

1. Guiding concept: Trade-offs
2. Guiding concept: Performance as integrative measure of individual fitness
3. Process I: Acquisition of environmental information
4. Process II: Internal communication and regulation of physiological function
5. Process III: Energy expenditure as one central hub for trade-offs
6. Process IV: Key innovations
7. Process V: Self-defense: Immunoecology
8. Application: Conservation physiology
9. Future challenges

Physiological ecologists study how animals live and function within environments that are constantly changing. Key guiding concepts in physiological ecology are that (1) individual animals are subject to trade-offs such that all (physiological) actions cannot be performed maximally at the same time. Trade-offs underlie the fact that "a jack of all physiological trades is a master of none," which in turn is the basis of the generalist-specialist continuum that brings about much of the niche differentiation in ecology. (2) A second guiding concept is that whole-organism performance provides an integrative measure of individual success in life. Quantifying individual performance allows physiological ecologists to assess the integration of traits within an organism and to determine how natural selection orchestrates not just one but all characteristics of an organism at the same time. Whereas in the past, physiological ecologists have also often studied animals in laboratory situations, technological advances now allow researchers to "go wild" and address individual physiological functions in the very environment where such functions have evolved. The importance of studying animal function in the wild cannot be overestimated because many organismal trade-offs are expressed only when food is scarce or predators are abundant.

GLOSSARY

constraints. These can absolutely limit certain actions of an organism. Even if all efforts in a trade-off scenario are devoted toward a particular action, this action is not sufficient to satisfy an organism's current needs.

energy. In biology, energy, which is essential for life, is gathered from the breaking of chemical bonds during metabolic processes. Energy is often stored by cells in the form of substances such as carbohydrate molecules (including sugars) and lipids, which release energy when reacting with oxygen.

hormones. These substances are chemical messengers that carry information from one part of the organism (e.g., the brain) to another (e.g., the gonads) often via the blood transport system. Hormones bind to receptors on target cells and thus regulate the function of their targets. Various factors influence the effects of a hormone, including its pattern of secretion, transport processes, the response of the receiving tissue, and the speed with which the hormone is degraded.

metabolic rate. Energy expenditure per unit time. Metabolic rate is normally expressed in terms of rate of heat production (kilojoules per time).

performance. This refers to whole-organism performance capabilities (e.g., how fast an organism can sprint) that are determined by physiological traits (e.g., composition of muscle fibers).

trade-offs. These attributes refer to the loss of one quality or aspect of something in return for gaining another quality or aspect.

Physiological ecology occupies a central role in the biological sciences and has a long tradition of integrating other biological disciplines. Physiological systems provide the interface between genomics at the lowest mechanistic level to organismal life history and evolution at the highest level of biological integration.

Every biological process linking genes to behavior will ultimately have to be understood mechanistically on the physiological level to truly provide a picture of how organisms function.

There are many levels at which physiological ecologists attempt to discern how organisms work. On the lowest level, physiological ecology meets genomics and proteomics. For example, Chi-Hing Chris Cheng and Art DeVries from the University of Illinois, working on the antifreeze protein in Antarctic fish, discovered that the protein is coded by a simple but frequent DNA repeat derived from a snippet in a trypsinogen-like protein gene, initially presumably by chance. This protein appeared to have just the right structure to recognize the surface structure of ice crystals that enter into the blood of the fish. Working up the physiological levels, because ice that enters into the fish's circulation always end up in the spleen, Cheng and DeVries hypothesized that the immune system, perhaps macrophages, of these fish living at subfreezing temperatures would take care of the nascent ice crystals encapsulated or presented by the antifreeze protein. Perhaps not unlike a pathogen, the immune system then either "kills" or lyses or excretes the nasty foreign body—a spiny ice crystal that would otherwise serve as a crystallization hotbed for more ice. What followed showed the true heuristic power of the physiological ecology approach. When Cheng and DeVries compared different antifreeze proteins among unrelated species of Antarctic and Arctic fish, they found that all of them use the same mechanism to deal with nascent ice in their blood and body fluids. It turned out that most fish can survive within the subfreezing, icy polar waters only if they have enough "antifreeze" in their circulation. Thus, Cheng and DeVries were able to integrate from a simple physiological innovation to explain a major ecological question: why there exist almost exclusively notothenioid, antifreeze fish around the Antarctic continent. Moreover, Cheng recently discovered that an unrelated innovation provides Arctic cod fishes with a near-identical antifreeze protein as the Antarctic notothenoids to brave the cold in the North.

However, organismal innovations rarely if ever come without a cost. It is not entirely clear what the cost is for Antarctic fish to have antifreeze protein, but we may soon find out if the Antarctic ocean circulation changes with global warming and the waters around the icy continent warm up. Such conditions could allow other, "nonantifreeze" fish to invade and challenge the old survivors, perhaps by bringing pathogens into a system that is not optimized to deal with anything else invading cells but ice crystals. If so, we may yet again see how physiological trade-offs govern ecological processes.

1. GUIDING CONCEPT: TRADE-OFFS

Physiological trade-offs are truly ubiquitous in nature. Everybody can immediately and intuitively understand them. If an organism puts too much energy into detoxifying ice crystals, other functions—perhaps predator defense, pathogen killing, or sperm maturation—lag behind (in fact, many notothenioid fish species are infested, often heavily, by parasites). Ecologists have discovered many pervasive life history trade-offs whose physiological underpinnings are currently under intensive investigation. For example, the more an animal reproduces, the more likely it is to lead a shorter life. The faster an animal grows, the more resources it needs, and again the more likely it is to lead a shorter life. However, there are circumstances when such trade-offs are not observed. In one, animals come in different qualities, with high-quality individuals within a species sometimes "living harder and dying older" than low-quality individuals. Such exceptions to the trade-off rule present considerable challenges and research opportunities for physiological ecologists. What mechanism(s) allow—at least in the short run of one or several generations—one individual to be more likely to survive or to live longer than others? Another challenge to the trade-off rule is presented by laboratory, domestication, and generally captive conditions. Under such circumstances, animals often appear to escape trade-offs. Again, it is yet unclear how animals can become "masters of all trades." The most likely physiological scenario is that the abundance of energy and nutrients provided in captivity allows individuals to obtain everything they need and thus to override physiological trade-offs. If confirmed and analyzed on the mechanistic level, this important distinction between feast and famine in the wild and almost pure feast in the laboratory could shed significant light onto one of the most pervasive principles in physiological ecology.

The question of how trade-offs come about immediately leads us to question how a multitude of organismal functions can be integrated and optimized. Physiological ecologists have found a simple, perhaps ingenious, way to ascertain how individual animals can deal with their environment.

2. GUIDING CONCEPT: PERFORMANCE AS INTEGRATIVE MEASURE OF INDIVIDUAL FITNESS

Instead of analyzing each physiological trait on its own in isolation, physiological ecologists resort to quantifying whole-organism function. Imagine the different ways in which one could answer whether lizard muscle fibers work well at low or at high temperatures. A valid reductionist approach could be to isolate each muscle fiber type, cultivate them all in vitro, expose the fibers to different temperatures, stimulate them electrically, and measure their energy expenditure and contraction rate and speed.

However, what matters for individual animals is how they use their entire complement of muscles to perform certain common tasks such as fast running. Maximum running speed may be related to male fighting ability, female nest-digging ability, insect-catching capacity, and agility to escape predators. Thus, all individuals in a lizard population are expected to rely on fast sprint speed. Ray Huey of the University of Washington made use of this experimental paradigm and showed in comparative studies of individual whole-body performance that most ectotherms are able to cope with a large range of low environmental temperatures. However, as individual performance reaches its maximum, it rapidly drops off toward even higher ambient temperatures. The physiological basis for this performance asymmetry is presumably found in temperature sensitivities of physiological or molecular processes.

Interestingly, individual performance is also subject to strong trade-offs. For example, although some species of ectotherms have large temperature ranges under which they can perform well, others have very narrow performance breadths (see below).

3. PROCESS I: ACQUISITION OF ENVIRONMENTAL INFORMATION

One of the most survival-relevant tasks of animals is to gather environmental information. Again, this task is subject to physiological trade-offs. Physiological ecologists working on bat echolocation determined that producing the ultrasound that bounces back from objects, i.e., provides bats with environmental information, is costly both in immediate energetic costs and in associated physiological costs. In addition to energetic costs, bats face the costs of producing organs and brain structures that enable them to expend energy on echolocation calls in the first place. Biologists actually exploit the fact that environmental information gathering is expensive. Bats spare the costs of echolocation when flying in known habitat and often do not echolocate there, allowing researchers to trap them with fine nylon nets.

High physiological costs of maintaining functioning tissue may also explain why juvenile migratory songbirds start out with a small hippocampus, a brain area involved in spatial memory and thus long-term information gathering. As individuals conduct their first transcontinental journeys, they add additional cells and connect their cells in more complex ways. However, because space in the brain capsule is presumably limited, the physically and physiologically expanded spatial memory for a life on the move may again be traded off against other brain functions that in turn deteriorate.

Energetic trade-offs between form and physiological function are particularly prominent in long-distance migratory songbirds that had to evolve streamlined foreheads for aerodynamic reasons, compared with their short-distance migrating relatives. Physiological ecologist Melissa Bowlin recently learned by studying heart rate in naturally migrating New World thrushes (songbirds) that even small morphological differences significantly affect costs of transport in the air.

4. PROCESS II: INTERNAL COMMUNICATION AND REGULATION OF PHYSIOLOGICAL FUNCTION

Once environmental information is gathered, it needs to be communicated most efficiently throughout the organism. Again it appears that cost minimization and trade-offs are key guiding physiological principles. Quick, practically immediate transfer of environmental information is achieved by costly electrical (neuronal) connections. However, for many types of information that either need to be communicated continuously or at least on the long term, electrical connections are by far too costly. Instead, animals use small and "cheap to produce" messenger chemicals (hormones) that bind to receptors in target tissues. The main advantage of a hormonal communication system is that it is inherently flexible at many levels, i.e., rates of physiological processes can be altered at production, at the chemically supported transport of hormones to target tissues, at the possible breakdown of messenger chemicals, and with respect to the number of receptors expressed at and by target tissues. Thus, for example, if a cell does not need (much) stimulation, it can degrade particular types of incoming hormone molecules (indicating particular, general environmental messages) in its periphery and/or provide only very few receptor sites as "mailboxes." Cells can also destroy the "mail" immediately so that it has no long-lasting effect.

Physiological ecologist John Wingfield showed that this cheap hormonal messenger system conveys both long-term and short-term environmental information and prepares the individual organism for certain activities. Many animals reproduce seasonally and grow reproductive organs in response to changes in day length, often mediated by the light-sensitive hormone melatonin. Because of physiological trade-offs, individuals do not allocate maximum efforts toward certain reproductive activities such as territorial defense from the outset. Instead, organisms often use behavior–physiology feedback loops to allow them to carefully regulate their efforts in response to environmental factors, in this case the actions of other members of the population. Thus, if population density is high in songbirds, individuals interact with others of the same species more often. Wingfield showed that individuals can ramp up reproductive hormones such as testosterone in response to a social, particularly reproductive, challenge. It is yet unclear whether this feedback is via increased physical activity (i.e., energy expenditure), increased neuronal stimulation (e.g., visual density), or a combination. In any case, it is clear that animals use hormones as a cheap means to communicate environmental information throughout the body.

5. PROCESS III: ENERGY EXPENDITURE AS ONE CENTRAL HUB FOR TRADE-OFFS

All along it has become obvious that organismal trade-offs can be expressed to a significant extent in terms of allocations in energy turnover. Energy is probably one of the physiological factors that are most limited under natural circumstances. It is thus not astonishing that physiological ecologists cast many of their discussions in energetic terms and consider energy as the central hub for physiological trade-offs.

Life follows the laws of thermodynamics, i.e., energy can neither be created nor destroyed (First Law). Furthermore, the disorder of a system (its entropy) increases over time as its energy content degrades to unusable heat. The only way animals can compensate for ever-increasing entropy is by constantly acquiring energy via food. However, foraging is again costly as well as time consuming, i.e., poses opportunity costs and is risky. The food then has to be broken down into chemicals usable by the organism, again a costly, damaging, time-consuming process.

Because animals will do anything to minimize costs, it should be obvious that environmental temperature is one of the most important habitat factors. Temperature has a hump-shaped influence on molecular processes such as enzyme activity. Coming from the low side, increasing temperatures enhance the rate of physiological processes and thus energy expenditure. Higher-than-optimum temperatures often show destructive effects and can result in serious structural damage.

Organisms incur costs at low environmental temperatures either because they are less agile (many ectotherms) or because they have to produce more internal heat (endotherms). Some animals have special tissues that help them produce heat very efficiently, such as brown fat in bats, which produces heat without shivering. Higher-than-optimum temperatures often become dangerous because organisms very rapidly lose performance and expend much energy in thermoregulatory activities, both behaviorally and physiologically (panting, activation of heat shock proteins).

Although most animals attempt to minimize energy expenditure for nonessential tasks, it has become clear that, across various types of animals, high energy expenditure has evolutionary benefits. Increased energy expenditure involving constantly high body temperatures with an associated constant interior milieu has perhaps been one of the key innovations in physiology.

6. PROCESS IV: KEY INNOVATIONS

Evolutionary key innovations give organisms access to new resources and cause rapid, sometimes spectacular adaptive radiation, as seen above in the case of antifreeze proteins. It has been postulated that a long sequence of key physiological innovations is responsible for the diversity of life forms present today.

For example, Michael Berenbrink and colleagues discovered that a key physiological innovation underlies the large adaptive radiation of fish. It is the unique ability of fish to secrete molecular oxygen into the swimbladder—a seemingly simple physiological process that had already been invented some 100 million years earlier in the eye. However, because certain fish were later able to regulate swimming behavior very cheaply using their new oxygen-filled swimbladders, they diversified hugely in form and function. The physiological key in this process was a change in the Na^+/H^+ exchange activity of red blood cells and a change in the content of surface histidine of hemoglobin (histidine is one of the 20 most common amino acids).

Another common key innovation—and again a highly efficient way of organisms to economize on physiological expenses—is to use special chemical components of other organisms. May Berenbaum demonstrated such a system very nicely in the interaction between the parsnip webworm and wild parsnips. Throughout the parsnip plant there exist a group of toxic chemicals called furanocoumarins that are the favorite food of the parsnip webworm. Furanocoumarins are so toxic that only very few herbivores can deal with them. However, webworms possess a highly efficient detoxification system involving cytochrome P450s, a very large and diverse superfamily of hemoproteins (iron-containing proteins) that simply insert one atom of oxygen into an organic substrate. Webworms use the toxic furanocoumarins to strongly deter predators from eating them.

Although animals can engage the help of others, perhaps through their chemicals, to defend themselves against predators, there are also more direct ways to fight pathogens and parasites.

7. PROCESS V: SELF-DEFENSE: IMMUNOECOLOGY

The study of the physiological ecology of immune reactions is a relatively new but fast growing and highly important field. In the past, immune biology has largely focused on very specific, fine-scale mechanisms of the immune defense. Immunoecology adds the systemic component to such detailed studies by addressing the integration of various immune responses on the individual level. In a key contribution, Kelly Lee and Kirk Klasing showed that the relative immune defense effort spent on either the innate or the adaptive arm of the immune system may be ecologically important. For example, such a differential allocation of efforts into different arms of the immune system may distinguish highly from poorly invasive species, such as the house sparrow and the tree sparrow, respectively.

Along the idea of whole-organism performance tests (see above), immunoecologists assess the reaction of individuals toward various immunological challenges simultaneously and as a composite measure. Physiological immune responses can be mediated by essentially two arms, the innate and the adaptive part of immune systems. The first line of defense is usually the innate arm. Specialized cells patrol tissues and have a superb ability to recognize an invader as foreign. As soon as the foreigner-recognition process starts, the first innate cells release signal molecules (cytokines) attracting bacteria- and virus-eating cells (scavenger macrophages, natural killer cells). Subsequently, the cells of the innate immune system send specialized signal molecules to the second (adaptive) arm of the immune system. The adaptive part of the immune system activates its machinery to produce antibodies that bind to and neutralize the foreign invaders. Whereas the innate system is costly to maintain and to activate, the adaptive system is costly to grow in the first place—once it is established, it appears fairly cheap to maintain.

It is important to note that organisms differ strongly in how much emphasis they put on the two arms of the system. Again it appears that because of omnipresent trade-offs, a jack of all immunological traits is a master of none. It is important to note in this context that some biomedical experimental subjects such as the house mouse do not necessarily provide systems that reflect the immune allocation in humans. Whereas humans are long-lived and invest heavily in the adaptive arm of the immune response (costly to develop but cheap to run), house mice are generally so short-lived and dependent on fighting each disease immediately that they invest much more strongly in the innate arm of the immune response. It will remain a challenge in physiological ecology to understand exactly how organisms allocate resources toward immune responses.

8. APPLICATION: CONSERVATION PHYSIOLOGY

Animals have always been sentinels for environmental changes and catastrophes. For example, when the causal (reproductive) effects of dichloro-diphenyl-trichloroethane (DDT) on top predators became clear, DDT-like substances were prohibited in large parts of the world. For conservation strategies to be successful, it is important to understand the physiological responses of organisms to their changed environment. Perhaps one of the most useful tools in conservation physiology is the rapid assessment of environmental stress via the measurement of glucocorticoid "stress" hormones. These steroid hormones are ubiquitous in vertebrates and occur at low (baseline) levels in all individuals. In many cases when individuals are experiencing increased environmental demands such as inclement weather or predation, glucocorticoids increase in the circulation and, subsequently, in the feces. Conservation physiologists often experimentally induce mild stress (capture and handling) to assess the capacity of an individual to react to environmental stress. The usefulness of conservation physiology is that it can reduce the complexity of conservation problems to highlighting a single set or small number of the most important stressors for organisms. New physiological techniques can enable a rapid assessment of the causes of conservation problems and the consequences of conservation actions.

9. FUTURE CHALLENGES

The biggest challenges in the future of physiological ecology will be to monitor, understand, and ultimately predict what animals do during their often long lives. Advanced biologging techniques of physiological parameters are at the brink of enabling field researchers to conduct studies that a few years ago were possible only in a laboratory situation. Furthermore, even small animals can perhaps soon be followed over large temporal and spatial scales in the wild. Such new data on physiological state and overall individual space use may ultimately allow researchers to understand the animal mind. Once we know in (almost) real time how individuals process environmental information (via hormonal mechanisms), and we know the environmental conditions in the vicinity of an individual (via animal-borne location loggers) in combination with the individual's physiological state, we may be able to predict decisions of animals mechanistically.

FURTHER READING

Cheng, Chris C., and Luise Chen. 1999. Evolution of an antifreeze glycoprotein. Nature 401: 443–444.

Janzen, Daniel H. 1967. Why mountain passes are higher in the tropics. American Naturalist 101: 233–249.

Lee, Kelly A., and Kirk C. Klasing. 2004. A role for immunology in invasion biology. Trends in Ecology and Evolution 19: 523–529.

Martin, Lynn B., Zac M. Weil, and Randy J. Nelson. 2007. Immune defense and reproductive pace of life in Peromyscus mice. Ecology 88: 2516–2528.

Wikelski, Martin, and Stephen Cooke. 2006. Conservation physiology. Trends in Ecology and Evolution 21: 38–46.

Wingfield, John C. 2003. Control of behavioural strategies for capricious environments. Animal Behaviour 66: 807–815.

I.3

Physiological Ecology: Plants

David D. Ackerly and Stephanie A. Stuart

OUTLINE

1. Introduction
2. Resource acquisition
3. Resource allocation and growth
4. Responses to environmental conditions
5. Ecophysiology, distributions, and global climate change

Plant physiological ecology addresses the physiological interactions of plants with the abiotic and biotic environment and the consequences for plant growth, distributions, and responses to changing conditions. Plants have three unique features that influence their physiological ecology: they are autotrophs (obtaining energy from the sun), they are sessile and unable to move, and they are modular, exhibiting indeterminate growth. Plant growth depends on acquisition of four critical resources: light, CO_2, mineral nutrients, and water. Light together with nitrogen-rich enzymes in the leaf drive photosynthetic assimilation of CO_2 into carbohydrates. Uptake of nitrogen and phosphorus, the elements most often limiting growth, is facilitated by symbiotic associations on plant roots with bacteria and fungi, respectively. Most water acquired by plants is lost in transpiration in exchange for CO_2 uptake through stomata. Water moves through a plant by cohesion-tension, drawn upward as a result of evaporation from leaves. Excessive tension can lead to embolism, in which air bubbles enter the water column and block water transport. Within the plant, allocation of resources to alternative functions creates important trade-offs that critically influence plant responses and performance in contrasting environments. Physiological ecology plays a critical role in understanding the distributions of individual species and of major biomes at a global scale and is vital to understand the potential impacts of global climate change on vegetation and biodiversity.

GLOSSARY

acquisition. The processes of acquiring resources from the environment, such as photosynthesis in leaves and nutrient uptake by roots.

allocation. The partitioning of resources among alternative structures or functions within a plant. The principle of allocation states that resources used for one purpose will be unavailable for other purposes, creating trade-offs that strongly influence plant growth and life cycles.

conditions. Factors of the environment that influence an organism but cannot be consumed or competed for (e.g., temperature, pH).

embolism (or cavitation). The blockage of water transport by air bubbles in the xylem (water-transporting cells), causing reduced water transport and, potentially, plant death.

leaf energy balance. The balance of energy inputs and outputs that influence leaf temperature. Solar radiation is the most important input, and transpirational cooling and convective heat loss are the most important outputs.

photosynthetic pathway. Plants exhibit three alternative photosynthetic pathways (C3, C4, and CAM) that differ in underlying biochemical and physiological mechanisms, resulting in contrasting performance depending on temperature and the availability of light, water, and nutrients.

resources. Aspects of the environment that are consumed during growth and that plants compete for. The most important are light, water, nutrients, and space.

water and nutrient use efficiency. The efficiency of photosynthesis relative to investment of water or nutrients, respectively.

1. INTRODUCTION

Physiological ecology examines how plants acquire and utilize resources, tolerate and adapt to abiotic conditions, and respond to changes in their environment. The study of physiological ecology considers plant physiology in relation to the physics and chemistry of the abiotic world on one hand, and a broad ecological and evolutionary context on the other. Plant physiological ecology provides the basic sciences with essential information about plant evolution, biodiversity, ecosystem productivity, and carbon and nutrient cycling. It also plays an instrumental role in a wide range of applied sciences, including agriculture, forestry, management of invasive species, restoration ecology, and global change biology.

In its early years, plant ecophysiology addressed two broad themes. One was the effort beginning in the mid-nineteenth century to understand the global distribution of major biomes and vegetation types, led by pioneering plant geographers such as A. von Humboldt, A.F.W. Schimper, and their followers. These workers recognized that similar vegetation types arise under similar climates in different parts of the world, and they developed basic principles of plant form and function that could explain these global patterns. This led to a subsequent emphasis, in the early twentieth century, on the question of how plants survive in extreme environments. Principles of physiology and biophysics were applied in natural settings to understand how plants can tolerate and even thrive from the heat of the desert to the extreme cold of the high arctic and the upper limits of vegetation on high mountains. Both of these traditions combined the mechanistic view of the physiologist with the idea of evolutionary adaptation to understand why species with different physiological characteristics dominate under contrasting environmental conditions.

In the United States, plant physiological ecology played a key role in the development of ecology as a discipline. *The Plant World*, published until 1919, was the forerunner of *Ecology*, the flagship journal of the Ecological Society of America. Ecology in the early twentieth century emphasized physiological, functional, and ecosystem ecology. Population and community ecology as we now know them had not yet emerged.

Three Important Things about Being a Plant

Plants share three important features that have profound consequences for their ecology and evolution, including physiological ecology. (1) Plants are autotrophs, converting sunlight to stored chemical energy that is the basis for terrestrial food webs and ecosystems. (Nonphotosynthetic and parasitic plants are an exception to this rule.) Photosynthesis is one of the outstanding products of evolution and is still more efficient than any photovoltaic mechanism for the capture and conversion of solar energy. (2) Plants are sessile—once a seed germinates and the seedling is established in the soil, plants cannot move. They cannot hide or escape from abiotic conditions or biotic enemies, and they cannot seek out mates, at least directly, for reproduction. Plants exhibit an enormous diversity of seed germination mechanisms that control the time and place of germination, thus shaping the environment the seedling and adult plant subsequently occupy. (3) Plants exhibit modular, indeterminate growth. They grow by cell division in regions known as *meristems*, located at the tips of growing branches, in axils at the base of leaves, beneath the bark of trees, and at the tips of roots. Meristematic cells are undifferentiated throughout the life of a plant, and most plants never reach a fixed, mature size. The combination of indeterminate growth and immobility means that growth and development are important mechanisms through which plants respond to the environment, and in this way growth in plants plays an analogous role to behavior in animals.

Conditions and Resources

At the core of physiological ecology is the study of how organisms respond to and are affected by the abiotic environment. In the case of plants, it is useful to divide the environment into conditions and resources. Conditions are factors that cannot be consumed or depleted by organisms, such as temperature, pH, or salinity. Resources are substances (or sources of energy) that are captured or consumed, can be depleted, and can be the focus of competitive interactions among individuals. The following sections address the acquisition and allocation of resources and the mechanisms by which plants respond to and tolerate a wide range of environmental conditions.

2. RESOURCE ACQUISITION

All plants require the same basic resources for growth and reproduction: carbon, light, mineral nutrients, and water. The essential challenge for terrestrial plants is that these resources are located in different places (above versus below ground) and have very different modes and rates of supply in the environment.

Carbon and Light

Carbon, in the form of atmospheric carbon dioxide, is available at a relatively constant concentration. CO_2 enters leaves through microscopic pores known as *stomata*. Stomata are formed by pairs of cells, known as guard cells, which are joined at either end, like two elongated balloons. When fluids move into the cells, they swell and bend, opening a small pore that allows gases to diffuse in and out of the leaf. The regulation of pressure within these cells is an intricate process influenced by chemical signals from the leaves and roots that depend on soil moisture availability and internal water status of the plant.

When stomata open, CO_2 diffuses from the atmosphere into the pores, where it crosses an air–liquid interface and dissolves in the interior fluids of the leaf as carbonic acid. At the same time, however, water evaporates from inside the leaves and diffuses through the stomata into the surrounding atmosphere. This exchange of water for CO_2 is one of the most fundamental trade-offs governing photosynthesis and plant growth. The concentration gradient driving the diffusion of water out is much steeper than the gradient for CO_2 coming in. As a consequence, plants lose 100 to 500 molecules of water for each molecule of CO_2 they absorb. Most of the water taken up by plants (see below) is used for this purpose. The ratio of water loss to CO_2 uptake is known as water use efficiency and represents a critical physiological trait that influences plant growth and distribution in contrasting climates.

Photosynthesis is a biochemical reaction that uses energy from sunlight to combine CO_2 and water to make carbohydrates (glucose, starch, and other sugars), releasing oxygen in the process. Photosynthesis involves two coupled processes, known as the light reactions and carbon reactions. The light reactions use solar energy to reduce $NADP^+$ to NADPH and phosphorylate ATP. These provide energy for the carbon reactions of the Calvin cycle. The most important of these reactions is the fixation of a CO_2 molecule to a five-carbon carbohydrate chain, followed by a rapid split into two three-carbon molecules, which gives this process the name C3 photosynthesis. The CO_2 fixation step is regulated by the enzyme RUBISCO, which, because of its relatively low efficiency, is present in very high levels in the leaf. RUBISCO represents up to 50% of all proteins in a leaf and is thought to be the most abundant protein on the planet.

The C3 pathway is dominant in plants of temperate and cool climates as well as in most trees. In deserts, grasslands, and other dry environments, two alternative photosynthetic pathways are found, each of which has evolved many times independently. C4 photosynthesis, common in grasses (including crops such as corn), utilizes the enzyme PEP-carboxylase instead of RUBISCO for the initial fixation of CO_2. This is a more efficient alternative, which can operate at lower internal CO_2 concentrations, so the plant can have fewer, smaller, or less open stomata and therefore lose less water. The first fixation step creates four-carbon compounds (hence the name), which are then shuttled to cells deeper inside the leaf where they are broken down, releasing the CO_2 for incorporation into the Calvin cycle. The extra steps of the C4 pathway require additional energy, so C4 plants occur primarily in warm and high-light environments.

The third pathway is known as CAM photosynthesis (Crassulacean acid metabolism), named for the plant family Crassulaceae where it was first discovered. CAM photosynthesis is widespread in cactus, tropical euphorbias, yuccas, and other succulents. CAM also utilizes PEP-carboxylase for the initial fixation step, but the stomata are opened only at night, allowing CO_2 to diffuse into the leaf with minimal water loss because of lower temperatures and higher relative humidity. The carbon is stored in carbon acids (hence the name) until daylight, when they are broken down and passed to the Calvin cycle. Almost all plants described as "succulents" have CAM photosynthesis, and the swollen and fleshy leaves or stems contain the expanded cells that are used to store the four-carbon compounds through the night. The nighttime uptake of carbon results in greatly enhanced water use efficiency because CAM plants lose only as few as 10 water molecules per CO_2 molecule acquired. However, overall photosynthetic rates are very low, limiting growth rates.

Photosynthesis in sun versus shade also presents trade-offs that are important for plants growing in heterogeneous light environments such as the forest understory. Plants with C3 photosynthesis exhibit a characteristic light response of photosynthesis. In complete darkness, photosynthesis is shut down, and leaves have a net loss of CO_2 as a result of background respiratory processes. With slight increases in light, photosynthetic rates increase until the light compensation point is reached, when photosynthesis balances respiration and there is no net loss or gain of CO_2 by the leaf. Net photosynthetic rates become positive above this light level and increase rapidly until they reach a point where the concentration and activity levels of RUBISCO and other enzymes become more limiting than the availability of light energy. At this point, photosynthetic rates reach a plateau known as the light-saturated photosynthetic rate.

In shade, photosynthesis is primarily limited by light energy rather than enzyme levels. As a result, shade leaves have lower nitrogen concentrations per unit leaf area, a lower saturated photosynthetic rate, and lower background respiration. The result is that the light compensation point is lower, and shade plants can maintain zero or positive carbon balance at lower light levels. The differences between sun and shade leaves are generally observed both within and between species.

Competition for light is largely asymmetric or one-sided: the highest leaves in the canopy capture the most light, and leaves lower down, on the same or other plants, receive much less. It has been argued that the evolution of plant height can be understood as an evolutionary game: if all the plants in a community "agreed" to reduce their height equally, they would all still receive the same amount of light. However, the community would be easily invaded by a taller "cheater" that received a disproportionate share of this critical resource. Taller strategies will continue to invade until the costs of additional height (in structural support, movement of water, etc.) outweigh the benefits, and an equilibrium is reached. This equilibrium will vary, depending on the availability of light, water, and nutrients and is thought to explain variation in the height of forests and other vegetation around the world.

Water

Water is central to the life of a plant. In addition to the water lost in exchange for carbon uptake (see above), water is needed for tissue hydration, nutrient uptake, long-distance signaling, and as a source of pressure for structural support and cell expansion.

Water transport begins in the soil, where root hairs provide a large surface area for water uptake. Water moves through and around cells until it reaches the endodermis, a root layer in which the cell walls are impermeable to water because of a waxy inclusion in the cell membrane known as the Casparian strip. To move beyond the Casparian strip, water must pass through living cells. This allows the plant to regulate how much water enters the active root tissue and can be used to generate root pressure. The main water transport tissue inside roots, trunk, branches, and leaves is the xylem, composed of hollow cells that are dead at maturity. Water travels through the xylem to the leaves and evaporates from air/water interfaces within the stomata; evaporation from the leaves is known as transpiration. Movement along this path occurs as water moves from areas of high water concentration (high water potential) to areas of lower water concentration (lower water potential). Under all but the most humid conditions, the concentration of water (water potential) is much lower in the atmosphere than it is within the plant. The difference in concentration drives evaporation into the atmosphere.

Within the xylem, water is connected by cohesion to form a single column between leaf and root. As a result, evaporation at the leaves' surface effectively pulls water out of the soil. Cohesion, which is the result of hydrogen bonding between water molecules, gives the water column the ability to withstand stretching, also known as tension or negative pressure. This scenario, first proposed in 1893, is known as the tension–cohesion theory. The underlying mechanisms have been questioned from time to time, but in general, it is considered to be well supported by empirical evidence.

Like a supersaturated or supercooled solution, water under tension is in a metastable state. As a result, it is vulnerable to disruption and can vacuum boil, spontaneously forming an air bubble or embolism. This process is known as cavitation. Because embolisms break the water column, they block the transport of water from root to leaf. MRI studies show that cavitation is constantly occurring, and being repaired, during transpiration. Stresses, such as drought and freezing, can cause more damaging levels of embolism. Xylem architecture is highly redundant, and plants can survive, and sometimes repair, many of these losses. However, if a sufficient proportion of the xylem is blocked, embolism can cause the death of distal branches and leaves. When a plant dies in a drought, embolism is likely the proximal cause of death.

Nutrients

Mineral nutrients, like water, are primarily acquired below ground. Roots are the essential foraging organs for below-ground resources. Many roots also sustain colonies of mycorrhizal fungi, which are important for the uptake of nutrients. The two soil nutrients that are most often limiting to plant growth are nitrogen, in the form of nitrate or ammonium, and phosphorus, usually taken up as phosphate. Other macronutrients required in relatively large quantities are potassium, calcium, magnesium, and sulfur; micronutrients, required in much smaller quantities, include chlorine, iron, manganese, boron, zinc, copper, nickel, and molybdenum and (in some plants) sodium, cobalt, and silicon.

Nitrogen is one of the most abundant elements in the biosphere but one that is often available in limited supplies. The primary source of nitrogen is dinitrogen gas from the atmosphere; however, plants are unable to assimilate this nitrogen directly. Instead, atmospheric nitrogen is assimilated by nitrogen-fixing bacteria; many of these bacteria enter into symbiotic relationships with plants and occupy nodules on plant roots. N-fixing symbioses are widespread among thousands of species in the legume family (Fabaceae) as well as at least nine other related groups. Once nitrogen has entered an ecosystem, decomposition of litter and mineralization of organic nitrogen by soil microbes makes nitrogen available for uptake by plants. *Photosynthetic nitrogen use efficiency* (PNUE) is the ratio of photosynthetic productivity to the concentration of nitrogen within the leaves (or the whole plant). On a short-term basis, PNUE is higher in faster-growing plants with short leaf lifespan, high tissue nitrogen concentrations, and high photosynthetic rates. However, over the lifespan of the leaf, PNUE tends to be higher in slower-growing plants with lower instantaneous rates of photosynthesis but long-lived leaves.

Phosphorus is primarily derived from weathering and soil formation and then is cycled within an ecosystem in parallel with nitrogen. Phosphorus is relatively immobile and diffuses very slowly in soil water. Symbiotic relationships between plants and mycorrhizal fungi play a critical role in phosphorus uptake, as the hyphae of the mycorrhizas greatly extend the foraging area of the root system. As in the N-fixing symbioses, plants provide carbohydrate as an energy source for the fungi. Plants may also leach organic acids into soil, and the reduction in soil pH increases the availability of phosphorus as cations are exchanged on clay particles and other surfaces. Recent studies have shown that 10% to 30% of net carbon gained by a plant may be lost into the soil either as leachates or carbohydrate supplied to symbiotic fungi or microorganisms.

3. RESOURCE ALLOCATION AND GROWTH

Resource acquisition is only a part of the story. To understand how plants grow, respond to the environment, and differ from each other, we must examine the allocation of resources. Allocation refers to the partitioning of acquired resources among different structures and functions within the plant. The principle of allocation underlies many of the fundamental trade-offs involved in plant growth: energy or materials can be allocated to only one structure or function at a time, so investment in one process will invariably entail trade-offs in others. Carbohydrates synthesized in the leaf by photosynthesis are loaded into the phloem and can be transferred to other parts of the canopy, to the branches and trunk, to flowers and fruits, and down into the roots. Nutrients, taken up in the roots, move upward in the xylem sap and are also divided among different parts of the canopy and utilized in the production of new leaves, stems, and roots; the provisioning of seeds; and the synthesis of enzymes and proteins throughout the plant.

Construction and maintenance of plant tissues require a significant amount of energy as well. For each gram of biomass used to construct new leaf tissue, approximately 0.5 g of carbohydrate will be required to provide the energy for biosynthesis. Biochemical reactions and maintenance of enzyme pools also require a continual input of energy. Leaves in particular, where the photosynthetic machinery are at work, will burn off 5–10% of their photosynthetic uptake as maintenance respiration.

Patterns of allocation are critically important for plant growth. In particular, allocation to leaves creates a positive feedback, as leaves can then capture more carbon, which can be used for additional growth, etc. Investment in leaves is like investment in a savings account, with the benefits of compound interest over time. For example, wild radish plants invest approximately twice as much of their carbon gain in new leaves, compared to the domestic radish, which has been selected for high allocation to the tuber. Although the leaves of the two types have identical photosynthetic rates, the wild type grows to three times larger than its domestic relative over the course of a season. On the other hand, there must be limits to the benefits of investing in leaves. A plant with too few roots would have lower growth rates as a result of insufficient water or nutrients or might simply fall over if it were not well rooted in the soil. A plant with too little above-ground structures (stems and petioles) would have a canopy packed with leaves, all shading each other, with very inefficient light capture. The principle of optimal allocation captures these trade-offs, as it is clear that there is an intermediate optimum in the allocation of resources to leaves at which the growth rate of the plant is maximized. In practice, it can be difficult to determine whether a plant is actually at its optimum, but the idea plays a central role as a guiding principle of ecology.

Over the entire life cycle, natural selection favors those genotypes that maximize their fitness, usually thought of in terms of lifetime reproductive output (see chapter I.14). In short-lived plants such as annuals, allocation must shift rapidly from vegetative growth to the production of flowers and the maturation of seeds and fruits. At the end of the life cycle, 50% or more of the biomass in many annual plants is allocated to reproductive structures. In longer-lived plants, a key component of lifetime fitness is survival through periods of adversity, including cold and drought, or recovery from disturbance or herbivory. Regrowth after disturbance relies on the mobilization of reserve energy in the form of nonstructural carbohydrates that can be stored in stems and roots. For example, shrubs of Mediterranean-type climate regions that are adapted to regrowth after fire have increased allocation to these energy reserves, with ensuing trade-offs in growth rate and annual reproduction.

4. RESPONSES TO ENVIRONMENTAL CONDITIONS

Because they are sessile and exothermic, plants must tolerate a wide range of conditions. They cannot seek shelter or migrate to a more hospitable habitat except through reproduction. As a result, almost anything that moves a plant away from its optimal conditions may be considered "stressful," including extremes of light levels (too much or too little), temperature (both hot and cold), water supply (drought or flooding), and soil composition. However, it is important to remember that what is "stressful" depends on what conditions the genotype has adapted to. For instance, salty conditions that would kill most crop or house plants may be those under which a salt-marsh or mangrove plant grows best and reproduces most efficiently. Nonetheless, it appears that plants that tolerate generally stressful habitats (the very cold, hot, dry, wet, salty, toxic, or nutrient-poor) do so through conservative, slow-growth strategies.

The study of leaf energy balance demonstrates the insights gained by combining ecophysiology with first principles of biophysics. The temperature of a leaf, like that of any other object, will reach an equilibrium when energy inputs and outputs are balanced. The primary input for plants is solar radiation, although the amount of radiation absorbed by a leaf may be reduced by reflective coverings such as hairs or a steeply angled leaf surface. The most important energy output is the heat loss that accompanies the evaporation of water lost in transpiration. As the temperature of a leaf increases, heat loss also goes up as a result of the steeper temperature gradient between the leaf and its surroundings, and this will eventually bring a leaf to equilibrium. Depending on radiation, wind, humidity, and other conditions, leaf temperatures may range from 5°C below to 15°C, or more, above ambient air temperatures. The size of a leaf has an important effect on leaf energy balance, as smaller leaves interrupt air circulation less, and are thus more closely coupled to air temperatures and less prone to overheat.

Both low- and high-temperature stresses are important determinants of photosynthesis, growth and distribution, and, by extension, of vegetation type and community composition. There is no one optimal temperature range for all plants; instead, optimal temperature ranges vary. In addition, the lowest or highest temperatures that a plant can tolerate can frequently be expanded by prior exposure to sublethal cold or hot temperatures, a process known as acclimation. Adaptations to heat stress include decreases in leaf size, increases in leaf reflectance, the production of molecular chaperones that stabilize proteins and membranes, and a shift toward saturated lipids in cell membranes. Adaptations to cold stress include narrow vessel diameters, the production of molecular chaperones that stabilize proteins and membranes, and a shift toward unsaturated lipids in cell membranes.

Drought stress occurs when the water potential of the soil drops below that of the plant and the atmosphere, and the plant cannot isolate itself from the soil or draw enough water to facilitate carbon gain. Flooding stress occurs when roots are deprived of oxygen and can no longer perform necessary functions such as water and nutrient uptake. The range of drought or flooding that can be tolerated varies widely across both clades and habitats. The composition of soil can also have important effects on the uptake of water and nutrients. Soil pH, salt concentration, and heavy metal concentration can all limit the uptake of water and nutrients and inhibit root growth. Some plants are adapted specifically to these stresses and thrive in alkaline, salty, or contaminated soils. Phytoremediation, the effort to remove soil contaminants through specially adapted vegetation, relies on such plants.

5. ECOPHYSIOLOGY, DISTRIBUTIONS, AND GLOBAL CLIMATE CHANGE

The science of ecology is often defined as the study of distribution and abundance of organisms. Ecophysiology clearly plays a central role in these broad questions, particularly in explaining distributional limits of species along environmental gradients. Species distributions often reflect intrinsic tolerance limits related to physiological traits. One particularly well-studied case is the distribution of the saguaro cactus in the southwestern United States, where the northern limits of the geographic range closely parallel the −7°C winter isotherm. More generally, the traits of species tend to change as one moves across environmental gradients because distribution patterns reflect the adaptations of plants to contrasting environments. A well-studied example is the relationship between the resistance to xylem embolism and water deficit and distributions, where less-resistant species either live closer to water sources or have deeper roots to maintain access to water through dry periods. However, species may employ very different mechanisms to survive in any particular environment, so there is no simple one-to-one relationship between any particular physiological trait and the environmental conditions where species live.

Understanding the physiological basis of species distributions is more important than ever in relation to global climate change. Paleoecological data demonstrate that plant distributions can track changing climate over centuries and millennia. Since the last glacial maximum, 10–20 thousand years ago, tree species in eastern North America and Europe have expanded their northern range limits by 1000 km or more. Rising CO_2 levels from burning of fossil fuels, deforestation, and other factors are expected to cause sharp increases in temperature in the next century, coupled with spatially and temporally variable shifts in precipitation. These changes are occurring much more rapidly than postglacial climate changes, raising significant concerns about whether plants and animals will be able to track favorable climates. Mechanistic models that incorporate physiological tolerances, as well as biotic interactions and dispersal capacity, are critical to improve these forecasts, especially for invasive species that may not occupy the full extent of their potential range in many parts of the world.

Physiological information has also been used to model the distribution of the world's major biomes. Vegetation modeling uses the idea of plant functional types, an idealized representation of a small number of physiological strategies. Carbon gain and growth of these life forms are simulated under mean climate characteristics of large grid cells that span the globe; the mix of types that prevail is then used to infer typical vegetation types, such as temperate deciduous forest, evergreen tropical forest, etc. These models have been calibrated with great success and are able to predict the broad patterns of global vegetation.

Within a region, vegetation type can be a critical determinant of energy, water, and nutrient cycles. Recent work suggests that understanding these cycles at an organismal level may be critical to understanding fluxes and cycles at the scale of landscapes, regions, or ecosystems. For instance, grasslands process, consume, and convert resources in different ways, and at different rates, than forests. This occurs, in part, because of the many physiological differences between grasses and trees. Combining information about the physiology and behavior of plants with an understanding of ecosystem-scale patterns and processes provides essential data for models of global climate, biogeochemistry, and atmospheric circulation. Interaction with these disciplines is essential to scaling up to landscape, biome, and global levels.

FURTHER READING

Ackerly, D. D., and R. Monson, eds. 2003. Evolution of plant functional traits. International Journal of Plant Sciences, 164, no. 3 (Special Issue). *Collection of twelve papers.*

Lambers, H., F. S. Chapin III, and T. L. Pons. 1998. Plant Physiological Ecology. New York: Springer-Verlag.

Mooney, H. A., R. W. Pearcy, and J. Ehleringer, eds. 1987. How plants cope: Plant physiological ecology. BioScience 37, No. 1. *Collection of six papers.*

Pearcy, R., J. Ehleringer, H. A. Mooney, and P. Rundel. 1989. Plant Physiological Ecology: Field Methods and Instrumentation. London: Chapman & Hall.

Tyree, M. T., and M. H. Zimmerman. 2003. Xylem Structure and the Ascent of Sap, 2nd ed. New York: Springer-Verlag.

Woodward, F. I. 1987. Climate and Plant Distribution. Cambridge: Cambridge University Press.

I.4

Functional Morphology: Muscles, Elastic Mechanisms, and Animal Performance

Duncan J. Irschick and Justin P. Henningsen

OUTLINE

1. Techniques and history
2. Examples
3. Future directions

Functional morphology is the study of relationships between morphology and organismal function. A simple inspection of animal diversity reveals a remarkable array of phenotypes and concomitant functions. For example, even within a single mammalian group (bats), one observes organisms consuming food of all types, such as blood, fruit, leaves, nectar, insects, and other animals. Accompanying this diversity in diet is a remarkable diversity in morphological structure ranging from vampire bats with fangs for making sharp incisions for drawing blood to leaf-eaters specialized for grinding and mastication. One also observes similar variation for different kinds of animal locomotion. Whereas some organisms have evolved wings for flight, such as in birds, bats, and flying insects, other species have evolved elongated hindlimbs for running or jumping, such as in some lizards and kangaroos. This diversity in form and function forms an essential template for functional morphologists because it provides the "menu" from which researchers can address how function relates to form.

GLOSSARY

biomechanics. A subfield of functional morphology that applies mathematical and biophysical theory to understand animal movement

function. The use, action, or mechanical role of phenotypic features

kinematics. Animal movement; the angles, velocities, and rates at which different body parts move throughout space and the study thereof

kinetics. Forces produced by organisms during dynamic movements and the study thereof

morphology. The descriptive features of the external and internal (anatomical) phenotype

performance. A quantitative measure of the ability of an organism to conduct an ecologically relevant task such as sprinting, jumping, or biting

structure. The configuration of muscles, bones, tendons, and other tissues that allow animals to achieve dynamic movements

Functional morphology is inherently mechanistic in that it seeks to understand the basic mechanical principles that explain organismal function. Thus, rather than focus purely on descriptive patterns of organismal function (i.e., the frog jumped 20 cm), functional morphology aims to understand the underlying physiological and morphological principles that allow organisms to conduct physical tasks such as swimming, running, flying, and feeding, among others. In contrast to reductionist research that studies living organisms from the biochemical or biophysical perspective (e.g., cell biology), functional morphology generally focuses on emergent functional properties arising from the whole organism. Whole-organism functional capacities represent the end output from integrated morphological, physiological, and behavioral attributes of organisms, and hence their study requires an integrative approach. For example, cheetahs are known for their remarkable sprinting capacities, and one can study how different aspects of their internal anatomy (i.e., lung and heart function, limb muscular morphology) allow cheetahs to sprint so quickly. However, functional morphology is less focused on functional capacities below the organismal level, such as the effectiveness of an enzyme at catalyzing reactions. This rule is not absolute, as, for example, many researchers study the function of individual muscle fibers to understand how larger muscle units function.

The field of functional morphology is built around several key ideas. First and foremost, the morphology of animals provides the foundation for all movement, such as the use of muscles and bones during locomotion. However, although descriptive studies of morphology are essential, by themselves they provide an incomplete picture of animal movement. Consequently, functional morphologists also aim to quantify animal function, such as feeding or locomotion. Before the advent of techniques for quantifying animal function, researchers assumed that function followed directly from morphology, but in fact, this relationship is complicated. Although the dimensions and configuration of nerves, bones, and muscle limit certain features of animal function (i.e., how fast an animal can run), they are rarely directly predictive. This is because the level of animal function (performance, see below) is driven not only by morphology but also by behavior, which is poorly understood in terms of anatomical bases.

Consider the example of Dick Fosbury, who pioneered the "Fosbury Flop" (jumping head first, with the back to the ground) to win the gold medal for the high-jump event at the 1968 Olympics. Before the advent of the technique, high-jumpers took off from their inside foot and swung their outside foot up and over the bar. By altering the "behavior" of jumping, Fosbury was able to achieve a significantly higher jumping capacity. This example also highlights an important concept in functional morphology, namely, whole-organism performance capacity, which is how well an organism completes an ecologically relevant task, such as maximum sprint speed or maximum bite force. Of course, performance capacities such as the high-jump today rarely matter to modern humans but were likely highly important during the evolution of modern humans, and similar performance traits remain important for animals. In the case of the high-jump, although the "function" is jumping, the performance metric is jump height. Therefore, alternations in behavior (how an athlete jumps) can greatly affect performance at a given function. A subtle aspect of this view is that performance as defined here is measured at the level of the whole organism and not at a level below the organism, such as in the case of an enzyme catalyzing a reaction. The reason for this distinction is that the actions of the whole organism are those that interact with the environment, and therefore, by understanding the dual nature of morphology and performance, we can acquire a reasonably complete view of how organisms operate in a natural environment.

We provide an overview of the state of functional morphology by first describing some of the techniques used along with a brief history of functional morphological studies. We then explain some general principles of functional morphological studies that have arisen from research over the last few decades. To provide an overview of the range of techniques used in functional morphological studies, we also provide four key examples of cutting-edge functional morphological research. We conclude by describing some promising future directions for this field.

1. TECHNIQUES AND HISTORY

The complex nature of functional morphology makes a comprehensive list of techniques impossible. Rather, we provide a brief and historical outline of common techniques and also describe recently emerging technologies that offer promise for the future.

Until relatively recent times, anatomical studies have been the mainstay of functional morphology. For centuries, scientists have used dissection to examine form–function relationships. Although form can be a poor predictor of function, the anatomical knowledge gained from these studies provided a foundation for many fields, including medicine and modern functional morphology. In the eighteenth and nineteenth centuries, scientists such as Dumeril, Couvier, and Mivart recorded many detailed anatomical drawings of a variety of vertebrates, many of which are still widely used today. The modern study of anatomy continues to employ basic dissection and description techniques but has remained robust by the incorporation of new imaging techniques. For example, the practice of clearing and staining enables researchers to examine patterns of bone, muscle, and tendon simultaneously and in situ. The advent of x-ray technology provided new insights into bone structure, which has now been enhanced with computerized tomography (CT) scans that enable the creation of three-dimensional models. Magnetic resonance imaging (MRI), first pioneered in 1977, has enabled unparalleled images of internal anatomy. Finally, increasingly sophisticated computer software capable of creating finite element models enables detailed reconstructions of internal structural components.

Quantification of movement has been a central goal of functional morphologists. The study of kinematics has progressed from the time of still cameras to modern high-speed cameras that can operate at framing rates of 1000 frames/sec. Before the beginning of the twentieth century, the study of movement was confined to still images and simple observation (e.g., stride length based on footprints from a horse in the dust). A major advance in imaging occurred when an artist, Eadweard Muybridge, took photographs of animal movement in rapid succession to allow visualization of basic gaits and locomotor patterns. The development of celluloid film and video technology has further improved the ability of functional morphologists to capture the movement of animals for quantification. Most recently, high-speed cameras have enabled biologists to digitally capture extremely rapid events that heretofore had remained largely mysterious. Examples include the snapping of appendages in snapping shrimp and the ballistic movements of the tongue in salamanders, both of which occur over a matter of milliseconds. A relatively new technological and conceptual advance has been the use of high-speed imaging equipment in the field, allowing biologists to capture behaviors in their natural environment. Such technology has been especially useful for gaining data on large marine mammals, which cannot be easily studied in a captive environment.

Studies of muscle anatomy and function have been another major focus of functional morphology. As with advances in imaging techniques for studying animal movement, technological innovations have continued to improve our ability to understand muscle function. Beginning in the 1940s, electromyography (EMG) has been used to detect and measure muscle activity. When muscles are activated, there is a measurable change in the electrical charge of the tissue. By placing wire electrodes into a muscle and measuring the current produced, one can determine when a muscle is activated. One of the early proponents of EMG work was Carl Gans, who wrote several books explaining how this method could be used to study muscle function in animals. A more recent development, sonomicrometry, enables researchers to study how individual muscle fibers change in length during movement. By inserting two tiny crystals into a muscle fiber at different points, one can measure the change in length of muscle fibers during contraction. When combined with EMG, sonomicrometry is particularly informative because these two techniques simultaneously provide information on the timing of muscle activation and length changes in individual muscle fibers. For vertebrates, bones serve as important anchors for movement. The process of locomotion, particularly in large animals (e.g., horses), imposes tremendous strains on bones, and relatively new methods using strain gauge technology enable researchers to measure the degree of torque (twisting force) and strain imposed on bones during locomotion. In a similar vein, force platforms, which amount to extremely sensitive three-dimensional scales, can effectively measure the forces that animals exert as they move across the ground. These kinetic techniques enable researchers to measure quantities such as power, force, acceleration, and energy use during locomotion. Finally, recent technological innovations have provided unparalleled ability to visualize animal movement in fluids. Unlike terrestrial movement, in which forces can be quantified by the use of force platforms, movement in aquatic environments is far more challenging to study. Methods such as digital particle image velocimetry (DPIV) use small reflective particles that interface with a laser beam to reconstruct the force vectors produced during aquatic movements of aquatic animals.

General Principles

Here we describe some general principles and recent findings for functional morphology based on three important areas of research, namely anatomy, energetics, and neuromuscular function.

Anatomy: Problems of Stress and Loading

Morphology provides a foundation for all movement. Many morphological structures are designed for withstanding peak stresses during normal activities, such as feeding, walking, or jumping, because a failure in morphological structure (e.g., a broken bone) could be catastrophic. Therefore, throughout the course of evolution, animals have experienced strong selection for morphological structures to withstand high peak stresses. An oft-used measure of the ability of bone to withstand stresses is the safety factor, which is calculated by dividing the maximum stress a bone can withstand without failing by the peak stress of normal locomotor activites.

The issue of withstanding high peak loads is particularly relevant during locomotion, when animals can exert high peak forces on limb joints, such as the knee or ankle. This problem is especially acute for very large animals (>300 kg) because as animals become larger, the ability of the musculoskeletal system to produce force and sustain stresses increases by a factor of 2, whereas body mass increases by a factor of 3. In other words, as animals become larger, the potential for catastrophic injuries (e.g., broken bones) increases dramatically. Consequently, one might also expect strong selection for constant "safety factors" for bones as animal size increases, and surveys and experiments with a variety of mammals show that safety factors are roughly similar (2–4) among mammals of different sizes. Large mammals have compensated for increased forces by adopting a more upright posture during movement, most obviously exhibited in elephants and horses, for example. This upright posture provides large mammals with greater mechanical advantage and decreases stress relative to smaller mammals, which adopt a more crouched limb posture.

The issue of peak stresses on morphological structures is also relevant during feeding, especially for animals that consume very hard prey. In this regard, animals that consume hard prey with morphological structures that seem ill suited for the task (e.g., cartilage) are especially intriguing. Cartilaginous animals, such as sharks and rays, consume large and hard prey without the benefit of hardened bones. One of the most spectacular examples of this phenomenon is the stingray, which can crush extremely hard clams and crabs in its jaws. Closer inspection of the internal anatomy of the jaws shows a device that operates much like a nutcracker, with prey (i.e., the "nut") being inserted on the "open" side; the triangle-like jaw structure then closes and effectively crushes the prey. This "nutcracker" design is far more effective at crushing prey than if the stingray used its parallel jaw parts to bite down "equally" (i.e., using the jaws in a parallel fashion). Further, the internal anatomy of stingray jaws also shows some intriguing convergent features with bone. Normal bone, especially the end of long bones where much of the weight loading occurs, is strengthened by the presence of a series of internal struts. MRI images of stingray jaws show a similar set of strengthening struts that enable stingrays to produce high forces without damaging their jaws.

Energetics: Elastic Elements as Drivers of Movement

Movement requires energy. For animals without muscles, movement is generated by a variety of mechanisms, such as hydrostatic pressure, but for most animals with muscles, a widely accepted model is that muscles drive locomotion using molecular motors, such as myosin, which converts chemical energy (in the form of ATP) to potential energy. According to this view, the speed or intensity of movement should correlate closely with the amount of energy spent. For the majority of movements driven by muscle, this simple prediction has been confirmed. For example, as reptiles run faster, their energetic expenditure increases linearly until the animal reaches muscular exhaustion, at which point muscle function rapidly decreases because of fatigue. However, this basic principle of "move faster, work harder" is not universal. When researchers began calculating levels of energy expenditure in large mammals during locomotion, they noticed that in some cases, expenditure would increase linearly with speed at low to modest speeds, but at higher speeds, energetic expenditure often increased at a far slower rate if at all. A notable example is jumping in kangaroos; when hopping at slow speeds, energetic expenditure increases linearly, but at high speeds, kangaroos can move as cheaply (from an energetic perspective) as if they were moving at slower speeds. In other words, these animals seem to be cheating; they increase speed without consuming additional energy. Extensive research into the anatomy of large mammals such as kangaroos and other large ungulates (deer, gazelles) provided a potential mechanism for this energetic savings. Many large ungulates possess elongated and enlarged tendons that act as springs during locomotion. When the animal's foot contacts the substrate during locomotion (particularly rapid locomotion), the tendon or ligament is compressed, storing elastic energy much like a compressed spring. As the foot leaves the ground, the pressure on the compressed tendons and ligaments is released, and elastic recoil from these springlike structures provides additional force to propel the animal, thus resulting in energetic savings.

The idea that locomotion is driven, at least in part, by compressive springlike mechanisms represents one example of a class of elastic mechanisms that can produce force for locomotion at relatively low energetic cost. Highly elastic structures such as tendons and ligaments are the primary culprits, but muscle also exhibits some elastic capacity. The use of elastic elements for saving energy appears to be ubiquitous in the animal kingdom and often enables higher performance than would be predicted based on simple calculations of muscle power alone. For example, some frogs can "store" energy in tendons and ligaments before jumping. On jumping, which is initiated by their hindlimb muscles, this stored elastic energy is released, providing additional force that enables frogs to jump longer distances than based on power output from their muscles alone. Other examples include a "catapult" mechanism in tiny leafhopper insects, in which elastic energy is stored and subsequently released by a catch mechanism during jumping. Fleas present perhaps the most spectacular example of elastic mechanisms for increasing performance; before a flea jumps, resilin, a highly elastic protein in the legs, is stretched, and this elastic energy is then released rapidly, enabling fleas to produce extremely high accelerations. This form of energy storage is energetically beneficial because the resilin is stretched slowly at low power outputs, yet released quickly, producing high power outputs over a very short time period, much like a catapult.

One well-studied form of elastic energy savings has been formulated into a predictive model, called the mass-spring model. The mass-spring model is most applicable to large terrestrial animals (e.g., kangaroos), and models a large mass (e.g., the body) attached to a large springlike structure (the limb, with its compressive tendons and ligaments). During locomotion, the mass compresses the spring during the middle of the stance phase, and then the body recovers the elastic energy at the end of the stance phase. The mass-spring model is especially useful for making energetic distinctions between walking and running in large mammals, such as humans. During walking, the hip follows an inverted-pendulum motion and is literally "vaulted" over the knee, with little compression of the elastic structures in the knee joint. By contrast, during running, the hip dips during midstance, and the elastic elements in the knee joint are compressed, allowing some elastic storage and recoil. This distinction implies that walking in humans is driven largely (if not entirely) by muscles, whereas running makes greater use of elastic elements. However, walking also derives some energetic savings from the manner in which the hip is first raised (to the "top" of the pendulum, which requires energy), and then lowered. The lowering of the hip is energetically cheaper than the raising of the hip because of the conversion of potential energy into kinetic energy. An important consideration is that the energetic "benefits" of such spring-like structures are probably most apparent for larger animals (e.g., horses) and are likely less important for very small animals such as insects. This is results from the simple fact that larger animals, because of their large mass, can exert much higher forces on tendons and ligaments during locomotion compared to smaller animals.

Neuromuscular Function: High-Frequency and Ballistic Movements

Despite the widespread influence of elastic mechanisms across the animal kingdom, muscles remain the primary driver of most movement, such as chewing, jumping, and running. The network of muscles, attached to bones via tendons, and the nerves innervating them make up the neuromuscular system. The neuromuscular system can be thought of as two primary parts: the muscles themselves, with their various structural components, and the nerves that innervate them and provide the wiring for effective control. Under "normal" conditions, the movement of most joints can be reasonably modeled, but certain extreme kinds of movement pose a challenge to conventional views of muscle dynamics.

High-frequency movements, such as tail rattling in rattlesnakes (~90 Hz) and vocalizations in some fish (e.g., the toadfish, ~200 Hz) pose a significant challenge to the neuromuscular system. At a first pass, one might guess that nature has already provided the solution to this challenge in the form of different muscle fiber types, each uniquely suited for different contraction speeds and force production. For example, fast-twitch muscles enable animals to produce rapid and powerful (high-force) movements over relatively short time periods, such as explosive jumps, whereas slow-twitch muscle fibers enable more sustainable and less powerful movements, such as maintaining a stationary posture. However, even fast-twitch muscle fibers would be hard-pressed to accomplish such high-frequency movements, such as observed in toadfish, especially over long time periods. Beyond the necessary coordination among the brain, nerves, and muscles, such high-frequency movements are challenging for other reasons, such as rates of calcium exchange (the driver of muscle function), ATP, and potential for rapid muscular fatigue. Inspection of the muscles used in high-frequency movements, such as the "boatwhistle" call of the toadfish and the rattle of rattlesnakes, show several key adaptations for rapid movements. These muscles exhibit improved release and sequestration of calcium from the sarcoplasmic reticulum, resulting in high rates of calcium movement within muscle fibers. These muscle fibers also show high rates of attachment and detachment of myosin cross-bridges and rapid removal of calcium from helper proteins. The rattle of rattlesnakes is especially confounding, as rattlesnakes can rattle at extremely high frequencies at relatively low energetic cost. Further, muscles involved in rattling show an interesting property: as the force of rattling increases, the energetic cost of muscular twitches remains constant. The rattler muscles seem to show a unique energy-saving property, namely, as the frequency of rattling increases, rattling force also increases, but the energetic cost of each twitch of the muscle remains constant.

A second significant challenge to the neuromuscular system comes in the form of extremely explosive movements. For example, many frogs rely on extremely rapid projections of their tongue (which can take place over time periods as short as 30 msec, or 30 one-thousandths of a second), to capture prey. In most animals that use tongues to feed, the tongue is moved by active muscle recruitment, and the overall length of projection is modest (< 2% of body length or less). In some salamanders, such as those within the genus *Hydromantes*, the tongue can be projected at up to 100% of body length over just a few milliseconds. Investigations of the anatomy surrounding this elongate tongue show some muscle fibers that essentially constrict the tongue, projecting it from the mouth at high speeds, much like a watermelon seed being squeezed between fingers. Other retractor muscles then "reel in" the long tongue after use. The result is an explosive ballistic projection of the tongue in which the tongue moves of its own accord, without any additional input of energy beyond the initial squeezing. A perhaps more spectacular example comes from chameleons, which also exhibit ballistic properties of their tongues, which can extend up to twice their body length, and which also adhere to prey via suction.

2. EXAMPLES

To provide an overview of the breadth of techniques and questions used by functional morphologists, we provide four brief examples of integrative research that spans a range of different animal taxa.

Gular Pumping during Locomotion in Monitor Lizards

The diversity of locomotor modes in animals is one of the most striking features of animal evolution. An oft-cited feature of this diversity is the dichotomy between reptiles and mammals in their aerobic capacity. Whereas mammals have substantial aerobic capacity and can run at relatively high speeds for long periods of time (> 30 min), reptiles quickly switch from aerobic to anaerobic locomotion during fast movements. However, the mechanism for this dichotomy has not been entirely well understood, but one possible explanation can be derived from the axial constraint hypothesis, which states that lizards face a respiratory trade-off when moving at moderate to high speeds. Lizards with a sprawling posture locomote with lateral undulations of the body trunk that are produced with alternating unilateral contractions of the intercostal muscles. The intercostal muscles are also used bilaterally to ventilate the lungs during resting respiration. Hence, the axial constraint hypothesis predicts that as the animal increases its speed, respiration will be inhibited. However, previous work with monitor lizards (*Varanus* spp.) and green iguanas (*Iguana iguana*) provided mixed support for this hypothesis. Researchers used videoradiography to observe locomotion in savannah monitors (*V. exanthematicus*) and green iguanas. This technique produces video x-rays that allow observation of internal structures, in this case the gular, or throat, cavity. Pressure transducers implanted in the lungs and gular cavity provided information about breathing cycles. The results were consistent with the axial constraint hypothesis in an interesting way. In green iguanas, lung ventilation decreased as locomotion speeds increased, as predicted. In the savannah monitors, however, lung ventilation was not impaired at higher speeds. The videoradiographs showed that the monitor lizards used gular pumping to force air into the lungs. That is, the hyobranchial apparatus compresses the gular cavity, creating positive pressure and causing air to be moved to the lungs (figure 1). This work provides an elegant example of how the proper application of technology allows functional morphologists to test hypotheses of organismal function in live animals. In this case, the research provides support for the axial-constraint hypothesis while showing that savannah monitors have circumvented the constraint through the evolution of a gular pump.

Fluid Dynamics during Swimming in Bluegill Sunfish

Many animals move in aquatic environments, such as fish, marine mammals, and invertebrates, among others. Unlike terrestrial environments, in which the quantification of forces is relatively straightforward (i.e., using force platforms; see below), detailing the forces involved during aquatic locomotion is far more difficult. Aquatic organisms typically move through fluids by use of specialized appendages (e.g., fins) that push against the fluid. The development of DPIV was an important advance for understanding aquatic locomotion because it allowed visualization of these more sophisticated force vectors. Research with sunfish has shown that complex fluid dynamics produced by a fish could be quantified in

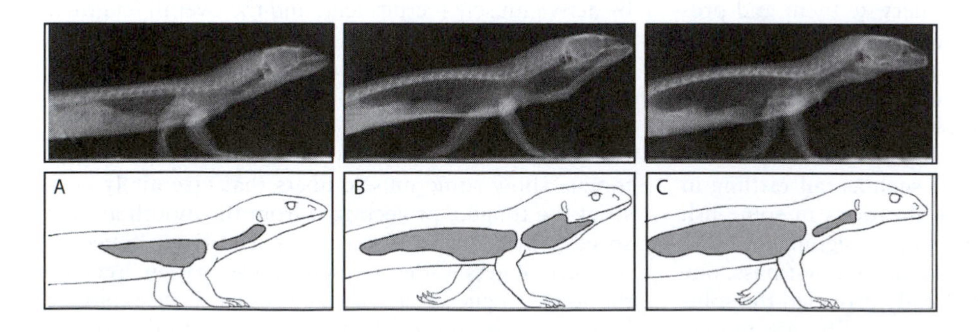

Figure 1. X-ray negative video images of a savannah monitor lizard (*V. exanthematicus*) walking on a treadmill at 1 km/hr, with corresponding drawings of the body, lungs, and gular cavity. The lizard is shown at three different stages of a single breath cycle: (A) end of exhalation, (B) end of costal and gular inspiration, and (C) end of gular pump. (From Owerkowicz et al., 1999)

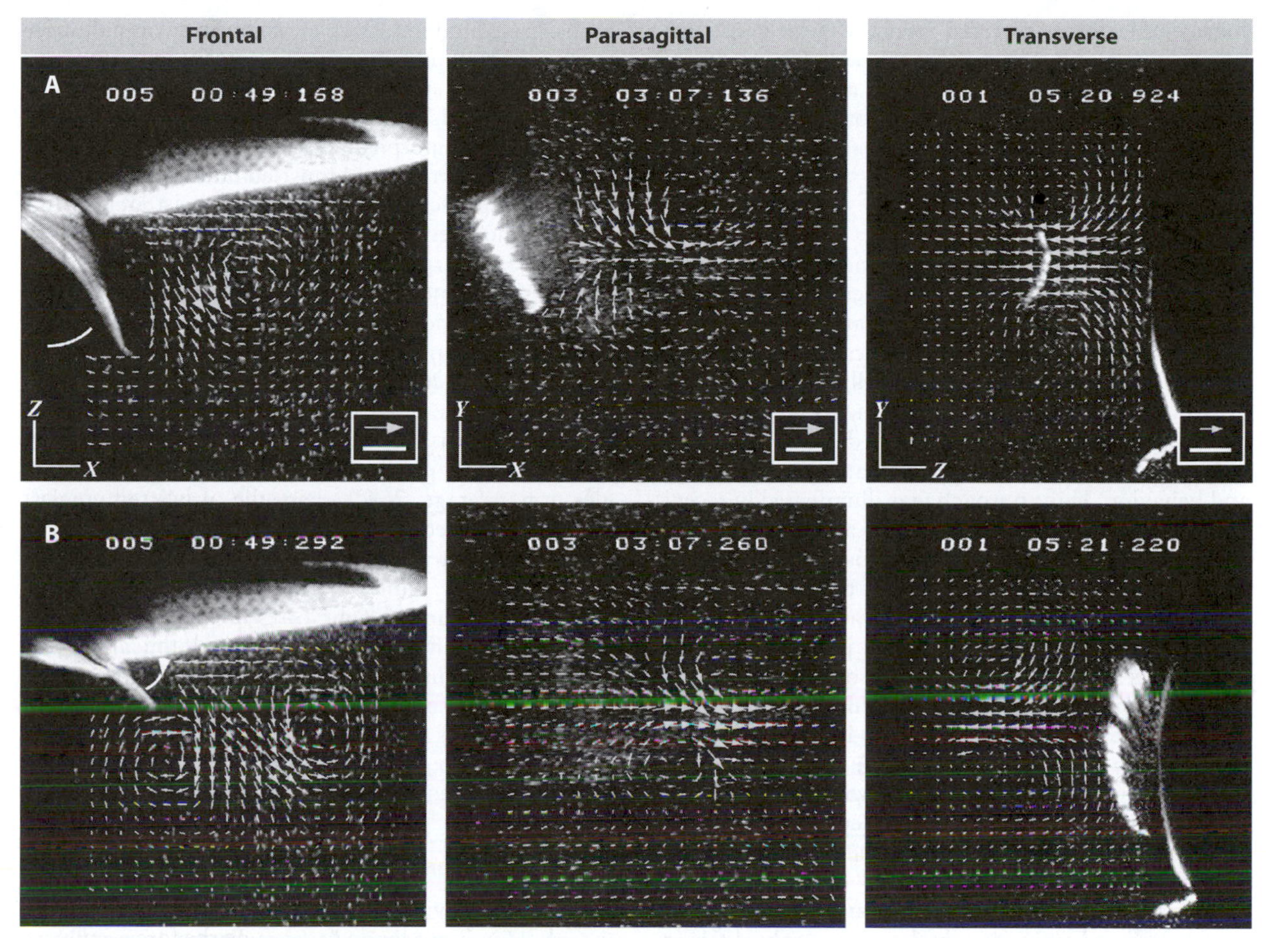

Figure 2. Water velocity vector fields calculated for orthogonal planar sections of the pectoral fin wake during swimming at 0.5 body length/sec. (Frontal Plane) Ventral view, anterior and upstream to the left. (Parasagittal Plane) Left lateral view, anterior and upstream to the left. (Transverse Plane) Posterior view showing the left side of body and the left fin; free-stream flow passes perpendicularly through the plane of the page toward the viewer. Flow patterns are shown at the time of (A) the stroke reversal, following fin downstroke in the direction of the upper curved arrow, and (B) mid- to late upstroke in the direction of the lower curved arrow. The last three digits in the numerical code in each panel denote time (in milliseconds). Mean free-stream flow velocity (10.5 cm/sec from left to right) has been subtracted from the frontal- and parasagittal-plane vector matrices to reveal wake structure. Note the spanwise component of flow in the frontal plane. By the middle to end of the upstroke, discrete pairs of counterrotating starting and stopping vortices are visible in each plane, each with a central jet of relatively high-velocity flow (B). Scales: arrow, 20 cm/sec; bar, 1 cm. (From Drucker and Lauder, 1999)

three dimensions. To accomplish this goal, researchers have used DPIV to measure the dynamics of the wake produced by swimming bluegill sunfish (*Lepomis macrochirus*). The water of a flow tank was seeded with small (approximately 12 μm diameter) silver-coated glass beads. These beads remain suspended in the water column in a relatively uniform pattern and are illuminated with a laser focused in a plane. As the fish swims, the beads move with the water. High-speed digital video records the movement of the particles, and these videos allow quantification of the forces exerted by the fish on the fluid medium. Measuring the change in position of a bead between consecutive frames allows calculation of a vector that can be used to determine quantities such as vorticity (rotational movement) and velocity. Filming from frontal, parasagittal, and transverse perspectives permits the quantities to be analyzed in three dimensions (figure 2). At slow speeds (0.5 body lengths/sec), when the bluegill sunfish are using labriform (with the pectoral fins only) swimming, the fins create vortices during the downstroke and stroke reversal. The propulsive forces generated by these fin strokes approximately balance the magnitude of drag and weight experienced by the fish. Researchers have also detected considerable forces directed medially, which may aid in maneuverability. This approach was significant not only in its individual discoveries but also in providing a roadmap for future research in locomotion in aquatic animals.

Jumping Performance in Anoles

One of the most exciting innovations in functional morphology has been the implementation of evolutionary principles, such as comparative methods. Because many functional morphological approaches are detailed in nature, studies are typically restricted to one or two species. However, improvements in technology have enabled researchers to capture detailed functional measurements on sets of species that differ in ecology and behavior. This approach was illuminated in recent research on jumping in Caribbean anole lizards. Caribbean anole lizards are both speciose and ubiquitous, and one of the primary features of their evolutionary radiation has been variation in their ability to jump and run. Some species are outstanding jumpers and runners, whereas other species display only modest capacities. Researchers examined the kinetics of jumping performance in 12 species of *Anolis* lizards that varied dramatically in morphology and jumping behavior. Anoles are particularly useful for examining evolution because they have undergone independent adaptive radiations on islands throughout the Greater Antilles, and the same morphological "types" (ecomorphs) have repeatedly and independently arisen. Some key questions addressed were (1) whether ecomorphs differ in jumping kinetics; (2) whether morphology correlates with jumping kinetics; and (3) whether anoles use takeoff angles that maximize jumping performance. To answer these questions, the authors measured the forces generated during jumping in 12 anole species using a force platform. This device uses a series of transducers (flexible and hollow metal rods) that are attached to a metal plate. When the lizard jumps from the plate, force is transmitted through the plate into the transducers. Bending in the transducers is then transmitted via attached wires to an amplifier and eventually a computer. When used in concert with high-speed digital cameras, force platforms allow researchers to examine the interaction between the forces used to propel locomotion (kinetics), and the movements used during locomotion (kinematics).

One surprising result was the lack of difference among morphologically divergent ecomorphs in jumping kinetics. This lack of a difference occurred because there were two ways to be a good jumper, namely having long hindlimbs versus short and stocky hindlimbs. Long hindlimbs accelerate the body over a longer time period, thus enhancing jump distance. By contrast, short and stocky hindlimbs provide extra muscle that provides high power outputs during jumping. As some anole ecomorphs have long hindlimbs, whereas others have shorter, more muscled hindlimbs, this two-pronged approach to jumping results in a lack of difference among ecomorph types.

A second surprising result was that anole lizards used a simple method to perform relatively long, yet short (in duration) and shallow (in height) jumps. Using simple jump equations, researchers have calculated the "expected" jump angles that anoles should use to maximize horizontal jump distance given their hindlimb length and takeoff velocity (figure 3). Interestingly, anoles jump using slightly lower takeoff angles (on average 2–4°) than the predicted angles. However, this reduction in angle had only a very slight effect (~1% on average) on horizontal distance traveled but greatly diminished jump duration (~7% on average) and jump height (~15% on average). The ecological significance of this biomechanical perturbation remains unresolved, but these jumps may allow anoles to jump more elusively in a cluttered arboreal environment, as shorter jump durations might allow an individual to change directions more rapidly when fleeing from predators. A nice biomechanical trick!

Flight Performance in Carpenter Bees

Despite the previous examples, the field of functional morphology is not limited to vertebrates. Invertebrates offer a nearly infinite palette of morphological form and behavior for functional morphologists to study, ranging from rapidly running spiders to slow and methodical sea worms. One exciting example of a functional morphological approach as applied to invertebrates is the hovering performance in carpenter bees (*Xylocopa varipunctata*). Functional morphologists are often interested in how morphology changes during the growth of an individual and how these changes affect function. In most animals, body shape changes dramatically from birth to adulthood, which is termed allometric growth. Because shape (morphology) is likely to dictate function and performance to some extent, one might expect differing selection pressures at different life stages.

Carpenter bees are large, robust bees that excavate nest cavities in wood. These insects have relatively high wing loading, meaning that they have a greater mass to wing area ratio than other bees. Researchers have used different mixtures of three gases to change the density of the air in a chamber in which a bee hovered. Oxygen

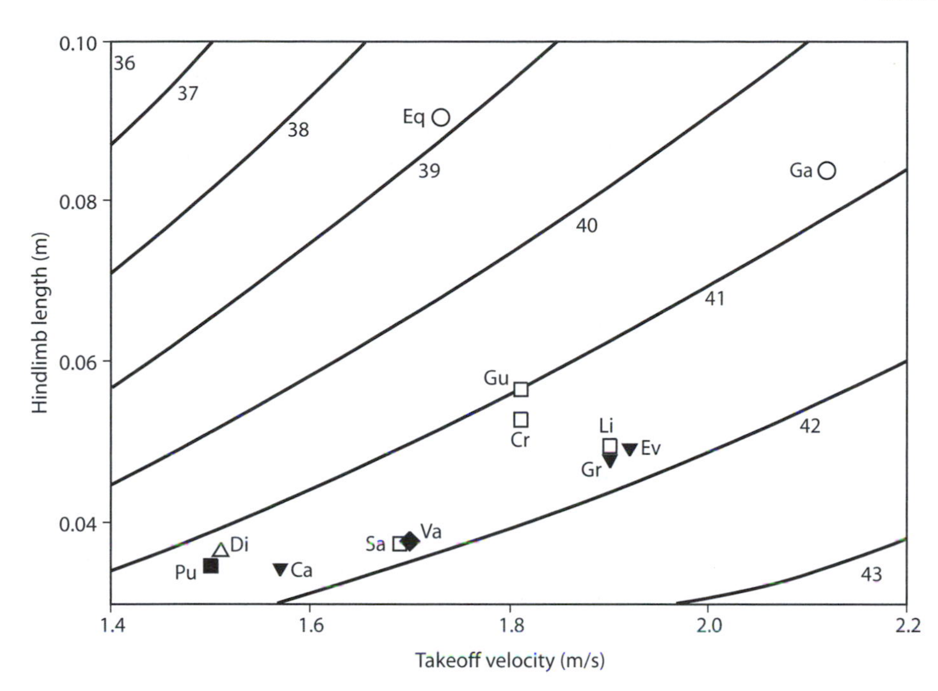

Figure 3. Landscape showing the theoretical optimal (i.e., distance-maximizing) takeoff angle as a function of hindlimb length (*H*) and takeoff velocity (*V*). Each species (N = 12) is represented by a single point noted by letters (e.g., Ev = *A. evermanni*) with its *H* and *V* coordinates. All predicted "optimal" takeoff angles fall within the 38.8°–41.8° range. (From Toro et al., 2004)

levels were kept constant (normoxic) so metabolic rates would remain unaffected while nitrogen (N_2), and helium (He) concentrations were varied. Because N_2 is denser than He, changing the proportion of these two gases in the chamber results in differing gas densities. By decreasing the density until the bee is unable to hover, the researchers were able to calculate maximum flight performance. Bees were filmed as they flew, and audio recordings of wingbeats were made simultaneously. Additionally, metabolic rates were measured with flow-through respirometry. Using bees with a wide range of body sizes allowed an ontogenetic examination of flight performance, kinematics, and energetics. In these bees, abdominal mass, thoracic mass, relative thoracic muscle mass, and wing loading all scaled allometrically with respect to body mass. The thoracic muscles are responsible for flight, and their relative mass decreases as body mass increases. Perhaps unsurprisingly, those bees with the highest relative thoracic muscle mass were able to hover in pure heliox. Interestingly, both wingbeat frequency and stroke amplitude correlated positively with body mass during hovering in normodense air but were mass independent during maximal flight. Power output and metabolic rate were significantly elevated during flight in hypodense air. In this case, large bees hover at near-maximal performance, whereas smaller bees have more power and kinematic reserves during hovering flight.

3. FUTURE DIRECTIONS

The field of functional morphology has emerged and evolved during the past century, and prospects for the future are exciting. We offer several areas that we feel constitute stimulating new frontiers for functional morphology. A conceptual advance that has arisen over the past several years has been the idea that researchers should aim to study organisms not only under controlled laboratory settings but also in natural surroundings. Ecological physiologists have long recognized the importance of understanding organismal function in an ecological setting, but the assumption that behaviors and functions obtained under laboratory settings are reasonable surrogates of natural behaviors persists. Recent work has shown that this is not always the case because animals will adjust their behavior as a function of their immediate environment. Technological advances are rapidly enabling researchers to take the "laboratory" into the field, thereby diminishing, or in some cases obviating, the traditional dichotomy between "field" and "laboratory" biology. For example, new advances in high-speed imaging now provide researchers with relatively inexpensive, lightweight, and portable high-speed cameras capable of filming events at up to 1000 frames/sec. Improved wireless technology enables researchers to use remote sensing devices to study changes in an animal's internal body temperature, hormone levels, or many other physiological parameters, all in a natural setting. An excellent example of the value of this approach is the study of movement in aquatic marine mammals. Unlike smaller aquatic animals, in which behavior and locomotion can be effectively studied in seminatural settings, such as large aquaria or tanks, human beings are not yet capable of recreating the habitat scale for marine mammals such as elephant seals or sea lions, many of which regularly travel dozens of kilometers in a single day. Further, many marine mammals regularly dive to depths greater than 100 m, and much of their behavior is closely tied to this daily rhythm of diving and returning to the surface for breaths of air. Advances in remote sensing technology have enabled researchers to measure swimming speeds, body temperatures, heart rates, and many other variables on free-swimming marine mammals. In the next several years, improvements in nanotechnology promise the exciting possibility of using similar remote-sensing devices on much smaller organisms, such as invertebrates, birds, reptiles, or small fish, for example.

Some of the most exciting recent developments have come from improvements in imaging. Until recently, researchers were able to gather only still shots of internal structures, such as bones and muscle, during their use, which begged the question as to how internal structures are used in normal circumstances. Researchers at Brown University, spearheaded by Elizabeth Brainerd, have developed an integrated 4-D MRI system that effectively visualizes the movement of internal structures in three dimensions (*X*, *Y*, and *Z*) in real time. Software being developed in conjunction with this MRI system allows the accurate reconstruction of bones and muscles during feeding, locomotion, and many other kinds of movement. Using this software, one can reconstruct movements of the skeleton in conjunction with key muscles, providing unparalleled views of dynamic movements. In addition to providing valuable basic information on organismal anatomy and function, these approaches offer exciting possibilities for sports medicine and rehabilitation after traumatic injuries (e.g., anterior cruciate ligament tears).

A surprising unification of efforts recently explored relationships between the morphology of sexual traits, such as peacock feathers, lizard horns, etc., and functional and physiological variables. Many animals exhibit elaborate and colorful sexual structures, and the reasons for which animals have evolved such seemingly deleterious structures constitute a point of contention within evolutionary biology. Functional morphological approaches offer promise for addressing this debate because the size, shape, and color of sexual signals may be "honest" signals of underlying functional variables that are valuable measures of male quality. Indeed, recent work has shown that the size and shape of such sexual structures often exhibit strong correlations with performance traits. For example, lizards with enlarged throat fans, or dewlaps, also exhibit high bite forces, despite no obvious functional link between the anatomy of head musculature (which dictates bite forces) and dewlap size and shape. In some species, male lizards with high bite forces are typically more dominant when competing against males with lower bite forces, which is reflected in their higher reproductive success. The imprint of sexual selection appears to be very strong in the realm of functional biology, and the subject promises increasing collaborations over the next decade between functional morphologists and researchers interested in sexual selection.

A final emerging area of growth concerns the use of animal models for the design of robots. Much has been written on the ability of organisms to execute difficult tasks with seeming alacrity, such as "cheap" (energetically) and fast jumping in kangaroos or fast and maneuverable swimming in fish and shark species. By collaborating with functional morphologists, engineers have constructed a wide variety of "animal robots" modeled after snakes, lobsters, and fish, among others, based on underlying principles of muscle anatomy and function. In an era of declining funding for functional morphologists from the federal government, many researchers are turning to alternative sources for funding, and although we applaud the applied use of functional morphological principles in general, we caution against overzealous devotion toward such applied purposes, as researchers should be aware of potentially destructive uses for the robots that they are helping to design.

FURTHER READING

Alexander, R. M. 1975. Biomechanics. Outline Studies in Biology. London: Chapman & Hall.

Arnold, S. J. 1983. Morphology, performance, and fitness. American Zoologist 23: 347–361.

Biewener, A. A. 2003. Animal Locomotion. Oxford: Oxford University Press.

Drucker, E. G., and G. V. Lauder. 1999. Locomotor forces on a swimming fish: Three-dimensional vortex wake dynamics quantified using digital particle image velocimetry. Journal of Experimental Biology 202: 2393–2412.

Dumont, E. R., J. Piccirillo, and I. R. Grosse. 2005. Finite element analysis of biting behavior and bone stress in the facial skeletons of bats. The Anatomical Record 293: 319–330.

Irschick, D. J., and T. Garland, Jr. 2001. Integrating function and ecology in studies of adaptation: Investigations of locomotor capacity as a model system. Annual Reviews of Ecology and Systematics 32: 367–396.

Irschick, D. J., A. Herrel, B. Vanhooydonck, and R. Van Damme. 2007. A functional approach to sexual selection. Functional Ecology 21: 621–626.

Lauder, G. V. 1996. The argument from design. In M. R. Rose and G. V. Lauder, eds., Adaptation. San Diego: Academic Press, 55–91.

Loeb, G. E., and C. Gans. 1986. Electromyography for Experimentalists. Chicago: University of Chicago Press.

Owerkowicz, T., C. Farmer, J. W, Hicks, and E. L. Brainerd. 1999. Contribution of gular pumping to lung ventilation in monitor lizards. Science 284: 1661–1663.

Roberts, S. P., J. F. Harrison, and R. Dudley. 2004. Allometry of kinematics and energetics in carpenter bees (*Xylocopa varipunctata*) hovering in variable-density gases. Journal of Experimental Biology 207: 993–1004.

Roberts, T. J., R. L. Marsh, P. G. Weyand, and C. R. Taylor. 1997. Muscular force in running turkeys: The economy of minimizing work. Science 275: 1113–1115.

Toro, E., A. Herrel, and D. J. Irschick. 2004. The evolution of jumping performance in Caribbean *Anolis* lizards: Solutions to biomechanical trade-offs. The American Naturalist 163: 844–856.

I.5

Habitat Selection

Judy Stamps

OUTLINE

1. Habitat and habitat selection at different spatial and temporal scales
2. Habitat selection: The behavior
3. Implications of habitat selection for basic and applied ecology

Separately and in combination, the terms *habitat* and *selection* mean different things to different audiences. This chapter focuses on habitat selection behavior at the level of individuals and considers how the processes that affect the choices made by organisms at different spatial scales affect the distributions at the population level. Because we initially focus on habitat selection at the level of individuals, habitat can be defined as a location in which a particular organism is able to conduct activities that contribute to survival and/or reproduction. That is, habitat is organism-specific rather than being determined by features that may be obvious to humans (e.g., vegetation type). Selection can be defined as a behavioral process by which an organism chooses a particular habitat in which to conduct specific activities. Hence, *habitat selection* implies that individual organisms have a choice of different types of habitat available to them and that they actively move into, remain in, and/or return to certain areas rather than others.

GLOSSARY

conspecific attraction. Attraction of individuals to conspecifics during the process of habitat selection

habitat selection. The process by which individuals choose areas in which they will conduct specific activities

heterospecific attraction. Attraction of individuals to other potentially competing species during the process of habitat selection

indirect cues. Stimuli that are produced by factors that are correlated with other factors with direct effects on intrinsic habitat quality

intrinsic habitat quality. The expected fitness of an individual when it uses or lives in a given habitat, after controlling for any effects of conspecifics on fitness

microhabitat. An area used for a specific type of activity (e.g., foraging, oviposition, nesting)

natal habitat preference induction. Exposure to cues in an individual's natal habitat increases the attractiveness of those cues during habitat selection

1. HABITAT AND HABITAT SELECTION AT DIFFERENT SPATIAL AND TEMPORAL SCALES

Habitats and habitat selection can occur at several different spatial and temporal scales. At larger scales, habitat refers to areas that are required for the long-term survival and reproduction of the members of a given population. In this case, habitat includes all of the areas required by all of the life stages of the members of that population, including areas that allow dispersers to travel among different patches of suitable habitat. For instance, from the perspective of a migratory bird, habitat includes breeding habitat, wintering habitat, and migratory stopovers that connect these venues. Habitat selection at large spatial and temporal scales occurs when individuals choose localities or regions that might be capable of supporting them, their offspring, and their descendents for an extended period of time.

At intermediate spatial and temporal scales, habitat refers to an area capable of supporting an individual for a biologically significant, finite period of its lifetime. Examples of habitat at this spatial scale include the selection of a feeding territory by a juvenile salmonid or an area suitable for feeding and oviposition by a female butterfly. Habitat selection at this scale is particularly important for sessile organisms such as barnacles because in this case a decision made early in life affects an individual's fitness for the rest of its life. In contrast, mobile organisms may select new habitats several times over the course of their lives as a result of changes in resource requirements, experience, or competitive ability during development, or as a consequence of seasonal movements from one area to another.

Finally, at even smaller spatial and temporal scales, habitat refers to an area in which an organism is able to conduct specific activities, such as foraging, resting, courtship, oviposition, or parental care. The term *microhabitat* is often used to refer to such areas. With the exception of sessile species, microhabitat selection typically occurs multiple times and involves many different types of habitats over the course of an individual's lifetime.

Recently, it has become apparent that scale matters and that models that predict behavior and distributions at small spatial scales may do a less satisfactory job at predicting them at larger spatial scales. In order to appreciate how scale affects habitat selection, it is helpful to consider one of the most influential general models of habitat selection, the ideal free distribution (IFD) (Fretwell and Lucas, 1970). The IFD predicts the area that an animal will select under the assumption that animals have accurate estimates of the intrinsic quality of the different areas that are available to them. In turn, *intrinsic quality* indicates the net fitness payoff that an individual can expect when it is using an area, after controlling for any effects of conspecifics on fitness. This particular model also assumes that animals compete with one another when they are using a habitat, such that fitness is inversely related to population density. Finally, it assumes that animals incur no costs when they are searching for a habitat, so they are always free to choose the habitat that maximizes their expected fitness. Under this simplified scenario, the probability that an individual will select a given area will be positively related to the relative intrinsic quality of that area and inversely related to the density of other individuals in that area.

Considerable empirical support for the IFD has been obtained in studies of habitat selection at small spatial scales, e.g., in studies of foraging patch selection conducted in tanks in the laboratory, or restricted areas in the field. This is not surprising because these are situations in which the assumptions of the ideal free distribution are most likely to be satisfied. Empirical support at microhabitat scales has also been obtained for modified versions of the IFD, e.g., models that assume that individuals differ with respect to competitive ability and sort themselves among habitat patches based on their competitive ability relative to the other individuals with whom they interact on a regular basis.

In contrast, studies of habitat selection at larger spatial and temporal scales often yield results at odds with the predictions of the IFD. These discrepancies have drawn attention to assumptions of this model that may not apply when animals select regions, localities, neighborhoods, home ranges, or territories for long-term use. One of the first assumptions of the IFD to be reevaluated was that individual fitness is inversely related to conspecific density. An alternative possibility is an Allee distribution, in which individual fitness increases as a function of density at low to intermediate densities and then declines at intermediate to high densities. Thus far, Allee distributions in the context of habitat selection have been documented for a wide array of taxa, including territorial animals, as well as species that live in colonies or groups. Of course, if individuals benefit from the presence of conspecifics at low to intermediate densities, then one would expect different patterns of habitat choice than if interactions between conspecifics were entirely competitive. For example, whereas the IFD predicts that empty patches of suitable habitat would be more attractive than a comparable habitat containing a moderate number of conspecifics, habitat selection under an Allee distribution predicts the reverse: newcomers should avoid an empty patch in favor of a patch that already contains members of their species.

2. HABITAT SELECTION: THE BEHAVIOR

Assessment of Habitat Quality

Indirect Cues and the Effects of Conspecifics and Heterospecifics on Habitat Selection

One of the assumptions of the IFD that is scale-dependent is that organisms have accurate estimates of the relative intrinsic quality of all of the habitats that are available to them. This assumption is most likely to be valid when individuals are able to extensively sample and evaluate many different habitats. For instance, an animal that has lived in a home range for an extended period of time probably has reasonable estimates of the relative quality of different foraging patches within its home range. However, extensive sampling of potential habitats is less feasible when animals are choosing habitats at larger spatial and temporal scales. In this situation, sampling can be constrained by many factors, including limits on the amount of time or energy that is available to search for and investigate novel habitats or elevated risk of mortality as animals explore unfamiliar areas. In addition, because large-scale habitat selection involves areas that individuals will use for extended periods of time, there is no guarantee that factors with major impacts on fitness will even be present when individuals choose a habitat. For instance, the larval dispersers of some benthic marine invertebrates settle in spring, when important attributes of the attachment site (e.g., exposure to hot temperatures in midsummer, or exposure to storm surges in winter) cannot be directly assessed.

When direct assessment of habitat quality is not an option, organisms may rely on indirect cues of intrinsic habitat quality. Indirect cues are stimuli that can be reliably detected when organisms are searching for habitats and that are correlated with biotic or abiotic factors that affect fitness after they have chosen a habitat. For many years, biologists have focused on indirect cues involving structural features of the habitat, e.g., shapes, colors, odors, sounds, or other cues that are likely to be correlated with other factors (food, predators, parasites, etc.) with direct impacts on fitness when animals are using a habitat. More recently, researchers have expanded the notion of indirect cues to include stimuli produced by conspecifics and heterospecifics. Thus, the conspecific cuing hypothesis argues that the presence or number of conspecifics in an area can provide information about other factors (predators, food supplies, parasites, etc.) that affect the intrinsic quality of that area. Similarly, the conspecific performance hypothesis argues that cues related to the reproductive performance of conspecifics in an area may provide information about factors that affect breeding performance in that area. Even other species that are potential competitors of a focal species may provide indirect cues to habitat quality, as is outlined in the heterospecific cuing hypothesis. These hypotheses have broadened the notion of indirect cuing to include the possibility that conspecifics, successful conspecific breeders, and heterospecifics affect habitat selection behavior because they provide information about other factors with major impacts on intrinsic habitat quality.

Indirect Cues versus Direct Benefits in the Effects of Cues on Habitat Selection

The issue of the effects of conspecifics and heterospecifics on habitat selection raises an important general question, namely, whether organisms respond to particular cues when selecting habitats because those cues are correlated with other factors that affect intrinsic habitat quality (an indirect cue), or whether they respond to those cues because they are produced by factors that directly affect fitness after they choose a habitat (direct benefits). That is, individuals may be attracted to cues from conspecifics or heterospecifics because these cues are associated with other factors that affect habitat quality (conspecific cuing, heterospecific cuing) and/or because individuals directly benefit from the presence of conspecific or heterospecific neighbors after settling in a habitat. As a result, two mutually nonexclusive hypotheses predict that individuals will be attracted to conspecifics (conspecific attraction) or to heterospecifics (heterospecific attraction) when choosing a habitat. In fact, a growing number of empirical studies indicate that individuals are attracted to conspecifics, to successful conspecifics, or to heterospecifics when choosing a habitat. However, it is not yet clear whether indirect cues, direct benefits, or both contribute to positive effects of cues from conspecifics or heterospecifics on habitat choice. Fortunately, this is a question that is currently under study, and we can expect more information on this topic in the coming years.

The Development of Responses to Cues during Habitat Selection

Over the years, it has become apparent that animals have two potential sources of information about associations between cues and intrinsic habitat quality: information from previous generations, via genes and maternal effects, and information from an individual's immediate past, via learned associations between cues and factors with direct effects on habitat quality. Information from the past is reflected by preexisting biases, which in the context of habitat selection are expressed when naive individuals with no previous exposure to a natural cue are attracted to that cue. For instance, in nature, larval anemonefish disperse from their natal habitat (a sea anemone of a particular species) and then attempt to locate and settle on a new host anemone. Naive larvae raised in the absence of sea anemones are more strongly attracted to the odor of their usual host anemone than to the odor of other sea anemones. Similar examples of differential attraction by naive individuals to cues produced by "ancestral" habitats have been reported in many different taxa.

An individual's personal experience can modify or alter preexisting biases for the attractiveness of cues from natural habitats. For instance, exposure to cues in an animal's natal habitat may increase the attractiveness of those cues to natal dispersers, a process termed "natal habitat preference induction" (NHPI). Thus far, NHPI has been reported for a number of taxa, including many insects and a scattering of vertebrates. The latter includes the anemonefish mentioned earlier, in which larvae raised in the presence of their typical host anemone are more strongly attracted to olfactory cues from their host anemone than are naive larvae. Preexisting biases and personal experience interact throughout development to affect preferences for cues for particular patches or types of habitat. As a result, nature and nurture both affect the ways that animals respond to indirect cues during the process of habitat selection.

Adding Search to Habitat Selection

By definition, habitat selection at larger spatial scales involves longer travel distances and higher travel costs than habitat selection at smaller spatial scales. Travel costs include increased risk of mortality from predators, accidents, infection, or other adverse conditions individuals face when traveling through unfamiliar, often inhospitable, terrain. These costs are compounded when individuals have difficulty detecting suitable habitats from a distance or when suitable habitat is sparsely distributed in the landscape. In such cases, individuals may travel considerable distances without finding any suitable habitat. Moreover, habitat selection at larger spatial scales is often time- or energy-limited. For instance, natal dispersal in brush mice is restricted to a 1- to 2-week period before sexual maturation, and natal dispersal times and distances in bark beetles are constrained by the fact that they do not feed en route and must rely on energy stored before they leave their natal habitat.

Long travel distances, high travel costs, and time- or energy-limited search constrain habitat selection behavior in ways that can often be safely ignored in studies of microhabitat selection. In the absence of these constraints, it is reasonable to assume that preference and choice will map onto one another and that organisms will select the habitat that they perceive (correctly or incorrectly) will yield the highest fitness. However, when organisms choose habitats at larger spatial scales, preference is only one of several factors that affect habitat choice. For example, when energy-limited individuals search for a habitat, theory predicts that individuals with higher energy reserves will be more selective than individuals with lower energy reserves. In that case, individuals in poor condition will be more likely to accept less-preferred habitats early in the search and less likely to end up settling in highly preferred habitats than individuals in good condition. This is an example of a "silver spoon effect" in which favorable conditions early in life increase the chances that an individual will be successful later in life. In fact, this type of silver spoon effect has been documented in studies of benthic marine invertebrates (bryozoans), in which larvae with large food reserves are more selective and more likely to settle in highly preferred habitats than larvae with lower food reserves.

Adding search to habitat selection highlights a number of other reasons why animals should accept less-preferred habitats, even though more-preferred habitats exist elsewhere in the same landscape. High travel costs, difficulty in detecting suitable habitats, long distances between suitable habitats, a shortage of time or energy available for search, and a scarcity of high-quality habitats are all factors that will favor individuals who are relatively nonselective when they are searching for a new habitat. Nonselective individuals still prefer some habitats to others and can express these preferences if provided with a choice of habitats located directly next to one another. However, under natural conditions, nonselective individuals will be more likely to accept any suitable habitat rather than incur the added costs of continuing to search for a more-preferred habitat. In turn, reduced selectivity during search increases the proportion of individuals that end up choosing habitats in proportion to their availability. Hence, individuals that are highly selective at smaller spatial scales (e.g., when choosing foraging patches within a home range or territory) may be considerably less so when choosing a habitat at larger spatial scales (e.g., when selecting a region in which to settle, or establishing a home range or territory within that region).

3. IMPLICATIONS OF HABITAT SELECTION FOR BASIC AND APPLIED ECOLOGY

Most ecologists and conservation biologists are not nearly as interested in the behavioral processes that generate habitat selection as they are in the effects of these processes on animal distributions and population viability. Indeed, in the ecological literature, the term *habitat selection* usually does not refer to the behavior of individual animals but rather to differential patterns of habitat use. In this literature, habitat selection is inferred when the density of individuals in a particular type of habitat is higher than predicted on the basis of a null model that assumes that individuals use different types of habitat in proportion to their availability. By extension, it is assumed that higher-than-predicted densities in a given type of habitat occur because organisms preferentially settle in, use, or remain in that type of habitat. However, active habitat choice is only one of several factors that can produce differential habitat use, so this assumption need not always be valid. For instance, newly settled larvae of benthic marine fish and invertebrates are often strongly associated with certain types of habitat. In the past, researchers assumed that differential patterns of habitat use by new recruits were a result of active habitat choice, but recent studies indicate that habitat-specific predation in the hours to days immediately following settlement contributes to these patterns. Because most researchers had assumed that the factors affecting the mortality of new arrivals were the same as the factors affecting the mortality of settled larvae, habitat-specific mortality early in the settlement period in larval recruits went undetected for many years.

A major goal of studies of habitat selection in the ecological literature is to identify the types of habitat that are most suitable for the members of a population or species. The notion that differential habitat use reflects differences in intrinsic habitat quality rests on yet another assumption, namely that organisms accurately estimate the relative intrinsic quality of different types of habitat, so that preference and performance are positively correlated across different types of habitat. If this assumption is valid, then the relative abundance of organisms in a given type of habitat may provide useful information about the quality of that type of habitat, relative to the quality of other types of habitat in the same area.

The assumptions outlined in the previous two paragraphs are actually quite similar to the underlying assumptions of the IFD. As a result, differential habitat use patterns are most likely to reflect habitat quality when the assumptions of the IFD are satisfied. Recall that the IFD assumes that individuals compete with one another while living in a habitat and that all of the individuals in a species are comparable with respect to their competitive ability. However, if individuals benefit from the presence of conspecifics after settling in a habitat, then the density of individuals in a given type of habitat need not reflect the relative quality of that type of habitat. For example, if individuals prefer to settle in the company of conspecifics, lower-quality patches that contain a moderate number of conspecifics may be more attractive to both local recruits and to potential immigrants than empty patches of higher-quality habitat. Alternatively, if individuals differ with respect to their competitive ability, then highly competitive individuals may be able to exclude less-competitive individuals from higher-quality habitats. In this case, there is no guarantee that population densities will be higher in habitats of higher intrinsic quality, even if these habitats are preferred by every member of the species.

The use of indirect cues in habitat selection can also disrupt relationships among population density, habitat preferences, and habitat quality. Even under the best of circumstances, the association between indirect cues and habitat quality is correlational rather than causal, so that organisms that rely on indirect cues will occasionally prefer lower-quality habitats by mistake. And because indirect cues provide only approximate estimates of habitat quality, organisms that rely on them are likely to have difficulty discriminating among habitats that do not differ very much with respect to habitat quality. Indeed, when organisms rely on indirect cues for habitat selection, ecologists with accurate estimates of habitat-specific mortality and reproductive rates probably have a better notion of the relative quality of different types of habitat than do the organisms that are selecting those habitats.

Although associations between indirect cues and habitat quality have always been imprecise, humans have contributed more than their share to the disruption of correlations between indirect cues and habitat quality. The recent literature on "ecological traps" considers cases in which a sudden environmental change (e.g., addition of a novel predator, altered habitat structure) has decoupled indirect cues from the true quality of the type of habitat that produces them. Most empirical studies of ecological traps have focused on situations in which humans are responsible for changing correlations between indirect cues and habitat quality. Examples include birds that preferentially settle in plantations of exotic trees rather than natural forests but that suffer lower nesting success in the former as a result of nest predation, or mayflies that prefer to lay their eggs on dry asphalt roads rather than ponds because asphalt reflects more of the polarized light that these animals use to choose oviposition sites. Hence, even if indirect cues used to be strongly correlated with habitat quality, there is no guarantee that this is still the case in today's altered world.

Another situation in which indirect cues can encourage mismatches between habitat preferences and relative habitat quality occurs when the attractiveness of indirect cues increases after exposure to those cues in the natal environment (NHPI). This is because NHPI encourages animals to select new habitats that are comparable to their natal habitat, even if other types of higher-quality habitats are available in the same landscape. Thus, NHPI may help explain situations in which animals raised in degraded habitats are reluctant to recruit to nearby patches of restored, high-quality habitat, or in which captive-raised or translocated animals fail to settle in habitats that are known to be of high quality for the members of their species.

Even if we are willing to assume that organisms have perfect estimates of the intrinsic quality and the density of conspecifics at every habitat that is available to them, and that every individual prefers the same type of habitat, adding search to habitat selection further complicates relationships among habitat preferences, habitat quality, and population density. When organisms have to search for a habitat, preference is no longer the only factor affecting habitat choice. Instead, the optimal behavior for a given individual depends not only on the benefits of finding a high-quality habitat but also on the costs of searching for it. Thus, if patches of high-quality habitat are rare and sparsely distributed, and if search is time- or energy-limited, then theory suggests that most of the individuals in a population will be relatively nonselective and, hence, likely to settle in habitats in proportion to their availability. As a result, habitat fragmentation and habitat degradation will not only reduce the amount of habitat that is available to support a population but also shift behavior in a direction that discourages individual selectivity and encourages individuals to accept habitats in proportion to their availability in the landscape. When this happens, an analysis of habitat use in relation to habitat availability might conclude (correctly) that individuals were not exhibiting habitat selection. However, it might also conclude (incorrectly) that individuals do not prefer some types of habitat to others or that all of the available habitats were of comparable intrinsic quality.

A number of other factors that occur when animals search for habitats can affect relationships among habitat choice, habitat preference, and the distribution of individuals. For instance, species with small perceptual ranges may have difficulty detecting suitable habitats. If searching individuals run a strong risk of not finding any suitable habitat, and if low-quality habitats produce cues that can be detected at longer distances than the cues from high-quality habitats, then individuals should be differentially attracted to, and differentially settle in, low- rather than high-quality habitats. This scenario may help account for the fact that pest species such as aphids recruit to large expanses of agricultural crops rather than to isolated patches of their native host plants, even though those crops are less suitable for feeding and oviposition than the native hosts. The condition of the individuals who are selecting habitats may also affect relationships between preference and choice because, as was noted above, when time- or energy-limited animals are searching for habitats, individuals in poor condition are expected to be less selective during search than individuals in good condition. Hence, habitat degradation may not only reduce the survivorship and reproductive success of individuals who live in those lower-quality habitats but also produce individuals who lack the stamina or stored resources necessary to locate and settle in patches of higher-quality habitat.

In conclusion, habitat selection behavior is scale-dependent. Although simple habitat-selection models do a reasonable job of predicting individual behavior and spatial distributions involving habitat selection at smaller spatial scales, more complex models may be required to predict patterns of habitat selection at larger spatial scales. Recent theoretical and empirical studies of habitat selection at larger spatial scales have expanded traditional models to consider situations in which organisms benefit from the presence of conspecifics or heterospecifics after settlement, rely on indirect cues to assess habitat quality, or incur costs when searching for potentially suitable habitats. On the debit side, this recent body of work provides a number of reasons why differential patterns of habitat use at larger spatial scales may not provide reliable estimates of either habitat preferences or intrinsic habitat quality. On the positive side, these recent studies have generated a number of new hypotheses about habitat-selection behavior, some of which have already been supported in studies of habitat selection at larger spatial scales. This new body of work provides possible explanations for distribution patterns that have been observed in nature and offers suggestions that may help applied biologists manage the habitat selection behavior of species of concern to humans.

FURTHER READING

The Ideal Free Distribution

Flaxman, Samuel M., and Christina A. de Roos. 2007. Different modes of resource variation provide a critical test of ideal free distribution models. Behavioral Ecology and Sociobiology 61: 877–886. *A recent study illustrating how the predictions of the IFD have been tested and validated for habitat selection behavior at small spatial scales.*

Fretwell, Steven D., and H. L. Lucas. 1970. On territorial behavior and other factors influencing habitat distribution in birds. I. Theoretical development. Acta Biotheoretica 19: 16–36. *The classic paper that presented the initial model.*

Trengenza, T. 1995. Building on the ideal free distribution. Advances in Evolutionary Research 26: 253–307. *A general review of the topic.*

Contributions that Consider Habitat Selection at Larger Spatial Scales

Jones, Jason. 2001. Habitat selection studies in avian ecology: A critical review. Auk 118: 557–562. *A review of habitat selection studies involving birds.*

Stamps, Judy A. 2001. Habitat selection by dispersers: Integrating proximate and ultimate approaches. In J. Clobert, E. Danchin, A. A. Dhondt, and J. D. Nichols, eds., Dispersal. Oxford: Oxford University Press, 230–242. *A review of factors affecting habitat selection behavior at larger spatial scales, with a focus on natal dispersers.*

Sutherland, William J. 1996. From Individual Behaviour to Population Ecology. Oxford: Oxford University Press. *A book that illustrates the ways that IFD models and modifications of these models can be used to study habitat selection at small to intermediate spatial scales.*

Underwood, Anthony J., Gee M. Chapman, and Tasman P. Crowe. 2004. Identifying and understanding ecological preferences for habitat or prey. Journal of Experimental Marine Biology and Ecology 300: 161–187. *A review of methods of studying habitat preferences with an emphasis on marine organisms.*

The Use of Indirect Cues in Habitat Selection

Johnson, Matthew D. 2007. Measuring habitat quality: A review. The Condor 109: 489–504. *A review of methods for estimating habitat quality with a focus on birds.*

Robertson, Bruce A., and Richard L. Hutto. 2006. A framework for understanding ecological traps and an evaluation of existing evidence. Ecology 87: 1075–1085. *A recent review of empirical studies of "ecological traps": situations in which indirect cues are no longer strongly correlated with habitat quality.*

Stamps, Judy A., and V. V. Krishnan. 2005. Nonintuitive cue use in habitat selection. Ecology 86: 2860–2867. *An overview of the use of indirect cues in habitat selection, with an emphasis of the use of indirect cues for habitat selection at larger spatial scales.*

I.6

Dispersal

Nicolas Perrin

OUTLINE

1. Definition, patterns, and mechanisms
2. Evolutionary causes
3. Demographic and genetic consequences of dispersal
4. Measuring dispersal

After a brief overview of the general patterns and the variety of mechanisms used for dispersal, this chapter delineates its evolutionary causes. Besides the spatial distribution and temporal dynamics of limiting resources, genetic structures resulting from mating or social systems play a role by affecting the potential for inbreeding and kin competition. Depending on conditions, however, dispersal may also have detrimental consequences at the population level, in terms of both demography and genetics. Finally, the chapter outlines recent developments in the way dispersal is measured.

GLOSSARY

coancestry. Probability that two alleles sampled from two different individuals are identical by descent.

F_{ST}**.** A measure of genetic differentiation among populations, expressing the proportion of variance within a set of demes that results from the differentiation among them.

genetic load. Decrease in average population fitness (relative to the fittest genotype) caused, e.g., by immigration of locally less-adapted immigrants (migration load), mating among relatives (inbreeding load), fixation of deleterious alleles (drift load), or any other population process.

heterosis. Increase in fitness resulting from matings among individuals from different populations (as a result, e.g., of superdominance or drift-load effects).

inbreeding depression. Drop in fitness resulting from the mating between relatives (caused, e.g., by recessive deleterious mutations).

local competition. Competition among relatives for limiting resources (including mates).

mass effects. Quantitative effects of dispersal on local population dynamics. Emigration from a population may have negative effects on its demography, whereas immigration may have positive (rescue) effects.

outbreeding depression. Drop in fitness resulting from the mating among distantly related individuals (from, e.g., the disruption of coadapted gene complexes).

phoresis. Mechanism of dispersal by attachment of the propagule to another, actively dispersing organism.

polygyny. Mating system in which a few males monopolize many females.

propagule. Any part of an organism used for the purpose of dispersal and propagation.

sink. Any population that consistently receives more immigrants than it sends emigrants.

source. Any population that consistently sends more emigrants than it receives immigrants.

1. DEFINITION, PATTERNS, AND MECHANISMS

There are many ways to define dispersal. The simplest and possibly most appealing one might be to define it as the movement of organisms away from their place of birth. The crucial feature here is that dispersers do not reproduce where they were born. This opposes dispersal to "philopatry," i.e., the tendency to reproduce at the natal place. Dispersal is referred to as "effective" when immigrants in a population contribute to local reproduction (i.e., when the rate of dispersal translates into a rate of gene flow among populations). Dispersal is a very ubiquitous feature throughout the living world, from bacteria to animals, including organisms that spend most of their life cycle in a sessile form (such as plants and fungi, but also many filter-feeding invertebrates).

Dispersal patterns can be described by a "dispersal kernel," which expresses settlement probability as a function of distance from the source. The shapes of such kernels are obviously bound to depend on the dispersal mechanisms involved, which might be passive (e.g., transport by wind or water) or more active (e.g., flight). In the latter case, the kernel will also depend on behavioral strategies (e.g., random walk versus directed movement) and cognitive abilities in interaction with landscape features. In its simplest form, the kernel is an exponential negative function of distance from the source. However, even slight departures from this simple function might be of importance. Long-distance dispersers have a disproportionate impact on population processes, in particular during colonization events, by determining the rate of spread and the establishment of long-lasting genetic structures. Whether dispersal kernels have thin or fat tails (i.e., decrease faster or slower than an exponential) thus becomes an important theoretical issue. It is also one difficult to address empirically because long-distance dispersal events are rare and therefore often missed in mark-recapture experiments.

Adaptations to dispersal are extremely diverse and often remarkably ingenious. Plants normally disperse passively by relying on currents (wind, water) or animals, both in the gametic (pollen) and zygotic (seed) dispersal phase. Pollen dispersal by animals involves complex interactions that are usually mutualisms (in which plants provide nectar to pollinators) but may include parasitism: more than one-third of orchid species do not provide their pollinators with either pollen or nectar rewards, relying on floral mimicry for pollination. Some species (e.g., the genus *Ophrys*) mimic the morphology and odors of female bees, inducing males to disperse their pollen through series of copulation attempts. As for the zygotic phase, adaptations to wind dispersal include seeds that resemble parachutes (e.g., the hairy expansions, or "pappus," on dandelion seeds), helicopters (maple trees), or gliders. The tiny sizes of some propagules are also an adaptation allowing efficient dispersal by winds (e.g., the seeds of orchids or the spores of fungi). Adaptations to water dispersal are commonly seen in littoral plants. Flotation of the fruit allows seeds to be carried away on the tide or ocean currents. Examples include *Rhizophora* spp. (mangrove trees) and *Cocos nucifera* (coconuts), which may successfully colonize very remote tropical islands (note that this mechanism of dispersal also implies specific adaptation for the long-term maintenance of germination ability).

Many plants have their seeds dispersed by animals and also build with them interactions that range from parasitism to comensalism to mutualism. The burs of burdock (*Arctium lappa*) and cocklebur (*Xanthium strumarium*) are covered with stiff, hooked spines that attach themselves to a passing animal's fur so that the animal will carry them away. The American devil's claw (*Proboscidea louisianica*) produces one of the largest hitchhiker fruits in the world, consisting of strange seed pods that attach to the feet of large herbivores. In mutualistic interactions, seeds are contained within a soft fruit adapted to animal consumption. Such fruits often present conspicuous bright colors when ripe (color vision in humans evolved as an adaptation to frugivory by our primate ancestors). These seeds have a tough protective outer coating so that although the fruit is digested, the seeds pass through their host's digestive tract intact and grow wherever they fall. The viscid berries of mistletoe (*Viscum album*), a hemiparasite on trees, are deposited on potential host plants by the European mistle thrush (*Turdus viscivorus*) when cleaning its bill on branches after a meal. Such fruits attractive to animals are among the most successful adaptations related to seed dispersal.

Other plant species have evolved mechanical means to overcome the tendency of a seed to drop close to its parent. Seedpods are built such that seeds are ejected away from the parent plant at maturation. Examples of explosive seed dispersal include cosmopolitan weeds such as *Oxalis corniculata* or *Impatiens* spp. As it dries, the capsule becomes sensitive to disturbance, ejecting tiny seeds in an explosive discharge. The fruit of the squirting cucumber (*Ecballium elaterium*) bursts when ripe, violently ejecting seeds together with a mucilaginous juice. Similar adaptations can be found in fungi: the spores of the dung-colonizing *Pilobolus* must go through a cow digestive system and come out with the dung to start the next generation. Because cows will not graze within a certain distance of dungs, spores have first to be dispersed away from their natal heap by an initial explosive event.

Sessile or low-mobility animals have developed larval stages that function as dispersing propagules. Marine invertebrates display a great variety of planktonic larvae (*planula* in cnidarians, *veliger* in mollusks, *trochophore* in polychaets, *zoe* in decapods, *pluteus* in echinoderms, etc.) whose morphological adaptations and body expansions (e.g., setae) increase floatability and allow long-distance dispersal by marine currents. Specific behavioral mechanisms allow settlement in suitable places before developing into the adult stage. The aerial plankton similarly contains a diversity of dispersing propagules, including the juveniles of many orb-weaver spiders, ballooning via silk lines. The wind lifts lines along with the spider and floats it off to a new area. Phoretic behavior is also widespread. Pseudoscorpions, for instance, use their claws to grasp the hair of mammals or legs of insects for a lift. Dispersal is often quite active in many evolved invertebrates (cephalopods, crustaceans, insects) because of their good swimming, walking, or flying abilities. The best skills in terms of mobility are to be found among vertebrates, whose cognitive capacities also allow dispersal and settlement decisions to be fine-tuned to environmental or social conditions prevailing locally.

Most animals, including vertebrates, tend to disperse first at the juvenile stage (natal dispersal), but many exceptions occur. A few others, such as aquatic insects, disperse as adults (the winged stage), flight being in this case the only way to disperse from pond to pond. Mayflies (Ephemeroptera), for instance, have very short adult lifespans (a few hours in some species), devoted entirely to mating and dispersal (adults do not feed). Dispersal is also often biased by sex. Males are usually the dispersing sex in mammals, whereas females tend to be the dispersing sex in birds. A few species have evolved a dispersal polymorphism, with morphologically distinct dispersing propagules. Within the Asteracea family, several genera (e.g., *Leontodon*, *Heterotheca*, *Senecio*) present two distinct types of seeds, the one aimed at long-distance dispersal being smaller, with a developed pappus. Some insects (mostly among Orthoptera and Hemiptera) also present a marked dispersal dimorphism with both long-winged and short-winged (or even wingless) individuals. This dimorphism is strongly marked in aphids (plant lice), which may display alternation of winged and wingless generations, depending on environmental conditions. Dispersal morphs have also been described in mammals (e.g., the naked mole rats *Heterocephalus glaber*).

2. EVOLUTIONARY CAUSES

Dispersal is costly. In addition to the fixed costs of building up specialized structures (including trade-offs involved), dispersers entail the energetic costs of crossing inhospitable habitats and associated mortality risks (up to 50% in small mammals). Settlers in new habitats may be at competitive disadvantage with residents, who benefit from a better knowledge of local areas and possibly help from relatives. Dispersal may also induce a migration load if immigrants are genetically maladapted to local conditions. Some benefits must therefore counteract these costs for dispersal to be evolutionary stable. Theoretical investigations in this field rely on the mathematical tools of game theory because the fitness returns of alternative strategies are frequency dependent (whether dispersing is the best decision also depends on what conspecifics are doing).

Some evident benefits to dispersal accrue in dynamic landscapes, where resources display significant spatial and temporal variance. This is most obvious when patch qualities present negative autocorrelations, as happens when suitable habitats have a determinate lifespan. *Pilobolus* fungi growing on a dung, or mycophagous *Drosophila* developing on the fruiting body of a mushroom, have to disperse their propagules because the patch currently exploited is bound to disappear soon.

But dispersal may also evolve in the absence of such negative autocorrelations. Under random extinction of local habitats, a purely philopatric lineage will eventually disappear because the survival probability of an occupied patch declines asymptotically to zero. Thus, to ensure long-term survival, lineages have to send away a proportion of migrants, the optimal value of which depends on extinction rate. The same is actually true in any sort of dynamic landscape, even in absence of extinction. When new areas become open to colonization, dispersing lineages enjoy a higher fitness than purely philopatric ones.

What about stable environments? Why leave a good local patch, even if crowded, if the other places you may reach are neither better nor less crowded, and you nevertheless have to endure the costs of dispersal? As it turns out, it may still pay to disperse in such conditions because of selective pressures imposed by local kin structures. Under complete philopatry, individuals would interact only with relatives. This might be disadvantageous in terms of inbreeding and also in terms of competition for mates or resources.

First, mating with relatives often induces some inbreeding depression, expressed as a reduced offspring fitness, as a consequence of the load of deleterious mutations that accumulate in all populations. Game-theoretical models show that, were inbreeding avoidance the only reason for leaving a natal patch, then one sex only, either male or female, should disperse. Indeed, if males disperse, females can avoid the costs of inbreeding without having to pay those of dispersal. Avoidance of inbreeding (mostly self-fertilization) is certainly the only force behind gametic dispersal in plants, and observed patterns match these expectations (only male gametes disperse). In animals (which normally do not show distinct gametic and zygotic dispersal phases), dispersal is quite commonly sex biased, but both sexes usually disperse to some extent, implying that inbreeding avoidance is rarely the only cause of dispersal. Numerical simulations suggest that inbreeding avoidance may account for about one-third of the level of dispersal in stable landscapes.

Second, a less obvious but potentially important cause of dispersal is the avoidance of local competition, which actually refers to competition with relatives. Competition is detrimental anyway, but competition with relatives is even worse. It affects not only the actor's direct fitness but also that of the related competitors, who share some of the actor's genes. If one has to compete anyway, it is better to do so with unrelated individuals. Kin selection thus promotes dispersal as a way to increase inclusive fitness (by dispersing, the actor leaves one breeding opportunity to a relative). The relevant limiting factors can be trophic resources, territories, or mates. Local competition for breeding partners is thought to be responsible for the male-biased dispersal of many mammals, because their polygynous habits and female-defense behavior potentially induce a strong local mate competition.

However, interactions with relatives may also bring benefits. Social structures usually emerge from cooperative interactions among kin. In many social mammals, including mice, helping among related females increases overall breeding success. These fitness benefits promote a strong female philopatry. In combination with polygyny, female philopatry boosts local relatedness because offspring within a group share the same father, and mothers are closely related. In turn, high relatedness increases the risks of inbreeding depression and local-mate competition, which promote male dispersal. Social species thus often display strong sex biases in dispersal. The combination of polygyny and heavily male-biased dispersal allows disentangling the dynamics of coancestry from that of inbreeding, which permits kin structures and social systems to develop without incurring inbreeding costs.

3. DEMOGRAPHIC AND GENETIC CONSEQUENCES OF DISPERSAL

When driven by the temporal fluctuations of local patch saturation, dispersal has the potential to positively affect regional demography, provided local fluctuations are not spatially correlated. This is truer when dispersal displays positive density dependence, being elicited by environmental cues linked to local saturation (e.g., resource depletion). In such a case, dispersal may dampen population fluctuations and prevent local extinctions through rescue effects. The positive effects of dispersal are obvious under metapopulation dynamics, when regional demographic equilibrium results from a balance between random extinctions and recolonizations of otherwise equivalent patches (see chapter II.4). Emigrating when resources become scarce also prevents local overexploitation of resources, with long-lasting negative effects. Impeding emigration from resource-depleted patches may induce dramatic density cycles that might lead to extinctions, as observed in ungulate populations introduced on small islands.

The positive effects of dispersal are less clear when local patches in the landscape show consistent and predictable differences in quality. Dispersal then becomes asymmetric, with a dominant flow from good patches (sources) to bad ones (sinks). Although immigration may fuel low-quality patches and maintain local populations above carrying capacity through mass effects, emigration also imposes a load on sources that may threaten their long-term existence. Once sources are extinct, the whole system will rapidly collapse. The levels of dispersal normally favored by natural selection are unlikely to be optimal in terms of population survival. This is by no means surprising because selection operates at the level of individuals, not populations. The possibility actually exists, at least theoretically, that individual selection drives dispersal patterns that ultimately lead to the collapse of populations (evolutionary "suicide").

As for genetic consequences, the main direct effect of dispersal is to homogenize gene pools. Here also, consequences may be positive as well as negative. The positive side consists of a genetic rescue of isolated populations stemming from two causes. First, isolated populations are threatened by deleterious mutations, which occur in all natural populations at a rate estimated to be one mutation per genome and per generation (order of magnitude). Mutations with large effects are easily purged (particularly in large populations) and do not last for long. The main threats actually come from small-effect mutations, which may accumulate in small populations, where drift is strong enough to counterbalance selection. Deleterious mutations segregating within populations are responsible for the inbreeding load, revealed by a decrease in the fecundity or fitness of offspring from matings among relatives. Once fixed in a population, deleterious mutations do not contribute any more to inbreeding load (because they occur in all mating partners, independent of their relatedness) but to the so-called drift load, revealed by the enhanced fecundity or fitness of offspring from matings among partners from different populations (heterosis).

Second, because of their small effective size and enhanced genetic drift, isolated populations are also threatened by a general loss of genetic variance, which jeopardizes their evolutionary potential. Connectivity with other populations may thus bring new genetic variance, thereby restoring their ability to respond adaptively to ecological changes. As a result of the interplay among drift, dispersal, and local selective pressures, a set of loosely interconnected populations might actually evolve more rapidly than a large panmictic population (Wright's shifting balance).

The negative genetic aspects of dispersal are referred to as *migration load*. This load may stem from the disruption of local adaptations, when migrants settle in habitats that differ from their natal habitat according to important environmental variables. This is more likely to happen when dispersal is passive and settlement is random (as opposed to active dispersal and targeting of favorable settlement sites). The migration load may also stem from the disruption of coadapted gene complexes (outbreeding depression). Small and isolated populations at the margin of species distribution have the potential to adapt to local conditions and, ultimately, to develop into new species. By homogenizing gene pools, dispersal may thus prevent the local adaptation of these marginal populations and their evolution toward the exploitation of differential ecological niches and, so, ultimately counteract the dynamics of speciation and the building of biodiversity.

The demographic and genetic consequences of dispersal obviously depend on specific features of both focal landscapes and study species, including their dispersal strategies and cognitive abilities. Individual-based modeling suggests that "blind" strategies (e.g., random walking) increase connectivity among patches, leading to low population structure (low F_{ST}), but, because of the high loss of propagules in the matrix, they are affordable only if propagules and dispersal are rather inexpensive. By contrast, "short-sighted" or "long-sighted" strategies (in which propagules are able to target favorable patches at some distance) are less costly in terms of matrix losses, but they induce stronger demographic (source-sink) and genetic structures because specific landscape features trap propagules into a restricted set of dispersal paths.

Efforts to integrate genetic and demographic consequences of dispersal into a common framework are still scarce despite their potential importance for applied ecology and should certainly be pursued. The potential negative aspects of dispersal (both demographic and genetic), in particular, have to be borne in mind when devising management strategies for conservation biology (e.g., when building dispersal corridors or reinforcing local populations).

4. MEASURING DISPERSAL

Our ability to measure dispersal depends on the choice of appropriate methods to study its occurrence in natural populations. Ideally, a combination of both field observations and genetic methods is required to obtain a comprehensive picture of dispersal patterns and to make inferences about its proximate and ultimate causes. Field data provide valuable insights into the species social and reproductive behavior, which are essential to better understand the potential causes to dispersal but usually do not allow quantification of how dispersal translates into gene flow because effective dispersal can be low even when there is high mobility. Moreover, for species that are particularly vagile, difficult to individually identify or to mark and recapture, estimating dispersal by direct observation is not always feasible. Genetic methods can be used to complement and reduce the invasiveness, effort, and expense of mark-recapture studies and give insights into how dispersal translates into effective dispersal and gene flow. An appreciation of the species life history is essential, at the very least to establish when dispersal is likely to occur. Because dispersal is often a juvenile trait, sampling juveniles provides access to predispersal individuals, whereas sampling adults provides a mixture of residents and immigrants. It is important to emphasize that these different cohorts should be analyzed separately. Doing otherwise would reduce our ability to gain insightful information from the contrast between pre- and postdispersal samples. Conventional genetic methods for measuring dispersal can be classified into those that measure either past gene flow or instantaneous dispersal. Both classically assume a population island model and simple population genetics framework.

The simplest approach to estimate past gene flow builds on Wright's formula for genetic differentiation between subpopulations,

$$F_{ST} = \frac{1}{4N_e m + 1},$$

which provides an indirect measure of effective dispersal rate ($N_e m$), a product of effective size (N_e) and dispersal rate (m). This, however, offers only an approximate solution because it relies on simplistic assumptions of island models, including stable populations of similar effective sizes and homogeneous dispersal rate. Sex biases in gene flow can similarly be estimated using sex-specific markers. The mitochondrial DNA (mtDNA) is inherited maternally in most animals. Plants also usually transmit mitochondria and chloroplasts uniparentally, either paternally or maternally. Similarly, the Y chromosome in male-heterogametic species (and the W chromosome in female-heterogametic ones) is transmitted uniparentally. Because these markers are nonrecombining, information on historical patterns of gene flow is maintained in successive generations, and sex-biased gene flow can be inferred most simply through patterns of haplotype distribution or from the relative estimate of gene flow in males compared to females (obtained, e.g., by contrasting F_{STY} and F_{STmt}). It is important to stress, however, that the difference in F_{ST}

between Y and mtDNA may stem from differences in male and female effective population size as well as dispersal rate. A female-biased dispersal may combine with a highly polygynous mating system to generate a particularly small male N_e and a strong contrast in sex-specific genetic structure ($F_{STY} \gg F_{STmt}$ in hamadryas baboons, for instance). It is worth noting, however, that a large sampling variance is associated with Y and mtDNA markers because of their lack of recombination, so differences may partly stem from stochastic events.

Recombining biparental markers (such as autosomal microsatellites) can also be used to estimate instantaneous dispersal, possibly in a sex-specific way. One approach builds on the contrast in F_{ST} between adults and predispersal juveniles. Assuming an island model of dispersal, the ratio of (sex-specific) F_{ST} estimated after dispersal over F_{ST} estimated before dispersal is a simple function of the (sex-specific) dispersal rate and can thus be used to estimate the proportion of immigrant individuals (males and females) in a subpopulation per generation. Alternative approaches rely on genetic assignment techniques, which calculate the probability of origin of a focal individual, given its genotype and the gene frequencies in potential source populations. Maximal power is usually achieved when dispersal rate is at an intermediate value (approximately 10% per generation). With high dispersal, a population will consist of a large proportion of immigrants, so that populations will not be differentiated enough. With low dispersal, immigrants constitute only a small proportion of the individuals sampled and may not be detected at all. Individual-based assignment tests based on likelihood or Bayesian principles offer several advantages over summary statistics and should be more powerful because they do not average over the population, allow immigrant individuals to be readily identified, are more geographically explicit, and in the latter do not require populations to be predefined. Although individual assignment techniques based on Bayesian principles applied to multilocus genotypes are becoming standard tools in molecular ecology, their potential for studying dispersal has perhaps yet to be realized.

FURTHER READING

Bowler, D. E., and T. G. Benton. 2005. Causes and consequences of animal dispersal strategies: Relating individual behaviour to spatial dynamics. Biological Reviews 80: 205–225.

Clobert, J., E. Danchin, A. A. Dhondt, and J. D. Nichols. 2001. Dispersal. Oxford: Oxford University Press.

Clobert, J., R. A. Ims, and F. Rousset. 2004. Causes, consequences and mechanisms of dispersal. In I. Hanski and O. Gaggiotti, eds., Ecology, Genetics, and Evolution of Metapopulations. Amsterdam: Academic Press, 307–335.

Levin, S. A., H. C. Muller-Landau, R. Nathan, and J. Chave. 2003. The ecology and evolution of seed dispersal: A theoretical perspective. Annual Review of Ecology and Systematics 34: 575–604.

Ronce, O. 2007. How does it feel to be like a rolling stone? Ten questions about dispersal evolution. Annual Review of Ecology, Evolution and Systematics 38: 231–253.

I.7

Foraging Behavior

Joel S. Brown

OUTLINE

1. Foraging behaviors, adaptations, and autecology
2. Finding food
3. Handling time
4. To eat or not to eat?
5. Patch use
6. Social foraging
7. Fear and foraging
8. Coadaptations between foraging behaviors and morphology
9. Nutrient foraging in plants

A need for energy and resources for survival, growth, and reproduction is a universal property of life. Hence, all organisms must forage. Even plants have noncognitive foraging behaviors. Life exhibits a wonderful diversity of feeding behaviors and associated morphological and physiological adaptations. Food must be found and handled. Letting the food come to the forager (sit and wait) or actively seeking food items (active pursuit) are two tactics for finding food items. Handling a food item may be as simple as absorption (endocytosis by a single cell organism) or a complex choreography of subduing, dismembering, and/or digesting a prey. Diet choice involves foragers deciding which food items to accept or reject. Patch use considers how thoroughly a forager should deplete the food from a spot before giving up and moving to a fresh spot. Foraging often occurs socially because groups permit sharing of information, scrounging, group hunting, task specialization, and, most often, safety in numbers. Predation risk and fear loom large in foraging, as animals balance the conflicting demands of finding food while avoiding becoming food themselves. All of these topics of foraging behavior become central to understanding an organism's ecology and evolution.

GLOSSARY

diet choice. The decisions made by foragers regarding which encountered food items to consume and which to reject. The abundances of different food types, their ease of finding and handling, and their value to the forager generally influence the decisions to eat or not to eat.

foraging games. The behavioral challenges facing both predator and prey when the prey can perceive and respond to the hunting tactics of the predator, and the predator can perceive and respond to the antipredator tactics of its prey. These can be as straightforward as pursuit-evasion games; or complex sets of decisions summed up by when and where to forage; or levels of prey vigilance and predator boldness. Finally, foraging games such as producer–scrounger games or behaviors involving territoriality and interference may occur between members of the same species.

nutrient foraging. The noncognitive foraging behaviors of plants as they adjust allocations to roots and shoots, alter uptake kinetics or growth forms to influence the uptake of water, light, nitrogen, and other nutrients.

patch use. The behaviors of foragers regarding how to deplete the food items of a given spot. Most importantly, when should the forager leave an area with food before moving to a fresh area? A forager should leave a depleted food patch when the benefits of continuing to harvest the patch no longer exceed the sum of metabolic, predation, and missed opportunity costs of foraging.

social foraging. When feeding occurs as groups of the same or different species. Social foraging may allow for information sharing, producer–scrounger games, group hunting, task specialization, and very often safety in numbers. Safety in numbers occurs through the many eyes, dilution, and confusion effects.

1. FORAGING BEHAVIORS, ADAPTATIONS, AND AUTECOLOGY

An animal's ecology can be summed up as follows: where does it live, what does it eat, and who eats it? For instance, on the sand dunes of Bir Asluj in the Negev

Desert of Israel, afternoon winds blow in from the Mediterranean, redistributing the sand and uncovering seeds. At sunset, as the wind abates, Greater Egyptian sand gerbils (*Gerbillus pyramidum*) emerge from their burrows and move under shrubs or across open spaces to search for seeds. A knowledge of where seeds likely aggregate guides their search paths. With a keen sense of smell, gerbils hone in on patches of seeds or seeds buried under the sand. With its forepaws, the gerbil recovers seeds and lifts them to its mouth, either transporting them in internal cheek pouches or deftly husking them with practiced coordination of forepaws and incisors. While the gerbil seeks food, predators seek the gerbil. A gerbil's ears and auditory system can detect the low-frequency sounds of a barn owl's wingbeat. A gerbil's quick reactions may save it from the strike of a horned viper, and erratic locomotion permits escape from a pursuing red fox. With cheek pouches full, the gerbil returns to its burrow and deposits the seeds underground in its larder, or it may save time by burying the seeds in a shallow depression, contributing another snack to its scatterhoard. Perhaps another gerbil will pilfer this cache before the owner returns for it. As the night draws on, the gerbils deplete the available seeds and conclude the night's foraging. Most will return to their burrows to await another wind and another night. A few will have fed the predators. Central to the gerbil's ecology are its foraging behaviors and the foraging behaviors of its predators. These behaviors have been engineered by natural selection through the circumstances of making a living as a seed-eating desert rodent. As a result, feeding behaviors are often the most frequent and tangible expression of an organism's ecology.

Feeding behaviors, as products of natural selection, emerge from all organisms' need for energy and resources to sustain life, permit growth, and allow for reproduction. Foraging behaviors are as diverse and varied as life itself. We may describe them in broad brushstrokes such as a sperm whale diving 400 m beneath the waves and submerging for over 40 min as it somehow seeks giant squid. Or these behaviors can be described in fine detail as the exact path and number of steps that a browsing white-tailed deer takes as it moves from one specific shrub to another, stopping to nibble particular leaves from particular branches. Each bite may necessitate some number of chews before the masticated mouthful is swallowed. The precise choreography of step, bite, and chew continues for hours. But in each case, the foraging behaviors can be seen as contingent responses to environmental opportunities and hazards. As behaviors, we can describe the animal's repertoire of specific actions that it uses to find and harvest food. As adaptations, we can ask why did the whale dive to a particular depth and spend a particular time? Why did the deer favor some leaves over others? Was it the leaves' nutrition, the presence of spines, or plant toxins?

In what follows, we explore both of these aspects of foraging behaviors—the sequence of actions required to get food and the adaptive nature of feeding behaviors. Change the organism, and the suite of available behaviors likely changes too. The exact behaviors available to a single-cell *Paramecium* are literally worlds apart from that of a web-building spider. Change the environment, and the behaviors of a given animal may change dramatically. Bream, a common fish of northern European lakes, can opt to snatch zooplankton from the water when available, or they can probe for tasty detritus in the muck of the lake's bottom.

Here, the focus is on the categories of foraging behaviors and scenarios and the concepts that permit general understanding of foraging behaviors as adaptations. Animals decide where to forage. This topic is covered under Habitat Selection (chapter I.5). Once an animal is searching for food, the sequences of activities alternate between finding food and then handling food. The forager, as it finds and harvests food, faces decisions of "to eat or not to eat" and when to give up a food patch (patch use) as it depletes. One often imagines solitary foragers going about their business peaceably, but social foraging and fear and foraging recognize how foraging can occur in groups and occur under the threat of predation. Through coadaptations of foraging and morphology, there is a wonderful evolutionary feedback between adaptive feeding behaviors and the other physiological (see chapter I.2) and morphological traits of the species. Finally, plant nutrient foraging examines the noncognitive behaviors of plants to access nitrogen, phosphate, light, and water. Throughout, foraging behaviors emerge as adaptations (MacArthur and Pianka, 1966; Emlen, 1966) that permit organisms to acquire food and resources quickly, efficiently, and/or safely.

2. FINDING FOOD

The chemical reaction involving the polymerization of atoms and molecules is limited by the concentration of molecular building blocks and the rate at which these building blocks can be "found" and "consumed" by the growing polymer. Such was the foraging behavior of the first replicators at the dawn of life on Earth, as partially autocatalytic reactions built combinations of proteins and/or nucleic acids. Brownian motion within an aqueous solution allowed these protolife forms to find food. Three billion years later, finding food looms large in all sequences of foraging behaviors. Active-pursuit and sit-and-wait tactics provide two evolutionary strategies for finding food. Single-cell archaebacteria and other prokaryotes likely evolved to be either free-floating or attached to stone surfaces.

A sit-and-wait strategy demands less energetically and physiologically of an organism. It generally takes the form of filter-feeding small food items or ambushing large prey items. For it to work, the food must move, either passively as food particles within a water current or actively as mobile prey. Caddisfly larvae attached to stream pebbles extrude a mucus to which food particles become stuck. Many clams buried in the mudflats of the intertidal create their own water current by "siphoning" water through a tube, past a filter-like organ, and then back into the water column. Because the gerbils thoroughly scour the sand dunes each night for seeds, the horned viper can remain motionless, coiled, and ready to strike. Sit-and-wait foraging tends to promote foraging efficiency (reward per unit cost) over foraging speed (captures per unit time).

Mobile foragers that actively seek and pursue their prey generally enhance speed at the expense of foraging efficiency. Sessile or slow-moving food strongly selects for actively moving and searching foragers. Gerbils must go find their seeds, as their seeds do not find them.

The encounter probability (units of per time) is a key foraging parameter. It describes the likelihood of a searching forager encountering a given food item. The encounter probability depends heavily on the forager's senses. Vision, smell, touch, pressure sensors, hearing, and even cuing in on electromagnetic distortions provide tools for encountering prey. Then there are the cues emitted by the food items or prey. Together, the senses of the forager and the cues of its food determine the forager's detection radius for food and its likelihood of accurately sensing the food item. For gerbils, smell may provide a detection radius of ~ 10 cm with touch concluding a successful encounter. Larger seeds are easier to detect than smaller seeds, and with humidity, seeds become more odiferous.

Random search is the simplest case where a feeding animal's likelihood of encountering any given food item is constant and independent of the total number of prey. This idealized condition becomes distorted when predators can observe the distribution and abundance of many food items in advance. A hummingbird moving among flowers or a black rhinoceros among acacia trees can be a "traveling salesman" and map a best route for collecting the food items—they can do better than random encounter. More prey items can enhance the encounter probability by drawing the forager's attention, or it may challenge the forager with a confusion effect as multiple prey flee haphazardly at the predator's approach. Finally, the prey themselves may alter and distort encounter probabilities through camouflage, deception, and even direct signals to the forager that it has been detected. With aposematic coloration, dangerous or unpalatable prey (bees, monarch butterflies, coral snakes) communicate their unsuitablility as food items. Conversely, red flowers and intensely colored fruits attract the attention of hummingbirds and robins, respectively.

3. HANDLING TIME

Handling time describes the effort and activities required to harvest an encountered food item. In its simplest form, handling time can be the fixed time required for a gerbil to husk and consume a seed. More generally, it includes all of the effort required to subdue (if necessary), transport (if not consumed on the spot), prepare, and ingest the food. In animals such as a python or a ruminating antelope, handling time can also include a digestive pause that precludes searching for and handling additional food items. In gerbils, handling time may include caching behaviors. An emperor penguin's handling effort can include marching to and from the colony to provision young.

For predators, encountering the prey may be much easier than actually capturing it. Stanley Temple (1987) recorded how a red-tailed hawk had success rates of 28%, 18%, and 12% when initiating a strike on eastern chipmunk, cottontail rabbit, and eastern grey squirrel, respectively. Predators have additional foraging behaviors of stealth, pursuit, and tactics for killing the prey while avoiding injury themselves. Barn owls attacking gerbils appear to use hearing to encounter and initiate a strike while using vision to enhance the accuracy of the final impact.

The handling behaviors of many foragers may include preparing the prey for consumption and choosing which bits of prey to consume. The gerbil can facilitate digestion by first husking and chewing each seed. Sparrows may "whirr" the wings off insects before ingesting them. If handling is time consuming, such as for a squirrel consuming a hazelnut, the forager may carry the food item to a safer, more comfortable setting. The animal recoups its preparation time by speeding digestion and increasing the efficiency of assimilation. Partial prey consumption, such as a scorpion consuming only the yummier parts of an isopod (sowbug), increases the quality of the ingested food.

If small glass beads are mixed into a pile of seeds, a gerbil will harvest and pouch some of these beads along with the seeds. When licking up termites, an aardvark may consume more dirt than termites. Situations arise where there may be no advantage to taking the time to discriminate between good and bad food items.

Foragers will forgo recognition time when undesirable items are few and far between, relatively harmless to consume, hard to discriminate, and time-consuming to separate. Otherwise, foragers will invest time to distinguish among potential food items.

Handling time can be as simple as the time taken to consume an item or a sophisticated choreography of time, effort, and risk. For a mosquito, handling time begins when the humming of its wings stops as it alights on your skin. She seeks a promising capillary bed within which to insert her proboscis. She injects a bit of anticoagulant (with luck, free of malaria!) and begins the process of gorging her stomach. All the while, she aims to avoid your wrath should you awake and claim her life.

4. TO EAT OR NOT TO EAT?

Diet choice is one of the fundamental consequences of adaptive feeding behaviors. Organisms do not consume different foods in direct proportion to their abundances in the environment. Feeding animals always appear more or less selective. Diet-choice studies show how foraging behavior results in a triaging of what is available to what is actually consumed. In all cases, the mapping of food availability into diet involves aspects of finding and handling food.

Ronald Pulliam (1974) developed a classic model of diet choice based on the simplest assumptions of random search (constant and fixed encounter probability) and constant handling time. Search is undirected in the sense that the forager does not know what food type it will find until it stumbles on a food item. While searching, the forager cannot alter its encounter probabilities for one food relative to another (search images allow foragers to do this). To the forager, a food type can be characterized by the encounter probability, a, its abundance in the environment, R, its handling time, h, and its energetic value, e.

Even this simple model suggests quite a bit. For instance, the likelihood that the next encountered food item is food 1 as opposed to food 2 is $a_1R_1/(a_1R_1+a_2R_2)$. The forager should prefer the food with the higher energy-to-handling-time ratio. So, food 1 is preferred if $e_1/h_1 > e_2/h_2$. To maximize its feeding rate, the forager should always consume its preferred item. But should it consume the less-preferred food? The answer is straightforward and simple. If the energy gain from handling an encountered item of the less-preferred food, e_2/h_2, is less than what could be gained from searching for and handling a preferred item, $e_1a_1R_1/(1+a_1h_1R_1)$, then the forager should be selective. Otherwise, the forager should be opportunistic and consume all encountered items.

This model and its many variants suggest how animal diets represent a biased sample of availability. If the forager actually rejects consuming less-preferred food items, then diet choice is an extreme all or nothing. Increasing the abundance of its preferred food should cause the forager to reject less preferred items. A bountiful environment encourages picky eaters.

When a forager is opportunistic and consumes all encountered food items, it will have a diet that appears to favor those foods that are easier to find (higher encounter probabilities). Cryptic foods will be underrepresented, conspicuous foods overrepresented in the diet. This is why gerbils will harvest a greater fraction of the large seeds than the small seeds from a given patch of sand. This effect of encounter probability on diet explains why flowers and fruits have evolved to be conspicuous (it is adaptive to be harvested) and why moths, stick bugs, and other prey have evolved camouflage (it is nonadaptive to be eaten).

Biases in diets can result from foods occurring in separate patches or habitats. When foods occur apart, search is no longer random with respect to food type. It is now directed toward one food or the other, but not both. The forager may appear to favor one food over another simply because that food occurs in particularly rich patches or safe habitats.

The state of the forager may alter its diet choice. For a mountain lion, mule deer are hard to encounter and successfully capture, but they pose minimal risk of injury to the mountain lion. Porcupines are the opposite. They are easier to encounter but they pose greater risks of injury to the mountain lion. Hence, a well-fed, successful mountain lion should eschew porcupines. But a down-and-out mountain lion should prefer to try its luck on capturing a porcupine rather than succumbing to the certainty of starvation. Bruce Patterson (2004) and others note this factor in the foraging behavior of large, man-eating cats. The man-eating lions of Tsavo likely switched diet as a consequence of prior crippling injuries.

Nutritional relationships—substitutable, complementary, antagosnistic, and essential—among foods can loom large in diet choice. Foods may offer different essential or complementary combinations of carbohydrates, fats, proteins, minerals, and vitamins. Shy on salt, moose of the northern Great Lakes of North America favor a salt-concentrated plant. Moose along coastal Scandinavia lose interest in this plant because much of their diet automatically includes plants impregnated with Baltic sea salt. A balanced diet means that foragers appear to favor the rarer food type or the food type with the scarcer nutrient. This balancing of nutrients can apply to plant toxins as well. Different plant species defend themselves with different chemical toxins such as tannins and oxalates. To an herbivore, it may be better to consume some tannins and some oxalates rather than a lot of just one—dose makes the poison, and feeding animals will often include this fact in their foraging behaviors.

5. PATCH USE

A jar of peanut butter or jam becomes increasingly frustrating and unsatisfying as the contents deplete. When full, a single swipe of the knife yields a bountiful spread. When mostly depleted, repeated strokes of the knife yield paltry returns. Eventually, we give up and discard the jar even though some contents remain. We share this dilemma of when to give up a depleted food patch and seek another with almost all feeding animals. Food items generally occur patchily, and the rate of food harvest declines as the patch becomes depleted. At what point should the forager abandon the patch, and how much unharvested food will it be leaving behind?

Eric Charnov (1976) proposed the Marginal Value Theorem for how long to remain in a food patch. The forager should leave its current patch when its harvest rate within the patch no longer exceeds what the forager's average harvest rate would be from leaving this patch, traveling to a fresh patch, and foraging that patch to the same quitting harvest rate. Put simply, leave a patch when the marginal rate of return (current harvest rate) drops to equal the forager's average harvest rate from the environment at large.

A forager should spend less time in a poor patch than a rich patch; a forager should spend less time in a patch of a rich environment than a poor environment; and a forager should leave patches sooner when travel time between patches is less. Foragers generally conform to these predictions, but with caveats. The costs and benefits of patch use may be more varied, and this has inspired variations and extensions of Charnov's model.

More generally, a forager should remain in a food patch until the benefits of harvesting resources, H, no longer exceed the sum of metabolic, C, predation, P, and missed opportunity, MOC, of foraging. Leave a patch when the harvest rate drops to $H=C+P+MOC$. If the forager's harvest rate within the food patch is directly related to the remaining abundance of food within the patch, the animal's patch use strategy also results in some amount of food being left behind. This remaining amount of food is referred to as the "giving-up density." The size of this giving-up density should be proportional to the animal's perceptions of foraging costs.

The gerbils of Bir Asluj demonstrate how giving-up densities change with these foraging costs and benefits. Gerbils have lower giving-up densities on foods that are more valuable, on foods that are easier to find, and within food patches that offer higher encounter probabilities on foods. Cold nighttime temperatures raise the gerbils' metabolic rates (C), and consequently they have higher giving-up densities on cold nights than warm nights. Gerbils feel safer (P) and have lower giving-up densities when seeds are under shrubs than a few meters away in the open, and they have lower giving-up densities on nights with no moon than with full moon. When resources are abundant within the environment (MOC), or when the gerbil has large stores of food, it will forage to a higher giving-up density and exaggerate its avoidance of the risky, open microhabitat even more. Well-off animals have more to lose from being killed by predators than animals in low states of energy or well-being.

The ability of foragers to detect and respond to variability in the distribution of food among patches is important for their ecology and their foraging behaviors. Perfect information on food availability allows the animal to perfectly balance its foraging time toward rich and/or safe food patches. Poor information on patch quality leaves the forager spending too much time in poor patches and too little time in rich patches. In reality, animals use sensory cues to "visualize" and assess patch qualities before investing time in the patch. Additionally, the foraging animal can use its experience within the patch to estimate patch quality. If the forager is having an easier time finding food than it expected, this may indicate a higher than average food abundance. Bayesian foraging studies how animals can use prior expectations and current experience to form an estimate of patch quality. The actual patch use behaviors of animals suggest that few have perfect information. Rather, foragers use a combination of preharvest sensory cues and sample information while foraging to form and update their estimate of patch quality.

The gerbils deplete their food patches by actually harvesting the seeds. But for foragers that have prey that can run, hide, or become vigilant, patch use takes the form of behavioral resource depression. The mere presence of the predator causes the "patch" to become less valuable as prey flee or become more wary. For predators with fearful prey, the catchability of their prey becomes as important as the number of prey. Cows in a pasture enjoy a very different proximity to birds than does the Cooper's hawk or Goshawk that aims to capture these birds.

Patch use behavior, through the giving-up density, has important implications for the distribution and abundance of the forager's food or prey. It may be that what we see in nature is simply the residue of feeding behaviors. What we see may often be what the foragers care not to eat, cannot eat, or cannot catch.

6. SOCIAL FORAGING

Leaf-cutter ants coordinate foraging as ants in the tree canopy drop their harvest to the ground where others transport the leaf discs back to the colony. Hyraxes post a sentinel that allows the other hyraxes to forage less fearfully. To counter these social foragers, one black eagle of a pair may circle in one direction from the colony, permitting the other eagle to fly in from elsewhere and surprise the otherwise distracted colony. Pelicans, seagulls, and cormorants are famous for forming noisy aggregations around promising patches of schooling fishes. These are all facets of foraging in groups. They reveal the competing interests associated with task specialization, predator detection, group hunting, information sharing (or dissembling), and shameless scrounging.

On first inspection, social foraging makes no sense. If a gerbil seeks to comb the sand dunes for seeds, doing so as a group simply means everyone has to walk farther for the same reward. When searching for food, better to divide the space and spread out. Hence, two critical factors loom large in social foraging—the forager's prey is behaviorally responsive, and the foragers fear their own predator. Advantages to social foraging as an antipredator adaptation accrue from having alarm calls, sentinels, many eyes, the dilution effect (better to catch my neighbor than me), and the confusion effect (many fleeing foragers may distract the predator from capturing any one forager).

When prey can flee or react, group foraging may permit task specialization (driving prey into an ambush), the ability to aggregate the prey (dolphins and whales corralling fish), beating the brush (banded mongooses moving abreast to scare up insects), or permitting the capture of large dangerous prey (army ants on vertebrates, wolves on a moose).

Information sharing looms as a benefit and consequence of group foraging. Spotting where others have found food may reduce the entire group's efficiency at finding food, but it may reduce the variance in food consumption. Less successful foragers join the feeding frenzy created by one forager stumbling on a particularly rich food patch. Some animals such as vampire bats and African hunting dogs will regurgitate and share food. Overly satiated members feed hungrier members. Such food sharing can even allow for task specialization where some individuals collect food even as others incubate a nest, protect a brood, or defend a territory from intruders. Of course, information sharing and gauging the successes of others introduces conflicts of interest where an individual may prefer to scrounge rather than produce its own harvest. Social groups encourage freeloading and producer–scrounger games. It may be that in some groups where siblings help their parents raise offspring (Florida scrub jays, Arabian babblers, and other birds), the balance of the relationship rests on the willingness of the parents to tolerate their "adult" offspring so long as they contribute food for their newest sibs.

Ant and bee colonies represent eusociality, the extreme of social foraging. These species exhibit caste systems, information sharing, group hunting or harvesting, and food sharing. What makes these systems special relative to a wolf pack or a naked mole rat colony? It may be the evolutionary objectives of the foragers that dictate the dividing line between a eusocial system and one that is merely a highly despotic social hierarchy. Individual worker ants and bees have been shown to forage in a way that completely subordinates themselves toward the fitness and success of the colony, whereas wolves and even individual naked mole rats seem to promote their own self-interests tempered by their need to be part of and treated well by the group.

7. FEAR AND FORAGING

Not a section of this chapter has gone by without some role for predators in shaping foraging behaviors. Foragers face a fundamental trade-off between food and safety. This trade-off becomes exacerbated and almost ensured by the adaptive behavior of having higher giving-up densities in risky than safe habitats. In most places and at most times, feeding animals face an environment in which background food abundance is high in risky habitats and low in safe places. A clever forager will use the tools of time allocation and vigilance to balance this trade-off. A clever predator will consider its prey's behaviors when doing its own foraging. The reciprocal behavioral responses of prey and predators lead to studies of predator–prey foraging games. Games of fear and stealth abound in nature across all taxa and ecosystems.

Steven Lima and William Mitchell have described the predator–prey shell game as prey seeking places free of predators and predators seeking to be where the prey are. The environment determines the form of the game. The afternoon winds at Bir Asluj ensure abundant seeds at dusk. This encourages gerbils to emerge early, which encourages clever owls to do the same. The responses of gerbils to seeds and of owls to gerbils create three temporal gradients. The seeds decline steadily as the gerbils deplete them. Gerbils start the night wary of owls and become increasingly less so as the night draws on. The owls modulate their behavior to track the seeds—busy early and less so later.

Foraging games can encompass several prey and predators. Owls encourage gerbils to forage more under shrubs than in the open. Snakes take advantage of this fear response by lying under shrubs in ambush. Owls and snakes create predator facilitation where the presence of one predator species makes it easier for the other to capture the shared prey. Furthermore, the nightly decline of seeds and risk promotes the coexistence of the Greater Egyptian sand gerbil with a smaller cousin, Allenby's gerbil (*G. andersoni allenbyi*). The size, temperament, and behavior of the large gerbil suits it for early in the night, whereas the little gerbil has adaptations and behaviors more suited to the resource poor but safer periods of the night. Burt Kotler (1984), through "fear and foraging," showed the role of predation risk in the foraging behaviors and coexistence of kangaroo rats and pocket mice at a desert site in Nevada.

The behaviors of the prey may facilitate the coexistence of diverse predators, and the behaviors of predators may similarly promote diverse prey. Nowhere is this more likely than the reciprocal radiation of insects and plants. The feeding behaviors of herbivorous insects select for plant defenses. The evolution of additional insect species to overcome these defenses simply encourages the evolution of additional and more diverse defenses among an increasing number of plants species. So the game of feeding and defending promotes other morphological adaptations and perhaps even speciation and adaptive radiations.

8. COADAPTATIONS BETWEEN FORAGING BEHAVIORS AND MORPHOLOGY

Coral reef fishes offer a bedazzling array of sizes, colors, and shapes. Many of these species feed on corals or the algae that grow in them. Close inspection of these fish reveals delicate differences in the mouthparts, mandibles, and teeth. Like a tray of dental instruments, these varied mouthparts permit the different species to scrape algae from diverse surfaces, chew coral, and probe interstices within the coral for food. The body sizes, fin dimensions, and body forms of the fish serve to stabilize and maneuver the fish within the water column to permit access to food and escape from predators. We see a fine-tuned coadaptation of feeding behaviors, mouthparts, and other morphological attributes. But, what came first—the behavior, the mouthparts, or the body form?

My doctoral advisor Michael Rosenzweig would tell us how "Natural selection can never adapt an organism to something it does not do." A feeding behavior must then precede coadaptive changes in physiology and morphology. But the species' prior physiology and morphology must at the very least allow for the behavior. This necessitates an important distinction between behaviors being selective versus opportunistic, and morphological adaptations as being specialist versus generalist (Rosenzweig, 1991).

A feeding animal may be more or less picky in its selection of foods and/or places to feed. A North American robin may choose to feed selectively on insects or fruits, or it may opportunistically feed on both as they are encountered. When the robin is feeding just on insects, its gut modulates to enhance the digestion of insects at the expense of fruit, and vice versa when robins feed primarily on fruits. Finally, the body size and morphology of a robin make it adept at probing for insects in the soil and leaf litter, moderately apt at picking insects from branches and leaves, and quite unable to collect insects from under bark or by "flycatching" insects from midair.

As natural selection engineers a fit between form and function, feeding behaviors or their absence can have profound consequences for the other traits of an organism. If a feeding opportunity arises, then a species previously nonadapted to this opportunity may acclimate by altering its foraging behavior. As this opportunity becomes an important part of its ecology, there will be selection on the species morphology and physiology to adapt. For instance, the cultivation of apples in the New World led to the apple-maggot fly evolving a new species. The precursor species inhabited native hawthorns. Those that switched to apples were now selected to fine-tune their breeding strategies to better match the flowering and fruiting phenology of apples. As the flip side of this same force, if a forager ceases to have a particular feeding opportunity, the absence of this behavior from its repertoire could lead to the loss of morphological and physiological adaptations aimed at improving the rewards from the now-absent behavior. Conserving a species may require us to preserve environments that maintain its full suite of feeding behaviors.

9. NUTRIENT FORAGING IN PLANTS

Plants forage too. They exhibit noncognitive behaviors and responses to light, nutrients, and water. Their "behaviors" represent allocation decisions and growth patterns. Their architecture and investment into roots contribute water and nutrients. Investment into aboveground leaves and stems influences carbon fixation. When viewed as nutrient foraging, most, if not all, of the principles and concepts of animal foraging behavior apply to plants—often with dramatic effect.

We often take wood for granted. Clearing trees created farmland and pastures. The wood itself could heat homes and power machines. As a building material it is sturdy, strong, and durable. The chair I sit in now is made from maple. Why is there wood? Competition for light. Nutrient foraging for light creates a special form of the tragedy of the commons. To be successful at having full sunlight, a plant need only be a bit taller than its neighbors. But if these neighbors respond in kind, an arms race ensues with ever greater and greater investment in sturdy, tall, woody trunks. What determines the canopy height? The costs and benefits of foraging. The benefit of being in the sunlight remains mostly constant because the available pool of light does not change with height (unless one gets demonstrably closer to the sun!). Yet the costs multiply with ever thicker trunks, greater surface area for pathogens and boring insect pests, greater mechanical challenges of transporting water to the canopy and photosynthates back to the roots, and greater chances of toppling over. As the trees play an evolutionary game of light competition, they achieve a canopy height at which no individual can benefit from being a bit taller and no individual is willing to concede light by being shorter. The environment-specific and tree species–specific adjustments of these costs and benefits produce 80-m-tall redwood forests and 30-m-tall European beech forests.

Other strategies for light foraging abound. Light gaps encourage the lateral growth of branches and strange bends in stalks or stems. Maple trees will produce "sun-loving" leaves for their canopy and "shade-tolerant" leaves for their subcanopy branches. Some plants will track the path of the sun with their leaves. Leaf size, morphology, greenness, and stem structure all contribute to the hugely diverse ways by which plants forage for light. A kind of producer–scrounger game happens when species of vines skip the investment in wood and simply achieve the canopy by growing up another's trunk.

The same holds for belowground nutrient foraging via roots. Plants may overproliferate roots with the goal of "stealing" nutrients from a neighbor. Of course, the neighbor is selected to respond in kind, and a belowground tragedy of the commons ensues. Roots show other varieties of noncognitive foraging behaviors. Plants will direct root proliferation toward areas of high nutrients. Plants may modulate root architecture (fineness of roots, density of root hairs) and root uptake kinetics (ability to actively transport nutrients) in response to nutrient opportunities.

The bargaining game between mycorrhizal fungi and plants presents an emerging frontier. Mycorrhizae are adept at concentrating nitrogen and phosphorus and then exchanging these with the roots of a plant for carbohydrates. To what extent is this symbiosis best modeled as a nutrient game? Elevated carbon dioxide levels in the atmosphere pose one of the greatest and most interesting challenges for the twenty-first century. Can nutrient foraging by plants play a role in understanding the dynamics of atmospheric CO_2 and the concomitant climate change? This author thinks so.

Whether animal or plant, universal aspects of feeding behaviors involve tactics for searching for and handling resources, foods, and prey. This process reaps rewards in terms of the value of the harvest and incurs costs that include the risk of injury or predation. The interplay between natural selection and the variety of environmental circumstances produces the myriad of foraging behaviors found among the millions of species inhabiting the planet. These behaviors allow foragers to seek and handle foods quickly, efficiently, and safely. The wind will blow, the seeds will redistribute, the sun will set, and the gerbils and owls will emerge to forage.

FURTHER READING

Charnov, E. L. 1976. Optimal foraging: The marginal value theorem. Theoretical Population Biology 9: 129–136.

Emlen, J. M. 1966. The role of time and energy in food preference. American Naturalist 100: 611–617.

Kotler, B. P. 1984. Predation risk and the structure of desert rodent communities. Ecology 65: 689–701.

MacArthur, R. H., and E. Pianka. 1966. On optimal use of a patchy environment. American Naturalist 100: 603–609.

Patterson, B. D. 2004. The Lions of Tsavo: Exploring the Legacy of Africa's Notorious Man-eaters. New York: McGraw-Hill Professional.

Pulliam, H. R. 1974. On the theory of optimal diets. American Naturalist 108: 59–75.

Rosenzweig, M. L. 1991. Habitat selection and population interactions: A search for mechanism. American Naturalist 137: S5–S28.

Stephens, D. W., J. S. Brown, and R. Ydenberg, eds. 2007. Foraging: Behavior and Ecology. Chicago: University of Chicago Press.

Temple, S. A. 1987. Do predators always capture substandard individuals disproportionately from prey populations? Ecology 68: 669–674.

I.8

Social Behavior

Eldridge S. Adams

OUTLINE

1. Ecological consequences of social behavior
2. The evolution of cooperation and altruism
3. Mechanisms of social behavior

Social life is a mix of cooperation, altruism, and selfishness. In species as diverse as slime molds, army ants, and great apes, individuals coordinate actions to achieve common goals. Yet competition and conflict are common within social groups and may lead to lethal altercations. Consider, for example, a well-integrated, long-lived society such as a colony of the honeybee *Apis mellifera*. Cooperative foraging is organized by communication among the worker bees, allowing the colony to allocate effort flexibly over a large region surrounding the hive. By acting in concert, nestmates build combs, raise young, and maintain a comfortable nest temperature even through snowy winters. When the hive is threatened by a vertebrate predator, workers sacrifice their lives to protect the queen and her offspring, perhaps the most celebrated example of altruism among the insects. Yet the benefits of this collective activity are not evenly shared. Although there may be well over 15,000 females in the society, one of them—the queen—lays the vast majority of eggs while eggs laid by other females are quickly eaten. Other conflicts are evident. As winter approaches, males, which do no work, are dragged out of the hive and left to die. When a new queen is reared, she may sting to death other prospective queens still developing in their royal cells, bringing reproductive competition to a deadly conclusion. Like other social species, the honeybee prompts two central questions. How do organisms benefit from group living? What prevents conflicts within the group from undermining cooperative aspects of social life?

GLOSSARY

Allee effect. An inverse relationship between population density and per capita population growth rate. Allee effects can accelerate the decline of a shrinking population.

altruism. Behavior that is costly to the individual performing it and is beneficial to one or more other individuals; costs and benefits are measured in terms of effects on fitness, which can be quantified by lifetime reproductive success.

coefficient of relatedness. The probability that one animal shares an allele carried by another as a result of descent from a common ancestor.

cooperation. Behavior that benefits two or more interacting individuals.

kin selection. Selection resulting from the effects of an organism on the fitness of relatives, as well as through the organism's own reproduction.

policing. Actions by group members that suppress or punish selfish behavior by other group members.

selfishness. Behavior that benefits the individual performing it at a cost to one or more other individuals.

self-organization. In social species, this refers to phenomena in which group organization arises spontaneously, without central control, because of the actions and interactions of multiple individuals.

1. ECOLOGICAL CONSEQUENCES OF SOCIAL BEHAVIOR

Why do so many organisms live in groups? To behavioral ecologists, the abundance and diversity of social species suggest that in many environments the benefits of group living outweigh its costs. The forces driving sociality vary, but field studies have revealed a few principal advantages, which recur in diverse taxa. One set of advantages emerges in the context of foraging. By acting in groups, animals can improve foraging success through enhanced search, ability to overcome prey defenses, or ability to outcompete other groups or individuals. On the other side of the hunt, potential prey often seek to escape capture by clustering and moving together. Group activity makes it more difficult or more dangerous for a predator to attack, and potential prey can maneuver for positions in which they are less vulnerable than other group members. Shared vigilance allows animals to spend more time foraging and less time watching for predators, increasing energetic efficiency. It is common for social predators to attack social prey, as when pods of dolphins hunt fish or squid, when packs of African wild dogs (*Lycaon pictus*) chase impala (*Aepyceros melampus*), or when columns of army ants overrun the nests of paper wasps.

Other advantages of group living arise in interactions between parasites and their hosts. Parasites and pathogens can overwhelm host defenses by acting in concert, using chemical signals to synchronize attacks in space and time. Some ant species specialize on parasitizing other ants, organizing raids in which brood is stolen to augment the worker force of the raiders. Potential hosts can also employ social behavior to defend themselves. Social grooming, common in primates and social insects, helps to reduce parasitism and even to improve immunity.

Coordinated groups are formidable competitors for limited resources, including food, nest sites, and opportunities to mate. Competition between social species is seen in faunas as diverse as social carnivores in African grassland and ant colonies in the canopies of tropical rainforests, where the number of allies may override individual fighting ability to determine who gains access to food. Moreover, by promoting repeated contacts among individuals competing for opportunities to mate, social life itself produces an environment favorable to establishment of competitive coalitions. Thus, male lions (*Panthera leo*), dolphins (*Tursiops* spp.), and wild stallions (*Equus caballus*), among others, form alliances to acquire or guard mates. Subordinate or bachelor males team up to gain access to mates defended by stronger individuals, and the dominant male in turn may recruit assistance to fend off challengers.

In other contexts, group living benefits animals by improving the efficiency of movement (e.g., the V-formation of geese), because of thermal advantages of clustering (e.g., social hibernation in marmots), and by improving the efficiency of nest construction (e.g., paper wasps). Much social behavior occurs in the context of rearing young. Cooperative breeding, with some individuals playing a supportive role to others, is seen in some birds, mammals, fish, and snapping shrimp and in thousands of species of social insects.

Similar principles underlie success in most of these examples: the collective effort of a group greatly exceeds that of a solitary animal; effectiveness of fighting is improved by outnumbering antagonists; sharing tasks permits animals to devote more time and energy to other needs; division of labor allows individuals to focus on complementary activities.

In animal conservation, the population consequences of sociality create a special concern. Social life can buffer a group against environmental change. Ants, for example, curtail colony growth when food is in short supply and can even eat their young, allowing the colony to live through the period of scarcity. However, species dependent on social strategies may be subject to Allee effects, which occur when the per capita rate of population growth is inversely related to density. When populations are sparse, animals may be unable to form cooperative associations of sufficient size, causing the population size to spiral downward. Field evidence shows that reduced group size can lead to colony failure. For example, in the highly social Damaraland mole rat (*Cryptomys damarensis*), small colonies perish during droughts because they lack the workforce needed to excavate underground tunnels leading to the storage roots that form the bulk of their diet. Allee effects can amplify the risk of extinction for social species when their populations suffer modest declines.

2. THE EVOLUTION OF COOPERATION AND ALTRUISM

Despite the advantages outlined above, group living often fosters competition for limited resources and opportunities to mate. There are five primary hypotheses for the evolution of helping behaviors, in which one animal acts to increase the survival or reproduction of others. Each hypothesis proposes a different viewpoint on the forces that keep competition in check.

Mutualism

By helping another individual, an organism can help itself. When group activity allows an outcome that cannot be achieved by acting alone, then there is little temptation for an individual to cheat by declining to participate. Doing so dooms the entire enterprise. This form of cooperation is sometimes called “by-product mutualism” because cooperation results from each individual acting in its own best interest. Yet if the collective action requires coordination and communication among group members, it is not merely an accidental consequence of selfish actions. Furthermore, even though cooperative behavior may produce mutual advantages, this does not guarantee that the benefits are evenly shared. Suppose, for example, that cooperation is essential for capture of large prey. Conflicts can still arise over how the food is divided or because individuals that did not participate in the chase seek a share of the catch. Such discord is seen in the spider *Amaurobius ferox*, in which young must act in groups to capture crickets, which are larger than the spiders themselves. Kil Won Kim and colleagues showed that success is very unlikely without participation by several spiders; nevertheless, conflicts and freeloading are common. When the cricket is comparatively small, the spiders are more likely to fight among themselves for opportunities to feed on captured prey, and when the cricket is much larger, spiders are more likely to join in feeding without taking part in the capture.

Kin Selection

William D. Hamilton reasoned that an organism can promote the spread of its genes in two ways: directly, by producing its own offspring, and indirectly, by helping relatives to survive and reproduce. The sum of these two components is referred to as the organism's *inclusive fitness*. The effectiveness of the indirect route is governed by the coefficient of relatedness, the probability that a randomly chosen allele carried by the helper is shared by the beneficiary as a result of descent from a common ancestor. Relatedness ranges from 0 for nonrelatives, to 1 for identical twins or members of the same clone. Changes in gene frequency resulting from both the direct and indirect pathways constitute kin selection.

The most famous effect of kin selection is that it can lead to the evolution of altruism, behavior by which an individual helps another at a cost to itself. Consider, for example, a bird that must decide whether to expend effort raising a son or daughter, or to help its parents to produce one more offspring—that is, a brother or a sister. Foregoing breeding to help parents is common in dozens of species of cooperatively breeding birds, such as the Florida scrub jay (*Aphelocoma coerulescens*). In birds, as in most familiar vertebrates, the coefficient of relatedness between parents and offspring is 0.5, because each parent has a 50% chance of passing a particular allele to a random son or daughter. Yet the coefficient of relatedness between siblings in these species is also 0.5. Therefore, a bird produces as many copies of its own genes by raising a sibling as it would by raising one of its own offspring. Which option is favored depends on their relative effectiveness. A shortage of mating opportunities for young adults or greater efficiency of groups can tip the balance in favor of helping relatives. In Florida scrub jays, for example, suitable breeding habitat is very limited, so young birds, especially males, profit from remaining in their parents' territories and helping to rear siblings. Eventually, a male may inherit all or part of the parental territory, so he derives both indirect benefits from raising siblings and direct benefits from acquiring a breeding territory.

Kin selection is demonstrated by adaptive evolution of the sterile castes of social insects. In the termite family Termitidae, for example, the soldiers have evolved varied morphological adaptations for defense, some with sickle-shaped mandibles, others with a nozzle-like extension at the front of the head used to spray chemicals onto ants or other attackers. But these soldiers do not reproduce, so how can their morphology or behavior evolve? Soldiers act to protect their parents, the queens and kings, which carry and transmit essentially all of the soldiers' genes. More effective forms of defense are favored by kin selection, entirely through the indirect route of helping relatives.

The calculus of kin-selected altruism is summarized by Hamilton's rule. An animal is favored to perform an altruistic behavior if c, the cost to itself, is exceeded by b, the benefit to the recipient, multiplied by r, the coefficient of relatedness ($c < br$). This condition is more easily satisfied for close relatives than for distant relatives. Appreciation of the importance of the indirect route of gene transmission has transformed the way social behavior is studied. Estimating genetic relatedness among group members is now a standard part of the analysis of animal social behavior.

Reciprocity

An animal may be favored to help another, at a cost to itself, if at a later time the roles are reversed. Robert Trivers termed this type of interaction "reciprocal altruism," noting that the behavior is altruistic in the short run but cooperative in the long run. The chief requirements are that the benefit to the recipient exceeds the cost to the donor and that some mechanism protects against "cheaters" that accept help from others but then do not reciprocate.

Game theoreticians have published hundreds of analyses of the evolution of reciprocity, yet there are few clear examples from nonhuman animals. Possible instances are seen in hermaphroditic species, such as the black hamlet, *Hypoplectrus nigricans*, a reef-dwelling fish. Eric Fischer argued that, in a mating between two hamlets, each individual is favored to supply the relatively inexpensive sperm rather than the larger and more costly eggs. However, if neither individual provides eggs, then reproduction cannot take place. Among black hamlets, this dilemma is solved by breaking mating into a series of bouts, in which the two fish alternate in the male and female roles. Black hamlets take longer to offer eggs to a mate that did not reciprocate in a previous encounter, a tendency that can protect against cheating.

Coercion and Policing

In some societies, coercion suppresses selfish actions and promotes helping behaviors. Dominance hierarchies govern reproductive rates in many social groups, with aggression by those at the top inhibiting reproduction by subordinates and sometimes inducing them to work. Helping behavior, in this case, is not entirely voluntary but rather is the best option remaining after choice of action has been restricted by dominant animals.

In some social insects, attempted selfish behavior by a group member is kept in check by responses of other group members, actions known as "policing." This has been best studied in honeybees. In a typical honeybee colony, the queen lays the vast majority of eggs. However, workers can lay unfertilized eggs, which, because of the unusual method of sex determination in honeybees, develop into reproductive males. Workers are more closely related to their own sons than to the queen's sons but are more closely related to the queen's sons than to the sons of other randomly chosen workers. Therefore, although each worker has an incentive to lay eggs herself, she should oppose egg-laying by other workers, preferring that reproduction be left to the queen. In fact, worker honeybees usually thwart attempts by other workers to produce male offspring. Worker eggs are distinguished from the queen's eggs, probably on the basis of odor, and are eaten by other workers before they develop. Effective policing lowers the incentive for workers to lay eggs in the first place and so can promote the evolution of worker sterility.

Group Selection

Cooperation and altruism can be promoted by the increased survival and reproduction of groups in which these behaviors are prevalent. This hypothesis invokes selection operating at the level of groups and in opposition to individual-level selection within groups. The group selection hypothesis fell into disfavor in large part because of George C. Williams' influential book *Adaptation and Natural Selection*, published in 1966. Williams critically reassessed former claims that particular behaviors evolved for the good of a group or the good of a species. He emphasized that individual selection is more powerful than group selection and that most putative examples of group-selected traits turn out, on closer inspection, to be advantageous to the individuals performing them.

However, further theoretical work showed that the conditions under which group selection can shape social behavior are not as restrictive as previously thought. Older models relied on differential survival of groups and required groups to be well separated, with very limited gene flow. The newer models rely more on differential reproduction and allow groups to be temporary. To highlight the differences in model structure, the newer models are said to represent "trait-group selection." Although claims of group selection continue to stimulate objections and misunderstanding, the controversy is resolved in part by recognizing that group selection can be formally equivalent to kin selection. In other words, in many cases where group selection works to promote altruism, altruists are on average related to the group members that they aid. The usefulness of the group selection perspective is that it provides a way to describe the effects of behaviors and selection at both the individual and the group level.

For an example of group selection overriding selection for selfishness within groups, we can turn again to honeybees. Recall that most worker-laid eggs are eaten by other workers before they develop. Beekeepers have discovered a small number of colonies in which this arrangement breaks down. In these "anarchic" colonies, some worker bees have more highly developed ovaries and can lay male eggs that are much less likely to be consumed by nestmates, presumably because they smell like the queen's eggs. These workers are able to escape policing and therefore to produce their own sons. This escape represents a selfish behavior by the anarchic bees, and one that could potentially spread rapidly because of the huge increase in individual reproduction that it allows. However, colonies with anarchic bees fare poorly, producing many males but few workers, and without active intervention by the beekeeper, most anarchic colonies perish. Thus, although evasion of policing leads to individual success within the colony, the colony itself is quickly doomed.

3. MECHANISMS OF SOCIAL BEHAVIOR

The mechanisms underlying sociality are as diverse as the animals and behaviors themselves, encompassing genes and development, endocrinology and neurobiology, communication and cognition. The ability to identify specific genes affecting behavior and to follow their action through the physiology and development of the animal has accelerated rapidly. To mention a single example, the behavior of mice is altered by insertion of the vasopressin receptor gene from a more social rodent, the prairie vole. Larry Young and colleagues showed that male transgenic mice respond to the hormone vasopressin by increasing social behavior toward females, a response normally seen in prairie voles but not in mice.

From the standpoint of understanding the ecology and evolution of sociality, two categories of mechanisms are particularly important.

Proximate Mechanisms for Coordinating Action

Coordinated social behavior originated among single-celled organisms. Among some existing species of bacteria, combined action is triggered by "quorum-sensing," in which high densities are detected by the buildup of signal molecules released into the external environment. When these signals reach a critical concentration, they stimulate profound changes in gene expression and cell behavior. This sensitivity to cell density allows the bacteria to secrete proteins only when their high concentration is likely to have a beneficial effect. For example, quorum-sensing in the pathogenic bacterium *Staphylococcus aureus* stimulates release of toxins at high population densities, allowing the bacteria to outcompete other strains and increasing their virulence toward the host.

In large groups of animals, much of the communication needed to organize collective action is achieved by simple, anonymous responses. The characteristic movement patterns of flocks, schools, and herds result largely from simple rules by which individuals adjust their spacing, alignment, and speed relative to other nearby animals. If the group-level behavior arises without central control, from the local decisions and interactions of numerous individuals, the pattern is said to be "self-organizing." Even when individual animals have incomplete information, self-organization can produce a collective intelligence allowing favorable decisions. Ant traffic, for example, can coalesce on the shortest of several available routes despite the fact that no individual directly compares alternative pathways. Use of the shortest route comes about as an automatic consequence of the way ants deposit and respond to chemical trails. The chemical signal is reinforced more rapidly on shorter routes simply because it takes ants less time to walk from one end to the other. Other ants are then drawn to the trail segments that are more strongly marked.

At the other extreme, group processes rely on tight feedback between particular individuals playing different roles. For example, Redouan Bshary and colleagues showed that two species of fish use signals to coordinate cooperative hunts. One species, the grouper *Plectropomus pessuliferus*, hunts in the open water over coral reefs, causing prey to seek cover. The other, the giant moray eel, *Gymnothorax javanicus*, readily moves through crevices, causing prey to flee into the open. A grouper initiates cooperative hunts by a visual signal, shaking its head back and forth rapidly while facing a moray eel. The two fish may then hunt together for more than 30 min, and both benefit from an increased rate of capture as a result of their complementary hunting styles.

Proximate Mechanisms Curtailing Cheating and Selfishness

The stability of some forms of social behavior, including reciprocity and altruism, requires that the choice of partners be restricted. As discussed above, altruism can evolve by kin selection if the donor and the beneficiary are related. Directing altruism preferentially toward kin does not necessarily require any special cognitive abilities. Instead, the spatial or group structure of the population may ensure that animals interact primarily with relatives. Alternatively, animals may learn the characteristics of group members during early development and then offer helping behavior only to those animals that resemble this learned template. Because individuals who are close by during juvenile stages are likely to be relatives, this type of learning allows helping behaviors to be directed toward kin. Many studies have sought evidence of a more refined ability to distinguish degrees of relatedness among equally familiar group members, but the evidence is scant. In principle, relatedness can be detected by shared heritable tags. This phenomenon is known as the *armpit effect* if the animal develops the standard for comparison by learning its own phenotype (e.g., by sniffing its own armpit), or the *greenbeard effect* if individuals recognize which other individuals carry copies of the same gene for altruism because the gene also codes for an identifying label (e.g., a green beard).

Reciprocal altruism does not require that animals be related, or even of the same species, but it does require protection against freeloaders, which accept help but do not offer it in return. Reciprocity is closely related to concepts of fairness, scorekeeping, and reputation building. The cognitive and emotional capacity to remember the past behavior of other animals and to respond with generosity or reprisal is well developed in some nonhuman primates. Frans de Waal and colleagues have documented these abilities in chimpanzees and capuchin monkeys, which share food with individuals other than offspring. Chimpanzees are more likely to share food with particular individuals from which they have recently received grooming and to respond aggressively toward those who have not groomed them. The evidence from capuchins goes even further. In controlled experiments, capuchins were offered food rewards for performing certain tasks, while also watching the rewards given to a paired monkey for the same performance. The monkeys were willing to perform the task for a low-value food item, such as a piece of cucumber, so long as both monkeys were given the same reward. However, when the partner received a food item of greater value, such as a grape, monkeys were less willing to perform the same task unless they too were given a grape. Social primates can base actions on the memory of previous encounters, comparing the reward received for a given effort to the rewards obtained by other group members. Like other social animals, they can at once struggle to ensure that the group succeeds and to improve success relative to others within the group.

FURTHER READING

Barron, Andrew B., Benjamin P. Oldroyd, and Francis L. Ratnieks. 2001. Worker reproduction in honey-bees (*Apis*) and the anarchic syndrome: A review. Behavioral Ecology and Sociobiology 50(3): 199–208.

Camazine, Scott, Jean-Louis Deneubourg, Nigel R. Franks, James Sneyd, Guy Theraulaz, and Eric Bonabeau. 2001. Self-organization in Biological Systems. Princeton, NJ: Princeton University Press.

Courchamp, Franck, Tim Clutton-Brock, and Bryan Grenfell. 1999. Inverse density dependence and the Allee effect. Trends in Ecology and Evolution 14(10): 405–410.

Crespi, Bernard J. 2001. The evolution of social behavior in microorganisms. Trends in Ecology and Evolutionary Biology 16(4): 178–183.

de Waal, Frans. 2006. Primates and Philosophers: How Morality Evolved. Princeton, NJ: Princeton University Press.

Dugatkin, Lee A. 1997. Cooperation among Animals: An Evolutionary Perspective. New York: Oxford University Press.

Krause, Jens, and Graeme D. Ruxton. 2002. Living in Groups. New York: Oxford University Press.

Seeley, Thomas D. 1995. The Wisdom of the Hive: The Social Physiology of Honey Bee Colonies. Cambridge, MA: Harvard University Press.

I.9

Phenotypic Plasticity

Joseph Travis

OUTLINE

1. Introduction
2. The spectrum of phenotypic plasticity
3. The evolution of adaptive plasticity
4. The ecological importance of phenotypic plasticity
5. Horizons for future ecological research on phenotypic plasticity

Phenotypic plasticity is the ability of an individual to express different features under different environmental conditions. Examples of plasticity surround us: plants have broader leaves when grown in shady conditions, and animals are smaller when they develop in crowded conditions. Although some of these changes reflect unavoidable consequences of adverse conditions, many of them are the product of natural selection molding an organism's ability to survive and reproduce in a world whose conditions vary from time to time and from place to place. Put another way, many examples of phenotypic plasticity reflect the evolution of a developmental system that attempts to produce different traits under different conditions because no single trait is best suited for all conditions. Plasticity facilitates a species' ability to occupy a variety of habitats, persist in uncertain environments, and stabilize its interactions with other species whose incidence and numbers change over time and across space.

GLOSSARY

carapace. The hard outer shell surrounding the bodies of small animals such as waterfleas and larger animals such as turtles.

diapause. A state of arrested development in which the animal can survive long periods of challenging conditions such as low temperatures or drought by lying dormant.

ectothermic animals. Animals that use external sources of heat for metabolism and whose rates of metabolism are closely linked to external temperatures, such as invertebrates, fish, amphibians, and reptiles.

fitness. The number of offspring an individual leaves behind for the next generation; fitness has two major components, survival (or length of life) and reproductive rate.

numerical stability. A steady-state equilibrium in population size, that is, numbers of individuals, to which a system will return if it is perturbed; stability in predator–prey systems refers to the numerical stability of both predator and prey that allows them to coexist indefinitely.

phenotypic plasticity. The ability of an individual to express different features under different environmental conditions.

1. INTRODUCTION

Phenotypic plasticity is the ability of an individual to express different features under different environmental conditions. This "adaptive plasticity" is one of the most remarkable products of Darwinian evolution. For adaptive plasticity to emerge, the developmental machinery to build different traits must be integrated with a sensory system that detects reliable cues about the prevailing environmental condition so that suitable traits are expressed in a timely manner. Adaptive plasticity is an interesting topic for evolutionary biology, but it is also an important topic in ecology. One reason is that plasticity can enable a species to cope with highly seasonal environments or occupy diverse habitats. But more subtly, plasticity can have a substantial effect on a variety of ecological processes and thereby act as an important influence on which species we see where and at what population sizes.

2. THE SPECTRUM OF PHENOTYPIC PLASTICITY

Phenotypic plasticity can be either reversible or irreversible. The most obvious examples of reversible changes are behavioral responses to environmental conditions. For example, tadpoles change their foraging patterns in response to the presence of predators. When predators are removed, the tadpoles adjust accordingly. Other well-known reversible responses include physiological changes such as the increase in mitochondrial density in terrestrial vertebrates in response to experiencing lower oxygen levels and the changes in specific fatty acids incorporated into animal cell membranes in response to changing thermal conditions. Morphological changes can also be reversible: the gills of aquatic salamanders increase or decrease in response to oxygen levels in the water, and vertebrate muscles change in form and density in response to the amount of use they receive.

As one might expect, reversible plasticity appears when environmental conditions change, often within an individual's lifetime. In most cases, individuals retain the ability to change their features for most of their lives. The exception to this rule is diapause in insects and other arthropods. Diapause is a state of arrested development in which the animal can survive long periods of challenging conditions such as low temperatures or drought by lying dormant. When conditions improve, the animal breaks diapause and resumes normal activity and development. A species can enter diapause in only one stage, for example, eggs in crickets and larvae in beetles, and once broken, diapause cannot be reentered.

Irreversible changes occur trivially when an organism adjusts the timing of a life history transition in response to environmental circumstances. Once an annual plant initiates flowering in response to its lighting conditions, there is no going back. Less trivially, irreversible changes are reflected in features that, once expressed, are not altered regardless of how conditions may change. For example, waterfleas in ponds develop spines and a thicker carapace in response to the presence of a predatory fly larva in the water; once developed, the carapace is not altered appreciably even if the predators disappear. A species of African acacia develops long spines on its stems in response to being browsed by giraffes and elephants; these spines remain for the lifetime of the tree, even if it never suffers from additional browsing.

Irreversible plasticity appears when environmental conditions are less volatile and less likely to change drastically within the lifetime of an individual. In many of these cases, there is a narrow window of development within which the individual is sensitive to the cues in the environment that trigger the expression of the feature. Outside of that window, the cues elicit no response. When these narrow windows of sensitivity exist, the individual is committing itself for the future in response to conditions in one relatively short period.

Whether reversible or irreversible, plasticity is expressed in response to a wide range of environmental factors. Some factors act ubiquitously; nearly all plants alter the expression of shoots, leaves, and flowers in response to variation in their lighting environments, and most animals alter development in response to variation in their thermal environments. Classes of biotic agents—predators, pathogens, potential competitors—also induce plastic responses. In some cases, the cue for the response is direct: the African acacia develops spines after it has been browsed. In others, the cue is indirect: waterfleas develop thicker carapaces in response to a chemical cue that alerts them to the presence of a larval midge predator, even before there is any attack on an individual waterflea.

The many examples of plasticity in nature might suggest that just about any feature of an organism can be phenotypically plastic and just about any environmental condition can induce a plastic response. This is true if one looks at all of nature's examples en masse; every trait responds to some environmental factor, and just about any environmental factor imaginable affects some trait in some species. But in a very important sense, it is not: plasticity can be quite specific. To be sure, there are general patterns of plasticity; nearly all ectothermic animals make larger eggs at lower temperatures. But the more striking observation is that the development of certain traits responds in specific species to specific cues; traits in a species that respond to one environmental agent may not respond to a different one, and the same features in different species may not respond to the same agent. Put another way, when one says "Trait X is plastic," one needs to specify in which species and in response to variation in which environmental condition.

There are several striking examples of this specificity. Damselfly species that coexist with fish behave differently in the presence of fish than in their absence, but species that do not coexist with fish fail to respond to their presence and are more likely to be eaten. Plasticity can even be specific at the population level; wild parsnip populations with a history of heavy herbivory respond to leaf damage by releasing compounds toxic to insect herbivores, whereas populations without a history of heavy herbivory do not.

Even more subtly, plasticity can be quite precise. That is, a trait may respond only to a particular range of variation in an environmental factor, and the same trait in different species may respond to a different range of variation in that same factor. Insect diapause is a classic example: populations of the same species at different latitudes enter diapause in response to different combinations of temperature and day length.

The specificity and precision of so much phenotypic plasticity suggest that it is not merely an ineluctable consequence of animal or plant physiology but a well-honed evolutionary response to variable environments of a particular kind.

3. THE EVOLUTION OF ADAPTIVE PLASTICITY

Adaptive plasticity should evolve whenever individuals with the capacity to adjust their development to the prevailing conditions outperform, in the long run, individuals that express the same trait values or features constitutively, that is, regardless of condition. By "outperform" we mean "have a higher fitness," that is, be more likely to survive or leave more offspring behind. The subtlety is in the phrase "in the long run." In any single circumstance, the individual with the capacity to adjust its development to express the most suitable feature will perform just as well as the individual who expresses the same feature constitutively. But it will outperform all of the individuals who express unsuitable features constitutively. Individuals with the capacity to adjust development have high fitness in all conditions, whereas individuals with constitutive development patterns for the same set of features have high fitness in some conditions but low fitness in most conditions. In the long run, over many generations or many locations, individuals with the capacity to adjust development have the highest average fitness.

To illustrate the argument, consider the waterfleas that develop a thicker carapace in response to the presence of a predatory fly larva. Developing a thicker carapace takes energy that would be used otherwise to accelerate maturation and reproduction. When predators are present, the thicker carapace repays the investment because it reduces the ability of the fly larva to capture and kill the animal before it reproduces. In the absence of the predator, the thicker carapace is a waste of energy because it detracts from the ability of the waterflea to get on with the business of maturing, mating, and reproducing. A waterflea that made a thin carapace regardless of conditions would do well in the absence of predators but poorly in their presence; conversely, a waterflea that made a thick carapace regardless of conditions would thrive in the presence of predators but do poorly in their absence. The waterflea with the plastic developmental system has the best of both worlds and, if predators are present at some times but not others, would, in the long run, have a higher average fitness than waterfleas that develop thick or thin carapaces constitutively.

If plasticity is such an obvious advantage over constitutive development, why would developmental systems be anything but plastic when different features are suited to different conditions? The apparently transparent advantage of phenotypic plasticity, as illustrated by the waterflea example, is based on three assumptions. The first assumption is that a reliable cue exists to inform the developing waterflea about the risk of predation from fly larvae. The second assumption is that there is no cost to plasticity; that is, the plastic developmental system produces a waterflea as fit as the constitutively thick carapace in the presence of fly larvae and as fit as the constitutively thin carapace in the absence of fly larvae. The third assumption is that each of the two conditions, presence or absence of flies, occurs with sufficient frequency that each constitutive development pattern often has the worse fitness.

Clearly, adaptive plasticity cannot evolve if the assumptions are blatantly false. For example, if there were no cue about the presence of predators, then there is no way to ensure the morphology appropriate for the condition, and the waterflea may as well guess which morphology to express. But what if we relax but do not void the assumptions? Suppose that a cue exists but is not perfectly reliable. Suppose that there is a fitness cost to plasticity; that is, the plastic system makes a slightly thinner carapace in the presence of the predator than does the unconditional "thick" system (and so is not quite as fit as "thick" when flies are present) and a slightly thicker carapace in the absence of the predator than does the unconditional "thin" system (and so is not quite as fit as "thin" when flies are absent). And suppose that the two conditions, presence or absence of predatory fly larvae, do not occur with equal frequency.

Now the prospects for the evolution of adaptive plasticity depend on complicated relationships among the reliability of the cue, the cost of plasticity, and the evenness in frequency of the two conditions. The waterflea example can illustrate this complexity. Consider what happens when only one condition is very common; perhaps predatory fly larvae are almost always abundant. In this case, the individuals expressing the thick carapace are likely to prevail because they are the fittest individuals nearly all of the time. For plasticity to persist, individuals carrying the plastic developmental system must have a tremendous fitness advantage over the individuals expressing thick carapaces constitutively when predatory flies are absent in order to make up for their comparative deficiency in fitness when flies are present. The greater the cost of plasticity when flies are present, and the more often flies are present, the greater the advantage the plastic waterfleas must have when flies are absent.

For a specific set of fitness relationships, the higher the variability in environmental circumstances, the more likely that plasticity in development will emerge as a successful adaptation to that variability. However, this rule of thumb is valid only to a point. When conditions change too quickly, cues become unreliable, and plasticity does not improve on constitutive development or even random expression of features. This is especially true when plasticity is irreversible and the sensitivity to cues is restricted to a short period during development. If the environment changes faster than the time between the sensitive period and the expression of the appropriate feature, then plasticity is actually deleterious because it will perform worse than random expression of features.

Adaptive phenotypic plasticity enables individuals to cope with circumstances that vary from time to time and place to place but are not so variable as to preclude reliable cues to guide development. This enabling of individuals propagates upward to the level of the population and beyond to produce some important ecological consequences.

4. THE ECOLOGICAL IMPORTANCE OF PHENOTYPIC PLASTICITY

The obvious ecological consequence of phenotypic plasticity is that it allows a species to expand its range to seasonal environments and diverse habitats. A seasonal environment is the ideal situation for the evolution of plasticity; seasons change frequently enough to promote reversible plasticity but not too frequently compared to the time scale of trait expression, reliable cues abound, and many of the features of different seasons are predictable. Nearly everyone is familiar with the many adjustments that plants and animals make to the changes of season in temperate regions from the physiological changes underlying migratory behavior in birds to those underlying the onset of winter dormancy in trees.

Phenotypic plasticity can also allow species to occupy very uncertain habitats. Temporary ponds offer an example; the regular drying of the pond precludes sustainable fish populations, but the duration of the pond is uncertain, depending on the amount and timing of local rainfall. Nonetheless, temporary ponds harbor a considerable diversity of aquatic animals. Ponds offer refuge from what would otherwise be devastating predation by fish. But the dry periods would seem to preclude continuous occupancy by completely aquatic animals, and a short pond lifetime can leave the aquatic stage of animals that spend only part of their time in the water, such as tadpoles and dragonfly nymphs, high and dry if they cannot metamorphose quickly enough. Species that inhabit temporary ponds show remarkable varieties of phenotypic plasticity in response to drying conditions. Some copepods produce diapausing eggs that rest in the soil, many of the frog and salamander larvae can accelerate their development as waters recede, and sirens (large, completely aquatic salamanders) burrow into the soil, secrete a waterproof cocoon around their bodies to prevent desiccation, and enter estivation until the waters return.

Habitats can also be uncertain in their biotic components, and plasticity in response to the risks of predation and parasitism enables a species to cope more effectively with varying levels of risk. Temporary ponds exemplify this situation as well. Not only is their duration uncertain, but so is the period between drying and refilling. When the pond refills soon after drying, it is colonized quickly by predaceous insects including dragonflies and backswimmers. The aquatic larvae can achieve very high densities by the time that tadpoles appear later in the season. But if the ponds are dry for a long time, tadpoles have little risk of predation because the insects are at very low densities and are very small in body size. Many tadpoles from temporary ponds display extensive phenotypic plasticity to the presence or absence of predators. Most species change their activity patterns to reduce their encounter rate with predators, and some species alter their tail coloration and morphology to avoid predator detection and escape predator attack.

But an example like this one raises an interesting question: if an organism evolves adaptive plasticity in response to variation in predation risk, does the advantage conferred by that plasticity have a reciprocal effect on the predator? This general question is at the heart of the close scrutiny that ecologists have been giving many examples of phenotypic plasticity. Indeed, reciprocal effects on predators or other biotic agents that induce plastic responses have been found in many studies and can ramify through a community and an ecosystem, with far-reaching consequences. To visualize this point, consider the tadpoles and dragonflies again. If the dragonflies are less able to procure tadpoles as food, they will increase their consumption of other prey such as aquatic invertebrates and cause their densities to decrease. Other predators in the system, which had been using aquatic invertebrates as their principal food resource, may then be forced into other trophic pathways. In effect, the adaptive plasticity in the tadpoles, once established, might drive a substantial change in species diversity, community structure, and perhaps even ecosystem processes such as nutrient cycling.

This kind of effect has been found in many cases, and the indirect effect of one species on another, mediated through the consequences of expressing a feature that is a response to a third species, is often called a *trait-mediated interaction*. In our example, the decreased density of aquatic invertebrates represents an indirect effect of the tadpoles as they express the tail morphology that reduces their mortality rate from dragonfly predation. Trait-mediated interactions have been shown to be responsible for some interesting patterns of species diversity. For example, the presence of spiders in a New England old field causes several of their potential insect prey species to find refuge and foraging substrate on different plants than they would exploit in the absence of spiders. The plant preferred in the presence of spiders is actually a dominant competitor, and grazing by the insects reduces its density sufficiently for a competitively inferior species to increase in its density. The end result is that the presence of the spider increases the species diversity of the plant community.

A growing body of mathematical theory has elaborated on these basic ideas, indicating potentially profound effects of plasticity on species interactions. Much of this theory has been inspired by a particular type of adaptive plasticity, the inducible defenses of plants. *Induced defenses* are morphological or chemical responses by plants in response to herbivore attack. The production of toxic chemicals in some populations of wild parsnip in response to herbivore damage is an example of an induced chemical defense. Induced chemical defenses are known in a wide variety of plants, from freshwater algae to trees. Although the defensive compounds produced by plants can be synthesized and deployed relatively quickly, they can be costly to manufacture, diverting energy away from other functions. If the risk of herbivory is high, plants that produce them have higher fitness than those that do not; if the risk is low, chemical defense production is a waste of energy. Analogous to the argument for the carapace thickness of waterfleas, inducible defenses are favored when herbivory is sufficiently variable and a reliable cue is available (and being chewed is usually a reliable signal that herbivores are active).

Models inspired by inducible defenses indicate that adaptive plasticity can stabilize the numerical relationship between predator and prey or herbivore and host. To see this without mathematics, remember that predator–prey systems are inherently unstable because predators tend to overconsume prey. Any feature that protects a minimum fraction of the prey population from the predator can stabilize the system and allow predator and prey to coexist. Consider a herbivore–host system in which a constitutive defense appears via mutation. When this defense is expressed in some of the plants, it will protect a minimum fraction of individuals and stabilize the system. But as it spreads so that nearly all plants are protected, the herbivore loses its food resource and is likely to suffer a serious drop in population size and perhaps even extinction. Now consider an inducible defense that is expressed only when the risk of herbivory is high. Initially, when the inducible defense is present in only a few plants, it stabilizes the interaction. As more individuals express the defense, the herbivores become food-limited, and their density starts to decrease. But as herbivore densities decrease, so does the risk of predation; fewer individuals will express the defense, leading to a greater opportunity for the herbivores, whose density can then increase. Eventually, the herbivore and plant populations reach equilibrium, and the proportion of plants expressing the defense also attains equilibrium. An experimental study of algae with and without inducible defenses has confirmed that inducible defenses can stabilize herbivore–host systems and even stabilize a system with three trophic levels: host, herbivore, and predator.

But theory shows that adaptive plasticity will stabilize a dynamic predator–prey or herbivore–host system only if prey respond to the cue—predation risk high or low—with just the right speed, compared to the rate at which predators or herbivores can change their consumption rate. Obviously, a response that is too slow will be ineffective at deterring predation. A response that is too fast introduces a time lag between the appearance of the defense and the effect on the predators that destabilizes the system. There are too many predators when prey are well defended and too few when they are not. Systems like this will start cycling in numbers to the point where either the prey or the predator becomes extinct. Whether rapid plastic responses actually destabilize species interactions is one of many empirical questions about adaptive plasticity that remain to be answered.

5. HORIZONS FOR FUTURE ECOLOGICAL RESEARCH ON PHENOTYPIC PLASTICITY

The most important of the longstanding unresolved issues is the cost of plasticity. This is a difficult problem. It is rare to find both constitutive and plastic expression of suitable features in one population, so it is usually not possible to make the appropriate comparisons of fitness. The most common experiments that attempt to measure the cost of plasticity compare families that differ in their levels of plasticity. The results have been equivocal; some experiments have detected apparent costs, but others have not. The tantalizing prospect of using genetic engineering to create constitutive expression offers considerable promise for resolving the magnitude of costs and whether those costs occur similarly in all environments.

The enthusiasm for studying trait-mediated interactions has produced an extensive documentation of their existence and immediate effects. But in most cases, we do not know enough about the precision with which the traits are expressed, the relative frequencies of the different circumstances that provoke different expressions, or the full extent of the indirect effects that emerge in the community. We know that plasticity can have profound effects, but we do not know whether the documented cases of profound effects are exceptional.

Although we know a great deal about which factors induce plastic responses, we know far less about the actual cues that organisms exploit. Delineating those cues is important for illuminating their reliability, which is a critical feature governing plasticity's evolution and persistence. But there is another reason to identify the cues. Global change, sensu lato, could make erstwhile reliable cues unreliable, perhaps by dissociating combinations of signals that had been serving as very reliable cues. There is some evidence that this is happening in diapausing insects and migratory animals that use combinations of temperature and day length as their cue.

The mysteries of what we do not know about phenotypic plasticity should not detract from the marvel of what is well known. Through adaptive plasticity, an organism can remake itself, within limits, to suit its circumstances. And the organism that remakes itself to suit its circumstances can also remake the ecological circumstances around it, creating myriad possibilities for itself and for those who would understand the distribution and abundance of organisms.

FURTHER READING

Bradshaw, A. D. 1965. Evolutionary significance of phenotypic plasticity in plants. Advances in Genetics 13: 115–155. *This article remains the single best essay on the entire subject. In this essay, Bradshaw describes phenotypic plasticity and distinguishes it from related ideas in the literature, tracing its intellectual history accurately from a letter of Charles Darwin in 1881 to the scientific literature of the early 1960s. Further, this article produced the technical terms still in use today, and Bradshaw's concluding section on research horizons helped determine the research on plasticity for several academic generations. Bradshaw cited a large number of examples, mostly but not entirely from plants, to support his claim that there were patterns in plasticity and that "plasticity is therefore a property specific to individual characters in relation to specific environmental influences." He discussed the types of variable environments in which one would expect to find plasticity, and his reasoning presaged the results of more sophisticated mathematical theory that would emerge over two decades later.*

DeAngelis, D. L., M. Vos, W. M. Mooij, and P. A. Abrams. 2007. Feedback effects between the food chain and induced defense strategies. In N. Rooney, K. McCann, and D. Noakes, eds., From Energetics to Ecosystems: The Dynamics and Structure of Ecological Systems. New York: Springer Verlag, 213–236. *This report is among the most recent mathematical investigations of how inducible defenses can affect the stability of predator–prey or herbivore–host systems and, in a larger context, the responses of individual species and the ecosystem to nutrient enrichment. The discussion section of the article offers an excellent introduction to the literature on mathematical models of the consequences of plasticity for those interested either in further reading or, especially, initiating research on the subject.*

DeWitt, T. J., and S. M. Scheiner, eds. 2003. Phenotypic Plasticity: Functional and Conceptual Approaches. New York: Oxford University Press. *This edited volume includes a broad range of papers that, together, cover every facet of the subject from the varieties of plasticity in nature to what we know (or knew in 2003) about the genetic control of plastic development. Readers who are considering initiating research in the broad area of phenotypic plasticity should use this volume as their road map to its current research horizons. A virtue of this collection is the significant number of essays by younger workers with fresh perspectives.*

Karban, R., and I. T. Baldwin. 1997. Induced Responses to Herbivory. Chicago: University of Chicago Press. *This is a very readable monograph that reviews and synthesizes the literature on the varieties of inducible defenses in plants. The text brings theory, as it existed at the time, to bear on the diversity of ways in which plants respond to herbivory, and its wealth of examples still serves as a readable and effective introduction to the topic.*

Kats, L. B., and L. M. Dill. 1998. The scent of death: Chemosensory assessment of predation risk by prey animals. Ecoscience 5: 361–394. *This is an underappreciated review paper that is focused on the diversity of chemical signals used by animals to assess predation risk and cue antipredator plasticity in a variety of traits. It is one of the few reviews in the ecological and evolutionary literature devoted primarily to a serious, thoughtful examination of specific cues and the all-important theoretical issue of their reliability.*

Miner, B. G., S. E. Sultan, S. G. Morgan, D. K. Padilla, and R. A. Relyea. 2005. Ecological consequences of phenotypic plasticity. Trends in Ecology and Evolution 20: 685–692. *This short paper is one of the few review papers focused specifically on the ecological consequences of plasticity and argues for its importance as an ecological topic, not merely a topic in evolutionary biology. It is focused primarily on the effects of plasticity on species interactions and less on how plasticity enables habitat breadth. The paper and its bibliography offer an introduction to the recent literature on the various aspects of trait-mediated interactions and the effects of inducible defenses.*

Pigliucci, M. 2001. Phenotypic Plasticity: Beyond Nature and Nurture. Baltimore, MD: The Johns Hopkins University Press. *This book is a recent synthesis of the evolution of plasticity, and Pigliucci's advocacy for thinking about integrated developmental systems is, in some ways, a modern counterpart to Schmalhausen's book. One of the book's strengths is its treatment of modern theory for the evolution of plasticity; the text offers lucid explications of some very complicated ideas, many of which have their origins in sophisticated mathematical theory, and clarifies the relationships among different theoretical approaches and the results of individual papers. Readers interested in a comprehensive introduction to the theory for the evolution of plasticity should read the treatment in this book.*

Schmalhausen, I. I. 1949. Factors of Evolution. Philadelphia: The Blakiston Company. Reprinted Chicago: University of Chicago Press, 1986. *This classic monograph emphasizes the evolution of integrated development systems for organisms. Schmalhausen took a unified view of evolutionary development, placing plasticity in the same conceptual context as its opposite, canalization, which is the process of minimizing the variation in development so as to produce the same features or trait values regardless of environmental conditions. He discussed how and when evolution might take each course and set these ideas firmly in the context of what were, at that time, modern ideas in evolutionary genetics. The book still offers a compelling argument that developmental systems are adaptive evolution's most breathtaking product.*

Shapiro, A. M. 1976. Seasonal polyphenism. Evolutionary Biology 9: 259–333. *This underappreciated review is a very thoughtful treatment of seasonal variation in morphology, coloration, and life history, with some close attention to insects. The text touches on the major themes in the evolution of plasticity and, despite its age, remains an excellent source of ideas and a laudable example of how to synthesize natural history, conceptual issues, and data.*

Sumner, F. B. 1932. Genetic, distributional, and evolutionary studies of the subspecies of deer mice (*Peromyscus*). Bibliographica Genetica 9: 1–106. *This is a classic paper that summarizes and synthesizes Sumner's decades of study of deer mouse ecology, genetics, and development. Sumner took the integrated approach to ecology and evolution that is so often proclaimed but so rarely practiced. The paper discusses how local adaptation (genetic differences produced by Darwinian adaptation to local conditions) and phenotypic plasticity combine to allow deer mice to occupy diverse habitats. His experimental dissections of phenotypic variation into its genetic, environmental, and interactive components remain models for modern emulation.*

Tollrian, R., and C. D. Harvell, eds. 1998. The Ecology and Evolution of Inducible Defenses. Princeton, NJ: Princeton University Press. *This volume offers a comprehensive look into its subject, and the papers included in the volume examine topics from the biochemistry of defensive compounds to trait-mediated interactions. Although the papers were written before the recent flowering of mathematical theory for the consequences of plasticity, the ideas that those theories examine are set out in several of these papers, and the volume clearly played a role in accelerating this area of research. For a reader interested in the variety of induced defenses, this volume offers a strong introduction to a very diverse literature. In addition, the authors of individual papers come from several schools of thought, and therefore the volume offers varied perspectives on its topic that some other edited volumes do not.*

Travis, J. 1994. Evaluating the adaptive role of morphological plasticity. In P. C. Wainwright and S. M. Reilly, eds., Ecological Morphology: Integrative Organismal Biology. Chicago: University of Chicago Press, 99–122. *This review paper was written for the scientist who is not a specialist in evolutionary biology or ecology and wishes to learn about phenotypic plasticity. It offers a synthetic examination of phenotypic plasticity, reviewing the conclusions of mathematical theory—but without the mathematics—for its evolution and matching a large number of examples, primarily drawn from animals, to the classes of theoretical treatments to which those examples apply. Although theory has advanced considerably since it was written, it is still a lucid introduction to the literature, especially the terminology, and its strength is in describing clear patterns in the vast array of examples.*

I.10

Life History

William F. Morris

OUTLINE

1. Variation in life history among species and the notion of trade-offs
2. Key life history patterns and associated trade-offs

The term *life history* summarizes the timing and magnitude of growth, reproduction, and mortality over the lifetime of an individual organism. Important features of an individual's life history include the age or size at which reproduction begins, the relationship between size and age, the number of reproductive events over the individual's lifetime, the size and number of offspring produced at each reproductive event, the sex ratio of offspring, the chance that the individual dies as a function of age or size, and the individual's lifespan or longevity (the time elapsed between the birth and death of the individual). Although all of these features (so-called life history traits) describe individuals, some are more easily understood when viewed as aggregate properties of a population of individuals. This is particularly true of mortality and lifespan. Each individual dies once, at a certain age. But in a population of identical individuals, some may die at a young age and some at an old age. By imagining that the fraction of this population that is still alive at a given age also represents the probability that an average individual survives to that age, we see that the chance of survival to a given age, which is the converse of the chance of dying, or mortality, is a property of an individual. Similarly, we can envision the average lifespan (or "life expectancy") even though each individual has a single age at death. All sunflowers and the vast majority of sequoia seedlings die before reaching one year of age. Yet in a sequoia population, individuals have the potential to live for several millennia, which distinguishes sequoias from sunflowers.

GLOSSARY

fertility. The number of daughters to which a female gives birth during a specified age interval

geometric mean. The *n*th root of the product of *n* numbers

iteroparity. A reproductive pattern in which individuals reproduce more than once in their lives

life table. A table summarizing age-specific survivorship and fertility used to calculate the net reproductive rate

net reproductive rate. The average number of daughters to which a newborn female gives birth over her entire life

semelparity. A reproductive pattern in which individuals reproduce only once in their lives

survivorship. The probability that a newborn survives to or beyond a specified age

1. VARIATION IN LIFE HISTORY AMONG SPECIES AND THE NOTION OF TRADE-OFFS

As for the difference in life expectancy between sunflowers and sequoias, each of the key life history traits varies 1000-fold or more among species, as illustrated in figure 1. Life history traits also vary among individuals of the same species. The fundamental question in ecological and evolutionary studies of life history is: why is there so much variation in life history traits among and within species?

To answer this question, we start by recognizing that life history features are traits just like any other (e.g., coloration, bill shape, cold tolerance, body size, etc.) that can be acted on by natural selection. Moreover, variation in life history traits among individuals in a population often has a genetic basis, so genotypes favored by natural selection can potentially increase in frequency from one generation to the next. If life history traits are genetically based and subject to selection, evolution of life history might be expected to lead to an organism that begins reproducing immediately after birth and produces a large number of well-provisioned offspring in a series of reproductive events throughout an infinitely long life (such an organism has been termed a "Darwinian monster" because it would quickly displace all other species from Earth). The reason we do not see Darwinian monsters even though

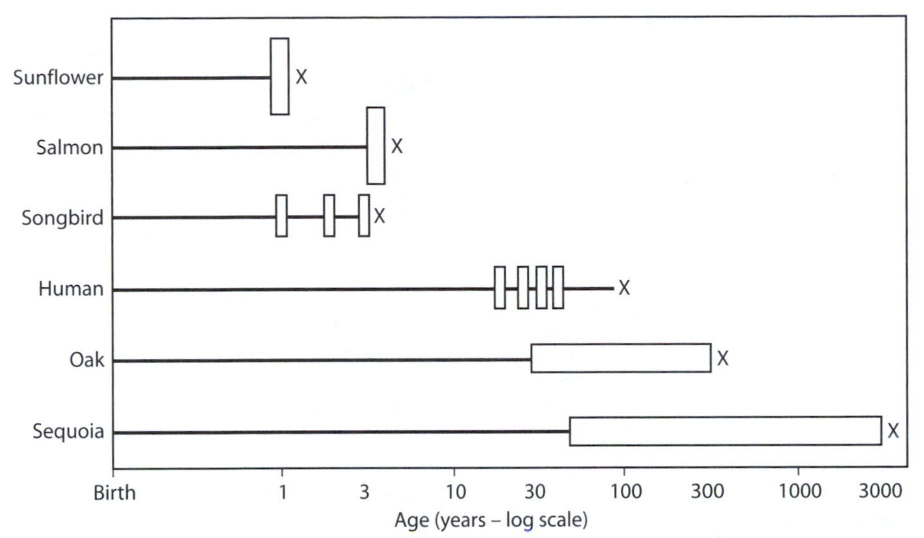

Figure 1. Diversity of life histories for six representative species, three plants (sunflower, oak, and sequoia trees) and three animals (salmon, songbird, and human). Rectangles show reproductive events; height of each rectangle indicates the magnitude of reproductive effort (separate reproductive events are merged for oaks and sequoias). An "x" marks the age at death of an adult. Note that age is on a logarithmic scale; sequoias can live 3000 times longer than sunflowers.

life history traits evolve is that different life history traits are not independent.

Because the resources that an organism has available to invest in maintenance and survival, in growth, and in reproduction are always limited, life history evolution is constrained by trade-offs: a greater investment in one life history trait must come at the expense of a smaller investment in one or more other life history traits. Trade-offs between many different pairs of life history traits have been documented, and we will see several examples in the following section of this article. In recognizing trade-offs, we no longer expect that evolution will produce Darwinian monsters, but rather that natural selection will balance, for example, improvements in reproduction with reductions in survival. On one hand, the optimal balance may depend on features of the environment the organism occupies. On the other hand, multiple combinations of life history traits may produce equally fit organisms. Both provide explanations for the diversity of life histories we see among Earth's biota.

2. KEY LIFE HISTORY TRAITS AND ASSOCIATED TRADE-OFFS

Age and Size at Reproductive Maturity

All else being equal, an organism should begin reproducing as soon as possible, for two reasons. First, because of factors such as predators and diseases, adverse weather, or genetic defects, the organism may not survive for long, so delaying reproduction carries the risk of dying before reproducing. This advantage of early reproduction is easily illustrated with a basic demographic tool, the life table (table 1). A life table has two principal columns, survivorship, usually denoted l_x, which is the probability that a newborn female survives to age x or older, and fertility, usually denoted m_x, which is the average number of daughters a mother of age x produces over the next age interval (note that life tables typically track females only). One use of a life table is to compute the average number of daughters a female will produce over her entire life, which is called the net reproductive rate and is usually denoted R_0. Natural selection can be expected to favor production of more daughters. As shown in table 1, for a fixed set of l_x values, fertility skewed toward earlier ages will lead to a higher R_0 simply because females will be more likely to survive to reproduce. The second reason why early reproduction is advantageous is that daughters produced earlier will themselves begin reproducing sooner than will daughters produced later in the mother's life. If we think of a mother and her female descendents as a lineage, a lineage founded by an early-reproducing mother will grow faster than will the lineage of a later-reproducing founder, even if both lineages have the same R_0 (figure 2).

However, the production of offspring costs resources that the parent could use for other purposes, such as growth. Individuals that reproduce early in life

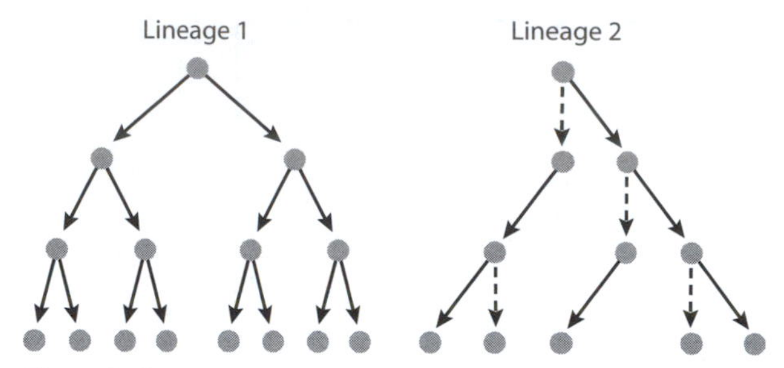

Figure 2. Growth of two female lineages. In Lineage 1, each mother produces two daughters in her first year and then dies. In Lineage 2, each mother produces one daughter in her first and one in her second year and then dies. For both lineages, $R_0=2$, but the first lineage grows faster. Circles are females, dashed vertical arrows show survival of the same female, and solid arrows show production of daughters.

may grow less rapidly, a trade-off between early reproduction and early growth that may prevent early reproducers from reaching a large final size. Moreover, because larger individuals often have both a greater chance of surviving and a greater reproductive potential, early reproduction may also be involved in trade-offs with later survival and later reproduction. If reproductive capacity increases rapidly as the size of an organism increases, delaying reproduction in order to achieve a larger size may allow an individual to produce more offspring over its entire life despite the advantages of early reproduction illustrated in table 1 and figure 2.

Delaying reproduction in order to grow more rapidly is especially important for males of species with territorial breeding systems. For example, the victors of fights between male elephant seals gain nearly exclusive reproductive access to harems of females. Because small males have little chance of winning fights, young males invest energy in growing larger rather than in futile attempts to breed. In contrast, females do not need to win fights to breed, so they begin reproducing several years earlier in life than do males. Thus, males and females of the same species may face different trade-offs between early reproduction and growth.

Interestingly, a breeding system in which males delay reproduction to grow large enough to win male–male contests may open the door for alternative mating strategies used by smaller or younger males. In many fish species, small males known as *satellites* may mimic females or use other methods to sneak into the territory of a mating pair in order to fertilize some of the female's eggs when she releases them into the water. Satellites also occur in other animal groups (including lizards). Satellites can increase in a population when they are rare, because there are many territorial males to "parasitize," but their success declines as they become a larger proportion of the population. They may achieve reproductive success similar to that of territorial males, or they may be making the best of a small size caused by genetic or environmental factors.

Number of Lifetime Breeding Events

Constant Environments

Once an individual becomes reproductively mature, it can breed only once (semelparous species), as in sunflowers and salmon, or breed multiple times (iteroparous species). Intuitively, we might think that if breeding once is good, breeding more than once is

Table 1. Hypothetical life tables

Daughters produced at ages 2 to 4				*Daughters produced at ages 1 to 3*			
Age, x	l_x	m_x	$l_x m_x$	*Age, x*	l_x	m_x	$l_x m_x$
0	1	0	0	0	1	0	0
1	0.6	0	0	1	0.6	1	0.6
2	0.4	1	0.4	2	0.4	2	0.8
3	0.2	2	0.6	3	0.2	1	0.2
4	0.1	1	0.1	4	0.1	0	0
5	0.0	0	0	5	0.0	0	0
R_0 = sum of $l_x m_x$ values			0.9	R_0 = sum of $l_x m_x$ values			1.6

Note: Hypothetical life tables with the same survivorship (l_x) schedules but with daughters produced at later (left) versus earlier (right) ages of the mother. The product $l_x m_x$ is the number of daughters a female is expected to produce at age x, accounting for the fact that she might not survive to age x. The sum of the $l_x m_x$ values over all ages is the net reproductive rate, R_0, which is higher when daughters are produced earlier in the mother's life.

better. But Lamont C. Cole pointed out in 1954 that the lineage of an immortal organism that reproduces every year would grow at the same rate as the lineage of an organism that reproduces only once at 1 year of age and then dies, but produces just one more offspring than does the iteroparous organism at each of its breeding events (it is easy to modify figure 2 to see how Cole's claim might be true). More than one additional offspring would give the advantage to the semelparous organism, and if the cost of investing in survival is high, an organism that reproduced once and then died might be able to achieve even higher reproduction (note that we have just assumed a survival–reproduction trade-off). Thus, Cole claimed that iteroparity was more paradoxical than semelparity. Other ecologists showed later that Cole's result requires that adults and newborn offspring have the same chance of surviving each year, which we now demonstrate with a simple mathematical model. If I_t and S_t are the numbers of individuals in the iteroparous and semelparous lineages in year t, then I_{t+1} and S_{t+1}, the numbers in the two lineages the following year, are predicted by

$$I_{t+1} = F_N \times B_I \times I_t + F_A \times I_t = (F_N \times B_I + F_A) \times I_t$$

$$S_{t+1} = (F_N \times B_S) \times S_t,$$

where F_N and F_A are the fractions of newborns and adults surviving the year, and B_I and B_S are the numbers of newborns produced by each iteroparous and semelparous organism each year. There is no F_A term in the equation for the semelparous lineage because individuals die after breeding. The two lineages will grow at the same rate if the terms in parentheses in the two equations are equal, that is, if

$$F_N \times B_I + F_A = F_N \times B_S$$

or, dividing both sides of the equation by F_N, if

$$B_I + F_A/F_N = B_S.$$

If instead the left side of the preceding equation is larger than the right side, the iteroparous lineage will grow faster. Note that if the same fraction of adults and newborns survive the year (i.e., if $F_A = F_N$), we obtain Cole's result (because the semelparous organisms are producing one more newborn than are the iteroparous organisms). However, because newborns are smaller and often more vulnerable than adults, for many species and environments, a smaller fraction of newborns than adults will survive. Because F_A/F_N may then be substantially greater than 1, semelparous reproduction may need to be a good deal greater to achieve a fitness equal to that of the iteroparous lineage. Thus, an important advantage of iteroparity is that it capitalizes on the greater value of adults, as measured by their higher survival rates, even at the expense of lower offspring production at each breeding event.

Variable Environments

In the preceding section, we assumed that newborn survival is the same every year. Long-lived adults are even more valuable in an environment in which newborn survival varies from year to year, as is likely to occur for many species and environments. Imagine that there are two kinds of years, good and bad, for newborn survival. Assume that in good years, $F_N = 0.5 + D$, and in bad years, $F_N = 0.5 - D$; by increasing the number D, we increase the contrast between good and bad years. If $D = 0$, 50% of newborns survive every year. If $D = 0.5$, all newborns survive in good years, and none survive in bad years (higher values of D are meaningless because the survival fraction cannot be less than 0 or greater than 1). If we assume that good and bad years are equally frequent but occur at random, then regardless of the value of D, the average newborn survival across years is 0.5.

If newborn survival varies from year to year, the terms in parentheses in the equations we used above to predict the growth of iteroparous and semelparous lineages, which represent the annual lineage growth rates, also vary from year to year.

Note that to get I_{t+2}, the size of the iteroparous lineage in year $t+2$, we would first compute I_{t+1} by multiplying I_t by the annual growth rate for the iteroparous lineage in year t [which would be $(0.5+D) \times B_I + F_A$ if year t were a good year and $(0.5-D) \times B_I + F_A$ if year t were a bad year] and then multiply the result by the annual growth rate for year $t+1$ (which again might be a good or a bad year). Thus, over a period of years, the size of a lineage (either iteroparous or semelparous) is determined by the product of the annual lineage growth rates. If good and bad years are equally likely, then over a long period of years, close to half of the years will be good and half will be bad, and in a typical 2-year period one will be good and one will be bad. Thus, the growth of the iteroparous lineage over a typical 2-year period will be determined by the product of the good-year and bad-year growth rates, $[(0.5+D) \times B_I + F_A] \times [(0.5-D) \times B_I + F_A]$, and the growth over a typical 1-year period will be the square root of this product. Similarly, the typical 1-year growth rate of the semelparous lineage will be the square root of the product $[(0.5+D) \times B_S] \times [(0.5-D) \times B_S]$. These typical growth rates represent the geometric means of the annual growth rates. (The geometric mean of two numbers is the square root of their product, whereas the more familiar average or arithmetic mean is one-half of their sum.) Because lineage growth is a multiplicative process, the geometric mean is a more appropriate measure of typical annual growth.

Now let us set F_A to 0.9 (90% of adults survive a year) and B_I and B_S to 0.5 and 2.5, respectively (the semelparous organism produces on average two more offspring per year than does the iteroparous organism). If $D = 0$, all years are the same, and the annual growth rate of the iteroparous lineage, $B_I \times F_N + F_A = 0.5 \times 0.5 + 0.9 = 1.15$, is less than the annual growth rate of the semelparous lineage $B_S \times F_N = 2.5 \times 0.5 = 1.25$. Thus, in a constant environment characterized by the survival and reproductive rates we have chosen, the semelparous lineage outperforms the iteroparous lineage. But what happens as we increase the contrast between good and bad years by increasing the value of D? As figure 3 illustrates, the semelparous lineage continues to grow faster than the iteroparous lineage when year-to-year variability in newborn survival (as determined by D) is low. However, the growth rates of both lineages decline as D increases, but much more so for the semelparous lineage, so that once D exceeds 0.2, the iteroparous lineage outgrows the semelparous lineage. Note that we have not changed the average newborn survival rate, so the switch in relative performance of the two lineages shown in figure 3 is driven entirely by the increase in variability of newborn survival. The presence of long-lived adults in the iteroparous lineage allows it to persist during years when few or no newborns survive. In contrast, persistence of the semelparous lineage requires that at least some newborns survive every year. That is why the geometric mean growth rate of the semelparous lineage is zero when $D = 0.5$ and its bad-year annual growth is zero; a single zero will cause the product of annual growth rates to be zero because a single year in which all newborns die will cause extinction of the lineage.

Thus, iteroparity, even at a cost of reduced annual reproduction, can be favored when newborn survival is low on average and/or is highly variable from year to year. Essentially, long-lived adults spread their reproductive efforts over multiple years, so their lifetime reproductive success is less sensitive to a single bad year. Although iteroparity is one life history adaptation to randomly varying environments, semelparous species also possess life history adaptations to environmental variability, namely dormancy and diapause, which we omitted from our simple model. Annual plants, by definition, reproduce only once, but many of them produce seeds of which a fraction lie dormant in the soil for one or more years. Because the offspring of a parent plant then germinate in different years, the parent is effectively spreading its reproduction over several years, just as an iteroparous organism does, thus reducing its sensitivity to a single bad year in which all seedlings die and increasing its geometric mean fitness. Similarly, many insects and other invertebrates can remain in an inactive diapause state as adults for more than a year, allowing the set of offspring of a single mother to sample different years even though each offspring reproduces only once.

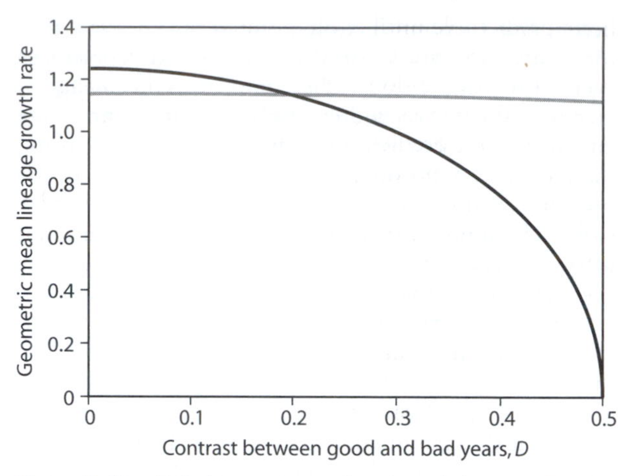

Figure 3. The typical annual growth rates of the semelparous lineage (shown as a black line) and of the iteroparous lineage (shown as a gray line) when newborn survival varies randomly from year to year. The typical growth rate is measured as the geometric mean, and the contrast in newborn survival between good and bad years increases as D increases.

We assumed above that only newborn survival varied from year to year, but in reality, the survival, growth, and reproduction of individuals of all ages are likely to vary. In a completely unpredictable environment, year-to-year variability in all of these life history traits will depress the geometric mean growth rate of a lineage, just as variability in newborn survival does in figure 3, but variability in traits such as adult survival that make large contributions to the lineage growth rate will be more detrimental than variability in less influential traits. Across many species, there is growing evidence that the most influential life history traits vary less from year to year than do less influential traits, suggesting that mechanisms have evolved that buffer the life histories of those organisms against the most detrimental types of variability.

However, not all year-to-year variability in life history traits is detrimental. Many environments are only partly unpredictable. For example, in fire-prone ecosystems, it may be difficult to predict whether a fire will occur in a given year, but once a fire has occurred, conditions of abundant resources, low competition, and low likelihood of additional disturbance may be quite predictable until enough fuel has accumulated to allow the next fire to occur. Many species inhabiting such disturbance-dominated systems have evolved timing of the phases in their life histories to exploit this environmental predictability. For example, many fire-adapted plants produce seeds that germinate only in the period soon after a fire, so their offspring can take advantage of abundant postfire resources. Conversely, reproduction of these plants is often restricted to late in the interfire interval, after plants have grown to reproductive size and when their seeds will be poised to exploit the next interfire interval. In these species, among-year variation in life history traits reflects adaptation to multiyear environmental cycles rather than the detrimental influence of environmental variability.

Lifespan and Aging

Even iteroparous organisms eventually die. Moreover, for many species, the chance of dying in a given interval of time may initially decline after birth as newborns grow to a less vulnerable size or as those with developmental defects die, but it eventually increases as individuals reach more advanced ages. The process of aging is defined as an increase in mortality risk late in life. As we would expect that the ability to continue living and reproducing would be favored by natural selection, why does aging occur? Peter B. Medawar argued, and William D. Hamilton showed mathematically, that the ability of natural selection to weed out genes that increase mortality or decrease fertility declines with an organism's age. The reason for this decline in the strength of selection is that even individuals with good genes are increasingly likely to have died from external causes, such as predators or bad weather, as age increases. Therefore, few individuals will still be alive to enjoy the advantage of decreased mortality risk or increased fertility at advanced ages, whereas most individuals will benefit from increases in early-life survival or fertility. Two types of genes may underlie an increase in mortality or decrease in fertility with age. Detrimental genes that are expressed only late in life would experience only weak selection against them and so would tend to accumulate in the genome. Genes that have beneficial effects on survival or fertility early in life but detrimental effects late in life would be maintained because positive selection for their early effects would overwhelm negative selection for their late effects. The former type of genes play a role in Peter Medawar's *mutation accumulation* theory of aging, whereas the latter type of genes are central to George C. Williams' *antagonistic pleiotropy* theory of aging. There is evidence that both types of genes may be present in the same organisms.

For many organisms, including fruit flies, some plants, and humans, the risk of mortality reaches a plateau rather than continuing to increase at very advanced ages. A mortality plateau could arise because individuals that are frail from genetic or environmental factors are increasingly likely to have already died as age increases. But as Hamilton's theory shows that natural selection will be powerless to eliminate detrimental genes expressed at all ages past the age at which reproduction ceases, detrimental late-acting genes could simply be maintained at intermediate frequencies by a balance between the nonselective evolutionary forces of mutation and genetic drift.

Although Hamilton's work predicts that there will be no selection to reduce mortality once reproduction ceases, that work only accounts for offspring directly produced by a mother. However, mothers can also contribute to the growth of their lineage by providing direct care to their granddaughters or by providing information that improves the quality of care their daughters provide to their own offspring. These direct and indirect transfers across two generations may explain why, in social species such as primates, mortality does not increase rapidly once a female ceases to give birth.

Number and Size of Offspring

Whether an organism reproduces once or more than once, the resources it invests in a single breeding event can be used to produce a single large offspring, or those resources can be divided to produce more than one, but smaller, offspring. The trade-off between the size and number of offspring is a fundamental constraint on life history. Producing more offspring will cause a lineage to grow faster, but not if each offspring is too small to have a good chance of surviving. Therefore, the best solution in the face of a size–number trade-off is to produce the number (and therefore size) of offspring at which the product of the offspring number and the size-dependent probability that each offspring survives is at a maximum. Other factors may skew the optimum solution toward more and smaller offspring. For example, trees with seeds dispersed by wind typically make many small seeds. Even though the small seedling emerging from each seed has a low chance of surviving, by increasing the area reached by its wind-blown seeds, a mother plant will be more likely to place at least some of its seeds in sites suitable for seedling growth and survival, even if those sites are few and far between.

A large number of studies have addressed the question of why most female birds lay fewer eggs in each nest than the maximum number observed for the species. Why do lineages that produce more eggs per nest not come to replace lineages that produce fewer eggs per nest? David L. Lack proposed that birds should maximize the number of offspring surviving to leave the nest rather than the number of eggs per nest. Because the parents can provide less food to each chick when a nest contains many chicks, the chance that each chick survives declines with the number of eggs (and therefore chicks) in a nest. As a result, the product of the number of eggs per nest and the probability that the chick hatching from each egg survives to leave the nest is usually highest at an intermediate egg number. However, this number is often higher than the average number of eggs actually observed in nest. Birds may lay fewer eggs than Lack's argument predicts because excessive investment in one nest reduces the parents' chance of survival or their future reproductive success. That is, trade-offs between current reproduction and future survival or future reproduction may constrain the amount invested in a single bout of reproduction.

See also chapters I.13, I.14, and II.1 in this volume.

FURTHER READING

General

Roff, D. A. 1992. The Evolution of Life Histories. New York: Chapman & Hall.

Stearns, S. C. 1992. The Evolution of Life Histories. Oxford: Oxford University Press. *Two excellent and comprehensive overviews of the field.*

Iteroparity

Cole, L. C. 1954. The population consequences of life history phenomena. Quarterly Review of Biology 103: 103–137. *A classic paper that spurred research into reproductive patterns and longevity.*

Aging

Hamilton, W. D. 1966. Moulding of senescence by natural selection. Journal of Theoretical Biology 12: 12–45. *Although rather mathematical, this paper is the basis for most evolutionary theories of aging.*

Partridge, L., and D. Gems. 2006. Beyond the evolutionary theory of ageing, from functional genomics to evo-gero. Trends in Ecology and Evolution 21: 334–340. *A recent review of genetic causes of aging.*

Rose, M. R. 1991. Evolutionary Biology of Aging. New York: Oxford University Press.

Rose, M. R., C. L. Rauser, L. D. Mueller, and G. Benford. 2006. A revolution for aging research. Biogerontology 7: 269–277. *A review of the evidence for and causes of mortality plateaus.*

Offspring Number

Godfray, H.C.J., L. Partridge, and P. H. Harvey. 1991. Clutch size. Annual Review of Ecology and Systematics 22: 409–429. *Emphasizes determinants of offspring number in organisms other than birds.*

Lack, D. L. 1954. Natural Regulation of Animal Numbers. Oxford: Oxford University Press. *This book includes Lack's argument for why birds should lay fewer than the maximum number of eggs per nest.*

I.11

Remote Sensing and Geographic Information Systems

Catherine H. Graham and Scott J. Goetz

OUTLINE

1. Basic concepts of remote sensing and geographic information systems
2. Applications of RS and GIS in ecology

Remote sensing (RS) and geographic information systems (GIS) provide data and tools that are used extensively across ecology, evolution, biogeography, and conservation biology. Some fields in particular, such as landscape ecology and biogeography, have relied heavily and increasingly on sophisticated analyses afforded by these data and tools.

GLOSSARY

electromagnetic energy. Energy or radiation in a wave in space with an electrical field that varies in magnitude in a direction perpendicular to the direction in which the radiation is traveling and a magnetic field oriented at right angles to the electrical field.

geospatial. The distribution of information in a geographic sense such that entities can be located by some coordinate of a reference system (i.e., latitude and longitude), which places these entities at some point on the globe.

global positioning system (GPS). This system is a set of 24 satellites that orbit the Earth and communicate their position to a ground receiving device providing the geographic location of that receiver.

1. BASIC CONCEPTS OF REMOTE SENSING AND GEOGRAPHIC INFORMATION SYSTEMS

Broadly, RS is the gathering and processing of data about the physical world by a device detecting electromagnetic energy that is not in contact with the object, area, or phenomenon under investigation. For example, the images in Google Earth (http://earth.google.com) are products of RS that provide information about vegetation and human uses of the land, such as agriculture, roads, and buildings. As such, RS is generally used to generate data that are often then imported into a GIS. A GIS is a collection of tools that provide the ability to capture, display, manage, and analyze most forms of spatial data that are geographically referenced to the Earth's surface (i.e., identified according to location).

An important feature in GIS is the ability to relate different types of information, such as remotely sensed vegetation images and maps of human population density, in a geospatial context to explore associations and relationships among these various types of information. Together, RS and GIS, along with other relatively recent tools, such as geographic positioning systems (GPS) and sophisticated spatial statistics, provide a powerful platform to advance our understanding of the natural world. In this chapter, we first describe RS and GIS in greater detail, and we then provide some examples of how they are used in ecology.

Remote Sensing

RS data provide real-time information about what is happening on our planet and can be used in a wide range of ecological studies. For example, it is now relatively easy to quantify deforestation rates across different ecosystems or even detect a fire in a remote area. RS uses theory developed in physics to measure electromagnetic energy emitted or reflected from distant objects. Electromagnetic radiation/energy can be described in terms of a stream of photons, which are massless particles, each traveling in a wavelike pattern and moving at the speed of light. This radiation varies across a spectrum of different amounts of energy in photons and size and frequency of waves.

RS applications typically use wavelengths that include the visible wavelengths (blue through red), the

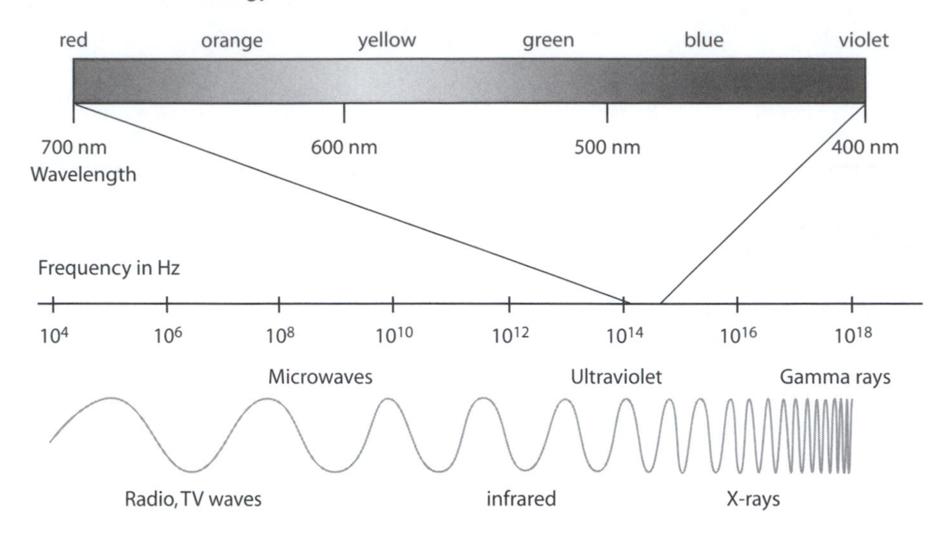

Figure 1. The electromagnetic spectrum. (From http://cache.eb.com/eb/image?id=73584&rendTypeId=35)

infrared, and microwave regions of the electromagnetic spectrum (figure 1). Different types of objects (such as grassland and forest canopies) reflect, absorb, or scatter electromagnetic energy differently and, as a result, emit electromagnetic waves at different magnitudes in different portions of the electromagnetic spectrum. The principle behind RS is to detect and identify these characteristic waves (i.e., varying energy levels) for different materials, which are often referred to as *spectral signatures*. This is done by documenting the reflectance across a range of electromagnetic wavelengths for one or more objects or vegetation types. This information is often displayed as a spectral reflectance curve (figure 2; for an interactive tool, visit http://geospatial.amnh.org/remote_sensing/widgets/spectral_curve/index.html). One of the first and still widely used sources of RS data is imagery from the Land Satellite (Landsat) series of sensors, first launched in 1973. The most recent Landsat images have seven different bands that cover a spectrum of 0.450 μm to 2.35 μm. Information about the reflectances in each of these bands can be used to classify different regions of an image (generally referred to as picture elements or pixels) as a certain type of land cover (figure 2). The eighth Landsat satellite in the series is scheduled for launch in 2011 and is called the Landsat Data Continuity Mission.

Satellite-based sensors have several unique characteristics, including how data reach the sensor, the number and width of spectral bands, and the spatial and temporal resolution of the data. There are two broad types of sensors that differ in how they sense and capture data. These are referred to as passive or active sensors. In passive sensors, radiation emitted from an object is simply measured as it reaches the sensor. In contrast, active sensors emit a pulse of energy and measure the portion of the energy that is returned or bounced back to the detector (figure 3). For example, land cover is often determined using a passive sensor, such as the Landsat images described above. Vegetation vertical structure or surface topography is more commonly measured with an active sensor, where returns of the energy pulses sent by the sensor are influenced by the complexity of object—a measure of its structure. For example, RADAR (radio detection and ranging) or LiDAR (light detection and ranging) returns from a tropical forest canopy may be sensitive to the density of biomass or the branching structure of trees.

The width of the bands of the electromagnetic spectrum that a sensor can detect is known as its spectral resolution. Some sensors, especially older ones, can detect information in only a few wide spectral bands (typically blue, green, red, and near-infrared), whereas other sensors can obtain information across a narrow range of spectra that are sensitive to absorption by atmospheric water vapor or other gases. If a sensor has many bands of narrow width (commonly known as hyperspectral), it may be easier to differentiate among different objects by using the more detailed information from its spectral signature.

Finally, sensors gather data at different spatial and temporal resolutions. Spatial resolution refers to the pixel size at which data are collected, whereas temporal resolution is the "revisit" time or how often a satellite flies over a given location (high temporal resolution = short revisit time). A high temporal resolution is important for some research questions, such as tracking algal blooms in the ocean or the leaf-out of trees in the northern hemisphere spring. Often there is a technological trade-off between spatial and temporal

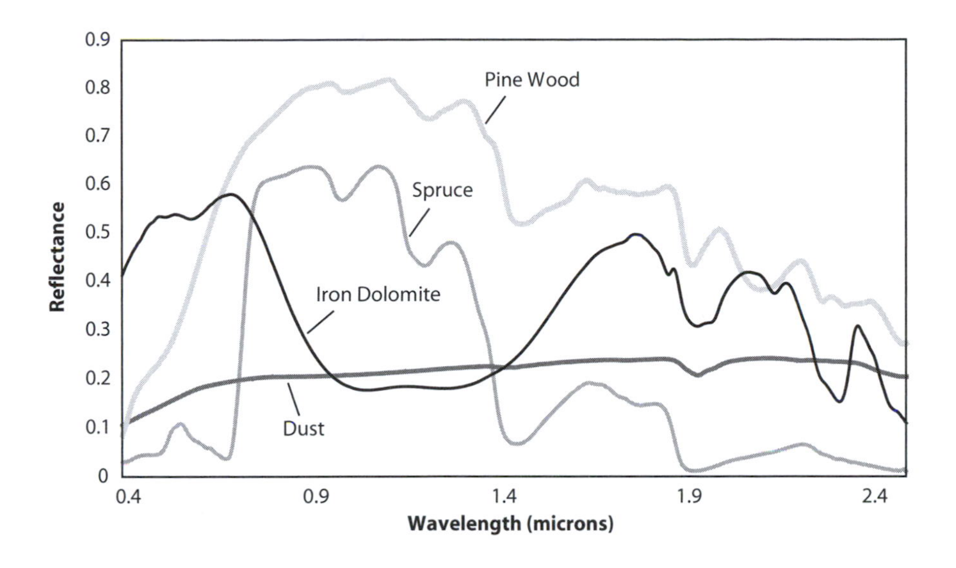

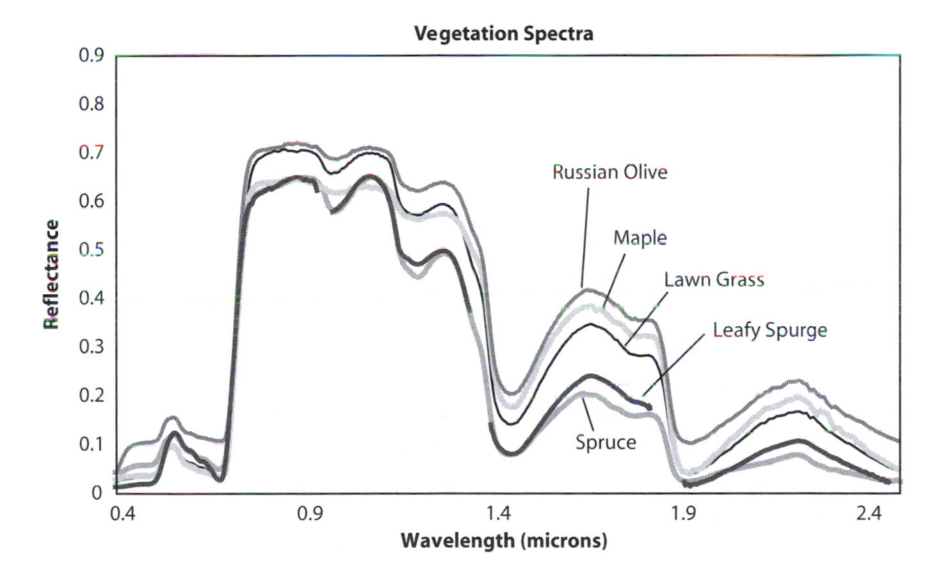

Figure 2. Spectral reflectance curves for different kinds of material. (From http://www.profc.udec.cl/~gabriel/tutoriales/rsnote/cp1/1-9-1.gif)

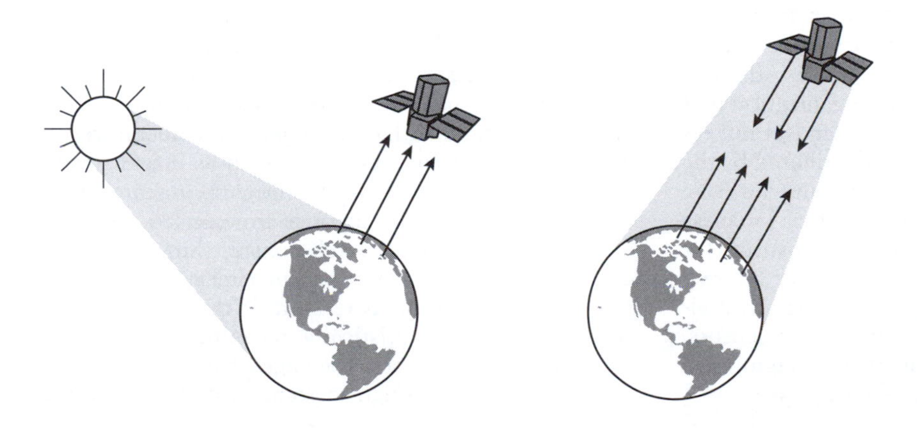

Figure 3. Passive (left) and active (right) sensors. (From http://www.csc.noaa.gov/products/nchaz/htm/ccap5.htm)

Table 1. Examples of ecological variables and data sources useful for ecological research

Ecological variable	*Sensor*	*Spatial resolution*	*Revisit time*	*Description*
Land cover	Moderate-resolution Imaging Spectrometers (MODIS)	250–1000 m	1–2 days	Can discriminate different land surfaces
	Landsat	30 m	16 days	
Chlorophyll	Sea-viewing Wide Field-of-view Sensor (SeaWiFS)	1000 m	1 day	Measure reflectance to assess presence/absence of vegetation and relative greenness of ocean and land chlorophyll to calculate productivity
	QuickBird	4 m	2–4 days	
Phenology	MODIS	250–1000 m	1–2 days	Information on leaf phenology, and in some cases flowering/fruiting cycles
	Advanced Spaceborne Thermal Emission and Reflection Radiometer (ASTER)	10 m	4–16 days	
Topography	Shuttle Radar Topography Mission (SRTM)	90 m	NA	Digital elevation models derived from radar singles (SRTM)
Vertical canopy structure	Laser Vegetation Imaging Sensor (LVIS)	1–10 m	NA	Provides 3D measurements via laser pulses; provides biomass estimation and information about vegetation structure

Source: Modified from Turner et al. (2003). There are many more data than listed here; for a more complete list, see Turner et al. (2003).

resolution in which sensors with a high revisit time have larger pixel sizes. An extreme example of this is a meteorological satellite that views the Earth every few minutes but only at very coarse spatial resolution.

This variation in sensor type (i.e., passive/active) and characteristics of images different sensors provide (i.e., bandwidth, number, pixel size, and revisit time) results in a broad range of images available for research in ecology. Table 1 lists some of these images, what sensor they come from, and their uses in ecological research. The National Aeronautics Space Administration (NASA) is a U.S. government organization that provides many RS images, such as the Landsat images mentioned earlier, although data are distributed by the U.S. Geological Survey (USGS). Other countries have similar organizations, such as Japan (the Japanese Space Agency, JAXA) or France (National Institute for Agricultural Research, INRA). Further, there are several commercial satellites that have very high-spatial-resolution (just 1–4 m pixels) multispectral sensors, such as the IKONOS system from Space Imaging and the QuickBird system from DigitalGlobe.

Geographic Information Systems

A GIS is a method for depicting relationships between different kinds of information, usually in the form of maps, each depicting different kinds of information. Because each layer has a specific geographic location, a GIS can be used to explore how these layers relate to each other or to a particular phenomenon, and this often provides new insights or analysis options for ecologists. For example, a researcher might want to study how the spatial pattern of American crow abundance might be influenced by a series of environmental variables such as rainfall, habitat type, and temperature, as well as socioeconomic variables such as housing type and density. Using GIS would allow the researcher to map all these variables at once in order to obtain a visual sense of potential relationships among them, generate hypotheses regarding their relationships, and run statistical tests to determine if the hypothesized correlations were statistically meaningful (plate 1). Currently, GIS software packages have an array of statistical tools, although in some cases GIS data are exported to statistical software packages for more complex analyses. Additional data, referred to as attributes, can be tied to the spatial data and used to address questions in ecology. For example, we might have information on how bird flu relates to American crow abundance.

2. APPLICATIONS OF RS AND GIS IN ECOLOGY

There are a myriad of ways that RS and GIS are currently used—and could be used—in ecology. These tools provide far more than the ability to make maps. Their uses are limited only by human imagination and technology. With technological capacity expanding at a phenomenal rate, it is our job as students and professionals to use these data to solve theoretical and applied problems in ecology. What follows are four examples of how RS and GIS are used in ecology.

Bird Abundance Patterns and Vegetation Structure

It has long been known that the diversity of habitat is related to the diversity of species that occupy that habitat. RS has the ability to provide information on both the two-dimensional (horizontal) and three-dimensional (vertical) aspects of habitat diversity. Satellite sensors like those on Landsat have been used in many studies linking some aspect of biodiversity (either species richness or abundance) to land cover or vegetation maps. This works in areas where there is a diversity of cover types, and species specialize on each of those cover types. So, for example, it has been possible to estimate the diversity of birds in a temperate area where there is a mixture of forest, shrub, and grassland vegetation. If the number of bird species or the abundance (total number of individuals of a given species) associated with each of the vegetation types is known (has been measured in the field), then one can approximate the diversity of species by simply multiplying all the pixels of each given type by the number of species that use that habitat (a "paint by numbers" approach). This approach tends not to work very well in densely forested areas, where habitat is more likely to be "partitioned" vertically and many species may be present in any given vegetation type. In this case, an active sensor such as RADAR or LiDAR can provide information on the vertical structure or diversity of habitat.

One such study was conducted in the dense temperate forest of the Patuxent Wildlife Refuge in Maryland. Wildlife biologists working at the refuge counted the number of birds they observed or heard within a 5-min period at a given location and then walked 400 m (1000 feet) and counted again. They did this 266 times over a regular grid, ultimately counting over 5000 birds from 88 different species. The number of bird species at any given grid location ranged from 2 to 27 (averaging 12), and abundance ranged from as few as 3 to over 100 (averaging 19).

RS scientists supported by NASA then flew over the refuge in an airplane equipped with an imaging LiDAR system. They flew at night because the instrument (known as the laser vegetation imaging system, or LVIS, pronounced "Elvis") is an active sensor that does not rely on light from the sun reflected by the vegetation canopy. The imagery derived from LVIS included an initial return from the top of the vegetation as well as multiple returns reflected from different elements in the canopy and ultimately a strong return from the ground surface. The difference between the first return (the top of the canopy) and the ground return was used to estimate canopy height (an oblique view of this canopy height is shown in plate 2). Other useful canopy metrics include the heights at which 25%, 50%, and 75% of the total summed energy was returned. The point at which half of the energy was returned is referred to as the height of median energy (HOME). This variable is of particular interest because, unlike canopy height, it provides some information on the vertical complexity of the canopy. When HOME and height are used together, one can get a sense for whether there is, for example, a dense understory or a shrub vegetation layer or very little of either. Because some birds specialize in foraging and/or nesting in these specific vegetation layers within the forest canopy, the LVIS metrics provide information that can help to estimate bird diversity (and possibly abundance, although that has yet to be tested because it can vary substantially from year to year for reasons unrelated to habitat diversity).

The results of this analysis indicate that LiDAR measurements of vegetation vertical structure provide a useful proxy for habitat diversity, and this, in turn, was related to bird species diversity. A number of different statistical analysis techniques showed that the LVIS metrics of habitat heterogeneity performed significantly better than those based on optical RS (i.e., Landsat), and use of both Landsat and LVIS improved little on use of LVIS alone. This is an important finding because it demonstrates not only that unique information can be derived from LiDAR but that this information has utility to biodiversity research. If a LiDAR sensor were placed on a satellite orbiting the Earth on a daily basis, it would provide valuable information on biodiversity and other aspects of ecosystems, including carbon stored in biomass, which is of interest for estimating carbon emissions to the atmosphere from deforestation and biomass burning. Just such a system is now under development at NASA and is expected to be launched in 2014. The scientists who conducted the bird diversity study described here are members of the science team in order to ensure that the data will be useful for both biodiversity and carbon-climate applications.

Determining Species Persistence in Dynamic Landscapes

A major goal for researchers, government agencies, and, often, local residents interested in conservation is to maintain viable populations of all the different kinds of species that live in a given region. A viable population is one that is likely to persist over a long period of time. Defined in terms of demographics, a population is viable when the new births in the population, combined with individuals colonizing from other areas (i.e., immigration), are greater than or equal to the deaths in the population and those individuals leaving the population through emigration, given human activities, variation in weather, and other natural disturbances (floods, hurricanes). Scientists have developed a series of tools to predict if a population is likely to remain viable and how changes in a species habitat (defined as places where the intrinsic rate of population growth is greater than 0) will affect its persistence.

Species habitats are continuously changing as a result of natural processes, such as storms or floods and human land use. More and more, humans have removed natural forests, meadows, or other vegetation types for housing developments, resources (wood harvesting), or recreation, and as a result they have reduced and changed the spatial configuration of habitat for species. GIS mapping and modeling can be combined with population viability analyses to determine how changes in land use might influence the probability that a species might persist for long periods of time. Further, different scenarios can be explored; for example, in a forest you could determine how different rates and types of forest extraction would influence the viability of a species dependent on forest.

A recent example of combining GIS mapping and prediction with viability analyses was used to make management recommendations for the sharp-tailed grouse (*Tympanuchus phasianells*), a pheasant-sized bird native to open and shrubby vegetation (steppe grasslands) in the Midwest (Akcakaya et al., 2004). Grouse populations have steadily declined because of conversion of its habitat to agriculture, housing developments, and forest plantations. Currently, much of the grouse population in the United States is contained within the Pine Barrens region of northwest Wisconsin. The habitat in the Pine Barrens is highly fragmented with patches of suitable habitat separated from other such patches. Further, the region is dynamic, mostly a result of natural fires and silviculture (forestry), so that the number, size, and location of patches change over time. This presents a modeling challenge because the only way to correctly predict population viability for the species is to do it in the context of a highly heterogeneous and dynamic landscape.

To address this challenge, Akcakaya and his colleagues created a series of different landscapes in a GIS, each simulated based on different silviculture and burning regimes. Basically, at a given time step they could predict the area and spatial configuration of grouse habitat and combine this with grouse demographics, such as how many young grouse were predicted to be born (fecundity) based on the size and quality of each patch. At the next time step, the area and spatial configuration of habitat changed based on the management regime being modeled, and as a result, the demographic responses also had to be recalculated. Using this approach, Akcakaya and his colleagues showed that the population viability of the sharp-tailed grouse depended both on landscape dynamics and on demographic variables such as fecundity and mortality. Further, demographic modeling in a static landscape provided overly optimistic results about the long-term viability of the species when compared to those obtained using dynamic landscapes.

For species like the grouse, which depends on temporary habitat patches in fragmented landscapes, and where the viability of the species is strongly influenced by the rate of appearance and spatial arrangement of patches in the landscape, complex GIS models are required to predict its persistence. This approach has also been used to determine the viability of Bell's sage sparrow, a species of special concern in California. This species relies on early-successional shrubland (chaparral) on the coast in California. Chaparral is a habitat that was extensive in the coastal dunes in California, which have been heavily impacted by human development. Further, chaparral is maintained by burning and is highly flammable. This presents an obvious and difficult issue for coastal California, where burning also causes devastation to housing developments. GIS modeling combined the species habitat, human land use changes, and demographic information on the sage sparrow to evaluate its long-term persistence. Without an integrated model run on a GIS platform, studies of this sort would not be possible.

Predicting Invasive Species (Weed) Spread Using Ecological Niche Modeling

Invasive species from many different taxonomic groups are increasing in abundance and distributional range at an alarming rate across many ecosystems on Earth. As humans and human products move around the globe at ever-increasing rates, nonnative species move with them, either intentionally or accidentally. Only a small proportion of nonnative species become invasive, but this small group of weeding species has had tremendous ecological effects on native species and ecosystems by outcompeting native species for resources, preying on native species, or changing cycling of important nutrients in ecosystems. Further, the estimated damage and control cost of invasive species in the United States alone amount to more than $138 billion annually (Pimentel et al., 2004). It is difficult to predict which nonnative species will become invasive, but once we have some reason to believe that a species will be invasive, we need tools to predict how and where it will spread. Anticipating future distributions of invasive species is essential for management prioritization, early detection, and control.

One way to predict where a species might spread is to evaluate the environmental conditions where a species exists on its native range and then use this information to predict where it might spread. Ecological niche modeling (also referred to as species distribution modeling) is a GIS-based model that can be used to predict geographic patterns of invasive species. Niche modeling requires two kinds of data: georeferenced occurrence records of where the species is in its native range and GIS-based maps of the environmental variables (e.g., temperature, precipitation) that are likely to influence the suitability of the environment for that species (plate 1). Using a GIS, we can extract the environmental information from environmental layers (maps) for each occurrence record of the species in the native range, establish a statistical relationship between species occurrence and this environmental information, and use this statistical relationship to predict where the species might exist in a different region. These models can be refined to include maps of human land use, such as roads across which people (and potentially invasive species) move.

The house crow, a common bird in Asia (India, Pakistan, Sri Lanka, southwest Thailand, and coastal southern Iran), has been expanding its range into new regions, such as East Africa, and it has been observed in Australia and parts of Europe. It is associated with human settlements in all of its range, from small villages to large cities. Recently, Nyari and colleagues (2006) used ecological niche modeling to determine globally where the species could exist. They used information on human land use, in this case a map of the human footprint (Sanderson et al. 2002), which includes information about cities, roads, population density, and satellite images of night lights (i.e., intensity of electricity use). Using GIS tools, maps, and RS images, they could identify regions that might be susceptible to an invasion by the house crow.

A similar project was conducted by Broennimann and colleagues (2007) with spotted knapweed (*Centaurea maculosa*), an invasive plant well established in North America that is native to Europe. As with the crow study, they used climate information and occurrences from its native range to predict where the plant could live in North America. They found a large correspondence between the model prediction and the actual range of the plant in North America, but certain populations of knapweed were not well predicted by the model. This result indicated that knapweed exists under novel environmental conditions in North America; the plant occupies a distinct climatic niche on its invaded range. Although this is somewhat disappointing from the perspective of accurate and complete model prediction of invasive species, it does provide fascinating insights into phenotypic plasticity and the potential for species to evolve into new environmental niches. Ongoing research is evaluating these different possibilities.

FURTHER READING

Akcakaya, H. R., J. Franklin, A. D. Syphard, and J. R. Stephenson. 2005. Viability of Bell's sage sparrow (*Amphispiza belli* ssp. *belli*): Altered fire regimes. Ecological Applications 15: 521–531. *An example of combining GIS data and viability modeling to study population persistence.*

Akcakaya, H. R., V. C. Radeloff, D. J. Mladenoff, and H. S. He. 2004. Integrating landscape and metapopulation modeling approaches: Viability of the sharp-tailed grouse in a dynamic landscape. Conservation Biology 18: 526–537. *An example of combining GIS data and viability modeling to study population persistence.*

Broennimann, O., U. A. Treier, H. Muller-Scharer, W. Thuiller, A. T. Peterson, and A. Guisan. 2007. Evidence of climatic niche shift during biological invasion. Ecology Letters 10: 701–709. *An example of using ecological niche modeling to study how species niches might shift during an invasion.*

Goetz, S., D. Steinberg, R. Dubayah, and B. Blair. 2007. Laser remote sensing of canopy habitat heterogeneity as a predictor of bird species richness in an eastern temperate forest, USA. Remote Sensing of Environment 108: 254–263. *An example of using remote sensing to study bird diversity.*

Graham, C. H., S. Ferrier, F. Huettman, C. Moritz, and A. T. Peterson. 2004. New developments in museum-based informatics and applications in biodiversity analysis. Trends in Ecology & Evolution 19: 497–503. *A review of spatial biodiversity data available and how to use it in a GIS.*

Kerr, J. T., and M. Ostrovsky. 2003. From space to species: Ecological applications for remote sensing. Trends in Ecology & Evolution 18: 299–305.

Nyari, A., C. Ryall, and A. T. Peterson. 2006. Global invasive potential of the house crow Corvus splendens based on ecological niche modelling. Journal of Avian Biology 37: 306–311. *An example of using ecological niche modeling and human landuse data to predict the spread of invasive species.*

Pimentel, D., R. Zuniga, and D. Morrison. 2004. Update on the environmental and economic costs associated with alien-invasive species in the United States. Ecological Economics 52: 273–288. *An evaluation of the economic cost associated with managing alien-invasive species.*

Sanderson E. W., M. Jaiteh, M. A. Levy, K. H. Redford, A. V. Wannebo, and G. Woolmer. 2002. The human footprint and the last of the wild. Bioscience 52: 891–904. *A quantification of the influence of human behaviors on wild areas.*

Turner, W., S. Spector, N. Gardiner, M. Fladeland, E. Sterling, and M. Steininger. 2003. Remote sensing for biodiversity science and conservation. Trends in Ecology & Evolution 18: 306–314. *A review of remote-sensing data available for ecological and conservation research.*

I.12

Geographic Range

Kevin J. Gaston

OUTLINE

1. Range size
2. Range edges
3. Range structure
4. Fundamental units

No species occurs everywhere. Indeed, most are absent from the vast majority of sites across the globe. Those areas in which a species does occur constitute its geographic range. As such, the geographic range is one of the fundamental units in ecology. The sizes and distribution of geographic ranges give rise to patterns of species richness and change in species composition from site to site, and combined with their abundance and trait structure give rise to other spatial patterns in assemblages. Likewise, temporal changes in assemblages on both short and long time scales follow from changes in the size, position, and structure of geographic ranges.

GLOSSARY

area of occupancy. The area within the outermost geographic limits to the occurrence of a species over which it is actually found

extent of occurrence. The area within the outermost geographic limits to the occurrence of a species

intraspecific species-abundance distribution. The frequency of areas within a species' geographic range in which it attains different levels of abundance

range edge or limit. The outermost geographic occurrences of a species, usually excluding vagrant individuals

species–range size distribution. The frequency of species with geographic ranges of different sizes

1. RANGE SIZE

The sizes of the geographic ranges of species vary dramatically and can be characterized in two fundamentally different ways. Extent of occurrence is the area within the outermost limits to the occurrence of a species, and area of occupancy is the area over which the species is actually found. The latter will tend to be consistently smaller because no species is distributed continuously across space even within the broad geographic limits to its occurrence. The finer the spatial resolution and the shorter the time period over which area of occupancy is measured, the smaller will be the area over which the species is documented to occur, and the greater this disparity will be. At one extreme lie those, predominantly freshwater or terrestrial, species that are currently found occurring in a single small habitat patch (often with only a very small number of individuals), which are thus narrowly distributed in terms both of extent of occurrence and area of occupancy. At the other extreme lie some marine organisms. Species of microorganisms may be widespread across the oceans both in terms of extent of occurrence and area of occupancy, whereas some large-bodied species of vertebrate may have large oceanic distributions in terms of extent of occurrence but, because of the relatively low numbers of individuals, not area of occupancy.

Species–Range Size Distributions

Both within and across major taxonomic groups, the geographic ranges of the majority of species are relatively small, and only a very few are widespread. Indeed, within such groups species–range size distributions, the frequency of species with ranges of different sizes, are almost invariably strongly right-skewed. One important consequence is that the vast majority of occurrence records result from a small number of species. For example, by one estimation, at a spatial resolution of approximately 100×100 km, the 10% most globally widespread extant species of birds account for 50% of occurrence records. Given that the ratio of extents of occurrence to areas of occupancy may often be proportionately larger for rare species than for widespread ones, that is, they occupy their ranges less densely, the dominance of occurrence records by widespread species may increase when documented at finer spatial resolutions. This dominance may explain why it is the more widespread rather than, as often assumed, the restricted species that contribute disproportionately to spatial variation in species richness and related macroecological patterns.

Phylogenetic Constraint

The average sizes of geographic ranges can vary markedly between species in different major taxonomic groups. Thus, among nonmarine vertebrates, species of fish and amphibians tend naturally to have smaller ranges than do mammals, and mammals smaller ranges than do birds. However, within taxonomic groups, the extent to which the geographic range sizes of species exhibit phylogenetic constraint is contentious. Certainly range size is not as strongly conserved as are body size and many life history traits. Even where significantly conserved, it typically remains impossible to predict with any accuracy the range size of a species from that of its sister species or other close relatives, suggesting that such heritability has limited practical value (e.g., in estimating the range sizes of species whose distributions have not been well documented). This would tend to follow if the range sizes of different species are determined by the variable outcomes that result from the combinations of individual traits and environmental conditions occurring at particular times and places.

Spatial Dynamics

The mean size of the geographic ranges of the species within a higher taxon tends to vary spatially. Most obviously, ranges are typically smaller in situations in which dispersal and environmental conditions are geographically highly constrained, such as on islands and at high elevations, and in specialized habitats (e.g., desert springs, deep sea vents). However, more systematic spatial patterns have also been argued to occur, in particular, increases in the latitudinal extent of ranges from low to high latitudes, in their altitudinal extent from low to high elevations, and in their depth extent from shallow to deep waters. The first of these is a phenomenon termed Rapoport's rule. The pattern appears to be most evident in the terrestrial northern hemisphere but may actually reflect a general trend for terrestrial ranges to increase from high southern to high northern latitudes. Although other factors may also have an influence, this trend is at least in part a result of changes in land area.

Temporal Dynamics

More difficult to establish than patterns of spatial variation in geographic range sizes are the long-term temporal trends. How the mean range sizes of the species in a higher taxon have changed over geologic time remains virtually unknown (although it is likely to have been marked, given changes in the distributions of land masses, water bodies, and climatic conditions). Little more is understood about how the range of an individual species changes in size between its origination and its extinction. However, best evidence suggests that geographic ranges typically undergo a rapid increase in size following speciation and then a slower subsequent, and perhaps prolonged, decline to extinction. This is supported by studies of species introduced into areas in which they previously did not occur, which have revealed that following an initial lag phase, during which a species tends to remain rather restricted to the locale of its introduction and densities there tend to build up, spread can occur across large areas very rapidly (both phases are extremely short in terms of evolutionary time).

When we focus on the events at the outset and conclusion of a species' lifespan, geographic range size influences both the likelihood of speciation and that of extinction. At least when allopatric, the likelihood of speciation appears to be related to range size by a hump-shaped function. As ranges increase from small to moderate sizes, the likelihood of speciation increases because the chance of the range being bisected by a barrier to dispersal increases. However, at some point ranges will become sufficiently large that they will tend to engulf all but the largest potential barriers, such that they do not engender speciation, and the probability of division will decline. In addition, widespread species may have well-developed dispersal abilities and greater numbers of individuals that both help to maintain range contiguity and reduce speciation rates.

By contrast, the likelihood of extinction is strongly negatively correlated with geographic range size. Indeed, abundance and range size are in general the two best predictors of the probability that a species will go extinct in the near future (although there are examples of previously very widespread species that, usually as a consequence of anthropogenic pressures, have rapidly declined to extinction). Larger ranges typically comprise greater numbers of individuals and thus have a smaller probability of a random walk to extinction, and because of their greater areal coverage, they have a reduced risk that adverse conditions in one region will affect all individuals at the same time. This raises the possibility of species selection acting on range sizes.

Traits

Within taxonomic groups, interspecific variation in geographic range sizes is often correlated with such variation in other traits, including dispersal ability, breadth of resource use, or environmental tolerance, local abundance, and body size. In at least some cases, these relationships seem likely to be mechanistic, although the paths of causality may be variable. Thus, although it seems intuitive that a greater dispersal ability will tend to lead to a species becoming more widespread, other barriers may prevent this from occurring (see below), and if the structures associated with good dispersal abilities are costly to build or maintain, they may be reduced or lost. Indeed, there is evidence that when other limits are removed, species can show rapid acquisition of improved dispersal abilities.

Species that are able to exploit a wider variety of resources or persist under a wider range of environmental conditions should, all else being equal, be able to attain larger geographic ranges. Of course, all else may not be equal (e.g., extent of different resource types, dispersal abilities), which will tend to weaken any correlations between the level of such generalism and range size. The variety of resources and the breadth of environmental conditions that species can use are influenced both by the variety and breadth that can be exploited by individual organisms and by the differences in resource usage and tolerance of environmental conditions among individual organisms. The latter is probably a much more important influence on the relative geographic range sizes of species in many taxonomic groups, given evidence for marked spatial variation in the realized and fundamental niches of individuals, particularly of those species that are more widespread.

The geographic range sizes of species within taxonomic groups tend commonly to be positively correlated with their local density, such that widespread species not only have more individuals but disproportionately more so than restricted species. A number of plausible mechanisms have been proposed to explain such a pattern, including variation in niche breadth (the range of resources or conditions a species can exploit), niche position (how typical are the resources or conditions that a species can exploit), habitat selection (the tendency for species to use more habitats as they become more abundant), and metapopulation dynamics (in which dynamics in local populations depend on those in other such populations). It seems likely that a variety of potentially mutually reinforcing processes may be at work and that the pattern is an almost inevitable consequence of the aggregated spatial distributions of the individuals of most species.

Also within taxonomic groups, the body sizes and geographic range sizes of species tend to exhibit an approximately triangular form, such that although species of all body sizes may have large geographic range sizes (the upper limit normally being imposed by the size of the land mass or ocean mass), the minimum range size observed tends to increase with body size. A positive relationship likely occurs because, on average, larger-bodied species have larger-sized home ranges than smaller-bodied ones. They may thus also require larger total geographic range sizes if range-wide populations are to exceed some minimum viable size, which would tend to result in a positive interspecific range size–body size relationship, and indeed a triangular one because there is no necessary upper constraint on the range size of small-bodied species. One consequence of this mechanism is that the body sizes of the largest species tend to increase with the areas of the land masses on which they occur.

2. RANGE EDGES

Regardless of their extents of occurrence or areas of occupancy, the geographic ranges of few species are entirely congruent. Rather, the bounds fall in different places and shift position on both ecological and evolutionary time scales. Why at any given time, or averaged over a particular period, the range edges of a particular species fall quite where they do has been much debated, and for surprisingly few species is it well understood. Part of the difficulty is that the question has a variety of answers, depending on the terms in which it is couched. First, one can determine whether there are abiotic and/or biotic factors that prevent further spread, and if so what these are. Second, one can consider how, in response to these factors, the population dynamics of a species change such that it is unable to persist beyond this point. Third, one can establish the genetic mechanisms that prevent a species from evolving capacities that would enable it to overcome any limiting abiotic and biotic factors that prevent it from expanding the limits to its geographic range and becoming more widespread.

Abiotic and Biotic Factors

Two principal groups of abiotic factors have been argued to limit geographic ranges, physical barriers and climate. In both cases, it is actually the interplay between these factors and the traits of the particular species that is of concern. Thus, given its dispersal abilities and behavioral tendencies, the spread of a species may be delayed or entirely prevented by expanses of inhospitable habitat, such as water for terrestrial organisms and land for marine ones, high elevations for lowland organisms and shallow water for deep-sea ones, or grasslands for forest species and forests for grassland species. In some cases, the role of behavior may be more significant than that of dispersal ability per se, with even quite small disjunctions in the distribution of suitable habitat greatly restricting spread. In many cases, rare long-distance dispersal events effectively define what does and does not constitute a barrier.

Climatic constraints on geographic ranges attract by far the majority of attention from ecologists, particularly because these may be modified by anthropogenic climate change, with implications for, among others, agriculture, forestry, fisheries, and human health. Climate doubtless limits the potential occurrence of all species because their physiological tolerances are constrained as a consequence of the costs of maintaining wide tolerances and the trade-offs associated with being adapted to particular conditions. However, demonstrating that climatic factors actually determine the position of the limits to the range of a species is more difficult. A variety of forms of evidence are strongly suggestive, including observations of systematic latitudinal variation in elevational and depth limits to species occurrences (suggesting elevational and depth responses to changes in climate with latitude), of spatial coincidence between the occurrence of range edges and particular climatic conditions, and of temporal covariation in the position of range edges and particular climatic conditions. However, such patterns could reflect covariation with some other factors, such as the distribution of a key resource, predator, or parasite, which itself is climatically limited. Rather more convincing are demonstrations of the more direct influence of climatic conditions at range edges on reproduction and mortality, perhaps most commonly reflected in the failure of species to be able to complete their life cycles beyond the range boundary.

A wide variety of biotic factors have been argued to limit the geographic ranges of species, including the absence of essential resources (e.g., nutrients, prey) and the presence of competitors, predators, or parasites. The role of resources in limiting ranges is perhaps most clearly demonstrated by specialist consumers, whose distributions must be contained within that of their host. Almost invariably, such species do not occur throughout the distribution of this resource, which in the absence of other factors presumably reflects spatial variations in the abundance and quality of the host. The roles of competitors, predators, and parasites in limiting ranges frequently involve interaction among three or more species, such that predators and parasites have alternative resources to exploit and competitors have their influence through predators and parasites (apparent competition).

Although it is easiest to consider them separately, in practice the limitation of the geographic ranges of individual species by abiotic and biotic factors may often be complex. Combinations of these factors may act synergistically, and different factors may be limiting on different parts of the range boundary and at different times, on both ecological and evolutionary time scales.

Population Dynamics

The edges of geographic ranges are formed at the point at which births and immigration in local populations are exceeded by deaths and emigration. If we assume that dispersal is sufficient for the establishment of a species in peripheral sites but does not otherwise influence numbers in those local populations, then the key issues are the factors that drive local extinction, which is commonly observed to be higher among populations at range edges. These factors can include demographic stochasticity (in the extreme, in small populations by chance during the same time interval all individuals may die, all may fail to breed, or sex ratios may become highly skewed), the mean environmental conditions (including resource availability, interspecific competition, predation), and the temporal variance in environmental conditions (even when conditions on average are suitable, high variance will increase the likelihood of local extinction).

Although such simple scenarios at range edges may occur, immigration may often play an important role in the dynamics of local populations, enabling them to persist even when the death rates of individuals exceed the birth rates. This leads to the notion of source–sink dynamics, in which the geographic range of a species can embrace unfavorable niche conditions, such that it occurs in environments at the range edges under which in isolation any local population would rapidly become extinct. This emphasizes the potential importance of thinking about range limits in terms of the interactions among multiple local populations. Indeed, under such a scenario, limits can be formed simply because the proportion of the landscape occupied by local populations becomes sufficiently low that the influence of dispersal on local populations becomes insufficient, and the ratio of population extinctions to colonizations too high.

Genetics

A number of constraints have been suggested that prevent populations at range edges from evolving the capacity to spread further. Most attention is, however, focused on the possibility that if gene flow is sufficient, the occurrence at range edges of alleles that would otherwise enable range expansion may be swamped by alleles from other populations. This is particularly likely to be the case if gene flow occurs predominantly to edge populations from the typically larger numbers of local populations and individuals that do not occur at range edges, pushing the latter away from adaptation to local optima. However, a diverse array of other mechanisms have been suggested that include low levels of genetic variation in peripheral populations, that traits show low heritability as a consequence of directional selection in marginal environments, that traits show low heritability because of environmental variability in marginal environments, that changes in several independent characters are required for range expansion and so favored genotypes occur too rarely, that genetic trade-offs between fitness in favorable and stressful environments prevent the increase of genotypes adapted to stressful conditions, that genetic trade-offs among fitness traits in marginal conditions prevent traits from evolving, and that the accumulation of mutations that are deleterious under stressful conditions prevents adaptation.

3. RANGE STRUCTURE

Aside from their size and boundaries, geographic ranges are structured in complex ways. This structure is determined by the distribution of the individuals of a species within the range boundaries and by variation in the traits exhibited by those individuals.

Abundance and Occupancy

Across its geographic range, a species is almost invariably rare in most of the places in which it occurs and relatively abundant in only a few. That is, intraspecific species-abundance distributions are strongly right-skewed. The notion has long prevailed that areas in which a species attains higher densities tend to lie toward the center of its geographic range, with the range edge being an area of lower density. This would seem likely to follow if conditions were most favorable in the range center and declined in all directions away from that core. However, whereas this might be a useful model in the abstract, the empirical evidence to support such a pattern of abundance is limited, and there are ample examples in which it does not occur, including cases in which high abundances are found close to range edges and of marked latitudinal trends in abundance across ranges. Overall, it seems that just as they take a wide diversity of shapes, the spatial abundance structures of ranges are also very varied. This is important from an applied perspective, as without local abundance data, it is difficult to target conservation or other activities at abundance hotspots.

Although local abundances may not tend consistently to decline toward range edges, there may be a greater likelihood that levels of occupancy do so. That is, although the densities of individuals within local populations, and therefore of intraspecific interactions, may not do so, the densities of local populations may change systematically. This would be consistent with a scenario in which range boundaries were more often determined by reductions in the availability of suitable habitat patches rather than in the quality of those patches where they do exist.

Just as there are interspecific abundance–range size relationships, there are intraspecific ones such that as the local density of a species increases through time, so does its occupancy. This has significant implications for understanding how the geographic ranges of species spread and decline and also for a variety of applied issues such as strategies for harvesting species. As geographic ranges change in size, they seem essentially to move back and forth along trajectories in abundance–range size space.

Traits

A number of systematic patterns of variation in the traits of individual organisms have been documented across the geographic ranges of species (not simply between the edges and the rest of ranges). These include spatial trends in morphology (principally body size), physiology, and life history. Indeed, some of these have been regarded as sufficiently general as to constitute ecogeographical rules. They include the neo-Bergmannian rule or James's rule (an increase in the size of a species toward higher latitudes or lower temperatures), Foster's or the island rule (smaller species become larger and larger species smaller on islands compared with mainland areas), Gloger's rule (a tendency for endothermic animal populations in warm and humid areas to be more heavily pigmented than in cool dry areas), Jordan's rule (fish species develop more vertebrae in cold environments than in warm ones), and one of Rensch's rules (an increase in litter sizes of mammals and clutch sizes of birds in colder climates). In the main, such patterns in geographic range structure appear to be driven by broad spatial trends in environmental conditions or by spatial trends in the temporal variation (between seasons or years) in those conditions. They tend also to be observed predominantly among the more widespread species, whose geographic ranges extend over a greater range of environmental conditions. The trait structures of the ranges of the majority of species may thus be a good deal more complex.

4. FUNDAMENTAL UNITS

The study of geographic ranges has been revolutionized by dramatic increases in available data on the occurrences of species and on the environments in which they occur (particularly from remote sensing), and in the technology available to handle those data. The broad-scale perspective that these have enabled has served to highlight the significance of geographic ranges as fundamental units in ecology, and much of that understanding of population and community ecology can usefully be cast in terms of the size, distribution, and structure of geographic ranges.

FURTHER READING

Brown, James H. 1995. Macroecology. Chicago: University of Chicago Press.

Gaston, Kevin J. 1994. Rarity. London: Chapman & Hall.

Gaston, Kevin J. 2003. The Structure and Dynamics of Geographic Ranges. Oxford: Oxford University Press.

Gaston, Kevin J., and Tim M. Blackburn. 2000. Pattern and Process in Macroecology. Oxford: Blackwell Science.

Lomolino, Mark V., Brett R. Riddle, and James H. Brown. 2006. Biogeography, 3rd ed. Sunderland, MA: Sinauer Associates.

Williamson, Mark. 1996. Biological Invasions. London: Chapman & Hall.

I.13

Adaptation

Allan Larson

OUTLINE

1. Adaptation and Darwinism
2. Adaptation as a hypothesis of evolutionary history
3. Molecular population genetics of adaptation
4. Adaptation and selfish genetic elements

Darwin's theory of natural selection explains how genetically variable populations gradually accumulate traits that enhance an organism's ability to survive and to reproduce. Calling a particular character an adaptation denotes the hypothesis that the character arose gradually by natural selection for a particular biological role, which is called the character's function. Any hypothesis of character adaptation is therefore a historical explanation that must specify the particular population, the interval of evolutionary time, the geographic conditions in which the relevant evolution occurred, and the nature of character variation that was sorted by natural selection. Empirical rejection of the hypothesis of character adaptation suggests the alternative hypotheses of exaptation (a character co-opted by natural selection for a biological role not associated with the character's origin), nonaptation (a character not discriminated from alternatives by natural selection), or disaptation (a character disfavored by selection relative to alternative forms). I illustrate the contrast between adaptationist and anti-Darwinian theories of character origination using a longstanding debate concerning evolution of mimicry of wing patterns among butterfly species. I describe adaptation as a molecular population-genetic process using as an example the medical syndrome of sickle-cell anemia in African populations; depending upon its genetic and environmental contexts, hemoglobin S may constitute an exaptation, a nonaptation, a disaptation, or a component of an adaptive complex of epistatically interacting genes. Evolutionary developmental modularity and phenotypic accommodation may enhance the role of phenotypically discontinuous changes in evolution by natural selection. Selfish genetic elements likely underlie most organismal characters that arise as disaptations and nonetheless persist despite natural selection against them. Suppression of selfish genetic elements is potentially a major source of evolution by natural selection. The explicitly historical approach to adaptation illustrated here contrasts strongly with a now largely discredited analogistic approach used in older ecological literature.

GLOSSARY

adaptation (as a process). Evolution of a population by natural selection in which hereditary variants most favorable to organismal survival and reproduction are accumulated and less advantageous forms discarded; includes character adaptation and exaptation.

balanced polymorphism. Occurrence in a population of a selective equilibrium at which two or more different allelic forms of a gene each have frequencies exceeding 0.05.

character adaptation. A character that evolved gradually by natural selection for a particular biological role through which organisms possessing the character have a higher average rate of survival and reproduction than do organisms having contrasting conditions that have occurred in a population's evolutionary history; adaptation in this usage contrasts with disaptation, exaptation, and nonaptation.

developmental constraint. A bias in the morphological forms that a population can express caused by the mechanisms and limitations of organismal growth and morphogenesis.

disaptation. A character that decreases its possessors' average rate of survival and reproduction relative to contrasting conditions evident in a population's evolutionary history; a primary disaptation is disadvantageous within the populational context in which it first appears; a secondary disaptation acquires a selective liability not present at its origin as a consequence of environmental change or an altered genetic context.

exaptation. Co-option of a character by natural selection for a biological role other than one through which the character was constructed by natural selection.

function. The biological role through which an adaptive character was constructed by natural selection.

gradualism. Accumulation of individually small quantitative changes in a population leads to qualitative change; contrasts with saltation, in which a single genetic change induces a large qualitative change in phenotype.

mimicry. Evolution by natural selection in which a character is favored because it closely resembles one present in a different species; the species whose character is copied by a "mimic" is called the "model."

modularity. Evolution of developmental constraints by which one of two or more alternative, qualitatively different suites of characters can be activated by particular genetic or environmental cues.

nonaptation. A character not selectively distinguishable from contrasting conditions present in the evolutionary history of a population.

saltation. Evolution of a large, qualitative change in phenotype in a single mutational step; contrasts with gradualism.

selfish genetic element. Genes that spread at a cost to the organism; stretches of DNA that act narrowly to advance their own proliferation or expression and typically cause negative effects on nonlinked genes in the same organism (modified from Burt and Trivers, 2006).

1. ADAPTATION AND DARWINISM

Among the various meanings given to the term *adaptation* in evolutionary ecology, synonymy with evolution by natural selection is probably the most common one. Darwin's theory of natural selection explains how genetically variable populations accumulate traits that enhance an organism's ability to survive and to reproduce by making resources more accessible (see chapter I.14). Less-favorable alternative traits decline in frequency and are lost from the population because their possessors lose the struggle for survival and reproduction. A population produces variant forms at random with respect to an organism's needs, and natural selection retains only the advantageous forms.

Closely allied with Darwin's theory of natural selection is his theory of gradual change. Darwin considered abrupt changes of organismal form or physiology likely to disrupt normal functioning and thereby to be discarded by natural selection. The favorable traits that natural selection accumulates across generations each contribute only small phenotypic effects in the traditional Darwinian hypothesis. Evolution of a qualitative change in organismal form, such as the origin of a new anatomical structure or color pattern, occurs gradually across many generations as natural selection increases the populational frequencies of many small component parts so that they come to reside in the same individuals. Natural selection acting on incremental variation thus provides Darwin's major explanation for evolution of novel organismal forms.

To call a particular character an adaptation denotes the hypothesis that the character arose gradually by natural selection for a particular biological role, which is termed the character's function (Gould and Vrba, 1980). Any hypothesis of adaptation is a historical explanation that must specify a particular population, interval of evolutionary time, and geographic conditions in which the relevant evolution occurred. Gould and Vrba (1980) use an extant population as the focal point for analysis and restrict the term *adaptation* to a character whose current utility matches the function for which the character arose by natural selection. They apply the contrasting term *exaptation* to a character co-opted by natural selection for a biological role not associated with the character's origin. One need not restrict hypotheses of adaptation versus exaptation to extant populations, but the historical frame of reference must make explicit the temporal and spatial dimensions across which the relevant evolutionary processes occurred.

I illustrate the contrast between adaptationist and anti-Darwinian theories of character origination using a longstanding debate concerning evolution of mimicry of wing patterns among butterfly species. Many cases are documented in which two or more butterfly species share the same potential avian predators and also share closely matched patterns of warning coloration on their dorsal wing surfaces. Because an avian predator learns to associate specific warning coloration with distastefulness, a distasteful "model" species often evolves characteristic warning coloration. Other species that share the same potential predators as the model can gain a selective advantage by "mimicking" the warning coloration of the model species. In some cases, the mimic species is a desired prey item that tricks its potential predator by adopting warning coloration deceptively (called Batesian mimicry). If the mimic is distasteful, sharing of the same warning coloration among species provides mutual benefit (called Müllerian mimicry). In each case, evolution of warning coloration deters avian predators, all of which seek their prey visually and learn to associate particular wing patterns with distasteful prey. The adaptationist and anti-Darwinian explanations of Müllerian mimicry concur that natural selection for a shared warning pattern benefits members of each species because a predator needs to learn only one warning pattern to reduce mortality in each species.

In the late 1800s, the anti-Darwinian orthogenetic evolutionist Theodor Eimer used butterfly mimicry among other empirical examples to support an argument that natural selection cannot construct complex morphological characters by accumulating gradual changes. He argued that butterfly species have inherited from their common ancestor similar mechanics of wing development and shared biases in production of new patterns; genetic changes that introduce pigments onto a wing surface are therefore likely to produce similar geometric patterns in all species that share a particular set of developmental mechanisms. Natural selection acts to preserve shared warning coloration in multiple species, but the specific pattern is formally an exaptation; it is a consequence of developmental mechanics, not something evolved gradually by natural selection acting on randomly produced variation in pigmentation.

Ronald Fisher in 1930 used butterfly mimicry to support the opposite, adaptationist hypothesis: a mimic species gradually evolves a sequentially improved match to its model by natural selection acting on many genes whose variation exerts random and incremental effects on pigment deposition across the wing surface. The detailed matching of the model's pattern and coloration by the mimic species therefore constitutes character adaptation.

In the 1980s, John Turner reported detailed genetic analyses of Müllerian mimicry among South American species of *Heliconius* butterflies to reconstruct the genetic histories of evolution of their mimicry patterns (figure 1). He concluded that genetic changes of major phenotypic effect were important for producing close matches in pigmentation pattern among geographically codistributed butterfly species and that subsequent improvement of the matching occurred by accumulating multiple genetic changes of smaller phenotypic effect. This interpretation supports Eimer's general hypothesis that shared developmental constraints explain evolution of shared patterns and that mimicry evolves by exaptation; only the detailed fine-tuning of the matched patterns, as explained by Turner, constitutes character adaptation as argued by Fisher.

2. ADAPTATION AS A HYPOTHESIS OF EVOLUTIONARY HISTORY

Each specific case of butterfly mimicry involves separate evolutionary histories of at least two species, and hypotheses of character adaptation versus exaptation therefore must be tested separately for each case. The prevailing pattern reported by John Turner for *Heliconius* populations might or might not prevail in other groups. In each separate test of a hypothesis of adaptation, one seeks evidence capable of rejecting the claim that a hypothetically adaptive character arose gradually through accumulation of many genetic variants, each of which gave its possessors a higher net rate of converting resources into survival, growth, and/or reproduction (= "Darwinian fitness," see chapter I.14) than did the alternatives with which it formed population-level polymorphisms.

I emphasize the importance of historical precision in formulating and testing hypotheses of adaptation because careless uses of adaptation have elicited condemnation of adaptationist studies. I agree with the critics that one must resist an analogistic tradition in which one equates as similar or equivalent the character variation and selection pressures described for distantly related species. For example, claims in sociobiological literature that one can use behavioral ecological studies of "helpers at the nest" in a bird species to explain analogous behaviors in human families must be rejected as having no historical equivalence. Evolution by natural selection depends as critically on the specific character variation produced in a population and the genetic structure of that variation as it does on environmental conditions. The kinds of phenotypic variation produced independently in different species are comparable only to the extent that homologous developmental mechanisms channel the morphological expression to a few major alternative forms in each case, as appears to occur in wing patterns of *Heliconius* butterflies.

Historical hypotheses of adaptation can be categorized as microevolutionary or macroevolutionary depending on the investigator's vantage point with respect to the historical process being studied (Rose and Lauder, 1996). A microevolutionary study measures dynamics of populational polymorphisms on a generational time scale (see chapter I.17). At this scale, an investigator must distinguish natural selection per se from the genetic response of a population to natural selection. Because the relationship between genotype and phenotype is complicated by genetic dominance and epistasis and phenotypic plasticity, natural selection on phenotypic variation does not guarantee a particular genetic response of the population to selection. Agricultural geneticists are well aware that selecting for a favorable characteristic in a crop species does not always cause a corresponding genetic improvement of the population in the following generation. A macroevolutionary study, by contrast, begins with the knowledge that a particular evolutionary

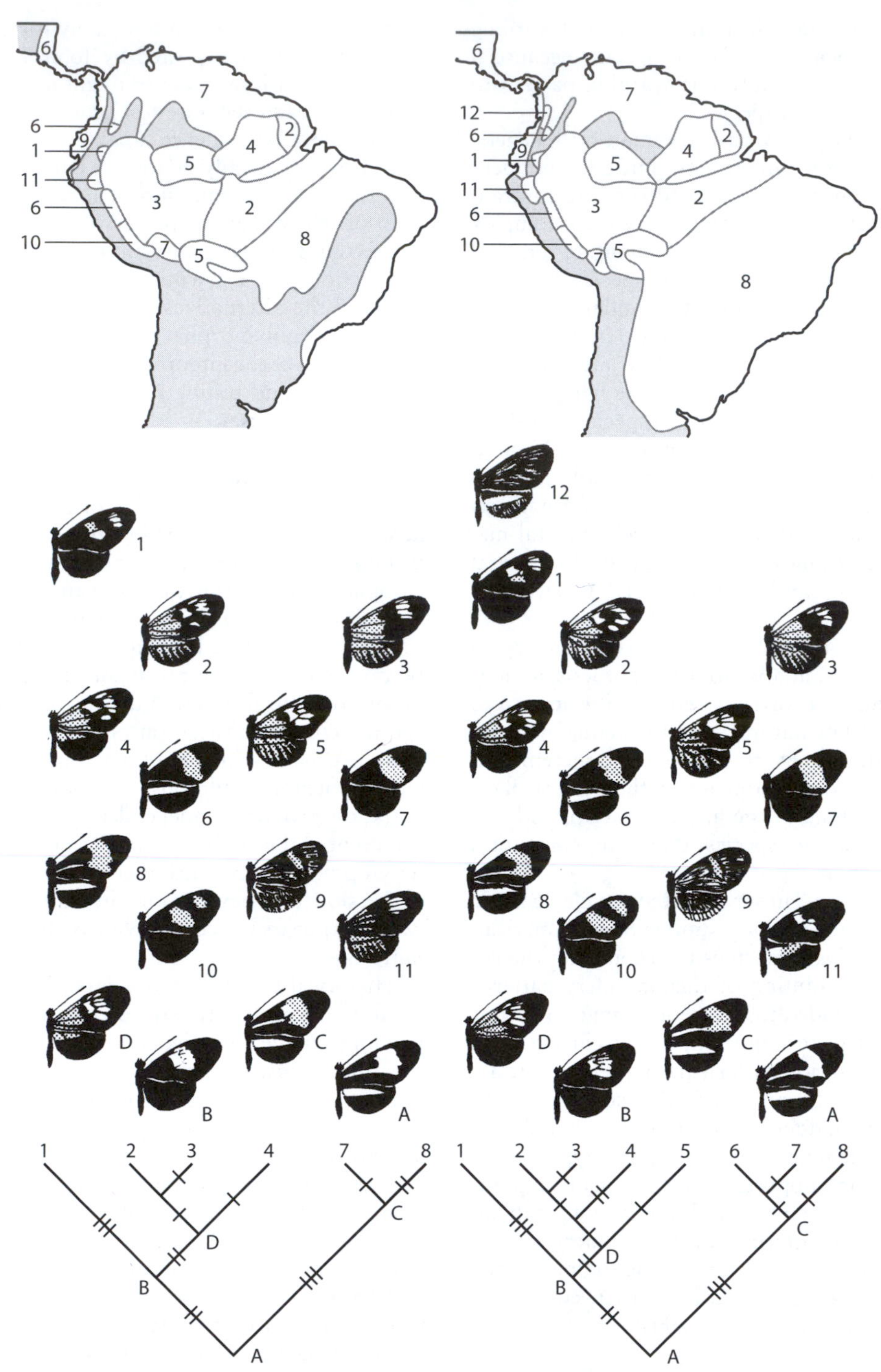

Figure 1. Evolution of shared antipredatory warning patterns by multiple geographic races of *Heliconius melpomene* (left) and *H. erato* (right) as interpreted by John R. G. Turner. Numbered areas on each map (top) indicate geographic distributions of corresponding numbered wing patterns below the map. Four inferred ancestral wing patterns (A–D) also are shown for each species. Tree diagrams show genetic substitutions of major effect that transform one color pattern to another by adding or subtracting large areas of pigmentation. Shared developmental constraints by these species likely underlie their parallel, saltational origin of the same major patterns; subsequent fine-tuning of the match between patterns of geographically codistributed races occurs by polygenic changes compatible with an interpretation of gradual adaptive evolution. (After Turner, J.R.G. 1981. Adaptation and evolution in *Heliconius*: A defense of neoDarwinism. Annual Review of Ecology and Systematics 12: 99–121. © 1981 by Annual Reviews, www .annualreviews.org. Used with permission.)

change, as inferred by phylogenetic analysis, has occurred (see chapter I.16). The unanswered question is whether the organismal variation and environmental contexts of a particular character transition are compatible with a specific selective explanation. For example, one can reject the macroevolutionary hypothesis that bird feathers evolved by natural selection for utility in flight because the fossil record shows that evolution of feathers preceded evolution of flight in birds. The utility of feathers for flight in living birds is therefore an exaptation, although details of the size and shape of wing feathers in particular species might constitute adaptations for flight evolved more recently in the species' evolutionary histories.

3. MOLECULAR POPULATION GENETICS OF ADAPTATION

I illustrate adaptation as a process with a strong empirical example of evolution by natural selection in the microevolutionary mode. Adaptation of human populations to resist malarial infection is perhaps the best case study in terms of documenting evolutionary change at the both the molecular genetic and phenotypic levels and in measuring critical environmental variables. My discussion draws on Templeton's (2006) synthetic analysis of relevant medical and epidemiological data with comments on the respective roles of character adaptation versus exaptation.

The medical syndrome of sickle-cell anemia in central African populations is perhaps the best-known case of evolution by natural selection at the molecular population-genetic level. Epidemic malaria was likely established in Central Africa as a consequence of agricultural practices introduced there about 2000 years ago (Templeton, 2006). The gene whose variation illustrates selectively guided change is the gene encoding β-hemoglobin. The most common and inferred ancestral allelic form of β-hemoglobin possesses as its sixth amino acid glutamic acid and is called "hemoglobin A." A single mutational change to hemoglobin A substitutes valine at this position to produce an alternative allele called "hemoglobin S." The *S* allele could have arisen from *A* more than once by mutation in different human populations. Hemoglobin S has a genetically dominant phenotype of malarial resistance. Hemoglobin S molecules form sickle-shaped aggregations within an erythrocyte under low-oxygen conditions; erythrocytes distorted by these aggregations are promptly destroyed by the spleen. Infection of an erythrocyte by a malarial parasite deprives the cell of oxygen, causing sickling; the spleen then destroys the infected erythrocyte and its malarial parasite before the parasite has completed its life cycle. A person heterozygous for the *A* and *S* alleles thereby gains resistance to malaria from hemoglobin S, and hemoglobin A has a genetically dominant phenotype for normal respiration under most environmental conditions. An individual homozygous for the *S* allele suffers the severe respiratory disability called "sickle-cell anemia." Severe anemia is therefore a recessive phenotype of hemoglobin S.

Before the introduction of epidemic malaria into Africa, the *S* allele would have been kept rare by natural selection because the *SS* homozygous individuals have greatly diminished chances of surviving to adulthood. The allele is preferentially removed from the gene pool by natural selection when it occurs in the *SS* genotype. If mating is random with respect to genotypic variation at the β-hemoglobin locus, the rare *S* allele occurs almost exclusively in heterozygous *AS* genotypes, and its selective consequences depend mainly on its phenotypic consequences in the *AS* genotype (a consequence of Hardy-Weinberg equilibrium; see chapter I.15).

In a malarial environment, selection favors *AS* individuals, thereby increasing the frequency of the *S* allele. Occurrence of the *S* allele at frequencies exceeding rarity is strongly geographically coincident with epidemic malaria (figure 2). An increase in frequency of *S* leads to more common occurrence of the selectively disfavored *SS* genotype producing a "balanced polymorphism" in which selection maintains both alleles in the population and moves allelic frequencies toward selective-equilibrium frequencies. The relative Darwinian fitnesses of the three genotypes in a malarial environment are *AA* (0.9), *AS* (1.0), and *SS* (0.3); the genotype with highest fitness is usually denoted 1.0 so that relative fitnesses of the other genotypes are expressed as a fraction of the optimal one. At selective equilibrium, the expected frequencies of the alleles in the population are *A* (0.89) and *S* (0.11). Because epidemic malaria in central Africa is a relatively recent introduction, many of these populations have not attained selective equilibrium for the hemoglobin ß-polymorphism.

Hemoglobin S arose by mutation before the establishment of epidemic malaria in central Africa; in the malarial environment, it was co-opted for fitness consequences in *AS* genotypes that are incidental to the mutational origin of hemoglobin S. Hemoglobin S is thus an exaptation in the evolutionary history of central African populations. The balanced polymorphism of alleles *A* and *S* constitutes a population-level adaptation. Eradication of malaria would convert the polymorphism from adaptation to what Baum and Larson (1991) call a secondary disaptation; an environmental change causes a character that formerly had a selective advantage over its evolutionary antecedent

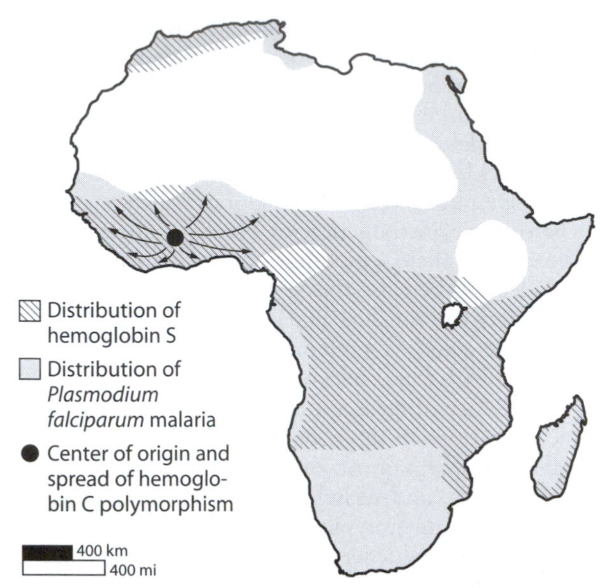

Figure 2. Evolution by natural selection at the β-hemoglobin locus in human populations of Africa. Polymorphisms for the hemoglobin C (circle and arrows) and hemoglobin S (crosshatching) forms of β-hemoglobin are associated geographically with occurrence of epidemic malaria (*Plasmodium falciparum*, gray shading). Hemoglobin A is the most common allelic form in all areas. The frequency of allele *C* is highest (~0.10) in the circled area and declines gradually with distance outside the circle through the region marked by arrows. Frequencies of *C* and *S* alleles are inversely correlated in West Africa because as *C* becomes more common, natural selection favors *C* and disfavors *S*. Where *C* is absent or rare, natural selection moves *S* toward an equilibrium frequency of approximately 0.11. Depending on its environmental and population-genetic contexts, hemoglobin S can be an exaptation, nonaptation, or disaptation for survival to adulthood.

(allele *A* close to fixation in this example) to one that is selectively disfavored relative to that condition. A polymorphism for the *A* and *S* alleles at the malarial selective-equilibrium frequencies is selectively disadvantageous relative to a population fixed for the *A* allele in an environment from which malaria has been eradicated; when *S* occurs at a frequency of ~ 0.1, approximately 1% of the polymorphic population suffers severe anemia (*SS* individuals), and the *S* allele no longer confers a selective advantage in *AS* genotypes. As the frequency of *S* drops to very low values in a nonmalarial environment, the *S* allele occurs strictly in heterozygous individuals, and presence of hemoglobin S ceases to generate natural selection. Alternative forms that have no selective consequences constitute nonaptation (a selectively "neutral" character lacking current utility; Vrba and Gould, 1986).

I illustrate the importance of geographic variation in adaptive evolution by extending this example from central African populations to western African ones (figure 2). A third form of β-hemoglobin allelic to hemoglobin A and hemoglobin S occurs in some African populations (Templeton, 2006): the hemoglobin C allele, derived by a single mutation from hemoglobin A, differs from both hemoglobin A and hemoglobin S in having lysine at the sixth amino-acid position. Allele C has a genetically recessive phenotype of malarial resistance and is associated with severe anemia only when heterozygous with allele *S*. In a malarial environment, the relative Darwinian fitnesses of the six genotypes formed by the various combinations of the *A*, *C*, and *S* alleles are *AA* (0.7), *AC* (0.7), *AS* (0.8), *CC* (1.0), *CS* (0.5), and *SS* (0.2). The *C* allele likely arose from *A* in Africa before the introduction of epidemic malaria, when the *A* allele was close to fixation and *S* was rare. Under these conditions, the *C* allele would occur entirely in *AC* genotypes and would be a nonaptation relative to the *A* allele. On malarial introduction, selection increased the frequency of *S* through the higher viability of *AS* genotypes over *AA* and *AC* genotypes. As the frequency of the *S* allele increases toward the selective equilibrium frequencies ($A = 0.89$, $S = 0.11$), the *C* allele would appear in *CS* genotypes as well as *AC* genotypes, and selection would decrease its frequency by disfavoring *CS* individuals. Selection acts to keep the *C* allele rare under these conditions, in which *C* is a disaptation relative to *A*. Although the *CC* genotype is the most favorable possible condition, the *CC* genotype is too rare under conditions of random mating to contribute a selective advantage.

Because alleles *A* and *C* are selectively equivalent in a nonmalarial environment, random genetic drift (see chapter I.15) might increase the frequency of *C* in a local population to a frequency high enough that *CC* genotypes occur regularly. Introduction of epidemic malaria into such a population would act very differently than it does in the population described in the preceding paragraph. Over many generations, the net effect of natural selection would be to increase frequency of the *C* allele toward fixation and to decrease frequencies of *A* and *S* ultimately to zero. Allele *C* constitutes an exaptation for malarial resistance in some western African populations (figure 2) in which allele *C* had drifted to sufficiently high frequencies before epidemic malaria that *CC* individuals were produced and subject to selective retention in a malarial environment. The contrasting selective consequences of the *A*, *C*, and *S* alleles of β-hemoglobin under slightly different starting frequencies of *C* and *S* in African malarial environments show that adaptive evolution depends critically on particular historical conditions.

Another dimension to adaptive evolution is gene exchange among different geographic populations of a species. As western African populations evolve by selection to increase the frequency of the *C* allele and central African populations evolve toward selective-equilibrium frequencies of the *A* (0.89) and *S* (0.11) alleles of β-hemoglobin, gene exchange occurs by interbreeding among these populations. Because fixation of allele *C* is the superior adaptive condition with respect to these three alleles, preferential contribution of *C* alleles from the favored west African genotypes should enable all populations eventually to undergo adaptive evolution in the manner of the west African populations by fixing *C* and eliminating *A* and *S*.

Further analysis of geographic variation in evolution of malarial resistance by Templeton (2006) reveals several cases in which genetic epistasis between the sickle-cell polymorphism at β-hemoglobin and variation at other loci produces different kinds of adaptive evolution. In Greek and Arabian populations, hemoglobin *S* occurs against a genetic background in which a mutation at a genetically linked locus causes fetal hemoglobin to be expressed throughout adulthood (called "persistence of fetal hemoglobin") rather than ceasing its expression after birth. This combination of alleles at two loci provides a simple example of a "coadapted gene complex"; hemoglobin S provides malarial resistance, while persistence of fetal hemoglobin alleviates the severe anemia associated with *SS* homozygotes. High fitness depends on a particular combination of alleles at these two genes.

One expects evolution of many advantageous phenotypes to occur by natural selection increasing the frequencies of selectively favored alleles at many genes. A consequence of such selection is that new mutations arising in different individuals and at different times in a population's history can be brought to high frequency independently by selection, thereby increasing their chances of occurring together in the same individuals by genetic recombination. If the combination of alleles thus achieved is favored by selection for the same biological role that brought the individual mutations to high frequency, selection gradually constructs an adaptive composite character. Such characters can become developmentally and genetically integrated into modules, whose expression or suppression during development can provide a store of potential exaptations, sometimes called a "toolkit" for constructing new organismal forms. A toolkit of adaptively evolved modules makes possible further evolution not confined to traditional Darwinian gradualism. A developmental module can be activated potentially by genetic or environmental factors or their interactions; "phenotypic accommodation" denotes a beneficial modification of organismal development made in response to a novel behavioral or environmental stimulus (West-Eberhard, 2005).

Hypotheses of developmental modularity and phenotypic accommodation share the expectation that evolutionary saltations are more likely than acknowledged by traditional Darwinism. A well-studied case of polymorphism for discrete developmental modules involves two constrasting feeding morphologies in tropical American fish of the genus *Cichlasoma*. The contrasting forms differ abruptly in the structures of jaws located in the pharynx and used to crush food (figure 3). The alternative states of the pharyngeal jaws are termed the "papilliform" morph versus the "molariform" morph; the molariform morph has hypertrophied skeletal and muscular components and greater ability to masticate hard food items, such as snails, which are often the less preferred food items. In some species, consuming snails early in life appears facultatively to trigger development of the molariform morph. The morphological contrast between these states in *C. minckleyi* of Mexico is so great that it has been called "intraspecific macroevolution." Widespread occurrence of molariform and papilliform feeding morphs among cichlid species indicates that they likely represent alternative developmental modules first assembled in the ancient history of cichlid fishes. The evolutionary origin of the alternative morphs is a macroevolutionary question, testable using interspecific phylogenetic analyses of how one form was constructed, perhaps gradually, from the other one. Microevolutionary studies of the alternative conditions as polymorphisms within species involve saltational changes governed by developmental switches and a prominent role for character exaptation in the adaptive trophic evolution of polymorphic populations.

4. ADAPTATION AND SELFISH GENETIC ELEMENTS

An important extension of the Darwinian evolutionary framework is to recognize semiautonomous selective processes occurring at the levels of genomic elements and species lineages in addition to the traditional level of varying organisms within populations. The abstract concepts of character sorting and selection have been expanded to encompass these levels (Gould, 2002; Vrba and Gould, 1986). At the genomic level, characteristic structures of retrotransposons include long terminal repeats and a coding region for reverse transcriptase, whose biological role is integral to the operation of the transposable element but whose origin cannot be explained as an organismal-level character adaptation. Unlike the genes encoding β-hemoglobin, mutational changes in transposable elements cannot be interpreted as having reached high evolutionary frequency to serve an organismal-level function, although their consequences can be co-opted as exaptations for

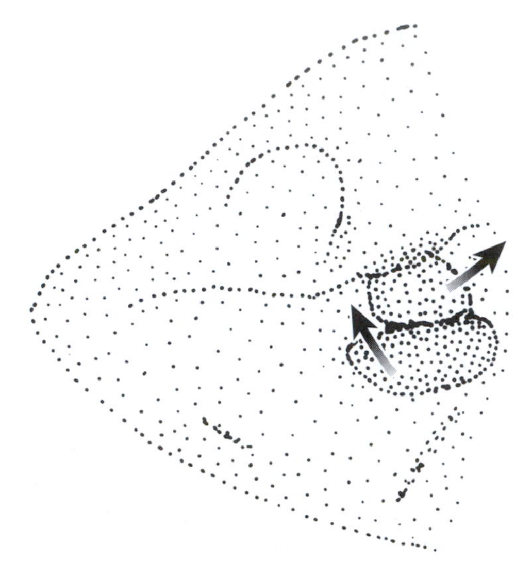

Figure 3. Cineradiographic tracing of the upper and lower pharyngeal jaws contacting to grind a snail in the molariform morph of *Cichlasoma minckleyi*. Arrows denote the directions of movement of the jaws as they make contact. This hypertrophied molariform morph and a much smaller papilliform morph represent alternative developmental modules in this and other species of *Cichlasoma*. The molariform morph can masticate hard food items, such as snails, not available to the papilliform morph. (After Liem, K. F., and L. S. Kaufman. 1984. Intraspecific macroevolution: Functional biology of the polymorphic cichlid species, *Cichlasoma minckleyi*. In A. A. Echelle and I. Kornfield, eds., Evolution of Fish Species Flocks. Orono: University of Maine Press, 196–217)

organismal-level roles. Burt and Trivers (2006) use "selfish genetic element" to denote "the minority of genes that spread at a cost to the organism" and "stretches of DNA . . . that act narrowly to advance their own interests" and "typically cause negative effects on nonlinked genes" in the same organism. For example, during spermiogenesis of male mice heterozygous for the *t*-allele, developing sperm containing the *t*-allele disable those containing the wild-type allele, permitting the *t*-allele to persist in populations despite selection against its detrimental effects on organismal phenotype, such as absence of a tail. Someone studying occurrence of the tailless phenotype in mice in a Darwinian context would conclude that this trait is a primary disaptation (Baum and Larson, 1991), a character disadvantageous relative to its ancestral alternative condition in the environmental context of its origin. Identification of primary disaptation using the adaptationist methodology described above would lead one to hypothesize association of such a phenotype with evolution of selfish genomic elements.

Selfish genomic elements that harm their "host" organism incur natural selection for their suppression by other genetic functions. One therefore expects to observe evolution of organismal-level adaptations whose function is to suppress selfish genomic elements. Given the phylogenetically widespread occurrence of the selfish genetic elements reviewed by Burt and Trivers (2006), mechanisms evolved to stabilize genomic structure and function probably constitute a large portion of the adaptive evolutionary diversity of life. The basic concepts of adaptation and exaptation and historical methods for testing specific hypotheses of adaptation as discussed above are directly applicable to studies of inherent conflicts between organismal character adaptation and the proliferative drives of selfish genomic elements.

FURTHER READING

Baum, D. A., and A. Larson. 1991. Adaptation reviewed: A phylogenetic methodology for studying character macroevolution. Systematic Zoology 40: 1–18.

Burt, A., and R. Trivers. 2006. Genes in Conflict: The Biology of Selfish Genetic Elements. Cambridge, MA: Harvard University Press.

Gould, S. J. 2002. The Structure of Evolutionary Theory. Cambridge, MA: Harvard University Press.

Gould, S. J., and E. S. Vrba. 1982. Exaptation: A missing term in the science of form. Paleobiology 8: 4–15.

Rose, M. R., and G. V. Lauder. 1996. Adaptation. San Diego: Academic Press.

Templeton, A. R. 2006. Population Genetics and Microevolutionary Theory. Hoboken, NJ: John Wiley & Sons.

Vrba, E. S., and S. J. Gould. 1986. The hierarchical expansion of sorting and selection: Sorting and selection cannot be equated. Paleobiology 12: 217–228.

West-Eberhard, M. J. 2005. Phenotypic accommodation: Adaptive innovation due to developmental plasticity. Journal of Experimental Zoology 304B: 610–618.

I.14

Phenotypic Selection

David W. Pfennig and Joel G. Kingsolver

OUTLINE

In this chapter, we describe the strength and patterns of natural selection in the wild. We focus on phenotypic selection because natural selection acts on the phenotypes of individual organisms. We begin by explaining what phenotypic selection is and how it works. We then explore how scientists study phenotypic selection in natural populations and discuss general patterns that have emerged from such investigations. Finally, we address common misunderstandings about selection and identify profitable avenues for future research.

GLOSSARY

fitness. The extent to which an individual contributes its genes to future generations relative to other individuals in the same population; a good operational definition of fitness is an individual's relative reproductive success.

heritability. In the broad sense, the fraction of the total phenotypic variation in a population that can be attributed to genetic differences among individuals; in the narrow sense, that fraction of the total phenotypic variation that results from the additive effects of genes.

natural (phenotypic) selection. A difference, on average, between the survival or fecundity of individuals with certain phenotypes compared with individuals with other phenotypes.

phenotype. The outward characteristics of organisms, such as their form, physiology, and behavior.

quantitative trait. A trait that shows continuous rather than discrete variation; such traits are determined by the combined influence of many different genes and the environment.

selection gradient. A measure of the strength of selection acting on quantitative traits: for selection on a single trait, it is equal to the slope of the best-fit regression line in a scatterplot showing relative fitness as a function of phenotype; for selection acting on multiple traits, it is equal to the slope of the partial regression in a scatterplot showing relative fitness as a function of all phenotypes.

sexual selection. A difference, among members of the same sex, between the average mating success of individuals with a particular phenotype and that of individuals with other phenotypes.

1. INTRODUCTION

In the introduction to *On the Origin of Species,* Darwin wrote, "a naturalist, reflecting on the mutual affinities of organic beings, on their embryological relations, their geographical distribution, geological succession, and other such facts, might come to the conclusion that each species had not been independently created, but had descended . . . from other species. Nevertheless, such a conclusion, even if well founded, would be unsatisfactory, until it could be shown *how* the innumerable species inhabiting this world have been modified . . ." (emphasis added). Thus, Darwin recognized that no theory of evolution would be complete if it failed to provide a plausible mechanism that could explain how living things change over evolutionary time. Darwin's theory of evolution by natural selection provided such a mechanism. Yet, Darwin's theory goes beyond explaining how living things change over time; it also explains the important concept of adaptation: the tendency for living things to evolve traits that make them so apparently well designed for survival and reproduction. Because of this broad explanatory power, Darwin's theory ranks among the most important ideas in the history of human thought.

Although the central concept of Darwin's theory is natural selection, Darwin never attempted to measure selection in nature. Moreover, in the century following the publication of *On the Origin of Species,* selection was generally regarded as too weak to be observed directly in natural populations. Partly for these reasons, some early evolutionists even questioned selection's efficacy in driving evolutionary change.

This view that selection is weak and cannot be measured has changed dramatically. Beginning in the 1930s, evolutionists demonstrated mathematically that natural selection alone could power evolutionary change and adaptation. Moreover, in the past three decades, selection has been detected and quantified in hundreds of populations in nature. These data demonstrate that not only does selection occur routinely in nature, but that it is often sufficiently potent to bring about substantial evolutionary change in a relatively short time period. Indeed, selection is now viewed as the cause of adaptive evolution within natural populations.

2. HOW PHENOTYPIC SELECTION WORKS

Phenotypic selection takes place when individuals with particular phenotypes survive to reproductive age at higher rates than do individuals with other phenotypes, or when individuals with particular phenotypes produce more offspring than do individuals with other phenotypes. In either case, selection results in differential reproductive success, where some individuals have more offspring than others. Thus, phenotypic selection requires phenotypic variation, where individuals differ in some of their characteristics, and differential reproduction, where some individuals have more surviving offspring than others because of their distinctive characteristics. Those individuals that have more surviving offspring are said to have higher fitness (note that an individual's fitness is measured as how well the individual performs relative to other individuals in the same population). Ultimately, phenotypic selection can lead to changes in the genetic makeup of populations over time—evolution. In particular, when the phenotypic characteristics under selection are heritable—that is, when the variations among individuals are, at least in part, passed from parents to offspring—selection will cause the population to change in these characteristics over time. Thus, evolution by natural selection requires three conditions: variation, differential reproduction, and heredity. Indeed, when these three conditions are satisfied, evolution by natural selection is a certain outcome.

Numerous factors in the environment can cause selection, including biological agents (such as an individual's competitors, predators, and parasites) and nonbiological agents (such as the weather). The specific phenotypic traits on which agents of selection act are termed targets of selection. As we will see, however, selection often acts on multiple traits simultaneously in the same individual, making it a challenge to determine precisely which trait represents the actual target of selection.

Although phenotypic selection always favors an increase in fitness, it does not invariably bring about the evolution of greater trait values. In particular, when selection acts on quantitative (i.e., continuously distributed) traits, three different modes of selection are possible, each of which produces a distinctive pattern of trait evolution (figure 1). With directional selection, fitness consistently increases (or decreases) with the value of the trait. When directional selection acts on a trait, it changes the value of that trait in the population. Directional selection also tends to reduce variation, although often not dramatically. With stabilizing selection, individuals with intermediate trait values have highest fitness. Stabilizing selection does not tend to change the mean trait value. It does, however, reduce variation by disfavoring individuals in the tails of the trait's distribution. Finally, with disruptive selection, individuals with extreme trait values have highest fitness. As with stabilizing selection, disruptive selection does not tend to change the mean trait value. Unlike stabilizing selection, however, disruptive selection increases variation by favoring individuals in the tails of the trait's distribution.

All three modes of selection drive evolution by eliminating individuals with low fitness and preserving individuals with high fitness. Moreover, as noted earlier, if the trait of interest is heritable, then evolution will result, but the trait distribution in the evolved population will differ depending on the mode of selection (see figure 1). In particular, for traits under positive directional selection, the population will evolve larger trait values (illustrated in figure 1), whereas for those under negative directional selection, the population will evolve smaller trait values. For traits under stabilizing selection, the population will evolve a smaller range of trait values as the average trait value becomes more common in the population. Finally, for traits under disruptive selection, the population will evolve a wider range of trait values, possibly leading to the evolution of discrete, alternative phenotypes (see figure 1).

Given this background, we now turn to the issue of how to measure the mode and strength of phenotypic selection.

3. MEASURING PHENOTYPIC SELECTION

Suppose we are interested in measuring possible selection acting on some trait in a population. The first

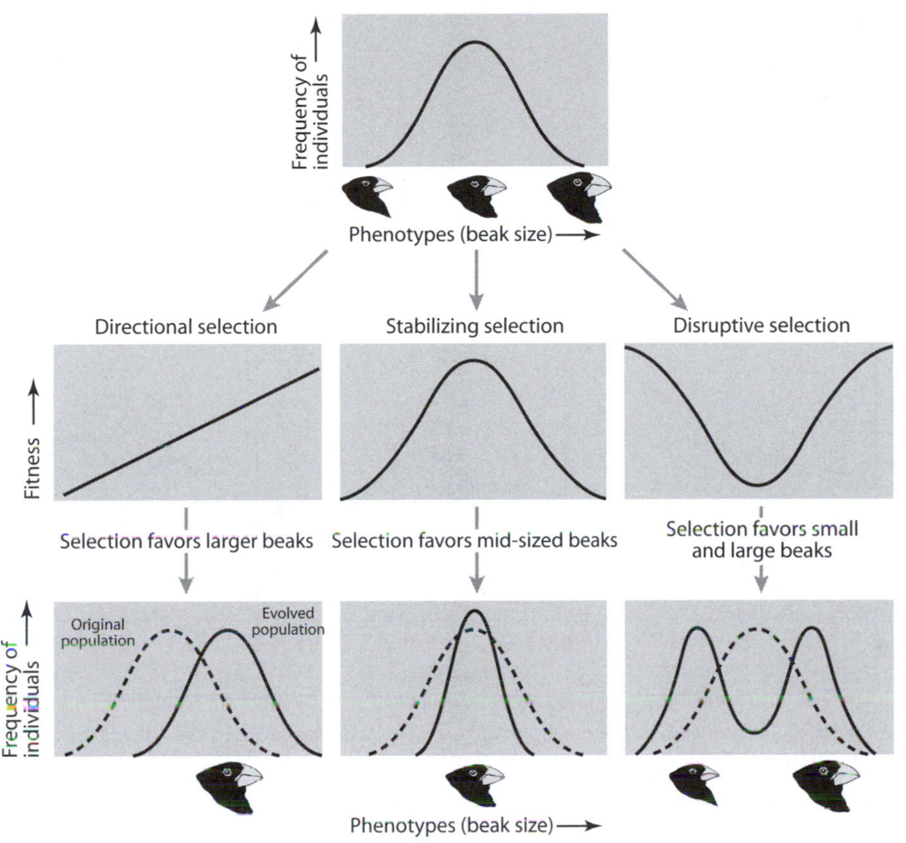

Figure 1. Three different modes of selection (directional, stabilizing, and disruptive) that may act on a quantitative trait (i.e., a trait that shows continuous rather than discrete variation). The top panel shows the distribution of beak sizes in a hypothetical population of birds before selection; the middle panels show fitness associated with different beak sizes during different modes of selection; and the bottom panels show the distribution of beak sizes following each form of selection. Note that for different modes of selection, the shape of the line relating fitness to phenotype varies (middle panels), as does the resulting pattern of trait evolution (bottom panels).

step is to estimate the fitness associated with different trait values. Ideally, we would identify individuals with different trait values and measure their overall fitness. In practice, however, most investigators measure only one component of fitness, such as survival, mating success, fecundity, or (even less directly) a trait that correlates with these fitness components, such as body size. Once we estimate fitness, we then fit a regression line (i.e., the best-fit line) through the data points relating fitness to phenotype. From the slope and shape of this regression line, we can determine the strength and mode of selection acting on our trait of interest. When this fitness function is described by a straight line (indicating directional selection; figure 1), the fitness (w) of the trait (z) can be estimated by the simple linear regression equation:

$$w = \alpha + \beta z,$$

where α is the y-intercept of the fitness function and β is the fitness function's slope. In this case, β measures the strength of directional selection. By contrast, when the fitness function has curvature (indicating stabilizing and disruptive selection; figure 1), quadratic regression is required to estimate the strength of selection. Here, fitness is estimated by:

$$w = \alpha + \beta z + (\gamma/2)z^2,$$

where γ measures the amount of curvature in the fitness function. In this case, γ measures the strength of quadratic selection. When $\beta = 0$ and γ is significantly negative (i.e., when the fitness function contains an intermediate performance maximum), we conclude that stabilizing selection is acting on the trait of interest. By contrast, when $\beta = 0$ and γ is significantly positive (i.e., when the fitness function contains an intermediate

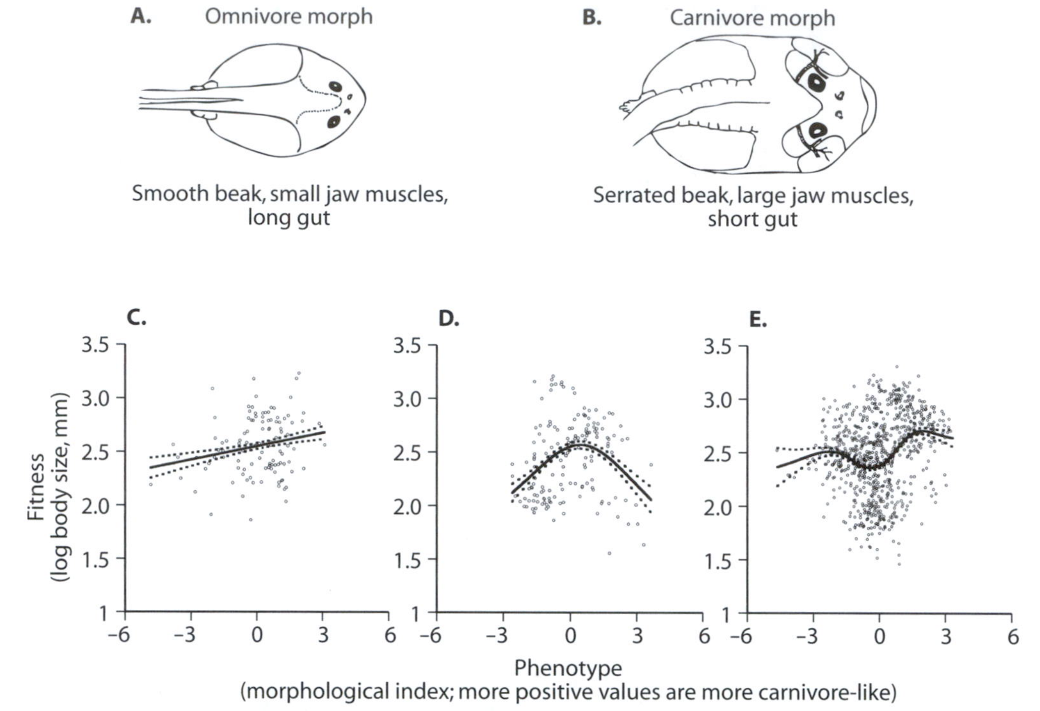

Figure 2. Spadefoot toad tadpoles (*Spea bombifrons* and *S. multiplicata*) are highly variable in resource use and feeding morphology as represented by two extreme morphotypes: (A) an omnivore morph, which feeds mostly on detritus, and (B) a carnivore morph, which specializes on fairy shrimp. The mode of selection operating on feeding morphology varies for different species and populations as revealed when the fitness of individual tadpoles is plotted on phenotype for (C) *S. bombifrons* from mixed-species ponds, (D) *S. multiplicata* from mixed-species ponds, and (E) *S. multiplicata* from single-species ponds. Each panel (C–E) shows cubic spline regression estimates bracketed by 95% confidence intervals.

performance minimum), we conclude that disruptive selection is acting.

To illustrate how each mode of selection may be manifest in natural populations, consider a recent study of spadefoot toad tadpoles by Pfennig and colleagues (2007). Tadpoles of two species from the southwestern United States, *Spea bombifrons* and *S. multiplicata*, are highly variable in resource use and feeding morphology as represented by two extreme morphotypes: an omnivore morph (figure 2A), which feeds mostly on the pond bottom on detritus (decaying organic material), and a carnivore morph (figure 2B), which feeds mostly in the water column on fairy shrimp. In some ponds, there is a clear dimorphism in feeding morphology; in other ponds, individuals with intermediate phenotypes may be most common.

The mode of selection operating on feeding morphology varies for different species and populations. In mixed-species ponds (i.e., ponds containing both species), the most carnivore-like *S. bombifrons* tadpoles are largest (figure 2C; body size serves as a suitable proxy for fitness because larger individuals have higher survival, mating success, and fecundity in this system). Thus, directional selection favors more carnivore-like *S. bombifrons*. Presumably, this pattern reflects selection on *S. bombifrons* to express resource-use phenotypes that minimize their overlap with *S. multiplicata* for food; *S. multiplicata* tend to be more omnivore-like than *S. bombifrons*.

A different mode of selection was detected among *S. multiplicata* in mixed-species ponds. In this species, stabilizing selection appears to favor individuals with intermediate phenotypes (figure 2D). Presumably, carnivore phenotypes in these individuals are selectively disfavored; earlier work had shown that *S. multiplicata* carnivores are competitively inferior to *S. bombifrons*. Yet why does selection not favor omnivores, which are as distinct as possible from *S. bombifrons*? Presumably, selection acts against *S. multiplicata* omnivores in mixed-species ponds because omnivores metamorphose later and at a smaller body size than carnivores. Because mixed-species ponds typically contain relatively high shrimp densities, those *S. multiplicata* that express an intermediate feeding morphology—and can thereby supplement their detritus diet with, but not specialize on, the more nutritious shrimp resource—may be selectively favored. Thus, in mixed-species ponds, selection appears to favor *S. multiplicata* that are as carnivore-like as possible, but that are not so carnivore-like that they overlap with *S. bombifrons* in resource use.

Finally, a third mode of selection was detected among *S. multiplicata* in single-species ponds (figure 2E). Here, disruptive selection favors extreme feeding morphologies. In these ponds, individuals expressing extreme phenotypes would most likely have fewer (and, in the case of extreme omnivores, perhaps lower-quality) resources available. Nevertheless, compared with the majority of the population that may be intermediate in phenotype (and in resource use), individuals expressing extreme phenotypes would also most likely have fewer competitors with which to share those resources. Thus, relative to intermediate individuals, the fitness of extreme omnivores and carnivores may be high.

Although the above example illustrates the general approach that is widely used for measuring phenotypic selection in the wild, a critical assumption behind this approach is that variation in the measured trait causes the observed variation in fitness. However, rather than acting directly on the trait of interest (through direct selection), selection may be acting on other, unmeasured traits that are correlated with the measured trait (through indirect selection), generating a spurious correlation between the focal trait and fitness. One way to reduce the problem of indirect selection is to experimentally alter the trait of interest and then evaluate the effects of the manipulation on subsequent fitness (*phenotypic engineering*).

To illustrate the latter approach, consider the following example. Male long-tailed widowbirds, *Euplectus progne*, are endowed with a half-meter-long tail. Malte Andersson hypothesized that these extraordinary tail feathers are selectively favored because females find them attractive; i.e., long tail feathers are favored by sexual selection. To test this hypothesis, Andersson predicted that experimentally augmenting a male's tail feathers should enhance the male's fitness. Andersson captured male widowbirds and then shortened the tails of some by removing a segment of tail feathers, only to glue them onto another bird's tail, thereby lengthening the latter bird's tail. He also had two control groups: one in which the male's tail feathers were cut off and glued back on, and another in which the males were handled in the same way but no tail feathers were removed.

The results of this phenotypic manipulation were dramatic. The tail-lengthened males were much more attractive to females than those that had suffered the loss of a portion of their tail feathers. Moreover, the tail-lengthened males also did better than controls. These data therefore indicate that tail length is a target of selection, with females acting as the selective agent.

Because it can expand the range of phenotypic values and reduce the problem of correlated traits, phenotypic engineering is especially useful for determining whether a trait is under direct selection and what mode of selection might operate on it. However, because phenotypic engineering often involves altering trait expression beyond the range of trait values observed in natural population, such manipulations do not help researchers estimate the strength of selection on natural populations in the wild, which is the topic we turn to next.

4. PHENOTYPIC SELECTION IN THE WILD

Numerous studies have used the above approaches to measure phenotypic selection in natural populations. Moreover, many of these studies measured selection acting on multiple traits on the same individual. Such data are particularly valuable because they allow us to distinguish direct selection on traits from the indirect effects of correlated traits. To estimate direct selection, we use a statistical approach known as multiple regression analysis. Multiple regression resembles simple linear regression (introduced in the previous section) except that fitness is regressed on multiple traits simultaneously, allowing us to measure the strength of direct selection acting on each trait after statistically controlling for the effects of correlations among other traits. Specifically, fitness (w) is estimated by:

$$w = \alpha + \beta_1 z_1 + \beta_2 z_2 + \beta_3 z_3 \cdots \beta_i z_i,$$

where β_i is the partial regression slope associated with trait z_i. This parameter, termed the linear selection gradient, measures the strength of direct selection acting on trait z_i. (Note that for selection on a single trait, the linear selection gradient is equal to the slope of the simple linear regression, as described in the previous section.) To allow comparisons among different types of traits and organisms, we can standardize the linear selection gradient by the amount of variation in the trait (e.g., by the standard deviation) to obtain a standardized measure of selection, β_s.

Kingsolver and colleagues recently reviewed studies that used these approaches to measure selection gradients in natural populations. They identified 993 estimates of directional selection (β_s), obtained from a diversity of organisms, ecological settings, and traits. Because positive and negative values of β_s occur with similar frequency, they used the absolute values, $|\beta_s|$, as an index of the magnitude of directional selection. The median value (50% of the values above and 50% below) of $|\beta_s|$ was 0.16, with a small fraction of values greater than 0.50, indicating strong selection. To put this strength of selection in perspective, imagine a population that experiences persistent directional selection of this magnitude ($\beta_s = 0.16$) on a trait that has a typical heritability of 0.3. In less than 70 generations, the population mean would exceed the initial range of variation in the population. In other words, phenotypic selection in many natural populations is sufficiently strong to cause substantial evolutionary change in a fairly short period of time on an evolutionary time scale.

Another important issue to resolve is the relative magnitude of natural selection (i.e., selection resulting from variation among individuals in survival or fecundity) compared to sexual selection (i.e., selection caused by variation among individuals in mating success). The available data on directional selection gradients suggest that sexual selection is typically stronger than natural selection. Indeed, the median magnitude of sexual selection is more than twice as great as that of natural selection. Thus, competition for mates may be important for rapid evolution in nature.

What are the patterns of quadratic selection in the wild? Kingsolver and colleagues (2001, 2007) identified 574 measures of the strength of quadratic selection, γ. They found that 50% of the values of γ are between −0.1 and +0.1, implying that the magnitude of quadratic selection is often modest. Moreover, the frequency distribution of γ is symmetric about zero, with negative and positive values equally common, which suggests that stabilizing selection is not more common than disruptive selection. Because disruptive selection is generally thought to be relatively rare in nature, this result is particularly surprising. It is possible that this result reflects sampling bias: only 16% of the values of γ in the literature are significantly different from zero. Thus, most studies do not have the sample size or statistical power to quantify quadratic selection of the magnitude that may be typical in natural populations. Alternatively, this result may reflect the true pattern of quadratic selection in nature; i.e., disruptive selection may actually be relatively common. The possible widespread occurrence of disruptive selection may reflect a ubiquitous agent of selection in nature: competition for resources, such as food. Because competition tends to decrease individual fitness, natural selection is generally thought to favor traits that lessen competition's intensity. One way for selection to do so is to favor evolutionary divergence between initially similar phenotypes through density-dependent or frequency-dependent disruptive selection (e.g., see figure 2E).

Thus, to summarize, phenotypic selection is common in nature, and it is often sufficiently strong to cause substantial evolutionary change in a relatively short time period. Moreover, sexual selection tends to be stronger than natural selection. Finally, stabilizing selection appears to be no more common than disruptive selection. However, because few studies have focused on quadratic selection specifically, it is difficult to say how common or how strong disruptive selection is relative to stabilizing selection in natural populations.

5. MISUNDERSTANDINGS ABOUT PHENOTYPIC SELECTION

Phenotypic selection is often misunderstood. We therefore highlight and clarify four common misunderstandings.

Misunderstanding 1: Phenotypic Selection Always Results in Evolution

Selection and evolution are not the same, although the two concepts are often incorrectly equated. Selection is a process that produces evolution, whereas evolution is the historical pattern of change through time. Phenotypic selection (the process) can lead to evolution (the pattern), but it is only one of several processes that can do so (the others are mutation, gene flow, nonrandom mating, and genetic drift). Moreover, if a trait lacks heritable variation, selection will not produce evolution.

Misunderstanding 2: Phenotypic Selection Causes Individuals to Change

A common misconception is that individual organisms evolve following selection. It is true that phenotypic selection acts on the phenotypes of individual organisms. However, after the selection event, none of the selected individuals are expected to change in any way. What does change are characteristics of the population. Thus, populations evolve; individual organisms do not.

Phenotypic selection may indirectly cause the phenotypes of individual organisms to change. Specifically, agents of selection often alter the developmental expression of traits through a process known as phenotypic plasticity. When phenotypes are plastic, individuals that are genetically identical may express radically different phenotypes if they develop in different environments. For example, the spadefoot toad tadpoles in figure 2 are born as omnivores but may develop into carnivores following a change in their diet. In many species, individuals often exhibit heritable variation in their tendency to respond to environmental cues through phenotypic plasticity, indicating that plasticity itself is subject to natural selection and evolutionary change. Indeed, adaptive phenotypic plasticity is thought to evolve because it enables organisms to produce the optimal phenotype for the various environments that they may experience during their lifetime. Thus, by favoring the evolution of phenotypic plasticity, agents of selection may indirectly change the phenotypes of individual organisms.

Misunderstanding 3: Selection Favors Individuals That Act for the Good of the Species

A common misunderstanding about phenotypic selection acting on behavior is that individual organisms will perform actions for the good of their species. However, if altruists survive and reproduce at lower rates than other individuals in the same population, then the tendency to behave altruistically should not evolve, unless the altruists receive some other benefit.

As it turns out, nearly every act of altruism that has been studied in detail increases the altruist's fitness, either because beneficiaries reciprocate or because the beneficiaries are genetically related to the altruist. Helping nondescendant kin (relatives other than offspring) can increase an altruist's fitness because relatives share genes. Moreover, fitness gained by personal reproduction (direct fitness) and fitness gained by helping nondescendant kin (indirect fitness) can both be expressed in identical genetic terms. We can sum up an individual's total contribution of genes to the next generation, creating a quantitative measure called *inclusive fitness*. Thus, altruism may be adaptive if it ultimately results in more shared genes being transmitted to the next generation. In general, natural selection should always favor traits that maximize an individual's inclusive fitness.

Misunderstanding 4: The Evolutionary Response to Phenotypic Selection Is Slow

It is often assumed that the evolutionary response to selection is slow. We have already seen, however, that phenotypic selection is often sufficiently strong to cause substantial evolutionary change in a relatively short time. Moreover, phenotypic selection may even produce substantial evolutionary change in only one generation. Consider a population that contains abundant phenotypic variation. If this variation has high heritability, and if there is strong truncating selection, in which individuals with a trait value above a certain threshold value survive or reproduce while those below this value do not, then the population will evolve dramatically in only one generation.

For example, Peter and Rosemary Grant recently documented character displacement—evolution in resource-acquisition traits stemming from competition between species—in a species of Galápagos finch that recently (i.e., in the last 25 years) confronted a novel competitor (Grant and Grant, 2006). Remarkably, their data suggest that the focal species may have evolved away from its competitor in beak morphology in only one generation. Thus, paradoxically, evolution may happen so rapidly that we may actually fail to detect it.

6. FUTURE DIRECTIONS

As we have seen, numerous recent studies have measured phenotypic selection in the wild. Many interesting patterns have emerged from these studies. However, a number of questions remain unanswered. Here, we list four such questions.

First, does phenotypic selection vary over time and space? In particular, does the fact that environmental conditions change frequently cause the magnitude and even the direction of selection to change also? Such fluctuating selection could explain why most organisms appear to be experiencing at least some directional selection. If environments vary frequently, then the organisms living in these environments will tend to possess trait values that are suboptimal for their particular environment. Consequently, directional selection would always be acting to drive the trait value toward the current optimum. We need many more long-term field studies of selection in the wild to determine if the magnitude, direction, or mode of selection varies in time and space.

Second, is disruptive selection relatively common in nature, and, if so, what agents drive it? Specifically, is disruptive selection often mediated by density- or frequency-dependent processes, such as competition? Resolving this issue is vital for understanding the origins and maintenance of alternative phenotypes in populations (e.g., see figure 2), and, possibly, the origin of new species.

Third, what measure of fitness provides the most complete picture of selection? An operational definition of fitness is that it is the total number of offspring that an individual produces in its lifetime. Yet, for practical reasons, most studies consider only components of fitness, such as survival. We need more studies that determine how reliably individual fitness components predict true lifetime fitness in natural populations. We especially need more studies that compare the relative magnitude of selection on survival or fecundity (natural selection) with selection on mating success (sexual selection). As noted in section 4 above, the available data indicate that sexual selection is typically significantly stronger than natural selection. Does this result generally hold across diverse taxa? Moreover, to develop a truly comprehensive view of how phenotypic selection drives trait evolution, we need more selection studies that determine how trait expression influences an individual's *inclusive* fitness.

Finally, what is the relative importance of evolution versus phenotypic plasticity in mediating rapid phenotypic responses to changing environments? Many organisms are currently undergoing rapid phenotypic change in response to ongoing human-mediated change in their environment. To what extent does such rapid phenotypic change reflect phenotypic plasticity as opposed to rapid evolution?

In sum, natural selection is the central organizing principle of evolutionary theory. This theory explains not only how living things diversify but also those features of living things that so wonderfully equip them for survival and reproduction. Although natural selection is a simple concept, modern research is only beginning to discover that it works in myriad and sometimes subtle ways.

FURTHER READING

Andersson, Malte. 1982. Female choice selects for extreme tail length in a widowbird. Nature 299: 818–820. *A classic example of the use of manipulative experiments to document phenotypic selection in the wild.*

Conner, Jeffrey K., and Daniel L. Hartl. 2004. A Primer of Ecological Genetics. Sunderland, MA: Sinauer Associates. *An excellent overview of ecological genetics with a clear summary of how to measure selection.*

Endler, John A. 1986. Natural Selection in the Wild. Princeton, NJ: Princeton University Press. *A seminal monograph that describes advantages and disadvantages of various approaches for measuring phenotypic selection in natural populations.*

Grant, Peter R., and B. Rosemary Grant. 2006. Evolution of character displacement in Darwin's finches. Science 313: 224–226. *An interesting example that describes rapid evolution in a classic system.*

Kingsolver, Joel G., Hopi E. Hoekstra, Jon M. Hoekstra, David Berrigan, Sacha N. Vignieri, Chris H. Hill, Anhthu Hoang, Patricia Gilbert, and Peter Beerli. 2001. The strength of phenotypic selection in natural populations. American Naturalist 157: 245–261. *Reviews numerous studies of selection in natural populations.*

Kingsolver, Joel G., and David W. Pfennig. 2007. Patterns and power of phenotypic selection in nature. BioScience 57: 561–572. *An overview of how phenotypic selection acts in natural populations. Portions of this chapter (especially parts of sections 3, 4, and 6) are adapted from this review.*

Losos, Jonathan B., Thomas W. Schoener, R. Brian Langerhans, and David A. Spiller. 2006. Rapid temporal reversal in predator-driven natural selection. Science 314: 1111. *Illustrates how directional selection can reverse direction rapidly.*

Pfennig, David W., Amber M. Rice, and Ryan A. Martin. 2007. Field and experimental evidence for competition's role in phenotypic divergence. Evolution 61: 257–271. *Describes how different modes of selection may act on the same species when confronted with different environmental circumstances.*

I.15

Population Genetics and Ecology

Philip Hedrick

OUTLINE

About 40 years ago, scientists first strongly advocated the integration of population ecology and population genetics into population biology (Singh and Uyenoyama, 2004). Even today these two disciplines are not really integrated, but there is a general appreciation of population genetic concepts in population ecology and vice versa. For example, the new subdiscipline molecular ecology, and many articles in the journal *Molecular Ecology*, use genetic markers and principles to examine both ecological and evolutionary questions. Although some aspects of population genetics have changed quickly in recent years, many of its fundamentals are still important for aspects of ecological study.

GLOSSARY

coalescence. The point at which common ancestry for two alleles at a gene occurs in the past.

effective population size. An ideal population that incorporates such factors as variation in the sex ratio of breeding individuals, the offspring number per individual, and numbers of breeding individuals in different generations.

gene flow. Movement between groups that results in genetic exchange.

genetic bottleneck. A period during which only a few individuals survive and become the only ancestors of the future generations of the population.

genetic drift. Chance changes in allele frequencies that result from small population size.

Hardy-Weinberg principle. After one generation of random mating, single-locus genotype frequencies can be represented as a binomial function of the allele frequencies.

neutral theory. Genetic change is primarily the result of mutation and genetic drift, and different molecular genotypes are neutral with respect to each other.

population. A group of interbreeding individuals that exist together in time and space.

selective sweep. Favorable directional selection that results in a region of low genetic variation closely linked to the selected region.

1. INTRODUCTION

The primary goals of population genetics are to understand the factors determining evolutionary change and stasis and the amount and pattern of genetic variation within and between populations (Hedrick, 2005; Hartl and Clark, 2007). In the 1920s and 1930s, shortly after widespread acceptance of Mendelian genetics, the theoretical basis of population genetics was developed by Ronald A. Fisher, J.B.S. Haldane, and Sewall Wright. Population genetics may be unique among biological sciences because it was first developed as a theoretical discipline by these men before experimental research had a significant impact.

The advent of molecular genetic data of populations in the late 1960s and DNA sequence data in the 1980s revolutionized population genetics and produced many new questions. Population genetics and its evolutionary interpretations provided a fundamental context in which to interpret these new molecular genetic data. Further, population genetic approaches have made fundamental contributions to understanding the role of molecular variation in adaptive differences in morphology, behavior, and physiology. A primary goal in determining the extent and pattern of genetic variation is to document the variation that results in selective differences among individuals, the "stuff of evolution."

The amount and kind of genetic variation in populations are potentially affected by a number of factors, but primarily by selection, inbreeding, genetic drift, gene flow, mutation, and recombination. These factors may have general or particular effects; for example, genetic drift and inbreeding can be considered to always reduce the amount of variation, and mutation to always increase the amount of variation. Other factors, such as selection and gene flow, may either increase or reduce genetic variation, depending on the particular situation. Combinations of two or more of these factors can generate many different levels and patterns of genetic variation. In 1968, Motoo Kimura introduced the important "neutral theory" of molecular evolution that assumes that genetic variation results from a combination of mutation generating variation and genetic drift eliminating it (Kimura, 1983). This theory is called neutral because allele and genotype differences at a gene are selectively neutral with respect to each other. This theory is consistent with many observations of molecular genetic variation (see below).

To understand the influence of these evolutionary factors, one must first be able to describe and quantify the amount of genetic variation in a population and the pattern of genetic variation among populations. In recent years, new laboratory techniques have made it possible to obtain molecular genetic data in any species, and a number of software packages have become available to estimate the important parameters in population genetics and related topics. In addition, the online Evolution Directory (EvolDir) is a source of information about different molecular techniques, estimation procedures, and other current evolutionary genetic information.

Let us first define the evolutionary or genetic connotation of the term *population.* As a simple ideal, a population is group of interbreeding individuals that exist together in time and space. Often it is assumed that a population is geographically well defined, although this may not always be true. Below we discuss the concept of effective population size, which provides a more explicit definition of population in evolutionary terms.

Many of the theoretical developments in population genetics assume a large, random-mating population that forms the gene pool from which the female and male gametes are drawn. In some real-life situations, such as dense populations of insects or outcrossing plants, this ideal may be nearly correct, but in many natural situations, it is not closely approximated. For example, there may not be random mating, as in self-fertilizing plants, or there may be small or isolated populations as in rare or endangered species. In these cases, modifications of the theoretical ideal must be made.

One of the basic concepts of population genetics is the Hardy-Weinberg principle (often called Hardy-Weinberg equilibrium [HWE]). It states that after one generation of random mating, single-locus genotype frequencies can be represented by a binomial (with two alleles) or multinomial (with multiple alleles) function of the allele frequencies. This principle allows great simplification of the description of a population's genetic content by reducing the number of parameters that must be considered. Furthermore, in the absence of factors that change allele frequency (selection, genetic drift, gene flow, and mutation), and in the continued presence of random mating, the Hardy-Weinberg genotype proportions will not change over time.

2. GENETIC DRIFT AND EFFECTIVE POPULATION SIZE

Since the beginning of population genetics, there has been controversy concerning the importance of chance changes in allele frequencies because of small population size, termed *genetic drift.* Part of this controversy has resulted from the large numbers of individuals observed in many natural populations, large enough to think that chance effects would be small in comparison to the effects of other factors, such as selection and gene flow. However, if the selective effects or amount of gene flow are small relative to genetic drift, then long-term genetic change caused by genetic drift may be important.

Under certain conditions, a finite population may be so small that genetic drift is significant even for loci with sizable selective effects, or when there is gene flow. For example, some populations may be continuously small for relatively long periods of time because of limited resources in the populated area, low tendency or capacity to disperse between suitable habitats, or territoriality among individuals. In addition, some populations may have intermittent small population sizes. Examples of such episodes are the overwintering loss of population numbers in many invertebrates and epidemics that periodically decimate populations of both plants and animals. Such population fluctuations generate genetic bottlenecks, or periods during which only a few individuals survive and become the only ancestors of the future generations of the population.

Small population size is also important when a population grows from a few founder individuals, a phenomenon termed *founder effect.* For example, many island populations appear to have started from a very small number of individuals. If a single female who was fertilized by a single male founds a population, then only four genomes (assuming a diploid organism), two from the female and two from the male, may start a new population. In plants, a whole population may be initiated from a single seed—only two genomes, if self-fertilization occurs. As a result, populations descended from a small founder group may have low genetic variation or by chance have a high or low frequency of particular alleles.

Another situation in which small population size is of great significance occurs when the population (or species) in question is one of the many threatened or endangered species (Allendorf and Luikart, 2007). For example, all approximately 500 whooping cranes alive today descend from only 20 whooping cranes that were alive in 1920 because they were hunted and their habitat destroyed. All 200,000 northern elephant seals alive today descend from as few as 20 that survived nineteenth-century hunting on Isla Guadalupe, Mexico. Further, all the living individuals of some species are descended from a few founders that were brought into captivity to establish a protected population, such as Przewalski's horses (13 founders), California condors (13 founders), black-footed ferrets (6 founders), Galápagos tortoises from Española Island (15 founders), and Mexican wolves (7 founders).

The population size that is relevant for evolutionary matters, the number of breeding individuals, may be much less than the total number of individuals in an area, the census population size, and is the appropriate measure for many population ecology studies. The size of the breeding population may sometimes be estimated with reasonable accuracy by counting indicators of breeding activity such as nests, egg masses, and colonies in animals or counting the number of flowering individuals in plants. But even the breeding population number may not be indicative of the population size that is appropriate for evolutionary considerations.

For example, factors such as variation in the sex ratio of breeding individuals, the offspring number per individual, and numbers of breeding individuals in different generations may be evolutionarily important. All these factors can influence the genetic contribution to the next generation, and a general estimate of the breeding population size does not necessarily take them into account. As a result, the effective population size, or N_e, a theoretical concept that incorporates variation in these factors and allows general predictions or statements irrespective of the particular forces responsible, is quite useful. In other words, the concept of an ideal population with a given effective size enables us to draw inferences concerning the evolutionary effects of finite population size by providing a mechanism for incorporating factors that result in deviations from the ideal.

The concept of the effective population size makes it possible to consider an ideal population of size N in which all parents have an equal expectation of being the parents of any progeny individual. In other words, the gametes are drawn randomly from all breeding individuals, and the probability of each adult producing a particular gamete equals $1/N$, where N is the number of breeding individuals. A straightforward approach that is often used to tell the impact of various factors on the effective population size is the ratio of the effective population size to breeding (or sometimes census) population size, that is, N_e/N. Sometimes, this ratio is only around 0.1 to 0.25, indicating that the effective population size may be much less than the number of breeding individuals. In general, the effect of genetic drift is a function of the reciprocal of the number of gametes in a population, $1/(2N_e)$, for a diploid population. If N_e is large, then this value is small, and there is little genetic drift influence. Or, if N_e is small, then this value is relatively large, and genetic drift may be important.

3. NEUTRAL THEORY

Neutral theory assumes that selection plays a minor role in determining the maintenance of molecular variants and proposes that different molecular genotypes have almost identical relative fitnesses; that is, they are neutral with respect to each other. The actual definition of selective neutrality depends on whether changes in allele frequency are primarily determined by genetic drift. In a simple example, if s is the selective difference between two alleles at a locus, and if $s < 1/(2N_e)$, the alleles are said to be neutral with respect to each other because the impact of genetic drift is larger than selection. This definition implies that alleles may be effectively neutral in a small population but not in a large population. Neutral theory does not claim that the relatively few allele substitutions responsible for evolutionarily adaptive traits are neutral, but it does suggest that the majority of allele substitutions have no selective advantage over those that they replace.

Kimura also showed that the neutral theory was consistent with a molecular clock; that is, there is a constant rate of substitution over time for molecular variants. To illustrate the mathematical basis of the molecular clock, let us assume that mutation and genetic drift are the determinants of changes in frequencies of molecular variants. Let the mutation rate to a new allele be u so that in a population of size $2N$ there are $2Nu$ new mutants per generation. It can be shown that the probability of chance fixation of a new neutral mutant is $1/(2N)$ (the initial frequency of the new mutant). Therefore, the rate of allele substitution k is the product of the number of new mutants per generation and their probability of fixation, or

$$k = 2Nu\left(\frac{1}{2N}\right) = u.$$

In other words, this elegant prediction from the neutral theory is that the rate of substitution is equal to the mutation rate at the locus and is constant over time. Note that substitution rate is independent of the effective population size, a fact that may initially be counterintuitive. This independence occurs because in a smaller population there are fewer mutants; that is, $2Nu$ is smaller, but the initial frequency of these mutants is higher, which increases the probability of fixation, $1/2N$, by the same magnitude by which the number of mutants is reduced. This simple, elegant mathematical prediction and others from the neutral theory provide the basis for the most important developments in evolutionary biology in the past half-century.

One of the appealing aspects of the neutral theory is that, if it is used as a null hypothesis, then predictions about the magnitude and pattern of genetic variation are possible. Initially, molecular genetic variation was found to be consistent with that predicted from neutrality theory. In recent years, examination of neutral theory predictions in DNA sequences has allowed tests of the cumulative effect of many generations of selection, and a number of examples of selection on molecular variants have been documented (see below).

Traditionally, population genetics examines the impact of various evolutionary factors on the amount and pattern of genetic variation in a population and how these factors influence the future potential for evolutionary change. Generally, evolution is conceived of as a forward process, examining and predicting the future characteristics of a population. However, rapid accumulation of DNA sequence data over the past two decades has changed the orientation of much of population genetics from a prospective one investigating the factors involved in observed evolutionary change to a retrospective one inferring evolutionary events that have occurred in the past. That is, understanding the evolutionary causes that have influenced the DNA sequence variation in a sample of individuals, such as the demographic and mutational history of the ancestors of the sample, has become the focus of much population genetics research.

In a determination of DNA variation in a population, a sample of alleles is examined. Each of these alleles may have a different history, ranging from descending from the same ancestral allele, that is, identical by descent, in the previous generation to descending from the same ancestral allele many generations before. The point at which this common ancestry for two alleles occurs is called coalescence. If one goes back far enough in time in the population, then all alleles in the sample will coalesce into a single common ancestral allele. Research using the coalescent approach is the most dynamic area of theoretical population genetics because it is widely used to analyze DNA sequence data in populations and species.

4. GENE FLOW AND POPULATION STRUCTURE

In most species, populations are often subdivided into smaller units because of geographic, ecological, or behavioral factors. For example, the populations of fish in pools, trees on mountains, and insects on host plants are subdivided because suitable habitat for these species is not continuous. Population subdivision can also result from behavioral factors, such as troop formation in primates, territoriality in birds, and colony formation in social insects.

When a population is subdivided, the amounts of genetic connectedness among the parts of the population can differ. Genetic connection depends primarily on the amount of gene flow, movement between groups that results in genetic exchange that takes place among the subpopulations or subgroups. When the amount of gene flow between groups is high, gene flow has the effect of homogenizing genetic variation over the groups. When gene flow is low, genetic drift, selection, and even mutation in the separate groups may lead to genetic differentiation.

It is sometimes useful to describe the population structure in a particular geographic framework. For example, within a watershed, there may be separated fish or plant groups that have a substantial amount of genetic exchange between them. On a larger scale, there may be genetic exchange between adjacent watersheds but in smaller amounts than between the groups within a watershed. On an even larger scale, there may be populations in quite separated watersheds that presumably have little direct exchange but may share some genetic history, depending on the amount of gene flow among the adjacent groups or occasional long-distance gene flow. This hierarchical representation is useful in describing the overall relationships of populations of an organism and in documenting the spatial pattern of genetic variation. Recently, there has been increasing interest in landscape and geographic approaches to estimating historical and contemporary gene flow. In addition, *phylogeography*, the joint use of phylogenetic techniques and geographic distributions, has been used to understand spatial relationships and distributions of populations within species or closely related species (Avise, 2000).

In general, the subdivision of populations assumes that the various subpopulations are always present. Another view assumes that individual population subdivisions at particular sites may become extinct and then later be recolonized from other subpopulations, resulting in a *metapopulation*. The dynamics of extinction and recolonization can make metapopulations quite different both ecologically and genetically from the traditional concept of a subdivided population.

Gene flow is central to understanding evolutionary potential and mechanisms in several areas of applied population genetics. First, the potential for movement of genes from genetically modified organisms (GMO) into related wild populations—that is, the gene flow of transgenes into natural populations—can be examined using population genetics. Second, the invasive potential of nonnative plants and animals into new areas may be affected by hybridization (gene flow) between nonnative and native organisms as well as adaptive change. Finally, a number of endangered species are composed of only one, or a few, remaining populations with low fitness. Gene flow from other populations of the same species can result in genetic rescue or genetic restoration of these populations by introducing new variation that allows removal of detrimental variation and restoration of adaptive change.

Estimating the amount of gene flow in most situations is rather difficult. Direct estimates of the amount of movement can be obtained in organisms where different individuals can be identified or individual marks are used. Many approaches have been employed to mark individuals differentially, such as toe clipping in rodents, leg banding in birds, coded-wire tags in fish, and radio transponders in many different vertebrates.

However, both movement of individuals and their incorporation into the breeding population are necessary for gene flow. Using highly variable genetic markers, it is now possible to identify parents genetically and thereby determine the spatial movement of gametes between generations without direct movement information of the parents. Or, individuals can be assigned to specific populations using genetic markers, thereby determining whether they are migrants or not.

Indirect measures of gene flow using genetic markers are useful to confirm behavioral or other observations or when these observations are inconclusive or impossible. Most commonly, the number of successfully breeding migrants between groups is measured using techniques to evaluate population structure. Theoretically, assuming finite subpopulations of size N_e and a proportion m migrants into each subpopulation each generation, then

$$F_{ST} \approx \frac{1}{4N_e m + 1}.$$

When $N_e m$ is large, the measure F_{ST} approaches 0, and when $N_e m$ is small, F_{ST} can approach 1. The value of F_{ST} for a group of populations can be estimated using the amount and pattern of molecular genetic variation over subpopulations.

The availability of molecular and DNA sequence data in many organisms provides a database to determine the relationships between populations or species that are not obvious from other traits. It is generally assumed that molecular genetic data better reflect the true relationships between groups than other data, such as morphology or behavior, because molecular data are less influenced by selective effects. Furthermore, differences between relationships generated from molecular data and from other traits provide an opportunity to evaluate the effect of selective effects on other traits.

Maternally inherited mitochrondrial DNA (mtDNA) data have been the workhorse for phylogeographic research because mtDNA does not recombine in most organisms and, as a result, shows a clearer phylogenetic record than many nuclear genes (chloroplast DNA and Y chromosomes are similarly useful). In addition, the effective population size for mtDNA (as well as for chloroplast DNA and Y chromosomes) is only about one-fourth that of nuclear genes so that divergence occurs about four times faster than for nuclear genes. However, this faster rate of divergence and potentially differential gene flow for the two sexes may cause the signal for these uniparentally inherited genes to be different from the phylogenetic pattern for nuclear genes, which constitute a very large proportion of the genome.

5. SELECTION

In the past few years, with the availability of extensive DNA sequence data for a number of organisms (particularly humans), there has been an intensive search for genomic regions exhibiting a signal of adaptive (positive Darwinian) selection. Many of the genomic regions identified have undergone a "selective sweep" because of favorable directional selection, as indicated by low genetic variation in genetic regions closely linked to the selected gene or regions.

An elegant example of a selective sweep is adaptive melanism in the rock pocket mouse of the southwestern United States (Hoekstra et al., 2004). The mouse is generally light-colored and lives on light-colored granite rocks, but it also has melanic forms that live on relatively recent black lava formations in several

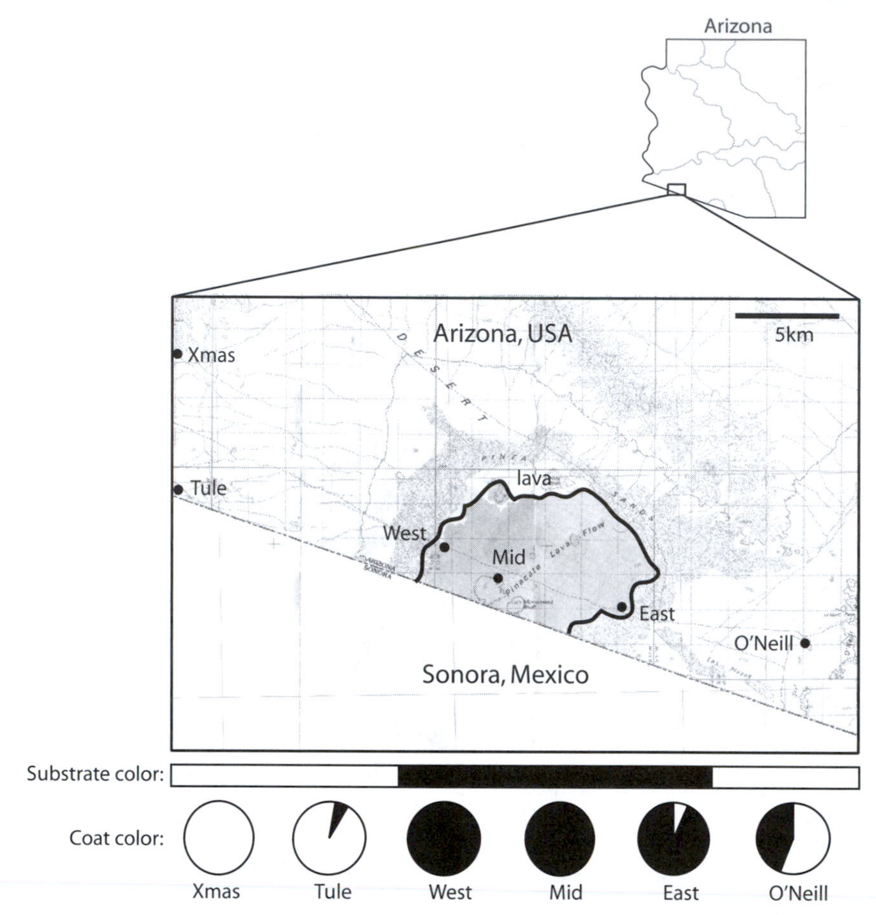

Figure 1. Six sampling sites (three on dark volcanic rock and three on light-colored substrate) and coat color frequencies (in pie diagrams) in rock pocket mice across a transect in the Sonoran desert. (From Hoekstra et al., 2004)

restricted sites. Figure 1 shows the frequencies of the normal recessive and dominant melanic forms from a 35-km transect in southwestern Arizona. Here the frequency of melanics is highly concordant with substrate color, that is, high frequencies of melanics in the center of this transect that has approximately 10 km of black lava and lower frequencies of melanics on the light-colored substrate sites at either end of the transect.

Investigation of molecular variation in the *Mc1r* gene, which is known to have variants that produce dark-colored house mice, was found to correlate nearly completely with the light and melanic phenotypes. The melanic and normal alleles were found to differ by four amino acids, and the nucleotide diversity for the melanic alleles was 1/20 that for light alleles. The lower variation among the melanic alleles is consistent with the expected pattern if selection has recently increased its frequency by a selective sweep in the area of black lava.

Some of best-documented examples of adaptive selection are those resulting from recent human changes in the environment (Hedrick, 2006), such as development of genetic resistance in insect pests to chemicals used to control them or genetic resistance in pathogens to antibiotics. The genetic basis of pesticide resistance may be the result of many genes, mutants at a single or a few genes, or expansion of gene families. Because the molecular basis of many of these adaptive changes is known, detailed genetic and evolutionary understanding is possible. For example, resistance to some insecticides among mosquitoes that are vectors for diseases such as malaria (*Anopheles gambiae*) and West Nile virus (*Culex pipiens*) is the result of a single amino-acid

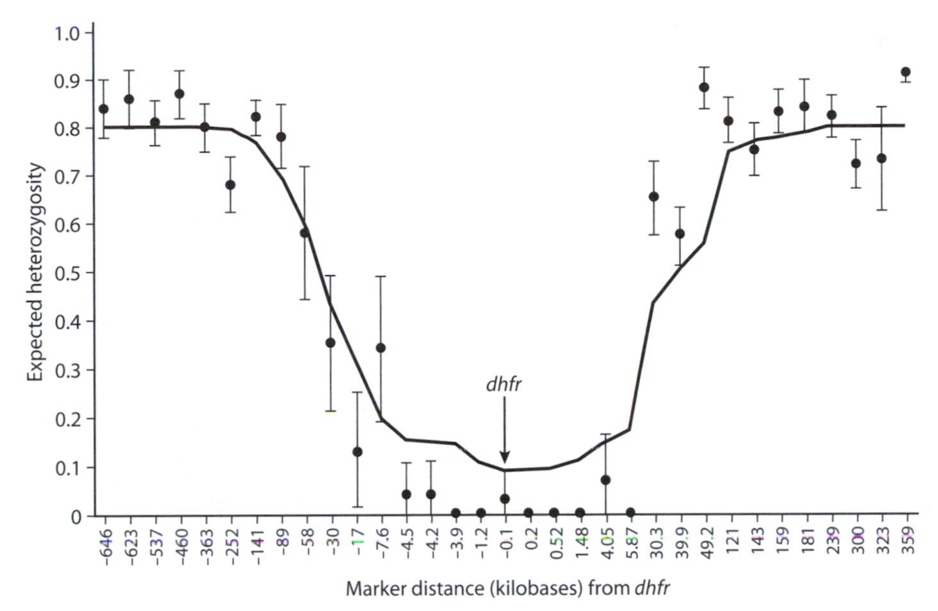

Figure 2. The observed (solid circles) and expected (lines) heterozygosity around the *dhfr* gene in the malaria parasite *Plasmodium falciparum*, which provides resistance to the antimalarial drug pyrimethamine. (From Nair et al., 2003)

substitution. In *C. pipiens*, a single nucleotide change, GGC (glycine) to AGC (serine) at codon 119 in the gene for the enzyme acetylcholinesterase (*ace*-1) results in insensitivity to organophosphates. A complete lack of variation within samples among resistant variants of this gene suggests that they have originated and spread quite recently.

Another set of examples of adaptive selection include those resulting from the development of host resistance to pathogens. For example, malaria kills more than one million children each year in Africa alone and is the strongest selective pressure in recent human history. As a result, selective protection from malaria by sickle cell, thalassemia, G6PD, Duffy, and many other genetic variants in the human host provide some of the clearest examples of adaptive variation and diversifying selection for pathogen resistance (Kwiatkowski, 2005). Genomic studies have demonstrated that selection for malarial resistance is strong, up to 10%, and that variants conferring resistance to malaria are recent, generally less than 5000 years old, consistent with the proposed timing of malaria as an important human disease. Often the resistant variants are in different populations, probably mainly in part because of their recent independent mutation origin.

Efforts to control the malaria parasite using antimalarial drugs have resulted in widespread genetic resistance to these drugs. For example, pyrimethamine is an inexpensive antimalarial drug used in countries where there is resistance in the malarial parasite to the widely used drug chloroquine. Pyrimethamine was introduced to the area along the Thailand-Myanmar border in the mid-1970s, and resistance spread to fixation in approximately 6 years. Resistance is the result of point mutations at the active site of the enzyme encoded by the gene *dhfr* on chromosome 4. Examination of genetic variation at genes near *dhfr* as shown in figure 2 showed remarkable reduced heterozygosity near *dhfr* and normal variation further away (Nair et al., 2003). This pattern of variation is consistent with a selective sweep, and the theoretical pattern expected from a selective sweep is shown by the curve in figure 2.

The major histocompatibility complex (MHC) genes are part of the immune system in vertebrates, and differential selection through resistance to pathogens is widely thought to be the basis of their high genetic variation (Garrigan and Hedrick, 2003). Variation in the genes of the human MHC, known as *HLA* genes, has been the subject of intensive study for many years because of their role in determining acceptance or rejection of transplanted organs, many autoimmune diseases, and recognition of pathogens. High *HLA* variation allows recognition of more pathogens, consistent with the fact that *HLA-B* is the most variable gene in the human genome. In recent years, there has been extensive research examining *R* (disease resistant) genes in plants, a system with similarities to MHC.

6. FUTURE DIRECTIONS

Because of the widespread availability of DNA sequence data in many organisms, the future application of population genetic data and principles in ecology appears almost unlimited. Many basic ecological questions, such as how many individuals there are in a population, what their relationships are, or whether they are immigrants, may be definitively answered in future years using genetic techniques. Such precise descriptions may then provide data to understand the evolutionary and ecological factors influencing population dynamics and distributions.

FURTHER READING

Allendorf, Fred, and Gordon Luikart. 2007. Conservation and the Genetics of Populations. Oxford: Blackwell Publishing. *A recent summary of the application of population genetics to conservation.*

Avise, John. 2000. Phylogeography: The History and Formation of Species. Cambridge, MA: Harvard University Press. *The joint use of phylogenetic relationships and geographic patterns to understand evolution.*

Garrigan, Daniel, and Philip Hedrick. 2003. Perspective: Detecting adaptive molecular evolution, lessons from the MHC. Evolution 57: 1707–1722. *A perspective on the strongest example of balancing selection, MHC, and the gain and loss of balancing selection signals.*

Hartl, Daniel, and Andrew Clark. 2007. Principles of Population Genetics, 4th ed. Sunderland, MA: Sinauer Associates. *A recent summary of the principles of population genetics.*

Hedrick, Philip. 2005. Genetics of Populations, 3rd ed. Boston: Jones and Bartlett Publishers. *A recent and thorough summary of the principles of population genetics.*

Hedrick, Philip. 2006. Genetic polymorphism in heterogeneous environments: The age of genomics. Annual Review of Ecology, Evolution, and Systematics 37: 67–93. *A summary of the recent empirical and theoretical examples of genetic polymorphism maintained by ecological variation.*

Hoekstra, Hopi, Kristen Drumm, and Michael Nachman. 2004. Ecological genetics of adaptive color polymorphism in pocket mice: Geographic variation in selected and neutral genes. Evolution 58: 1329–1341. *A discussion of adaptive melanism in pocket mice living on dark lava results from amino-acid changes in the* Mc1r *gene.*

Kimura, Motoo. 1983. The Neutral Theory of Molecular Evolution. Cambridge: Cambridge University Press. *A summary of the neutral theory from the view of its major architect, Motoo Kimura.*

Kwiatkowski, Dominic. 2005. How malaria has affected the human genome and what human genetics can teach us about malaria. American Journal of Human Genetics 77: 171–192. *A review of the many human genes that confer resistance to malaria, the strongest selective factor in recent human history.*

Nair, Shalini, Jeff Williams, Alan Brockman, Lucy Paiphun, Mayfong Mayxay, P. N. Newton, J. P. Guthmann, F. M. Smithuis, T. T. Hien, N. J. White, F. Nosten, and T. J. Anderson. 2003. A selective sweep driven by pyrimethamine treatment in southeast Asian malaria parasites. Molecular Biology and Evolution 20: 1526–1536. *An example of the fast development of antimalarial drug resistance by a selective sweep of new resistant variants in Plasmodium.*

Singh, Rama, and Marcy Uyenoyama, eds. 2004. The Evolution of Population Biology—Modern Synthesis. Cambridge: Cambridge University Press. *A summary of the contributions to population biology over recent decades.*

I.16

Phylogenetics and Comparative Methods

David D. Ackerly

OUTLINE

1. The role of phylogenetics in ecology
2. Phylogenies and the analysis of trait correlations
3. Phylogenetic signal: Pattern and significance
4. Phylogenetics and community ecology
5. Prospects for the future

The study of ecology frequently draws on comparative observations and experiments that rely on the similarities and differences among species and the correlations among species traits and the environment. In such studies, consideration of the phylogenetic relationships among species provides valuable information for statistical inference and an understanding of evolutionary history underlying present-day ecological patterns. From a statistical perspective, related species do not necessarily provide independent data points for hypothesis tests, due to inheritance of shared characteristics from common ancestors. This similarity can be addressed through a variety of statistical techniques, including the widely used method of phylogenetic independent contrasts. Independent contrasts play a particularly valuable role in the analysis of trait and trait–environment correlations and may point toward alternative interpretations of comparative data. In community ecology, measures of the phylogenetic clustering or spacing of co-occurring species provide a useful tool to test alternative processes underlying community assembly. Co-occurrence of close relatives most likely reflects ecological filtering, in which related species with similar traits share the ability to tolerate local conditions. The reverse pattern of phylogenetic spacing of co-occurring species may reflect a variety of processes, and additional observations of species traits in relation to environment and interacting taxa will be necessary to address underlying processes. Use of comparative methods has increased dramatically with the rapid growth in phylogenetic information and computing power and will continue to play an important role in ecological research.

GLOSSARY

See figure 1 for illustrations of main terms.

branch lengths. These may indicate either the number of inferred character changes or a measure of relative or absolute time along any particular branch connecting two nodes. If the molecular data underlying a phylogeny do not violate a molecular clock, a single rate may be imposed such that branch lengths will represent *relative time*, and contemporaneous taxa will be placed at the same distance from the root (i.e., the same age). If a molecular clock is violated, *rate-smoothing methods* have been developed to obtain the best-supported estimate of relative time. Fossils and biogeographic or paleoecological information may then be used to calibrate these branch lengths and convert them to units of *absolute time*. Rate-smoothing and calibration methods are fraught with difficulty, and branch lengths should be treated with caution. (Note that branch lengths may also be set arbitrarily for convenience when one is drawing trees, in which case they have no intrinsic biological meaning.)

character states. Phylogenetic trees are reconstructed based on analysis of a matrix of *characters*, where each character can take on one of two or more *states* (binary or multistate, respectively) for each taxon in the group. Phylogenies can be reconstructed from molecular and/or morphological data, although the former are now much more common. Analyses that include morphological data are advantageous as they make it possible to incorporate taxa or fossils for which molecular data are not available.

lineage. This refers to a single line of ancestor–descendant relationship, connecting nodes within a phylogeny.

most recent common ancestor (MRCA). The MRCA is the most recent node that is shared by any two taxa in a tree.

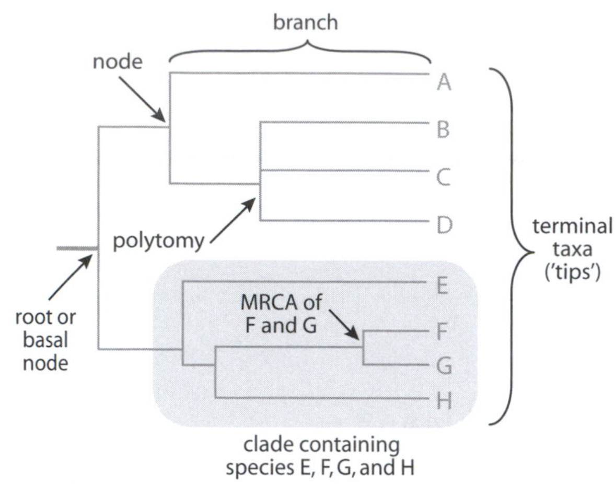

Figure 1. Example of a phylogenetic tree for eight taxa (A–H), illustrating some of the terms in the glossary. This tree is ultrametric, meaning that all terminal taxa are equidistant from the root of the tree.

phylogenetic distance. The phylogenetic distance between two nodes or taxa refers to the sum of branch lengths from one tip (or internal node) down to the MRCA and back up to another tip (or node) of a tree. The phylogenetic distance matrix is an $n \times n$ matrix (for n taxa) of such distances among all pairs of taxa, with 0s in the diagonal.

phylogeny. A phylogeny, or phylogenetic tree, is a branching diagram showing the hierarchy of evolutionary relationships among a group of taxa (extant and/or extinct). Terminal taxa or *tips* are connected by *branches* to internal *nodes* that indicate a hypothesized ancestor. A *clade* includes all of the taxa (extant and extinct) that descend from a node. Phylogenies can be either *rooted* or *unrooted*, where the root represents the hypothesized ancestor of all taxa on the tree.

polytomy. This refers to a node with three or more daughter nodes. A *soft polytomy* indicates uncertainty, where the true bifurcating relationships among the daughters are unknown. A *hard polytomy* represents a hypothesis of near simultaneous divergence where the sequence of individual speciation events cannot be meaningfully resolved. Most phylogenetic comparative methods treat polytomies as either hard or soft but do not always make the distinction explicit.

ultrametric. An ultrametric tree is one in which all terminal taxa are contemporaneous; more precisely, the sum of the branch lengths from the root to each tip is the same for all tips. Phylogenies of extant taxa will be ultrametric if branch lengths have been adjusted to represent relative or absolute time.

> In Great Britain there are 32 indigenous trees[:] of these 19 or more than half . . . have their sexes separated—an enormous proportion compared with the remainder of the British flora: nor is this wholly owing to a chance coincidence in some one Family having many trees & having a tendency to separated sexes: for the 32 trees belong to nine Families, & the trees with separate sexes to five Families.
>
> —Charles Darwin, manuscript for *Natural Selection* (unpublished)

In the quote above, Darwin observes an interesting pattern among plant species of Great Britain. He notes that among trees, the proportion of species that have individuals of separate sexes (as in humans and most vertebrates) is much higher than among the flora as a whole, most of which is composed of shrubs and herbaceous plants. He explained the high frequency of separate sexes as an adaptation to promote cross-fertilization in trees: because trees are large and have many flowers, the chance that an insect would carry pollen from one flower to another of the same individual is quite high. If all the flowers on a tree are of the same sex, these repeated visits by pollinators will not lead to high levels of self-fertilization.

Darwin's observations provide a nice example of what we now call comparative biology, which draws on comparisons of the similarities and differences among species to test ecological and evolutionary hypotheses. In addition, what Darwin recognized intuitively is that a simple count of the number of species exhibiting different characteristics might not be adequate to support his argument. If many of the species are drawn from the same family (that is, closely related in evolutionary terms), they are likely to share many ecological characteristics. Thus, a group with many tree species may also contain many species with separate sexes, reflecting their descent from a common ancestor. But if the evolutionary argument is sound—that trees should evolve separate sexes because of the problem of self-fertilization—then this combination of traits should evolve independently in many different taxonomic groups, and this is indeed what Darwin observed.

Throughout the past 150 years, since the publication of Darwin's *On the Origin of Species*, comparative biology has played a central role in ecology and evolutionary biology. In essence, each species alive today (or in the past) represents the outcome of a long, natural experiment. The results reflect the contemporary ecology of a species—interactions with the abiotic environment and with other forms of life—as well as the cumulative legacy of the past. Evolution works slowly, and most features are passed down from ancestor to descendant with little change. A penguin appears beautifully adapted to the challenges of surviving and reproducing under the extreme conditions of Antarctic life. But these adaptations must be understood in historical context: penguins are birds, and this experiment in polar living started with very specific initial conditions, including egg-laying, a feathered pelt, forelimbs modified into wings, and so on. Comparative research, placing penguins in the broader context of other birds and viewing them side by side with their closest relatives (loons, albatrosses, petrels, and shearwaters) is critical to an understanding and appreciation of their contemporary ecology and behavior.

In the past 30 years, comparative biology has grown rapidly as a new generation of methods emerged, combining the historical perspective outlined above with the quantitative tools of experimental statistics. The emergence of modern phylogenetics triggered these developments. The word *phylogeny* refers to the evolutionary relationships among a group of organisms, illustrated as a branching tree where the tips (or leaves) may represent individuals, populations, species, or groups of species, and the internal branching points are their common ancestors. The study of phylogenetics has been revolutionized by the combination of molecular biology (providing a trove of data), conceptual advances (the theory of cladistics), and the availability of high-speed computers. Together, these advances have made it possible to infer highly resolved phylogenies for many groups of organisms. With continuous improvements in methods and the availability of data, the tree of life is taking shape and revealing the hierarchy of evolutionary relationships among living (and extinct) organisms.

1. THE ROLE OF PHYLOGENETICS IN ECOLOGY

The science of ecology studies the interactions of organisms with their environment and the consequences of these interactions for where species live and how they interact. To address these questions, it is often useful to compare different species, either through observations or experiments. The similarities and differences in how species respond to their environment or interact with each other can provide important ecological insights. When data are gathered on different species, understanding how they are related to each other (i.e., their phylogenetic relationships) contributes valuable information that can affect data analysis and interpretation. In this chapter, I focus on two areas of ecological research where phylogenies play a particularly important role: the analysis of correlations among species traits and environmental conditions (like Darwin's example above) and the study of community ecology. In addition, I provide a brief discussion of the concept of *phylogenetic signal*, a general term for the similarity among close relatives.

In the discussion below, it is assumed that a phylogeny is available for each group under consideration. Most phylogenies are based on molecular data, particularly DNA sequences, sometimes combined with morphological or other characteristics. The computational methods used to search for the best-supported phylogeny are continually being improved and are beyond the scope of this chapter. Regardless of the method used, it is important to recognize that every phylogeny is a hypothesis of relationships, and like any scientific hypothesis, it is subject to revision and improvement. Phylogenies may also contain different degrees of uncertainty, both in terms of the topology (the pattern of who is related to whom) and the lengths of the branches, which represent the amount of evolutionary change or the amount of time elapsed between different nodes of the tree. This uncertainty can be incorporated into comparative analyses; in many cases, the results are quite robust across a range of possible alternatives, so strongly supported and fully resolved phylogenies are not a prerequisite for comparative analysis. An overview of some terminology used to describe phylogenies is provided in the Glossary.

2. PHYLOGENIES AND THE ANALYSIS OF TRAIT CORRELATIONS

Research in functional ecology, life history strategies, and related areas of ecology often addresses questions of interspecific trait–trait and trait–environment associations, such as: Do mammals with larger body sizes have larger home ranges? Do plant species of open habitats tend to have smaller seeds? How are the traits of invasive species different from those of native species in a community? The answers to these questions help us to understand how species traits influence distribution, abundance, and interactions with other species in a community. They also have important applications in conservation biology, restoration ecology, and the management of invasive species.

A variety of statistical methods can be applied to test hypotheses of trait associations, depending on the type of data available and the nature of the hypothesis. These include correlation, regression, analysis of variance, contingency table analysis, and others. One of the basic assumptions of virtually all statistical tests is that each data point represents an observation that is independent with respect to the underlying null hypothesis. This assumption is not required in order to calculate the various statistics; rather, it is essential to deriving the statistical significance of the outcome. For conventional statistics, this significance value (or *p*-value) represents the probability of observing the data if the underlying null hypothesis is true. When that probability is too low (conventionally, we use a cutoff value of 5%), we reject the null hypothesis and accept that there is a significant effect or relationship. For maximum-likelihood tests, which are playing an increasingly important role in ecology and comparative methods, the assumption of independence is used to assess the likelihood of the best-fit model relative to alternative models or hypotheses, given the observed data.

The fundamental argument underlying the development of many comparative methods arises from the observation that related species are ecologically and phenotypically similar to one another. This will not hold for every trait, as instances of rapid divergence and of convergent evolution are widespread and important. But on average, species resemble their close relatives more than they do more distant taxa, and this similarity reflects descent from recent common ancestors. Because of this inherited similarity, it is argued that in statistical terms species do not represent independent data points, violating this basic assumption of significance testing. One can also approach this problem in terms of the underlying historical processes. Trait associations among extant species arise through a historical sequence of correlated changes occurring along each branch of the phylogeny; ideally, we would like to estimate the correlation between these changes to more directly measure evolutionary linkages between the traits. It is now well established that the correlations observed among living species (at the tips of the phylogeny) do not provide a reliable estimate of this historical pattern of correlated evolutionary changes that have occurred along the branches of the phylogeny. Although some researchers are strongly motivated by the statistical arguments, and others more by the historical questions, both perspectives lead one to the use of phylogenetic comparative methods.

The selection of a comparative method to conduct associational analyses depends on the nature of the data and the hypothesis. One of the most common problems is the correlation (a measure of the strength of association) between two traits measured on a continuous scale (e.g., body size or seed size). Correlation coefficients range from 0 for two traits with no association up to 1 for traits that are very tightly linked (–1 if it is a negative association). In 1985, Joseph Felsenstein introduced the method of phylogenetic independent contrasts (often referred to as PICs) to address this question in a phylogenetic context; more than 20 years later, his method remains one of the most robust and widely used of all comparative methods. The method of independent contrasts rests on the assumption that the evolutionary change in a trait that occurs along each lineage leading up to present-day species represents an independent event with respect to the changes occurring in other branches. Independence, in this context, refers to the statistical notion that the changes are independent manifestations of underlying processes, although the same processes (e.g., natural selection as a result of climate change) may be affecting multiple lineages in a group. If the trait changes that occur in two lineages arising from a common ancestor are independent, then, as Felsenstein demonstrated based on statistical theory, the difference between the trait values of the two descendants will also represent a statistically independent observation. These differences are calculated by subtracting the trait value of one species from the value of its closest relative, and they are referred to as PICs (there is an additional step involving the branch lengths on the phylogeny, which I do not describe here). In addition, Felsenstein showed that one can continue to calculate contrasts at deeper nodes of the tree, based on an iterative process of averaging the trait values at successively deeper nodes. In a fully resolved phylogeny, N species will be connected by $N - 1$ common ancestors, so N trait values measured on the species will provide $N - 1$ contrasts; these contrasts can be used as the variables in correlation, regression, and multivariate statistical analyses.

A study that I conducted with Peter Reich in 1999 illustrates the application of independent contrasts and how they can impact the analysis of trait associations. We examined correlations among several functional attributes of leaves, including leaf size, leaf lifespan (the length of time a leaf persists on a plant), and specific leaf area (SLA, the ratio of leaf area to leaf dry mass; higher values indicate thinner or less dense leaves). Global studies of leaf function have found that leaves with higher SLA tend to have faster metabolic rates and shorter leaf lifespan, and this strategy is favored in more fertile habitats. The opposite set of traits is observed in leaves with low SLA. In addition, it is sometimes observed that leaves with low SLA and long leaf lifespan are smaller, and small leaves are often viewed as an adaptation to low-water or high-temperature environments. In particular, the needles of conifers (pines, spruces, etc.) are smaller in area and have a longer lifespan than the leaves of most flowering plants.

In a data set of about 100 species, including both conifers and flowering plants, there are negative correlations of leaf lifespan with both SLA and leaf size. However, when we apply independent contrasts, the results change dramatically. The evolutionary correlation

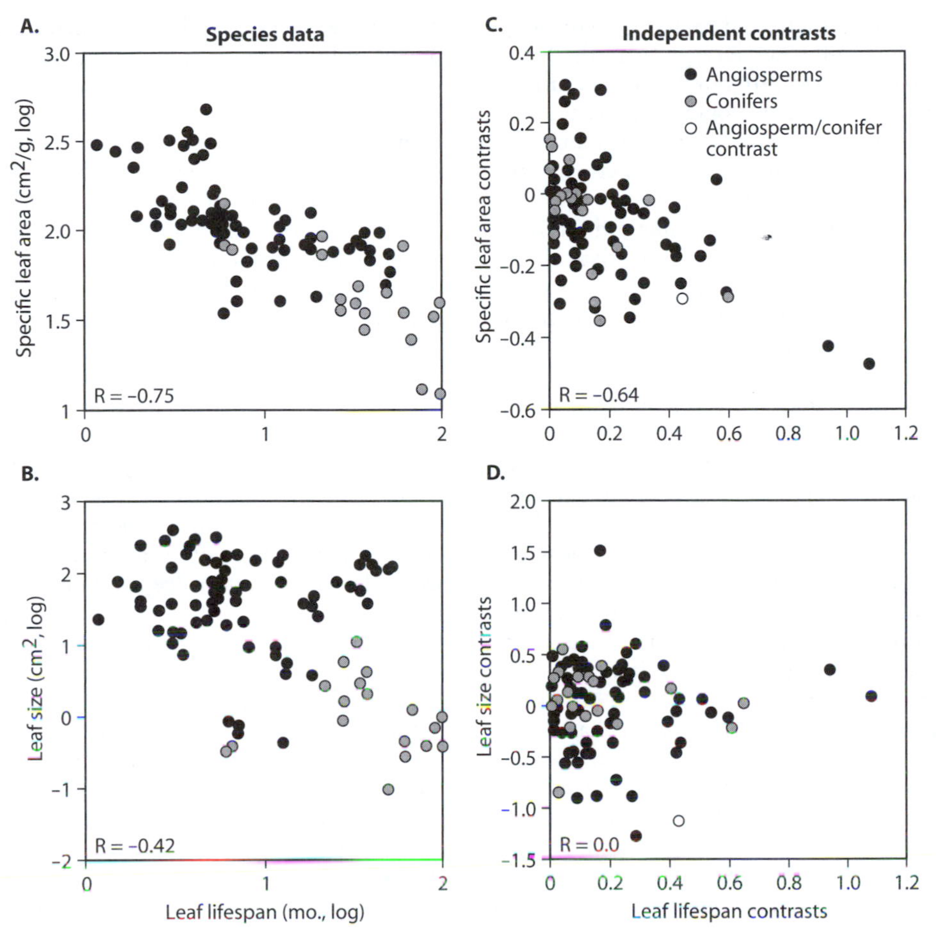

Figure 2. Analysis of interspecific correlations among leaf traits, using independent contrasts. Panels A and B show the correlations of leaf lifespan with leaf size and specific leaf area, respectively. Black circles are data for flowering plant species; gray circles are for conifers. The strength of the associations is indicated by the correlation coefficients in the lower left corner of each panel. Panels C and D show the corresponding relationships analyzed with independent contrasts. Black circles are contrasts between nodes within the flowering plant phylogeny, and gray circles are contrasts among conifers. The white circles represent the contrast at the basal node between the two groups. For convenience, the subtraction at each node is arranged such that the contrast for leaf lifespan is positive, and then the contrast for the other trait is positive or negative, depending on the trait values (subtraction must be in the same direction for both traits). (From Ackerly et al., 2000, Bioscience; copyright American Institute of Biological Sciences)

between SLA and leaf lifespan, based on contrasts, is similar to the pattern observed without using independent contrasts. But the evolutionary correlation between leaf lifespan and leaf size is essentially zero (figure 2). Why does this result shift so dramatically? As noted above, most of the correlation observed between leaf lifespan and size results from the marked difference between these traits in conifers and flowering plants, the deepest split in the phylogeny for this group of plants. Independent contrasts capture this shift as a single contrast. The rest of the contrasts, calculated among species of flowering plants or among species of conifers, exhibit no correlation in the shifts occurring in these two traits. In essence, the pattern observed if each species is treated as an independent data point reflects the influence of a single event deep in the evolution of these groups; when this single event is represented as one data point in the analysis (based on the one contrast), its influence is diminished, and we see that there is not a consistent evolutionary tendency for correlated changes between these two traits. Other lines of evidence are consistent with this result: there is no evidence that leaf lifespan and leaf size are functionally or evolutionary linked to each other, so the result from independent contrasts proves more reliable. We are still left with an important pattern in the present day: it is true that conifers have small, long-lived needles, which differ on average from the leaves of flowering plants. These differences may be important to understanding the ecological differences between these two groups of plants, but they should not be taken as evidence of an ongoing functional and evolutionary linkage between these traits.

As shown in this example, the method of independent contrasts addresses both the statistical and historical issues associated with the analysis of interspecific trait correlations. The contrasts are statistically independent, so significance values are reliable. The correlation or regression coefficients between the contrasts provide a much more precise measure of the underlying evolutionary pattern compared to a correlation of trait values from present-day species. However, like all statistical methods, independent contrasts invoke key assumptions, and these assumptions have been the source of some controversy. The most important assumption is that trait evolution conforms to a pattern of change known as a constant-variance random walk or Brownian motion. This model assumes that the changes occurring in each unit of time are equally likely to be positive or negative and are drawn from a normal distribution, such that small changes are more likely than large ones. Because these changes accumulate across multiple time steps, the total change along a branch is also expected to be proportional to the length of the branch. On the one hand, simulations have shown that statistical tests based on independent contrasts are quite robust to a variety of deviations from these basic assumptions, particularly if appropriate steps are taken to transform data or branch lengths in advance of analysis. In addition, the Brownian motion model is a reasonable first approximation of a model of evolutionary change based on our knowledge of quantitative genetics and the inheritance of continuous traits. On the other hand, a variety of other models of trait evolution may be considered; under some of these alternatives, species trait values are relatively independent of each other, and independent contrasts (or other comparative methods) do not necessarily provide a reliable measure of historical patterns.

There are several other classes of comparative methods that can be used for questions of trait associations. One of the most important is known as the phylogenetic regression, introduced by Alan Grafen in 1989, or phylogenetic general linear models. These approaches utilize statistical methods in which the user can specify the degree of independence among observations. The phylogeny is used to generate what is known as a variance–covariance matrix, which captures the expected degree of dependence among each pair of species in a study. This then opens up the full power of linear models, including multifactorial analysis of variance or covariance, with appropriate adjustment of significance tests reflecting the phylogeny. Although this facilitates a much broader range of hypothesis tests, one drawback is that the interpretation of results in terms of underlying historical processes is generally not as straightforward. A related class of methods uses maximum-likelihood approaches to find the best-fit model for a given set of interspecific trait data, given the phylogeny and alternative hypotheses of how the traits may be associated with each other. Maximum-likelihood approaches (and related Bayesian methods) have the general advantage that it is easier to invoke alternative underlying models of trait evolution. Further discussion of these methods, and the issues of branch lengths and evolutionary models, is beyond the scope of this chapter; researchers who will be using contrasts or other methods discussed here are well advised to seek a deeper understanding of these issues.

It is important to note that discrete characters, such as presence/absence of a trait or different states of a morphological character, usually require different approaches. Traditional tests of association for discrete characters involve chi-square or G-tests, based on contingency tables showing the frequency of different pairs of states. Phylogenetic approaches can be used to reconstruct historical transitions from one state to the other and then to test for associations between these transitions or between transitions in one character and the background state of the other character. Maximum-likelihood models, such as the DISCRETE program introduced by Mark Pagel, provide powerful solutions to this problem by testing whether the probabilities of transitions in different characters are associated with each other (see box 1).

BOX 1. SOFTWARE FOR PHYLOGENETIC COMPARATIVE METHODS

Phylogenetic comparative methods are computationally intensive, and a variety of software packages have been introduced that implement different tests. A few of the most important are briefly summarized here.

MacClade, first introduced by David and Wayne Maddison in 1987, set the standard for graphical elegance and ease of use in phylogenetic software. It is primarily used for reconstructing the evolution of discrete characters, based on parsimony methods, and also has limited capabilities for continuous characters.

Mesquite, also developed by the Maddisons, is a cross-platform and open-source program (http://www.mesquiteproject.org) with most of the features of MacClade plus a broader array of methods, including independent contrasts.

R is a freely distributed program for statistical analysis and programming; individual users develop and contribute libraries that implement different methods (http://www.r-project.org). Several libraries are now available (*ape*, *ade4*, *geiger*, *PHYSIM*, *PHYLOGR*) that implement numerous phylogenetic comparative methods. *R* is a very powerful program that is being adopted by many researchers in ecology (although it is difficult to learn at first).

COMPARE, written by Emilia Martins, is a Web-based program that implements independent contrasts, phylogenetic linear models, and related methods (http://www.indiana.edu/~martinsl/compare/).

Phylocom is a freely distributed program (http://www.phylodiversity.net/phylocom) that is widely used for phylogenetic analysis of community structure and also conducts independent contrasts and analyses of phylogenetic signal.

DISCRETE and ***Continuous***, both written by Mark Pagel and colleagues, implement several maximum-likelihood methods for the analysis of trait correlations, modes of trait evolution, and related methods. Both of these programs are now included in the BayesTraits program (http://www.evolution.rdg.ac.uk/BayesTraits.html).

3. PHYLOGENETIC SIGNAL: PATTERN AND SIGNIFICANCE

The fact that closely related species resemble each other—in ecological, morphological, behavioral, and other attributes—comes as no surprise to students of natural history. Evolution is generally a conservative process, and traits will usually change slowly, if at all, from one generation to the next. Adaptive radiations, in which species may diverge rapidly and take on novel adaptive traits and ecological lifestyles, are of interest precisely because they are unusual: at moments of ecological opportunity, following mass extinctions or the arrival of colonists on uninhabited islands, we see the potential for rapid evolutionary change. But most of the time, evolution is slow, and few changes accumulate, even over long periods of time. The lack of change is referred to as evolutionary stasis. The importance of understanding stasis in evolution has been highlighted by paleontologists, especially Steven Jay Gould, based on their study of the fossil record. When stasis, or at least a slow rate of change, plays out across the phylogeny, the result is that close relatives will be very similar.

Many terms have been used to describe this pattern of slow change: *phylogenetic inertia, phylogenetic constraint,* and *phylogenetic effects*. Often, these terms convey a sense that the phylogeny itself is the cause of ancestor–descendant resemblances. I find it useful to use the term *phylogenetic signal,* advocated in a recent essay by Simon Blomberg and Ted Garland, to emphasize that the similarity among relatives is a pattern and by itself does not reveal the underlying processes. An understanding of the causes of phylogenetic signal, and why it may vary in different groups and for different traits, draws on genetics, developmental biology, and ecology. We know that evolutionary change requires heritable, genetically based variation in a trait for selection to act on. Recent advances in the field of "evo-devo" are shedding light on how the process of development can influence the expression of genetic mutations, explaining why some traits vary more than others and why certain attributes may appear repeatedly in different lineages. On the other hand, even if ample genetic variation is available, natural selection may act to maintain traits in their current condition if an organism is well adapted to its current conditions. This process is known as stabilizing selection and may be pervasive in nature, although for a variety of technical reasons it can be quite hard to detect. The ability of plants and animals to migrate during episodes of climate change and track the environments to which they are well adapted may also be a process that reduces the rate of evolutionary change. There is no general consensus on the relative importance of these different factors that contribute to the phylogenetic signal in different traits, and it is very difficult to obtain all the relevant data in any particular case study.

In the context of ecological research, it can be useful to quantify the pattern of phylogenetic signal and compare observed patterns to those expected under alternative evolutionary models. The Brownian motion model, in particular, provides an important point of comparison because it is the foundation of many comparative methods. Although Brownian motion represents a random model of evolutionary change, it does generate a fairly high degree of phylogenetic signal, as sister taxa diverge gradually from their common ancestors. In contrast, null models in which trait values are randomly rearranged among the species in a study provide a baseline measure for the complete absence of phylogenetic signal. Two closely related measures, Pagel's λ and Blomberg's K statistic, are particularly useful, as they take on a value of 1 when patterns of trait similarity conform to expectations of Brownian motion and greater than or less than 1 when close relatives are more or less similar than expected, respectively. Another class of methods known as Mantel tests is based on the correlation between the phylogenetic distances between species (the distance down the branches of the phylogeny to the common ancestor and back up to another species) and the ecological or phenotypic differences between them. These methods are useful for ecological characteristics such as niche overlap and co-occurrence where the degree of similarity or dissimilarity between species is quantified directly.

Phylogenetic information can play an important role in the prediction of ecological traits when there is strong phylogenetic signal. For example, in a recent study, Jérôme Chave and colleagues demonstrated that wood density tends to be very similar among closely related tree species. Wood density is important for carbon storage, a critical factor in the global carbon cycle, but it has only been measured on a small proportion of tree species in the tropics. Knowledge that close relatives have similar wood density will allow more accurate prediction of carbon storage in diverse tropical forests, even for species for which wood density has not been measured directly.

4. PHYLOGENETICS AND COMMUNITY ECOLOGY

Phylogenetics is playing an increasingly important role in community ecology as a tool to gain insight into the processes that influence community structure. One of the earliest theoretical principles of ecology was the competitive exclusion theorem, formalized by Gause in the 1930s, which states that two species that utilize identical resources cannot coexist in a community. In the 1950s, this idea, together with the knowledge that closely related species are usually ecologically similar and therefore utilize similar resources, led to the prediction that species from the same genus should co-occur infrequently. This prediction was tested by calculating the average number of species per genus in isolated communities, such as islands, compared to the overall biota of the surrounding region. In the past 10 years, phylogenetic approaches to community ecology have been revitalized by the availability of highly resolved phylogenetic trees and new methods. In addition, developments in community assembly theory have emphasized an alternative view that co-occurring species may be more similar to each other than expected because similar traits may promote ecological success under particular environmental conditions. These two perspectives provide contrasting predictions regarding whether communities will be composed of more or less closely related species.

Three steps are required to quantify the phylogenetic structure of ecological communities and test hypotheses about whether this structure is significantly different than may be expected. First, the degree of relatedness among co-occurring species needs to be quantified, based on the best available phylogeny. Cam Webb and others have introduced several related methods to accomplish this. The simplest approach is simply to calculate the average phylogenetic distance between all pairs of species within the community. Other approaches take into account species abundance or measure the distance between each species and its closest relatives in the community, as opposed to more distant relatives. The second step is to specify a broader pool of species from which a particular community has been assembled. This provides the source pool to construct hypothetical communities that serve as a point of comparison with observed patterns. Ideally, the spatial scale defining this pool is large enough so that it includes all of the species that could, in a reasonable span of time, arrive at the community of interest. However, in practice, it is very difficult to determine exactly what this scale should be, and researchers rely on a variety of practical solutions to address this problem. Finally, one needs to construct an appropriate null model by which random communities can be drawn from this regional pool to determine whether the observed communities diverge from random expectations. Simple null models include a random draw of species, where each species is equally likely to be chosen. More complex models can be constructed, in which the probability of a species being chosen is proportional to its frequency of occurrence in the landscape. The construction and analysis of these null models are continuing points of discussion and development in this field.

Many studies of phylogenetic community structure have appeared in recent years, and some generalizations are beginning to emerge. First, empirical and theoretical studies suggest an asymmetry in the interpretation of phylogenetic community data. It appears that clustering of close relatives within a community arises primarily from an ecological filtering process, in which similar species are favored as they share adaptations that are appropriate for the particular conditions. On the other hand, many different processes can lead to the opposite pattern in which communities are composed of more distant relatives than expected. These include competition, small-scale habitat heterogeneity, facilitative interactions among functionally disparate species, and even a filtering process when the traits that promote success have evolved independently in different clades. Theoretical studies also suggest that it is much harder to detect patterns in which coexisting species are distantly related, compared to the opposite pattern.

A second result is the realization that communities will not be structured either by filtering or by competition or by any other single process. Many processes are likely at work, mediated by different sets of traits. For example, Jeannine Cavender-Bares and colleagues studied the composition of oak-dominated forests in Florida and found that local communities were generally composed of distantly related species. These species tended to share physiological traits affecting their water relations, with drought-adapted species occurring together on drier sites. Moreover, these hydraulic traits exhibited low phylogenetic signal, so similar species tended to be distantly related for these characteristics. On the other hand, co-occurring species displayed a high disparity of trait values related to acorn maturation and wood density. These traits exhibited a high degree of phylogenetic signal, but closely related species with similar values were distributed across different communities. Thus, it is critical to specify the traits that may be relevant to community assembly and examine their distribution on the phylogeny carefully before interpreting patterns of phylogenetic community structure in terms of particular underlying processes.

Finally, there is a fascinating pattern in plant communities of a shift from the co-occurrence of more distant relatives when studies focus on a narrow clade (e.g., oaks) to a pattern of clustering of close relatives in broader studies that encompass the full spectrum of flowering plants or all seed plants. A similar shift occurs moving from smaller to larger spatial scales. Both of these patterns are consistent with a stronger role for resource partitioning among closer relatives and at smaller spatial scales, whereas habitat filtering becomes more apparent at larger spatial and phylogenetic scales.

5. PROSPECTS FOR THE FUTURE

The potential role of phylogenetics in ecology was heralded by several articles and books published in the late 1980s to mid-1990s. In the relatively short interval since then, many methods have been introduced or improved, and growth in research has been rapid. The number of citations in the scientific literature under the keywords *phylogen* and *ecology* rose from 4 in 1990 to 87 in 1995, 130 in 2000, and 275 in 2006. An important engine of this growth has of course been the constantly expanding availability and improved resolution of phylogenies for diverse groups of taxa, accompanied by new methods, fast computers, and easy-to-use software. This chapter highlights two areas that are most relevant to ecological research. Measures of phylogenetic diversity are also used as criteria to help prioritize taxa and habitats in conservation biology, and a wide variety of comparative methods are in use in evolutionary biology, including the study of adaptation, diversification, adaptive radiations, and related topics.

Several important areas of challenge and opportunity lie ahead. One is the improved resolution of branch lengths and node ages on phylogenies, which will be provided by including more species and more genes and improvements in fossil calibration. Time-calibrated phylogenies are opening the door to linkages between comparative methods and paleoecology and will facilitate investigation of a new generation of questions. A second area is the development of global databases for ecological traits. This will allow us to assess questions of phylogenetic signal and ecological trait correlations across the entire phylogeny of major clades and to understand how the assembly of local floras and faunas relate to global patterns of ecological diversity. Third, phylogenetic methods are providing new insights into ecology and biogeography of microbes, fungi, and other groups that are difficult to study directly in the field. These are but a few of the growth areas at the intersection of phylogeny and ecology—the most exciting advances will be those that at this point are not even anticipated.

FURTHER READING

Blomberg, S. P., and T. Garland, Jr. 2002. Tempo and mode in evolution: Phylogenetic inertia, adaptation and comparative methods. Journal of Evolutionary Biology 15: 899–910.

Felsenstein, J. 1985. Phylogenies and the comparative method. American Naturalist 125: 1–15.

Harvey, P. H., and M. Pagel. 1991. The Comparative Method in Evolutionary Biology. Oxford: Oxford University Press.

Maddison, W. P., and D. R. Maddison. 1992. MacClade: Analysis of Phylogeny and Character Evolution. Sunderland, MA: Sinauer Associates.

Pagel, M. D. 1999. Inferring the historical patterns of biological evolution. Nature 401: 877–884.

Webb, C. O., D. D. Ackerly, M. McPeek, and M. J. Donoghue. 2002. Phylogenies and community ecology. Annual Review of Ecology and Systematics 33: 475–505.

I.17

Microevolution

Michael A. Bell

OUTLINE

1. Evolution: Micro versus macro
2. "The ecological theater and the evolutionary play"
3. Microevolutionary mechanisms
4. Contemporary microevolution
5. The unintended consequences of human technology
6. Geographic variation
7. Phylogeography
8. Genomics and microevolution
9. Prospects

Microevolution occurs within and among populations of a species and usually involves changes in the mean value or relative frequencies of alleles and phenotypes that are shared by most populations of the species. Divergence among populations of a species (i.e., conspecific populations) is often associated with habitat differences, and such divergence often has important ecological consequences. Population genetics deals with evolution in terms of allele and genotype frequencies within populations, so it provides the theoretical foundation to study microevolution. Widespread species typically exhibit geographic variation, which has generally been thought to take thousands of generations to evolve. However, recent research on contemporary evolution suggests that geographic variation can evolve within a few generations after species colonize new habitats or experience environmental change. The high rate at which microevolution can occur is important because it means that pathogens, pests, and harvested natural populations can rapidly evolve traits that adversely affect people. DNA variation within and among conspecific populations can be studied as a product of microevolution, and it also provides powerful tools to tease apart the contributions of common ancestry and local adaptation to the evolution of geographic variation. Thus, previously intractable problems in microevolution and its applications to natural resource management can now be studied using the emerging technologies of molecular biology and genomics.

GLOSSARY

character displacement. This is the evolution of enhanced differences between species where they occur together as a result of selection against members of one or both species that use the same resources as members of the other species (i.e., ecological character displacement) or against individuals that tend to hybridize with members of the other species (i.e., reproductive character displacement).

cline. A cline is a geographic gradient in the frequency or mean value of a phenotype or genotype.

monophyletic group. This is a group of species that are more closely related to each other than any is to species outside the group.

phenotypic plasticity. A change in an individual phenotype that does not alter its genetic constitution and is not inherited by its offspring.

random walk. In population genetics, this is a change in allele frequencies from their initial values as a result of repeated episodes of genetic drift.

taxon. A taxon (including higher taxon) is any named group (e.g., Vertebrata, Mammalia, *Homo sapiens*) at any taxonomic rank (e.g., Kingdom, Class, Species); higher taxa are more inclusive.

1. EVOLUTION: MICRO VERSUS MACRO

Biological evolution is change through time in the heritable properties of a lineage or monophyletic group (clade). Microevolution is generally confined to evolution within and among conspecific populations, and it occurs within relatively short time spans. In contrast, macroevolution involves changes in the number or characteristic properties (e.g., average body size) of the species of a clade. It depends on the variation among species generated by microevolution and unfolds over longer periods. Nevertheless, the definitions of microevolution and macroevolution have been controversial, and there is disagreement about their mechanistic relationships and even the value of the terms.

The division between microevolution and macroevolution is usually placed at speciation because members of different species do not routinely interbreed, and the evolutionary fates of separate species are largely independent. Microevolution involves changes in the frequencies of alleles and genotypes and of interactions between different genes. These changes are manifested as recognizable changes in the mean values or frequencies of biochemical, physiological, behavioral, developmental, and morphological phenotypes. A separate set of macroevolutionary mechanisms influences the probability of speciation and extinction. Thus, properties of species that promote speciation or impede extinction will tend to increase in a monophyletic group over time. Both microevolution and macroevolution contribute to biodiversity, but microevolution affects individuals and changes the properties of populations, whereas macroevolution alters the relative frequencies of species with different properties.

There are also practical reasons to distinguish microevolution and macroevolution. Microevolution can be studied using comparative, observational, or experimental methods to study individuals and populations over a few generations in the laboratory and field. Existing genetic properties and ecological conditions can be used to interpret microevolution. In contrast, macroevolutionary studies focus on differences among species. Careful species description, characterization of clades, and investigation of phylogenetic relationships among taxa are paramount in macroevolutionary research. The environmental factors and genetic properties that influenced speciation and extinction have typically been lost in the dim past and are hard to infer. Consequently, microevolution and macroevolution are generally studied using different methods.

2. "THE ECOLOGICAL THEATER AND THE EVOLUTIONARY PLAY"

G. Evelyn Hutchinson's famous 1965 book, from which the title of this section was borrowed, emphasized that evolution occurs within an ecological context. Although existing genetic properties of a population (e.g., presence of an advantageous allele) influence its microevolutionary response to natural selection, ecology is a major factor in microevolution and a crucial source of information to interpret it. Furthermore, if environmental differences cause microevolutionary divergence among conspecific populations, they will exhibit differences that must be taken into account in ecological studies. Studies of microevolution and ecology are intimately associated and reciprocally illuminating.

3. MICROEVOLUTIONARY MECHANISMS

Because microevolution involves changes in the relative frequencies of heritable traits within populations of a species, it can be analyzed in terms of the behavior of alleles and genotypes within populations. This is the subject of population genetics, and the Hardy-Weinberg equilibrium is the starting point to develop the genetic theory of microevolution. The Hardy-Weinberg equilibrium describes the distribution of alleles among diploid genotypes in a population in the total absence of evolution. It will be sketched here only briefly, but most textbooks on evolutionary biology develop it in detail (see chapter I.15).

The Hardy-Weinberg Equilibrium

If no evolutionary forces impinged on a population, the relative frequencies of alleles and genotypes in the population would reach equilibrium values that would never change after one generation of random mating. Genotype frequencies under these Hardy-Weinberg equilibrium conditions can be calculated using the simple equation, $1 = (p + q)^2$, where p and q are the relative frequencies of two alleles of a gene and must sum to 1. Of course, no real population ever conforms to Hardy-Weinberg equilibrium conditions, although deviations from equilibrium frequencies are often so small that they are undetectable. Detectable deviations from equilibrium frequencies, however, indicate that microevolution is taking place and may suggest its causes. Potential causes for deviations from equilibrium frequencies include mutation, meiotic drive, assortative mating, gene flow, genetic drift, and natural selection, the most important of which are mutation, gene flow, genetic drift, and natural selection.

Mutation

Mutations are heritable changes in DNA and are the ultimate source of variation for microevolution. However, mutation rates are so low (10^{-4}–10^{-9} mutations/generation/trait/individual) that they do not usually produce measurable departures from expected Hardy-Weinberg equilibrium frequencies. Phenotypic changes caused by mutation are not biased (i.e., random) to produce adaptation.

Gene Flow

Gene flow occurs when an individual is born in a source population and reproduces after migrating to a recipient population. Its effects depend on the magnitude of genetic differences and rate of migration between the donor and recipient populations. Gene flow is frequently a more important source of genetic variation than mutation, but it also tends to homogenize populations that otherwise would evolve differences.

Genetic Drift

Genetic drift is the change of allelic frequencies between generations just by chance (i.e., sample error). The magnitude of genetic drift is inversely related to effective population size (N_e), which, in turn, increases with the number of breeding individuals and evenness of the sex ratio and decreases as variation in the number of offspring per pair increases. When N_e is large, genetic drift causes very small differences between successive generations, but if N_e is small for even one generation, it can cause major changes in the genetic composition of a population. Even if N_e is consistently large, there will always be some drift, and its effects will accumulate, causing a "random walk" of small deviations that can add up to major changes in allelic frequencies over many generations. In populations with small N_e, rare alleles tend to be lost by drift, and even traits that are disfavored by natural selection can drift to high frequencies. Because there is initially only one copy of a new mutant allele, even advantageous mutations tend to be lost by drift.

Natural Selection

Natural selection depends on three components: survival selection, fecundity selection (ability to produce offspring), and sexual selection (mating success of individuals compared to other members of the same sex). Each component results from differences in the relative rates of success of different phenotypic classes, and selection can be quantified as Darwinian fitness or a selection coefficient. Darwinian fitness depends strongly on the interaction of the phenotype with the environment, and environmental changes can cause changes in fitness associated with a phenotype. If a phenotype with high fitness is heritable, alleles that produce it will tend to increase through time (i.e., evolve). New phenotypes may appear in a population by means of gene flow, mutation, and sexual recombination of existing alleles, and natural selection can increase their frequencies and cause a population's phenotype to evolve beyond its previous range of variation. Thus, natural selection is the major cause for evolutionary adaptation and phenotypic divergence among conspecific populations.

Genetic drift also causes divergence among conspecific populations, and it is necessary to distinguish the effects of drift and selection. Phenotype–environment correlations indicate selection but are not sufficient to identify the environmental variable that causes it. Further evidence based on differences in function or Darwinian fitness of phenotypes is necessary to confirm inferences based on phenotype–environment correlations. It is surprisingly difficult to establish that natural selection has caused microevolution of a specific trait.

Nonheritable Change

Not all phenotypic variation among conspecific populations represents microevolution. Phenotypic plasticity may result from conditions experienced by the individual. Genetically identical individuals or the same individual at different times may differ because of phenotypic plasticity. For example, muscles may grow larger from exercise, skin may become darker from exposure to sunlight, and learning may alter behavior, but such changes do not affect the genetic constitution of the individual that experiences the phenotypic plasticity or that of its progeny. Similarly, maternal effects, phenotypic differences caused by a female's nongenetic contributions (e.g., messenger RNA, yolk, parental care) to its progeny, may influence the offspring's phenotype but not be inherited. However, the individual's ability to exhibit phenotypic plasticity (i.e., show a phenotypic response to environmental variation) may be heritable, and thus plasticity may evolve. Phenotypic plasticity may cause nonheritable but ecologically important phenotypic variation among conspecific populations. Much of the phenotypic change caused by human environmental disturbance apparently results from phenotypic plasticity and not microevolution.

4. CONTEMPORARY MICROEVOLUTION

It is widely believed that microevolution is rarely rapid enough to be observed in progress. For many years, industrial melanism in the UK stood as the lone well-confirmed example of contemporary evolution. The peppered moth, *Biston betularia*, and other moths and beetles evolved dark pigmentation where soot from industrial pollution darkened tree bark and killed light-colored lichens on which the moths rest during the day. Although the speed with which industrial melanism evolved was never in doubt, questions arose about evidence that bird predation selects against moths that contrast with bark color. Recent results seem to confirm this effect, and many other cases of rapid evolution in response to human-induced environmental change have been reported in recent years.

Initially, most of these cases involved evolution of resistance to insecticides by insects and to antibiotics by bacteria. It seemed possible that these cases of

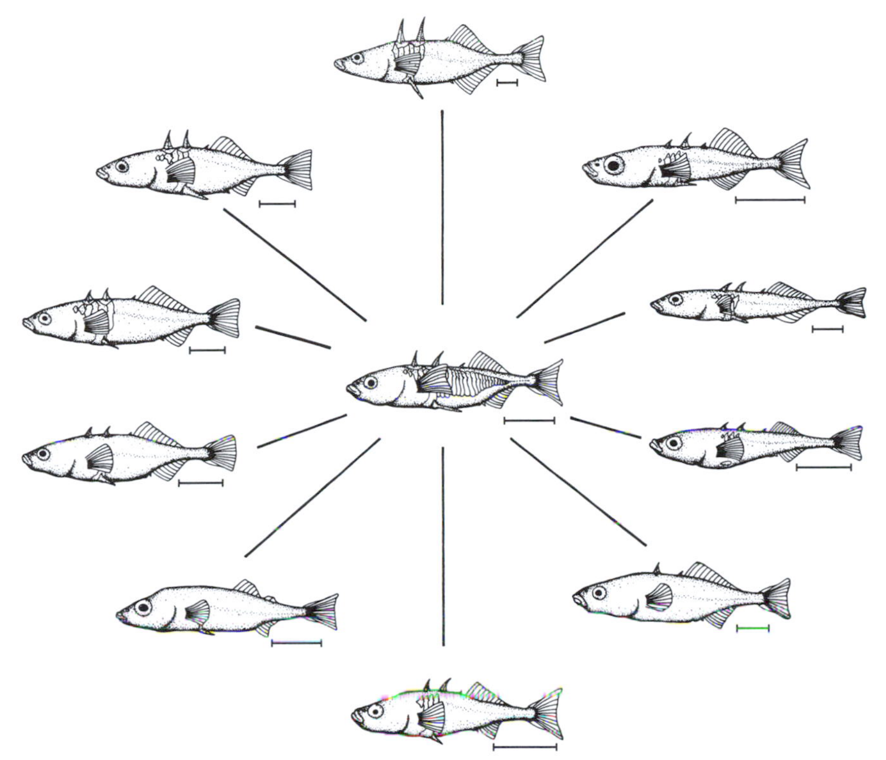

Figure 1. Variation in armor, size, and shape of threespine stickleback. The specimen in the middle is a completely plated (high *Eda* allele), anadromous (sea-run) stickleback, and those around the periphery are low-plated (low *Eda* allele), but otherwise phenotypically diverse, freshwater stickleback from western North America. All specimens were drawn to be the same size, and the scale bars equal 1 cm. Variable armor traits include the length and number of dorsal spines, expression of the pelvis (including absence), and number (including zero) and distribution of lateral plates. (Reprinted with permission from Bell and Foster, 1994)

contemporary evolution might differ from typical microevolution. Insects have large populations and short generation times, both of which favor rapid evolution, and bacteria have even larger populations and shorter generation times. Additionally, selection imposed by human technology might be more severe than selection under natural conditions. However, it is also possible that evolution of resistance in insects and bacteria may not be atypical; it may just be more conspicuous because it has serious consequences for people. Consequently it is quickly noticed and carefully studied.

Darwin's finches are the classic case of adaptive radiation, and they have been closely observed for decades (see chapter I.19). They occur on the Galápagos Islands of Ecuador, which are relatively undisturbed. Dramatic evolutionary changes in body and bill size and in bill shape evolved in the medium ground finch, *Geospiza fortis*, and the cactus finch, *G. scandens*, in response to climate-induced changes in food availability during a 30-year period. Similarly, field mustard, *Brassica rapa*, grown from seeds collected after a 5-year drought in California exhibited higher tolerance to drying and flowered earlier than seeds collected before the drought.

Sea-run (anadromous) threespine stickleback fish, *Gasterosteus aculeatus*, which colonized a lake in Alaska, also evolved rapidly under relatively natural conditions. Freshwater stickleback are phenotypically diverse and differ strikingly from sea-run populations (figure 1). Within a dozen generations after the lake was colonized, several armor, body shape, and feeding traits evolved, and this young population has become indistinguishable from adjacent lake populations.

Contemporary evolution is not restricted to life history and anatomical traits. The well-known fruit fly, *Drosophila melanogaster*, originated in the Old World but has been transported to every continent except Antarctica by humans. It lays its eggs in rotting fruit, where ethanol may reach lethal concentrations. Alcohol dehydrogenase (ADH) is one of the enzymes that detoxifies ethanol. There are two common alleles for the *alcohol dehydrogenase* gene, *Adh-F* and *Adh-S*. *Adh-F* is less stable at high temperatures and has higher activities at lower temperatures. It gradually increases in frequency (i.e., clinally) going away from the equator in Australia, Asia, and North America. The functional attributes of enzymes encoded by these enzyme alleles (i.e., allozymes) and other evidence suggest that these clines result from natural selection, and they must have evolved since *D. melanogaster* was introduced to these continents. Moreover, within the last 20 years, the position of an *Adh* cline in Australia has shifted southward by 400 km so that high *Adh-F* allele frequencies now occur farther from the equator than before, as would be expected from global warming. Thus, this recent microevolutionary change can be used to monitor global climate change.

Numerous additional examples of contemporary evolution have been described in recent years (see Hendry and Kinnison, 2001), indicating that natural environmental change and colonization of new habitats frequently cause detectable microevolutionary change on a human time scale and that microevolution is often so fast that close monitoring is necessary to catch it in the act.

5. THE UNINTENDED CONSEQUENCES OF HUMAN TECHNOLOGY

In 1962, Rachel Carson's *Silent Spring* sounded the alarm that insecticides, which had been in use since the late 1940s to limit crop pests and disease vectors, were also eliminating many desirable, nontarget species. She did not know that insecticides also caused rapid microevolution of resistance in the insect populations that were their targets. Of course, large population size, short generation time, and intense selection by insecticides should cause rapid evolution of resistance. By 1997, more than 500 insect species had evolved resistance to insecticides. Weeds were no different, and herbicide resistance had evolved in more than 200 weed populations by 2001. It has also become painfully clear that pathogenic bacteria, fungi, protozoans, and metazoans routinely evolve resistance to drugs that had previously produced cures. These microevolutionary responses adversely affect human health and agriculture, but they can sometimes be mitigated by using management strategies based on microevolutionary theory.

For example, genes from the bacterium *Bacillus thuringiensis* (Bt) have been inserted into the genome of cotton plants, conferring on them the ability to express Bt toxin, which kills pink bollworm, *Pectinophora gossypiella*. After 8 years of planting Bt-transgenic cotton, bollworm populations in Arizona still had not evolved resistance to Bt toxin. Failure of *P. gossypiella* to evolve resistance was apparently achieved by growing small plots of nontransgenic cotton, where nonresistant bollworms thrived. Because the nonresistant bollworms far outnumbered rare, resistant bollworms that survived in the Bt-transgenic-cotton fields, and they tend to disperse into the Bt-transgenic fields, most bollworms the following season are either nonresistant or hybrids between resistant and nonresistant parents, both of which are killed by Bt toxin. By sacrificing small plots of nonresistant cotton to bollworm infestation, natural selection favoring resistance to Bt toxin has been retarded by gene flow from nonresistant bollworm populations.

Experience with the evolution of antibiotic resistance has been far less encouraging. By 1943, penicillin production was under way, and other antibiotics would soon follow. Even as the first antibiotics became available for clinical use, penicillin-resistant bacteria had already appeared, and resistance soon occurred in one bacterial pathogen after another. For example, penicillin could easily cure *Staphylococcus aureus* (staph) infection in the early 1950s, but by the late 1960s it had become ineffective. Methicillin still worked, but it became ineffective by the 1990s. Most staph infections can still be cured by vancomycin, but resistance to this "antibiotic of last resort" is spreading. New drugs continue to be developed, but this is an expensive arms race with tragic consequences and no end in sight.

In retrospect, microevolution of antibiotic resistance in bacteria is not surprising. Genetic variation for antibiotic resistance is common in bacterial populations, and their large size (i.e., N_e) and short generation time both facilitate the appearance of new mutants. Genes for resistance may protect bacteria from multiple antibiotics, and they can be transferred in plasmids between bacterial species. Natural selection for antibiotic resistance has been hastened by the misuse of antibiotics for diseases against which they are ineffective, failure of patients to complete antibiotic treatment, and widespread, chronic, low-dose antibiotic treatment to increase livestock productivity. All of these practices selectively eliminate less resistant bacterial clones, leaving behind more resistant ones to found new bacterial populations.

Under favorable conditions, it is also possible to select for reduction in the severity of disease (i.e., virulence). Many pathogens rely on their host's social interactions to spread to new hosts before it dies or mounts an immune response that eliminates the infection. Consequently, if hosts with the most severe infections are isolated from other susceptible individuals, the most virulent pathogen strains will fail to spread. For example, installation of window screens in the southeastern United States during the first half of the twentieth century prevented mosquitoes from biting people who stayed indoors with the most serious cases of malaria, contributing to evolution of lower virulence in *Plasmodium*, the malaria pathogen. By separating malaria victims with the most severe symptoms from mosquitoes that spread *Plasmodium*, the most virulent strains could not spread, and the disease became less serious.

Human-induced microevolution may also play a crucial role in the loss of valuable commercial fish populations. Commercial fishing gear selectively captures larger fish, but smaller individuals slip through the net. Fisheries policies are intended to allow young fish to escape and grow to a larger size, at which they both reproduce and become more valuable as food. However, selective fishing for larger fish also favors individuals that stop growing at a smaller size and reproduce earlier in life. Because body size and reproductive schedules are heritable, size-selective fishing should cause evolution of smaller adult body size and earlier reproduction. After fishing is halted, the survivors should have genotypes for smaller body size and early reproduction. Many commercially fished populations never recover numerically, and those that do are often descendants of small individuals from which they inherit small body size. The conclusion that size-selective fishing has caused evolution of smaller body size in commercial fishes has been controversial because the quality of the marine habitats in which declining fish populations live has also deteriorated. Nevertheless, a growing minority advocates the incorporation of microevolutionary principles into fisheries' management policy.

6. GEOGRAPHIC VARIATION

Variation that has evolved among populations in response to local ecological differences is a common phenomenon and an important source of evolutionary insight. Comparison of mainland or large central populations to populations on islands or peripheral habitat patches (e.g., mountain tops, desert springs) often reveals variation that is associated with environmental differences. Island populations may be isolated from predators, competitors, parasites, and pathogens that occur on the mainland, or they may encounter resources that are unavailable on continents, leading to evolution of unusual traits. The divergent properties of insular populations must be interpreted with care because insular populations are often small (i.e., low N_e) or were bottlenecked in the past, allowing genetic drift to influence their evolution.

Ecological character displacement may occur when closely related species that are usually allopatric occur sympatrically. In sympatry, the members of each species that most closely resemble those of the other one may compete poorly with it, and natural selection will tend to eliminate intermediate individuals and favor evolution of enhanced differences between them. Inference of ecological character displacement has been controversial, but some cases are well supported.

Clines, which were mentioned in passing before, are geographic gradients in the frequencies of genotypes or phenotypes or in phenotypic means. They may evolve in an initially homogeneous population that experiences an environmental gradient or even sharp differences in natural selection in different parts of its range. Clines may also form where previously isolated contrasting populations come into secondary contact and hybridize. Populations separated by a cline may eventually merge, or selection against hybrids owing to ecological or genetic differences may cause the populations to retain their differences. Similar clines may occur in multiple species and are recognized as *biogeographic rules*. For example, in endothermic vertebrates, body size tends to increase (Bergmann's rule), extremities tend to be shorter (Allen's rule), and coloration tends to be paler (Gloger's rule) toward the poles. Similarly, the number of vertebrae increases toward the poles in fishes (Jordan's rule). Clinal variation among conspecific populations is a conspicuous and ubiquitous manifestation of microevolution.

7. PHYLOGEOGRAPHY

DNA sequence variation is most strongly influenced by genetic drift, and its evolution should be largely independent of phenotypic microevolution. It should carry a strong signal of evolutionary relationships or phylogeny. Heritable phenotypic differences among populations, however, may reflect both phylogeny and local natural selection (adaptation). Although gene flow complicates the analysis, it is possible to reconstruct the phylogeny of conspecific populations, which is called phylogeography, using DNA sequence data to distinguish the effects of phylogeny and adaptation. It is possible that different genes will indicate different relationships among populations of a species because one gene may have been present when one population split into two, and another may have entered one population by gene flow long after the two populations split. Variation of allozymes and restriction fragment length polymorphisms (RFLP) in mitochondrial DNA (mtDNA) were used in phylogeography until the late 1980s, when development of the polymerase chain reaction (PCR) enabled widespread use of DNA sequences from the nuclear genome.

Phylogeographic analysis of sockeye salmon, *Oncorhynchus nerka*, illustrates the value of this approach. Sea-run sockeye salmon are widespread throughout the north Pacific, but rare, isolated lake-resident populations of *O. nerka*, called kokanee, also exist.

Anadromous sockeye are about twice the size of kokanee but spend only 1 to 3 years at sea before spawning in fresh water. The smaller kokanee remain in fresh water and spawn after 2 to 7 years. The phenotypic similarity of isolated kokanee populations throughout their range suggested that they evolved in one place and spread from there, but their wide distribution suggested that they evolved repeatedly from local anadromous sockeye. Analyses of DNA sequence variation showed decisively that kokanee populations are genetically similar to adjacent migratory sockeye populations. They must have evolved many separate times from sockeye, and the similarity of isolated kokanee populations is a result of repeated (convergent) microevolution of similar adaptations to similar habitats.

8. GENOMICS AND MICROEVOLUTION

Development of large DNA-sequence databases, including whole, sequenced genomes, laboratory methods to inexpensively obtain DNA data, and statistical methods to analyze them have created exciting opportunities to study microevolutionary processes, phylogeography, and the genetics of microevolution. Molecular markers, including RFLP, single-nucleotide polymorphisms (SNPs), and short tandem repeats (microsatellites) can be used to study the number of genes, the relative importance of different genes, their location in the genome, and even which parts of genes underlie phenotypic evolution. In the threespine stickleback, for example, at least four genes on separate chromosomes control variation in the number of lateral armor plates (see figure 1), but the *Ectodysplasin* (*Eda*) gene accounts for more than 75% of the variation in plate number. *Eda* affects development of other vertebrate traits (e.g., teeth, fish scales, mammal hair, and sweat glands), so a change in the structure of the EDA protein would probably have adverse effects on other stickleback traits. As expected, the protein-coding region of *Eda* does not differ consistently between alleles for high- and low-plate-number phenotypes, implicating altered expression of *Eda* in evolution of plate number. Insertion of an *Eda* gene from a mouse into the genome of a low-plated stickleback caused an increase in the number of plates, confirming *Eda*'s role in plate number evolution. Remarkably, an ancestral *Eda* allele for low-armored phenotypes originated by mutation more than 10 million years ago, and alleles that have evolved from it have spread across the Pacific, Arctic, and Atlantic oceans from a single point of origin, providing the genetic variation on which natural selection acted to cause evolution of low plate number throughout this huge range. However, different genes with smaller effects may cause plate number differences in neighboring populations.

Genomics has also contributed an entirely new method to study natural selection. If a novel allele enters a population by mutation or gene flow, it may initially be represented by one copy. If this allele experiences strong positive selection, it will quickly be fixed (i.e., reach 100%), and the DNA sequence surrounding it will also be fixed before recombination with DNA surrounding other alleles for the same gene can occur. Until this region accumulates mutations, heterozygosity will tend to be depressed around the positively selected allele. Such depressed heterozygosity is the signature of a recent rapid *selective sweep*. Studies of the human and other genomes have detected numerous selective sweeps, and extended DNA sequences surrounding genes already believed to have experienced selective sweeps tend to be surrounded by regions of depressed heterozygosity. For example, an allele for the *Lct* gene that confers the ability in adult humans to digest milk sugar (lactose) is surrounded by a long stretch of DNA with reduced heterozygosity, suggesting a recent selective sweep when humans became consumers of raw milk.

9. PROSPECTS

A wide range of methods continues to be used to develop new insights into the mechanisms for evolution within species. Since the mid-1960s, application of molecular biology to microevolutionary problems has revolutionized the field. Molecular methods and genomic data will continue to be used in phylogeography, and they will increasingly be applied to research on the genetics and development of phenotypes that vary within and among conspecific populations. As the power of molecular methods and number of sequenced genomes increase, new opportunities to use geographic variation to investigate basic problems in genetics and development will appear and create additional tools for research in microevolution. Increasingly, microevolutionary theory and molecular biology will be combined to address problems related to human health, agriculture, and natural resource management.

See also chapters I.9 and I.13–I.16.

FURTHER READING

Avise, John. 2000. Phylogeography: The History and Formation of Species. Cambridge, MA: Harvard University Press. *A review of methods to use molecular traits to infer microevolutionary history of conspecific populations.*

Bell, Michael A., and Susan A. Foster, eds. 1994. The Evolutionary Biology of the Threespine Stickleback. Oxford:

Oxford University Press. *A review of microevolutionary phenomena and mechanisms in this microevolutionary model species.*

Colosimo, P. R., K. E. Housemann, S. Balabhadra, G. Villarreal, Jr., M. Dickson, J. Grimwood, J. Schmutz, R. Myers, D. Schluter, and D. M. Kingsley. 2005. Widespread parallel evolution in sticklebacks by repeated fixation of *Ectodysplasin* alleles. Science 307: 1928–1933. *An extraordinary molecular analysis demonstrating that a single mutation created the variation on which natural selection has acted to produce a global pattern of geographic variation in a conspicuous skeletal trait.*

Endler, John A. 1986. Natural Selection in the Wild. Monographs in Population Biology 21. Princeton, NJ: Princeton University Press. *An encyclopediac review of natural selection.*

Endler, John A. 1995. Multiple-trait coevolution and environmental gradients in guppies. Trends in Ecology and Evolution 10: 22–29. *An overview of the microevolution of multiple phenotypes in guppies.*

Ewald, Paul W. 1994. Evolution of Infectious Disease. New York: Oxford University Press. *A ground-breaking book on pathogen microevolution.*

Grant, Peter R. 1986. Ecology and Evolution of Darwin's Finches. Princeton, NJ: Princeton University Press. *An account of microevolution and speciation placed within exceptionally thorough biological and environmental contexts.*

Hendry, A. P., and M. T. Kinnison. 2001. Microevolution: Rate, Pattern, Process. Dordrecht: Kluwer Academic Publishers. (Also published as Genetica, vol. 112–113.) *An excellent collection of review papers concerning rates, patterns, mechanisms, and ecological contexts for microevolution in contemporary populations.*

Lederberg, Joshua, and Esther M. Lederberg. 1952. Replica plating and indirect selection of bacterial mutants. Journal of Bacteriology 63: 399–406. *A classic study showing that mutations are random with respect to adaptation.*

Majerus, Michael E. N. 1998. Melanism: Evolution in Action. Oxford: Oxford University Press. *An excellent review of the classic case of contemporary microevolution, it exposed some methodological flaws, which, however, do not invalidate this case.*

Palumbi, Stephen R. 2001. The Evolution Explosion. New York: W. W. Norton & Company. *An excellent popular review of microevolution in response to human technology.*

Wilson, Edward O., and William H. Bossert. 1971. A Primer of Population Biology. Sunderland, MA: Sinauer Associates. *A concise quantitative treatment of basic population genetics.*

I.18

Ecological Speciation: Natural Selection and the Formation of New Species

Patrik Nosil and Howard Rundle

OUTLINE

1. Ecological speciation: What it is and how to test for it
2. Forms of divergent selection
3. Forms of reproductive isolation
4. Genetic mechanisms linking selection and reproductive isolation
5. Geography of ecological speciation
6. Generality of ecological speciation
7. Remaining questions in the study of ecological speciation

Understanding how new species arise is a central goal of evolutionary biology. Recent years have seen renewed interest in the classic idea that adaptive evolution within species and the origin of new species are intimately linked. More specifically, barriers to genetic exchange between populations (termed *reproductive isolation*) are the hallmark of species, and evolutionary biologists have been asking whether ecologically based divergent natural selection, the process that is responsible for adaptive divergence between populations, may cause such reproductive barriers to evolve. Convincing examples of this process, termed *ecological speciation*, are accumulating in the literature, and comparative approaches suggest that it may be a widespread phenomenon taxonomically. Attention is now being given to understanding details of the process and uncovering generalities in its operation. Three main components of ecological speciation can be recognized: a source of ecologically based divergent selection, a form of reproductive isolation, and a genetic mechanism linking the two. Current research is focused on understanding these components during the various stages of ecological speciation from initiation to completion.

GLOSSARY

ecologically based divergent selection. Selection arising from environmental differences and/or ecological interactions (e.g., competition) that acts in contrasting directions on two populations (e.g., large body size confers high survival in one environment and low survival in the other) or favors opposite extremes of a trait within a single population (i.e., disruptive selection)

linkage disequilibrium. A statistical association between alleles at one locus and alleles at a different locus, the consequence of which is that selection on one locus (e.g., a locus affecting an ecological trait such as color pattern) causes a correlated evolutionary response at the other locus (e.g., a locus affecting mating preference)

pleiotropy. Multiple phenotypic effects of a gene (e.g., a gene affecting color pattern also affects mating preferences)

postmating isolation. Barriers to gene flow that act after mating (e.g., intermediate trait values of hybrids that make them poor competitors for resources, reducing their fitness)

premating isolation. Barriers to gene flow that act before mating (e.g., divergent mate preferences that prevent copulation between individuals from different populations)

reproductive isolation. A reduction or lack of genetic exchange (gene flow) between taxa

sympatric speciation. A geographic mode of speciation whereby a single population splits into two species in the absence of any geographic separation, often via disruptive selection

1. ECOLOGICAL SPECIATION: WHAT IT IS AND HOW TO TEST FOR IT

The idea that the macroevolutionary phenomenon of speciation is the result of the microevolutionary process of adaptation dates back at least to Charles Darwin. However, it was not until the popularization of the biological species concept in the middle of the last century, whereby speciation was defined as the process by which barriers to genetic exchange evolve between populations, that the study of Darwin's "mystery of mysteries," the origin of species, became empirically tractable. The past two decades have witnessed an explosion of speciation research, with much attention being given to understanding the role of divergent selection in speciation.

As defined by Dolph Schluter and others, *ecological speciation* is the process in which barriers to genetic exchange evolve between populations as a result of ecologically based divergent natural selection. Selection is ecological when it arises from differences in the environment or from interactions between populations over resource acquisition. Ecologically based selection can thus arise, for example, from an individual's quest to obtain food and other nutrients, attract pollinators, or avoid predators. It can also arise from an individual's interaction with other organisms in its attempt to achieve these goals (e.g., resource competition, predation). Selection is divergent when it acts in contrasting directions in the two populations. Included here is the special case of disruptive selection on a single population, in which selection favors opposite extremes of the same trait. During ecological speciation, populations experience divergent selection between environments or niches and thus differentiate in ecologically important traits. If these traits, or ones that are genetically correlated with them, affect reproductive isolation, then speciation occurs as a consequence.

Ecological speciation is distinguished from other models of speciation in which the evolution of reproductive isolation involves processes other than ecologically based divergent selection. These include models in which chance events play a central role, including speciation by polyploidization, hybridization, genetic drift, and population bottlenecks (i.e., drastic reductions in population size). Nonecological speciation also includes models in which selection is involved, but it is nonecological (e.g., sexual conflict, in which selection arises from an evolutionary conflict of interest between the sexes over traits related to reproduction), or it is not divergent between environments.

An alternative definition of ecological speciation would restrict it to situations in which the reproductive barriers themselves are ecological in nature, such as reduced hybrid fitness arising because intermediate hybrid phenotypes cause them to perform poorly in either parental environment (i.e., see the third point below). In contrast, incompatibilities between the parental genomes, expressed when they are brought together in hybrids, is an example of a nonecological barrier. However, when the goal is to understand mechanisms of speciation, it is of interest when both ecological and nonecological forms of reproductive isolation evolve through a specific evolutionary process (e.g., ecologically based divergent selection). Ecological speciation can therefore involve the evolution of any type of reproductive barrier, so long as ecologically based divergent selection is responsible. Ecological speciation can also occur under any spatial arrangement of populations, with population pairs being geographically separated (*allopatry*), contiguous (*parapatry*), or in complete contact (*sympatry*). Under any of these geographic scenarios, if divergent selection drives the evolution of reproductive isolation, then speciation is classified as ecological.

Laboratory evolution experiments using *Drosophila* fruit flies have shown that ecological speciation is feasible: when replicate populations are independently adapted to one of two environments, reproductive isolation tends to arise between populations from different environments, but not between populations evolved in similar environments. Classic examples of such experiments come from the work of Diane M. B. Dodd and of George Kilias and colleagues (figure 1). Convincing examples of ecological speciation in nature are also accumulating, with empirical tests tending to focus on three forms of evidence.

First, ecological speciation predicts that the strength of reproductive isolation between pairs of populations will be positively related to the magnitude of their ecological differentiation, independent of any correlation with divergence time (figure 1). This has been shown in *Timema* walking-stick insects studied by Patrik Nosil, Bernie J. Crespi, and Cristina P. Sandoval, in which pairs of populations adapted to different host-plant species exhibit stronger reproductive isolation than do pairs of populations adapted to the same host-plant species (figure 1). A special case of this scenario, termed parallel speciation, occurs when the same reproductive barriers evolve in independent populations experiencing similar environments. The *Drosophila* laboratory experiments described above demonstrate the initial stages of parallel speciation: independent populations adapted to one environment were reproductively isolated from populations adapted to the other environment but not from one another. A prime example of parallel speciation in nature comes from freshwater stickleback (*Gasterosteus*) fishes studied by

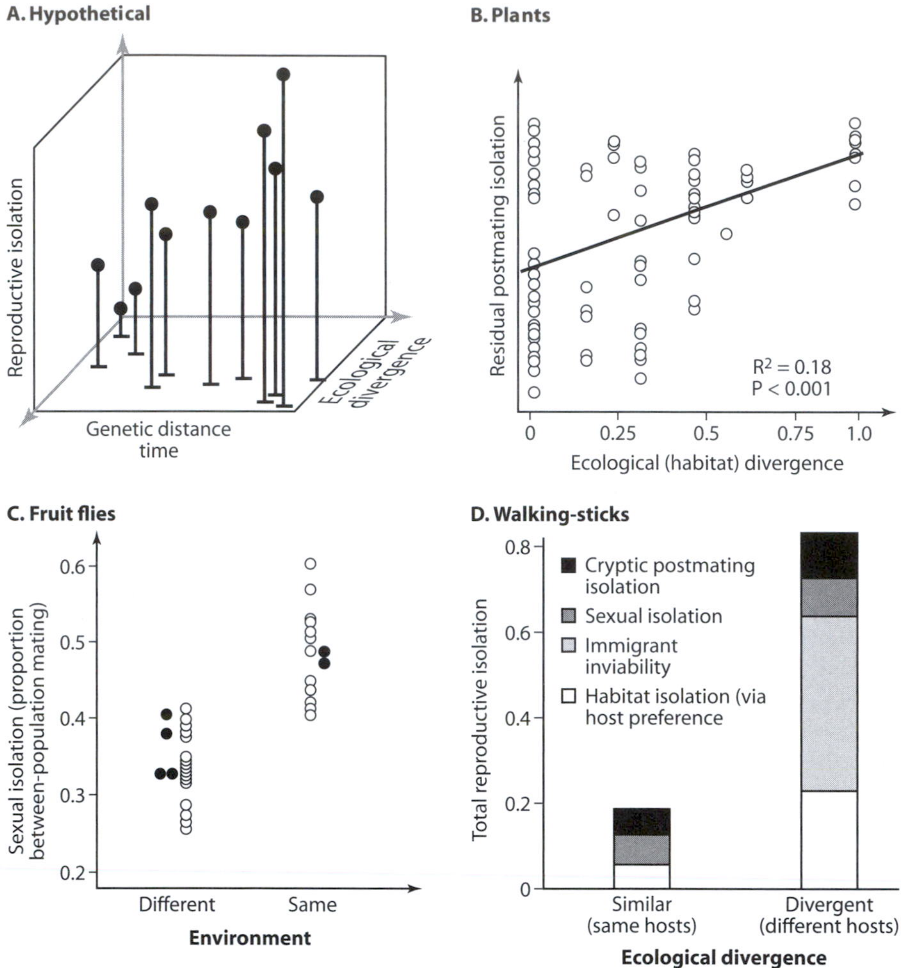

Figure 1. Tests for ecological speciation, where the premise is to isolate an association between ecological divergence and levels of reproductive isolation, independent of the amount of time that population pairs have had to diverge from one another via non-ecological processes such as genetic drift. (A) A hypothetical scenario in which reproductive isolation increases with both genetic distance (a proxy for time) and ecological divergence. (B) The pattern predicted by ecological speciation, in this case a positive association between habitat divergence and residual postmating isolation (the effects of time have been statistically removed) between angiosperm species. The data come from a comparative study by Funk and colleagues, and the figure is reprinted with permission of the National Academy of Sciences U.S.A. (C) Evidence for ecological speciation from laboratory evolution studies. Shown is the proportion of matings occurring between independently evolved lines of *Drosophila* as a function of the similarity of their environments. Between-population mating is less common when populations have been adapted to different environments. Open circles are from work by Dodd, and closed circles are from work by Kilias and colleagues. (D) In *Timema cristinae* walking-stick insects, multiple forms of reproductive isolation are stronger between pairs of populations using different host-plant species (i.e., pairs with divergent ecologies) than between similar-aged pairs of populations using the same host-plant species (i.e., pairs with similar ecologies). The pattern was documented in a series of studies by Nosil, Crespi, and Sandoval, and the figure is reprinted with permission of the American Society for Naturalists.

Dolph Schluter, Howard D. Rundle, and colleagues. These fish come in two main forms, a slender limnetic that feeds primarily on plankton in the open water of a lake and a more robust benthic, which feeds on invertebrates in the shallows. Sympatric limnetic–benthic pairs occur in a number of lakes in western Canada, and molecular genetic evidence suggests that the pairs have arisen independently (i.e., the present-day phenotypic similarity of limnetics, and of benthics, from separate lakes is the result of parallel evolution and not shared ancestry). Mating trials demonstrate that reproductive isolation between limnetics and benthics has likewise evolved in parallel: premating isolation is strong between limnetics and benthics, even when they

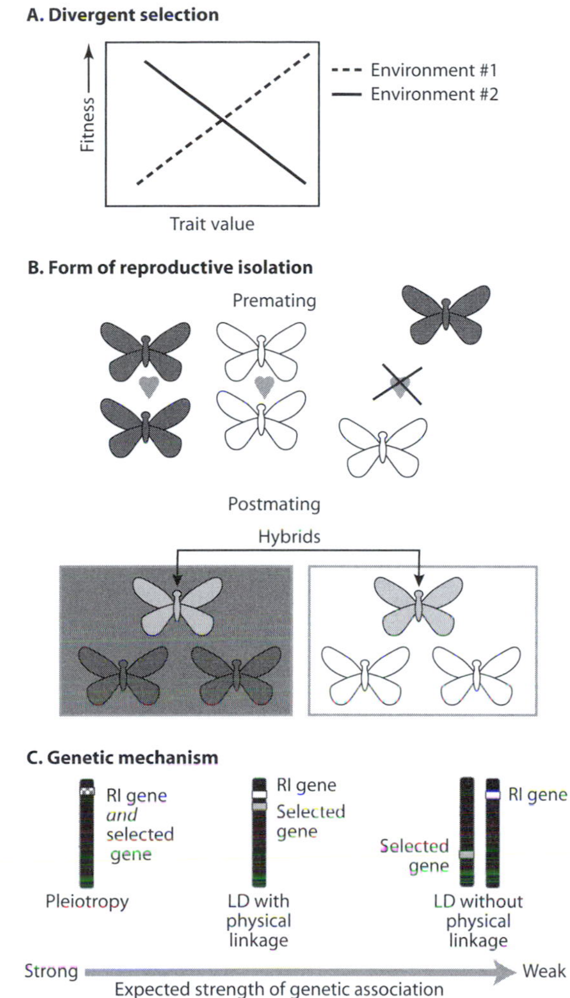

Figure 2. A schematic illustration of the three components of ecological speciation. (A) A form of divergent selection is required, where selection is divergent when it acts in contrasting directions in two populations. (B) Forms of reproductive isolation are numerous and can act either before or after mating (premating and postmating isolation, respectively). Depicted are butterflies from populations adapted to different habitats (dark gray versus white-winged individuals) and hybrids between these parental forms (light gray). The form of premating isolation depicted is sexual isolation, where individuals prefer to mate with individuals that are the same color as themselves. The form of postmating isolation shown is one where hybrids suffer reduced fitness because their intermediate phenotype renders them unfit in either parental environment. Specifically, the hybrids do not match the substrate in either parental environment and suffer increased rates of visual predation as a result of this lack of crypsis. (C) A genetic mechanism is required to transmit selection on genes conferring local adaptation to genes causing reproductive isolation. Reproductive isolation can evolve because the genes under selection are the same as those conferring reproductive isolation (i.e., pleiotropy). Alternatively, reproductive isolation might evolve via the statistical association between genes under selection and those conferring reproductive isolation. This statistical association is termed *linkage disequilibrium* and is facilitated by proximity of the different genes on the same chromosome (i.e., physical linkage).

are taken from separate lakes, whereas premating isolation is absent within a form, even when they derive from different lakes (e.g., premating isolation is lacking between limnetic forms from different lakes). The repeated evolution of the same barriers to gene flow, in correlation with ecological divergence, is unlikely to occur via nonecological processes (e.g., genetic drift) and thus provides strong comparative evidence for ecological speciation.

Second, ecological speciation is facilitated when the traits under divergent selection also cause reproductive isolation pleiotropically; such pleiotropy is therefore predicted to be common in cases of ecological speciation. A clear example comes from the work of Jim Mallet, Chris Jiggins, and colleagues on butterflies in the genus *Heliconius*. In this group of tropical butterflies, natural selection acts on mimetic coloration to reduce visual predation. Geographic variation in the phenotype of the comimic generates divergent selection among populations to match the local form. Because these color patterns are also used in mate choice, divergence in coloration generates premating isolation (i.e., sexual isolation) as a side effect.

Third, ecologically based divergent selection predicts that if hybrids can be formed between the parental populations, their fitness should be reduced for ecological reasons. This normally occurs because hybrid phenotypes are intermediate between the two parental forms, making them ill-suited for various tasks (e.g., acquiring resources, avoiding predation, finding a mate) in either parental environment. This type of hybrid fitness reduction is unlikely to arise via nonecological mechanisms of speciation. Ecologically dependent reductions in growth rate have been shown in limnetic–benthic hybrids, and reduced survival of *Heliconius* hybrids is likely to arise from their intermediate coloration that fails to mimic either parental form.

As the above studies highlight, the case for ecological speciation is compelling: it clearly can, and does, occur. Attention is now being given to understanding the details of the process, including the three main components (a source of divergent selection, a form of reproductive isolation, and a genetic mechanism linking the two; figure 2), to testing for generalities, and to uncovering the geographic context of ecological speciation.

2. FORMS OF DIVERGENT SELECTION

The first component of ecological speciation is a source of divergent selection (figure 2A). Three main sources have been recognized: differences between environments, ecological interactions, and sexual selection.

Differences between Environments

Divergent selection can arise because of differences between populations in their environments, including, for example, habitat structure, climate, and resource availability. As populations adapt to different environments, they may diverge from one another in many ways, evolving to look different, smell different, and behave differently. Such differences will contribute to speciation if, for example, they reduce the likelihood of between-population mating (perhaps because individuals from different populations no longer recognize one another as potential mates as in the *Heliconius* butterflies discussed above), or they reduce the fitness of any hybrids that are formed (perhaps, as mentioned earlier, because such hybrids are ill-suited to either parental niche, as in the stickleback example discussed above). It is highly unlikely that any two environments are identical, and it is not surprising, therefore, that environmental differences appear to be a common cause of divergent selection during ecological speciation.

Ecological Interactions

Divergent selection may also arise between populations as a result of their ecological interactions with one another, most notably competition for shared resources. Divergent selection arising from such interactions is frequency dependent because individual fitness depends on the frequency of the various phenotypes within the population. Although interspecific competition appears common in nature and may play a key role in sympatric divergence between taxa, its consequences for the evolution of reproductive isolation are poorly understood. In the threespine sticklebacks discussed earlier, for example, resource competition has been strongly implicated in the morphological divergence of limnetics and benthics (i.e., ecological character displacement), as has adaptation to their different environments. Although the latter form of divergent selection has been implicated in the evolution of reproductive isolation, unambiguous evidence that the former promotes reproductive isolation is lacking.

Interbreeding (hybridization), another type of interaction between populations, can also contribute to ecological speciation via a process known as *reinforcement*. Reinforcement occurs when hybrids have reduced fitness such that selection favors parental individuals that are less likely to hybridize, thereby strengthening premating isolation. Although it features prominently in many models of speciation, reinforcement is difficult to categorize because it can complete a speciation process initiated by any mechanism, ecological or not. However, if hybrid fitness is reduced by ecological means, then reinforcement can be considered a component of ecological speciation. Reinforcement has been implicated in the ecological speciation of limnetic and benthic threespine sticklebacks: postmating isolation is ecological in nature, and premating isolation in sympatry appears to have been strengthened in response.

Sexual Selection

Sexual selection has long been hypothesized to be a powerful mechanism of speciation because it involves communication between a signaler and a receiver, thereby creating the potential for rapid coevolutionary diversification of mating signals and preferences that may generate reproductive isolation. Divergent sexual selection arises when mate preferences differ between populations. Such selection is considered a component of ecological speciation when it is initiated by divergent selection between environments. This can occur, for example, if two habitats vary in their signal transmission properties such that different signals are most detectable (i.e., favored by natural selection) within each. Different mating signals and preferences may then evolve in populations occupying either habitat. For example, Manuel Leal and Leo J. Fleishman studied populations of *Anolis* lizards that occupy different (mesic versus xeric) habitats. They found that light conditions differ between these habitats and that the dewlap spectral traits of the lizards, which are important for social and mating communication, have diverged between populations using different habitats in ways that increase signal detectability within the native habitat, potentially generating premating isolation between these populations. Divergence of mating signals can also occur if sensory or communication systems adapt to their specific environment, even outside of the mating context (e.g., to facilitate resource acquisition or predator avoidance). Several examples of divergence in display traits, sensory systems, or preferences in correlation with environment have now been reported.

3. FORMS OF REPRODUCTIVE ISOLATION

The second component of ecological speciation is the form of reproductive isolation, of which many are possible, and speciation may involve one or more of them. Forms of reproductive isolation are commonly classified according to whether they occur before or after mating (premating and postmating isolation, respectively; figure 2B). The role of these reproductive barriers in speciation was thoroughly reviewed in a recent book on speciation by Jerry Coyne and H. Allen Orr (2004).

Premating Isolation

Premating isolation can arise when populations are separated in space (habitat) or time. Habitat isolation occurs when populations exhibit genetically based preferences for separate habitats, reducing the likelihood of between-population encounters and thus of interbreeding. For example, divergent host-plant preferences cause partial reproductive isolation between many herbivorous insect populations that mate on the plant on which they feed. Temporal isolation occurs when populations exhibit divergent developmental schedules such that mating happens at different times in each. A classic empirical example of such forms of reproductive isolation comes from the apple and hawthorne host races of *Rhagoletis* flies studied by Guy Bush, Jeffery L. Feder, and colleagues. Differences in host-plant preferences and developmental schedules cause substantial reproductive isolation between these host races. Additionally, if individuals immigrating into a foreign habitat are maladapted and die before mating, this *immigrant inviability* will also act to reduce interbreeding. All these forms of reproductive isolation are inherently ecological and thus are expected to commonly play a role in ecological speciation.

Another barrier that acts before mating is sexual isolation, arising from differences between populations in their mating signals and preferences. Sexual isolation is considered by many to be the main component of reproductive isolation between recently evolved taxa. Consistent with this, studies in cichlids and *Drosophila* have shown that sexual isolation appears necessary for species to coexist in nature. Likewise, the laboratory *Drosophila* experiments discussed above demonstrate that sexual isolation can evolve relatively rapidly when populations are subjected to divergent natural selection (figure 1). A number of examples from nature also exist in which adaptation to different environments has been implicated in the evolution of sexual isolation, including beetles, walking sticks, butterflies, and stickleback fish. For example, in the stickleback fish discussed previously, adaptation to their different habitats (open water versus shallows) causes divergence in body size, and because mate choice is assortative with respect to size, sexual isolation arises as a by-product.

Postmating Isolation

Postmating isolation can arise when hybrid fitness is reduced because of an ecological mismatch between intermediate hybrid phenotypes and the environment, as was discussed earlier in the evidence for ecological speciation. An example of such ecologically dependent reductions in hybrid fitness stems from work on limnetic–benthic sticklebacks. Hybrids between the limnetic and benthic forms exhibit high fitness in the laboratory. In contrast, the fitness of hybrids in the wild is reduced relative to parental forms. Use of various types of hybrid crosses has shown that this reduction was a direct result of their intermediate phenotype and was not caused by genetic incompatibilities between the two forms (that could arise via any mechanism of speciation). Hybrids can also suffer reduced fitness because their sexual display traits and/or mate preferences reduce their mating success, in effect generating sexual selection against them. This has been shown in hybrid male sticklebacks in work by Steven Vamosi.

Postmating isolation can also result from genetic incompatibilities between divergent genomes, caused by negative interactions between genes that differ between populations, when these genes are brought together in hybrids. These incompatibilities reduce the fitness of hybrids and do not depend on an ecological interaction between phenotype and environment. However, it is still possible that such incompatibilities evolve as a by-product of ecologically based divergent natural selection, for example, if alleles favored by selection within each population are incompatible with one another when brought together in the genome of a hybrid.

4. GENETIC MECHANISMS LINKING SELECTION AND REPRODUCTIVE ISOLATION

The final component of ecological speciation is the genetic mechanism by which selection on ecological traits is transmitted to the genes causing reproductive isolation, thereby driving the evolution of the latter. There are two ways this can occur, distinguished by the relationship between the genes under divergent selection (i.e., those affecting ecological traits) and those causing reproductive isolation (figure 2C). In the first, these genes are the same (e.g., a gene affecting color pattern pleiotropically affects mate preference). In this case, reproductive isolation is said to evolve by direct selection because the alleles responsible for reproductive isolation are themselves under selection, albeit for another reason. In the second, the genes under divergent selection are physically different from those causing reproductive isolation. In this case, reproductive isolation is said to evolve by indirect selection because selection acts on genes causing reproductive isolation only to the extent that they are nonrandomly associated (i.e., in linkage disequilibrium) with the genes directly under selection. When selection acts on genes affecting ecological traits, such nonrandom associations will cause a correlated evolutionary response in genes conferring reproductive isolation. The nature of these genetic associations is important because it affects the strength of selection transmitted to the genes affecting reproductive isolation.

Pleiotropy and Direct Selection

Speciation is facilitated when genes under divergent selection cause reproductive isolation pleiotropically, and there are numerous ways this can occur. These include, for example, habitat isolation that evolves as a direct consequence of selection on genes affecting habitat choice. Selection might also act on ecological traits that incidentally affect mate preferences; the *Drosophila* lab experiments discussed previously suggest that this is not an unlikely occurrence, and the previously discussed mimetic color patterns in tropical butterflies of the genus *Heliconius* provide a classic example from nature. In plants, adaptation to different pollinators can cause premating isolation as a side effect. For example, work by Douglas Schemske and Toby Bradshaw has shown that divergent natural selection acts on a flower color gene in *Mimulus* monkeyflowers via the effects of color on attractiveness to pollinators. In *Mimulus lewisii*, pink-colored flowers are favored by bumblebees and discriminated against by hummingbirds. In contrast, *M. cardinalis* has red flowers, which are favored by hummingbirds and discriminated against by bumblebees. Adaptation to different pollinators via divergence in this flower color gene therefore directly affects the probability of cross-pollination (i.e., hybridization), a form of sexual isolation.

Other forms of reproductive isolation could also involve pleiotropy. For example, temporal isolation, caused by differences in flowering time could arise as a pleiotropic effect of adaptation to different environments, whereas postmating isolation can arise pleiotropically if alleles favored by selection within each population contribute to genetic incompatibilities in hybrids.

Linkage Disequilibrium and Indirect Selection

Indirect selection is less effective than direct selection in the evolution of reproductive isolation. This is because the genetic association between the two sets of genes (i.e., linkage disequilibrium) is not perfect, thereby reducing the strength of selection transmitted to the genes causing reproductive isolation (a phenomenon that has been likened to the slipping of a car's clutch in that the wheels experience only a fraction of the power provided by the engine). The amount of linkage disequilibrium that exists can be affected by three factors. The first is the genetic basis of reproductive isolation, of which there is an important distinction between what are termed one-allele and two-allele mechanisms. In a one-allele mechanism, reproductive isolation arises from the fixation of the same allele in both populations (e.g., an allele causing individuals to prefer mates phenotypically similar to themselves, for example, individuals similar in body size). In a two-allele mechanism, different alleles fix in each population (e.g., a preference allele for large individuals in one population and a preference allele for small individuals in the other). This distinction is important because, in a two-allele mechanism, recombination will tend to break down linkage disequilibrium between the genes under divergent selection and those causing reproductive isolation. In contrast, recombination creates no such problem for a one-allele mechanism, and it is therefore a more powerful mechanism of speciation. The prevalence of these genetic mechanisms in nature is unknown.

The second factor is physical linkage. The maintenance of linkage disequilibrium is greatly facilitated by the physical linkage of genes on a chromosome because the likelihood of a recombination event declines with decreasing genetic map distance. Chromosomal inversions may thus play a role in ecological speciation by suppressing recombination, thus physically linking large regions of the genome.

The third factor is the strength and form of selection. Linkage disequilibrium can be maintained by strong selection that favors specific combinations of genes (i.e., correlational selection), and such selection may be important during sympatric speciation.

In general, data examining the relationship between genes under divergent selection and those causing reproductive isolation are sparse. In practice, separating pleiotropy from close physical linkage will be a difficult task, although their effects may ultimately be very similar. Important questions are how common pleiotropy and tight physical linkage are and how often they are of the form that would facilitate ecological speciation. Finally, we note that almost nothing is known about the types of genes involved in ecological speciation. Information on the genetics underlying ecological speciation may improve our mechanistic understanding of its operation in nature, including the type of genes involved and how they cause reproductive isolation.

5. GEOGRAPHY OF ECOLOGICAL SPECIATION

Traditionally, speciation has been classified not by the mechanisms responsible (e.g., ecological speciation) but rather by the geographic context under which it occurs. These include allopatric, parapatric, and sympatric, with the latter being especially controversial and garnering much attention. Ecological speciation, however, can occur under any of these geographic contexts. The divergence of allopatric populations is unimpeded by the constraining effects of gene flow, so reproductive isolation is eventually expected to arise between them from chance events (e.g., genetic drift). Ecologically based divergent selection, however, can greatly accelerate this process and may commonly do so because allopatric pairs of populations often occupy different environments and are therefore subject to divergent selection. Sympatric speciation, in contrast, represents the opposite extreme in which speciation occurs in the absence of any geographic isolation. Strong disruptive selection is therefore required to overcome the homogenizing effects of gene flow, and such selection is expected to be ecological in origin. Ecological speciation is therefore a likely mechanism of sympatric speciation. Parapatric speciation represents an intermediate scenario in which gene flow is reduced but not eliminated by geographic barriers (including distance). Divergent selection is again required to overcome the effects of gene flow, making ecological speciation a likely mechanism.

Although attractive, the classification of speciation into these distinct geographic contexts may be overly simplistic and fail to capture the complexity of some speciation events in nature. Ecological speciation, for example, may often occur in stages that involve different geographic contexts (figure 3). The idea is that speciation begins when populations are allopatric, with reproductive isolation accumulating as a by-product of adaptation to their different environments. The second stage is initiated on secondary contact (parapatry or sympatry), with genetic exchange becoming possible at this point. Although the resulting gene flow is generally thought to constrain adaptive divergence and hamper speciation, ecological interactions are added as a source of divergent selection, and reinforcement also becomes possible. The amount of reproductive isolation that evolves during each stage indicates the primary geographic context of speciation, with the classic scenarios of allopatric and sympatric speciation representing the extremes in which, before secondary contact, reproductive isolation was either essentially complete or absent, respectively. Intermediate scenarios may be more common in nature, however, as suggested in the speciation of limnetic and benthic sticklebacks, which appears to have involved some reproductive isolation evolving during both phases. More complex scenarios are also possible, as suggested by recent molecular data from the apple and hawthorn races of *Rhagoletis* flies,

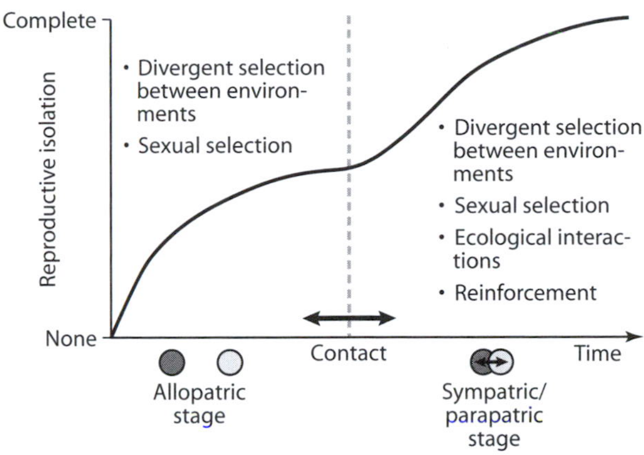

Figure 3. A scenario for ecological speciation under various geographic contexts. Reproductive isolation between two populations is absent at the beginning of the speciation process (at left) and evolves to completion (at right), as indicated by the solid line. The geographic context of speciation is indicated by the position of the dashed vertical line dividing the allopatric stage (left) from the sympatric/parapatric stage (right). This division can occur at any point in time, thus accommodating a range of geographic contexts and either a one- or two-stage process. For example, the extreme case of fully sympatric speciation occurs when the allopatric stage is absent (division line coincides with the *y*-axis). The opposite extreme of fully allopatric speciation occurs when reproductive isolation has evolved to completion before secondary contact occurs (division line falls at the right-hand extreme). Depicted is an intermediate, two-stage scenario in which partial reproductive isolation evolves in allopatry, but reproductive isolation does not evolve to completion until after secondary contact has occurred. The ecological causes of divergent selection by which reproductive isolation may evolve are listed within the panel for each stage. The figure is modified from a review of ecological speciation by Rundle and Nosil and reprinted with the permission of the British Ecological Society.

traditionally put forward as a classic case of sympatric speciation. Feder and colleagues (2004) have shown that inversion polymorphisms, containing genetic variation affecting ecologically important diapause traits, trace their origins to allopatric populations in Mexico.

6. GENERALITY OF ECOLOGICAL SPECIATION

Comparative approaches can be used to investigate generalities of ecological speciation, including which, if any, forms of reproductive isolation are common, in what order they tend to arise, and the forms of divergent selection that drove their evolution. Likewise, the taxonomic generality of ecological speciation can also be explored using comparative approaches, as was done in a recent study by Daniel J. Funk, Patrik Nosil, and William B. Etges (2006). Using published data involving more than 500 species pairs of plants, invertebrates, and vertebrates, they found a positive association between ecological divergence and reproductive isolation for seven of eight groups. By controlling for divergence time using published genetic data, their results suggest that ecological speciation may be taxonomically widespread.

7. REMAINING QUESTIONS IN THE STUDY OF ECOLOGICAL SPECIATION

There is convincing evidence from the laboratory and nature that divergent natural selection can drive the evolution of reproductive isolation. Ecological speciation therefore most certainly occurs. Current research is aimed at determining the ecological sources of divergent selection, the types of reproductive barriers involved, and the genetic mechanisms linking them. Attention is also being given to the generality of the findings associated with these components.

There is much work remaining. Insufficient attention has been given to understanding the contribution of ecological interactions and sexual selection to the evolution of reproductive isolation. The relative importance of various forms of reproductive isolation and the likelihood that each evolves via divergent selection are also not well resolved. Direct tests of the genetic link between traits under selection and those conferring reproductive isolation are lacking too. Finally, the factors affecting the degree of progress toward the completion of ecological speciation are poorly understood. We hope that studies using a diversity of taxa and examining a wide range of divergence, from incipient to established species, will shed light on how ecological speciation unfolds from beginning to end.

FURTHER READING

Bradshaw, Toby, and Douglas Schemske. 2003. Allele substitution at a flower colour locus produces a pollinator shift in monkeyflowers. Nature 426: 176–178.

Coyne, Jerry, and H. Allen Orr. 2004. Speciation. Sunderland, MA: Sinauer Associates.

Feder, Jeffery, Stewart Berlocher, Joe Roethele, Hattie Dambroski, James Smith, William Perry, Vesna Gavrilovic, Kenneth Filchak, Juan Rull, and Martin Aluja. 2003. Allopatric genetic origins for sympatric host-plant shifts and race formation in *Rhagoletis*. Proceedings of the National Academy of Sciences U.S.A. 100: 10314–10319.

Funk, Daniel, Patrik Nosil, and William Etges. 2006. Ecological divergence exhibits consistently positive associations with reproductive isolation across disparate taxa. Proceedings of the National Academy of Sciences U.S.A. 103: 3209–3213.

Jiggins, Chris, Russell Naisbit, Rebecca Coe, and James Mallet. 2001. Reproductive isolation caused by colour pattern mimicry. Nature 411: 302–305.

Leal, M., and Leo J. Fleishman. 2004. Differences in visual signal design and detectability between allopatric populations of Anolis lizards. American Naturalist 163: 26–39.

Mayr, Ernst. 1963. Animal Species and Evolution. Cambridge, MA: Harvard University Press.

Nosil, Patrik. 2007. Divergent host-plant adaptation and reproductive isolation between ecotypes of *Timema cristinae*. American Naturalist 169: 151–162.

Rundle, Howard, Laura Nagel, Janette Boughman, and Dolph Schluter. 2000. Natural selection and parallel speciation in sympatric sticklebacks. Science 287: 306–308.

Rundle, Howard, and Patrik Nosil. 2005. Ecological speciation. Ecology Letters 8: 336–352.

Schluter, Dolph. 2000. Ecology of Adaptive Radiation. Oxford: Oxford University Press.

I.19

Adaptive Radiation

Rosemary Gillespie

OUTLINE

1. Conditions promoting adaptive radiation
2. Are certain taxa more likely to undergo adaptive radiation?
3. Examples of current adaptive radiations
4. Initiation of adaptive radiation
5. Speciation in adaptive radiation
6. Community assembly
7. Molecular basis for adaptive change
8. Testing adaptive radiation

Adaptive radiation is generally triggered by the appearance of available niche space, which could result from (1) *intrinsic factors* or key innovations that allow an organism to exploit a novel resource, and/or (2) *extrinsic factors*, in which physical ecological space is created as a result of climatic changes or the appearance de novo of islands. There are no general rules as to what taxa are more likely to undergo adaptive radiation, although some lineages may have certain attributes that facilitate adaptive radiation in the appropriate setting. The process of adaptive radiation is described below, as well as some prime examples of the phenomenon. Adaptive radiation is generally initiated by expansion of ecological amplitude of a taxon into newly available ecological space, followed by specialization, the process possibly facilitated through adaptive plasticity. Speciation associated with adaptive radiation may involve one or more of the following: founder events, divergent natural selection, sexual selection, and hybridization. Competition is generally implicated in divergent natural selection and in dictating the communities of species formed during the course of adaptive radiation. Current research is focused on (1) examining the molecular underpinnings of apparently complex morphological and behavioral changes that occur during the course of adaptive radiation, and (2) experimental manipulation of bacteria to assess the conditions under which adaptive radiation occurs.

GLOSSARY

adaptive radiation. Rapid diversification of an ancestral species into several ecologically different species, associated with adaptive morphological, physiological, and/or behavioral divergence

attenuation. Decline in number of species represented on islands with distance from a source of colonists

divergent natural selection. Selection arising from environmental forces acting differentially on phenotypic traits (morphology, physiology, or behavior) resulting in divergent phenotypes; reproductive isolation may occur as a side effect, either in sympatry or allopatry

ecological character displacement. Divergence in ecological traits (which may lead to reproductive isolation as a by-product) caused by competition for shared resources

ecological release. Expansion of habitat or use of resources by populations into areas of lower species diversity with reduced interspecific competition

ecological speciation. Process by which barriers to gene flow evolve between populations as a result of ecologically based divergent natural selection

ecomorph. A group of populations, species, etc., whose appearance is determined by the environment

escalation/diversification. Diversification of a herbivore/parasite in concert with its host in which the adaptations of the host to counter exploitation by the herbivore or parasite build one on each other, and vice versa

escape and radiation. Diversification of a herbivore/parasite in concert with its host in which the host is generally considered to radiate before exploitation and subsequent radiation by the herbivore or parasite, and vice versa

founder event. Establishment of a new population with few individuals that contain a small, and hence unrepresentative, portion of the genetic diversity relative to the original population, potentially leading to speciation

key innovation. Any newly acquired structure or property that permits the occupation of a new environment, or performance of a new function, which, in turn, opens a new adaptive zone

nonadaptive radiation. Elevated rate of speciation in the absence of noticeable ecological shifts

sexual selection. Form of natural selection based on an organism's ability to mate such that individuals with attributes that allow them greater access to the opposite sex, either through (1) combat with the same sex or (2) attributes that render them more attractive to the opposite sex, mate at higher rates than those that lack these attributes

taxon cycle. Temporal sequence of geographic distribution of species from (1) colonizing to (2) differentiating to (3) fragmenting and to (4) specializing

Adaptive radiation is the rapid diversification of a lineage into multiple ecologically different species, generally associated with morphological or physiological divergence. The phenomenon can be characterized by four criteria: common ancestry, a phenotype–environment correlation, trait utility, and rapid speciation. The concept of adaptive radiation, and particularly diversification of ecological roles by means of natural selection, has had a long history, beginning with the observations of Charles Darwin on the Galápagos Islands, and has played a pivotal role in the development of the Modern Synthesis (see Givnish and Sytsma, 1997, for a detailed history of the concept).

1. CONDITIONS PROMOTING ADAPTIVE RADIATION

The most familiar present-day adaptive radiations are known from isolated archipelagoes or similar island-like settings (e.g., lakes). However, it is quite likely that much of the diversity of life originated through episodes of adaptive radiation during periods when ecological space became available for diversification. Within this context there are two primary mechanisms through which ecological space can become available: (1) intrinsic changes in the organism often associated with key innovations, and (2) extrinsic effects, including environmental change and colonization of isolated landmasses. The two situations may be linked; for example, an intrinsic change may allow an organism to exploit a new environment. In either case, individuals exploiting the newly available niche space must be isolated to some extent from the remainder of the population to allow for genetic divergence associated with adaptive radiation. If a new habitat becomes available in close proximity to other such habitats (not isolated), it will be colonized repeatedly by taxa from those habitats, and patterns of species diversity will be governed by ecological processes of immigration and extinction, rather than by evolutionary processes.

Intrinsic Factors: Key Innovations

A key innovation is a trait, or a suite of traits, that allows an organism to exploit a novel resource or increase the efficiency with which a resource is used, thereby allowing a species to enter a "new" adaptive zone; the ecological opportunity thus provided may permit diversification. For example, the evolution of C4 photosynthesis, which enhanced rates of carbohydrate synthesis in open environments, likely served as a key innovation preceding radiation of most of the major lineages of grasses. Among recent adaptive radiations, one of the best-known key innovations is that of the pharyngeal jaw apparatus of cichlid fish (figure 1). Although most bony fishes have pharyngeal gill arches modified to process prey, the cichlid pharyngeal jaw is unique in having upper pharyngeal jaw joints, a "muscular sling," and suturing that functionally fuses the lower pharyngeal jaw. The features of the jaw appear to have allowed cichlids to exploit a diversity of prey types, including large fish and hard-shelled prey that are unavailable to most other aquatic vertebrates. In flowering plants, floral spurs have evolved at least seven times, each time resulting in higher rates of diversification, perhaps the best known radiation being in the columbines (*Aquilegia*, Ranunculaceae).

Interacting species, such as herbivores or parasites and their hosts, may themselves create ecological opportunity and provide some notable examples of key innovations. Here, the host may develop a defense to the herbivore or parasite (e.g., a plant may develop toxicity), but in due course the herbivore or parasite may develop resistance to the defense of the host. Paul Ehrlich and Peter Raven (1964) examined such coevolutionary responses and hypothesized that when plant lineages are temporarily freed from herbivore pressure via the origin of novel defenses, they enter a new adaptive zone in which they can undergo evolutionary radiation. However, if a mutation then arises in a group of insects that allows them to overcome these defenses and feed on the plants, the insect would then be free to diversify on the plants in the absence of competition. The major radiations of herbivorous insects and plants may have arisen through repeated steplike opening of novel adaptive zones that each has presented to the other over evolutionary time. Often referred to as escape and radiation, the host is generally considered to radiate before exploitation and subsequent radiation by the herbivore or parasite. This idea has been supported by

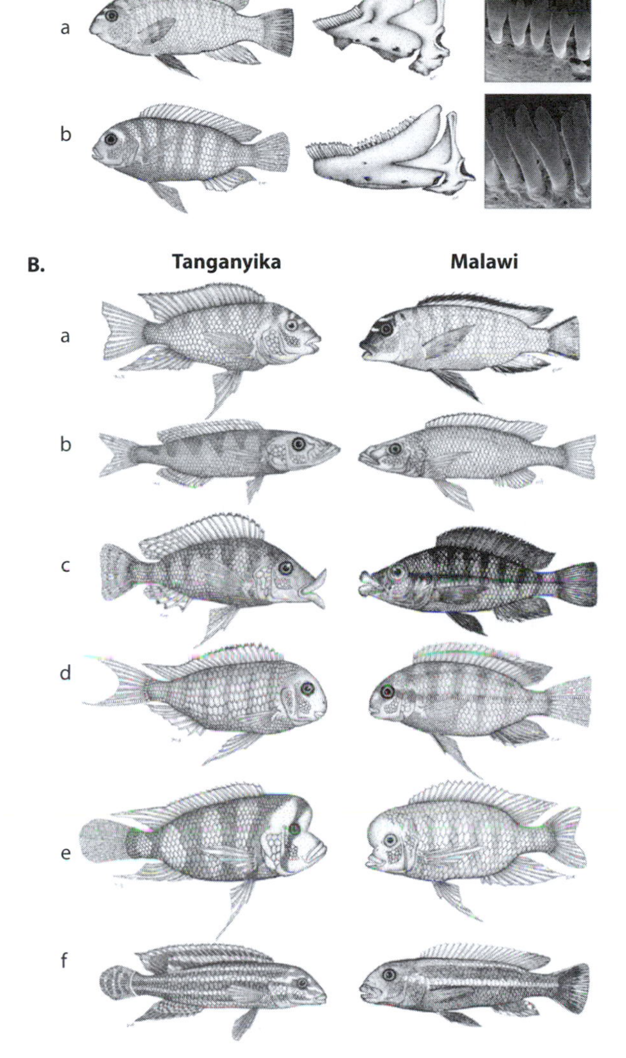

Figure 1. Adaptive radiation in cichlid fish. The pharyngeal jaw morphology, which allows dietary specialization, appears to have served as a key innovation in facilitating adaptive radiation in this group. (A) Biting and sucking species exhibit distinct morphologies. *Labeotropheus fuelleborni* (top) is a specialized biting species characterized by a short, robust lower jaw and an outer row of closely spaced tricuspid teeth. *Metriaclima zebra* (bottom) forages with a sucking mode of feeding and has a more elongate jaw and an outer series of larger bicuspid teeth. (B) Cichlids exhibit remarkable evolutionary convergence. Similar ecomorphs have evolved repeatedly within different cichlid assemblages. All of the cichlids in the left-hand column are from Lake Tanganyika. All of the cichlids in the right-hand column are from Lake Malawi and are more closely related to one another than to any species within Lake Tanganyika. Note the similarities among color patterns and trophic morphologies. (From Albertson, R. C., and T. D. Kocher. 2006. Genetic and developmental basis of cichlid trophic diversity. Heredity 97: 211–221, http://www.nature.com/hdy/journal/v97/n3/full/6800864a.html)

more recent studies of insect diversification in the context of their host plants, with repeated evolution of angiosperm feeding in phytophagous beetles associated with an increased rate of diversification. There is a consistently greater diversity of beetles among plants in which latex or resin canals have evolved as protection against insect attack. In the same way, adaptive radiation of parasites may occur as a consequence of host switching to a new lineage of hosts.

The development of a symbiotic association can also serve as a key innovation, providing a possible avenue through which taxonomic partners can enter into a new set of habitats unavailable to one or both of the symbiotic partners alone. For example, the development of gut endosymbionts and the concomitant ability to digest cellulose in ruminants appear to have led to the radiation of bovids in the African savannas. Likewise, the presence of algal endosymbionts has played a major role in the evolution and diversification of certain clades of Foraminifera.

Extrinsic Factors

Rates of speciation are frequently accelerated with the physical appearance of new habitat (figure 2). In particular, changes in the temperature or humidity of the environment over various eras have repeatedly resulted in mass extinctions coupled with the opening of ecological space (see also figure 4). For example, the Cretaceous–Paleocene boundary event resulted in numerous extinctions of plants and insects and set the stage for the subsequent adaptive radiation of other groups. Environmental changes appear to form the basis of the Phanerozoic revolutions. Rising temperature and nutrient supplies as a result of submarine volcanism may have triggered later Mesozoic and perhaps early Paleozoic diversification episodes. Similar factors may underlie the iterative radiations of ammonoids through the geological record in which each radiation appears to have originated from a few taxa, which then went on to give rise to a wealth of morphological diversity.

The emergence de novo of isolated habitats can be considered a kind of environmental change, as, for example, the formation of islands in the ocean, and many adaptive radiations are associated with such situations. For example, the adaptive radiation of finches in the Galápagos Islands appears to have been triggered by the appearance of land through volcanic activity over the last 3 million years. As the number of islands increased, so did the number of finch species. In the same way, adaptive radiations in the Hawaiian Islands are mostly associated with the volcanic activity that resulted in the formation of the current high islands that date back approximately 5 million years. Likewise, the

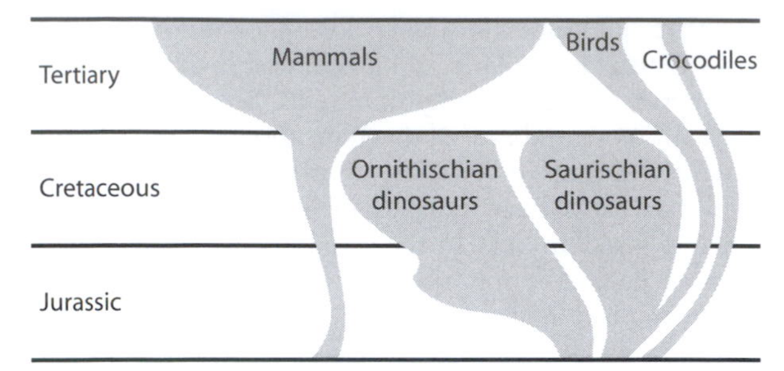

Figure 2. Open niches vacated by dinosaur extinctions at the end of the Cretaceous may have created empty ecological space and allowed mammals to radiate into these positions. Likewise, lineages that colonize isolated islands may give rise to adaptive radiations because the colonists are free from competition with other species. (From Understanding Evolution, http://evolution.berkeley.edu/evolibrary/article/side_o_0/adaptiveradiation_01)

evolution and adaptive radiation of the African cichlids appears to have been initiated in Lake Tanganyika approximately 5–6 million years ago when rivers in the area became progressively more swampy, with diversification of fish being initiated when river species became isolated in the deepening lake. In each of these cases, the new habitats were extremely isolated, resulting in infrequent colonization, thus giving the few successful colonists sufficient time to "explore" the ecological space available and diversify into multiple species.

2. ARE CERTAIN TAXA MORE LIKELY TO UNDERGO ADAPTIVE RADIATION?

Whether some species are predisposed to undergo adaptive radiation because of a broad environmental tolerance, generalized feeding patterns, or perhaps some proclivity to develop novel associations has been the subject of much debate. For example, birds have undergone extensive adaptive radiation in the insular Pacific, whereas butterflies have not. This has led some authors to suggest that speciation in butterflies may be constrained by the mechanics of insect–plant coevolution preventing rapid diversification in the insects. However, this argument is not well supported, as other insects with similar coevolutionary ties have undergone some of the most spectacular insular adaptive radiations known. It appears that, given conditions of isolation and time, almost any group of organisms is capable of undergoing adaptive radiation given ecological opportunity that it can exploit. However, certain groups do appear to be predisposed to adaptive radiation. For example, the occurrence of parallel radiations of sister clades of plants, the Hawaiian silverswords and California tarweeds, suggests that this lineage has certain attributes that facilitate adaptive radiation in the appropriate settings.

3. EXAMPLES OF CURRENT ADAPTIVE RADIATIONS

Adaptive radiation has now been documented in every kingdom of life and in a large number of phyla and in many different circumstances. However, the original concepts were developed through studies on islands, as these systems are discrete and amenable for studying the basis of adaptive radiation. The earliest groups to be examined in this context were vertebrates, in particular birds, with perhaps the best-known example being that of the Galápagos finches (figure 3) initially described by John Gould, and used by Charles Darwin as a key demonstration of his theory of evolution by natural selection. Currently, there are 13 recognized species in three lineages: ground finches, tree finches, and warbler-like finches, with sympatric species occupying distinct ecological roles. Research by Peter and Rosemary Grant and colleagues has shown that the finches have considerable genetic variation within populations, which is intermittently subject to both natural and sexual selection, with the final community of finches on an island dictated by food resources and interspecific competition for these resources. In the Hawaiian Islands, the endemic honeycreepers, which show even more extraordinary morphological and ecological differentiation than the Galápagos finches, comprise 56 species in a single lineage. However, only 22 species of honeycreeper are currently extant, with others known only from historical or fossil collections, making it difficult to develop hypotheses regarding processes underlying the adaptive radiation. Other well-known vertebrate radiations include lizards (anoles) in the Greater Antilles of the Caribbean in which diversification has allowed species to occupy a range of ecological roles, with as many as 11 species occurring sympatrically. Different species live, for example, on twigs, in the grass, or on tree trunks near the ground. Jonathan Losos and colleagues recognize six types of habitat specialists on the basis of morphological measurements (see plate 4). Among frogs, a remarkable example of adaptive radiation has recently been found in Madagascar. Among mammals, the best-known adaptive radiation is also from Madagascar, where lemurs constitute a spectacular diversification of more than 65 species, although at least 15 of these are now extinct. In addition, some striking radiations of small mammals have been documented, including the rodents on the islands of the Philippines and bats in southeast Asia.

Additional spectacular examples of adaptive radiation in vertebrates come from lacustrine fish, with the best known being cichlids (mentioned above), which

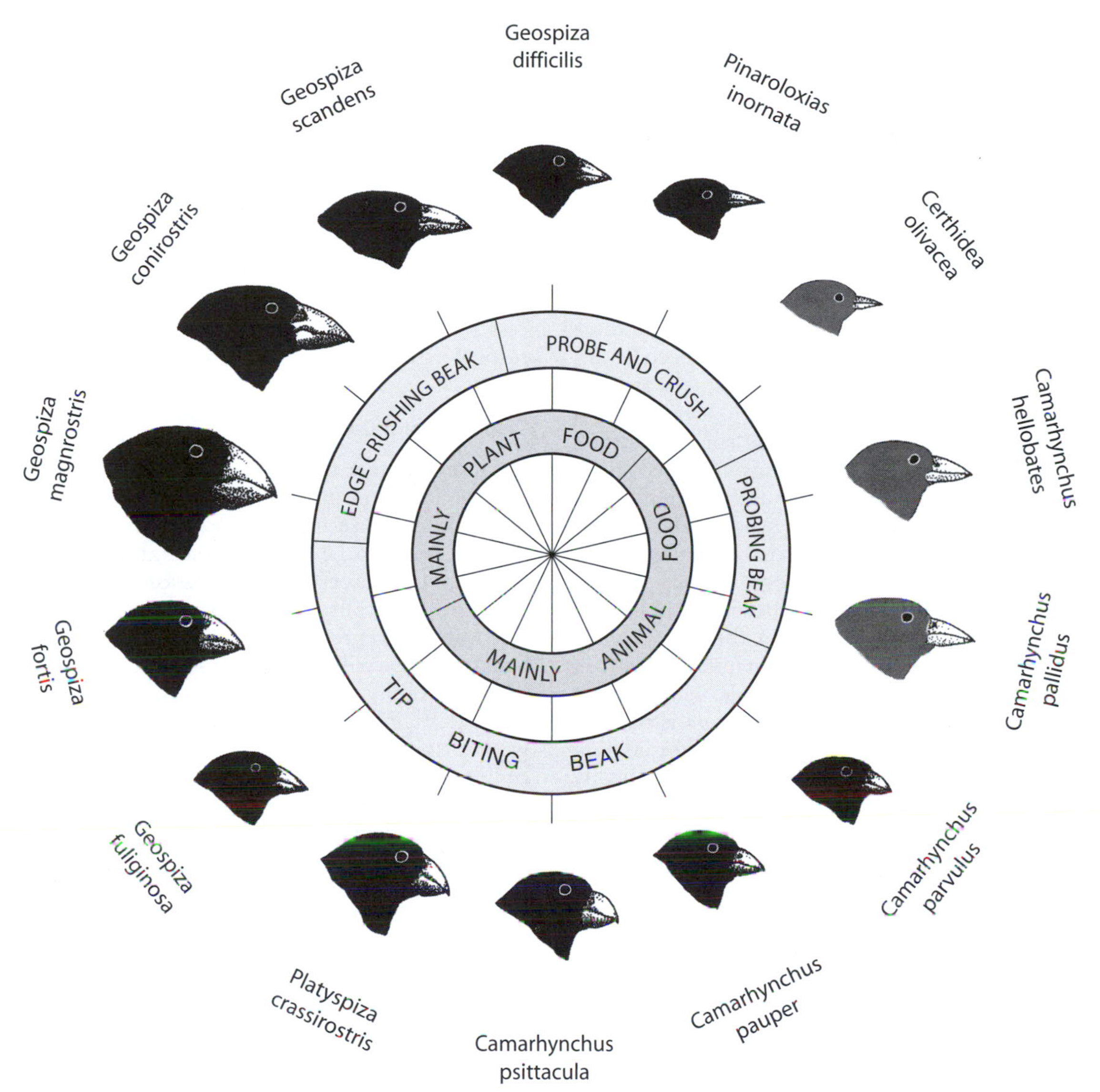

Figure 3. Adaptive radiation in Darwin's finches. Diagram illustrating the morphological and associated ecological diversity among the radiation of Darwin's finches in the genus *Geospiza* (Emberezidae). The 14 species evolved from a common ancestor about 3 million years ago. (From Grant, P. R., and B. R. Grant. 2008. How and Why Species Multiply: The Radiation of Darwin's Finches. Princeton, NJ: Princeton University Press)

reach their highest diversity in the East African lakes of Victoria (over 400 species), Malawi (300–500 species), and Tanganyika (approximately 200 species). In each of these lakes, the fish exhibit diversity in trophic morphology, including specialist algal scrapers, planktivores, insectivores, piscivores, paedophages, snail crushers, and fin biters. In addition to their trophic diversity, they display a striking array of color patterns, which appear to be involved in courtship and recognition. Other fish radiations include darters of the Central Highlands of eastern North America and threespine sticklebacks in deglaciated lakes in Canada. The latter in particular have now been heralded as an outstanding example of recent and ongoing adaptive radiation, within which it is possible to study processes involved. No more than two species occur in any one lake, but pairs of species in different lakes appear to have evolved independently of other pairs as a result of parallel bouts of selection for alternate trophic environments.

Among invertebrates, there are now a number of examples that show adaptive radiation associated with historic climatic change (figure 4). One of the best known ongoing radiations in insects is that of the Hawaiian *Drosophila* flies, in which courtship behavior can be very elaborate; the males of the so-called picture winged species often have ornately patterned wings as well as unusual modifications of the mouthparts and legs. At the same time, clades are characterized according to whether they breed on fungi, leaves, fruit, or bark, suggesting a role of both sexual selection and ecological shifts in allowing diversification of these flies. Hawaiian swordtail crickets have become increasingly recognized over recent years as a striking example of a very rapid island radiation. In common with other crickets, courting males "sing" to attract females by rubbing their forewings together. Each species has a unique song, and females respond preferentially to the song of the same species. Differentiation appears to occur through sexual selection on genetically well-structured populations. Among spiders, several radiations have been described in the Hawaiian Islands, one of the largest being in the genus *Tetragnatha*, where ecomorphs have arisen independently in much the same way as Caribbean lizards. In other parts of the Pacific, the land snail genus *Partula* is particularly well known for its radiation on different islands in the South Pacific. Like other land snails, they are highly polymorphic with respect to the color, banding, and chirality of the shell; competition appears to be important in dictating the array of species at a site. The Canary Islands are also well known for adaptive radiations of insects (in particular, beetles and psyllids) and spiders. Here, although most groups show evidence of competition in shaping communities, there is little biogeographic congruence between groups; stochasticity in species arrival patterns plays a prominent role in dictating species composition on any one island.

The Hawaiian silversword alliance (Asteraceae) has been considered a prime example of adaptive radiation in plants (plate 3). It consists of 28 species, which exhibit a large diversity of life forms, including trees, shrubs, mat-plants, monocarpic and polycarpic rosette plants, cushion plants, and vines, that occur across a broad environmental spectrum, from rainforests to desert-like settings. Additional plant radiations include columbines (mentioned above) in North America and numerous other island radiations, such as *Argyranthemum* in the Canary Islands and *Psiadia* in the Mascarenes, both of which are in the same family as the Hawaiian silverswords (Asteraceae) and have diversified in a parallel fashion.

One of the most recently described adaptive radiations is that of the soil bacterium *Pseudomonas fluorescens*, which, over a short period of time, can develop from an isogenic population under conditions of environmental heterogeneity into several somewhat predictable ecomorphs. Accordingly, this system has been hailed as one within which it is possible to conduct experimental studies of adaptive radiation.

4. INITIATION OF ADAPTIVE RADIATION

Although there have been many studies that describe different adaptive radiations, the initiation of the process remains poorly understood. Yet it is possible to recognize some general patterns.

Dispersal and Colonization

Because adaptive radiation requires colonization and differentiation in an ecologically available space, the taxa that colonize must necessarily be few. Although there appears to be a substantial random element to colonization, successful colonization of very isolated locations generally requires high dispersal abilities. Accordingly, representation of taxa within biotas in isolated areas will be skewed toward those that can disperse readily. For example, as one ventures farther into the Pacific Ocean from west to east (i.e., toward greater isolation), there is an attenuation in the number of lineages of terrestrial groups that have colonized by over-water dispersal. In less isolated archipelagoes, such as Fiji, the fauna is relatively rich with numerous continental affinities. Farther east, Samoa is less rich at higher taxonomic levels than Fiji but still has many families and orders that are lacking from the native fauna of more remote Polynesian islands. East of Samoa, the number of floral and faunal groups that have been able to reach the remote islands diminishes, and here the few colonists have frequently undergone adaptive radiation, accentuating the unrepresentative nature of the biota. The Hawaiian archipelago (4000 km from the nearest continent, North America; 3200 km from nearest island group) demonstrates this pattern most acutely: among insects, the native fauna is represented by only 50% of insect orders and 15% of known families and exhibits extraordinarily high levels of endemism (95–99% in invertebrates) with numerous cases of adaptive radiation.

High dispersal of colonists clearly contrasts with the apparently much restricted ranges and dispersal abilities that are frequently associated with members within an adaptive radiation. In general, there appears to be a dramatic loss of dispersal ability and/or attainment of a more specialized habitat at the outset of adaptive radiation. Indeed, a tendency toward reduced dispersal following colonization of new ecological space

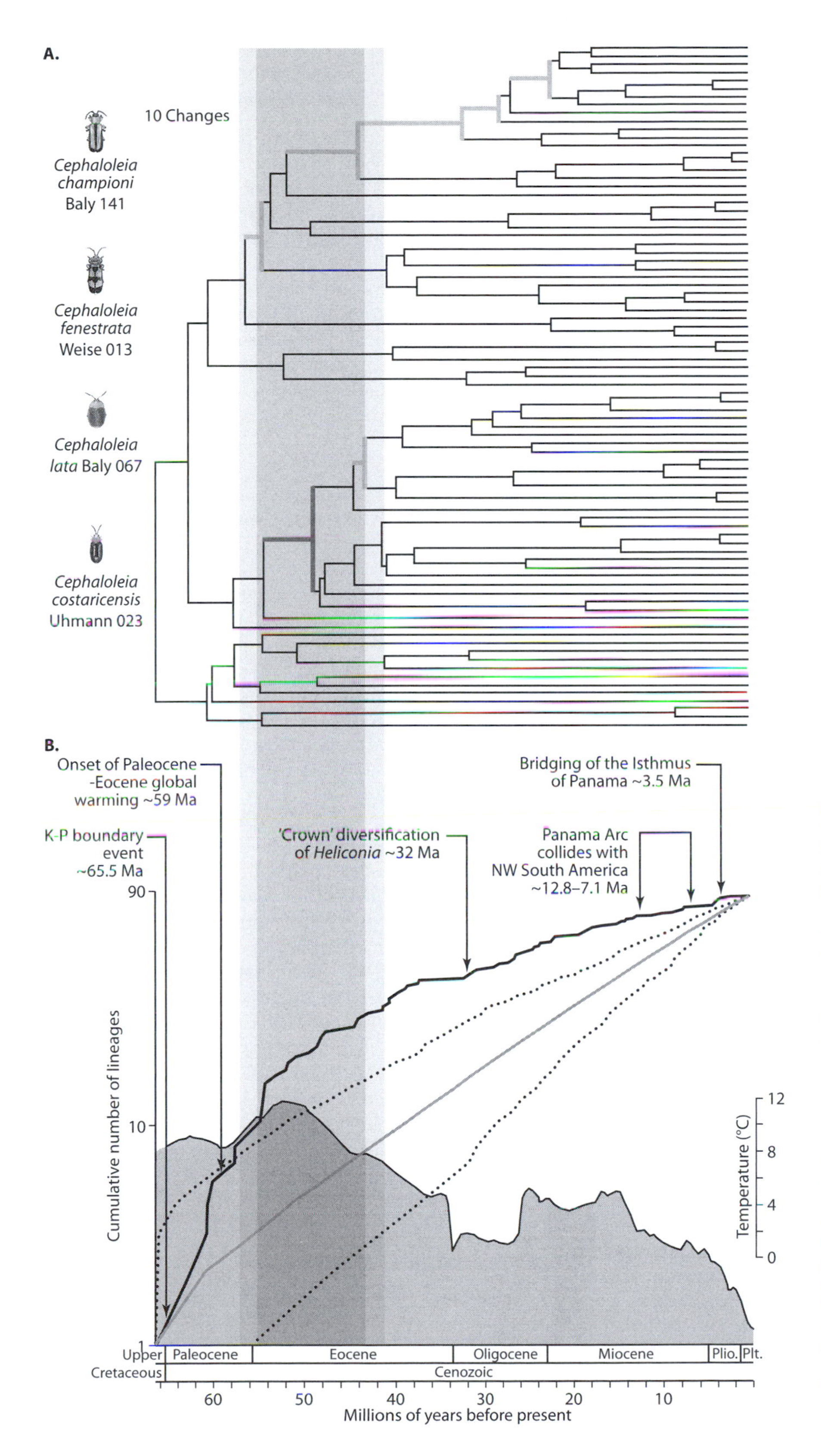

Figure 4. Tempo in the adaptive radiation of *Cephaloleia* (Coleoptera: Chrysomelidae). (A) Chronogram showing significant diversification rate shifts (each terminal corresponds to a species) and source(s) of support. Shading indicates the timeframe over which significant diversification rate shifts occurred. (B) Semilogarithmic plot of lineages through time (LTT) for *Cephaloleia* (LTT plot; heavy upper line) superimposed on a time-averaged record of high-latitude sea-surface temperatures (lower irregularly descending line), a proxy for global climate. The constant diversification rate model is rejected if the empirical LTT curve falls outside the 95% confidence interval generated by simulation (middle ascending lines). Upturns or downturns in the empirical LTT plot reflect changes in rates of speciation or extinction. (From McKenna, Duane D., and Brian D. Farrell. 2006. Tropical forests are both evolutionary cradles and museums of leaf beetle diversity. Proceedings of the National Academy of Sciences U.S.A. 103: 10947–10951)

may be an important factor in allowing diversification to proceed.

Ecological Release and Specialization

When a taxon first colonizes a new area, it frequently expands its ecological range, a phenomenon referred to as ecological release. Regular cycles of ecological and distributional change following colonization of islands are well known—the phenomenon was described first in detail for Melanesian ants and subsequently for Caribbean birds. The idea is that widespread, dispersive populations or species (Stage I) give rise to differentiating (Stage II) and then fragmented (Stage III) and ultimately specialized endemic species (Stage IV). This pattern is consistent with the early stages of adaptive radiation. However, once local endemics have formed, there appears to be little evidence to suggest that species become progressively more specialized over time. Phylogenetic analysis of some radiations suggests that specialized species, when they colonize new habitat, may be able to expand their range and accordingly may reinitiate the "cycle" to give rise to other specialist species.

Behavioral Plasticity

Although it has been suggested that behavioral and ecological plasticity may impede adaptive diversification as it allows a single taxon to exploit a broad environmental range, recent research suggests otherwise. Mary Jane West-Eberhard (2003) has argued that adaptive plasticity (including behavior) promotes evolutionary diversification, in particular when the environment is variable. Recent theoretical work has supported this idea, and plasticity has now been suggested as playing a role in a number of radiations. For example, in Caribbean anoles, plasticity may allow species to occupy a new habitat in which they otherwise might not be able to survive. Once in these habitats, their behavior may become modified, and, as new mutations arise, selection may act to accentuate the initial, relatively minor, morphological changes.

Conclusive demonstration that phenotypic plasticity precedes, and then permits, subsequent evolution has been difficult to obtain because once the ancestral populations have evolved, they may lose the pattern of plasticity present at the start of adaptive differentiation. In the threespine stickleback, however, where ancestral oceanic sticklebacks likely have changed little since colonization and diversification of freshwater species, the pattern of behavioral plasticity in the ancestral species supports the argument that phenotypic plasticity can guide subsequent evolutionary change and facilitate adaptive radiation.

5. SPECIATION IN ADAPTIVE RADIATION

Rapid speciation coupled with phenotypic diversification are key features of adaptive radiation, and accordingly, many studies have examined the basis for genetic diversification. Because adaptive radiation involves ecological shifts, divergent natural selection has been most broadly implicated across a spectrum of lineages, with founder effects and sexual selection also playing a role in certain situations. However, the mechanism through which species form during adaptive radiation is still only very poorly understood, although it appears that geographic barriers to gene flow are generally involved at least in the initiation of speciation.

Founder Events

A founder event refers to the establishment of a new population by a few individuals that carry only a small sample of the genetic diversity of the parent population. Many studies have suggested that founder events can play a role in adaptive radiation, as taxa within a radiation often have very small population sizes, with ample opportunity for isolation. Because of the effects of random sampling, a founder event will lead to differences in allele frequencies at some loci as compared to the parent population. However, considerable debate has focused on the nature of genetic changes that occur subsequent to the founder event, during the period of population growth, with some traditional arguments suggesting that founder events trigger rapid species formation, although more recent studies have largely refuted a role for founder events in reproductive isolation. Genetic drift can lead to changes during the bottleneck, but the effect becomes weaker as the population starts to grow. At the same time, a large proportion of alleles are lost during a bottleneck, and few new mutations can occur while the population size is small. The effect of these opposing forces is that the number of beneficial mutations fixed per generation remains largely unchanged by the bottleneck. Nevertheless, selection subsequent to a genetic bottleneck can preserve alleles that are initially rare and that would otherwise tend to be lost as a result of stochastic events in populations of constant size.

Divergent Natural Selection

As with other factors considered to play a key role in adaptive radiation, competition plays a dual role. On the one hand, reduced competition is generally associated with the presence of open resource niches, either as organisms move into new habitats or through the acquisition of a key innovation, thereby providing increased opportunities for diversification. On the other hand, competition, often with ecological character displacement, is frequently implicated in promoting adaptive change between close relatives. However, there are no known direct tests that link the evolution of reproductive isolation to interspecific competition, and the role of competition in the early stages of adaptive radiation remains unclear. Nevertheless, divergent natural selection driven by interspecific competition appears to have shaped current phenotypic differences in many radiations ranging from sticklebacks in the Canadian lakes to lizards on the islands of the Caribbean, finches in the Galápagos, and spiders in the Hawaiian Islands.

Predation has also been suggested as a possible operative that may work together with (or instead of) competition to allow adaptive differentiation, but this has been difficult to test. However, the specific role of predation in facilitating adaptive radiation has recently been demonstrated in both walking-stick insects and bacteria.

Sexual selection (see below) has been shown in some cases to act in concert with divergent natural selection. For example, sexual dimorphism in Caribbean anole lizards allows a species to exploit different niches, thereby serving as an alternative means of ecological diversification.

Hybridization and Gene Flow

Gene flow among populations in the process of diverging, or hybridization between incipient species, may slow the process of diversification and homogenize populations. However, interspecific hybridization may also be a possible source of additional genetic variation within species: Hybridization increases the size of the gene pool on which selection may act and therefore may be a significant process in the adaptive radiation of some species. Such effects have now been shown in Darwin's finches, African cichlids, Lake Baikal sculpins, and several lineages of Hawaiian plants.

Sexual Selection

Sexual selection has been implicated in the diversification of species within some of the most explosive adaptive radiations. In particular, it has been suggested that sexual selection drives species proliferation in Hawaiian *Drosophila* flies and *Laupala* crickets. The mechanism for this, as proposed by Kenneth Kaneshiro and colleagues for *Drosophila,* is that, when a newly founded population is small, female discrimination is relaxed; accordingly, sexual behavior becomes simpler with more intraspecific variability. Divergence of sibling species may then occur during isolation as a result of a shift in the distribution of mating preferences. In Hawaiian *Laupala* crickets, closely related species are morphologically similar with no ecologically recognizable features and distinguishable only by the pulse rate of the male courtship song, a secondary sexual trait used in mate attraction. These crickets demonstrate the highest rate of speciation recorded so far in arthropods.

Other taxa in which extreme diversification has been attributed in part to sexual selection include haplochromine cichlids, in which sexually dimorphic breeding coloration, with brightly colored males and often dull females, appears to have arisen through female mate choice for male coloration. Disruptive sexual selection on male coloration can result in genetic isolation between fish that exhibit small differences in male coloration and female preference for male coloration. Likewise, diversification in jumping spiders in the sky islands of the western United States appears to be the product of female preference for greater signal complexity or novelty.

In all of these radiations, ecological differentiation still occurs, but sexual selection may act somewhat independently and accelerate the rate of differentiation. Mate choice is the primary isolating mechanism, and hybrids between rapidly diverging sibling species are often fully viable and fertile.

6. COMMUNITY ASSEMBLY

During the course of adaptive radiation, speciation plays a role similar to that of immigration, although over an extended time period, in adding species to a community. It appears that when a lineage diversifies in a community, it may adapt to multiple different ecological settings, with the development of specific sets of attributes, or ecomorphs, to match a given microhabitat. In an archipelago situation, where similar habitats occur on different islands, it has been found that similar sets of ecomorphs can arise independently within the same lineage through convergent evolution, a phenomenon first demonstrated in anole lizards in the Caribbean (plate 4) and now shown in a wide range of other adaptive radiations including Himalayan birds, Galápagos finches, cichlid fish of the African rift lakes, Canadian lake sticklebacks, ranid frogs in Madagascar, spiders in the Hawaiian islands, and snails in the Bonin Islands of Japan. These results point to a model of ecological community assembly that incorporates evolutionary effects of interspecific competition.

The central importance of competition in shaping communities has recently been challenged by Stephen Hubbell's "neutral theory," which postulates that differences between members of a community of ecologically equivalent species are "neutral" with respect to their success. One outcome of neutral theory has been to prompt investigation of when, and to what extent, ecological equivalence might play a role. In the course of adaptive radiation, neutral processes may govern the identity of taxa initially colonizing a new area, potentially resulting in a transient period with multiple ecological equivalents. Likewise, communities formed through sexual selection may include some ecologically similar species. However, the end product of subsequent ecological speciation appears inevitably to be a set of taxa that are ecologically distinct.

7. MOLECULAR BASIS FOR ADAPTIVE CHANGE

That repeated evolution of similar forms has occurred in many species undergoing adaptive radiation has led to considerable research on the molecular underpinnings of such apparently complex changes. With the development of molecular techniques, we now have the opportunity to analyze the genetic architecture of species differences and the role of new mutations versus standing genetic variation in adaptation and divergence in adaptive radiation. For example, in fish, it appears that gene duplication provided a genomic mechanism for adaptive radiation of teleosts, with lineages arising after duplication being much more species-rich than the more basal groups. Studies on the body plating in a radiation of sticklebacks have shown that related alleles are responsible for a given phenotype, indicating that even though a given phenotype (ecomorph) may have multiple independent origins, the same ancestral allele appears to be involved in producing the same phenotypes.

In groups in which sexual selection has been implicated in diversification, investigation of rapidly evolving genes for traits associated with sex and reproduction have been the target of research on the process of speciation. For *Drosophila*, results show differential patterns of evolution of genes expressed in reproductive and nonreproductive tissues, supporting the role of sexual selection as a driving force of genetic change between species.

8. TESTING ADAPTIVE RADIATION

Experimental tests of the processes underlying adaptive radiation are now possible using microbial systems to probe the relative roles of niche space, competition, predation, and time in dictating when and how adaptive radiation occurs. These studies, although still in their infancy, have shown many parallels between adaptive diversification of bacterial genotypes over the space of a few weeks and the presumed early stages of adaptive radiation of macroorganisms over millions of years. The challenge is to apply the knowledge gained from these rich bacterial systems to a more general understanding of adaptive radiation.

FURTHER READING

Benton, Michael J., and Brent C. Emerson. 2007. How did life become so diverse? The dynamics of diversification according to the fossil record and molecular phylogenetics. Palaeontology 50: 23–40.

Carlquist, Sherwin, Bruce G. Baldwin, and Gerald D. Carr, eds. 2003. Tarweeds and Silverswords: Evolution of the Madiinae (Asteraceae). St. Louis: Missouri Botanical Garden Press.

Ehrlich, Paul R., and Peter H. Raven. 1964. Butterflies and plants: A study in coevolution. Evolution 18: 586–608.

Gillespie, R. G. and D.A. Clague, eds. 2009. Encyclopedia of Islands. Berkeley: University of California Press.

Givnish, T.J.K., and K. J. Sytsma, eds. 1997. Molecular evolution and adaptive radiation. New York: Cambridge University Press.

Grant, Peter R., and B. Rosemary Grant. 2007. How and Why Species Multiply: The Radiation of Darwin's Finches. Princeton, NJ: Princeton University Press.

Jessup, C. M., R. Kassen, S. E. Forde, B. Kerr, A. Buckling, P. B. Rainey and B.J.M. Bohannan. 2004. Big questions, small worlds: Microbial model systems in ecology. Trends in Ecology and Evolution 19: 189–197.

Losos, Jonathan B. 2001. Evolution: A lizard's tale. Scientific American 284: 64–69.

Ricklefs, Robert E., and Eldredge Bermingham. 2007. The causes of evolutionary radiations in archipelagoes: Passerine birds in the Lesser Antilles. American Naturalist 169: 285–297.

Schluter, Dolph. 2000. The Ecology of Adaptive Radiation. Oxford: Oxford University Press.

Seehausen, Ole. 2006. African cichlid fish: A model system in adaptive radiation research. Proceedings of the Royal Society B–Biological Sciences 273: 1987–1998.

Wagner, Warren L., and Vicki A. Funk, eds. 1995. Hawaiian Biogeography: Evolution in a Hotspot Archipelago. Washington, DC: Smithsonian Institution Press.

West-Eberhard, Mary Jane. 2003. Developmental Plasticity and Evolution. New York: Oxford University Press.

II

Population Ecology

H. Charles J. Godfray

Understanding what determines the average abundance of species, why their numbers fluctuate, and how they interact with each other is a major part of modern ecology often united under the term *population ecology*. Of course, the boundaries of population ecology are ill-defined and porous: on the one hand the field grades into physiological ecology—how individuals interact with the environment—and on the other hand into community ecology—the study of large assemblages of species. Population ecology is part of the larger subject of population biology that encompasses both the evolutionary and the ecological processes affecting populations.

The human race has always been concerned with the abundance and fluctuations of the plants and animals that share its environment, not least because they provide its food. But the modern study of populations begins with Thomas Malthus (1798), who, in his *Essay on the Principle of Population*, realized that if birth and death rates remain constant with the former greater than the latter, then population size will grow geometrically until some extrinsic factor comes into play. The conclusions that Malthus, an upper-class English vicar, drew from his insights were of the importance of doing something about the "irresponsibly fecund lower orders" (as well as the need to attend to other "problems" such as "liberal women" and the French!). Fortunately, Malthus is not remembered as a politician, but his writing hugely influenced the first generation of biologists to think about animal populations, and in particular Charles Darwin, who realized that geometric population growth implied massive mortality and hence a huge advantage to any heritable trait that helped individuals in the struggle for survival. Today we use the Malthusian parameter, the population's rate of geometric growth assuming demographic parameters remain the same, as an index of the state of the population. A closely related parameter, the growth rate of a rare mutation, is intimately connected to notions of evolutionary fitness. Calculating population growth rates (population projection) is quite straightforward for some species, for example, those with discrete generations. It can be much more complicated when there are overlapping generations and where the population is composed of individuals of different classes (differing in age, size, or other variable), issues discussed in chapter II.1.

But demographic rates do not remain the same forever, and in particular, as population densities increase, birthrates go down or death rates go up. It is these density-dependent effects that are critical in determining the typical range of abundance of different organisms, as discussed in chapter II.2. Density-dependent effects may increase smoothly as population size gets larger but may also be much more capricious, cutting in only above a threshold, the latter itself possibly varying from year to year. The chief factor determining observed population densities at any particular time is often a density-independent process such as the weather, and the densities of some populations may fluctuate in a random way for many generations before they become large enough for density-dependent processes to come into play. However, no population can be regulated, that is, persist indefinitely within certain bounds, without density dependence occurring.

Where density-dependent processes act instantaneously and increase gently with population size, the outcome of population regulation will be a stable equilibrium (though in nature random perturbations will mean that an absolutely constant population density is unlikely to be observed). But if there is a time lag between population increase and the impact of density dependence, or if density dependence is very strong, then overcompensation may occur, and the population will show cycles. As was first realized by ecological theoreticians, particularly by Robert May, in the 1970s, stronger density dependence and larger time lags may lead to population fluctuations that are chaotic—purely deterministic yet impossible to predict in detail. Hastings (chapter II.3) explores these issues and discusses recent findings about how deterministic population dynamics and random environmental effects may produce a complex array of fascinating dynamics.

Animal and plant populations do not exist in one place but occur in a normally complex spatial landscape. The first generation of ecological studies tended to ignore this spatial component, but their considerations are now central to much theoretical and practical ecology and are discussed in several of the chapters in this section. One particularly fruitful line of inquiry is discussed in greater depth by Hanski (chapter II.4). Consider a species that inhabits a constellation of habitat patches: some may be empty, and others may be occupied by subpopulations with substantially independent dynamics, which may become extinct or send out colonizers to form new populations. This is a metapopulation, and the theoretical study of their dynamics, coupled with some superb long-term field studies, in particular by Hanski himself, has greatly enriched the subject and proved very important in applied ecology, especially in conservation biology.

Turning from single populations to pairs and small collections, a number of authors explore the different ways in which species may interact. This is important in its own right but also as the building blocks from which communities and ecosystems are composed. Populations can flourish only if they have resources for growth and reproduction, and interspecific resource competition is a potent force that is thought to structure many communities. We have two entries on competition: chapter II.5 focuses on plants and chapter II.6 on animals. Although there are commonalities, the facts that plants are (literally) rooted to the ground and compete for a relatively small range of essential quantities (space, light, water, nutrients) has meant that plant competition ecology has developed rather differently from its zoological cousin.

Competition is sometimes referred to as a "– –" interaction because each species involved suffers from the presence of the other. In a "+ –" interaction, one species gains at the expense of the other. We have four chapters on such interactions. Chapter II.7 discusses predator–prey dynamics including issues such as the circumstances under which predator–prey cycles may occur and what processes may tame the intrinsic tendency for instability in such interactions. Such concerns are also explored in chapter II.8, which considers parasitoid–host dynamics. Parasitoids are insects, typically wasps and flies, with a life history somewhat intermediary between predators and parasites. They are numerically abundant and important mortality factors affecting a wide variety of insect hosts and are also very important as biological agents. Chiefly because of this last reason their population dynamics has been closely studied by ecologists. Both the dynamics of true parasites and the dynamics of pathogens are discussed in chapter II.9. Although the importance of disease for humans and their livestock and crops has long been realized, it is only relatively recently that the importance of disease in natural ecosystems has been appreciated, a shift in focus that has been greatly helped by the molecular biology revolution that has furnished new tools for studying microbial pathogens. Interestingly, mathematical tools and techniques developed by ecologists are now used routinely by epidemiologists studying human, farm animal, and crop pathogens. The final "+ –" interaction chapter is by Morris (chapter II.10) on plant–herbivore interactions. The major difference here is that plants are frequently able to tolerate varying degrees of herbivory in a way that, say, an individual zebra cannot tolerate predation by a lion. Morris explores the consequences of these differences and asks the degree to which herbivory may influence plant population dynamics.

Mutualism and symbiosis, "+ +" interactions, are rather the Cinderella of this type of population ecology and have received far less attention than competition and consumer–resource interactions. In chapter II.11, Bronstein discusses why ecologists may have underestimated their importance, drawing examples from some wonderful model systems, in particular involving plants and their insect pollinators. Many partners in mutualisms and symbioses are microorganisms, and we also include a chapter discussing some of the special issues concerning microbial population ecology. Chapter II.12 explores how all the types of ecological processes that affect larger organisms also operate at this smaller scale but shows how factors such as the size of microbial populations and their particular growth dynamics require special treatment. It also illustrates how microbial systems, with their very short generation times, provide fabulous models systems for exploring processes that affect all living organisms. Finally, Thompson (chapter II.13) explores how ecological interactions such as competition, predation, and mutualism result in evolutionary pressures that are experienced by both the interacting species. The resultant coevolution has shaped the morphology, biochemistry and behavior of many, perhaps most, organisms on Earth, including ourselves.

II.1

Age-Structured and Stage-Structured Population Dynamics

Mark Rees and Stephen P. Ellner

OUTLINE

1. Age-structured models: Life tables and the Leslie matrix
2. Stage-structured matrix models
3. Integral projection models
4. Continuous-time models with age structure
5. Applications and extensions
6. Coda

When all individuals in a population are identical, we can characterize the population just by counting the number of individuals. However, the individuals within many animal and plant populations differ in important ways that influence their current and future prospects of survival and reproduction. For example, larger individuals typically have greater chances of survival, produce more and sometimes larger offspring, and often have slower growth rates. In such cases, characterizing the population structure—the numbers of individuals of each different type—is critical for understanding how the population will change through time. In this chapter, we examine some of the main types of models used for describing and forecasting the dynamics of structured populations. Age-structured models in discrete time, appropriate for populations in seasonal environments, were developed centuries ago by the great mathematician Leonhard Euler (1707–1783). These are considered first, before moving to models where individuals are characterized by their stage in the life cycle (e.g., seed versus flowering plant, larva versus adult). Next we look at how to incorporate differences among individuals that vary continuously, such as size. Having explored discrete-time models, we briefly turn to continuous-time models and then present some applications and extensions.

GLOSSARY

age structure. Distribution of ages in a population

matrix. A rectangular array of symbols, which could represent numbers, variables, or functions

1. AGE-STRUCTURED MODELS: LIFE TABLES AND THE LESLIE MATRIX

The simplest age-structured models assume that each individual's chance of survival and reproduction depends only on its age; there are no effects of population density. The standard model counts only females (assuming no shortage of mates) and assumes that all births occur in a single birth pulse immediately before the population is censused (a so-called postbreeding census). The population dynamics is then summarized by the following equations:

$$\begin{aligned} n_0(t+1) &= f_0 n_0(t) + f_1 n_1(t) + f_2 n_2(t) + \cdots \\ &= \sum_{a=0}^{A} f_a n_a(t) \\ n_a(t+1) &= p_{a-1} n_{a-1}(t), \end{aligned} \tag{1}$$

where $n_a(t)$ is the number of individuals of age a at time t, f_a, and p_a are, respectively, the average fecundity and the probability of survival to age $a+1$ of age a individuals, and A is the maximum possible age (or the maximum age at which reproduction occurs, if postreproductives are omitted from the population count). Because births occur just before the next census, $f_a = p_a m_{a+1}$, where m_{a+1} is the number of offspring produced by an age $a+1$ female.

Another way of formulating the model is to assume a prebreeding census, so the population is censused immediately before the birth pulse. This has two important consequences: (1) all individuals are at least age 1, and (2) in this case $f_a = p_0 m_a$, so fecundity depends on the number of offspring produced now, m_a, and the chance that they survive to be censused at age 1, p_0.

The simple age-structured model can be written as a matrix, commonly known as a Leslie matrix after British ecologist P. H. Leslie. Expressing equation 1 in matrix form simply means putting the *f*s and *p*s in the right places:

$$\begin{bmatrix} n_0(t+1) \\ n_1(t+1) \\ \vdots \\ n_A(t+1) \end{bmatrix} = \begin{bmatrix} f_0 & f_1 & f_2 & \cdots & f_A \\ p_0 & 0 & 0 & \cdots & 0 \\ 0 & p_1 & 0 & \cdots & 0 \\ \vdots & \vdots & \ddots & & \vdots \\ 0 & 0 & & p_{A-1} & 0 \end{bmatrix} \begin{bmatrix} n_0(t) \\ n_1(t) \\ \vdots \\ n_A(t) \end{bmatrix} \tag{2}$$

or, more compactly,

$$\mathbf{n}(t+1) = \mathbf{L}\mathbf{n}(t), \tag{3}$$

where **L** is the matrix is equation 2. When **L** is a matrix with n columns and $\mathbf{n}(t)$ a column vector of length n, then $\mathbf{Ln}(t)$ is a vector whose ith element is

$$[\mathbf{Ln}(t)]_i = \sum_{j=1}^{n} \mathbf{L}_{ij}\mathbf{n}(t)_j, \tag{4}$$

where $\mathbf{L}_{ij}$ is the number in the ith row and jth column of **L**. Matrix multiplication expresses equation 1 as a single operation; it also means that the tools of linear algebra can be used to study how the population varies through time.

Now that we have formulated the model, how does it behave? To answer this question we need to solve equation 3. Starting with some initial age distribution, $\mathbf{n}(0)$, we find:

$$\mathbf{n}(1) = \mathbf{Ln}(0)$$
$$\mathbf{n}(2) = \mathbf{Ln}(1) = \mathbf{L}[\mathbf{Ln}(0)] = \mathbf{L}^2\mathbf{n}(0),$$

and so on; so the general solution is

$$\mathbf{n}(t) = \mathbf{L}^t\mathbf{n}(0). \tag{5}$$

It is difficult to intuit what $\mathbf{L}^t$ is doing, but some insight is gained by solving the model numerically. For an example, setting

$$\mathbf{L} = \begin{bmatrix} 0 & 3 & 5 \\ 0.2 & 0 & 0 \\ 0 & 0.5 & 0 \end{bmatrix} \tag{6}$$

and iterating the model (figure 1) shows that after some initial fluctuations, the total population size grows at a constant rate, and the proportion of individuals in each age class becomes constant. This suggests that, rather remarkably, an age-structured population will behave very much like a simple unstructured population undergoing exponential growth, in which the numbers at time t are given by $n(t) = n(0)\lambda^t$. This fact allows us to derive one of the most important equations in age-structured dynamics. We know that $n_a(t) = l_a n_0(t-a)$, where l_a is the probability an individual survives to age a ($l_0 = 1$, $l_a = p_0 p_1 p_2 \ldots p_{a-1}$). Substituting this into the first line of equation 1 gives

$$n_0(t+1) = \sum_{a=0}^{A} f_a l_a n_0(t-a). \tag{7}$$

Assuming that n_0 grows exponentially at some rate λ, substituting $n_0(t) = c\lambda^t$ into equation 7 and simplifying gives the famous Euler-Lotka equation,

$$1 = \sum_{a=0}^{A} \lambda^{-(a+1)} l_a f_a = \sum_{a=1}^{A+1} \lambda^{-a} l_a m_a. \tag{8}$$

This equation shows how the long-term population growth rate λ is determined by the age-specific survival and mortality. Critically, when $\lambda > 1$ the population increases, and when $\lambda < 1$ it decreases. Consequently, λ is of great importance in applied contexts: for control of pest species we would like to make $\lambda < 1$, whereas for species of conservation interest we would like to ensure that $\lambda > 1$. As λ gets larger, the right-hand side of equation 8 gets smaller; using this fact and substituting $\lambda = 1$ into equation 8, we find that $\lambda > 1$ if and only if $R_0 > 1$, where

$$R_0 = \sum_{a=0}^{A} l_a f_a.$$

R_0 is the average number of offspring produced by a newborn female over her lifetime, given by summing the chance of surviving to each age times the number of offspring produced at that age. Thus, the population will increase ($\lambda > 1$) in the long run only if each female more than replaces herself.

Understanding the model further requires some results from matrix algebra. Eigenvalues turn out to be key quantities and are defined as follows: λ is an eigenvalue of **L** if there is a nonzero vector **w** such that

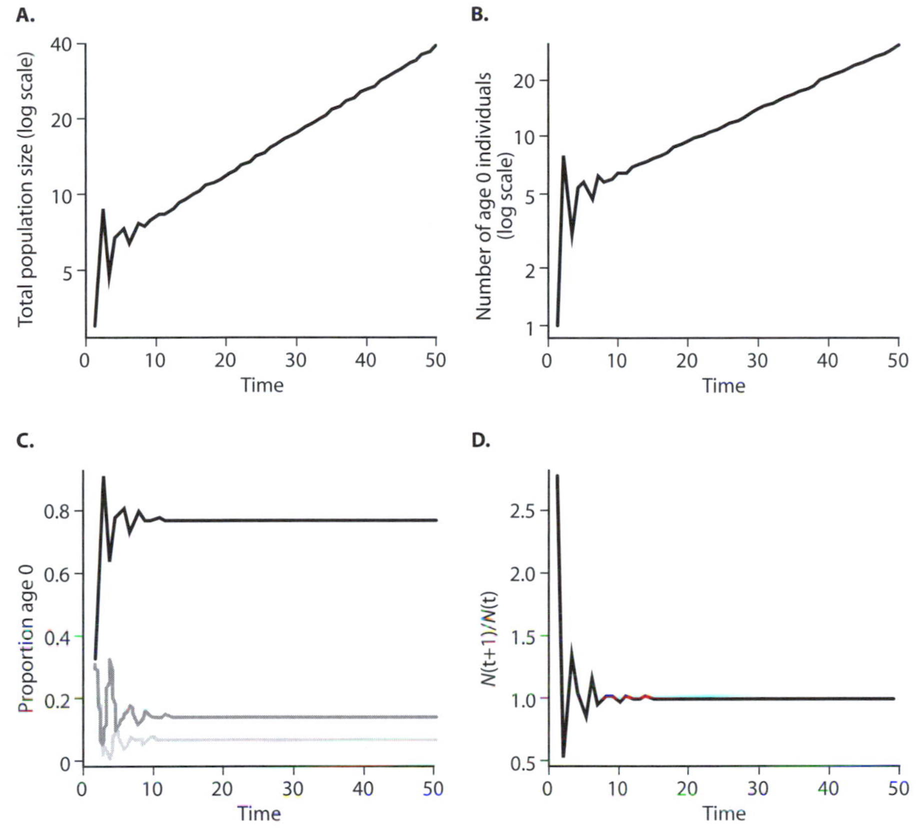

Figure 1. Numerical solution of the Leslie matrix model (equation 6) assuming $\mathbf{n}(0) = (1, 1, 1)$. The panels show time series plots of (A) total population size $N(t)$, (B) the number of newborn individuals, (C) the proportion of individuals of each possible age (black = 0, dark gray = 1, and light gray = 2), and (D) the population growth rate $N(t+1)/N(t)$.

$\mathbf{L}\mathbf{w} = \lambda\mathbf{w}$, and $\mathbf{w}$ is called the corresponding eigenvector. If there are $A+1$ distinct eigenvalues, then the corresponding eigenvectors are *linearly independent,* and so any population vector can be expressed as $\mathbf{n}(0) = c_0\mathbf{w}_0 + c_1\mathbf{w}_1 + c_2\mathbf{w}_2 + \cdots + c_A\mathbf{w}_A$. Then we can rewrite equation 3 as

$$\begin{aligned}\mathbf{n}(1) &= \mathbf{L}\mathbf{n}(0)\\ &= \mathbf{L}(c_0\mathbf{w}_0 + c_1\mathbf{w}_1 + c_2\mathbf{w}_2 + \cdots + c_A\mathbf{w}_A)\\ &= c_0\mathbf{L}\mathbf{w}_0 + c_1\mathbf{L}\mathbf{w}_1 + c_2\mathbf{L}\mathbf{w}_2 + \cdots + c_A\mathbf{L}\mathbf{w}_A\\ &= c_0\lambda_0\mathbf{w}_0 + c_1\lambda_1\mathbf{w}_1 + c_2\lambda_2\mathbf{w}_2 + \cdots + c_A\lambda_A\mathbf{w}_A,\end{aligned}$$

so moving one year forward corresponds to multiplying the coefficients c_i by the corresponding λ_i. Thus, the model solution (equation 3) can be written as

$$\mathbf{n}(t) = \sum_{i=0}^{A} c_i\lambda_i^t\mathbf{w}_i. \qquad (9)$$

So as t becomes large, the solution will be determined by the largest-magnitude λ_i, termed the dominant eigenvalue and its eigenvector; hence,

$$\mathbf{n}(t) \approx c_1{\lambda_1}^t\mathbf{w}_1, \qquad (10)$$

where $\approx$ means approximately with a relative error decreasing to zero as t becomes large. This explains the numerical results (figure 1) that after an initial period of transients, the population grows at a constant rate, given by λ_1, and the proportion of individuals in each age class becomes constant and is proportional to $\mathbf{w}_1$. For this reason $\mathbf{w}_1$ is called the *stable age distribution.*

The existence of a dominant eigenvalue is guaranteed so long as $\mathbf{L}$ is power positive: some power $\mathbf{L}^m$ has all entries greater than zero. $\mathbf{L}$ will be power positive providing $f_A > 0$ and any two consecutive *f*s are positive, or more generally if all the entries of $\mathbf{L}^{A^2+1}$ are positive. For a nonnegative $\mathbf{L}$ that is power positive, the Perron-Frobenius Theorem implies that $\mathbf{L}$ has a unique dominant eigenvalue that is real, positive, and strictly larger in magnitude than any other eigenvalue, guaranteeing convergence to the stable age distribution and stable growth rate λ_1.

When the matrix lacks power positivity, we can get more exotic behavior. For example, consider a population where all the reproduction is concentrated in the last age class, such as

$$\mathbf{L} = \begin{bmatrix} 0 & 0 & 11 \\ 0.2 & 0 & 0 \\ 0 & 0.5 & 0 \end{bmatrix}. \qquad (11)$$

In this case, the age structure continually cycles with a cycle length of 3, and population size never settles into growing at a constant rate (figure 2). These population waves were initially explored by Harro Bernardelli in relation to oscillations in the age structure of the Burmese population. For a matrix like that in equation 11, with all reproduction in the final age class, each individual of age a at time t gives rise, after m time steps (where m is the number of age classes) to R_0 age-a individuals at time $t+m$, and R_0 is the product of the nonzero matrix entries. Consequently, any initial age structure gives rise to a cycle of age structures that repeats indefinitely with period A, and the long-term population growth rate is $R_0^{1/m}$.

2. STAGE-STRUCTURED MATRIX MODELS

In a stage-structured model, individuals are divided into discrete categories conventionally called "stages" or "stage classes." These sometimes represent discrete stages in the life history, say eggs, larvae, pupae, and adults of an insect species, but very commonly stages are categories imposed on a continuously varying trait such as size. For example, a plant population might be characterized by small, medium, and large individuals, and all between-stage transitions may be possible as a result of growth and shrinkage. Despite this, many of the ideas developed for age-structured populations carry over.

In place of the Leslie matrix, reproduction and the movements of surviving individuals between stages are governed by a population projection or Lefkovitch matrix, **M**. The dynamics are then given by

$$\mathbf{n}(t+1) = \mathbf{M}\mathbf{n}(t). \qquad (12)$$

The Perron-Frobenius Theorem still applies provided that **M** is power positive, so the long-term growth rate is given by the dominant eigenvalue, λ_1, of **M**, and the stable stage distribution by the corresponding eigenvector, $\mathbf{w}_1$.

To give a concrete example, here is the (slightly rounded) projection matrix used by Katriona Shea and David Kelly (1998) to explore the dynamics of *Carduus nutans*, an invasive thistle:

$$\mathbf{M} = \begin{array}{c} \\ SB \\ S \\ M \\ L \end{array} \begin{array}{c} \begin{array}{cccc} SB & S & M & L \end{array} \\ \begin{bmatrix} 0.04 & 8.25 & 179.41 & 503.14 \\ 0.19 & 1.09 & 22.18 & 62.18 \\ 0 & 0.01 & 0 & 0 \\ 0 & 0.01 & 0.02 & 0 \end{bmatrix} \end{array}. \qquad (13)$$

SB is the number of seeds in the seedbank, and S, M, L refer to thistle rosettes that are small, medium, and large in size. The matrix has the following simple interpretation: each column gives the expected contribution of a particular stage to each of the other stages. So the first column says that 4% of the seeds in the seedbank will stay there, and 19% will become small rosettes; the second column says that each small rosette will give rise to 8.25 seeds in the seedbank, 1.09 small rosettes, and a small number of medium and large rosettes, and so on.

Constructing the matrix **M** for a real population requires selecting appropriate stages. If the life cycle is divided into discrete stages, this is straightforward. Otherwise things become more complicated, as it is necessary to (1) decide on the appropriate measure of individual state and (2) set the boundaries between stages. Practical issues of data collection and the ability to predict an individual's fate may determine how to measure an individual's state. Typically a single variable is used (e.g., longest leaf length or rosette diameter as a measure of plant size), but more complex classifications, say by age and size, are also possible. Setting boundaries may be problematic. Ideally there should be many categories, so all individuals within a category really behave in a similar way, as the model assumes. However, the more categories there are, the fewer observations there are on each category, so estimates of the elements of **M** become less reliable. Integral projection models, discussed in the next section, provide an elegant way around these problems.

An enormous amount of work has been done analyzing the properties of projection matrices and using those properties to study real populations, much of it summarized in the landmark monograph by Hal Caswell (Caswell, 2001; first edition 1989). For example, we can use elasticity analysis to explore how fractional changes in matrix elements affect the long-term population growth rate λ_1. Specifically, defining

$$e_{ij} = \frac{\text{fractional change in } \lambda_1}{\text{fractional change in } m_{ij}} = \frac{\partial \lambda_1 / \lambda_1}{\partial m_{ij} / m_{ij}}, \qquad (14)$$

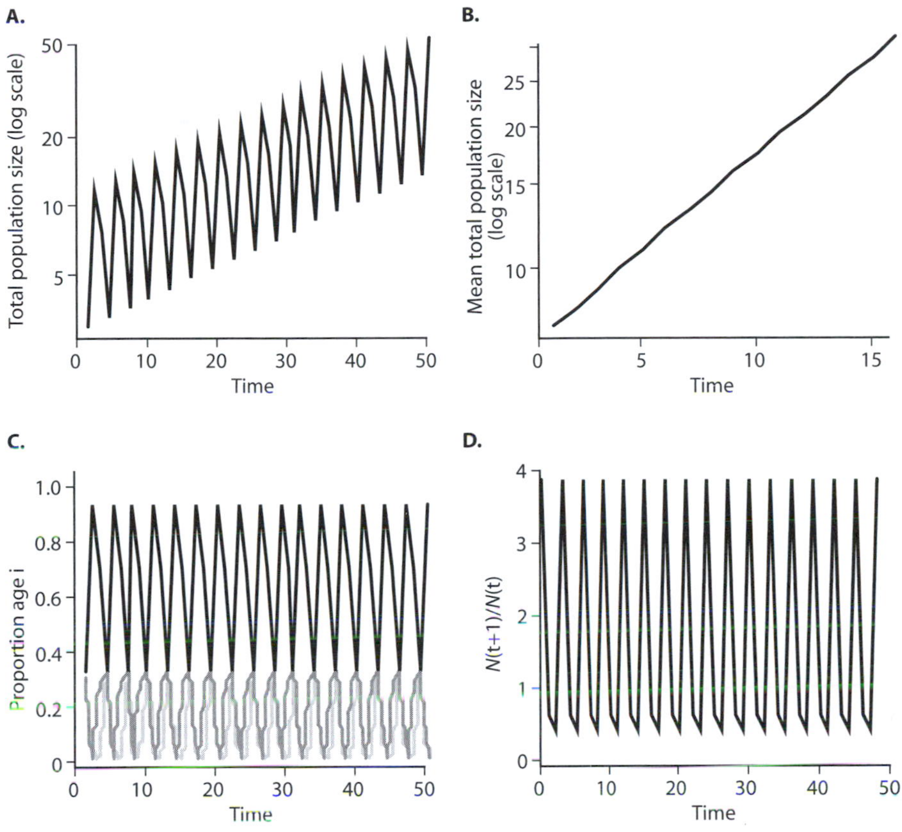

Figure 2. Numerical solution of the Leslie matrix model (equation 11) assuming $\mathbf{n}(0) = (1, 1, 1)$ in each panel; we have a time series plot of (A) total population size $N(t)$, (B) total population size averaged over the cycle, (C) the proportion of individuals in each age class (black = 0, dark gray = 1, and light gray = 2), and (D) the population growth rate $N(t+1)/N(t)$.

it can be shown that

$$e_{ij} = \frac{m_{ij}\ v_i w_j}{\lambda_1 \mathbf{v}\cdot\mathbf{w}}, \tag{15}$$

where $\mathbf{v}\cdot\mathbf{w}$ is the dot-product ($\mathbf{v}\cdot\mathbf{w} = v_1w_1 + v_2w_2 + \cdots v_nw_n$), and $\mathbf{v}$ is the left eigenvector of $\mathbf{M}$ ($\mathbf{vM} = \lambda\mathbf{M}$). For the thistle matrix (equation 13), the elasticities are

$$\mathbf{e} = \begin{array}{c} \\ SB \\ S \\ M \\ L \end{array} \begin{array}{c} \begin{array}{cccc} SB & S & M & L \end{array} \\ \begin{bmatrix} 0.004 & 0.198 & 0.017 & 0.031 \\ 0.247 & 0.308 & 0.025 & 0.045 \\ 0 & 0.044 & 0 & 0 \\ 0 & 0.075 & 0.001 & 0 \end{bmatrix} \end{array}.$$

These results suggest that the transitions $SB \to S$, $S \to SB$, and $S \to S$ are critical for population growth, and therefore, management strategies should focus on reducing these transitions. The unintuitive prediction that it will be far more effective to concentrate on small plants rather than large ones is made apparent only by computing the elasticities.

Matrix models can be generalized in many ways, such as by adding density dependence and/or stochastic variation from one time step to the next. Exploring all these would require an entire large book, which, fortunately, Caswell (2001) has already written.

3. INTEGRAL PROJECTION MODELS

Plant and many other types of organisms do not just come in small, medium, and large sizes. For example, consider Platte thistle, with individual size measured by the root crown diameter (figure 3). If individuals are divided into three size classes (indicated by the vertical lines in figure 3), then from the fitted curves, it is clear that some categories contain very different individuals; for example, individuals in the "large" category have probabilities of flowering that vary systematically from ≈0.2 to over 0.8. A matrix projection model with three size classes ignores these differences and treats all "large" individuals as identical.

To avoid this problem, in 2000 Michael R. Easterling, Stephen P. Ellner, and Philip M. Dixon proposed

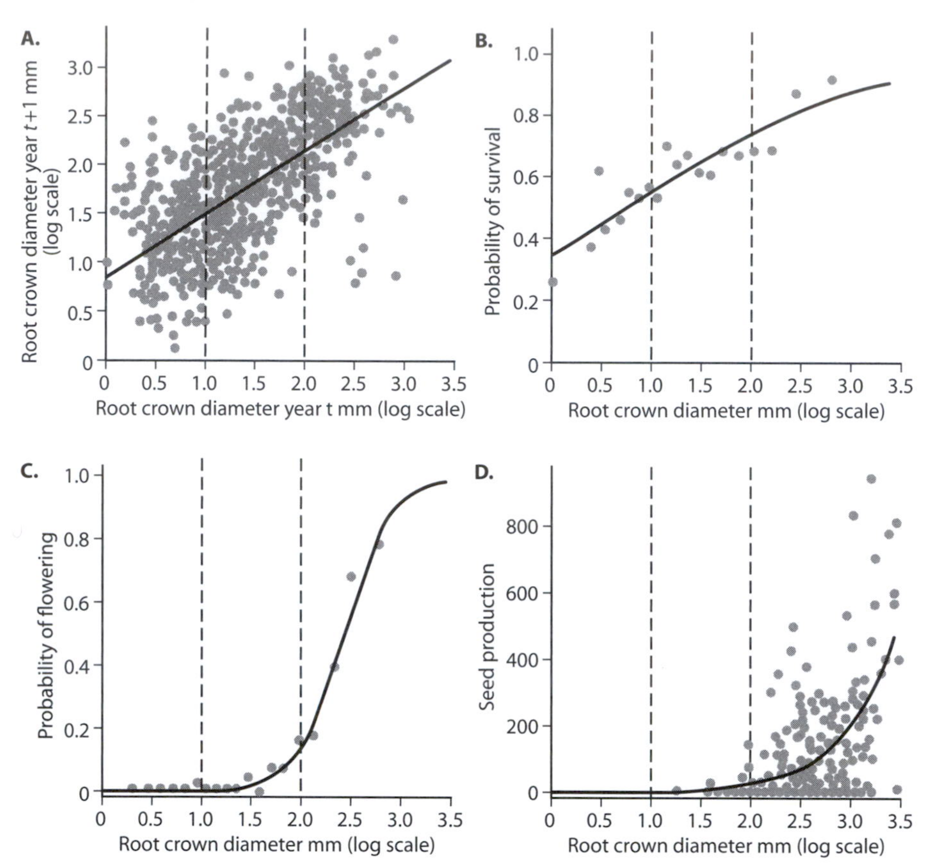

Figure 3. Size-structured demographic rates for Platte thistle, *Cirsium canescens*. (A) Growth (as characterized by plant size in successive years), (B) survival , (C) the probability of flowering, and (D) seed production all vary continuously with size and can be described by simple regression models. (Redrawn from Rose et al., 2005) In panels B and C, the data were divided into 20 equal-sized categories, and the plotted points are fractions within each category, but the logistic regression models (plotted as curves) were fitted to the binary values (e.g., flowering or not flowering) for each individual.

the integral projection model (IPM) in which individuals are characterized by a continuous variable x such as size. The state of the population given by $n(x,t)$, such that the number of individuals with sizes between a and b is $\int_a^b n(x,t)dx$. Instead of the matrix **M**, the IPM has a projection kernel $K(y,x)$, so that

$$n(y,t+1) = \int_s^S K(y,x)n(x,t)dx,$$

where s and S are the minimum and maximum possible sizes. The integration is the continuous version of equation 4, adding up all the contributions to size y at time $t+1$ by individuals of size x at time t. Providing some technical conditions are met (see Ellner and Rees, 2006, for details), the IPM behaves essentially like a matrix model, and so the results described above carry over.

Constructing the projection kernel $K(y,x)$ is straightforward using the regressions shown in figure 3. For an individual of size x to become size y, it must (1) grow from x to y, (2) survive, and (3) not flower (flowering is fatal in monocarpic plants like Platte thistle). These probabilities are calculated from the fitted relationships in figures 3A, 3B, and 3C, respectively. The use of regression models to construct the projection kernel brings some advantages: (1) accepted statistical approaches can be used for selecting an appropriate regression model; and (2) additional variables characterizing individuals' states can be included by adding explanatory variables rather than having to select a single best state variable. For example, in some thistles the probability of flowering depends on both an individual's size and age and is often described by a logistic regression such as $\text{logit}\, p_f(a,x) = \exp(\beta_0 + \beta_s x + \beta_a a)$. Extending a size-structured model to include age-dependent flowering therefore requires the estimation of a single additional parameter rather than estimation of many age- and size-class-specific flowering probabilities in the analogous matrix model.

4. CONTINUOUS-TIME MODELS WITH AGE STRUCTURE

The simplest starting point is the continuous-time analog of the Leslie matrix, in which vital rates depend only on individual age a, ignoring effects of population density and environmental factors. The state of the population (as usual counting only females) is then characterized by $n(a,t)$, so that as in IPMs

$$\int_a^b n(s,t)ds = \text{Number of individuals of age } a \text{ to } b. \tag{16}$$

The dynamics of $n(a,t)$ are generated by the age-specific per capita birthrate $b(a)$ and death rate $\mu(a)$. To be age a at time t, an individual must have been age $a - dt$ at time $t - dt$ and have not died; that is,

$$n(a,t) = n(a - dt, t - dt)[1 - \mu(a - dt)dt].$$

Rearranging and letting $dt \to 0$, we obtain the McKendrick–von Foerster equation

$$\frac{\partial n}{\partial t} + \frac{\partial n}{\partial a} = -\mu(a)n(a,t), \tag{17}$$

which describes the dynamics of $n(a,t)$ for $a > 0$. The boundary condition, describing the birth of new individuals, is

$$n(0,t) = \int_0^\infty b(a)n(a,t)da. \tag{18}$$

As with the previous models we expect exponential solutions, so $n(a,t) \approx Cn^*(a)e^{rt}$. By arguments analogous to those for the discrete-time age-structured model, the long-term instantaneous growth rate r can be shown to satisfy the continuous Euler-Lotka equation

$$1 = \int_0^\infty e^{-ra} l(a)b(a)da, \tag{19}$$

where $l(a)$ is the chance of surviving to age a given by

$$l(a) = \exp\left(-\int_o^a \mu(s)ds\right).$$

The fate of the population then depends on r, with increasing populations having $r > 0$ and decreasing ones $r < 0$. So not surprisingly r, like λ, plays a key role in population management and life-history theory. Just as in the discrete case, we can compute expected lifetime fecundity

$$R_0 = \int_0^\infty l(a)b(a)da,$$

so $r > 0$ if and only if $R_0 > 1$, as expected.

Age has the unforgiving property that one year on, you will always be one year older, and this property was exploited when studying age-structured models. In a size-structured model, we must specify how size changes over time. It is then possible, by looking at the flows of individuals into and out of some small size range, to derive the dynamics of the size structure. If individuals grow deterministically, growth can be described by an equation $dm/dt = g(m)$. This leads to the McKendrick–von Foerster equation for size-structured dynamics (although it was almost surely known to Euler as the equation for passive particles carried by a moving fluid):

$$\frac{\partial n}{\partial t} + \frac{\partial (gn)}{\partial m} = -\mu(m)n(m,t). \tag{20}$$

Specifying appropriate boundary conditions is less straightforward for equation 20 than for equation 17. If we assume that individuals are size m_0 at birth, then equation 20 applies for $m > m_0$, and the boundary condition is

$$n(m_0,t) = \frac{1}{g(m_0)} \int_{m_0}^\infty b(m)n(m,t)dm; \tag{21}$$

the prefactor before the integral is needed to convert the birthrate (individuals per unit time) into the resulting contribution to the size distribution (individuals per increment of size). Much as for age-structured models, the size-structured model has a long-term size structure and population growth rate, which can be derived using the (nonlinear) relationship between age and size entailed by the deterministic growth pattern. The basic model can be extended in many ways, such as allowing a random component to growth (including a chance of shrinkage), variable offspring size, reproduction by fission, and characterization of individuals by multiple measures of size (e.g., lipid and nonlipid body mass) or by size and age.

Extending the basic models to include density dependence and interspecific interactions is difficult, and indeed, William S. C. Gurney and Roger Nisbet (1998) described continuous-time models in which individuals are distinguishable by age *and* size as "a traditional source of mathematical headaches." To make some progress, Gurney, Nisbet, and John Lawton (1983) suggested grouping individuals into stages and assuming constant vital rates within each stage. However, unlike the stage-structured matrix model, individuals within a stage may have different states. For example, even if all juveniles have the same growth rate, younger juveniles may be smaller and less likely to mature soon into adults.

To see what this means, consider Gurney, Nisbet, and Lawton's model for laboratory blowfly populations based on the classic experimental studies of A. J. Nicholson. The model assumes two stages, Juvenile and Adult.

1. Ages 0 to τ are Juveniles with a constant per capita mortality rate $\mu(a,t)=\mu_J$ and birthrate $b(a,t)=0$.
2. Ages τ and above are Adults with a constant per capita mortality rate $\mu(a,t)=\mu_A$ and per capita birthrate $b(a,t)=qe^{-cN_A(t)}$, where $N_A(t)$ is adult population size, so the birthrate is density dependent.

These assumptions imply a set of differential equations describing the dynamics,

$$\begin{aligned} dN_J/dt &= R_J(t) - R_A(t) - \mu_J N_J \\ dN_A/dt &= R_A(t) - \mu_A N_A, \end{aligned} \tag{22}$$

where $R_J(t)$ and $R_A(t)$ are the recruitment rates into the Juvenile and Adult stages. By definition, $R_J(t) = qN_A(t)e^{-cN_A(t)}$. Because the Juvenile stage lasts exactly τ time units, $R_A(t)$ equals the recruitment into the Juvenile stage $t-\tau$ time units ago times the survival through the Juvenile stage $S_J = e^{-\tau\mu_J}$; hence, $R_A(t) = S_J R_J(t-\tau)$. Substituting $R_J(t)$ into $R_A(t)$ gives a single equation for the dynamics

$$dN_A/dt = S_J N_A(t-\tau)e^{-cN_A(t-\tau)} - \mu_A N_A. \tag{23}$$

The key simplifying assumption in this model is that all Juveniles have the same demographic rates. Juveniles do differ in their state though: some are nearly mature and will soon become adults, whereas others are recently born and will not become mature for some time. Although the final model involves only the total numbers in each class, its structure reflects the fact that newborns all wait τ time units before maturing into Adults. The presence of the time delay τ is the price for allowing individuals within a stage to differ in state. In this model, stages correspond exactly to a range of ages, but similar models can be constructed in which stages correspond to a range of sizes, or in which there is no exact correspondence of stages to age or size; rather, each individual within a stage has a probability (potentially depending on age, size, etc.) of making the transition to another stage. These models may require state variables in addition to the population counts for each stage to track the within-stage state dynamics and its consequences.

5. APPLICATIONS AND EXTENSIONS

Here we discuss some applications of the ideas presented in the previous sections: a stochastic density-dependent Leslie matrix model for an Asiatic wild ass; stage-structured models used to understand the dynamics of laboratory populations; and use of integral projection models to make evolutionary predictions of life-history strategies.

Climate Variability and Persistence of Asiatic Wild Ass

David Saltz, Daniel I. Rubinstein, and Gary C. White (2006) developed a Leslie matrix-type model for the population of an Asiatic wild ass (*Equus hemionus*) reintroduced into the Makhtesh Ramon Nature Reserve in Israel, based on long-term monitoring (1985–1999). Their main goal was to explore possible effects of increased rainfall variability on the population's risk of extinction because increased variability is predicted under some global climate change scenarios even in areas where no changes in mean rainfall are predicted.

Their model includes a number of important extensions to the basic Leslie matrix model described above. Saltz et al. used their data to fit a model predicting an adult (age ≥ 3) female's chance of successful reproduction as a function of total rainfall in the current and previous years. Their model also included a negative effect of adult female density, but reproductive success was unrelated to age. Age-specific survivorship was based on a published survivorship curve for zebra, with additional mortality of 30% or higher during drought years (rainfall < 40 mm) based on data for other ungulates.

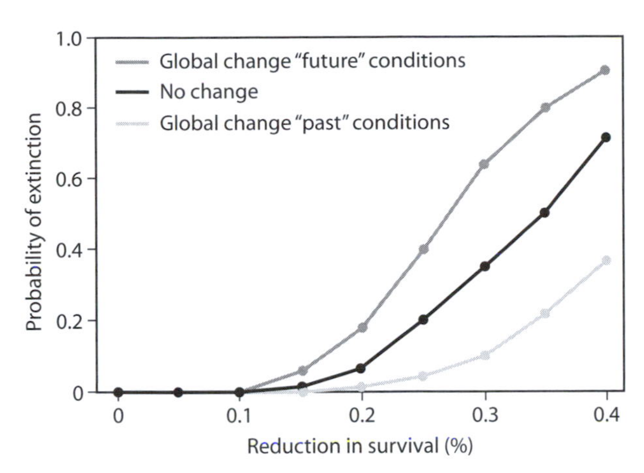

Figure 4. Probabilities of extinction of an Asiatic wild ass population under various climate change scenarios.

Because the model was explicitly linked to variation in rainfall, model simulations could be based on the 41 years of rainfall data for the study area. In particular, simulations to assess extinction risk over a 100-year time period were run by bootstrapping from either the first 20 years of rainfall data (when variance was lower), the second 21 years (when variance was higher), or the complete data set. They also incorporated demographic stochasticity: for example, rather than having 30% of adults die in a drought year, they did a simulated "coin toss" to determine whether each individual lived or died.

The strong effect of drought years on survival and reproduction produced a strong impact of rainfall variability on population persistence. At the low-end estimate of drought-induced mortality (30%), the increase in variance between the first and second halves of the rainfall data produced a more then fivefold increase in the probability of population extinction within 100 years (figure 4).

Nicholson's Blowflies: Continuous-Time Stage Structured Models and Density-Dependent Leslie Matrices

Gurney et al. (1983) used Nicholson's data to estimate the parameters for the model in equation 23. Nicholson conducted a series of long-term experiments, using sheep blowflies, designed to explore the effects of resource limitation at different life stages. The blowfly has four distinct life stages—eggs, larvae, pupae, and adults—but feeds only in the larval and adult stages. In the experiments considered by Gurney et al., larvae were given unlimited resources, whereas the adults received protein (in the form of ground liver) at a fixed rate, leading to competition. To apply the stage-structured model, Gurney et al. (1983) estimated the stage-specific mortalities, fecundities, and durations:

- Using the duration of each stage and the stage-specific mortality to estimate the egg-to-adult delay time as $\tau \approx 15.6$ days and egg-to-adult survival $S_J \approx 0.91$.
- Estimating the egg-production rate by combining data on egg production versus food supply with the assumption that food is divided evenly among adults, to get $b(N_A) \approx 8.5e^{-N_A/600}$.
- Using the rate of decline in adult population when no recruitment is occurring to estimate the adult mortality rate $\mu_A \approx 0.27/\text{day}$.

With these estimates, the model produces sustained cycles with a period of about 37 days (compared to an average observed period of about 38 days), and adult population varying between a minimum of 150 and a maximum of 5400 (compared to observed minima and maxima of 270 ± 120 and 7500 ± 500; figure 5). This is remarkable given that no model parameters were adjusted to fit the experimental time series, and perhaps even more remarkably the model solutions exhibit the "double peak" that usually occurred in the data. The population cycles occur because egg production is overcompensating, and the period of the cycles is determined by τ; analysis of the model suggests the period will be in the range $(2\tau, 4\tau)$, in good agreement with the numerical solution.

To simulate the model without the difficulties of solving delay-differential equations, it can be expressed as an age-structured model, similar to equation 2. Because the maturation time is 15.6 days, it is convenient to use time and age increments of 0.1 days. The instantaneous mortality and fecundity rates in the continuous-time model can be converted into rates per 0.1 days. For example, if juveniles in the discrete-time model become mature adults when they exit the 15.6-day-old age class, the juvenile survival probability in the discrete-time model is $S_J^{1/157}$ per time increment. Then 157 age classes are needed for the juveniles, but only one for the adults, giving a Leslie matrix that is large (of size 158×158) and density dependent but straightforward to implement on a computer.

Integral Projection Models for Plants: Linking Evolution and Ecology

Population dynamics and evolution are intimately linked because the fate of new genetic mutants depends on their ability to spread in a population. Coupling evolutionary ideas with demographic models for the

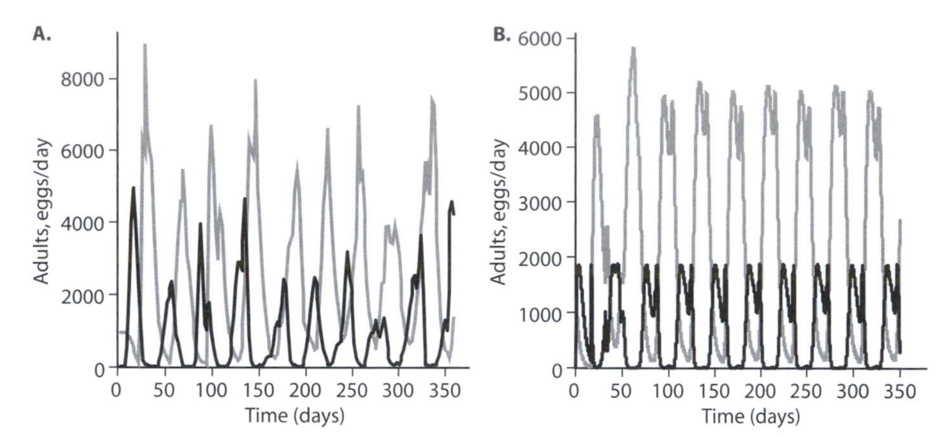

Figure 5. (A) Experimental time series of blowfly adult (gray) and egg (black) dynamics (from Nicholson, 1954) and (B) predicted dynamics from the simple stage-structured model (equation 23).

growth of the mutant subpopulation thus allows predictions of how natural selection shapes individual behaviors and life histories. The key evolutionary idea is John Maynard Smith's concept of an evolutionary stable strategy (ESS): a strategy that cannot be displaced by a rare mutant if it has become fixed in a population. In a population at demographic equilibrium ($\lambda = 1$), the established strategy cannot be displaced if λ is less than 1 for a rare mutant strategy with some other strategy. In this way, ESSs in real populations can be identified.

As an illustration, an integral projection model for *Oenothera biennis* (evening primrose) can be used to predict the size dependence of flowering probability (figure 3C). This relationship is determined by balancing the benefits of flowering at a large size (increased seed production, figure 3D) against the mortality costs of growing large. Using published data, Rees and Rose (2002) produced a fully parameterized IPM for this species. The probability of flowering was size dependent and described by a logistic regression, $\text{logit}p_f(x) = \beta_0 + \beta_s x$, where x is log rosette diameter, β_0 and β_s and the fitted intercept and slope. Making β_0 smaller reduces the probability of flowering for all sizes and so increases the mean size at flowering. In *Oenothera,* density dependence acts only at the recruitment stage, so the ESS is characterized by maximizing R_0, the total reproductive output of an individual that survives through the recruitment stage (Mylius and Diekmann, 1995). Numerical evaluation of R_0 as a function of β_0 shows that estimated value is very close to the predicted ESS (figure 6). This example illustrates how structured population models, coupled with evolutionary ideas, provide a general framework for understanding the evolution of organisms' life cycles subject to trade-offs and constraints, a vast subject known as life-history theory.

6. CODA

All populations are structured: by age, size, genotype, social status, and so on. Structured population models have arguably become the core theoretical framework for population ecology, and a modern course on

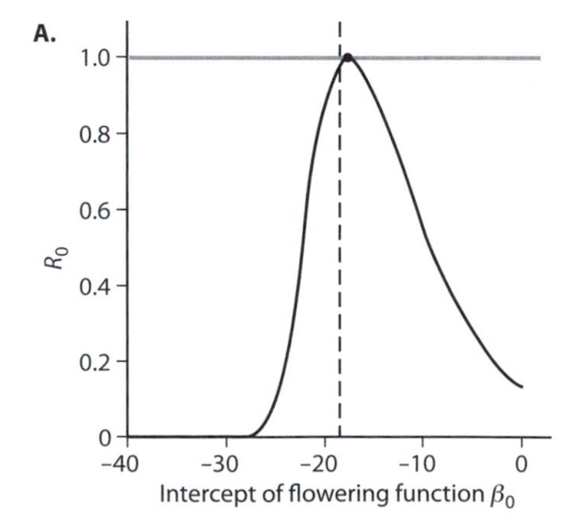

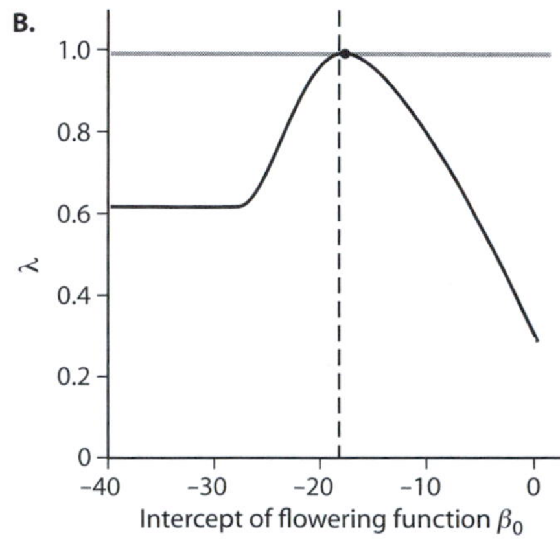

Figure 6. Relationships between β_0 and (A) R_0 and (B) λ for the *Oenothera* IPM. The ESS is marked with a dot, and the estimated β_0 is indicated by the vertical line.

population ecology would be in large part a course on structured population modeling. The scope of theory and applications vastly exceeds the space available here. Read on!

FURTHER READING

Caswell, Hal. 2001, Matrix Population Models: Construction, Analysis and Interpretation. Sunderland, MA: Sinauer Associates (1st edition 1989). *This volume is the classic and comprehensive monograph on matrix models for structured populations, clear, authoritative, and amusing. This book and the ones below by Metz and Diekmann, Tuljapurkar, and Tuljapurkar and Caswell are essential reading for the structured population modeler.*

Ellner, Stephen P., and Mark Rees. 2006. Integral projection models for species with complex demography. American Naturalist 167: 410–428. *This article generalized the size-structured IPM introduced by Easterling, Ellner, and Dixon by allowing individuals to be cross-classified by several traits. The article includes model construction, elasticity analysis, stable distribution theory for density-independent models, evolutionary optimality criteria, and local stability analysis for density-dependent models.*

Gurney, William S. C., and Roger M. Nisbet. 1998. Ecological Dynamics. Oxford: Oxford University Press. *This volume is a very readable text covering a wide range of structured population models, with many case studies that illustrate the concepts and processes involved in constructing structured models.*

Gurney, W.S.C., R. M. Nisbet, and John. H. Lawton. 1983. The systematic formulation of tractable single-species population models incorporating age structure. Journal of Animal Ecology 52: 479–495. *This paper introduced continuous-time models with discrete stage structure and showed how they could explain qualitative differences between the dynamics of two laboratory insect populations.*

Metz, Johannes A. J., and Odo Diekmann, eds. 1986. The Dynamics of Physiologically Structured Populations. Berlin: Springer. *This is the volume that moved continuous-time size-structured models and their relatives into the mainstream of population modeling, with a mix of mathematical theory and applications. Often mathematically challenging; the article by de Roos in Tuljapurkar and Caswell (1997) provides a "gentle introduction" that may be useful to read first.*

Murdoch, William M., Cheryl J. Briggs, and Roger M. Nisbet. 2003. Consumer–Resource Dynamics. Princeton, NJ: Princeton University Press. *This definitive volume presents systematic development and real-world applications of stage-structured models in the style of Gurney-Nisbet-Lawton for predator–prey and host–parasitoid population interactions, summarizing the fruits from two decades of focused effort. If we could all work like this, ecology would be much the better for it.*

Mylius, Sido D., and Odo Diekmann. 1995. On evolutionarily stable life-histories, optimization and the need to be specific about density-dependence. Oikos 74: 218–224. *This represents an elegant theoretical paper analyzing the properties that characterize ESSs in structured populations.*

Tuljapurkar, Shripad. 1990. Population Dynamics in Variable Environments. New York: Springer. *Although sadly now out of print, this volume is an essential reference on stochastic matrix models by the author of many fundamental articles on this topic.*

Tuljapurkar, S., and H. Caswell, eds. 1997. Structured-Population Models in Marine, Terrestrial, and Freshwater Systems. New York: Chapman & Hall. *This book provides a wealth of applications, and some accessible reviews of basic theory, derived from a summer course on structured population models at Cornell University.*

Applications discussed in this chapter

Rees, M., and K. E. Rose. 2002. Evolution of flowering strategies in *Oenothera glazioviana*: An integral projection model approach. Proceedings of the Royal Society 269: 1509–1515.

Rose, K. E., S. M. Louda, and M. Rees. 2005. Demographic and evolutionary impacts of native and invasive insect herbivores: A case study with Platte thistle, *Cirsium canescens*. Ecology 86: 453–465.

Saltz, David, Daniel I. Rubenstein, and Gary C. White. 2006. The impact of increased environmental stochasticity due to climate change on the dynamics of Asiatic wild ass. Conservation Biology 20: 1402–1409.

Shea, K., and D. Kelly. 1998. Estimating biocontrol agent impact with matrix models: *Carduus nutans* in New Zealand. Ecological Applications 8: 824–832.

II.2

Density Dependence and Single-Species Population Dynamics

Anthony R. Ives

OUTLINE

1. Three questions about the dynamics of single species
2. Density dependence
3. Endogenous population variability
4. Exogenous population variability
5. Returning to the three questions

In ecology, population dynamics refers to how populations of a species change through time. The study of single-species population dynamics encompasses three general questions: (1) What explains the average abundance of a population? (2) What explains the fluctuations in abundance of a population through time? and (3) How do average abundances and fluctuations in abundance vary among populations in different geographic locations? Any of these questions can be asked of any population of any species, yet some populations pose particularly interesting challenges for one or more of the questions. Thus, ecologists often focus on populations that are remarkably large (pests) or small (endangered species), that have dramatic fluctuations through time, or that vary markedly from one location to another.

GLOSSARY

density dependence. Density-dependent population growth occurs when the per capita population growth rate changes as the population density changes. Because it changes with population density, density-dependent growth is not exponential.

dynamics. The dynamics of a population consists of the changes through time in the population size or a related measure such as density.

endogenous variability. Endogenous population variability is driven by density-dependent factors that involve interactions among individuals in the system specified by a researcher. The system could consist of a single population or populations of interacting species.

exogenous variability. Exogenous population variability is driven by factors outside the system that are not themselves influenced by population fluctuations within the system. Examples include not only environmental factors such as weather but also the abundances of other species if the dynamics of these species is not affected by the focal species within the system.

exponential population growth and decline. When the per capita population growth rate remains constant, the population experiences exponential growth or decline. Exponential population growth can also occur when the per capita population growth rate varies through time provided its average remains constant.

intrinsic rate of increase. The intrinsic rate of increase is the maximum per capita population growth rate for a population with a stable age structure (i.e., the proportions of the population in different age groups remain the same). The intrinsic rate of increase is often achieved when the population is at low density.

per capita population growth rate. The per capita population growth rate is the rate at which a population changes per individual in the population. It is often expressed as the natural logarithm of the ratio of population densities at consecutive sample times, $\log_e x(t+1)/x(t)$.

population. A population is a group of individuals of the same species occupying a specified geographic area over a specified period of time. The area may be ecologically relevant (an island) or irrelevant (political districts), and the boundaries may be porous, with individuals immigrating to and emigrating from the population.

stability. Stability is defined in many ways in ecology. In models of population dynamics, stability is generally used in two ways. First, when there is no environmental stochasticity, stability describes how populations change when they are around points or cycles. A stable point, for example, is one in which, if the population density is initially near the point, it will move generally closer to the point through time. Second, when there is environmental stochasticity, a more stable system is one in which population variability is small for a given level of environmental variability in the per capita population growth rate. There are additional ways that stability can be defined in model and real systems, which necessitate care in using the word *stability*.

stochasticity. Stochasticity is random (unpredictable) variability that is described by a probability distribution giving the mean, variance, and other properties of the random process.

1. THREE QUESTIONS ABOUT THE DYNAMICS OF SINGLE SPECIES

The three broad questions about single-species population dynamics boil down to: What explains the abundance of a population and changes in abundance through time and space? As an example of the first question, we could ask why, unfortunately, there are more mosquitoes than moose in Wisconsin. The answer might seem simple; moose are so much bigger than mosquitoes, they simply occupy more space and need more food. The question becomes more difficult, however, when asking why the roughly 60 species of mosquitoes in Wisconsin differ hugely in abundance. One species, *Aedes vexans*, is many times more common than most other species. Is this because *A. vexans* is more flexible in its breeding requirements and capable of breeding in more diverse habitats than other species? Is it because the females have more catholic tastes for the hosts that unwillingly give up a blood meal that the females convert to eggs? Or is it because *A. vexans* somehow is more adept at avoiding the many predators that turn mosquitoes into lunch? This set of questions poses a real challenge to ecologists, and a challenge of possible practical importance. Along with being common, *A. vexans* females also include humans in their range of suitable hosts, and if we understood why it is so common, we might also be able to change this situation.

The second question is best illustrated with a figure showing two example populations (figure 1). The first, the moose population on Isle Royale in Lake Superior, fluctuates over the 45 years of data, showing a peak of 2500 individuals followed by a crash in the late 1990s to 500 individuals. This fivefold variation, however, is small in comparison to the 500,000-fold variation in the abundance of midges in Lake Myvatn, Iceland. The root causes of the fluctuations of both populations are the same: a combination of depleted food resources (balsam fir trees for moose, algae for midges) and predation (by wolves on moose and a variety of species on midges). Despite having the same general causes, however, why are the fluctuations in midges so much more dramatic?

Finally, an example of the third question is posed by the pattern of population dynamics shown by many small rodent species such as voles. Many populations at high latitudes show strong fluctuations, often exceeding

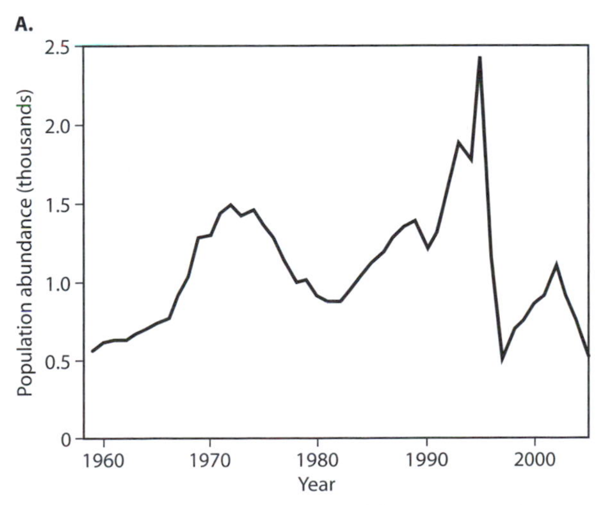

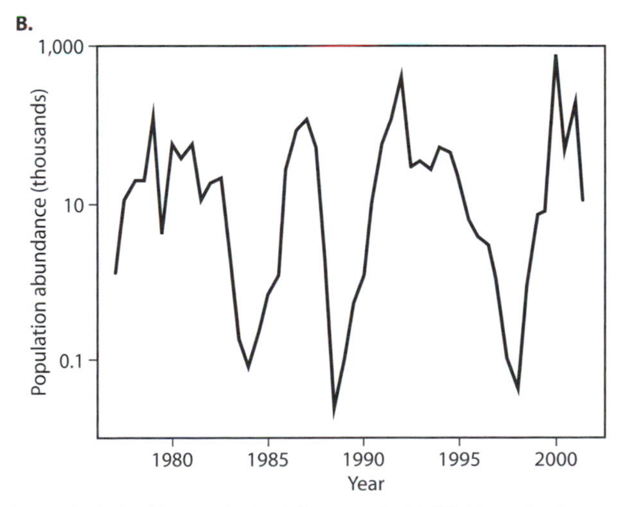

Figure 1. Population abundances of (A) moose on Isle Royale (http://www.isleroyalewolf.org) and (B) the midge *Tanytarsus gracilentus* in Lake Myvatn, Iceland (Ives et al., 2008). Note the $\log_{10}$ scale in B.

two orders of magnitude, whereas populations of the same species fluctuate much less dramatically at lower latitudes. A tempting explanation for this pattern is simply that populations at higher latitudes have to contend with a more severe climate, in particular hard winters, and the severe climate drives greater population fluctuations. This answer cannot be the sole explanation, however, because the fluctuations in rodent populations at high latitudes do not match the fluctuations in weather conditions. In fact, some peaks in rodent abundances occur in winter instead of summer. Ecologists suspect that predators are involved in the high population fluctuations of rodents at high latitudes and that the importance of these predators for some reason diminishes for populations closer to the equator.

This chapter is about single-species population dynamics, so a discussion of what is a single-species population is in order. In ecology, what is meant by a population is often given by the context of the discussion or study. In the examples above, the moose population on Isle Royale is clearly delineated; it is the number of individuals on the island, and Lake Superior gives a clear, ecologically relevant boundary through which moose do not easily pass (despite being excellent swimmers). In other cases, the boundaries might be clear but ecologically irrelevant. For example, the population of moose in Wisconsin is delineated by political, not ecological boundaries. For issues of population management, political boundaries make sense, but not so for ecological concerns. Furthermore, the boundaries are porous, with moose obliviously crossing from Michigan to Wisconsin and back again. Nonetheless, the moose population of Wisconsin can still be delineated and counted, making it clear what the population is.

Obviously, single-species populations consist of a single species, although no species lives on an island unto itself. Population dynamics of any species will be affected by other species—species that it consumes, species that consume it, species that compete, and species that might in some way help (such as pollinators helping plants). Although ecologists recognize the importance of interactions among species, very often studies are conducted on a single focal species, with other species considered only to the extent that they affect the dynamics of the focal species. In this way, other species are treated somewhat like the weather. Often, the reason for focusing on a single species is simply pragmatic. If there is a single species that is of applied or academic interest, it is necessary to limit the ecological extent of the investigation to what can sensibly be studied with the resources available.

The other topic of this chapter is density dependence, and explaining this requires an entire section.

2. DENSITY DEPENDENCE

The easiest way to explain density dependence is to consider when it is absent. In particular, consider European rabbits. Rabbits have remarkable reproductive proclivity. At 3 months of age, the females start to breed, and a single pair can produce up to 40 offspring per year. The consequence of this breeding ability was seen when Thomas Austin introduced 24 rabbits for sport hunting onto his property in the state of Victoria, Australia, on Christmas Day, 1859. Within 10 years, the descendants of these rabbits reached a population numbering in the millions that spread throughout much of eastern Australia. By the early twentieth century, the population plateaued at about 200 million.

The population growth of rabbits can described by a simple mathematical formula,

$$x(t+1) = e^{r}x(t),$$

where $x(t)$ is the population size of rabbits in year t, and r is a biological parameter called the intrinsic rate of increase, which gives the maximum rate at which the rabbit population can increase. Studies have shown that r for invading European rabbits is roughly 2.5 year^{-1}, which when plugged into this equation means the population can increase by a factor of $e^{2.5} = 12$ each year. This reproductive potential is impressive, the more so when the equation above is used to predict the growth of the rabbit population. Starting with 24 rabbits, assuming they and their descendants maintain the intrinsic rate of increase of 2.5 year^{-1}, in 5 years the population would be over 250,000, and in 10 years it would be almost 100 billion (10^{11}). Following this exponential growth a little longer, in about 21 years the mass of rabbits would exceed that of the Earth (6×10^{24}kg).

This did not happen. The reason is that rabbits experienced density-dependent reductions in their per capita population growth rate. The per capita population growth rate is the natural logarithm of the number of individuals in a population at some time $t+1$ divided by the number at time t, $\log_e x(t+1)/x(t)$, where time is measured in units appropriate for the species (years for rabbits, minutes for bacteria, etc.). The per capita population growth rate integrates both reproduction and survival, and when the per capita population growth rate is density dependent, the birth and/or death rates change with density. For rabbits in Australia, when densities became very high, birthrates declined and death rates increased as rabbits suffered food shortages.

Like rabbits, all populations cannot maintain their intrinsic rate of increase indefinitely. Eventually, densities will become high enough that birthrates decline and/

or death rates increase. Eventually, the per capita population growth rate will drop to zero. Populations that persist for long periods of time must have negative per capita population growth rates at high densities that stop unbounded increases and positive per capita population growth rates at low densities to stop population extinction. In fact, in the long run, a population must have an average per capita population growth rate of zero.

Although all persisting populations must have density-dependent population dynamics, the factors leading to density dependence are often multiple, complex, and not easily identified and understood. For example, the exotic dynamics shown by midges in Lake Myvatn (figure 1B), with fluctuations of over five orders of magnitude, involves density dependence that causes populations to crash from very high densities. But what explains the timing of the crashes, why do the midges crash for several generations in a row, and what saves the population at low density so that the species does not become extinct? Detailed answers to these questions about midges, and similar questions for other species, are often extremely hard to answer. Much of the study of population ecology revolves around explaining the factors causing density dependence and the consequences they have for population dynamics.

3. ENDOGENOUS POPULATION VARIABILITY

Density dependence not only bounds a population above and below but also sets the character of the population dynamics. To illustrate this, it is easiest to use another mathematical model that includes a density-dependent per capita population growth rate, specifically

$$x(t+1) = e^{r(1 - x(t)/K)}x(t),$$

where, as before, r is the intrinsic rate of increase, and K is often called the carrying capacity. Here, the per capita population growth rate, $r(1 - x(t)/K)$, decreases as the population size $x(t)$ increases, reaching zero when $x(t) = K$. Thus, K for the rabbits in Australia would be around 200 million. Because the per capita population growth rate is positive when $x(t)$ is less than K and negative when $x(t)$ is greater than K, it seems reasonable to expect that the population will eventually settle close to K. Although this is the case sometimes, it is not always so. This is because density dependence itself can generate population variability.

Before we proceed, a disclaimer is needed. Simple models such as the one above are extremely helpful in understanding basic ecological phenomena, such as the possible consequences of density dependence. Nonetheless, they are not very realistic and do not necessarily do a good job describing the dynamics of any real species. But it is in fact their unrealistic simplicity that makes these models didactically valuable; the point is not that real populations act exactly like the model but instead that the general types of phenomena shown by the model may in fact have counterparts in real systems.

Figure 2 illustrates the population dynamics generated by the simple model with density-dependent per capita population growth rates for three values of the intrinsic rate of increase, r. When r is low ($r = 0.1$), the population can increase only slowly, so a graph of $x(t+1)$ versus $x(t)$ is nearly a straight line (figure 2A). Nonetheless, the line curves down slightly, showing that the population is increasing when $x(t)$ is below the carrying capacity K and decreasing when $x(t)$ is above K. When r is intermediate ($r = 1$), there is a higher per capita population growth rate when densities are low, yet the per capita population growth rate declines more rapidly so that $x(t+1) = x(t)$ again when $x(t) = K$ (figure 2B). Finally, when r is high ($r = 2.2$), the population increases very rapidly from low densities and declines rapidly from high densities (figure 2C). In fact, it increases so rapidly that the relationship between $x(t+1)$ and $x(t)$ is strongly hump-shaped; when the population density starts at some intermediate value, say $K/2$, the resulting population at the next time step is higher than K, whereas if the population starts at high values, say $2K$, the population subsequently crashes to very low levels.

Plots of the model populations over time reflect these patterns. When r is low, the population rises slowly toward the carrying capacity K (figure 2D), whereas when r is intermediate, the population attains K rapidly (figure 2E). However, when r is high, the population experiences perpetual booms and busts; when populations are initially low, they bounce to high densities in the next time step and then drop down again in the time step after that. In this case, the formerly stable carrying capacity becomes unstable; even if the population started very close to K, it would exhibit cycles of increasing amplitude until it settled down to the perpetual cycle. Thus, although the carrying capacity K is unstable, the cycle is stable. Such stable boom-and-bust cycles are rarely seen so starkly in real populations, but this in no way diminishes the lesson from the simple model that density dependence can itself generate population fluctuations. There is no environmental variability in the model, so the only factor creating these cycles is endogenous.

Do purely endogenously driven population fluctuations occur in nature? Certainly, although not necessarily as clearly and simply as in the model. In laboratory systems, sustained fluctuations have been

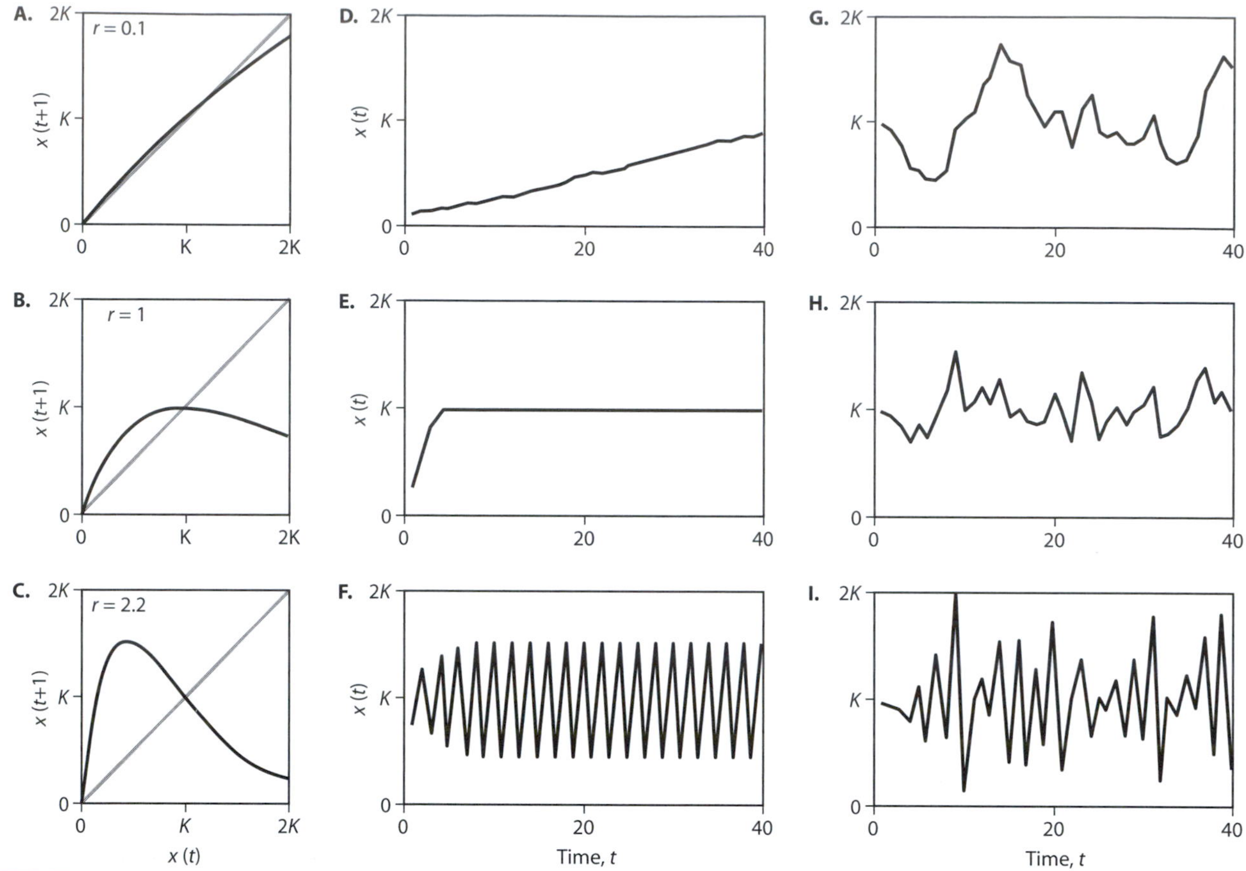

Figure 2. Hypothetical population dynamics generated by the model $x(t+1) = e^{r[1-x(t)/K+\varepsilon(t)]}$ for values of the intrinsic rate of increase r = 0.1, 1, and 2.2. Panels A–C plot the population abundance at time step $t+1$ against the abundance in the previous time step t. The gray line gives the one-to-one line; therefore, $x(t+1)$ crosses the dashed line at K, because at K, $x(t+1) = x(t)$. Panels D–F give trajectories of population abundance, $x(t)$, through time t when there is no environmental stochasticity, $\varepsilon(t) = 0$. Panels G–I give population trajectories with environmental stochasticity starting from K; each trajectory is subjected to the same sequence of environmental fluctuations given by values of $\varepsilon(t)$.

created for flies and beetles that involve either populations that reach high enough densities to consume all their food, or cannibalism in which high densities of adults consume large numbers of juveniles. In nature, sustained cycles consistent with a single-species population model have been observed for fish species in which a few, large adult individuals dominate the population by consuming most of the juveniles. This domination is punctuated by bursts of juvenile success when the current large adults senesce and die, and the burst of juvenile recruitment establishes the next dominant cohort of adults. These types of endogenous fluctuations involve species that have distinct stage structures, a topic discussed in chapter II.1. Even more complex possibilities occur when there are strong interactions among two or more species, another topic described in chapter II.3.

4. EXOGENOUS POPULATION VARIABILITY

In the models so far, the only source of population fluctuations has been density dependence, but real populations are buffeted by purely environmental forces. The consequences of these exogenous sources of population variability on population dynamics depend on density dependence. To illustrate the importance of density dependence, the population model can be modified to include environmental stochasticity—variability that is unpredictable. Specifically, let the per capita population growth rate be $r(1 - x(t)/K) + \varepsilon(t)$, where $\varepsilon(t)$ is selected at random from a normal probability distribution. Thus, the per capita population growth rate includes density dependence and environmental stochasticity.

Figure 2G–I illustrates examples of population trajectories generated by the new stochastic model.

Even though the environmental stochasticity [the values of $\varepsilon(t)$] is the same for all examples, the impact it has on population fluctuations is markedly different. These differences can be understood by comparing populations with and without environmental stochasticity (figure 2D–F). When the intrinsic rate of increase, r, is low, population densities are brought very slowly toward the carrying capacity K (figure 2D). Therefore, in the presence of environmental stochasticity (figure 2G), the weak endogenous force of density dependence does little to counteract the environmental fluctuations that buffet populations away from their initial abundance at K. In contrast, for intermediate r (figure 2E), population densities away from K are returned rapidly, so in the stochastic case, the environmentally driven fluctuations are more rapidly damped out (figure 2H). Finally, for the high value of r there are sustained fluctuations driven by endogenous processes alone (figure 2F). Environmental stochasticity adds to this variability, creating cycles that tend to have higher booms and lower busts (figure 2I).

This simple model with environmental stochasticity illustrates the dual nature of density dependence. Density dependence can reduce the magnitude of population fluctuations if it acts to bring the population rapidly to some stable point (K in the simple model). Conversely, density dependence can itself generate endogenous population variability by driving a point unstable. Thus, understanding population fluctuations requires understanding the interactions between density-dependent per capita population growth rates and exogenous stochasticity.

5. RETURNING TO THE THREE QUESTIONS

The three questions concerning single-species population dynamics—what explains the abundance of a population, and how its abundance changes through time and space—intimately involve density dependence. Density dependence sets the average population abundance because the average abundance is where the average per capita population growth rate is zero. Density dependence also determines the characteristics of the fluctuations in population abundance through time because it can generate endogenous population fluctuations. Furthermore, density dependence determines the impact of environmental stochasticity on population dynamics, either damping out environmental stochasticity rapidly or allowing it to produce large fluctuations in population abundance. Finally, when populations in different geographic locations differ in either average abundance or the characteristics of their dynamics, then density dependence somehow differs among populations.

Although density dependence is fundamental to population dynamics, determining how per capita population growth rates depend on density for a specific species is often very difficult. To measure how per capita population growth rates depend on density, it is necessary to observe populations at both very low and very high abundances. Some populations fluctuate sufficiently violently that they naturally occur at both very low and very high abundances, but other populations do not. Experiments designed to perturb population abundances are often the best way to measure density dependence, although for many species, such manipulative experiments are impossible or unethical. Another difficulty is that species dynamics are generally affected by those of other species, such as species that are eaten by the focal species, compete with the focal species, or eat the focal species. For species with strong and close interactions, the dynamics of one cannot be separated from those of the other species.

Despite these difficulties, we know a lot about the dynamics and density dependence of a large number of plant and animal species. This is not a random selection of species. Some species were studied because they have practical significance, either being pests that we want to eliminate or endangered species that we want to protect. Other well-studied species were selected because they are particularly easy to study, at least relative to related species in other areas. For example, the study of moose on Isle Royale has the advantage that the population is well defined and isolated from factors (such as deer and elk populations) that could complicate its dynamics. A final set of well-studied species consists of those whose dynamics is sufficiently dramatic to beg an explanation that an ecologist cannot resist, for example, the hugely fluctuating midge population in Lake Myvatn. To understand the population dynamics of any species requires long-term study into the many facets of the species ecology that affect its per capita population growth rate.

FURTHER READING

Gotelli, Nicholas J. 2001. A Primer of Ecology, 3rd ed. Sunderland, MA: Sinauer Associates.

Gurney, W.S.C., and R. M. Nisbet. 1998. Ecological Dynamics. New York: Oxford University Press.

Ives, Anthony R. 1998. Population ecology. In S. I. Dodson, T.F.H. Allen, S. R. Carpenter, A. R. Ives, R. L. Jeanne, J. F. Kitchell, N. E. Langston, and M. G. Turner, eds. Ecology. New York: Oxford University Press, 235–314.

Turchin, Peter. 2003. Complex Population Dynamics: A Theoretical/Empirical Synthesis. Princeton, NJ: Princeton University Press.

II.3

Biological Chaos and Complex Dynamics

Alan Hastings

OUTLINE

1. Fluctuations in populations
2. Brief guide to dynamic systems
3. Chaotic dynamics in models in ecology and population biology
4. Search for chaos in data
5. Resolution as noisy clockwork
6. Other complex dynamics

The cause of fluctuations in ecological populations has long been the subject of study, with the goal of understanding the relative importance of exogenous versus endogenous forces in explaining observed dynamics. The discovery of the likelihood of chaotic dynamics in simple discrete-time models that could be used to describe single-species population dynamics spurred much research focused on understanding chaos and its importance and likelihood in ecological systems. To understand the importance of chaos, we consider the role of fluctuations in ecological systems, the generation of chaotic dynamics in models, and the determination of chaos from time series. This naturally leads to more general questions on the role of complex dynamics in ecology and to a more synthetic view of the causes of observed fluctuations.

GLOSSARY

asymptotically stable solution. A solution that is approached by all nearby solutions is asymptotically stable. This is also known as an attractor.

chaos. Chaos is a property of an attractor in a dynamic system that can be roughly characterized as sensitive dependence on initial conditions and can be detected by the presence of a positive Lyapunov exponent.

cycle. A cycle is a solution that repeats at regular intervals.

equilibrium. An equilibrium of a model is a solution that does not change in time.

Lyapunov exponent. A Lyapunov exponent represents the exponential rate of divergence (if positive) or convergence (if negative) of (two) solutions started on or near an attractor.

1. FLUCTUATIONS IN POPULATIONS

A key observation that is central in ecology is that populations fluctuate in time. These fluctuations can exhibit some regularity or can be irregular. Periodical cicadas emerge with great regularity, whereas outbreaks of other insects such as locusts are both dramatic and irregular. The cause of fluctuations in populations in ecology has been a central question in ecology for many years. Early in the history of ecology, Volterra and Lotka focused on the regular oscillations produced by interactions between predator and prey in their models. Shortly thereafter, Gause attempted to reproduce these oscillations in laboratory systems using microorganisms and found that sustained oscillations were difficult to reproduce. In the simple laboratory systems, either the predator ate up all the prey and then starved or the predator could not find enough food and starved with only the prey surviving. This set up a problem that remains until today, namely, what allows predator and prey to coexist. Also, many of the mechanisms that might allow coexistence of species might lead to more complex dynamics, and more often coexisting species fluctuate.

In any examination of natural populations, fluctuations in numbers have been found to be the almost universal outcome. These fluctuations could range from relatively regular cycles, such as those observed in small mammal populations, or more dramatic changes, such as outbreaks of insect populations. A classic debate in ecology has focused on the causes of these fluctuations. One potential source of fluctuations could be external influences, such as changes in weather or climate. These exogenous forces could be responsible for changes in the dynamics of populations, producing cycles that were either regular or irregular. Another cause of changes in the numbers of populations would be endogenous forces within the population that would lead to cycles or more complex dynamics. Enlarging the question to look at fluctuations not just in time but also over space was thought of as one way to decide. Population fluctuations that are synchronous over space would either have to be caused by exogenous forces that were synchronous over space (the Moran effect) or endogenous forces such as dispersal that would synchronize populations.

One particular kind of fluctuating population that we will return to later that deserves special attention is the incidence of disease. In particular, the numbers of individuals with various childhood diseases (measles, mumps, rubella, and others) in the prevaccination era have been intensely studied. These particular fluctuations have played a special role for several reasons. The data are of much higher quality, with more, and more accurate, observations than for the numbers of many organisms. Another important aspect is that the underlying interactions producing the dynamics of diseases are relatively simple and relatively easily described. Finally, because data are known for multiple diseases in multiple locations, deeper understanding is possible.

2. BRIEF GUIDE TO DYNAMIC SYSTEMS

The idea I now consider is that the primary cause of the fluctuations in ecological systems is interactions within and between populations rather than primarily external influences. The role of internal dynamics should be carefully examined in model systems using ideas from the mathematical theory of dynamics. This theory has undergone long development and can be traced back to attempts by physical scientists to understand the motion of the planets and other systems. A brief review of the theory of dynamic systems can elucidate many implicit assumptions ecologists make when using models framed in differential or difference equations or partial difference equations in attempts to understand what drives dynamics of ecological systems.

First, dynamic systems are systems that incorporate changes through time. In many engineering or physical examples, the systems, initial conditions, and parameters controlling the systems are well defined. However, in ecology, this is rarely the case. Ecological systems are so complex that only rough descriptions are often possible, the measurements of population sizes are notoriously difficult, and parameters such as birth and death rates are known only imprecisely. Thus, the way to understand the behavior of ecological systems is not to look for exact solutions of fully specified dynamic systems through either numerical or analytic means. An understanding of what is known as the qualitative behavior is much more appropriate and important.

Fortunately for ecologists, advances in understanding dynamic systems have focused primarily on the qualitative behavior of these systems. One focus is on long-term or asymptotic behavior of dynamic systems. Several definitions are needed. The simplest long-term dynamic is an equilibrium, or a solution that remains constant in time. The next most complex dynamic behavior is a cycle, or a solution that repeats regularly in time, so $x(t+T) = x(t)$ for some cycle period T for all times t. Before we turn to other more complex asymptotic behaviors, we need a definition of stability.

I use heuristic, rather than mathematically rigorous, definitions. An asymptotic solution to a dynamic system is stable if it is approached from all nearby initial conditions. Such a solution is also known as an attractor. Attractors can be as simple as equilibrium points but can also be more complex. Cyclic attractors are also known as limit cycles. Quasicyclic attractors oscillate but with two (superimposed) incommensurate periods so the solutions do not exactly repeat. Attractors can also be chaotic.

A chaotic attractor is a solution that is still stable in the sense that it is approached by nearby solutions. A chaotic attractor has the property that even solutions that start on the chaotic attractor diverge from each other at an exponential rate, so all solutions have very sensitive dependence on initial conditions. The understanding and appreciation of the concept of chaotic dynamics can be traced back to work by Lorenz over 40 years ago and further back to work of Poincaré and Andronov and others more than a century ago. The importance of chaotic dynamics is that it challenges notions of predictability. However, note that even solutions that exhibit chaotic dynamics are predictable over short time scales, even if prediction is not possible over long time scales.

3. CHAOTIC DYNAMICS IN MODELS IN ECOLOGY AND POPULATION BIOLOGY

Within ecology, the first appreciation of chaotic dynamics arose in studies of single-species models with overcompensatory density dependence and nonoverlapping generations. The key idea of overcompensatory density dependence refers to the fact that not only does the per capita production of individuals in the next generation decline as the number of individuals in the current generation goes up, but additionally, the total population size of the next generation eventually declines as the number of individuals in the current generation is increased. These models take the general form

$$N(t+1) = N(t)f[N(t)],$$

where $N(t)$ is the population size of a single species at time t, and $f[N(t)]$ is the mean number of individuals in the next generation left by an individual in the current generation. The dependence of the function f on the population size $N(t)$ is used to describe the action of density-dependent factors in determining the population size of the following generation. Thus, if $f[N(t)]$ is a declining function of the population size, the model exhibits density dependence. And if $N(t)f[N(t)]$ eventually declines, then the system is said to incorporate overcompensatory density dependence. Classic density-dependent terms used in fisheries can be incorporated this way, with the Ricker function exhibiting overcompensatory density dependence, and the Beverton-Holt form not exhibiting this property. It is the functional forms that include overcompensatory density dependence that can lead to complex dynamics and chaos.

One important observation is that chaotic dynamics is essentially a generic property of discrete-time models with strong enough overcompensatory density dependence, so it makes sense to study simple models. Some of the first investigations of chaotic dynamics used an idealized form to describe the population dynamics, namely,

$$x(t+1) = r\,x(t)[1 - x(t)],$$

where $x(t)$ is a scaled (between 0 and 1) measure of population size at time t, and r (between 0 and 4) is a measure both of the growth rate when rare and degree of overcompensation in the density dependence. This seemingly simple model, the quadratic or logistic population model, was extensively investigated by Robert May (1974b) and others in the 1970s and since.

A description of the dependence of the dynamic behavior of the quadratic model as a function of the parameter r provides insight into the possible aspects of chaotic dynamics in populations, and similar behavior is found in other models. For small values of r, the population cannot survive. For larger values of r, the population inevitably approaches a stable equilibrium. As the parameter r is increased, the asymptotic behavior of the model is a two-point cycle with the population alternating between two values. This period-doubling behavior continues as r increases, with asymptotic behavior of four-point cycles, then eight-point cycles, then 16, and so on. The ranges of values of the parameter r over which the cycles of period 2^n occur get smaller and smaller as n gets bigger. A limit is reached at a critical value of r, beyond which the dynamics is much more complex, and chaotic solutions are found. The presence of this period-doubling sequence is one of several "routes to chaos."

Although an emphasis on the study of the discrete-time model for a single species seemed to imply that chaos was strictly a property of ecological systems with nonoverlapping generations, this is not the case. In continuous-time models, there need to be at least three interacting quantities (e.g., species) for the system to exhibit chaotic dynamics. Since the discovery of chaos in discrete-time models, studies have emphasized that chaotic dynamics is likely to be found in simple models of food webs (or food chains) with three or more species with nonlinearities as would arise from functional responses used to describe predation. Essentially, chaos seems to arise when dynamics is characterized by interactions among oscillating systems with different periods as is the case with predator–prey systems. Even systems with enough competitive relationships can produce chaos.

Chaos and other forms of complex dynamics also arise very naturally in descriptions of ecological systems that include different population levels at different spatial locations. Chaotic behavior can arise in models as simple as those describing two coupled predator–prey oscillators.

One additional class of models that can exhibit chaotic dynamics comes from epidemiology. Among the simplest epidemiological models are those phrased in terms of susceptible, infectives, and removed individuals, the SIR model first discussed by Kermack and McKendrick in the 1920s. Because this is a continuous-time model that can be reduced to two quantities, it cannot produce chaos. However, if the contact rate describing the transition from susceptible to infective varies seasonally (periodically), the model does have chaotic solutions.

4. SEARCH FOR CHAOS IN DATA

The behavior of the simplest population models that could exhibit chaotic dynamics, the single-species discrete-time models, seemed to provide a possible way of explaining irregular fluctuations in natural populations. However, the possibility of chaotic dynamics is not the same as the existence of chaotic dynamics. Thus, great efforts have been made to uncover evidence for chaotic dynamics in ecological systems, both natural and in the laboratory.

One procedure for uncovering chaotic dynamics begins with the collection of a time series of population abundances or disease incidence or similar data. At first, the possibility of uncovering chaotic dynamics would seem to be doomed because typical time series of natural populations in ecology or population biology focus on only a single species or the incidence of a single disease. Here a powerful idea that shows how to study a system with many dimensions (e.g., many species) from a single dimension (time series for a single species) comes to the rescue. The idea is based on a powerful mathematical argument (though the conditions that justify the procedure are almost never checked) that says that the full dynamic behavior of a system can be understood and replicated by a reconstruction procedure beginning with a single time series. Instead of focusing on the time series itself, say $x(t)$, consider representing the data with lags and plotting $x(t)$ versus $x(t-T)$ and $x(t-2T)$. The actual number of lags chosen (which may often be more than two) and the length of the lag (T) can be critical, and unfortunately there is no well-established procedure for their choice. However, by using this procedure, one can clearly focus on the search for chaos.

At this point, there is still the problem of looking for a signature of chaos, such as a positive Lyapunov exponent. For data sets much larger than any found in ecology, there would be direct procedures for estimating this Lyapunov exponent. In ecology, given the limited data sets, an approach based on choosing a model and fitting this model to the data and then looking at properties of the best-fitting model must be used. There are relatively standard approaches for estimating a Lyapunov exponent from a model. There are, however, two different kinds of approaches for making models fit, based either on choosing a very general functional form, which is purely phenomenological, or on choosing from among a set of much more mechanistic descriptions of the biological processes. The former approach has the advantage of flexibility while perhaps ignoring important biological constraints. The latter approach has the advantage of biological realism but can be difficult to apply to data from a single time series of a more complex system. The latter approach also depends critically on choosing a good set of candidate models that can exhibit an appropriate range of dynamic behavior.

The general approach of using time series to look for chaos has been applied to a variety of data sets, including childhood diseases (Finkenstadt and Grenfell, 2000) and laboratory systems such as flour beetles (Costantino et al., 1997). The evidence for chaos in childhood diseases such as measles is not completely clear-cut, but there at least seems to be a very strong possibility of chaotic dynamics.

For the flour beetle system studied by Costantino and collaborators (1997), the evidence for chaos in this highly controlled system is clearer. The calculations of the Lyapunov exponent for a model that fits the system well yield a positive Lyapunov exponent, one hallmark of chaos. Moreover, both the model (clearly) and experimental results (somewhat less clearly) also exhibit the period-doubling behavior that is one of the signatures of chaos.

For systems outside the laboratory other than childhood diseases, the evidence for chaotic dynamics is much weaker. This may be because natural systems are not chaotic, but part of this may be a result of the difficulty of obtaining high-quality data. One point that clearly is important is that stochastic forces must play a role, as environmental fluctuations and demographic heterogeneity inevitably influence all populations. Stochasticity clearly plays an important role.

5. RESOLUTION AS NOISY CLOCKWORK

To some extent, so far we have focused on the causes of observed fluctuations in population levels as a dichotomy: either endogenous or exogenous. It is clear that stochastic forces must be important for the dynamics of natural populations, and it is equally clear that there are strong interactions within and between species affecting population dynamics. This sets up what is essentially a false dichotomy between two forces. Instead, it is much more realistic to consider the interplay between endogenous and exogenous forces.

This idea that the dynamics of natural populations must depend on both stochastic and deterministic forces has been referred to as "noisy clockwork" (Bjornstad and Grenfell, 2001), although the idea has a longer history. One important aspect is that the stochastic aspects of population dynamics cannot be thought of as small perturbations of a deterministic population trajectory. Instead, the complex endogenous aspects of population dynamics and the exogenous forces are inexorably intertwined. These two aspects together produce the complex population trajectories we observe.

One can study populations from this point of view and obtain new insights. One can define a Lyapunov exponent for a stochastic system in terms of expectations and therefore sensibly ask whether chaotic dynamics exist in natural systems. However, difficulties of limited data still make detection of chaos a challenge.

6. OTHER COMPLEX DYNAMICS

There is one other way that the emphasis on chaotic dynamics in ecology may have led investigators away from important ecological behavior. Although it is possible to sensibly define chaotic behavior on shorter time scales, much of the study of chaotic behavior, especially in model systems, has emphasized asymptotic behavior. However, many ecological systems may best be understood by studying transient dynamics rather than asymptotic behavior.

The modern approaches to understanding dynamic systems and new statistical approaches for understanding time series that have been used in the study of chaos in ecology can also be used to shed light on other dynamic behavior. Transient dynamics can be studied using ideas from dynamic systems. There are mathematical tools for understanding spatial systems and, in particular, systems of coupled oscillators (e.g., predator–prey systems and epidemiological systems) and synchrony, even in the presence of stochasticity.

New and novel statistical approaches will also likely prove useful in understanding the forces producing observed population fluctuations. Model-based frequentist approaches and Bayesian methods that truly incorporate different kinds of stochasticity will both contribute to a deeper understanding of population dynamics.

FURTHER READING

Bjornstad, O. N., and B. T. Grenfell. 2001. Noisy clockwork: Time series analysis of population fluctuations in animals. Science 293: 638–643.

Costantino, R. F., R. A. Desharnais, J. M. Cushing, and B. Dennis. 1997. Chaotic dynamics in an insect population. Science 275: 389–391.

Finkenstadt, B. F., and B. T. Grenfell. 2000. Time series modelling of childhood diseases: A dynamical systems approach. Journal of the Royal Statistical Society Series C–Applied Statistics 49: 187–205.

Hastings, A. 2004. Transients: The key to long-term ecological understanding? Trends in Ecology and Evolution 19: 39–45.

Hastings, A., C. Hom, S. Ellner, P. Turchin, and H.C.J. Godfray. 1993. Chaos in ecology: Is mother nature a strange attractor? Annual Reviews of Ecology and Systematics 24: 1–33.

Huisman, J., and F. J. Weissing. 2001. Fundamental unpredictability in multispecies competition. American Naturalist 157: 488–494.

May, R. M. 1974a. Stability and Complexity in Model Ecosystems. Princeton, NJ: Princeton University Press.

May, R. M. 1974b. Biological populations with nonoverlapping generations: Stable points, stable cycles, and chaos. Science 186: 645–647.

II.4

Metapopulations and Spatial Population Processes

Ilkka Hanski

OUTLINE

Most landscapes are complex mosaics of many kinds of habitat. From the viewpoint of a particular species, only some habitat types, often called "suitable habitat," provide the necessary resources for population growth. The remaining landscape, often called the (landscape) matrix, can only be traversed by dispersing individuals. Often the suitable habitat occurs in discrete patches, an example of which is a woodland in the midst of cultivated fields—for forest species, the woodland is like an island in the sea. The woodland may be occupied by a local population of a forest species, but many such patches are likely to be temporarily unoccupied because the population became extinct in the past and a new one has not yet been established. At the landscape level, woodlands and other comparable habitat patches comprise networks in which local populations living in individual patches are connected to each other by dispersing individuals. A set of local populations inhabiting a patch network is called a metapopulation. In other cases, the habitat does not consist of discrete patches, but even then, habitat quality is likely to vary from one place to another. Habitat heterogeneity tends to be reflected in a more or less fragmented population structure, and such spatially structured populations may be called metapopulations. Metapopulation biology addresses the ecological, genetic, and evolutionary processes that occur in metapopulations. For instance, in a highly fragmented landscape, all local populations may be so small that they all have a high risk of extinction, yet the metapopulation may persist if new local populations are established by dispersing individuals fast enough to compensate for extinctions. Metapopulation structure and the extinction–colonization dynamics may greatly influence the maintenance of genetic diversity and the course of evolutionary changes. Metapopulation processes play a role in the dynamics of most species because most landscapes are spatially more or less heterogeneous, and many comprise networks of discrete habitat patches. Human land use tends to increase fragmentation of natural habitats, and hence, metapopulation processes are particularly consequential in many human-dominated landscapes.

GLOSSARY

connectivity. An individual habitat patch in a patch network and a local population in a metapopulation are linked to other local populations, if any exist, via dispersal of individuals. Connectivity measures the expected rate of dispersal to a particular patch or population from the surrounding populations.

dispersal. Movement of individuals among local populations in a metapopulation is dispersal. Migration is often used as a synonym of dispersal.

extinction–colonization dynamics. Local populations in a metapopulation may go extinct for many reasons, especially when the populations are small. New local populations may become established in currently unoccupied habitat patches. Local extinction and recolonization are called turnover events.

extinction debt. If the environment becomes less favorable for the persistence of metapopulations through, e.g., habitat loss and fragmentation, species' metapopulations start to decline. For some metapopulations, the new environment may be below the extinction threshold. Extinction debt is defined as the number of species for which the extinction threshold is not met and that are therefore predicted to become extinct but have not yet had time to become extinct.

extinction threshold. Classic metapopulation may persist in a habitat patch network in spite of local extinctions if the rate of recolonization is sufficiently high. Long-term persistence is less likely the smaller the habitat patches are (leads to high extinction rate) and the lower their connectivity (leads to low recolonization rate). Below extinction threshold, recolonizations do not occur fast enough to compensate for local extinctions, and the entire metapopulation becomes extinct even if some habitat patches exist in the landscape.

local population. This is an assemblage of individuals sharing common environment, competing for the same resources, and reproducing with each other. In a fragmented landscape, a local population typically inhabits a discrete habitat patch.

metapopulation. A classic metapopulation is an assemblage of local populations living in a network of habitat patches. More generally, spatially structured populations at landscape scales are often called metapopulations.

metapopulation capacity. This is a measure of the size of the habitat patch network that takes into account the total amount of habitat as well as the influence of fragmentation on metapopulation viability.

network of habitat patches. In a fragmented landscape, habitat occurs in discrete patches, each one of which may be occupied by a local population, and which together compose a network that may be occupied by a metapopulation.

source and sink populations and habitats. A local population that has negative expected growth rate, and that therefore would go extinct without immigration, is called a sink population, and the respective habitat is a sink habitat. A population that has sufficiently high growth rate when small to persist even without immigration is called a source population, and the respective habitat is source habitat.

1. METAPOPULATION PATTERNS AND PROCESSES

Different Kinds of Metapopulations

Many classifications of metapopulations have been proposed, and they serve a purpose in facilitating communication, but it should be recognized that in reality there exists a continuum of spatial population structures rather than discrete types. The following terms are often used.

Classic metapopulations consist of many small or relatively small local populations in patch networks. Small local populations have high or relatively high risk of extinction, and hence, long-term persistence can occur only at the metapopulation level, in a balance between local extinctions and recolonizations (further discussed under the Levins Model below).

Mainland–island metapopulations include one or more populations that are so large and live in sufficiently big expanses of habitat that they have a negligible risk of extinction. These populations, called mainland populations, are stable sources of dispersers to other populations in smaller habitat patches (island populations). The MacArthur-Wilson model of island biogeography is an extension of the mainland–island metapopulation model to a community of many independent species.

Source–sink metapopulations include local populations that inhabit low-quality habitat patches and would therefore have negative growth rate in the absence of immigration (sink populations), and local populations inhabiting high-quality patches in which the respective populations have positive growth rates (source populations; further discussed under Source and Sink Populations below).

Nonequilibrium metapopulations are similar to classic metapopulations, but there is no stochastic balance between extinctions and recolonizations, typically because the environment has recently changed and the extinction rate has increased, the recolonization rate has decreased, or both (this is further discussed under Transient Dynamics and Extinction Debt below).

Dispersal and Population Turnover

Three ecological processes are fundamental to metapopulation dynamics: dispersal, colonization of currently unoccupied habitat patches, and local extinction. Dispersal has several components: emigration, departure of individuals from their current population; movement through the landscape; and immigration, arrival at new populations or at empty habitat patches. All three components depend on the traits of the species and on the characteristics of the habitat and the landscape, but they may also depend on the state of the population, for instance, on population density.

Dispersal may influence local population dynamics. In the case of very small populations, a high rate of emigration may reduce population size and thereby increase the risk of extinction. Conversely, immigration may enhance population size sufficiently to reduce the risk of extinction. Immigration to a currently unoccupied habitat patch is particularly significant in potentially leading to the establishment of a new local

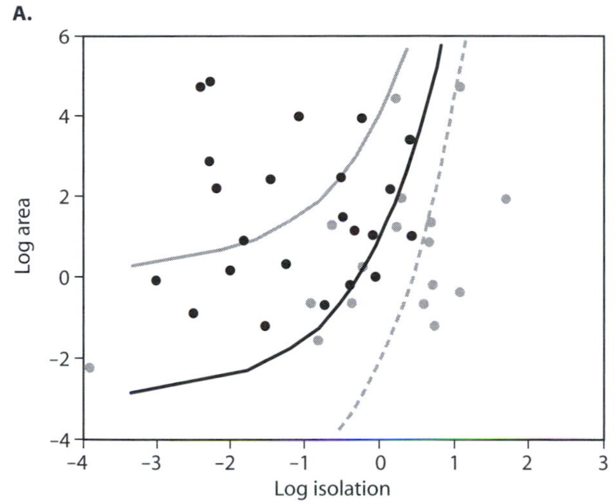

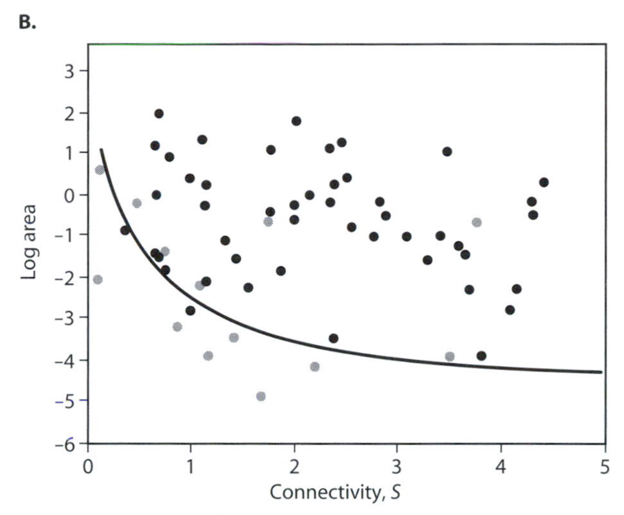

Figure 1. Two examples of the common metapopulation pattern of increasing incidence (probability) of habitat patch occupancy with increasing patch area and connectivity. Black circles represent occupied, gray circles unoccupied, habitat patches at the time of sampling. (A) Mainland–island metapopulation of the shrew *Sorex cinereus* on islands in North America. Isolation, which increases with decreasing connectivity, is here measured by distance to the mainland. The lines indicate the combinations of area and isolation for which the predicted incidence of occupancy is greater than 0.1, 0.5, and 0.9, respectively. (From Hanski, I. 1993. Dynamics of small mammals on islands. Ecography 16: 372–375) (B) Classic metapopulation of the silver-spotted skipper butterfly (*Hesperia comma*) on dry meadows in southern England. The line indicates the combinations of area and connectivity above which the predicted incidence of occupancy is greater than 0.5. (From Hanski, I. 1994. A practical model of metapopulation dynamics. Journal of Animal Ecology 63: 151–162)

population. From the genetic viewpoint, two extreme forms of immigration and gene flow have been distinguished, the "migrant pool" model, in which the dispersers are drawn randomly from the metapopulation, and the "propagule pool" model, in which all immigrants to a patch originate from the same source population. The latter is likely to reduce genetic variation in the metapopulation.

Colonization of a currently unoccupied habitat patch is more likely the greater the numbers of immigrants, which is measured by the connectivity of the patch (see next section). All populations have smaller or greater risks of extinction from demographic and environmental stochasticities (see Stochasticity in Metapopulation Dynamics below) and other causes. Typically, the smaller the population the greater the risk of extinction.

Assuming that a local population in patch i has a constant risk of going extinct, E_i, and that patch i, if unoccupied, has a constant probability of becoming recolonized, C_i, the state of patch i, whether occupied or not, is determined by a stochastic process (Markov chain) with the stationary (time-invariant) probability of occupancy given by

$$J_i = \frac{C_i}{C_i + E_i}. \qquad (1)$$

J_i is often called the incidence of occupancy. This formula helps explain the common metapopulation patterns of increasing probability of patch occupancy with patch area (which typically decreases E_i and hence increases J_i) and with decreasing isolation (which typically increases C_i and hence increases J_i). Figure 1 gives two examples.

Connectivity

Local populations and habitat patches in a patch network are linked via dispersal. Connectivity measures the strength of this coupling from the viewpoint of a particular patch or local population. Connectivity is best defined as the expected number of individuals arriving per unit time at the focal patch. Connectivity increases with the number of populations (sources of dispersers) in the neighborhood of the focal patch; with decreasing distances to the source populations (making successful dispersal more likely); and with increasing sizes of the source populations (larger populations send out more dispersers). A measure of connectivity for patch i that takes all these factors into account may be defined as

$$S_i = A_i^{\zeta im} \Sigma_{j \neq i} \exp(-\alpha d_{ij}) J_j A_j^{\zeta em}. \qquad (2)$$

Table 1. Four types of stochasticity affecting metapopulation dynamics

Type of stochasticity	*Entity affected*	*Correlation among entities*
Demographic	Individuals in local populations	No
Environmental	Individuals in local populations	Yes
Extinction–colonization	Populations in metapopulations	No
Regional	Populations in metapopulations	Yes

Here, A_j is the area and J_j is the incidence of occupancy of patch j, d_{ij} is the distance between patches i and j, $1/\alpha$ is species-specific average dispersal distance, and ζ_{im} and ζ_{em} describe the scaling of immigration and emigration with patch area. This formula assumes that the sizes of source populations are proportional to the respective patch areas; if information on the actual population sizes N_j is available, the surrogate $J_j A_j$ in equation 2 may be replaced by N_j.

Stochasticity in Metapopulation Dynamics

Metapopulation dynamics are influenced by four kinds of stochasticity (types of random events; table 1): demographic and environmental stochasticity affect each local population, and extinction–colonization and regional stochasticity affect the entire metapopulation. Both local dynamics and metapopulation dynamics are inherently stochastic because births and deaths in local populations are random events (leading to demographic stochasticity) and so are population extinctions and recolonizations in a metapopulation (extinction–colonization stochasticity). Environmental stochasticity refers to correlated temporal variation in birth and death rates among individuals in local populations, whereas regional stochasticity refers to correlated extinction and colonization events in metapopulations. Metapopulations are typically affected by regional stochasticity because the processes generating environmental stochasticity, including temporally varying weather conditions, are typically spatially correlated.

It can be demonstrated mathematically that all metapopulations with population turnover caused by extinctions and colonizations will eventually become extinct. It is a certainty that, given enough time, a sufficiently long run of extinctions will arise by "bad luck" and extirpate the metapopulation. However, time to extinction can be very long for large metapopulations inhabiting large patch networks (see the Levins Model below), and the metapopulation settles for a long time to a stochastic quasiequilibrium, in which there is variation but no systematic change in the number of local populations.

Source and Sink Populations

Populations may occur in low-quality sink habitats if there is sufficient dispersal from other populations living in high-quality source habitats. Therefore, the presence of a species in a particular habitat patch does not suffice to demonstrate that the habitat is of sufficient quality to support a viable population. Conversely, a local population may be absent from a habitat patch that is perfectly suitable for population growth when a local population happened to become extinct for reasons unrelated to habitat quality.

In a temporally varying environment, sink populations may, counterintuitively, enhance metapopulation persistence. This may happen when source populations exhibit large fluctuations leading to a high risk of extinction. The habitat patches supporting such sources may become recolonized by dispersal from sinks, assuming this happens before the sink populations have declined to extinction. In general, dispersal among local populations that fluctuate relatively independently of each other (weak regional stochasticity) enhances the metapopulation growth rate. This happens because when a population has increased in size in a good year, and the offspring are spread among many independently fluctuating populations, subsequent bad years will not hit them all simultaneously. It can be shown that this spreading-of-risk effect of dispersal may be so substantial that it allows a metapopulation consisting of sink populations only to persist without any sources.

2. LONG-TERM VIABILITY OF METAPOPULATIONS

Mathematical models are used to describe, analyze, and predict the dynamics of metapopulations living in fragmented landscapes. A wide range of models can be constructed differing in the forms of stochasticity they include (see Stochasticity in Metapopulation Dynamics above), in whether change in metapopulation size occurs continuously or in discrete time intervals, in how many local populations the metapopulation consists of, in how the structure of the landscape is represented, and so forth. A minimal metapopulation includes two local populations connected by dispersal. At the other extreme, assuming infinitely many habitat patches leads to a particularly simple description of the classic metapopulation, which is discussed next.

Levins Model

The Levins model has special significance for metapopulation ecology, as it was with this model that the American biologist Richard Levins introduced the metapopulation concept in 1969. The Levins model captures the essence of the classic metapopulation concept—that a species may persist in a balance between stochastic local extinctions and recolonization of currently unoccupied patches. For mathematical convenience, the model assumes an infinitely large network of identical patches, which have two possible states, occupied or empty. The state of the entire metapopulation can be described by the fraction of currently occupied patches, p, which varies between 0 and 1.

If each local population has the same risk of extinction, and each population contributes equally to the rate of recolonization, the rate of change in the size of the metapopulation is given by

$$\frac{dp}{dt} = cp(1-p) - ep, \tag{3}$$

where c and e are colonization and extinction rate parameters. This model ignores stochasticity, but it is a good approximation of the corresponding stochastic model for a large metapopulation inhabiting a large patch network. The Levins model is structurally identical with the logistic model of population growth, which can be seen by rewriting equation 3 as

$$\frac{dp}{dt} = (c-e)p\left(1 - \frac{p}{1-e/c}\right). \tag{4}$$

$c - e$ gives the rate of metapopulation growth when it is small, and $1 - e/c$ is the equilibrium metapopulation size ("carrying capacity"). The ratio c/e defines the basic reproductive number R_0 in the Levins model. A species can increase in a patch network from low occupancy if $R_0 = c/e > 1$. This condition defines the extinction threshold in metapopulation dynamics. In reality, in a finite patch network, a metapopulation may become extinct because of stochasticity even if the threshold condition $c/e > 1$ is satisfied. When a diffusion approximation is used to analyze the stochastic Levins model, the mean time to extinction T can be calculated as a function of the number of habitat patches n and p^*, the size of the metapopulation at quasiequilibrium. Figure 2 shows the number of patches that the network must have to make T at least 100 times as long as the expected lifetime of a single local population (given by the inverse of the extinction rate). For metapopulations with large p^*, a modest network of $n = 10$ patches is sufficient to allow long-term persistence, but for rare species (say $p^* < 0.2$), a large network of $n > 100$ is needed for long-term persistence.

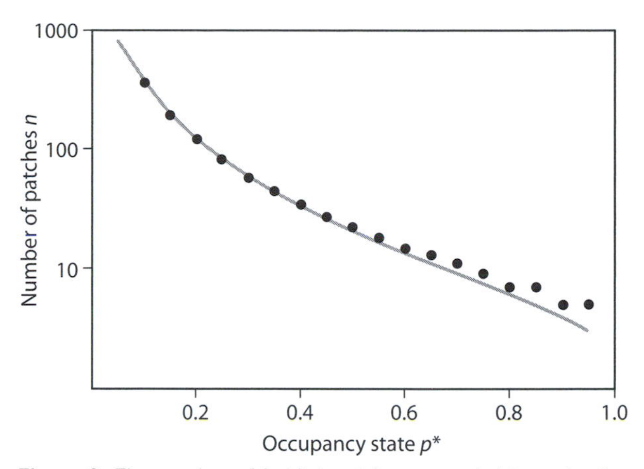

Figure 2. The number of habitat patches n needed to make the mean time to metapopulation extinction T at least 100 times longer than the mean time to local extinction. The dots show the exact result based on the stochastic logistic model, and the line is based on the following diffusion approximation:

$$T = \sqrt{\frac{2\pi}{n}} \frac{e^{-(n-1)p^*}}{p^{*2}(1-p^*)^{n-1}}.$$

(From Ovaskainen and Hanski, 2003)

The stochastic Levins model includes extinction–colonization stochasticity but no regional stochasticity, which leads to correlated extinctions and colonizations. In the presence of regional stochasticity, the mean time to metapopulation extinction does not increase exponentially with increasing n as in figure 2 but as a power function of n, the power decreasing with increasing correlation in extinction and recolonization rates, reducing long-term viability. This result is analogous to the well-known effects of demographic and environmental stochasticities on the lifetime of single populations.

Spatially Realistic Metapopulation Models

There is no description of landscape structure in the Levins model; hence, it is not possible to investigate with this model the consequences of habitat loss and fragmentation. Real metapopulations live in patch networks with a finite number of patches; the patches are of varying size and quality, and different patches have different connectivities, which affect the rates of immigration and recolonization. These considerations have been incorporated into spatially realistic metapopulation models. The key idea is to model the effects of habitat patch area, quality, and connectivity on the processes of local extinction and recolonization. Generally, the extinction risk decreases with increasing patch area because large patches tend to have large populations with a small risk of extinction, and the colonization rate increases with connectivity to existing populations.

The theory provides a measure to describe the capacity of an entire patch network to support a metapopulation, denoted by λ_M and called the metapopulation capacity of the fragmented landscape. Mathematically, λ_M is the leading eigenvalue of a "landscape" matrix, which is constructed with assumptions about how habitat patch areas and connectivities influence extinctions and recolonizations. The size of the metapopulation at equilibrium is given by

$$p_\lambda{}^* = 1 - e/(c\lambda_M), \quad (5)$$

which is similar to the equilibrium in the Levins model, but with the difference that metapopulation equilibrium now depends on metapopulation capacity, and metapopulation size p_λ is measured by a weighted average of patch occupancy probabilities. The threshold condition for metapopulation persistence is given by

$$\lambda_M > e/c. \quad (6)$$

In words, metapopulation capacity has to exceed a threshold value, which is set by the extinction proneness (e) and colonization capacity (c) of the species, for long-term persistence. To compute λ_M for a particular landscape, one needs to know the range of dispersal of the focal species, which sets the spatial scale for calculating connectivity (parameter α in equation 2), and the areas and spatial locations of the habitat patches. The metapopulation capacity can be used to rank different fragmented landscapes in terms of their capacity to support a viable metapopulation: the larger the value of λ_M, the better the landscape. Figure 3 gives an example in which metapopulation capacity explains well the size of butterfly metapopulations in dissimilar patch networks.

3. METAPOPULATIONS IN CHANGING ENVIRONMENTS

A fundamental question about metapopulation dynamics concerns long-term viability, which has great significance for the conservation of biodiversity in fragmented landscapes (see Metapopulations, Spatial Population Processes, and Conservation below). A

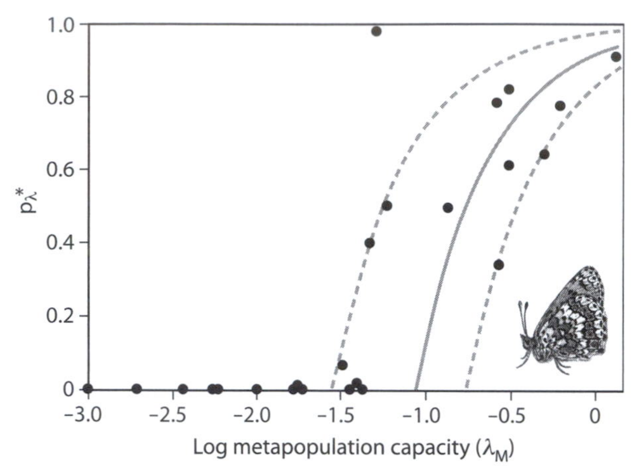

Figure 3. Metapopulation size of the Glanville fritillary butterfly (*Melitaea cinxia*) as a function of the metapopulation capacity λ_M in 25 habitat patch networks. The vertical axis shows the size of the metapopulation based on a survey of habitat patch occupancy in 1 year. The empirical data have been fitted by a spatially realistic model (continuous line; the broken lines give the model fit to the second smallest and the second largest positive values). The result provides a clear-cut example of the extinction threshold. (From Hanski and Ovaskainen, 2000)

patch network will not support a viable metapopulation unless the extinction threshold is exceeded, and even if it is, a metapopulation may become extinct for stochastic reasons (see Levins Model above). Long-term viability is further reduced by environmental change.

Ephemeral Habitat Patches

Innumerable species of fungi, plants, and animals live in ephemeral habitats such as decaying wood. A dead tree trunk may be viewed as a habitat patch for local populations of such organisms. The trunk is not permanent, largely because of the action of the organisms themselves, and local populations necessarily become extinct at some point. Parasites living in a host individual can be similarly considered as comprising a local population, which necessarily becomes extinct when the host individual dies. This example reflects fundamental similarities between metapopulation biology and epidemiology.

Regular disappearance of habitat patches increases extinctions, but the metapopulation may still persist in a stochastic quasiequilibrium. Equation 1 may be extended to include patch extinction and the appearance of new patches:

$$J_i = \frac{C_i - [C_i(1 - C_i - E_i)^{age}]}{C_i + E_i}, \quad (7)$$

where *age* is the age of patch *i*. Following its appearance, a new patch is initially unoccupied, $J_i = 0$ when $age = 0$. When the patch becomes older, the incidence of occupancy approaches the equilibrium $[C_i/(C_i + E_i)]$ given by equation 1 and determined by the extinction–colonization dynamics of the species. The precise trajectory is given by equation 7, where the term in square brackets declines from C_i when $age = 0$ to zero as *age* becomes large and when equation 1 is recovered.

Transient Dynamics and Extinction Debt

Human land use often causes the loss and fragmentation of the habitat for many other species. Following the change in landscape structure, especially if the change is abrupt, it takes some time before the metapopulation has reached the new quasiequilibrium, which may be metapopulation extinction. Considering a community of species, the term *extinction debt* refers to situations in which, following habitat loss and fragmentation, the threshold condition is not met for some species, but these species have not yet become extinct because they respond relatively slowly to environmental change. More precisely, the extinction debt is the number of extant species that are predicted to become extinct, sooner or later, because the threshold condition for long-term persistence is not satisfied for them following habitat loss and fragmentation.

How long does it take before the metapopulation has reached the new quasiequilibrium following a change in the environment? The length of the transient period is longer when the change in landscape structure is greater, when the rates of extinction and recolonization are lower, and when the new quasiequilibrium following environmental change is located close to the extinction threshold. The latter result has important implications for conservation. Species that have become endangered as a result of recent changes in landscape structure are located, by definition, close to their extinction threshold, and hence, the transient period in their response to environmental change is predicted to be long. This means that we are likely to underestimate the level of threat to endangered species because many of them do not occur presently at quasiequilibrium with respect to the current landscape structure but are slowly declining because of past habitat loss and fragmentation. On the positive side, long transient time in metapopulation dynamics following environmental change gives us humans more time to do something to reverse the trend.

4. EVOLUTION IN METAPOPULATIONS

The hierarchical structure of metapopulations, from individuals to local populations to the entire metapopulation, has implications for evolutionary dynamics. In addition to natural selection occurring within local populations, different selection pressures may influence the fitness of individuals during dispersal and at colonization. Individuals that disperse from their natal population and succeed in establishing new local populations are likely to comprise a nonrandom group of all individuals in the metapopulation. Particular phenotypes and genotypes may persist in the metapopulation because of their superior performance in dispersal and colonization even if they would be selected against within local populations. This is often called the metapopulation effect.

The most obvious example relates to emigration rate and dispersal capacity. The most dispersive individuals are selected against locally because their local reproductive success is reduced by early departure. However, these individuals may find a favorable habitat elsewhere, which increases their fitness in the metapopulation. Which particular phenotypes and genotypes are favored in particular metapopulations depends on many factors. Local competition for resources and competition with relatives for mating opportunities selects for more dispersal, and so does temporal variation in fitness among populations, but mortality during dispersal selects against dispersal. Because of many opposing selection pressures, habitat fragmentation may select for more or less dispersive individuals depending on particular circumstances.

5. SPATIAL DYNAMICS IN NONPATCHY ENVIRONMENTS

The classic population concept and corresponding population models assume that all individuals interact equally with all other individuals in the population. This is called panmictic population structure. On the other extreme is the classic metapopulation concept, which assumes a set of dynamically independent local populations, within which individual interactions take place. Many real populations have intermediate spatial population structures: individuals do not occur in discrete local populations, but the population is more continuous across a large area, yet ecological interactions and dispersal are more or less localized; hence, the population is not panmictic at a large spatial scale. In this case, what matters most is the local density experienced by individuals rather than the overall density of individuals in the large population as a whole.

Population dynamics across a landscape is determined by the opposing forces of localized interactions, which tend to increase differences in local dynamics and dispersal among neighboring population units, which reduces their dynamic independence. In single panmictic populations, strongly nonlinear dynamics may lead to population cycles and other complex dynamics. In large populations distributed across a landscape, nonlinear dynamics and dispersal may generate complex spatially structured dynamics. For instance, the dynamics of measles in human populations may exhibit traveling waves, in which epidemics initiated in large core populations (cities) lead, with some time lag, to epidemics in the surrounding smaller communities (see chapter II.9). Comparable complex spatiotemporal patterns of population density have been detected in some animal populations.

6. METAPOPULATIONS, SPATIAL POPULATION PROCESSES, AND CONSERVATION

Loss and fragmentation of natural habitats are the most important reasons for the current catastrophically high rate of loss of biodiversity on Earth. The amount of habitat matters because long-term viability of populations and metapopulations depends, among other things, on the environmental carrying capacity, which is typically positively related to the total amount of habitat. Additionally, the spatial configuration of habitat may influence viability because most species have limited dispersal range, and hence, not all habitat in a highly fragmented landscape is readily accessible, giving rise to the extinction threshold (see Levins Model above).

Consequences of Habitat Loss and Fragmentation

Metapopulation models have been used to address the population dynamic consequences of habitat loss and fragmentation. In the Levins model, where there is no description of landscape structure, habitat loss has been modeled by assuming that a fraction $1-h$ of the patches becomes unsuitable for reproduction, while fraction h remains suitable. Such habitat loss reduces the colonization rate to $cp(h-p)$. The species persists in a patch network if h exceeds the threshold value e/c. At equilibrium, the fraction of suitable but unoccupied patches $(h-p^*)$ is constant and equals the amount of habitat at the extinction threshold $(h=e/c)$.

The spatially realistic metapopulation model (see above) combines the metapopulation perspective of the Levins model with a description of the spatial distribution of habitat in a fragmented landscape. In the model described by equation 5, the metapopulation capacity λ_M replaces the fraction of suitable patches h in the Levins model, and the threshold condition for metapopulation persistence is given by $\lambda_M > e/c$. λ_M takes into account not only the amount of habitat in the landscape but also how the remaining habitat is distributed among the individual habitat patches and how the spatial configuration of habitat influences extinction and recolonization rates and hence metapopulation viability. The metapopulation capacity can be computed for multiple landscapes, and their relative capacities to support viable metapopulations can be compared: the greater the value of λ_M, the more favorable the landscape is for the particular species (figure 3).

Reserve Selection

Setting aside a sufficient amount of habitat as reserves is essential for conservation of biodiversity. Reserve selection should be made in such a manner that a given amount of resources for conservation makes a maximal contribution toward maintaining biodiversity. In the past, making the optimal selection of reserves out of a larger number of potential reserves was typically based on their current species richness and composition, without any consideration for the long-term viability of the species in the reserves. More appropriately, we should ask which selection of reserves maintains the largest number of species to the future, taking into account the temporal and spatial dynamics of the species and the predicted changes in climate and land use. Metapopulation models can be incorporated into analyses that aim at providing answers to such questions.

7. CONCLUSION

Metapopulations are assemblages of local populations inhabiting networks of habitat patches in fragmented landscapes. The local populations have more or less independent dynamics because of their isolation, but complete independence is prevented by large-scale similarity in environmental conditions and by dispersal, which occurs at a spatial scale characteristic for each species. Metapopulation models are used to describe, analyze, and predict the dynamics of metapopulations. Important questions include the conditions under which metapopulations may persist in particular patch networks and for how long, how landscape structure influences metapopulation persistence, and the response of metapopulations to changing landscape structure. Metapopulation dynamics in highly fragmented landscapes involves an extinction threshold, a critical amount, and spatial configuration of habitat that is necessary for long-term persistence of the metapopulation.

FURTHER READING

Hanski, I. 1999. Metapopulation Ecology. Oxford: Oxford University Press.

Hanski, I., and O. E. Gaggiotti, eds. 2004. Ecology, Genetics, and Evolution of Metapopulations. Amsterdam: Elsevier.

Hanski, I., and O. Ovaskainen. 2000. The metapopulation capacity of a fragmented landscape. Nature 404: 755–758.

Hanski, I., and O. Ovaskainen. 2002. Extinction debt at extinction threshold. Conservation Biology 16: 666–673.

Ovaskainen, O., and I. Hanski. 2002. Transient dynamics in metapopulation response to perturbation. Theoretical Population Biology 61: 285–295.

Ovaskainen, O., and I. Hanski. 2003. Extinction threshold in metapopulation models. Annales Zoologici Fennici 40: 81–97.

Tilman, D., R. M. May, C. L. Lehman, and M. A. Nowak. 1994. Habitat destruction and the extinction debt. Nature 371: 65–66.

Verheyen, K., M. Vellend, H. Van Calster, G. Peterken, and M. Hermy. 2004. Metapopulation dynamics in changing landscapes: A new spatially realistic model for forest plants. Ecology 85: 3302–3312.

II.5

Competition and Coexistence in Plant Communities

Ray Dybzinski and David Tilman

OUTLINE

Numerous species commonly compose natural plant assemblages from the poles to the equator, and a wealth of classic ecological experiments have demonstrated that these often compete strongly with one another for resources such as nutrients or light. Theoretical ecologists have demonstrated that competing species are only expected to stably coexist (i.e., coexist in the long run) when each is protected from local extinction by density- or frequency-dependent processes that benefit it when rare, or equivalently, when the net negative effects of intraspecific (within-species) competition exceed those of interspecific (between-species) competition. Competition for nutrients, such as nitrogen and phosphorus, is "size symmetric" because smaller and larger individuals are potentially equal competitors on a per-biomass basis. In contrast, shoot competition for light is "size asymmetric" because taller individuals are advantaged irrespective of biomass. We describe the different modeling approaches that this difference requires. However, in either case, stable coexistence requires trade-offs such that species that are better competitors for one limiting nutrient or in one light environment are necessarily worse competitors for a second limiting nutrient or in a second light environment. As one important example of how these models might account for the stable coexistence of numerous species across landscapes, we consider the effects of habitat heterogeneity in mean growing season temperature coupled with trade-offs in performance among species at different mean temperatures. Finally, given our theoretical understanding, we close with a discussion of the current challenges to and opportunities for advancing our empirical understanding of competition and coexistence in the real world.

GLOSSARY

coexistence. The indefinite persistence of two or more species. The empirically relevant sort of coexistence is termed "stable coexistence" in which species will continue to persist in the face of perturbations in their abundances. It is important to note that species that co-occur may or may not be stably coexisting; it is possible that one or more species are on their way to local extinction at a time scale that might appear slow to a casual observer.

competition. Most broadly, an interaction between individuals in which neither benefits. Here, we are considering exploitation competition for limiting resources in which the resource consumed or intercepted by one individual is no longer available to the second individual, thereby decreasing its fitness.

exclusion. A condition in which a species is driven to local extinction as a consequence of a competitive interaction.

founder control. A condition in which the dominant species in a competitive interaction is the species that is initially most abundant.

interspecific competition. Competition among individuals of different species.

intraspecific competition. Competition among individuals of the same species.

invader/invasion. In the context of theoretical ecology, an invader is a species introduced at arbitrarily small abundance to a habitat of a resident species at equilibrium. The question is asked: will the invader increase in abundance? Note that this use of the term is different from the sense in which an "invader" may be a foreign species with negative ecological consequences.

local extinction. A condition in which a species is no longer present within a defined habitat area. Local extinction is very different from the common use of the term "extinction," in which a species is no longer present anywhere.

resource. Broadly, something that may be consumed by one individual such that it is no longer available to another organism. Relevant resources for plants are nutrients, such as nitrogen, phosphorus, potassium, and micronutrients, along with light, water, and carbon dioxide.

1. INTRODUCTION TO COMPETITION AND STABLE COEXISTENCE

Plant species, be they trees in tropical or temperate forests, forbs in prairies or tundra, or algae in lakes or oceans, frequently compete with other plant species for various limiting resources, such as nitrate, phosphate, and light (Harper, 1977). Plants are also impacted by interactions such as mutualism, predation, herbivory, and disease. Each of these interactions has the potential to influence both the types of habitats in which species occur and their abundances in those habitats. Here we focus on mechanisms of competition for nutrients and light and their influence on coexistence and competitive exclusion among competing plant species.

As developed in the pioneering work of Lotka and Volterra, competition between two species, or interspecific competition, is defined as an interaction in which an increase in the abundance of one species causes a decrease in the growth and abundance of the other species and vice versa. By removing neighboring plants from around a target plant and documenting an increase in the target plant's growth rate, numerous experimental studies have shown that interspecific competition can be a strong force in plant communities. Competition between two individuals of the same species, intraspecific competition, is similarly defined and is almost certainly as strong as or stronger than interspecific competition.

Competition occurs because all vascular plants require light, water, and 20 or so mineral nutrients, including N, P, K, Ca, Mg, and various trace metals to survive and grow (e.g., Harper, 1977). As plants grow and consume these resources in a particular habitat, one or more resources become limiting because their rates of supply are insufficient to meet the demand from consumption. The outcome of the resultant resource competition may be exclusion or coexistence and depends on the dynamics of resource supply, on the resource requirements of the various species, and on the dependence of these factors on the physical attributes of the habitat, including its temperature, humidity, soil pH, slope, and aspect. The outcome of competition may also depend on initial species abundances. If each species maintains dominance whenever it is initially dominant, competition is said to be "founder controlled."

Species will coexist if each species can increase when it is rare and other species are at equilibrial abundances. This generally requires that intraspecific competition be stronger than interspecific competition. The mere co-occurrence of two or more competing species does not necessarily mean that they are coexisting; some of these species may be on their way to local extinction, albeit too slowly to be detected by casual observation. Ecological theory provides a way to formally distinguish stable coexistence from competitive exclusion at any time scale. First, if a competition model has a multispecies equilibrium point at which competing species have abundances greater than zero, the stability of this point can be assessed mathematically. If species abundances return to the equilibrium point after abundances are perturbed away from that point, this indicates that coexistence is stable. However, it is also possible for two or more competing species to persist indefinitely even when they do not have a stable multispecies equilibrium point. In these cases, one way to test for stable coexistence is to determine if there is "mutual invasibility" of species within the model (note that "invasion" in this chapter has no connotation of processes related to problematic nonnative species). In the simple case of two species, A and B, one asks if the geometric mean growth rate of a relatively small number of As in a monoculture of Bs is positive. If so, species A is said to be able to invade species B. If B can also invade A, then each species increases when rare and is protected from local extinction. The two species stably persist (coexist), even though their abundances are not constant. If A can invade B but B cannot invade A, then species A will exclude species B. If neither species can invade an equilibrial population of the other, then there is founder control; the initial dominant will maintain dominance.

2. COMPETITION FOR NUTRIENTS

Although it is possible for many nutrients to limit growth, it is easiest to begin by considering a single consumer species and a single limiting nutrient (hereafter referred to more generically as a "resource"). A resource is considered limiting if increases/decreases in its supply lead to increases/decreases in the growth rate of the species (Tilman, 1982). To understand dynamics, we must consider the factors that control both the rates of growth and loss of a species and the rates of supply and consumption of the resource. Because many plants have indeterminate growth and must increase in mass by two to six or more orders of magnitude when growing from a seed to an adult plant, it is customary to consider the dynamics of plant biomass rather than the dynamics of the number of individuals (as often done for animal populations). Thus, the specific rate of growth or loss of plant biomass is represented by $dB/dt\ 1/B$, where B is biomass per unit area (for terrestrial plants) or per unit volume (for marine and freshwater planktonic algae).

The specific growth rate of a species is often a monotonically increasing but asymptotically saturating function, $f[R]$, of the concentration or level of the limiting resource, R (figure 1A). Plant species experience a variety of sources of loss, including mortality and tissue loss from senescence, damage, herbivory, and disease. These losses, m, can be expressed in terms of density-independent changes in plant biomass, giving equation 1:

$$\frac{1}{B}\frac{dB}{dt} = f[R] - m. \tag{1}$$

As a plant population grows, it will consume the limiting resource. To see this, consider the dynamics of the resource:

$$\frac{dR}{dt} = a(S - R) - QBf[R], \tag{2}$$

where S is the supply rate of all forms of resource R, a is the rate of conversion of currently plant-unavailable forms of this resource (such as, for example, soil organic compounds containing nitrogen) into a plant-available form (such as nitrate), and Q is the quota of this resource needed to make a unit of new biomass (e.g., the ratio of nitrogen in plant tissues to biomass). Equation 2 thus states that the rate of change in available resource, R, depends on its rates of supply, $a(S - R)$, and consumption, $QBf[R]$.

In combination, equations 1 and 2 state that, at equilibrium, a species growing by itself will reach a biomass at which growth and loss balance each other, at which resource supply and consumption balance each other, and thus, at which $dB/dt = dR/dt = 0$. This occurs when the balance between supply and consumption has reduced the level of the resource down to R^* (figure 1A; Tilman, 1982). R^* is the resource requirement of this species in the sense that it is the amount of resource at which resource-dependent growth balances all sources of loss (figure 1A). If the

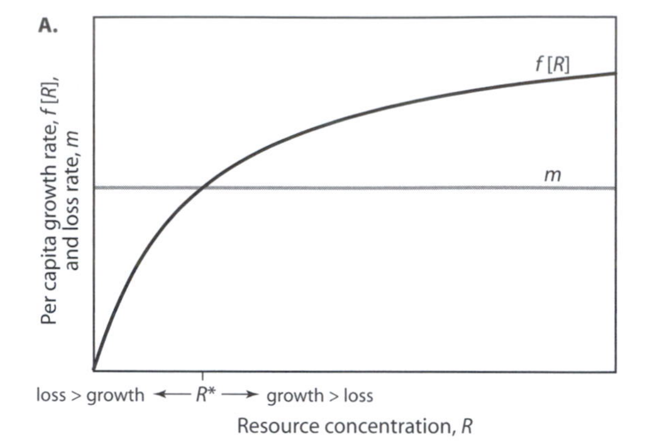

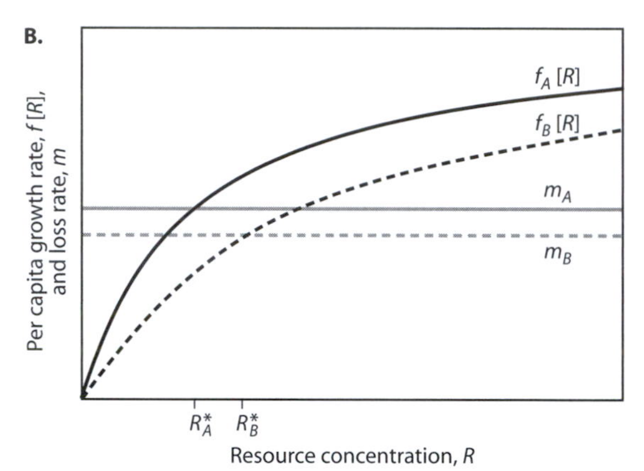

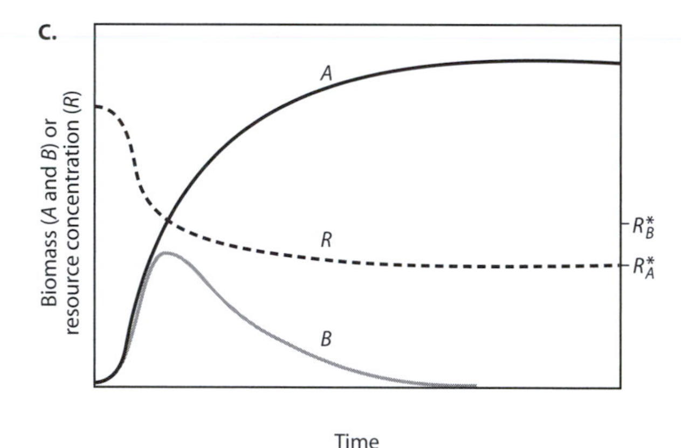

Figure 1. Details of resource consumption and competition for a single nutrient. (A) A single plant species growing on a single resource will reach equilibrium at resource concentration R^*, where per capita growth, $f[R]$, exactly balances per capita losses, m. (B) Depending on their physiology, morphology, and ecology, different species (e.g., Species A and Species B) are expected to have different resource-dependent growth rates and loss rates. As a consequence, different species are expected to have different R^* values for a given resource. (C) When species compete for a single resource, the species with the lowest R^* value is expected to win. The species with the lowest R^* value reduces equilibrial resource concentrations to a level where the loss rates of its competitors exceed their growth rates, and they are driven locally extinct.

level of the resource were greater than R^*, the abundance of the species would increase ($dB/dt > 0$) because growth would be greater than loss. Greater plant abundance would ultimately reduce the level of the resource to R^*. If the level of the resource were less than R^*, the abundance of the species would decrease ($dB/dt < 0$) because loss would be greater than growth. Less plant abundance would decrease resource consumption and ultimately increase the level of the resource to R^*.

When two or more species compete for a single limiting soil resource, the long-term outcome of their competition, should an equilibrium be reached, is that the species with the lowest R^* value (as determined by its physiology, morphology, and environmental conditions and their combined effects on $f[R]$ and m; figure 1B) would win and exclude all other competitors (figure 1C). This occurs because the species with the lowest R^* value would continue to increase in abundance until it had reduced the resource down to its R^*. At this resource level, the growth rates of all other species would be less than their loss rates, causing them to decline in abundance until locally extinct. Thus, if there is only one limiting resource, coexistence is not possible at equilibrium; only a single species will persist.

There are two important points to be made about this result. First, note that the model is predictive in the sense that a species trait measured in monoculture (R^* or physiological or morphological traits, such as root mass and fine root diameter, correlated with R^* in a given environment) allows one to predict the outcome of competition. This is much different from phenomenological models, such as the Lotka-Volterra competition equations, in which the competition coefficients that might allow one to predict the outcome of competition are measurable only after a competition experiment has been performed (thereby making a prediction of the outcome unnecessary!).

Second, note that the model's prediction that one species should displace all is at odds with the coexistence of numerous competing species in almost all natural ecosystems. This discrepancy suggests that one or more of the simplifying assumptions used in the model are violated in nature. This would not be surprising because the model implicitly assumes that there is only one limiting resource; that sites within a habitat do not differ in physical factors that may also limit growth, such as temperature, soil pH, and humidity; that plant abundances are not limited by dispersal; that interspecific interactions go to equilibrium; and that organisms on other trophic levels that might exert density- or frequency-dependent loss rates, such as herbivores, seed predators, pathogens, and mutualists, are unimportant. More complex and realistic theories have shown that addition of any one or more of these complexities of nature can allow many species to coexist, but only if species have appropriate trade-offs, as discussed below.

Competition for Two Limiting Nutrients

Few ecosystems have but a single limiting nutrient. Nutrient addition experiments in freshwater and marine ecosystems have shown that two or more nutrients often limit phytoplankton, especially nitrate, phosphate, silicate, and/or iron. The most commonly limiting nutrients in most terrestrial ecosystems are soil nitrogen and phosphorus. Other nutrients, especially potassium, calcium, or trace metals, can be limiting depending on the age and origin of the soil. Although not a nutrient per se, water is also limiting in many terrestrial ecosystems. In all but very nutrient poor or dry terrestrial ecosystems and in many freshwater and marine ecosystems, light also limits at least some plant species. We address the special case of light limitation in a separate section below.

Although a single number, R^*, can summarize the competitive ability of a plant for a single limiting resource, resource-dependent growth isoclines are needed to do so for two or more limiting resources. A resource-dependent growth isocline defines the concentrations of all limiting resources for which a species has a given growth rate. The isocline defining the growth rate that is equal to the loss rate of a given species (i.e., $dB/dt = 0$), which is called the nullcline, is directly analogous to R^* in its ability to predict the outcome of competition for two or more limiting resources. Most commonly, nullclines are plotted on graphs in which the axes represent the possible concentrations of limiting resources (figure 2A). When resource concentrations fall between the origin and a species' nullcline, the species will experience greater loss than growth. When resource concentrations fall beyond the nullcline relative to the origin, the species will experience greater growth than loss (figure 2A).

The shape of the nullcline reflects the type of resources that are limiting. Most plant resources are nutritionally essential; individual plants must have N, P, K, etc., and they cannot eliminate their need for one of these elements by substituting increased amounts of a different element. Nutrients tend to be perfectly essential for morphologically simple plants, such as single-celled phytoplankton species. This means that the nullclines of algae tend to have right-angle corners (figure 2A), with each branch of the isocline representing the R^* of the species for that resource. However, because higher

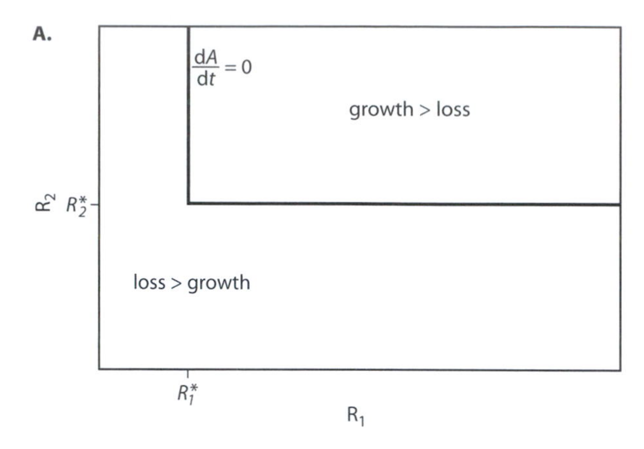

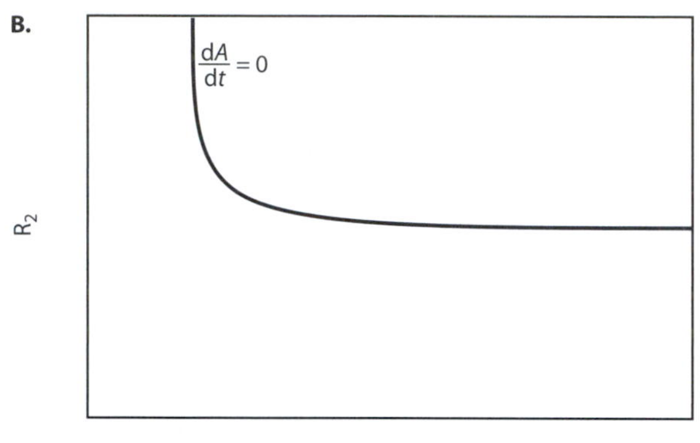

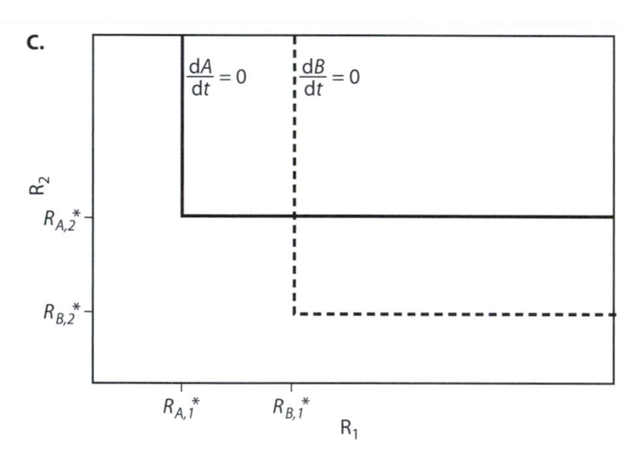

Figure 2. Details of resource consumption for two nutrients. (A) A nullcline defines the concentration of two resources for which a plant species will be in equilibrium, i.e., where $dA/dt = 0$ and per capita growth is exactly balanced by per capita loss. The nullcline shown depicts the response of a species to perfectly essential resources, such as nitrogen and phosphorus, where no amount of one resource can compensate for a deficit of the other resource. (B) A bent nullcline depicts the response of a species to interactive essential resources, where plants can somewhat shift investment toward physiological, morphological, or ecological structures that capture the most limiting resource. For example, a plant may invest less photosynthate in mycorrhizal associations and more photosynthate in the construction of fine roots when nitrogen is more limiting than phosphorus. Characterizing resources as interactive essential instead of perfectly essential does not change the qualitative outcome of competition. (C) The intersection of the nullclines of two species is necessary, but not sufficient, for their coexistence. Their intersection implies that one species has a lower R^* value for one resource and that the other species has a lower R^* value for the other resource. If coexistence does occur (see figure 3), equilibrial resource concentrations will be fixed at the intersection point. When the nullclines do not intersect (not shown), the species with lower R^* values for both resources will always win in competition, as in the case for a single resource.

plants can vary their morphology and the physiology of different tissue types, they have some ability to substitute one resource for the other when both are approximately equally limiting. Thus, they have interactive-essential resources: their nullclines tend to have rounded corners (figure 2B). The difference between perfectly essential and interactive-essential resources has no major qualitative effects on the outcomes of interspecific resource competition described below.

Two species can stably coexist at equilibrium in a homogeneous habitat only if their nullclines cross, which occurs only if the species have an interspecific trade-off. For instance, consider species A and B of figure 2C. Species A has a lower requirement (i.e., a lower R^* value) for R_1 than species B, but species B has a lower requirement for R_2 than species A. The two-species equilibrium point at which the nullclines cross shows the levels to which these two species would reduce R_1 and R_2 if they were to stably coexist at equilibrium. Although nullclines must cross for two species to coexist at equilibrium, that alone is not sufficient to ensure coexistence. The outcome of competition is also determined by their rates of consumption of the resources.

Assuming that each species forages optimally for these two resources, such that no resources are consumed that do not increase growth (i.e., no "luxury consumption," a reasonable assumption given that resource consumption requires energy), each species will consume resources in the ratio defined by the slope of the line drawn from the corner of its nullcline to the origin. Hence, consumption may be represented by vectors, c_A and c_B, that take those slopes (figure 3A). Consider the habitat represented by the supply point (S_1, S_2) in figure 3A. Such habitats have intermediate rates of supply of the two limiting resources. At equilibrium, the consumption by the two competitors would reduce resources down to the two-species equilibrium point. Indeed, coexistence would occur for any habitats with supply points within the area bounded by the shaded triangle that extend the slopes of the consumption vectors of the two species away from the

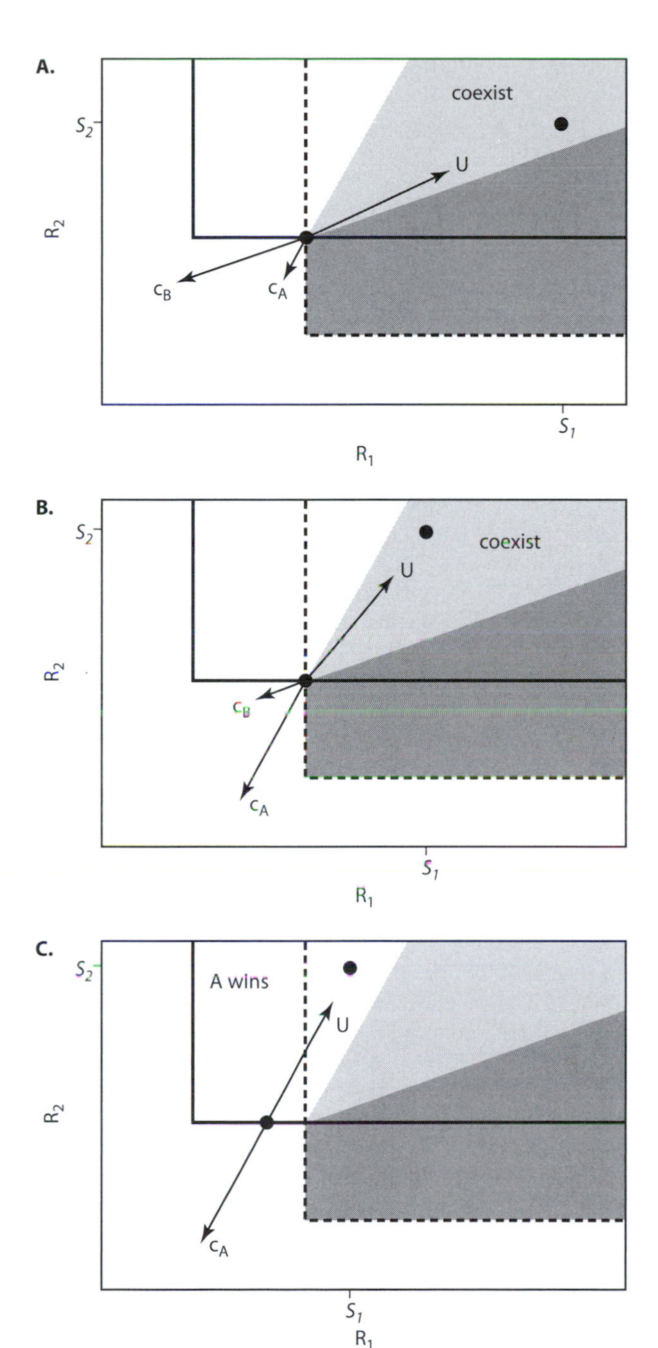

Figure 3. The outcome of competition for two nutrients. For all three panels: Species A's nullcline is solid and Species B's nullcline is dashed. The vector c_A represents Species A's rate-of-consumption vector. By assumption of optimal foraging, its slope is fixed by the ratio of resource concentrations that limit Species A, i.e., the slope between the origin and the bend in Species A's nullcline. Its magnitude is proportional to Species A's equilibrial biomass. The vector c_B is analogous for Species B. The vector U represents the rate of resource replenishment. The point (S_1, S_2) represents the supply concentration of the two resources in a given habitat. At equilibrium, U will point toward (S_1, S_2) and will be equal and opposite to the vector sum of the consumption vectors c_A and c_B. Because their slopes are fixed, coexistence will occur at equilibrium only when it is possible to add c_A and c_B from the nullcline intersection point such that their sum is equal and opposite to U. This will occur for all (S_1, S_2) inside the medium gray triangle (panels A and B). For all (S_1, S_2) inside the lightest gray area (panel C), Species A will win ($c_B = 0$). For all (S_1, S_2) inside the darkest gray area (not shown), Species B will win ($c_A = 0$). For all (S_1, S_2) inside the white area (not shown), both species will become locally extinct.

two-species equilibrium point (figure 3). This is because it is possible to create a linear combination of $\boldsymbol{c_A}$ and $\boldsymbol{c_B}$ that is exactly equal in magnitude and opposite in sign to the rate of resource replenishment, represented by the vector $\boldsymbol{U}$, such that resource supply is exactly balanced by consumption (compare figures 3A and 3B) and $dR/dt = 0$.

In habitats that have low rates of supply of R_1 relative to R_2, such as for the habitat depicted in figure 3C, both species are most limited by R_1, and species A, which is the better competitor for R_1 because it has a lower R_1^*, would competitively displace species B at equilibrium. Notice that because their slopes are fixed by assumption of optimal foraging, there is no linear combination of $\boldsymbol{c_A}$ and $\boldsymbol{c_B}$ in figure 3C that could exactly balance $\boldsymbol{U}$, as in figures 3A and 3B. Similarly, species B would win for habitats that have low relative rates of supply of R_2.

If species have trade-offs in their abilities to compete for two or more limiting resources, these trade-offs would mean that each species would have habitats, defined by supply rate of these resources, in which it would persist in competition with any other species. This is illustrated for four species (C, D, E, and F) competing for two limiting resources in figure 4A. Note that, with two limiting resources, only two species can stably coexist, at equilibrium, in any given habitat (i.e., any specific S_1, S_2) but that all four species can persist together in a landscape that has spatial heterogeneity in the supply rates of these resources as shown. Even more species could coexist with a fixed amount of heterogeneity if new species invaded with intermediate requirements for R_1 and R_2, an example of which is shown using a dashed nullcline in figure 4A. Alternatively, more species could coexist if the habitat encompassed even greater heterogeneity (not shown). The concept of trade-offs can easily be expanded to three or more resources.

3. COMPETITION FOR LIGHT

The model of competition described above applies to competition for nutrients but not to competition for light (Tilman, 1988). To see why, first realize that the

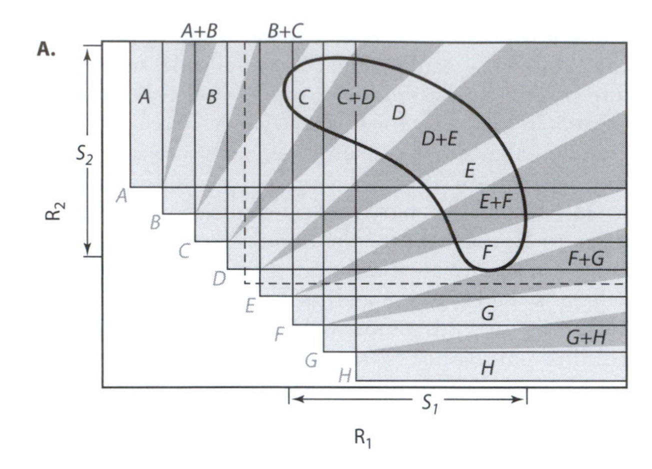

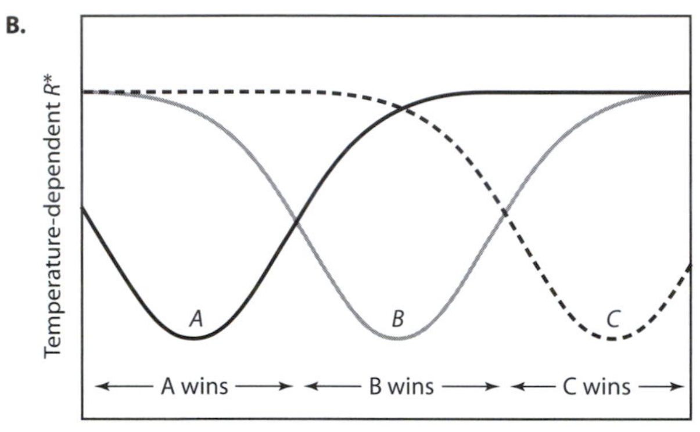

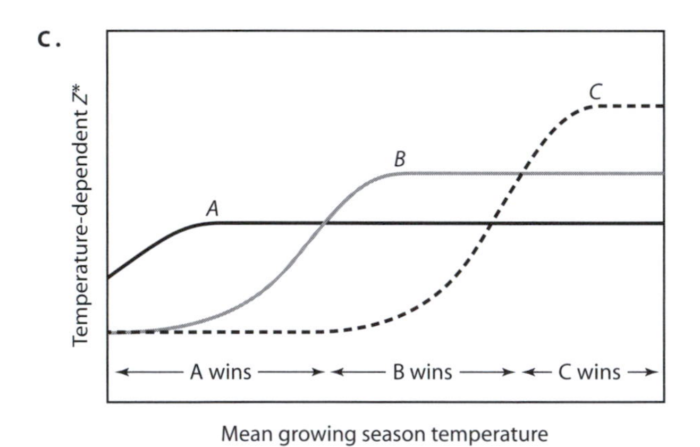

Figure 4. The effect of heterogeneity in resource supply points and temperature on coexistence. (A) With two limiting resources, at most two species can stably coexist at a single resource supply point (S_1, S_2). However, if supply points shift across a landscape (as depicted by the circle and the range of S_1 and S_2), many species can coexist, provided a trade-off causes a lowered R^* value for one resource to necessarily lead to an increased R^* value for the other resource. Light gray regions depict those for which supply points will cause a single species to dominate. Dark gray regions depict those for which supply points will cause two species to stably coexist. Gray letters represent species nullclines. Black letters represent the supply point zones in which a species will persist. The dashed nullcline shows an example of a new species that would be able to invade and coexist in this landscape. (B and C) Hypothetical dependence of R^* and Z^* values on mean growing season temperature. Ecophysiological constraints suggest that those characteristics that will lead to low R^* values or relatively high Z^* values at one temperature should prevent a species from having a low R^* value or relatively high Z^* value at a different temperature. Thus, heterogeneity in growing season temperature across a landscape should lead to the coexistence of multiple species.

concentration to which a plant reduces a nutrient through consumption (R) is the same concentration at which it must subsequently extract the nutrient (R). In other words, the nutrient concentration it "gets" is the same as the nutrient concentration it "creates." This is fundamentally different from competition for light: a tall tree intercepts full sunlight and reduces the intensity of light below its crown, such that trees below it receive less than full sunlight. Thus, the light level that the tall tree "gets" is distinct from the light level it "creates."

Weiner (1990) described differences in competitive interactions in terms of symmetry and asymmetry. "Size symmetric" competition occurs when larger individuals have no advantage over smaller individuals on a per-biomass basis, i.e., when the resource consumption of an individual is proportional to the biomass of its resource-capturing structures, such as roots for soil nutrients. "Size asymmetric" competition occurs when larger individuals take a disproportionate share of the resource, as is expected when taller plants preempt the light of shorter plants below them. Indeed, empirical studies have generally supported the idea that competition for soil resources is size symmetric, whereas competition for light is size asymmetric. Given size asymmetry, it is challenging to create a realistic mathematical model of light competition that is as simple and readily interpretable as the nutrient model described above.

The Perfect Plasticity Approximation

Recently, S. W. Pacala and co-workers discovered an approach that allowed them to create a simple yet mechanistic and predictive analytical model of plant competition for light. Although their model is formulated explicitly for single-canopy forests, it is extendable to other plant communities. The model treats trees as morphologically plastic, as being able to "lean" toward gaps by concentrating their growth in areas of higher light, where the photosynthetic return on structural investment is greater. To a first approximation, termed the "perfect plasticity approximation" (PPA), the result of each tree's individual plasticity and phototropism is canopy closure via tessellation, where each crown fills as much space as it can without growing into adjacent crowns. Moreover, under the PPA, the height above the ground at which each canopy crown touches an adjacent canopy crown tends toward a constant value across a forest canopy with a given set of species at a given stage of development.

Thus, forest trees may be classed as either in the canopy or in the understory and consequently subject to either full sunlight or the mean light remaining after transmission through the canopy. This has the dual advantage of simplifying the mathematics of a light competition model and yet making it more realistic (Adams et al., 2007). Consider a forest monoculture of species A at equilibrium. The forest consists of reproductive canopy trees and nonreproductive understory trees. The canopy trees reduce light to a level, L_A, in the understory. Because the forest is assumed to be in equilibrium, each canopy tree exactly replaces itself over a lifetime of reproductive effort, on average. Understory trees grow at rate $G_{D,A}[L]$ and die at rate $\mu_{D,A}[L]$. Canopy trees grow at rate $G_{L,A}$, die at rate $\mu_{L,A}$, and reproduce at rate F_A per unit area of crown. Crown area, $\pi(\alpha_A D)^2$, and height, $H_A(D)^\wedge\beta_A$, scale allometrically with trunk diameter, D. The equilibrial canopy height is $\hat{Z}_A{}^*$.

Invasion of species A's monoculture by a small number of species B individuals will be successful only if, on average, each invader produces more than one successful (i.e., reproductive) offspring. By assumption of the Adams et al. (2007) model, individuals reproduce only once they have reached the canopy. Thus, if some of the invaders die before reaching the canopy, successful invasion will depend on whether those individuals that do reach the canopy can produce enough successful offspring to make up for their loss. Resident species A can influence this process only by affecting the fraction, ζ, of invaders that survive to reach the canopy:

$$\zeta = e^{\frac{-\mu_{D,B}[L_A]}{G_{D,B}[L_A]}\left(\frac{\hat{Z}_A{}^*}{H_B}\right)^{1/\beta_B}}. \quad (3)$$

Adams et al. (2007) point out that invasion will be less likely if L_A is smaller (by decreasing the invader's understory growth rate, $G_{D,B}[L_A]$, and increasing its mortality rate, $\mu_{D,B}[L_A]$) or if $\hat{Z}_A{}^*$ is greater (by increasing the time required to reach the canopy and thus the likelihood of mortality before reaching the canopy).

If the crown transmissivities of the two species are equivalent, i.e., $L_A = L_B$, then Adams et al. (2007) demonstrate that light competition among trees is analogous to competition for a single soil resource, in the sense that one species is always expected to win, and the winner may be determined by examining the traits of the species in equilibrial monoculture. Specifically, the winner will be the tallest species, with the canopy height of species i, $\hat{Z}_i{}^*$, approximated by:

$$\hat{Z}_i{*} \approx H_i\left(\left(\frac{G_{D,i}[L_i]}{\mu_{D,i}[L_i]}\right)\ln\left[\frac{2\pi\alpha_i^2 F_i G_{L,i}^2}{\mu_{L,i}^3}\right]\right)^{\beta_i}. \quad (4)$$

Foresters have long recognized that forest succession, which may be viewed as prolonged competitive exclusion, generally proceeds from shade-intolerant species to shade-tolerant species. Purves et al. (2008) showed that shade-intolerant species tend to have low $\hat{Z}^*$, that shade-tolerant species tend to have high $\hat{Z}^*$, and that succession in the forests near the Great Lakes predictably proceeded from species with low $\hat{Z}^*$ to species with high $\hat{Z}^*$. This suggests that, relative to the importance of other factors that affect light competition during succession, the crown transmissivities of different tree species may be effectively equivalent.

Clearly, numerous tree species stably coexist in most forests, an observation that is at odds with the prediction that the single species with the highest $\hat{Z}^*$ should exclude all competitors. Again, this suggests that some of the model's assumptions, which largely overlap with the assumptions of the simple model of competition for a single soil resource described in the previous section, are false. However, before introducing exogenous factors into the model to explain diversity, Adams et al. (2007) note that at least some coexistence may be explained if we relax the assumption that the transmissivities of different tree species are equal, i.e., let $L_A \neq L_B$.

Adams et al. (2007) showed that it is possible for two species to coexist when competing for light if each species performs less well in the understory light environment that its canopy creates. Specifically, two species will coexist via competition for light when the species that casts more shade performs relatively better under less shade and the species that casts less shade performs relatively better under more shade. Each species creates the conditions that disfavor it. As discussed earlier, species are expected to stably coexist in resource-limited conditions when intraspecific competition is stronger than interspecific competition. Further theoretical and experimental work is needed to determine how likely this mechanism of coexistence might be relative to mechanisms that involve having both light and one or more additional factors limit plant growth.

4. COMPETITION AND TEMPERATURE

The rates of photosynthesis, respiration, and other physiological processes of all plants are dependent on temperature during the growing season. Temperatures below freezing are a separate but equally important limiting factor for plants because a host of chemical and mechanical adaptations are needed for cells to avoid the damage that freezing can cause. Moreover, the magnitude of such adaptations increases for temperatures further below freezing. Here we consider just the effects of the mean temperature of the growing season on plant growth rates.

The net effect of the dependence of plant physiology on temperature is that each plant species should have a mean growing season temperature at which it performs best. For habitats in which a single nutrient is limiting, the R^* of each species would be a function of mean growing season temperature, with each species having its lowest R^* value at its optimal temperature (figure 4B). For habitats in which light is limiting, equilibrial canopy height, $\hat{Z}^*$, will vary with temperature because of the dependence of growth, mortality, and fecundity rates on temperature.

Habitats have spatial variation in mean growing season temperature, such as along elevational gradients or among slopes with different aspects (north- versus south-facing slopes), and temporal variation, such as year-to-year or decade-to-decade differences. Such variation can allow numerous species to coexist. For example, the temperature-dependent R^* values of figure 4B define the temperature ranges at which each species would be the superior competitor (figure 4B). Species A would dominate the coolest sites/times, species B would dominate warmer sites/times, and species C would dominate the warmest sites/times. Strictly interpreted, only a single species would persist in a given site, but all three species would stably coexist on the larger landscape that encompassed the temperature variation illustrated in figure 4B. Similarly, the $\hat{Z}^*$ values of different species are expected to be maximal at a temperature for which they are well adapted, relative to other species at that temperature. Thus, across a landscape with variation in mean growing season temperature, the relative position of $\hat{Z}^*$ values would shift such that species A might win at lower temperatures, whereas species B might win at higher temperatures (figure 4C).

This coexistence occurs because plants, like all organisms, are constrained by trade-offs. No species has the lowest R^* value or highest $\hat{Z}^*$ value for all temperatures. Rather, each species has a range of temperatures at which it is the superior competitor and a range for which it is inferior. Just such a trade-off is expected based on the temperature dependence of both the activation energy and the denaturation of enzymes. If such a trade-off were an unavoidable aspect of plant life, there could be many other species, each with its own optimal temperature, that could persist between species A and B (figure 4B,C). Indeed, there would be no simple limit to the number of competing species that could coexist on a single limiting resource, be it nutrients or light, in a landscape with spatial variation in growing season mean temperature.

5. FROM MODELS TO REALITY: FUTURE CHALLENGES

The models described above provide several mechanisms, each of which, by itself, can maintain a potentially infinite number of stably coexisting species. The ecological literature is replete with other coexistence mechanisms that incorporate effects of space; disturbance and other nonequilibrium dynamics; or natural enemies on coexistence (see Further Reading). Given the appropriate trade-offs between species, most of these mechanisms can also maintain a potentially infinite number of stably coexisting species. The current and future challenge for ecology is determining which of these possible mechanisms actually maintain plant species diversity in nature. This is difficult for at least two reasons. First, feedbacks between plants and their abiotic and biotic environment and between soil biota and their abiotic and biotic environment make it difficult to isolate or manipulate only those aspects of the community or environment that theory suggests should impact some observable characteristic. For instance, it is difficult to remove specialized insect herbivores to test for their importance in diversity maintenance without also removing other insects, some of which may be mutualists.

The second difficulty lies in the long time scales over which many mechanisms are expected to operate, either because the dynamics are slow or because the mechanism relies on periodic but rare events. These defy detection in short-term field studies. As has happened in geology, substantial progress might be made if theoretical ecologists apply themselves to articulating short-term signatures of long-term mechanisms that field ecologists could then look for or test for. However, this appears more challenging in ecology than in geology because many of the most obvious signatures (e.g., particular rank abundance relationships) may be generated by multiple mechanisms. Alternatively, ecologists might turn to long-term data. For example, Purves et al. (2008) used long-term Forest Inventory and Analysis data (available from the U.S. Department of Agriculture) to parameterize and then test a variant of the light competition model described above and showed that the model largely predicted changes in total basal area, species composition, and the size structure of forest stands over 100 years of secondary succession. Unfortunately, such relevant long-term data are rare in ecology.

Despite these difficulties, many studies have demonstrated that particular mechanisms of coexistence may operate in particular habitats (Fargione and Tilman, 2002), typically by altering some aspect of the habitat and observing the predicted response. Such studies provide important insights into the ways particular mechanisms of coexistence operate and, to the extent that publications are not biased against negative results, provide data for broader conclusions via meta-analysis of multiple studies across a range of habitats and conditions. Although it might be intellectually gratifying to learn that just one mechanism maintains diversity throughout all of the varied biomes of the world, it seems much more likely that multiple mechanisms operate in any given habitat and that the relative importance of different mechanisms shifts across biomes and environmental gradients. In all cases, though, theory suggests that interspecific trade-offs will be both a root cause of coexistence of competing plant species and a useful signature of the mechanisms responsible for coexistence.

FURTHER READING

Adams, T. P., D. W. Purves, and S. W. Pacala. 2007. Understanding height-structured competition in forests: Is there an R^* for light? Proceedings of the Royal Society B: Biological Sciences 274: 3039–3047.

Chesson, P. 2000. Mechanisms of maintenance of species diversity. Annual Review of Ecology and Systematics 31: 343–366. *This review differentiates between the "equalizing" and "stabilizing" components of coexistence mechanisms. Chesson's distinction has received considerable attention in the recent literature, especially in the debate between niche and neutral processes. See, for instance: Adler et al. 2007. A niche for neutrality. Ecology Letters 10: 95–104.*

Fargione, J., and D. Tilman. 2002. Competition and coexistence in terrestrial plants. In U. Sommer and B. Worm, eds. Competition and Coexistence. Berlin: Springer-Verlag, 165–206. *This review has a good discussion of the limited number of studies that have successfully identified mechanisms of coexistence operating in natural systems.*

Harper, J. L. 1977. Population Biology of Plants. London: Academic Press. *This volume is a classic synthesis of early work on plant competition.*

Huisman, J., and F. J. Weissing. 1999. Biodiversity of plankton by species oscillations and chaos. Nature 402: 407–410. *This article is an elegant account of how the number of coexisting species may greatly exceed the number of limiting resources. The authors use a variant of the soil resource model presented above and find that the dynamics of competition lead to chaos and, as a result, species coexistence.*

Miller, T. E., J. H. Burns, P. Munguia, E. L. Walters, M. M. Kneitel, P. M. Richards, N. Mouquet, and H. L. Buckley. 2005. A critical review of twenty years' use of the resource-ratio theory. The American Naturalist 165: 439–448. *See also a reply by Wilson et al. and rejoinder by Miller et al. in 169: 700–708.*

Purves, Drew W., Jeremy Lichstein, Nikolay Strigul, and Stephen W. Pacala. 2008. Predicting and understanding forest dynamics using a simple tractable model. Proceedings of the National Academy of Sciences 105: 17018–17022. *This parameterized version of the light competition model presented above predicts the 100-year outcome of secondary succession in the forests of the Great Lakes states.*

Tilman, D. 1982. Resource Competition and Community Structure. Princeton, NJ: Princeton University Press. *This volume presents a detailed treatment of the soil resource competition model and its implications.*

Tilman, D. 1988. Plant Strategies and the Dynamics and Structure of Plant Communities. Princeton, NJ: Princeton University Press. *This volume presents a detailed treatment of a simulation model of plant allocation to leaves, roots, etc. and its effect on competition and coexistence.*

Tilman, D., and S. Pacala. 1993. The maintenance of species richness in plant communities. In R. E. Ricklefs and D. Schluter, eds. Species Diversity in Ecological Communities. Chicago: The University of Chicago Press, 13–25. *This is a review of coexistence mechanisms.*

Weiner, J. 1990. Asymmetric competition in plant populations. Trends in Ecology and Evolution 5: 360–364. *This article presents a clear presentation of the difference between "symmetric" competition for soil resources and "asymmetric" competition for light.*

II.6

Competition and Coexistence in Animal Communities

Priyanga Amarasekare

OUTLINE

1. Introduction
2. Basic principles of competitive coexistence
3. Coexistence under a single limiting factor
4. Coexistence under multiple limiting factors
5. Multiple coexistence mechanisms
6. Summary and conclusions

Competition is the most ubiquitous of species interactions. It occurs any time a resource that is essential to growth and reproduction (e.g., food, shelter, nesting sites) occurs in short supply. The acquisition of the resource by one individual simultaneously deprives others of access to it, and this deprivation has a negative effect on both the fitness of individuals and the per capita growth rates of populations. Competition is thus an interaction that has mutually negative effects on its participants. Coexistence results when populations of several species that utilize the same limiting resources manage to persist within the same locality. This chapter focuses on mechanisms that allow competitive coexistence in animal communities. Animals have two characteristics that determine the kinds of resources they can use and the mechanisms by which they can tolerate or avoid competition for these resources. First, animals are heterotrophs and have to ingest other organisms to obtain the energy required for growth and reproduction; competition thus involves biotic resources. Second, most animals are mobile and hence able to avoid or reduce competitive effects through dispersal.

GLOSSARY

density dependence. Dependence of the per capita growth rate on the abundance or density of the organism in question.

exploitative competition. Individuals have indirect negative effects on other individuals by acquiring a resource and thus depriving others of access to it.

functional response. The relationship between per capita resource consumption and resource abundance.

interference competition. Individuals have direct negative effects on other individuals by preventing access to the resource via aggressive behaviors such as territoriality, larval competition, overgrowth, or undercutting.

per capita growth rate. Per-individual rate of increase as a result of reproduction, mortality, emigration, and immigration.

stable coexistence. Competing species maintain positive abundances in the long term and are able to recover from perturbations that cause them to deviate from their long-term or steady-state abundances.

1. INTRODUCTION

A thorough understanding of the mechanisms of coexistence requires a thorough understanding of the mechanism of competition. Because animals rely on biotic resources which themselves grow and reproduce, the appropriate theoretical framework is one in which the resource dynamics are considered explicitly. Tilman's resource competition theory, although motivated by plant competition, provides such a framework for animal communities as well. When two or more species are limited by the same resource, the species that can maintain a positive per capita growth rate at the lowest resource level will exclude all other species. This is called the R^* rule in exploitative competition (Tilman, 1982). Coexistence mechanisms are the processes that counteract the R^* rule. They do so by increasing the strength of intraspecific competition relative to that of interspecific competition. The exact means by which this is achieved is obvious in some cases and quite subtle in others. There are several basic principles that underlie all coexistence mechanisms, and a clear grasp of these principles is necessary to understand the more subtle coexistence mechanisms.

2. BASIC PRINCIPLES OF COMPETITIVE COEXISTENCE

First, stable coexistence requires species to exhibit ecological differences. These differences are typically thought of as the species' niches. Following Chesson (2000), a species' niche has four dimensions: resources, natural enemies, space, and time. Species could differ in terms of (1) which resources or natural enemies they are limited by, (2) *when* they use the resource or encounter the natural enemy, or (3) *where* they use the resource or encounter the natural enemy. Niche differences are essential to coexistence because they allow species to depress their own per capita growth rates more than they do the growth rates of their competitors (Chesson, 2000). Elucidating exactly how this occurs is often difficult, but such an understanding is vital for a mechanistic understanding of coexistence.

A useful starting point is the idea of a negative feedback loop. Such a loop can cause a species' per capita growth rate to decrease when the population size is large and to increase when the population size is small. Negative feedback processes arise naturally when individuals compete with conspecifics (other individuals of the same species) for a limiting resource. Coexistence requires that this self-limiting negative feedback be stronger than the negative effect that the species has on the per capita growth rate of another species it competes with; i.e., when the focal species' population size is large, effects of resource limitation should affect its survival and reproduction more than it affects the survival and reproduction of its competitor. In what follows, I discuss the various ways in which this can be achieved. I begin with the case of species competing for a single limiting factor, which is the most restrictive case for coexistence. I then discuss cases where species compete for multiple limiting factors.

3. COEXISTENCE UNDER A SINGLE LIMITING FACTOR

The salient point to keep in mind is that coexistence requires mechanisms that counteract the R^* rule. When two or more species compete for a single limiting resource, this can occur via two basic classes of mechanisms. The first class of mechanisms enables coexistence via density-dependent negative feedback processes that operate within local communities. These mechanisms can operate in the absence of spatial or temporal variation in the environment. The second class of mechanisms enables coexistence by allowing organisms to avoid or minimize interspecific competition in space or time. These mechanisms rely on spatial or temporal variation in the abiotic environment. I next discuss these two classes of mechanisms.

Mechanisms That Are Independent of Environmental Variation: Intraspecific Interference

Intraspecific interference mechanisms are inherently density-dependent phenomena that have little or no effect on a species' per capita growth rate when it is rare and a strong negative effect when it is abundant. Such mechanisms therefore enable species to depress their per capita growth rates more than they would the per capita growth rates of other species they compete with. Intraspecific interference mechanisms are widespread in animal communities. In species where mates and/or nest sites are in short supply, territoriality limits per capita reproductive success when population sizes are large. Overgrowth and undercutting in sessile marine organisms have a similar self-limiting effect. In insect parasitoids, direct interference via aggression during oviposition or superparasitism (several females oviposit within the same host, leading to larval competition via direct combat or physiological suppression) reduces the per capita reproductive success of individual female parasitoids and decreases the per capita population growth rate at high density. A similar self-limiting effect can arise even in the absence of direct interference: per capita oviposition success declines with increasing parasitoid density because females keep rediscovering already parasitized hosts. This phenomenon, termed pseudointerference, can allow coexistence if parasitoid species that compete for a common host species have aggregated distributions, or attacks, that are independent of one another (Murdoch et al. 2003). This allows higher encounter rates with conspecifics than with heterospecifics, thus leading to stronger intraspecific than interspecific competition.

Mechanisms Dependent on Environmental Variation

When species compete for a single limiting factor, coexistence can occur via mechanisms that depend on environmental variation. Temporal variation enables coexistence by allowing species to differ in terms of *when* they use a limiting resource, and spatial variation enables coexistence by allowing species to differ in terms of *where* they use a limiting resource. In both cases, coexistence results because intraspecific competition is concentrated, in time or space, relative to interspecific competition. Below I discuss specific examples of spatial and temporal coexistence mechanisms.

Coexistence Mechanisms Driven by Temporal Variation

Nonlinear Competitive Responses

Most animal species exhibit nonlinear functional responses. The most common of these is the Type II functional response where the per capita consumption rate saturates at high resource abundances because of long handling times in predators or egg limitation in insect parasitoids. Type II functional responses cause the species' per capita growth rates to depend on resource abundance in a nonlinear manner. If the resource abundance fluctuates over time, and species that compete for the resource differ in the degree of nonlinearity in their functional responses, stable coexistence is possible if the species with the more nonlinear response is more disadvantaged, when abundant, by resource fluctuations than the species with the less nonlinear response (Armstrong and McGehee, 1980). Large resource fluctuations depress the per capita growth rate of the species with the more nonlinear response. This in turn reduces competition on the species with the less nonlinear response and allows it to invade a community where the former species is resident. The species with the more nonlinear functional response, which has the lower R^*, is better at exploiting the resource when resource abundance is lower, and the species with the less nonlinear functional response is better at exploiting the resource when resource abundance is higher. Thus, temporal fluctuations in the resource allow coexistence via temporal resource partitioning. Resource fluctuations can arise via abiotic environmental variation (e.g., seasonal variation in temperature and/or humidity) or via the resource's interaction with the consumer species with the more nonlinear functional response (Armstrong and McGehee, 1980).

Temporal Storage Effect

A second type of temporal coexistence mechanism, termed the temporal storage effect (Chesson, 2000), occurs when competing species differ in their responses to abiotic environmental variation. For instance, a species' developmental period or resource consumption rate may vary depending on seasonal variation in temperature or humidity. Such species-specific differences, termed the environmental response (Chesson, 2000), modify the nature of competitive interactions between species. For example, two competing species that differ in their temperature sensitivity may be active at different times of the year, and the resulting reduction in temporal overlap will reduce the intensity of interspecific competition between them. This effect is quantified as the covariance between the environmental response and competition. When species-specific responses to environmental variation modify competition, intraspecific competition is the strongest when a species is favored by the environment, and interspecific competition is the strongest when a species' competitors are favored by the environment. Thus, the covariance between the environmental response and competition is negative (or zero) when a species is rare and positive when it is abundant. This relationship can be understood as follows. When a species is favored by the environment, it can respond with a high per capita population growth rate (i.e., positive or strong environmental response) and reach a high abundance. When abundant, the species experiences mostly intraspecific competition because the environment is unfavorable to the other species it competes with. Thus, a strong environmental response increases intraspecific competition, resulting in a positive covariance between the environmental response and the strength of competition. In contrast, when the species is not favored by the environment, its per capita population growth rate is low (weak environmental response), and it will remain rare. It will experience mostly interspecific competition (because the environment is now favorable to the species' competitors), but because it is rare, the strength of competition it experiences will be low. Thus, a weak environmental response reduces interspecific competition, resulting in a zero or negative covariance between the environmental response and the strength of competition.

Stable coexistence via the temporal storage effect requires a third ingredient, buffered population growth (Chesson, 2000). Buffered population growth occurs when the decrease in population growth when the abiotic environment is unfavorable is offset by an increase in population growth when the abiotic environment is favorable. This follows from the mathematical phenomenon called Jensen's inequality, i.e., when the per capita growth rate is a nonlinear function of a life history trait or a vital rate that is subject to environmental variation, the growth rate averaged over the range of environmental variation experienced by the species is different from the average of the growth rates it experiences at different points of the environmental gradient (Ruel and Ayres, 1999). Buffered population growth allows species to tide over unfavorable periods resulting from strong interspecific competition and/or abiotic conditions that are not conducive to growth and reproduction. Life history traits that enable buffered population growth in animals include resting eggs in zooplankton species (e.g., *Daphnia*), high adult longevity in species in which competition occurs at the juvenile stage (e.g., coral reef fish), and dormancy and diapause (e.g., desert rodents) that allow species to be inactive during harsh environmental conditions.

Thus, species-specific responses to the abiotic environment ensure that species experience mostly intraspecific competition when they are favored by the environment, and buffered population growth ensures that species experience minimal interspecific competition when they are not favored by the environment. The overall outcome is an increase in the strength of intraspecific competition relative to interspecific competition, and this ensures stable coexistence.

The difference between nonlinear competition and a temporal storage effect is that, in the former, the competing species differ in the nonlinearity of their responses to a limiting resource that fluctuates over time (where fluctuations are a response to abiotic environmental variation or a result of the resource's interaction with a consumer species), whereas in the latter, the species differ in their responses to an abiotic environmental factor that fluctuates over time, which alters the species' responses to the limiting factor. In the first case, the differences in species' responses affect competition directly because they respond to the limiting factor itself, whereas in the latter, differences in species' responses affect competition indirectly because they respond to an environmental factor whose variation determines *when* species engage in competition.

The most obvious biological example of temporal storage effect comes from temporal refuges. Such refuges are common in invertebrates, particularly insects, whose developmental periods, life history traits, and vital rates are all strongly influenced by temperature variation.

Such temporal refuges typically arise from seasonal variation in the environment, with species differing in their tolerance of harsh environmental conditions. In insects in particular, species-specific differences are typically manifested as differences in activity periods. This situation is well documented in pest–enemy systems where multiple natural enemy species attack the same pest species. For instance, the leaf-feeding beetle, *Galerucella calmariensis*, is a successful biological control agent of the invasive plant pest purple loosestrife (*Lythrum salicaria*), but predation by the omnivorous mirid bug (*Plagiognathus politis*) disrupts control; however, the beetle emerges earlier in the season than the bug, and hence has a predation-free window that allows it to establish and inflict severe damage on its host plant. The parasitoids of the olive scale (*Parlatoria oleae*) *Aphytis maculicornis* and *Cocophagoides utilis* engage in intraguild predation and also exhibit differential temperature sensitivity: *A. maculicornis* is intolerant of warmer temperatures, whereas *C. utilis* is intolerant of colder temperatures; *C. utilis* thus has a temporal refuge from intraguild predation during the warm summer months, which promotes coexistence as well as complementary pest control.

Coexistence Mechanisms Driven by Spatial Variation

Spatial Storage Effect

A spatial storage effect can arise if competing species differ in their responses to spatial variation in the abiotic environment (Chesson, 2000). This can occur if spatial variation changes the species' R^* values such that a species that is the superior competitor in one locality may be an inferior competitor in another locality. When species differ in their responses to spatial variation in the abiotic environment, they grow to high abundances in the habitat patches most favorable to their growth and reproduction and experience strong intraspecific competition. The advantage of favorable environmental conditions is counteracted by strong intraspecific competition, leading to a negative covariance between the environmental response and strength of competition. In contrast, species tend to be rare in habitat patches unfavorable to their growth and reproduction, and they experience mostly interspecific competition (because the habitat patches unfavorable to the focal species tend to be favorable to its competitors). However, the species' rarity minimizes the impact of interspecific competition, leading to a weak or negative covariance between strength of competition and the quality of the habitat patch. In a spatially varying environment, the fact that per capita growth rates will depend on the habitat patch quality ensures buffered population growth. For instance, the reduction in the per capita growth rate in unfavorable patches (because of low habitat quality and strong interspecific competition) is compensated for by an increase in the per capita growth rate, when the species is rare, in favorable patches. Together, the environmental response (and the resulting covariance between competition strength and habitat quality) and buffered population growth ensure that intraspecific competition is stronger than interspecific competition, and this promotes stable coexistence.

Dispersal-Mediated Coexistence: Source–Sink Dynamics

A spatial storage effect automatically implies the existence of sources and sinks, localities where a species, because of its lower R^*, can escape competitive exclusion (sources) as opposed to localities where a species, because of its higher R^*, is subject to competitive exclusion (sinks). In such situations, dispersal from source habitats can prevent the exclusion of inferior competitors from sink habitats. Competitive exclusion is prevented because dispersal creates a negative density-dependent effect that increases the strength of intraspecific competition relative to interspecific competition (Amarasekare, 2003). Such dispersal-mediated coexistence is, however, not robust to increases in the dispersal rate. High rates of dispersal eliminate the spatial variation in competitive rankings such that the R^* rule operates at the metacommunity scale, and the species that has the lowest R^* when averaged across all habitat patches excludes the other species. Moreover, if emigration rates are sufficiently large, source communities themselves can experience negative growth rates, resulting in region-wide exclusion of the species. A key point to appreciate is that coexistence via source–sink dynamics is contingent on there being a mechanism within the source communities that ensures the persistence or coexistence of a given species; i.e., the species should exhibit a self-limiting negative feedback loop in localities that are favorable to its growth and reproduction.

The Role of Trade-offs in Coexistence

The role of trade-offs (energetic and other constraints that prevent species from simultaneously doing well on all aspects of growth and reproduction) in coexistence has spawned many a misconception. When two or more species compete for a single limiting resource, trade-offs cannot allow coexistence unless they involve a density- or frequency-dependent negative feedback mechanism. Trade-offs that lack such negative feedback mechanisms can reduce the differences between species in their R^* values but cannot provide the kind of negative feedback processes that are necessary for stable coexistence (Chesson, 2000). For instance, a species that is efficient at acquiring resources may do so at the cost of higher density-independent mortality and hence have a higher R^* than it would otherwise, but this trade-off will not allow coexistence because it will still competitively exclude (or be excluded by) a less efficient species subject to lower density-independent mortality that has a higher (lower) R^*. Trade-offs that involve negative feedback mechanisms arise naturally when species compete for multiple limiting factors.

4. COEXISTENCE UNDER MULTIPLE LIMITING FACTORS

Trade-offs in Resource Use

When species compete for multiple limiting factors, trade-offs in resource use can allow coexistence by increasing intraspecific competition relative to interspecific competition. This occurs because strong dependence of a species on a particular resource (which also leads to resource depletion) creates a negative feedback loop that leads to self-limitation (Chesson, 2000). For instance, a species will enjoy a high per capita growth rate when it is rare and the resource is abundant but suffer a low or negative per capita growth rate when it is abundant and the resource is rare. Thus, if species 1 has a negative feedback loop with resource A and species 2 has a similar feedback loop with resource B, they can coexist even if species 2 does utilize resource A because their overlap in the use of resource A is less that it would be had they not been utilizing two resources. Each species exhibits a trade-off in resource use (i.e., it is a superior exploiter of one resource but at the cost of being an inferior exploiter of another resource), and as long as such trade-offs cause species to limit themselves more than they would their competitors, stable coexistence is possible.

Trade-offs between Competitive Ability and Susceptibility to Natural Enemies

Most animals are limited by resources and natural enemies alike. Coexistence is possible if species that share common resources and natural enemies exhibit a trade-off between resource exploitation ability (as determined by their R^*s) and susceptibility to a natural enemy (as determined by an equivalent P^* rule, i.e., the prey species with the greatest resistance/tolerance to a natural enemy will exclude all other prey species). For instance, two competing prey species can coexist if one is more strongly limited by the resource and the other is more strongly limited by the natural enemy. Coexistence becomes possible because each species has a negative feedback loop with the resource or natural enemy that enables it to depress its own per capita growth rates more than it depresses the per capita growth rates of the other species.

5. MULTIPLE COEXISTENCE MECHANISMS

Species coexistence in most animal communities is likely to result from multiple mechanisms. For instance, parasitoid species that attack the same host species often engage in interference mechanisms, but they also exhibit temporal refuges resulting from differential sensitivities to abiotic environmental variation. The key issue in elucidating the operation of multiple mechanisms is the potential for interactions between different types of coexistence mechanisms. Such interactions often involve nonadditive effects and lead to outcomes that are often unexpected and counterintuitive. Thus, it is crucial that empirical investigations of coexistence mechanisms be guided by a theoretical framework that can both distinguish between different coexistence mechanisms and predict the outcomes of interactions between mechanisms.

6. SUMMARY AND CONCLUSIONS

Coexistence in animal communities can occur via a variety of mechanisms, all of which share the property that negative feedback loops between a given species and the resource (or natural enemy) that it is most limited by leads to strong self-limitation, which overwhelms the effects of interspecific competition. The key issue in elucidating coexistence mechanisms in natural communities is to understand the mechanisms by which such self-limitation occurs. It is equally important to be able to distinguish between different types of mechanisms and to understand the nature of interactions between mechanisms that operate simultaneously. The axes provided here, competition for single versus multiple limiting factors and coexistence mechanisms that are independent of versus dependent on abiotic environmental variation, provide the basis for developing comparative predictions that can guide empirical investigations of multiple coexistence mechanisms in animal communities.

I want to acknowledge Peter Chesson, whose ideas have shaped my thinking about coexistence and whose work I have drawn heavily from in writing this chapter.

FURTHER READING

Amarasekare, P. 2003. Competitive coexistence in spatially structured environments: A synthesis. Ecology Letters 6: 1109–1122. *This article synthesizes recent theory on spatial coexistence mechanisms and develops comparative predictions that can be used to distinguish between different mechanisms.*

Armstrong, R. A., and R. McGehee. 1980. Competitive exclusion. American Naturalist 115: 151–170. *A classic paper that provides a rigorous mathematical analysis of coexistence via nonlinear competitive responses.*

Chesson, P. 2000. Mechanisms of maintenance of species diversity. Annual Review of Ecology and Systematics 31: 343–366. *An important paper that is essential reading for those who want a rigorous and comprehensive understanding of coexistence mechanisms.*

Murdoch, W. W., C. J. Briggs, and R. M. Nisbet. 2003. Consumer–Resource Dynamics. Princeton, NJ: Princeton University Press. *The definitive work on consumer–resource dynamics, with a comprehensive analysis of coexistence mechanisms in parasitoids and predators.*

Ruel, J., and M. Ayres. 1999. Jensen's inequality predicts effects of environmental variation. Trends in Ecology and Evolution 14: 361–366. *A discussion of Jensen's inequality that illustrates how the mathematical concept applies to ecological phenomena.*

Tilman, D. 1982. Resource Competition and Community Structure. Princeton, NJ: Princeton University Press. *A comprehensive treatment of resource competition theory, based on the* R* *rule, that has provided the foundation for much of the later theory on competitive coexistence.*

II.7

Predator–Prey Interactions

Robert F. Denno and Danny Lewis

OUTLINE

1. Evidence that predators reduce prey populations
2. Reciprocal density effects and predator–prey cycles
3. Mathematical models of predator–prey interactions
4. Factors stabilizing predator–prey interactions and promoting their persistence
5. Predation in complex food webs
6. Predation, biodiversity, and biological control
7. Evolutionary interactions between predators and prey
8. Epilogue

In natural food webs, consumers fall victim to other consumers such as predators, parasitoids, parasites, or pathogens. Predators kill and consume all or parts of their prey and do so either before or after their catch has reproduced. A lynx stalking, attacking, and consuming a snowshoe hare is an example from the vertebrate world. Spiders snaring moths in their webs, assassin bugs lancing caterpillars with their beaks, and starfish ravaging mussel beds in rocky intertidal habitats are all instances of invertebrate predation. By contrast, parasitoids such as small wasps and flies usually attack only the immature stages of their arthropod hosts, thus killing them before they reproduce. Parasites live on (e.g., fleas and lice) or in (e.g., tape worms) host tissues, often reducing the fitness of their host but not killing it. Pathogens (e.g., viruses, bacteria, and fungi) induce disease and either weaken or ultimately kill their hosts. Although this chapter focuses on predators, there are many similarities among predator–prey, host–parasitoid, and host–pathogen interactions.

GLOSSARY

food web. Network of feeding relationships among organisms in a community

functional response. The relationship between prey density and the number of prey consumed by an individual predator

intraguild predation. A predation event in which one member of the feeding guild preys on another member of the same guild (predators consuming predators)

keystone species. A species that has a disproportionate effect on its environment relative to its abundance

mesopredator. A predator that is fed on by another predator, usually a top carnivore

numerical response. The relationship between the number of predators in an area and prey density

omnivory. Feeding at more than one trophic level such as occurs when a predator consumes herbivores as well as other predators

top carnivore. A predator at the top of the food chain feeding on organisms at lower trophic levels (e.g., mesopredators and herbivores)

trophic cascade. Reciprocal predator–prey effects that alter the abundance, biomass, or productivity of a community across more than one trophic link in the food web (e.g., removing predators enhances herbivore density, which in turn diminishes plant biomass)

Unlike many consumers, predators are often generalized in their feeding habits, consuming a diversity of prey species that can even represent different trophic groups. For instance, coyotes (top carnivores) feed on other predators such as foxes (mesopredators), both of which consume rabbits (herbivores) and opossums (omnivores). When predators consume other predator species, the act is called intraguild predation, whereas cannibalism occurs when predators consume members of their own species. Wolf spiders (*Lycosa* and *Pardosa*), for example, are notoriously cannibalistic, consuming smaller individuals in the population and even their own offspring. Although many predators are generalists, feeding on a diversity of prey species, there are some very specialized feeders. Desert horned lizards (*Phrynosoma platyrhinos*) are ant specialists, and ground beetles in the genus *Scaphinotus* feed selectively on mollusks and have a long head and mandibles adapted for reaching deep into snail shells.

Predation can have widespread ecological, evolutionary, and economic effects on biological communities in both natural and managed habitats. Predation, for instance, can be a powerful evolutionary force with natural selection favoring more effective predators and less vulnerable prey. In an ecological sense, predators can dramatically affect the abundance and distribution of their prey populations. Moreover, the diverse feeding habits of predators form linkages that are responsible for the flow of energy through food webs, thus affecting food-web dynamics. Predators can act as keystone species, preventing superior competitors from dominating the community and promoting biodiversity at lower trophic levels. In contrast, the invasion of native ecosystems by exotic predators often has very negative effects on resident prey species. On a more positive note, invertebrate predators have been used as effective control agents of agricultural pests, increasing crop yields without the adverse consequences of pesticides. Thus, in both theoretical and applied contexts, it is imperative to understand the process of predation and its complex effects on species interactions, food-web dynamics, and biodiversity. In the remainder of this article, we explore critical elements of predator–prey interactions, namely how predators and prey influence each other's population size and dynamics, what factors stabilize predator–prey interactions and promote their persistence, how predation promotes complex species interactions and stabilizes food webs, and how predators and prey have reciprocally influenced each other's evolution.

1. EVIDENCE THAT PREDATORS REDUCE PREY POPULATIONS

Excluding or adding predators to natural prey populations provides support that predators indeed can reduce populations of their hosts, very significantly in some cases. A classic example involves the mule deer herd on the Kaibab Plateau on the north rim of the Grand Canyon in Arizona. Before 1905, the deer herd numbered about 4000 individuals, but it erupted more than 10-fold over the course of the next 20 years when a bounty resulted in the demise of native deer predators such as wolves, coyotes, and cougars. At a much larger spatial scale in eastern North America, white-tailed deer populations have erupted following the extinction of top predators. These examples and others from the vertebrate world suggest that predators impose natural controls on prey populations and diminish the chances for so-called prey release.

The biological control of crop pests following the release of natural enemies provides further evidence that predators suppress prey populations. With the accidental introduction of Cottony cushion scale (*Icerya purchasi*) from Australia, the California citrus industry became seriously threatened by this severe insect pest. In the late 1800s, a predaceous ladybug beetle (*Rodolia cardinalis*) was collected in Australia and subsequently released into California citrus groves. Shortly after the release of this efficient predator, it completely controlled the scale insect and saved the citrus industry from financial ruin (Caltagirone and Doutt, 1989). Since this classic case, the encouragement or release of arthropod predators has frequently resulted in reduced pest populations (Symondson et al., 2002).

Manipulative experiments also show that invertebrate predators impose controls on prey populations in natural habitats. For example, herbivorous planthoppers (*Prokelisia marginata*) and their wolf spider predators (*Pardosa littoralis*) co-occur on the intertidal marshes of North America (Döbel and Denno, 1994). When spiders are removed from habitat patches, planthopper populations erupt to very high levels. If spiders are removed but are then added back into habitat patches at natural densities, planthopper populations remain suppressed. The question arises as to how predators reduce prey populations. In planthopper–spider systems, spiders can reduce prey directly by consuming them (consumptive effect), or they can indirectly affect prey populations via nonconsumptive effects (Cronin et al., 2004). For example, the mere presence of spiders promotes the local dispersal of planthoppers. Similarly, when grasshoppers are exposed to nonlethal (i.e., defanged) spiders, they undergo a feeding shift from grasses to poor-quality forbs, where they avoid spiders but incur increased mortality from starvation (Schmitz et al., 1997). Notably, the mortality arising from this antipredator behavior rivals that seen when grasshoppers are killed directly by spiders with their fangs intact. In fact, evidence is building from many systems that predators adversely affect prey populations via both consumptive and nonconsumptive effects.

It should be evident that predators often inflict high mortality on prey populations and that in the absence of predation prey populations often erupt. There are cases, however, in which predator removal does not result in increased prey density. Often, such cases involve compensatory mortality whereby the mortality inflicted by predation is replaced by mortality from another limiting factor such as food shortage. In the sections that follow, we consider how predators and prey interact to affect each other's long-term population dynamics, specifically address the role of predation in population cycles, and explore factors that promote the persistence of predator–prey interactions in nature.

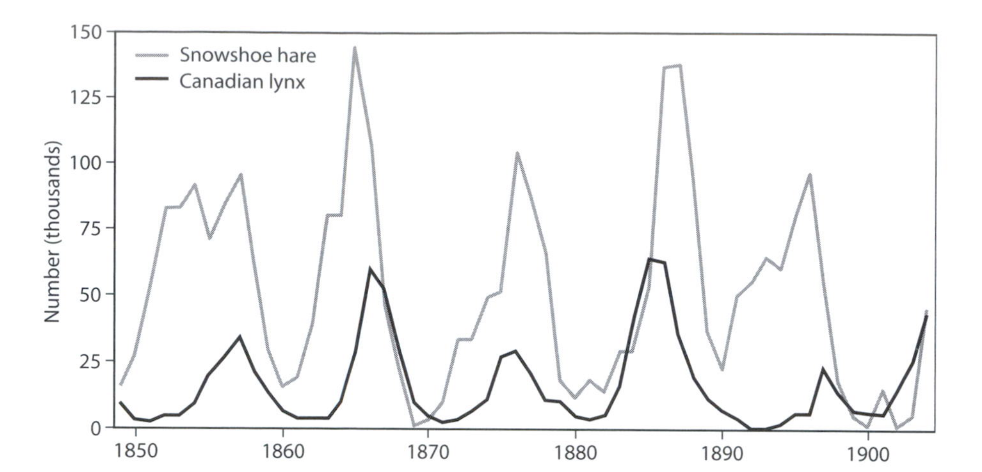

Figure 1. Fluctuations in lynx and snowshoe hare populations based on the number of pelts purchased by the Hudson Bay Company between 1845 and 1930 (data from NERC, 1999).

2. RECIPROCAL DENSITY EFFECTS AND PREDATOR–PREY CYCLES

Population cycles occur in a diverse array of animals ranging from arctic mammals to forest insect pests. Traditionally, predation is one factor thought to induce such cycles (Gilg et al., 2003). Historic support for the view that a coupled predator–prey interaction can drive population cycles came from an analysis of about 100 years of fur-trapping records by the Hudson Bay Company in boreal Canada. An analysis of the number of lynx and snowshoe hare pelts showed spectacular cyclicity with peaks and valleys of abundance occurring at roughly 10-year intervals (figure 1). When hares were numerous, lynx increased in numbers, reducing the hare population, which in turn caused a decline in the lynx population. With predation relaxed, the hare population recovered, and the cycle began anew. It should be noted, however, that there is controversy over the singular role of predation in driving population cycles in boreal mammals.

3. MATHEMATICAL MODELS OF PREDATOR–PREY INTERACTIONS

The first ecologists to model the cyclic dynamics of predator–prey interactions were Alfred Lotka (1925) and Vito Volterra (1926), who independently derived the "predator–prey equations" (Lotka-Volterra equations), a pair of differential equations describing the coupled dynamics of a single specialized predator and one prey species. Both ecologists based their models on observations of reciprocal predator–prey cycles in nature. Volterra's ideas were motivated by watching the rise of fish populations in response to decreased fishing pressure during World War I, whereas Lotka was inspired by observing parasitoid–moth cycles. The Lotka-Volterra equations demonstrate the inherent propensity for predator–prey populations to oscillate, in what is called "neutral stability" (Begon et al., 1996).

For the prey or host population, the rate of population change through time (dH/dt) is represented by the equation:

$$\frac{dH}{dt} = r_h H - \alpha HP,$$

where H is prey density, r_h is the rate of increase of the prey population (birthrate), α is a constant that measures the prey's vulnerability and predator's searching ability, and P is predator density. Thus, exponential growth of the prey population ($r_h H$) is countered by deaths from predation (αHP). Change in the predator population through time (dP/dt) is shown by:

$$\frac{dP}{dt} = c\alpha HP - d_p P,$$

where c is a constant, namely the rate that prey are converted to predator offspring, and d_p is the rate of decrease in the predator population (death rate). The death rate of the predator population ($-d_p P$) is offset by the rate at which predators kill prey and convert them to offspring ($c\alpha HP$). The two equations provide a periodic solution in that predator and prey populations oscillate in reciprocal fashion through time (figure 2A). When the dynamics of predator and prey populations resulting from the Lotka-Volterra equations is plotted in two-phase space (predator density versus prey density), a neutral limit cycle results whereby both predator and prey populations cycle perpetually in time (figure 2B).

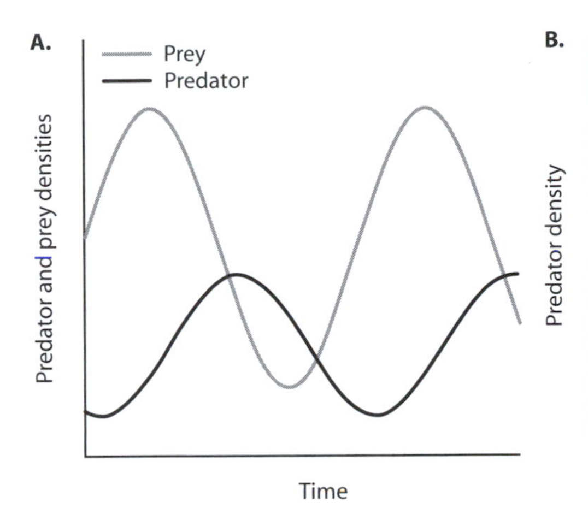

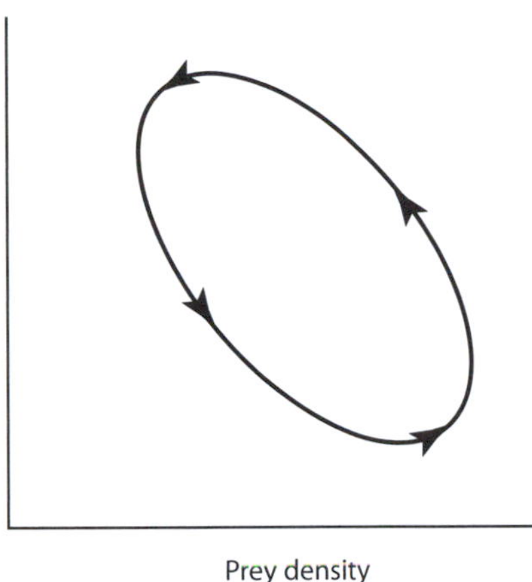

Figure 2. (A) Oscillating predator and prey populations and (B) a neutrally stable predator–prey limit cycle generated by the Lotka-Volterra equations.

Seeing that simple models could generate predator–prey oscillations prompted numerous researchers to duplicate such persistent cycles under simple laboratory conditions. However, these attempts often failed. A representative example involved the predaceous ciliate protozoan *Didinium nasutum* and its prey, another ciliate, *Paramecium caudatum*. Five *Paramecium* were placed in laboratory cultures, and after 2 days three predators were added. Initially, prey populations exploded in the absence of predators, but with the addition of predators, *Paramecium* populations were quickly driven to extinction. Moreover, in the absence of prey, predators subsequently perished (figure 3A).

These unexpected results, and those from many other laboratory attempts, raised the question of why predator–prey cycles could not be easily reproduced in the laboratory, why such simple systems were inherently unstable, and why the predator–prey interaction did not persist. Simply stated, more is needed to understand why prey is not driven to extinction at high predator densities and why predators persist when focal prey are rare. As the following sections demonstrate, ecologists have since identified multiple factors missing from the Lotka-Volterra model that introduce realism into predator–prey interactions and lend accuracy in predicting real-world dynamics. As a result, more recent models incorporate biological features such as saturating predator functional responses (inability of predators to capture all available prey when prey are abundant), nonlinear reproductive responses, predator interference, refuges, spatial processes such as immigration, alternative prey, and multiple trophic levels (Canham et al., 2003; Grimm and Railsback, 2005).

4. FACTORS STABILIZING PREDATOR–PREY INTERACTIONS AND PROMOTING THEIR PERSISTENCE

Although the Lotka-Volterra equations generate coupled predator–prey oscillations, they are inherently oversimplified. It is unrealistic to expect predators and prey to cycle as predicted (figure 2). For instance, several assumptions of the early predator–prey models are not met by real organisms. The models assume exponential growth of prey in the absence of predation and exponential decline of the predator population in the absence of prey. Prey, however, are often resource limited, and their population growth can be slowed independent of predation. As shown in the discussion of functional responses below, predators are rate limited in their ability to capture and process prey, which constrains their ability to suppress prey populations at high densities. Also, predators often interfere with one another at high predator densities, further relaxing predation pressure on prey. There is also the unrealistic model expectation that predators and prey respond instantaneously to changes in each other's densities. For microorganisms with high reproductive rates, this expectation is not far fetched. However, for larger predators, their reproductive response to increased prey density is lagged, providing the opportunity for prey to escape predator control, ultimately leading to an unstable dynamic.

There are a multitude of other reasons why simple models inadequately predict predator–prey dynamics and do not capture the complexity of predator–prey interactions in nature. Foremost is that predator–prey interactions do not take place in closed systems in the absence of spatial processes such as emigration and

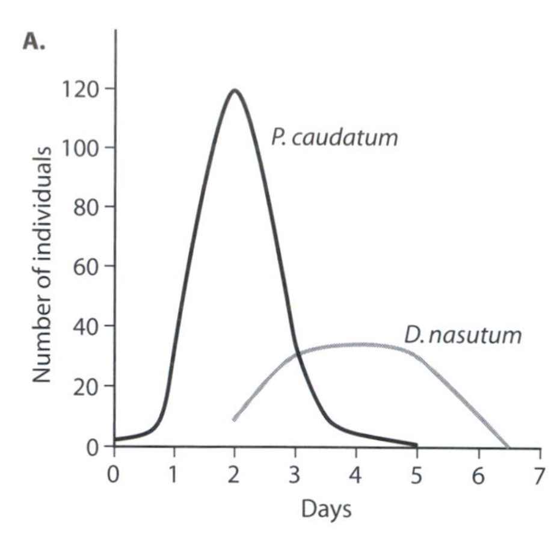

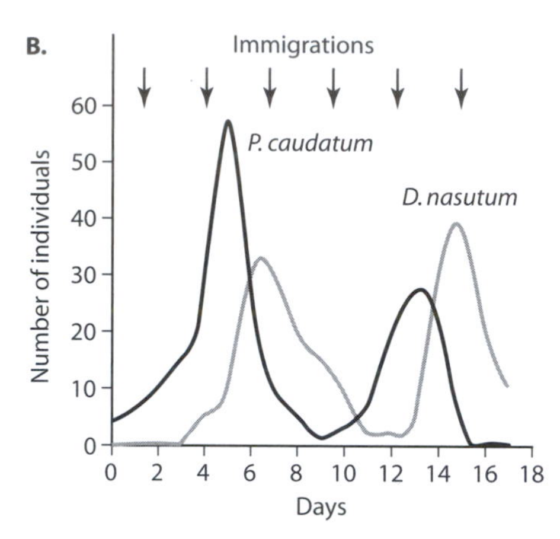

Figure 3. Interaction between predator (*Didinium nasutum*) and prey (*Paramecium caudatum*) in laboratory microcosms (A) without immigration and (B) with immigration (the addition of a single individual of the predator and prey once every 3 days as indicated by arrows). Low-level immigration of predators and prey into the system promoted the cycling and persistence of predator and prey populations. (From Gause, 1934)

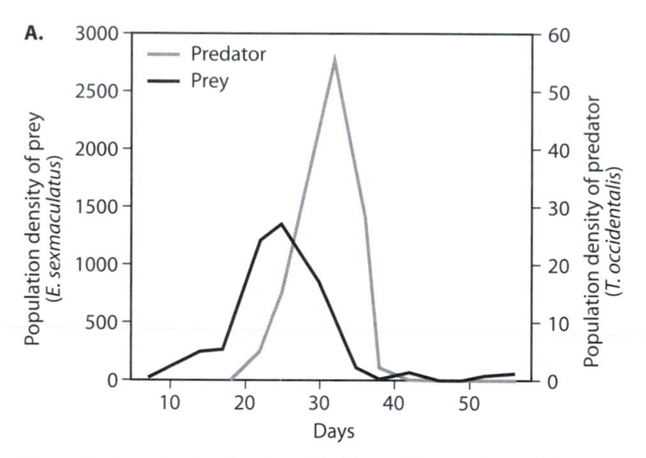

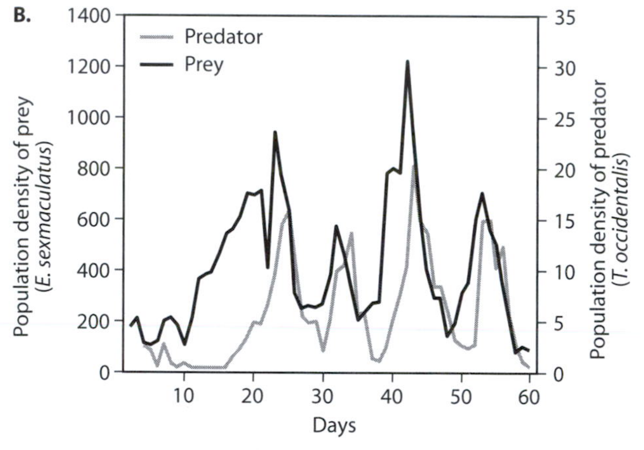

Figure 4. In a simple-structured habitat without refuges (A), predacious mites drive herbivorous mites to low densities, leading to the starvation and ultimate extinction of predatory mites. In a complex-structured habitat (B), herbaceous mites find refuge from predation, and the predator–prey oscillation persists until food (oranges) quality for the prey deteriorates (Huffaker, 1958).

immigration. At low prey densities, predators often disperse to areas of higher prey density, thus relaxing predation on the local prey population rather than driving it to extinction. Moreover, immigration from neighboring patches can rescue declining local populations. Even in very simple lab settings, immigration can encourage the persistence and cycling of predators and prey. Returning to the *Didinium–Paramecium* system, the addition of a single individual predator and prey every third day of the experiment resulted in a persistent predator–prey cycle (figure 3B).

In addition to spatial processes, complex habitat structure and the refuge it provides for prey from predation also lend persistence to predator–prey interactions. A classic example involves interactions between the citrus-feeding mite *Eotetranychus sexmaculatus* and its predatory mite *Typhlodromus occidentalis* (Huffaker, 1958). The population dynamics of the mites was compared between two experimental habitats: a simple habitat consisting of a monoculture of oranges arranged on trays and a complex-structured habitat where oranges were interspersed among rubber balls and little posts from which prey could disperse. In the simple universe, predaceous mites easily dispersed throughout the habitat, prey were driven to a threateningly low density, and the predator then became extinct (figure 4A). In the complex habitat, prey dispersed and found refuges from predation, and three complete predator–prey oscillations resulted before the food quality of oranges deteriorated and the system collapsed (figure 4B). This study highlights the importance of refuges in promoting the coexistence of predators and prey, but it also emphasizes that other factors such as food quality bear on the persistence of the interaction.

Prey species also escape predation as a result of constraints on the ability of predators to catch and handle prey (functional response) and increase their population size (numerical response) as prey densities rise (Holling, 1959, 1965). In his component analysis of predation, Holling described three types of functional response (figure 5).

For predators exhibiting a Type I functional response, the consumption rate of a single individual is limited only by prey density (figure 5A). Thus, over a wide range of densities, per capita consumption and prey density are linearly related. Many filter feeders (e.g., rotifers and sponges) that consume suspended zooplankton exhibit this response. For predators showing a Type I response, the proportion of prey captured of the total number offered remains constant and independent of prey density (figure 5B).

Most invertebrate predators (e.g., hunting spiders, preying mantises, ladybug beetles) exhibit a Type II functional response, in which consumption rate levels off with increasing prey density to an upper plateau (saturating response) set by handling time (the time required to subdue and consume each prey item) and satiation (figure 5C). At high prey densities, most of a predator's time is spent handling captured prey, and little time is spent searching for additional prey. Notably, for predators with a Type II response, the fraction of prey captured of the available total decreases with increasing prey density (figure 5D). As prey density increases, such predators are less able to reduce prey population growth, thus providing prey with an ever-growing opportunity to escape from predation.

Many vertebrate predators (e.g., birds and mammals) and some invertebrate predators show a sigmoidal or Type III functional response (figure 5E). For such predators, consumption rate responds slowly to increases in prey density when prey is scarce. At somewhat higher prey densities, consumption rate rises rapidly, and at very high prey densities, consumption rate saturates and is limited by handling time as in a Type II response. The rapid rise in consumption rate at intermediate prey densities occurs because predators learn to discover and capture prey with increased efficiency or they simply increase their searching rate as they encounter more prey. In some cases, predators respond to volatile chemicals emitted by their prey and thereafter rapidly increase and direct their searching rate accordingly.

Polyphagous predators often switch to alternative prey when the density of their preferred prey species falls below a certain threshold. Cases of prey switching can transform a Type II response into a Type III because the consumption rate of focal prey is relaxed at low prey densities. Regardless of the mechanism (learning, increased search rate, or prey switching), predators exhibiting a Type III response are thought to stabilize predator–prey interactions. Stability is imposed because the fraction of prey consumed by a predator is low at low densities, preventing predators from driving prey to extinction (figure 5F). Yet, with an increase in prey density, the fraction of prey consumed increases (is density dependent), thus reducing the opportunity for prey to escape predator control. Only at very high densities are predators satiated such that the fraction of prey consumed decreases and the prey population escapes.

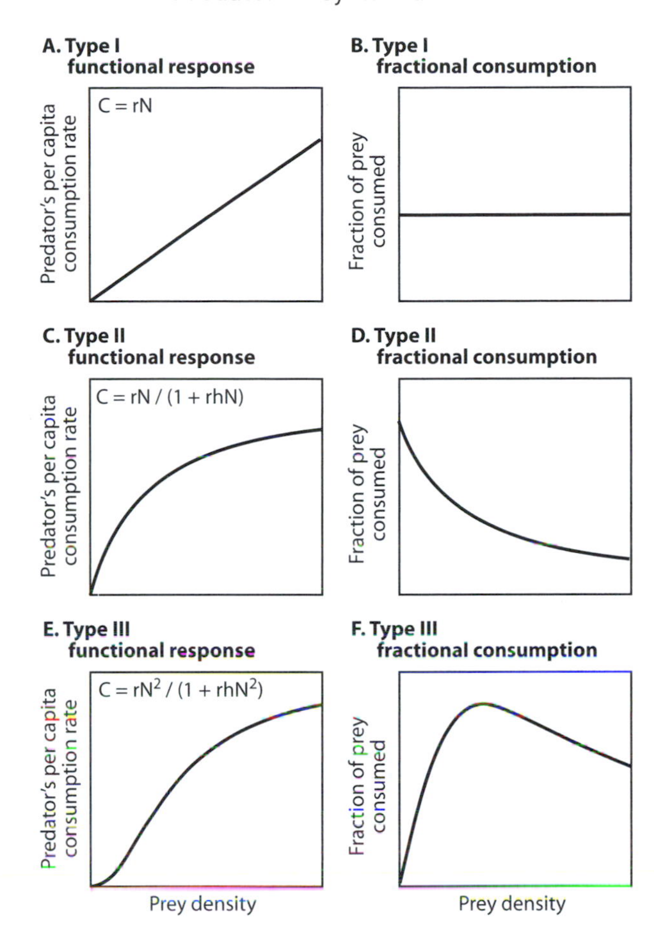

Figure 5. Consumption rate of predators when offered prey at increasing densities. Per capita consumption rates are shown for predators exhibiting Type I, II, and III functional responses (A, C, and E, respectively), as is the fractional consumption rate (proportion of prey taken of the total number offered) for each functional response (B, D, and F). Associated equations describe the specific functional response: C = per capita consumption rate, N = prey density, r = "risk of discovery" (a measure of prey vulnerability), and h = handling time.

So far, we have considered only the consumption rate of an individual predator under conditions of increasing prey density. To gain a complete picture of how predators might control prey populations, we also need to know how many predators are present in the population and how they respond to increasing prey densities. Most predators exhibit a numerical response by becoming more abundant as the density of their preferred prey increases. Two independent mechanisms underlie this pattern. First, predators often aggregate in areas where prey abound, a response that results from a short-term change in the spatial distribution of predators. For instance, the local density of the wolf spider *Pardosa littoralis* can be dramatically enhanced over a 3-day period when prey is experimentally added to its habitat (Döbel and Denno, 1994). Second, if prey density remains high for an extended period of time, predator populations often build as a consequence of increased reproduction. The density of Arctic foxes, for instance, increases greatly during peak lemming years because of elevated breeding success (Gilg et al., 2003). Thus, predator aggregation and enhanced reproduction can both account for the numerical response of a predator to increased prey density. Like functional responses, numerical responses level off at intermediate prey densities because continued increases in prey densities do not result in a higher predator density because of reproductive limitations or interference among conspecific predators. Nonetheless, strong aggregative responses are often thought to stabilize predator–prey interactions because predation is relaxed in vacated habitats where prey density is low, whereas predation is increased in colonized habitats where prey is more abundant.

Combining a predator's functional and numerical responses into its total response predicts a predator's overall response to increased prey density and thus its overall impact on the prey population. A predator's total response (number of prey consumed/unit area) can be calculated by multiplying its functional response (number of prey consumed/predator) by its numerical response (number of predators/unit area). Because both functional and numerical responses level off at intermediate prey densities, further increases in prey density result in an increasingly smaller proportion of the prey population that is killed by predators. Thus, the opportunity for escape exists at high prey densities, and one defensive strategy is for prey to satiate the predator population by emerging synchronously at very high densities. Periodical cicadas (*Magicicada*) appear to employ this strategy in that the proportion of mortality attributable to predation is drastically reduced during times of peak emergence when cicadas reach incredibly high densities.

Other life history traits of prey, such as dispersal capability, stage class, and body size may also provide escape from predation and thus contribute to the persistence of predator–prey interactions. Regarding dispersal, a highly mobile lifestyle appears to promote escape from predator control. Planthopper species, for example, vary tremendously in their dispersal ability with both highly mobile and extremely sedentary species represented. In a survey of species, invertebrate predators inflicted significantly more mortality on immobile species than on their migratory counterparts (Denno and Peterson, 2000). Invulnerable stage classes also provide a refuge from predation. The true bug *Tytthus* attacks only the eggs of planthoppers; thus, once planthopper eggs have hatched, emerging nymphs are immune to predation from this specialist predator. Also, because the act of predation requires overpowering victims, predators usually profit by attacking smaller or weaker individuals in the prey population. Size-selective predation has been observed across a wide range of vertebrate (snakes, fish, birds, and mammals) and invertebrate predators (insects, spiders, starfish). Importantly, because predators often focus their attacks on smaller prey, larger prey obtain a partial refuge from predation.

5. PREDATION IN COMPLEX FOOD WEBS

So far, our focus has been on interactions between a single predator and prey species and how inherent limitations imposed by a predator's functional and numerical responses and size-selective predation can offer prey a partial escape from predation. However, predators and prey are nested into food webs and do not occur in isolation from other players. In fact, refuges for prey exist because of other species in the system. We have already seen how the presence of alternative prey species can relax predation on focal prey when its density drops to low levels. Moreover, mesopredators are also subject to predation themselves from top carnivores. Thus, interactions among predators can result in intraguild predation, which often relaxes predation on shared prey. A good example involves heteropterans bugs (*Zelus* and *Nabis*) and lacewing larvae (*Chrysoperla*), all of which prey on cotton aphids (Rosenheim et al., 1993). In the absence of heteropterans predators, lacewing predation drives the aphid population to a low level. When heteropterans are added to the system, they focus their attack on the more vulnerable lacewings, aphids experience a partial refuge from predation, and aphid populations rebound. Thus, the presence of multiple predator and prey species in the community can alter the interaction between a specific predator–prey pair and often lends stability to any specific interaction.

By now it should be clear that many factors influence the dynamics of a specific predator–prey interaction. Even in the simplest of systems, it is difficult to draw solely on the focal pair of players to explain each other's population fluctuations. For instance, the 4-year population cycles of lemmings in Greenland are driven by a 1-year delay in the numerical response of stoat and stabilized by density-dependent predation imposed by arctic foxes, snowy owls, and skuas (Gilg et al., 2003). Moreover, recent assessments of snowshoe hare population dynamics (figure 1), including a large-scale predator exclusion and food enhancement experiment, suggest that hare population cycles result from interactions among three trophic levels (Krebs et al., 1995, 2001). When lynx and other predators were experimentally excluded, hare populations doubled. Hare populations tripled when plant biomass was increased via fertilization. Strikingly, hare populations increased 11-fold when predators were excluded and plant resources were enhanced. This finding highlights the view that predators and prey can indeed affect each other's abundance, but the dynamic of the interaction is complex and can not be studied in isolation from other factors.

Visiting an invertebrate system further underscores why understanding predator–prey interactions requires a multitrophic approach (Finke and Denno, 2004, 2006). *Spartina* cordgrass is the only host plant for *Prokelisia* planthoppers, which in turn are consumed by the mesopredator *Tytthus* and the intraguild predator *Pardosa*. In this intertidal system, there is considerable variation in leaf litter as a result of elevational differences in tidal flushing and decomposition. In litter-rich habitats, *Pardosa* spiders abound and readily aggregate in areas of elevated planthopper density. In these structurally complex habitats, *Tytthus* finds refuge from *Pardosa* predation, the predator complex effectively suppresses planthoppers, and cordgrass flourishes. By contrast, in litter-poor habitats, spiders are less abundant, *Tytthus* experiences intraguild predation, and overall predation on planthoppers is relaxed, leading to planthopper outbreak and reduced plant biomass. Thus, both vegetation structure and the predator assemblage interact in complex ways to influence the strength of the spider–planthopper interaction and the probability for a trophic cascade, namely the extent to which predator effects cascade to affect herbivore suppression and plant biomass. This example and many others further emphasize that alternative population equilibria exist for prey and that release from predation is dependent on spatial refuges and the composition of other players in the system.

Although there is evidence that cycling does occur in some simple predator–prey systems in the boreal north, coupled cycling is not often characteristic of predator–prey dynamics in more complex food webs. Here, plant-mediated effects, alternative prey, intraguild predation, and refuges collectively dampen predator–prey cycles. Moreover, such multitrophic interactions are the rule and are thought to lend stability to food webs, making them more resistant to environmental disturbance and invasion by other species (Fagan, 1997).

6. PREDATION, BIODIVERSITY, AND BIOLOGICAL CONTROL

Predators can act as keystone species influencing the species composition and biodiversity of the prey community. A classic case involves starfish, which graze mussels and barnacles in intertidal habitats, precluding them from dominating the community, allowing other invertebrate species to persist, and enhancing the overall diversity of the benthic community (Paine, 1974). From a conservation perspective, however, exotic predators that invade natural habitats can have very negative effects on resident prey species, effects that can cascade throughout the food web. When the brown treesnake was accidentally introduced to Guam, its population erupted in the absence of native predators, ultimately leading to the widespread extirpation of many native vertebrate species including birds, mammals, and lizards. Similarly, rainbow trout have been purposefully introduced throughout the world, often with devastating effects on native stream communities. In parts of New Zealand, trout incursion resulted in a trophic cascade, whereby this efficient predator reduced populations of native invertebrates that graze on benthic algae, which in turn promoted dramatic increases in algal biomass.

Another alarming consequence of our rapidly changing world is the loss of biodiversity as a result of habitat disturbance, fragmentation, and loss. In particular, consumers at higher trophic levels such as predators are at risk of extinction. In coastal California, for example, urbanization and habitat fragmentation have promoted the disappearance of coyote, the historic top carnivore in this sage–scrub habitat (Crooks and Soulé, 1999). Its disappearance has fostered increased numbers of smaller mesopredators (e.g., foxes and skunks), which in turn are contributing to the extinction of scrub-breeding birds.

A practical extension of the consequences of multiple-predator interactions is whether single or multiple predators are more effective in suppressing agricultural pests. The effect of increased predator diversity on

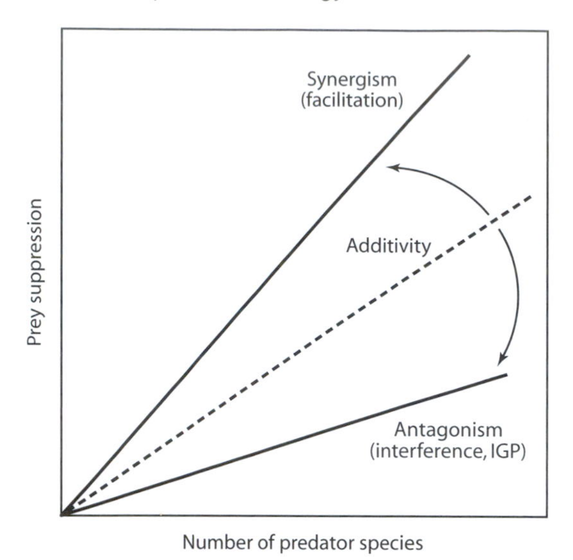

Figure 6. Relationships between the number of predator species in the system and prey suppression. Synergism results from facilitation, additivity arises from predator complementarity, and antagonism occurs from intraguild predation (IGP) or interference.

biological control, and ecosystem function at large, depends on how predator species interact and complement each other (figure 6). We have already seen how increasing predator diversity by adding an intraguild predator to the enemy complex can relax predation on shared herbivore prey. However, not all predator–predator interactions are antagonistic. In stream systems, predators interact synergistically, whereby stonefly predators drive mayflies from under stones making them more susceptible to fish attack (Soluk and Collins, 1988). Thus, predators that interact synergistically can enhance prey suppression beyond additive expectations. Likewise, if predators complement each other by attacking prey at different developmental stages or during different times of the year, increasing predator diversity can enhance prey suppression. The key to elucidating the relationship between predator diversity and ecosystem function or manipulating the composition of the predator assemblage for more effective biological control rests on the nature of predator–predator interactions in the system. The issue remains open in biological control because there is system-specific evidence that increasing predator diversity can either increase or decrease pest suppression.

7. EVOLUTIONARY INTERACTIONS BETWEEN PREDATORS AND PREY

There is little doubt that predators have exerted selection on prey that has resulted in evolutionary change. For instance, prey species have evolved a wide range of defenses in response to selection from predation. Such defenses can be categorized as primary, secondary, or tertiary depending on when in the predation sequence (detection, capture, or handling) they operate. Primary defenses (e.g., crypsis and reduced activity when predators forage) operate before prey is detected by a predator. Secondary defenses operate to deter capture after prey is detected by the predator. Examples include escape mechanisms (aphids dropping from leaves in the presence of a foraging ladybug), startle behaviors (moths displaying wings with eye spots to frighten away birds), and evasive behaviors (moths detecting the sonar of bats and initiating strategic flight-avoidance tactics). Tertiary defenses interrupt predation after capture and during the handling phase. Such defenses include mechanisms that deter, repel, or even kill the predator directly (contact toxins, venoms, or morphological structures such as spines). The consequences of some tertiary defenses for predators can be severe. The neurotoxin injected by the death-stalker scorpion (*Leiurus quinquestriatus*) can cause rapid paralysis and death to an attacking small mammal.

Clearly, predation has promoted a wide array of prey defenses, and the question arises as to whether there have been counteradaptations in predators. Have predators and prey engaged in an "evolutionary arms race" such that reciprocal selection has promoted a continuing escalation of predator offense and prey defense? Some evidence is consistent with this hypothesis. For instance, the drilling abilities of predaceous gastropods and the shell thickness of their bivalve prey have increased through geologic time (Vermeij, 1994). Similarly, marine snails have become more heavily armored, while the correlated response in predaceous crabs has been the evolution of larger claws for crushing the better-defended snails. In both of these instances, predators may have evolved greater weaponry in reciprocal response to improved prey defense (coevolution hypothesis), or predator armaments may have evolved in response to other predators or competitors (escalation scenario).

Overall, however, evidence suggests that reciprocal selection on predators may be weaker than that on prey, thus precluding a classic evolutionary arms race (Brodie and Brodie, 1999). In part, the asymmetry arises because many predators are generalist feeders, and selection imposed by any one of its prey options is likely small. In general, coevolution between exploiter and exploited is unlikely when the intimacy of the interaction is low. Moreover, selection on predators from effective primary and secondary prey defenses is probably weak. For instance, if a predator fails to detect cryptic prey or catch a stealthy individual, it simply searches for another, without penalty. The exception occurs when predators interact with dangerous prey, prey that possess tertiary defenses such as toxins that can kill the attacker. In such cases, predators experience strong selection from prey and are expected to evolve either innate avoidance behavior or traits that allow them to exploit dangerous prey. Such a case of coevolution has likely occurred between the toxic newt *Taricha granulosa* and its garter snake predator *Thamnophis sirtalis*. The skin of the newt contains one of the most potent neurotoxins known, one that kills all other predators outright. Across populations, geographic variation in the level of newt toxin covaries with levels of resistance in the snake. Thus, garter snakes are evolving in response to newt defenses, and the "arms race" is apparently taking place (Brodie and Brodie, 1999).

8. EPILOGUE

Predation is a central process in community ecology because it is responsible for energy flow among multiple trophic levels. Moreover, the many reticulate linkages resulting from predation across multiple trophic levels (omnivory) are an important stabilizing force in food-web dynamics. Predators, however, by virtue of their precarious position at the apex of the food chain, are often at greater risk of extinction when natural systems are disturbed, often with dire consequences for the diversity and functioning of the community at large. Also, predators are important members of natural-enemy complexes that can provide effective pest control in agroecosystems. From the perspective of the consequences of predation, however, it is crucial to realize that the objectives of conservation biology and biological control seek very different ends. Conservation biologists seek to retain trophic diversity, preserve trophic linkages (e.g., intraguild predation) that stabilize food-web interactions, and reduce the probability that predator effects will cascade to affect plant productivity. Reconstructing stable food-web dynamics and ecosystem function are particularly important in habitat restoration projects. By contrast, biological control aims to induce trophic cascades, whereby antagonistic interactions among predators are minimized, pests are collectively suppressed by the enemy complex, and crop yield is enhanced. This contrast in objectives provides an ideal impetus for exploring the complex role of predation in community dynamics from both applied and theoretical perspectives.

FURTHER READING

Begon, M., J. A. Harper, and C. R. Townsend. 1996. Ecology, 3rd ed. London: Blackwell Science.

Brodie, E. D. III, and E. D. Brodie, Jr. 1999. Predator–prey arms races. BioScience 49: 557–568.

Caltagirone, L. E., and R. L. Doutt. 1989. The history of the vedalia beetle importation to California and its impact on the development of biological control. Annual Review of Entomology 34: 1–16.

Canham, Charles D., Jonathan J. Cole, and William K. Lauenroth. 2003. Models in Ecosystem Science. Princeton, NJ: Princeton University Press.

Cronin, J. T., K. J. Haynes, and F. Dillemuth. 2004. Spider effects on planthopper mortality, dispersal and spatial population dynamics. Ecology 85: 2134–2143.

Crooks, K. R., and M. E. Soulé. 1999. Mesopredator release and avian extinctions in a fragmented habitat. Nature 400: 563–566.

Denno, R. F., and M. A. Peterson. 2000. Caught between the devil and the deep blue sea, mobile planthoppers elude natural enemies and deteriorating host plants. American Entomologist 46: 95–109.

Döbel, H. G., and R. F. Denno. 1994. Predator–planthopper interactions. In Planthoppers: Their Ecology and Management. R. F. Denno and T. J. Perfect, eds. New York: Chapman & Hall, 325–399.

Fagan, W. F. 1997. Omnivory as a stabilizing feature of natural communities. American Naturalist 150: 554–567.

Finke, D. L., and R. F. Denno. 2004. Predator diversity dampens trophic cascades. Nature 429: 407–410.

Finke, D. L., and R. F. Denno. 2006. Spatial refuge from intraguild predation: Implications for prey suppression and trophic cascades. Oecologia 149: 265–275.

Gause, G. F. 1934. The Struggle for Existence. Baltimore: Williams & Wilkins. Reprinted 1964, New York: Hafner.

Gilg, O., I. Hanski, and B. Sittler. 2003. Cyclic dynamics in a simple vertebrate predator–prey community. Science 302: 866–868.

Grimm, V., and S. F. Railsback. 2005. Individual-based Modeling and Ecology. Princeton, NJ: Princeton University Press.

Holling, C. S. 1959. The components of predation as revealed by a study of small mammal predation of the European pine sawfly. Canadian Entomologist 91: 293–320.

Holling, C. S. 1965. The functional response of predators to prey density and its role in mimicry and population regulation. Memoirs of the Entomological Society of Canada 45: 5–60.

Huffaker, C. B. 1958. Experimental studies on predation: Dispersion factors and predator–prey oscillations. Hilgardia 27: 343–383.

Krebs, C. J., R. Boonstra, S. Boutin, and A.R.E. Sinclair. 2001. What drives the 10-yr cycle of snowshoe hares? BioScience 51: 25–36.

Krebs, C. J., S. Boutin, R. Boonstra, A.R.E. Sinclair, J.N.M. Smith, M.R.T. Dale, K. Martin, and R. Tarkington. 1995. Impact of food and predation on the snowshoe hare cycle. Science 269: 1112–1115.

NERC Centre for Population Biology, Imperial College. 1999. The Global Population Dynamics Database. http://www.sw.ic.ac.uk/cpb/cpb/gpdd.html.

Paine, R. T. 1974. Intertidal community structure: Experimental studies of the relationship between a dominant competitor and its principal predator. Oecologia 15: 93–120.

Rosenheim, J. A., L. R. Wilhoit, and C. A. Armer. 1993. Influence of intraguild predation among generalist insect predators on the suppression of an herbivore population. Oecologia 96: 439–449.

Schmitz, O. J., A. P. Beckerman, and K. M. O'Brien. 1997. Behaviorally mediated trophic cascades: Effects of predation risk on food web interactions. Ecology 78: 1388–1399.

Soluk, D. A., and N. C. Collins. 1988. Balancing risks? Responses and nonresponses of mayfly larvae to fish and stonefly predators. Oecologia 77: 370–374.

Symondson, W.O.C., K. D. Sunderland, and M. H. Greenstone. 2002. Can generalist predators be effective biocontrol agents? Annual Review of Entomology 47: 561–594.

Vermeij, G. J. 1994. The evolutionary interaction among species: Selection, escalation, and coevolution. Annual Review of Ecology and Systematics 25: 219–236.

II.8

Host–Parasitoid Interactions

Cheryl J. Briggs

OUTLINE

1. Parasitoid terminology and taxonomy
2. Parasitoids and behavioral ecology
3. Parasitoids in theory: The quest for persistence and stability
4. Heterogeneity in risk of parasitism
5. Large-scale spatial dynamics
6. Stage structure in systems with overlapping generations
7. Parasitoids in biological control: Case study of the successful control of California red scale

Parasitoid–host interactions have been popular topics of study in the areas of population biology and behavioral ecology because they represent potentially simple, tightly coupled interactions in which the oviposition behavior of the adult female parasitoid searching for hosts translates directly into fecundity and therefore fitness. The Nicholson-Bailey model, which predicts that the interaction between a single host and a single parasitoid species, in its simplest form, will result in the extinction of one or both species, spawned several decades of research into uncovering the mechanisms that allow real host–parasitoid interactions to persist. Work in this area has focused on the potential stabilizing effects of heterogeneity across the host population in the risk of parasitism, large-scale spatial processes, and stage-structured interactions.

GLOSSARY

classical biological control. The purposeful release of natural enemies of a pest (often from the pest's area of origin) with the hope that the enemy will both suppress the density of a pest species and also persist to suppress future outbreaks of the pest.

oviposition/ovipositor. The act of laying an egg on or in a host/the specialized structure that many adult female parasitoids have that allows them to lay an egg on or in a host.

parasitoid. Parasitoids are insects in which the adult female lays one or more eggs on, in, or near the body of another insect (the host), and the resulting parasitoid offspring use the host for food as they develop, killing the host in the process.

population regulation. In the history of ecology, this has been a surprisingly difficult term to define; the tendency of a population to persist within bounds.

pseudointerference. A form of temporal density dependence in which the parasitoid efficiency decreases at high parasitoid densities because an increasing fraction of parasitoid attacks are wasted on already-parasitized hosts.

stability. A population equilibrium is stable if the population returns to the equilibrium following a perturbation.

1. PARASITOID TERMINOLOGY AND TAXONOMY

Parasitoids are insects the adult female of which lays one or more eggs on, in, or near the body of another insect (the host), and the resulting parasitoid offspring use the host for food as they develop, killing the host in the process. Some authors have used the term *parasitoid* more generally to describe parasitic species that spend the majority of their life in close association with a single host individual, ultimately resulting in the death of that host; however, the term has been used mainly in reference to insects with this type of life history. Parasitoids are distinguished from parasites in that parasitoids kill their host in the process of completing their life cycle. They differ from predators because they require only a single host to complete their development. The majority of insect parasitoids are Hymenoptera (wasps) or Diptera (flies), but the parasitoid life history is also present in the Coleoptera (beetles) and occasionally in representatives of other orders of insects.

Most parasitoids attack the juvenile stages of their host, and the parasitoid literature is filled with specialized terminology to describe their mode of attack.

Parasitoids are often characterized by the stage of host that is attacked, e.g., in egg parasitoids, the adult female parasitoid lays her egg in the egg stage of the host, and in larval parasitoids, it is the larval host stage that is initially attacked. Some parasitoids (termed *idiobionts*) immediately kill or permanently paralyze their host at the time of attack, whereas others (*koinobionts*) permit their host to continue to feed, grow, and develop for some time before it is killed, allowing the developing parasitoid offspring to gain more resources from the host beyond those present at the time of oviposition. In solitary parasitoids, a single egg is laid on a host, whereas in gregarious parasitoids, a few to several hundred eggs can be laid on the same host. In some species, the female parasitoid gains all of the protein needed to produce all of her eggs from the host on which she developed (*proovigenic* parasitoids), whereas in others, the female continues to develop eggs during her adult life (*synovigenic* parasitoids), and the adult female can feed on additional hosts to gain the necessary nutrients. When host feeding, the adult female parasitoid can pierce the host with her ovipositor and then turn around and consume the host fluids, rather than laying an egg. Host feeding generally results in the death of the host.

2. PARASITOIDS AND BEHAVIORAL ECOLOGY

Parasitoid–host interactions have been highly influential in the development of ecological theory pertaining to population dynamics, population regulation, and species interactions, in part because, at least conceptually, the parasitoid–host interaction represents a simple, tightly coupled consumer–resource interaction in which each host attacked can lead to one or a clutch of new parasitoids. Because of this direct link between the oviposition behavior by searching female parasitoids and their fecundity, and therefore fitness, parasitoids are also the subjects of intensive study in behavioral ecology. Parasitoids display a staggering array of interesting behaviors. Parasitoids have been shown to assess the quality and size of the hosts that they encounter and modify their oviposition strategy accordingly. Larger hosts contain more resources for the developing parasitoids and therefore can produce larger, or more, parasitoid offspring. In many cases, parasitoids have been shown to lay larger clutches of eggs in larger or higher-quality hosts. Hymenopterous parasitoids are haplodiploid, such that females result from fertilized eggs and males result from unfertilized eggs. Therefore, the adult female can determine the sex of her offspring through fertilization. Female parasitoids (which produce costly eggs) generally benefit more than males from being large, and as such, many parasitoid species have been shown to lay female eggs in large hosts and male eggs in small hosts. In synovigenic parasitoid species, females tend to host-feed on smaller hosts and oviposit in larger, higher-quality, hosts.

In many cases, parasitoids have been shown to be able to discriminate between unparasitized and parasitized hosts, and they can distinguish between hosts that have been parasitized by members of their own species (conspecifics) versus other parasitoid species (heterospecifics). Superparasitism describes the situation in which the adult female parasitoid lays additional eggs on a host individual that has already been parasitized by a conspecific, and in multiparasitism, the adult female parasitoid oviposits on a host that has been parasitized by a heterospecific competitor parasitoid. In either case, the adult parasitoid may kill any eggs present on the host (ovicide) by probing them with her ovipositor, or the resulting parasitoid larvae may compete within the host resulting in the death of one or both larvae. A further type of interaction occurs when a heterospecific parasitoid larva that is already present is itself used as a host. Some parasitoids are facultative hyperparasitoids in that they can attack either unparasitized hosts or hosts that have already been parasitized by individuals of another species, whereas other species are obligate hyperparasitoids in that they attack only parasitized hosts. Autoparasitism (heteronomous hyperparasitism) is one of the more unusual life history strategies exhibited by parasitoids in which female eggs are laid on an unparasitized host while male eggs are laid on juvenile parasitoids, either of the same species or another species.

In addition to being of interest to behavioral and evolutionary ecologists, all of these oviposition behaviors have the potential to influence the dynamics of the host–parasitoid interaction. For example, superparasitism can lead to density-dependent juvenile parasitoid survival and can have a potentially stabilizing effect. Facultative hyperparasitism has the same dynamic effect as intraguild predation (in which competing predator species also eat each other) and can therefore affect the outcome of competition between parasitoid species.

3. PARASITOIDS IN THEORY: THE QUEST FOR PERSISTENCE AND STABILITY

The Nicholson-Bailey model lies at the heart of most of the theory of host–parasitoid population dynamics. This model considers the host–parasitoid interaction in its simplest form: the population of a single host species is attacked by a single parasitoid that searches for hosts at random within the closed population. The hosts and parasitoids have synchronized, nonoverlapping generations, making the model most relevant for temperate insect systems with a single generation per year. The host population is not regulated by any factor other than the parasitoid (i.e., no other sources of density dependence). The model predicts the density of hosts (N) and parasitoids (P) in year $t+1$ based on their densities in year t:

$$\text{Hosts}: \quad N_{t+1} = R\, N_t f(N_t, P_t)$$
$$\text{Parasitoids}: \quad P_{t+1} = cN_t[1 - f(N_t, P_t)]$$

where R is the net reproductive rate of hosts in the absence of the parasitoids (the host population would increase by a factor of R each year if no parasitoids were present). The function $f(N_t, P_t)$ is the fraction of hosts that avoid being parasitized in year t. The Nicholson-Bailey model assumes that this function does not depend on the density of hosts but decreases exponentially as the density of parasitoids in year t increases: $f(N_t, P_t) = f(P_t) = exp(-a\, P_t)$. This function is the zero term of a Poisson distribution and can be thought of as the result of parasitoids searching randomly for hosts with the average number of encouters (aP_t) increasing linearly with parasitoid density. The constant, a, can be interpreted as the fraction of the total area searched by each parasitoid during year t. If $f(N_t, P_t)$ is the fraction of hosts that escape parasitism, then the remaining fraction $[1 - f(N_t, P_t)]$ is the fraction of hosts in year t that are parasitized. Each of these parasitized hosts results in c new parasitoids in the following year.

As long as R is greater than 1, this system always has a positive equilibrium with both host and parasitoid present. The equilibrium, however, is always unstable. For all parameter combinations, the host and parasitoid populations oscillate, with the magnitude of the oscillations rapidly increasing in amplitude through time until one or both populations become extinct. The parasitoid population lags behind the host, and the only two potential outcomes are extinction of the parasitoids followed by geometric growth of the host and extinction of the host followed by extinction of the parasitoid. In this simple model, persistence of the host–parasitoid interaction is not possible. The instability occurs because the parasitoids overexploit the host population, causing them to become rare, which in turn leads to a crash in parasitoid numbers, allowing the host population to recover; because of the time lags in the system, the successive cycles of overexploitation and recovery increase progressively in amplitude until one party becomes extinct.

The inherent instability and lack of persistence of the Nicholson-Bailey model fly in the face of the fact that parasitoid–host interactions do exist in nature and are therefore apparently persistent. Numerous examples of persistent and sometimes stable dynamics have been observed in real laboratory and field host–parasitoid interactions. In many cases, parasitoid species that have been released to control pest insect species have resulted in successful classical biological control, which relies on persistence of the parasitoid with its host. This disconnect between the predictions of the Nicholson-Bailey model and the observed persistence of real host–parasitoid systems spawned decades of research into uncovering the mechanisms that could potentially regulate and stabilize host–parasitoid interactions.

The Nicholson-Bailey model makes very simple assumptions about how the host–parasitoid interaction works. Much of the theoretical research in this area has investigated the effects of violating these assumptions to determine their effects on the persistence and stability of the host–parasitoid interaction. In most cases, the theoretical developments have been inspired by empirical studies, either the results of laboratory experiments or observations on the life history features of specific species.

Key assumptions of the Nicholson-Bailey model are:

- No density dependence in the host reproduction or survival
- Random searching by parasitoids in a well-mixed host population
- Constant parasitoid search rate that does not depend on parasitoid or host density
- Efficiency at converting parasitized hosts to new searching parasitoids does not depend on host or parasitoid density
- No density dependence in parasitoid survival
- Synchronized, nonoverlapping generations
- Closed populations of host and parasitoid
- One host/one parasitoid: the host is attacked by only a single parasitoid species, and the parasitoid species specializes on a single host

Models have been developed to determine the effects on persistence and stability of all of these biological mechanisms. Three types of potentially stabilizing mechanisms are highlighted here: heterogeneity in risk of parasitism, large-scale spatial dynamics (e.g., metapopulation dynamics), and stage structure in systems with overlapping generations.

4. HETEROGENEITY IN RISK OF PARASITISM

The Nicholson-Bailey model assumes that all hosts in the population at any given time are at equal risk of being attacked by parasitoids and that the parasitoids search at random within the host population. Much attention has been directed to the different ways that these assumptions can be violated and to the effects of heterogeneity in risk of parasitism. Robert May showed that allowing nonrandom attack by the parasitoids can drastically alter the predictions of the Nicholson-Bailey model. He did this in a phenomenological way by replacing the function that describes the fraction of hosts that escape parasitism, $f(P_t)$, in the Nicholson-Bailey model with the zero term of a negative-binomial distribution [$f(P_t) = (1 + aP_t/k)^{-k}$, rather than the zero term of a Poisson distribution as in the original model]. The negative-binomial model assumes that there is heterogeneous risk of parasitoid attack across the host population, according to the clumping parameter, k. For high values of k, the negative-binomial model approaches the case of homogeneous parasitoid attack, as in the original model. As k is reduced, the distribution of risk across the host population becomes more skewed, with high risk of parasitism concentrated in a smaller fraction of the population. This could come about, for example, if certain hosts are more exposed to parasitoids (e.g., near the plant surface) or are on plants that attract more parasitoids. Adding heterogeneous attack to the Nicholson-Bailey model dramatically alters the dynamics. With small values of k, hosts and parasitoids are not only able to persist, they persist at a stable equilibrium. The condition for stability is $k < 1$. However, as k gets smaller, and the risk of parasitism gets concentrated on a smaller fraction of the host population, a larger fraction of the host population escapes parasitism, and the host density at equilibrium gets larger. Therefore, as with many factors that can stabilize host–parasitoid dynamics, increasing stability also leads to increasing host equilibrium density, potentially presenting biological control practitioners with a dilemma.

The negative-binomial model is a phenomenological description of heterogeneity of risk of parasitism across the host population, which could represent a number of different biological mechanisms, including heterogeneity in host susceptibility caused by behavioral or physiological differences among individuals in the population. Some individuals in the population may be better able to fight off parasitoid attack because of their behavioral traits or physiological condition. Alternatively, as a result of differences in phenology, some individuals in the population might overlap temporally with the adult searching parasitoids for a longer period of the year than others. The far greatest amount of attention in this area, however, has been devoted to the potential for spatial heterogeneity in parasitoid search behavior and parasitoid aggregation to stabilize host–parasitoid interactions.

Michael Hassell and colleagues have explored a range of different variants of a discrete-time model in which each generation the host population is distributed across a number of patches (e.g., host plants), and at the start of each generation the parasitoids distribute themselves across these host patches according to some distribution. The parasitoid distribution can either be independent of host density or aggregated such that parasitoids are concentrated in areas of high host density. At the start of each generation, the hosts and parasitoids from all of the patches mix and redistribute themselves. Hassell and colleagues proposed a general rule, the "$CV^2 > 1$ Rule," which states that the host–parasitoid equilibrium will be stable if the coefficient of variation squared of the density of searching parasitoids around each host is greater than 1. This rule is a reasonable approximation for the stability criterion for a range of different models of this type. Both host density-dependent and host density-independent heterogeneity in parasitoid attack can be stabilizing.

The stabilizing effect of spatial heterogeneity, however, results not directly from the spatial distribution of parasitoid attack but from how this translates into temporal density dependence. In these models, the parasitoid attack is concentrated on the hosts at high risk (i.e., those in patches with high parasitoid density), whereas those at low risk (low parasitoid density) may escape parasitism. This leads to a form of temporal density dependence, termed pseudointerference, in which the parasitoid efficiency decreases at high parasitoid densities because an increasing fraction of parasitoid attacks are wasted on already-parasitized hosts. In this type of discrete-time model, the parasitoids are assumed to effectively choose a patch at the start of the generation and not redistribute as the density of hosts in the high-risk patches are depleted. A general finding of subsequent models appears to be that spatial heterogeneity that maintains this heterogeneity of risk to the hosts across the generation is stabilizing, but the stabilizing effect is lost if redistribution of parasitoids homogenizes the risk.

5. LARGE-SCALE SPATIAL DYNAMICS

An alternative to persistence of host–parasitoid interactions through local processes is the possibility envisioned by A. J. Nicholson in the 1930s that the dynamics of hosts and parasitoids in any single location might be unstable and characterized by frequent extinctions, but the collection of host and parasitoid populations across a larger (metapopulation) scale might persist if the local populations are loosely connected by dispersal. If each of the local populations has an unstable equilibrium, then the key to long-term persistence of a host–parasitoid metapopulation (or metacommunity) is that the dynamics on patches in different parts of the environment remain asynchronous to some degree, such that not all populations of hosts are becoming extinct or reaching outbreak densities at the same time. There must be some intermediate level of dispersal of hosts and parasitoids between the patches: too much dispersal and the patches will become synchronized; too little and they will revert to isolated, nonpersistent populations.

Over the last two decades, there has been an explosion of models describing spatial host–parasitoid interactions. The structure of these models varies greatly, including patch-occupancy models that characterize each patch only as being occupied or unoccupied by hosts and/or parasitoids, patch models in which the within-patch models are described explicitly, grid or lattice models in which each cell of the grid is either an individual or a population, continuous-time reaction-diffusion models that follow the density of the populations distributed across continuous space, interacting particle models in which discrete individuals bump into each other in continuous space, etc. In general, this vast array of models has shown that spatial host–parasitoid interactions can frequently persist for longer than their nonspatial counterparts, and in many cases, they can persist indefinitely. In some cases, long-term persistence results from only "statistical stabilization," in which the variability observed in the sum of a number of unregulated population trajectories will be less than the variability of individual trajectories (e.g., diversifying your stock portfolio). In other cases, linking together populations that would be unstable in isolation can actually stabilize the dynamics at both the local population and metapopulation levels (for this to occur, there must be some mechanism that maintains asynchrony between the local populations). In models in which space is modeled explicitly, persistence in space is sometimes accompanied by various types of spatial pattern formation, where parts of the environment have high densities of hosts and/or parasitoids and others have low densities. The spatial patterning can either be static in space or can move through the environment (e.g., spiral waves).

Theoretical studies of spatial host–parasitoid interactions far outnumber empirical studies. Work by John Maron and Susan Harrison on western tussock moths on lupines in California provides one potential example of spatial pattern formation caused by a parasitoid. Patches of lupine with high densities of tussock moth are surrounded by apparently habitable lupine plants with low moth densities, and these high-density patches can remain in place for many years. Moths experience lower parasitism rates within the patches than in the areas immediately surrounding the patches. Maron and Harrison hypothesized that the parasitoids, which have higher dispersal ability than their hosts, maintain the spatial patterning by spilling out of the high-density patches, causing a "halo of death" around the patch. In another study, Jens Roland and Philip Taylor showed that the degree of spatial fragmentation, such as that caused by deforestation, can alter the ability of parasitoids to find and potentially control their host populations. For three of four of the parasitoid species attacking a single species of forest tent caterpillar, the parasitism rate decreased as the degree of fragmentation increased. But for the smallest of the parasitoid species, the pattern was reversed, and the highest parasitism rates were achieved in highly fragmented forests.

6. STAGE STRUCTURE IN SYSTEMS WITH OVERLAPPING GENERATIONS

The Nicholson-Bailey model and its descendants assume discrete host and parasitoid generations, as in many temperate insect systems in which there is a single generation of each species per year. In many tropical, subtropical, and Mediterranean (e.g., California) climates, insects can have many generations per year with all life stages present at any time, and because all life stages co-occur, there is the potential for interesting stage-dependent interactions that can affect the dynamics of the host–parasitoid interaction. Continuous-time host–parasitoid models with stage structure have been written to describe this type of situation. In this type of model, frequently written as delayed differential equations, the life cycle of each species is divided into a number of discrete stages (e.g., eggs, larvae, pupae, adults) that can have different demographic rates. In the case of the parasitoid, only the adult female stage searches for hosts, and in the case of the host, generally only a subset of the juvenile stages is attacked by the parasitoids (e.g., the larval stage). The dynamic effects of a range of stage-dependent parasitoid oviposition behaviors have also been investigated through this type of model.

Stage-structured models can produce new types of interesting dynamics (e.g., generation cycles) caused by time lags and interactions between stages, and details of the stage-structured interaction can affect the persistence and stability of the host–parasitoid interaction. Because many parasitoids attack only juvenile stages of their host, one key finding that is likely to be relevant to a number of real systems is that a relatively long-lived adult stage that is invulnerable to parasitism can have a strong stabilizing effect on the host–parasitoid equilibrium. The long-lived adult stages act as a life-history refuge for the host when parasitoid populations are high and would otherwise overexploit the host. Additionally, the dynamics of a system in which the parasitoid has a relatively short development time tends to be more stable than one in which the parasitoid has a long development time (the yearly time lag between parasitism of the host and the production of new parasitoids inherent in the Nicholson-Bailey model is central to the instability of its dynamics). The reason for this is that the parasitoid then acts more as a direct rather than delayed density-dependent cause of mortality, and the reduction of the time lag tends to prevent the cycles of overexploitation, crash, and recovery.

In many species, parasitoids produce female offspring from attacks on older (larger) juvenile hosts but continue to attack and kill younger (smaller) hosts, either to produce male offspring or for host feeding. The attacks on younger host stages lead to a form of delayed feedback in the parasitoid recruitment rate (because attacks on the young stages now will result in fewer female parasitoid-producing larger hosts later). Murdoch and co-workers found that stage-dependent parasitoid attacks can have a stabilizing effect but also can lead to a new type of instability (delayed-feedback cycles) if this effect is too strong.

7. PARASITOIDS IN BIOLOGICAL CONTROL: THE CASE STUDY OF THE SUCCESSFUL CONTROL OF CALIFORNIA RED SCALE

Parasitoids have been commonly used as in classical biological control in which natural enemies are introduced (often from the pest's area of origin) with the hope that the enemy will both suppress the density of a pest species and also persist to quell future outbreaks of the pest. Although most biological control efforts fail, either because the natural enemy does not become established or because it is not successful at suppressing the target pest population, there are many examples of phenomenal successes. These successes are dramatic illustrations of the ability of parasitoids to exert top-down control on their host populations and to persist with their host, sometimes in very stable interactions.

One of the most detailed efforts to determine the factors that allow persistence of a host–parasitoid system is the work of William Murdoch and colleagues in attempting to understand the successful biological control of California red scale, *Aonidiella aurantii*, by a parasitoid. California red scale is a pest of citrus that is controlled at a small fraction of its potential density by the action of the parasitic wasp, *Aphytis melinus*. The interaction is not only persistent but remarkably stable, with host and parasitoid densities fluctuating within very narrow bounds through time. Murdoch and co-workers tested and rejected many of the potential stabilizing mechanisms through field observations and experiments, including parasitoid aggregation, a refuge from parasitism, and large-scale spatial dynamics. Stability in this case appears to arise from details of the stage structure of the interaction. Red scale has a long-lived adult stage that is invulnerable to attack by the parasitoid. *A. melinus* has a short development time relative to that of the host, allowing it to respond rapidly to increases in host abundance. Additionally, in situations in which the parasitoid population reaches high densities (e.g., following high densities of the host), high levels of reattack (both parasitism and host feeding) on already parasitized hosts occur, leading to density dependence in the juvenile parasitoid survival.

This system is not only an excellent example of top-down control of an insect by a natural enemy, it is also a classic example of competitive displacement. *A. melinus* was one of many parasitoid species introduced. Following its introduction, *A. melinus* rapidly displaced the earlier-released and less-effective parasitoids by suppressing the host abundance to lower levels than that achieved by the earlier parasitoids.

FURTHER READING

Briggs, Cheryl J., and Martha F. Hoopes. 2004. Stabilizing effects in spatial parasitoid–host and predator–prey models: A review. Theoretical Population Biology 65: 299–315. *Summarizes different ways that parasitoid–host interactions can be stabilized by spatial processes.*

Godfray, H. Charles J. 1994. Parasitoids: Behavioral and Evolutionary Ecology. Princeton, NJ: Princeton University Press. *Summary of parasitoid natural history, reproductive strategies, and the relevant evolutionary and ecological theory.*

Hassell, Michael P. 1978. The Dynamics of Arthropod Predator–Prey Systems. Princeton, NJ: Princeton University Press. *The classic summary of early host–parasitoid models.*

Hassell, Michael P. 2000. The Spatial and Temporal Dynamics of Host–Parasitoid Interactions. Oxford: Oxford University Press. *An update on more recent work, including host density-dependent and host density-independent parasitoid aggregation, and patch models. Concentrates on discrete-time models.*

Hassell, Michael P., and Steven W. Pacala. 1990. Heterogeneity and the dynamics of host–parasitoid interactions. Philosophical Transactions of the Royal Society: Biological Sciences 330: 203–220. *More on the CV^2 Rule.*

Hawkins, Bradford A., and Howard V. Cornell, eds. 1999. Theoretical Approaches to Biological Control. Cambridge, UK: Cambridge University Press. *Edited volume with more information on parasitoids in biological control situations.*

Mills, Nicholas J., and Wayne M. Getz. 1996. Modeling the biological control of insect pests: A review of host–parasitoid models. Ecological Modeling 92: 121–143.

Murdoch, William W., Cheryl J. Briggs, and Roger M. Nisbet. 2003. Consumer Resource Dynamics. Princeton, NJ: Princeton University Press. *Discusses more continuous-time models than Hassell's books and has a heavy emphasis on stage-structured interactions.*

Murdoch, William W., Cheryl J. Briggs, and Susan Swarbrick. 2006. Biological control: Lessons from a study of California red scale. Population Ecology 48: 297–305. *Recent summary of the efforts to understand the factors leading to stability in the California red scale biological control example.*

Nicholson, A. J. 1933. The balance of animal populations. Journal of Animal Ecology 2: 131–178. *Foresees most of the major developments in population ecology of the twentieth century.*

Tayor, Andrew D. 1993. Heterogeneity in host–parasitoid interactions: "aggregation of risk" and the "$CV^2 > 1$ Rule." Trends in Ecology and Evolution 8: 400–405. *A good description of how heterogeneity of risk is stabilizing in discrete-time models through leading to pseudo-interference and temporal density dependence in the parasitoid recruitment rate.*

II.9

Ecological Epidemiology

Michael Begon

OUTLINE

1. Parasites, pathogens, and other definitions
2. The importance of ecological epidemiology
3. The dynamics of parasites within populations: Transmission
4. The population dynamics of infection
5. Parasites and the dynamics of hosts
6. Shared parasites—zoonoses

Strictly speaking, *epidemiology* is the study of the dynamics of disease in a population of humans. In ecology, however, the term takes on a slightly different meaning. Ecologists tend to expand the usage to cover populations of any species, animal or plant, but they then restrict it to infectious diseases (as opposed to, say, cancers or heart disease). Studies of human epidemiology usually treat the host (human) population as fixed in size and focus on the dynamics of disease within this population. What distinguishes "ecological" epidemiology is an acknowledgment that the dynamics of the parasite and the host populations may interact. Hence, we are interested in the dynamics of parasites in host populations, that may themselves vary substantially in size, and also in the effects of the parasites on the dynamics of the hosts.

GLOSSARY

basic reproductive number. Usually denoted R_0, for microparasites, the average number of new infections that would arise from a single infectious host introduced into a population of susceptible hosts; for macroparasites, the average number of established, reproductively mature offspring produced by a mature parasite throughout its life in a population of uninfected hosts

critical population size. The population size of susceptible hosts for which $R_0 = 1$, where R_0 is the basic reproductive number, and which must therefore be exceeded if an infection is to spread in a population

density-dependent transmission. Parasite transmission in which the rate of contact between susceptible hosts and the source of new infections increases with host density

frequency-dependent transmission. Parasite transmission in which the rate of contact between susceptible hosts and the source of new infections is independent of host density

herd immunity. Where a population contains too few susceptible hosts (either because of natural infection or immunization) for infection to be able to establish and spread within a population

macroparasite. A parasite that grows but does not multiply in its host, producing infective stages that are released to infect new hosts; the macroparasites of animals mostly live on the body or in the body cavities (e.g., the gut); in plants, they are generally intercellular

microparasite. A small, often intracellular parasite that multiplies directly within its host

transmission threshold. The condition $R_0 = 1$, where R_0 is the basic reproductive number, which must be crossed if an infection is to spread in a population

vector. An organism carrying parasites from one host individual to another, within which there may or may not be parasite multiplication

zoonosis. An infection that occurs naturally and can be sustained in a wildlife species but can also infect and cause disease in humans

1. PARASITES, PATHOGENS, AND OTHER DEFINITIONS

A parasite is an organism that obtains its nutrients from one or a very few *host* individuals, normally causing harm but not causing death immediately. This distinguishes parasites from predators, which kill and consume many prey in their lifetime, and from grazers, which take small parts from many different prey. If a parasite infection gives rise to symptoms that are clearly harmful, the host is said to have a disease. *Pathogen*, then, is a term that may be applied to any parasite that gives rise to a disease (i.e., *is pathogenic*).

The language used by plant pathologists and animal parasitologists is often very different, but for the ecologist, these differences are less striking than the resemblances. One distinction that *is* useful is that between microparasites and macroparasites. Microparasites are small, often intracellular, and they multiply directly within their host where they are often extremely numerous. Hence, it is usually impossible to count the number of microparasites in a host: ecologists normally study the number of infected hosts in a population. Examples include bacteria and viruses (e.g., the typhoid bacterium and the yellow net viruses of beet and tomato), protozoa infecting animals (e.g., the *Plasmodium* species that cause malaria), and some of the simpler fungi that infect plants.

Macroparasites grow but do not multiply in their host. They produce infective stages that are released to infect new hosts. The macroparasites of animals mostly live on the body or in the body cavities (e.g., the gut) of their hosts. In plants, they are generally intercellular. It is often possible to count or at least to estimate the numbers of macroparasites in or on a host. Hence, ecologists study the numbers of parasites as well as the numbers of infected hosts. Examples include parasitic helminths such as the intestinal nematodes and tapeworms of humans, the fleas and ticks that are parasitic in their own right but also transmit many microparasites between their hosts, and plant macroparasites such as the higher fungi that give rise to the mildews, rusts, and smuts.

Cutting across the distinction between micro- and macroparasites, parasites can also be subdivided into those that are transmitted directly from host to host and those that require a vector or intermediate host for transmission, i.e., are either simply carried from host to host by another species (aphids carrying viruses from plant to plant) or need to parasitize a succession of two (or more) host species to complete their life cycle (both mosquitoes and humans being parasitized by the malaria *Plasmodium*).

2. THE IMPORTANCE OF ECOLOGICAL EPIDEMIOLOGY

Parasites are an important group of organisms in the most direct sense. Millions of people are killed each year by various types of infection, and many millions more are debilitated or deformed. When the effects of parasites on domesticated animals and crops are added to this, the cost in terms of human misery and economic loss becomes incalculable. Parasites are also important numerically. A free-living organism that does *not* harbor several parasitic individuals of a number of species is a rarity.

Thus, ecological epidemiology is important from an entirely practical point of view. If we wish to control the diseases that have afflicted us and our domesticated species historically—malaria, tuberculosis—and those that have emerged recently or threaten us—HIV-AIDS, SARS, avian influenza—then we must seek to understand their dynamics. But it is also the case that a major question in ecology that not only remains unanswered but has only recently been seriously addressed is: To what extent are animal and plant populations and communities in general affected by parasitism and disease? Ecologists have long been concerned with the effects of food resources, competitors, and predators on their focal species; only recently have parasites and pathogens been afforded similar attention.

3. THE DYNAMICS OF PARASITES WITHIN POPULATIONS: TRANSMISSION

Transmission dynamics, in a very real sense, is the driving force behind the overall population dynamics of pathogens. Different species of parasite are of course transmitted in different ways between hosts, the most fundamental distinction being between parasites that are transmitted directly (either through close contact between hosts or via an environmental reservoir to which infectious hosts have contributed) and those that require a vector or intermediate host for transmission.

Irrespective of these distinctions, the rate of production of new infections in a population depends on the per capita transmission rate (the rate of transmission per susceptible host "target") and also on the number of susceptible hosts there are. That per capita transmission rate depends on the infectiousness of the parasite, the susceptibility of the host, and so on, but it also depends on the *contact rate* between susceptible hosts and whatever it is that carries the infection.

For directly transmitted parasites, we deal with the contact rate between infected hosts and susceptible (uninfected) hosts; for hosts infected by long-lived infective agents, it is the contact rate between these and susceptible hosts; with vector-transmitted parasites it is the contact rate between host and vector. But what determines this contact rate? Essentially, two factors are determinative: the contact rate between a susceptible individual and *all* other hosts, and the proportion of these that are actually infectious.

For the first of these, ecologists have tended to make one of two simplifying assumptions: either that this contact rate increases in direct proportion to the density of the population (density-dependent transmission) or that it is utterly independent of population density (frequency-dependent transmission). The former imagines individuals bumping into one another at random: the more crowded they become, the more contacts they make. The latter, by contrast, assumes that the number of contacts an individual makes is a fixed aspect of its behavior. Frequency-dependent transmission has therefore conventionally been assumed for sexually transmitted diseases—the frequency of sexual contacts is independent of population density—but it is increasingly recognized that many social contacts, territory defense for instance, may come into the same category. It has also become increasingly apparent that real contact patterns usually conform to neither of these simplifying assumptions exactly, but they nonetheless represent two valuable benchmarks through which real data sets can be understood.

There has also often been an assumption that the "infectious proportion" can be calculated from, and also applies throughout, the whole host population. In reality, however, transmission typically occurs locally, between adjacent individuals. Thus, there are likely to be hot spots of infection in a population, where the infected proportion is high, and corresponding cool zones. Transmission, therefore, often gives rise to spatial waves of infection passing through a population rather than simply an overall, global rise.

4. THE POPULATION DYNAMICS OF INFECTION

We begin by looking at the dynamics of disease within host populations without considering any possible effects on the total abundance of hosts. We then take the more "ecological" approach of considering the effects of parasites on host abundance in a manner much more akin to conventional predator–prey dynamics (see chapter II.7).

The Basic Reproductive Number and the Transmission Threshold

In the study of the dynamics of parasites, there are a number of particularly key concepts. The first is the basic reproductive number, usually denoted R_0. For microparasites, this is the average number of new infections that would arise from a single infectious host introduced into a population of susceptible hosts. For macroparasites, it is the average number of established, reproductively mature offspring produced by a mature parasite throughout its life in a population of uninfected hosts.

The transmission threshold, which must be crossed if an infection is to spread, is then given by the condition $R_0 = 1$. An infection will eventually die out for $R_0 < 1$ (each present infection or parasite leads to fewer than one infection or parasite in the future), but an infection will spread for $R_0 > 1$. Insights into the dynamics of infection can be gained by considering the various determinants of the basic reproductive number. We do this in some detail for directly transmitted microparasites with density-dependent transmission (see above) and then deal more briefly with related issues for other parasites.

Directly Transmitted Microparasites and the Critical Population Size

For such microparasites, R_0 can be said to increase (1) with the average period of time over which an infected host remains infectious, L; (2) with the number of susceptible individuals in the host population, S, because greater numbers offer more opportunities for transmission; and (3) with the transmission coefficient, β, the strength or force of transmission. Thus, overall:

$$R_0 = S\beta L. \qquad (1)$$

Note immediately that by this definition, the greater the number of susceptible hosts, the higher the basic reproductive number of the infection. But in particular, the transmission threshold can now be expressed in terms of a critical population size, S_T, where, because $R_0 = 1$ at that threshold:

$$S_T = 1/\beta L. \qquad (2)$$

In populations with numbers of susceptibles less than this, the infection will die out ($R_0 < 1$). With numbers greater than this, the infection will spread ($R_0 > 1$). These simple considerations allow us to make sense of some very basic patterns in the dynamics of infection.

Consider first the kinds of population in which we might expect to find different sorts of infection. If microparasites are highly infectious (large βs), or give rise to long periods of infectiousness (large Ls), then they will have relatively high R_0 values even in small populations and will therefore be able to persist there (S_T is small). Conversely, if parasites are of low infectivity or have short periods of infectiousness, they will have relatively small R_0 values and will be able to persist only in large populations. Many protozoan infections of vertebrates, and also some viruses such as herpes, are persistent within individual hosts (large L), often because the immune response to them is either ineffective or short lived. A number of plant diseases, too, such as club-root, have very long periods of infectiousness. In each case, the critical population size is therefore small, explaining why the diseases can and do survive endemically even in small host populations.

On the other hand, the immune responses to many other human viral and bacterial infections are powerful enough to ensure that they are only very transient in individual hosts (small L), and they often induce lasting immunity. Thus, for example, a disease such as measles has a critical population size of around 300,000 individuals and is unlikely to have been of great importance until quite recently in human biology. However, it has generated major epidemics in the growing cities of the industrialized world in the eighteenth and nineteenth centuries, and in the growing concentrations of population in the developing world in the twentieth century.

The Epidemic Curve

The value of R_0 itself is also related to the nature of the *epidemic curve* of an infection. This is the time series of new cases following the introduction of the parasite into a population of hosts. Assuming there are sufficient susceptible hosts present for the parasite to invade (i.e., the critical population size, S_T, is exceeded), the initial growth of the epidemic will be rapid as the parasite sweeps through the population of susceptibles. But as these susceptibles either die or recover to immunity, their number, S, will decline, and so too, therefore, will R_0. Hence, the rate of appearance of new cases will slow down and then decline. And if S falls below S_T and stays there, the infection will disappear—the epidemic will have ended. Not surprisingly, the higher the initial value of R_0, the more rapid will be the rise in the epidemic curve. But this will also lead to the more rapid removal of susceptibles from the population and hence to an earlier end to the epidemic: higher values of R_0 tend to give rise to shorter, sharper epidemic curves. Also, whether the infection disappears altogether (i.e., the epidemic simply ends) depends very largely on the rate at which new susceptibles either move into or are born into the population because this determines how long the population remains below S_T. If this rate is too low, then the epidemic will indeed simply end. But a sufficiently rapid input of new susceptibles should prolong the epidemic or even allow the infection to establish endemically in the population after the initial epidemic has passed.

Cycles of Infection

This leads us naturally to consider the longer-term patterns in the dynamics of different types of endemic infection. As described above, the immunity induced by many bacterial and viral infections reduces S, which reduces R_0, which therefore tends to lead to a decline in the incidence of the infection itself. However, in due course, and before the infection disappears altogether from the population, there is likely to be an influx of new susceptibles into the population, a subsequent increase in S and R_0, and so on. There is thus a marked tendency with such infections to generate a sequence from many susceptibles (R_0 high), to high incidence, to few susceptibles (R_0 low), to low incidence, to many susceptibles, etc., just as in any other predator–prey cycle. This undoubtedly underlies the observed cyclic incidence of many human diseases, with the differing lengths of cycle reflecting the differing characteristics of the diseases: measles with peaks every 1 or 2 years, whooping cough every 3 to 4 years, and so on.

By contrast, infections that do not induce an effective immune response tend to be longer lasting within individual hosts, but they also tend not to give rise to the same sort of fluctuations in S and R_0. Thus, for example, protozoan infections tend to be much less variable (less cyclic) in their prevalence.

Immunization Programs

Recognizing the importance of critical population sizes also throws light on immunization programs in which susceptible hosts are rendered nonsusceptible without ever becoming diseased (showing clinical symptoms), usually through exposure to a killed or attenuated pathogen. The direct effects here are obvious: the immunized individual is protected. But by reducing the number of susceptibles, such programs also have the indirect effect of reducing R_0. Indeed, seen in these terms, the fundamental aim of an immunization program is clear: to hold the number of susceptibles below S_T so that R_0 remains less than 1. To do so is said to provide "herd immunity."

In fact, a simple manipulation of equation 2 gives rise to a formula for the critical proportion of the population, p_c, that needs to be immunized in order to provide herd immunity (reducing R_0 to a maximum of 1, at most). This reiterates the point that in order to eradicate a disease, it is not necessary to immunize the whole population—just a proportion sufficient to bring R_0 below 1. Moreover, this proportion will be higher the greater the "natural" basic reproductive number of the disease (without immunization). It is striking, then, that smallpox, the only known disease where in practice immunization seems to have led to eradication, has unusually low values of R_0 (and hence p_c).

Frequency-Dependent Transmission

Suppose, however, that transmission is frequency dependent. Then there is no longer the same dependence for spread on the number of susceptibles, and hence, no threshold population size. Such infections can therefore persist even in extremely small populations, where, to a first approximation, the chances of sexual contact, say, for an infected host are the same as in large populations.

Vector-Borne Infections

For microparasites that are spread from one host to another by a vector, the life-cycle characteristics of both host and vector enter into the calculation of R_0. In particular, the transmission threshold ($R_0 = 1$) is dependent on a ratio of vector:host numbers. For a disease to establish itself and spread, that ratio must exceed a critical level; hence, disease control measures are usually aimed directly at reducing the numbers of vectors and are aimed only indirectly at the parasite. Many virus diseases of crops, and vector-transmitted diseases of humans and their livestock (malaria, onchocerciasis, etc.), are therefore controlled by insecticides rather than chemicals directly targeting the parasite.

Directly Transmitted Macroparasites

The effective reproductive number of a directly transmitted macroparasite (no intermediate host) is directly related to the length of its reproductive period within the host (i.e., again, to L) and to its rate of reproduction (rate of production of infective stages). Most directly transmitted helminths have an enormous reproductive capability. For instance, the female of the human hookworm *Necator* produces roughly 15,000 eggs per worm per day. The critical threshold densities for these parasites are therefore very low, and they occur and persist endemically in low-density human populations, such as hunter–gatherer communities.

Indirectly Transmitted Macroparasites

Finally, for macroparasites with intermediate hosts, the threshold for the spread of infection depends directly on the abundance of both (i.e., a product as opposed to the ratio, which was appropriate for vector-transmitted microparasites). This is because transmission in both directions is by means of free-living infective stages. Thus, because it is inappropriate to reduce human abundance, schistosomiasis, a helminth infection of humans for which snails are intermediate hosts, is often controlled by reducing snail numbers with molluscicides in an attempt to depress R_0 below unity (the transmission threshold). The difficulty with this approach, however, is that the snails have an enormous reproductive capacity, and they rapidly recolonize aquatic habitats once molluscicide treatment ceases.

5. PARASITES AND THE DYNAMICS OF HOSTS

It is part of the definition of parasites that they cause harm to their host, and although it is not always easy to demonstrate this harm, there are numerous examples in which all sorts of parasites have been shown to affect directly the key demographic rates: birth and death.

Parasites Interact with Other Ecological Processes

On the other hand, the effects of parasites are often more subtle than a simple reduction in survival or fecundity. For example, infection may make hosts more susceptible to predation. For example, postmortem examination of red grouse (*Lagopus lagopus scoticus*) carried out by Peter Hudson and colleagues showed that birds killed by predators carried significantly greater burdens of the parasitic nematode *Trichostrongylus tenuis* than the presumably far more random sample of birds that were shot. Alternatively, the effect of parasitism may be to weaken an aggressive competitor and so allow weaker associated species to persist. For example, a study by Pennings and Callaway showed that dodder (*Cuscuta salina*), a plant parasitic on other plants, which has a strong preference for *Salicornia* in a southern Californian salt marsh, is highly instrumental in determining the outcome of competition between *Salicornia* and other plant species within several zones of the marsh.

Thus, parasites often affect their hosts not in isolation but through an interaction with some other factor: infection may make a host more vulnerable to competition or predation; or competition or shortage of food may make a host more vulnerable to infection or to the effects of infection. This does not mean, however, that the parasites play only a supporting role. Both partners in the interaction may be crucial in determining both the overall strength of the effect and which particular hosts are affected.

Parasites Affect Host Abundance/Dynamics

What role, if any, do parasites and pathogens play in the dynamics of their hosts? Data sets showing reductions in host abundance by parasites in controlled, laboratory environments, in which infected and uninfected populations are compared, have been available for many years. However, good evidence from natural populations is extremely rare.

Red Grouse and Nematodes

The difficulties of finding such evidence are illustrated by further work on the red grouse—of interest both because it is a "game" bird, and hence the focus of an industry in which British landowners charge for the right to shoot at it, and also because it is a species that often, although not always, exhibits regular cycles of abundance (repeated increases and crashes): a pattern demanding an explanation. The underlying cause of these cycles has been disputed, but one mechanism receiving strong support, especially from Hudson and his colleagues, has been the influence of the parasitic nematode *Trichostrongylus tenuis* occupying the birds' gut caeca and reducing both survival and fecundity.

Mathematical models for this type of host–macroparasite interaction are supportive of a role for the parasites in grouse cycles (i.e., the results of the models are consistent with field observations), but they fall short of the type of "proof" that can come from a controlled experiment. Hudson and others therefore carried out a field-scale experimental manipulation in the late 1990s designed to test the parasites' role. In two populations, grouse were treated with an anthelmintic (worm killer) in the expected years of two successive population crashes; in two others, grouse were treated only in the expected year of one crash; and two further populations were monitored as unmanipulated controls. Grouse abundance was measured as "bag records": the number of grouse shot. The anthelmintic had a clear effect in the experiment—population crashes were far less marked—and it is therefore equally clear that the parasites have an effect normally: that is, the parasites affected host dynamics. The precise nature of that effect, however, remained a matter of some controversy. Hudson and his colleagues believed that the experiment demonstrated that the parasites were responsible for both the existence of the host cycles and their amplitude.

Others, however, felt that rather less had been fully demonstrated, suggesting for example that the cycles may have been reduced in amplitude in the experiment rather than eliminated, especially as the very low numbers normally observed in a trough are a result of there being no shooting when abundance is low. That is, the worms may normally have been important in determining the amplitude of the cycles but not for their existence in the first place. Redpath, Hudson, and others therefore carried out a very similar field experiment, almost 10 years after the first, but with a greater proportion of experimental birds treated for worms and with abundances measured more accurately. This time, the demonstrable effect of worms was far less profound, and the conclusion drawn was that the parasites appear to be part of a much larger web of interactions, molding host dynamics. As evidence accumulates, this seems likely to be a much more general conclusion: that parasites may play a neglected but important role in determining host dynamics, alongside and interacting with many other factors.

6. SHARED PARASITES—ZOONOSES

Finally, we turn from systems comprising one host to those with more than one host species, and to a subset of these interactions that is of particular importance to humans. For any species of parasite (be it tapeworm, virus, protozoan, or fungus), the potential hosts are a tiny subset of the available flora and fauna. The overwhelming majority of other organisms are quite unable to serve as hosts.

The delineation of a parasite's host range, moreover, is not always as straightforward as one might imagine. Species outside the host range are relatively easily characterized: the parasite cannot establish an infection within them. But for those inside the host range, the response may range from a serious pathology and certain death to an infection with no overt symptoms. What is more, it is often the "natural" host of a parasite, i.e., the one with which it has coevolved, in which infection is asymptomatic. It is then often "accidental" hosts in which infection gives rise to a frequently fatal pathology. (*Accidental* is an appropriate word here because these are often dead-end hosts that die too quickly to pass on the infection, within which the pathogen cannot therefore evolve, and *to* which it cannot therefore be adapted.)

These issues take on not just parasitological but also medical importance in the case of *zoonotic infections*: infections that circulate naturally, and have coevolved, in one or more species of wildlife but also have a pathological effect on humans. Good examples of zoonotic infections are bubonic and pneumonic plague, the human diseases caused by the bacterium *Yersinia pestis*. *Y. pestis* circulates naturally within populations of a number of species of wild rodent: for example, in the great gerbil, *Rhombomys opimus*, in the deserts of Central Asia, and probably in populations of kangaroo rats, *Dipodomys* spp. in similar habitats in the southwestern United States. (Remarkably, little is known about the ecology of *Y. pestis* in the United States despite its widespread nature and potential threat.) In these species, there are few if any symptoms in most cases of infection. There are, however, other species where *Y. pestis* infection is devastating. Some of these are closely related to the natural hosts. In the United States, populations of prairie dogs,

Cynomys spp., also rodents, are regularly annihilated by epidemics of plague, and the disease is an important conservation issue. But there are also other species, only very distantly related to the natural hosts, where untreated plague is usually, and rapidly, fatal. Among these are humans. Why such a pattern of differential virulence so often occurs—low virulence in the coevolved host, high virulence in some unrelated hosts, but unable even to cause an infection in others—is an important unanswered question in host-pathogen biology. But it is a question that urgently requires an answer as the list of zoonotic infections threatening us—HIV-AIDS, Ebola, avian influenza—grows ever larger.

FURTHER READING

Anderson, Roy M., and Robert M. May. 1992. Infectious Diseases of Humans. Oxford: Oxford University Press. *Although focused on humans, the key reference in the field to the fundamentals of the topic, written by the two main pioneers of ecological epidemiology.*

Daszak, Peter, Andrew A. Cunningham, and Alex D. Hyatt. 2000. Emerging infectious diseases of wildlife—threats to biodiversity and human health. Science 287: 443–449. *A review of wildlife diseases covering both medical and conservation perspectives.*

Gage, Kenneth L., and Michael Y. Kosoy. 2005. Natural history of plague: Perspectives from more than a century of research. Annual Review of Entomology 50: 505–528. *A review that throws light not only on plague, but also on the complexities of studying zoonotic diseases in their wildlife reservoirs generally.*

Hudson, Peter J., Annapaola Rizzoli, Bryan T. Grenfell, Hans Heesterbeek, and Andy P. Dobson, eds. 2002. The Ecology of Wildlife Diseases. Oxford: Oxford University Press. *A valuable collection of multiauthored reviews, resulting from an international workshop attended by many of the most important scientists working in the field.*

Redpath, Stephen M., Francois Mougeot, Fiona M. Leckie, David A. Elston, and Peter J. Hudson. 2006. Testing the role of parasites in driving the cyclic population dynamics of a gamebird. Ecology Letters 9: 410–418. *The most recent study on the experimental approach to the role of parasites in red grouse dynamics, which therefore also provides an entrée to the work that preceded it.*

II.10

Interactions between Plants and Herbivores

Rebecca J. Morris

OUTLINE

1. The diversity of herbivores
2. Herbivore–plant population dynamics
3. The impact of herbivores on plant populations
4. Plant responses to herbivory
5. The impact of plants on herbivore populations
6. Herbivore–plant interactions at the community level

Herbivores are animals that feed on living plants. Herbivory is one of most common ecological interactions and is exhibited by species ranging from microscopic mites to giant pandas. Herbivore–plant interactions have features in common with all other consumer–resource interactions, although there are significant differences. Notably, plants do not necessarily die when they have been attacked by herbivores. Although there is no compulsory link between herbivore and plant dynamics, herbivores can affect the population dynamics of the plants on which they feed, and plants can affect herbivore population dynamics. Herbivore–plant interactions have been studied through a combination of observational time series data, mathematical modeling, and experimentation, and here a variety of examples are discussed.

GLOSSARY

functional response. Results from switching behavior when the herbivore alters the composition of its diet as a result of short-term changes in relative food availability

herbivore. An animal that feeds solely on living plant tissue

herbivory. The consumption of living plant material

host plant. The plant on which an insect herbivore feeds

numerical response. Acts by dispersal with mobile herbivores aggregating in regions of high food availability, or in the longer term by increasing reproductive success

population cycles. Changes in the numbers of individuals in a population repeatedly oscillating between periods of high and low density

population dynamics. The variation in time and space in the size and density of a population

resource. An environmental factor that is directly used by an organism and that potentially influences individual fitness; plants are a resource for herbivores

1. THE DIVERSITY OF HERBIVORES

Herbivores are animals that feed solely on living plant material. They are taxonomically and ecologically diverse and range from single-celled zooplankton to wildebeest, and from leaf-mining moths to marine iguanas. They can be found in terrestrial, marine, and freshwater ecosystems. Insects and mammals are the most well-known groups of herbivores and have been studied most intensively, but there are many other types of herbivore including some species of birds, fish, reptiles, crustaceans, and molluscs.

Herbivores can feed on all the different types of living plant tissue including leaves, fruits, pollen, flowers, and seeds. Each herbivore, however, tends to specialize on a particular type of plant tissue. Herbivores exhibit a variety of feeding methods including chewing, sucking, boring, and galling. Folivores, which feed on leaves, are some of the most common herbivores and include mammals such as deer and insects such as grasshoppers. Frugivores are fruit eaters ranging from monkeys to wasps; and granivores are the seed eaters, or seed predators, including squirrels and weevils. Herbivores remove approximately 10% of net primary production, at least in terrestrial ecosystems. Herbivory typically does not kill the plant but influences the fitness of plants by reducing growth and reproduction and potentially increasing mortality. However, seed eaters (and some species that feed on seedlings) do have a significant effect on seed abundance and can directly influence plant populations, assuming that the plant is seed limited.

There is such a wide variety of herbivore species that it is useful to consider the differences between the two most studied groups of herbivores: insects and mammals. Insect herbivores differ from mammalian herbivores in their size, metabolic rate, population density, numbers, and the kinds of damage they cause. Insects tend to be more specialized than mammalian herbivores and are more likely to have an intimate lifelong association with their host plant. Mammalian herbivores are likely to have a more immediate and, in the long term, more profound impact on plant populations than invertebrates because of their greater body size, polyphagy, individual bite size, mobility, and tolerance of starvation. A relatively high proportion of mammalian herbivore populations are food limited, whereas a comparatively high proportion of insect herbivore populations are regulated by predators, parasites, and diseases.

2. HERBIVORE–PLANT POPULATION DYNAMICS

The population dynamics of herbivore–plant systems shares features exhibited by all consumer–resource interactions. Consequently, the mathematical models for these systems have the same logical foundations, largely based on the Lotka-Volterra model and its variations. However, herbivore–plant systems differ from other consumer–resource relationships, for example, predator–prey relationships, in several important ways. Classifying consumer–resource relationships according to the closeness and duration of the relationship (intimacy) and the probability that the interactions will result in the death of the organism concerned (lethality) can highlight these differences. Many herbivores, in particular grazers, score low on both intimacy and lethality. Other herbivores, such as sapsuckers like aphids, are functional parasites and score highly on the intimacy scale but are still low on the lethality scale.

Herbivores rarely kill the individuals on which they feed and throughout their lives will eat parts of many individuals. Whereas predators tend to kill their prey, grazers consume only part of a resource individual, perhaps concentrating on the young leaves or the flowers. This has important implications for theoretical models. Simple theoretical models employ a logistic model for vegetative regrowth after herbivory, but in fact, a linear initial regrowth model is more apt because the plant biomass is not usually reduced to near zero. The primary productivity of plant communities is also an important factor affecting herbivore–vegetation dynamics. Models of plant growth include the logistic hyperbolic functional response, in which the dynamics of the system becomes increasingly less stable as plant standing biomass increases, and the globally stable regrowth–herbivory–regrowth model.

The functional response of herbivores is measured as units of plant biomass removed, whereas for predators, it is the number of individuals eaten. The functional response of herbivores is difficult to quantify because it must take into account the parts of the plant that are available to the herbivores as well as differences in nutritional quality. A hyperbolic functional response is used as a reasonable approximation of herbivore foraging, at least for grazers. The spatial immobility of plants also sets them apart from other resources such as prey. The spatial arrangements of plants can influence herbivores by affecting their average density of bites and average bite size.

Another factor that distinguishes herbivory from predation is that dynamic changes in resource quality are much more likely. Plant quality may change if herbivores consume the better-quality resources first, thus decreasing the average quality of the remaining vegetation; or it may alter if the plant increases defense of its remaining biomass or increases the amount of new biomass produced. Consequently, herbivory can modify the frequency distribution of plant qualities.

Despite the distinguishing features of herbivore–plant dynamics, there is no necessary link between fluctuations in the numbers of herbivores and the plants on which they feed. Fluctuations in plant populations may have nothing to do with herbivore feeding and may be caused by extrinsic factors such as the weather or by competitive interactions with neighboring plants. Similarly, herbivore numbers may be determined by natural enemies or shortage of breeding sites and may have nothing to do with plant numbers.

Stabilizing Influences

There is an inherent tendency for consumer–resource systems to oscillate. However, the dynamics of herbivore–plant interactions is stabilized by the fact that many plants are long-lived. Plants also have an absolute refuge, their below-ground biomass, which acts as a powerful stabilizing influence. Other mechanisms also stabilize the herbivore–plant systems, for example, the switching behavior of herbivores as well as mechanisms that depend on fluctuations in population densities and environmental factors in space and time. The latter can generate heterogeneity in the plant or herbivore distributions, resulting in refuges. Territorial behavior, for example, can have a stabilizing effect by affecting the population densities of the herbivores in space and thus providing refuges for the plant.

3. THE IMPACT OF HERBIVORES ON PLANT POPULATIONS

It is well accepted that herbivores can have a negative impact on the growth, reproductive output, and survival of many plant species. However a strong herbivore effect on plant performance does not automatically imply an equally intense effect on population dynamics. Even if herbivores reduce plant abundance, it does not necessarily follow that this mortality will lead to a reduction in the number of individuals in subsequent generations. For example, if herbivore-induced mortality of plants reduces plant density, the survival or fecundity of plants that escape herbivory may be increased as a result of reduced intraspecific competition. Herbivores can have a big effect on plant populations if they have no natural enemies or other limiting resources. The impact of herbivory on plant abundance is determined in part by whether or not the herbivore is capable of mounting a response to changes in plant abundance. The response must be large and rapid enough to check or reverse change in plant abundance and could be a functional or a numerical response. In the following sections, a mixture of observational data and experiments are described to explore the effects of herbivores on plant populations.

Insect Outbreaks: The Spruce Budworm

Insect herbivore outbreaks provide good evidence that herbivores can have a significant impact on plant population dynamics, with the best examples coming from forest pests. When herbivore and plant populations interact, there can be multiple equilibria or alternative stable states. These result in a large change in abundance (an outbreak and subsequent crash) as a result of either a small change in carrying capacity or a small environmental perturbation when carrying capacity is close to a critical threshold value. Eruptive or cyclic outbreak can be caused by a sudden increase in availability of high-quality food. For example, the Spruce budworm (*Choristoneura fumerana*) feeds on balsam fir and white spruce in Canada and experiences outbreaks correlated with a dramatic increase in the availability of high-quality food. In the 1940s, it killed these two tree species over an area of 52,000 km^2 because of a dramatic increase in high-quality food, as a large area of balsam fir in monoculture matured at the same time.

The Release of Insect Biological Control Agents—Cactoblastis and the Prickly Pear Cacti

The best observational evidence of the impact of insect herbivores on plant populations comes from the release of specialist insect biological control agents against target weeds. Biological control involves the introduction of herbivores outside their native range, where they may be unrestricted by natural enemies and competitors, and consequently they may exhibit stronger effects than those on plant populations in natural systems. For example, the South American moth *Cactoblastis cactorum* was introduced to control the prickly pear cacti, *Oputia inermis* and *O. stricta*, in Australia. Before the introduction of the moth, vast areas of Australia were covered in the cacti, but the moth wiped out the cacti, which has since remained at low population levels and only in small patches. Several factors have affected the success of *Cactoblastis* in Australia. First, heterogeneity in the distribution of the herbivores may exert a regulatory effect. The refuge for the cacti lies in the highly aggregated egg-laying behavior of the moth. If larval densities are too high, the insects will not survive and will perish along with the plant itself, thus stabilizing the interaction at low densities. Second, cacti growing under conditions of water or nutrient stress have thick mucilaginous segments that suppress the development of the *Cactoblastis* population in certain areas, such as the coastal areas of Queensland. Finally, high temperatures are also regarded as being of major importance in reducing the fecundity of the summer generation of *Cactoblastis*. *Cactoblastis* is now considered a pest and a serious threat to the high diversity of *Opuntia* species, in particular in the southwestern United States and Mexico.

Experimental Exclusion of Insect Herbivores

The best experimental evidence for the effect of herbivory on plant dynamics comes from exclusion experiments using insecticides in natural communities. Using insecticides to remove insects has its drawbacks, but it is the best way of doing so experimentally. A classic experiment from the 1960s studied the effects of insect exclusion using insecticides on the hemiparasitic woodland herb cow wheat (*Malampyrum lineare*), demonstrating a dramatic increase in plant abundance. More recently, the exclusion of flower-head-feeding insects using insecticides on thistle (*Cirsium canescens*) in Nebraska has shown higher densities of seedlings on sprayed sites as well as higher mature plant densities in the next generation.

Introducing or Excluding Mammalian Herbivores

Observational evidence of mammalian herbivores affecting plant populations comes from the profound effect of introduced mammals on native vegetation, particularly on islands or island continents. Rabbits introduced to Australia have reduced plant biomass, altered plant communities, and caused erosion, and introduced red deer (*Cervus elaphus*) browsing native forests have had significant effects in New Zealand. Exclusion experiments for mammals are easier to carry out than those on insects and typically involve fencing plots. There have been many large herbivore exclusion experiments showing that the abundance of plants is affected by removing herbivores. For example, the gazelle (*Gazella dorcas*) feeds on the bulbs, leaves, and flowers of the desert lily (*Pancratium sickenbergeri*) in the Israeli desert. In exclosure experiments, fenced populations of lilies had twice as many plants as unfenced, showing a significant negative impact of herbivory on the plant population. More recently, an effect of ungulates on the population dynamics of two montane herbs has been demonstrated during a 7-year experiment in Spain. Spanish ibex (*Capra pyrenaica*) and domestic sheep (*Ovis aires*) had a significant negative effect on the population dynamics of *Erysimum mediohispanicum* and *E. baeticum*. As well as having a direct impact on the abundance of plants, mammalian herbivores [including black-tailed prairie dogs (*Cynomys ludovicianus*) in North American savanna and mule deer (*Odocoileus hemionus*) and elk (*Cervus elaphus*) in Arizona pine forest] can also prevent the transition from grassland to woodland.

4. PLANT RESPONSES TO HERBIVORY

The tolerance of plants to herbivory is a major plant strategy, much more important in herbivore–plant relationships than in other consumer–resource interactions. Plants can compensate for attack by herbivores by increasing the amount of new growth or by changing the effectiveness of existing plant parts. For example, partial defoliation may result in a more effective use of existing leaves rather than the production of new leaves. Plants also show a diversity of defensive adaptations against herbivores. These include physical barriers such as tough leaves, thorns, or hairs, and chemical defense (using secondary compounds such as cucurbatacin or nickel, which may be either deterrents or toxic compounds). Other organisms may also be used for defense; for example, in the ant–acacia mutualism, acacia ants (*Pseudomyrmex ferruginea*) actively defend the acacia plant from herbivores, and in return the ants are supplied with shelter in the form of modified thorns, protein-rich Beltian bodies, and carbohydrates from extrafloral nectaries. Some defenses are already present on the plant, and others are induced after the plant has been attacked. In turn, herbivores have found ways of circumventing plant defenses by changing their behavior through avoiding the plants or disabling the defense or by detoxifying or excreting the compounds. In the case of the chrysomelid beetle (*Labidomera clivollis*) feeding on milkweed (*Asclepias syriaca*), disabling the defense involves the beetle biting into lateral leaf veins near the midvein in order to cause the veins to leak before feeding.

5. THE IMPACT OF PLANTS ON HERBIVORE POPULATIONS

Plants are an important food resource for herbivores. In the 1960s, Hairston, Smith, and Slobodkin hypothesized that herbivores could not be resource limited because "the world is green," and plants are obviously abundant and intact. Their hypothesis was, and still is, criticized because the world is not always green and because green plants are not necessarily edible or of high enough quality for the herbivores to eat. It is now widely accepted that variation in quantity and quality of plant resources can have a fundamental effect on herbivore abundance and population dynamics. Changes in plant quality caused by herbivore feeding also have the potential to feed back and influence subsequent growth, reproduction, and mortality of herbivores. However, predation and parasitism are still thought to be important regulatory agents as well, particularly for insect herbivores.

Tracking Host Plant Abundance: The Cinnabar Moth

The cinnabar moth (*Tyria jacobaeae*) and its food plant, tansy ragwort (*Senecio jacobeae*), studied by Dempster in the 1970s, provides a classic example of a food-limited specialist insect herbivore tracking the abundance of its host plant. Larval numbers are dependent on the amount of food present, but plant numbers are not determined by the amount of larval feeding. The moth larvae build up in numbers until the limit in its food supply is reached, and the population then crashes as a result of starvation. The plants survive defoliation, and may in fact multiply as a result, because damaged plants can produce new shoots during the year of defoliation, which develop and produce mature seeds later the same year. Consequently, after the crash, the population of the moth is allowed to rise once more. Increased larval dispersal as a result of food shortage also leads to overexploitation of the host resulting in large population fluctuations. Recent analyses of long-term data sets have shown that the cinnabar moth–ragwort system can show fundamentally different dynamics at two different sites. In coastal dunes in the Netherlands, the moth has a delayed density-dependent effect on plant growth rate, and ragwort has a positive effect on changes in moth density when at low density but inhibits moth population growth rate at high densities. In contrast, in southern England, there is no evidence for either a direct or delayed effect of the moth on ragwort, or vice versa. In southern England, direct density-dependent mortality and mass plant recruitment resulting from soil disturbance (by rabbits) stabilize the plant–herbivore system, but plant and herbivore dynamics are uncoupled. In the Netherlands, time delays, and in particular delayed density dependence acting on the insect, determine the plant–herbivore population dynamics. These differences are caused by differences in the importance of seed limitation in plant recruitment in the two locations.

Experimental Effects of Plant Quality and Quantity on Insect Populations

It is difficult to find experimental studies that focus solely on the bottom-up effects of the plant on insect herbivores and do not investigate the top-down effects of predators concurrently. However, the greater abundance of pests attacking agricultural crops, compared to wild vegetation, provides indirect evidence that an increase in plant quantity and quality (as a result of fertilization) leads to an increase in insect abundance. Manipulative experiments have been used to investigate the effect of plant resources on insect herbivore dynamics, with an assemblage of sap-feeding phytophagous insects inhabiting Atlantic-coast salt marshes. The population density of all herbivores increased when the quality of the plant (in this case leaf nitrogen content) was elevated. Other field experiments have found that the density of a native perennial shrub, *Gossypium thurberi*, in the United States had a significant effect on survival and cumulative numbers over several generations of its most abundant herbivore, the lepidopteran caterpillar, *Bucculatrix thurberiella*.

Food-Limited Mammalian Herbivores: Population Cycles of Arctic Lemmings

A strong positive correlation between the biomass of large vertebrate herbivores (migratory African buffalo and Serengeti wildebeest) in African grasslands and the rate of plant productivity suggests that the herbivores may be food limited. However, the current view is that rather than being regulated by resources in general, animal numbers are regulated in a density-dependent manner by the forage available in key resource areas, especially during droughts. Population cycles, where some species exhibit fairly regular density cycles, are often evident in mammalian herbivore populations, and there is evidence that some may be caused by herbivore–plant interactions. For example, arctic lemmings (*Lemmis* and *Dicrostonyx* spp.) show population cycles linked to their resources. There are many other examples of population cycles in mammals, but although food may be a contributory factor influencing the population cycling, there are other more important causative factors, including predators. For example, boreal forest populations of snowshoe hares (*Lepus americanus*) go through 10-year population cycles in Arctic Canada. Heavy exploitation during a period of high-quality food is followed by an extended period when the regrowth foliage is of low quality (because of increased levels of induced defenses). The rate of increase in herbivores becomes positive only once the plant quality recovers its initial high levels. The quality and quantity of food do have important effects on population dynamics but cannot explain cycles completely, and predation is thought to be the most important factor, as confirmed by an 8-year experiment involving the manipulation of supplemental food and predator abundance.

6. HERBIVORE–PLANT INTERACTIONS AT THE COMMUNITY LEVEL

Studying the causes of herbivore population dynamics is complex. It is often difficult to establish whether a species is tracking or causing the population changes in another species. Experiments are difficult to carry out because stopping the cycles is not always possible and because when one causative factor is removed, another may come into play. Searching for single-factor explanations for herbivore or plant dynamics will often be futile, not to mention extremely difficult to establish unambiguously. The huge range of explanations even among rodents, such as lemmings and voles, is testimony to the fact that no single mechanism can explain population cycles in every individual case, especially given that even the same species in different locations can exhibit very different dynamic patterns.

Although many examples of herbivores affecting plant populations, and vice versa, are described here, there are also many studies that have shown no effect at all, strongly suggesting that the effects are system specific. This chapter has focused on simple herbivore–

plant interactions between two species, but plant and herbivore populations do not exist in isolation—they exist in multispecies systems. Although understanding the dynamics of herbivore–plant interactions is essential, it is crucial to remember that other factors may also play an important role in the community dynamics. Direct interactions with predators, competitors (plants or herbivores), and the environment, as well as a multitude of indirect interactions, can all affect the interactions between plant and herbivore species.

FURTHER READING

Crawley, Michael J. 1983. Herbivory. The Dynamics of Animal–Plant Interactions. Oxford: Blackwell Scientific Publications. *A comprehensive overview of herbivore–plant interactions research up to the early 1980s, but considerable progress has been made since.*

Crawley, Michael J. 1997. Herbivore–plant dynamics. In M. J. Crawley, ed. Plant Ecology, 2nd ed. Oxford: Blackwell Science, 401–474. *An invaluable and comprehensive review of interactions between herbivores and plants.*

Danell, Kjell, Patrick Duncan, Roger Bergstrom, and John Pastor, eds. 2006. Large Herbivore Ecology, Ecosystem Dynamics and Conservation. Cambridge: Cambridge University Press. *A recently published book with chapters addressing the impact of large herbivores on plant population dynamics and vice versa.*

Herrera, Carlos M., and Olle Pellmyr, eds. 2002. Plant–Animal Interactions. An Evolutionary Approach. Oxford: Blackwell Scientific Publications. *Provides a broad and basic background to insect and mammalian herbivory and granivory, including small sections on population dynamics.*

Maron, John L., and Elizabeth Crone. 2006. Herbivory: Effects on plant abundance, distribution and population growth. Proceedings of the Royal Society London Series B 273: 2575–2584. *A recent review focusing on when and where herbivores have the greatest effect on plant populations and suggesting future directions for the field of herbivore–plant dynamics.*

Turchin, Peter. 2003. Complex Population Dynamics: A Theoretical/Empirical Synthesis. Princeton, NJ: Princeton University Press. *A detailed synthesis of principles and models of population dynamics considering herbivore–plant interactions as a special case.*

II.11

Mutualism and Symbiosis

Judith L. Bronstein

OUTLINE

1. Interspecific interactions and mutualism
2. Types of mutualism
3. Major ecological features of mutualisms
4. Conservation of mutualisms

Mutualisms are interactions between two species that benefit both of them. Individuals that interact successfully with a mutualist experience greater success than those that do not. Behaving mutualistically is therefore of direct benefit to the individual itself. As Charles Darwin first pointed out in *On the Origin of Species*, mutualism does not require any special concern for the well-being of the partner. Although knowledge of mutualism lags behind that of other interspecific interactions, some important generalizations have emerged: nearly all mutualisms involve costs, not only benefits; outcomes of mutualisms are often context dependent; and mutualisms are often beset with cheaters that take advantage of rewards without conferring benefits in return. Mutualisms are increasingly recognized as fundamental to patterns and processes within ecological systems and are of growing concern in a conservation context. Persistence of individual species may frequently depend on preservation of the organisms, not only the habitats, on which they depend.

GLOSSARY

context dependency. Spatial and temporal variation in the strength and/or outcome of mutualism that can be attributed to the local environmental context; also referred to as conditionality

cooperation. Mutually beneficial interactions among individuals of the same species, often involving social interactions such as foraging or parental care

facilitation. Modification of some component of the abiotic or biotic environment by one organism that enhances colonization, recruitment, and establishment of another

facultative mutualism. A mutualism that increases an organism's success but that is not absolutely required for its survival and/or reproduction

mutualism. A two-species interaction that confers survival and/or reproductive benefits to both partners

obligate mutualism. A mutualism without which an organism will fail to survive and/or reproduce

symbiosis. An interaction (positive, negative, or neutral) in which two species exist in intimate physical association for most or all of their lifetimes and are physiologically dependent on each other

1. INTERSPECIFIC INTERACTIONS AND MUTUALISM

Interactions between species influence ecological processes at the level of the population, the community, and the ecosystem. Virtually all species on Earth are involved in multiple interspecific interactions at any one time. For example, an individual plant may simultaneously interact with pollinators, seed dispersers, root symbionts, herbivores, seed predators, and plant competitors.

Interspecific interactions are most commonly classified according to their effects on the two species. The effect of any given interaction on a population attribute (usually either population growth or fitness) of a given species can be positive (+), negative (−), or neutral (0). Thus, there are six possible pairwise outcomes, commonly referred to as mutualism (+,+), competition (−,−), commensalism (+,0), neutralism (0,0), amensalism (−,0), and predation, parasitism, and herbivory (+,−). This classification is based on discrete [+, −, 0] effects on each of the interacting populations. As will be discussed in what follows, however, divisions among different forms of interspecific interactions are not nearly so black and white: effects actually range continuously from positive to negative in interesting and important ways.

Ecologists have given deep and prolonged attention to two interspecific interactions: predation and competition, relationships that are negative for either one or both of the participants. Mutualisms, i.e., mutually positive interactions, are more poorly understood. However, they are increasingly recognized to be fundamental to patterns and processes within ecological systems. Mutualisms occur in habitats throughout the world, and ecologists now recognize that almost every species on earth is involved directly or indirectly in one or more of these interactions. In tropical rainforests, for example, the large majority of plants depend on animals for pollination and seed dispersal. Over 80% of all flowering plants are involved in mutualisms with beneficial fungi (mycorrhizae) that live on and in their roots. In the ocean, both coral reef communities and deep-sea vents are exceptionally rich with mutualisms. In fact, corals themselves obligately depend on the photosynthetic algae that inhabit them. Influences of mutualism transcend levels of biological organization from cells to populations, communities, and ecosystems. They are now thought to have been key to the origin of eukaryotic cells, as both chloroplasts and mitochondria were once free-living microbes. Mutualisms are crucial to the reproduction and survival of many organisms as well as to nutrient cycles in ecosystems. Moreover, the ecosystem services that mutualists provide (e.g., seed dispersal, pollination, and carbon, nitrogen, and phosphorus cycles resulting from plant–microbe interactions) are leading mutualisms to be increasingly considered a conservation priority.

2. TYPES OF MUTUALISM

Mutualisms usually involve exchanges of two qualitatively different kinds of benefit. In fact, mutualisms have recently been modeled as economic exchanges that take place within a "biological marketplace." Within this marketplace, organisms offer their mutualists commodities that are relatively cheap to acquire or produce. In exchange, they receive commodities that would otherwise be difficult or impossible for them to acquire. Three kinds of commodities are traded: transportation either of the partner itself or of its gametes; protection of the partner from the biotic or abiotic environment; and provision to the partner of limiting nutrient(s).

One particularly significant mutualistic exchange is between the commodities of transportation and nutrition. Perhaps the premier example is biotic pollination, in which certain animals visit flowers to obtain resources (usually food in the form of nectar) and return a benefit by transporting pollen among the plants they visit. As one indication of the importance of plant/pollinator mutualisms, Steven Buchmann and Gary Nabhan (1996) have estimated that half the foods humans consume are the products of one of these interactions. Biotic seed dispersal, in which animals disperse the seeds of the plants whose fruits they consume, involves a similar exchange of nutrition for transportation. In this case, seeds are transported away from the parent plant, toward more suitable sites for germination and growth.

Protection from biological enemies is another commodity that mutualists can provide their partners. For example, ants commonly feed on sugar-rich substances (honeydew) excreted by aphids and related insects; the ants vigorously defend these resources, providing those insects with protection against predators and parasites. Ants similarly guard caterpillars in the family Lycaenidae as well as plants in over 90 families; like aphids, these organisms produce nutritional rewards that serve to attract and then retain a corps of defenders. In all of these examples, protection is exchanged for a nutritional commodity. In other cases, protection is exchanged for the benefit of transportation. For example, certain hermit crabs actively place anemones on their shells. When the crabs are attacked, the anemones effectively come to their defense, stinging the attacker. The anemones, in turn, benefit by being transported among rich feeding sites.

Finally, a large class of ecologically and economically important mutualisms involve the exchange of nutrients. Although such mutualisms can be found in marine and terrestrial habitats worldwide, the two best-studied nutritional mutualisms take place in association with plant roots. *Rhizobium* bacteria found in nodules on the roots of many legume (bean) species fix atmospheric nitrogen into NH_3, a form that can be easily taken up by plants. In return, the bacteria receive carbon fixed by photosynthesis by their hosts. Consequently, legumes can thrive in nitrogen-poor environments such as deserts where few other plant species can persist. Many plant species also transfer fixed carbon to mycorrhizal fungi. The primary benefit of mycorrhizae is to vastly increase plants' access to soil phosphorus, a nutrient severely limiting to plant growth. This mutualism is so crucial that many researchers believe that its evolution was a critical step allowing plants to invade land around 400 million years ago.

Dividing mutualisms up according to exchanged benefits is a valuable way of recognizing parallels among interactions that superficially seem quite different. Mutualisms can be grouped in other ways that highlight common features. Some of these systems are based on mutualism's diverse evolutionary origins. Some but not all mutualisms, for instance, appear to have originated as predatory interactions. They evolved into mutualisms as the victim acquired traits or behaviors that benefited its predator so much that it became advantageous for the predator to retain rather than to consume it. Another approach to grouping mutualisms is to divide them into those in which benefits are received either in the form of direct exchanges or as indirect effects mediated by yet additional species. For example, pollination and seed dispersal mutualisms involve direct benefits (an exchange of food for transportation between the two partners). In contrast, an ant–aphid protection mutualism is indirect, as aphids benefit from ants only insofar as they reduce the abundance or change the behavior of other species, the aphid's enemies.

Mutualisms are also commonly grouped by the level of dependency between partners and by the degree of species specificity of the relationship. Facultative mutualists are ones whose populations persist (although less successfully) in the absence of a partner, whereas obligate mutualists are ones whose populations become extinct if a mutualist is absent. In species-specific mutualisms, only a single partner species confers mutualistic benefits, whereas in generalized mutualisms, an array of species can provide the necessary benefit. Individual types of mutualism can vary greatly in obligacy and specificity. For example, a plant that can reproduce only through the actions of a single pollinator species is engaged in a species-specific, obligate pollination mutualism; in contrast, a plant that can self-fertilize to some extent and that can be pollinated by multiple flower visitors is involved in a facultative, generalized pollination mutualism. One recent insight is that the degree of obligacy and specificity often differs greatly between mutualistic partners. In interactions in which one side of the mutualism is highly specialized and obligate, the other is commonly looser and much more open.

A final, very important division depends on whether or not mutualism is symbiotic. Two species that exist in intimate physical association for most or all of their lifetimes and that are physiologically dependent on each other are considered to be in symbiosis. *Symbiosis* is often taken to mean the same thing as mutualism, but in fact, that term confers no information about the outcome of the interaction: a symbiosis can benefit both, one, or neither of the two partner species, and only in the first case is it a mutualism. Put another way, only a subset of symbioses are mutualistic, and only a subset of mutualisms are symbiotic. The beneficial exchange in the vast majority of symbiotic mutualisms is nutritional, in either one or both directions; often, one species (the symbiont) is not free living, but inhabits the body of another species (the host). Of those mutualisms mentioned above, only plant–*Rhizobium* and plant–mycorrhizae associations are generally considered to be symbiotic. In some cases, symbionts become so closely integrated that their individuality becomes difficult and ultimately impossible to distinguish. The eukaryotic cell is now believed to have originated as a symbiosis between a primitive cell and bacteria that were ancestors to modern-day mitochondria.

The terminology used to describe beneficial interactions has been confusing and inconsistent since the terms *mutualism* and *symbiosis* were first coined, in different contexts, in the 1870s. Other interactions are often confused with mutualism as well. Although mutualism is an interspecific interaction, cooperation is usually used to describe mutually beneficial interactions among individuals of the same species, often involving social interactions such as foraging and parental care. Facilitation typically refers to the modification of some component of the abiotic or biotic environment by one organism that then enhances colonization, recruitment, and establishment of another, such as occurs during succession. It can involve two different species as well as a two-way exchange of benefit, but does not necessarily do so.

3. MAJOR ECOLOGICAL FEATURES OF MUTUALISMS

Even when mutualisms are grouped according to characteristics such as type of benefit, evolutionary origin, degree of intimacy, or specificity, one is struck by the tremendous variation in natural history that they exhibit. At first glance, it can be difficult to see any similarities between (for example) a pollination mutualism involving hummingbirds and plants and a nutritional mutualism involving cows and the symbiotic bacteria that live in their guts. This diversity can give the impression that mutualisms are nothing more than a grab-bag of pairwise interactions that happen to share a few common features. However, in recent years researchers have begun to recognize a variety of ecological features that unite mutualisms regardless of their divergent natural histories.

Mutualisms Have Costs as Well as Benefits

Although the benefits that mutualists confer on one another have been exhaustively documented, it has only recently become recognized that there are costs associated with these interactions as well. Most of the commodities that organisms provide their mutualists involve some kind of investment. For example, up to 20% of a plant's total carbon budget can be allocated to supporting mycorrhizae, and over 40% of their total energy investment may be devoted to producing nectar for pollinators. In some cases, these investments are fixed; that is, they do not vary with the number of mutualists in attendance and are experienced even if no mutualists are attracted. For example, flowers of many plants fill with nectar only once, before the arrival of pollinators. In other cases, investments are more modulated. Some lycaenid caterpillars, for instance, secrete more rewards if more ant guards are in attendance or if they sense the presence of predators.

In the act of conferring benefits on their partners, mutualists can inflict collateral damage. For instance, certain pollinators lay eggs on the plants they visit, and their offspring consume either the leaves or some of the seeds of the same plants. Fig trees are pollinated by minute wasps that exhibit the latter behavior. This remarkably costly mutualism is, oddly enough, both species-specific and obligate for both partners: that is, a given fig species can be pollinated by only one fig wasp species, and that wasp can develop successfully only in the fruits of that fig. Somewhat more abstract costs are associated with an organism's inability to succeed in the absence of its mutualists. Plants may reproduce poorly in parts of their range where their most effective pollinators are unable to persist, for example. In the case of fig trees, their distributional limits appear to be primarily determined by their obligate mutualists' inability to survive and fly in cold weather. As discussed in section 4, dependencies like this have important consequences for conservation.

It has become increasingly apparent that natural selection has acted on mutualisms not only in the context of their benefits but of their costs as well. Traits are being recognized that reduce the costs that mutualists inflict on their partners while not impeding the mutualism itself. For example, ant-tended plants need to keep their mutualistic guards away from their flowers: the ants can both destroy pollen and deter pollinators by their aggressive actions (which are desirable when directed against the plant's enemies rather than friends). The flowers of several ant-tended species have recently been shown to contain chemical repellents that keep the ant guards away.

Mutualisms Frequently Show Context-Dependent Outcomes

An interaction is mutualistic only when the benefits that each of the two species receives from the interaction exceed the costs that each experiences. Yet, the magnitudes of both the costs and benefits of mutualism are known to be highly variable in space and time. This variation in the strength and outcome has become known as conditionality or context dependency of mutualism. Mutualistic outcomes can vary depending on numerous factors, including the abundance of predators and competitors, the supply of resources such as nutrients, the density and distribution of mutualists, and the size, stage, or age classes of interacting species.

In some circumstances, a single interaction can vary all the way from mutualism to commensalism (i.e., benefiting one partner and neutral to the other) to antagonism (i.e., benefiting one partner and harmful to the other). For example, certain ants defend plants from most herbivores while at the same time guarding another herbivore because it too produces an attractive nutritional reward. When honeydew-producing herbivores are locally rare, the net effect of the interaction is likely positive for the plants so long as other herbivores are present that the ants remove. However, when honeydew producers are abundant, the plants might well suffer more by the ants' presence than by their absence (because the ants are fostering their enemies). Under these circumstances, one would have to call the ant–plant interaction antagonistic rather than mutualistic. Another example of context-dependent outcomes involves a marine protection mutualism involving the hermit crab and a small hydroid that it frequently places onto its shell. Hydroids are able to exclude most other organisms that attach to and damage hermit crab shells. This behavior clearly benefits the hermit crab. However, shells with hydroids are preferentially colonized by a common marine worm. These worms weaken the shells so much that they can easily be crushed by the hermit crab's own predators (blue crabs). The net effect of the association with hydroids thus varies greatly for the hermit crab: it is mutualistic when the worms or the predatory blue crabs are scarce but antagonistic when both worms and blue crabs are common. This case clearly demonstrates how dependent the outcome of an interaction can be on the particular environment in which that interaction occurs. Note that the discrete classification of interactions presented in section 1 of this chapter, as well as in most textbooks, fails to capture this rich dynamism of outcome.

Context-dependent variation can affect both distributions and population sizes of species that rely on mutualists. It is also likely to have important implications for how these interactions evolve, although work on this question is just beginning. John Thompson has argued that variation in outcome is the raw material for the evolution of interactions, just as variation in traits in populations is the raw material for the evolution of species.

Mutualisms Are Afflicted by Cheating

Mutualisms are not altruistic interactions. Rather, each partner is attempting to maximize the benefits that it receives, independent of the costs this might inflict on its partner. Thus, for example, ants defend honeydew-producing insects because they are valuable food sources, not in order to guarantee their well-being; similarly, most pollinators visit flowers in order to feed upon nectar, not in order to assure that the plants successfully reproduce. If a certain feeding behavior led pollinators to feed more efficiently but to pick up or deposit less pollen, that behavior would be favored by natural selection. In general, unless an organism has an immediate interest in the well-being of its partner—something thought to be extremely rare, at least in nonsymbiotic mutualisms—natural selection can be expected to favor individuals that obtain what they require from their partners while investing as little as possible in them. For this reason, mutualisms can profitably be thought of as "reciprocal parasitisms." Taken to its logical conclusion, this means that a partner could end up exhibiting behaviors that do not benefit the partner at all and that would in fact be outright detrimental to it. Thus, mutualism seems quite fragile evolutionarily, prone to breaking down as one or the other partner evolves into a better exploiter of its partner. Understanding why mutualisms are so common in spite of this apparent threat is a major goal of evolutionary research on these interactions.

In addition to the "threat from within," mutualisms of every type are assaulted by species that take advantage of rewards produced expressly for the mutualists. These organisms have been branded with a variety of not-very-flattering names, including cheaters, exploiters, robbers, thieves, parasites, interlopers, and *aprovechados* (Spanish for "one who takes advantage"). For example, many insects access floral nectar through holes chewed through the corolla. These "nectar robbers" are efficiently exploiting a commodity relied on by mutualistic pollinators while conferring no benefit in return because these feeding behaviors generally do not bring the robber into contact with the pollen. Furthermore, many nectar robbers damage the flower to the point where it either falls off or is avoided by true pollinators. Another example involves ants associated with ant–plant protection mutualisms that live and feed on the plant but ignore herbivores that the mutualists attack. Some of these ants inflict additional damage. For example, some clip off flowers the way a gardener might. Pruning increases leaf production, and because the reward-producing organs are at the tips of the leaves, their food availability increases as a result of their actions. This occurs to the clear detriment of the plant's reproduction.

Can mutualisms evolve in ways that minimize the impact of such exploitation? A few cases suggest that this may be possible. For example, chili fruits are infamous for the fiery sensation they produce in our mouths, an effect of the chemical capsaicin. Ecologist Joshua Tewksbury and his collaborators have compared the fates of the common edible chili (*Capsicum annuum*) and fruits of a nonpungent, capsaicin-free mutant of the closely related *Capsicum chacoense*. Rodents readily fed on the capsaicin-free, palatable chilies and destroyed the seeds when they did. They left the fiery, capsaicin-producing ones alone. In contrast, birds acted as mutualistic seed dispersers, feeding on chilies in a way that left the seeds intact and germinable. Birds treated fiery and palatable chilies identically because they lack the taste receptors for capsaicin. These observations suggest that the function of capsaicin in nature may be to deter exploiters (mammals that feed on chilies and destroy their seeds) while not disrupting the seed dispersal mutualism. It is an open question, however, whether traits like this are the rule or the exception in nature. Certainly, the ubiquity of cheaters suggests that mechanisms of deterring them, even if they exist, may not be foolproof.

4. CONSERVATION OF MUTUALISMS

Mutualisms have come to be seen as essential for reproduction and survival of species in every habitat on Earth. There is an alarming corollary of this realization: if efforts are not made to secure the well-being of an organism's mutualists, then conservation efforts devoted at that organism may well be in vain. Yet, at this point, remarkably little is known about either the processes that disrupt mutualisms or how easily they might be reassembled.

One of the more striking effects of human land use, and one that has increased dramatically in recent decades, is habitat fragmentation. Fragmentation creates small populations from large ones by weakening or severing their linkage through dispersal. Reductions in population size of one species caused by fragmentation can lead to failure of its mutualists as well. Habitat loss and alteration may reduce habitat quality for mutualists and thus mutualist population sizes. Furthermore, habitat patches may become so isolated that mobile mutualists become unable or unwilling to travel between them. For example, the loss of native bee pollinators from forest fragments in Argentina has reduced seed production of almost three-quarters of the plant species within those fragments, and reproduction of some species has ceased almost entirely.

If the loss of partners can raise a major ecological threat to mutualisms, the reverse phenomenon—the addition of new species—can be equally problematic. For example, the Argentine ant (*Linepithema humile*), a particularly successful invader worldwide, can decimate populations of ground-dwelling insects. In Hawaii, these ants are reducing the abundance of certain native insect pollinators, with potentially disastrous consequences for the persistence of native plants. Invaders can sometimes outcompete and displace native mutualists, often to the detriment of their partners. The Argentine ant has replaced native ant species as seed dispersers of a hyperdiverse plant family, the Proteaceace, in South Africa, leading to reduced seedling establishment and a shift in composition at the scale of entire plant communities. Interestingly, however, not all introduced species have negative impacts. Certain invaders join native mutualist assemblages with minimal effects on the residents. Furthermore, invaders can fill the gap created when a native mutualist has been driven to extinction, saving its partner from a similar fate. For example, an introduced opossum is now an effective pollinator of a New Zealand liana whose bat pollinator has been driven to extinction.

The studies described here indicate the potential magnitude of the conservation challenge. Importantly, they suggest that the problem of mutualism loss is not one restricted to specific species deprived of their partners; rather, the effects can scale up to entire communities through a cascade of altered interactions. In addition, these studies emphasize that the consequences of mutualism loss are complex and difficult to predict. What features predict whether a given mutualism will be resilient to change or susceptible to breakdown? When do invasive species displace native mutualists (for better or for worse), when are they excluded by natives, and when do invasives join native mutualists in robust but altered mutualistic communities? Finding answers to these questions depends on acquiring a deeper understanding of the basic ecology of mutualism.

FURTHER READING

Bascompte, Jordi, and Pedro Jordano. 2007. Plant-animal mutualistic networks: The architecture of biodiversity. Annual Review of Ecology, Evolution, and Systematics 38: 567–593. *A community perspective on specialization and generalization in plant/animal mutualisms, using network theory. Although rather technical, this work represents an important leading edge in mutualism research.*

Boucher, Douglas S., ed. 1985. The Biology of Mutualism. New York: Oxford University Press. *Chapters in this now-classic compiled volume span a diversity of topics, including the natural history, ecology, population dynamics, and evolution of mutualisms. Some but not all chapters require a technical background.*

Bronstein, Judith L. 1994. Our current understanding of mutualism. Quarterly Review of Biology 69: 31–51. *An overview of the scientific literature on mutualism that highlights areas in which understanding is particularly thorough and points to areas in need of further research. The patterns it highlights on which aspects we understand most and least thoroughly are equally relevant today.*

Bronstein, Judith L., Ruben Alarcón, and Monica Geber. 2006. Tansley Review: Evolution of insect/plant mutualisms. New Phytologist 172: 412–428. *This presents a detailed review of some of the best-understood mutualisms from an ecological and evolutionary perspective. A general outline is set out of the major angles of current research on the evolutionary biology of mutualism.*

Buchmann, Stephen L., and Gary P. Nabhan. 1996. The Forgotten Pollinators. Washington, DC: Island Press. *This popular, passionately written work details the ecological and economic importance of pollination mutualisms and the need for conservation measures to protect them.*

Sapp, Jan. 1994. Evolution by Association: A History of Symbiosis. New York: Oxford University Press. *This fascinating work on the history of research into symbiosis is as informative about the sociology of science as it is about the basic biology of these interactions.*

Schwartz, Mark W., and Jason D. Hoeksema. 1998. Specialization and resource trade: Biological markets as a model of mutualisms. Ecology 79: 1029–1038. *Although rather technical, this paper presents a novel approach to mutualism based on simple analogies taken from the field of economics.*

Tewksbury, Joshua J., and Gary P. Nabhan. 2001. Directed deterrence by capsaicin in chillies. Nature 412: 403–404. *As discussed in the text, this is a clever study demonstrating that mutualistic seed dispersers are not deterred by a fruit chemical that seed predators cannot tolerate. It demonstrates the value of experiments for studying mutualism and incidentally gives insight into humans' own relationship with these fruits.*

Thompson, John N. 2005. The Geographic Mosaic of Coevolution. Chicago: University of Chicago Press. *A lengthy, detailed, but easy to read discussion of the evolution of interspecific interactions, this volume is particularly valuable for the information it presents on context-dependent mutualisms.*

II.12

Ecology of Microbial Populations

Thomas Bell

OUTLINE

1. Introduction
2. Microbial diversity
3. Growth in culture
4. Constraints on growth
5. Multiple resources
6. Multiple populations
7. Outside the laboratory

Laboratory studies of microbial populations have informed many early population models, and there is now an enormous literature describing the dynamics of microbial populations under controlled conditions. This chapter outlines the major developments in the study of microbial populations, from the simplest-case scenario of a single population feeding on a single substrate to situations where there are interactions among multiple populations. Currently, the greatest difficulty is in extrapolating the results of the laboratory studies to understand natural microbial communities.

GLOSSARY

batch/continuous culture. In batch culture, strains are grown for a fixed period (e.g., a few days) before being transferred to fresh medium. In continuous culture, there is a continuous input of nutrients and output of spent medium, resulting in constant environmental conditions. The rate at which nutrients are input (and output) into the microcosm is called the dilution rate. Continuous culture experiments are conducted in a chemostat.

cometabolism. Simultaneous metabolism of two substrates such that the metabolism of one substrate occurs only in the presence of a second substrate.

culturability. The ability to grow strains in the laboratory in pure culture. For example, it is estimated that as few as 1% of bacteria species are culturable. Surveys of microbial communities therefore often rely on *culture-independent* techniques. For example, it is possible to construct a *clone library* of amplified DNA sequences to characterize a particular microbial community.

diauxie. Literally "double growth"; diauxie describes the way in which bacterial populations feed on mixtures of substrates (usually sugars). Diauxic growth is characterized by an initial growth phase, followed by a lag where the strain switches from the first to the second substrate, which is in turn followed by a second growth phase as the second substrate is utilized.

microbe. Here defined as an organism that is small (<1 mm) and unicellular. The current discussion is also restricted to free-living microbes (i.e., excluding parasites).

Monod equation. Named after the microbiologist Jacques Monod, the equation describes the relationship between substrate concentration and the growth rate of a microbial population. The form of the equation is equivalent to the Michaelis-Menten equation of enzyme kinetics.

syntrophy. A mutualistic interaction where two strains can utilize a substrate that neither could utilize when the other is absent.

yield. The number of microbial cells produced per unit of substrate.

1. INTRODUCTION

There is a popular conception of the microbial world as an unseen host of germs hiding in unwashed corners, intent on infecting people and crops, contaminating water and food. However, microbial populations are intrinsic to the ecology of animal and plant communities and play a vital role in the flow of nutrients and energy in ecosystems. In aquatic ecosystems, phytoplankton are often the principal source of primary production, thereby controlling the quantity of organic material available to higher trophic levels. Bacteria and fungi control the rate of decomposition in most ecosystems and, therefore, the amount of inorganic matter (e.g., inorganic nitrogen and phosphorus) that is recycled to primary producers. Clearly there is great interest in understanding the dynamics of microbial populations to elucidate their wider influence on the ecology of animal and plant communities. There are also pressing economic reasons for understanding microbial populations because microbes are the most important pathogens of humans. There has therefore been a great investment in understanding how microbial disease populations proliferate and spread. Finally, microbial populations can be harnessed to help produce useful products such as beer and wine.

Microbes cannot usually be observed directly, so our intuition and understanding of how microbial populations operate are often misinformed or misconstrued. Although we might casually observe that sparrows are abundant this year, or that the tulips have not flowered, these kinds of observational records are lacking for most microbial species. In fact, it is only relatively recently that microbiologists have developed the tools to identify and track the vast majority of microbial populations that cannot be directly observed. It is for this reason that the era of molecular biology has brought renewed excitement to microbial ecology research.

There is no strict definition of a "microbe," but typical definitions require a unicellular organism with an arbitrary upper size limit of around 1 mm. Many of the basic tenets of population biology also apply to microbial communities, but there are a few notable ways in which microbial populations exhibit some fundamental differences compared to larger organisms. For example, many microbial species have the ability to transform (incorporate) genetic material (DNA) directly from the surrounding environment, which has obvious implications for how populations adapt over evolutionary time scales. The colossal size of microbial populations and the rapidity (in absolute terms) with which they are able to grow and evolve are additional reasons for devoting a chapter to microbial populations. The purpose of the current chapter is first to give some background on how microbial populations in particular have been used in carefully designed laboratory studies and then to outline our understanding of how this knowledge translates to real-world ecosystems.

2. MICROBIAL DIVERSITY

Even under powerful microscopes, there are often few physical differences that distinguish closely related microbial species. Most bacteria, for example, have historically been described by gross morphological features, such as shape or colony morphology. However, the perception of microbes as a uniform group of simple organisms is wildly inaccurate. Microbes are an eclectic mix of organisms that differ significantly in their biology and encompass an enormous range of biological diversity. In fact, we now know that the vast majority of the diversity of life on Earth lies with the microbes. Phylogenies of conserved DNA sequences have revealed that the genetic differences among microbial taxa actually exceed the genetic differences between microbes and larger, multicellular organisms. In other words, there are microbes as genetically different from *Escherichia coli* (the bacterium found, among other places, in the human gut) as *E. coli* is from humans. Because of this enormous diversity, any discussion of general principles in microbial population biology will be beset with exceptions.

Even identifying microbial populations in situ is problematic. The principal method for identifying microbial species has been to measure the genetic distance between strains by sequencing a converved locus. For example, identification of prokaryotic populations has in the past relied on sequencing of a stretch of ribosomal DNA specific to bacterial cells. This has become refined in recent years by using multiple loci or even sequencing whole genomes from environmental samples. However, because species definitions often rely on whether individuals can interbreed, there are some substantial and ongoing issues in defining asexual microbial populations in particular.

Perhaps as a consequence of the difficulty in tracking microbial populations in a natural setting, much of microbiological research has split into two approaches. The first is to use controlled laboratory conditions to understand the dynamics of well-described model species. This is an approach that has been extremely successful in other areas of biology and has produced great advances in developmental biology and genetics in particular. For the model system tactic to be successful, it must be assumed that the results obtained using the model systems can be extrapolated to microbial populations in general, and it is not clear whether this is the case. The principal alternative is to study microbial populations in their natural setting using exclusively culture-independent techniques. For example, many of the most abundant organisms in the world have never been observed except as sequences in clone libraries. Although it is possible to track populations over time using these methods, it is more difficult to infer causal relationships between environmental factors (e.g., temperature, pH) and the rise and fall of the populations that are under observation.

3. GROWTH IN CULTURE

Microbiologists have been tracking the dynamics of pure cultures of specific strains in test tubes since the inception of modern microbiology and for hundreds of years for the goods and services they produce. Familiar examples include alcohol fermentation by yeasts to produce beer and wine or the production of penicillin by fungi for use as an antibiotic. To optimize the rate at which these goods and services are produced, there is great interest in describing the growth of microbial populations in pure culture, as well as in understanding the basic processes that prevent population crashes. Thus, Jacques Monod (1949), one of the founding fathers of the field, wrote that "[t]he study of the growth of bacterial cultures does not constitute a specialized subject or branch of research: it is the basic method of Microbiology."

Importantly, experiments investigating the growth of a population of microbes are relatively easy to conduct because it is straightforward to isolate a species by inoculating an environmental sample onto nutrient agar. Individual cells are able to grow on the nutrient agar into visible colonies as the cells proliferate. The colonies represent a single genotype and can be picked off the agar and stored indefinitely for future experiments. Alternatively, environmental samples can be serially diluted until only a single cell remains, which is then allowed to proliferate. In the archetypal experiment, a single strain is placed in a growth medium, and the population size is tracked over time.

If each cell divides asexually into two daughter cells, the simplest possible model of microbial population growth contains only parameters for the size of the population and the rate at which the cells divide:

$$\frac{dN}{dt} = rN,$$

where dN/dt is the rate of change in a population containing N individuals, and r is the population growth rate. The model assumes that there are no restrictions on population growth, so the population grows exponentially without limit, and:

> A single cell of the bacterium *E. coli* would, under ideal circumstances, divide every twenty minutes. That is not particularly disturbing until you think about it, but the fact is that bacteria multiply geometrically: one becomes two, two become four, four become eight, and so on. In this way it can be shown that in a single day, one cell of *E. coli* could produce a super-colony equal in size and weight to the entire planet Earth. (Michael Crichton. 1969. *The Andromeda Strain*, New York: Dell, p. 247)

Scary stuff, but in reality, there must be restrictions on growth for any reasonable biological system. The

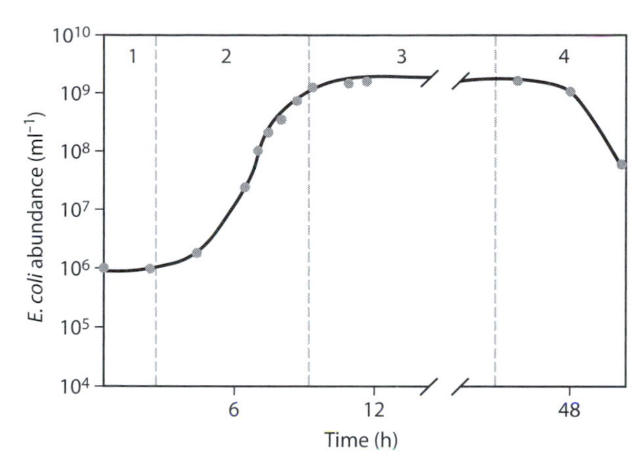

Figure 1. Population dynamics in batch culture showing the four phases of bacterial population growth: (1) lag, (2) exponential growth, (3) stationary, (4) death. (From Finkel, S. E. 2006. Long-term survival during stationary phase: Evolution and the GASP phenotype. Nature Reviews Microbiology 4: 113–120)

logistic equation was developed as a more reasonable model of population growth that took into account these constraints on growth. Although the logistic equation is now used throughout population biology, much of the original research was to describe the growth of yeast populations in culture and then to extrapolate the model to predict the growth of human populations. In the logistic equation, the change in population size over time (dN/dt) is modulated by the degree to which the population size (N) differs from the carrying capacity (K):

$$\frac{dN}{dt} = rN\left(1 - \frac{N}{K}\right).$$

When N is very different from the carrying capacity, there is a rapid change in population size (because $1 - N/K$ is a large negative or positive number). As the population size approaches the carrying capacity, $1 - N/K$ approaches zero, so the change in population size also approaches zero. In other words, the growth rate is zero ($dN/dt = 0$) when $N = K$ (and also, less interestingly, when $N = 0$). These properties of the logistic equation capture many of the characteristics of the growth of microbial population in laboratory cultures (figure 1). In particular, empirical observations showed that there was generally an initial lag before any observable population growth where there is no apparent change in population size even over relatively long periods of time. This is then followed by an exponential increase in population size (the exponential or log phase), which rapidly flattens into an asymptotic population size that is maintained for prolonged periods (the stationary phase). The logistic equation does not account for population decreases (the mortality phase), which can follow the stationary phase.

The purpose of the logistic equation is to describe the form of population growth, which it does reasonably well (excluding the mortality phase). The main difficulty with this description of microbial population dynamics is that it does not identify the biological mechanisms that underlie the parameters in the equation. Here, "carrying capacity" and "growth rate" are abstract phenomena that cannot be derived from first principles, and it is left to explain why there is a certain carrying capacity or growth rate. Monod was among the first to describe these in terms of the specific mechanisms that modulated the growth rate of microbial populations under controlled conditions.

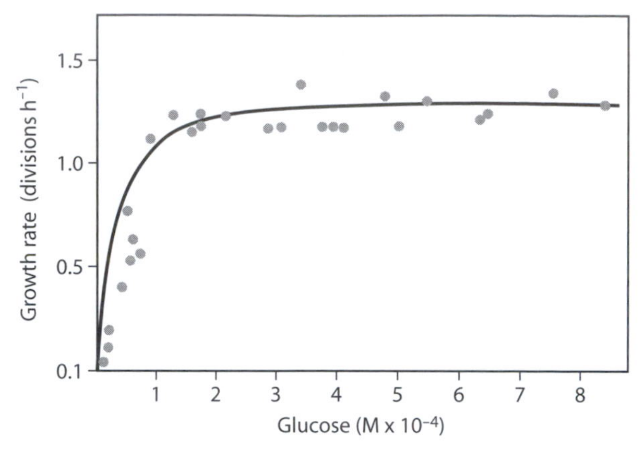

Figure 2. Growth rate of *E. coli* as a function of glucose concentration. (From Monod, 1949)

4. CONSTRAINTS ON GROWTH

When microbes are grown in batch culture (i.e., in closed microcosms with no inputs or outputs), several processes are occurring simultaneously as cells proliferate. First, microbial cells are utilizing an energy source (a substrate) to create more cells. If the substrate is finite, it will be depleted as it is utilized. Second, byproducts are created as the substrate is metabolized, which alters the environmental conditions. Both of these provide the opportunity for the population growth rate to be altered.

Monod was among the first to demonstrate convincingly that substrate concentration influenced population growth rate. To do so, it was necessary to maintain substrates at a constant concentration. Because substrate is used during population growth, he used continuous culture methods that allowed substrates to be replaced as quickly as they were used. When cells are grown in continuous culture, substrate is continuously added, and an equal volume of medium is taken from the microcosm. This allows direct measurements of growth rate for a set substrate concentration. In an unstructured population where all cells are equivalent, it was found that the specific growth rate was directly related to substrate concentration according to:

$$r = \frac{k_{\max}[S]}{k_s + [S]},$$

where $K_{\max}$ is the maximum growth rate, K_s is the substrate concentration at half the maximum growth rate (called the half-saturation constant), and $[S]$ is the actual concentration of the substrate. The relationship is saturating because the population growth rate reaches an asymptote at high substrate concentrations (figure 2). The consequence for the growth of microbial populations is that small increases in substrate concentration can have a substantial effect on growth rates when substrates are at low concentration. In contrast, even large alterations to substrate concentration will have little impact on growth when substrates are abundant.

Because we can now describe the effect of substrate concentration on microbial population size, it is also possible to estimate the depletion of the substrate in terms of microbial population growth. We would write:

$$\frac{dS}{dt} = -\gamma \frac{dN}{dt},$$

where γ is the microbial yield; for every gain of N bacteria, there is a γ depletion of the substrate. The yield is thought to represent a fundamental life history trade-off between the rate at which substrate is utilized (i.e., $K_{\max}$ and K_s) and the efficiency with which the substrate is utilized (γ). The yield is roughly constant for a particular strain but does vary to some extent with changes in substrate concentration. Although in theory microbes either deplete resources rapidly but wastefully (resulting in a low yield) or slowly and efficiently (resulting in a high yield), the experimental evidence remains circumstantial. For example, long-term experiments with *E. coli* have found little evidence for such a trade-off because genotypes with high yield were also characterized by high growth rates.

Once the parameters of this model of growth have been estimated in continuous culture, it is possible to understand what was happening in batch culture. First, substrate is abundant, and cells are dividing at nearly their maximum capacity. Substrate is depleted equally rapidly, and there is a breaking point at which substrates rapidly become rare; population growth tails off equally rapidly. Although the model successfully predicted growth dynamics, it was deficient in some significant areas. Namely, there was no explanation for differences in the length of the lag phase and no account of the death phase. Finally, the model does not account for adaptation to changing environmental conditions, which might result in increases in growth rates or yield over time.

In addition to depleting the substrate, population growth also alters environmental conditions as the by-products of metabolism accumulate. For example, under aerobic conditions, oxygen is depleted during respiration. Even if oxygen is allowed to enter the microcosm (e.g., if there is no lid on the flask), the rate at which oxygen diffuses into the culture medium might be insufficient to maintain ambient oxygen levels, in which case oxygen concentrations will decrease until the inputs and outputs of oxygen are balanced. In aquatic ecosystems, dissolved oxygen concentration is particularly important in situations where population growth rates need to be maintained at high levels (e.g., in sewage treatment plants). In a similar fashion, hydrogen molecules are transported into microbial cells during metabolism. If productivity is sufficiently large, this results in a decrease in the hydrogen ion concentration (pH) of the medium. Finally, only a fraction of the energy source is converted from organic carbon to ATP, and the rest is lost as heat. Especially in nutrient-rich environments, high levels of productivity can raise temperatures to an extent that a different suite of microbes is selected. Oxygen concentration, temperature, and pH represent important environmental variables for most microbial populations, and most strains can maintain positive population growth only within a limited range of pH, temperature, and oxygen concentration. Even if substrates are never depleted, the by-products of metabolism have the capacity to constrain growth rates.

5. MULTIPLE RESOURCES

So far we have considered only situations where populations are feeding on a single resource, but a more complicated dynamics is possible when there are multiple resources. The challenge is to extend the Monod equation describing the rate of population growth to m substrates ($S_1, S_2, \ldots, S_m$). In general, there are two possibilities. The first is that all of the resources are essential (i.e., required for growth). For example, algal cells require at least some inorganic nitrogen and phosphorus in order to grow. If there is a great surplus of nitrogen, phosphorus concentration will determine

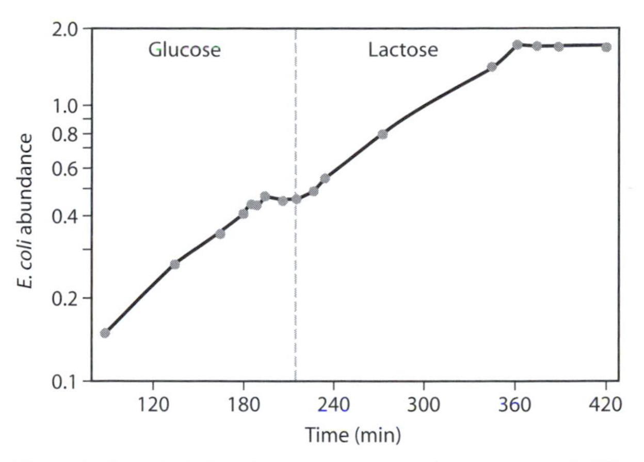

Figure 3. Growth of *E. coli* on a mixture of 0.05% glucose and 0.15% lactose. Glucose is consumed during the first phase, followed by the diauxic lag; lactose is then consumed, followed by the final stationary phase. *E. coli* abundance is measured as optical density at 600 nm. (From Traxler, M. F., D.-E. Chang, and T. Conway. 2006. Guanosine 3′,5′-bispyrophosphate coordinates global gene expression during glucose-lactose diauxie in *Escherichia coli*. Proceedings of the National Academy of Sciences U.S.A. 103: 2374–2379)

the growth rate, and vice versa. In this scenario, the principal question is to discover which substrate is limiting (e.g., iron is often limiting in marine environments, whereas phosphorus is often limiting in freshwater lakes) and then to estimate the population growth rate from the Monod equation based on the concentration of the limiting substrate.

When there are several substrates but no specific substrate is required for growth, a couple of different scenarios are possible. First, microbial cells encounter substrates at random and metabolize substrates as they are discovered. This model would predict growth in the population roughly equal to the average energetic "value" of the mixture of resources. However, this does not seem to occur for most well-studied cases. The second scenario is that the microbial cells concentrate on just one of the substrates as a result of either physiological constraints or optimal foraging behavior. Experiments have shown that microbial cells will often specialize on the most profitable resource until the resource reaches a sufficiently low concentration, at which point it switches to the next-most-profitable resource. For example, when *E. coli* are grown in a mixture of glucose and lactose, they will first feed only on glucose until it is exhausted, after which they will switch to the sugar that is the next most efficient for their growth. The growth of the population proceeds in a series of steps, where each lag, log, and stationary phase occurs sequentially as each sugar is used, a phenomenon called diauxie (figure 3). Extended diauxic lags between growth phases appear to occur particularly when there is a need to significantly alter the metabolic machinery before processing the next substrate; when *Pseudomonas* sp. is grown on a mixture of glucose and phenol, there is a lag of 2 to 3 days between the initial growth phase (when glucose is consumed) and the second growth phase (when phenol is consumed). This represents a pause of tens of generations (hundreds of years in human terms) while the population adjusts to the new conditions.

6. MULTIPLE POPULATIONS

When more than one population is present, any combination of negative (−), positive (+), or neutral (0) interactions is possible, including predation (+, −), competition (−, −), mutualisms (+, +), commensalism (+, 0), and so forth. An exhaustive summary of all types of interactions is beyond the scope of the current chapter, but some interactions have received particular attention from microbiologists. For example, some of the pioneering research on competition for resources was conducted on microbial communities, notably by G. F. Gause (using protozoa) and later by David Tilman (using algae). In Gause's classic experiments using *Paramecium caudatum* and *P. aurelia*, he showed that both grew adequately when cultured separately, but *P. aurelia* drove *P. caudatum* extinct in mixture (figure 4). This illustrates one of the simplest kinds of indirect interaction where two populations are competing for a single substrate. The substrate will be depleted to the point where only one of the populations is able to persist. All else being equal, the population that is able to subsist on the lowest ration of substrate will be the eventual winner. In well-mixed microcosms, the two populations can coexist only if they are competing for more than a single substrate, and if there is also a trade-off between performance on one substrate and performance on the other substrate. The theory of competition in the microbial literature provides the template for competition theory in other fields of ecological research and demonstrates the utility of microbial populations in providing general tests of ecological theory.

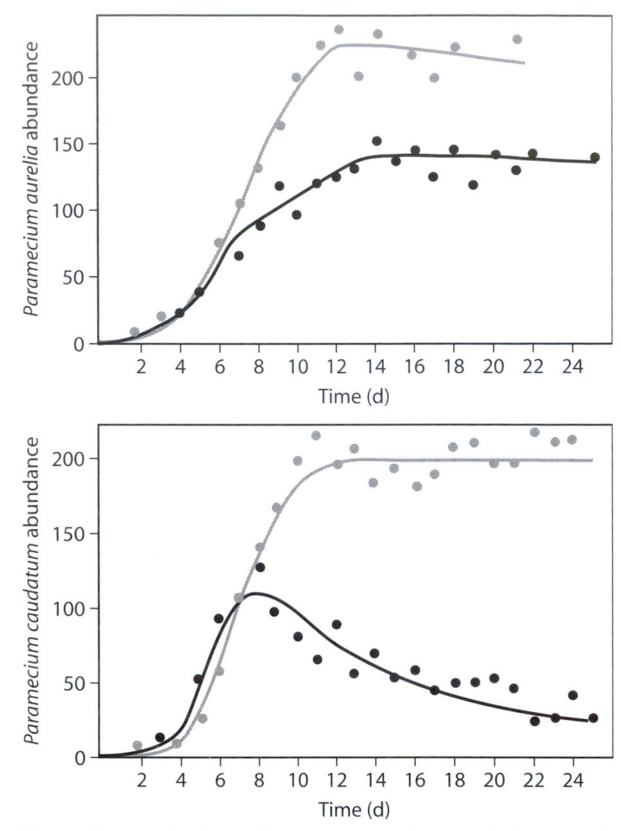

Figure 4. Growth of two ciliates, *Paramecium aurelia* (top panel) and *P. caudatum* (bottom panel) singly (gray circles) and in mixture (black circles). *P. caudatum* is outcompeted in mixture. (From Gause, G. F. 1934. The Struggle for Existence. Baltimore: Williams & Wilkins)

Much of the research on interactions among microbial populations has concentrated on commensal (+, 0) and mutualistic (+, +) interactions, especially in applied microbiology. However, it is unclear whether this reflects the importance of these kinds of relationships in natural microbial communities or whether they are simply picked up as interesting case studies. It might also be the case that the compounds being used for many industrial applications are so exotic that a single strain is unlikely to contain the machinery necessary to completely metabolize the substrate. In microbiology, *syntrophy* is the term used to describe a situation where populations provide the substrates required for each others' growth. One well-studied example is the interaction between *E. coli* and *Enterococcus faecalis* in the human gut. Putrescine is an important metabolite for both species, but neither can produce putrescine from arginine on their own. Rather, *E. faecalis* converts arginine to ornithine, and *E. coli* produces putrescine from ornithine; putrescine is then available to both populations. Similarly, *Lactobacillus arabinosus* and *Streptococcus faecalis* can only grow on glucose minimal medium when they are grown together. *L. arabinosus* produces the folic acid that *S. faecalis* requires for growth, while *S. faecalis* provides the phenylalanine that *L. arabinosus* cannot produce (figure 5). Another well-studied synergistic relationship exists between the soil fungi *Penicilium piscarium* and *Geotrichum candidum*, which results in the breakdown of propanil (an agricultural herbicide) into a

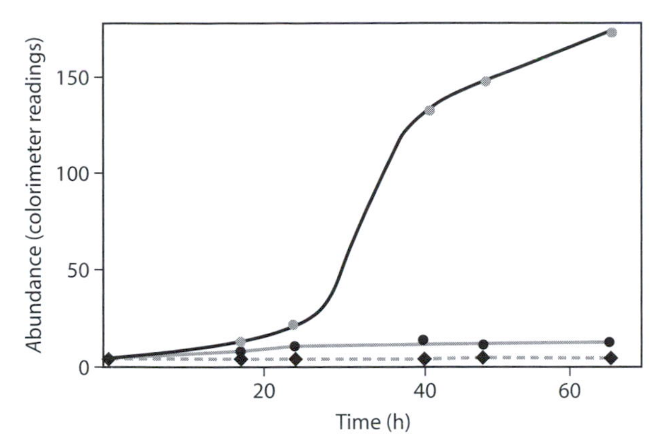

Figure 5. Syntrophy between *Lactobacillus arabinosus* and *Streptococcus faecalis*. Both *L. arabinosus* (black diamonds) and *S. faecalis* (black circles) fail to grow when grown singly but are able to grow when grown together (gray circles). The mechanism is described in the text. (From Nurmikko, V. 1956. Biochemical factors affecting symbiosis among bacteria. Experientia 12: 245–284)

substance that is less toxic to both species. Clearly, there are economic incentives to understanding the extent of such interactions in natural microbial communities.

Commensal relationships (where one species benefits and the other is unaffected by the interaction) might be even more common. For example, many microbial species excrete excess vitamins and amino acids into the environment, and other microbial populations can be reliant on these freely available excreta. Recalcitrant substrates are often metabolized along a processing chain, where a series of microbes break down the substrate to simpler molecules that are available to the next strain along the chain. In such processing chains, each population relies on upstream processing but is unaffected by downstream feeding. In cometabolism, a second (unused) substrate is oxidized as a by-product of metabolism of a substrate. For example, *Myxobacterium vaccae* cometabolizes cyclohexane to cyclohexanol when grown on propane. Other bacterial populations (that are unable to oxidize cyclohexane) are then dependent on *M. vaccae* for the production of cyclohexanol, but *M. vaccae* is unaffected by producing the cyclohexanol.

7. OUTSIDE THE LABORATORY

There are practical reasons for studying laboratory cultures, for example, because they are used to produce useful goods and services. However, it is also hoped that the laboratory studies will have some bearing on what is occurring in microbial populations, communities, and ecosystems under more realistic circumstances. Although microbial ecology has been hampered by methodological difficulties in identifying what constitutes a species, tools are rapidly becoming available for observing natural microbial assemblages. Despite recent methodological advances, extrapolating laboratory findings to natural environments remains a formidable task. Many microbial strains are unculturable (i.e., cannot be isolated and grown in the laboratory), so environmental microbiologists are often forced to conduct observational surveys and so draw their conclusions from exclusively observational data. Many of the artificial conditions imposed on microbial communities in laboratory settings are rarely seen in natural ecosystems; for example, culture media are often much more resource-rich than would be found in a natural ecosystem, and population dynamics at high nutrient concentrations might be largely irrelevant for understanding how natural populations operate. In addition, many of the factors that are clearly important in natural communities, such as predation and long-distance dispersal, are not accounted for in studies of single populations in the laboratory. Much current research is therefore devoted to reconciling laboratory experiments with observations of natural communities. As with the classic laboratory experiments, studies of natural communities have begun by concentrating on the degree to which substrate availability, competition, and predation affect local population dynamics.

Microbial populations are clearly often limited by substrate availability in nature, perhaps to the extent that the appropriate conditions for growth are rare for most microbial populations at most times. In marine environments, microbes in open-water environments are often iron limited, so iron fertilization has a dramatic impact on the carrying capacity. Similarly, addition of substrates to freshwater and soil environments leads to sustained increases in microbial standing crop and productivity. One important outstanding question is whether microbes exhibit trade-offs in their preference for individual substrates, which would explain why influxes of particular substrates have differential effects among genotypes or among species. Such specialization is often a prerequisite for coexistence but has not been shown convincingly in microbial communities, at least for the detailed comparisons that have been conducted comparing across genotypes of *E. coli*.

Predation and lysis by viruses are thought to be the principal forms of microbial mortality. As with the effect of substrate additions, the addition of a novel predator can have a variety of effects, some of which are counterintuitive; nutrients released from lysed cells can actually have positive effects on the growth of some microbial populations. Although laboratory studies have indicated rapid resistance to virus predators, it is unclear whether interactions between a specific virus and a specific host are sufficiently tightly coupled for coevolutionary interactions to be common in natural environments. Along with viruses, protozoa are the principal microbial predators. However, it is similarly unclear whether protozoa select specific prey items (i.e., whether they specialize on particular prey species), although they plainly specialize on a restricted range of cell sizes. Consequently, very small microbes may avoid predation and often increase in abundance when protozoan predators are present. Large cells also avoid predation, so many microbes form filaments when protozoan predators are common.

If predators are rare, local extinction appears to occur rarely in microbial communities. Although nutrients might frequently be rare in nature, many microbial strains can avoid significant mortality by transforming from an active to a dormant state by lowering their metabolic rate and halting cell division (e.g., during stationary phase, see figure 1). Thus, temperature fluctuations lead to cycles of dormancy and revival in *Vibrio vulni*; similarly, cyanobacteria can transform to a dormant state during periods of water stress. The consequence of such a resting state is that past population growth is stored when the environment is unfavorable, making populations largely resistant to local extinction except under exceptional circumstances. This leads to the opportunity for many microbial strains to have a cosmopolitan distribution because newly founded populations are able to cling to unfavorable environments. There is, consequently, a longstanding suggestion that, for most microbial communities, "everything is everywhere; the environment selects." Although there is mounting molecular evidence that there is a degree of endemism in microbial communities, the degree to which local (population growth and mortality) versus regional (immigration and extinction) effects influence the dynamics of natural microbial populations remains largely unresolved.

FURTHER READING

Atlas, Ronald M., and Richard Bartha. 1998. Microbial Ecology: Fundamentals and Applications. Reading, MA: Benjamin Cummings. *Currently the authoritative textbook on microbial ecology.*

Gause, G. F. 1934. The Struggle for Existence. Baltimore: Williams & Wilkins. *Classic experiments of predator–prey interactions and competition for shared resources using microbial microcosms. Reprinted 2003 by Dover Phoenix Editions.*

McArthur, J. Vaun. 2006. Microbial Ecology: An Evolutionary Approach. London: Academic Press. *A much-needed undergraduate-level textbook that briefly outlines the main themes in microbial ecology. Importantly, the textbook tries to integrate ecological ideas with microbiological examples.*

Monod, J. 1949. The growth of bacterial cultures. Annual Review of Microbiology 3: 371–394. *Seminal review of the early period of microbial population biology, summarizing the major developments in the field.*

Panikov, Nikolai S. 1995. Microbial Growth Kinetics. London: Chapman & Hall. *Authoritative overview of the development of ideas in modeling microbial populations. There are admirable attempts to extrapolate these results to soil communities in particular. Unfortunately, the book can be difficult to track down. The chapter on the historical development of the kinetic model of population growth is particularly recommended.*

Pernthaller, Jakob. 2005. Predation on prokaryotes in the water column and its ecological implications. Nature Reviews Microbiology 3: 537–546. *An accessible overview of the importance of predators in controlling bacterial populations in natural environments.*

II.13

Coevolution

John N. Thompson

OUTLINE

1. All complex organisms depend on coevolved mutualistic interactions
2. Coevolution shapes defenses and counterdefenses
3. Coevolution of competitors further structures the web of life
4. Species coevolve as a geographic mosaic
5. Coevolution may sometimes foster speciation
6. Coevolution may result in predictable webs of interaction
7. The coevolutionary process is pervasive in human endeavors

Coevolution is reciprocal evolutionary change among interacting species driven by natural selection. It is the evolutionary process by which many predators and prey, parasites and hosts, competitors, and mutualists adapt to each other in the constant struggle for life. It is also a process that can sometimes lead to new species, as different populations of interacting species coevolve in different ways in different geographic regions. Through its effects on adaptation and speciation, coevolution continually reshapes the web of life. Moreover, human society is increasingly altering the coevolutionary process through manipulation of ecological relationships among species within and among ecosystems, alteration of the genetic structure of crop plants, and development of novel strategies for mitigation of human diseases.

GLOSSARY

coevolution. Reciprocal evolutionary change in interacting species driven by natural selection

coevolutionary cold spot. Geographic regions in which one of a set of interacting species does not occur or in which the interaction, although occurring, does not result in reciprocal evolutionary change

coevolutionary hot spot. Geographic regions in which interactions between two or more interacting species result in reciprocal evolutionary change

local adaptation. Adaptation of populations to the local physical environment or to the local populations of other species with which they interact

1. ALL COMPLEX ORGANISMS DEPEND ON COEVOLVED MUTUALISTIC INTERACTIONS

Coevolution has been a major part of the process of evolution at least since the beginnings of complex life on Earth. In fact, many of the major events in the history of life are a direct result of the coevolutionary process that has created mutualistic symbioses among species. All complex organisms rely on mitochondria for cellular respiration. Those mitochondria are ancient bacteria that coevolved with their hosts and eventually became obligate organelles within the cells of all eukaryotic life. Every multicellular organism therefore has two genomes, a nuclear genome and a mitochondrial genome, as a direct result of this ancient coevolutionary process. Most animal species harbor one or more other coevolved symbionts that are necessary for their survival and reproduction. Among the most common are gut symbionts that aid digestion and nutrition. In many insect species, for example, coevolved symbionts provide one or more essential amino acids that are missing in the diet.

Plants harbor yet other ancient coevolved partners: obligate symbionts called chloroplasts. These organelles drive photosynthesis, and few plant species can survive without them. Many plants also rely on mycorrhizal fungi that attach to the roots of plants and aid in nutrition. Legumes and a few other plant taxa have coevolved relationships with rhizobial bacteria that convert atmospheric nitrogen into a form usable by the plant. In addition, the leaves of some herbaceous plants and some trees are laced with endophytic fungi, whose coevolved relationships with plants are only now being explored in depth. A majority of plants also rely on animal pollination for reproduction. Hence, survival and reproduction in most plant species require interactions with multiple other species, and many of these interactions are highly coevolved.

In general, all major biological communities are based on coevolved mutualistic relationships that form the underpinnings for community structure and succession. Most terrestrial communities rely on lichens (which are mutualistic symbioses between fungi and algae), mycorrhizal fungi, rhizobial bacteria, and chloroplasts to create the organic base on which other microbial life and animal species rely. Take away those coevolved mutualisms, and terrestrial life as we know it would disappear.

The same central ecological role for coevolved interactions holds for marine communities. Coral reefs, which harbor so much of the diversity of marine life, are a result of coevolved mutualistic symbioses between corals and the dinoflagellates (sometimes called zooanthellae) they harbor. The phenomenon called "coral bleaching," which has increasingly devastated coral reefs worldwide, results from the loss of the dinoflagellate symbionts brought about by multiple forms of environmental change. Much deeper in the ocean, deep-sea vents harbor species that rely on complex symbioses that oxidize sulfide to produce energy, much as plants use the sun's energy for photosynthesis. As new molecular tools continue to be developed to study microbial species, it is becoming increasingly evident that coevolved symbioses with microbial taxa permeate oceanic and terrestrial communities worldwide.

In addition to the intimate associations found in symbiotic mutualisms, many species are involved in short-term mutualisms that are fundamental to the structure and maintenance of biological communities. The majority of plants in terrestrial communities from the polar regions to the tropics rely not only on pollinators to fertilize their ovules but also on seed dispersers to distribute their seeds to new sites for germination. In marine communities, some fish species are involved in mutualisms with other fish species, called cleaner fish, that groom parasites from the skin of host fish.

As these mutualisms coevolve, natural selection hones the complementarity of the traits involved in the interaction (e.g., deep floral corollas in some flowers and long tongues in some pollinators), and other species evolve to converge on that common set of traits. The result is the formation of webs of mutualistic species. For example, in interactions between plants and pollinators, plants have evolved to provide pollen, nectar, or resins as rewards to attract pollinators, and a wide range of animal species have converged in the shapes of their mouthparts to extract the rewards from flowers. Plant species in different families have often converged on traits that attract particular groups of pollinators, just as some birds, bats, or insects have converged on traits that allow them to visit flowers with particular shapes.

Hence, mutualisms often accumulate groups of phylogenetically related and unrelated species over time, as species converge through natural selection on similar traits. Similarly, many unrelated plant groups have converged on similar fruit sizes, shapes, and colors to attract frugivorous birds, and birds from multiple avian families have converged on the traits that allow them to exploit these fruits as a major part of their diet. The combination of convergence and complementarity of traits therefore sometimes creates a coevolutionary vortex that continues to collect new species into the interaction over millions of years and shapes the structure of ecological communities.

2. COEVOLUTION SHAPES DEFENSES AND COUNTERDEFENSES

Not all interactions, of course, are mutualistic. Almost all species are attacked by parasites and predators. These interactions drive yet other forms of coevolution as species evolve defenses and counterdefenses against each other. Ongoing coevolution holds true for even the simplest forms of life. Bacteria are attacked by a great diversity of viruses known as bacteriophages. Estimates of the molecular diversity of oceanic bacteria and phages indicate that phage diversity may be at least as great as bacterial diversity. Multiple experiments have shown that bacteria and phages undergo very sophisticated forms of coevolution involving molecular changes in cell walls of bacteria, driven by a relentless battle between the ability of bacteria to prevent phages from breaking through the cell wall and the ability of phages to thwart those defenses and penetrate the cell wall. In experimental studies within laboratory ecological microcosms, bacteria and phages have been observed repeatedly to undergo rapid coevolution through natural selection on new mutations within the period of only a few weeks. These experiments suggest that even the simplest forms of life undergo continual coevolution with enemies.

Coevolution with enemies is equally common among multicellular species, and the signature of that process can be seen in the ecological lifestyles and the traits of species. The most common way of life on Earth is parasitism. There are more known species of parasites than there are of all other kinds of species. In turn, every species that has been studied in detail has been shown to have traits that have evolved to thwart attack by parasites or predators. Tens of thousands of chemical compounds found in plants are thought to have evolved as defenses against enemies. Similarly, all animals have physiological defenses that thwart attack by the many parasitic species that are constantly attempting to gain access to animal tissues. In most cases, we do not know many of the details about how coevolution shapes these interactions, but we do know that they are pervasive. Humans, for example, are subject to at least 1400 known diseases caused by pathogens and parasites.

Coevolution with predators or parasites can sometimes lead to multiple rounds of increasingly high levels of investment in defenses and counterdefenses through the process of directional natural selection. Such coevolutionary arms races are not sustainable in the long term but can nevertheless result in highly exaggerated traits. One of the most visual examples is that of camellia fruits that have evolved to be much larger in some populations than in others as a defense against camellia weevils that must eat through the fruit to get to embedded seeds on which they lay eggs. In response, some camellia weevils have evolved a highly elongated head that is longer than the entire length of the rest of the body, allowing the insects to penetrate the fruit to the seeds.

Alternatively, some parasites and hosts coevolve through natural selection that favors rare genetic forms of the host species, resulting in fluctuating genetic polymorphisms in the coevolving species. Parasites evolve adaptations to the most common genotypes in a host population, and natural selection therefore favors hosts that have rare genotypes to which the parasite is not adapted. This process of frequency-dependent natural selection (i.e., selection favoring the least frequent genotype) results in the maintenance of multiple genetic forms within the coevolving populations. The remarkable allelic diversity found in the mammalian immune system is thought by biologists to be maintained by this form of coevolutionary selection acting on hosts and their parasites.

Genetic novelty among individuals within populations can be enhanced by the recombination of genes that accompanies sexual reproduction, and one of the leading hypotheses for the evolution of sexual reproduction is that it is favored by coevolution between parasite and hosts. According to this hypothesis, sexual females are more likely than asexual females to produce offspring with rare genotypes, and these genetically different offspring of sexual females are, on average, more likely to be resistant to parasites than the genetically identical offspring of asexual females. Hence, one of the most important aspects of the ecology of organisms, the process of sexual reproduction, may result at least in part from the process of coevolution with fast-evolving parasites.

Some parasites or predators interact locally with multiple host species, creating more complex forms of coevolution. One major hypothesis for this form of multispecific coevolution is called coevolutionary alternation. According to the hypothesis, natural selection favors parasites or predators that attack the currently least-defended host or prey species. Selection acting on a host or prey population that is being strongly attacked favors individuals that have higher levels of defense, leading eventually to an overall increase in the level of defense in that population. Those increased defenses in turn favor parasites or predators that attack other, less-defended, species. The result is a constantly changing mix over time of which host or prey species are attacked.

The best-known potential example of this form of coevolution is that between European cuckoos and the birds they parasitize. Cuckoos lay their eggs in the nests of other bird species, and the unsuspecting hosts raise the cuckoo nestlings. Cuckoos differ across Europe in the bird species they parasitize, and there appear to be geographic differences in preference for host species. The bird species that serve as hosts differ among populations in the defenses they harbor. In addition, some host populations are heavily attacked but have low defenses, suggesting that they are locally new hosts for the cuckoos. Some populations of other bird species have high defenses but are rarely attacked, suggesting that these populations may have formerly been hosts but have now been abandoned in favor of local host species with lower defenses. In general, these interactions appear to coevolve through a continual reshuffling of preference in the cuckoos and levels of defense in the host species.

Not all interactions with parasites lead either to escalating arms races, fluctuating polymorphisms, or coevolutionary alternation. In some interactions between parasites and hosts, the relationship may coevolve toward attenuation of the level of antagonism, especially if parasites are transmitted directly from mothers to daughters at birth. This mode of transmission can favor parasites that are less virulent than other parasites, because selection favors parasite genotypes that do not kill their host. If a female host dies before she can reproduce, that lineage of parasite is not transmitted and becomes extinct. Over time, then, less virulent lineages tend to spread in populations under these conditions of transmission. Although some biologists in the past have assumed that parasites will tend to evolve toward decreased virulence over time, much empirical work and mathematical theory in coevolution has shown that assumption to be false. There is nothing inevitable about the direction of evolution in interactions between parasites and hosts. The trajectory of coevolutionary selection depends on the ecological and genetic conditions in which the interaction occurs.

3. COEVOLUTION OF COMPETITORS FURTHER STRUCTURES THE WEB OF LIFE

The web of life gains additional structure from coevolution among competing species. Wherever species share resources that are insufficient to support all individuals, competition favors those that are either more efficient than others in garnering limited resources or those individuals that use alternative resources. Through competition for limited resources, coevolutionary selection favors divergence among species in the use of those resources. The divergence can take multiple forms, including the use of different foods, the use of different habitats, or the use of the same foods or habitats but at different times of year.

The outcome of such competition is called character displacement. Its effects can be seen through studies of geographic variation in the morphologies of species that co-occur with competitors in some parts of their geographic range but not others. Competing animal species often differ in body size or other morphological characters to a greater degree in regions where the species occur together than in regions where they do not overlap. In fact, most published studies of competition have focused on how species diverge in morphological characters associated with food choice when in competition with other, closely related species.

Although competition for resources may be the cause of most instances of character displacement, some instances of displacement may result from other ecological causes. For example, if two prey species are attacked by the same predator species, then natural selection could favor character displacement among the prey species. A prey species that is different from other coexisting species may have a selective advantage because the local predators may become adapted to only the most common species.

In plants, some instances of character displacement are driven by selection to avoid hybridization with other plants that share the same pollinators. In such instances of reproductive character displacement, natural selection favors plants with flowers that minimize the chance that pollen from another species will reach their stigmas and produce hybrid offspring with lower Darwinian fitness. Reproductive character displacement may involve the evolution of flowering at different times of the year from closely related plants, or the evolution of floral shapes that exclude pollinators that visit other plant species.

4. SPECIES COEVOLVE AS A GEOGRAPHIC MOSAIC

As species coevolve through mutualistic and antagonistic interactions throughout their geographic ranges, natural selection may come to differ among environments. Traits of interacting species that are favored by selection in one environment may be ineffective in other physical or biotic environments. Over time, then, interacting species may exhibit complex geographic patterns in the traits shaped by this selection mosaic. One species may evolve to be larger than other competitors in one geographic region but evolve to be smaller than other competitors in other regions. Similarly, parasites and hosts, or predators and prey, may evolve different arsenals of defenses and counter-defenses in different populations. For example, some plant species are known to have evolved different mixes of chemical defenses in different geographic areas, and the specialist insects that attack these plants have evolved, in response, mixes of detoxification compounds that are customized to their local plant populations. The resulting geographic selection mosaic in local adaptation is the raw material that fuels the process of coevolution across the Earth's constantly changing landscapes.

In addition, natural selection may sometimes be intensely reciprocal in only some environments. These coevolutionary hot spots may be embedded in a matrix of coevolutionary cold spots, where the interaction does not occur or is detrimental or beneficial to one species but not to other interacting species. A parasite, for example, may commonly cause death to its hosts in some environments but have relatively little effect on host fitness in other environments. Coevolving species may therefore commonly be a complex mix of populations that differ in the degree to which coevolution is driving the evolution of the interacting populations.

Gene flow, random genetic drift, and metapopulation dynamics can add to the geographic mosaic of coevolution by continually changing the geographic areas in which different combinations of coevolving traits occur. A trait evolved in one population may spread to other, but not necessarily all, populations of a species. In addition, traits can be lost in some local populations through random genetic drift. The continual remixing of genes and traits among populations adds to the geographic mosaic of coevolving species.

The geographic mosaic theory of coevolution argues that selection mosaics, coevolutionary hotspots, and trait remixing are common features of coevolving interactions. Evidence for the geographic mosaic of coevolution has been shown in an increasingly wide array of interacting species, including herbaceous plants and insect herbivores, conifers and seed-eating birds, and snakes and toxic salamanders. The persistence of some interactions over millions of years may be a result of the geographic mosaic of coevolution, which allows coevolution to proceed simultaneously in different directions and involve different traits in different populations.

5. COEVOLUTION MAY SOMETIMES FOSTER SPECIATION

In some cases, the geographic mosaic of coevolution may lead to speciation. As coevolving populations adapt locally to each other, they may diverge to such an extent that they become reproductively isolated from other populations. In fact, one of the major hypotheses for the remarkable diversification of life is that coevolution has repeatedly favored starbursts of speciation between interacting lineages as the species involved continue to evolve novel defenses and counterdefenses.

This process of diversifying coevolution may result in groups of closely related species that are specialized to interact with only one or a few other species. For example, crossbills are a group of birds specialized for extracting seeds from conifers before the cones open. The birds use their crossed bills to pry apart the scales of conifers so that they can then extract the seeds with their tongues. Different crossbill populations have evolved different beak shapes, each adapted to a different conifer species. Some are specialized on particular pine species, others on particular fir or spruce species. In response to attack by these specialist populations, some conifer species have evolved novel cone shapes and sizes that increase the difficulty for the crossbills when they are attempting to extract seeds. Moreover, populations of crossbills and conifers even within the same species differ geographically in their patterns of coadaptation. Overall, crossbills appear to have diverged into a group of specialist species each adapted to different conifer species, and sometimes even to particular populations of a conifer species, as a direct result of their ongoing coevolution with conifers.

6. COEVOLUTION MAY RESULT IN PREDICTABLE WEBS OF INTERACTION

Although it is relatively easy to understand how coevolution may shape interactions between a pair of species, it is a greater challenge to understand how coevolution shapes more complex webs of interacting species. In species-rich biological communities, some species may interact with many other species, whereas others may interact with few species, creating tangled webs of interaction. Recent studies, however, suggest that coevolution may shape webs of dozens or even a hundred or more interacting species in predictable ways. Research on food webs of predators and prey has shown that these interactions are commonly compartmentalized into multiple smaller subwebs of species that interact more often with one another than with other species within the larger web. This is expected from coevolutionary theory because coevolution between antagonistic species can lead to groups of species that share similar defenses and counterdefenses, and natural selection favors prey that interact with fewer rather than more predators.

In contrast, mutualistic interactions between free-living species such as plants and their pollinators or frugivores tend to group into few subwebs. Instead, they form more of what is called a nested structure, where a core group of generalist species interacts with each other and a group of specialist species interacts preferentially with the most generalist species. This structure of specialization is consistent with coevolutionary theory because natural selection on mutualistic interactions among free-living species can result in a coevolutionary vortex that draws more species into the interaction over time without breaking it up into many subwebs. Hence, natural selection often favors mutualists that interact with multiple other mutualists, and it appears to favor a few species that specialize on the core species within the mutualist network.

These mechanisms by which natural selection shapes large webs of coevolving species are among the least-understood aspects of coevolution. Studies of coevolving webs, however, are becoming an increasingly important part of evolutionary biology and ecology because they are central to understanding how coevolution has shaped the overall organization of biodiversity within and among ecosystems.

7. THE COEVOLUTIONARY PROCESS IS PERVASIVE IN HUMAN ENDEAVORS

Human society has increasingly altered the coevolutionary process as we have transformed all the major ecosystems on Earth. The wholesale movement of species among continents by humans has created new interactions among species at a global scale unprecedented in the Earth's history. Those species are now coevolving with native species, but we have little understanding of how these biological communities will stabilize over the coming centuries.

Even more directly, we have increasingly co-opted the coevolutionary process in our attempt to increase our food supplies and minimize diseases. Much of the development of agricultural crops has been driven by cycles of selective breeding for new varieties that are resistant to pests, followed by the evolution of new pests that can overcome that resistance, which is then followed by subsequent selective breeding for yet newer resistant varieties. These same kinds of cycles of human-induced coevolution are occurring in the development of new antibiotics against diseases that afflict humans and domesticated animals.

Some aspects of human-induced coevolution closely match those found in the natural process of coevolution, whereas others do not. We can, however, use studies of coevolution within natural populations as a guide for the development of new strategies to slow down our coevolutionary arms races with agricultural pests and human diseases. Hence, a science of applied coevolutionary biology is beginning to emerge and is likely to grow in importance as human society continues its attempt to manipulate the ecological structure of biodiversity worldwide.

FURTHER READING

Brodie, E. D. III, and E. D. Brodie, Jr. 1999. Predator–prey arms races. Bioscience 49: 557–568. *This article presents a concise explanation of the complex ways in which natural selection shapes the evolution of defenses and counter-defenses through coevolution.*

Bronstein, J. L., R. Alarcón, and M. Geber. 2006. The evolution of plant–insect mutualisms. New Phytologist 172: 412–428. *This article presents an overview of some of the most common mutualisms found in terrestrial environments and the role of coevolution in shaping these interactions.*

Pennisi, E. 2007. Variable evolution. Science 316: 686–687. *A leading science writer interviews multiple coevolutionary biologists on our current understanding of the coevolutionary process.*

Sotka, E. E. 2005. Local adaptation in host use in marine invertebrates. Ecology Letters 8: 448–459. *This summary of studies discusses local adaptation and the potential for coevolution in a diverse group of marine species.*

Thompson, J. N. 2005. The Geographic Mosaic of Coevolution. Chicago: University of Chicago Press. *This volume presents a synthesis of current understanding of the coevolutionary process, including a wide range of examples of how coevolution continues to reshape interactions among species.*

III

Communities and Ecosystems

Michel Loreau

Ecology is the science of the interactions between living organisms and their environment. What makes ecology so fascinating, and at the same time so disturbing for the layperson, is the extraordinary diversity and complexity of these interactions, which create a wide range of nested complex systems from the scale of a droplet of water to that of the entire planet. Anything that happens here and now is almost certain to have an effect elsewhere and later. And this also concerns us as humans. Just as any other species, we transform the world around us by the mere act of living—consuming resources, releasing waste products, changing land and sea for our purposes. But are the effects elsewhere and later predictable? Can we make sense of this complexity, or is it better to just ignore it and to attend to our affairs without worrying about their consequences? Questions of this type are crucial, both for ecology as a science to understand the world in which we live and for society at large to cope with the unforeseen consequences of its past and current actions.

Ecology has approached complex ecological systems from two different angles, which have gradually led to two distinct subdisciplines, community ecology and ecosystem ecology. A *community* is a set of species that live together in some place. The focus in community ecology has traditionally been on species diversity: What exogenous and endogenous forces lead to more or less diverse communities? How do species interactions constrain the number of species that can coexist? What patterns emerge from these species interactions? An *ecosystem* is the entire system of biotic and abiotic components that interact in some place. The ecosystem concept is broader than the community concept because it includes a wide range of biological, physical, and chemical processes that connect organisms and their environment. However, the focus in ecosystem ecology has traditionally been on the overall functioning of ecosystems as distinct entities: How is energy captured, transferred, and ultimately dissipated in different ecosystems? How are limiting nutrients recycled, thereby ensuring the renewal of the material elements necessary for growth? What factors and processes control energy and material flows, from local to global scales? Community ecology and ecosystem ecology, then, provide two complementary perspectives on complex ecological systems.

Community ecology has been a very lively subdiscipline during the last decades because of a growing interest in biodiversity and the stability of complex systems. The first chapters in this section summarize some of the main findings and debates in this area. Robert Colwell (chapter III.1) begins by defining the concept of biodiversity and examining how it can be measured. *Biodiversity* is a relatively new term that has gained wide popularity in the general public because of current concerns about its loss as a result of growing human environmental impact; but the concept of biological diversity from which it arose is much older and has been studied abundantly in ecology since its inception. Measuring species diversity from local to regional scales is a much more challenging task than might appear at first sight. Robert Colwell provides a very accessible digest of decades of research on this issue. Jérôme Chave (chapter III.2), Robert Holt (chapter III.3), and Ragan Callaway (chapter III.4) then move from patterns to processes and explore the respective roles of the three types of elementary species interactions that hold communities together. *Competition* has long been regarded as the primary factor that governs species coexistence and, hence, species diversity and other community-wide patterns. Recent research has both confirmed its prevalence and relativized its role in the organization of natural communities (chapter III.2). *Predation* is another interaction that affects species persistence and diversity. But it can either enhance or hamper prey species coexistence depending on the details of species interactions, which argues for a thorough investigation of its mechanics if we are to predict its impacts on natural communities (chapter III.3). Positive species interactions, such as *facilitation* and *mutualism*, have received comparatively much less attention until recently. Yet there is ample evidence that facilitation is widespread and plays an important role in the organization of plant communities (chapter III.4).

Competition, predation, and facilitation are simple pairwise species interactions that can be viewed as the building blocks of more complex communities. But whenever three or more species are engaged in such interactions, we see the emergence of *indirect effects* in which one species affects another through a shared, intermediary species. Oswald Schmitz (chapter III.5) explores some of the myriad fascinating ways that indirect effects emerge in, and affect, communities and ecosystems. An indirect effect of special interest is that which emerges from a simple food chain in which carnivores eat herbivores, and herbivores eat plants. Predation by carnivores tends to suppress herbivore populations, which in turn tends to release plants from control by herbivores. This cascade of effects as a result of *top-down control* of the food chain, which is known as the *trophic cascade*, can strongly affect the biomass of the various trophic levels and ecosystem functioning. Elizabeth Borer and Dan Gruner (chapter III.6) assess its importance relative to bottom-up processes in the regulation of ecosystems. When a large number of species prey on each other, however, they do not always fall into well-defined trophic levels. They then constitute a *food web* rather than a food chain. The persistence and stability of these enormously complex food webs are one of the main enigmas of ecology. Kevin McCann (chapter III.7) reviews the history of thought on this issue and suggests that the solution to this enigma may lie in the stabilizing role of variability in space and time. Spatial dynamics has received increasing attention in ecology during the last decades. Many of the patterns and processes within local communities are strongly influenced by movements of organisms and materials at the landscape or regional scale. Mathew Leibold (chapter III.8) examines the consequences of these movements on the maintenance of biodiversity in *metacommunities*, i.e., in sets of communities connected by dispersal.

Compared with community ecology, ecosystem ecology generally offers a more macroscopic, integrated view of ecological systems. A second set of chapters in this section summarizes current knowledge on the flows of energy and materials in ecosystems, from local to global scales. Julien Lartigue and Just Cebrian (chapter III.9) provide a comparative analysis of patterns of *carbon flows* and productivity across a wide range of ecosystems, both terrestrial and aquatic. In particular, they examine the factors that determine the movements of carbon through the activity of herbivores and decomposers, two of the main agents of carbon flows in ecosystems. Carbon flows are usually closely associated with energy flows: both energy and carbon are typically captured by primary producers through photosynthesis and ultimately released through respiration. In contrast, other elements that often limit primary production, such as nitrogen and phosphorus, are tightly recycled within ecosystems. Peter Vitousek and Pamela Matson (chapter III.10) present an overview of *nutrient cycling* in ecosystems, the factors that control it, and its implications for global *biogeochemical cycles*. Although much research has focused on the global cycle of carbon, in part because of the importance of carbon dioxide in the climate system, they show that humanity has altered the cycles of other elements to a much greater extent than that of carbon. The following three chapters by Richard Houghton (chapter III.11), Darren Bade (chapter III.12), and Paul Falkowski (chapter III.13) examine the biogeochemical cycles of carbon and other elements in more detail in the three great types of ecosystems on Earth, i.e., terrestrial, freshwater, and marine ecosystems. Collectively, these chapters summarize the fundamental ecological knowledge that is necessary to understand and predict human impacts on the biosphere.

Although community ecology and ecosystem ecology provide different perspectives on ecological systems, these perspectives should ultimately be compatible with each other and inform each other. A recent trend in ecology has been the emergence of unifying approaches that seek to lay bridges among these subdisciplines through common principles or topics that lie at their interface. One of these topics, which has attracted a great deal of attention in the last decade, is the *relationship between biodiversity and ecosystem functioning*. Interest in this issue arose from the realization that biodiversity is currently being lost at an accelerating rate globally. But the impacts of biodiversity loss on the functioning of ecosystems and on their ability to provide ecosystem services to human societies were largely unknown until recently. Andy Hector and Andy Wilby (chapter III.14) summarize the progress achieved by more than a decade of active research in this area. Linking biodiversity and ecosystem functioning clearly requires working at the intersection between community ecology and ecosystem ecology. But there are other ways to work toward integration across subdisciplines. *Ecological stoichiometry* examines how the nutrient content of organisms shapes their ecology. Robert Sterner and Jim Elser (chapter III.15) show how stoichiometric constraints play a role in a wide range of ecological phenomena, from the growth rate of animals through population dynamics to the rates of recycling of elements in food webs, thus providing a unifying theme through ecology.

Macroecology is another emergent research program in ecology, which Pablo Marquet reviews in chapter III.16. Macroecology studies ecological patterns and processes at large spatial and temporal scales. One of its goals is to identify statistical regularities that might reflect the operation of elementary principles underlying both community structure and ecosystem functioning.

As do all biological systems, communities and ecosystems change through time as a result of exogenous environmental changes, endogenous ecological interactions, and evolution. Dynamic and evolutionary constraints play a significant role in shaping ecological systems, whether these are viewed from a community or an ecosystem perspective. In particular, complex systems can have tipping points, where the slightest disturbance can trigger rapid change until they reach an alternative stable state. Marten Scheffer (chapter III.17) presents the theory of these *regime shifts* and the empirical evidence that supports them. He also discusses how insights into regime shifts can be used in ecosystem management. Erika Zavaleta and Nicole Heller (chapter III.18) examine how communities and ecosystems respond to current *global environmental changes* driven by human activities, in particular in the Earth's climate, atmosphere, and biogeochemistry. A picture emerges of natural systems altered drastically by the accumulating effects of multiple global changes. They conclude that safeguarding the capacity of ecosystems to adapt to change is a minimum requirement to preserve options for the future. *Evolution* is increasingly recognized as an important determinant of community structure and ecosystem functioning. Nicolas Loeuille (chapter III.19) summarizes its effects on community and ecosystem properties along a gradient of increasing evolutionary complexity, from evolution of single species to coevolution of a large number of interacting species. He shows how complex adaptive systems emerge from the evolutionary dynamics of organisms that compose them.

The chapters in this section vividly illustrate the vital importance of ecology as a science that provides a rigorous body of knowledge on the complex natural systems of which we humans are part. They show that the diversity and complexity of the interactions between living organisms and their environment can be understood. Even though ecological systems are intrinsically too complex to be amenable to a complete description by a few simple laws yielding simple predictions, some principles, rules, and general trends do emerge and can usefully guide the way we manage ecosystems and cope with environmental changes. Ecology provides a unique perspective on diversity and complexity that will be critically important as humankind faces the most formidable environmental challenges of its history.

III.1

Biodiversity: Concepts, Patterns, and Measurement

Robert K. Colwell

OUTLINE

1. What is biodiversity?
2. Relative abundance: Common species and rare ones
3. Measuring and estimating species richness
4. Species diversity indices
5. The spatial organization of biodiversity
6. Estimating β and γ diversity from samples
7. Species–area relations

Life on Earth is diverse at many levels, beginning with genes and extending to the wealth and complexity of species, life forms, and functional roles, organized in spatial patterns from biological communities to ecosystems, regions, and beyond. The study of biodiversity encompasses the discovery, description, and analysis of the elements that underlie these patterns as well as the patterns themselves. The challenge of quantifying patterns of diversity at the species level, even when the organisms are known to science, is complicated by the problem of detecting rare species and the underlying complexity of the environmental template.

GLOSSARY

α, β, and γ diversity. The species diversity (or richness) of a local community or habitat (α), the difference in diversity associated with differences in habitat or spatial scale (β), and the total diversity of a region or other spatial unit (γ)

biodiversity. The variety of life, at all levels of organization, classified both by evolutionary (phylogenetic) and ecological (functional) criteria

diversity index. A mathematical expression that combines species richness and evenness as a measure of diversity

evenness. A measure of the homogeneity of abundances in a sample or a community

functional diversity. The variety and number of species that fulfill different functional roles in a community or ecosystem

rarefaction curve. The statistical expectation of the number of species in a survey or collection as a function of the accumulated number of individuals or samples, based on resampling from an observed sample set

relative abundance. The quantitative pattern of rarity and commonness among species in a sample or a community

richness estimator. A statistical estimate of the true species richness of a community or larger sampling universe, including unobserved species, based on sample data

species accumulation curve. The observed number of species in a survey or collection as a function of the accumulated number of individuals or samples

species–area relation. The generally decelerating but ever-increasing number of species as sampling area increases

species richness. The number of species in a community, in a landscape or marinescape, or in a region

1. WHAT IS BIODIVERSITY?

Although E. O. Wilson first used the term *biodiversity* in the literature in 1988, the concept of biological diversity from which it arose had been developing since the nineteenth century and continues to be widely used. Biodiversity encompasses the variety of life, at all levels of organization, classified both by evolutionary (phylogenetic) and ecological (functional) criteria. At the level of biological populations, genetic variation among individual organisms and among lineages contributes to biodiversity as both the signature of evolutionary and ecological history and the basis of future adaptive evolution. Species that lack substantial genetic variation are thought to be more vulnerable to extinction from natural or human-caused changes in their environment.

It is at the species level that the term *biodiversity* is most often applied by ecologists and conservation biologists, although higher levels of classification (genera, families, orders) or patterns of evolutionary diversification are sometimes also considered, especially in paleontology. Species richness is the number of species of a particular taxon (e.g., birds or grasses) or life form (e.g., trees or plankton) that characterize a particular biological community, habitat, or ecosystem type. When data are not available at the community, habitat, or ecosystem level, political units (counties, states or provinces, countries) are often used as the basis of statements about species richness.

Within biological communities and ecosystems, functional diversity refers to the variety and number of species that fulfill different functional roles. A *food web* and some measure of its complexity and connectivity is one way to depict the functional diversity of a community. Another is the classification and enumeration of species representing different *functional groups*, such as primary producers, herbivores, and carnivores. Within forest communities, for example, plant functional groups that are often distinguished include fast-growing pioneer species that quickly colonize disturbed habitats, slower-growing species that characterize mature forests, and plants that fill special functional roles, such as those that fix atmospheric nitrogen. A marine biologist working on soft-bottom communities might categorize benthic organisms by the physical effect they have on the substrate as well as by source of nutrients. In microbial communities, microbial taxa that depend on and transform different chemical substrates represent distinct functional groups.

At the level of landscapes, marinescapes, or ecosystems, biodiversity is conceived on a landscape or larger scale, often in terms of the number, relative frequency, and spatial arrangement of distinguishable ecosystem types, or *ecoregions*.

2. RELATIVE ABUNDANCE: COMMON SPECIES AND RARE ONES

The species that characterize any natural community differ in relative abundance, usually with a few species quite common and most species much less so. Another way of looking at it is that most individuals belong to the few common species in a typical community. For example, in a study of the soil "seed bank" in a Costa Rican rainforest, by B. J. Butler and R. L. Chazdon, the 952 seedlings that germinated from 121 soil samples included 34 species. The most common single species was represented by 209 seedlings, and the next most common had 109. In contrast, the least common 15 species each had 10 or fewer seedlings.

One way to plot such species abundance data (an approach originated by R. H. Whittaker) is a *rank-abundance curve*, in which each species is represented by a vertical bar proportional to its abundance. Figure 1 shows such a plot for the seed bank data. Notice the long "tail" of rarer species. A community with such striking disparities in abundance among species is said to have low evenness. A rank-abundance plot for a hypothetical community with perfect evenness would be flat instead of declining, indicating that every species had the same abundance.

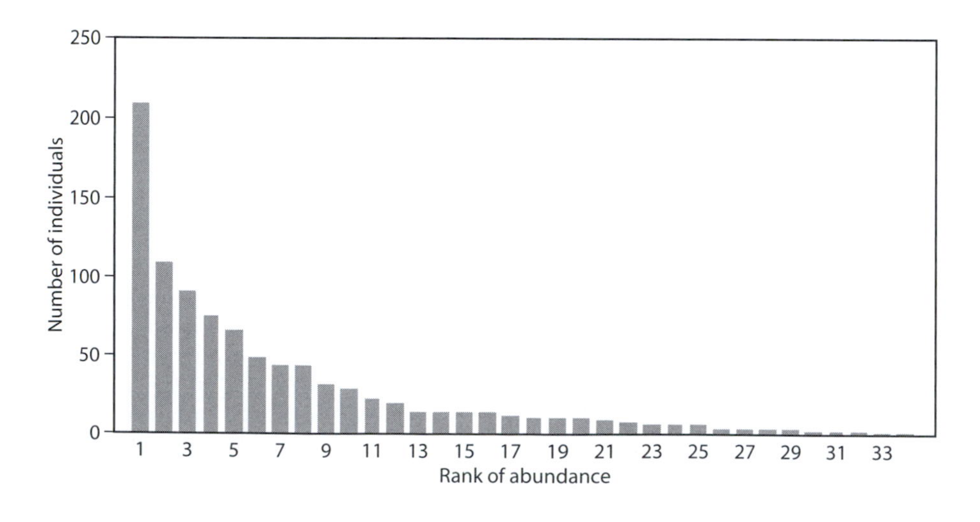

Figure 1. A rank-abundance curve.

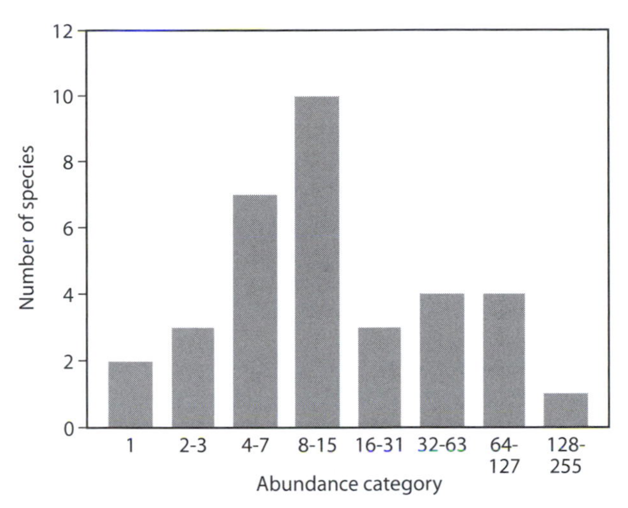

Figure 2. A log abundance plot.

Another way to plot the same species abundance data is to count up the number of species in each abundance category, starting with the rarest species, and plot these frequencies against abundance categories, as in figure 2. It is customary to use abundance categories in powers of two, which gives a *log abundance plot* (originated by F. W. Preston). When relative abundance distributions approximate a *normal* (bell-shaped) curve in a log abundance plot (the seed bank data in figure 2 come close), the statistical distribution is called *lognormal*. Lognormal distributions of relative abundance are common for large, well-inventoried natural communities. Many other statistical distributions have been used to describe relative abundance distributions, including the *log-series* distribution, which is described later in the context of diversity indices.

Conservation biologists are concerned with relative abundance because rare species are more vulnerable to extinction. Some species that are rare in one community are common in another (e.g., gulls are rare in many inland areas, but common along coasts), but some species are scarce everywhere they occur (e.g., most large raptors). In a classic paper, D. Rabinowitz classified species by three factors: (1) size of geographic range (not localized versus localized); (2) habitat specificity (not habitat specific versus habitat specific); and (3) local population density (not sparse versus sparse). She pointed out that there are seven ways to be rare, by this classification, but only one way to be common: not localized, not habitat specific, not sparse. Species that are rare by all three criteria (localized, habitat specific, and sparse), such as the ivory-billed woodpecker in the United States, are the most vulnerable to extinction.

3. MEASURING AND ESTIMATING SPECIES RICHNESS

On first consideration, measuring species diversity might seem an easy matter: just count the number of species present in a habitat or study area. In practice, however, complications soon arise. With the exception of very well-known groups in very well-known places (for which we already have good estimates of total richness anyway), species richness must generally be estimated based on samples. First of all, even for groups as well known as birds or flowering plants, not all species that are actually present are equally easy to detect. Although size, coloration, and—for animals—behavior can affect the detectability of individuals, relative abundance is the most important influence on the effort required to record a species. As every beginning stamp or coin collector soon discovers, the common kinds of coins or stamps are usually the first to be found. As the collection grows, the rate of discovery of kinds new to the collection declines steadily, as rarer and rarer kinds remain to be found.

For species richness, this process can be depicted as a *species accumulation curve*, sometimes called a *collector's curve*. The jagged line in figure 3 shows a species accumulation curve for the seed bank data of figure 1, as the 121 soil samples were added one at a time to the total. Because the order in which the soil samples were added to the collection was arbitrary, a smoothed version of such a curve, called a rarefaction curve, makes more sense. Conceptually, a rarefaction curve can be produced by drawing *1, 2, 3,...N* samples (or individuals) at a time (without replacement) from the full set of samples, then plotting the means of many such draws. Fortunately, this is not necessary, as the mathematics of combinations allows rarefaction curves to be computed directly, along with 95% confidence intervals (the dashed lines in figure 3), based on work by C. X. Mao and colleagues. Rarefaction curves are especially useful for comparing species richness among communities that have not been fully inventoried or have been inventoried with unequal effort.

Richness estimation offers an alternative to rarefaction for comparing richness among incompletely inventoried communities. Instead of interpolating "backward" to smaller samples as in rarefaction, richness estimators extrapolate beyond what has been recorded to estimate the unknown asymptote of a species accumulation curve. Simple (regression-based) or sophisticated (mixture model) curve-fitting methods of extrapolation can be used, or nonparametric richness estimators can be computed. The latter depend on the frequencies of the rarest classes of observed species to

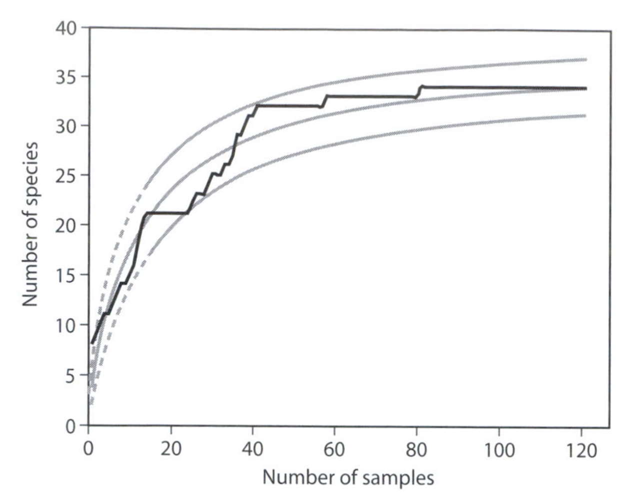

Figure 3. Species accumulation and rarefaction curves.

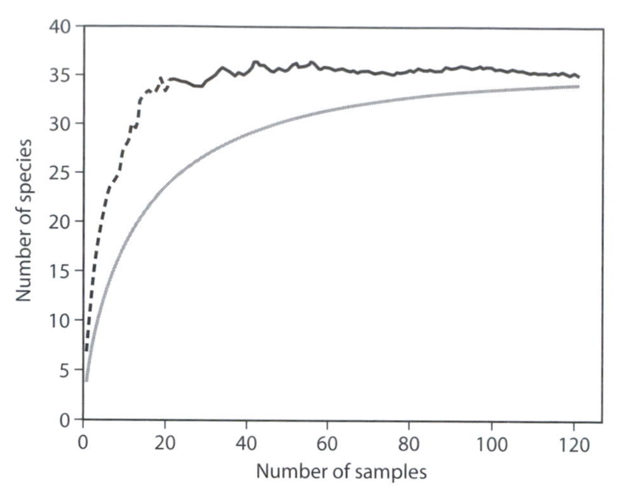

Figure 4. Estimated species richness and rarefaction curves.

estimate the number of species present but not detected by the samples. The simplest nonparametric estimator, *Chao1*, augments the number of species observed (S_{obs}) by a term that depends only on the observed number of *singletons* (a, species each represented by only a single individual) and *doubletons* (b, species each represented by exactly two individuals):

$$S_{est} = S_{obs} + \frac{a^2}{2b}.$$

For the seed bank example of figures 1 and 2, when all samples are considered, 34 species were observed. Of these species, two were singletons, and two were doubletons, so that the estimated true richness is 35 species, confirming the visual evidence from the rarefaction curve that the inventory was virtually complete. The real utility of estimators, however, lies in their potential to approximate asymptotic species richness from much smaller samples. Figure 4 shows the same rarefaction curve (solid line) as in figure 3, with the estimated (asymptotic) species richness (shown by the dashed line) for the *Chao1* estimator, which begins to approximate true richness with as few as 20 samples. (The estimator curve shows the mean of 100 random draws for each number of samples.) It should be noted that richness estimators are not a panacea for problems of undersampling. Hyperdiverse communities with large numbers of very rare species, such as tropical arthropods, have so far resisted efforts to provide reliable nonparametric richness estimators.

4. SPECIES DIVERSITY INDICES

The concept of diversity, including biodiversity itself as well as the narrower concept of species diversity, is a human construct without any unique mathematical meaning. The simplest measure of species diversity is species richness, but a good case can be made for giving some weight to evenness as well. For example, the subjective sense of tree species richness is likely to be greater for a naturalist walking through a forest composed of 10 species of trees, each equally represented, than a forest of 10 species in which one species contributes 91% of the individuals and the others each 1%.

Diversity indices are mathematical functions that combine richness and evenness in a single measure, although usually not explicitly. Although there are many others, the most commonly used diversity indices in ecology are *Shannon diversity, Simpson diversity,* and *Fisher's* α. If species i comprises proportion p_i of the total individuals in a community of S species, the Shannon diversity is

$$H = -\sum_{i=1}^{s} p_i \ln p_i \text{ or, preferably, } e^H$$

and Simpson diversity is

$$D = 1 - \sum_{i=1}^{s} p_i^2 \text{ or, preferably, } D' = \left(\sum_{i=1}^{s} p_i^2\right)^{-1}.$$

Both Shannon and Simpson diversities increase as richness increases, for a given pattern of evenness, and increase as evenness increases, for a given richness, but they do not always rank communities in the same order. Simpson diversity is less sensitive to richness and more sensitive to evenness than Shannon diversity, which, in turn, is more sensitive to evenness than is a simple count of species (richness, S). At the other extreme, a third index in this group, the *Berger-Parker index*, depends exclusively on evenness; it is simply the inverse of the proportion of individuals in the community that belong to the single most common species, $1/p_i(\max)$. Because rare species tend to be missing from smaller samples, the sensitivity of these indices to sampling effort depends strongly on their sensitivity to richness. In practice, which measure of diversity to use depends on what one wishes to focus on (pure richness or a combination of richness and evenness), the relative abundance pattern of the data, comparability to previous studies, and the interpretability of the results. These four diversity measures (richness, the exponential form of Shannon diversity, the reciprocal form of Simpson diversity, and the Berger-Parker index) can be shown to be specific points on a diversity continuum defined by a single equation based on the classical mathematics of Rényi entropy, as first shown in the ecology literature by M. O. Hill in 1972 and periodically rediscovered since then. L. Jost, in 2005, reviewed these relationships and provided compelling arguments for preferring the exponential version of Shannon index and the reciprocal (D') version of the Simpson index.

Fisher's α is mathematically unrelated to the Rényi family of indices. It is derived from the log-series distribution, proposed by R. A. Fisher as a general model for relative abundance:

$$\alpha x, \alpha x^2/2, \alpha x^3/3, \alpha x^4/4, \ldots \alpha x^n/n,$$

where successive terms represent the number of species with 1, 2, 3,...n individuals, and α is treated as an index of species diversity. Estimating α from an empirical relative abundance distribution, however, depends only on S (the total number of species) and N (the total number individuals) but nevertheless requires substantial computation because iterative methods must be used. Fisher's α is relatively insensitive to rare species, and the relative abundance distribution need not be distributed as a log-series.

5. THE SPATIAL ORGANIZATION OF BIODIVERSITY

Imagine walking through a forest into a grassland or snorkeling across a coral reef beyond the reef edge toward the open sea. The testimony of our own eyes confirms that the biosphere is not organized as a set of smooth continua in space but rather as a complex "biotic mosaic" of variably discontinuous assemblages of species. On land, the discontinuities are driven in the shorter term by topography, soils, hydrology, recent disturbance history, dispersal limitation, species interactions, and human land use patterns, and in the longer term and at greater spatial scales by climate and Earth history. The same or analogous factors structure biodiversity in the sea.

If you were to keep track of the plant or bird species encountered, in the form of a species accumulation curve, during a long walk in a forest followed by a long walk in an adjacent grassland, the curve would first rise quickly, as the common forest species were recorded, leveling off (if the walk is long enough) as the rarest forest species are finally included. The number of species accumulated at that point (or a species diversity index computed for the accumulated data) is called the α diversity (or local diversity) for a habitat or community, a concept originated by R. H. Whittaker. (Note that α diversity has nothing to do with Fisher's α, in terms of the names, although the latter may be used as one measure of the former.) As you leave the forest and enter the grassland, the curve will rise steeply again, as common grassland species are added to the list. Once rarer grassland species are finally included, the curve begins to level off at a new plateau. The increment in total species (or the change in a diversity index) caused by the change in habitat is one measure of β diversity, in Whitaker's terminology (sometimes called *differentiation diversity*), although there are many ways to quantify β diversity and little agreement about which is best. The total richness or diversity for both habitats combined (the second plateau in the species accumulation curve) is the γ diversity (regional diversity) for this hypothetical forest–grassland landscape.

The forest-to-grassland example presents a classic illustration of β diversity, as originally conceived by Whittaker, but the concept has been generalized to include spatial differentiation of biotas within large expanses of continuous, environmentally undifferentiated habitat as well as between isolated patches of similar habitat. Within expanses of homogeneous habitat, β diversity is usually considered to be the result of *dispersal limitation*—the failure of propagules (fruits, seeds, juveniles, dispersive larval stages, migrants, etc.) to mix homogeneously over the habitat—but in practice, it is often hard to rule out subtle differences in environment as a cause of biotic differentiation.

6. ESTIMATING β AND γ DIVERSITY FROM SAMPLES

Estimating β or γ diversity for a region or landscape, from samples, is a daunting prospect for any but the best-known groups of organisms. Over larger spatial or climatic scales, the "patches" of the mosaic can be better viewed as ordered along gradients, in either physical or multivariate environmental space. Unfortunately, the geometry of the biotic mosaic is remarkably idiosyncratic (although it may be properly *fractal* for some organisms at some scales), which means that designing a scheme for estimating richness at large spatial scales is likely to require many ad hoc decisions—it is more like designing trousers for an elephant than finding yourself a hat that fits.

A common approach to coping with idiosyncratic biotic patterns is to take advantage of biotic discontinuities to define "patch types" in the mosaic for sampling purposes. For example, the vegetation of treefalls in a forest might be distinguished from the riparian (streamside) vegetation and from the mature forest matrix. Or the fish fauna of isolated patch reefs might be distinguished from the fish fauna of fringing reefs. An alternative is to select sampling sites along explicit gradients, such as elevational transects on land or depth and substrate gradients in the sea. Both strategies represent forms of stratified sampling in which the strata are the patch types or gradient sites, and multiple samples within them are treated as approximate replicates, meaning, in practice, that samples within patch types or gradient sites are expected to be more similar than samples from different types or sites.

Any particular definition of patch types and the scale that underlies them is inevitably somewhat arbitrary. A seemingly less arbitrary alternative would be spatially random sampling over the entire region of interest, analyzed using a multivariate approach to assess the relationship of richness and species composition to underlying environmental and historical factors. But, given limited resources (are they ever otherwise?), random sampling over heterogeneous domains is often highly inefficient because of the uneven relative abundance of patch types: the biota of common patch types are oversampled compared to the biota of rarer patch types, which may even be missed entirely. If one accepts a within- and between-patch-type design framework, the definition of patch types (or sample spacing on gradients) is best made at the design phase based on expert advice and whatever prior data exist, with the possibility of later iterative adjustment.

Although comparisons of α diversity among patch types by rarefaction are interesting in their own right, they fail to provide the information needed to estimate γ diversity because some species are likely to be shared among patch types and some species may be missed by the sampling in all patch types. If we had full knowledge of the biota (complete species lists) for all patch types within a region, it would be simple to determine the total biota for two, three,...all types combined, computing some measure of (average or pair-specific) β richness (species turnover) along the way. For sampling data, the problem is much more difficult. Undetected species within patch types are not only undetected, they are unidentified, so that that we do not know whether the same or different species remain undetected in different patch types.

Nonetheless, it is possible in principle to estimate lower and upper bounds for γ (regional) richness. The union of detected species lists for all patch types, pooled, provides a lower-bound estimate of total domain richness, on the assumption that every species undetected in one patch type is detected in at least one other patch type. The sum of total richness estimates over all patch types (including undetected species from each patch type, using nonparametric estimators or extrapolation techniques), adjusted for the number of observed shared species, is an approximate upper-bound estimate of total regional richness, assuming that undetected species included in the estimates are entirely different for each patch type and were detected in none.

The truth inevitably lies between these bounds, for data from nature. To estimate the true regional richness, we need information about the true pattern of shared species among patch types. Statistical tools for estimating the true number of species shared by two sample sets, including species undetected in one or both sets, are scarce, and this is an area in which much more work is needed. Many studies have attempted to address the problem of estimating β diversity, or pooling samples (between patch types or random samples) by using similarity indices, such as the Sørensen or Jaccard indices. Unfortunately, the number of observed, shared species is almost always an underestimate of the true number of shared species because of the undersampling of rare species. This means that species lists based on samples generally appear proportionally more distinct than they ought to be, similarity indices are routinely biased downward, and slope estimates for the decline in similarity with distance ("distance decay of similarity") are likely to be overestimated. Recently, A. Chao and others have developed estimation-based similarity indices that greatly reduce undersampling bias and promise to help correct this longstanding dilemma. These indices are based on the probability that two randomly chosen individuals, one from each of two samples, both belong to species shared by both samples (but not necessarily to the same shared species). The estimators for these indices take into account the contribution to the true value of this probability made by species actually present at both sites but not detected in one or both samples.

7. SPECIES–AREA RELATIONS

Ecologists and biogeographers have long documented a striking regularity in the pattern of increase in the species count as larger and larger geographic areas are considered. When the number of species or its logarithm (depending on the case) is plotted against the logarithm of area, an approximately linear relationship is revealed. With either plot (a log-log power curve or a semilog exponential curve), the pattern on arithmetic axes is a decelerating but ever-increasing number of species as area increases. This pattern, known as the species–area relation (SAR), has been called one of the few universal patterns in ecology, but its causes are not simple.

There are many variants on SARs, but the primary dichotomy separates plots based on nested sampling schemes from plots in which the areas of increasing size are distinct places, such as islands in lakes or seas, habitat islands on land, or simply political units (states, countries) of different areas. There are two important causes for the increase in species count with increasing area. The first cause is undersampling. Especially in the case of nested sampling schemes, in which smaller areas lie within larger ones, the smaller units may be too small or too poorly sampled to reveal all species characteristic of the habitat(s) they represent. In this case, the supposed SAR for the smaller areas is better described as a species accumulation curve or rarefaction curve. B. D. Coleman and colleagues pointed out that, even for a completely homogeneous species pool, larger areas will have more species because they contain more individuals; the model they proposed is virtually indistinguishable from a rarefaction curve.

The second cause of increasing species count with area is β diversity, in all its varieties. (1) Within large expanses of homogeneous habitat, species composition may vary spatially simply because of dispersal limitation, so that larger areas contain more species. (2) Larger areas are more likely to include a greater number of habitat types or ecoregions, each with its own distinct or partially distinct biota. (3) For very large areas, on continental scales, ecologically similar biotas may have very different evolutionary histories. For example, the lizard fauna of coastal Chile and coastal California share many ecological similarities but have no species (or even genera) in common. Such cases could be viewed as an extreme form of dispersal limitation, as we discover to our dismay when alien species from similar biomes on other continents become local invasives (e.g., California poppy, *Eschscholzia californica*, in Chile, and the Chilean ice plant, *Carpobrotus chilensis*, in California).

FURTHER READING

Chao, A. 2004. Species richness estimation. In N. Balakrishnan, C. B. Read, and B. Vidakovic, eds., Encyclopedia of Statistical Sciences. New York: Wiley. *A comprehensive review of the statistical methods for estimating richness.*

Magurran, A. E. 2004. Measuring Biological Diversity. Oxford: Blackwell Publishing. *The standard reference for conceptual and quantitative aspects of diversity measurement.*

Rosenzweig, M. 1995. Species Diversity in Space and Time. New York: Cambridge University Press. *An overview and analysis of the geography of species richness, especially species–area relations.*

Soulé, M. E., ed. 1986. Conservation Biology: The Science of Scarcity and Diversity. Sunderland, MA.: Sinauer Associates. *A classic collection of papers focusing on the conservation of biodiversity.*

Wilson, E. O., and F. M. Peter, eds. 1988. Biodiversity. Washington, DC: National Academy Press. *An important collection of papers that launched public awareness of biodiversity and its importance.*

III.2

Competition, Neutrality, and Community Organization

Jérôme Chave

OUTLINE

Competition has long been thought to play a foremost role in the organization of ecological communities, and this has been a core concept in the building of niche theory. However, many observed patterns in nature are difficult to reconcile with the predictions of niche theory, and they reflect the historical nature of community assembly. The neutral theory of biodiversity has recently been developed to provide an alternative interpretation of patterns in community organization in the absence of competitive difference among coexisting species.

GLOSSARY

Hutchinsonian ratio. Body size ratio of the larger species over the smaller species in a pair of species; niche theory predicts that co-occurring species should have larger body size ratio than expected by chance

neutrality. Assumption of equivalence in individuals' prospects of reproduction or of death, irrespective of the species they belong to

phylogenetic overdispersion and clustering. The tendency of species to be on average more or less (respectively) evolutionarily related in a sample than in the larger species pool

species abundance distribution $\Phi(n)$. Number of species with exactly n individuals in a sample

1. INTRODUCTION

Ecological communities are complex assemblages of organisms shaped by environmental constraints and interacting through a variety of ecological processes but also reflecting historical contingencies. Understanding these processes has been a central goal of animal and plant ecology for almost a century. Early on, researchers put forward the idea that communities were tightly organized associations of species, with sharp boundaries, and that they were amenable to classification, just as species can be classified taxonomically. This idea may be related to the famous work of the Russian experimentalist Georgyi Gause. He and others developed research projects on simple species assemblages easily amenable to experiments, such as yeast and paramecia (Gause, 1934). Because these systems included only a few species, typically two, and controlled environmental conditions, it was possible to find the theoretical conditions under which species may coexist stably in association. Nicely, these experiments could be reframed into the mathematical theory developed by the Italian mathematician Vito Volterra (1931). The major finding of both theory and experiments was that, of two species with identical ecological requirements and competing for limited resources in a stable environment, one will eventually exclude the other. This result later developed into a fundamental principle, called the *principle of competitive exclusion.*

The view that ecological communities should be the result of tight species association did not remain uncriticized over these years. In 1926, the plant ecologist and taxonomist Henry Gleason, for instance, suggested that species do not, as a rule, live their lives in tight associations, a view that was echoed in 1935 by one of the most prominent British ecologists of the first half of the twentieth century, Sir Arthur Tansley. Some authors even suggested the extreme view that species interactions play no role in community organization (Andrewartha and Birch, 1954). This longstanding debate in ecology was admirably synthesized by Robert MacArthur (1972). Although he acknowledged the role of history in shaping species assemblages, he noted "unravelling the history of a phenomenon has always appealed to some people and describing the machinery of the phenomenon to others." He further suggested that the ecologist tends to be "machinery oriented," whereas the biogeographer tends to be "history oriented." To MacArthur, useful patterns of species diversity are shaped by repeatable phenomena, not by chance events.

An important issue here is: How do experimental findings generalize to natural communities and to more than a couple of species? This question was most clearly addressed in 1959 by the American limnologist G. Evelyn Hutchinson, one of the most influential ecologists of the twentieth century. Central to the understanding of species coexistence is the concept of the ecological niche of a species, the biological equipment of a species in relation to competition and to environmental conditions. For animals, niches may be defined in terms of the food of a species, as developed early on by Charles Elton. For plants, an operational definition of the niche is more problematical, even though much research has been devoted to quantifying plant life-history strategies. Although the concept of the niche was developed historically earlier, it was Hutchinson and his students, including MacArthur, who proposed an operational definition of this concept to most ecologists (for a historical overview, see Sharon Kingsland, 1995).

Following Gause's principle, one would expect that competition plays a role in species coexistence if *interspecific* competition is more intense than *intraspecific* competition. This implies that one could assess the role and nature of competition experimentally. Studies on this topic are almost innumerable. In 1983, Thomas Schoener and Joseph Connell both reviewed this huge literature (over 200 published studies at that time and many more since then), and they found that well over half of the published studies convincingly demonstrated some effect of competitive exclusion. However, they also found that studies with more species tended to have fewer species evidencing competition. This suggests that interspecific competition may be more diffuse in species-rich communities. Also important is the fact that although numerous researchers have claimed to quantify competition, as had been reviewed by Connell and Schoener, few have actually addressed the importance of competition in shaping ecological communities. Deborah Goldberg and Andrew Barton (1992) found that, in the plant community, no more than 89 studies had addressed patterns of competition, 63% of which actually focused on a single species in a background of competing species. Fewer than half a dozen experiments had suitably addressed the consequences of plant competition on community-wide patterns by the early 1990s.

After almost a half-century of research on the role of competition on community organization, one would think that a clear theory had emerged and that quantitative and predictive theories are by now available. Not quite so. Predictions of niche theory have proven difficult to test. As a result, the original quest for a universal role of competition in community assembly rules has turned into a more ramified research program, including a search for the relative importance of processes aside from competition.

Dispersal limitation has long been known to limit the distribution of species and hence to contribute to the makeup of communities. Just by chance, a species may be absent from a site where it would have thrived. This simple remark has resulted in an elegant model by John G. Skellam in a famous paper (1951). Likewise, demographic processes may cause a rare species to be present in a community by chance rather than because it is a superior competitor (see figure 1). These two processes have long been considered a nuisance by ecologists but are in fact essential to explain many patterns of community organization. In 1987, Robert Ricklefs called this bias against nondeterministic interpretations of community assembly the "eclipse of history" in ecology.

This view has recently been reinforced by the neutral theory of biodiversity. The concept of neutrality is over three decades old in ecology: it was developed in 1976 by Hal Caswell, who used it as a null model for testing the importance of competition. Caswell constructed his model by assuming that co-occurring species were not interacting. This approach remained largely unnoticed, with the notable exception of work by Stephen Hubbell (1979, 1986, 1997), who tested theories of tree species coexistence in tropical rain forests. Hubbell's thinking eventually matured into an important book (Hubbell, 2001), which offered a novel interpretation of the neutral theory. In his book, Hubbell assumes that individuals in a community all have the same prospects of birth and death and that species interaction is fully symmetrical. In this sense, Hubbell's theory is *neutral*. As Hubbell remarked, many people are confused by the word "neutral," which, to many, is equivalent to "nothing is going on." However, the neutral assumption does not imply that individuals or species do not compete with each other, only that competition should be

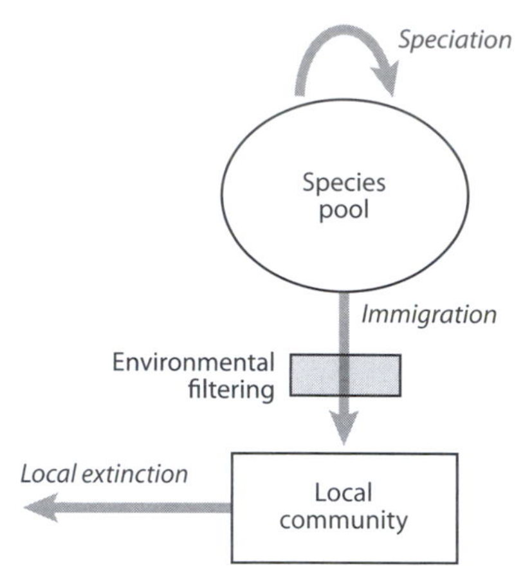

Figure 1. Local community assembly from a regional species pool. At the regional scale, the evolutionary processes dominate, in particular speciation. A local community is a filtered sample of this regional species pool because not all species can reach the site (dispersal limitation) and because not all species are able to settle in the local environment (environmental filtering). The immigration of new species is balanced by the local extinction of species through competitive displacement, demographic drift, or catastrophic events.

symmetrical: an individual of species A has the same influence on an individual of species B as an individual of species B has on an individual of species A. Most importantly, Hubbell constructed a theory in which a local community is seen as a dispersal-limited sample of a regional species pool. In the regional species pool, species arise through speciation, a process that is summarized by a single parameter θ, equal to the product of the per capita speciation rate and the number of individuals in the region. At the local community level, the intensity of dispersal limitation is quantified by the immigration rate m, the probability that a new individual is an immigrant.

Beyond Caswell's seminal work, Hubbell's theory finds its mathematical inspiration in the neutral theory of molecular evolution, proposed in 1964 by Motoo Kimura, itself rooted in population genetics theory of the first half of the twentieth century. Kimura, impressed by the variability of patterns in the genetic data that were just becoming available at that time, suggested that most of these patterns could be caused by random changes in the course of molecular evolution rather than by deterministic processes. That most of the genetic variability was a result of chance rather than selection was utterly counterintuitive, and Kimura's ideas were harshly criticized by the selectionist school. A direct consequence of this theory, however, was that if molecular evolution is neutral, then the cumulative amount of molecular divergence between groups could be a useful measure of the time since the split occurred.

The molecular clock concept, although much debated, has become a standard tool in population genetics and in phylogenetics. Moreover, powerful statistical methods have been developed from the neutral theory, and these were used to test a selective departure from neutrality (see Warren Ewens' book on mathematical population genetics, republished in 2004). As in molecular evolution, Hubbell's theory was bluntly criticized, which led to a series of tests of the hypotheses or the predictions of this theory. Brian McGill et al. (2006) offered an interesting overview of these critiques and concluded that there is an "overwhelming weight of evidence against neutral theory." Despite this view, these authors and many others still believe that this approach provides an interesting framework for testing ecological hypotheses.

The next section summarizes evidence for the role of competitive and neutral processes in the organization of communities.

2. SPECIES ABUNDANCE

One of the most classic ecological patterns in community ecology is the species abundance distribution, the number of species of a given abundance in a biodiversity survey. Let us call $\Phi(n)$ the number of species with exactly n individuals in such a survey. With this definition, the total number of individuals in the survey is $N = \sum_n n\Phi(n)$, and the total number of species is $S = \sum_n \Phi(n)$. In a classic work, the geneticist Ronald Fisher (1943) provided a simple model for the species abundance distribution. The Fisher species abundance distribution was predicted to be a so-called log-series distribution, $\Phi(n) = \alpha x^n/n$, where α and x are simple functions of S and N defined above (see for instance Evelyn C. Pielou's 1975 book for a derivation of this result). Interestingly, the following relation: $S = \alpha \ln(1 + N/\alpha)$ holds in Fisher's model. This suggests that α, also known as Fisher's α in the ecological literature, is an unbiased index of diversity, if the model's assumptions are met. Fisher used his model to explain data on Lepidoptera accumulated over years by two colleagues of his, Steven Corbet and C. B. Williams. The data suggested that most of the species in a survey should be represented by a few individuals. Later, Preston (1948) revisited this question, suggesting that empirical samples of a community would be better explained by the so-called log-normal distribution $\Phi(n) \sim A \exp(-B[\ln(n/n_0)]^2)$, truncated at $n=1$ because only species with at least one individual can be observed in a sample (figure 2). Hence, the left part of the species abundance distribution is "veiled" because

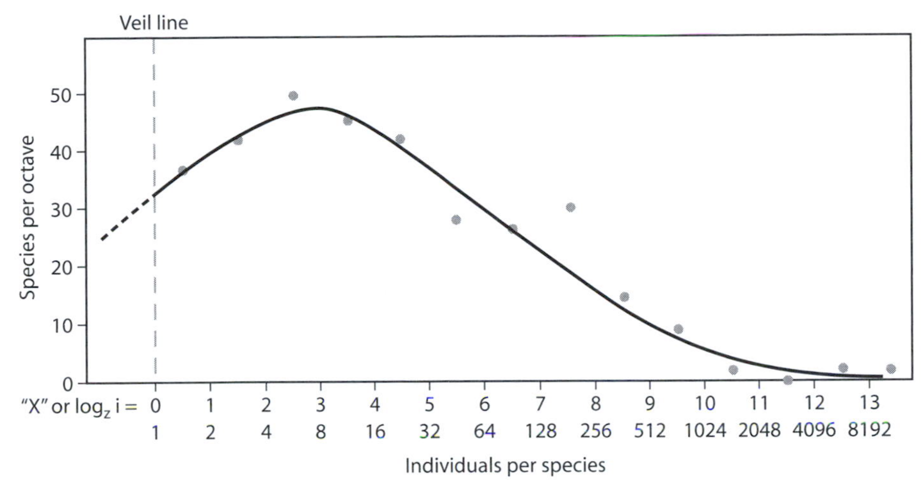

Figure 2. Species abundance distribution plotted in logarithmic classes of abundance. The data correspond to a moth survey performed by C. O. Dirks in 1931–1934 at Orono, Maine (56,131 collections for 344 species). (Reprinted with permission from Preston, F. W. 1948. The commonness, and rarity, of species. Ecology 29: 254–283)

of a limited sampling, a phenomenon sometimes referred to as Preston's "veil line."

How can these patterns of species abundance be explained by the processes of community organization? In 1957, Robert MacArthur proposed a connection between niche theory and the species abundance distribution. He reasoned that if the total niche space in a community could be graphically represented by a segment of unit length, then the niche of each species would be represented by a fraction of this segment, further assuming nonoverlapping niches. This is sometimes called a "niche preemption model" in the literature. He then suggested that the niche segment was partitioned at random into S subsegments ("broken stick"), each representing the niche of a species. Finally, assuming that the size of the niche space for a species was proportional to the abundance of this species, he was able to derive a mathematical formula for the species abundance distribution under this model. Although MacArthur's "broken stick" model generally failed to reproduce empirical species abundance distributions, as he himself acknowledged, it eventually became a great source of inspiration for his colleagues, as it was a notable attempt to relate empirical patterns to the processes of niche partitioning at play in the organization of ecological communities.

It was subsequently discovered by Joel E. Cohen (1968) that the same result as MacArthur's could be obtained simply by drawing S random variables from an exponential distribution and normalizing these numbers such that their sum is equal to 1. Hence, exactly the same mathematical form of the species abundance distribution can be obtained based on a niche-based argument, or based on a purely probabilistic argument. This line of reasoning also offered a useful connection with Fisher's species abundance model. Let us assume that the species abundances are represented by S random variables x with the same two-parameter distribution,

$$P(x) = \frac{r^k}{\Gamma(k)} x^{k-1} \exp(-rx).$$

This is called the Γ distribution in probability theory. Then Cohen's exponential model holds if $k=1$, and Fisher's model is recovered in the limit when k tends to zero. More biologically relevant models were proposed later on, notably in 1996 by Steinar Engen and Russell Lande (see Lande et al., 2003) and by Hubbell (2001).

There is no shortage of competing models predicting species abundance distributions. However, a proper statistical theory able to assess how well a model fits data is more difficult to obtain. Because the species abundance data are far from normal, a more general method is based on likelihood theory. The *likelihood function*—a term coined by Fisher—quantifies how likely a model is for a set of model parameters and given the sample at hand. The model parameters may be estimated by the value that maximizes the likelihood function. It is a remarkable fact that for Hubbell's dispersal-limited neutral theory, a likelihood function can be computed exactly based on the species abundance data of a local sample, as was shown by Rampal Etienne and Han Olff, and the parameters θ and m can therefore be estimated directly (Alonso et al., 2006).

They constructed an alternative niche preemption model, for which they also constructed a likelihood theory. Using tropical forest tree species abundance data, they showed that the neutral model was more likely than the niche preemption model. Although much has been made of the fact that nonneutral models provide a better fit to empirical species abundance distributions, very few have actually based their results on Etienne and Olff's exact statistical theory (McGill et al., 2006). For instance, Igor Volkov and his collaborators (2005) suggested that a community model including density-dependent regulation always outperformed the neutral model's fits of empirical tropical tree species abundance distributions. With David Alonso and Rampal Etienne (2006), we reassessed this result based on an exact likelihood-based comparison, and we showed that their conclusion was inaccurate. These studies, as well as others, suggest either that nonrandom ecological processes are indeed much less important for community organization than chance events or that the species abundance distribution is an uninformative pattern to detect nonneutrality in ecological communities.

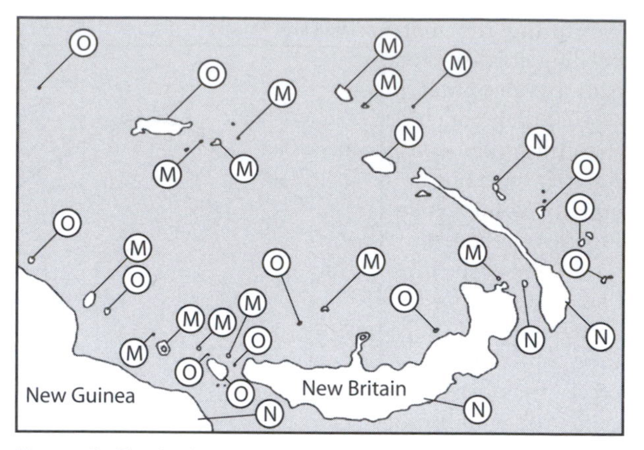

Figure 3. Checkerboard pattern of distribution for two related small cuckoo-doves in the Bismark archipelago. Symbol M stands for islands where only *Macropygia mackinlayi* occurs, N for islands where only *Macropygia nigrirostris* occurs, and O for islands where neither occurs. (Redrawn from Diamond, J. M. 1975. Assembly of species communities. In M. L. Cody and J. M. Diamond, eds. Ecology and Evolution of Communities. Cambridge, MA: Harvard University Press, 342–344.)

3. SPECIES CO-OCCURRENCE

Species abundance distributions are useful patterns in a single ecological community. However, a most powerful test of the role of competition in community organization makes use of the variation in the species composition across local communities. Charles Elton, in his famous paper "Competition and the Structure of Ecological Communities" published in 1946, sought to provide a simple pattern to assess whether island bird community assemblages were organized by competition. He reasoned that if competition is prevalent in islands, then closely allied species that share many niche traits should not be observed together. He used this remark to devise an original test of the competition theory: if competition is at work, then congeneric species should more rarely co-occur than expected by chance because they usually tend to share the same niche. Hence, the species to genus ratio (or S/G ratio) should decrease as competition intensity increases. He measured many S/G ratios in islands of the United Kingdom archipelago and showed that they were indeed smaller than observed in the corresponding source pools of species (the main islands). One problem with Elton's test, however, was that many genera typically tend to have a single species, although a few have many species. In other words, the S/G statistic is difficult to measure with a great reliability. Daniel Simberloff (1970) pointed out another deficiency in Elton's reasoning. He showed that if larger areas are sampled, the number of both species and genera are expected to increase, but the S/G ratio should also increase. The larger S/G ratio in mainland areas compared with islands was found by Simberloff to be almost solely a result of this bias.

The study of species co-occurrence on archipelagoes was to become a very active field of research. In 1975, Jared Diamond published a large analysis of bird communities in archipelagoes east of the island of New Guinea (figure 3). Of special interest to him was to test the existence of assembly rules for these bird communities. One of these assembly rules stipulated that "some pairs of species never coexist, either by themselves or as part of a larger combination"; that is, one species excludes competitively the other if they were occurring on the same island. Although Diamond found evidence for this pattern in his data, it generated a critical evaluation by Connor and Simberloff (1979), who pointed out that Diamond had not tested them against a proper null assumption. As in the S/G ratio case, the suitable way of testing that some species combinations never exist in nature is to provide a suitable "space of the possible" resulting solely from random migration events into the island and to compare these possible configurations to the observed one. One way of testing Diamond's rule, Connor and Simberloff proposed, would be to construct a matrix of species co-occurrences, such that an entry i, j is nonzero if species i and j co-occur, and zero otherwise. The next step would be to construct a "randomized" community by permuting the matrix of species co-occurrences according to different protocols, which has been made easy by computer programs. Connor and Simberloff (1979) then concluded, based on the graphic match of the null model compared with observed data, that allopatric speciation and limited dispersal would be equally valid explanations of the observed patterns. However, there are several ways of randomizing the data. One of the most fundamental critiques of Connor and Simberloff's (1979) solution is that, as Jonathan Roughgarden (1983) puts it, "Islands do not reach into urns and draw out their species. There are real processes that bring species to islands." Thus, the choice of a proper null model is a difficult one, and, following Roughgarden's prescription, one should seek null models with a better biological motivation than random assortment models. A historical review of the alternative solutions developed to test Diamond's fifth assembly rule was nicely summarized in 1996 by Nick Gotelli and Gary Graves (see their chapter 7), and a definitive test of the hypothesis was provided by Gotelli and McCabe (2002).

One puzzling problem with niche theory is that it can lead to contradicting predictions about the spatial distribution of species. The classic prediction is that related species should exclude each other competitively, as Elton suggested. A newer prediction, however, is that related species have similar abilities to withstand a given environment and hence are more likely to be found in the same habitat. This latter prediction suggests implicitly the existence of species "associations." Similar species may compete, but they are found in the same environments simply because they have overlapping ecological requirements. Many studies have sought evidence for such habitat associations in plants by relating spatial patterns of species distribution to abiotic environmental factors. Another prediction, derived from the neutral theory, is that changes in species composition across samples should not depend on the environment or on species identities but only on pairwise geographic distance between samples. Richard Condit et al. (2002) tested this hypothesis using networks of tropical forest tree plots in Panama, Ecuador, and Peru. They found that the prediction of the neutral theory that species similarity should decline logarithmically with geographic distance was consistent with field observations. However, working on plants abundant in the understory of tropical forests (ferns and Melastomataceae), Hanna Tuomisto and her colleagues (2003) found that the predictions of the neutral theory were not met. Instead, they suggested that environmental variation best explains variation in diversity in their data. Ordination techniques have also been used to quantify the relative role of geographic distance and environmental variation in explaining patterns of species similarity. Karl Cottenie (2005) published a meta-analysis of distributional data on 158 species, and he found that most of the variation in biodiversity was explained by a combination of both factors rather than a single factor. This body of evidence suggests again that even though the neutral theory cannot alone predict patterns of species co-occurrence, neither can pure niche-based theories.

4. BODY SIZE

Niche theory makes predictions not only for the spatial distribution and co-occurrence of species but also for the appearance and behavior of the species that do co-occur. For instance, in 1945 David Lack reported that beak size for Darwin finches in the Galápagos islands depended on the whether the species were or were not co-occurring with other finch species. This fact had also been remarked upon by G. Evelyn Hutchinson (1959) in two "water boatman" species, aquatic insects in genus *Corixa*, found in sympatry throughout Europe (*C. affinis* and *C. punctata*). Although these two species are difficult to separate on the basis of their morphology or ecology, a striking difference between them is that one species was much smaller than the other. He reasoned that these differences in size were not a fact of chance but that they enable the species to partition the food web structure although their ecological requirements are the same. Because Hutchinson provided as a clear niche-theoretical prediction that the body size of co-occurring species should differ, the ratio of larger body size over smaller body size was later referred to as the "Hutchinsonian ratio."

This study of body size ratios in co-occurring species was to become one of the most classic tests of competition theory in ecological communities. Two remarkable examples are here singled out. Thomas Schoener (1970) in a study of the *Anolis* lizards of the Lesser Antilles archipelago, a classic model in island biogeography, found that of the 27 Lesser Antillean islands, nine had two *Anolis* species, and these varied in size by a factor greater than 1.5. In contrast, in the single-species islands, species tend to be intermediate in size. Abbott et al. (1977), in a monograph on the ecology of Darwin finches, also confirmed this pattern, showing that in islands where two species are found together, their beak size ratios tend to be large. This prediction was also tested in continental environments, for instance by Jim Brown (1973), who explored body size ratios in desert rodents in the United States.

As for previous ecological patterns, the relation between size ratios and competition was critically reassessed by Daniel Simberloff and William Boecklen

(1981). To properly test whether size ratios in co-occurring species may be a signal of competition, they reasoned, a rigorous statistic of size overdispersion should be defined. When they ran this improved statistical test on published studies that had claimed to establish Hutchinsonian ratios in empirical data sets, they found that only a third of these claims were valid, the remaining two-thirds being not statistically different from a random assortment of species. For instance, even though both Schoener and Brown claimed their patterns of species co-occurrence should be explained by competition, Simberloff and Boecklen showed that chance would equally well explain their results.

Where do we stand today on Hutchinson's body size ratios? Simberloff and his collaborators have paved the way for more rigorous tests, and they offered the alternative hypothesis that chance, rather than deterministic factors, might be responsible for at least some of the body size patterns observed in nature. Since then, however, tests of Hutchinson's hypothesis have been mostly carried out by evolutionary ecologists, with hypotheses and methods that were largely inaccessible to community ecology during the 1970s and the 1980s. The next section summarizes these results.

5. EVOLUTION AND COMPETITION

That two different species should have a different size to partition the biotic environment was a remarkable enough hypothesis because it was amenable to testing. W. L. Brown and E. O. Wilson (1956) made an even more surprising remark: in the same species, the size may vary depending on whether or not they co-occur with a related species. As an illustration, they provided the striking example of two Old World bird species (genus *Sitta*). The species overlap only in Iran, and where they co-occur, one species tends to have much smaller beak and body sizes than when it is found alone, whereas the other species tends to have much larger beak and body sizes. This phenomenon was referred to by Brown and Wilson as "character displacement." This topic has attracted a number of ecologists and evolutionary biologists. Tamar Dayan and Daniel Simberloff (2004) recently reviewed published empirical evidence, and they concluded that research over the past two decades has found convincing evidence in favor of character displacement. If one were to extirpate one of the two bird species in Brown and Wilson's example, the other would likely return to its normal size.

In Hutchinson's example, however, body sizes are realized differences, and they are maintained irrespective of whether species co-occur or not. What is the difference? The evolutionary time scales involved in the two processes differ; Hutchinson's body size differences refer to realized evolutionary divergences. On the other hand, Brown and Wilson's intraspecific differences demonstrate spectacularly their potential for evolutionary divergence. In 1983, Ted Case offered a new terminology to classify possible causes of large body size ratios: character *adjustment* is the adaptive change in species caused by character displacement, whereas character *assortment* is the mechanism of ecological competitive exclusion in species that have already evolved divergent characters.

Is there a relation between character adjustment and character assortment, or, in other words, do plastic differences trigger events of speciation? To employ a wording more consistent with evolutionary theory, the issue is to know whether character displacement evolved before or after speciation was completed. If the former, then it is to be expected that character displacement may have played a role in the early stages of population divergence. It has long been thought that adaptive changes alone are in general insufficient to give rise to new species when the types are in sympatry. Instead, the primary cause of speciation is thought to be geographic isolation, followed by character divergence, as illustrated in figure 4. This question has been a central focus of evolutionary biology for several decades, and it remains a fertile point of contact between ecology and evolution, as demonstrated in Dolph Schluter's (2001) review.

In a landmark study of Schoener's *Anolis* lizards, Jonathan Losos (1990) proposed a new look at the causes of morphological divergence in species. Use should be made of the facts that co-occurring species have a shared evolutionary history and that this evolutionary history is reflected in the species phylogeny. If competitive exclusion is the prime mechanism for large body size ratios in two species, then it should be expected that the ancestral species should have an intermediate body size. Losos mapped events of body size changes onto the best phylogenetic hypothesis for genus *Anolis* available at the time, and he found that body size change did account for some of the observed body size ratios in two-species islands, but these were entirely caused by a single evolutionary event. He then suggested that most of the remaining differences in body size should be ascribed to character assortment rather than to character adjustment.

Such phylogeny-based analyses in community ecology have been scarce even after the heated debate over community organization processes in the 1980s. One explanation is simply that reliable phylogenetic

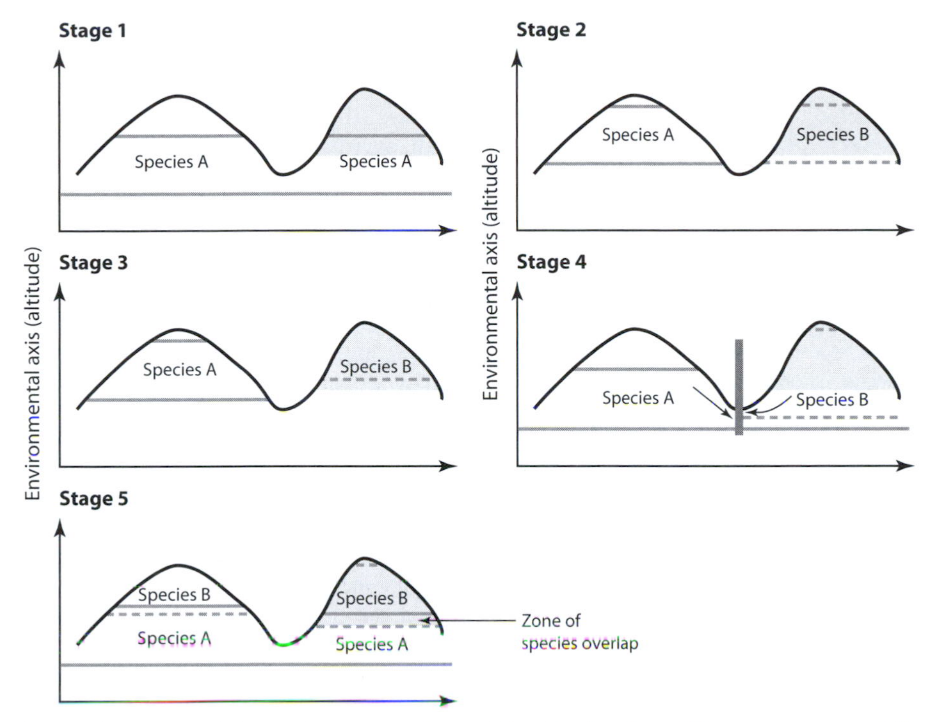

Figure 4. Stages of species formation. In this hypothetical example, species A is initially widespread (stage 1), but it becomes isolated as a result of a change in climatic conditions leading to its isolation in populations in two mountaintops (stage 2). After isolation, the two populations are exposed to different environmental conditions (gray zone on the right mountaintop versus white zone on the left one), leading to allopatric differentiation in the incipient species (stage 3). When the climatic conditions return to normal, the species come into contact again (stage 4), and they compete for space, potentially leading to character displacement and eventually to niche shift (stage 5). (Adapted from MacArthur, 1972; and from Wiens, J. J. 2004. Speciation and ecology revisited: Phylogenetic niche conservatism and the origin of species. Evolution 58: 193–197)

hypotheses for groups of co-occurring taxa are difficult to construct. It is only with the rise of modern molecular phylogenetics and robust phylogeny reconstruction methods, based either on molecular sequences or on supertree methods, that these questions came to the fore. Ancestral state reconstructions and comparative methods became standard in the ecologist's toolbox as a result of the book by Paul Harvey and Mark Pagel (1991). Then, ecologists began to include the historical information contained in phylogenies in tests of community organization. The underlying idea is to factor out lineage effects to assess the role of environment or of competition without potentially confounding historical factors. Losos (1996) and Webb et al. (2002) provided a comprehensive overview of this new line of thinking.

A fruitful use of these ideas consists in using the structure of the tree subtended by a local community and comparing it to that of the whole species pool. In the same way as in Elton's (1946) test, the idea is that if competition is really shaping ecological communities, then species should be more overdispersed phylogenetically than any subtree of the same size drawn at random in the species pool. Suppose that we are working with trees of the Borneo rainforest, as in Cam Webb's (2000) study, and that we have at hand a number of permanent 1-ha plots in which all trees have been censused. Assuming that the species in the same lineage share part of the same evolutionary history, they should have many similar niche features. Hence, they should be more prone to competitive exclusion. This implies that in any local community organized through competition, co-occurring species should be more phylogenetically dispersed than expected by chance in the regional pool's phylogeny (phylogenetic overdispersion, figure 5). If, on the other hand, whole lineages have evolved some sort of habitat specialization, e.g., through adaptive speciation, then we should observe that species tend to be more phylogenetically clustered. This alternative is similar to Ted Case's

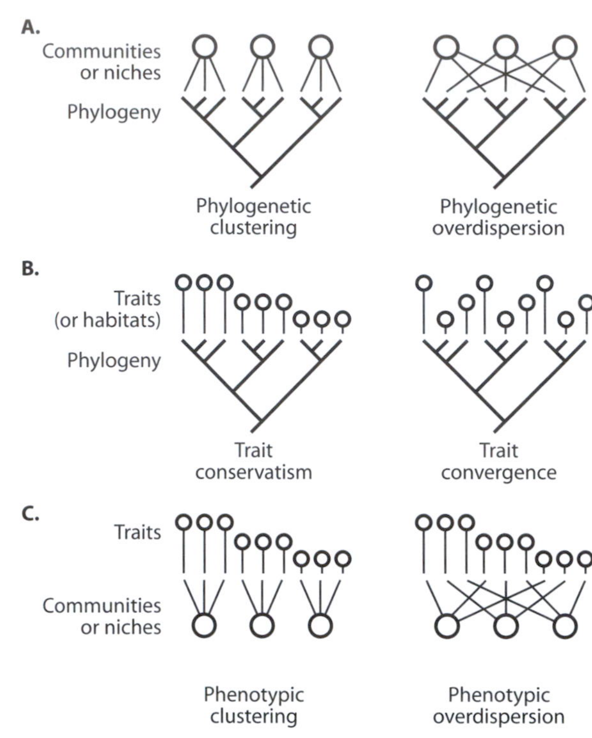

Figure 5. This graphic representation illustrates different possible outcomes of tests based on species distribution data paired to a phylogenetic hypothesis (A), species traits paired to a phylogenetic hypothesis (B), or species traits paired to species distribution (C). The left panels show a case of conservatism in co-occurring species (A), in traits (B), or of phenotypic clustering in communities (or niches). The right panels show the opposite case of overdispersion. The idea that phylogenetically related species have similar niche traits and hence are more likely to compete is represented here by panel B, left. If this hypothesis holds, then phylogenetically related species should not co-occur, which is depicted here by panel A, right. (C) Classic tests of community ecology, such as Elton's island models, that mostly ignore the phylogenetic relatedness among species. (Modified from Cavender-Bares, J., D. D. Ackerly, D. A. Baum, and F. A. Bazzaz. 2004. Phylogenetic overdispersion in Floridian oak communities. American Naturalist 163: 823–843)

concepts of adjustment versus assortment, but for niches rather than for single characters.

It was possible for Webb to rely on a molecular phylogeny of flowering plants put together by the Angiosperm Phylogeny Group 2 years earlier. He then constructed several statistics that measured whether the co-occurring tree species tended to be more clumped in the phylogeny or more overdispersed. He computed these statistics for the plot samples and compared his results with a null model of a community randomly assembled from the species pool. He found that the samples were phylogenetically clustered, suggesting that trees species assemblages are organized through habitat specialization and that interspecific competition plays a minor role during community assembly. This approach has been influential in community ecology over the past few years (see Webb et al., 2002). The neutral theory makes the prediction of neither under- nor overdispersion in the co-occurring species. However, because species assemblages are dispersal limited, their topology differs from random trees.

6. RECONCILING HISTORICAL AND MECHANISTIC INTERPRETATIONS

Returning to MacArthur's original remark that it is difficult to see all at once the "machinery" and the "history" of a phenomenon, ecologists now agree that both viewpoints are valuable, depending on the question being addressed. Quite likely, the recent development of the neutral theory of biodiversity has helped reconcile these viewpoints. In some cases, deterministic interpretations of a phenomenon hold valuable knowledge, and the hope of the ecologist is that, having unveiled this causal link, she will be able to make robust predictions. In other cases, the interpretation of a phenomenon cannot ignore historical contingencies. Indeed, for most complex systems, we know that it is illusory to search for a fully mechanistic interpretation of a phenomenon that will also give rise to robust and long-term predictions. Although this has been known by meteorologists for almost five decades, it has not discouraged them from constructing increasingly sophisticated methods of weather forecasting.

During decades of debate over the processes that control the organization of ecological communities, the prevailing view has matured. The original dogma that competition was the only mechanism able to explain patterns of community organization has drifted to tests including stochastic processes. As a result, recent studies tend to be considerably more mathematically sophisticated than those in the past. This reflects the fact that more ecological data, and also more computer power, are available today. Researchers have also opened the door to new approaches that not only are inspired by evolutionary theory but make full use of this body of knowledge. Evolution questions are historical in nature, and they call for a proper historical framework, for instance, a robust phylogenetic hypothesis. Although it has long been remarked that ecological questions should be properly interpretable in light of evolution, it is only with the advent of more easily accessible methods in molecular biology that this idea has swept through the field of ecology. Meanwhile, classic studies in community ecology also benefited from more interaction with physiologists. They gained from serious attempts to develop synthetic analyses involving field studies and theory as well. On the modeling side, much progress remains ahead of us, as increasingly complex objects (phylogenies, large-scale trait databases, regional species distribution) are being assembled for testing ecological hypotheses.

FURTHER READING

Alonso, David, Rampal S. Etienne, and Alan J. McKane. 2006. The merits of neutral theory. Trends in Ecology and Evolution 21: 351–356. *This article summarizes the mathematical foundations of the neutral theory of biodiversity.*

Dayan, Tamar, and Daniel Simberloff. 2005. Ecological and community-wide character displacement: The next generation. Ecology Letters 8: 875–894. *This article provides a comprehensive overview of 50 years of research on the consequences of competition in closely related species.*

Gause, Georgyi F. 1934. The Struggle for Existence. Baltimore: Williams & Wilkins. Available online at http://www.ggause.com/Contgau.htm. *This is a foundation of model testing and experimental methods in ecology. This book paved the way for much theoretical ecology until today.*

Gotelli, Nicholas J., and Gary R. Graves. 1996. Null Models in Ecology. Washington, DC: Smithsonian Institution Press. *This book reviews many of the important patterns of community ecology as well as tests of ecological theories using these patterns.*

Hubbell, S.P. 2001. The Unified Neutral Theory of Biodiversity and Biogeography. Princeton, NJ: Princeton University Press. *In only a few years, this book has become a classic in ecology. It provides a new perspective on patterns of community organization and offers a theoretical interpretation of these patterns that is an alternative to classic niche theory.*

Hutchinson, G. Evelyn. 1959. Homage to Santa Rosalia or why are there so many kinds of animals? American Naturalist 93: 145–159. *This is a classic in community ecology that formulates an interpretation for body size differences among closely related species.*

Kingsland, Sharon E. 1995. Modeling Nature: Episodes in the History of Population Ecology, 2nd ed. Chicago: University of Chicago Press. *This wonderful book provides a concise history of key concepts in ecology and of the lives of those who forged them.*

Losos, Jonathan A. 1990. A phylogenetic analysis of character displacement in Caribbean Anolis lizards. Evolution 44: 558–569. *This presents the first phylogenetically controlled test of Hutchinson body size ratio. This paper provided a much needed new light on an old heated debate in ecology.*

MacArthur, Robert H. 1972. Geographical Ecology: Patterns in the Distribution of Species. Princeton, NJ: Princeton University Press. *This is one of the tipping points in the history of ecology. It is a deep and easily accessible text for general readers.*

McGill, Brian J., Brian A. Maurer, and Michael D. Weiser. 2006. Empirical evaluation of neutral theory. Ecology 87: 1411–1423. *In this article, the authors provide an overview of several tests of the neutral theory of biodiversity. This is an excellent introduction to the recent literature on this topic, although some of their conclusions are still a matter of debate and should therefore be taken with a grain of salt.*

Schluter, Dolph. 2001. Ecology and the origin of species. Trends in Ecology and Evolution 16: 372–380. *This easily accessible overview of speciation theory summarizes the different modes of speciation and evidence for them. See also Jerry Coyne and Allen E. Orr's book on speciation.*

Webb, Campbell O., David D. Ackerly, Mark A. McPeek, and Michael J. Donoghue. 2002. Phylogenies and community ecology. Annual Review of Ecology and Systematics 33: 475–505. *This is the best overview to date of how the dialogue between phylogeny and ecology may lead to progress in both disciplines.*

III.3

Predation and Community Organization

Robert D. Holt

OUTLINE

Acts of predation are among the most dramatic events one can see in nature, but the impact of predation on ecological communities goes well beyond the effect of direct mortality on the prey species itself. Because species are embedded in complex food webs, predation on one species can lead to chains of indirect interactions affecting many other species. Predation can sometimes enhance diversity, for instance, if it is differentially inflicted on dominant or abundant competitors. This can free up space or resources, thus permitting inferior or scarce competitors that are better able to withstand predation to persist. However, indiscriminate predation can instead shift the relative competitive rankings of species without enhancing coexistence. Prey species that are highly productive can sometimes sustain predators at levels where less productive prey species are vulnerable to exclusion. There are many complexities in predator–prey interactions that have implications for community organization, including behavioral games between predators and prey and interactions among predators themselves, altering their net effects on their prey. A deeper understanding of all these dimensions of predator–prey interactions is essential for developing wiser policies of conservation and resource management in our rapidly changing world.

GLOSSARY

apparent competition. An indirect interaction between prey species where a given prey species experiences more intense predation because of the presence of the alternative prey as a result of changes in either predator abundance or predator behavior.

community. The assemblage of species that are found together at one place at one time that can potentially interact.

community module. A small number of species involved in a clearly defined pattern of interactions, such as two consumers competing for a shared resource, or two prey species interacting indirectly via their impacts on a shared predator.

community organization. A term that broadly encompasses the number of species found in a community, their relative abundances, and their pattern of interconnections via competition, exploitation, and mutualism.

indirect interactions. When there are three or more species, a given pair of species may influence each other via changing the abundance, activity, or traits of other species (one to many).

keystone predator. A predator that strongly interacts with its prey and facilitates their coexistence.

natural enemy. A species that utilizes another species (the "victim") as a resource and harms that other species in so doing. Natural enemies include "true" predators, parasitoids, pathogens, and herbivores.

predator. A natural enemy that kills its victim in order to utilize resources contained in that victim.

switching. A behavioral response by predators to relative prey abundance such that common prey are disproportionately attacked.

trophic cascade. A chain reaction in a community across trophic levels in which predation on one species relaxes consumption by that species of its own resource population.

1. INTRODUCTION

Few things in the world thrill the nature lover as much as the sight of a predator in action or repose—a lion lazing in the sun, a killer whale gamboling in the waves, a diamondback rattler coiled menacingly under a desert shrub, a spider spinning a silken coffin around its quivering moth prey. Predators provide intellectual thrills too, for some of the most dramatic and intellectually rich stories in ecology involve elucidating the impact of predators on communities. Before recounting some of these tales, which I will use to illustrate principles governing how predators influence communities, it is useful to clarify some terms and to provide a sense of the overall complexity of this issue.

Community organization denotes the number of species that co-occur and their relative abundances, along with those processes that control these structural features, including in particular interspecific interactions such as predation, competition, and facilitation. Predators as sources of mortality directly reduce prey and, if sufficiently severe, can cause extinction. Via such direct impacts, predation can profoundly influence community organization. Yet direct mortality (albeit dramatic and compelling) is just one of the many causal pathways by which predators govern community organization. Individual prey can respond adaptively to avoid predation, for instance by changing habitats, feeding rates, or even morphological traits. These responses may reduce predation, but at the cost of hampering resource acquisition, competitive ability, dispersal, or stress tolerance. Because predators and their prey are almost always embedded in complex communities where they interact with other species, the direct impact of predators on prey numbers, behaviors, and traits can set in motion many chains of indirect interactions across the community (see chapter III.5). Beyond the direct and indirect effects of predators on the average abundance of species, predators also provide feedback effects in communities, which sometimes stabilize interactions but, in other settings, can be dramatically destabilizing. Most communities have multiple predator species, and interactions among the predators can strongly influence the role of predation in community organization. Many species are both predator and prey. A *top predator* has no obvious predator consuming it (although top predators do have parasites and pathogens).

Understanding these ramifying impacts of predation presents an immense challenge. One avenue that has proven useful in analyzing the role of predation in communities is to examine community modules, small sets of species (or well-defined functional groups) interacting in configurations such as those shown in figure 1. In this figure, each species shown is assumed to be dynamically responsive to the other species connected to it by a feeding relationship; usually more species are present than the few shown here, but for some purposes these additional species can be ignored, either because they have negligible effects or because their action can be subsumed in some manner. Theoretical and experimental analyses of modules allow one to think clearly about important processes without getting bogged down in a proliferation of details.

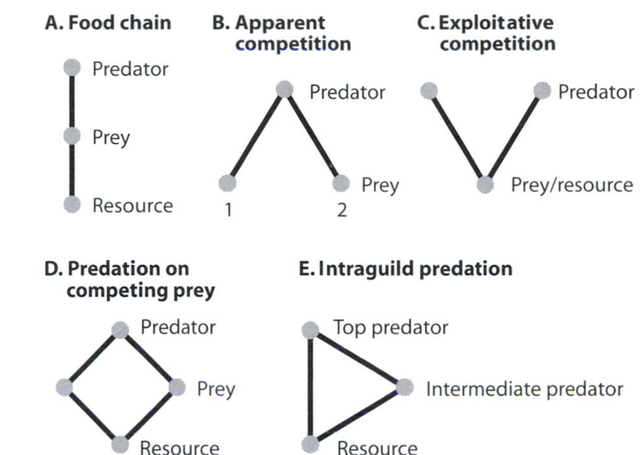

Figure 1. Community modules. These are simple subwebs drawn out of more complex community webs. In intraguild predation (E), top predators and intermediate predators utilize the same resources, and the top predator feeds as well on the intermediate predator.

2. PREDATORS CAN ENHANCE SPECIES COEXISTENCE

One theme that has received considerable attention is the role of predators in governing coexistence within guilds of their prey, where a *guild* is a set of species that utilize resources in similar ways and thus potentially compete (i.e., the prey are engaged in exploitative competition, comparable to figure 1C but one trophic level down). If predators are inefficient consumers or swamped by a surfeit of prey, there may be little effect of predation on prey species richness. But in other circumstances, predators can be essential in permitting species to coexist.

A celebrated experiment by Robert Paine in the intertidal of eastern Washington exemplifies the power of experiments to reveal the key role of predators in communities and highlights several important features of predation in a community context. A thick band of mussels (*Mytilus californianus*) and several species of barnacles dominate the rocky midintertidal. The lower intertidal has lower biomass and much higher species diversity, including immature individuals of these species as well as other space occupiers such as macroscopic benthic algae and a sponge, which in turn sustain browsers such as chitons and limpets. The top predator in the system is the starfish *Pisaster ochraceus*. Paine's experimental protocol was elegantly simple: he systematically removed *Pisaster* from a strip of the lower intertidal, with appropriate controls nearby. Within a few years, diversity on the rock surface was collapsing into a mussel monoculture. Mussels are superior competitors for space, crowding out algae and barnacles and indirectly forcing abandonment by browsing invertebrates.

This experiment crisply shows that predation can sustain components of diversity in a community, and do so dramatically. Note I said "components" of diversity. Mussel beds support hundreds of small invertebrate and plant species, living on and among the shells, a dependent community that might disappear when the beds are demolished by the starfish. It is unclear without further study if *Pisaster* enriches the entire community or just the guild of species that directly compete with mussels for space (and of course any species that largely depend on them). *Pisaster* in the lower rocky intertidal is the canonical keystone species, a species with such a large impact on the community that, in its absence, the community radically changes in species composition. We can draw out some key lessons of this study that pertain to many systems.

First, for predators to enhance diversity, it must hold that in their absence there is strong competition leading to exclusion. In the intertidal, space is contended for by species differing in their ability to colonize and monopolize space, and the mussel is clearly the dominant competitor. This broadly fits the exploitative competition module (figure 1C). Models of exploitative competition for a single resource predict that in the absence of mitigating factors (e.g., temporal or spatial heterogeneity), a single species should persist at the lowest level of the shared resource (here, empty space), excluding all others.

Second, the impact of *Pisaster* was particularly strong on a dominant competitor—the mussel. Predator selectivity helps determine whether or not predators enhance, or instead reduce, prey diversity. For selectivity to promote prey coexistence, there should be a negative trade-off in prey traits across species, so that those prey species superior in competition are more vulnerable to the predator. However, predator selectivity alone does not permit competitive coexistence. Imagine that the numbers of *Pisaster* are determined at a broader spatial scale than Paine's study site and that their attack behavior is fixed, independent of prey numbers. *Pisaster* predation might then act, to a reasonable approximation, as a fixed, density-independent mortality factor, but with different magnitudes on different prey. Figure 2 shows a simple graphic model for exploitative competition with added predation. The solid lines denote the birthrate of each competitor on the resource. The dotted line shows background mortality experienced by both prey species, when the predator is absent. Species 1 wins in competition because it persists at a lower resource level (its so-called R^*; note that $R_1^* < R_2^*$). When the predator is present, it selectively feeds on prey species 1, and so increases the level of mortality experienced by that species to the level shown by the dashed line. The resource levels now required by species 1 to sustain itself in the face of predation (increased to $R_1^{*\prime}$) exceed those needed by species 2, so species 2 will win. Density-independent selective predation clearly is important because it determines which species dominates. But on its own, it will not maintain prey diversity.

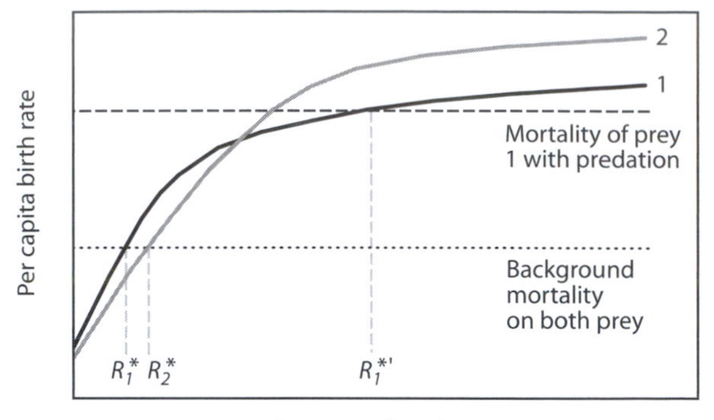

Figure 2. A graphic model of reversal of dominance in exploitative competition as a result of predation. The curved solid lines show how each consumer's birthrate increases with resource R in the absence of predation. The dotted line is background mortality experienced equally by both species. Species 1 has a lower maximal birthrate than does species 2 at high resource levels but can reproduce more effectively when resources are scarce. If species 1 is present at equilibrium, it sustains the resource at level R_1^*, and species 2 becomes extinct at this low resource level (because it needs at least R_2^* to persist). The dashed line includes the additional mortality suffered by species 1 when a selective predator is present. Consumer species 1 now requires a much higher resource level than R_1^* to persist (indicated by $R_1^{*\prime}$), and this level is higher than R_2^*, so now species 2 can exclude species 1—predation has reversed competitive dominance. (In this example, the same outcome holds even if the predator is an indiscriminate generalist, because prey 2 could still persist at a somewhat lower level of the shared resource, and so will win in competition.)

Now imagine that the predator is a generalist imposing mortality uniformly on both prey species. Prey numbers will decline, and resource levels will rise. In the example shown, such uniform predation shifts dominance from species 1 to species 2, but again without coexistence. In general, in competition for a single resource, predation that acts as a simple source of additional mortality on either or both species does not permit coexistence (although it may determine the winner).

This is not what is implied by Paine's study: reduction in mussel abundance leads to a large increase in diversity rather than a simple replacement of one competitor by another. For a diversifiying effect of predation, one of several mechanisms must come into play. In general, for predation to enhance coexistence, interspecific density dependence must be weakened relative to intraspecific density dependence.

A wide array of mechanisms can permit predators to facilitate coexistence, operating over different temporal and spatial scales. Space precludes a full treatment of these, so I will mention just a few highlights. The predator itself can show density-dependent and discriminatory feedback responses to its prey, in effect providing independent modes of regulation controlling prey numbers in addition to resource competition, and so stabilize competitive coexistence. Sometimes, a lowered overall abundance of a prey guild caused by predation may permit coexistence mechanisms to operate that are simply less effective at higher overall abundance. For instance, the starfish does not uniformly reduce mussel numbers but instead clears out patches. These patches can be colonized by species that are rapid dispersers but poor competitors. Alternatively, if refuges are present and in limited supply, prey may compete for refuges to avoid predation. If different prey species use different refuges (e.g., because of body size differences), the presence of predators leads to *competition for enemy-free space*. This may provide an axis for niche differentiation that does not exist in the absence of predation.

If a dominant competitor is attacked by a specialist predator, the added density-dependent feedback needed to facilitate coexistence can come from a strong numerical response of this specialist predator. The predator keeps the dominant below its carrying capacity, thus freeing resources for other species; when the dominant declines, so too (with a lag) does the predator, and the dominant can then rebound from low numbers. Specialization by predators on competitive dominants requires a *trade-off* between competitive ability and vulnerability to predation. Trade-offs emerge from morphological, behavioral, life history, or phenological constraints. For instance, turtles are protected from predators by a thick shell, but this reduces their ability to move rapidly over the landscape in search of food. Understanding trade-offs is fundamental to tying the details of organismal biology to the dynamics of predator–prey interactions. With a multiplicity of prey species, each with its own specialist predator, in principle an entire species-rich ensemble could be supported despite strong potential competition among the prey. Many parasitoids are relatively specialized and effective at limiting their insect hosts, so this speciose group of natural enemies may help explain the hyperdiversity of herbivorous insects. But such tight specialization is much less common in terrestrial and aquatic food webs with vertebrate predators and their prey.

3. PREDATORS CAN SOMETIMES HAMPER PREY SPECIES COEXISTENCE

In many circumstances, predation makes coexistence more difficult. This can happen in several ways. For instance, if prey respond behaviorally to predators by crowding into a limited supply of refuges, they may compete for those refuges directly or may compete more intensely for food resources inside refuges. If prey on their own do not strongly compete, predation can reduce species richness. There are many examples. For instance, experimental studies have examined the impact of *Anolis* lizards on web-building spider communities on Staniel Cay in the Bahamas. The lizards sharply reduced total spider abundance and species richness. The latter effect arose because spiders that were rare in the controls were differentially absent in the lizard sites; spiders that were relatively more common were less strongly affected. The lizards preferentially feed on species that are rare even in the absence of lizard predation, so added mortality from lizards pushes these spider species to the point of local extinction.

Predators have particularly devastating impacts on prey communities when the prey species have not had an evolutionary history equipping them with appropriate defenses. A tragic example comes from the island of Guam. During the last 40 years, Guam's native forest bird species have plummeted in abundance, and several species are now believed extinct or nearly so. The culprit is an introduced predatory snake, the brown tree snake *Boiga irregularis*, which was apparently transported to Guam in the mid-twentieth century. *Boiga*, a major predator on bird nests in its native range, has become very abundant on Guam, with declines in bird numbers paralleling increases in its range and abundance on the island. One reason the brown tree snake has so thoroughly decimated the native avifauna of Guam is that its numbers soared and remained high even after its bird prey largely disappeared. This reflects two features of Guam. First, *Boiga* has neither competitors nor predators that could keep its own numbers in check. It is the top predator in the community. Second, and crucially, *Boiga* is a generalist and also consumes small lizards such as skinks and geckos. These lizards are abundant on Guam. Because of their high reproductive potential, they withstand predation more readily than do species with lower reproductive potential (many island birds have notably low clutch sizes). These alternative prey species sustain a high snake density; the snake can then overexploit bird populations to the point of extinction without endangering its own persistence.

This kind of negative indirect interaction between prey is called apparent competition (figure 1B) because in many respects its outcome resembles the impact of direct or exploitative competition. If one could do a radical experiment and remove alternative prey species from Guam, the working hypothesis would be that snake numbers would be held in check, weakening predation pressure on the native birds. In general, the intensity of predation on any particular prey species indirectly depends on alternative prey in the diet of the shared predator. By supporting a predator, some prey species may indirectly limit the abundance of a dominant competitor, thereby facilitating their own persistence. Models of keystone predation (figure 1D) show that coexistence can occur if one prey species is superior at competing for a shared resource, and the other prey species is superior at withstanding—and sustaining—the shared predator, i.e., at apparent competition. But this mechanism of coexistence depends on a "Goldilocks effect." If prey productivity is low, few predators will occur, and so what mainly matters is competition among prey. If productivity is high, many predators will be sustained, and the ability to withstand predation looms large. Coexistence involving a balance between exploitative and apparent competition is most likely at intermediate productivity.

4. THE IMPACT OF PREDATOR AND PREY BEHAVIOR ON COMMUNITY ORGANIZATION

Predators and prey are not automatons but can respond behaviorally to shifts in each others' abundance and behavior, with important consequences for community organization and dynamics that are still being elucidated by ecologists.

A basic feature of predation is that the capacity of individual predators to consume prey is limited: the rate of predation saturates as prey density becomes large. Over short time spans, increases in one prey species can benefit another, simply because the predator's capacity to capture and process prey becomes saturated. This indirect mutualism between prey species may explain a number of phenomena such as mixed-species aggregations of ungulates or forest birds. But saturation on its own does not explain the maintenance of prey diversity because basically it just permits prey to escape predation and grow to numbers where they are likely to be limited by food or other resources.

Flexible predatory behaviors can lead to frequency-dependent predation that both keeps prey numbers low and helps maintain diversity. If rare species are relatively ignored by predators, they may enjoy a kind of protection at low numbers. For instance, predatory fish in coral reefs concentrate on cardinalfish when they are abundant, allowing recruits of many other fish species to escape unharmed. Such switching behaviors can permit stable coexistence of prey on a single resource. This has been long known as a theoretical possibility, but there are surprisingly few (if any) rigorously documented examples. In the coral reef example, there does not appear to be direct competition between the cardinalfish and beneficiary species such as butterfly fish. Many examples of indirect mutualism between prey involve predators shifting their attention between habitats with different prey species, relaxing predation when prey numbers in a habitat are temporarily low. In such cases, the spatial segregation of the prey that allows predator switching also means the prey may not be strongly competing in the first place. Moreover, because additional prey species should often boost predator numbers, positive interactions among prey species via predator saturation or switching may be outweighed by longer-term numerical responses leading to apparent competition. Wolves, for instance, are well known for flexible hunting behaviors, including switching. But having moose as an abundant and productive alternative prey for wolves has been implicated in the decline and local extirpation of woodland caribou in parts of Canada, because wolf numbers are substantially boosted by the availability of moose.

Beyond the issue of modulating species coexistence, flexible predator behaviors can strongly affect system stability. Theoretical studies suggest that mobile predators in heterogeneous landscapes that respond adaptively to changes in local prey numbers can have strong stabilizing effects. This is a compelling idea and surely helps explain the persistence of some complex ecological communities. However, flexible predator behaviors, including switching, can also at times destabilize systems. The North Pacific has seen a sequential collapse of marine mammals—first seals, then sea lions, and finally sea otters—over the last few decades. One might at first suspect that this has been caused by global change, such as shifts in climate, but recent evidence implicates switching by a top predator—the killer whale. Historically, killer whales focused their predation on the great whales, such as sperm whales, but whale numbers were strongly reduced by an upsurge of whaling after World War II, particularly by the Japanese whaling fleet. This sharp reduction in their primary food source led killer whales to start feeding on seals, and then when seal numbers were depleted, on smaller-bodied sea lions. When these in turn had been sufficiently cleaned out, the killer whales started feeding on the even smaller-bodied sea otters. Calculations suggest switching behavior by a relatively small number of killer whales suffice to explain the observed collapse of sea otter populations in the Aleutian Islands. Thus, flexible predator switching behavior has amplified an initial disturbance caused by humans, with reverberating impacts across an enormous oceanic ecosystem.

Behavioral responses by prey to predators are also important. Prey individuals faced with predators often reduce their foraging. There is increasing evidence that these nonlethal effects of predation can be quantitatively large, at times even more important than the direct lethal effects of predation on prey communities. Analyses of such trait-mediated indirect interactions between species is a very active area in community ecology (see chapter III.5).

5. PREDATORS CAN INITIATE TROPHIC CASCADES

If predators reduce the abundance or shift the behavior of their prey, this can have strong second-order effects on the resources consumed by those prey. This chain of indirect interactions is called a trophic cascade. There are an increasing number of examples of dramatic trophic cascades in both terrestrial and aquatic systems.

In North America, European colonists eliminated wolves over wide areas, but in recent years the wolf has returned as a result of growing sympathy by the public for such predators. Wolves prey on large ungulates such as elk. Researchers have compared areas of Banff National Park where wolves colonized to areas avoided because of human activities. Elk were more numerous by an order of magnitude in low-wolf areas, with higher survival and calf recruitment. This led to substantially lower aspen recruitment and willow production because of intense browsing.

This recent study is a microcosm of a widespread shift in trophic interactions that plausibly occurred over huge landscapes during the settlement of the American West. Large carnivores are highly vulnerable to direct elimination by humans and indirect impacts via habitat fragmentation. In the Great Plains, settlers, as they spread across the prairie in the late nineteenth century, substituted cattle for native migratory bison and systematically exterminated top predators such as wolves and grizzly bears. In protected areas such as Wind Cave National Park in southwestern South Dakota, in the twentieth century, there was an upsurge in the abundance of wild ungulates such as deer and elk, freed from predation by both wild and human predators. Analyses of cottonwood and bur oak stands reveal essentially zero recruitment for more than a century because of high levels of browsing. (Inside exclosures, there has been substantial recruitment.) Thus, a trophic cascade initiated by human decimation of top predators may have had profound consequences for Great Plains ecosystems.

Simple models of trophic cascades consider unbranched food chains (figure 1A) with one species at each trophic level. These models show that effective predators can free the basal species of control by the intermediate prey species. One worrisome effect of human impacts on the Earth's ecosystems is that top predators are particularly vulnerable to anthropogenic disturbance, leading to disrupted control of herbivore numbers. For instance, an accidental introduction of the pathogen canine parvovirus onto Isle Royale around 1980 caused a dramatic crash in wolves, which in turn permitted an upsurge of moose numbers and more intense browsing.

The topic of trophic cascades has been the focus of an intense and continuing debate among ecologists (see also chapters III.5 and III.6). Many feel that simple models matching the food chain of figure 1A leave out critical complicating factors. The strength of trophic cascades can vary with many factors, such as the overall complexity of the food web and the magnitude of direct plant defenses against herbivory. If a long coevolutionary struggle between plants and herbivores has created an armory of defenses, such as toxins, structural defenses, and low-quality tissues, herbivores may be more a dynamic annoyance than a prime driver, even in the absence of predators. Moreover, increases in herbivory can drive shifts in plant community composition so that the final community is dominated by species that the herbivores cannot readily eat. Short-term manipulative experiments may not capture such compositional changes (which may require colonization from external species pools).

John Terborgh has argued that few natural plant communities are immune to the strong effect of herbivores unchecked by predation. The huge artificial Lake Guri in Venezuela isolated hundreds of islands of varying sizes. On smaller islands, predators such as jaguars were absent, and folivorous howler monkeys increased to high numbers. Likewise, anteaters and armadillos disappeared, and the colonies of their prey, leafcutting ants, burgeoned. These hyperabundant herbivores turned to plants not normally part of their diets and devastated their preferred forage plants, with dramatic effects on tree recruitment. In effect, the absence of the original top predator means that the herbivores can grow until limited by their own resources. This sets up the opportunity for apparent competition between different plant species in the herbivore's diet, and plants with low tolerance of herbivory are vulnerable to extinction. The removal of the top predator then results in plant extinctions, two trophic levels down.

Turning this process around, we can return to the original theme of this chapter—identifying ways predation can facilitate prey species coexistence—and link this issue to both trophic cascades and apparent competition. If prey species themselves depend on living resources and are left unchecked, they can overexploit some of those resources to the point of extinction, driven by apparent competition (in this case, via the numerical response of the prey to its own resources). This can preclude niche partitioning among the prey. Conversely, a reduction in prey abundance or activity caused by predation can permit a wider range of resource species to persist, opening up avenues for potential niche partitioning and weakening interspecific competition relative to intraspecific competition. In effect, trophic cascades can mediate coexistence via niche partitioning at intermediate trophic levels, which is permitted because the top predator indirectly relaxes apparent competition at the basal trophic level.

6. THE DIVERSE EFFECTS OF PREDATOR DIVERSITY

Relatively few systems have just a single top predator, matching the simple modules of figure 1. Predator diversity has a wide variety of impacts on community organization and functioning, many of which have been poorly explored either theoretically or empirically.

Experimental studies in kelp forests have shown that increasing predator diversity strengthens the trophic cascade when both predator and herbivore trophic levels are diverse. The reason is that herbivores respond behaviorally to predation by reducing foraging rates, and different herbivores respond to different predators. With more predator species, fewer avenues of escape are open to the prey. Predator diversity thus has a synergistic effect on limitation of herbivore numbers, indirectly boosting kelp biomass. Conversely, in a study of invertebrate predation on planthoppers feeding in salt marshes, increased predator diversity dampened the trophic cascade. The reason is that this system included *intraguild predation* (figure 1E), a kind of omnivory where some predators eat other predators as well as a shared resource. For instance, spiders eat ladybugs, and both prey on planthoppers. Models of intraguild predation suggest that in the absence of factors such as alternative prey, for the predators to coexist, the top predator has to be less efficient at utilizing the shared resource. A mixture of predators will thus reduce the total impact of predation on the basal species and weaken trophic cascades. However, these models have yet to incorporate complexities such as trait-mediated interactions, spatial heterogeneity, and other realistic factors, and so this conclusion should be viewed as a tentative hypothesis

7. CONCLUSIONS

In this essay, I have just scratched the surface of the rich topic of predation and community organization. There are many important issues, such as the role of life history variation and population structure, the implications of spatial dynamics such as metapopulation processes, and the relationship between predation and the classic complexity–stability debate, which deserve much further scrutiny. The examples sketched above show that it is vital to understand the impact of predators on community organization, not just in terms of basic science but to inform conservation and management policies. It is difficult to understand the origin and maintenance of the diversity of life without paying attention to the profound impact of predation. It is even more difficult to imagine that we can preserve the biota with which we share the planet without paying due attention to the compelling drama of predation as a key driver in ecological systems.

FURTHER READING

Conceptual Papers and Reviews

Chase, J. M., P. A. Abrams, J. P. Grover, S. Diehl, P. Chesson, R. D. Holt, S. A. Richards, R. M. Nisbet, and T. J. Case. 2002. The interaction between predation and competition: A review and synthesis. Ecology Letters 5: 302–315.

Holt, R. D. 1997. Community modules. In A. C. Gange and V. K. Brown, eds., Multitrophic Interactions in Terrestrial Ecosystems, 36th Symposium of the British Ecological Society. Oxford: Blackwell Science, 333–349.

Holt, R. D., and J. H. Lawton. 1994. The ecological consequences of shared natural enemies. Annual Review of Ecology and Systematics 25: 495–520.

Levin, S. A. 1970. Community equilibria and stability, and an extension of the competitive exclusion principle. American Naturalist 104: 413–423.

Rosenheim, J. A. 2007. Special feature—intraguild predation. Seven papers on this topic. Ecology 88: 2679–2728.

Key Case Studies

Paine, R. T. 1966. Food web complexity and community stability. American Naturalist 100: 65–75.

Ray, J. C., K. H. Redford, R. S. Steneck, and J. Berger, eds. 2005. Large Carnivores and the Conservation of Biodiversity. Washington, DC: Island Press.

Ripple, W. J., and R. L. Beschta. 2007. Hardwood tree decline following large carnivore loss on the Great Plains, USA. Frontiers in Ecology and the Environment 5: 241–246.

Savidge, J. A. 1987. Extinction of an island forest avifauna by an introduced snake. Ecology 68: 660–668.

Spiller, D. A., and T. W. Schoener. 1998. Lizards reduce spider species richness by excluding rare species. Ecology 79: 503–516.

Springer, A. M., J. A. Estes, G. B. van Vliet, T. M. Williams, D. F. Doak, E. M. Danner, K. A. Forney, and B. Pfister. 2003. Sequential megafaunal collapse in the North Pacific Ocean: An ongoing legacy of industrial whaling? Proceedings of the National Academy of Sciences, U.S.A. 100: 12223–12228.

Terborgh, J., K. Feeley, M. Silman, P. Nunez, and B. Balukjian. 2006. Vegetation dynamics of predator-free land-bridge islands. Journal of Ecology 94: 253–263.

III.4

Facilitation and the Organization of Plant Communities

Ragan M. Callaway

OUTLINE

1. Introduction
2. What mechanisms cause positive interactions
3. Can we predict when positive or negative interactions may be important?
4. What do positive interactions mean for community theory?

Current plant community ecology, as presented in most textbooks, often promotes the perspective that communities are produced only by the traits of populations and that assemblages of different plant species exist primarily because each shares adaptations to particular abiotic conditions. To some degree, this perspective leads to the conclusion that plant communities are simply a handy typological construct. However, a large body of research accruing during the last 30 years demonstrates that many if not most plant communities have fascinating interdependent characteristics, and although they are not "organic entities," it is clear that many species create conditions that are crucial for the occurrence and abundance of other species. This research is the focus of this chapter.

GLOSSARY

continuum. A distribution of many species along a gradient in which each species appears to be distributed randomly with respect to other species

facilitation. The positive effect of one species on another

holistic communities. The idea that species within a community are highly interdependent, forming organism-like units

hydraulic lift. The process by which some plant species passively move water from deep in the soil profile, where water potentials are high, to more shallow regions where water potentials are low

indirect interactions. Interactions between two species that are modified by a third species

individualistic communities. The idea that communities are fundamentally groups of populations that occur together primarily because they share adaptations to the same abiotic environment; communities do not have organism-like qualities

niche complementarity. The condition in which different niches result in variation in the utilization of resources or space

1. INTRODUCTION

As the discipline of ecology emerged from its biogeographic origins in the early 1900s, two strikingly polar views on the nature of plant communities vied for recognition, and the conflict established a precedent for ecological thought today. Initially, the view of Frederic Clements was ascendant with most ecologists accepting the idea that

> [T]he community is an organic entity. As an organism the community arises, grows, and dies. Furthermore, each community is able to reproduce itself, repeating with essential fidelity the stages of its development . . . comparable in its chief features with the life history of an individual plant. (Clements, F. E. 1916. Plant Succession. Washington, DC: The Carnegie Institution, Publication 242)

This holistic perspective, however, was replaced in the middle of the 1900s by new ideas promoted by Henry Gleason. In this new individualistic world view, the community "is merely the resultant of two factors, the fluctuating and fortuitous immigration of plants and an equally fluctuating and variable environment . . . not an organism, scarcely even a vegetational unit, but merely a coincidence" (Gleason, H. A. 1917. The structure and development of the plant association. Bulletin of the Torrey Botanical Club 44: 463–481). It would be hard to dream up two more diametrically contrasting perspectives for how species are organized into groups.

Texts have a strong individualistic flavor, but most ecologists are fully aware that the nature of plant communities is more nuanced than the hyperdichotomy of individualistic versus organismal communities. Perceiving such nuances is important, but the dominant individualistic perception of plant community organization has probably left lingering but strong effects on the way we conduct research, leading to a great deal of information on negative interactions such as predation, competition for resources, and allelopathy. However, this dominant perception has probably impeded the progress of empirical research on facilitation and indirect interactions among plants.

Understanding the nature of communities is not just an academic issue. Whether or not communities have weak or strong tendencies toward independent or interdependent assembly has strong implications for conservation. For example, the view that plant species are fully individualistic and interchangeable in communities has been used to advocate active human involvement in "shaping and synthesizing *new* ecosystems, even in the 'natural' environment" (italics mine; Johnson and Mayeux, 1992). This may be reasonable if maintaining functional plant communities is simply a matter of finding a suite of populations that can grow in a particular set of conditions. But if interactions among plants are more complex and interdependent, as suggested by research on facilitation, indirect effects of herbivores and mycorrhizae, and networks of direct and indirect interactions within the plant community, shaping and synthesizing new communities may not work. Conservationists typically assume a high degree of interdependence in communities when they argue for the preservation of natural systems and biological diversity. The Ecological Society of America recommends the following for conservation priorities: Does the species play an especially important role in the ecosystem in which it lives? Do other species depend on it for their survival? Will its loss substantially alter the functioning of the ecosystem? These priorities assume interdependence.

In this chapter, I focus on several general questions: What mechanisms cause positive interactions? Can we predict when positive or negative interactions may be important? What do positive interactions mean for community theory?

2. WHAT MECHANISMS CAUSE POSITIVE INTERACTIONS?

Positive interactions can be direct, the effect of one species on one other species, or positive interactions can be indirect, requiring an intermediate species in order to occur (see chapter III.5). There are many direct and indirect facilitative mechanisms, probably far more than mechanisms for resource competition, and they can be difficult to separate experimentally or conceptually. I present a brief overview of mechanisms here, but for more detail on mechanisms see Callaway (2007).

Shade from other species can keep plant tissues below lethal temperatures, decrease respiration costs and transpiration loss, reduce ultraviolet irradiation, and increase soil moisture. Shade is one of nature's most important facilitative mechanisms. In the Sonoran Desert, the grouping of saguaro (*Carnegia gigantean*) seedlings under other desert perennials has been studied intensively and coined the "nurse plant" syndrome. Raymond Turner and colleagues (1966) studied several mechanisms with the potential to cause the associations between young saguaro cacti and nurse trees by experimentally transplanting young saguaros in factorial treatments of shade, supplemental water, and protection from herbivores. Turner and colleagues found that that predation was important, but all nonshaded saguaro seedlings died regardless of water addition.

In nature, the trade-offs between the facilitative and competitive effects of shade are complex because many species reach their maximum photosynthetic rates at light levels far below the natural maximum ($\sim 2000 \cdot \mu mol \cdot m^{-2} \cdot s^{-1}$). These species may benefit from the effects of shade from neighbors without any cost of decreased carbon gain. For example, *Arnica cordifolia* is a perennial herb that is common in conifer understories in the northern Rocky Mountains. Donald Young and Bill Smith found that a 30% decrease in light on the forest floor during cloudy days in the Medicine Bow Mountains of Wyoming resulted in a 37% *increase* in carbon gain for *Arnica* and an 84% reduction in transpiration (Young, 1983). They found that the photosynthetic rates of *Arnica* remained near saturation even on very cloudy days. In other words, *Arnica* gained from the lower transpiration rates associated with increased shade without an accompanying cost of lower photosynthetic rates.

Water relations of plants can also be facilitated in many different ways. Facilitators can build up litter, decrease soil bulk density, intercept rain or fog by canopies, snow accumulation, or hydraulic lift. Soil beneath canopies is commonly wetter than that in nearby open areas, and the difference in soil moisture has been correlated with facilitative effects in many systems.

Vegetation is a fundamental driver of soil development, and shrubs and trees add nutrients to the soil in ways that favor some species over others. In the 1950s, J. D. Ovington noted that "whilst the trees cannot alter primary site factors such as bedrock or topography, they may modify some secondary factors. Nutrients are removed from the soil and are returned in part as litter fall so that the trees influence those soil processes which affect the physical and chemical condition of the soil." Ovington also noted that deeply rooted perennials take up nutrients that are unavailable to more shallowly rooted understory plants and deposit them on the soil surface via litter fall and throughfall. Understory plants may eventually acquire these nutrients after the litter fall from the overstory decays.

Wind can damage plants by decreasing tissue temperatures, increasing vapor pressure differences between leaf and air, or simply by damaging plant parts. In environments where wind is extreme, many plants may be facilitated by sheltering beneath or behind other plants. This form of facilitation creates subalpine tree islands and ribbon forests with seedling regeneration restricted to the leeward side of the islands or ribbons. As another example, *Pinus flexilis* (limber pine) shades other species and protects them from high winds at the ecotone of the Rocky Mountains and the Great Plains.

Other mechanisms that have been shown to play important roles in facilitation include soil oxygenation, substrate building, protection from disturbance, and forms of chemical communication among plants. Most examples of chemical communication among plants involve herbivores and thus are indirect interactions requiring intermediate species, such as herbivores, pollinators, mycorrhizal fungi, soil microbes, or other competing plant species in order to occur.

The seminal paper on indirect defense interactions among plants was published by Peter Attsat and Dennis O'Dowd in 1976. They argued that many plant species were "functionally interdependent with respect to their herbivores." Soon afterward, Sam McNaughton published a paper demonstrating lower mortality rates of the highly palatable grass *Themeda triandra* when associated with unpalatable species. Since then many other studies have shown similar processes of indirect facilitation involving herbivores. Other indirect interactions can involve reproductive feedback determined by the density of individuals, enhanced sharing of pollinators or dispersers, feedback involving soil microbial communities, shared mycorrhizal networks, and intriguing indirect interactions among competing plant species.

Competitive interactions between two species can be altered by simultaneous competitive interactions with additional species or through cumulative diffuse effects that occur when numerous species have different kinds of direct effects that act on a single species. In all cases the facilitative effect is produced by something analogous to an alliance—an enemy of my enemy is my friend. This has been understood theoretically for a long time, but the first experiment designed explicitly to quantify indirect effects among interacting plants was conducted by Tom Miller. Miller found that direct and indirect effects were common and strong and that interactions sorted themselves out so direct negative effects among particular species were balanced by positive indirect effects. Strong direct inhibition by dominant competitors was consistently ameliorated by the presence of other competitors. For example, Miller found that *Ambrosia artemisiifolia* reduced the biomass of *Chenopodium album* by 94–98% in two-species experiments. In multispecies communities, however, *Ambrosia* reduced the biomass of *Chenopodium* by only 17%. Ecologists have tried to understand coexistence and species diversity in the context of niche partitioning, variation in particular resource requirements and uptake, shifts in competitive hierarchies in different microenvironments, and nonequilibrium processes; however, Miller's results suggest that the balance of competitive interactions may create facilitative effects that are crucial for sustaining coexistence among species in communities.

3. CAN WE PREDICT WHEN POSITIVE OR NEGATIVE INTERACTIONS MAY BE IMPORTANT?

In the 1950s, ecologists from the Intermountain Research Station in Utah found that herbaceous species were much smaller when grown under *Populus tremuloides* (quaking aspen) canopies than in open meadows near the trees, indicating that the trees had competitive effects. However, when they also trenched plots to exclude *P. tremuloides* root systems under canopies, the biomass of some of the herbs was greater than that in the open, demonstrating that strong facilitative and competitive effects were functioning at the same time in their system.

Since the 1950s a great deal of other research has also shown that facilitative and competitive interactions often operate in balance. For example, *Quercus douglasii* (blue oak) in California facilitates the growth of understory species through nutrients in litter and throughfall but often competes with the same species at the same time through its lateral root system.

Such co-occurring positive and negative interactions were given a strong element of predictability in the early 1990s when Mark Bertness experimentally demonstrated competition among salt marsh plants in relatively moderate abiotic conditions, showed facilitation in abiotically stressful conditions, and then eliminated facilitation by experimentally eliminating the abiotic stress. Bertness's field studies led to a general conceptual model for the relationship between stress and the relative importance of competition and facilitation, the "stress gradient hypothesis" proposed by Bertness and Callaway in 1994. In part derived from J. P. Grime's hypotheses about the relative importance of competition in plant communities, they postulated that competitive interactions would be most important to the organization of plant communities when abiotic stress does not strongly limit the ability of plants to acquire and exploit resources. Under relatively benign abiotic conditions that permit rapid resource acquisition, competition can be intense. However, if severe physical conditions restrict resource acquisition, amelioration of severe stress by a neighbor may be more likely to favor growth than competition with the same neighbor is to reduce growth (figure 1).

Strong support for the stress gradient hypothesis comes from many studies, but two groups in particular have conducted experiments in many places and under a wide range of abiotic conditions. Lorena Gómez-Aparicio and colleagues at the University of Granada in Spain conducted the largest-scale study to date of facilitation and competition in semiarid environments using 18,000 replicates of 11 different potential beneficiary species with 18 different species of potential nurse shrubs. The work was carried out over 4 years at many different sites. These results convincingly demonstrated that pioneer shrubs facilitate the establishment of woody, late-successional Mediterranean species and that nurse shrubs had a stronger facilitative effect on seedling survival and growth at low altitudes and sunny, drier slopes than at high altitudes or shady, wetter slopes. The second study was a series of experiments conducted in subalpine and alpine plant communities with 115 species in 11 different mountain ranges around the world (Callaway et al., 2002). Callaway and colleagues found that competition generally dominated interactions at lower elevations where productivity was higher and abiotic conditions are less physically stressful. In contrast, at high elevations where abiotic stress is high, the interactions among plants were predominantly positive.

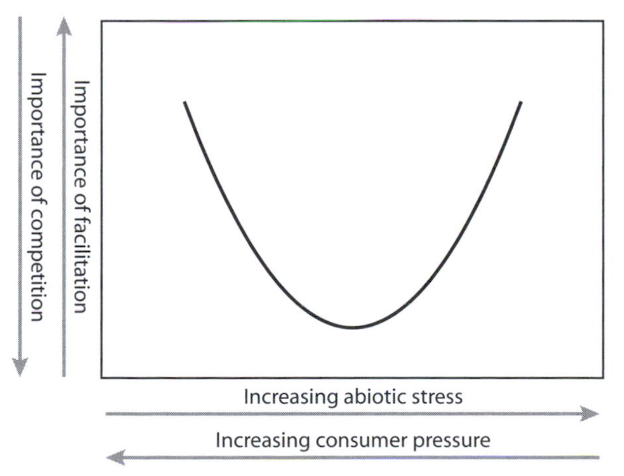

Figure 1. Conceptual model for variation in the importance of competition and facilitation in communities along gradients of abiotic stress and herbivory. (Redrawn from Bertness and Callaway, 1994)

Not all tests of the stress gradient hypothesis have supported it, but the idea is rooted in an effort to understand in what conditions we might expect to find strong facilitative effects. Facilitation virtually always occurs through the alleviation of some kind of stress experienced by a plant; thus, common sense suggests facilitation would be more common in relatively stressful conditions. Without some kind of stress there is nothing for a neighbor to facilitate.

4. WHAT DO POSITIVE INTERACTIONS MEAN FOR COMMUNITY THEORY?

Facilitation has important implications for several key concepts in ecology. Jon Bruno proposed that incorporating facilitation into ecological theory "will fundamentally challenge some of our most cherished paradigms" and "that current theory emphasizing competition or predation paints an incomplete, and in some cases misleading picture of our understanding of the structure and organization of ecological systems." Facilitation suggests new perspectives on the realized niche, diversity–community attribute relationships, the role of interactions in evolution, and, as noted at the beginning of the chapter, the nature of communities.

Facilitation and the Niche

Implicit in the process of facilitation is the idea that the realized niche can be *increased* by other species. It has been assumed that the performance of a species along a set of relevant environmental variables is sufficient to explain its fundamental niche, and competitive and consumer interactions have been incorporated into definitions of the realized niche. Discrepancies between realized and fundamental niches are virtually always attributed to resource competition. However, research on facilitation clearly demonstrates that both positive and negative interactions must be incorporated into the concept of the niche. Competition and facilitation may act somewhat symmetrically at different margins of a species' distribution with competition as a limiting factor at one extreme and facilitation an expanding factor at the other extreme.

Facilitation, Species Richness, and Ecosystem Function

A substantial amount of research has found a positive relationship between the number of species in a community and aspects of ecosystem function of the community (see chapter III.14). In general, the effects of facilitation have been combined conceptually with niche differentiation within the broad idea of complementarity as possible drivers of the effects of species richness in diversity–ecosystem function theory. This is an effective operational approach because the two processes are quite difficult. For example, legumes facilitate by fertilizing soil though N fixation, but the use of atmospheric N can also be considered a form of niche differentiation. *Complementarity* describes the divergence in niche space among species in a community that allows for an increase in total utilization of resources, something different from facilitation. Complementarity occurs when neighbors do not substantially infringe on each other's resource requirements; in other words, there is reduced competition. Facilitation occurs when a species benefits from the presence of a neighbor.

In a substitutive experimental design, the total number of individuals is constant. If interspecific competition is weaker than intraspecific competition because of niche differentiation, individuals on average will experience weaker competitive effects and should perform better. On the other hand, superior performance of individuals in species-rich mixtures compared to monocultures suggests the possibility of facilitation. As in many other studies, Maria Caldeira and colleagues found that productivity in species-rich plots was significantly higher than that in monocultures. However, species performed better as a community than as individuals. Caldeira found that some individuals performed better in mixtures. Plants were sown into Caldeira's plots, and thus the experiment was not clearly substitutive, but measurements of soil moisture and stable carbon isotope ratios in the leaves of several species indicated that plants growing in species-rich mixtures improved their water relations, which suggests that either complementarity or facilitation may have been responsible for the increase in productivity with diversity. Future experiments that separate these different conceptual mechanisms may allow a better understanding of the role of diversity on community and ecosystem functioning.

Indirect facilitation may also affect diversity–ecosystem function relationships. Johannes Knops and colleagues quantified foliar disease severity in plots varying in species richness. They found that for each of the four target plant species, foliar disease was significantly negatively correlated with plant species richness. However, they found that disease severity was also dependent on host plant density. Although not yet investigated in the context of diversity–ecosystem function, negative feedback that has been documented between plants and soil microbes could also enhance community diversity by increasing species turnover rates. As the number of species increases, turnover among species in a particular place may be much greater. Alternatively, high species richness may reduce strong negative feedbacks in general.

Facilitation and Evolution

Thymus vulgaris is one of the most widespread plant species in Europe, and the species is composed of several different chemotypes that differ in the biochemical composition of the essential oils produced in the leaves. Bodil Ehlers and John Thompson found that six different *Thymus* chemotypes could be identified using the dominant monoterpene in the essential oil, which is either phenolic or nonphenolic. The grass, *Bromus erectus*, is often spatially associated with all of the different *Thymus* chemotypes, suggesting facilitation. Ehlers and Thompson found that *B. erectus* from nonphenolic *Thymus* patches performed significantly better on its home soil than on soil from a different nonphenolic or phenolic *Thymus* patch. This superior performance of matched local *B. erectus* to familiar *Thymus* chemotypes was observed only for soil collected directly underneath *Thymus* plants and not on soil collected away from *Thymus*. These results suggest that *B. erectus* may be genetically adapting to soil modifications mediated by different *Thymus* chemotypes, and importantly, this may occur only because *Thymus* facilitates the growth of *B. erectus* in the chemically modified environment.

The scenario described by Ehlers and Thompson is unique in the ecological literature for plants, but it describes a reasonable way that facilitation may drive evolutionary changes. By pulling other species into an expanded niche (figure 2), benefactors may expose beneficiaries to new abiotic or biotic environments to which they may adapt.

In a similar example, Manuel Figueroa and colleagues examined facilitation and evolution in the

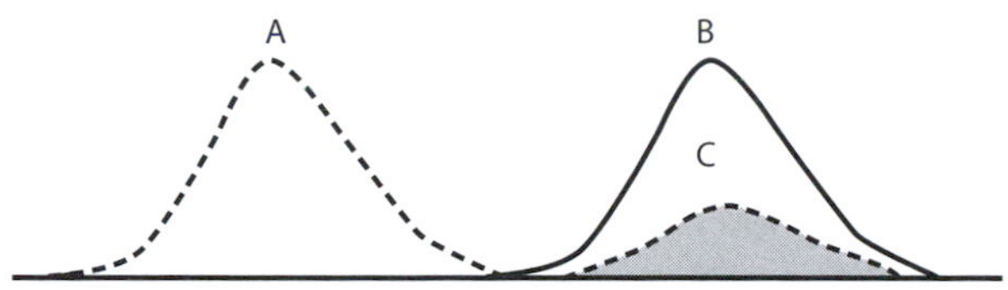

Figure 2. Illustration of how facilitation might create novel evolutionary opportunities for beneficiaries (e.g., Ehlers, B. K., and J. Thompson. 2004. Do co-occurring plant species adapt to one another? The response of *Bromus erectus* to the presence of different *Thymus vulgaris* chemotypes. Oecologia 141: 511–518.) Peak A represents the niche of a beneficiary species (e.g., *Bromus erectus*) in the absence of a benefactor species (e.g., *Thymus vulgaris*). Peak B represents the niche of the benefactor species. Peak C represents the realized niche of the beneficiary species in the presence of the benefactor, with the shaded area representing the new environment to which the beneficiary can now evolve. (Reprinted from Callaway, 2007)

Odiel Marshes in southwest Spain. They found that a *Sarcocornia* hybrid grows on the raised centers of *Spartina martima* patches. These patches are invaded by *Sarcocornia perennis*, a species common to lower parts of the marsh; however, once established in *Spartina* patches, an opportunity is provided for hybridization with *Sarcocornia fruticosa*, a species that occurs in higher parts of the marsh. Hybrids occur only on *Spartina* patches with *S. perennis*. Figueroa called this scenario with the hybrid *Sarcocornia* genetic facilitation and suggested that succession might be facilitated genetically through the establishment of conditions leading to hybridization.

Recently, Alfonso Valiente-Banuet and colleagues explored the facilitative effects of plant taxa that evolved during the drying climate of the more recent Quaternary (within the last 2 million years) period on more ancient plant taxa that evolved during the wetter Tertiary ($\sim$60 million years ago). Most global deserts and semiarid environments developed during the Quaternary. The development of desert corresponded with the evolution of new species, but, interestingly, many mesic-adapted Tertiary species did not become extinct in the drier climate. Valiente-Banuet found that "modern" species, those that arose during the Quaternary, currently facilitate ancient Tertiary species. In fact, very few ancient species recruited in any microhabitat other than beneath the canopies of other species. In other words, species that rapidly evolved to new stressful abiotic conditions appeared to be pulling ancient species, which were not adapted to xeric conditions, into modern communities by creating appropriate regeneration niches. These results have profound implications for the processes that sustain global biodiversity and for the nature of plant communities. Communities such as these are clearly not individualistic.

Facilitation and the Nature of Communities

As discussed at the beginning of this chapter, most ecologists probably do not perceive plant communities as fully individualistic. However, the presentation of communities as individualistic is almost the rule in general and specialized textbooks. Moreover, this classic and often artificial historical dichotomy in viewpoints is one with lingering impacts on the way we think and conduct research.

There are many arguments for the individualistic paradigm, but descriptions of continuous distributions of species along environmental gradients have been a central component of the argument for individualistic communities since the idea was first articulated by Robert Whittaker in 1951. Most gradient analyses show a continuum of apparently randomly overlapping species, and this is used to argue for individualistic communities. If species depended on each other would they not always occur together? However, it is now clear that species can facilitate each other in some conditions but compete with each other in different conditions. Furthermore, most gradient analyses do not quantify spatial relationships at a scale appropriate to detect positive associations, and some gradient analyses are at odds with the continuum. For example, the presence of *Prosopis velutina* on desert and terrace landforms and *Olneya tesota* throughout the Sonoran Desert is strongly associated with particular understory communities. The distributions of species on these gradients and many others are not continuous but grouped into nodes, and these nodes suggest facilitation and some degree of interdependence among species.

Robert Whittaker died in 1979, but in a story with an ironic twist, he posthumously published a paper in 1981 in which virtually perfect correlations occur between understory communities and the presence of different desert or chaparral shrubs—the nodal distribution of species along gradients indicative of holistic communities (figure 3). It is hard to imagine tighter correlations among the distributions of different plant species along gradients, yet even with these "strongly differentiated patterns," and despite the lack of evidence for a continuum in their results, there was no discussion of how these findings might be reconciled with Whittaker's paradigm; nor was there any discussion of rejecting the continuum as universal in plant communities.

Chris Lortie argued that recent experimental efforts to understand the relative importance of positive or negative interactions in many different communities allows ecologists "to explicitly reconsider what most ecologists appear to have done implicitly: our formal conceptual theory of the fundamental nature of

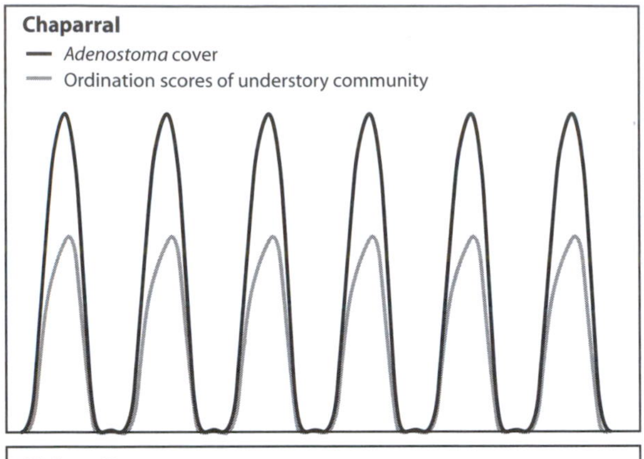

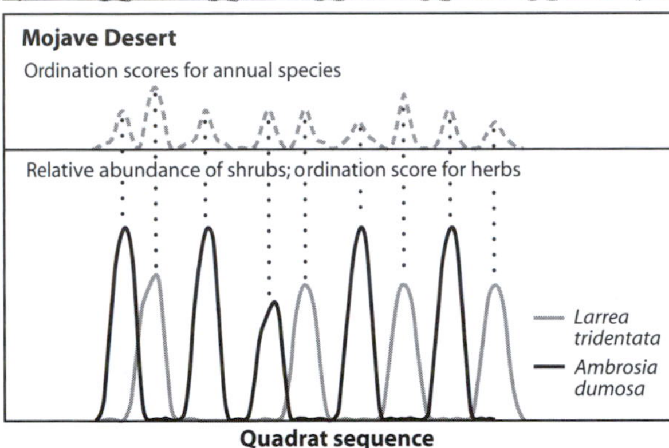

Figure 3. Stylized representation of graphic results presented by Schmida and Whittaker demonstrating strong nodality among species in plant communities. In the chaparral, high ordination scores for understory herbs correlate strongly with the presence of shrubs, and in the Mojave Desert high ordination scores correlate strongly with the presence of *Larrea* or *Ambrosia*. (Reprinted from Callaway, 2007)

communities." Lortie also proposed a conceptual model for the nature of plant communities termed the "integrated community concept." The integrated community concept is based on evidence that community composition is determined by (1) stochastic processes, (2) species-specific tolerances to local abiotic conditions, (3) positive and negative direct and indirect interactions among plants, and (4) direct interactions with other organisms. If communities are determined by complex interactions among all of these processes, including facilitation, communities should vary from those that act very much as collections of independent individual species to others that act as highly interdependent groups of species. As stated by Lortie, "communities (and even a single community) will encompass a range of different dependencies among species—or degrees of integration—determined by the relative importance, and variation in space and time, of each of the processes we proposed."

The integrated community concept, incorporating facilitation, offers a different and more mechanistically inclusive understanding of the organization of plant communities than the individualistic paradigm. Community composition, biological diversity, and coexistence are determined not only by competitive and consumer interactions but also by powerful and ubiquitous facilitative effects. Because facilitative effects suggest a substantial component of interdependence in plant communities, communities cannot be understood only by studying populations.

FURTHER READING

Archer, S., C. Scifres, and C. R. Bassham. 1988. Autogenic succession in a subtropical savanna: Conversion of grassland to thorn woodland. Ecological Monographs 58: 111–127.

Atsatt, P. R., and D. O'Dowd. 1976. Plant defense guilds. Science 193: 24–29.

Bertness, M. D., and R. M. Callaway. 1994. Positive interactions in communities. Trends in Ecology and Evolution 9: 191–193.

Brooker, R., Z. Kikvidze, F. I. Pugnaire, R. M. Callaway, P. Choler, C. Lortie, and M. Michalet. 2005. The importance of importance. Oikos 109: 63–70.

Bruno, J. F., J. J. Stachowitcz, and M. E. Bertness. 2003. Inclusion of facilitation into general ecological theory. Trends in Ecology and Evolution 18: 119–125.

Callaway, R. M. 2007. Positive Interactions and Interdependence in Plant Communities. Dordrecht: Springer.

Holmgren, M., M. Scheffer, and M. A. Huston. 1997. The interplay of facilitation and competition in plant communities. Ecology 78: 1966–1975.

Miller, T. E. 1994. Direct and indirect species interactions in an early old-field plant community. American Naturalist 143: 1007–1025.

Turner, R. M., S. M. Alcorn, G. Olin, and J. A. Booth. 1966. The influence of shade, soil, and water on saguaro seedling establishment. Botanical Gazette 127: 95–102.

Valiente-Banuet, A., A. V. Rumebe, M. Verdú, and R. M. Callaway. 2006. Modern Quaternary plant lineages promote diversity through facilitation of ancient Tertiary lineages. Proceedings of the National Academy of Sciences U.S.A. 103: 16812–16817.

III.5

Indirect Effects in Communities and Ecosystems: The Role of Trophic and Nontrophic Interactions

Oswald J. Schmitz

OUTLINE

1. Introduction
2. Mechanisms causing nontrophic effects
3. The nature of indirect effects in communities
4. The nature of indirect effects in ecosystems
5. Direct and indirect effects in context

Species in ecological communities interact directly with another species through consumer–resource, competitive, or mutualistic interactions. Whenever three or more species are engaged in such interactions, we see the emergence of indirect effects in which one species affects another through a shared, intermediary species. Indirect effects can reinforce or counter direct effects and lead to interesting emergent properties. This chapter explores some of the myriad ways that indirect effects emerge in communities and ecosystems. Through the use of selected examples, it shows why consideration of indirect effects is critical to a complete understanding of species interactions in ecological systems.

GLOSSARY

direct effect. The immediate impact of one species on another's chance of survival and reproduction through a physical interaction such as predation or interference

food chain. A descriptor of an ecological system in terms of the feeding linkages and energy and materials flows among major groups of species (plants, herbivores, decomposers, carnivores)

indirect effect. The impact of one species on another's chance of survival and reproduction mediated through direct interactions with a mutual third-party species

nontrophic interaction. A direct interaction that changes the behavior, morphology, or chemical composition of a species in response to the threat of being consumed

trophic interaction. A direct interaction involving the consumption of a resource species by a consumer species

1. INTRODUCTION

Imagine a herd of wildebeest grazing on a Serengeti plain. Imagine now that a prowling lion—a threat to their life—comes into their vicinity. This causes them all to stop feeding and look up in vigilance to see what the approaching predator will do. The wildebeest are nervous and tense, ready to flee at any sign of attack. Yet they are reluctant to flee because that would mean giving up feeding in a highly nutritious patch of forage, one of a few such high-quality patches currently available within a vast landscape. The resources in the patch are especially favored because they will enable the wildebeest to maximize their resource intake for growth, survival, and reproduction. The wildebeest face a critical choice: do they flee from the predator and give up the valuable food resource or do they stay and risk being captured? This choice is faced by individuals of every species of animal during the course of their daily existence. Nevertheless, the fear factor motivating this choice surely must be short-lived. After all, things will go back to normal once the predator has left or it has subdued the one victim out of the many comprising the herd, right? But the reality is, "No, not exactly."

The critical question here is: What is considered "normal"? Often the presumption is that once the predator has left, the threat disappears, and animals can resume feeding on their resources with little worry. But ecological science has revealed that this tends not to be the normal case. Instead, many individuals live in a chronic state of vigilance brought about by the fear of being captured. Ambush predators can lie in wait for long periods of time. Individuals that let down their guard and move within the vicinity of any predator lying in wait have a high likelihood of being the predator's next victim. Prowling predators can hunt in groups, so foraging individuals that do not regularly scan their surrounding environment may find themselves trapped. Normal, in many cases, means living in situations that pose continuous risks of being a predator's next victim.

The above vignette of wildebeest daily life on the plain encapsulates several key ecological concepts. First, because wildebeest are both consumers of their plant resources and at the same time resources for other consumers—their predators—they are inherently part of an ecological food chain. Their role in that food chain is identified by the kind of consumptive interaction, or technically trophic interaction, in which they are engaged. Because they eat plants, they belong to the herbivore trophic group. Their predators, because they eat herbivore prey, belong to the carnivore trophic group. By extension, species that consume mineralized nutrients and CO_2 in order to photosynthesize carbohydrates belong to the plant trophic group. To the victim (i.e., plants fed on by herbivore; herbivores fed on by carnivores), these direct trophic interactions are detrimental because it means loss of tissue or life.

But fascinating things happen when one considers how the trophic interactions play themselves out along the full length of the food chain. For instance, carnivores can lower the population abundance of herbivores through direct trophic interactions. This in turn means that there are fewer herbivores feeding on plants than in cases where carnivores are absent. Fewer herbivores mean more plants. Thus, by feeding on herbivores, carnivores provide a benefit to the plants. It is, however, an indirect benefit because carnivores do not interact directly with the plants. Rather, their effect is mediated by changes in herbivore abundance. Ecologists call this an indirect effect. Indirect effects emerge in all ecological systems whenever three or more species or trophic groups interact.

In the example of the Serengeti plain, we also see two different mechanisms causing the indirect effect of predators on plants. By engaging in a direct trophic or consumptive interaction with wildebeest, lions lower the numbers of wildebeest feeding on the plants. By scaring the wildebeest away from the resource, they also lower the number of wildebeest feeding on the plants. Moreover, by posing a constant threat that causes wildebeest to remain vigilant, they alter the rate at which wildebeest consume plants. This latter interaction between lion and wildebeest is called a nontrophic or nonconsumptive effect. Counterintutitively, by scaring prey within any given time period, predators can have a greater beneficial effect on plants through nontrophic interactions with their prey than through trophic interactions. By scaring *all* individuals within a herd, all herbivores stop feeding. In contrast, by directly killing prey, predators may only lower the number of individuals feeding on plants by the small fraction that is subdued within a given time period. Clearly, predators influence their prey populations through both trophic and nontrophic interactions, but a recent synthesis by Preiser et al. (2005) shows that nontrophic interactions can often have the stronger effect in ecological systems.

2. MECHANISMS CAUSING NONTROPHIC EFFECTS

One reason why nontrophic effects may be highly important in ecological systems is that, unlike trophic effects, which simply involve capturing and subduing prey, they come about by a variety of mechanisms involving changes in any of the morphological, behavioral, and chemical traits.

Morphological

Prey species may undergo defensive morphological changes that are induced by persistent cues of predation risk. For example, water fleas (Daphnia) that are exposed to a persistent predation threat develop spurs on their head and long, sharp tail spines. When tadpoles of many amphibian species are exposed to predation cues, they develop thick muscular tails that often are conspicuously colored. Thicker tails allow for greater acceleration to evade predator attacks, and bright tail coloration deflects the predator's attack from the vital head region of the prey to more expendable body parts. Mussels are often preyed on by snails that penetrate their shell by drilling through it. Cues from predaceous snails thus cause the mussels to develop thicker shells.

Behavioral

Predators often home in on their prey by looking for signs of prey activity. Vigilance and avoidance of risky habitat by prey are two behavioral mechanisms that can lower the risk of being captured. Becoming vigilant and decreasing the speed of movement decreases conspicuousness to predators. In addition, prey may switch their habitat use to areas devoid of predators or areas that afford greater cover. For example, in aquatic systems, species of mayflies avoid their fish predators by crawling off the surface of rocks that are covered with food resources and hiding under rocks. Snails vulnerable to crayfish predators that hunt on pond bottoms crawl up and feed on emergent vegetation in the water column. In grasslands, grasshoppers facing hunting spider predators reduce feeding on nutritious but highly risky grass and switch to leafier herbs that are less nutritious but serve as refuges.

Chemical

When fed on by herbivores, plants are often induced to produce chemicals aimed at deterring herbivore impacts. These chemicals can be quite volatile and hence be diffused into the air to attract species that are enemies of the herbivores. They can also chemically signal to neighboring plants that they have been impacted, thereby causing the neighbor plants to induce the production of chemicals that are nauseating or toxic to herbivores as a preemptive measure.

These morphological, behavioral, and chemical changes, however, come with costs. In a world with finite resources, reducing resource intake because of vigilance or allocating valuable resources toward defenses means that fewer resources are available for life-cycle development, growth, and reproduction. As a consequence reproductive output may be diminished or eliminated altogether if individuals fail to develop fully in ways that overcome seasonal environmental bottlenecks. For example, tadpoles need to develop into legged frogs or salamanders to escape their natal pond environment before it dries up in summer. Investing resources into thick tails to facilitate burst swimming to escape predators may delay body growth.

Inasmuch as these traits are properties of individuals, then, understanding indirect effects in ecological systems necessarily requires scaling from the level of the individual to the level of communities and ecosystems. And just as evolutionary history shapes individuals' ability to flexibly change these traits to balance the trade-offs, understanding indirect effects in ecological systems necessarily requires blending principles of evolutionary ecology with community and ecosystem ecology.

3. THE NATURE OF INDIRECT EFFECTS IN COMMUNITIES

Fundamental Direct and Indirect Effects

Predation, competition, and mutualism are often considered to be fundamental direct interactions that determine the structure and functioning of ecological communities. But whenever more than two species are linked together by such interactions, we see the emergence of indirect effects in which the middle species mediates the nature and strength of the indirect effect of the first species on the third species.

Perhaps the most familiar indirect effect is the one described above for the Serengeti food chain. In this kind of system, the two predators (herbivores feeding on plants, carnivores feeding on herbivores) can have negative direct effects on their respective prey through direct consumptive effects. But, once being linked in a feeding chain, the top predators have an indirect positive effect on plants that counteracts the herbivore effect by virtue of lowering the abundance of herbivore prey.

Two species may also have a negative indirect effect on each other's abundance by interacting with an intermediary species. For example, in systems in which species share a common resource but never interact directly with each other for access to that resource, their own trophic interaction with the resource reduces the availability of the resource for the other species. Here, both species are competitors (they have mutually negative effects on each other), but their effect on each other is indirect.

Three species may also compete directly with each other by preempting each others' access to space or to important resources through direct physical struggles or territorial interactions. Here, direct competitors can cause indirect interactions that are again opposite in sign to their direct effects. One species, by competing directly with a second species, relieves competitive pressure of the second species on a third species. In this case, the first species will have a positive indirect effect on the abundance of the third species.

The nature of indirect effects can also be quite different even within the same kind of ecological system, depending on whether trophic or nontrophic interactions are dominant.

Indirect Effects Caused by Morphological Changes

In a rocky intertidal system, barnacles, mussels, and algae compete directly for space on rock surfaces to which they affix themselves. Barnacles tend to be competitively dominant to mussels and thus usurp most of the space. This enables algae to fill in the interstices between the barnacles. However, a species of whelk (snail predator) prefers to prey on the barnacles and thus opens up space for mussels to become established. The mussels in turn exclude the algae. In other words, barnacles, mussels, and algae are direct competitors for space, and barnacles have a beneficial indirect effect on algae by precluding mussels from becoming established. The whelk in turn has an indirect beneficial effect on mussels and hence an indirect negative effect on algae by directly consuming barnacles. However, this outcome occurs only when snails feed on patches containing adult barnacles that are fully developed. The predatory snails have an alternative effect when they try to feed on patches of younger barnacles. Here, young barnacles develop a predation-resistant morphology when faced with predation cues. This in turn reduces the ability of the predator to suppress the barnacle's competitive dominance. This nontrophic effect leads to predator indirect effects on mussels and algae that are opposite in sign to those found when trophic interactions dominate.

Some insect species lay their eggs in autumn on the ends of plant shoots. This causes the shoots to die back because of mechanical damage. In the following spring, the plant compensates by produce longer shoots that tend to be very leafy. This change in plant morphology in turn provides new habitat promoting the population sizes of many caterpillar species that would normally not exist on the plant. The caterpillars eat the plants, but they also roll leaves to form shelters. Once the caterpillars abandon the shelters to develop into adults, the leaf rolls are inhabited by aphids and three species of ants that tend the aphids for their honeydew production and in turn protect the aphids from predators. Here, short-term plant damage can induce plant morphological responses that lead to nontrophic indirect effects that enhance the diversity and abundance of a series of insect species.

Indirect Effects Caused by Behavioral Changes

In an old-field meadow community, a species of generalist grasshopper faces spider predator species with different hunting modes. It turns out that predator hunting mode has important implications for the nature and sign of the indirect effects on plants. In the absence of predators, grasshoppers prefer to consume grass. Mortality from a species of predator with a sit-and-wait ambush hunting mode is comparatively low, but mortality risk caused by chronic predator presence in the upper vegetation canopy induces grasshoppers to switch from feeding on grass to seeking refuge in and foraging on a less nutritious goldenrod species. Thus, the sit-and-wait predator has a net positive indirect effect on abundance of grass and a net negative indirect effect on the abundance of the goldenrod induced by a nontrophic (habitat shift) interaction with the grasshopper. This happens because the spider predator presents a persistent point-source cue of presence within the habitat. The outcome is much different when the grasshopper faces an actively hunting spider species that wanders widely throughout the vegetation and thus presents a diffuse cue of presence. In this case, grasshoppers respond only to imminent predation risk when they directly encounter the predator. Actively hunting predators tend to capture many grasshoppers and thus have a strong effect on the numerical abundance of grasshoppers that overrides the nontrophic effect. This translates into a greatly reduced total numbers of grasshoppers feeding on both grasses and herbs than in the absence of predators. Such a trophic interaction leads to positive indirect effects on both grass and goldenrod.

Indirect Effects Caused by Chemical Changes

Plants can also take their defense into their own hands. In some cases, plants produce extrafloral bodies that produce nectar to attract ants. In exchange for this reward from the plants, ants defend the plant against attack by herbivorous insects. Here the plant and ant species engage in a mutualistic interaction in which the plants change their morphology to provide a direct benefit to the ant; and the ant in turn provides an indirect benefit to the plant through either trophic (eating herbivore pests) or nontrophic (scaring pests off the plant) effects. In an arid system, a species of herb (coyote tobacco) is attacked by three herbivores (the hornworm caterpillar, a beetle, and a leaf bug) that either eat plant tissue or suck plant sap. On attack by the herbivores, the plant releases volatile chemicals into the air, and the chemical plumes attract predatory insects that in turn prey on the herbivores. In this case, a trophic effect—herbivores feeding on the plant—induces a chemical response by the plant that leads to a nontrophic effect—attraction of predators—that in turn precipitates an indirect positive effect of predators on the plants.

4. THE NATURE OF INDIRECT EFFECTS IN ECOSYSTEMS

An ecosystem is a conceptualization of nature that considers the communities of species comprising a location, the rate and efficiency of energy and materials transfer among species within the community, and vital ecosystem processes such as plant production.

Ecosystems are often viewed as being organized into chains of feeding dependencies, comprised of at least three trophic levels. There are fundamentally two kinds of food chains that determine the pathway of energy and material flow throughout a system. The plant-based chain involves live plant biomass, herbivores, and carnivores. The detritus-based chain involves nonliving plant matter, decomposers, and carnivores. In both cases, plants draw up water and nutrients from the soil and carbon dioxide from the air and are stimulated by sunlight to convert those different chemicals into tissue. In the plant-based chain, herbivores eat that plant tissue and are themselves eaten by their predators. In the detritus-based chain, decomposers eat the dead plant matter and are themselves eaten by their predators. In both chains, old individuals die, and the chemical constituents of their body are also broken down by decomposers and are recycled back through the system by nourishing plants, etc.

The multilevel trophic structure of the plant-based and detritus-based chain also means that indirect effects can propagate within an ecosystem. These newly discovered indirect effects involve a combination of trophic and nontrophic interactions and influence not only the abundance of plants and plant matter but also the rate and efficiency of material cycling.

Effects on Material Flows and Production

A deeply ingrained view in ecology is that herbivores have direct negative effects on plant abundance and production by consuming plants. However, this view has been challenged in light of some observations that modest levels of herbivory might indeed enhance plant production. Such a direct and mutually beneficial effect of herbivory to both herbivores and plants is known as the grazing optimization hypothesis. It turns out, however, that this direct mutualism may only be apparent. Instead, the enhanced production may be driven by an indirect interaction in which herbivores alter cycling of nutrients that are essential for plant growth. If herbivores return nutrients back to the soil in the vicinity of their grazing locations, through urination and defecation, then higher levels of grazing may translate into proportionately higher rates at which herbivores return those nutrients to the soil than lower levels of grazing. Thus, it is the indirect effect of herbivores on plants mediated by nutrient cycling that enables plants to compensate better for loss of tissue to herbivores at intermediate levels of herbivory than at lower levels of herbivory.

Alteration of nutrient cycling and alteration of plant production have also been observed to occur across the entire food chain. For example, the nontrophic indirect effects of predators in the old-field meadow system described above lead to important and ramifying indirect effects on plant productivity, plant species diversity, and the biophysical properties of the whole ecosystem. In the absence of predators, herbivores have a comparatively weak effect on highly productive goldenrod, which allows it to grow rapidly into tall, dense stands that shade the surrounding soil. In the presence of predators, herbivore consumption both thins goldenrod stands and stunts the height of the remaining stems, thus suppressing the most productive plant species in this ecosystem. This creates a more open and patchy environment, enabling more photosynthetically active solar radiation to reach the soil surface. This in turn facilitates the proliferation of other, less-productive herb species, which are intolerant of shady conditions caused by goldenrod. The altered plant species composition of this ecosystem further causes changes in the rate of nitrogen cycling because dead goldenrod plants are more difficult to decompose than other herb species.

Introduction of foxes onto the Aleutian Island chain has had a hugely transformative effect on some of these arctic island ecosystems because the foxes substantially reduced abundant seabird populations that breed in colonies on these islands. Seabirds normally provide an important nutrient subsidy to the islands by feeding on marine organisms and then excreting nitrogen- and phosphorus-rich guano onto the islands. Nutrient-subsidized fox-free islands supported lush, thick plant communities dominated by grasses. Fox-infested islands tended to be composed of less lush low-lying herbs and dwarf shrubs. These different plant communities and their associated productivity supported different compositions and abundances of arthropod species. Thus, foxes indirectly influenced plant productivity and composition, and animal species composition, by directly disrupting a major source of nutrients to the islands. This was achieved through a combination of trophic and nontrophic effects. Devastation of seabird populations through trophic interactions causes the loss of a major vector of nutrients. The threat of future predation (a nontrophic effect) also discourages surviving seabirds from returning to the breeding colonies in later years. Over the long term, this can eliminate the offshore nutrient subsidy altogether.

The detritus-based chain is a major pathway of energy and material flow in a tropical river system. In this system, a detritivorous fish species has major effects on the cycling of carbon, an important building block of living organisms. By consuming detritus, the fish lower the abundance of dead organic matter particles in the river. This in turn indirectly lowers the abundance of algal and bacterial biofilms and enables the establishment of nitrogen-fixing bacteria that contribute to live-biomass production that serves as an important food resource for other species. In addition, consumption of water-borne particulate matter by the fish clears the water column. This enables solar radiation to penetrate deeper into the water column, thereby indirectly enhancing the production of nitrogen-fixing bacterial biomass. The fish, in turn, redistribute organic material more evenly by excreting organic matter throughout the river system, which in turn indirectly enhances the ability of other detritus-consuming organisms to exist within the river system.

Effects on Trophic Transfer Efficiencies

Trophic and nontrophic interactions can have qualitatively different effects on the efficiency of energy and nutrient transfer up food chains. In rocky intertidal ecosystems, predatory green crabs influence the behavior and foraging rate of one of its principal prey, a carnivorous snail that feeds on barnacles. When faced with predation risk, the snail becomes increasing vigilant and therefore feeds less. But it also becomes stressed. Such stress, in turn, elevates the snail's metabolic costs, thus leaving less resource available for growth and development. In other words, the efficiency at which barnacle tissue is converted into snail tissue—called secondary production—becomes diminished relative to conditions in which predation risk is absent. This finding questions the classical view that transfer efficiencies between trophic levels in ecosystems tend to be fixed. In addition, the poorer nutritive quality of snails stressed by predation risk means that the total amount of energy transferred further up the food chain to the snail's predators will ultimately be reduced. In this situation, predators indirectly harm their own welfare through nontrophic interactions with their prey. The increased attenuation of energy transfer and secondary production caused by nontrophic interactions may bolster the idea that lack of energy flow up food chains is why so many food chains in nature are short.

5. DIRECT AND INDIRECT EFFECTS IN CONTEXT

Species in communities and ecosystems are wholly dependent on other species for their survival and reproduction. These dependencies can be direct, as, for example, a predator capturing and subduing a prey species, or indirect, where a carnivorous predator may benefit a plant species by consuming the plant species' herbivore enemies. Direct and indirect effects can come about through a variety of mechanisms including trophic interactions and myriad forms of nontrophic interactions. The number of indirect effects in ecological systems rises in direct proportion to the number of species that are directly linked together in a chain of dependencies. These myriad direct and indirect interactions are what contribute to the fascinating complexity of ecological systems. A complete understanding of species interactions in communities and ecosystems therefore requires explicit consideration of direct interactions in tandem with indirect interactions.

FURTHER READING

Agrawal, Anurug A. 2001. Phenotypic plasticity in the interactions and evolution of species. Science 294: 321–326. *This article presents an important synthesis showing how species traits change flexibly as a consequence of the nature and strength of species interactions in communities.*

Hairston, Nelson G., and Nelson G. Hairston Jr. 1993. Cause–effect relationships in energy flow, trophic structure and interspecific interactions. American Naturalist 142: 379–411. *This key paper formalizes the concept of ecosystem in terms of species interactions and their influence on energy and materials cycling and transfer efficiencies.*

Loreau, M. 2001. Linking community, evolutionary and ecosystem ecology: Another perspective on plant–herbivore interactions. Belgian Journal of Zoology 131: 3–9. *This seminal article elaborates the need to link evolutionary ecology with ecosystem ecology and uses the case example of grazing optimization in plant–herbivore interactions to show how indirect effects can control processes that were previously thought to be driven by direct effects.*

Maron, John L., James A. Estes, Donald A. Croll, Eric M. Danner, Sarah C. Elmendorf, and Stacey L. Buckelew. 2006. An introduced predator alters Aleutian Island plant communities by thwarting nutrient subsidies. Ecological Monographs 76: 3–24. *This article presents a detailed analysis tracing how introduced foxes affect the source and fate of nitrogen and phosphorus and its implications for island ecosystem production.*

Ohgushi, Takayuki. 2005. Indirect interaction webs: Herbivore-induced effects through trait changes in plants. Annual Review of Ecology Evolution and Systematics 36: 81–105. *This review presents a synthesis of the ways that herbivore attack on plants can propagate morphological changes in plants that then alter the composition and abundance of arthropod species inhabiting the plants.*

Preisser, Evan L., Daniel L. Bolnick, and Michael F. Benard. 2005. Scared to death? The effects of intimidation and consumption in predator–prey interactions. Ecology 86: 501–509. *This article presents a comprehensive meta-analysis that quantifies the relative strength of trophic and nontrophic effects observed in a host of experimental studies of predator–prey interactions.*

Price, Peter W., Carl E. Bouton, Paul Gross, Bruce A. McPheron, John N. Thompson, and Arthur E. Weiss. 1980. Interactions among three trophic levels: Influence of plants on interactions between insect herbivores and natural enemies. Annual Review of Ecology and Systematics 11: 41–65. *This classic, foundational paper spells out a working hypothesis for the direct and indirect effects of plant antiherbivore defenses in ecological food chains.*

Schmitz, Oswald J. 2006. Predators have large effects on ecosystem properties by changing plant diversity not plant biomass. Ecology 87: 1432–1437. *This experimental study traces the direct and indirect effects that top predators have on plant diversity and nitrogen cycling in a meadow ecosystem.*

Skelly, David K. 1997. Tadpole communities. American Scientist 85: 36–45. *This article presents an empirical synthesis in narrative form that illustrates how amphibian community structure is influenced by trophic and nontrophic effects of predators.*

Strauss, Sharon Y. 1991. Indirect effects in community ecology: Their definition, study and importance. Trends in Ecology and Evolution 6: 206–210. *This landmark article clarifies and defines direct and indirect effects. The paper stimulated much formal experimental analysis on indirect effects in ecological systems.*

Taylor, Brad W., Alexander S. Flecker, and Robert O. Hall. 2006. Loss of harvested fish species disrupts carbon flow in a diverse tropical river. Science 313: 833–836. *This experimental study examines how a dominant fish species influences ecosystem structure through consumption and translocation of carbon in plant detritus.*

Trussell, Geoffrey C., Patrick J. Ewanchuk, and Catherine M. Matassa. 2006. The fear of being eaten reduces energy transfer in a simple food chain. Ecology 87: 2979–2984. *This article discusses the experimental study that quantifies the extent to which nontrophic effects of predators on prey limit secondary production of prey in a rocky intertidal system.*

III.6

Top-Down and Bottom-Up Regulation of Communities

E. T. Borer and D. S. Gruner

OUTLINE

1. What are "top-down" and "bottom-up" processes?
2. A history of converging views
3. A few system vignettes
4. Theory: Seeking generality
5. Moving beyond vignettes: Empirical generality and tests of theory
6. Where do we go from here?

In this chapter we briefly trace the historical debate and outline the theoretical and empirical evidence for factors controlling the biomass of predators, herbivores, and plants within and among ecosystems.

GLOSSARY

autotroph. Organisms that make their own food by synthesizing organic compounds from inorganic chemicals, usually via photosynthesis (e.g., algae, vascular plants).

biomass. The total mass of living biological material.

consumer. See heterotroph.

food web. Network of feeding relationships among organisms in a local community.

heterotroph. Organisms that must consume organic compounds as food for growth (e.g., animals, most bacteria, and fungi).

primary producer. See autotroph.

trophic. From Greek, "food," this term refers to feeding of one species on another, as in "trophic interactions" or "trophic links."

trophic level. Feeding position in a food chain: autotrophs form the basal trophic level, herbivores represent the second trophic level, and so on.

1. WHAT ARE "TOP-DOWN" AND "BOTTOM-UP" PROCESSES?

Humans are dramatically altering the global budgets of elemental nutrients that limit the growth and biomass of autotrophs, or primary producers. Through activities such as fossil fuel combustion and application of agricultural fertilizers, global pools of nitrogen and phosphorus have doubled and quintupled, respectively, relative to preindustrial levels. The impacts of these nutrient fertility bonanzas are most obvious in surface waters of lakes and coasts. Nutrient eutrophication often causes rapid and explosive blooms of algae and microorganisms and equally rapid death, decomposition, and ecosystem-wide oxygen starvation, or hypoxia. The Gulf of Mexico hypoxic "dead zone" at the mouth of the Mississippi River annually swells over areas exceeding 18,000 km^2, larger than the U.S. state of Connecticut. Nutrient eutrophication is a jarring example of a *bottom-up* process, resource supply, that can dramatically alter autotrophs and the food webs that rely on them for energy and nutrition.

Concurrently, humans are changing the role and composition of consumers in food webs via species removals and additions. Habitat loss and degradation and selective hunting and fishing deplete consumers disproportionately from food webs; many top predators such as tigers, wild dogs, wolves, and sharks have been hunted to near ecological extinction. At the same time, humans are adding consumers to food webs for endpoints such as conservation, recreation, and agriculture as well as accidentally introducing invasive consumer species. In a dramatic example, the brown tree snake (*Boiga irregularis*), a nocturnal predator, was accidentally introduced to Guam after World War II. This single species has eaten its way through

Guam's native food web, causing direct reductions or complete extinctions of dozens of native birds, bats, and reptiles, and indirect negative impacts to native arthropods, forest tree seed dispersal, and recruitment. This example highlights an extreme change in *top-down* processes, or consumption of organic biomass, that can have dramatic effects throughout food webs.

Management of algal blooms, crop fertilization, agricultural insecticide use, and wildlife conservation are prime examples in which complex interactions between bottom-up processes (i.e., fertility) and top-down processes (i.e., consumption) challenge us to better understand the critical processes that bridge communities and ecosystems. Thus, understanding the ways in which altered resources and consumer community structure interact to control the biomass of predators, herbivores, and plants is not simply a problem for basic science but one that has an immediate impact on humans. Biological control of crop pests and control of lake clarity are two management realms that draw on knowledge about these interacting processes to bring about planned changes in whole ecosystems.

Thousands of scholarly studies report on the implications of fertility manipulations and biological weed or pest control introductions for applied endpoints such as agricultural yield. We focus here on the basic science underlying such bottom-up and top-down applications. Although such factors as genetics, disease, nutrition, dispersal, and spatial structure can be critically important in structuring communities, we focus primarily on fertility and consumer controls of communities, as even this more restricted literature is quite vast. We refer to an extremely simplified theoretical community, or "module," describing one predator, one herbivore, one plant, and one nonbiological resource (e.g., nitrogen; figure 1). Most common mathematical descriptions of this module treat each level as a single species; however, these "species" often are conceptualized as unified "trophic levels," each containing multiple interacting species. We will provide empirical examples of strong top-down and bottom-up control and examine evidence across the literature for whether these are special cases or represent general patterns in ecosystems. We will end by outlining a few of the most fruitful future directions for this vibrant and rapidly progressing field of community ecology.

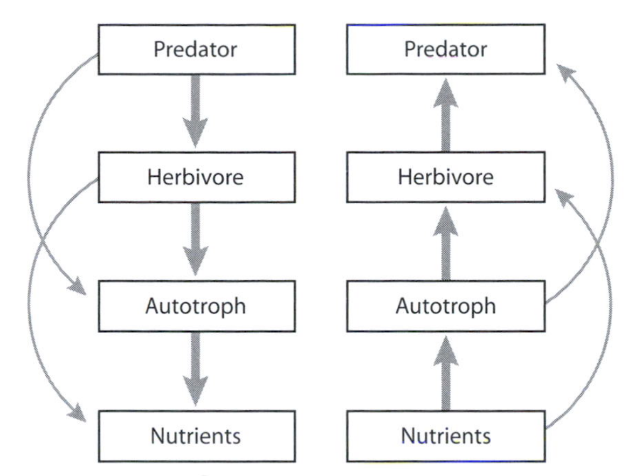

Figure 1. Community modules illustrate top-down and bottom-up direct and indirect interactions. In the first scenario, direct consumption (thick arrows) leads to indirect increases or reductions (thin arrows) in lower trophic levels and resource pools (e.g., soil nitrate). This scenario represents top-down consumer control of a community. In the second scenario, exemplifying bottom-up community regulation, fertilization directly and indirectly alters the abundance of autotrophs and consumers. These modules may define an entire community with only a single species at each level, may be composed of several species interacting within each level (e.g., several herbivore or plant species), or may be part of a larger, more complex food web.

2. A HISTORY OF CONVERGING VIEWS

First, let us not lose sight of the forest for the trees. Terrestrial biomes (e.g., tropical savannah, desert, arctic tundra) are defined by their dominant plants, which, in turn, grow in these regions primarily because of the regional combination of solar radiation, temperature, and precipitation. Aquatic systems have parallel, broadly defined regions (e.g., kelp forest, coral reef), with their location determined, in large part, by the regional combination of light availability and temperature. These broadly defined regions in both aquatic and terrestrial systems tend to have characteristic nutrient availability, but even the most nitrogen-rich oceans contain only about 1/10,000 the nitrogen of topsoil. At regional and local scales in all systems, nutrient supply and consumption may be virtually irrelevant for determining community biomass in habitats with extreme physical disturbance, such as those exposed to heavy surf or volcanic activity.

In addition, although herbivores such as sea urchins, locusts, or rabbits can decimate plant biomass in some circumstances, as a group, herbivores consume less than an estimated 20% of annual terrestrial plant production, less than 40% of ocean production, and approximately 50% of freshwater production. Although herbivores such as crop pests can consume large proportions of crop biomass, herbivores substantially control the biomass of all producers in a region in only infrequent and notable cases. Thus, consumers can be important drivers of producer biomass patterns only if producers have sufficient light, precipitation, and nutrients, and physical disturbance is not extreme or frequent. Given this global context, in this chapter we step a bit closer to examine the roles of altered fertility and consumption *within* communities.

Our current understanding of the interactive effects of fertility and consumption in controlling the relative abundance of plants and consumers rests on the refinement of historical debates. As in most fields of research, this debate began with two opposing views. Both views were well supported by empirical evidence, and both had grounding in mathematical theory.

1. Bottom-up: Fertility is the key to understanding plant biomass, which, in turn, controls the biomass of consumers.
2. Top-down: Consumers control prey, such that predators reduce herbivore biomass and release plant biomass from herbivore control.

Charles Elton first proposed a "pyramid of numbers" in which primary producers dominate and consumer biomass decreases as trophic levels become more remote from the base of production. This generality seems to agree with our observations of terrestrial systems, but aquatic ecosystems often violate Elton's rule with inverted biomass pyramids, or ratios of heterotroph-to-autotroph biomass (H:A) greater than 1. By observing successional transitions in producer quality from lakes to bogs to terrestrial communities, Raymond Lindeman reconciled this aquatic–terrestrial contrast by hypothesizing systematic and taxonomic differences in trophic conversion and assimilation efficiencies. This hypothesis explained both the increasing domination by plants in terrestrial habitats and, ultimately, the limitation of energy reaching the top consumers across all food webs.

This bottom-up view largely prevailed until Hairston, Smith, and Slobodkin (HSS) introduced the classic top-down alternative, that predators protect the "green world" from rabbits and sea urchins by regulating their densities below outbreak levels that could decimate producers. Robert Paine later coined the term *trophic cascade,* as an indirect effect of predators on plant biomass via consumption of herbivores, to describe this phenomenon. Numerous empirical examples, notably from aquatic or relatively simple terrestrial systems (e.g., monoculture crops), confirmed that experimental removal of predators could cause reductions in plants via increased herbivory.

Early criticism of HSS noted that primary producers are neither uniformly edible nor immediately available to consumers. Physical attributes, such as spines, trichomes, or tough and thickened leaf tissues, as well as a variety of constitutive or rapidly inducible chemical compounds, protect producers from herbivore consumption. A body of optimality theory predicted (1) that plants should develop adaptive defenses in proportion to the risk of herbivore consumption and (2) that available resources should constrain plant defenses, with trade-offs to other plant functions such as growth or reproduction. Some authors argued forcefully that consumers were irrelevant in tropical and temperate forests, deserts, and many other terrestrial habitats. Traits that confer strong competitive ability for limited resources on land (e.g., woody stems to grow for light) also reduce consumption and assimilation of plant biomass to herbivores and higher trophic levels.

With theoretical advances and increasing empirical data, the dialectic of top-down versus bottom-up forces yielded to a nuanced view acknowledging the dual role of these pressures within communities. Oksanen and colleagues used a simple mathematical model to propose that the total community biomass, number of trophic levels, and strength of top-down pressure in food webs should depend ultimately on the productivity of a system, thus melding top-down and bottom-up paradigms into a single hypothesis (see box 1). This hypothesis added the twist that food chain length can predict whether predators should have positive or negative indirect effects on producers. In three-level examples envisioned by HSS, predator presence should increase plant production relative to the same community lacking predators. In four-level food chains, the top predator releases herbivores indirectly by consuming third-level predators, thereby indirectly controlling producers (box 1). This theory predicts that the effects of productivity and predation should remain strong across all trophic links; however, Menge and Sutherland suggested an alternative model in which the effects of nutrients and predation attenuate as they travel through a food web. In this case, resource supply and species competition should most strongly control autotrophs, but at higher trophic levels, predation should increase in its controlling effect.

BOX 1

$$\frac{dR}{dt} = rR\left(1 - \frac{R}{K}\right) - aRH$$

$$\frac{dH}{dt} = caRH - aHC - dH$$

$$\frac{dC}{dt} = caHC - dC$$

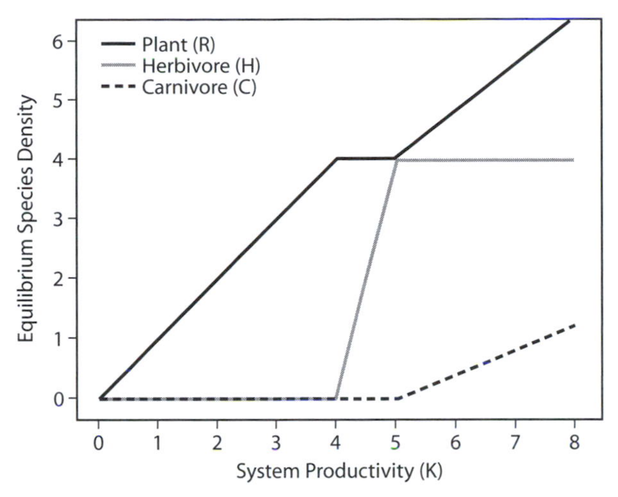

Box Figure 1. This simple mathematical model of a plant resource (R), an intermediate herbivore (H), and a top carnivore (C) produces a variety of predictions across a gradient of increasing productivity, including increased species richness, increased numbers of trophic levels, and changing relative abundances of each species.

These equations provide a dynamic mathematical description of an extremely simplified linear trophic web with a plant resource (R), an intermediate herbivore (H), and a top carnivore (C). Here, the plant increases via logistic growth to a fixed carrying capacity (K). Plants and herbivores are consumed only by the adjacent trophic level; there is no omnivory in this description. For simplicity, the attack (a), conversion efficiency (c), and death (d) rates are the same for both consumers. These equations allow us to make predictions for a community module with this structure at equilibrium with increasing system carrying capacity, K. The plot in box figure 1 shows the equilibrium predictions for this community module when $r = 10$, $a = 0.5$, $c = 0.5$, and $d = 1$.

System productivity controls total community biomass and the number of trophic levels, whereas consumption controls the distribution of biomass among trophic levels. Higher trophic levels can easily be added to this mathematical description; communities with an odd number of trophic levels (e.g., three: plant, herbivore, and predator) are predicted to release plants, and those with an even number of trophic levels (e.g., four: plant, herbivore, intermediate predator, and top predator) suppress plant growth.

3. A FEW SYSTEM VIGNETTES

The classic paradigm of a top-down trophic cascade is exemplified by lake systems with phytoplankton as the dominant primary producers. Phytoplankton are typically grazed by zooplankton (e.g., *Daphnia* spp.), which are consumed by small planktivorous fish (e.g., minnows). Whole-lake manipulations of piscivorous fish (e.g., bass) have shown that three-level chains lacking piscivorous fish result in green lakes because planktivorous fish (third trophic level) limit zooplankton (second level) and release phytoplankton (primary producers), whereas blue lakes result where piscivorous fish are added as the fourth trophic level. A great number of empirical examples show similar top-down scenarios from lake systems, and an entire class of restoration techniques focuses on the removal or addition of trophic levels ("biomanipulation") to achieve desired states of primary productivity and water clarity in lakes.

A rare example of a terrestrial top-down trophic cascade comes from Isle Royale National Park, Michigan. Time series data revealed that the width of annual tree growth rings on balsam fir (*Abies balsamea*) correlated positively with the annual abundance of wolves (*Canis lupus*) and negatively with moose (*Alces alces*), the prey of wolves and the dominant browsing herbivore on the island. This example also demonstrates that pathogens or parasites may act as effective consumers in food webs. An outbreak of canine parvovirus decimated the wolf population in the 1990s, releasing the moose population from top-down control, again in the manner expected from theory for four-trophic-level chains (figure 2A). Controversy for years raged over whether terrestrial trophic cascades were empirically rare because they are less prevalent and powerful than those in aquatic systems or because they are simply more difficult to study at appropriate scales of time and space.

Top-down and bottom-up forces can shift in their relative importance over large spatial and temporal scales, as exemplified by studies from several marine communities. Along the Pacific coast of the Americas, kelp forests dominate hard substrates in the shallow subtidal zone, providing resources and habitat structure for a complex food web of microbes, algae, invertebrates, fish, and mammals. Seminal work from Alaska has shown that sea otters (*Enhydra lutris*) consume invertebrate grazers, such as sea urchins, which in the absence of otters can completely eliminate kelp forests and shift the community to an alternative state ("urchin barren"). Historical declines in large marine mammals recently have induced orca whales (*Orcinus orca*) to increase their predation intensity on otters, thus reducing the protective role of otters for kelp in a manner predicted by food chain theory (figure 2B; box 1). Geographic variation in seawater temperature may ultimately drive the relative importance of top-down and bottom-up forces in this kelp forest system. In the

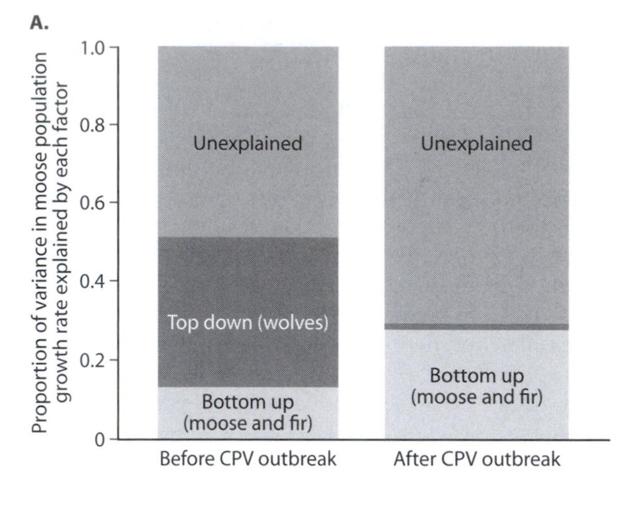

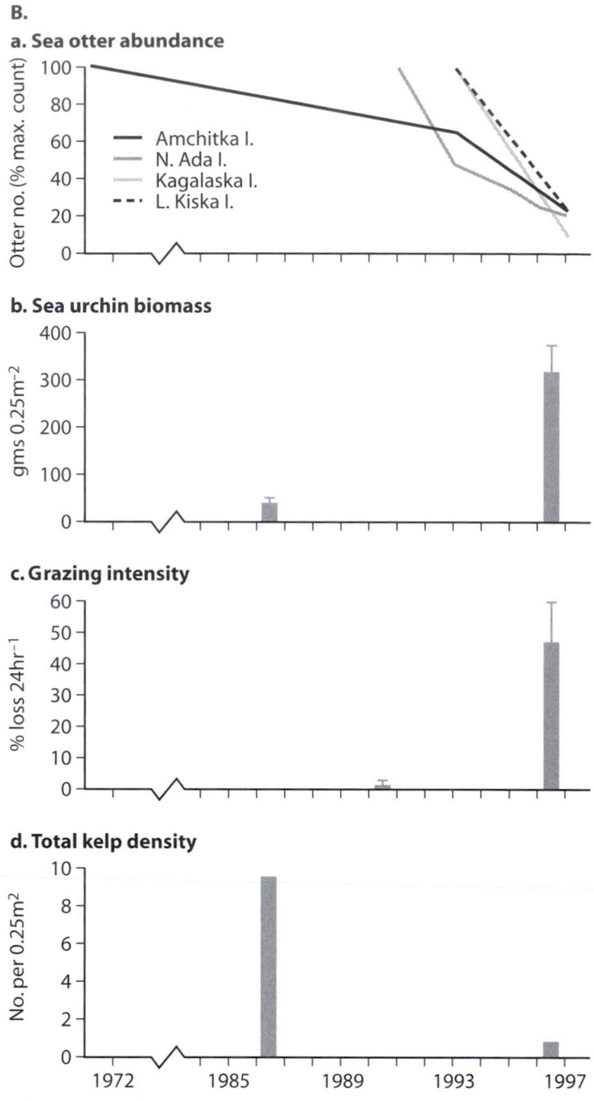

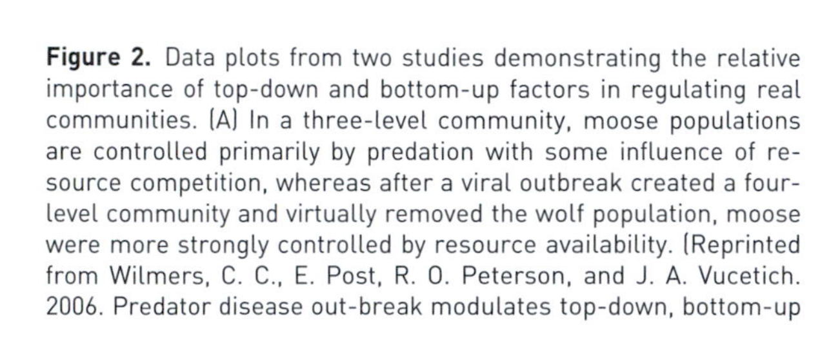
Figure 2. Data plots from two studies demonstrating the relative importance of top-down and bottom-up factors in regulating real communities. (A) In a three-level community, moose populations are controlled primarily by predation with some influence of resource competition, whereas after a viral outbreak created a four-level community and virtually removed the wolf population, moose were more strongly controlled by resource availability. (Reprinted from Wilmers, C. C., E. Post, R. O. Peterson, and J. A. Vucetich. 2006. Predator disease out-break modulates top-down, bottom-up and climatic effects on herbivore population dynamics. Ecology Letters 9: 383–389) (B) Evidence for both three- and four-level top-down control. (Reprinted from Estes, J. A., M. T. Tinker, T. M. Williams, and D. F. Doak. 1998. Killer whale predation on sea otters linking oceanic and nearshore processes. Science 282: 473–476) In the 1980s, sea otter (a) and kelp (d) biomass was high, whereas sea urchin biomass (b) was low. In contrast, in the late 1990s, when otter abundance declined, urchins reached high biomass, and kelp was rare.

cold, nutrient-rich waters of Alaska, the nitrate needed for kelp growth is seldom limiting, and predators instead control food web dynamics. In the warmer waters along California's southern coast, by contrast, otters apparently play a diminished role, and nitrate concentrations decline precipitously unless replenished by upwelling events that churn nutrients off the sea floor. Warm El Niño–Southern Oscillation (ENSO; coupled oceanic and atmospheric phenomena in the Pacific Ocean causing altered sea surface temperature, winds, and rainfall) events further intensify the nutrient limitation of kelps and can cause forest diebacks on broad scales. A similar pattern of geographic variation has been shown on the Northern Atlantic shelf, where food webs in colder waters were more sensitive to heavy exploitation of cod—and were more likely to be controlled by top-down trophic cascades—than the more resilient, species-rich food webs in warmer subtropical, nutrient-limited waters.

4. THEORY: SEEKING GENERALITY

A relatively simple mathematical description of single-species interactions (box 1) incorporates both system fertility and consumptive interactions. This mathematical description predicts that, at equilibrium, predators should benefit plants indirectly via a trophic cascade. Thus, communities lacking predators should have less plant biomass and more herbivore biomass than a similar community with predators present. But this same mathematical description also predicts that alterations of system fertility or productivity should affect all trophic levels of the community such that more enriched communities should contain relatively more predator biomass than comparable, but less enriched, communities. This mathematical description allows us to develop logical, if quite general, hypotheses about the resulting biomass of plants and herbivores when we examine or manipulate fertility or consumers in real communities.

As born empiricists observing our world, we know that most communities are composed of many species of interacting plants, herbivores, and predators, so this description of a whole community using a three-species module (box 1), with a single predator species specializing on a single herbivore species, which in turn specializes on a single plant with a single limiting resource, offends our empirical sensibilities. Obviously, this mathematical description is an extreme simplification of any natural community. Given its simplicity, it is surprising how often this exceedingly simplified mathematical description matches our observations of communities, as in the lake, Isle Royale, and kelp forest communities described earlier in this chapter (figure 2).

There are many examples of communities in which fertilization and consumer addition or removal do not support the predictions of our simple model. In these cases, complexities of the community appear to either enhance (rarely) or reduce (commonly) the predicted direct and indirect effects of fertility and consumption. Most community ecologists have a pet hypothesis about when and where top-down or bottom-up effects should be most apparent (box 2), and the simple theory presented here (box 1) has been extended in many ways to produce testable predictions. In many cases, a slight addition of complexity can aid our logic and increase our predictive power. For example, modules of single-species consumptive interactions (figure 1) can exist within more complex food webs, but even species-poor communities commonly have several species in each trophic level. In fact, compensation within the plant community caused by a trade-off between abilities to resist herbivory and compete for limited resources tends to reduce both direct and indirect top-down effects. Top-down and bottom-up effects can still interact, even within complex communities; in communities featuring both competition within a trophic level and consumption among levels, top-down effects are predicted to be strongest at low to moderate productivity levels. Similarly, herbivore avoidance of predators or preference among plant species, for example, can cause a variety of community responses to enrichment, including altered composition and biomass of both plants and herbivores.

BOX 2

When is consumption most likely to indirectly affect plant biomass? When should we expect fertility to control community composition? A plethora of hypotheses exist in the literature to address these questions, some based in models, others based in logical arguments. These hypotheses range from biological to methodological, falling into broad categories that in many cases are interlinked:

Spatial heterogeneity. Predators will be ineffective if herbivores can hide. Overall, predators will control the biomass of herbivores and plants in less complex habitats with few herbivore refuges.

Food webs that deviate from a linear food chain. Real communities are often better described as a food *web* than a food *chain*. Trophic interactions such as omnivory, competition, or intraguild predation, or behavioral interactions such as avoidance or territoriality, tend to increase interference and reduce the overall impact of each consumer level on the next. Predators should have little control over the biomass of subsequent trophic levels in complex communities, leading to stronger bottom-up control.

Predation risk, nonconsumptive effects, and flexible foraging. At intermediate levels in food chains, herbivores must balance the trade-off between eating and being eaten. Predator cues or the threat of predation may shift the quality and quantity of herbivore dietary intake—effects that can cascade to change plant community composition and ecosystem processes without consumptive changes in herbivore densities.

Species turnover. Changes in nutrient and consumer regimes can precipitate turnover in community species composition. Herbivore pressure can transform plant communities to favor less edible, more defended plants or may select for abundance of tolerant plants that rapidly recover to replace less-tolerant species. These changes may depend on nutrient resource supply.

Communities at equilibrium. When they experimentally examine the effects of predation on herbivore and plant biomass, studies measuring treatment effects before plants have had time to regenerate are likely to measure only the effects of consumption, not the effects of long-term alterations in equilibrium biomass. In addition, because herbivory can lead to a change in the composition of the plant community, short-term studies may produce a biased estimate of the true long-term effects of predator removal on plant and herbivore biomass. Finally, communities subject to frequent or intense disturbance may never reach equilibrium, reducing the importance of top-down or bottom-up controls.

Resource availability and quality. Where plant resources are nutritious and easy to consume, consumption rates by herbivores should also be higher. Thus, herbivores should strongly control the biomass of nutritious plants, and reduction of herbivores by predation should cause greater increases in plant biomass in a community with nutritious plants than in a similar community with less nutritious plants.

Predator or herbivore efficiency. Consumers with low metabolic costs (e.g., many invertebrates) can convert a higher percentage of what they consume into reproduction than can those with higher metabolic costs (e.g., mammals). In general, consumers that are extremely effective at converting food into progeny should effectively control the biomass of their resource, leading to greater control by predators over the biomass of herbivores and plants.

5. MOVING BEYOND VIGNETTES: EMPIRICAL GENERALITY AND TESTS OF THEORY

Historically, empiricists have approached ecology with a system-specific mindset, tending to think and work within a single ecosystem type. Ecological generalities, such as the factors determining the relative importance of consumers and fertility, are indeed interesting and important within systems but are not constrained by system boundaries. In fact, examining and manipulating consumers and fertility both within and among ecosystems is a promising avenue for confirming, refuting, or suggesting new hypotheses about when fertility or consumers should control community biomass distribution (see box 2). Clearly an important step in discovering generalities about communities is to quantify the relative importance of consumers and fertility across gradients in fertility and disturbance, species composition, and community complexity as well as among ecological systems. But a single, replicated, manipulative experimental study to successfully accomplish this goal is virtually impossible to envision.

The literature examining single cases of consumer addition or removal under a variety of fertility regimes has grown in the decades since the publication of HSS and spans a broad range of systems, taxa, spatial and temporal scales, and food web complexity. Drawing from this diverse literature, recent quantitative syntheses have examined the emergent evidence for the relative influence of top-down and bottom-up control of communities across numerous empirical case studies. Most of these reviews and meta-analyses have focused on within-system comparisons, but several also have examined the evidence for fertility and consumer controls among systems.

These quantitative analyses of the existing empirical data show a great deal of variation in the relative importance of consumers and fertility in regulating H:A ratios and community composition. However, some consistent patterns emerge from these analyses. Both top-down and bottom-up forces can have substantial food web effects; the direct effects of nutrients and herbivores on plants and predators on herbivores are often quite strong. However, predators can impart strong top-down effects across entire food chains, whereas nutrients primarily affect plant productivity. When responses are compared among systems, plant biomass tends to increase with fertilization in all systems, whereas herbivory consistently suppresses plants in oceans and lakes but has variable effects on land. The indirect effects of predators on plants tend to be strongest in some marine and lake communities and weakest in streams and on land. Overall, the herbivore–plant link is weak; in contrast to the predictions of simple community theory (box 1), herbivores in predator-free communities appear unable to take advantage of increases in ecosystem productivity. Regulation of community biomass by consumers appears to be associated with consumer metabolism, an intriguing finding that deserves further experimental examination across a variety of taxa (see Underrepresented Taxa and Systems in the next section).

6. WHERE DO WE GO FROM HERE?

After decades of exhaustive theoretical, empirical, and synthetic investigation into the roles of top-down and bottom-up forces in diverse ecosystems, our understanding of the relative strengths of these processes has come a long way since Elton and Lindeman. Even so, there remain many rapidly progressing areas in which major questions remain. Here, we outline a few exceptionally promising avenues for future research.

- Community composition. Most communities are made up of many interacting species at each trophic level, and we are often as interested in which species are present as we are in the total biomass of a trophic group. Many experiments manipulating nutrients or consumers have found that the composition of a community (e.g., dominant species, invasive species) can change dramatically while biomass at each trophic level remains relatively constant. As one species within a trophic level declines in abundance, another may increase, leading to a minimal effect of consumers or resources on biomass but a dramatic effect on which species are present. In Western European heath/moorlands, for example, increased nitrogen deposition has been blamed for the decline in cover of native *Calluna* heather, bryophytes, sedges, and rushes and their replacement by nonnative grasses. Grazing by sheep can increase the negative feedback on native heath and moorlands and accelerate species turnover without appreciable changes in overall biomass. In this case, top-down and bottom-up effects do not interact to affect overall producer biomass, but they do shift the composition and nature of this plant community in substantial ways. In addition, the degree to which discrete trophic levels exist can determine the response of the entire community to altered nutrients or consumers. Thus, a more general predictive understanding of the ways in which top-down and bottom-up perturbations alter the number and identities of species in trophic groups and whole communities will make an important contribution to this field.
- Stoichiometric constraints. Food is not uniformly nutritional. Stoichiometric models show that explicit incorporation of chemical mass balance into mathematical descriptions of consumption can produce predictions that differ from those of population-level models. For example, one algae–herbivore stoichiometric model predicts that algal biomass should increase with nutrient addition; however, algal quality should decline, causing herbivore nutritional limitation in spite of increased algal quantity. This example clearly demonstrates the importance of nutrition and chemical mass balance in making predictions about biomass distribution in communities. Further research into defining and examining the consequences of nutritional quality of resources will foster a more thorough understanding of the conditions under which nutrition matters to biomass distribution throughout communities.
- Factorial experiments. Every recent quantitative synthesis of empirical data has bemoaned the paucity of factorial manipulations of consumers and resources. Whereas hundreds of studies to date have concurrently manipulated nutrients and herbivores, only a few dozen studies have simultaneously manipulated predators and nutrient concentrations and measured the effect on both plants and herbivores. Factorial experiments in terrestrial systems, particularly nonherbaceous communities, lag behind those in lakes and oceans. Well-replicated, large-scale full factorial manipulations are necessary to quantitatively assess the strength and direction of control for community biomass.
- Underrepresented taxa and systems. Our understanding of generalities in community regulation among marine, terrestrial, lake, and stream communities is hampered by the limited suites of species that have been studied in these systems. For example, the vast majority of studies of terrestrial communities examine aboveground arthropods in herbaceous-dominated systems, whereas lake, marine, and stream studies are dominated by those with vertebrate predators. Resource and consumer controls of belowground terrestrial communities remain virtually unstudied. Vertebrate herbivores are rarely examined, particularly in studies combining nutrient and predator manipulations. The majority of terrestrial studies enumerate the producer species, whereas lake and stream studies tend to report only total producer biomass. The bottom line is that our understanding of when, how, and where communities are regulated by nutrients or consumers is limited by the combinations of species that have been studied, to date.

We thank all our past and current mentors, colleagues, and collaborators who have shaped our thinking on this topic. We also thank Angela Brandt, Matt Bracken, and Michel Loreau, who read and commented on drafts, and Eric Seabloom, who assisted with *R* coding for box 1.

FURTHER READING

Borer, E. T., B. S. Halpern, and E. W. Seabloom. 2006. Asymmetry in community regulation: Effects of predators and productivity. Ecology 87: 2813–2820. *This quantitative meta-analytic review examines predator removal experiments across productivity gradients and in factorial fertilizations in marine, freshwater, and terrestrial systems.*

Gruner, D. S., J. E. Smith, E. W. Seabloom, S. A. Sandin, J. T. Ngai, H. Hillebrand, W. S. Harpole, J. J. Elser, E. E. Cleland, M.E.S. Bracken, E. T. Borer, and B. M. Bolker. 2008. A cross-system synthesis of herbivore and nutrient resource control on producer biomass. Ecology Letters 11: 740–755. *This synthesis uses meta-analyses to test predictions from simple mechanistic models and demonstrates strong and independent effects of herbivores and nutrient resources on plant community biomass that are broadly similar in relative magnitude across ecosystem types.*

Hairston, N. G., F. E. Smith, and L. B. Slobodkin. 1960. Community structure, population control, and competition. American Naturalist 94: 421–425. *This foundational paper argues via a verbal model for top-down control maintaining high plant biomass; often referred to as the Green World Hypothesis.*

Hillebrand, H., D. S. Gruner, E. T. Borer, M. E. Bracken, E. E. Cleland, J. J. Elser, W. S. Harpole, J. T. Ngai, E. W. Seabloom, J. B. Shurin, and J. E. Smith. 2007. Community structure and ecosystem productivity mediate the intrinsic control of producer diversity across major ecosystem types. Proceedings of the National Academy of Sciences, U.S.A. 104: 10904–10909. *This meta-analysis examines results from hundreds of published experiments and demonstrates how top-down and bottom-up forces can interact to change producer composition even when biomass is not affected.*

Leibold, M. A., J. M. Chase, J. B. Shurin, and A. L. Downing. 1997. Species turnover and the regulation of trophic structure. Annual Review of Ecology and Systematics 28: 467–494. *This is a review of models and evidence for fertility-driven shifts in species composition that alter the strength of trophic interactions.*

Lindeman, R. L. 1942. The trophic–dynamic aspect of ecology. Ecology 23: 399–418. *This article presents a classic synthesis, published posthumously, that anchored the bottom-up perspective as the dominant paradigm for a generation of ecologists. This paper merged the ecosystem-level ideas of nutrient and energy flux with community-level feeding relationships to explain long-term dynamics of plant community succession.*

Matson, P. A., and M. D. Hunter, eds. 1992. Special feature: The relative contributions to top-down and bottom-up forces in population and community ecology. Ecology 73: 723–765. *This special feature included four seminal contributions (authored by Hunter and Price, Power, Strong, and Menge) that surged this topic to the forefront of community ecological thinking. All contributions argued for a pluralistic view that embraces the complexity and heterogeneity of real ecosystems and communities.*

Menge, B. A., and J. P. Sutherland. 1976. Species diversity gradients: Synthesis of the roles of predation, competition, and temporal heterogeneity. American Naturalist 110: 351–369. *This foundational verbal model argues that the relative importance of nutrients and consumption should vary with environmental disturbance: consumption should structure communities that experience low physical disturbance but should be relatively unimportant in physically disturbed habitats.*

Oksanen, L., S. D. Fretwell, J. Arruda, and P. Niemela. 1981. Exploitation ecosystems in gradients of primary productivity. American Naturalist 118: 240–261. *This article melded the bottom-up and top-down paradigms into a single mathematical framework with predictions for food chain length and strength of trophic cascades.*

Polis, G. A. 1999. Why are parts of the world green? Multiple factors control productivity and the distribution of biomass. Oikos 86: 3–15. *This article presents an engaging narrative review of broad-scale patterns and case studies showing the multiplicity of potential abiotic and biotic limiting factors that alter the relative strengths of top-down and bottom-up control.*

Polis, G. A., and D. R. Strong. 1996. Food web complexity and community dynamics. American Naturalist 147: 813–846. *This article presents a critical appraisal of simple food chain models of trophic dynamics that argues that strong indirect trophic interactions in complex food webs are the exception, not the rule.*

III.7

The Structure and Stability of Food Webs

Kevin McCann

OUTLINE

1. Introduction
2. Diversity and stability: The early years
3. Robert May and the limits to diversity
4. Diversity and food web structure
5. Structure, variability, and stability
6. Future directions: Food webs across space and time

The role of diversity and structural complexity in the dynamics and stability of ecosystems is a longstanding and unresolved issue in ecology. Here, I review the history of this major ecological problem and highlight three relatively distinct historical periods in thought. The first period was one of mostly intuitive belief that suggests nature's diversity gives rise to stability. This period was followed by a second that arose with the rigorous application of mathematics and dynamic systems theory that, more or less, puts this intuitive belief to the test. This theoretical result ultimately pushed ecologists to look beyond diversity to understand the dynamics of these complex natural entities. In response to this theory, a group of intrepid empirical ecologists began to map real food webs and so begin the search for patterns in food web structure. More recently, conceptual developments in ecology have begun to consider how specific food web modules (i.e., common natural food web structures) and variability in space and time govern the stability of ecological systems. The emerging answer appears to suggest that the variability itself may ultimately be responsible for the persistence of these enormously complex entities.

GLOSSARY

food web compartment/channel. A highly and strongly connected set of species (i.e., subweb) that connect with much lower frequency and much lower strength to other species in the larger web.

food web connectance (*C*). Given *S* species in a food web, then connectance is the number of actual links or interactions (*L*) divided by the maximum possible links (S^2), so $C=L/S^2$.

food web modules/motifs. All possible topologies of sub–food webs of *n*-species; thus, a specific module or motif consists of a given two-species interaction (e.g., predator–prey, mutualism), three-species interaction (food chain, omnivory, etc.).

food web or ecological network. A set of species that are connected to one another via trophic interactions (i.e., fluxes of matter and energy).

food web pathways. A directed set of interactions from any one species to another (e.g., a resource to consumer to a predator of the consumer).

food web structure. At its most general level, nonrandom patterns in the food web topology, interaction strengths, densities, and other ecological traits (e.g., age structure). As one example, some authors have argued that omnivory is ubiquitous and so is found in real food webs more than expected in randomly constructed food web networks. Network analysts use "motifs" to ask if there is a specific topology that is significantly overrepresented relative to random networks.

interaction strength (IS). The dynamic influence of one species on another. This is measured in a variety of metrics, but some standard measures have emerged. (1) Direct metrics: these measures estimate the direct influence of one species on another. Energy or biomass flux has been frequently employed (e.g., the IS of predator on prey is equivalent to the amount of biomass consumed by the predator). Another similar measure is the elements of the Jacobian matrix that assesses the instantaneous rate of change of one species with respect to a very small change in density of another species. (2) Indirect metrics: often employed in the field, these metrics assess the change of one species with respect to the change in density of another species (often the complete exclusion of the species). These are not instantaneous measures and so include both direct (one species immediate effect on the other species) and indirect consequences (e.g., the excluded species has effect on other species, which in turn changes the focal species density).

stability. There exist several common dynamic measures of stability that generally assess the food web's rate at which it returns to some defined aspect of its food web structure or a food web's ability to retain some defined aspect of its structure. (1) Resilience: dynamic response to a temporary perturbation (e.g., equilibrium or nonequilibrium stability; effectively tracks how fast densities return to their original values). (2) Variability: the variability in population dynamics of individual species or groups of species. Large variability is assumed to mean that there is significant chance of losing food web structure in a "noisy" world (a species at low density may readily experience local extinction). (3) Resistance: the ability to retain structure in the face of a perturbation (e.g., a community is resistant to an invasive species).

1. INTRODUCTION

The role of diversity and structural complexity in the dynamics and stability of ecosystems is a longstanding and unresolved issue in ecology. Here, I review the history of this major ecological problem and highlight three relatively distinct historical periods in thought. A period of mostly intuitive belief that suggests nature's diversity gives rise to stability was followed by a second period that arose with the rigorous application of mathematics and dynamic systems theory that, more or less, put this intuitive belief to the test. This theoretical result ultimately pushed ecologists to look beyond diversity to understand the dynamics of these complex natural entities. In response to this theory, a group of intrepid empirical ecologists began to map real food webs and so begin the search for patterns in food web structure. More recently, conceptual developments in ecology have begun to consider how specific food web modules (i.e., common natural food web structures) and variability in space and time govern the stability of ecological systems. The emerging answer appears to suggest that the variability itself may ultimately be responsible for the persistence of these enormously complex entities.

2. DIVERSITY AND STABILITY: THE EARLY YEARS

The early ecological interest in diversity and stability revolved largely around intuitive interpretations. As far back as Charles Darwin's tangled bank, scientists have tended to suggest that diverse, highly interconnected systems might be responsible for the persistence and consistency of natural systems. As such, early ecologists attempted to create a logical basis for this belief. E. P. Odum, in his 1953 book, *Fundamentals of Ecology*, simply came out and defined community stability as "the amount of choice which the energy has in following the paths up through the food web." The specific reason for the choice of this definition was not explicitly discussed, but it is arguable that Odum's definition came from the intuitive idea that portioning up the energetic pie ought to stabilize an organism's food supply. The precise assumptions behind this intuitive argument, however, remained a mystery.

Not long after Odum's definition, Robert MacArthur (1955) published a well-known paper that attempted to give some rigor to Odum's earlier statement. In this paper, MacArthur sketched a series of simplified food webs and discussed the ramifications of energy partitioning for stability using elementary arguments from information theory. His idea, although intriguing, still reads more like an intuitive appeal masqueraded in mathematics. MacArthur's theory, for example, completely sidestepped population and community dynamics. By ignoring internal dynamics MacArthur was effectively allowed to define stability in a manner identical to Odum. Sadly, the approach gives us little additional mechanistic insight, though, because many of the uncertainties behind diversity and its relationship to dynamic stability depend on the dynamics of the interacting species (e.g., predators can make their prey oscillate). It is important to point out that, as with Odum, the notion that multiple pathways in a food web might stabilize ecological systems remained a possibility. The logic, however, was still far from complete.

Shortly after MacArthur's semiformal treatment, Charles Elton (1958) took an entire chapter in his famous book, *The Ecology of Animal and Plant Invasions*, to explore the relationship between diversity and stability. His approach, like the others, was largely intuitive, although it drew from earlier simple theoretical models and anecdotal empirical evidence pertaining to the influence of invasions on ecological communities. Elton presented six lines of reasoning that can be summarized using three more broadly defined themes. Although his definition of stability vacillated, he went to great length to emphasize that he was generally considering dynamic instabilities (see definitions above) that drove "destructive oscillations" and "population explosions" in food webs.

First, Elton felt that because simple model systems were subject to extraordinary instabilities, then increased stability would likely accompany increased model complexity and diversity. Here, Elton was drawing from the early theoretical work of Alfred Lotka and Vito Volterra that produced neutrally stable dynamics and the fact that simple model laboratory experiments (i.e., microcosms) had proven to be the very definition of instability (e.g., microcosms are frequently so unstable that stability is measured as time to extinction). Second, he argued that simplified food webs were more vulnerable to invaders. In this case, he relied on accumulating empirical evidence that suggested that pest outbreaks occur readily in monocultures and other habitats greatly simplified by human impact. Third, he argued that island food webs were notorious for extensive impact incurred by invasive species. Here, the logic was that island food webs are less diverse than mainland webs. Because diversity was related to all aspects of these arguments, this implied diversity might be positively related to stability.

All these lines of reasoning are interesting and together compelling; however, they also clearly included a suite of other factors that obfuscate the exact role of diversity. The mere fact that diversity correlates with all Elton's arguments does not necessarily imply that diversity is the governing force behind his three generalizations. As an example, each of the three generalizations offered by Elton is also clearly related to spatial scale. Microcosms, monocultures, and models take place in a spatially homogeneous arena, whereas island food webs are clearly systems limited in spatial extent relative to their continental counterparts. One could just as easily argue that space is the actual driver behind Elton's arguments.

Elton's line of reasoning attempted to be dynamic (e.g., Elton's destructive cycles); however, these arguments, like Odum and MacArthur's arguments, were clearly not yet rigorous theoretical applications. The mathematics of dynamic systems was not yet a part of the ecologist's toolbox. Soon things were to change as a number of mathematicians and physicists were entering ecology and clearing the way to challenge some of these more intuitive ideas. They were to find that this early reasoning, although seemingly sound, had some logical holes.

3. ROBERT MAY AND THE LIMITS TO DIVERSITY

In the early 1970s, mathematical ecologists began to wrestle with the diversity–stability problem for large model communities. This was the start of a time that did much to point out the limits of diversity as a stabilizer. The development of computer technology was beginning to open up the area of dynamic systems to new avenues previously unexplored by the pencil-and-paper mathematical techniques of the 1950s. Previously intractable systems, or intractable questions, were fair game, and the analysis of large and/or nonlinear systems became a focus of intense scientific interest in many scientific disciplines. Scientists from numerous realms made significant contributions to these mathematical developments, including Robert May's finding that simple, discrete ecological systems beget chaotic dynamics (May, 1976).

The hallmark of chaos is an extreme sensitivity to initial conditions (e.g., the butterfly flaps its wings, and it causes a storm elsewhere) and dynamics that twist out noisy patterns. The curious property of chaos is that it contains elements of both pattern (e.g., a geometrically defined attractor) and random behavior (unless the initial value is precisely known, the dynamics is relatively unpredictable), all of which can arise from a deterministic system. Chaos gave scientists the sudden recognition that a nonlinear feedback process of a defined signature interacts and mixes with other feedback processes with different temporal signatures. Perhaps the most amazing aspect of chaotic dynamic systems is that these underlying signals can mix in a way that the addition of these feedback sums to be greater than the individual feedbacks themselves. This can be seen on examination of the spectral signature of a chaotic time series. Although there are spikes at some characteristic frequencies (the signatures of the major oscillatory drivers responsible for the pattern in the chaotic dynamics), there are also spikes pretty much everywhere else. Curiously, and almost magically, the major driving processes mix together in an amazing blend of pattern and noise.

It is frequently argued that the emergence of chaos had a significant influence on the perspective of science in general. It seemed clear that if even simple systems can generate chaotic dynamics, then nature's complex palette may inspire an even more delirious and unpredictable form of chaos. From a diversity–stability perspective, the culmination of this intense period of activity can be seen in Robert May's seminal book (1973). May's work, and others, found in no uncertain terms that unconstrained complexity and diversity readily and rapidly drive inspired amounts of instability. In other words, diversity begets instability. The logic behind this is not far off that which I just suggested above. In May's own words:

> A variety of explicit counterexamples have demonstrated that a count of food web links is no guide to stability. This straightforward fact contradicts the intuitive verbal argument often invoked, to the effect that the greater the number of links, and alternative pathways in the web, the greater the chance

of absorbing environmental shocks, thus damping down incipient oscillations. The fallacy in this intuitive argument is that the greater the size and connectance of the web, the larger the number of characteristic modes of oscillation it possesses: since in general each mode is as likely to be unstable as stable (unless the increased complexity is of a highly special kind), the addition of more and more modes simply increases the chance for the total web to be unstable. (May, 1973)

At first one may wish to dismiss May's model experiments on the grounds that they are based on overly simplified constructs. To do so, though, is to miss the point. May intentionally constructed a very clear test of the early intuitive ideas behind diversity and stability. His models, although simplified, directly assessed the premise that diversity and complexity (where these are equivalent to increased connectance) positively influenced stability. To do this, he generated diversity and connectance using a random statistical universe (i.e., there were no other implicit assumptions about diversity). Thus, he had set up a clear theoretical test of the hypotheses that diversity and increased connectance, in and of themselves, beget stability. The formal mathematics suggests that diversity and complexity do not make systems more stable on average.

Nonetheless, May was aware of the fact that these theoretical results did not resonate with what most empiricists were finding—that diversity and complexity tended to correlate with more stable communities. May's results do not suggest that diversity is uncorrelated to critical ecological structure, but, rather, the results are a strong argument against diversity as the major driver of stability. His work meant that ecologists must seek to understand what critical underlying biological structures (e.g., food web structure) impart stability in real systems. Interestingly, although underplayed, these early investigations made some suggestions for the resolution of this problem. May and others suggested that patterns in interaction strength and compartmentation may play an important role in stabilizing complex dynamic systems (May, 1973; Pimm, 1991). Both of these are aspects of the way interactions are patterned. As a result, ecologists were poised to explore patterns in the structuring of food webs.

4. DIVERSITY AND FOOD WEB STRUCTURE

The theoretical work of the early 1970s thus inspired a factory-like production of researchers interested in documenting and revealing patterns in food webs. Shortly after the 1970s, a statistical network approach to food web structure was inspired by some of the seminal work of Joel Cohen and Frederic Briand's synthesis of 33 early food webs. The early work suggested a suite of patterns that were robust across a number of webs (well reviewed by Pimm, 1991). Further, early theoretical and empirical results suggested that omnivory and compartmentation ought to be rare, somewhat counter to May's suggestion. Nonetheless, a number of researchers began to question the resolution and meaning of these early food web statistics and their theories. Most webs were naturally biased toward specific taxa, and all webs were certainly incomplete. The question remained whether or not these early patterns were meaningless artifacts of incomplete and poorly resolved data. Surprisingly, although it appears as though some things have changed (discussed below), many of these early patterns remain with the more resolved webs (see Dunne, 2005, for a thorough review).

A crowd of empiricists studying food webs began to find recurrent structures in some of the food webs that had been well documented. Gary Polis, for example, pressed hard to show scientists that omnivory was replete in desert food webs and, with Don Strong, championed the role of spatial subsidies and multichannel pathways in food webs. Indeed, Polis and Strong (1996) argued that ecologists needed to expand their spatial scale and recognize that many of the focal webs we were studying actually were coupled through generalist consumers via both top-down (consumption) and bottom-up (nutrient transfer) mechanisms. At a level, the multichannel arguments of Polis and Strong are intimately related to the early notion of compartments put forward by May and others. These empirically motivated ideas resonated with many ecologists in that they emphasized that food webs operate over vast spatial scales, coupling many ecological systems together. At the same time, other researchers pointed out that food webs were not only variable in space but also highly variable in time (reviewed extensively by Polis and Winemiller, 1996). Empirical studies were expanding the spatial and temporal scale of food web ecology in their search for important biological structure mediating stability.

During this same period, while many ecologists were attempting to piece together the topology of the food web, some ecologists were also beginning to estimate the strength of these interactions. Robert Paine, for example, pioneered animal exclusion experiments to empirically estimate indirect or functional interaction strength. Paine would exclude a species and measure how much that changed the abundance of the remaining players in the food web (i.e., an indirect measure of interaction strength). Interestingly, Paine found that most of these indirect measures of interaction strengths in a food web are weak with a few very strong interactions between species. More recently, ecologists have also begun to look at the distribution of direct measures of interaction strength (e.g., energy flux), and here too it seems that many interactions are weak, but more empirical work is required.

5. STRUCTURE, VARIABILITY, AND STABILITY

All in all, this surge of empirical work placed ecologists in a good position to revisit food web theory. Much theory up to this point had emphasized a theory for whole food webs (May, 1973). This whole-systems approach, however, is complimented by theory that looks more explicitly at the dynamic implications of specific food web structures that comprise the whole food webs (e.g., predator–prey interactions, food chain, omnivory). Robert Holt has coined these important smaller subsystems food web modules (network theorists call these subsystems motifs) and argued that ecological theory would do well to expand beyond the well-established theories of the base modules (e.g., predation, competition) to food web modules that embody simple ubiquitous natural food web structures (e.g., omnivory). In an attempt to map empirical structure to food web theory, ecologists have followed Holt and begun to develop a theory for specific food web modules (e.g., omnivory, spatial subsidies).

Most of the early food web theory relied heavily on equilibrium assumptions. This assumption has some obvious mathematical advantages in that it allows theoreticians to use some relatively simple linear techniques to derive stability. This has been a powerful theoretical tool and will undoubtedly remain so, but a number of ecologists have begun to relax these equilibrium assumptions to consider how variability in space and time may influence food web stability and structure (DeAngelis and Waterhouse, 1987).

One result that followed out of these emerging perspectives was the averaging effect (Tilman and Downing, 1994). The basic idea behind the averaging effect is that variable population dynamics can sum to give relatively stable (i.e., less variable) community dynamics as long as the different populations show differential responses to variable conditions. This differential response can be randomly driven, competively driven, or induced by life history differences that drive differential response to varying abiotic conditions. Thus, although each individual population can show significant variation through time, the sum of all these differentially responding organisms produces a relatively stable aggregate community biomass.

Another emerging theory, viewed from either an equilibrium or a nonequilibrium perspective, explored the implication of interaction strength within Holt's food web module perspective. Here, researchers asked how interaction strength influenced the dynamic behavior and stability of some ubiquitous food web modules (reviewed in McCann et al., 2000). The results indicated that interaction strength can indeed play a potent role in stabilizing food webs. Specifically, weak interactions (in the sense of per capita energy flow; direct interaction strength) can be easily positioned within a food web such that they mute potentially strong and oscillatory interactions. This stabilization appears to result from two primary mechanisms:

1. Weak interactions, properly positioned, can redirect energy away from a potentially oscillatory consumer–resource interaction. In essence, the weak interaction reverses the paradox of enrichment as energy/productivity is shunted away from a potentially oscillatory interaction. Note, however, a strong interaction placed similarly would only contribute to more intense and complex oscillations (sensu May, 1973).
2. Generalist consumers can, through preferential feeding, drive out-of-phase resource or prey dynamics. These out-of-phase dynamics sum to give a relatively stable resource or prey community biomass and thus enable a relatively stable response of the consumer to the resource variability (i.e., the out-of-phase dynamics gives the consumer the option to respond to low resource densities by shifting its attention toward a resource that is not at low resource densities).

This latter mechanism again suggested that interesting stabilizing structures may be unfolding because of variability itself. In fact, this mechanism is clearly closely aligned to the bottom-up ideas of the averaging effect (Tilman and Downing, 1994). Intriguingly, certain trade-offs readily generate out-of-phase prey dynamics. For example, if organisms that are more competitive are less tolerant to predation, then this readily produces a situation with differential prey responses. As an example, the sudden increase in predators drives the suppression of the dominant competitor (because it is most preferred), and thus, the weaker competitor is released to flourish (i.e., they are out of phase). Similarly, a decrease in predators releases the strong competitor to flourish, which suppresses the weak competitor (i.e., again out of phase). Both life history theory and empirical data have found abundant evidence for such trade-offs.

These new theories still assume a homogenized spatial world, and the empiricists' results spoke again and again to the importance of space (reviewed in Polis and

Winemiller, 1996; Polis and Strong, 1996). We now turn to some very recent directions and briefly anticipate some future directions in this important area of research.

6. FUTURE DIRECTIONS: FOOD WEBS ACROSS SPACE AND TIME

Although space has played a large role in population ecology, consideration of the role of space on food web dynamics is relatively recent. Robert Holt, Michel Loreau, Mathew Liebold, and others have begun to tie metapopulation theory to community and ecosystem perspectives (dubbed metacommunity and metaecosystem, respectively). They have argued cogently that this larger perspective has the potential to unite population, community, and ecosystem perspectives. More specifically, they have argued that expanding the spatial scale of food webs may allow us to more completely understand such longstanding issues as food chain length, trophic control, island biogeography, and food web stability/instability.

Along a similar research theme, some ecologists have begun to consider empirical arguments to frame a more general spatial theory of food webs. Polis and Strong (1996) emphasized that different habitats contained different primary producers and that these tended to be coupled by higher-ordered generalist consumers. This result is consistent with two empirical generalizations: one, that generalist foraging tends to increase with higher-order consumers, and two, that higher-order organisms tend to be larger and more mobile than their prey. Recent theory has begun to consider the implications of such spatial coupling on the dynamics and stability of coupled food webs (McCann et al., 2005). The results suggest that in spatially extended systems with differentially responding resources or prey, behavior (i.e., movement) by the larger, more mobile organism can act as a potent stabilizing force, especially when considered in a nonequilibrium context.

The result is intuitively easily presented and consistent with earlier theory emerging from spatial population ecology. Effectively, larger organisms can respond to variation in space by moving away from areas where prey or resource densities are low and toward areas where prey or resource densities are high. The outcome is the release of predatory pressure on prey when prey species are at low densities and increasing predatory pressure when prey species attain high densities—precisely the arrangement that is needed to reduce extreme variation in density. From the consumer perspective, their rapid behavioral response makes it possible to track variable resource or prey densities at a larger spatial scale. Again, the result relies on the underlying idea that resources in different habitats are responding differentially through time. It turns out that this variation can be abiotically driven or driven by the top-down predatory pressure of generalist consumers if the consumer tends to prefer one organism significantly more than other organisms (this is a manifestation of the weak interaction effect discussed above). So again, like the averaging effect described for a single trophic level (Tilman and Downing, 1994), the notion of differential responses within a nonequilibrium perspective suggests that food web stability may unfold from variability in space and time.

Following May's suggestion, ecologists also looked into food web compartmentation (Pimm, 1991). Early work found little evidence for compartments in food webs except at huge spatial scales. Recent analysis of food webs by Anna Krause and others, using flux-based measures of interaction strength, found that compartments might be more ubiquitous than early investigations suggested. Soil ecologists have argued for such compartmented structure in their soil food webs. They have suggested that bacterial energy channels tend to break down more labile detritus and also turn over much more rapidly than fungal energy channels that tend to arise out of more recalcitrant detrital sources. In a 2006 *Nature* paper, Neil Rooney and collaborators have extended this argument to the benthic detrital pathways versus pelagic pathways in aquatic ecosystems. Benthic invertebrates tend to turn over on a much longer time scale than the rapid turnover of zooplankton on phytoplankton. Finally, it has been suggested for some time that detrital webs are slower, more donor-controlled than grazing webs. If compartments (like species) tend to respond differentially in time and space, then mobile higher-order consumers capable of coupling these distinct subwebs can average across these variable out-of-phase subsystems. Hence, strong and weak compartments may be a very important form of food web structure that contributes to the persistence of ecological systems.

It appears that this variability in biological structure may, in fact, lead us to understand intriguing structural changes in food webs as a response to dynamic conditions. Thus, embracing the "noise" in space and time might contribute to significant empirical and theoretical advances in understanding. Importantly, a number of common threads (nonequilibrium perspective, differential response, scale) in the above discussion suggest that there may be some ways to unfold this vast complexity in a manageable way (e.g., compartments in space, compartments in time). Similarly, these advances suggest that ecologists must continue the recent trend of crossing longstanding scientific boundary lines (community versus ecosystem ecology; evolutionary dynamics versus ecological dynamics).

FURTHER READING

DeAngelis, D. L., and J. C. Waterhouse. 1987. Equilibrium and non-equilibrium concepts in ecological models. Ecological Monographs 57: 1–21.

Dunne, J. 2005. The network structure of food webs. In M. Pascual and J. A. Dunne, eds., Ecological Networks: Linking Structure to Dynamics in Food Webs. Oxford: Oxford University Press.

Elton, C. 1958. The Ecology of Invasions by Animals and Plants. Chicago: University of Chicago Press.

MacArthur, R. 1955. Fluctuations of animal populations, and a measure of community stability. Ecology 36: 533–536.

May, R. 1973. Stability and Complexity in Model Ecosystems. Princeton, NJ: Princeton University Press.

May, R. 1976. Simple mathematical models with very complicated dynamics. Nature 261: 459–467.

McCann, K. S. 2000. The diversity–stability debate. Nature 405: 228–233.

Pimm, S. L. 1991. The Balance of Nature? Ecological Issues in the Conservation of Species and Communities. Chicago: University of Chicago Press.

Polis, G. A., and D. R. Strong. 1996. Food web complexity and community dynamics. American Naturalist 147: 813–846.

Polis, G. A., and K. O. Winemiller, eds. 1996. Food Webs: Integration of Patterns and Dynamics. New York: Chapman & Hall.

Tilman, D., and J. Downing. 1994. Biodiversity and stability in grasslands. Nature 367: 363–365.

III.8

Spatial and Metacommunity Dynamics in Biodiversity

M. A. Leibold

OUTLINE

1. Two important consequences of dispersal in metacommunities
2. The four paradigms of metacommunity ecology
3. Synthetic efforts
4. Application of metacommunity thinking to food webs and ecosystems
5. A critique of metacommunity thinking

Spatial dynamics presents some of the biggest challenges in modern ecology. These occur when the movement of organisms in space affects their populations and consequently affects how they interact with other species. It has long been known that spatial dynamics can be very important in regulating species interactions. For example, Huffaker (1958) found that spatial structure in the form of patchy resources with limited dispersal was important in allowing coexistence of the predatory mite *Tylodromus occidentalis* with its prey, the six-spotted mite *Eotetranychus sexmaculatus*. In a different context, Watt (1947) recognized that a spatial "mosaic" of patches was key in regulating the process of succession in communities because patches at different stages of succession were key sources of colonists during the process as patches underwent successional cycles. Despite this long recognition that spatial effects were important in community ecology, however, a satisfying conceptual, theoretical, and experimental understanding of spatial dynamics is still in development (Tilman and Kareiva, 1997; Hanski, 1999; Chesson et al., 2005).

GLOSSARY

mass effects. Variation in community composition determined by source–sink relations among patches

metacommunity. A set of local communities connected by the dispersal of at least one component species

neutral dynamic. Variation in community composition determined by stochastic effects of dispersal and demography among species with equivalent niches

patch dynamics. Variation in community composition determined by extinctions of species in patches and colonization among patches

species sorting. Variation in community composition determined by the optimization of fitness among species across patches

Spatial dynamics is intimately linked with the principle of dispersal. Much of the work has examined passive dispersal in which organisms do not have much control over where they go (cases of dispersal where there is such control are mostly studied in behavioral ecology, where they often involve habitat selection behavior). There are numerous approaches to understanding how dispersal affects community interactions, and some of these are outlined in table 1. These approaches vary (Durrett and Levin, 1994; Bolker and Pacala, 1997) in whether they view space as consisting of discrete patches or a continuous landscape, whether they view dispersal as a local process or a global one, and whether they account for space explicitly (having a "map" of locations) or implicitly (just taking into account that there are distinct areas but not keeping track of where they are) as well as whether they account for the discrete nature of individuals. Generally, the simpler approaches are easier to understand but are more likely to oversimplify the situations than the more complex ones. These approaches also differ in their goals, with some of them focused on accounting for how population density varies in space and time, some focused on understanding coexistence, and some focused on understanding diversity or other questions. Although different approaches often give somewhat different answers, there are many common insights that can result (Durrett and Levin, 1994).

Table 1. Some spatial approaches to community ecology

Approach	*Continuous or discrete levels of spatial structure*	*Local or global dispersal*	*Spatially explicit or implicit*	*Representative and/or recent publication*
Metacommunity ecology	Discrete	Global	Implicit	Leibold et al., 2004
Moment closure	Continuous	Local	Either	Bolker and Pacala, 1997
Scale transition	Continuous	Global	Implicit	Chesson et al., 2005
Partial differential equations	Continuous	Local	Either	Holmes et al., 1994
Individual-based models	Continuous	Local	Explicit	DeAngelis and Mooij, 2005
Lattice models	Discrete	Local	Explicit	Durrett and Levin, 1994

A useful and simple organizing framework for thinking about the some of the basic elements of these approaches is the metacommunity. A metacommunity is defined as a set of local communities that are linked by dispersal of at least one component species (plate 5). It thus views spatial structure in a simple hierarchical way with local communities existing at a distinct and lower level than the metacommunity itself. The advantage of "metacommunity thinking" is that it captures many of the salient features of spatial ecology in a way that is reasonably accessible for verbal modeling, for guiding our intuition, and for generating more precise theoretical models. And although there are a number of important challenges for future work (some of these discussed below) and limitations, it also serves as a useful way to explore more complex spatial dynamics that does not match the strict hierarchy of spatial organization assumed in the metacommunity concept.

1. TWO IMPORTANT CONSEQUENCES OF DISPERSAL IN METACOMMUNITIES

Current work on metacommunity thinking has focused on two effects that dispersal plays in such a simple hierarchy. First, dispersal is key in allowing new species to colonize local communities from which they were previously absent (Hanski, 1999). Thus, in a closed community (no dispersal), changes in community composition are limited to extinction (and possibly sympatric speciation), but this will be very different in communities that can receive colonizing immigrants from other communities. Thus, dispersal within a metacommunity is a key process affecting local community assembly (the process of colonizations and extinctions that determines which species are present in a community such as occurs during succession), and this is one way that the composition of a regional biota can influence local communities. If dispersal among local communities is very slow, the process of community assembly will also be slower and likely to be more stochastic and less predictable than if dispersal is high, and this may have many consequences to patterns of biodiversity and community composition.

The second effect of dispersal among local communities in a metacommunity is to homogenize differences among the local communities. This effect is particularly true when the dispersal rates are sufficient to maintain "sink" populations in some local communities that are supported by immigration from "source" populations in other local communities (these are also sometimes termed mass effects). As dispersal gets higher and higher, any intrinsic local differences in the fitness (per capita production) of local populations will affect these local population densities less and less because they are increasingly overwhelmed by the composition of the migrants. If such homogenization simultaneously affects many species in the metacommunity, then community composition will be homogenized among the local patches. Taken to an extreme, if the dispersal rate is extremely high, such homogenization will mean that the spatial patchiness that might be identified at the lower level (the local community) is actually irrelevant to the organisms involved. Instead, these organisms view such an assemblage of patches as a single patch with properties that are some sort of weighted average of the component patch attributes. At this point, what we call the metacommunity is effectively just a local community as far as the organisms involved are concerned.

These two consequences of dispersal can interact with each other if different species have very different dispersal rates. It is possible, for example, that one set of species will be strongly subjected to the homogenization effect, whereas another is more strongly affected by the dispersal-limited community assembly. Work is only now beginning to understand the consequences of such variability in dispersal.

2. THE FOUR PARADIGMS OF METACOMMUNITY ECOLOGY

Metacommunity thinking is still in the early stages of development. Historically, a number of approaches to

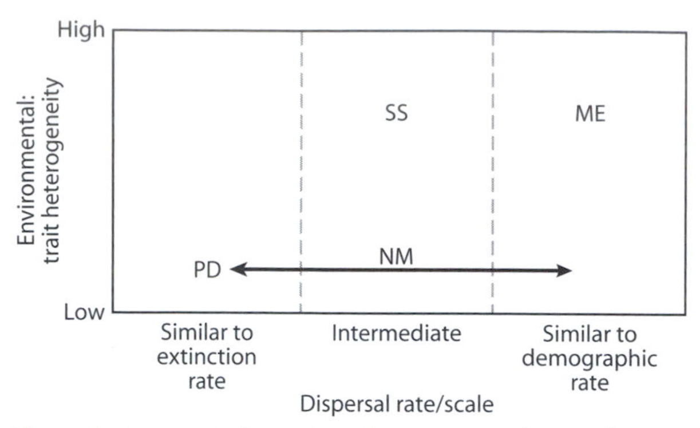

Figure 1. A conceptual overview of metacommunity paradigms in relation to the amount of dispersal and the amount of environmental heterogeneity that affect local traits of component species. SS refers to species sorting, ME refers to mass effects, PD refers to patch dynamics, and NM refers to neutral models. The arrows that point from NM indicate that the neutral models can account for any of the time scales involved. One important distinction not shown in this figure is that NM assumes all species have similar dispersal rates, whereas PD does not.

thinking about metacommunity dynamics have been developed in relative isolation from one another (Leibold et al., 2004; Holyoak et al., 2005). These approaches differ in the assumptions they make about dispersal rates and about the amount of trait environmental heterogeneity (figure 1). Current work is trying to synthesize these paradigms into a common framework, but they still illustrate the dominant views that guide much of the thinking about metacommunities.

The patch dynamics paradigm is closely related to metapopulation models in population biology (Hanski, 1999) and has mostly focused on patch occupancy (whether a species is present in a patch rather than its density). Much of the work done in this area has not adequately evaluated how environmental heterogeneity among patches affects results. Instead, the focus has been on colonization–extinction dynamics, often under a possible trade-off among species between their colonizing ability and their competitive ability. A unique feature of these models is that extinctions within patches occur for stochastic reasons (either demographic stochasticity in small populations or disturbance/environmental change; Lande, 1993). If colonization events are on time scales that are similar to or slower than these extinctions, patch dynamics can explain some of the variation in community assembly that results. One sometimes confusing issue is that a number of authors have used the same mathematical formalism to address sessile organisms by assuming that patches (which might be better called "microsites") consist of single individuals (e.g., Tilman, 1994; Hubbell, 2001; Mouquet and Loreau, 2003) so that the death of individuals is equivalent to the extinction rate, and the establishment of an individual is equivalent to the rate of colonization and/or competitive exclusion. These microsite patch dynamic models have in some cases then served to model theories about the patches (containing many such individuals that occur in a metacommunity at a yet higher spatial scale subject to mass effects dispersal; see below). Thus, the full model is really addressing mass effects even though the way individuals are modeled corresponds to the patch dynamics approach.

The species sorting paradigm is perhaps the most intuitively obvious (see Chase and Leibold, 2003). Here, dispersal is seen as the fuel for community assembly, and local interactions determine how this assembly proceeds. Most of the work has focused on the assembly of either competitive assemblages (e.g., Tilman, 1982) or of food webs (e.g., Holt et al., 1994; Leibold, 1996), and much of this work has studied how environmental context (i.e., environmental heterogeneity among patches) alters expected patterns of community structure. Relatively little work in this area has considered how regional communities are regulated by the cumulative effects of these processes. This framework usually ignores any effects of dispersal on local population sizes or their dynamics and is thus more appropriate when dispersal is small relative to demographic rates but still large relative to local population extinction rates. The bulk of equilibrium population ecology theory can easily be related to this approach (especially relevant are mechanistic approaches such as those described by MacArthur, 1972, and Tilman, 1982, reviewed by Chase and Leibold, 2003), but only limited work has examined how regional community structure is regulated under this view (Leibold, 1998; Shurin et al., 2004).

The mass effects paradigm (Shmida and Wilson, 1985) has probably received the most attention even if it is probably the most complicated case. Here, dispersal is sufficient to have consequences on local population persistence, size, and dynamics. Numerous approaches fall into this general framework including much of what is also considered spatial ecology (see Tilman and Kareiva, 1997; table 1). The most important phenomena in this perspective occur when dispersal can allow populations to persist in local communities as sink populations that are supported by immigration from other populations that are source populations for them (Holt, 1985; Pulliam, 1988). Obviously, this may allow more species to coexist in local populations than might be predicted by niche-based models of local community structure (i.e., species sorting), but the results can also be more complex if the mass effects are sufficient to alter the likelihood of persistence of local populations (e.g., Amarasekare and

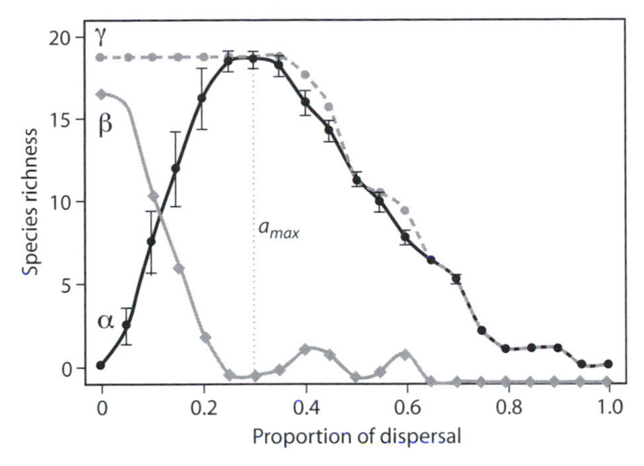

Figure 2. Species richness as a function of the proportion of dispersal between communities. Black circles, local (α) scales; diamonds, between-community (β) scales; gray circles, regional (γ) scales. a_{max} is the dispersal value at which species diversity is maximal. (From Mouquet and Loreau, 2003)

Nisbet, 2001; Amarasekare et al., 2004). Thus, immigration of organisms from elsewhere can overwhelm local populations even if these are otherwise better suited to local conditions and drive down diversity instead.

The neutral dynamics paradigm is the most recently developed one (Bell, 2001; Hubbell, 2001). It is based on the premise that differences among species in their ecological traits are negligible and that stochastic forces of demography and migration among local communities are more important in regulating some aspects of community ecology. This premise is somewhat controversial, and ongoing theoretical and empirical debates are still unresolved (see the special feature on this topic in *Ecology*, volume 87, issue 6, 2006). At a minimum, however, this approach does two things. First, it can serve as a null hypothesis for conclusions made by the other approaches (Bell, 2001). Second, and perhaps more importantly, it is also important in drawing attention to the stochastic demographic processes that tend to be ignored by other approaches but likely interact with the more deterministic ones described by the other paradigms.

3. SYNTHETIC EFFORTS

Although these four paradigms have been developed in reasonably independent ways, they can be viewed as a continuum depending on time scales and on the degree of environmental and trait heterogeneity (figure 1). Several studies have explored this continuum. Law and Leibold (2005) explore the relationship between species sorting and patch dynamics in a simple model of nontransitive competition among three species and show that the assumption of stochastic extinctions plays a critical role in regulating metacommunity structure at both the regional and local scales. Shurin et al. (2004) examine how patch dynamics and species sorting interact to affect the likelihood of alternate stable states in competitive metacommunities. They show that local priority effects do not always lead to likely existence of alternate stable states because coexistence among the species at the regional level can be strongly constrained (i.e., species that would produce local alternate stable states at a local scale do not always coexist easily at the metacommunity scale).

More work has been done at the interface of species-sorting and mass effects (e.g., Amarasekare and Nisbet, 2001; Mouquet and Loreau, 2003). The study by Mouquet and Loreau (2003) is illustrative of this continuum and shows how dispersal affects local (α) and regional (γ) diversity as well as among-community composition turnover (β diversity) as one goes from a situation that more closely matches species sorting to one that ranges through the various effects that mass effects have on diversity (figure 2). At very low dispersal, each distinct patch is inhabited by a species that is a specialist on that patch type (so that local diversity is minimal and β diversity is maximal). As dispersal increases, there is a point at which local diversity is increasingly enhanced by mass effects (immigrants from other patches survive long enough to maintain increasingly large populations). Thus, α diversity increases, and β diversity decreases. However, at yet higher dispersal rates, some species become extinct in the metacommunity as a whole because their average fitness (over all the patches they inhabit) is less than that of surviving species given the increasing homogenization of the metacommunity. This then means that local diversity also declines. Finally, when the dispersal rate is so high that the metacommunity is effectively one homogeneous patch (from the point of view of the interacting species), there is only one species present in the metacommunity (and in any local patch as well).

Predictions of this model for biodiversity have now begun to be successfully tested in microcosms (Cadotte, 2006a; Matthiesen and Hillebrand, 2006; figure 3), in mesocosms (Forbes and Chase, 2002), and even in some convenient natural systems such as the biota that inhabit pitcher plants (Kneitel and Miller, 2003), but the issue is still somewhat unresolved in broader-scale meta-analyses (Cadotte, 2006b).

On the empirical side, synthetic studies have also begun to seek ways to identify which of these paradigms is more apparent based on census data from metacommunities. The most common approach has been to recognize that species sorting predicts that

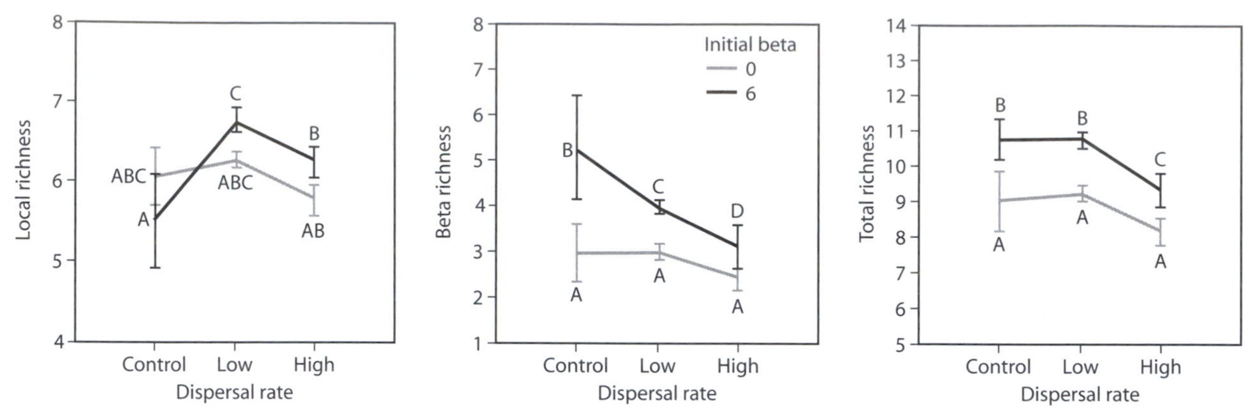

Figure 3. Effects of dispersal rate on local (α in figure 2) and total (γ in figure 2) diversity as well as on β diversity (i.e., community differentiation, β in figure 2) in experimental protist metacommunities in the experiments of Cadotte (2006). In addition to manipulating dispersal, the initial β diversity was manipulated to either 0 (all initial local communities were identical with 13 species present) or 6 (each local community had 7 of the 13 species present, whereas the entire metacommunity had 13 species present). Points identified with different letters (A–D) in each panel identify points that do not differ significantly from each other.

differences in composition among local communities reflect environmental differences. And conversely, it predicts that there should be no purely spatial effects (those that are unrelated to how environmental conditions change in space; of course, it is always difficult to establish that all relevant environmental variables are understood, and it is known that environmental variation covaries with space, but some of this can be accounted for in statistical analyses). Alternatively, both mass effects and patch dynamics would predict that there should be some resemblance between communities if they are near each other regardless of whether they resemble each other environmentally. Thus, similarity of community composition can be related to environmental similarity and/or spatial proximity using Mantel tests (Manly, 1986) or similar methods (e.g., Borcard et al., 2004). To date, such studies have almost universally shown that there is substantial similarity in composition that can be explained by similarity of environments (consistent with species sorting), but they also show that there is often some additional degree of similarity related to proximity (see Cottenie, 2005), which could be consistent with any of the other paradigms. There is also often statistical interaction between the two, although these studies have not adequately described this interaction or interpreted it. Beisner et al. (2006; table 2) show that variation in such patterns can be related to the general dispersiveness of different taxa with slowly dispersing taxa showing more effects of proximity and less effect of environment than more rapidly dispersing taxa. This would suggest that, in that study at least, the proximity effects result from patch dynamic processes rather than mass effects. Cottenie et al. (2003) also show that there can be proximity effects in the composition of zooplankton species in ponds that are interconnected by stream flow (table 3), but this effect seems most important when the flow rates in the streams are high (Michels et al., 2001). In this case, the proximity effects would thus be more likely to depend on mass effects rather than patch dynamics.

4. APPLICATION OF METACOMMUNITY THINKING TO FOOD WEBS AND ECOSYSTEMS

The ideas described above focus on how metacommunity dynamics (involving any or all the paradigms described above) affect coexistence and diversity of competing species. However, metacommunity thinking is also being used to better understand other aspects of communities involving not just competition but other aspects of community–ecosystems ecology including food-web structure and features of ecosystems.

The implications of metacommunity dynamics for understanding food webs are only beginning to be investigated, but an intriguing set of phenomena have begun to emerge. First, following on the initial work on predator–prey interactions (Huffaker, 1958), spatial structure can stabilize otherwise unstable interactions between predators and their prey in more complex food webs (Holyoak, 2000; Holt, 2002). Second, community assembly of complex food webs that involve cyclical recurrent patterns of community composition are facilitated by dispersal and may produce different patterns of biodiversity at local versus regional scales (Steiner and Leibold, 2004). Finally, there are complex, novel, and sometimes counterintuitive mechanisms of interactions in food webs subject to mass effects in metacommunities (e.g., Holt, 2002; Callaway and

Table 2. Variance partitioning of community composition for bacteria (fastest dispersal), zooplankton (intermediate dispersal), and fish (slowest dispersal) in lakes

Component of variance explained	*Taxa*		
	Bacteria	*Zooplankton*	*Fish*
Environment	0.14*	0.18*	0.08
Space	0.06	0.17*	0.11*
Space-by-environment	0.04	0.04	0.00
Residual	0.76	0.64	0.81

Source: Modified from Beisner et al. (2006).

Note: The proportion of variance in the similarity of communities attributable purely to measured aspects of the environment (Environment), distance along watercourses (Space), and their possible interaction as well as the remaining variation are tabulated. Asterisks (*) denote statistically significant effects.

Hastings, 2002; Brose et al., 2004). And although it has been suggested that food web architecture may vary consistently with spatial scale (e.g., Brose et al., 2004), it is not yet clear if the links between theory and data are conclusive in helping us to understand these processes.

Perhaps more intriguing are the ways that metacommunity thinking is changing the interpretation of ecosystem attributes in ecology. Several recent studies illustrate this. First, Mouquet et al. (2002) show that the effect of biodiversity on emergent aggregate properties of ecosystems (e.g., their productivity or standing crop of plants) depends on how the initial biodiversity was maintained. If many of the species are maintained as relatively poorly locally adapted sink populations, then the effects of changes in diversity on ecosystem attributes may be different than if all the species are self-maintained at the local scale. Additional work by Thebault and Loreau (2006) shows that this may additionally depend on food web dynamics. Second, Leibold et al. (1997) have argued that the scaling of plant and herbivore abundances with productivity results from the ways that different plant species with different defense versus exploitation traits are selected at different levels of productivity and that this occurs only when communities are interconnected by dispersal.

Table 3. Variance partitioning of community composition for zooplankton

Component of variance explained	*Zooplankton*
Environment	0.20*
Space	0.17*
Space-by-environment	0.02
Residual	0.62

Source: Modified from Cottenie et al. (2003).

Note: The proportion of variance in the similarity of communities attributable purely to measured aspects of the environment (Environment), distance along watercourses (Space), and their possible interaction as well as the remaining variation are tabulated. Asterisks (*) denote statistically significant effects.

The idea that metacommunity dynamics has important implications for ecosystems is even more developed in the concept of "meta-ecosystems" (Loreau et al., 2003), where the movement of materials in space (either passively via diffusion or flow, or actively, via the movement of individuals through dispersal) is also considered (Polis et al., 1997). Work in this area is just beginning.

5. A CRITIQUE OF METACOMMUNITY THINKING

The above discussions illustrate a rich array of ways that taking into account the dispersal of organisms influences community and ecosystem thinking. Many of these insights are based on the simplest version of the concept of metacommunity and ignore numerous details that might matter a lot. Thus, for example, it may not be enough to consider the simple hierarchy of local–regional community that is implied by the discussion above; it may matter that some local communities are more isolated than others; it may matter that some species disperse more than others; the particular arrangement of patches may matter; and what about spatial dynamics in more continuous (less discrete) situations such as landscapes and gradients? What about organisms that disperse actively and selectively rather than passively? What about organisms that evolve in response to their environments? These and a number of other complications are barely addressed by the simple metacommunity concept outlined above. They indicate that metacommunity thinking can open new ways of thinking about community ecology.

Addressing these issues more satisfyingly, however, will require more sophisticated approaches (Chesson et al., 2005). Some of these are already ongoing as outlined in table 1, but many are not. Work to date shows that many of these issues do modify our expectations to some degree. However, this work also demonstrates that the broad insights provided by simple metacommunity thinking described above can be quite general. Just how they resolve themselves and how important these issues are present an exciting current direction in ecology both on the theoretical and empirical fronts. Overall, these studies show that many aspects of ecology are strongly modified by dispersal so that previous ecological work that is strongly limited to closed communities is likely to be of limited use in understanding larger-scale patterns in biodiversity and other aspects of community ecology.

FURTHER READING

Amarasekare, P., and R. M. Nisbet. 2001. Spatial heterogeneity, source–sink dynamics, and the local coexistence of competing species. American Naturalist 158: 572–584.

Beisner, B. E., P. R. Peres, E. S. Lindstrom, A. Barnett, and M. L. Longhi. 2006. The role of environmental and spatial processes in structuring lake communities from bacteria to fish. Ecology 87: 2985–2991

Bell, G. 2001. Neutral macroecology. Science 293: 2413–2418.

Bolker, B., and S. W. Pacala. 1997. Using moment equations to understand stochastically driven spatial pattern formation in ecological systems. Theoretical Population Biology 52: 179–197.

Borcard, D., P. Legendre, C. Avois-Jacquet, and H. Tuomisto. 2004. Dissecting the spatial structure of ecological data at multiple scales. Ecology 85: 1826–1832.

Brose, U., A. Ostling, K. Harrison, and N. D. Martinez. 2004. Unified spatial scaling of species and their trophic interactions. Nature 428: 167–171.

Cadotte, M. W. 2006a. Metacommunity influences on community richness at multiple spatial scales: A microcosm experiment. Ecology 87: 1008–1016.

Cadotte, M. W. 2006b. Dispersal and species diversity: A meta-analysis. American Naturalist 167: 913–924.

Callaway, D. S., and A. Hastings. Consumer movement through differentially subsidized habitats creates a spatial food web with unexpected results. Ecology Letters 5: 329–332.

Chase, J. M., and M. A. Leibold. 2003. Ecological Niches. Chicago: University of Chicago Press.

Chesson, P., M. J. Donahue, B. A. Melbourne, and A.L.W. Sears. 2005. Scale transition theory for understanding mechanisms in metacommunities. In M. Holyoak, M. A. Leibold, and R. D. Holt, eds., Metacommunities: Spatial Dynamics and Ecological Communities. Chicago: University of Chicago Press, 279–306.

Cottenie, K. 2005. Integrating environmental and spatial processes in ecological community dynamics. Ecology Letters 8: 1175–1182.

Cottenie, K., E. Michels, N. Nuytten, and L. De Meester. 2003. Zooplankton metacommunity structure: Regional vs. local processes in highly interconnected ponds. Ecology 84: 991–1000.

DeAngelis, D. L., and W. M. Mooij. 2005. Individual-based modeling of ecological and evolutionary processes. Annual Review of Ecology and Systematics 36: 147–168.

Durrett, R., and S. A. Levin. 1994. The importance of being discrete and spatial. Theoretical Population Biology 46: 363–395.

Forbes, A. E., and J. N. Chase. The role of habitat connectivity and landscape geometry in experimental zooplankton metacommunities. Oikos 96: 433–440.

Hanski, I. 1999. Metapopulation Ecology. Oxford: Oxford University Press.

Holmes, E. E., M. A. Lewis, J. E. Banks, and R. R. Veitt. 1994. Partial-differential equations in ecology—spatial interactions and population dynamics. Ecology 75: 17–29.

Holt, R. D. 1985. Population dynamics in two-patch environments: Some anomalous consequences of an optimal habitat distribution. Theoretical Population Biology 28: 181–208.

Holt, R. D. 2002. Food webs in space: On the interplay of dynamic instability and spatial processes. Ecological Research 17: 261–273.

Holt, R. D., J. Grover, and D. Tilman. 1994. Simple rules for interspecific dominance in systems with exploitative and apparent competition. American Naturalist 144: 741–777.

Holyoak, M. 2000. Habitat subdivision causes changes in food web structure. Ecology Letters 3: 509–515.

Holyoak, M., M. A. Leibold, and R. D. Holt, eds. 2005. Metacommunities: Spatial Dynamics and Ecological Communities. Chicago: University of Chicago Press.

Hoopes, M. F., N. Mouquet, and M. Holyoak. 2004. Mechanisms of coexistence in competitive metacommunities. American Naturalist 164: 310–326.

Hubbell, S. 2001. The Unified Neutral Theory of Biodiversity and Biogeography. Princeton, NJ: Princeton University Press.

Huffaker, C. B. 1958. Experimental studies on predation: Dispersion factors and predator–prey oscillations. Hilgardia 27: 343–383.

Kneitel, J. M., and T. E. Miller. 2003. Dispersal rates affect species composition in metacommunities of *Sarracenia purpurea* inquilines. American Naturalist 162: 165–171.

Lande, R. 1993. Risks of population extinction from demographic and environmental stochasticity and random catastrophes. American Naturalist 142: 911–927.

Law, R., and M. A. Leibold. 2005. Assembly dynamic in metacommunities. In M. Holyoak, M. A. Leibold, and R. D. Holt, eds. Metacommunities: Spatial Dynamics and Ecological Communities. Chicago: University of Chicago Press, 263–278.

Leibold, M. A. 1996. A graphical model of keystone predators in food webs: Trophic regulation and the abundance, incidence, and diversity patterns in communities. American Naturalist 147: 784–812.

Leibold, M. A. 1998. Similarity and local coexistence of species in regional biotas. Evolutionary Ecology 12: 95–110.

Leibold, M. A., J. M. Chase, J. B. Shurin, and A. L. Downing. 1997. Species turnover and the regulation of trophic structure. Annual Review of Ecology and Systematics. 28: 467–494.

Leibold, M. A., M. Holyoak, N. Mouquet, P. Amarasekare, J. M. Chase, M. F. Hoopes, R. D. Holt, J. B. Shurin, R. Law, D. Tilman, M. Loreau, and A. Gonzalez. 2004. The metacommunity concept: A framework for large scale community ecology? Ecology Letters 7: 601–613.

Loreau, M., N. Mouquet, and R. D. Holt. 2003. Metaecosystems: A theoretical framework for a spatial ecosystem ecology. Ecology Letters 6: 673–679.

Manly, B.F.J. 1986. Randomization and regression methods for testing for associations with geographical, environ-

mental and biological distances between populations. Researches in Population Ecology 28: 201–218.
Matthiessen, B., and H. Hillebrand. 2006. Dispersal frequency affects local biomass production by controlling local diversity. Ecology Letters 9: 652–662.
McArthur, R. H. 1972. Geographical Ecology: Patterns in the Distribution of Species. New York: Harper & Row.
Michels, E., K. Cottenie, L. Neys, and L. De Meester. 2001. Zooplankton on the move: First results on the quantification of dispersal of zooplankton in a set of interconnected ponds. Hydrobiologia 442: 117–126.
Mouquet, N., and M. Loreau. 2003. Community patterns in source–sink metacommunities. American Naturalist 162: 544–557.
Mouquet, N., J. L. Moore, and M. Loreau. 2002. Plant species richness and community productivity: Why the mechanism that promotes coexistence matters. Ecology Letters 5: 56–65.
Polis, G. A., W. B. Anderson, and R. D. Holt. 1997. Toward an integration of landscape and food web ecology: The dynamics of spatially subsidized food webs. Annual Review of Ecology and Systematics 28: 289–316.
Pulliam, H. R. 1988. Sources, sinks, and population regulation. American Naturalist 132: 652–661.
Shmida, A., and M. V. Wilson. 1985. Biological determinants of species diversity. Journal of Biogeography 12: 1–20.
Shurin, J. B., P. Amarasekare, J. M. Chase, R. D. Holt, M. F. Hoopes, and M. A. Leibold. 2004. Alternative stable states and regional community structure. Journal of Theoretical Biology 227: 359–368.
Steiner, C. F., and M. A. Leibold. 2004. Cyclic assembly trajectories and scale-dependent productivity–diversity relationships. Ecology 85: 107–113.
Thebault, E., and M. Loreau. 2006. The relationship between biodiversity and ecosystem functioning in food webs. Ecological Research 21: 17–25.
Tilman, D. 1982. Resource Competition and Community Structure. Princeton, NJ: Princeton University Press.
Tilman, D. 1994. Competition and biodiversity in spatially structure habitats. Ecology 75: 2–16.
Tilman, D., and P. Kareiva. 1997. Spatial Ecology: The Role of Space in Populations Dynamics and Interspecific Interactions. Princeton, NJ: Princeton University Press.
Watt, A. S. 1947. Pattern and process in the plant community. Journal of Ecology 35: 1–22.

III.9

Ecosystem Productivity and Carbon Flows: Patterns across Ecosystems

Julien Lartigue and Just Cebrian

OUTLINE

1. Nature of carbon budgets
2. Rationale and approach for studying patterns of ecosystem productivity and carbon flow
3. Patterns in ecosystem productivity and carbon flow
4. Conclusion

The characterization and understanding of carbon flows in aquatic and terrestrial ecosystems are topics of paramount importance for several disciplines, such as ecology, biogeochemistry, oceanography, and climatology. Scientists have been studying such flows in many diverse ecosystems for decades, and sufficient information is now available to investigate whether any patterns are evident in how carbon flows in ecosystems and to determine the factors responsible for those patterns. In particular, a wealth of information exists on the movement of carbon through the activity of herbivores and consumers of detritus (i.e., decomposers and detritivores), two of the major agents of carbon flows in ecosystems. This chapter analyzes the transference of carbon through herbivory and decomposition in aquatic and terrestrial ecosystems, documents the nature and implications of salient patterns, and explains why those patterns emerge.

GLOSSARY

absolute decomposition. The amount (in g $C{\cdot}m^{-2}{\cdot}year^{-1}$) of detritus consumed by microbial decomposers (e.g., bacteria, fungi) and detritivores, which range from detritivorous micro-, macro-, and gelatinous zooplankton in pelagic systems to micro- (<100 μm), meio- (100–500 μm), and macrofauna (>500 μm) in benthic and terrestrial systems

decomposition rate. The proportion of detrital mass decomposed per unit time (e.g., day^{-1}), often estimated by fitting the following single exponential equation to the pattern of detritus decay observed in experimental incubations, $DM_t = DM_{t_0} e^{-k(t-t_0)}$, where k is the decomposition rate, DM_t is the detrital mass remaining in the experimental incubation at time t, DM_{t_0} is the initial detrital mass, and $(t-t_0)$ is the incubation time

detrital production. The amount (in g $C{\cdot}m^{-2}{\cdot}year^{-1}$) of net primary production not consumed by herbivores, which senesces and enters the detrital compartment

detritus. Dead primary producer material, which normally becomes detached from the primary producer after senescence

herbivory. The amount (in g $C{\cdot}m^{-2}{\cdot}year^{-1}$) of net primary production ingested or removed, including primary producer biomass discarded by herbivores

net primary production. The amount (in g $C{\cdot}m^{-2}{\cdot}year^{-1}$) of carbon assimilated through photosynthesis and not respired by the producer

nutrient concentration (producer or detritus). The percentage of nitrogen and phosphorus within producer biomass or detritus on a dry weight basis

1. NATURE OF CARBON BUDGETS

Carbon enters the biotic component of an ecosystem when inorganic carbon, often carbon dioxide, is taken up and converted into organic compounds. With the rare exception of chemosynthetic organisms, the energy for this conversion comes from photosynthesis. Once inorganic carbon has been converted into organic compounds, it is considered fixed. This production of fixed carbon is known as primary production, and those organisms that can fix carbon are primary producers. Gross primary production is the entire amount of carbon fixed by a primary producer. Net primary production is gross primary production minus

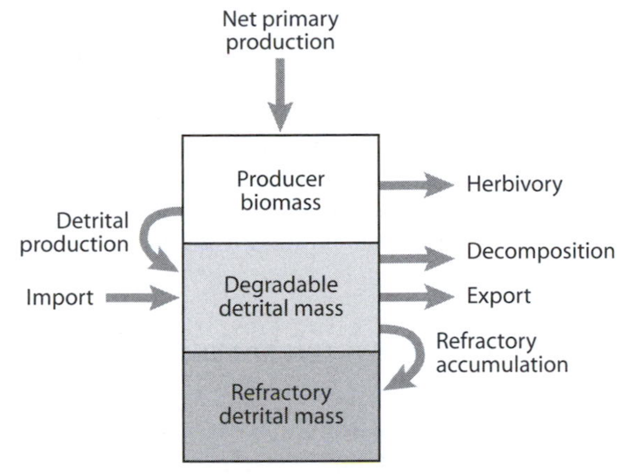

Figure 1. Diagram of carbon flow into and out of the producer and detrital pools in an ecosystem. (Adapted from Cebrian, 1999)

the organic compounds that have been broken down during respiration to fuel cellular processes within the primary producer.

It is the fixed carbon measured by net primary production that becomes primary producer biomass and part of the producer carbon pool (figure 1). This fixed carbon will then either remain as producer biomass, be consumed by herbivores, or enter the detrital pathway and become part of the detrital carbon pool. The import or export of detritus can also alter the amount of carbon in the detrital pool, but regardless of the source of the detritus, detrital carbon will either be recycled by decomposers and detritivores or stored as refractory carbon.

In both aquatic and terrestrial ecosystems, the transfer of fixed carbon from primary producers to herbivores and decomposer/detritivores provides major pathways for the flow of energy and nutrients. As a result, these transfers have consequences not only for carbon storage but also for nutrient recycling and herbivore and decomposer/detritivore populations.

In assessing these transfers, it is important to recognize that they can be viewed in absolute as well as proportional terms. Absolute size refers to the amount or magnitude of the transfer measured in units of producer carbon often over space and time (i.e., $g\,C\,m^{-2}\,year^{-1}$), whereas proportional size refers to the percentage of net primary production consumed by herbivores or the percentage of detrital mass consumed per unit time by decomposers and detritivores.

When regarded as an absolute flux, herbivory sets limits to the level of herbivore production maintained in an ecosystem. Because of herbivore respiration and herbivore egestion of nonassimilated producer biomass, the transfer of fixed carbon from primary producers to herbivores is not complete, and only a fraction of producer biomass ingested becomes herbivore biomass. When herbivory is considered as the percentage of net primary production removed, its implications for the impact of herbivores on carbon and nutrient recycling and storage as producer biomass in the ecosystem become apparent. If herbivores remove a large percentage of net primary production, only a small percentage of the carbon fixed and nutrients taken up by producers is available for accumulation as producer biomass. In such cases, herbivores have the potential to exert significant control on carbon and nutrient storage by producers, which is commonly referred to as top-down regulation (see chapter III.6). Likewise, as the percentage of net primary production consumed increases, so does consumer-driven recycling of carbon and nutrients in the ecosystem. It is important to mention that, when diverse ecosystems are compared, absolute consumption and percentage of net primary production consumed are not always related. Ecosystems with high net primary production may support large absolute consumption by herbivores, which may still represent a small percentage of that high net primary production, in comparison with ecosystems with lower net primary production supporting less absolute consumption but a larger percentage of net primary production lost to herbivores.

As is the case for herbivory, decomposition can also be viewed as an absolute flux or as a proportion of detrital mass decomposed per unit time (i.e., decomposition rate). When considered as an absolute flux, decomposition corresponds to the amount of detritus consumed by microbial decomposers and detritivores. This consumption leads to the reduction of particulate and dissolved detritus into simpler and simpler constituents and, ultimately, to nutrient mineralization. Much like herbivory, decomposition, when regarded as an absolute flux, is indicative of the potential levels of decomposer and detritivore production maintained in the ecosystem because only a fraction of the carbon ingested by decomposers and detritivores is metabolized into biomass of these organisms. When decomposition is viewed as the proportion of detrital mass decomposed per unit time, its implications for how fast carbon and nutrient flow through the detrital pathway become apparent. Ecosystems whose decomposition rate is high tend to have faster nutrient and carbon recycling rates and store less carbon in their detrital pools regardless of any large differences in detrital production. It is worth mentioning that, when diverse ecosystems are compared, higher values of absolute decomposition do not always equate to higher decomposition rates. Ecosystems with low detrital production may have high decomposition rates, yet small absolute decomposition, when compared with other ecosystems with high detrital production, low decomposition rates, and large absolute decomposition.

2. RATIONALE AND APPROACH FOR STUDYING PATTERNS OF ECOSYSTEM PRODUCTIVITY AND CARBON FLOW

The first studies that measured productivity and the flow of carbon focused on individual ecosystems. These studies sought to characterize the transfer of carbon between trophic levels with the goal of understanding how energy moved through an ecosystem. The pioneering studies of Howard T. Odum in freshwater springs in Florida and John M. Teal in the salt marshes of Georgia are classic examples of this early work.

Later studies sought to understand what factors limited net primary production and decomposition by investigating differences in these processes across environmental gradients within the same type of ecosystem. These studies led to now-well-known patterns being established. Annual precipitation is a major determinant of net primary production in grassland ecosystems in arid regions across the Great Plains of the United States. In eastern deciduous forests in the United States, net primary production increases as the length of the growing season increases. In the mountains of Hawaii, net primary production and decomposition rates are positively associated with temperature along an elevation gradient. In aquatic ecosystems, light and nutrient availability frequently limit net primary production.

These studies and their successors have led to a growing body of work measuring net primary production, herbivory, detrital production, decomposition, and the nutritional content of both producers and detritus across a variety of aquatic and terrestrial ecosystems. Such a wealth of data can be extremely useful for detecting more general trends in productivity and how carbon flows through ecosystems.

Researchers have compiled published values of net primary production, herbivory, decomposition, and producer and detrital nutrient content for aquatic and terrestrial ecosystems into large data sets. In assembling such comprehensive data sets, the researchers need to ensure that the values compiled reflect adequately the ecosystems examined. To do so, the values entered into the data set must include the most abundant species of producers and consumers in the ecosystem and encompass at least a year or the entire growing season. In addition, when making comparisons of productivity and carbon flow across a wide range of ecosystems and using data from multiple studies, we must deal in a common currency or unit. The most common unit of choice is grams of carbon or "g C." Last, researchers need to ensure that the conclusions obtained from multistudy data sets are not compromised by the uncertainty that results from compiling values from studies that use different methods, assumptions, and sample sizes. Meta-analysis and estimation of error propagation are two examples of techniques that allow researchers to test the robustness of conclusions obtained from literature comparisons.

3. PATTERNS IN ECOSYSTEM PRODUCTIVITY AND CARBON FLOW

Having discussed the nature of carbon budgets and the rationale behind developing ecosystem carbon budgets as well as assembling this information into larger data sets, we now consider the general patterns in productivity and carbon flow that emerge from the analysis of these larger data sets. In this section, we analyze these patterns and flow by first exploring the overall differences between aquatic and terrestrial ecosystems and then exploring patterns within each type of ecosystem.

Aquatic and Terrestrial Ecosystems: General Differences

Net primary production in aquatic and terrestrial systems is highly variable, but production in both is similar (figure 2A). Aquatic ecosystems, however, do support greater carbon flow to herbivores, both as an absolute carbon flux and as a percentage of net primary production (figure 2D,E). Because herbivore production efficiency, the ratio of herbivore growth to carbon ingested, does not seem to vary significantly between aquatic and terrestrial systems, this greater absolute flux of producer carbon to herbivores implies that aquatic systems should support higher levels of herbivore production compared to terrestrial systems, as has recently been demonstrated.

The higher percentages of net primary production removed by herbivores in aquatic ecosystems suggest that herbivores are more influential in carbon and nutrient recycling and accumulation of producer biomass in aquatic ecosystems relative to their role in terrestrial ecosystems. Indeed, many of the examples of herbivores controlling producer biomass (i.e., top-down control) are from aquatic ecosystems, although on occasion herbivores are found to regulate producer biomass in terrestrial ecosystems as well. The evidence for herbivores as important agents of nutrient recycling in aquatic ecosystems is also abundant, whereas there is considerably less evidence for such a role for herbivores in terrestrial ecosystems, which tend to channel a

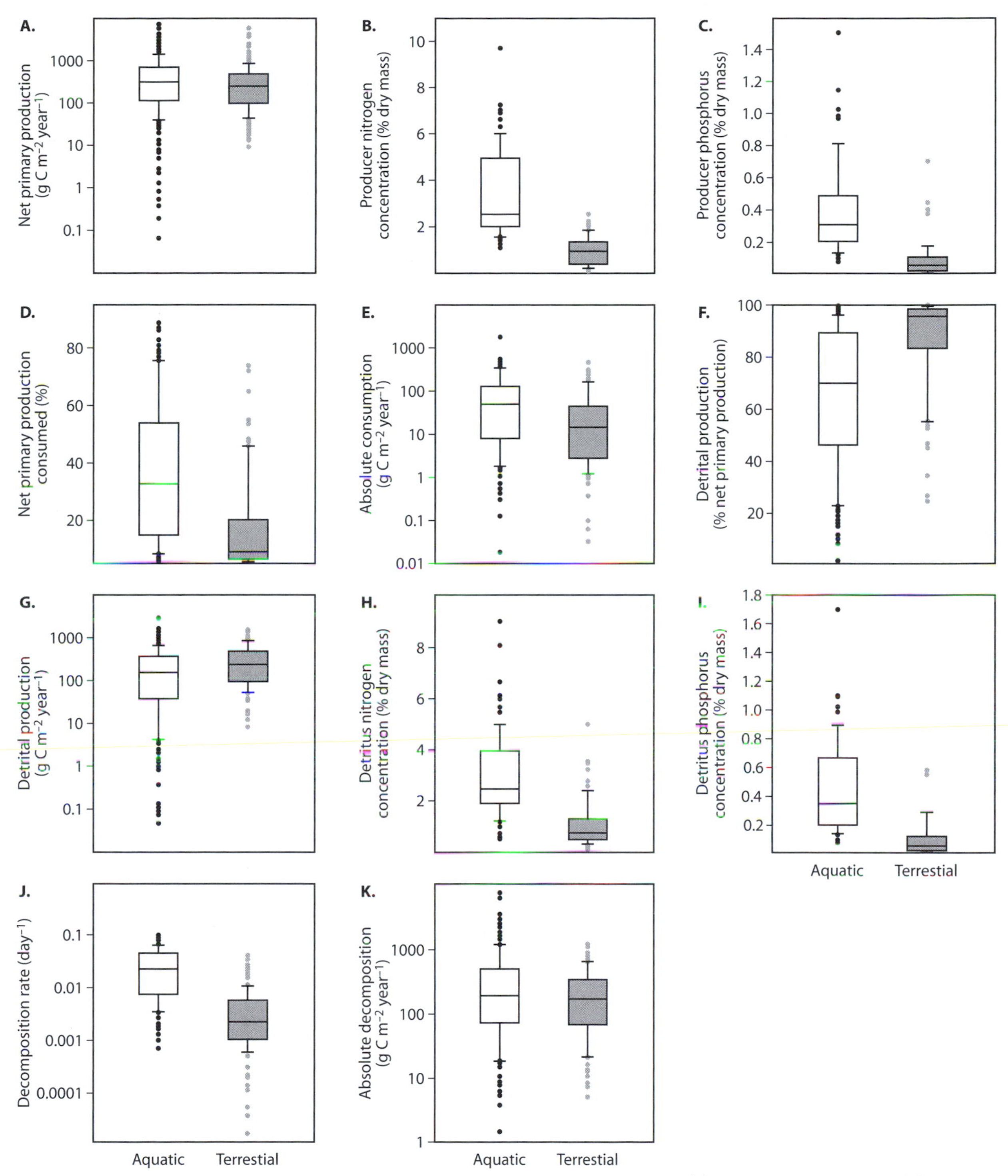

Figure 2. Box plots comparing aquatic and terrestrial ecosystems: (A) net primary production, (B) producer nitrogen concentration, (C) producer phosphorus concentration, (D) percentage of net primary production consumed, (E) absolute consumption, (F) detrital production as a percentage of net primary production, (G) detrital production, (H) detritus nitrogen concentration, (I) detritus phosphorus concentration, (J) decomposition rate, and (K) absolute decomposition. Boxes encompass 25th and 75th percentiles, and the central line is the median. Bars are 10th and 90th percentiles with measurements outside of these percentiles indicated by closed circles. Data set used to generate the box plots is from Cebrian and Lartigue (2004).

higher percentage of net primary production into the detrital pathway.

There are several possible explanations for the greater herbivory measured as an absolute flux of carbon or as a percentage of net primary production consumed in aquatic versus terrestrial ecosystems. One explanation is that primary producers in aquatic ecosystems tend to have higher nutrient concentrations than those in terrestrial ecosystems (figure 2B,C). There is growing evidence that the growth rates of aquatic and terrestrial herbivores are limited by the nutrient content of their diets (see chapter III.15). Under such a premise, herbivore metabolism and growth in aquatic ecosystems are promoted by a diet of higher nutritional quality, and higher rates of absolute consumption and larger percentages of net primary production removed by herbivores result. Indeed, aquatic ecosystems support greater herbivore standing stocks than do terrestrial ecosystems. Higher concentrations of structural, refractory compounds, such as lignin, in terrestrial producers could also lead to lower rates of herbivory in terrestrial ecosystems. Other compounds in the producers, such as fatty acids and digestible carbohydrates, and differences in the availability of nutrients in the producer for herbivore digestion may also affect the growth rates of herbivores and the intensity of herbivory, but their role in explaining the differences in herbivory observed between aquatic and terrestrial ecosystems requires further research. Herbivore behavior, size, energy demands (endothermy versus ectothermy), and predation intensity as well as other factors may also have an impact on herbivory intensity and supersede the expected effects of producer nutritional quality, especially when only few ecosystems are being compared.

Because net primary production differs little between aquatic and terrestrial ecosystems, and herbivory is greater in aquatic ecosystems, aquatic ecosystems tend to transfer a smaller flux of producer carbon, both in absolute terms and as a percentage of net primary production, to the detrital pathway than do terrestrial ecosystems (figure 2F,G). However, most net primary production in both types of ecosystem is not consumed by herbivores and enters the detrital compartment.

The higher nutrient concentrations of aquatic producers compared to terrestrial producers carry over into the detrital compartment as well, where aquatic detritus has higher nutrient concentrations than terrestrial detritus (figure 2H,I). Aquatic detritus also decomposes at a faster rate than does terrestrial detritus (figure 2J), possibly because microbial decomposers and invertebrate and vertebrate detritivores, like herbivores, appear limited in their metabolism and growth rate by the nutrient concentrations of their diets. With higher-quality detritus to consume, aquatic decomposers and detritivores should generally have higher metabolic and growth rates than their terrestrial counterparts leading to faster decomposition rates in aquatic than in terrestrial ecosystems. In other words, the greater nutritional detritus found in aquatic ecosystems leads to more active decomposers and detritivores and faster decomposition rates in comparison with terrestrial ecosystems. And indeed, comparisons have often found faster decomposition rates in aquatic than in terrestrial ecosystems despite substantial environmental variability between the two types of ecosystem.

Faster decomposition rates in aquatic systems indicate faster rates of nutrient recycling through the detrital pathway. This is supported by evidence of faster turnover rates of nutrients through the detrital pool in aquatic ecosystems compared to terrestrial ecosystems. In addition, faster decomposition rates in aquatic ecosystems along with lower detrital production imply the accumulation of smaller standing stocks of detritus in comparison with terrestrial ecosystems.

However, aquatic and terrestrial ecosystems show similar values of absolute decomposition despite the higher decomposition rates found in the former systems (figure 2K). The reason for this lies in the interplay between detrital production and decomposition rates; aquatic ecosystems produce less detritus than terrestrial ecosystems, but it decomposes faster. As a consequence, absolute decomposition, which corresponds to the product between detrital production and decomposition rate, remains similar between the two types of ecosystem. Because the efficiency of decomposer and detritivore production does not differ between aquatic and terrestrial ecosystems, aquatic and terrestrial ecosystems should support similar amounts of decomposer and detritivore production. Interestingly, aquatic ecosystems feature lower standing stocks of decomposers and detritivores than do terrestrial ecosystems, pointing to higher rates of predation on decomposers and detritivores in the former ecosystems.

Patterns within Aquatic and Terrestrial Ecosystems

We now move from discussion of differences between aquatic and terrestrial ecosystems to consideration of patterns within each type of ecosystem. Net primary production and producer nutrient concentrations are uncorrelated within aquatic and within terrestrial ecosystems. The identification of such a general independence may seem surprising at first (figure 3A,B). Indeed, many fertilization experiments have shown that increased nutrient availability often leads to increased nutrient concentrations in producer biomass

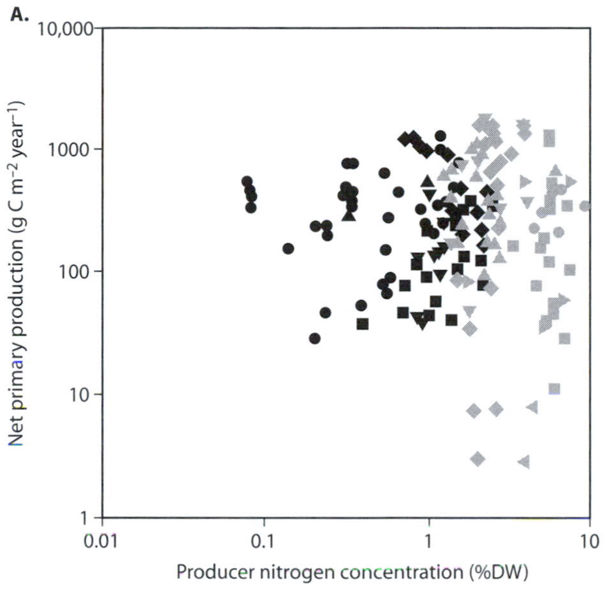

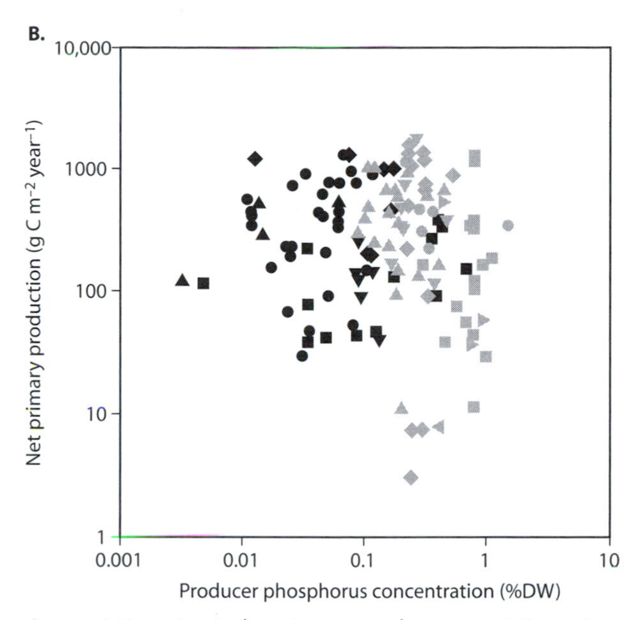

Figure 3. Net primary production from aquatic (gray symbols) and terrestrial (black symbols) ecosystems plotted against producer (A) nitrogen concentration and (B) phosphorus concentration. The aquatic communities represented are freshwater phytoplankton (gray circles), marine phytoplankton (gray squares), freshwater benthic microalgae (gray left-pointing triangles), marine benthic microalgae (gray right-pointing triangles), marine macroalgae (gray diamonds), freshwater submerged macrophytes (gray down-pointing triangles), and seagrass (gray up-pointing triangles). The terrestrial communities represented are tundra shrubs and grasses (black down-pointing triangles), freshwater and marine marshes (black diamonds), temperate and tropical shrublands and forests (black circles), temperate and tropical grasslands (black squares), and mangroves (black up-pointing triangles). (Adapted from Cebrian and Lartigue, 2004)

and higher levels of net primary production in aquatic and terrestrial ecosystems. This independence likely stems from the large environmental variability encompassed when a large range of ecosystems are compared within the aquatic or terrestrial realm. Growth limitation by light and temperature, water availability (in terrestrial ecosystems), wave action (in aquatic ecosystems), and other types of environmental stress may very well prevent a positive association between producer nutrient concentrations and net primary production over a broad range of ecosystems.

Aquatic and terrestrial ecosystems composed of producers with higher nutrient concentrations do tend to have a greater percentage of net primary production removed by herbivores (figure 4A,B) despite contrasting natures (invertebrate versus vertebrate), metabolic patterns (ectothermy vesus endothermy), behavior (migratory versus resident), and feeding specificity (specialized versus generalized) of the herbivore populations in the ecosystems compared. As is the case for the general comparison between aquatic and terrestrial ecosystems, this relationship within each type of ecosystem is likely fueled by the growth rates of aquatic and terrestrial herbivores, which are often limited by the nutrient content of their diets. On this basis, producers with higher nutrient concentrations lead to faster herbivore growth rates and larger percentages of net primary production consumed. It follows that, regardless of whether aquatic or terrestrial ecosystems are considered, herbivores should exert a greater control on producer biomass accumulation and carbon and nutrient recycling in ecosystems composed of producers with higher nutrient concentration.

Although there is a positive association between producer nutrient concentration and the percentage of net primary production consumed by herbivores in aquatic ecosystems, herbivory measured as an absolute flux of producer carbon to herbivores is only poorly associated with producer nutrient concentration (figure 4C,D). Absolute consumption is, however, strongly associated with the absolute magnitude of net primary production (figure 4E) in aquatic ecosystems. In other words, more productive aquatic ecosystems, but not aquatic ecosystems having producers with higher nutrient concentrations, support greater absolute consumption by herbivores. This pattern stems from the interaction between the variability in net primary production and the variability in the percentage consumed within aquatic ecosystems (figure 4F). A higher percentage of net primary production is lost to herbivores

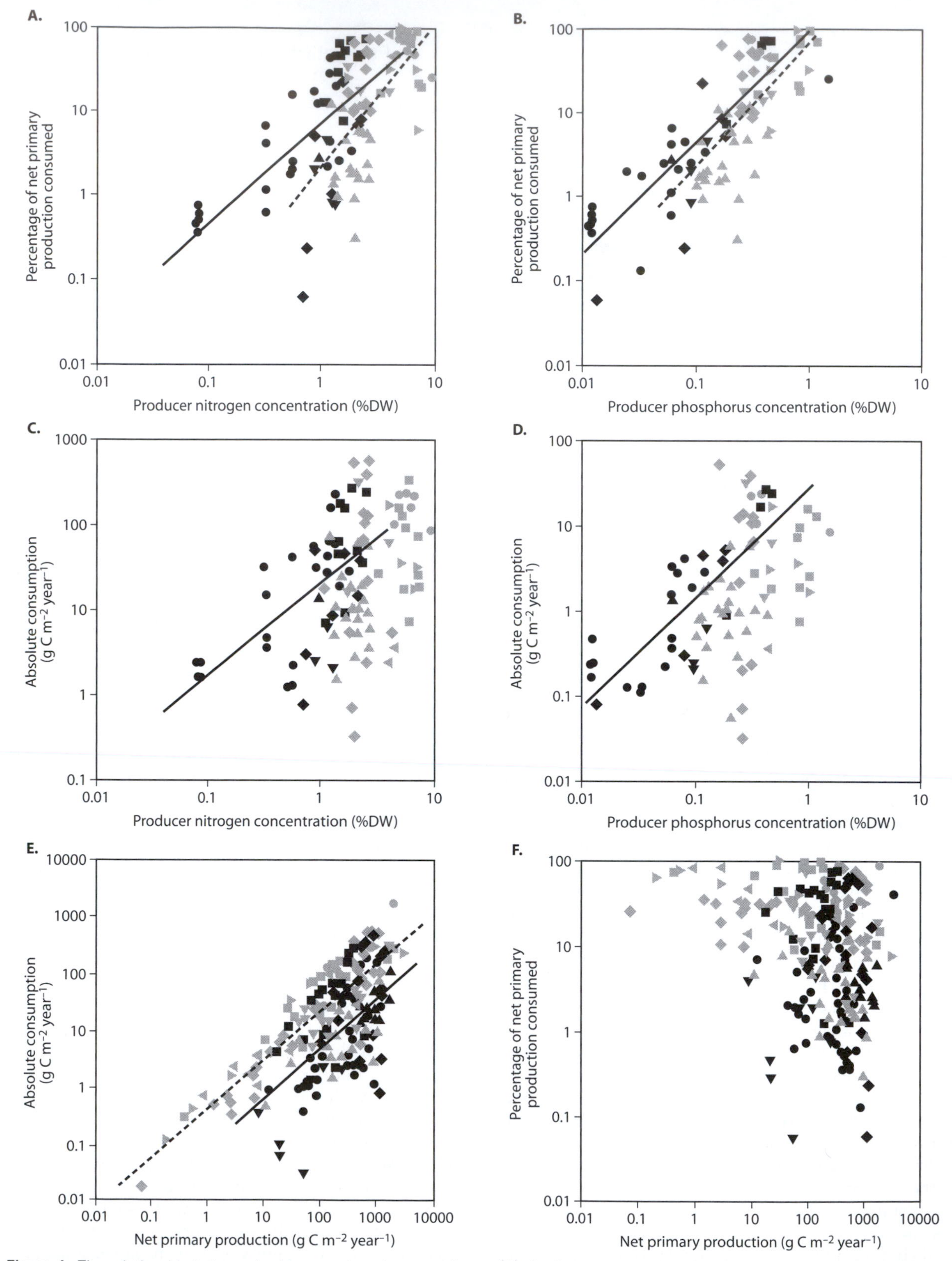

Figure 4. The relationship between herbivory and producer nutrient concentrations or net primary production in aquatic and terrestrial ecosystems: (A) percentage consumed versus producer nitrogen concentration (dashed line, aquatic ecosystems regression, $R^2 = 0.37$; solid line, terrestrial ecosystems regression, $R^2 = 0.40$), (B) percentage consumed versus producer phosphorus concentration (dashed line, aquatic ecosystems regression, $R^2 = 0.44$; solid line, terrestrial ecosystems regression, $R^2 = 0.65$), (C) absolute consumption versus producer nitrogen concentration (solid line, terrestrial ecosystems regression, $R^2 = 0.38$), (D) absolute consumption (solid line, terrestrial ecosystems regression, $R^2 = 0.64$), (E) absolute consumption versus net primary production (dashed line, aquatic ecosystems regression, $R^2 = 0.66$; solid line, terrestrial ecosystems regression, $R^2 = 0.25$), and (F) absolute consumption versus net primary production. Symbols are the same as in figure 3. (Adapted from Cebrian and Lartigue, 2004)

in aquatic ecosystems comprised of producers with higher nutrient concentrations. However, this percentage varies little compared to the much larger differences in net primary production within aquatic ecosystems. As a result, absolute consumption, which is the product of net primary production and the percentage consumed by herbivores, remains more closely associated with net primary production and only poorly associated with the percentage consumed and producer nutrient concentrations. An implication of these patterns is that aquatic ecosystems with higher net primary production, and not those composed of more nutritional producers, transfer more producer carbon to herbivores and, because the efficiency of herbivore production does not seem to vary consistently across ecosystems, also support higher herbivore production.

A different situation exists within terrestrial ecosystems, where absolute consumption is positively associated with producer nutrient concentration but less so with net primary production. Again, the explanation lies in the interaction between the variability in net primary production and the variability in the percentage consumed within terrestrial ecosystems (figure 4F). Within terrestrial ecosystems, net primary production varies to a lesser degree than does the percentage consumed, and, as a result, absolute consumption remains more closely associated with the percentage consumed and, by extension, with producer nutrient concentration than with net primary production. Therefore, terrestrial ecosystems composed of producers with higher nutrient concentrations, in addition to supporting a greater impact by herbivores on the accumulation of producer biomass and carbon and nutrient recycling, transfer a greater flux of producer carbon to herbivores and should have higher levels of herbivore production. Recent work, however, has shown that herbivore production is positively related to net primary production in terrestrial ecosystems because absolute consumption and net primary production are positively related, albeit not strongly, within these ecosystems.

Because most net primary production enters the detrital compartment in both aquatic and terrestrial systems, detrital production is strongly associated with net primary production within each type of system. Conversely, detrital production is unrelated to detritus nutrient concentration within each type of ecosystem. This lack of association stems from the independence between net primary production and producer nutrient concentration. Within both aquatic and terrestrial ecosystems, the nutrient concentration of producers changes little through senescence in relation to the variability among producers. This, along with the strong association between detrital production and net primary production within each type of ecosystem, explains why the independence between net primary production and producer nutrient concentration drives the independence between detrital production and detritus nutrient concentration.

Detritus with higher nutrient concentrations tends to exhibit faster decomposition rates within both aquatic and terrestrial ecosystems, although the trend is not always strong (figure 5A,B). Yet this association is relevant given the substantial environmental variability that may exist among ecosystems and the contrasting effects on decomposition rates that result from differing levels of temperature, soil or sediment reduction–oxidation reaction conditions, and, in terrestrial systems, moisture.

The association between faster decomposition rates and more nutritional detritus found within ecosystems, regardless of whether these are aquatic or terrestrial, probably results from the limitation exerted by the nutrient content of the detritus on the metabolic and growth rates of decomposers and detritivores; higher nutrient concentrations in the detritus stimulate the metabolic and growth rates of these organisms, resulting in faster decomposition rates. Two important corollaries follow. First, ecosystems with more nutritional detritus, regardless of whether they are aquatic or terrestrial, should feature faster nutrient recycling rates through the detrital pathway. Second, ecosystems with more nutritional detritus should also accumulate smaller detrital pools provided the differences in decomposition rates among ecosystems exceed the differences in detrital production.

Despite the association between faster decomposition rates and higher detritus nutrient concentrations within both aquatic and terrestrial ecosystems, decomposition when viewed as an absolute flux is independent of detritus nutritional quality (figure 5C,D). Instead, absolute decomposition is strongly associated with detrital production and net primary production within either type of ecosystem (figure 5E). The reason for this pattern lies in the interaction between the variability in detrital production and the variability in the percentage of detrital production decomposed within a year among ecosystems (figure 5F). In aquatic ecosystems, and to a lesser degree in terrestrial ecosystems, detrital production varies to a larger extent among ecosystems than does the percentage decomposed. As a consequence, absolute decomposition, which corresponds to the product of detrital production and the percentage decomposed, remains closely associated with detrital production and net primary production and independent of the percentage of detrital production decomposed. Because the percentage of detrital production decomposed is a surrogate for decomposition rates, absolute decomposition also remains unrelated to decomposition rates

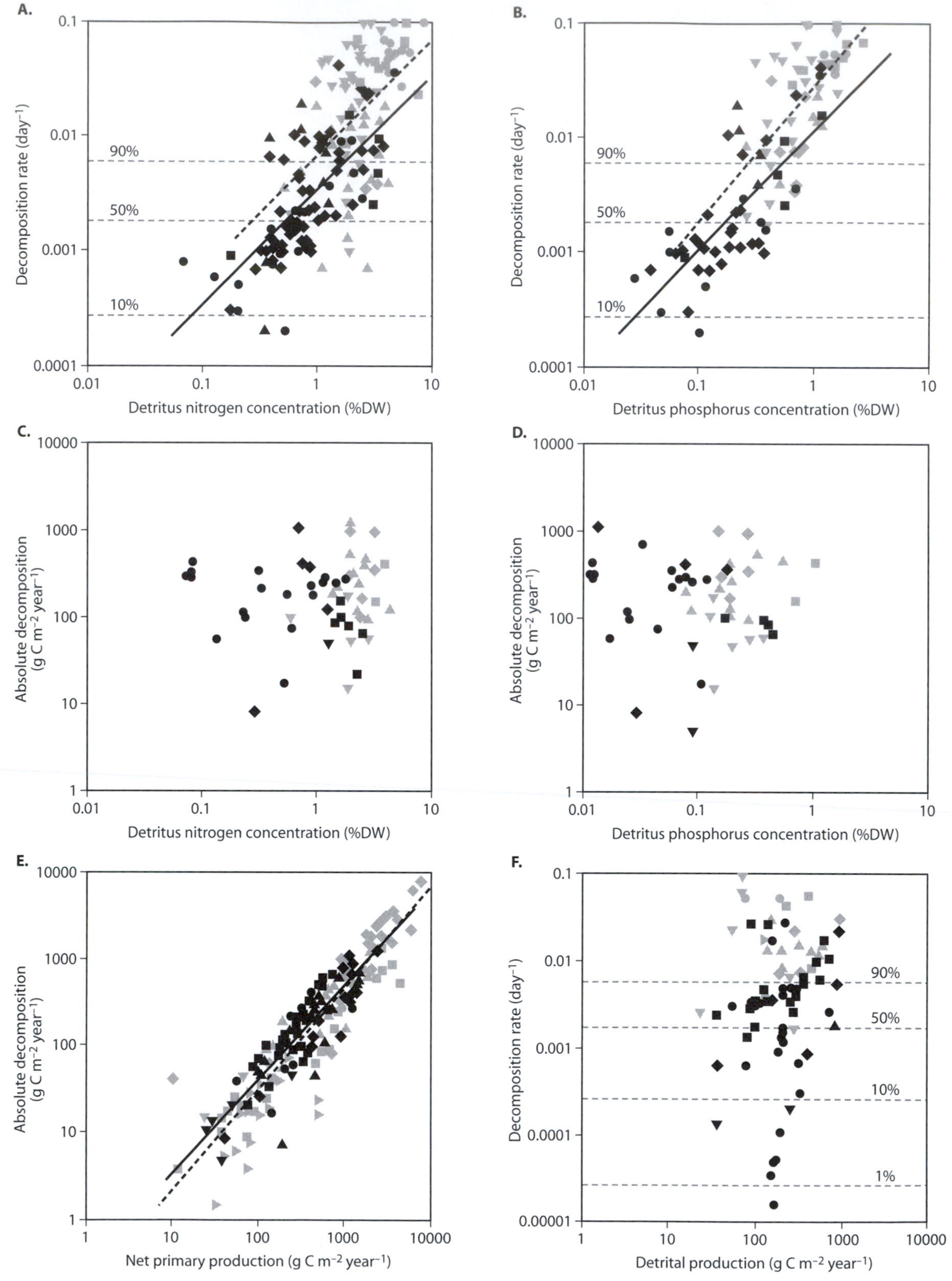

Figure 5. The relationship between decomposition and detritus nutrient concentration or detrital production in aquatic and terrestrial ecosystems: (A) decomposition rate versus detritus nitrogen concentration (dashed line, aquatic ecosystems regression, $R^2 = 0.21$; solid line, terrestrial ecosystems regression, $R^2 = 0.46$), (B) decomposition rate versus detritus phosphorus concentration (dashed line, aquatic ecosystems regression, $R^2 = 0.34$; solid line, terrestrial ecosystems regression, $R^2 = 0.54$), (C) absolute decomposition versus detritus nitrogen concentration, (D) absolute decomposition versus detritus phosphorus concentration, (E) absolute decomposition versus net primary production (dashed line, aquatic ecosystems regression, $R^2 = 0.84$; solid line, terrestrial ecosystems regression, $R^2 = 0.76$), and (F) decomposition rate versus detrital production. Horizontal dashed lines and percentages indicate the percentage of detrital production that would be decomposed within a year at the given decomposition rate. Symbols are the same as in figure 3. (Adapted from Cebrian and Lartigue, 2004)

and, by extension, to detritus nutrient concentration when either aquatic or terrestrial ecosystems are compared. Therefore, aquatic and terrestrial ecosystems with higher primary and detrital production, and not those having more nutritional detritus, transfer more detrital carbon to decomposers and detritivores and, because the efficiency of decomposer and detritivore production varies little across ecosystems, support higher decomposer and detritivore production.

4. CONCLUSION

Producer nutritional quality and net primary production are two independent predictors of herbivory and decomposition in aquatic and terrestrial ecosystems. Herbivory, expressed as the percentage of net primary production consumed by herbivores, and decomposition, expressed as the proportion of detrital mass consumed per day by decomposers and detritivores, are positively associated with producer nutrient concentration but independent of net primary production, regardless of whether the comparison is done between aquatic and terrestrial ecosystems or within each type of ecosystem. Thus, producer nutrient concentration, and not net primary production, stands out as a potential indicator of top-down regulation of the pools of producer biomass and detritus and nutrient and carbon recycling rates by first-order consumers in ecosystems. The reverse situation is often found when herbivory and decomposition are expressed as absolute fluxes, which are then positively associated with net primary production but independent of producer nutrient concentration because net primary production often varies to a larger extent than does the percentage consumed by herbivores or decomposed across a broad range of ecosystems. Therefore, net primary production, and not producer nutrient concentration, is often the indicator of secondary production (i.e., production of herbivores and consumers of detritus) in ecosystems.

FURTHER READING

Cebrian, Just. 1999. Patterns in the fate of production in plant communities. American Naturalist 154: 449–468.

Cebrian, Just. 2004. Role of first-order consumers in ecosystem carbon flow. Ecology Letters 7: 232–240. *Investigates herbivore and decomposer and detrivore biomass and their impact on the turnover of producer-fixed carbon in aquatic and terrestrial ecosystems.*

Cebrian, Just, and Julien Lartigue. 2004. Patterns of herbivory and decomposition in aquatic and terrestrial ecosystems. Ecological Monographs 74: 237–259. *A more detailed and technical discussion of the patterns in productivity and decomposition discussed in the chapter.*

Odum, Howard T. 1957. Trophic structure and productivity of Silver Springs, Florida. Ecological Monographs 27: 55–112.

Sterner, Robert W., and James J. Elser. 2002. Ecological Stoichiometry: The Biology of Elements from Molecules to the Biosphere. Princeton, NJ: Princeton University Press. *An introduction to the discipline of ecological stoichiometry—the study of the balance of elements and energy in ecological interactions.*

Teal, John. 1962. Energy flown in the salt marsh ecosystem of Georgia. Ecology 43: 615–624.

III.10

Nutrient Cycling and Biogeochemistry

Peter M. Vitousek and Pamela A. Matson

OUTLINE

1. Element cycles in terrestrial ecosystems
2. Global element cycles
3. Illustration: Nutrient cycling in practice

Studies of nutrient cycles involve integrating information from very fine spatial and temporal scales (the dynamics of enzymes in the neighborhood of microbes) to very coarse scales (the global biogeochemical cycles); they involve integrating the dynamics of organisms with those of the environment that they inhabit and help to shape. Some of the finest-scale, most biological of processes (e.g., the growth of microbial populations on chemically recalcitrant plant litter) control important aspects of the Earth system (e.g., the persistence of nitrogen limitation to primary production, as in the example above). Nutrient cycles cannot be studied effectively in isolation, whether that means isolation from a consideration of both biological and geochemical processes or isolation from understanding the substantial and increasing influence of human activity on the Earth system.

GLOSSARY

biological nitrogen fixation. The enzyme-mediated reduction of atmospheric dinitrogen (N_2) to chemical forms that can be used by most organisms.

eutrophication. Overenrichment of ecosystems resulting from excessive additions of nutrients; eutrophication may create anaerobic conditions ("dead zones") in aquatic ecosystems.

mineralization. With reference to phosphorus and nitrogen, mineralization is the microbially mediated conversion of organically bound nutrients to soluble, biologically available inorganic forms.

mycorrhizae. Mycorrhizae are a symbiosis between the roots of most higher plants and several groups of fungi, in which the fungal partner typically derives energy from the plant and the plant receives nutrients from the fungus.

nitrification. The biologically mediated oxidation of ammonium (NH_4) to nitrate (NO_3); specialized microorganisms derive their energy from this transformation.

nutrient limitation. Nutrient limitation occurs where the rate of a biological process like productivity or decomposition is constrained by a low supply of one or more biologically essential elements.

weathering. The breakdown of rocks and minerals, at least partly into soluble and biologically available components.

within-system cycle. Transfers of nutrients among plants, animals, microorganisms, and soil and/or solution, within the boundaries of an ecosystem.

We define a "nutrient" as an element that is required for the growth of some or all organisms—and one that plants typically acquire from soil or solution (as opposed to the uptake of carbon from gaseous forms). The cycles of nutrients are interesting to ecologists for many reasons, including the following:

- A low supply of a nutrient can constrain the growth and populations of organisms and the productivity, biomass, diversity, and dynamics of entire ecosystems.
- Losses of nutrients from terrestrial ecosystems represent inputs to aquatic systems and to the atmosphere. In the atmosphere, reactive nitrogen gases influence atmospheric chemistry and climate; in freshwater and marine systems, inputs of N and P can drive eutrophication (overenrichment). Element losses thus represent a useful currency for evaluating land–water and land–atmosphere interactions.
- The cycles of multiple elements are altered on regional and global scales by human activity. Much research in this area has focused on the global cycle of carbon, in part because of the importance of CO_2 in the climate system, but

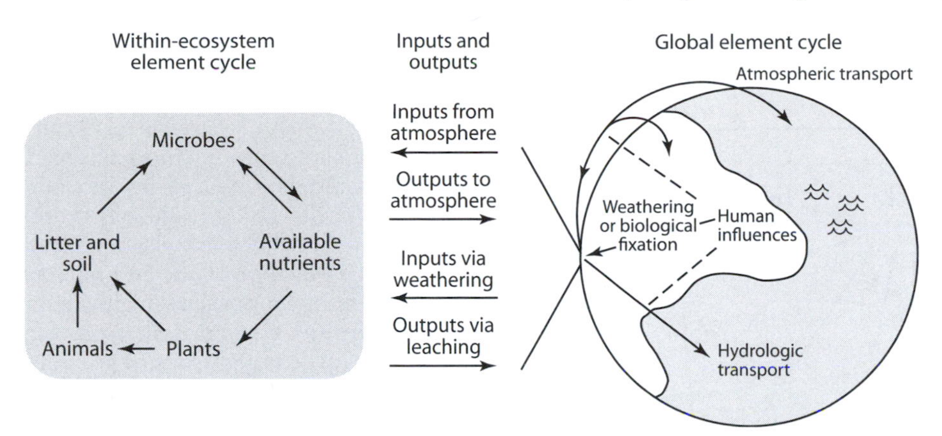

Figure 1. Overview of nutrient cycling. Elements cycle within ecosystems (on left), from biologically available forms through plants, animals, soils, and microbes, and eventually back to available forms. This within-system cycle interacts with pathways of element inputs into and outputs from the ecosystem; those inputs and outputs in turn connect to element cycles on regional and global scales.

humanity has altered the cycles of nitrogen, phosphorus, and sulfur to a much greater extent than that of carbon.

- Substantial differences in the biology, geology, and chemistry of the cycles of different elements make attempts to integrate cycles on local, regional, and global scales both challenging and rewarding.
- Just as element supply can shape the growth and distribution of organisms, organisms can affect the supply of elements by affecting inputs, outputs, or rates of cycling of nutrients.

Ecological research in nutrient cycling and biogeochemistry has focused strongly on nitrogen and phosphorus for the good reason that of the many elements that plants and animals require, these two most often control plant growth, community diversity, and ecosystem-level processes such as productivity. They are not the only such controls; the supply of iron controls the growth and biomass of algae in large areas of the ocean, a low supply of potassium or sulfur can constrain the growth of plants (especially after nitrogen and phosphorus requirements are met), silica supply often regulates the growth of diatoms in lakes and ocean, and calcium availability has long been recognized as an excellent predictor of the distribution of plants and plant communities in many regions. Nevertheless, nitrogen and phosphorus control organisms and ecosystems across a very broad range of sites and conditions, and we focus on them here.

1. ELEMENT CYCLES IN TERRESTRIAL ECOSYSTEMS

The dynamics of elements with strong biological affinities (e.g., nitrogen and phosphorus) can be viewed in terms of a within-ecosystem cycle that is nested inside a set of inputs to and outputs from an ecosystem; that input–output cycle is in turn nested within a global cycle (figure 1). The within-ecosystem cycle is characterized by biological uptake of elements and their incorporation into organisms, ultimately followed by breakdown of organic material and release of the elements it contains into forms available for subsequent uptake. Inputs and outputs represent fluxes of elements across ecosystem boundaries, to and from the atmosphere, rock, or hydrologic systems; these inputs and outputs ultimately connect any particular ecosystem to the world.

Nutrient Cycling within Ecosystems

For both nitrogen and phosphorus, the quantity taken up by organisms and released from organic matter each year substantially exceeds (usually by at least 10-fold) the amount of these elements that enters or leaves ecosystems. We will discuss exceptions to this generalization, but it holds in most intact ecosystems. The predominance of within-ecosystem cycling relative to inputs and outputs means that the quantities of nitrogen and phosphorus that are available to organisms on a year-to-year basis are determined proximately by the rate at which nutrients cycle within ecosystems rather than by nutrient inputs to ecosystems.

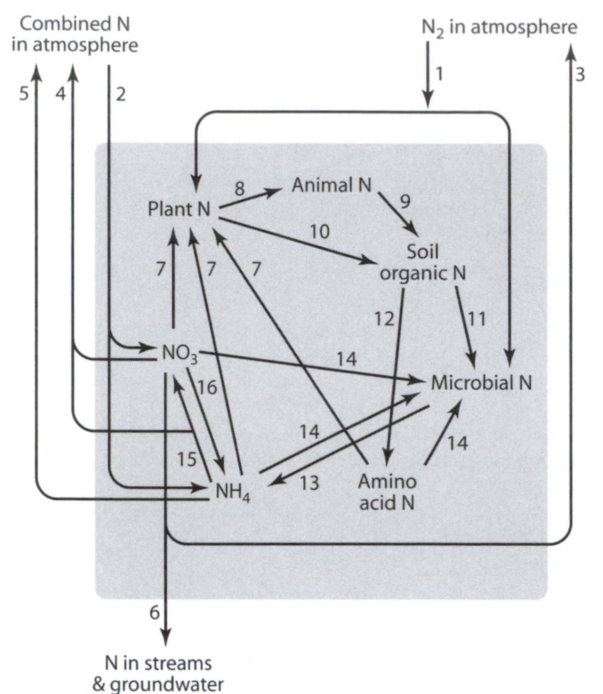

Figure 2. Pathways of nitrogen cycling in terrestrial ecosystems. The nitrogen cycle is complex, with multiple transformations and oxidation/reduction reactions; indeed, this depiction is a substantial simplification of the N cycle in that it leaves out many of the microbial populations that carry out nitrogen transformations. Major fluxes of nitrogen shown here include (1) biological nitrogen fixation; (2) atmospheric deposition of combined nitrogen; (3) denitrification to dinitrogen; (4) fluxes of oxidized trace gases of nitrogen that occur during nitrification and denitrification; (5) ammonia volatilization; (6) solution losses of nitrogen; (7) plant uptake of available nitrogen; (8) consumption of plant nitrogen by animals; (9) flux of nitrogen to soil from excretion or animal death; (10) flux of nitrogen to soil in plant litter; (11) uptake of organic nitrogen by microbes that carry out decomposition; (12) mobilization of amino acids from soil organic nitrogen through the action of extracellular enzymes; (13) release of ammonium by microbes; (14) microbial uptake of available nitrogen; (15) nitrification of ammonium to nitrate; and (16) dissimilatory reduction of nitrate to ammonium.

Although nitrogen and phosphorus are alike in the importance of internal cycling relative to inputs and outputs, they differ in many important respects. We will compare and contrast them, starting with the dynamics of inorganic nitrogen and phosphorus in soil, then follow their cycles through plants (and sometimes animals) back to forms of nitrogen and phosphorus in plant litter and organic matter in soil, and through the microorganisms that decompose organic material and (often) release nitrogen and phosphorus back to inorganic forms in soil.

Biologically, the dynamics of soil nitrogen is more complex than the dynamics of phosphorus (figure 2). Most inorganic nitrogen initially becomes available in soils in the reduced ammonium (NH^+_4) form. This ammonium cation can be exchangeably bound to soil clays and organic matter; it also can be acquired directly by plants and microorganisms. In addition, ammonium can be oxidized by specialized microorganisms to nitrite (NO_2^-) and nitrate (NO_3^-). Nitrite is generally short-lived in soils, but nitrate can be utilized by a wide range of plants and microorganisms. Especially as nitrate, nitrogen is highly mobile in soils; it can diffuse readily through soils to roots, but it can also leach rapidly through soils to groundwater or streamwater. As discussed below, nitrate also can be utilized as an electron acceptor (replacing oxygen) by many microorganisms.

In contrast, inorganic phosphorus in soils is geochemically more complex than nitrogen (figure 3). Although it remains in the form of oxidized phosphate (PO_4^{3-}) through its cycle, without substantial oxidation/reduction reactions, it is strongly adsorbed by soil colloids (much more so than ammonium), and it forms very sparingly soluble complexes with calcium in basic soils and with aluminum and iron in acid soils. Because of these geochemical reactions, the mobility of phosphate in soils is very low; it diffuses to roots much more slowly than nitrate or ammonium, but it is also retained within soils more efficiently against losses via leaching. The mobility of phosphate is so low that most plants require (or at least utilize) help in obtaining it; their roots are colonized by mycorrhizal fungi that derive their energy from the plant and that produce fungal hyphae that explore larger volumes of soil than are accessible to plant roots, passing much of the phosphate they obtain to their hosts. Mycorrhizae can be useful in nitrogen acquisition as well, but they are particularly important for phosphate.

Within plants (and animals and microorganisms), both nitrogen and phosphorus are integral contributors to the basic biogeochemical machinery of life. Nitrogen is part of proteins, including the enzymes that catalyze most of the biochemical reactions within organisms. Phosphorus makes up the backbone of nucleic acids (DNA and RNA) and serves as the fundamental energy currency of living organisms. Accordingly, nitrogen and phosphorus occur in a characteristic ratio (or stoichiometry; see chapter III.15). This ratio is not fixed—organisms can use nitrogen and phosphorus differently, for structure and defense as well as growth and metabolism, but these elements are closely linked.

Animals typically acquire their nitrogen and phosphorus from the food they consume rather than from the environment. Most consumers of terrestrial plants (herbivores) are much richer in nitrogen and phosphorus than the plants they eat; accordingly, low

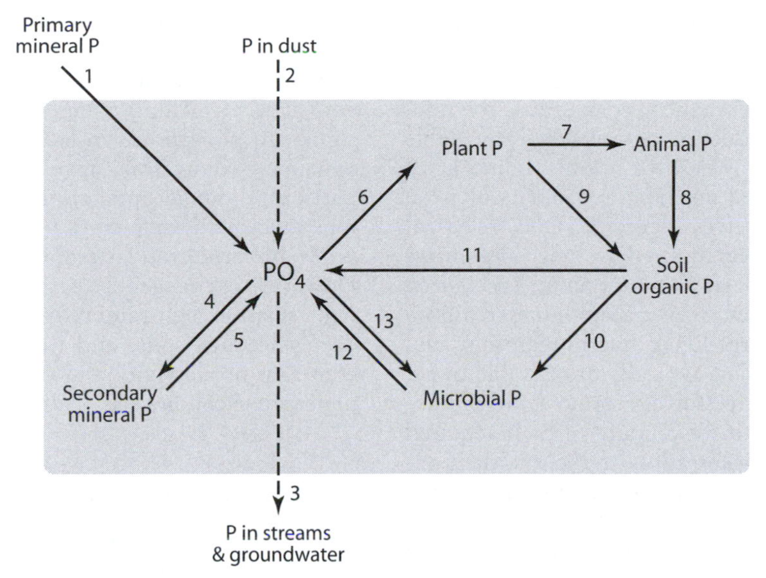

Figure 3. Pathways of phosphorus cycling in terrestrial ecosystems. Although the phosphorus cycle lacks some of the biological complexity of the nitrogen cycle, it participates in a broader array of important geochemical reactions. Fluxes of phosphorus shown here include (1) weathering of primary minerals; (2) inputs of phosphorus from the atmosphere, mostly as dust; (3) leaching of phosphorus to streams and groundwater; (4) formation of secondary minerals within the soil; (5) weathering of secondary minerals; (6) plant uptake of phosphate; (7) consumption of plant phosphorus by animals; (8) flux of phosphorus to soil from excretion or animal death; (9) flux of phosphorus to soil in plant litter; (10) uptake of organic phosphorus by microbes that carry out decomposition; (11) release of phosphate from soil organic phosphorus through the action of extracellular enzymes; (12) release of phosphate by microbes; and (13) uptake of phosphate by microbes.

concentrations of nitrogen (protein) and phosphorus in plants can constrain the growth of herbivores and/or force substantial inefficiencies in their feeding and energy metabolism. Predators typically consume animals with nitrogen and phosphorus contents that align more closely with their own requirements.

Most nitrogen and phosphorus return to soil contained in organic compounds when plants or animals die or when plants shed leaves and roots. Much of the nitrogen and phosphorus in leaves is pulled back from those leaves as they senesce, so that leaf litter (the senescent leaves dropped by plants) has substantially lower concentrations of nitrogen and phosphorus than do live leaves. The soluble and readily decomposable fractions that remain in leaf litter when it reaches the soil break down relatively quickly, and the nutrients they contain are utilized by microorganisms. However, the more recalcitrant fractions of litter—and the still more recalcitrant organic compounds that make up most of the dispersed organic matter in soils—are broken down by extracellular enzymes produced by soil bacteria and fungi. These enzymes break complex and insoluble organic compounds into smaller, more labile forms.

The consequences of decomposition via extracellular enzymes differ for nitrogen versus phosphorus. Most nitrogen in organisms and organic matter is bonded directly to carbon, and recalcitrant compounds containing organic nitrogen have a wide array of structures and in many cases the ability to precipitate soluble proteins (such as enzymes). In contrast, the majority of organic phosphorus is bonded to carbon compounds through ester linkages (C-O-P). Consequently, much of the phosphorus in litter and soil organic matter can be released into solution by phosphatase enzymes that cleave their ester linkage, releasing (or "mineralizing") phosphate ions without decomposing the organic compound that contained them. In contrast, the coordinated action of several enzymes may be required to release nitrogen from recalcitrant organic compounds, and nitrogen mineralization generally is linked to the breakdown of the organic compounds that contain it. Moreover, all extracellular enzymes contain substantial quantities of nitrogen; a plant or microorganisms with sufficient nitrogen but inadequate phosphorus can straightforwardly invest some of its nitrogen in enzymes that could mobilize phosphorus, but if the organism's supply of nitrogen is insufficient, acquiring more by producing extracellular enzymes represents a greater challenge. For these reasons, the cycling and supply of biologically available nitrogen are less flexible (relative to biological demands) than those of phosphorus.

Just as the rate-limiting step in the decomposition of recalcitrant organic matter is the enzyme-catalyzed breakdown of complex and insoluble compounds to soluble ones that microorganisms can utilize, the rate-limiting step in the production of biologically available nitrogen in soil is the release of soluble amino acids from their complex and insoluble precursors or from microorganisms themselves (figure 2). Once released, amino acids can be metabolized by microorganisms that may release nitrogen as ammonium; specialized microorganisms can then oxidize ammonium to nitrate as described above. Free-living microorganisms, mycorrhizae, and plant roots typically are capable of acquiring any of these three forms (amino acids, ammonium, and nitrate), with the amount of each acquired depending on their concentrations in the soil, their relative mobility, and the affinity of organisms' uptake systems.

The fact that free-living microorganisms and plants/mycorrhizae draw on the same sources of nitrogen and phosphorus raises the possibility of competition for these potentially limiting resources, and the contrast in C-N-P stoichiometry between microorganisms and terrestrial plants accentuates this possibility. Plants use carbon to build structure, in their cell walls and support tissues, and senesced leaves typically have C:N:P ratios that range from 750–2000:15:1 (on a mass basis). (As described above, N:P ratios are not constant, but our focus here is on the interaction between carbon and the embodied energy it represents on the one hand and nitrogen and phosphorus on the other.) These ratios are even wider in woody tissues. In contrast, the most important microorganisms that carry out decomposition typically have C:N:P ratios in the range of 90–220:15:1, with fungi on the upper end of that range. Accordingly, soil organisms live in a carbon-rich, nitrogen- and phosphorus-poor world in comparison to their own requirements for growth and metabolism, and under many circumstances they can be highly effective competitors with plants for nitrogen and phosphorus, reducing the availability of these nutrients to plants.

Multiple elements in addition to nitrogen and phosphorus are required by organisms, and under some circumstances insufficient supplies of these elements can constrain biological activity. Among these elements, sulfur, calcium, magnesium, and potassium are required in the largest quantities by terrestrial plants. Sulfur goes through oxidation-reduction reactions similar to those of nitrogen; it is used within organisms in reduced form, but specialized microorganisms derive their energy by oxidizing reduced sulfur to sulfate (SO_4^{2-}) in soils, and most organisms acquire their sulfur as sulfate and then reduce it. Soils often contain a relatively large adsorbed pool of sulfate that can buffer sulfur availability within terrestrial ecosystems.

In contrast to sulfur and nitrogen, the macronutrient cations (calcium, magnesium, and potassium) are similar to phosphorus in lacking oxidation–reduction dynamics. Many soils have relatively large (in comparison to annual uptake) pools of exchangeable cations associated with clays and organic matter; these pools buffer cation availability in the soil solution. However, exchangeable cation pools can be vanishingly small in high-rainfall tropical ecosystems on old, deeply leached soils, and the supply of cations may constrain productivity and other ecosystem processes under these conditions.

Inputs and Outputs of Nutrients

On a year-to-year basis, the cycling of nitrogen and phosphorus within ecosystems is far more important as a source of these elements than are element inputs from outside the ecosystem, for most ecosystems most of the time. However, in the long term, the balance between ecosystem-level element inputs and element losses controls the quantities of nitrogen and phosphorus within an ecosystem and ultimately their availability to organisms. As with internal nutrient cycles, the dominant sources, pathways, and regulation of inputs and outputs of nitrogen and phosphorus differ substantially.

Inputs

The most important input pathways for nitrogen in terrestrial ecosystems are atmospheric deposition and biological nitrogen fixation (figure 2); some sedimentary rocks release nitrogen when they weather, and lower-slope or wetland ecosystems can receive dissolved inputs of nitrogen from upslope ecosystems. Atmospheric inputs of nitrogen involve the deposition of inorganic (mostly ammonium and nitrate) and organic nitrogen dissolved in rainfall, sedimentation of particles from the atmosphere or their impact on plant surfaces, and the adsorption/absorption of nitrogen-containing gases including ammonia, nitric acid, and nitrogen dioxide. Under background atmospheric conditions (those little altered by human activity), all of these forms of combined nitrogen (defined as nitrogen that is bonded to carbon, hydrogen, or oxygen) get into the atmosphere primarily through the emission of reactive nitrogen-containing gases from soils and vegetation (the latter particularly during fires). As discussed below, these emissions (and subsequent depositions) have been vastly increased by human activity.

Biological nitrogen fixation is the enzyme-mediated reduction of triple-bonded dinitrogen gas (N_2) to ammonium and carbon-bonded amines; it is carried out by a number of microorganisms, some of which have established symbiotic relationships with certain groups of plants (notably the legumes). Biological nitrogen fixation is energetically demanding, but the microorganisms and symbioses that express it have access to the vast pool of dinitrogen in the atmosphere. The conundrum of why nitrogen remains in short supply in many ecosystems despite the near-ubiquitous distribution of organisms with the capacity to fix nitrogen from the atmosphere is discussed below. Typically, neither atmospheric deposition under background conditions nor biological nitrogen fixation adds large amounts of nitrogen to terrestrial ecosystems in any given year, although in circumstances where symbiotic nitrogen fixers dominate plant communities, rates of biological nitrogen fixation can be substantial. However, their cumulative effects lead to the accumulation and maintenance of combined nitrogen within terrestrial ecosystems.

In contrast to nitrogen, phosphorus has no meaningful gas phase, and the main source of phosphorus to terrestrial ecosystems is rock. Most phosphorus in rocks is in the mineral apatite; in the presence of water and acidity, apatite breaks down and releases soluble phosphate (and calcium). Over time, much of the phosphate in soils comes to reside in secondary minerals that form within soils—the calcium, iron, and aluminum compounds that slow the mobility of phosphorus in soil—and these secondary minerals themselves can weather and release phosphorus (figure 3).

The quantity of apatite in soils is limited, and so the supply of phosphorus via weathering can be depleted. New apatite can be added to ecosystems by geological disturbances (volcanic eruptions, glaciations) that bring fresh rock to near the surface, or by tectonic uplift that in essence represents a flux of unweathered rock into ecosystems from below. Deposition of dust from the atmosphere also can bring phosphorus into terrestrial ecosystems. Phosphorus inputs via dust generally are small relative to weathering in ecosystems where rock-derived elements are present in soil. However, where these minerals have been depleted through many millennia of weathering, dust can become a predominant source of phosphorus.

Outputs

In keeping with its dynamic cycle and its high mobility, losses of nitrogen are more rapid and occur by more diverse pathways than those of phosphorus. One of the most important pathways of loss is as mobile nitrate in solution. Because organisms readily take up nitrate, large nitrate losses generally occur at times and places where combined nitrogen is abundant enough to saturate the demand of organisms, either continuously or episodically during disturbances. Ammonium and dissolved organic nitrogen also can be lost in solution, although retention via cation exchange makes the former substantially less mobile than is nitrate. Losses of dissolved organic nitrogen are interesting in that some forms are not utilized by organisms rapidly (if at all), and losses of these forms may continue even when nitrogen is in short supply.

Nitrogen also can be lost to the atmosphere by multiple pathways (figure 2). Ammonium is in pH-dependent equilibrium with ammonia gas, and in neutral to basic conditions ammonia can volatilize from soils or plants. Some nitric oxide (NO) and nitrous oxide (N_2O) is produced during nitrification, the oxidation of ammonium to nitrate, and a fraction of these gases escape to the atmosphere. Under anaerobic conditions, nitrate can serve as a terminal electron acceptor in respiration in place of oxygen, and this reaction (termed denitrification) can release substantial quantities of dinitrogen to the atmosphere (as well as some nitrous oxide; a related pathway can produce some ammonium that is retained within ecosystems). In addition, fire can represent a substantial pathway of nitrogen loss to the atmosphere; most of the nitrogen in plant and litter material that is consumed by fire is emitted to the atmosphere, much of it in the form of nitric oxide.

Nitrogen exists in two isotopic forms—^{14}N (with seven protons and seven neutrons) is the more abundant isotope, but ^{15}N (with seven protons and eight neutrons) represents about one atom in 300 of nitrogen. These isotopes behave very similarly, but some pathways of element loss (particularly ammonia volatilization and denitrification) fractionate in favor of the lighter ^{14}N isotope. Accordingly, it is often possible to see the legacy of these processes in the relatively ^{15}N-enriched nitrogen that remains within ecosystems after the lighter isotope is lost preferentially by these pathways.

Losses of phosphorus are much slower than those of nitrogen and occur via fewer pathways. The mobility of dissolved phosphorus in soils is very low, but over many millennia, losses of phosphorus via leaching in the forms of phosphate or dissolved organic phosphorus can remove much of the phosphorus that originally was present in apatite when the soil began to develop. Most of the phosphorus that remains in old, highly leached soils is in highly insoluble and physically protected forms, leading to what the New Zealand soil scientist T. W. Walker has described as a terminal steady state of profound phosphorus depletion.

Phosphorus (and nitrogen) can also be lost to terrestrial ecosystems via erosion, where the soil particles that contain them are removed via wind erosion in dry areas (the complement of inputs via dust deposition) and through slope and hydrological processes anywhere.

Other Elements

Inputs of the other nutrient elements that plants require in abundance (sulfur, calcium, magnesium, and potassium) differ from those of nitrogen and phosphorus in that all of them are relatively abundant in sea salt and thus also in the marine aerosols that transfer elements from the oceans to the land. These marine aerosols constitute a variable but significant component of atmospheric deposition. Nitrogen and phosphorus both are in short supply in most of the surface ocean, and little of either is transferred to land in marine aerosol. Sulfur and the cations also enter terrestrial ecosystems through rock weathering (which is quantitatively the most important source of cations, at least in soils that have not been depleted by multiple millennia of intense leaching), dust deposition, and (for sulfur) as a consequence of anthropogenic emissions of reactive sulfur-containing gases upwind.

The mobility of these other elements is intermediate between those of nitrogen and phosphorus. Sulfur is lost from terrestrial ecosystems by as diverse an array of processes as is nitrogen—via leaching as sulfate or dissolved organic sulfur, through emissions of a wide range of gases, and during fires. Most cation losses occur via leaching and, secondarily, erosion, although potassium in particular is volatilized during vegetation fires.

2. GLOBAL ELEMENT CYCLES

Just as the internal element cycles of ecosystems are embedded in a system of input–output dynamics, those inputs to and outputs from any particular ecosystem are embedded in regional and global element cycles. Inputs to any given ecosystem often reflect outputs of nutrients from other ecosystems that lie upwind or upstream, and similarly, losses of nutrients from any ecosystem can influence other ecosystems downwind and downstream. The substantial and remarkably recent human-caused alterations of global element cycles make it particularly difficult to evaluate nutrient cycling in any ecosystem without considering its regional and global context. Moreover, systematic changes in nutrient outputs from terrestrial ecosystems represent a component of as well as a response to human alteration of global element cycles.

The anthropogenic increase in atmospheric carbon dioxide is perhaps the most familiar component of human-caused global change, and this increase of over 30% since the beginning of the industrial revolution represents a substantial alteration to the global cycle of carbon. This increase influences the cycles of other nutrients; any increases in plant growth that it fuels cause increases in plants' demands for nutrients and/or a dilution in the concentrations of elements within plants.

The cycles of nitrogen and phosphorus (and sulfur) have been altered to an even greater extent than that of carbon, although the changes are not distributed as evenly across the globe as for carbon. Before substantial human influence, the global cycle of nitrogen could be summarized by considering biological nitrogen fixation—the flux from dinitrogen in the atmosphere into combined nitrogen on land and in the ocean (plus a small contribution from fixation associated with lightning); this biological fixation probably amounted about 100 million metric tons of N into terrestrial ecosystems (table 1) and about the same amount into marine systems. Most of the N that was fixed in terrestrial ecosystems eventually returned to the atmosphere through denitrification; about a quarter of it moved through hydrologic pathways to the ocean.

Humanity has more than doubled the quantity of dinitrogen fixed on land, most importantly through an

Table 1. Global fluxes of nitrogen and phosphorus from the major reservoir of each (the atmosphere for nitrogen, rocks for phosphorus) into potentially biologically available forms

Nitrogen	
Background	
Biological nitrogen fixation on land	90–140
Fixation by lightning	3–5
Year 2000	
Industrial N fixation for fertilizer	90
Crop N fixation	40
Fossil fuel fixation and mobilization	20–32
Phosphorus	
Background	
Weathering of P-bearing minerals	10
Year 2000	
Mining and distribution of P-rich rock	19.8

Note: These fluxes are summarized under background conditions (without extensive human influence), and in the human-influenced Earth around the year 2000; the latter fluxes are in addition to the background fluxes. All values in millions of metric tons per year.

Source: From the Millennium Ecosystem Assessment Current Status and Trends Assessment (2005).

increase in the production of industrial nitrogen fertilizer that is designed to overcome nitrogen limitation to crop productivity. Industrially fixed nitrogen fertilizers barely existed in 1950, but they now account for about 90 million tons per year of fixed nitrogen. Other human-associated production of combined nitrogen occurs through the cultivation of symbiotic nitrogen-fixing crops and forages (such as soybean and alfalfa) and the inadvertent fixation and mobilization of nitrogen through fossil fuel combustion. Fossil-fuel-derived combined nitrogen is emitted directly to the atmosphere in reactive forms (much of it as nitric oxide), and a substantial fraction of fertilizer nitrogen is emitted to the atmosphere as well, mostly as ammonia. Consequently, atmospheric deposition of combined nitrogen has increased greatly (by 5- to 10-fold or even more) above background levels in urban, industrial, and intensive agricultural areas of the world. Initially, these effects were most pronounced on a regional level in eastern North America and Western Europe, but they are increasing in importance in rapidly industrializing areas of the world, most notably in eastern and southern Asia. Most of the nitrogen emitted into the atmosphere by human activities is reactive; typically it is transported through the atmosphere for hundreds of kilometers, but not globally. Some emissions of N occur as the greenhouse gas nitrous oxide, which is increasing in concentration globally and contributing to the radiative forcing that underlies global warming.

About a quarter of the anthropogenic fixed nitrogen added to terrestrial ecosystems moves via hydrologic pathways to rivers and the coastal ocean, where it increases fluxes of nitrogen regionally and drives substantial estuarine and coastal eutrophication. The extensive "dead zone" that has formed near the mouth of the Mississippi River in recent decades is one example (among many) of this phenomenon.

The background cycle of phosphorus (before these substantial human influences) can be summarized by considering the flux from rock to potentially accessible forms via weathering. Humanity has altered that flux substantially by mining phosphorus-rich rocks (table 1), often extracting the phosphorus they contain in purified forms, and then either applying that phosphorus as fertilizer or using it in industrial processes. This mining and mobilization of phosphorus has more than doubled the background flux via weathering. Because of its relative immobility, phosphorus does not spread as widely as does nitrogen from the sites where it is applied, although the relatively small fraction of applied phosphorus that makes its way into aquatic systems has caused extensive eutrophication of freshwater ecosystems and contributes to the alteration of the coastal ocean. Some of the phosphorus mobilized by human activity is distributed more widely in dust. Globally, the fluxes of dust as well as the concentration of phosphorus in dust have increased as cultivation and overgrazing have reduced plant cover in semiarid regions and enhanced rates of erosion by wind.

Among the other major nutrients, the global cycle of sulfur has been altered substantially by the emission of the sulfur contained in fossil fuel. This global-scale change is notable for having been mitigated substantially since 1980 through the choice of lower-sulfur fuels and the capture of sulfur from the emission stream of fossil fuel power plants.

3. ILLUSTRATION: NUTRIENT CYCLING IN PRACTICE

One of the more intriguing challenges to our understanding of nutrient cycling and biogeochemistry is the regulation of biological nitrogen fixation. It seems logical to expect that if the supply of available nitrogen is so small as to constrain the growth of plants and/or microorganisms substantially, then organisms with the capacity to fix nitrogen biologically should gain a substantial advantage by drawing on the vast supply of dinitrogen in the atmosphere. Over time, the activity of biological nitrogen fixers should bring combined nitrogen into ecosystems, ultimately increasing its overall availability. Why, then, is the supply of nitrogen so low that it constrains primary productivity and other ecosystem processes in many ecosystems? It would seem that this constraint, termed nitrogen limitation, would be merely a transitory or marginal phenomenon. Why do biological nitrogen fixers not respond to nitrogen limitation and reverse it as a by-product of their own production of combined nitrogen?

Just that happens in many lake ecosystems. Where most lakes receive abundant phosphorus relative to combined nitrogen, nitrogen-fixing cyanobacteria soon dominate among the primary producers, and, in the course of their growth and metabolism, they add enough combined nitrogen to bring the lake's nitrogen: phosphorus ratio roughly into stoichiometric balance.

In contrast, nitrogen limitation to primary production and other ecosystem processes is widespread and apparently persistent in many terrestrial and coastal marine ecosystems—both fertilizer experiments and the results of regional-scale human alterations of the nitrogen cycle demonstrate that. Why? What keeps biological nitrogen fixers from responding to nitrogen deficiency in many nitrogen-limited ecosystems?

Several constraints to biological nitrogen fixation have been identified; these include (1) its greater energetic cost relative to acquiring combined nitrogen, (2) a greater demand for phosphorus and trace metals (particularly molybdenum and iron) relative to nonfixers, and (3) preferential grazing on protein- and nitrogen-rich symbiotic plants in terrestrial ecosystems. Each of these mechanisms is known to be important in particular ecosystems, but few studies have evaluated their relative importance.

One analysis by Peter Vitousek and Sarah Hobbie focused on constraints to heterotrophic nitrogen fixation in Hawaiian montane rainforest. Forests growing on young Hawaiian soils are nitrogen limited; weathering of their relatively fresh lava substrate releases phosphorus and other elements, but no nitrogen. Native plants with symbiotic nitrogen fixation do not colonize these sites; their absence here may simply reflect their sparse representation in the native flora because symbiotic nitrogen-fixing plants are known to colonize young soils in many other areas of the world. However, heterotrophic nitrogen fixers—bacteria that draw energy from decomposing organic matter and in the process fix dinitrogen from the atmosphere—are present here as they are everywhere, and in some Hawaiian forests (and elsewhere) they fix substantial quantities of nitrogen. They add very little nitrogen to young nitrogen-limited forests in Hawaii, though. It is reasonable to ask "Why not more?"

Two potential mechanisms that could constrain heterotrophic nitrogen fixers have been identified. First, the supply of elements other than nitrogen could be insufficient to support nitrogen fixation. Concentrations of phosphorus in leaf litter are relatively low in these sites, even though experiments demonstrate that nitrogen supply limits tree growth. Alternatively, the structure and biochemistry of leaf litter could be so recalcitrant and/or toxic as to constrain nitrogen fixation. Trees growing in low-nutrient sites (including these) use the nutrients that they do obtain more efficiently than do those in high-nutrient sites, largely by increasing the residence time of nutrients in their tissues. In practice, this means that trees retain their leaves longer, and they may reabsorb nutrients from senescing leaves more effectively. Longer leaf retention times require greater investment by the plants in the structure and defense of those leaves, so that when those leaves finally do drop, they are packed with nasty carbon-rich compounds such as lignin and soluble polyphenols. In such situations, leaf litter contains abundant total energy but very little available energy, and biological nitrogen fixation could be constrained by this lack of available energy.

Experimental studies on the controls of heterotrophic nitrogen fixation in Hawaii were facilitated by the distribution of the native dominant tree, ohia (*Metrosideros polymorpha* in the Myrtaceae), across a very broad range of sites differing in climate and soil fertility and so in tissue biochemistry. Ohia dominates sites that are rich and poor, wet and dry, and by selecting among them, it is possible to find populations with leaf litter that represents all of the relevant combinations of nutrient (especially phosphorus) concentrations and tissue biochemistry: high phosphorus and high lignin, high phosphorus–low lignin, low phosphorus–high lignin, and low phosphorus–low lignin. That leaf litter was collected, transported to a nitrogen-limited young site, and embedded in an ongoing forest fertilization experiment in which different sets of the litter received additions of nitrogen, phosphorus, and other nutrients individually and in complete factorial combinations. Rates of decomposition and biological nitrogen fixation were measured for the various leaf litters in the various nutrient treatments.

Neither additions of phosphorus and other nutrients nor high initial concentrations of those elements in leaf litter had significant direct effects on rates of nitrogen fixation in decomposing leaf litter. However, rates of nitrogen fixation were significantly reduced in the presence of high lignin (and soluble polyphenol) concentrations in leaf litter. Heterotrophic nitrogen fixation added more than 1 mg of combined nitrogen per gram of initial litter mass to leaf litter from four sites with lignin concentrations less than 15%, representing a substantial contribution of combined nitrogen to decomposing leaves, and (if all leaves in a site fixed N at these rates) to the forests in which the litter was produced. Litter from four sites with lignin concentrations greater than 20% averaged nitrogen fixation of only 0.16 mg of nitrogen fixed per gram of initial litter mass—more than sixfold less.

Further, added nitrogen stimulated the breakdown of low-lignin litter but had little or no effect on the decomposition of high-lignin litter. Vitousek and Hobbie concluded that nitrogen limitation caused trees in young sites to retain leaves and nutrients longer, leading to leaf litter in those sites being high in lignin and polyphenols. As a consequence, decomposition of those leaves was constrained by the complex and recalcitrant carbon chemistry of the litter, not by nitrogen or other nutrients. Not surprisingly, heterotrophic nitrogen fixers largely were inactive in this litter; they would have received little benefit in carrying out the energetically expensive process of nitrogen fixation if nitrogen supply were not a controlling factor in decomposition (figure 4). All of these results make sense—but note that the trees are unproductive and produce recalcitrant litter because their growth is constrained by nitrogen. Further, because the trees' litter is recalcitrant, microorganisms do not fix nitrogen that would

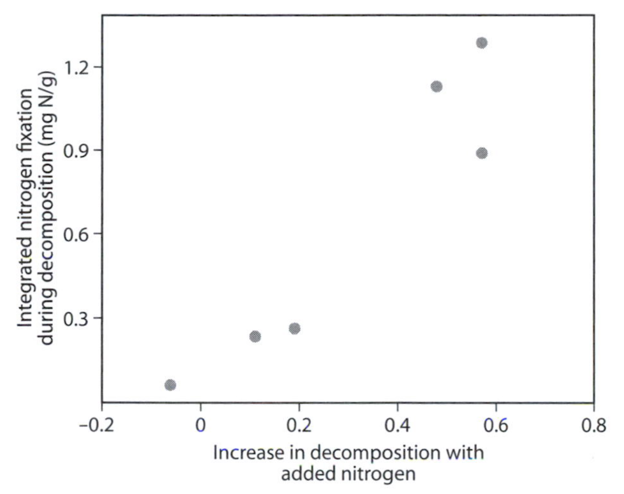

Figure 4. Heterotrophic nitrogen fixation in decomposing leaf litter from Hawaiian forests. Leaf litter of a single tree species was collected in multiple sites and decomposed in a common site, with and without nitrogen fertilization. Trees in some sites produced leaf litter that was high in lignin; decomposition of that litter was unresponsive to added nitrogen (*x*-axis), and rates of nitrogen fixation (in the absence of fertilizer, shown on the *y*-axis) were low. Where lignin concentrations were low, litter decomposition was stimulated by added nitrogen, and rates of biological nitrogen fixation (in the absence of fertilizer) were low. Microbes fixed substantial quantities of nitrogen only where nitrogen supply limited the rate of decomposition, but, as discussed in the text, the trees produced leaf litter with high lignin concentrations in large part because their growth rate was limited by nitrogen. (From Vitousek and Hobbie, 2000)

alleviate limitation to the trees and ultimately could cause them to produce less recalcitrant litter. Interactions of this sort may be widespread in nutrient cycling and biogeochemistry.

FURTHER READING

Aber, J. D., and J. M. Melillo. 2001. Terrestrial Ecosystems, 2nd ed. New York: Academic Press.

Aerts, R., and F. S. Chapin. 2000. The mineral nutrition of wild plants revisited: A reevaluation of processes and patterns. Advances in Ecological Research 30: 1–67.

Chapin, F. S., III, P. A. Matson, and H. A. Mooney. 2002. Principles of Terrestrial Ecosystem Ecology. New York: Springer.

Millennium Ecosystem Assessment. 2005. Global Status and Trends Assessment. Washington, DC: Island Press.

Schindler, D. W. 1977. Evolution of phosphorus limitation in lakes. Science 195: 260–262.

Schlesinger, W. H. 1997. Biogeochemistry: An Analysis of Global Change. New York: Academic Press.

Sterner, R., and J. J. Elser. 2002. Ecological Stoichiometry: The Biology of Elements from Molecules to the Biosphere. Princeton, NJ: Princeton University Press.

Vitousek, P. M. 2004. Nutrient Cycling and Limitation: Hawaii as a Model System. Princeton, NJ: Princeton University Press.

Vitousek, P. M., and S. E. Hobbie. 2000. The control of heterotrophic nitrogen fixation in decomposing litter. Ecology 81: 2366–2376.

III.11

Terrestrial Carbon and Biogeochemical Cycles

R. A. Houghton

OUTLINE

1. The production equations (carbon)
2. To what extent do C, N, and P limit photosynthesis in terrestrial ecosystems?
3. To what extent do C, N, and P limit NPP? (What determines Rs_A?)
4. To what extent do C, N, and P limit NEP? (What determines Rs_H?)
5. To what extent do C, N, and P limit the amount of carbon in vegetation and soil?
6. Disturbances limit C accumulation

Two modes of explanation account for the accumulation of carbon in terrestrial ecosystems: metabolism and demography. Carbon, nitrogen, and phosphorus (as well as temperature and moisture) affect the metabolic processes that control the rate at which ecosystems fix and accumulate carbon in organic matter. Whether they also control the total amount of carbon that can be held in the biomass and soils of ecosystems is less clear. An alternative explanation is that maximum carbon storage is limited, at least in forests, by disturbances, both natural and anthropogenic, that initiate changes in demography.

GLOSSARY

autotrophic respiration. The metabolic process by which primary producers (green plants) convert sugars to carbon dioxide, releasing energy.

denitrification. A process by which nitrate and nitrite are reduced to ammonium.

global carbon balance. The total sources of carbon from fossil fuels and land-use change must sum to the total sinks (accumulations) of carbon in the atmosphere, oceans, and land.

gross primary production. The amount (or rate) of organic matter (sugars) produced from CO_2 by green plants through photosynthesis.

heterotrophic respiration. The metabolic process by which consumers (heterotrophs) convert sugars to carbon dioxide, releasing energy.

net ecosystem production. The amount (or rate) of organic material produced by green plants after both autotrophic and heterotrophic respiration.

net primary production. The amount (or rate) of organic material produced by green plants after (autotrophic) respiration.

nitrification. A process by which ammonium is oxidized to nitrite, and nitrite to nitrate.

nitrogen fixation. A process by which molecular nitrogen is reduced to form ammonium.

A major impetus for studying carbon and biogeochemical cycles has always been propelled from the applied sciences of agriculture and forestry: How can yields be increased? Ecologists have also investigated the cycles of C, N, and P to understand how ecosystems function. Most recently, a third motive has emerged for studying the biogeochemistry of ecosystems: carbon management. How much carbon is (can be) sequestered in the vegetation and soil of ecosystems? Which ecosystems are best at sequestering carbon? How rapidly does this process occur? And what factors limit both the rate of accumulation and the total amount?

The global carbon balance suggests that terrestrial ecosystems have been accumulating carbon for several decades, but the reasons for this are not entirely clear. Candidate explanations include (1) ecosystem responses to changing environmental conditions [for example, increasing concentrations of CO_2 in the atmosphere, increased mobilization of nutrients (nitrogen, phosphorus) from human activities, changes in climate] and (2) recovery of forests from past harvests, abandonment of farmlands, and fire exclusion.

The mechanisms responsible for current carbon sinks on land are important to understand, first, because the environmental variables driving these mechanisms may be different in the future, thereby either enhancing or diminishing current rates of sequestration, and, second, because an understanding of the mechanisms should indicate the types of management likely to increase terrestrial carbon sequestration. What is it that limits the rate of carbon sequestration? The focus of this chapter is to explore these limits in the context of biogeochemical cycles.

The concept of limiting factors is an extension of the Law of the Minimum, attributed to Liebig but recognized by others before Liebig's formulation in 1840 (Browne, 1942). The Law of the Minimum states that, if all of the mineral nutrients but one are available in the quantities required for the growth of a plant, the deficiency of that one nutrient will prevent growth. The concept of limiting factors in ecology includes not only biogeochemical factors but all resources, including water and light, as well. This chapter focuses on the biogeochemical factors. Furthermore, the concept of limiting factors applies not only to plant growth but also to the decomposition of organic matter.

1. THE PRODUCTION EQUATIONS (CARBON)

The processes governing carbon accumulation occur over minutes to hours as leaves fix atmospheric CO_2 and over thousands to millions of years as soils develop during primary succession. The exchanges of carbon between terrestrial ecosystems and the atmosphere are described by the following chemical equation:

$$6CO_2 + 6H_2O \leftrightarrows C_6H_{12}O_6 + 6O_2. \qquad (1)$$

In photosynthesis, carbon dioxide is fixed into glucose, releasing oxygen. In respiration, CO_2 is released as a result of the oxidation of glucose. Respiration occurs in both the autotrophic and heterotrophic components of ecosystems.

Carbon cycling in ecosystems is best illustrated by the production equations (Woodwell and Whittaker, 1968). The rate at which atmospheric CO_2 is fixed through photosynthesis is called gross primary production (GPP). Some the fixed carbon is respired by the leaves (both light and dark respiration), and some is respired by leaves and other tissues for maintenance and growth. Respiration by the plant (and leaves) is autotrophic respiration (Rs_A). What is left after plant respiration (GPP $- Rs_A$) is net primary production (NPP):

$$\mathrm{NPP} = \mathrm{GPP} - Rs_A. \qquad (2)$$

NPP is the rate of accumulation of carbon in plants (leaves, stems, and roots). It is the amount of organic matter available (over some time interval) for those organisms (heterotrophs) that consume preformed organic matter rather than make their own (from sunlight and CO_2). The heterotrophs include not only grazers and browsers (secondary producers) but the microorganisms (bacteria and fungi) that decompose preformed organic matter. Heterotrophs respire or mineralize organic carbon back to CO_2.

What is left after autotrophic and heterotrophic respiration (Rs_H) is called net ecosystem production (NEP):

$$\mathrm{NEP} = \mathrm{GPP} - Rs_A - Rs_H, \qquad (3)$$

$$\mathrm{NEP} = \mathrm{NPP} - Rs_H. \qquad (4)$$

NEP is the rate of accumulation of organic matter in an ecosystem. The accumulation may be either in the biomass of autotrophs and heterotrophs or in soil organic matter. Notice that NEP may be negative if more organic matter is consumed or respired than fixed over some time interval.

The production equations implicitly include environmental (light, water, temperature) and biogeochemical constraints. Although numerous cations (e.g., calcium, magnesium, potassium, sodium) and anions (e.g., phosphate, nitrate) are required for plant and animal growth and metabolism, this discussion focuses on the elements nitrogen (N) and phosphorus (P) because they are thought to be most limiting to the productivity of terrestrial ecosystems.

The nitrogen cycle involves many forms of N. Plants take up inorganic N [either ammonium (NH_4^+) or nitrate (NO_3^-)] and incorporate it in organic matter (most of the uptake is of nitrate, as ammonium is toxic at high concentrations). Mineralization of organic matter releases not only CO_2 but the inorganic forms of N back into the environment. Unlike the carbon cycle, the N cycle involves primarily "aquatic" flows (nutrients in solution) rather than gaseous flows, although the primary source of terrestrial N is from the atmosphere. Certain forms of bacteria can fix atmospheric N_2 (nitrogen fixation) into organic matter. Other forms convert ammonium to nitrite and nitrite to nitrate (nitrification); and still other forms convert nitrate and nitrite to ammonium (denitrification). Nitrification occurs in aerobic environments; denitrification, in anaerobic environments. Nitrification and denitrification may cause N to be lost from terrestrial ecosystems because both processes may release NO or N_2O to the atmosphere.

The phosphorus cycle is simpler, as phosphorus does not have a gaseous form. Phosphorus is made available through the weathering of rock. It also becomes unavailable through binding with clays and complex organic matter.

This chapter concerns the roles N and P, as well as C, play in limiting the rates of carbon accumulation identified in the production equations. The approach is hierarchical, dealing with the effects of N and P on photosynthesis, NPP, and NEP, and, finally, carbon stocks.

It should be clear, at least theoretically, that the rate of carbon accumulation depends not only on the input of carbon through photosynthesis but on the rate of respiration as well. Decreasing rates of decay or respiration could have as large an effect on increasing terrestrial carbon storage as increasing rates of production. The sections below explore the biogeochemical limits to the following processes:

- Do C, N, and P limit the rate of photosynthesis?
- Do they limit the rate of NPP? (What determines Rs_A?)
- Do they limit the rate of NEP? (What limits Rs_H?)
- Do they limit the amount of carbon in wood and soil?

To some extent, the different limitations to carbon storage explored in the following sections result from differences in temporal and spatial scales. By and large, however, they result from different ecological scales, that is, the inclusion of more and more processes as observations move from the leaf to the ecosystem, or as explanatory mechanisms move from physiological ecology to age structure. Uncertainty surrounding the capacity of terrestrial ecosystems to function as carbon sinks, and the limiting factors, results in part from the different scales used for explanation.

2. TO WHAT EXTENT DO C, N, AND P LIMIT PHOTOSYNTHESIS IN TERRESTRIAL ECOSYSTEMS?

Experiments with elevated CO_2 concentrations generally stimulate photosynthesis, at least in the short term, indicating that photosynthesis is limited by carbon (Koch and Mooney, 1996). Plants often, but not always, acclimate (or down-regulate) to elevated CO_2 concentrations, such that their photosynthetic capacity is not increased in the long term (weeks to years). Nevertheless, the net effect of higher concentrations of CO_2 in the short term is generally to increase the rate at which carbon is fixed in photosynthesis. The fate of the "extra" organic carbon is less clear. Some of it may be respired by the plant; some may increase the growth rate of the plant (leaves, stem, roots) (NPP) (see section 3, below); and some may accumulate as biomass, litter, and soil organic matter, thus increasing the storage of carbon in the ecosystem (see sections 4 and 5, below).

Over a range of plant species and ecosystems, rates of photosynthesis are correlated with the nitrogen content of leaves. Much of this nitrogen is in enzymes, including the major photosynthetic enzyme, ribulose bisphosphate carboxylase. Leaf phosphorus is also implicated in determining the photosynthetic capacity of some species.

Thus, photosynthesis may be limited by C, N, or P, although neither the CO_2 nor the N effect on photosynthesis is linear. Photosynthesis eventually saturates at high CO_2 concentrations (800–1000 ppm) and at high concentrations of leaf nitrogen.

3. TO WHAT EXTENT DO C, N, AND P LIMIT NPP? (WHAT DETERMINES Rs_A?)

Not all of the carbon taken up through photosynthesis (GPP) turns into plant biomass. Some is respired by the plants (Rs_A). The "average" fraction respired (NPP: GPP) appears to be approximately 0.5, but the range is wide, and the processes that determine the ratio are not well understood. A recent review suggests that the ratio NPP:GPP in trees is inversely related to age. Further, neither GPP nor NPP continues to increase as forests age. Rather, they reach a maximum and then decline. The decline in GPP, or photosynthesis, may result from N or P limitation, as described in section 2, above, or from other factors. One explanation is that tree height limits the amount of water that can be drawn up from the soil to support evapotranspiration, but other factors seem to be involved. The limitations discussed in this section are different from those discussed in section 2, above, in part because photosynthesis and growth are often observed at different scales.

Despite the initial increases in biomass observed in crops, annual plants, and tree seedlings under elevated concentrations of CO_2, experiments at the level of ecosystems, and experiments longer than a few years, suggest greatly reduced responses. Where CO_2 fertilization experiments have been carried out for more than a few years, they often show an initial CO_2-induced increment in biomass that diminishes over time. The diminution of the initial response occurred after 2 years in an arctic tundra and after 3 years in a rapidly growing loblolly pine forest. The pine forest was chosen in part because CO_2 fertilization was expected to be greatest in a rapidly growing forest. The decline after 3 years is attributed to some factor other than CO_2 becoming limiting.

Nitrogen and phosphorus are likely candidates. That is, the initial increase in NPP uses up the available N or P, incorporating (immobilizing) it in new biomass. Nitrogen is thought to limit NPP because, when N is added to terrestrial ecosystems, NPP often increases, at least in temperate zone and boreal regions. Elevated CO_2 and nitrogen fertilizer, together, have been shown to have a greater effect on forest growth than the sum of their individual effects. Is N limitation the reason why CO_2 effects on NEP decline with time (the so-called progressive N limitation hypothesis)? Several studies have investigated whether the initial enhancement to growth is diminished when the N initially available becomes bound in biomass. The results are mixed. In two forests, CO_2-enhanced productivity continued for 6 years, in part, because the C:N ratio increased in the high CO_2 treatment and in part because the elevated CO_2 stimulated root growth, which tapped a larger volume of soil for N. In these cases, N did not become limiting to NPP. In another forest and in two grasslands, productivity declined, although changes in the allocation of N between plants and soil served to delay the nutrient-induced decline.

Progressive N limitation may also explain the negative synergy sometimes observed between elevated CO_2 and other factors. Alone, increases in temperature, precipitation, nitrogen deposition, and atmospheric CO_2 concentration each increased NPP in a California grassland. When the treatments were combined, however, elevated CO_2 decreased the positive effects of the other treatments. That is, elevated CO_2 increased productivity under "poor" growing conditions but reduced it under favorable growing conditions. The most likely explanation is that some soil nutrient became limiting, either because of increased microbial uptake or decreased root allocation.

The observation that N is often limiting is puzzling because there is a vast reservoir of nitrogen (as N_2) in the atmosphere, and certain leguminous species and micorrhizae are capable of fixing this atmospheric N_2 into biological available forms, such as ammonium or nitrate. Why should ecosystems not be able to replenish low stocks of fixed N? Observations suggest that P may ultimately limit not only NPP but N availability: nitrogen fixation by free-living bacteria appears inversely related to the N:P ratio in soil, and the rate of accumulation of N is greatest in soils with high P content. More recent studies suggest that low P availability does indeed limit the response of N fixers to elevated CO_2.

Age is also important: the nutrient limiting NPP appears to vary with age of the ecosystem. In general, P is thought to be the nutrient limiting terrestrial productivity in the long term. Forests growing on old, highly weathered soils are conservative of P, whereas forests growing on young soils are more conservative of N. The observation is consistent with the fact that N is derived from the atmosphere through N fixation and thus accumulates in biomass and soil over time, whereas rock-derived P is obtained from weathering and becomes bound up in unavailable forms through time. Because tropical soils are generally more highly weathered (i.e., old), tropical forests are thought to be limited by P, whereas temperate zone and boreal forests, generally growing on younger soils, are more limited by N. The generality has been observed in Hawaiian chronosequences over thousands of years and in secondary forests in the Amazon over decades. The generality is also consistent with the observation that the N:P ratio of foliage and litter in mature forests increases along the latitudinal gradient from boreal to tropical regions (i.e., conservation of the limiting nutrient).

Human activities, including fertilizer use and combustion, have increased the amount of available N in the environment and may have increased NPP. However, the story of N deposition is complicated because much of the N is in the form of acid precipitation, and it is difficult to distinguish the fertilization effects of N from the adverse effects of acidity. Other factors, such as tropospheric ozone and sulfur (acid rain), have been shown to reduce productivity. Interestingly, regions where N inputs are high are often regions where ozone concentrations are also high, and the effects may be largely offsetting in terms of productivity. Finally, high levels of N may saturate ecosystems, interfere with other processes of metabolism, and eventually reduce NPP.

In summary, it is unclear from results to date whether NPP in terrestrial ecosystems will remain higher at elevated concentrations of CO_2, or whether other factors will limit the response. Because the availability of N and P is controlled in large part by mineralization of organic matter, the response of NPP requires an understanding of what limits or controls the supply of nutrients; that is, what controls the remineralization or decomposition of litter and dead organic matter.

4. TO WHAT EXTENT DO C, N, AND P LIMIT NEP? (WHAT DETERMINES Rs_H?)

Although adding N to forests may increase NPP and the carbon stored in plants, it may not increase the carbon stored in soil. All else being equal, an increase in NPP would be expected to lead to an increase in carbon stocks as the pools of carbon in biomass and soil reach a new steady state at which inputs balance outputs. However, if the increase in productivity is in tissues with a rapid turnover (fine roots, foliage), the enhanced growth may be respired within a year or two, yielding little or no gain in carbon storage.

Furthermore, adding N to forests may modify soil organic matter so as to increase or decrease its residence time, thereby altering carbon storage. The effect of added N on the rate of decomposition depends on the predominant forms of N present in the organic material; that is, "substrate quality." Rs_H generally consumes the young, just-fixed organic matter first. As this labile material is decomposed, the residual organic matter becomes progressively refractory over time, including complex humic structures in which N occurs within phenols and other ring structures of large compounds. In contrast, young labile organic matter includes readily available N in the form of amino acids and amino sugars as well as substrates that contain little or no N, such as cellulose. Decomposition of this wide array of substrates depends on the soil microorganisms and the enzymes they are capable of producing. Synthesis of microbial biomass and enzymes requires N. If N is in short supply to the microbes, then adding N may enhance their growth and their production of enzymes, thus increasing decomposition. This is often the case for cellulase because the decomposition of cellulose does not release N, and so the bacteria must obtain an alternate source of N. In contrast, in the case of phenol oxidases, the fungi that are the dominant producers of this enzyme obtain most of their N from the metabolism of the N-containing phenolic substrates. In fact, addition of exogenous N can inhibit the production of phenol oxidases and hence slow the rate of decomposition of these complex forms of soil organic matter.

5. TO WHAT EXTENT DO C, N, AND P LIMIT THE AMOUNT OF CARBON IN VEGETATION AND SOIL?

The largest stocks of carbon occur in (1) forests, such as those in the Northwestern United States, where there exists an especially favorable combination of light, seasonal temperatures, and moisture; and (2) in deep peatlands, where the rate of NPP may be low, but the rate of decomposition is even lower because of waterlogged conditions and, sometimes, permafrost.

What factors limit the accumulation of carbon in terrestrial ecosystems? It is difficult to make the argument that either C or N is limiting to the amount of carbon that can be sequestered in terrestrial ecosystems, given the large reservoirs of these elements in the atmosphere. Under elevated CO_2, the *rate* of plant growth is often stimulated, but the biomass of mature forests may not increase even if photosynthesis and growth are increased through elevated CO_2. Nutrients, as well, are unlikely to play a role in maximum biomass (carbon storage) despite the influence they have on NPP. The most productive sites often have the highest turnover (mortality), such that standing biomass is not greater than in low-nutrient sites. The classic paper by Jordan and Herrera (1981) argued that forests growing on fertile and nonfertile soils might have the same high levels of biomass. The difference between the two forests is the rates at which they recover following a disturbance. The one on the fertile soil would accumulate biomass more rapidly.

One potential mechanism for increasing the amount of carbon held in an ecosystem without increasing the amount of N is to change the distribution of carbon (and N) stocks from soil (where the C:N ratio is low) to wood (where the C:N ratio is high). An increase in the rate of mineralization of soil organic matter, through warming for example, might lead to such a shift. If the mineralization of carbon (and its release to the atmosphere) is less than the amount accumulated in new wood, carbon will have been accumulated. However, N released through mineralization may also be immobilized in soils or lost from the ecosystem, becoming largely unavailable in either case. Evidence from an experiment in arctic tundra showed that, although warming led to an increased mineralization of N and an increased storage of carbon in woody plants, it led to an even greater decrease in soil carbon storage and, hence, a net reduction in carbon stocks for the ecosystem.

The *rate* at which carbon, N, and P are incorporated in organic matter may also limit the sequestration of carbon because terrestrial ecosystems do not have an infinite amount of time to accumulate carbon. Rather, they are periodically disturbed, some frequently, by fires, storms, insects, as well as human activity. These disturbances transfer living carbon to dead carbon, perhaps losing C, N, and P from the ecosystem in the process.

6. DISTURBANCES LIMIT C ACCUMULATION

The focus of this chapter has been on the biogeochemical mechanisms by which C, N, or P limits production, growth, and carbon storage in ecosystems. There is a process of a different scale, however, that limits carbon storage in terrestrial ecosystems, namely, the process of disturbance and recovery. Most ecosystems, particularly forests, are not at maximum biomass because they are recovering from past disturbances, including harvests, abandonment of farmlands, and fire exclusion.

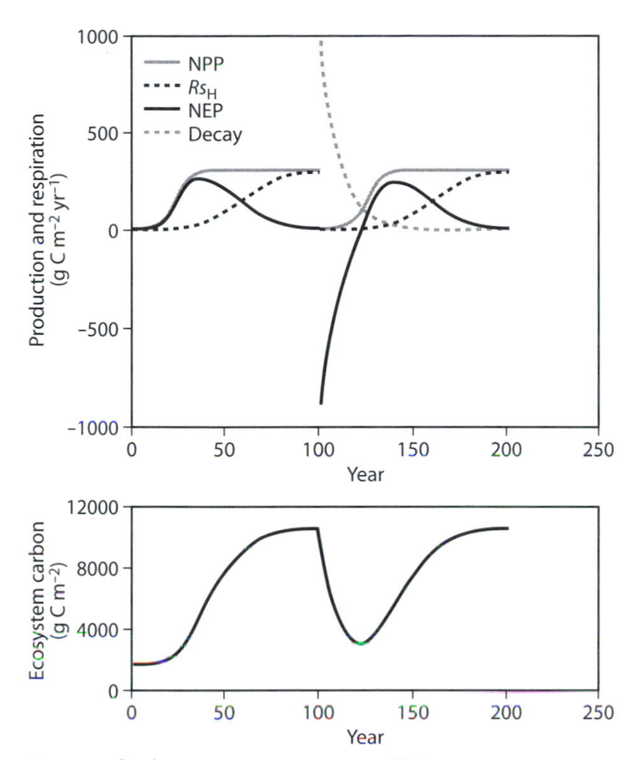

Figure 1. (top) Annual NPP, Rs_H, and NEP for an idealized forest over 100 years. A disturbance in year 101 kills all of the living biomass, and the forest recovers over the next century. NEP is negative for several years following the disturbance because the rate of carbon loss from decay of accumulated biomass is greater than the rate of carbon gain through NPP. (bottom) Total amount of carbon in the ecosystem.

Responses to changing environmental conditions involve largely physiological processes (photosynthesis, allocation, and respiration), whereas the recovery (or degradation) of forests involves demographic processes (growth, mortality). The two categories of explanation differ in underlying mechanisms. Clearly, the explanations overlap, as changes in environmental conditions may alter rates of disturbance and mortality as well as rates of photosynthesis. Nevertheless, the physiological and demographic mechanisms are different, as are the data used to investigate these mechanisms (short-term measurement of CO_2 flux by eddy covariance, for example, as opposed to longer-term changes in wood volumes from forest inventories).

Recovery from disturbance is the dominant factor in explaining NEP in U.S. forests, and in five chronosequences in Europe, the age effect accounted for 92% of the total variability in NEP. When the same five chronosequences were standardized for age (by considering the entire rotation period, disturbance to disturbance), the factor most important in explaining variability was N deposition. Because all of the flux data were obtained in recent years, the effect of CO_2 on NEP could not be evaluated.

Organic carbon accumulates in the vegetation and soils of ecosystems as they age. The production equations describe the rate of accumulation of organic matter (NEP) as a forest grows and matures (shown in idealized form in figure 1, top, left). NPP increases more rapidly than Rs_H, and thus, NEP is highest in year 30 before declining to zero in year 100. A disturbance occurs in year 101 (figure 1, top) and kills any living material, after which NPP and Rs_H repeat as in figure 1 (top, left). After the disturbance, however, the NEP that had accumulated in the first 100 years begins to decay. The decay of this dead material is at first much larger than NPP, and thus, NEP is negative. After year 150, this dead material is largely gone, and NEP for years 150–200 looks very much like NEP for years 50–100. The "decay" in figure 1, top, should be included as part of Rs_H. It is separated in the figure for clarity.

From the perspective of the ecosystem, the first 100 years show a gradual accumulation of carbon to a steady state of ~10,000 gC/ha (figure 1, bottom panel); the second century shows a rapid decline before a reaccumulation to the steady-state level. The atmosphere sees the inverse of the carbon accumulation curve; i.e., slow withdrawals as forests grow and a rapid increase following disturbance.

The cycles of N and P and other biogeochemical cycles in terrestrial ecosystems follow these changes in carbon and organic matter in some respects and act independently in other respects. As carbon accumulates during succession, or forest growth, N and P also accumulate in tree biomass. Once a forest approaches maturity, however, and further accumulation of C in biomass is negligible, N and P, although still limiting, must either accumulate in soil or be lost from the ecosystem. As a forest matures, a greater fraction of the CO_2 fixed in GPP is released through Rs_A, and less of the N and P entering the ecosystem is taken up by plants. The logic leads to the hypothesis that young ecosystems are better at retaining limiting nutrients; old ecosystems are leakier.

The relative importance of biogeochemical and demographic explanations for today's carbon sink is of more than academic interest. If environmentally enhanced growth explains the current terrestrial carbon sink, the sink may persist longer into the future than if regrowth is the explanation: the accumulation of carbon declines as forests age, whereas growth enhancement as a result of higher levels of CO_2, N, and P is likely to continue or even increase. The uncertainty contributes significantly to projections of future climate, and, thus, partitioning the current carbon sink between these competing explanations is an important topic for further research.

FURTHER READING

Browne, C. A. 1942. Liebig and the law of the minimum. In F. R. Moulton, ed. Liebig and after Liebig: A Century of Progress in Agricultural Chemistry. Washington, DC: AAAS, 71–82.

Davidson, E. A., and I. A. Janssens. 2006. Temperature sensitivity of soil carbon decomposition and feedbacks to climate change. Nature 440: 165–173.

Jordan, C. F., and R. Herrera. 1981. Tropical rain forests: Are nutrients really critical? American Naturalist 117: 167–180.

Koch, G. W., and H. A. Mooney (eds.). 1996. Carbon Dioxide and Terrestrial Ecosystems. San Diego, CA: Academic Press.

Woodwell, G. M., and R. H. Whittaker. 1968. Primary production in terrestrial communities. American Zoologist 8: 19–30.

III.12

Freshwater Carbon and Biogeochemical Cycles

Darren Bade

OUTLINE

1. Freshwater ecosystems
2. Reduction–oxidation reactions
3. Metal cycling: Fe and Hg
4. Phosphorus cycling
5. Nitrogen cycling
6. Sulfur cycling
7. Carbon cycling

Freshwater lakes provide an ideal example for considering the carbon cycle and other biogeochemical cycles. A range of redox conditions exists in lakes that allows observation of numerous chemical and biochemical processes. The processes are not limited to freshwater lakes, and similar examples can be found in marine and terrestrial systems. The cycles of carbon and other elements are closely linked. Production of organic carbon depends on cycling of nutrients such as nitrogen and phosphorus. Respiration of organic carbon alters the redox condition, which in turn influences the cycling of nutrients. Many other elements can be influenced by redox conditions (e.g., sulfur, iron, and mercury). The elements can have indirect effects on carbon cycling or can be deleterious to organisms present in the ecosystem.

GLOSSARY

airshed. A region sharing a common flow of air

biogeochemistry. The scientific study of the physical, chemical, geological, and biological processes and reactions that govern the cycles of matter and energy in the natural environment

ecosystem. A natural unit consisting of all plants, animals and microorganisms (biotic factors) in an area functioning together with all of the nonliving physical and chemical (abiotic) factors of the environment.

lake. A body of water of considerable size surrounded entirely by land

micronutrient. A chemical element necessary in relatively small quantities for organism growth

nutrient. A chemical element necessary for organism growth

watershed. The area of land where all of the water that is under it or drains off of it goes into the same place

1. FRESHWATER ECOSYSTEMS

Streams, rivers, lakes, and wetlands constitute some of the most obvious natural freshwater ecosystems. Additionally, groundwater and intermittent pools can be considered in this context. Artificial ecosystems, such as small impoundments, large reservoirs, and engineered wetlands, also are labeled as freshwater ecosystems.

There are many physical, chemical, and biological differences among these varied ecosystems. Hydrology offers one brief indication of the differences. Water residence time, the average time a molecule of water spends in an ecosystem, varies from minutes to years to millennia in streams, lakes, and ancient groundwater. Related to the water residence time is the movement of water. At one end of the spectrum are streams, which have strong unidirectional flow governed by gravity. At the other end of the spectrum, lakes typically show little directional flow.

There is one key commonality that links all the freshwater ecosystems. Indeed, this commonality also links marine and terrestrial counterparts. The cycling of elements in all ecosystems is mainly controlled by aqueous chemical reactions, including those reactions mediated by organisms. In that vein, the cycles examined in this chapter are considered mainly in the context of freshwater lakes.

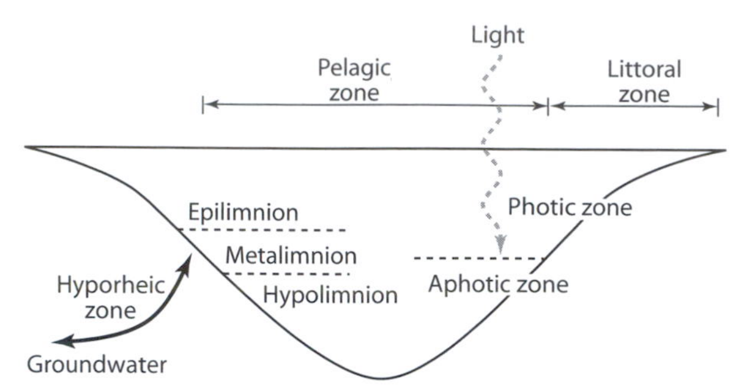

Figure 1. Typical zones described in lakes.

Lakes provide an ideal example to consider element cycling for several reasons. Because lakes often have easily defined ecosystem boundaries and fairly long water residence times, studying the internal cycling of elements becomes more tractable. Also, lakes contain many habitats that are similar to habitats in other freshwater ecosystems. A brief description of these habitats is warranted (figure 1).

Two important zones, the photic and aphotic zones, are present in many lakes with sufficient depth. The photic zone represents the surface water of a lake where light is sufficient for photosynthesis to occur. The aphotic zone is defined as the volume of water where photosynthetically available radiation is less than 1% of that present at the surface. Because light limits photosynthesis in the aphotic zone, respiratory processes usually dominate. Processes dominant in the aphotic zone of a lake are also likely to occur in groundwater. In streams, the hyporheic zone, an area below the streambed that contains the interaction between surface and groundwater, has similar processes as the aphotic zone. Although relatively unstudied, lakes, like streams, also contain an area where lake and groundwater interact.

Lake habitats can be further divided into littoral and pelagic zones. The littoral zone is the region of the lake, near shore, where there is significant interaction between bottom substrates and the photic zone. Surface water in streams and wetlands is often subject to similar light and temperature regimes as the littoral zone of a lake. Because rooted plants can occupy the littoral zone of a lake, the littoral zone provides many similarities to wetlands, and many times there is little distinction between the two. The pelagic zone is the region of the lake beyond the littoral zone.

One unique feature of lakes is the potential for layers of water to stratify because of differences in density. In most freshwater lakes, temperature causes the difference in density. For brevity, only considering lakes warmer than about 4°C, the warmer the water, the less dense it is. Just as oil floats on vinegar in a salad dressing, less-dense water floats above more dense water. Shaking the bottle of salad dressing to mix it is akin to the energy that must be applied in order to mix a stratified water column. This energy usually comes from wind, which causes waves and currents to be set in motion. Density differences and stratification caused by temperature depend on seasonal cycles. A common seasonal cycle for lakes is as follows. (1) In spring, the lake is isothermal, there is little difference in density throughout the water column, and the lake is easily mixed by wind. (2) In summer, increased solar radiation warms the surface waters, decreasing their density to the point where wind energy is insufficient to completely mix the water column. (3) In the fall, the combination of surface water cooling and bottom water warming causes the density difference in lakes to decrease to the point where wind energy can mix the lake again.

Three zones can be identified with respect to summer thermal stratification. The epilimnion is the surface water that has relatively higher, and fairly uniform, temperatures. The metalimnion is the layer of transition between the warm surface water and the colder deep water. It is characterized by a strong thermal and density gradient. The hypolimnion consists of colder water, and temperature tends to decrease with depth. Stratification is important because the hypolimnion can effectively be sealed off from interactions with the atmosphere, because diffusion of gases (e.g., O_2) or solutes through the metalimnion is extremely slow. Additionally, the depth of the top of the hypolimnion may fall below the photic zone, resulting in an absence of photosynthesis. With limited supply of O_2 to the hypolimnion but continued respiration, O_2 can become depleted in the hypolimnia of some lakes.

Oxygen is of key importance to biogeochemical cycling in all freshwater environments. Oxygen is only sparingly soluble in water and diffuses much more slowly in water than in air. Not only can O_2 be extremely low in wet environments, but strong gradients of O_2 can exist. Gradients from O_2 saturation to depletion can exist within centimeters in sediments of lake, streams, and wetlands. A similar gradient might exist over the distance of several meters in the metalimnion of a lake or reservoir. These oxic/anoxic interfaces are important sites of chemical transformations because of the unique reduction/oxidation potential (see below) that exists in some of these environments.

Over the course of this chapter we explore the cycles of iron and mercury (Fe, Hg), phosphorus (P), nitrogen (N), sulfur (S), and finally carbon (C). Oxygen and reduction/oxidation potentials will be important in many of these sections.

The broad concepts captured by this chapter are that freshwater ecosystems receive fluxes of solutes and nutrients from the surrounding watershed and airshed. These solutes and nutrients can be significantly cycled and transformed in freshwater ecosystems. In addition, these nutrients provide substrate for primary production and respiration, ultimately impacting the carbon cycle of the recipient ecosystem. In a landscape, lakes and wetlands can store significant amounts of C, P, and N. However, because of terrestrial subsidies, a large amount of carbon can also be lost from freshwaters in the form of carbon dioxide (CO_2) and methane (CH_4). Nitrogen can also be lost from aquatic ecosystem in gaseous form (N_2). Cycling of metals plays a significant part in the functioning of freshwaters as micronutrients, as potential toxins, and because of their interactions with other cycles. Finally, humans have significantly altered the cycling of many of these elements, not only by altering their input to freshwater ecosystems but also by creation or destruction of freshwater ecosystems.

2. REDUCTION–OXIDATION REACTIONS

Just as the flow of electrons from a battery provides the energy to drive an electric motor, a flow of electrons is needed to drive the metabolic machinery of organisms. For this flow to occur, there must be an electrochemical gradient. In aqueous environments, this gradient derives from the mixture of reduced and oxidized compounds that exist in a particular location. For instance, the most basic equation of heterotrophic respiration,

$$CH_2O + O_2 \rightarrow CO_2 + H_2O,$$

is the coupled oxidation of organic carbon (CH_2O) to CO_2 and reduction of O_2 to H_2O. Biological enzymes are needed to catalyze this reaction, but there is an overall release of energy for metabolism. Just as the electrons flow in the battery, in the process of respiration, electrons (coupled with hydrogen) flow from the organic carbon to the O_2. In all redox reactions there must be a source of electrons and an electron acceptor. Alternatively, the creation of organic matter through photosynthesis,

$$CO_2 + H_2O \rightarrow CH_2O + O_2,$$

reduces CO_2 and oxidizes water. Photosynthesis requires the input of energy as photons from the sun and is catalyzed by enzymes in the cell. In another analogy, this input of energy to create organic carbon is similar to the input of energy that occurs when a battery is charged. The organic matter then stores this potential energy for later release in reactions such as respiration, above. When oxygen is not present, other elements or compounds can act as electron acceptors and become reduced during anaerobic respiration. For example, iron in its oxidized form [Fe(III)] can be used as an electron acceptor and reduced to Fe(II). Iron is extremely electroactive (easily exchanges electrons), and the reduced form of iron [Fe(II)] is oxidized in the presence of dissolved O_2 fairly rapidly, even without the presence of biological enzymes. During the oxidation of Fe(II), energy is released, and although the amount of energy is modest, and the reaction does not need to be catalyzed by enzymes, there are bacteria that take advantage of this flow of electrons for energy. Generally, if significant energy can be gained from a redox reaction, microorganisms have evolved to enzymatically exploit it. These reduction and oxidation reactions are central to the cycling of several elements including C, N, S, Fe, and Hg.

The ability of an environment to donate or accept electrons is the redox potential (E_h; expressed in millivolts, mV). This potential is derived from the oxidation state of the constituents in the mixture. Recall from general chemistry the oxidation state of some familiar nitrogen molecules (e.g., $NH_4^+ : N = -III$, $H = +I$; $N_2 : N = 0$; $NO_3^- : N = +III, O = -II$). In practice, redox potential is measured with a platinum electrode attached to a voltmeter. Because the environment contains a mixture of many different reducing and oxidizing species, not always in equilibrium, measurements from electrodes only give a relative indication of the true redox potential. These measurements can still shed light on reactions that are likely to occur under certain situations. For instance, the surface waters of lakes or streams generally have a large positive redox potential (e.g., above +400 mV), indicative of an oxidizing environment (oxygen is the most energetically favorable electron acceptor), whereas water in organic sediments can have very large negative redox potential (e.g., more negative than −400 mV), indicative of a strongly reducing environment (oxygen is not present, and many of the other electron acceptors have also been reduced).

In general, sunlight provides the energy to create large amounts of reduced organic carbon and a reservoir of oxygen. The presence of these two products is a prime example of the nonequilibrium conditions that often exist with respect to redox conditions in natural waters. Respiratory, fermentative, and other nonphotosynthetic organisms take advantage of the energy

Table 1. Sequence of electron acceptors used for respiratory oxidation of organic carbon

Reaction
Reduction of O_2 to H_2O
$O_2 + 4H + 4e^- \Leftrightarrow 2H_2O$
Reduction of NO_3^- to N_2
$2NO_3^- + 12H^+ + 10e^- \Leftrightarrow N_2 + 6H_2O$
Reduction of Mn^{4+} to Mn^{2+}
$MnO_2 + 4H^+ + 2e^- \Leftrightarrow MN^{2+} + 2H_2O$
Reduction of Fe^{3+} to Fe^{2+}
$Fe(OH)_2 + 3H^+ + e^- \Leftrightarrow Fe^{2+} + 3H_2O$
Reduction of SO_4^{2-} to H_2S
$SO_4^{2-} + 10H^+ + 8e^- \Leftrightarrow H_2S + 4H_2O$
Reduction of CO_2 to CH_4
$CO_2 + 8H^+ + 8e^- \Leftrightarrow CH_4 + 2H_2O$

Note: The reactions are listed in decreasing order of amount of energy that can be obtained from each redox reaction.

stored in the reduced carbon and enzymatically catalyze a myriad of reactions that tend to restore equilibrium. Table 1 lists several of these potential reactions, many of which are further explored in this chapter. Therefore, the carbon cycle, which depends not only on the cycles of other nutrients, can alter the cycle of these nutrients through their reduction–oxidation reactions.

3. METAL CYCLING: Fe AND Hg

The cycles of two metals, iron and mercury, are of special interest in freshwater ecosystems. Iron is required as an important micronutrient in many cellular functions of both autotrophs and heterotrophs. Iron cycling has additional ramifications because of its interaction with other cycles, namely phosphorus and sulfur. Mercury concentrations have been increasing in freshwater ecosystems as a result of human activities. Because of its toxicity, the cycling of Hg warrants special attention. All of these metals are highly electroactive, so their cycling is closely linked to redox conditions.

Iron exists in freshwater environments in both the oxidized [Fe(III)] and reduced [Fe(II)] forms. The primary factor controlling the cycling of Fe is the difference in solubility between the oxidized and reduced moieties. The reduced forms tend to be much more soluble than the oxidized forms. In addition to the cycling within an aquatic ecosystem, the difference in solubility also determines how these metals are transported to the ecosystem. Dissolved organic matter (DOM), because of its ability to bind with metals, can also influence the transport and cycling of Fe. Although it is not considered in this chapter, manganese has very similar properties to Fe, and the cycles are nearly similar.

Rust is a familiar form of oxidized iron. Oxidized iron, also called ferric iron, has relatively low solubility in most surface freshwaters and is generally in the solid form of $Fe(OH)_3$. Therefore, it is mainly transported to lakes via streams or atmospheric deposition in particulate or colloidal form or bound or chelated to DOM. On the other hand, reduced iron (ferrous iron) is much more soluble and can be transported in its ionic form (Fe^{2+}), as long as the water is anoxic or strongly acidic. Therefore, groundwater with low concentrations of oxygen can carry reduced iron to lakes. The solubility of Fe^{2+} is mainly controlled by $FeCO_3$ and FeS.

In well-oxygenated waters such as the epilimnion, ferrous iron is oxidized quickly to form the precipitate, $Fe(OH)_3$. Other ions can be precipitated with the ferric ion or become adsorbed onto the ferric hydroxide, most notably orthophosphate (see section 4). Surface waters of lakes thus tend to have low concentrations of iron because of its low solubility and the tendency for particulate iron to leave the surface water via sedimentation. In some instances, the low concentration of iron in the surface water can directly limit algal primary production, as is the case in some parts of the ocean (chapter III.13).

Oxygen consumption by heterotrophic bacteria usually creates substantial redox gradients in the sediments or hypolimnia of lakes. This redox gradient affects the internal cycling of Fe significantly. When oxygen becomes depleted, iron can act as an electron acceptor for respiration, and iron is reduced. Because reduced iron tends to be more soluble, Fe^{2+} can diffuse upward through the sediments toward the overlying water column. When it reaches an oxygenated strata, the solid $Fe(OH)_3$ is again formed. When this reaction occurs in the water column, the precipitate again sediments out of the water column. Whether the location of the zone of iron oxidations is within the sediments or in the water column has large ramifications for the cycling of P. The repeated cycling of iron oxidation states is sometimes referred to as the "ferrous wheel" (figure 2). Some bacterial assemblages take advantage of the ferrous wheel to continually supply a source of energy to fix carbon from the oxidation of reduced iron and, conversely, to supply an electron acceptor for respiration during the reduction of oxidized iron. In the presence of reduced forms of sulfur, ferrous iron can also form amorphous FeS, which is fairly insoluble and tends to remove iron from further cycling. The

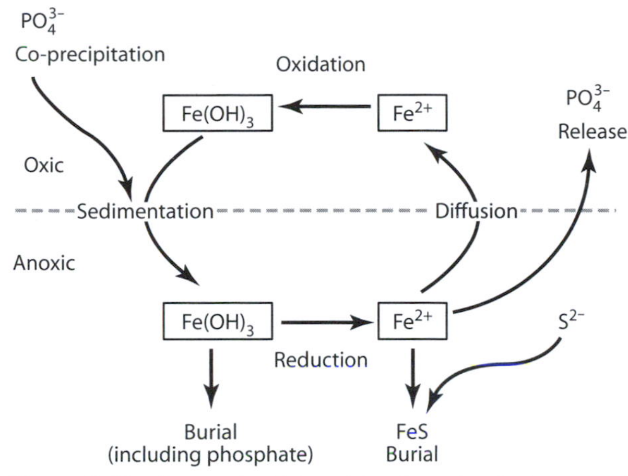

Figure 2. The iron cycle. External inputs and outputs are not considered. The impact of the iron cycle on both orthophosphate and sulfide is noted in this figure.

production of FeS can be a significant factor in the cycling of both Fe and S.

Mercury mainly exists in two different oxidation states in the environment: elemental mercury (Hg^0) and the more oxidized mercuric mercury (Hg^{2+}) (figure 3). Mercury occurs naturally in the environment, most often as inorganic salts [e.g., $HgCl_2$, $Hg(OH)_2$, HgS]. Elemental mercury, the familiar silver-colored liquid in thermometers, is rare in the environment. However, the presence of total mercury in the environment has been increasing mainly because of the combustion of fossil fuels and other industrial processes. In the atmosphere mercury can exist in both oxidation states; however, Hg^0 has a longer residence time in the atmosphere and can be transported large distances. Mercuric Hg is more quickly removed from the atmosphere in precipitation, and both forms can be deposited on land and water as dry deposition. The solubility and movement of Hg from land to water are strongly controlled by its complexation with DOM. Once in a lake or wetland, mercury can undergo changes in oxidation state, similar to Fe. Elemental mercury is not very soluble in water and has a tendency to volatilize back into the atmosphere. This can be one significant pathway of loss in aquatic ecosystems. Another significant loss of mercury from a lake is via sedimentation of organic or inorganic materials that have complexed with Hg. Neither of the inorganic forms of Hg is extremely toxic to biota.

One peculiar facet of mercuric mercury, however, is its ability to form covalent bonds as opposed to ionic bonds. One covalently bonded molecule in particular, methyl mercury (CH_3Hg^+), is extremely toxic. In addition, methyl mercury tends to bioaccumulate in organisms. This has led to advisories suggesting limited consumption of fish from certain areas. The balance between inflows and losses, and the processes that methylate and demethylate mercury, control the concentrations of toxic methyl mercury in fish. Two correlates with high fish methyl mercury concentrations are high dissolved organic carbon concentrations and low pH (although these two parameters are often correlated).

Methylation of mercury is thought to be carried out predominately by sulfate-reducing bacteria (see section 6, below). Once it has been methylated, methyl mercury still has a large affinity for complexing with particles, such as bacteria and algae. This may be an important mechanism for methyl mercury to enter the food web. Once in an organism, methyl mercury is lipophilic and is not easily excreted. This leads to the bioaccumulation that poses increasing threats for large predators, including avian predators that rely on a large portion of fish for their diet. Several bacteria are also able to demethylate mercury, returning it to either Hg^0 or Hg^{2+}. Additionally, reactions with UV light can lead to photodegradation of methyl mercury. There are still large uncertainties in the specific details of many aspects of mercury cycling. The uncertainty in these processes and the need for sound policy regarding mercury emissions have created contention among state, federal, and international authorities.

4. PHOSPHORUS CYCLING

Phosphorus is often found to be limiting to primary production in many freshwater ecosystems, especially lakes. Nitrogen, another key limiting nutrient is discussed below. The phosphorus cycle has been greatly augmented by man. The use of phosphorus as a fertilizer and detergent has led to increased loading into freshwater ecosystems. The consequence of this excess phosphorus is cultural eutrophication (see chapter VII.5).

Although the P cycle is closely linked to redox conditions, phosphorus in water is mainly found at one oxidation state in the form of orthophosphate (PO_4^{3-}). At this oxidation state, P exists in several different inorganic and organic forms, either dissolved or particulate. The primary P-containing geologic mineral is apatite. Sedimentary rocks such as $CaCO_3$ also can contain significant amounts of P. Inorganic forms include dissolved orthophosphate or particulate orthophosphate adsorbed to clay, carbonates, ferric hydroxides, or organic particles. Phosphorus is found in cellular organic material such as nucleic acid, phosphoproteins, phospholipids, phosphate esters, or nucleotide phosphates (e.g., ATP).

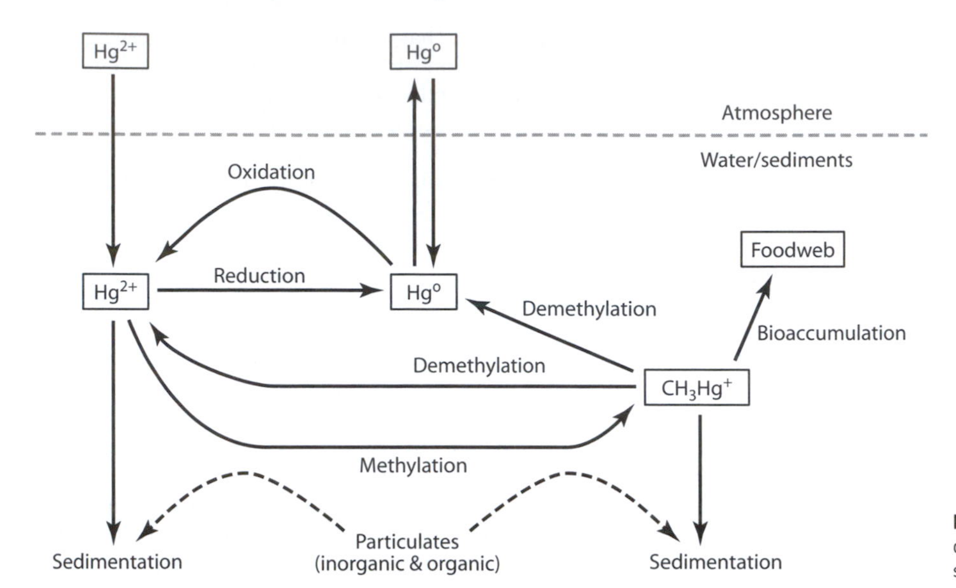

Figure 3. The mercury cycle. Additional stream and groundwater sources are not displayed.

In the absence of human influence, most terrestrial landscapes substantially retain P. Phosphorus is largely held in the vegetation, in soil organic matter, or adsorbed to inorganic particles. In these situations, the primary flux of P to aquatic ecosystems comes from atmospheric deposition or fluvial inputs of dissolved organic P. In the presence of human activity, the forms and quantities of P loading change. Use of P fertilizer can increase the amount of dissolved orthophosphate in streams and groundwater. Land use practices that cause erosion increase the proportion of P that enters aquatic ecosystems adsorbed or contained in particulates.

The simplest example of the aquatic P cycle occurs in lakes with high levels of oxygen throughout the water column. These conditions are usually found in oligotrophic lakes, lakes with low levels of nutrients that limit primary productivity. In oligotrophic lakes, very little phosphorus is found in the dissolved form, and most is contained in particulate forms. Phosphorus is rapidly and efficiently cycled in the biota. There is strong competition between algae and bacteria for limited P resources. Algae and bacteria also produce enzymes to release bound phosphorus (phosphatases). Losses of particulate P via sedimentation constitute the dominant loss term in this oligotrophic P cycle.

Several conditions such as increased productivity, large external loading of organic carbon, or a high ratio of sediment surface area to hypolimnetic volume can lead to oxygen depletion in the hypolimnion. The change in redox conditions and zone of Fe oxidation in the sediments can significantly alter the P cycle (figure 2). Under oxic conditions, orthophosphate can be adsorbed with iron hydroxide or precipitated as $FePO_4$. When Fe is reduced in anoxic sediments, soluble Fe^{2+} is produced, and PO_4^{3-} is also released. If the zone of Fe oxidation occurs within the sediments, PO_4^{3-} is effectively "trapped" in the sediments as it readsorbs with iron precipitates. Under extremely anoxic conditions, there is no zone of oxidation in the sediments, and the PO_4^{3-} can be returned to the water column. If reduced iron is buried as FeS, PO_4^{3-} can escape, but there will be little iron available to be reoxidized, allowing the PO_4^{3-} to remain in solution. Therefore, P cycling can be linked to both Fe and S cycling.

The recycling of orthophosphate from the sediments represents an internal feedback mechanism that has significant impact on the mitigation of human-caused eutrophication. The recycling of P from the sediments sustains high levels productivity and reinforces the conditions for P recycling. In these cases, mitigation of the original P pollution will have only a limited effect on reducing the problems of eutrophication (see chapter VII.5).

5. NITROGEN CYCLING

Nitrogen exists in several different oxidation states, and the variation in redox conditions in freshwater ecosystems creates an environment where transformations between oxidation states are significant. Most nitrogen exists in its molecular form, N_2, as a gas. Biological fixation of N_2 to organic forms is therefore a significant pathway in the N cycle. Humans have invented methods for N_2 fixation, and these methods have contributed more fixed nitrogen to the environment than forms of biological fixation. The effect of this added N has had large ramifications for many environments including freshwater ecosystems (see chapter III.10).

The forms of nitrogen found in water include dissolved gases such as N_2, N_2O (nitrous oxide), and (ammonia), dissolved ions of NH_4^+ (ammonium), NO_2^- (nitrite), and NO_3^- (nitrate), and dissolved and particulate organic forms. The reactive inorganic forms of N (species of N, excluding N_2) most prominent in the environment are NH_4^+ and NO_3^-. The organic forms range from amino acids, amines, and proteins to complex organic compounds with low nitrogen content.

Nitrogen is often considered a key limiting nutrient for terrestrial ecosystems. Therefore, N tends to be highly retained in vegetated landscapes. As with P, this terrestrial retention can cause N to be limiting (or co-limiting with P) to aquatic primary production. For lakes in pristine environments, terrestrial landscapes do contribute some dissolved and particulate organic nitrogen. Atmospheric deposition is a significant source of reactive inorganic nitrogen (NH_4^+ or NO_3^-). Biological fixation of N_2 by cyanobacteria and other heterotrophic bacteria can also be important.

Several sources of N pollution significantly alter the flux of N to recipient freshwater ecosystems. Fertilizer use can increase the flux of both NH_4^+ and NO_3^- from streams and groundwater. Agricultural practices also tend to increase the amount of gaseous NH_3, which can be deposited on the lake surface. Of particular interest is the emission of nitrogen oxides (NO_x) emanating from high-temperature combustion of fossil fuels, mainly attributed to automobiles. Nitrogen oxides are transformed in the atmosphere or in the lake and become nitric acid, an important component of acid rain. All these inputs can increase the productivity of the lake, with the side effect of increased respiratory loss of O_2 in the aphotic zone.

Fixation of N_2 by cyanobacteria is a key process in supplying nitrogen to freshwater ecosystems when external sources are small. Nitrogen fixation by cyanobacteria requires nitrogenase enzyme, organic carbon, and light energy. In most cyanobacteria, nitrogen fixation takes place in specialized cells called heterocysts. Because nitrogenase enzyme is inactivated by O_2, the heterocysts provide a microenvironment of low oxygen to allow for nitrogen fixation. Heterocysts are only found in filamentous cyanobacteria; however, not all species of filamentous cyanobacteria have heterocysts. Also, some unicellular cyanobacteria have the ability to fix N_2 without the presence of heterocysts. Nitrogen fixation generally increases when inorganic nitrogen concentration is low or when the N:P ratio is decreased, sometimes because of P pollution.

Within a lake, the major transformations of N are largely maintained by microbial processes (figure 4). Algae can use both NH_4^+ and NO_3^- to form N-containing organic compounds, most notably amino acids. However, NO_3^- must first be reduced to NH_4^+ before it can be incorporated as an amine group. This process is known as assimilatory (because the nitrogen is incorporated in the biomass) nitrate reduction. Nitrate assimilation is therefore more energetically costly than ammonium assimilation, and ammonium is preferentially used, if available. Because of the preferential uptake of ammonium by plants and algae, it generally

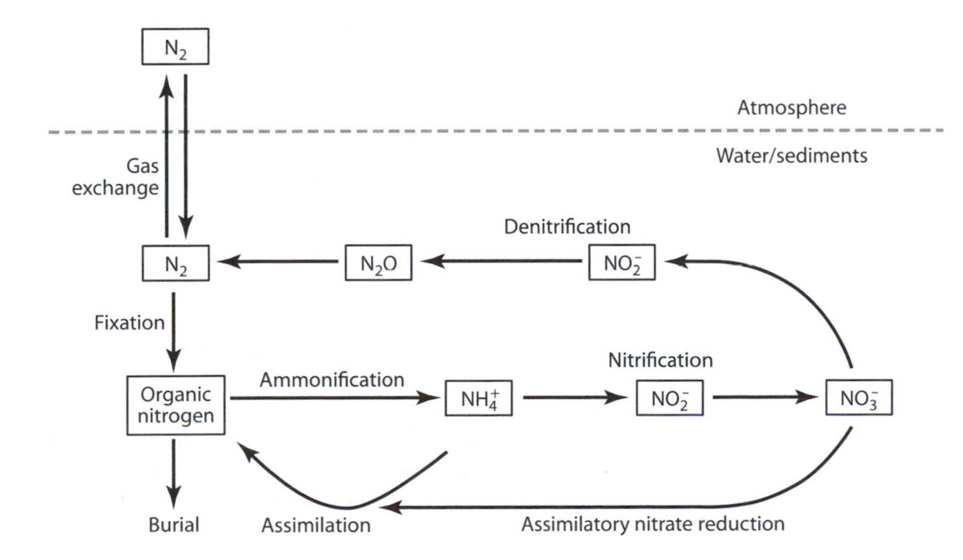

Figure 4. The nitrogen cycle. External inputs of inorganic and organic nitrogen from streams, groundwater, and precipitation are not displayed; however, they are often much more significant than N_2 fixation.

has very low concentrations in surface waters. Another process that keeps ammonium at relatively low concentrations is nitrification.

Nitrification is the process of NH_4^+ oxidation. The oxidation is biologically mediated by a relatively small number of bacterial species and has two distinct steps. These reactions are a form of chemoautotrophic production. The first step,

$$NH_4^+ + 1.5O_2 \leftrightarrow 2H^+ + NO_2^- + H_2O,$$

is carried out mainly buy *Nitrosomonas* bacteria. The second step,

$$NO_2 + 0.5O_2 \leftrightarrow NO_3^-,$$

is carried out by *Nitrobacter*. The overall reaction,

$$NH_4^+ + 2O_2 \leftrightarrow 2H^+ + NO_3^- + H_2O,$$

points to some interesting aspects of nitrification. First, nitrification is an aerobic process, and although it is autotrophic, it actually consumes 2 moles of O_2 per mole of N oxidized. In the absence of O_2, nitrification can not proceed, and it is possible for relatively higher concentrations of NH_4^+ to accumulate (e.g., in sediments or hypolimnia that experience anoxia). Second, nitrification is an acidifying process; 2 moles of H^+ are produced per mole of N oxidized. Therefore, NH_4^+ pollution can indirectly lead to acidification after nitrification.

Particulate organic nitrogen has two significant fates. First, the particulate organic nitrogen (also dissolved organic nitrogen) can be decomposed by heterotrophic bacteria. If the N in the detrital organic matter is in excess of the needs of the bacteria, it will be released as NH_4^+. The release of NH_4^+ from organic material decomposition is called mineralization or ammonification. At this point the NH_4^+ can again be assimilated by microorganisms or be nitrified. The second major fate of particulate organic nitrogen is sedimentation. As bacteria preferentially decompose simpler compounds or molecules where N is more accessible, there is a decrease in the quality of the remaining organic matter for further degradation. The N bound in this organic matter is then permanently buried in the sediments.

One additional process is key to understanding N cycling in freshwater ecosystems. Denitrification is the reduction of NO_3^- with the concomitant oxidation of organic matter. Nitrate is used as a terminal electron acceptor when oxygen becomes depleted (table 1) and yields nearly as much free energy as aerobic respiration. Denitrification is carried out by many facultative anaerobic bacteria. During denitrification, NO_3^- is sequentially reduced to N_2 following these intermediate steps:

$$NO_3^- + \rightarrow NO_2^- \rightarrow N_2O \rightarrow N_2.$$

The process can be interrupted at any of the intermediate steps, but generally there is little accumulation of either NO_2^- or N_2O. Denitrification tends to be limited to certain areas because it requires a source of NO_3^-, organic carbon, and low oxygen. Sediments of lakes and wetlands often have high rates of denitrification, as do the hyporheic zones of streams and rivers. The loss of N_2 to the atmosphere after denitrification is a permanent loss of nitrogen from the system. Because denitrification decreases concentrations of inorganic nitrogen, wetlands are often constructed with the specific purpose of denitrifying large amounts of nitrate in order to reduce the load of nitrate to downstream ecosystems. Denitrification is also important in regard to the impacts of acid deposition because the process consumes hydrogen ions.

6. SULFUR CYCLING

A large interest in the sulfur cycle of lakes developed after the recognition of the impacts of acid precipitation, which is often dominated by sulfuric acid. Sulfur dioxide (SO_2), mainly emitted by coal combustion and ore smelting, is oxidized to form sulfuric acid (H_2SO_4) in the atmosphere or in the lake. Acidic precipitation has increased the flux of sulfate ions (SO_4^{2-}) from both direct inputs to a lake. Some regions can have naturally high concentrations of SO_4^{2-}. Sedimentary deposits often contain significant amounts of calcium sulfate. Weathering and oxidation of sulfur (S^0) or sulfide (S^{2-})-containing minerals will also produce sulfate ions and acidity.

Lakes have some natural ability to mitigate the inputs of SO_4^{2-}. Just as the reduction of nitrate consumes H^+, sulfate reduction also consumes acidity (2 moles of H^+ per mole of SO_4^{2-} reduced). The reduction of SO_4^{2-} also has important connections to Fe, Hg, and P cycling. Sulfate can be used as a terminal electron acceptor for anaerobic respiration by sulfate-reducing bacteria (table 1). Sulfate-reducing bacteria produce hydrogen sulfide (H_2S). Hydrogen sulfide is quickly oxidized if it diffuses into oxygenated waters. Or, the reduced sulfide can react with reduced Fe to form fairly insoluble precipitates such as FeS. The formation of FeS, if it occurs beyond the zone of oxidation in the sediments, can represent a permanent burial of

S. Sulfur-containing organic matter can be formed by reduction of sulfate and by abiotic reactions of sulfide and organic matter. This organic bound sulfur tends to be less prone to reoxidation in the sediments than FeS.

Regions with high loading of SO_4^{2-} ions and significant sulfate reduction can precipitate a large quantity of iron as FeS. The relationship between Fe and S cycling also impacts P cycling (figure 2). With large amounts of sulfide and FeS formation, very little Fe^{2+} diffuses to the water column, but PO_4^{3-} still escapes. It has been speculated that increased flux of SO_4^{2-} from acid rain could exhaust the supply of iron in some lakes. The ramifications of this are twofold. First, if FeS can no longer be formed because of limited Fe, sulfide will remain as H_2S. Because H_2S can be easily reoxidized, there will be little permanent loss of sulfate, and the consumption of acidity by sulfide burial will be reduced. Second, the loss of Fe^{2+} that can be oxidized could possibly increase the release of PO_4^{3-} from sediments, affecting eutrophication. Additionally, sulfate-reducing bacteria play a significant role in the methylation of Hg, and increased SO_4^{2-} concentrations could alter the cycling of Hg.

Two groups of sulfur-oxidizing bacteria can use H_2S. Chemosynthetic sulfur-oxidizing bacteria derive energy from the H_2S and oxidize it to elemental sulfur or sulfate. Photosynthetic sulfur bacteria also oxidize H_2S. Both the green and purple photosynthetic sulfur bacteria are anaerobes and require light as a source of energy. Hydrogen sulfide is used as an electron donor, with the concomitant production of elemental sulfur, just as water is used as an electron donor in oxygen-producing photosynthesis.

7. CARBON CYCLING

Inorganic Carbon

Central to inorganic C cycling is the role of carbonate chemistry. Carbonate chemistry describes which species of inorganic carbon are present in water. To start the explanation of carbonate chemistry, it is probably easiest to consider a very simplistic example of water in a beaker. This removes some of the complexities of biogeochemistry that are considered later.

Carbon dioxide (CO_2) from the atmosphere is soluble in water. Once dissolved in water CO_2 is hydrated by water to yield carbonic acid:

$$CO_2 + H_2O \leftrightarrow H_2CO_3.$$

The equilibrium concentration of carbonic acid is considerably less than that of dissolved CO_2, such that the amount of carbonic acid is negligible compared to the other constituents. Carbonic acid can dissociate to form bicarbonate ions (HCO_3^-) and carbonate ions (CO_3^{2-}).

$$H_2CO_3 \leftrightarrow H^+ + HCO_3^-,$$

and

$$HCO_3^- \leftrightarrow H^+ + CO_3^{2-}.$$

The sum of these three carbonate species—dissolved CO_2, bicarbonate, and carbonate—are often described as the dissolved inorganic carbon (DIC).

The speciation of the different carbonate compounds is pH dependent, and if the pH of the beaker is altered, the amount of DIC and the relative proportion of each of the species will change. At an approximate pH of 6.4, the concentrations of bicarbonate and dissolved CO_2 are equal, and the relative contribution of carbonate ions to the amount of DIC is negligible. At an approximate pH of 10.4, the concentrations of bicarbonate and carbonate are equal, and the relative contribution from dissolved CO_2 is negligible.

The range of pH observed in freshwater systems is generally between 4 and 10. For a simple approximation, one can assume that if the pH of the system is below 5, dissolved CO_2 dominates the DIC. As pH increases, the proportion of CO_2 declines, and bicarbonate dominates between pH 7 and 9. As pH further increases, carbonate begins to become appreciable above pH 9.5.

Adding to the complexity of the beaker example, the addition of some other ions can influence the carbonate chemistry. Calcium has a large interaction with carbonate chemistry. In the presence of high concentrations of calcium and high pH, calcium carbonate ($CaCO_3$) can precipitate from the system:

$$Ca^{2+} + 2HCO_3^- \leftrightarrow CaCO_3 + CO_2 + H_2O$$

The reverse of this reaction can also take place, with $CaCO_3$ becoming soluble with increased acidity or the presence of increasing CO_2 (which increases the acidity through carbonic acid).

The final layer of complexity includes the biological processes that make a lake infinitely more interesting than the beaker example (figure 5). The main biological processes of interest are those that consume or generate CO_2, namely, photosynthesis and respiration. Photosynthesis consumes CO_2, and this can alter the carbonate chemistry in several ways. As CO_2 is consumed, pH increases, and at very high rates of photosynthesis pH can increase to greater than 9. In the presence of

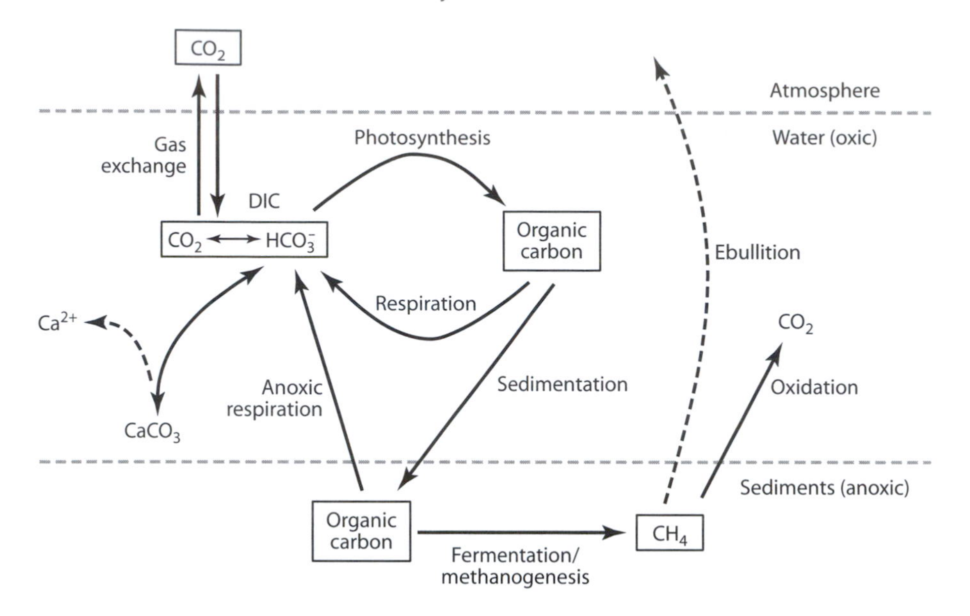

Figure 5. The carbon cycle. This very generalized diagram does not consider other forms of autotrophic production besides photosynthesis, but in general, these processes are much less significant.

enough calcium, during these events of high productivity, $CaCO_3$ can precipitate. This tends to restore the system to lower pH, as CO_2 is released during the precipitation of calcium carbonate. The precipitation of calcium carbonate can be an important sedimentary process in a lake. The sedimentation of $CaCO_3$ can increase the burial of organic carbon as well as orthophosphate, which can coprecipitate with the $CaCO_3$. Respiration, on the other hand, releases CO_2 and will cause the pH to decline.

We must keep in mind that the patterns of dissolved gases, such as O_2 and CO_2 are also modulated by the physics of gas exchange. If the concentration is higher than equilibrium with the atmosphere, gas will be lost to the atmosphere, and vice versa. The rate of this flux is dependent on the concentration gradient between air and water and the amount of turbulence, most often caused by wind, at the air–water interface.

The source of inorganic carbon in lakes is often much greater than just the amount coming from atmospheric exchange, however. Inflowing streams and groundwater can import large quantities of inorganic carbon. Groundwater is unique in that it can often be much more supersaturated in CO_2 than surface waters because of a predominance of respiration and very slow atmospheric gas exchange.

Organic Carbon

There are two sources of organic carbon for lakes: primary production of carbon within the lake (autochthonous carbon) and inputs of external sources of carbon, primarily from the terrestrial component of the watershed (allochthonous carbon). Organic carbon can either be dissolved or particulate (DOC and POC). Particulate carbon can be further categorized as living organisms or detritus (nonliving organic carbon). Once present in the lake, organic carbon has several fates (figure 5). The organic carbon can be exported downstream or buried in the sediments of the lake. Also, the organic carbon can be respired to inorganic compounds (CO_2 or CH_4) and lost via gaseous exchange with the atmosphere. Therefore, freshwater ecosystems can have an interesting role in the landscape, capable of both burying organic carbon and releasing CO_2 back to the atmosphere.

The balance of organic carbon burial and loss of CO_2 to the atmosphere is represented by net ecosystem production (NEP). NEP is the difference of total respiration (R_T) from gross primary production (GPP):

$$NEP = GPP - R_T.$$

Positive NEP means that GPP exceeds R_T and that CO_2 is taken from the atmosphere and buried as organic carbon. Alternatively, negative NEP means that R_T exceeds GPP. In order for R_T to exceed GPP, there must be a subsidy of allochthonous carbon. With sufficient external inputs of organic carbon, there can be burial of organic carbon and release of inorganic carbon from aquatic ecosystems.

Respiration of organic carbon in aerobic conditions produces CO_2 and is the most energetically favorable. Under anaerobic conditions, other electron acceptors besides O_2 must be used with declining energetic benefit (table 1). In many freshwater ecosystems that experience oxygen depletion, fermentation and methanogenesis represent the final steps in organic matter decomposition. During fermentation, organic carbon acts as both an electron donor and acceptor. A common example of fermentation is the breakdown of glucose, which produces acetic acid, CO_2, and hydrogen (H_2). Fermentation produces very little energy compared to other forms of respiration, but it is a significant process because fermentative organisms can degrade many organic compounds that are not accessible to other nonfermentative organisms.

The final step in organic matter decomposition is methanogenesis, which is closely tied to fermentation. Two pathways of methanogenesis are prevalent. The first pathway, CO_2 reduction, is a chemoautotrophic process. In this pathway,

$$4H_2 + CO_2 \rightarrow CH_4 + 2H_2O,$$

hydrogen, supplied from fermentation, is a source of energy, and CO_2 is both an electron acceptor and the source for cellular carbon production. In the second pathway, acetate is decomposed to CO_2 and methane:

$$CH_3COOH \rightarrow CO_2 + CH_4.$$

Methane can be further cycled in the water column. Methane can be oxidized by chemoautotrophic organisms. Also, methane is not very soluble in water, and in sediments where production of methane is high, methane comes out of solution to form bubbles. When the bubbles escape the sediments, they bypass the opportunity for oxidation and escape to the atmosphere (ebullition). The contribution of freshwater ecosystems, especially wetlands, to the global flux of methane is significant. Because methane is about 20 times more effective than CO_2 as a greenhouse gas, the cycling of methane in freshwaters is of special interest.

The cycling of organic carbon in freshwater ecosystems is closely tied to the cycling of other elements through changing redox conditions in different environments. Conversely, the supply of many elements is critical for the production of new organic carbon. An important question is whether the carbon and biogeochemical cycling of freshwater ecosystems is significant on a large, even global, scale. This question is especially interesting in the light of understanding the global carbon budget. In the case of lakes, many are supersaturated in CO_2, implying that R_T exceeds GPP, and NEP is negative. A significant amount of allochthonous carbon must be respired for this to occur. This appears to be a common occurrence in many freshwater ecosystems. As organic carbon moves from land to the ocean through a series of aquatic ecosystems, a significant portion of the carbon is returned to the atmosphere as CO_2.

But freshwater ecosystems can also bury significant amounts of carbon produced either internally or externally. As redox conditions decrease in the anoxic zones of freshwater ecosystems, decomposition of organic carbon becomes less and less energetically favorable, and carbon can be effectively buried. As part of a working group at the National Center for Ecological Analysis and Synthesis, Cole and colleagues (2007), estimated globally that aquatic ecosystems received about 1.9 Pg C $year^{-1}$ from adjacent terrestrial ecosystems. Of this amount, about 0.8 Pg C $year^{-1}$ was lost to the atmosphere as CO_2, and about 0.2Pg C $year^{-1}$ was buried as sediments. The remainding 0.9 Pg C $year^{-1}$ was supplied to the oceans from the worlds rivers as organic or inorganic carbon. Evidence is mounting that carbon and biogeochemical cycling of elements in freshwater ecosystems matters, even at the large, global scale.

FURTHER READING

Allan, J. D., and M. M. Castillo. 2007. Stream Ecology: Structure and Function of Running Waters. Dordrecht: Springer.

Cole, J. J., Y. T. Prairie, N. F. Caraco, W. H. McDowell, L. J. Tranvik, R. G. Striegl, C. M. Duarte, P. Kortelainen, J. A. Downing, J. J. Middelburg, and J. Melack. 2007. Plumbing the global carbon cycle: Integrating inland waters into the terrestrial carbon budget. Ecosystems 10: 171–184.

Mitsch, W. J., and J. G. Gosselink. 2000. Wetlands. New York: John Wiley & Sons.

Schlesinger, W. H. 1997. Biogeochemistry: An Analysis of Global Change. San Diego: Academic Press.

Wetzel, R. G. 2001. Limnology: Lake and River Ecosystems. San Diego: Academic Press.

III.13

The Marine Carbon Cycle

Paul Falkowski

OUTLINE

1. The two carbon cycles
2. Chemoautotrophy
3. The evolution of photoautotrophy
4. Selective pressure in the evolution of oxygenic photosynthesis
5. Primary production
6. Who are the photoautotrophs?
7. Carbon burial
8. Carbon isotope fractionation in organic matter and carbonates
9. Concluding remarks

> The system of scholastic disputations encouraged in the Universities of the middle ages had unfortunately trained men to habits of indefinite argumentation, and they often preferred absurd and extravagant propositions, because greater skill was required to maintain them; the end and object of such intellectual combats being victory and not the truth.
>
> —Charles Lyell, Principles of Geology, 1830

Approximately 50% of all the primary production on Earth occurs in the oceans, virtually all by microscopic, single-celled organisms that drift with the currents, the phytoplankton. On ecological time scales of days to years, the vast majority of the organic matter produced by phytoplankton is consumed by grazers such that the turnover time of marine organic carbon is on the order of 1 week, compared with over a decade for terrestrial plant ecosystems. On geological time scales of millions of years, however, a small fraction of the carbon fixed by phytoplankton organisms is buried in marine sediments, thereby both giving rise to oxygen in Earth's atmosphere and providing fossil fuel in the form of petroleum and natural gas. In this chapter, we examine the factors controlling the marine carbon cycle and its role in the ecology and biogeochemistry of Earth.

GLOSSARY

acid–base reactions. A class of (bio)chemical reactions that involve the transfer of protons without electrons.

chemoautotrophy. A mode of nutrition by which an organism can reduce inorganic carbon to organic matter in the absence of light using preformed bond energy contained in other molecules.

isotopic record of carbon. The changes in the ratio of ^{13}C to ^{12}C over geological time in marine carbonates or in organic matter in sediments or sedimentary rocks.

net primary production. The organic carbon that is produced by photosynthetic organisms and becomes available for other trophic levels in an ecosystem.

photoautotrophy. A mode of nutrition by which an organism can reduce inorganic carbon to organic matter using light energy.

phytoplankton. Microscopic, mostly single-celled photosynthetic organisms that drift with the currents.

redox reactions. A class of (bio)chemical reactions that involve the transfer of electrons with or without protons (i.e., hydrogen atoms). Addition of electrons or hydrogen atoms to a molecule is called "reduction"; removal of electrons or hydrogen atoms from a molecule is called "oxidation." "Redox" is a contraction of the terms *reduction* and *oxidation*.

1. THE TWO CARBON CYCLES

All life on Earth is critically dependent on the fluxes of six elements: H, C, N, O, S, and P. Of these, the flux of C is unique. Not only is C used to make the substrates of key biological polymers, such as lipids, carbohydrates, proteins, and nucleic acids, the oxidation and reduction of C provide the major conduit of energy supply for life itself. The biological carbon cycle is based on electron transfer (i.e., redox) reactions in which the formation and utilization of the bond energy of C–H and, to a lesser extent, C–C molecules provide the major driving force of life. However, and perhaps paradoxically, the overwhelming majority of C on Earth is contained in a relatively immobile pool in the lithosphere in the form of carbonate rocks (table 1). This oxidized pool of carbon contains no biologically available energy. To sustain a flux of carbon (and hence, an essential biological building block and energy supply) on geological time scales, the lithospheric, oxidized carbon in carbonates must enter one of two mobile pools, either the atmosphere or the ocean, from which biological processes can access the carbon, reduce it to organic matter, and transfer the organic matter through metabolic processes. Hence, there are two parallel carbon cycles on Earth. One cycle is slow and abiotic, and its chemistry is based on acid–base reactions. The physical processes that drive this cycle play a key role in Earth's climate. The second is fast and biologically driven; its chemistry is based on electron transfer reactions. The biological processes that drive this cycle play a key role in sustaining ecosystems. Let us briefly consider the two carbon cycles and then focus on the unique role of the ocean as the conduit where both cycles meet.

The "Slow" Geological Carbon Cycle

The slow geological carbon cycle operates on multimillion-year time scales and is dictated by tectonics, which is itself related to the amount of radiogenic heat produced in the Earth's interior. In this cycle, CO_2 is released from Earth's mantle to the atmosphere and oceans via volcanism and sea floor spreading. CO_2 is a unique gas, however. In aqueous solution, it reacts with water to form carbonic acid (H_2CO_3), which is a weak acid. In the atmosphere, this acid is formed in precipitation. When rain falls on silicate-rich rocks such as granite, the carbonic acid reacts with the silicates, and Ca^{2+} and Mg^{2+} ions are extracted and solublilized. Orthosilicic acid and the two cations are carried to the sea via rivers. In the ocean, Ca^{2+} and CO_3^{2-} are generally present in excess of the concentrations in equilibrium with $CaCO_3$, meaning that the solution is supersaturated. In the contemporary ocean, many organisms including corals, shellfish, and several taxa of plankton catalyze the precipitation of carbonates as $CaCO_3$ and $Ca(Mg)CO_3$, which ultimately are sources of marine sediments and sedimentary and metamorphic rocks (e.g., marble). On geological time scales, most of the carbonates are subsequently subducted into the mantle, where they are heated, and their carbon is released as CO_2 to the atmosphere and ocean to carry out the cycle again. This cycle would operate whether or not there were life on the planet.

Absent organisms to catalyze the precipitation of carbonates, ultimately the ocean would become highly supersaturated, and carbonates would spontaneously precipitate. Indeed, such a situation almost certainly occurred over the first 3 billion years of Earth's 4.5-billion-year history. This slow carbon cycle is a critical determinant of the concentration of CO_2 in Earth's atmosphere and oceans on time scales of tens and hundreds of millions of years. CO_2, in turn, has an infrared absorption cross section, making it one of the most important greenhouse gases on Earth.

Once carbon, derived either from volcanism or from sea floor spreading, enters the atmosphere or oceans, it becomes mobile. In the atmosphere, virtually all of the carbon is in the form of gaseous CO_2. In the ocean, however, carbonic acid (H_2CO_3) forms a buffer system, which can be described by the following equations:

$$H_2O + CO_2 \leftrightarrow H_2CO_3 \leftrightarrow H^+ + HCO_3^- \leftrightarrow 2H^+ + CO_3^{2-} \leftrightarrow Ca(Mg)CO_3. \quad (1)$$

The equilibrium reactions are shifted toward the right at high pH and toward the left at low pH. Specifically, in seawater at 20°C, the p*K* for the first deprotonation reaction is about 6, and that for the second deprotonation is about 9. Thus, in the ocean, with an average pH of 8.2, virtually all (>95%) of the inorganic

Table 1. Carbon pools in the major reservoirs on Earth

Pools	*Quantity* ($\times 10^{15}$ g)
Atmosphere	720
Oceans	38,400
Total inorganic	37,400
Surface layer	670
Deep layer	36,730
Dissolved organic	600
Lithosphere	
Sedimentary carbonates	>60,000,000
Kerogens	15,000,000
Terrestrial biosphere (total)	2000
Living biomass	600–1000
Dead biomass	1200
Aquatic biosphere	1–2
Fossil fuels	4130
Coal	3510
Oil	230
Gas	140
Other (peat)	250

carbon is present in the form of bicarbonate. This buffer system is the major determinant of the pH of the ocean.

The ionic forms of CO_2 do not contribute to the vapor pressure of the gaseous form; thus, the concentration of the sum of all the dissolved inorganic carbon (Tco_2) can greatly exceed the atmosphere/water equilibrium concentration of gaseous CO_2 (Pco_2). The vapor pressure is predicted from Henry's law:

$$[CO_2] = K_H Pco_2, \tag{2}$$

where $[CO_2]$ is the concentration of CO_2 in moles/liter, K_H is the Henry's law constant of about 10–1.5 and is a weak function of temperature and ionic strength, and Pco_2 is the partial pressure of the gas in atmospheres. In the surface ocean, for example, total dissolved inorganic carbon (i.e., Tco_2) is approximately 2 mM, whereas $[CO_2]$ is only about 10 μM. This $[CO_2]$ is close to that of the atmosphere (corresponding at present to approximately 380 parts per million by volume) and resulting in approximately a 50-fold higher concentration of dissolved inorganic carbon in the ocean than of CO_2 in the atmosphere (table 1). Indeed, on time scales of thousands of years, the concentration of atmospheric CO_2 is determined by oceanic processes that control the dissolved CO_2 concentrations in surface waters.

Because of the partitioning into the three phases (equation 1), the inorganic carbon system in aquatic environments has very little chance to reach equilibrium. On the left side of equation 1, CO_2 in solution tends to equilibrate with the gas phase (i.e., the CO_2 in the overlying atmosphere under natural conditions), whereas on the right side of the equation, the CO_3^{2-} tends to equilibrate with the solid phase of $CaCO_3$ or $MgCa(CO_3)_2$. Furthermore, the equilibrium constants for the various inorganic carbon reactions are temperature and salinity dependent. The partitioning of CO_2 between aqueous solution and gas phase increasingly favors the gas phase as temperature or salinity increases.

Calcification leads to a loss of one Ca^{2+} for each atom of carbon precipitated. The loss of Ca^{2+} is compensated by the formation of H^+, which shifts the equilibrium of the inorganic carbon system, described in equation 1, to the left. Thus, calcification potentiates the formation of CO_2, leading to higher Pco_2 while simultaneously reducing the concentration of total dissolved inorganic carbon. It should be noted that although the biological formation of $CaCO_3$ requires metabolic energy, the energy is not stored in the chemical bonds of the product; that is, calcification is not a chemical reduction of CO_2. Rather, the energy is used to reduce the entropy in formation of the crystalline carbonate.

The "Fast" Biological Carbon Cycle

The second carbon cycle is dependent on the biologically catalyzed reduction of inorganic carbon to form organic matter, the overwhelming majority of which is oxidized back to inorganic carbon by respiratory metabolism. This cycle, which is observable on time scales of days to millennia, is driven by redox reactions that evolved over about 2 billion years, first in microbes and subsequently in multicellular organisms. A very small fraction of the reduced carbon escapes respiration and becomes incorporated into the lithosphere. In the process, some of the organic matter is transferred to the slow carbon cycle.

To form organic molecules, inorganic carbon (CO_2 and its hydrated equivalents) must be chemically reduced, a process that requires the addition of hydrogen atoms (not just protons but protons plus electrons) to the carbon atoms. Broadly speaking, the biologically catalyzed reduction reactions are carried out by two groups of organisms: chemoautotrophs and photoautotrophs, which are collectively called primary producers. The organic carbon they synthesize fuels the growth and respiratory demands of the primary producers themselves and all remaining organisms in the ecosystem.

All oxidation–reduction reactions are coupled sequences. Reduction is accomplished by the addition of an electron or hydrogen atom to an atom or molecule. In the process of donating an electron to an acceptor, the donor molecule is oxidized. Hence, oxidation–reduction reactions require pairs of substrates and can be described by a pair of partial reaction, or half-cells:

$$A_{ox} + n(e^-) \leftrightarrow A_{red}, \tag{3a}$$

$$B_{red} - n(e^-) \leftrightarrow B_{ox}. \tag{3b}$$

The tendency for a molecule to accept or release an electron is therefore relative to some other molecule being capable of conversely releasing or binding an electron. Chemists scale this tendency, called the redox potential, *E*, relative to the reaction:

$$H_2 \leftrightarrow 2H^+ + 2e^-, \tag{4}$$

which is arbitrarily assigned an *E* of 0 at pH 0 and is designated E_0. Biologists define the redox potential at pH 7, 298 K (i.e., room temperature), and 1 atmosphere pressure (= 101.3 kPa). When so defined, the redox potential is denoted by the symbols E'_0 or sometimes E_{m7}. The E'_0 for a standard hydrogen electrode is –420 mV.

2. CHEMOAUTOTROPHY

Organisms capable of reducing sufficient inorganic carbon to grow and reproduce in the dark without an external organic carbon source are called chemoautotrophs (literally, "chemical self-feeders"). Genetic analyses suggest that chemoautotrophy evolved very early in Earth's history and is carried out exclusively by procaryotic organisms in both the Archea and Bacteria superkingdoms.

Early in Earth's history, the biological reduction of inorganic carbon may have been directly coupled to the oxidation of H_2. At present, however, free H_2 is scarce on the planet's surface. Rather, most of the hydrogen on the surface of Earth is combined with other atoms, such as sulfur or oxygen. Activation energy is required to break these bonds in order to extract the hydrogen. One source of energy is chemical bond energy itself. For example, the ventilation of reduced mantle gases along tectonic plate subduction zones on the sea floor provides hydrogen in the form of H_2S. Several types of microbes can couple the oxidation of H_2S to the reduction of inorganic carbon, thereby forming organic matter in the absence of light.

Ultimately all chemoautotrophs depend on a nonequilibrium redox gradient, without which there is no thermodynamic driver for carbon fixation. For example, the reaction involving the oxidation of H_2S by microbes in deep sea vents described above is ultimately coupled to oxygen in the ocean interior. Hence, this reaction is dependent on the chemical redox gradient between the ventilating mantle plume and the ocean interior that thermodynamically favors oxidation of the plume gases. Maintaining such a gradient requires a supply of energy, either externally, from radiation (solar or otherwise), or internally, via planetary heat and tectonics, or both.

The overall contribution of chemoautotrophy in the contemporary ocean to the formation of organic matter is relatively small, accounting for less than 1% of the total annual primary production in the sea. However, this process is critical in coupling reduction of carbon to the oxidation of low-energy substrates and is essential for completion of several biogeochemical cycles.

The oxidation state of the ocean interior is a consequence of a second energy source: light, which drives photosynthesis. Photosynthesis is an oxidation–reduction reaction of the general form:

$$2H_2A + CO_2 + \text{light} \xrightarrow{\textit{Pigment}} (CH_2O) + H_2O + 2A, \tag{5}$$

where A is, for example, an S atom. In this formulation, light is specified as a substrate, and a fraction of the light energy is stored as chemical bond energy in the organic matter. Organisms capable of reducing inorganic carbon to organic matter by using light energy to derive the source of reductant or energy are called photoautotrophs. Analyses of genes and metabolic sequences strongly suggest that the machinery for capturing and utilizing light as a source of energy to extract reductants was built on the foundation of chemoautotrophic carbon fixation; i.e., the predecessors of photoautotrophs were chemoautotrophs. The evolution of a photosynthetic process in a chemoautotroph forces consideration of both the selective forces responsible (why) and the mechanism of evolution (how).

3. THE EVOLUTION OF PHOTOAUTOTROPHY

Reductants for chemoautotrophs are generally deep in the Earth's crust. Vent fluids are produced in magma chambers connected to the Athenosphere. As a result, the supply of vent fluids is virtually unlimited. Although the chemical disequilibria between vent fluids and bulk seawater provide a sufficient thermodynamic gradient to continuously support chemoautotrophic metabolism in the contemporary ocean, in the early Earth the oceans would not have had a sufficiently large thermodynamic energy potential to support a pandemic outbreak of chemoautotrophy. Moreover, magma chambers, volcanism, and vent fluid fluxes are tied to tectonic subduction and spreading regions, which are transient features of Earth's crust and hence only temporary habitats for chemoautotrophs. In the Archean and early Proterozoic oceans, the chemoautotrophs would have had to have been dispersed throughout the oceans by physical mixing in order to colonize new vent regions. This same dispersion process would have helped ancestral chemoautotrophs exploit solar energy near the ocean surface.

Although the processes that selected photosynthetic reactions as the major energy transduction pathway remain obscure, central hypotheses have emerged based on our understanding of the evolution of Earth's carbon cycle, the evolution of photosynthesis, biophysics, and molecular phylogeny. Photoautotrophs are found in all three major superkingdoms; however, there are very few known Archea capable of this form of metabolism. Efficient photosynthesis requires harvesting of solar radiation and hence the evolution of a light-harvesting system. Although some Archea and Bacteria use the pigment protein rhodopsin, by far the most efficient and ubiquitous light harvests are based on chlorins; no known Archea has a chlorin-based photosynthetic metabolic pathway. The metabolic pathway for the synthesis of porphyrins and chlorins is one of the oldest in biological evolution and is found in all chemoautotrophs. It has been proposed that the chlorin-based photosynthetic energy conversion apparatus originally arose from the need to prevent UV radiation from damaging essential macromolecules such as nucleic acids and proteins. The UV excitation energy could be transferred from the aromatic amino acid residues in the macromolecule to a blue absorption band of membrane-bound chlorins to produce a second excited state, which subsequently decays to the lower-energy excited singlet. This energy dissipation pathway can be harnessed to metabolism if the photochemically produced charge-separated primary products are prevented from undergoing a back reaction but rather form a biochemically stable intermediate reductant. This metabolic strategy was selected for the photosynthetic reduction of CO_2 to carbohydrates, using reductants such as S^{2-} or Fe^{2+}, which have redox potentials that are too positive to reduce CO_2 directly.

The synthesis of reduced (i.e., organic) carbon and the oxidized form of the electron donor permits a photoautotroph to use "respiratory" metabolic processes but operate them in reverse. However, not all of the reduced carbon and oxidants remain accessible to the photoautotrophs. In the oceans, cells tend to sink, carrying with them organic carbon. The oxidation of Fe^{2+} forms insoluble Fe^{3+} salts that precipitate. The sedimentation and subsequent burial of organic carbon and Fe^{3+} remove these components from the water column. Without replenishment, the essential reductants for anoxygenic photosynthesis would eventually become depleted in the surface waters. Thus, the necessity to regenerate reductants potentially prevented anoxygenic photoautotrophs from providing the major source of fixed carbon on Earth for eternity. Major net accumulation of reduced organic carbon in Proterozoic sediments implies local depletion of reductants such as S^{2-} and Fe^{2+} from the upper ocean. These limitations almost certainly provided the evolutionary selection pressure for an alternative electron donor.

4. SELECTIVE PRESSURE IN THE EVOLUTION OF OXYGENIC PHOTOSYNTHESIS

Water (H_2O) is a potentially useful biological reductant with an effectively unlimited supply on Earth. Water contains about 100 kmol/m^3 of H atoms, and, given $>10^{18}$ m^3 of water in the hydrosphere and cryosphere, more than 10^{20} kmol of reductant is potentially accessible. Use of H_2O as a reductant for CO_2, however, requires a larger energy input than does the use of Fe^{2+} or S^{2-}. Indeed, to split water by light requires 0.82 electron volts at pH 7 and 298 K. Utilizing light at such high energy levels required the evolution of a new photosynthetic pigment, chlorophyll *a*, which has a red (lowest singlet) absorption band that is 200 to 300 nm blue shifted relative to bacteriochlorophylls. Moreover, stabilization of the primary electron acceptor to prevent a back reaction necessitated thermodynamic inefficiency that ultimately led to the evolution of two light-driven reactions operating in series. This sequential action of two photochemical reactions is unique to oxygenic photoautotrophs and presumably involved horizontal gene transfer through one or more symbiotic events. As discussed below, oxygenic photosynthesis appears to have arisen only once in a single clade of Bacteria (the cyanobacteria).

In all oxygenic photoautotrophs, equation 3 can be modified to:

$$2H_2O + CO_2 + \text{light} \xrightarrow{Chl\,a} (CH_2O) + H_2O + O_2, \quad (6)$$

where Chl *a* is the pigment chlorophyll *a* exclusively utilized in the reaction. Equation 6 implies that somehow chlorophyll *a* catalyzes a reaction or a series of reactions whereby light energy is used to oxidize water:

$$2H_2O + \text{light} \xrightarrow{Chl\,a} 4H^+ + 4e + O_2, \quad (7)$$

yielding gaseous molecular oxygen. Hidden within equation 7 are a complex suite of biological innovations that have not been yet successfully mimicked by humans. At the core of the water-splitting complex is a quartet of Mn atoms that sequentially extract electrons, one at a time, from two H_2O molecules, releasing gaseous O_2 to the environment and storing the reductants on biochemical intermediates.

The photochemically produced reductants generated by the reactions schematically outlined in equation 7 are subsequently used in the fixation of CO_2 by a suite of enzymes that can operate in vitro in darkness, and, hence, the ensemble of these reactions is called the dark reactions. At pH 7 and 25°C, the formation of glucose from CO_2 requires an investment of 915 calories per mole. If water is the source of reductant, the overall efficiency for photosynthetic reduction of CO_2 to glucose is approximately 30%; i.e., 30% of the absorbed solar radiation is stored in the chemical bonds of glucose molecules.

5. PRIMARY PRODUCTION

When one subtracts the costs of all other metabolic processes by the chemoautotrophs and photoautotrophs, the organic carbon that remains is available for the growth and metabolic costs of heterotrophs. This remaining carbon is called *net* primary production (NPP). NPP provides an upper bound for all other metabolic demands in an ecosystem. If NPP is greater than all respiratory consumption of the ecosystem, the ecosystem is said to be net autotrophic. Conversely, if NPP is less than all respiratory consumption, the system must either import organic matter from outside its bounds, or it will slowly run down—it is net heterotrophic.

It should be noted that NPP and photosynthesis are not synonymous. On a planetary scale, the former includes chemoautrophy; the latter does not. Moreover, photosynthesis per se does not include the integrated respiratory term for the photoautotrophs themselves. In reality, that term is extremely difficult to measure directly; hence, NPP is generally approximated from measurements of photosynthetic rates integrated over some appropriate length of time (a day, a month, a season, or a year), and respiratory costs are either assumed or neglected. From satellite data used to estimate upper-ocean chlorophyll concentrations, satellite-based observations of incident solar radiation, atlases of seasonally averaged sea-surface temperature, and models that incorporate a temperature response function for photosynthesis, it is possible to estimate global net photosynthesis in the world oceans. Although estimates vary among models based on how the parameters are derived, for illustrative purposes we use a model based on empirical parameterization of the daily integrated photosynthesis profiles as a function of depth. The physical depth at which 1% of irradiance incident on the sea surface remains is called the euphotic zone. This depth can be calculated from surface chlorophyll concentrations and defines the base of the water column at which net photosynthesis can be supported. Given such information, net primary production can be calculated following the general equation:

$$PP_{eu} = C_{sat} \cdot Z_{eu} \cdot P^{b}_{opt} \cdot DL \cdot F, \qquad (8)$$

where PP_{eu} is daily net primary production integrated over the euphotic zone, C_{sat} is the satellite-based (upper water column, derived from table 2) chlorophyll concentration, P^{b}_{opt} is the maximum daily photosynthetic rate within the water column, Z_{eu} is the depth of the euphotic zone, DL is the photoperiod, and F is a function

Table 2. Comparison of marine and terrestrial net primary productivity across biomes

	Ocean NPP		Land NPP
Seasonal			
April to June	10.9		15.7
July to September	13.0		18.0
October to December	12.3		11.5
January to March	11.3		11.2
Biogeographic			
Oligotrophic	11.0	Tropical rainforests	17.8
Mesotrophic	27.4	Broadleaf deciduous forests	1.5
Eutrophic	9.1	Broadleaf and needleleaf forests	3.1
Macrophytes	1.0	Needleleaf evergreen forests	3.1
		Needleleaf deciduous forests	1.4
		Savannas	16.8
		Perennial grasslands	2.4
		Broadleaf shrubs with bare soil	1.0
		Tundra	0.8
		Desert	0.5
		Cultivation	8.0
Total	48.5		56.4

Source: From Field, C. B., M. J. Behrenfeld, J. T. Randerson, and P. Falkowski. 1998. Primary production of the biosphere: Integrating terrestrial and oceanic components. Science 281: 237–240.

Note: Units are in Pg (10^{15}g) per annum. See Field et al. (1998) for complete discussion of how these data were derived using satellite observations of the ocean on terrestrial ecosystems.

describing the shape of the photosynthesis depth profile. This general model can be both expanded (differentiated) and collapsed (integrated) with respect to time and irradiance; however, the global results are fundamentally similar. The models predict that NPP in the world's oceans amounts to 40–50 Pg per annum.

6. WHO ARE THE PHOTOAUTOTROPHS?

In the oceans, oxygenic photoautotrophs are a taxonomically diverse group of mostly single-celled photosynthetic organisms that drift with currents. In the contemporary ocean, these organisms, called phytoplankton (derived from the Greek, meaning to wander), are comprised of approximately 20,000 morphologically defined species distributed among at least eight taxonomic divisions or phyla. By comparison, higher plants are comprised of more than 250,000 species, almost all of which are contained within one class in one division. Thus, unlike terrestrial plants, phytoplankton appear to be represented by relatively few morphological species, but they are phylogenetically extremely diverse. This deep taxonomic diversity is reflected in their evolutionary history and ecological function.

Within this diverse group of organisms, three basic evolutionary lineages are discernible. The first contains all prokaryotic oxygenic phytoplankton, which belong to one class of bacteria, namely the cyanobacteria. Cyanobacteria are the only known oxygenic photoautotrophs that existed before about 2.4 billion years ago. These prokaryotes numerically dominate the photoautotrophic community in contemporary marine ecosystems, and their continued success bespeaks an extraordinary adaptive capacity. At any moment in time, there are approximately 10^{24} cyanobacterial cells in the contemporary oceans. To put that in perspective, the number of cyanobacterial cells in the oceans is two orders of magnitude more than all the stars in the sky. The other two groups are eukaryotic. One, broadly speaking, contains chlorophylls *a* and *b* and is called the "green" line. These organisms, which are the progenitors of terrestrial plants, are not as abundant as a third group, which contains chlorophylls *a* and *c* and is often called the "red" line. The red line includes diatoms, coccolithophorids, and most dinoflagellates. All three groups are extremely important players in NPP and carbon burial in the contemporary ocean.

7. CARBON BURIAL

On geological time scales, there is one important fate for NPP, namely burial in the sediments. By far the largest reservoir of organic matter on Earth is locked up in rocks. Virtually all of this organic carbon is the result of the burial of exported marine organic matter in coastal sediments over literally billions of years of Earth's history. On geological time scales, the burial of marine NPP effectively removes carbon from biological cycles and places most (not all) of that carbon into the slow carbon cycle. A small fraction of the organic matter escapes tectonic processing via the Wilson cycle and is permanently buried, mostly in continental rocks. The burial of organic carbon effectively removes reducing equivalents from the atmosphere and ocean and thereby allows oxygen to accumulate in Earth's atmosphere.

Carbon burial is not inferred from direct measurement but rather from indirect means. One of the most common proxies used to derive burial on geological time scales is based on isotopic fractionation of carbonates. The rationale for this analysis is that the primary enzyme responsible for inorganic carbon fixation is ribulose 1,5-bisphosphate carboxylase/oxygenase (RuBisCO), which catalyzes the reaction between ribulose 1,5-bisphosphate and CO_2 (not HCO_3^-), to form two molecules of 3-phosphoglycerate. The enzyme strongly discriminates against ^{13}C, such that the resulting isotopic fractionation amounts to approximately 27 parts per thousand relative to the source carbon isotopic value. The extent of the actual fractionation is somewhat variable and is a function of carbon availability and of the transport processes for inorganic carbon into the cells as well as the specific carboxylation pathway. However, regardless of the quantitative aspects, the net effect of carbon fixation is an enrichment of the inorganic carbon pool in ^{13}C, whereas the organic carbon produced is enriched in ^{12}C.

8. CARBON ISOTOPE FRACTIONATION IN ORGANIC MATTER AND CARBONATES

The isotopic fractionation in carbonates mirrors the relative amount of organic carbon buried. It is generally assumed that the source carbon, from volcanism (so-called mantle carbon) has an isotopic value of about −5 parts per thousand. Because mass balance must constrain the isotopic signatures of carbonate carbon and organic carbon with the mantle carbon, then, in the steady state, the fraction of buried organic matter of the total carbon buried (f_{org}) can be calculated from the relationship,

$$f_{org} = (\delta_w - \delta_{carb})/\Delta_B, \quad (9)$$

where f_{org} is the fraction of organic carbon buried, δ_w is the average isotopic content of the carbon weathered,

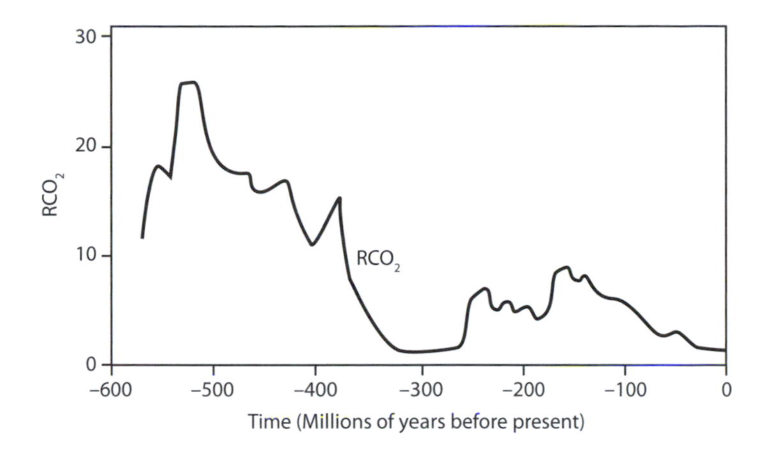

Figure 1. The change in the CO_2 concentrations over the past 550 million years.

δ_{car} is the isotopic signature of the carbonate carbon, and Δ_B is the isotopic difference between organic carbon and carbonate carbon deposited in the ocean. Equation 9 is a steady-state model that presumes the source of carbon from the mantle is constant over geological time. This basic model is the basis of nearly all estimates of organic carbon burial rates.

Carbonate isotopic analyses reveal positive excursions (i.e., implying organic carbon burial) in the Proterozoic and more modest excursions throughout the Phanaerozoic (figure 1). Burial of organic carbon on geological time scales implies that export production must deviate from the steady state on ecological time scales. Such a deviation requires changing one or more of (1) ocean nutrient inventories, (2) the utilization of unused nutrients in enriched areas, (3) the average elemental composition of the organic material, or (4) the "rain" ratios of particulate organic carbon to particulate inorganic carbon to the sea floor.

Over the past 200 million years, the carbon isotopic record indicates that a significant amount of organic carbon has been buried in the lithosphere. The burial of organic carbon denotes the burial of reductants. As a consequence, oxidized molecules must have accumulated in some other domain. The oxidized molecule is O_2. Hence, to a geochemist, the burial of organic matter formed by oxygenic photosynthetic organisms requires the oxidation of the atmosphere. Quantitative analysis of isotopic record of carbonates suggests that oxygen rose from about 11% 200 million years ago to the contemporary value of 21% as a result of the burial of the organic matter, largely in marine sediments. The removal of a small fraction of the buried carbon by humans to fuel the current industrialization of the world represents a reversal of this process, namely the consumption of oxygen and the reoxidation of organic matter by machines rather than by biological respiration.

9. CONCLUDING REMARKS

The evolution of primary producers in the oceans profoundly changed the chemistry of the atmosphere, ocean, and lithosphere of Earth. Primarily through the utilization of solar radiation, the biological (fast) carbon cycle allows for a disequilibrium in geochemical processes, such that Earth maintains an oxidized atmosphere and ocean. This disequilibrium prevents atmospheric oxygen from being depleted, maintains a reduced atmospheric CO_2 concentration, and simultaneously imprints the ocean interior and the lithosphere elemental compositions that reflects that of the bulk biological material from which it is derived. Although primary producers in the ocean comprise only about 1% of Earth's biomass, their metabolic rate and biogeochemical impact rival those of the much larger terrestrial ecosystem. On geological time scales, these organisms are the little engines that are essential to maintaining life as we know it on this planet.

Over the past 200 years, the fossil remains of marine photosynthetic organisms have been extracted from the lithosphere by humans at a rate approximately 1 million times faster than they accumulated. The extraction and subsequent combustion of fossil fuels have temporarily inverted the carbon cycle; the oceans are not in equilibrium with the atmosphere, and the excess CO_2 potentially will alter Earth's climate rapidly and dramatically. The rise of CO_2 is not debatable; it is a scientific fact. Unfortunately, however, the fundamental scientific facts pertaining to the carbon cycle on Earth are still debated, obscuring a sustainable path forward. Were he alive today, Charles Lyell might think we had not yet left the philosophy encouraged by universities in the Middle Ages.

FURTHER READING

Berner, R. A. 2004. The Phanerozoic Carbon Cycle: CO_2 and O_2. Oxford: Oxford University Press.

Blankenship, R. E., and H. Hartman. 1998. The origin and evolution of oxygenic photosynthesis. Trends in Biochemical Sciences 23: 94–97.

Falkowski, P. G., E. A. Laws, R. T. Barber, and J. W. Murray. 2003. Phytoplankton and their role in primary, new, and export production. In M.J.R. Fasham, ed., Ocean Biogeochemistry: A JGOFS Synthesis. Berlin: Springer-Verlag, 99–121.

Falkowski, P. G., and J. A. Raven. 2007. Aquatic Photosynthesis. Princeton, NJ: Princeton University Press.

Falkowski, P., R. J. Scholes, E. Boyle, J. Canadell, D. Canfield, J. Elser, N. Gruber, K. Hibbard, P. Hogberg, S. Linder, F. T. Mackenzie, B. Moore, T. Pedersen, Y. Rosenthal, S. Seitzinger, V. Smetacek, and W. Steffen. 2000. The global carbon cycle: A test of our knowledge of earth as a system. Science 290: 291–296.

Hedges, J. I., and R. G. Keil. 1995. Sedimentary organic matter preservation: An assessment and speculative synthesis. Marine Chemistry 49: 81–115.

III.14

Biodiversity and Ecosystem Functioning

Andrew Hector and Andy Wilby

OUTLINE

1. Background and history
2. Biodiversity and ecosystem functioning relationships
3. Mechanisms
4. Multitrophic systems
5. Diversity and stability
6. Ecosystem multifunctionality
7. Ecosystem service provision
8. The next phase of research

Forecasts of ongoing biodiversity loss prompted ecologists in the early 1990s to question whether this loss of species could have a negative impact on the functioning of ecosystems. *Ecosystem functioning* is an umbrella term for the processes operating in an ecosystem, that is, the biogeochemical flows of energy and matter within and between ecosystems (e.g., primary production and nutrient cycling). The first general phase of research on this topic addressed this question by assembling model communities of varying diversity to measure the effects on ecosystem processes. The results of the meta-analyses of this first wave of studies show that biodiversity generally has a positive but saturating effect on ecosystem processes that is remarkably consistent across trophic groups and ecosystem types. These relationships are driven by a combination of complementarity and selection effects with complementarity effects nearly twice as strong as selection effects overall. However, diverse communities rarely function significantly better than the best single species, at least in the short term. In the longer term, biodiversity can provide an insurance value similar to the risk-spreading benefits of diverse portfolios of financial investments. The effects of biodiversity on ecosystem functioning may have been underestimated by the first phase of research because of the short duration of many studies and the focus on single-ecosystem processes in isolation rather than a consideration of all important ecosystem functions simultaneously. The next phase of research will focus, in part, on whether the benefits of biodiversity seen in experiments translate to real-world settings.

GLOSSARY

biodiversity. A contraction of biological diversity that encompasses all biological variation from the level of genes through populations, species, and functional groups (and sometimes higher levels such as landscape units)

complementarity effect. The influence that combinations of species have on ecosystem functioning as a consequence of their interactions (e.g., resource partitioning, facilitation, reduced natural enemy impacts in diverse communities)

ecosystem functioning. An umbrella term for the processes operating in an ecosystem

ecosystem processes. The biogeochemical flows of energy and matter within and between ecosystems, e.g., primary production and nutrient cycling

ecosystem service. An ecosystem process or property that is beneficial for human beings, e.g., the provision of foods and materials or sequestration of carbon dioxide

selection effects. The influence that species have on ecosystem functioning simply through their species-specific traits and their relative abundance in a community (positive selection effects occur when species with higher-than-average monoculture performance dominate communities)

1. BACKGROUND AND HISTORY

Darwin, in *On the Origin of Species*, initially proposed that changes in biodiversity could affect ecosystem functioning if niche space is more fully occupied in more diverse communities than depauperate ones. We use *ecosystem functioning* as an umbrella term to embrace all the biogeochemical processes that operate within ecosystems, primary production for example. This early work was apparently forgotten until the early 1990s, but the same reasoning was around in the mid-twentieth century, when it was proposed that more diverse mixtures of fish species should lead to greater productivity: "Presumably fish production will increase as the number of niches increases . . . [and] probably the proportion of occupied niches increases as the number of species of fishes increases." Indeed, both of these early studies even presented data in support of this relationship (figure 1).

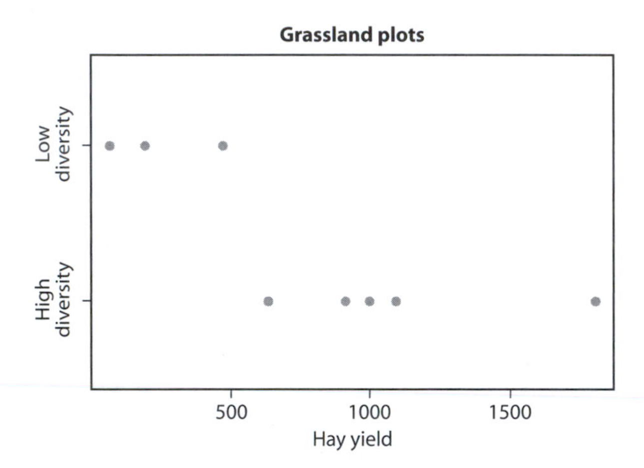

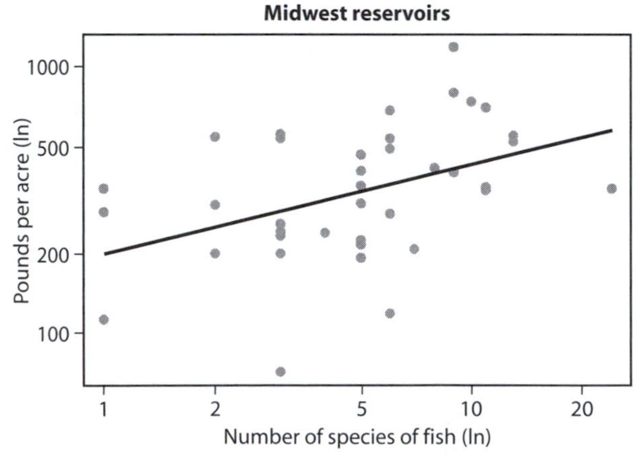

Figure 1. Early evidence for a link between biodiversity and ecosystem functioning from (top panel) an early-nineteenth-century large-scale experimental garden at Woburn Abbey, UK, mentioned by Darwin in *On the Origin of Species* (after Hector and Hooper, 2002), and (lower panel) relation between standing crops and numbers of species of fish present in Midwestern reservoirs (after Carlander, 1955).

General concern about the impact of anthropogenic biodiversity was voiced at the Rio Earth Summit in 1992, at which time the Convention on Biological Diversity (CBD) was launched with the signatures of 150 heads of government. This international treaty designed to promote sustainable development and the protection of biodiversity was evidence of political acceptance that anthropogenic biodiversity loss may have serious detrimental effects on humankind. Concerns highlighted at Rio and in the Convention also led to renewed scientific interest and a concerted effort by ecologists to understand the effects of changes in biodiversity on ecosystem functioning and the likely significance of such changes for humankind. More than a decade's worth of research has now been published, accompanied by a debate that focused in large part on the mechanisms underlying the relationship between biodiversity and functioning. Synthesis of the first decade of results through meta-analysis is helping to reveal both pattern and mechanism.

2. BIODIVERSITY AND ECOSYSTEM FUNCTIONING RELATIONSHIPS

The main approach that has been used to investigate the relationship between biodiversity and ecosystem functioning is the direct manipulation of biodiversity by the assembly of synthesized model communities in the laboratory or field. An alternative approach is to remove species from natural communities. A third, nonmanipulative approach is to infer the relationship between biodiversity and ecosystem functioning by seeing how the two are correlated across habitats. All three approaches have strengths and weaknesses. In this chapter, we focus on the assembly of model communities of varying diversity.

Meta-analysis of the first decade of research clearly shows a positive relationship between biodiversity and ecosystem functioning (e.g., figure 2), a pattern that is remarkably consistent across trophic groups (producers, herbivores, detrivores, and predators) and present in both terrestrial and marine ecosystems. However, the relationship between biodiversity and ecosystem functioning is generally saturating, suggesting that the effect of random biodiversity loss on ecosystem functioning will be initially weak but will accelerate.

The first phase of research on biodiversity and ecosystem functioning was focused on identifying general patterns (whether biodiversity change can affect ecosystem functioning or not), and species were therefore removed at random to generate experimental diversity gradients. Another key result of these studies reveals that there is considerable variation among species or

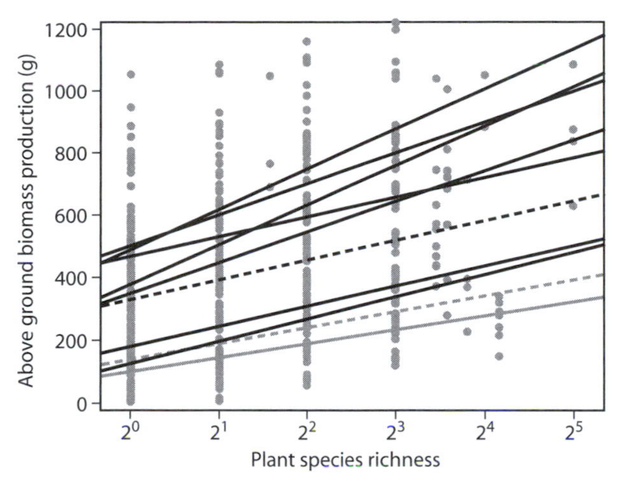

Figure 2. Effects of plant species richness on the production of aboveground plant biomass for 11 plant biodiversity experiments. Lines are linear regression slopes as a function of the number of plant species ($\log_2$ scale) for the eight BIODEPTH project experiments (solid black lines; see Hector et al., 1999), the Cedar Creek biodiversity experiment (solid gray line; see Tilman et al., 2001), the Jena biodiversity experiment (dashed black line; Roscher et al., 2005), and Van Ruijven and Berendse (2005; dashed gray line).

species assemblages in their impact on functioning. This suggests that the actual effect of biodiversity loss on ecosystem functioning seen in real-world situations will depend strongly on which species are lost. Moving from random to more realistic real-world situations is a key goal for the next phase of research.

3. MECHANISMS

The early studies mentioned in the introductory background section identify only one way in which biodiversity changes can affect ecosystem functioning, namely by affecting the degree of species complementarity (basically by affecting the number of underutilized or vacant niches). That is, more diverse communities utilize a greater proportion of available niche space. However, as mentioned above, biodiversity changes can also affect ecosystem functioning by the simple presence or loss of particular species with strong intrinsic effects on ecosystem processes (so-called sampling or selection effects); more diverse communities are more likely to contain those species or assemblages that strongly affect functioning. There has been widespread debate over the last decade about whether the positive relationships reviewed above were explained by complementarity or selection effects.

Additive partitioning methods are one approach that allows separation of the overall net effect of biodiversity on ecosystem functioning into complementarity effects that arise from species interactions and selection effects that are species specific. Meta-analysis reveals that almost all studies are driven by a combination of these effects but that overall complementarity effects were nearly twice as strong as selection effects (figure 2). However, even though complementarity effects have a greater impact than selection effects, they are not strong enough to cause mixtures to do significantly better than monocultures in most cases (figure 3). In summary, although the relationship between biodiversity and ecosystem functioning is positive, and complementarity effects contribute approximately twice as much as selection effects in generating these relationships, diverse communities do not generally perform better than the best individual species. However, this result is influenced by the short duration of many of the experiments performed to date because the relationship between biodiversity and ecosystem functioning grows stronger over time (figure 3) as a result of increasing complementarity. Nevertheless, it appears that diverse communities are rarely able to do substantially better than a monoculture of the best-performing species that they contain.

This appears to be in part because communities are often not dominated by the most productive species but by species with a lower performance. In fact, in over 40% of the reviewed studies, communities were dominated by a species with a lower-than-average monoculture biomass, leading to a negative selection effect with a negative influence on the performance of the ecosystem as a whole. An important implication of this meta-analysis for future research is that studies must be longer term if they are to reveal the full effects of biodiversity on ecosystem functioning; experiments to date have, if anything, underestimated the effects of random loss of species on ecosystem functioning.

4. MULTITROPHIC SYSTEMS

Alongside the work on biodiversity and ecosystem functioning, there has also been significant interest in the functional importance of biodiversity in the context of multitrophic interactions. Here the focus has been more on the impact of diversity at one trophic level on the population density at the trophic level below. Most commonly this has involved studies of predator species diversity and impact on prey populations. Recently, attempts have been made to link the considerable bodies of work on biodiver-

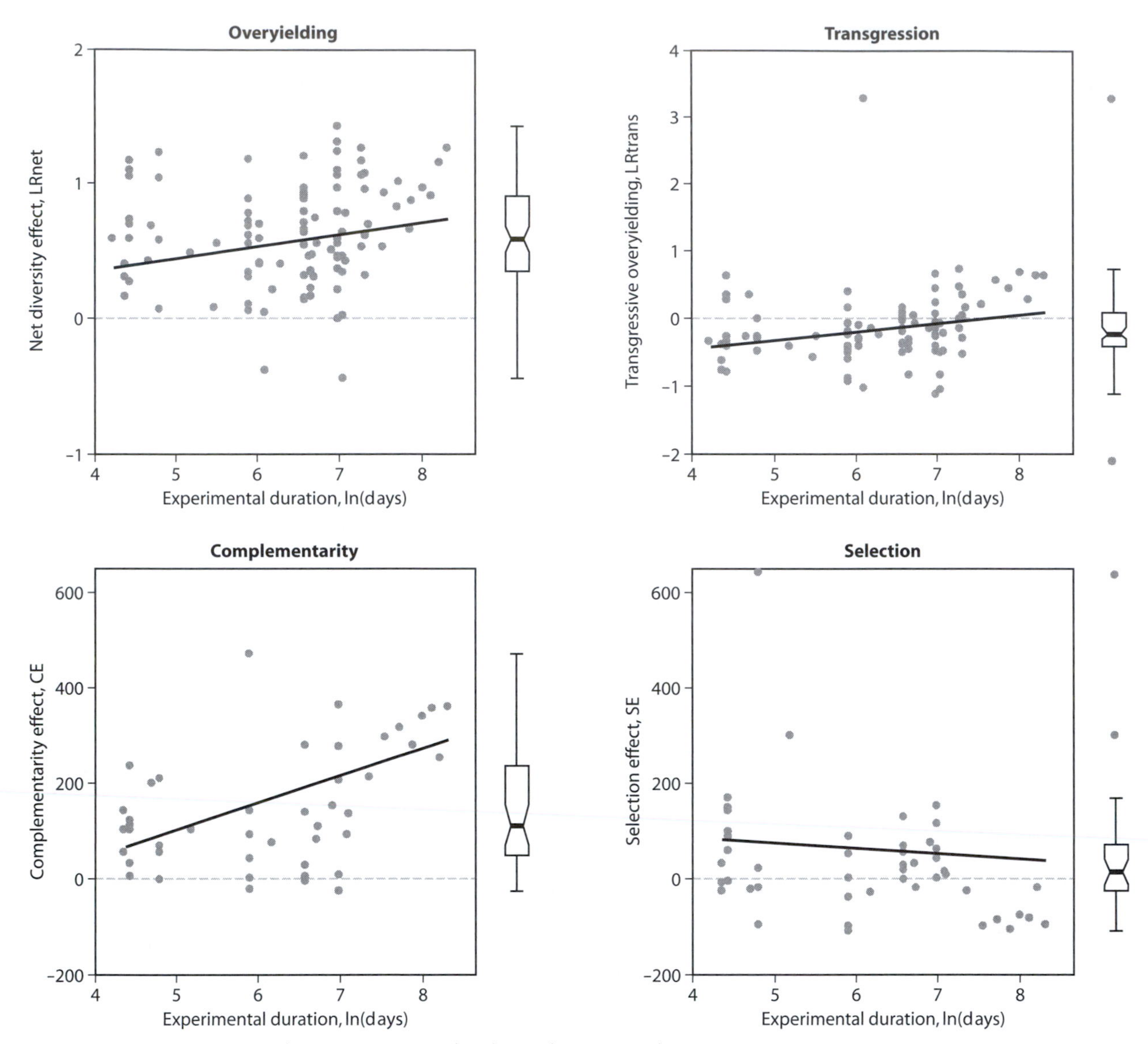

Figure 3. Meta-analysis results (Cardinale et al., 2007) for (top left) overyielding, (top right) transgressive overyielding, (lower left) complementarity effects, and (lower right) selection effects. The data for the upper and lower panels are for different (but overlapping) sets of studies. Note the different *y*-axis scales in the two upper panels. Tukey box and whisker plots to the right of each scatter plot summarize the variation averaged over time (notches provide an approximate 95% interval for the medians).

sity and ecosystem functioning and predator–prey interactions.

One striking difference between the predator–prey and biodiversity–ecosystem functioning perspectives is the relative importance ascribed to interspecific interactions among target species. Interactions among species are not explicitly considered in biodiversity–ecosystem functioning studies, whereas predator–prey theory has a long history of investigating direct and indirect interactions among predator species and how these affect the population size of the prey species. For example, intraguild predation where one predatory species preys on another is a common interaction in nature and has the capacity to reduce the joint impact of the predator species on the original prey species. The opposite outcome can occur when facilitative interactions occur among predator species. One commonly reported example of this is when the avoidance behavior of the prey to one predator makes them more susceptible to predation by a second. Aphids, for example, commonly drop from the plant when approached by a foliar predator, but this can leave them susceptible to ground-foraging predators so that the functioning of ground and foliar predators together is greater than the sum of the functioning of each alone.

Facilitative and negative interactions among constituent species can occur in basal trophic levels, such as primary producers or detritivores. Plants are known to take part in allelopathic interactions in which they impact each other negatively via the production of toxic chemicals. There is also strong evidence that plant species may facilitate each other by enriching the soil by nitrogen fixation or by moderating harsh abiotic environments, for example. A key question is whether the predictive power of biodiversity–ecosystem functioning theory would be improved by the incorporation of such species interactions. Generally, meta-analyses of biodiversity–ecosystem functioning reveal consistent positive effects of diversity on functioning, but results from terrestrial predator–prey systems are more equivocal, with almost half of the studies reporting negative or neutral effects of increasing species diversity on prey suppression. Where significant species interactions occur, it may be useful to think of observed relationships between biodiversity and ecosystem functioning as the net effect of co-occurring positive mechanisms (resource use differentiation and facilitation) and negative mechanisms (intraguild predation, interference). Experimental evidence from predator–prey systems suggests that, at least in some cases, negative interactions among species outweigh the positive mechanisms and cause reduced functionality in more diverse communities.

5. DIVERSITY AND STABILITY

Ecological stability commonly refers to one of three general properties of ecosystems: the temporal variation in a property of the ecosystem (e.g., primary production) or the response (resistance) or recovery (resilience) of these properties following perturbation. One possible value of biodiversity to humans is its potential to increase stability by buffering ecosystem processes such as production against environmental variation and making them more resistant and resilient to perturbations. This insurance value of biodiversity has most often been considered in the context of fluctuations over time, where it has been likened to the risk-spreading benefits of diverse portfolios of investments in financial markets but could also apply to spatial environmental variation. For this insurance effect to occur requires only that fluctuations in the abundances of a guild of species not be perfectly synchronized because under perfect synchrony, an entire guild or trophic level would effectively behave as one species. When species responses are not perfectly positively correlated, changes in some species can be compensated by others, and the averaging of their asynchronous fluctuations smoothes the collective productivity of the whole community (figure 4).

One potentially confusing or counterintuitive aspect of the insurance hypothesis is that diversity has a stabilizing effect on aggregate community or ecosystem properties (such as primary productivity) at the same time as the fluctuations of the constituent species may be destabilized as a result of interactions with greater numbers of species (although destabilization is not inevitable). The key consequence to understand is that it is the lack of perfect synchrony of individual species fluctuations that leads to the stabilizing effect of diversity on ecosystem processes. This asynchrony through independent or compensatory species responses can be interpreted as a form of temporal niche differentiation among species.

A recent review of the diversity–stability literature emphasizes its breadth and complexity because of the many different types of stability and the range of different variables that stability measures can be calculated for (e.g., stability of population abundance versus total community biomass as introduced above; figure 5). For experiments where diversity was directly manipulated, there are reports of two positive effects of plant species diversity on the stability of biomass production and three positive effects of microbial diversity on the stability of biomass or carbon dioxide production. There are no reports of negative or neutral effects of diversity on temporal stability of ecosystem processes from grassland experiments but one negative effect of increased multitrophic diversity on the temporal stability of biomass production in seagrass beds and one neutral and one negative effect of microbial diversity on the stability of microbial biomass production. Observational studies have also looked at stabilizing effects of biodiversity on ecosystem processes producing five positive effects of plant diversity on temporal stability and one neutral effect. In summary, evidence from both natural and experimental systems of plants and microbes suggests that insurance effects of biodiversity on temporal stability may be relatively widespread.

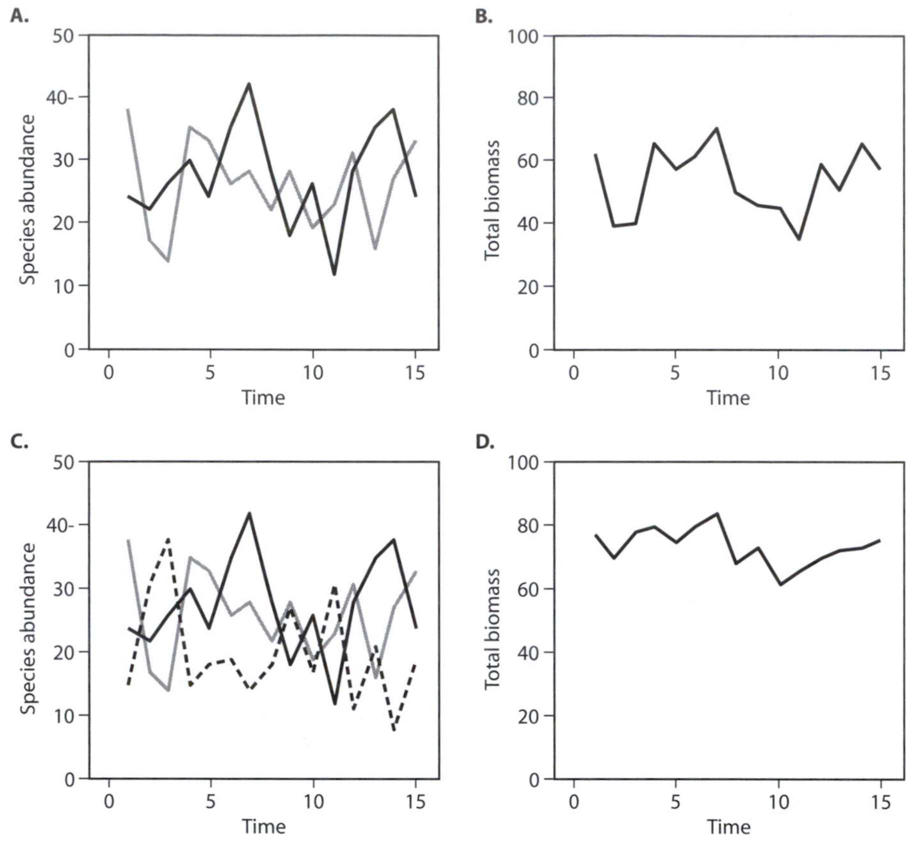

Figure 4. Asynchronous population fluctuations buffer total community biomass. In this hypothetical example, the total biomass of the three-species community (D) is less variable than that of the two-species community (B) because of the asynchrony of individual species biomasses (A and C).

6. ECOSYSTEM MULTIFUNCTIONALITY

As summarized above, meta-analysis of the results of the first generation of experimental research on biodiversity and ecosystem functioning has revealed that individual ecosystem processes generally show a positive but saturating relationship with increasing diversity. The saturating relationship suggests that some species are redundant with respect to a single function. However, nearly all studies to date have been short term and address the effect of biodiversity on ecosystem functioning only at a given point in time and under a relatively narrow set of conditions. Much of the other work reviewed above suggests that biodiversity can sometimes have an insurance value by buffering ecosystem-level processes in a way analogous to that in which diverse investment portfolios spread financial risk and improve average performance in the longer term. Nevertheless, all of the research to date considers ecosystem processes ex-

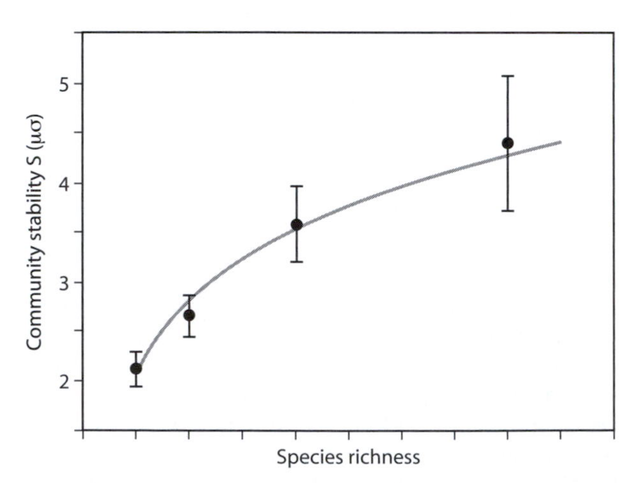

Figure 5. Diversity increases stability (S) of ecosystem primary production (μ = mean and σ = standard deviation of biomass through time). (Reproduced from van Ruijven and Berendse, 2007)

amined individually despite the fact that most ecosystems are managed or valued for several ecosystem services or processes: so-called ecosystem multifunctionality. If it is the case that a single species, or group of species, controls ecosystem functioning, then the remaining species are functionally redundant. Although it seems unlikely that a single species could control all ecosystem processes, it is possible that a single group of species may. However, if there is appreciable lack of overlap in the groups of species that influence different ecosystem processes, then higher levels of biodiversity will be required to maintain overall ecosystem functioning than indicated by analyses focusing on individual ecosystem processes in isolation. Only one study of ecosystem multifunctionality exists to date, but this analysis of seven ecosystem processes measured in a network of grassland biodiversity experiments supports the ecosystem multifunctionality hypothesis: the greater the number of ecosystem processes included in the analysis, the greater the number of species found to affect overall functioning (figure 6).

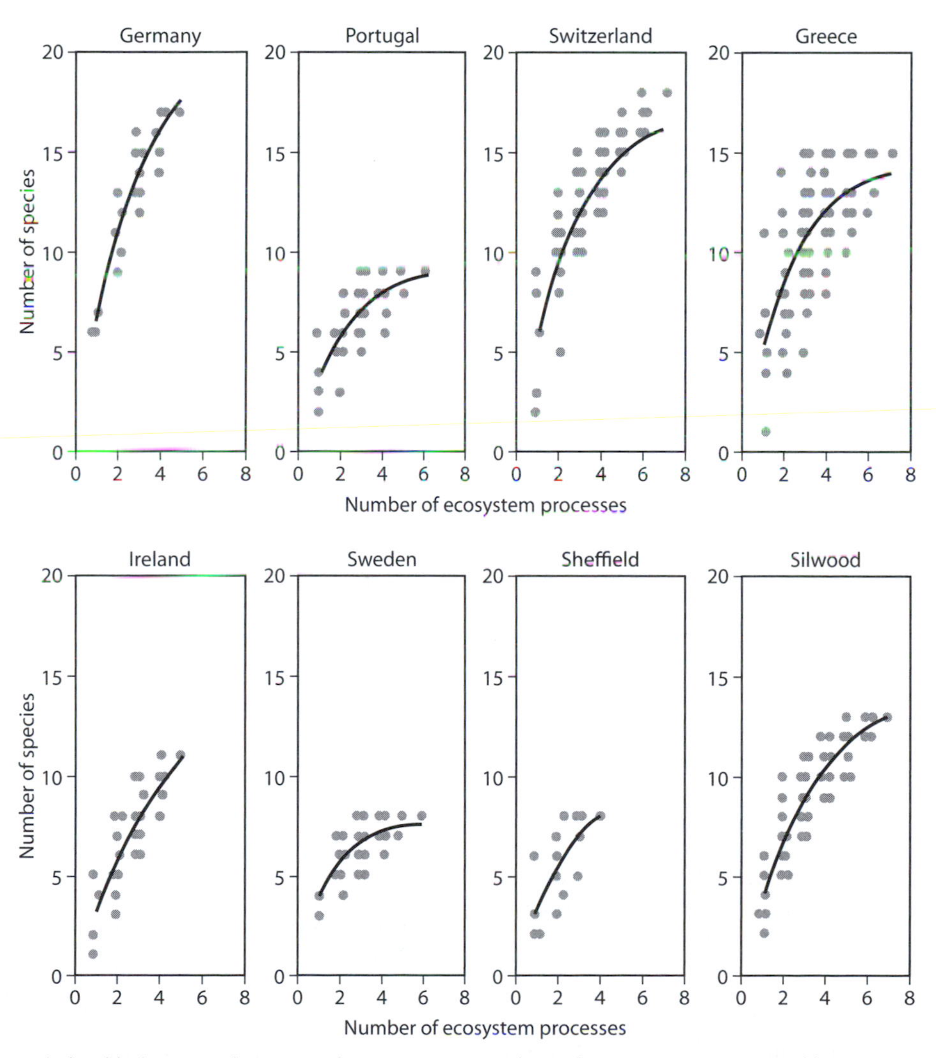

Figure 6. Positive relationship between the range of ecosystem processes considered and the number of species that affect one or more aspects of ecosystem functioning. The points (jittered for clarity) show numbers of species required for all possible combinations of ecosystem processes. Lines are average predictions based on the mean number of species required for a single process and the average overlap in the sets of species required for each pair of processes. (After Hector and Bagchi, 2007)

7. ECOSYSTEM SERVICE PROVISION

The Millennium Ecosystem Assessment defines ecosystem services as the benefits provided by ecosystems to humans. Ecosystem services include the provision of materials (food, genetic resources, water, etc.), cultural and psychospiritual well-being, supporting services (nutrient cycling, soil formation, etc.), and regulating services (pest and disease control, pollination, erosion control, climate regulation, etc.). The ecosystem processes that we have covered in this chapter are closely aligned with both supporting and regulating services. The evidence from the meta-analyses discussed above suggests that, in general, we expect the provision of such services to be compromised because of anthropogenic declines in biodiversity. Direct evidence of impacts of biodiversity loss on ecosystem functioning is accumulating. For example, increased diversity of wild host species has been shown to lead to dilution effects that reduce the probability of human infection by zoonotic diseases. Loss of biodiversity is also implicated in causing reduced carbon sequestration and therefore a net release of carbon into the atmosphere, where it contributes to global climate change. However, just as in experimental studies of biodiversity and ecosystem functioning, effects will depend strongly on which species are lost. The provision of ecosystem services has been of particular concern in agricultural systems both because of their spatial extent (they are estimated to cover a quarter of the terrestrial Earth surface, rising to almost three-quarters in some developed regions) and the severe losses of biodiversity they endure. Intensification of production in many parts of the globe has resulted in extreme declines in biodiversity in agricultural systems, in terms of both homogenization of production systems (simplified landscapes and fewer breeds/varieties grown) and declines in the wild species inhabiting agricultural ecosystems. Such simplification requires that compromised services such as pest regulation and maintenance of soil fertility and condition be replaced by synthetic pesticides and fertilizers, which are inherently unsustainable because of their reliance on externally derived energy and materials and their negative impacts on nontarget taxa, including humans. Enhancement and utilization of ecosystem services in agriculture are seen as one route to increased sustainability of food production.

8. THE NEXT PHASE OF RESEARCH

The first phase of research on biodiversity and ecosystem functioning primarily used experimental communities to investigate the effects of random species loss. Recent meta-analysis suggests that there generally are effects of species loss on ecosystem functioning in these experiments and that these effects are generally positive but saturating. Both complementarity and selection effects play a role in generating these relationships, with the effects of complementarity being nearly twice as strong as those of selection. Nevertheless, diverse mixtures rarely perform better than the best-performing species, at least in the short term (complementarity effects grow stronger over time in these studies).

A key goal for the next phase of research is a move away from artificial experimental systems toward more realistic settings to see if the biodiversity effects seen in the experiments translate to real-world situations. This will also necessitate a move away from the random loss of species used in the first phase of research toward more realistic scenarios of species loss and the incorporation of multiple trophic levels. The move from experimental to real-world settings will also require a move to larger field-scale study systems and, as suggested by the recent meta-analyses, to longer-term research. The first phase of research reviewed here has demonstrated the potential for biodiversity to have positive effects on ecosystem functioning. The question now is whether these experimental results will translate into positive effects of biodiversity on the provision of ecosystem services in the real world. The value of ecosystem services to humans is enormous (see part VI), and it is now critical to ascertain what role biodiversity plays in the provision of these services to human societies.

FURTHER READING

Balvanera, P., A. B. Pfisterer, N. Buchmann, J.-S. He, T. Nakashizuka, D. Raffaelli, and B. Schmid. 2006. Quantifying the evidence for biodiversity effects on ecosystem functioning and services. Ecology Letters 9: 1146–1156.

Cardinale, B. J., D. S. Srivastava, J. Emmett Duffy, J. P. Wright, A. L. Downing, M. Sankaran, and C. Jouseau. 2006. Effects of biodiversity on the functioning of trophic groups and ecosystems. Nature 443: 989–992.

Cardinale, B. J., J. P. Wright, M. W. Cadotte, I. T. Carroll, A. Hector, D. S. Srivastava, M. Loreau, and J. J. Weis. 2007. Impacts of plant diversity on biomass production increase through time due to species complementarity: A meta-analysis of 44 experiments. Proceedings of the National Academy of the U.S.A. 104: 18123–18128.

Hector, A., and R. Bagchi. 2007. Biodiversity and ecosystem multifunctionality. Nature 448: 188–190.

Ives, A. R., and S. R. Carpenter. 2007. Stability and diversity of ecosystems. Science 317: 58–62.

Kinzig, A., D. Tilman, and S. Pacala, eds. 2002. The Functional Consequences of Biodiversity: Empirical Progress and Theoretical Extensions. Princeton, NJ: Princeton University Press.

Loreau, M., and A. Hector. 2001. Partitioning selection and complementarity in biodiversity experiments. Nature 413: 548.

Loreau, M., S. Naeem, and P. Inchausti, eds. 2002. Biodiversity and Ecosystem Functioning: Synthesis and Perspectives. Oxford: Oxford University Press.

Loreau, M., S. Naeem, P. Inchausti, J. Bengtsson, J. P. Grime, A. Hector, D. U. Hooper, M. A. Huston, D. Raffaelli, B. Schmid, D. Tilman, and D. A. Wardle. 2001. Biodiversity and ecosystem functioning: Current knowledge and future challenges. Science 294: 804–809.

Worm, B., E. B. Barbier, Nicola Beaumont, J. E. Duffy, C. Folke, B. S. Halpern, J.B.C. Jackson, H. K. Lotze, F. Micheli, S. R. Palumbi, E. Sala, K. A. Selkoe, J. J. Stachowicz, and R. Watson. 2006. Impacts of biodiversity loss on ocean ecosystem services. Science 314: 787–790.

III.15

Ecological Stoichiometry

R. W. Sterner and J. J. Elser

OUTLINE

1. What is ecological stoichiometry?
2. Major patterns of nutrient content in organisms
3. Influence of stoichiometry on animal growth and community structure
4. Nutrient cycling in ecosystems
5. Influence of stoichiometry on species dynamics
6. Whole-lake food web experiments
7. Light:nutrient ratios and the ecology of Australia

Ecological stoichiometry examines how the nutrient content of organisms shapes their ecology. Although the chemistry of living things is constrained by their need to have a certain representation of major biomolecules such as DNA, RNA, proteins, lipids, etc., there is enough flexibility in these allocations that different species have nonidentical chemical contents. Thus, community structure is related to the portioning of elements in ecosystems. Stoichiometric considerations play a role in the rate of growth of animals, in the rates of recycling of elements by food webs, in the rate of mineralization of nutrients from organic matter, and in many other ecological phenomena. Stoichiometric models often have complex dynamics not seen in models lacking explicit treatment of stoichiometry, which suggests that stoichiometry is an important force shaping ecological dynamics.

GLOSSARY

autotroph. An organism that converts inorganic carbon to organic carbon and thus does not need to ingest or absorb other living things. Green plants (including certain algae and cyanobacteria) are photoautotrophs because they use light energy to make this conversion.

ecological stoichiometry. The balance of multiple chemical substances in ecological interactions and processes, or the study of this balance.

geophagy. The eating of dirt. This behavior may be used to balance mineral intake for animals living in low-food-quality environments.

growth rate hypothesis. Differences in organismal C:N:P ratios are caused by differential allocations to RNA necessary to meet the protein synthesis demands of rapid biomass growth and development.

heterotroph. An organism that relies on organic carbon for energy. Heterotrophs include herbivores, carnivores, and detritivores as well as omnivores that may feed on more than one trophic level.

homeostasis. Maintenance of constant internal conditions in the face of externally imposed variation. In ecological stoichiometry, homeostatic regulation of organism nutrient content causes some species to have narrower bounds to their chemical content than others.

nullcline. A set of points in an ecological model where the rate of change of one species is zero (it is at equilibrium). In community models, intersections of nullclines indicate points where more than one species is at equilibrium.

nutrient content. The quantity of an element in an organism's biomass. May be measured as moles or grams per organism, as the percentage of mass made up by a given element, or as the X:C ratio, where X is a nutrient such as N or P.

threshold element ratio. The nutrient ratio of an organism's food where that organism switches from limitation by one of those elements to limitation by another. For example, in the case of C:P, when food is above the TER, that organism will be limited by P, and when food is below the TER, that organism will be limited by C.

1. WHAT IS ECOLOGICAL STOICHIOMETRY?

Some branches of ecology are oriented toward understanding the dynamics of individual species, and others focus on the fluxes of matter and energy among collections of species in ecosystems. Ecological stoichiometry fits between these two approaches because it deals with the patterns and processes associated with the chemical content of species. Numerous ecological phenomena from the success or failure of populations to the carbon storage of whole ecosystems have a stoichiometric component. The term *ecological stoichiometry* is relatively recent, but the field is based on some of the most classic of ecological studies.

Formally defined, ecological stoichiometry is "the balance of multiple chemical substances in ecological interactions and processes, or the study of this balance." In addition, ecologists interested in stoichiometry often consider the availability of solar or chemical energy relative to the availability of one or more chemical substances.

Ecological stoichiometry is concerned with the contents of multiple elements in living and dead organic matter. There are approximately 90 naturally occurring elements, of which 11 predominate in living organisms. Only four of these (C, H, O, and N) make up about 99% of living biomass; the other seven (Na, K, Ca, Mg, P, S, and Cl) are essential to all living things. About 10 others, metals and nonmetals, are required by most but not necessarily all species. Finally, about eight other elements are required by more limited numbers of species. Some elements, especially C, H, O, and N, provide the atomic-level skeletons for biomolecules. Others are involved in materials providing structure at the organismal level, for example, the Ca and P in vertebrate bone. These elements all are generally required in high amounts. Other elements are used in energy transduction processes, where electrons are energized and deenergized. These elements, such as Fe and Mg, are just as necessary for life, but they are required in lower quantities. Although the theories and tools of ecological stoichiometry could be applied to any of these elements, most studies to date concern C, N, and P.

In the abiotic world (air, water, rocks, etc.), elements can be combined in almost limitless proportions. Living things, however, are based on a much more restricted chemistry utilizing carbon-containing organic molecules combined using more-or-less defined proportions of nucleic acids, lipids, proteins, and carbohydrates. But, and this is a crucial point in ecological stoichiometry, in spite of a commonality to the chemistry of living things, species throughout the tree of life do not have precisely identical chemistry. Nor do all species regulate their chemical content to the same degree. Ecological stoichiometry considers the many phenomena that emerge from the patterns of chemical content in organisms as well as the proximate and ultimate reasons they have the chemical composition they do. Some aspects of ecological stoichiometry arise because of a first-order commonality to the chemistry of life, whereas others arise as a result of the differences in chemistry among living things that one observes when one pays careful attention to the patterns of element content of different species.

Inorganic chemistry teaches us about the different characteristics of the chemical elements, their tendency to ionize, the number of covalent bonds they may form, etc. According to these physical principles, biochemistry makes use of the different elements in different ways. A special element is phosphorus. P makes up 1% of the soft tissues of most living things and a much higher percentage of some hard tissues such as bone. P has many biochemical functions. It is central to metabolism in the ATP/ADP energy capture and utilization system. It helps form the backbone of the information-carrying and protein-assembling nucleic acids (DNA and RNA; more on this role of P below). It combines with lipids to make cell membranes and therefore is involved in cell structure. Each of these different roles is crucial to the living cell. P also is a large component of the skeletal system of vertebrate animals. P is critical in these different ways, but not all cells or organisms combine these separate functions in the same ways; therefore, cellular or organismal P content can be species specific. Although necessary in all these ways and others, P is not always greatly abundant in nature relative to biological demand.

Living organisms concentrate certain elements while rejecting others. One of the hallmarks of life is its ability to maintain relatively constant internal conditions in the face of external variability. A well-known example is the narrow range of temperature that a healthy endothermic vertebrate such as a human being maintains in spite of being exposed to wide environmental temperature fluctuations. The negative feedback associated with maintaining relatively constant internal conditions is called homeostasis. Homeostatic regulation of element content is a key aspect of ecological stoichiometry and requires a formal definition. The degree of stoichiometric homeostasis varies for different organisms and different elements. Because homeostasis is a resistance to change, we measure it by relating the elemental content of an organism to the elemental content of its food or its neighboring environment, as the case may be; homeostasis is indicated with the Greek letter eta (H) (figure 1).

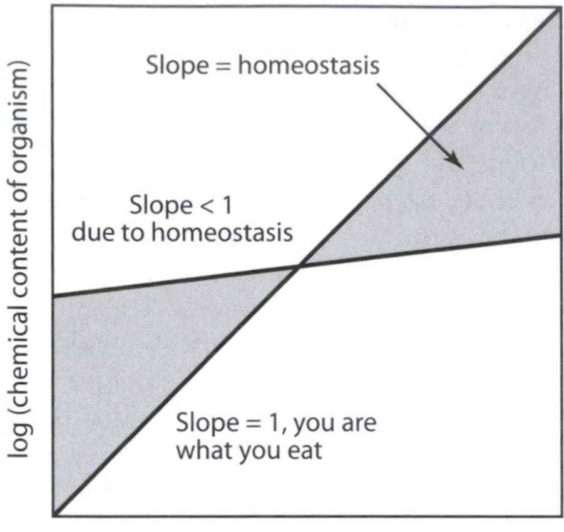

Figure 1. Homeostasis in ecological stoichiometry. If one plots the logarithm of the element content of an organism versus the logarithm of the chemical content of the resources it consumes (both measured on a common scale), the slope is referred to as homeostasis. A slope of 1 indicates that the organism's chemistry changes in lockstep with its food, or "it is what it eats." Shallow slopes indicate a resistance to change. Homeostasis results from the tendency of living things to shape their own chemistry to their needs.

2. MAJOR PATTERNS OF NUTRIENT CONTENT IN ORGANISMS

Homeostasis varies with the organism and with the chemical of interest. Photoautotrophic organisms (e.g., cyanobacteria, algae, and plants) often display a wide range of variation in C:N:P ratios according to conditions of light, nutrient, and growth rate as well as across different species and functional groups. Under severe nutrient limitation, autotrophs produce biomass with extremely low nutrient content (high C:nutrient ratio). This connection links ecological dynamics to stoichiometric patterns. In contrast, animals are more homeostatic and generally regulate their C:N:P ratios around stage- or species-specific values that are more nutrient rich than those for nutrient-limited autotrophs.

The wide range of autotroph C:N:P ratios in ecosystems reflects contrasts in the abiotic and biogeochemical conditions that supply CO_2, light, and nutrients to photoautotrophs in different ecosystems. For example, in broad cross sections of both North American and Norwegian lakes, seston C:P ratio has been shown to be positively correlated with ecosystem light:nutrient ratio, which itself is determined by local conditions affecting light intensity (mixed layer depth, light attenuation) and external P supplies. In terrestrial ecosystems, local soil conditions, canopy development, and water supply interact to affect plant C:N:P stoichiometric ratios.

It is also increasingly recognized that various anthropogenic perturbations, such as atmospheric N deposition and increased CO_2 concentrations can also affect autotroph C:N:P ratios in both aquatic and terrestrial ecosystems. For example, a doubling of CO_2 concentration reduces plant N content by about 16%, on average. There are also broad-scale patterns in the N:P ratio of plant biomass in which N:P ratio decreases moving toward the poles. This pattern may reflect differences in edaphic conditions (e.g., differences in soil age that affect soil P supply) or effects of selection operating on plant growth rate (see growth rate hypothesis below). Despite the wide intraspecific variation in C:N:P ratios that can be produced by differences in growth conditions, there also are significant differences in plant stoichiometry that derive from phylogenetic affiliation. For example, legumes that harbor N-fixing symbionts generally have higher N:P ratios than other taxa.

Heterotrophs such as bacteria and metazoans also exhibit substantial variation in C:N:P ratios, but physiological variation caused by growth or dietary conditions is thought to be relatively minor compared to such effects in autotrophs. Heterotrophs are much more homeostatic in their element content than are autotrophs. Variation in element content in different heterotrophs reflects differences in organismal allocation to major biochemical and structural components. For microbes and small invertebrates (figure 2), C:N:P variation is tied to growth-related allocation to P-rich ribosomal RNA (the growth rate hypothesis), as the content of rRNA generally increases with growth rate, comprising a significant fraction of overall biomass and containing 8.6% P by mass. Indeed, in the bacteria, zooplankton, and insects shown in figure 2, on average about 50% (and sometimes over 90%) of total organismal P was contributed by the P contained in RNA. However, because growth rate decreases with increasing body size, the contribution of P in RNA to overall body C:N:P stoichiometry also declines with body size, becoming relatively insignificant (less than 10% of total P) for animals larger than about 0.1 g dry mass.

C:N:P ratios among larger vertebrate animals also vary considerably as a result of differential allocation to structural P in bones (the mineral apatite that makes up bone is 17.6% P by mass). In terrestrial vertebrates, the percentage of whole-organism mass devoted to skeleton increases with body size. In aquatic vertebrates, because of the suspension action of water, gravity exerts less of a selective pressure, and differences in skeletonization reflect other biological aspects; bony fish are well protected from predators, for example. These intra- and interspecific variations in vertebrate P requirements have important implications for

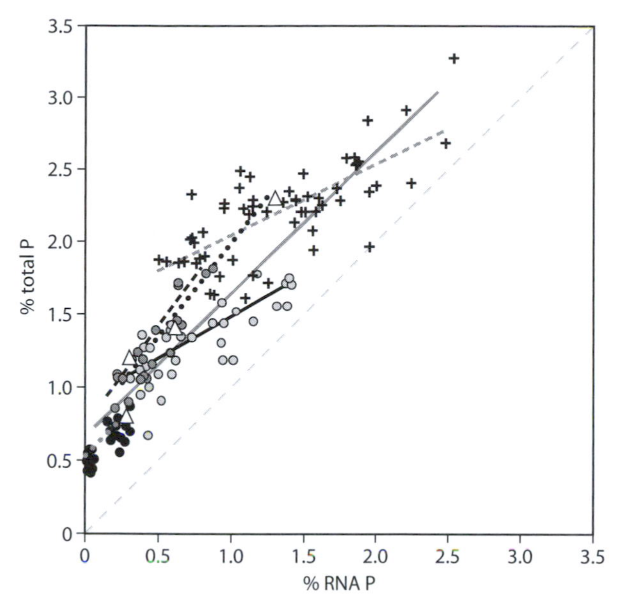

Figure 2. Intra- and interspecific relationships between total body P content (percentage of dry mass) and total body RNA P content. (RNA P content is the percentage of body mass contributed by the P contained in RNA; it is calculated by multiplying the RNA content by 0.086, the mass fraction of P in RNA.) The dashed 1:1 line indicates the condition of having all the cellular P made up by RNA-P. The figure shows P-limited *Escherichia coli* in chemostats (crosses), various freshwater crustacean zooplankton under different food conditions (dark gray circles and open triangles), larval *Drosophila melanogaster* during ontogenetic development (light gray circles), and field-collected mesquite-feeding weevils sampled during dry and wet years (black circles). Various shorter lines indicate significant intraspecific relationships. The longer solid line is a fit to the entire data set ($R^2 = 0.87$); its slope is 0.97 (~ 1), indicating that, across the entire data set, variation in P content is directly and quantitatively attributable to variation in RNA content. In all cases shown, P- and RNA-rich organisms have higher growth rates than low-P and low-RNA organisms.

their mineral nutrition and for their role in cycling of limiting nutrients in the ecosystem.

3. INFLUENCE OF STOICHIOMETRY ON ANIMAL GROWTH AND COMMUNITY STRUCTURE

An animal's niche is defined by numerous ecological factors, including climate, physical habitat structure, and presence or absence of predators. Some important niche dimensions are stoichiometric. These stoichiometric dimensions are likely most significant for detritivores, which consume nonliving food that may have very low nutrient content but also are often very important for herbivores because of the aforementioned differences in C:P and C:N ratios between plant and animal biomass.

In theory, there is only one particular composition containing elements in precisely the right proportions to optimize growth and maintenance of any given consumer at a particular moment in its life. Deviations from this optimum reduce growth rate and fitness, sometimes severely. Across terrestrial and aquatic habitats, there are numerous studies showing that one or more elements, including Na, N, P, Fe, or Ca, may be low enough in autotroph biomass to shape the foraging decisions or limit the growth of individual herbivore species. As a corollary, under these conditions, other resources including energy are in relative excess. Elements in surplus are not held with great efficiency by the metabolism of the consumer and instead are recycled to the environment, whereas dividends are paid on consumers holding that element most limiting their growth with high efficiency. Ecologists define a particular ratio in an organism's food where it switches from limitation by carbon to limitation by another element as a "threshold element ratio" (TER). The actual value of the TER reflects the taxonomic identity and nutrient ratio of the consumer itself as well as its efficiency at processing carbon or nutrients.

When an animal cannot locate sufficient quantities of an element in its environment, that lack may limit the animal's growth. C:N:P ratios in a variety of ecosystems indicate that animals, especially herbivores and detritivores, must often subsist on food with low element content. For example, in a large compilation of published values of C:N:P ratios for foliage in terrestrial plants, the average C:N and C:P ratios (moles: moles) were 36 and 968, respectively (N:P 28), in contrast to average C:N and C:P ratios of 6.5 and 116 for herbivorous insects (N:P 26.4). Similarly, freshwater seston (suspended organic matter containing phytoplankton and other microscopic biota on which filter-feeding zooplankton depend) also has high C:N and especially C:P ratios (10.2 and 307; N:P 30.2) compared to the average freshwater zooplankton species (6.3 and 124; N:P 22.3). Marine seston, however, generally has lower C:N and C:P ratios, much more in line with the elemental composition of marine zooplankton themselves. These contrasts suggest that, based on stoichiometric imbalance alone, marine food webs should operate more efficiently in processing organic matter than freshwater and especially terrestrial food webs. Indeed, existing data do indicate that a considerably greater fraction of the low-nutrient primary production of terrestrial food webs enters detrital food chains (that is, is not consumed by herbivores) than does so in freshwater and especially marine food chains.

A well-studied example is the waterflea *Daphnia* and the concentration of P in its algal food. This is a freshwater, herbivorous zooplankter and an important keystone species in aquatic food webs. *Daphnia* is considered a high-P zooplankter, consistent with its rapid-growth life history as explained by the growth rate hypothesis described above. Laboratory studies in which *Daphnia* have been raised on algal foods of varying C:P ratio have consistently shown that unless food quantity is very low, *Daphnia* growing on algal foods with C:P above about 300 or so show reduced rates of biomass gain, longer times to first reproduction, and reduced fecundity. Low-P zooplankton are much less susceptible to effects such as these. Similarly, studies examining stoichiometry across multiple lakes have shown that high-P *Daphnia* are less likely to be abundant in habitats where the potential food base has a high C:P ratio. Hence, the population-level dynamics that are well studied in the laboratory correctly predict certain aspects of community structure in the field.

4. NUTRIENT CYCLING IN ECOSYSTEMS

Above, we focused on the single element most limiting for consumer growth. In an ecological interaction, mass must balance, and any matter that is ingested but not incorporated into consumer biomass must be returned to the environment in a solid, liquid, or gaseous state. The existence of stoichiometric mismatches between food and consumer, as well as differences in homeostatic regulatory ability of elements, therefore, has considerable bearing on the patterns of nutrient recycling in ecosystems.

One place this is observed is in litter decomposition. As described above, leaves of higher plants exhibit a wide range in C:N:P ratios. When leaves die, there is a range in C:N:P of the corresponding detritus that is as wide or wider than that in living leaves. Nutrient content in detritus also is lower than in leaves because of nutrient resorption before abscission. This potentially severe elemental imbalance between detritus and the organisms that consume it generates a strong stoichiometric component to litter decomposition. Litter of relatively high N or P content (C:N $\approx$ 10 or C:P $\approx$ 100) breaks down rapidly, as quickly as 1% loss of litter mass per day. In contrast, litter of low nutrient content (C:N $\approx$ 100 or C:P $\approx$ 1000) breaks down much more slowly, about 0.01% of mass per day. This contrast in litter breakdown rates and nutrient remineralization rates has a strong bearing on many aspects of terrestrial nutrient cycling, carbon storage, and other phenomena.

Stoichiometric constraints on nutrient cycling in ecosystems are well illustrated by consideration of the N:P ratio resupplied by foraging, homeostatic consumers. Theoretical analysis has predicted several patterns. First, the N:P recycled should generally increase with the N:P of the food consumed (but not in a linear way, see below). Consumers ingesting high-N:P food should tend to recycle high N:P ratios. Second, the N:P recycled should be a decreasing function of the N:P of the consumers themselves. High N:P consumers must retain N and lose P in order to meet their growth requirements. Finally, and perhaps most interestingly, the N:P recycled by homeostatic consumers should generally be a more extreme version of the N:P they eat. Consumers eating high-N:P food are best served by keeping P but losing N, and vice versa. This accentuation of nutrient ratios, if repeated over and over again in a relatively closed system while consumer biomass builds up, can cause N:P limitation patterns to diverge. This is an ecosystem instability generated by the homeostasis of the foraging consumer. We return to this subject below.

One example where such recycling effects have been studied is in lake communities where the zooplankton herbivores may be limited by high-N:P copepods or low-N:P *Daphnia*. Theory says that species shifts between these two groups should result in different nutrient limitation patterns, with recycling by copepods generating low-N:P conditions and recycling by *Daphnia* generating high-N:P conditions. Experimental studies of lake food webs generally have borne out those predictions; some of these results are described in an upcoming section.

5. INFLUENCE OF STOICHIOMETRY ON SPECIES DYNAMICS

The large stoichiometric imbalance between nutrient-limited autotrophs and the herbivorous animals consuming them suggests that food quality may play an important role in regulating herbivore dynamics in nature. However, nutrient recycling by the consumers themselves may ameliorate the low nutrient content of autotrophs via this important feedback. Thus, nutrient–autotroph–herbivore systems have the potential for interesting and complex feedbacks that may affect population dynamics and food-web structure. These interactions have been analyzed from a theoretical perspective in a number of mathematical models.

These models, "stoichiometrically explicit" versions of the famous Lotka-Volterra equations, generally contain several key components that distinguish them from nonstoichiometric models: variable and growth-rate-dependent nutrient content of the autotrophic prey; strict homeostasis of nutrient content in the herbivorous consumer; and overall mass conservation of multiple elements in the system. We analyze these models by using nullclines. A grazer–autotroph model will have two nullclines, one for each trophic level. A

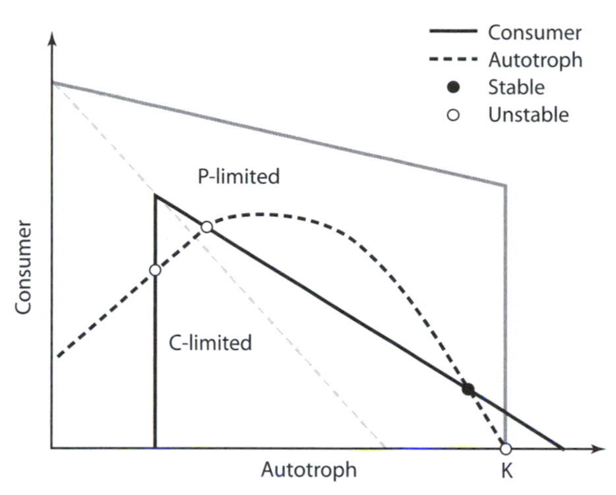

Figure 3. An example of nullcline analysis for a stoichiometrically explicit predator–prey model. In the figure, the gray box delineates the entire domain within which population fluctuations are confined because of the fixed amount of limiting nutrient (in this case, P) present in the system. The point on the autotroph axis labeled *K* represents the carrying capacity for the autotrophs. As discussed in the text, *K* reflects overall light intensity and is varied to examine potential dynamic impacts of light on the consumer–autotroph system. The interior of the gray box is delineated into two regions indicating situations where individual consumer growth is limited by C (low food quantity) or by P (low food quality). Intersections of the consumer (dark, solid) and autotroph (dashed) nullclines indicate potential equilibria. In this case, only one intersection is a stable equilibrium (solid circle) in which autotrophs have high biomass but consumer biomass is low and consumer growth is P-limited.

nullcline is the set of all consumer and autotroph biomass values for which the rate of biomass change for a given population is zero. In most nonstoichiometric models, the shape of the nullclines is such that only one intersection is possible, meaning there is only one combination of autotroph and grazer densities where there is no change with time. Imposition of stoichiometric food quality effects, which generally manifest when autotrophs have produced high biomass with low nutrient content, forces the consumer's nullcline (where its rate of change is zero) to be hump shaped (figure 3). The hump means that the nullcline can intersect the autotroph's nullcline to form more than one equilibrium point.

Stoichiometric models predict a much richer range of population dynamics than nonstoichiometric models. For example, increasing the parameter (*K*) for autotroph carrying capacity (to simulate the effects of increased light intensity) causes an intriguing series of bifurcations. As light intensity ($\propto K$) increases, the dynamics shifts from a single stable point (with consumers limited by total food quantity) to a limit cycle to a second stable point (with consumers limited by poor food quality; situation shown in figure 3) to a stable point involving deterministic extinction of the consumer. Finally, a stoichiometric model containing one autotroph species together with two consumer species has also been analyzed to show that the two consumers can stably coexist with each other indefinitely under certain conditions of high light intensity and poor food quality. Under such coexistence, the single food type acts as two different resources, expanding the niche space.

The predictions of these stoichiometrically explicit models have been borne out in experimental studies. In an artificial ecosystem experiment involving a green alga and two species of *Daphnia*, increased light intensity led to very slow production of *Daphnia* biomass because of the poor food quality along with decreased trophic transfer efficiency from algae to *Daphnia*. As predicted by theory, near *Daphnia* extinction occurred in one mesocosm with highest light intensity, whereas high light also resulted in sustained coexistence of the two *Daphnia* species. In fact, the data showed that, under low light intensity (P-rich algae), there was normal density dependence in the herbivores: individual female fecundity was negatively correlated with *Daphnia* population size, indicating strong intra- and interspecific competition. However, under high light intensity (low-P algae), fecundity was positively correlated with *Daphnia* abundance, indicating intra- and interspecific facilitation. This latter relationship reflects the indirect effects of consumer-driven nutrient recycling: as the animal population built up, it cropped the low-P algae while excreting some P, thus increasing the P content of the remaining algae and improving its quality. A variety of additional studies have extended this work to field situations and have shown that modifications of light intensity and nutrient supply can significantly affect the C:N:P ratios of seston and thus alter the production and dynamics of zooplankton consumers under natural conditions.

Stoichiometric effects on trophic interactions and food web dynamics also occur in terrestrial ecosystems, as in the well-studied example of the role of low dietary N content (high C:N ratio) in limiting herbivore production. More recently, it has been shown that food P content can also play a role in limiting consumer performance in terrestrial settings. For example, experimental manipulation of the P content of the plant *Datura wrightii* significantly increased the growth rate of caterpillars of the moth *Manduca sexta*.

6. WHOLE-LAKE FOOD WEB EXPERIMENTS

Previous sections explored the dual roles of stoichiometric mechanisms in affecting trophic dynamics: first, effects of food quantity and quality, and second, internal nutrient processing via consumer-driven nutrient recycling. In natural ecosystems, these mechanisms may come together in complex and interesting ways. We now consider a pair of whole-lake food web manipulations that were designed to evaluate stoichiometric dimensions to nutrient cycling related to food web structure. The experiments were based on predicted differential C:N:P stoichiometry of major herbivores that dominate under different food web conditions. Lakes with three dominant trophic levels (phytoplankton, zooplankton, zooplanktivorous fish) tend to have low zooplankton biomass dominated by copepods but lacking *Daphnia* because *Daphnia* are especially susceptible to predation by small fish. Such a zooplankton community will have high N:P ratio. In contrast, lakes with four trophic levels (the previous three plus piscivorous fish such as pike or bass at the top) tend to have high zooplankton biomass dominated by low-N:P *Daphnia* because the piscivores hold the zooplanktivores in check. The experiment thus relied on couplings of community structure and stoichiometry.

One experiment was performed in Lake 227 (L227), a lake that was experimentally eutrophied for more than two decades by addition of P-rich fertilizer, making this one of the longest-running whole-lake experiments in ecology. Its fish community lacked piscivores, and the lake supported dense populations of planktivorous minnows and a zooplankton community dominated by copepods and rotifers. Consistent with the low N:P ratios of the lake's nutrient loading, the dense

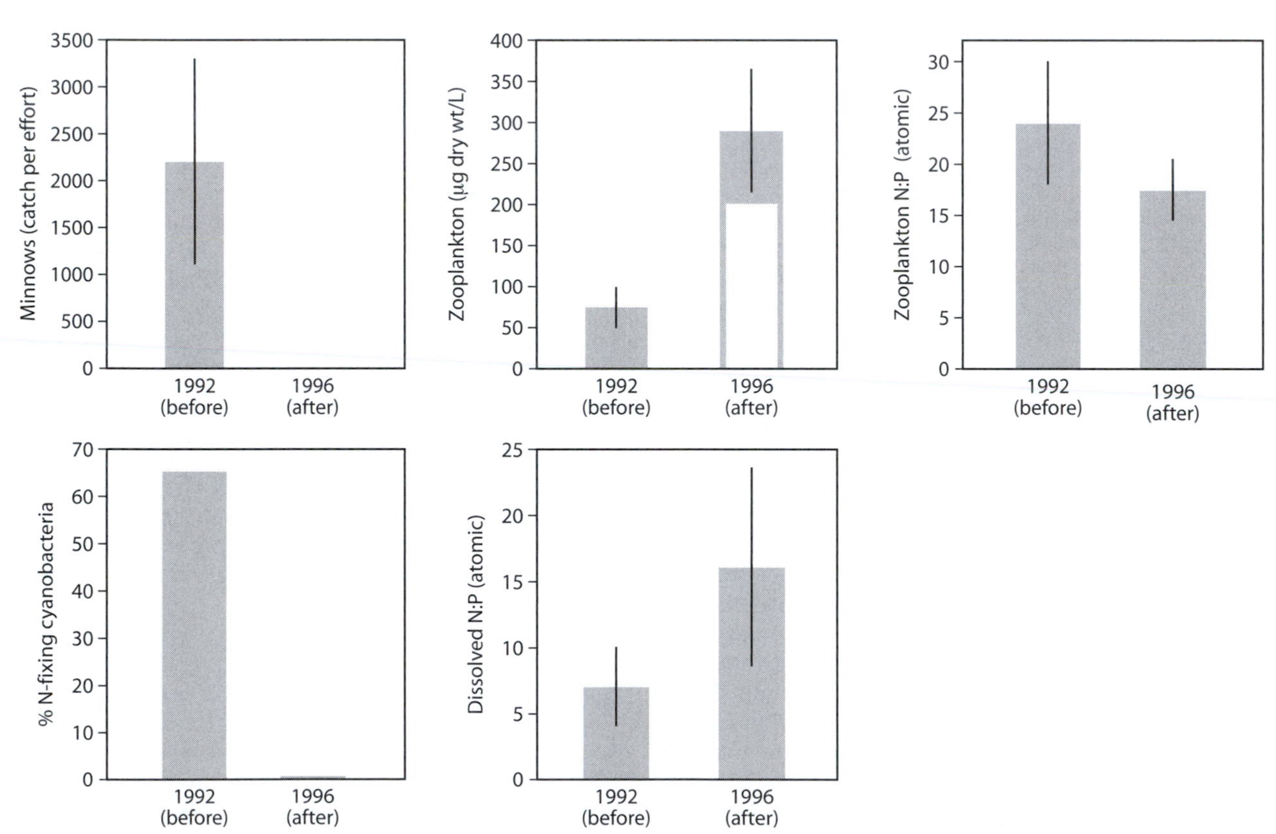

Figure 4. Food web dynamics after food web manipulation of experimentally eutrophied Lake 227 at the Experimental Lakes Area, Ontario, Canada. Two hundred piscivorous pike were introduced into the lake in 1992 and 1993, leading to elimination of planktivorous minnows (first panel). Consistent with a trophic cascade, zooplankton biomass increased dramatically (second panel), especially *Daphnia* (white portion of the zooplankton graph). Consistent with increased predominance of *Daphnia* in the zooplankton, the N:P ratio of total zooplankton biomass declined (third panel). Increased *Daphnia* and overall zooplankton biomass in 1996 was associated with a massive shift in phytoplankton community structure away from previous dominance by N-fixing cyanobacteria (fourth panel). This is consistent with an altered internal nutrient-recycling regime that disproportionately increased the availability of N relative to P (fifth panel), likely reflecting differential retention of P in the low-N:P *Daphnia* biomass. Error bars indicate +1 standard error.

phytoplankton community was dominated by N-fixing cyanobacteria, as it had been for much of the period since low-N:P fertilization had begun in the mid-1970s. Into this configuration, 200 piscivorous pike were introduced to the lake over the course of 2 years, thus adding a fourth trophic level where it had not existed.

Within a year of pike introduction, minnow densities had decreased dramatically, and during the third year following introduction of pike, a massive increase in zooplankton biomass (largely *Daphnia*) and a corresponding decrease in zooplankton N:P ratio were observed (figure 4). These changes were associated with large declines in algal biomass and overall increases in concentrations of dissolved nutrients, especially dissolved nitrogen, resulting in an increased N:P ratio of available nutrients. Most striking was the nearly complete absence of previously dominant cyanobacteria when *Daphnia* dominated the lake. These results provide support for a strong stoichiometric component of the trophic cascade, in which differential recycling of limiting nutrients by zooplankton taxa with different body N:P ratios affects phytoplankton community structure by altering the competitive arena.

A similar experiment was performed in Lake 110, a lake with a similar food web structure (dense minnow populations, no *Daphnia* in the zooplankton), but a lake still in its natural oligotrophic state. One crucial difference between L227 and L110 (and indeed other unfertilized lakes of the region) was that the C:P ratio of seston in L110 was considerably higher (>500 compared to <200 for L227). As for L227, piscivorous pike were introduced, and the response of the system

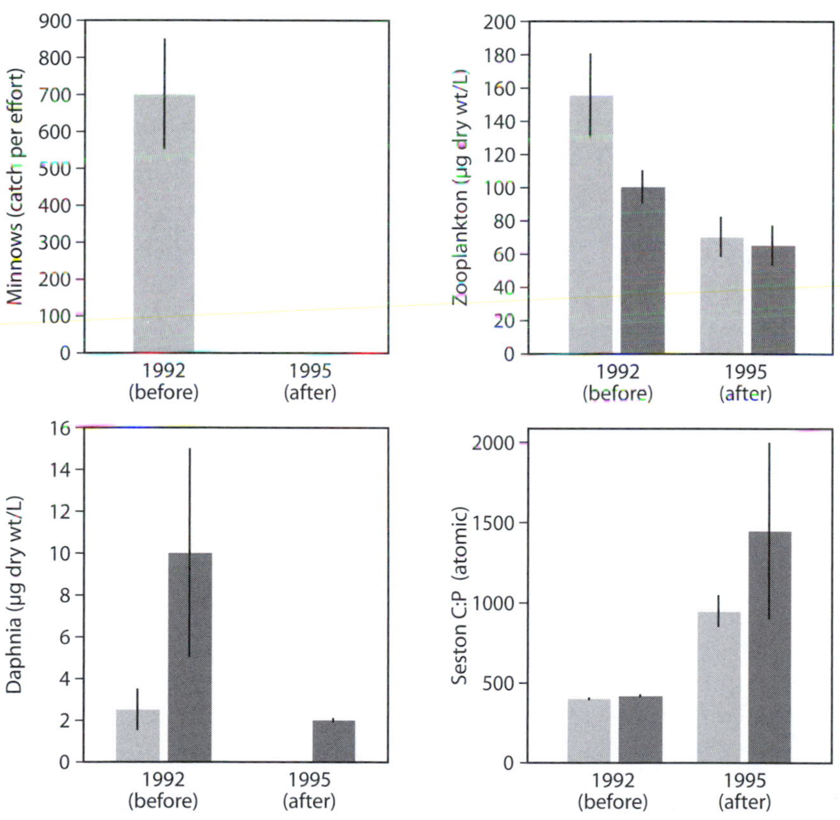

Figure 5. Food web dynamics after food web manipulation of naturally oligotrophic Lake 110 at the Experimental Lakes Area, Ontario, Canada. One hundred fifty-three piscivorous pike were introduced into the lake in 1992 and 1993, leading to elimination of planktivorous minnows (first panel). Inconsistent with a trophic cascade, zooplankton biomass decreased dramatically (second panel), including *Daphnia* (third panel), in both the manipulated lake (L110, light gray bars) and the reference lake (L239, dark gray bars). The zooplankton declines were associated with major increases in seston C:P ratio (fourth panel) in both lakes. These dynamics suggest that regional climatic effects altered phytoplankton growth conditions and worsened phytoplankton P limitation, raising seston C:P ratio and imposing a stoichiometric constraint on zooplankton and thereby truncating the expected trophic cascade. Error bars indicate ±1 standard error.

was monitored over ensuing years along with dynamics of a similar oligotrophic lake (L240).

As in L227, pike introduction greatly reduced the activity and abundance of minnows (figure 5). However, instead of an increase of zooplankton biomass and especially of *Daphnia*, zooplankton biomass in L110 declined substantially, and *Daphnia* populations became nearly extinct. Interestingly, zooplankton in the reference lake underwent similar dynamics. These anomalous responses become understandable by noting that the year of *Daphnia* collapse corresponded to a year in which seston C:P ratios were unusually high in both L110 and L239. This suggested that regional climatic changes had altered nutrient supply to the lakes, accentuating phytoplankton P limitation and thus increasing seston C:P ratio and worsening food quality for zooplankton and especially for *Daphnia*. This "stoichiometric constraint" on the trophic cascade was supported by later dietary P supplementation studies that showed that *Daphnia* in nearby, similar lakes do experience direct P limitation of their growth because of high seston C:P ratios.

Taken together, these two whole-lake experiments illustrate the ecosystem-scale operation of the two sides of the stoichiometric coin: food quality and nutrient recycling.

7. LIGHT:NUTRIENT RATIOS AND THE ECOLOGY OF AUSTRALIA

To further illustrate the scope and potential explanatory power of ecological stoichiometry, we close with a specific application that ties many observations together. Recently proposed, the "Nutrient Poverty/Intense Fire Theory" purports to explain a host of aspects of the ecology of Australia, the flattest, driest, and geologically oldest vegetated continent. Australia has a uniquely large proportion of highly nutrient-deplete soils. These high-light:nutrient conditions promote several notable features in the autotrophs. The surplus carbon (over nutrient availability) is used to produce foliage that is well defended from herbivores as well as large quantities of lignified tissues. These factors negatively impact foliovores. However, many of the plants also produce unusually high quantities of readily digestible exudates, and Australia is unusual in the number of vertebrates (mostly birds) that pollinate its flowers, attracted and supported, it would seem, by the high exudation rates.

In regard to the low-quality forage, note that large amounts of high-carbon plant biomass not consumed by herbivores build up and provide fuel for intense fire. Frequent fires characterize much of Australia, and these generate a positive feedback because fire volatilizes potentially limiting elements for animals; these elements include N, S, I, and Se, further depleting these potentially limiting elements.

The unusual plant biomass, caused by the positive feedback of high-light:nutrient conditions, presents challenges for the herbivores of the continent. The Nutrient Poverty/Intense Fire theory suggests that it is no coincidence that folivorous vertebrates of Australia are unusually small relative to other continents and have low metabolic rates. Another unusual aspect of the animals of the continent is the absence of geophagy (soil eating). Though geophagy is common on other continents, the extremely nutrient-poor soils of Australia may not provide adequate supplementation of trace elements to promote this style of foraging.

By examining the ratios of available energy to nutrients, particularly scarce nutrients, ecologists may identify processes not previously recognized as important for life forms or biotic adaptation on other continents.

FURTHER READING

Andersen, T., J. J. Elser, and D. O. Hesson. 2004. Stoichiometry and population dynamics. Ecology Letters 7: 884–900. *This article provides an up-to-date review of the complex population dynamics that is characteristic of stoichiometrically explicit models.*

Cebrián, J. 1999. Patterns in the fate of production in plant communities. American Naturalist 154: 449–468. *This article summarizes a large amount of scientific literature on the processing of carbon in ecosystems, for example, how much of primary production is grazed versus enters into detrital food chains. The article shows how autotroph nutrient content (and growth rate) relates to carbon processing.*

Elser, J. J., K. Acharya, M. Kyle, J. Cotner, W. Makino, T. Markow, T. Watts, S. Hobbie, W. Fagan, J. Schade, J. Hood, and R. W. Sterner. 2003. Growth rate–stoichiometry couplings in diverse biota. Ecology Letters 6: 936–943. *This article examines the stoichiometric signatures of growth rate in diverse heterotrophs. P, RNA, and growth are closely coupled in small heterotrophs when P is limiting their growth.*

Elser, J. J., W. F. Fagan, R. F. Denno, D. R. Dobberfuhl, A. Folarin, A. Huberty, S. Interlandi, S. S. Kilham, E. McCauley, K. L. Schulz, E. H. Siemann, and R. W. Sterner. 2000. Nutritional constraints in terrestrial and freshwater food webs. Nature 408: 578–580. *This article reviews the patterns of C:N:P ratios in autotrophs and herbivores from terrestrial and aquatic food webs. It shows that there often are large differences in the C:nutrient ratio across the plant–animal interface in terrestrial and freshwater systems.*

Hessen, D. O. 2006. Determinants of seston C:P ratio in lakes. Freshwater Biology 51: 1560–1569. *This article explores the reasons why organisms in lakes have such widely varying nutrient ratios. Lakes with high light but low nutrients have high seston C:P ratios and low numbers of P-rich herbivores.*

Kay, A. D., I. W. Ashton, E. Gorokhova, A. J. Kerkhoff, A. Liess, and E. Litchman. 2005. Toward a stoichiometric framework for evolutionary biology. Oikos 109: 6–17. *Like any phenotypic variation, stoichiometric differences among species are dependent on evolutionary history and subject to natural selection. This review considers a number of evolutionary facets to stoichiometric ecology.*

Orians, G. H., and A. V. Milewski. 2007. Ecology of Australia: The effects of nutrient-poor soils and intense fires. Biological Review 82: 393–423. *This thought-provoking recent article suggests that many of the unique aspects to the ecology of Australia are caused by high light and low nutrients, with a positive feedback of fire involved.*

Reich, P. B., B. A. Hungate, and Y. Luo. 2006. Carbon–nitrogen interactions in terrestrial ecosystems in response to rising atmospheric carbon dioxide. Annual Reviews of Ecology and Systematics 37: 611–636. *This article is a thorough examination of many relevant stoichiometric linkages in global change.*

Sterner, R. W. 1995. Elemental stoichiometry of species in ecosystems. In C. Jones and J. Lawton, eds. Linking Species and Ecosystems. Boca Raton, FL: Chapman & Hall, 240–252. *An early, concise examination of stoichiometric reasoning applied to ecological systems.*

Sterner, R. W., and J. J. Elser. 2002. Ecological Stoichiometry: The Biology of Elements from Molecules to the Biosphere. Princeton, NJ: Princeton University Press. *This book is the most comprehensive examination of stoichiometry in ecology. Topics range from the molecular to the global.*

III.16

Macroecological Perspectives on Communities and Ecosystems

Pablo A. Marquet

OUTLINE

1. The road to macroecology
2. Macroecology: Toward a definition
3. Macroecological patterns
4. Neutral macroecology
5. Metabolic theory

Macroecology is an emergent research program in ecology that examines patterns and processes in ecological systems at large spatial and temporal scales. It acknowledges the complexity of ecological systems and the limitation of reductionistic approaches, emphasizing a statistical description of patterns in ensembles of multiple species. One of its goals is the identification of regularities that might eventually unveil the general principles underlying the structure and functioning of communities and ecosystems.

GLOSSARY

energetic equivalence. Concept that denotes the equivalence of species in terms of the amount of energy that their populations use within natural communities

metabolism. Network of chemical reactions that take place in living entities and by which energy and materials are taken up from the environment, transformed into the component of the network that sustains it, and allocated to perform specific functions

metacommunity. Set of local communities that are linked by the dispersal of their components and potentially interacting species

metapopulation. Set of local populations of one species linked through dispersal

reductionism. Scientific approach by which understanding of complex systems can be obtained by reducing them to the interactions among their constituent parts

scaling. Name given to the existence of a power–law relationship between two variables of the form $y=ax^{\theta}$, where θ is the scaling exponent and is normalization constant

species–area relationship. Relationship that describes how the number of species increases with the area sampled or with the size of the system under analysis (e.g., lake, habitat fragment, or island)

Theory of Insular Biogeography. Equilibrium theory proposed by MacArthur and Wilson in 1963 that proposes that the number of species in a given island results from the dynamic equilibrium of the opposite processes of immigration from a source and local extinctions

1. THE ROAD TO MACROECOLOGY

As do most research programs in science, macroecology represents the crystallization of a line of inquiry that started two centuries ago with the discoveries of the German naturalist Alexander von Humboldt, published in 1807, and his remarks on the latitudinal distribution of biodiversity (the pole-to-tropic gradient) and continued, with different intensity, in the works of Olof Arrhenius, Carrington Bonsor Williams, John Christopher Willis, Frank Preston, Leigh Van Valen, George Evelyn Hutchinson, Robert MacArthur, Eduardo Rapoport, and several others. One can ask in retrospective, What makes the work of these authors macreocological? The common theme in all of them was the usually large spatial extent (i.e., regional to continental) of the patterns they reported and the use of statistical descriptions of species ensembles with regard to attributes such as abundance, richness, geographic distribution, or body mass, with an emphasis on the emerging patterns rather than on the component species. Before macroecology, these patterns were studied in isolation and interpreted as resulting from evolutionary processes and/or ecological or biogeographic dynamics. Macroecology provided a synthetic and common framework for all of them by explicitly recognizing the importance of, and the links among, ecological, evolutionary, and biogeographical processes and scales in the understanding of ecological phenomena.

Three major events contributed to the consolidation of macroecology as a research program in ecology.

1. First is the recognition of the role played by regional factors in affecting the local dynamics of populations and communities. The importance of regional effects became recognized thanks to the analysis of the degree of coupling between local and regional diversity championed by Robert Ricklefs and the development of metapopulation theory, which, although formally introduced in 1969 by Richard Levins, started to flourish in the 1980s, most notably through the work of Ilkka Hanski.
2. As we elaborate in greater detail below, macroecological work is usually concerned with patterns occurring at regional to global scales where experiments are not feasible and data are difficult to obtain. However, this changed during the last two decades with the explosive development and/or availability of data such as atlases on the distribution and abundance of different taxa (e.g., Breeding Bird Survey, Gentry Plots) and the development of new technological tools to deal with and to generate data on environmental variables at large spatial scales (e.g., satellite imagery, remote sensing, and geographic information systems).
3. Finally, one the main drivers of the macroecological approach was the growing recognition of the limitation inherent to the reductionistic, microscopic approaches that became dominant in ecology since the 1970s, which try to understand ecological communities from detailed knowledge on between-species interactions through manipulative experiments of short duration and limited spatial extent.

Reductionistic approaches, although powerful in characterizing the outcome of pairwise species interactions at a given locale, cannot deal appropriately with the vexing complexity of ecological systems composed of networks of many species, linked through direct and indirect paths of different strengths and degrees of nonlinearity, and subjected to processes acting at different temporal and spatial scales (e.g., species extinction and speciation and individual birth, death, and dispersal). In this context, it comes as no surprise that communities, under the microscopic paradigm, were considered as highly variable and idiosyncratic with regard to the relative importance of specific biotic interactions (e.g., competition, predation, mutualism) and their effect on local coexisting populations. Two representative quotations from major figures in the field of ecology can help us to clarify this point. Lord Robert May, in his MacArthur Award address published in the journal *Ecology* in 1986, wrote: "Ecology is a science of contingent generalizations, where future trends depend (much more than in the physical sciences) on past history and on the environmental and biological setting." This view was also sponsored by two prominent community ecologists, Jared Diamond and Ted Case, who in the introduction to the edited volume *Community Ecology,* wrote:

> The answers to general ecological questions are rarely universal laws, like those of physics. Instead, the answers are conditional statements such as: for a community of species with properties A_1 and A_2 in habitat B and latitude C, limiting factors X_2 and X_5 are likely to predominate.

2. MACROECOLOGY: TOWARD A DEFINITION

Most people are unaware that the first use of the word *macroecology,* curiously, appeared in a small monograph published in Spanish in 1971 by Guillermo Sarmiento and Maximina Monasterio, two Venezuelan researchers working in tropical savanna ecosystems. They used the word macroecology to mean the analysis of broad patterns in vegetation that resulted from the interaction between geomorphology and soil properties at large spatial scales (between regional and landscape scales). They compared this approach with what they called microecology (by analogy with the distinctions made in economics between macro- and microeconomics), which they characterized as a detailed inventory of species abundance and composition at small (plot) scales. The research program we usually call macroecology, however, was formally introduced in 1989 in a seminal paper by James H. Brown and Brian A. Maurer. They defined macroecology as the study of how species divide resources (energy) and space at large spatial scales, with its goal the study of the assembly of continental biotas. By undertaking analyses at this large spatial scale, they expected that local idiosyncratic noise would tend to cancel out so they would be able to pick up the fingerprint of general

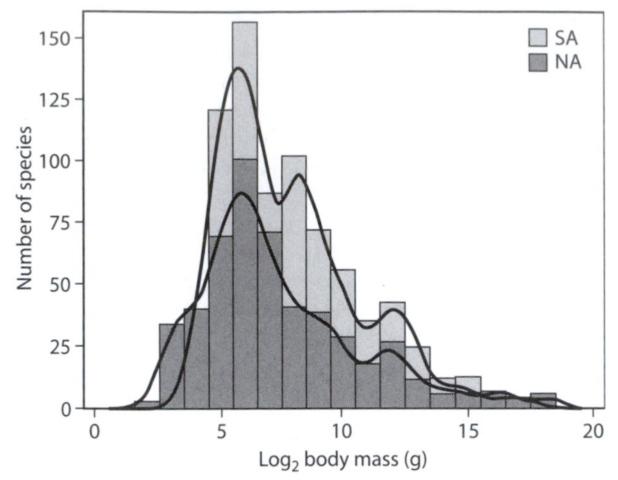

Figure 1. Body size distribution for North American (NA) and South American (SA) terrestrial mammals. The continuous lines represent the result of applying a kernel smoothing to reveal the complex structure of these distributions with lumps and gaps in body size. (After Marquet, Pablo A., and Hernán Cofré. 1999. Large temporal and spatial scales in the structure of mammalian assemblages in South America: A macroecological approach. Oikos 85: 299–309)

patterns and principles affecting ecological systems (figure 1).

After its original definition, the term *macroecology* has taken different meanings. The two most commonly in use are (1) macroecology as the study of biodiversity patterns and processes at large spatial and long temporal scales, a sort of large-scale community ecology, and (2) macroecology as a sort of statistical mechanics, where the emphasis is on the statistical regularities that emerge from the study of ensembles or large collections of species, about which it tries to make the fewest possible assumptions, the same as with particles in statistical mechanics. Under this definition, macroecology is concerned with the existence of statistical patterns in the structure of communities that seemingly reflect, or might provide some clues on, the operation of general principles or natural laws. This makes of macroecology, to some extent, a radical (from the Latin radix = root) attempt aimed at understanding ecological systems from first principles.

3. MACROECOLOGICAL PATTERNS

Macroecologists have analyzed a myriad of patterns in biodiversity. As would be expected in any research program at its early stages, large efforts have been placed in the documentations of patterns, which serve as the empirical foundations on which deductive and prediction-rich theories rest. It is beyond our scope here, however, to describe and provide an in-depth discussion of each of the patterns that macroecologists most commonly work with, but for the sake of simplicity, those patterns can be separated into three categories.

Patterns in the Frequency Distribution of Ecological Attributes

As is usually emphasized in descriptive statistics, knowledge of the shape of the frequency distribution of the variables under study is of paramount importance, as it not only informs us if the data fit the assumptions required by statistical tests but may also shed light on the type of processes that underlie its emergence. Macroecologists have been deeply concerned with the shape of frequency distributions of traits such as body size, abundance, and distribution and how these change across time and space (figure 1). Important questions that macroecologists have tried to answer are related to, for example: What determines the size of the largest- and the smallest-sized species found in a given biota? Are there gaps in body size distributions? Why are most species of small to medium size? How much does this distribution vary across time and/or from local to global scales? And do other taxa, such as trees, bacteria, or birds show similar distributions? So far, very few of these questions have a definitive answer. We know, for example, that the area of a landmass determines the size of the largest species that can potentially evolve there and that the shape of the body size distribution is highly variable across space and taxonomic groups. Just as with other macroecological patterns, body size distributions are affected by how speciation and extinction rates vary with body size and by how the strength and direction of these relationships are affected by environmental factors such as temperature and area. However, as of yet, macroecologists are far from achieving a general explanation for body size patterns, one that is not only able to explain the patterns we already know but can at the same time predict new patterns in size distributions across time, space, and taxa.

Patterns in the Covariation of Attributes

Patterns in covariation have been widely analyzed by macroecologists. Among them are the famous relationship between local and regional species richness and that between density and body mass (figure 2), among many other ecological and life history traits that covary with individual size, such as geographic range, home range area, population variability, and lifespan, which have been thoroughly documented in several books such as Robert Henry Peters (*The Ecological Implications of Body Size*) and William Alexander Calder III (*Size, Function and Life History*), both published in 1983, and more recently, in 1994, Karl Joseph Niklas (*Plant Allometry*).

The relationship between population density and body size has a long history in ecology. Carl O. Mohr originally proposed it in 1940, in the context of analyzing wildlife census techniques. He reasoned that a plot in which different estimates of density are plotted for each species would allow one to assess what he called economic densities (i.e., the number of individuals per unit of habitat actually used by the species) and thus compare the estimates of alternative census techniques. Mohr, however, did not pay too much attention to the biology in this relationship. It was not until 1981 that John Damuth revisited it and showed that population density of herbivore mammals decreased with body size and that density was related reciprocally to individual metabolic requirements, implying that different species, regardless of their size, tend to use similar amounts of energy within communities. We revisit and expand on this relationship in a later section.

Patterns of Change in Attributes along Time or Space

Probably the best-known macroecological patterns in time and space are those between species richness and latitude and the well-known relationship between species richness and area. These patterns, although well documented for a variety of taxa, still remain an active area of inquiry. For example, the latitudinal pattern in species richness (figure 3) has been reported in a great diversity of organisms, from microbes to trees and vertebrates, and it is known to occur in terrestrial, marine, and freshwater environments. Ultimately, the gradient should reflect latitudinal variations in rates of speciation and extinction. The prevailing view is that most species originated in tropical areas, which served as the "cradle of diversity," and that a large fraction of them have remained there, meaning that it has also served as a "museum of diversity." The reasons why speciation is high and extinction low in the tropics, however, are still much debated, as many biotic and abiotic factors, such as temperature, productivity, and area, can affect these rates.

Macroecologists have also focused on some of the so-called ecogeographic rules such as Bergmann's rule, which refers to the tendency for individuals of a given species to increase in size toward the cooler areas of its geographic range. Other well-studied patterns are the tendency for lineages to increase their size over geological time (i.e., Cope's rule) and the tendency for small species to evolve toward larger size (gigantism) and large species to evolve toward smaller size (dwarfism) in islands (the island rule). But not only patterns in the temporal and spatial variation of morphological traits are within the domain of macroecology; the way population density changes across the geographic range of a species has also been the focus of much research, as has the temporal dynamics of ranges themselves (i.e., their expansion and collapse) as well as geographic patterns in the size and shape of geographic ranges.

There is no doubt that macroecology is rich in patterns. The discovery of patterns and their statistical description, however, are the beginning of a process whose end is the proposition of a hypothesis about mechanisms that could potentially give rise to the observed phenomenon. The history of ecology tells us that the time it takes to traverse this path, from the identification of a general pattern to the proposition of generative mechanisms, is usually long and is marked by bursts of activity reflected in the generation and coexistence of several alternative models and hypotheses that can explain the same phenomenon. And this is true for most macroecological patterns. Two recent theoretical developments [the Metabolic Theory of Ecology (MTE) and the Neutral Unified Theory of Biodiversity (NUTB)], however, have shown that several macroecological patterns are interconnected and can, as a first approximation, be understood as resulting from the action of simple and general principles. As we will see in the next section, these theories have revolutionized the field, highlighting the simplicity underlying complex ecological systems.

4. NEUTRAL MACROECOLOGY

Neutral theories have a long history in biology. Motoo Kimura introduced the first neutral model within biology in the 1960s. This neutral model was intended as a vehicle for understanding the forces affecting allelic variation in the context of population genetics. The neutral theory of population genetics asserts that most allelic changes in a population are selectively neutral, or nearly so, and driven by mutation, migration, and genetic drift. Although originally viewed as an antiselectionist theory, it is now appreciated as an important complement to our understanding of the factors driving adaptive evolution. The application of these ideas to understand patterns in ecological systems dates back mainly to the work of Hal Caswell, who, in 1976, developed a neutral theory in the context of community ecology with the goal of understanding the role of

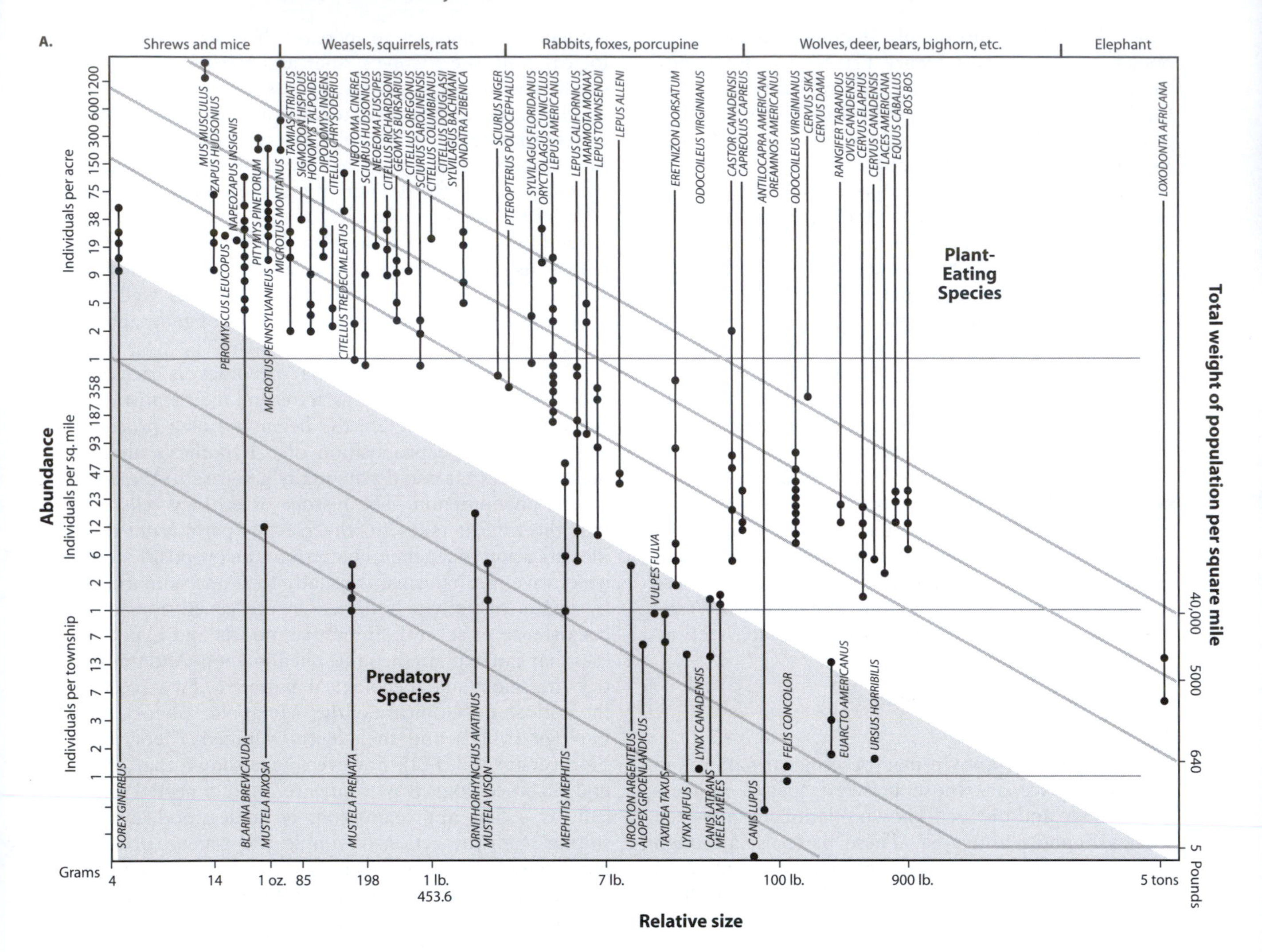

Figure 2. The relationship between population density and body size for (A) mammalian herbivore and carnivore species (from Mohr, Carl O. 1940. Comparative populations of game, fur and other mammals. American Midland Naturalist 24: 581–584) and for (B) 307 species of mammalian primary consumers. (From Damuth, J. 1981. Population density and body size in mammals. Nature 290: 699–700)

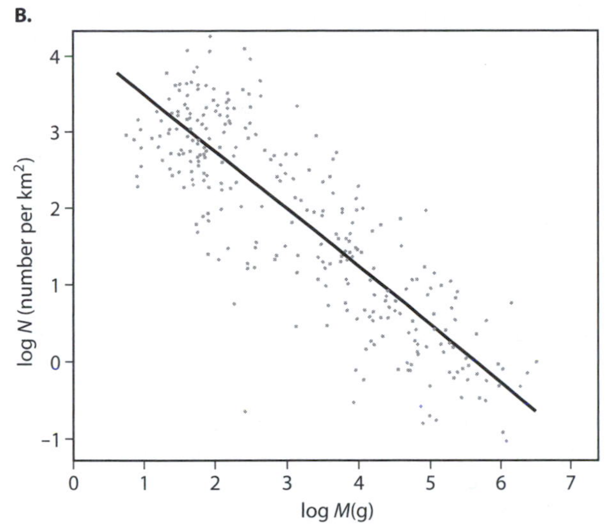

Figure 2. (*cont.*)

biotic forces in affecting diversity regulation. His reasoning was that one approach to assess the importance of biotic factors (such as competition and predation) was to compare empirical patterns against the results of a stochastic model that do not assume their existence. To do this, he used stochastic models first developed in the context of population genetics under neutrality. Twenty-five years later, in 2001, Stephen P. Hubbell expanded this approach and developed what he originally called the unified neutral theory of biodiversity and biogeography, which is also known as neutral macroecology because it is capable of generating several of the patterns usually studied by macroecologists, such as species abundance and species–area relationships.

The quantitative nature and predictive potential of neutral macroecology facilitated the development of testable null hypotheses for macroecological patterns under the assumption that individuals are equivalent (neutral) in terms of their vital rates of death, birth, migration, and the probability of becoming a new species. In practice, neutral macroecology builds on the theory of insular biogeography proposed by MacArthur and Wilson in 1963 by assuming a source area of immigrants or metacommunity, which in the context of neutral theory represents a biogeographic unit within which most member species originate, live, and become extinct, and which in turn contains several local communities embedded in it. However, unlike island biogeography theory, it assumes that individuals, instead of species, are equivalent. This allows neutral macroecology theory to provide precise predictions of the shape of the distribution of species' relative abundance among other patterns.

In its current form, neutral macroecology applies to trophically similar species, which, in local communities,

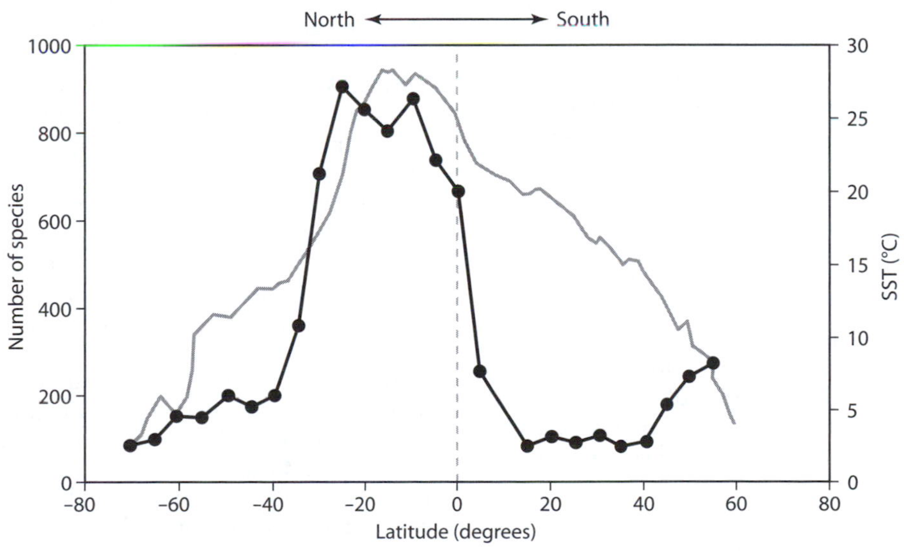

Figure 3. North–south view of the latitudinal diversity gradient of marine gastropod prosobranchs along the north- and southeastern Pacific shelves, from north Alaska to Cape Horn (gray line). Mean sea surface temperature (SST) along the continental margin is also shown (segmented line). (From Valdovinos, Claudio, Sergio A. Navarrete, and Pablo A. Marquet. 2003. Mollusk species diversity in the Southeastern Pacific: Why are there more species toward the south? Ecography 26: 139–144)

undergo random fluctuations in abundance, a process called ecological drift, as a consequence of stochastic birth, death, and immigration processes. Diversity in the local community is maintained by immigration from the metacommunity, where in addition to death and birth processes, speciation occurs. Thus, under neutral macroecology, community assembly is a process resulting from stochastic immigration only (i.e., dispersal-assembled communities) instead of resulting from adaptive divergence in species niches (i.e., niche-assembled communities). According to neutral macroecology, the relative species abundance distribution and the shape of the species–area relationship can take different forms, depending on the average rate of immigration from the metacommunity and the value of the so-called biodiversity number, $\theta = 2J_M\upsilon$, where J_M is the total number of individuals in the metacommunity and the speciation rate.

The dynamic, individual-based, quantitative and stochastic character of neutral macroecology theory and its ability to make predictions that can be compared to empirical patterns have led to its rapid development as a null hypothesis for macroecological patterns.

5. METABOLIC THEORY

Ever since Alfred Lotka, energy availability, acquisition, and apportionment have been thought to be essential for understanding biodiversity and ecosystem functioning. The metabolic theory of ecology provides a simple framework to analyze the role of energy flows from individuals to ecosystems. The core of metabolic theory rests on the understanding of individual metabolism and deriving its consequences for population, community, and ecosystem patterns and dynamics.

The efficiency with which energy is captured, delivered, and transformed by an organism for survival, growth, and reproduction is critical to individual fitness. This being the case, one might expect to find some general principles and regularities associated with energy acquisition and transformation. One such regularity, referred to as Kleiber's rule (after the Swiss physiologist Max Kleiber), describes how body size, M (mass in grams), is constrained by individual metabolic rate, B (watts, W), the total rate of energy transformation by an organism. This relationship is described by a simple mathematical function called a scaling law of the form

$$B = B_0 M^b, \tag{1}$$

where B_0 is a normalization constant independent of body size, and b is an "allometric" or scaling exponent. Kleiber was first to recognize that this scaling exponent takes a characteristic value of $b \approx \frac{3}{4}$. Despite the simplicity and ubiquity of Kleiber's rule, the mechanisms underlying it eluded biologists for more than 70 years.

A general, mechanistic explanation for Kleiber's rule was first proposed 1997 by Geoffrey West, Brian Enquist, and James H. Brown. The fundamental assumption of their model is that natural selection has resulted in the optimization of biological distribution networks in order to minimize the costs of transporting energy and materials within organisms. They demonstrate, theoretically, that the optimal solution to this problem is a hierarchical, fractal-like distribution network with space-filling geometry and size-invariant terminal metabolic units (e.g., mitochondria, chloroplasts). One of the consequences of these geometric constraints is the $\frac{3}{4}$ power exponent discovered by Kleiber. The model provides a parsimonious explanation for why a large number of functional and structural characteristics of organisms, which had been well known to ecologists and physiologists, such as growth rate, lifespan, and home range, relate to body size with scaling exponents that are simple multiples of $\frac{1}{4}$ and under which circumstances they might deviate from this theoretical expectation. The existence of such universal scaling laws implies that dynamically and organizationally, all mammals, for example, are on the average scaled manifestations of a single idealized mammal, whose properties are determined as a function of its size. That is, in terms of almost all biological rates, times, and internal structure, an elephant is approximately a blown-up gorilla, which is itself a blown-up mouse, all scaled in an appropriately nonlinear, predictable way. This work paved the way to the MTE outlined in 2004 by James H. Brown and collaborators.

By explicitly focusing on the causes and consequences of individual metabolic rate, the MTE provides new opportunities to deal with the inherent complexity of ecological systems at different levels of biological organization. We exemplify this approach by showing how metabolic theory can be used to derive predictions on species abundance within communities, as derived by these authors.

Because the maintenance and reproduction of an individual require energy, everything else being equal, the maximum number of individuals per unit area, N, that a species can achieve will be proportional to the ratio of the rate of resource supply per unit area in the environment, R, and the individual metabolic rate, B

$$N \propto \frac{R}{B}. \tag{2}$$

This expression leads to

$$N \propto M^{-\frac{3}{4}}. \quad (3)$$

Indeed, as we saw earlier, John Damuth had already reported in 1981 that the relationship between abundance and body size typically yields exponents of approximately $-3/4$, which has since been shown to hold across a wide variety of organisms from microbes to trees. Equation 3 also implies that that total energy flux by a population (i.e., $N \times M^{\frac{3}{4}}$) is independent of body size, meaning that species populations within communities are equivalent in the amount of energy that they control, the so-called Energetic Equivalence Rule. However, this relationship should hold for species using the same resource, which is not the case if we work with complete communities where species occupy different trophic positions and feed on different resources. To extend the theory to local communities of trophically dissimilar species, we need to consider that species occupying different trophic positions usually differ in size (e.g., predators are usually larger than their prey) and that the efficiency of energy transfer between trophic levels is usually low ($\sim 10\%$). Once these characteristics are taken into account, metabolic theory predicts that population density (N) across trophic levels should scale with mass as M^{-1} and that biomass (i.e., $N \times M$) should be independent of body mass, which happens to be a well-known pattern in aquatic ecosystems. Empirically, analyses of freshwater and marine communities across trophic levels agree with the expectations of metabolic theory.

A novel contribution of MTE has been to derive a new expression that characterizes the combined effects of body size and temperature on metabolic rate:

$$B \approx b_0 M^{3/4} e^{-E/kT}. \quad (4)$$

In statistical mechanics, the term $\exp(-E/kT)$, often referred to as the Boltzmann factor, is proportional to the fraction of molecules of a gas that attain an energy state of E at an absolute temperature T. To react, the molecules must possess *activation energy*; that is, they must collide with one another with sufficient energy to change their state. Temperature increases the proportion of molecules that attain sufficient energy to react. Hence, the Boltzmann factor can be used to describe individual level biochemical kinetics and metabolic rate. Recent expansions of metabolic theory range from the analysis of the effects of body size and temperature on nucleotide substitutions, speciation rates, and the latitudinal diversity gradient to ecosystem respiration and the carbon cycle.

Macroecology likely represents a moment in the investigation of complex ecological systems, an attempt to come to terms with ecological complexity. Much remains to be done to advance the macroecological research program and to cope with some of the criticisms against it. Macroecology needs more and better data of ecological systems and dynamics from local to regional, continental, and global scales. Although for many questions the reliance on published data sets and compilation studies will be inevitable, macroecology needs to back up some of its key empirical patterns and claims with experiments and field studies specially designed to assess them or to test their predictions. In addition to this, there are two exciting scientific paths whose developments are of paramount importance to macroecology. The first is the need to better link reductionistic (microecological) with macroecological patterns and explanations. At the end of the day, fundamental principles and laws should also be at work no matter the scale of analyses and thus can help to design and interpret small-scale experiments involving few species or the response of ecosystems to human-driven perturbations. Notice that this is not a purely academic exercise. A general theory of ecological systems and dynamics will help to generate new and better management strategies and policies to ameliorate the impact of humans on ecosystems and to restore their functioning and the services they provide to humanity. This link has not been sufficiently explored. The second is the daunting task to unify neutral macroecology and metabolic theory. Energy and stochasticity are essential components of living systems from cells to ecosystems. Their unification in a stochastic theory of energy flow will be a great scientific achievement.

FURTHER READING

Alonso, David, Rampal S. Etienne, and Alan J. McKane. 2006. The merits of neutral theory. Trends in Ecology and Evolution 21: 451–457. *An excellent synthetic presentation of neutral macroecology, its development, current status, and future projections.*

Brown, James H. 1995. Macroecology. Chicago: The University of Chicago Press. *The first book introducing macroecology with emphasis on the existence of general patterns and principles in ecological systems and the role of energy as a fundamental currency in macroecology.*

Brown, James H., James F. Gillooly, Andrew P. Allen, Van M. Savage, and Geoffrey B. West. 2004. Toward a metabolic theory of ecology. Ecology 85: 1771–1789. *Outstanding presentation of the Metabolic Theory of Ecology as a general theory of ecological systems.*

Brown, James H., and Brian A. Maurer. 1989. Macroecology: The division of food and space of species on continents. Science 243: 1145–1150. *Seminal paper introducing and defining the term macroecology.*

Gaston, Kevin J. 2003. The Structure and Dynamics of Geographic Ranges. Oxford: Oxford University Press. *A magnificent overview of patterns in the structure of geographic ranges of species and their implications.*

Gaston, Kevin J., and Tim M. Blackburn. 2000. Pattern and Processes in Macroecology. Oxford: Blackwell Science. *A comprehensive presentation and synthesis of the scope and aims of macroecology understood as large-scale community ecology and its importance for understanding patterns from local to global scales.*

Gaston, Kevin J., and Tim M. Blackburn, eds. 2003. Macroecology: Concepts and Consequences. Boston: Blackwell Publishing. *First compilation of the different approaches to macroecological research written by the major figures in the field.*

Hubbell, Stephen P. 2001. The Unified Neutral Theory of Biodiversity and Biogeography. Princeton, NJ: Princeton University Press. *A landmark book introducing, in a comprehensive way, the stochastic theory of neutral macroecology, with emphasis on its explicative and synthetic character in comparison to alternative approaches inspired in niche differences among species.*

Maurer, Brian A. 1999. Untangling Ecological Complexity. Chicago: University of Chicago Press. *A fascinating introduction to the attributes of complex ecological systems and the merits of the macroecological approach to the understanding of patterns and processes in ecological communities.*

Storch, David, Pablo A. Marquet, and James H. Brown. 2007. Scaling Biodiversity. Cambridge: Cambridge University Press.

West, Geoffrey B., and James H. Brown. 2005. The origin of allometric scaling laws in biology from genomes to ecosystems: Towards a quantitative unifying theory of biological structure and organization. Journal of Experimental Biology 208: 1575–1592. *A wonderful and lucid review of the origin and implications of scaling laws in biology.*

III.17

Alternative Stable States and Regime Shifts in Ecosystems

Marten Scheffer

OUTLINE

1. Introduction
2. The theory in a nutshell
3. Shallow lakes as an example
4. Mechanisms for alternative stable states in ecosystems
5. How to know if a system has alternative stable states
6. Using alternative stable states in management

Complex systems ranging from cells to ecosystems and the climate can have tipping points, where the slightest disturbance can cause the system to enter a phase of self-propagating change until it comes to rest in a contrasting alternative stable state. The theory explaining such catastrophic change at critical thresholds is well established. In particular, an early influential book by the French mathematician René Thom catalyzed the interest in what he called "catastrophe theory." Although many claims about the applicability to particular situations were not substantiated later, catastrophe theory created an intense search for real-life examples, much like chaos theory later. In the 1970s, C. S. Holling was among the first to argue that the theory could explain important aspects of the dynamics of ecosystems, and an influential review by Sir Robert May in *Nature* promoted further interest among ecologists. Nonetheless, not until recently have strong cases for this phenomenon in ecosystems been built.

GLOSSARY

alternative stable states. A system is said to have alternative stable states if under the same external conditions (e.g., nutrient loading, harvest pressure, or temperature) it can settle to different stable states. Although genuine "stable states" occur only in models, the term is also used more liberally to refer to alternative dynamic regimes.

attractor. A state or dynamic regime to which a model asymptotically converges, given sufficient simulation time. Examples are a stable point, a cycle, or a strange attractor.

catastrophic shift. A shift to an alternative attractor that can be invoked by an infinitesimal small change at a critical point known as catastrophic bifurcation.

hysteresis. The phenomenon that the forward shift and the backward shift between alternative attractors happen at different values of an external control variable.

regime shift. A relatively fast transition from one persistent dynamic regime to another. Regime shifts do not necessarily represent shifts between alternative attractors.

resilience. The capacity of a system to recover to essentially the same state after a disturbance.

1. INTRODUCTION

The idea of catastrophic change at critical thresholds is intuitively straightforward in physical examples. Suppose you are in a canoe and gradually lean over to one side more and more. It is difficult to see the tipping point coming, but eventually leaning over too much will cause you to suddenly capsize and end up in an alternative stable state from which it is not easy to return. Still, people have been hesitant to believe that large complex systems such as ecosystems or the climate would sometimes behave in a similar way. Indeed, fluctuations around gradual trends rather than "tipping over" seem the rule in nature. Nonetheless, occasionally sudden changes from one contrasting fluctuating regime to another one are observed. Such abrupt changes have been termed regime shifts.

As we shall see in this chapter, regime shifts are indicative of the existence of tipping points and alternative stable states but by no means a proof. Indeed, rigorous experimental proofs are possible only in small controlled systems. Such a difficulty of proving that a mechanism is at work in nature is common in ecology. For instance, it has proven remarkably hard to demonstrate unequivocally the role of a mechanism as basic as competition. Nonetheless, the role of alternative stable states in driving some ecosystems dynamics has become an important focus of research. I first briefly show the key aspects of the theory and elaborate the case of shallow lakes as an example. Subsequently, I briefly highlight some other mechanisms and discuss how one may find out if a system has alternative stable states. Finally, I reflect on how insights in such stability properties can be used in managing ecosystems.

2. THE THEORY IN A NUTSHELL

Smooth, Threshold, and Catastrophic Response to Change

In most cases, the equilibrium of a dynamic system moves smoothly in response to changes in the environment (figure 1A). Also, the system is often rather insensitive over certain ranges of the external conditions, although it responds relatively strongly around some threshold conditions (figure 1B). For instance, mortality of a species usually drops sharply around a critical concentration of a toxicant. In such a situation, a strong response happens when a threshold is passed. Such thresholds are obviously important to understand. However, a very different, much more extreme kind of threshold than this occurs if the system has alternative stable states. In that case, the curve that describes the response of the equilibrium to environmental conditions is typically "folded" backward (figure 1C). Such a *catastrophe fold* implies that, for a certain range of environmental conditions, the system has two alternative stable states separated by an unstable equilibrium (dashed lines), which marks the border between the basins of attraction of the alternative stable states.

This situation is the root of catastrophic shifts or, with less negative connotation, critical transitions (figure 2). When the system is in a state on the upper branch of the folded curve, it can not pass to the lower branch smoothly. Instead, when conditions change sufficiently to pass the threshold (F_2), a "catastrophic" transition to the lower branch occurs. Clearly this point is a very special point. In the exotic jargon of dynamic systems theory it is called a *bifurcation point*. The point we have in our picture marks a so-called *catastrophic bifurcation*. Such bifurcations are characterized by the fact that an infinitesimally small change in a control parameter (reflecting, for instance, the temperature) can invoke a large change in the state of the system if it crosses the bifurcation. The bifurcation points in a catastrophe fold (F_1 and F_2) are known as *fold bifurcations*. (They are also called "saddle–node" bifurcations because in these points a stable "node" equilibrium meets an unstable "saddle" equilibrium.)

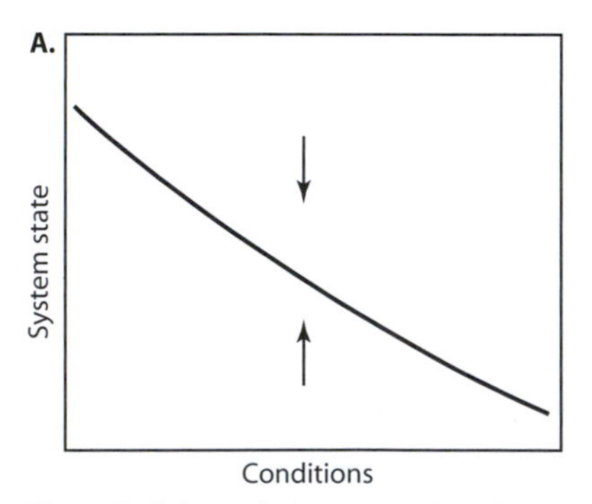

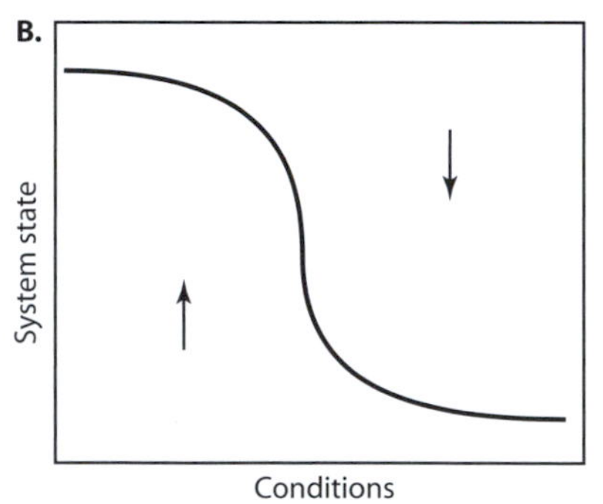

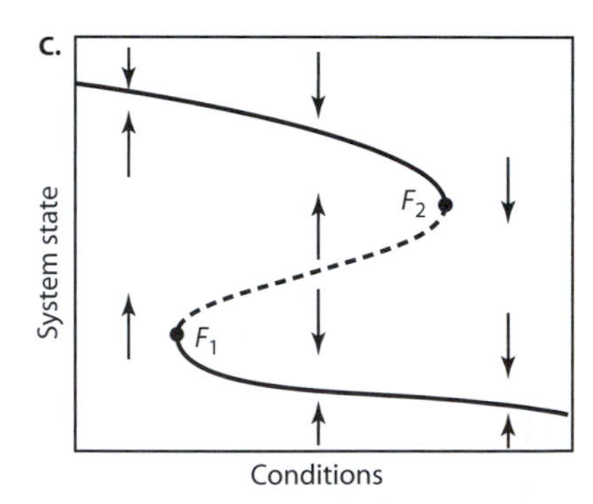

Figure 1. Schematic representation of possible ways in which the equilibrium state of a system can vary with conditions such as nutrient loading, exploitation, or temperature rise. In panels A and B only one equilibrium exists for each condition. However, if the equilibrium curve is folded backward (panel C), three equilibria can exist for a given condition. The arrows in the graphs indicate the direction in which the system moves if it is not in equilibrium (i.e., not on the curve). It can be seen from these arrows that all curves represent stable equilibria except for the dashed middle section in panel C. If the system is pushed away a little bit from this part of the curve, it will move further away instead of returning. Hence, equilibria on this part of the curve are unstable and represent the border between the basins of attraction of the two alternative stable states on the upper and lower branches.

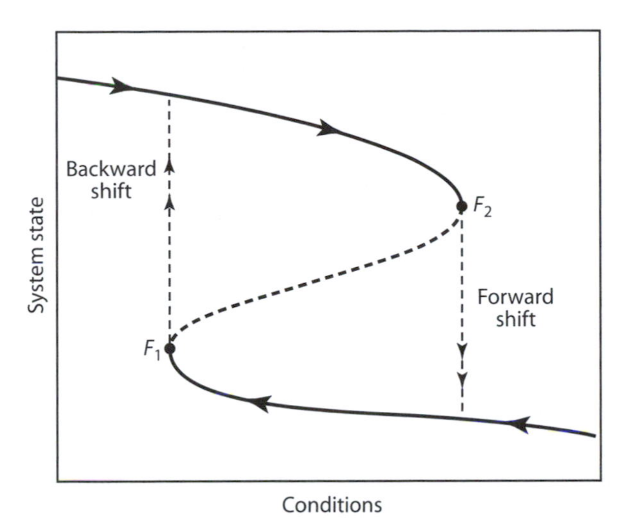

Figure 2. If a system has alternative stable states, critical transitions and hysteresis may occur. If the system is on the upper branch but close to the bifurcation point F_2, a slight incremental change in conditions may bring it beyond the bifurcation and induce a critical transition (or catastrophic shift) to the lower alternative stable state ("forward shift"). If one tries to restore the state on the upper branch by means of reversing the conditions, the system shows hysteresis. A backward shift occurs only if conditions are reversed far enough to reach the other bifurcation point F_1.

Hysteresis

The fact that a tiny change in conditions can cause a major shift is not the only aspect that sets systems with alternative stable states apart. Another important feature is the fact that in order to induce a switch back to the upper branch, it is not sufficient to restore the environmental conditions that existed before the collapse (F_2). Instead, one needs to go back further, beyond the other switch point (F_1), where the system recovers by shifting back to the upper branch. This pattern in which the forward and backward switches occur at different critical conditions (figure 2) is known as *hysteresis*. From a practical point of view, hysteresis is important because it implies that this kind of catastrophic transition is not so easy to reverse.

The idea of catastrophic transitions and hysteresis can be nicely illustrated by stability landscapes. To see how stability is affected by change in conditions, we make stability landscapes for different values of the conditioning factor (figure 3). For conditions in which there is only one stable state, the landscape has only one valley. However, for the range of conditions where two alternative stable states exist, the situation becomes more interesting. The stable states occur as valleys, separated by a hilltop. This hilltop is also an equilibrium (the slope of the landscape is zero). However, this equilibrium is unstable. It is a *repellor*. Even the slightest movement away from it will lead to a self-propagating *runaway change* moving the system toward a stable equilibrium. Such a stable equilibrium is an *attractor*.

To see the catastrophic transitions and hysteresis in this representation, imagine what happens if you start in the situation of the landscape up front. The system will then be in the only existing equilibrium. There is no other attractor, and therefore, this state is said to be *globally stable*. Now, suppose that conditions change gradually, so that the stability landscape changes to the

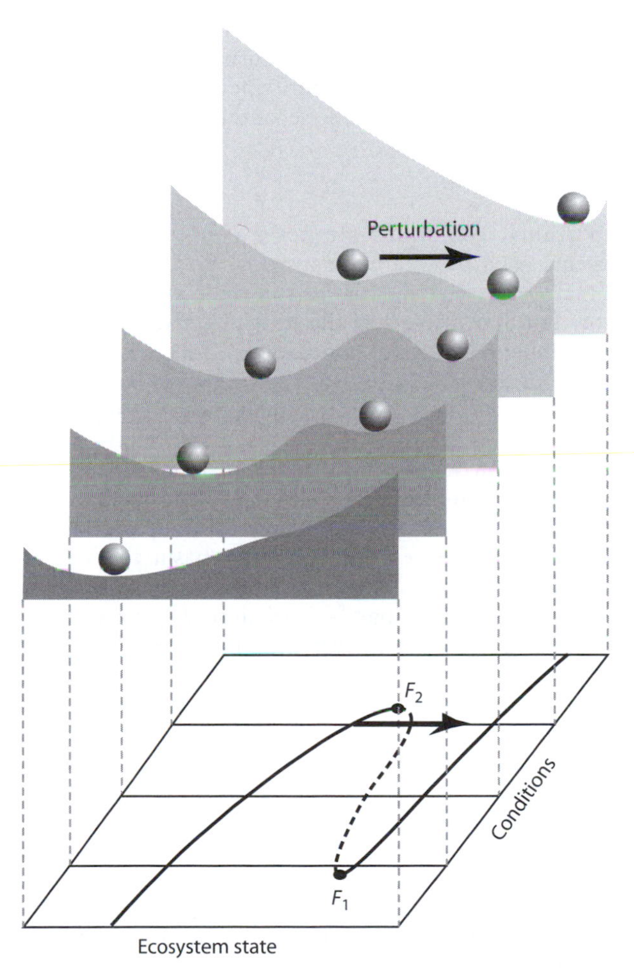

Figure 3. External conditions affect resilience of multistable systems to perturbation. The bottom plane shows the equilibrium curve as in figure 2. The stability landscapes depict the equilibria and their basins of attraction at five different conditions. Stable equilibria correspond to valleys; the unstable middle section of the folded equilibrium curve corresponds to hilltops. If the size of the attraction basin is small, resilience is small, and even a moderate perturbation may bring the system into the alternative basin of attraction.

second or third one in the row. Now, there is an alternative attractor, implying that the state in which we were has become locally (rather than globally) stable. However, as long as no major perturbation occurs, the system will not move to this alternative attractor. In fact, if we would monitor the state of the system, we would not see much change at all. Nothing would reveal the fundamental changes in the stability landscape. If conditions change even more, the basin of attraction around the equilibrium in which the system rests becomes very small (fourth stability landscape) and eventually disappears (last landscape), implying an inevitable catastrophic transition to the alternative state. Now, if conditions are restored to previous levels, the system will not automatically shift back. Instead, it shows hysteresis. If no large perturbations occur, it will remain in the new state until the conditions are reversed beyond those of the second landscape.

Resilience

In reality, conditions are never constant. Stochastic events such as weather extremes, fires, or pest outbreaks can cause fluctuations in the conditioning factors but may also affect the state directly, for instance by wiping out parts of populations. If there is only one basin of attraction, the system will settle back to essentially the same state after such events. However, if there are alternative stable states, a sufficiently severe perturbation may bring the system into the basin of attraction of another state. Obviously, the likelihood of this happening depends not only on the perturbation but also on the size of the attraction basin. In terms of stability landscapes (figure 3), if the valley is small, a small perturbation may be enough to displace the ball far enough to push it over the hilltop, resulting in a shift to the alternative stable state. Following Holling, I use the term *resilience* to refer to the size of the valley or basin of attraction around a state that corresponds to the maximum perturbation that can be taken without causing a shift to an alternative stable state. Note that gradually changing conditions may have little effect on the state of the system but nevertheless reduce the size of the attraction basin. This loss of resilience makes the system more fragile in the sense that it can be easily tipped into a contrasting state by stochastic events.

The Continuum between Catastrophic and Smooth Response

A tricky and often overlooked problem is that we cannot generalize stability properties of a system. For instance, we cannot make statements such as: the critical nutrient level for a lake to collapse into a turbid state is 0.1 mg/L phosphorus. In fact, we cannot even say that "lakes have alternative stable states." In technical terms, the problem is that the positions of critical bifurcation points (e.g., F_1 and F_2) always depend on various parameters of a model. In practice, this means that the corresponding thresholds are not fixed values. For instance, the critical nutrient level for a shallow lake to flip from a clear to a turbid state depends on its size. In a wider sense, it means that safe limits to prevent critical transitions will usually not have universal fixed values.

A corollary is that the degree of hysteresis may vary strongly. For instance, shallow lakes can have a pronounced hysteresis in response to nutrient loading (figure 1C), whereas deeper lakes may react smoothly (figure 1B). Often a parameter can be found that can be changed such that the bifurcation points move closer together and eventually merge and disappear. This "bifurcation of bifurcations" is known as *cusp bifurcation*, after the hornlike shape produced by the fold bifurcations moving together (figure 4). It marks the change from a situation in which the system can respond in a catastrophic way to a situation in which the response to a control parameter (parameter 2 in figure 4) is always smooth, as there are no alternative attractors. Thus, the panels with distinct possible responses to external conditions (figure 1) do in fact represent snapshots of a continuum of possible behavior that may be displayed by a single system. A positive feedback is usually responsible for causing a threshold response. A moderate feedback may turn a smooth response (figure 1A) into a threshold response (figure 1B), and a stronger feedback may cause the response curve to turn into a catastrophe fold (figure 1C).

Another point to note is that there is in principle no limit to the number of alternative attractors in a system. In general, complex systems may have complex stability landscapes, with numerous smaller or larger attraction basins. In analogy to the scenarios for two alternative stable states, gradually changing conditions (e.g., temperatures) may alter the landscapes, making some attraction basins larger and causing others to disappear. Meanwhile, disturbances occasionally flip the system out of an attraction basin that has become small, allowing it to settle into a more resilient state. Obviously, this is still a very stylized world view, but before elaborating it, I will give an example of an ecosystem with alternative stable states.

From Simple Theory to Complex Reality

It is obvious that the elegant and simple models described in dynamic systems theory can capture only

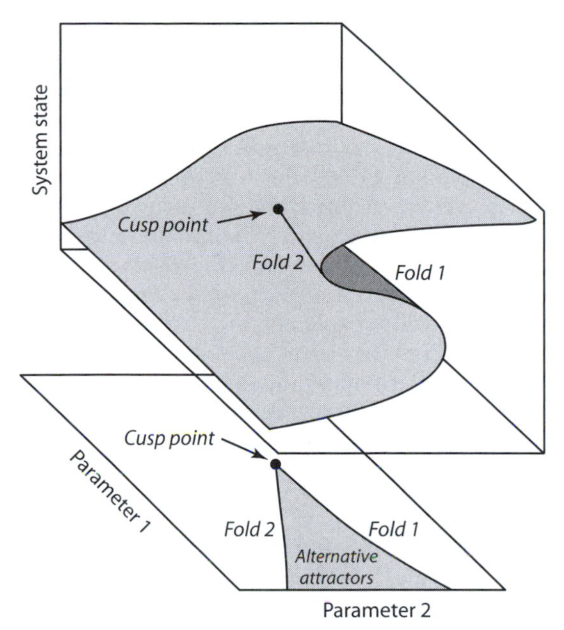

Figure 4. The cusp point where two fold bifurcation lines meet tangentially and disappear marks the change from a system with catastrophic state shifts in response to change in parameter 2 to one that responds smoothly to that parameter.

some aspects of what happens in complex reality. For instance, the idea of "stable states" is a gross simplification of the dynamic regimes observed in nature. There are always fluctuations, driven by a combination of stochastic forcing and seasonal cycles in the weather, mixed in many systems with intrinsically generated chaotic dynamics or cycles. Therefore, it is perhaps more appropriate to speak about alternative dynamic regimes rather than alternative stable states if we are thinking about real systems. Nonetheless, I shall stick to the simple "stable states" terminology in this chapter. On very long time scales, there is obviously no stability either, as eventually lakes become land, and the Earth system evolves inevitably. Another complication is the heterogeneity of nature, as compared to the simple models used by theoreticians. The models usually assume homogeneous environments with only a few key players. In practice, heterogeneity of environmental conditions may complicate the picture, and so may the great variety of species that are usually involved in ecosystem dynamics. Some places may be better for growth than others, and some species may replace others depending on conditions. Spatial exchange may then help to smooth patterns at larger scales. I cannot elaborate on all those aspects here, but the bottom line is that although fluctuations, heterogeneity, and spatial processes may smooth things, there are also situations where the mechanism leading to alternative stable states is strong enough to dominate the dynamics, and real systems show a behavior that can be quite well explained by the simple theory presented above.

3. SHALLOW LAKES AS AN EXAMPLE

A particularly well-studied example of bistability of an entire ecosystem is the development of submerged vegetation in turbid shallow lakes. Submerged plants can enhance water clarity by a suite of mechanisms such as control of excessive phytoplankton development and prevention of wave resuspension of sediments. However, the submerged plants also need low turbidity in order to get sufficient light. As a consequence, there is a positive feedback: plants promote water clarity and vice versa. It seems intuitively straightforward that such a positive feedback may lead to alternative stable states: one vegetated and another one without plants. However, things are more complex than that. First, as argued, alternative equilibria arise only if the feedback effect is strong enough. Second, stability of one of the states can be lost if external factors such as climate or nutrient input change (cf. figure 3).

To see how such loss of stability can happen, consider a simple graphic model of the response of shallow lakes to nutrient loading (figure 5). An overload with nutrients such as phosphorus and nitrogen derived from waste water or fertilizer use tends to make lakes turbid. This is because the nutrients stimulate growth of microscopic phytoplankton, which makes the water greenish and turbid. Although this *eutrophication* process can be gradual, shallow lakes tend to jump abruptly from the clear to the turbid state. This behavior can be explained from a simple graphic model based on only three assumptions: (1) Turbidity increases with the nutrient level because of increased phytoplankton growth. (2) Vegetation reduces turbidity. (3) Vegetation disappears when a critical turbidity is exceeded.

In view of the first two assumptions, equilibrium turbidity can be drawn as two different functions of the nutrient level: one for a macrophyte-dominated and one for an unvegetated situation. Above a critical turbidity, macrophytes will be absent, in which case the upper equilibrium line is the relevant one; below this turbidity the lower equilibrium curve applies. The emerging picture shows that, over a range of intermediate nutrient levels, two alternative equilibria exist: one with macrophytes and a more turbid one without vegetation. At lower nutrient levels, only the macrophyte-dominated equilibrium exists, whereas at

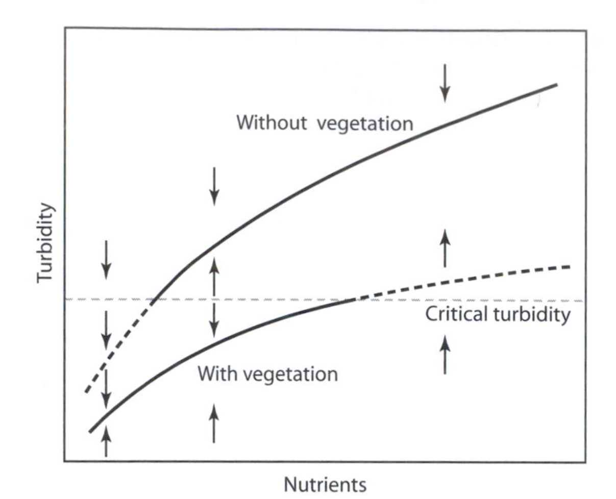

Figure 5. Alternative equilibrium turbidities caused by disappearance of submerged vegetation when a critical turbidity is exceeded (see text for explanation). The arrows indicate the direction of change when the system is not in one of the two alternative stable states. (From Scheffer, M., S. H. Hosper, M. L. Meijer, B. Moss, and E. Jeppesen. 1993. Alternative equilibria in shallow lakes. Trends in Ecology and Evolution 8: 275–279)

the highest nutrient levels, there is only a vegetationless equilibrium.

The zigzag line formed by the stable and unstable equilibria in this graphic model corresponds to the folded line in figure 1C and the panel below the stability landscapes (figure 3). However, this simple example may serve to convey a better feeling for the way in which a facilitation mechanism may cause the system to respond to environmental change showing hysteresis and catastrophic transitions. Gradual enrichment starting from low nutrient levels will cause the lake to proceed along the lower equilibrium curve until the critical turbidity is reached at which macrophytes disappear. Now, a jump to a more turbid equilibrium at the upper part of the curve occurs. In order to restore the macrophyte-dominated state by means of nutrient management, the nutrient level must be lowered to a value where phytoplankton growth is limited enough by nutrients alone to reach the critical turbidity for macrophytes again. At the extremes of the range of nutrient levels over which alternative stable states exist, either of the equilibrium lines approaches the critical turbidity that represents the breakpoint of the system. This corresponds to a decrease of resilience. Near the edges, a small perturbation is enough to bring the system over the critical line and to cause a switch to the other equilibrium.

Small lakes have the advantage that they can be experimentally manipulated relatively well. Whole-lake experiments in which the fish stock is strongly reduced for a brief period can induce a shift from the turbid to the clear state. The short-term effect of fish removal on turbidity is explained by a trophic cascade (fewer fish results in more zooplankton and therefore less phytoplankton) and the role of bottom-dwelling fish in recycling of nutrients and resuspending sediments. However, fish quickly reproduce, allowing the populations to recover from the brief fishing campaign. The fact that the result of such "shock therapy" can be long term as well is explained by the positive feedback between water clarity and submerged vegetation. The whole-lake experiments and other lines of evidence have made the shallow lake case one of the best-documented examples of alternative stable states in ecosystems.

4. MECHANISMS FOR ALTERNATIVE STABLE STATES IN ECOSYSTEMS

By use of models, it can be shown that alternative stable states may plausibly arise from a range of mechanisms. The shallow lakes case in its simplest representation is an example of *facilitation*. The submerged plants facilitate their own growth by "engineering" the environmental conditions, making the water clearer. A similar thing may happen in semiarid environments where plants may enhance the water conditions in their environment. This can happen on a very local scale. For instance, an adult plant canopy can ameliorate hot and dry conditions, thereby facilitating growth of smaller plants. However, vegetation may also lead to more cloudy and rainy conditions on a regional scale in some places. For instance, in Northwest Africa, the monsoon circulation that brings rains from the ocean to the land is promoted by vegetation cover, implying a positive feedback. Such positive feedback between plants and water conditions can create alternative stable states on local as well as regional states.

Another mechanism that is well known to create alternative stable states in some situations is *competition* between two species or functional groups. In classical competition models, the necessary and sufficient condition is that intraspecific competition is weaker than interspecific competition. Thus, both species should do better if they are in a monoculture than when they are together with the other species. If the two competing species or functional groups are dominant in the ecosystem, such a competitive play may dominate the entire ecosystem dynamics. An example is the shift from a diverse phytoplankton community to dominance by particular groups of cyanobacteria in lakes. The cyanobacteria are more shade tolerant but also create more shade given the same amount of nutrients. They thus create conditions in which they are also better competitors. This can lead to a runaway shift toward cyanobacterial dominance under some conditions.

A third mechanism for alternative stable states that should be mentioned is *overexploitation*. Classical work by Noy-Meir in the 1970s has shown elegantly how, under some conditions, a population can come into free fall if it is exploited beyond a certain critical level. The mechanism is that at low population numbers the overall production declines with population density. Therefore, if exploitation levels do not decline proportionally or more, there will be a self-propagating further decline in the population.

Numerous, more complicated mechanisms for alternative stable states in ecology exist. For instance, metapopulations in scattered habitats may collapse beyond a certain critical level of habitat fragmentation. Also, alternative stable states may occur if a predator controls the natural enemies of its offspring. For instance, it has been hypothesized that Nile perch could come to dominate the Lake Victoria fish community only after the cichlids that can be important egg predators were driven to low enough densities by overfishing and eutrophication. Now the abundant Nile perch helps in keeping the cichlid populations low through predation.

5. HOW TO KNOW IF A SYSTEM HAS ALTERNATIVE STABLE STATES

Evidence from Field Data

Jumps in Time Series or Regime Shifts

Sudden changes in a system are always an interesting feature, and not surprisingly, various statistical techniques have been developed to check whether a shift in a time series can be explained by chance. However, it is important to keep in mind that even if a critical transition in a time series is significant, this does not necessarily imply that it was a jump between alternative attractors. Probably the most common cause of a sudden change is a sudden change in the conditions. For instance, the closure of a dam for a major reservoir may cause a drastic shift in the downstream river ecosystem. Another possible explanation of a sudden shift is that conditions changed gradually but exceeded a limit at which the system changes drastically although not catastrophically (i.e., not a stability shift related to a catastrophic bifurcation). For example, the onset and termination of a period of ice-cover in a lake can be quite sudden, even if temperature changes develop gradually. Thus, sudden shifts in a time series of some indicator of the state of a system may often simply be caused by a sudden drastic change in an important control parameter (figure 6A) or a control parameter reaching a range where the system responds strongly, even though there is no bifurcation (figure 6B). On the other hand, they can be true critical transitions caused by a tiny but critical change in conditions (figure 6C) and/or a disturbance pushing the system across the border of a basin of attraction (figure 6D). I should stress again, that although real stability shifts (lower two panels) are a distinctly different occurrence (e.g., they can be triggered by infinitely small change and have some irreversibility), there is in fact a continuum of possibilities in the range from linear to catastrophic system responses (figure 4).

It may seem impossible to detect from a time series whether the system went through a real stability shift or, rather, jumped to one of the mechanisms illustrated in the upper two panels of figure 6. However, at least in theory, there are some options to sort that out. First, there is a statistical approach to infer whether alternative attractors are involved in a shift based on the principle that any attractor shift implies a phase in which the system is speeding up as it is diverging from the repelling border of the basin of attraction. Another approach is to compare the fit of contrasting models with and without attractor shifts or compute the probability distribution of a bifurcation parameter. Unfortunately, all such tests require extensive time series of good quality and containing many shifts. Thus, although jumps in time series are an indication that something interesting is happening, they are usually not enough to determine if we are dealing with true stability shifts.

In ecology, much discussion has been devoted to the question of how to map effects of random massive colonization events to stability theory. These happen, for instance, in marine fouling communities that, once established, can be very persistent and hard to replace until the cohort simply dies of old age. It seems inappropriate to relate such shifts to alternative stable regimes unless the new state can persist through more generations by rejuvenating itself. The latter might be the case, for instance, in dry forests, where an adult plant cover is essential for survival of juveniles except in very rare wet years, which trigger initial massive seedling establishment.

Sharp Boundaries and Multimodality of Frequency Distributions

The spatial analog to jumps in time series is the occurrence of sharp boundaries between contrasting states. For instance, lush kelp forests on rocky coasts

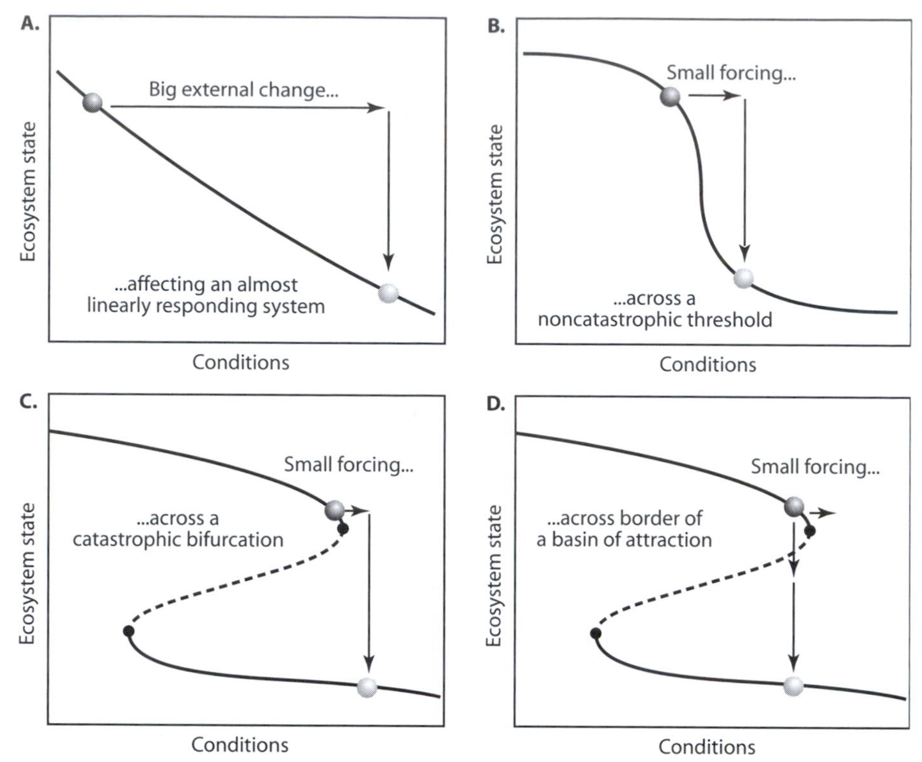

Figure 6. Four ways in which a sudden jump in a time series can be explained: (A) a sudden big change in conditions, (B) a small change in conditions in a range in which the system is very sensitive, (C) a small change in conditions passing a catastrophic bifurcation, and (D) a disturbance of the system state across a boundary of a basin of attraction causing a stability shift.

can be interrupted by remarkably distinct *barrens* where grazers prevent development of the macroalgae. Similarly, if one samples many distinct systems such as lakes, one may find them to fall into distinct contrasting classes. Statistically, the frequency distributions of key variables should be multimodal (e.g., figure 7B) if there are alternative attractors. Sophisticated tests are available for multimodality, but again these require rich data sets and have low power for the limited data sets often available for ecological studies. Therefore, there is a good chance of concluding that one mode is sufficient even when the data are truly multimodal. Importantly, significant multimodality does not necessarily imply alternative attractors. There may often be alternative explanations, analogous to those described in the previous section on shifts in time series: a conditioning factor may itself show a sharp change along a spatial gradient or be multimodally distributed. Also, the system may show a threshold response to a spatially varying factor without having alterative basins of attraction.

Dual Relationship to Control Factor

Part of the difficulty in interpreting jumps in time series and spatial patterns as indicators of alternative stability domains stems from the problem that we do not know how conditioning factors vary. If one has sufficient data and insight in the role of driving factors, one can push the diagnosis a step further by directly plotting the system state against the value of an important conditioning factor (such as temperature). Ideally, this produces plots that are directly comparable to the lower panels in figure 6. Statistically, this is not completely straightforward, but one may, for instance, test whether the response of the system to a control factor is best described by two separate functions rather than one single regression (e.g., figure 7C). Such tests for multiplicity of regression models are easily conducted using likelihood ratios, the extra sum of squares principle, or information statistics. Dual relationships, if they exist, are suggestive of an underlying hysteresis curve. Still, it may be that a shift in some unknown other control factor has simply taken the system to a different state in

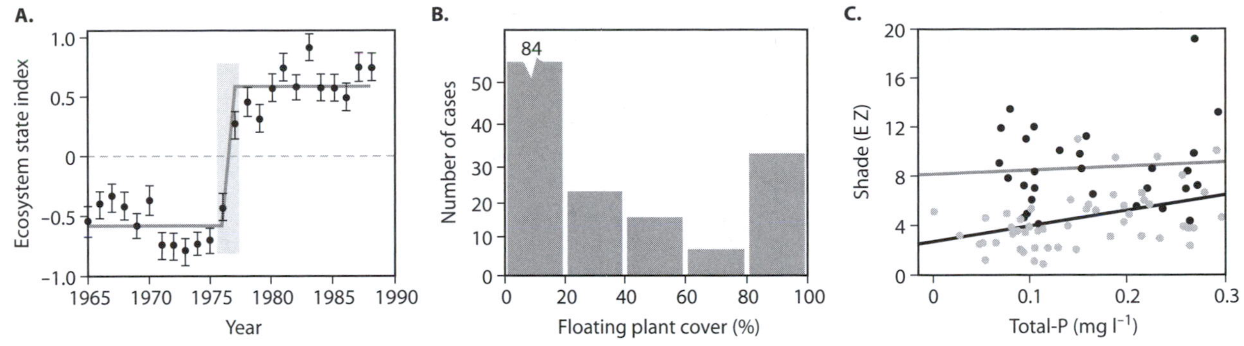

Figure 7. Three types of hints of the existence of alternative attractors from field data: (A) a shift in a time series, (B) multimodal distribution of states, and (C) dual relationship to a control factor. The specific examples are (A) a regime shift in the Pacific Ocean ecosystem (Modified with permission from Hare, S. R., and N. J. Mantua. 2000. Empirical evidence for North Pacific regime shifts in 1977 and 1989. Progress in Oceanography 47: 103–145); (B) bimodal frequency distribution of free-floating plants in a set of 158 Dutch ditches (Modified with permission from Scheffer, M., S. Szabo, A. Gragnani, E. H. van Nes, S. Rinaldi, N. Kautsky, J. Norberg, R.M.M. Roijackers, and R.J.M. Franken. 2003. Floating plant dominance as a stable state. Proceedings of the National Academy of Sciences U.S.A. 100: 4040–4045); and (C) different relationships between underwater shade and the total phosphorus concentration for shallow lakes dominated by Cyanobacteria (black circles) and lakes dominated by other algae (gray circles). (Modified with permission from Scheffer, M., S. Rinaldi, A. Gragnani, L. R. Mur, and E. H. Van Nes. 1997. On the dominance of filamentous cyanobacteria in shallow, turbid lakes. Ecology 78: 272–282)

which all kinds of relationships between variables and environmental factors look different.

In conclusion, one may obtain indications for the existence of alternative attractors from descriptive data, but the evidence can never be conclusive. There is always the possibility that discontinuities in time series or spatial patterns result from discontinuities in some environmental factor. Alternatively, the system might simply have a threshold response that is not related to alternative stability domains. The latter possibility is, of course, still very interesting. First, it helps to know that the system can change sharply if it is pushed across a threshold. Second, it often implies that under different conditions (as represented, for instance, by parameter 1 in figure 4), true alternative attractors could arise in the same system.

Experimental Evidence

Experiments can be difficult to perform on relevant scales. However, they are much easier to interpret than field patterns. I discuss three major ways in which experiments can provide evidence for the existence of alternative attractors.

Different Initial States Lead to Different Final States

By definition, systems with more than one basin of attraction can converge to different attracting regimes depending on the initial state. In ecosystems, several sets of field observations suggest such so-called path dependency. For instance, similar excavated gravel pit lakes in the same area of the United Kingdom stabilized in either a clear or a turbid state in which they persisted for decades depending on the excavation method. Wet excavation created initially murky conditions and left the lakes turbid. By contrast, if the water was pumped out during excavation and the lake was allowed to refill only afterward, the initial state was one of clear water, and such lakes tended to remain clear over the subsequent decades. As always, there might be various alternative explanations for convergence to different endpoints.

However, path dependency can well be explored experimentally. The requirement is that one can study a set of replicates of a system that start their development from slightly different states and follow their evolution over time. An example is a study on the competition between floating and submerged aquatic plants (figure 8A). The development in a series of buckets incubated with different initial densities of the two plant types was followed. Although the set of initial states represented a gradual range of plant densities, all buckets developed toward dominance by one of the two types, indicating that the mix of the two types was unstable and that dominance by either of the two species represented alternative stable states. Another example of the experimental detection of path dependency comes from a study of plankton

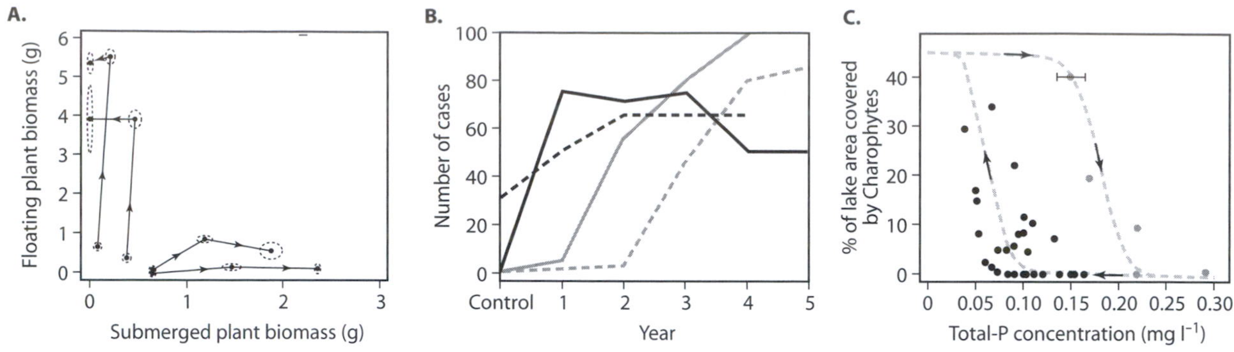

Figure 8. Three types of experimental evidence for alternative attractors: (A) different initial states leading to different final states, (B) disturbance triggering a shift to another permanent state, and (C) hysteresis in response to forward and backward change in conditions. Specific examples are the following: (A) path dependency in growth trajectories from competition experiments of a submerged plant (*Elodea*) and a floating plant (*Lemna*), which tend to different final states depending on the initial plant densities (Reproduced with permission from Scheffer, M., S. Szabo, A. Gragnani, E. H. van Nes, S. Rinaldi, N. Kautsky, J. Norberg, R.M.M. Roijackers, and R.J.M. Franken. 2003. Floating plant dominance as a stable state. Proceedings of the National Academy of Sciences U.S.A. 100: 4040–4045); (B) shifts of different shallow lakes to a vegetation-dominated state triggered by temporary reduction of the fish stock (Modified with permission from Meijer, M. L., E. Jeppesen, E. Van Donk, B. Moss, M. Scheffer, E.H.R.R. Lammens, E. H. Van Nes, J. A. Berkum, G. J. De Jong, B. A. Faafeng, and J. P. Jensen. 1994. Long-term responses to fish-stock reduction in small shallow lakes—interpretation of five-year results of four biomanipulation cases in the Netherlands and Denmark. Hydrobiologia 276: 457–466); and (C) hysteresis in the response of charophyte vegetation in the shallow Lake Veluwe to an increase and subsequent decrease in the phosphorus concentration. (Modified with permission from Meijer, M. L. 2000. Biomanipulation in the Netherlands—15 Years of Experience. Wageningen, the Netherlands: Wageningen University)

communities in small aquaria. Here it was shown that different orders of colonization from a common species pool may result in alternative endpoint communities, which are all stable in the sense that they are resistant against colonization by other species from the pool.

Disturbance Can Trigger a Shift to Another Permanent State

Another feature of systems with alternative attractors that can be tested experimentally is the phenomenon that a single stochastic event might push the system to another basin of attraction where it converges to an alternative persistent regime (figure 8B). This is often more practical in a real-life situation than trying to set a large set of "replicate" systems to a slightly different initial condition. As mentioned earlier, a good example from ecosystem management is that a temporary drastic reduction in the fish stock (biomanipulation) can move lakes from a turbid state to a stable clear condition. Lasting effects of single disturbances have also been studied in ecotoxicological research, where the inability of the system to recover to the original state after a brief toxic shock has been referred to as *community conditioning*. Such experiments should be interpreted cautiously. If one wants to demonstrate that the new state is stable, the return of the original species should not be prevented by isolation. Another problem is the potentially long return time to equilibrium, which may suggest an alternative stable regime even if it is just a transitional phase. For instance, biomanipulated lakes may remain clear and vegetated for years until they start slipping back to the turbid state.

Hysteresis in Response to Forward and Backward Change in Conditions

Demonstration of a full hysteresis in response to slow increase and subsequent decrease in a control factor also comes close to proving the existence of alternative attractors (figure 8C). Examples of hysteresis are seen in lakes recovering from acidification or eutrophication and in hemlock–hardwood forests responding to change in disturbance intensity. However, a hysteretic pattern may not indicate alternative attractors if the response of the system is not fast enough relative to the rate of change in the control factor. Indeed, one will always see some hysteresis-like pattern unless the system response is much faster than the change in the control variable.

In conclusion, experiments are potentially a powerful way to test whether a system may have alternative attractors, but there are important limitations to exploring large spatial scales and long time spans.

Summary

In summary, there is no silver-bullet approach to find out if a system has alternative basins of attraction separated by critical thresholds. Observations of sudden shifts, sharp boundaries, and bimodal frequency distributions are suggestive but may have other causes. Experiments that demonstrate hallmarks such as "path dependency" and hysteresis are much more powerful but can only be done on small, fast systems. Models that formalize mechanistic insights are essential to help improve our understanding of complex systems but remain difficult to validate. Clearly, our best approach is to build a case carefully, using all possible complementary approaches, and interpret the results wisely.

6. USING ALTERNATIVE STABLE STATES IN MANAGEMENT

There are two ways in which insight into possible alternative stable states is useful when it comes to management of ecosystems. First, we try to manage the ecosystem in such a way that the risk of collapsing into another unwanted state is reduced. Equally important, and certainly more rewarding at first sight, is the possibility of using the insight for novel approaches to restoration, invoking a shift from an unwanted into a preferred alternative state.

Managing Resilience

From a management point of view, a crucially important phenomenon in systems with multiple stable states is that gradually changing conditions may have little effect on the state of the system but nevertheless reduce the size of the attraction basin (figure 3). This loss of resilience makes the system more fragile in the sense that it can easily be tipped into a contrasting state by stochastic events. This is also one of the most counterintuitive aspects. Whenever a large transition occurs, the cause is usually sought in events that might have caused it: The collapse of some ancient cultures may have been caused by droughts. A lake may have been pushed to a turbid state by a hurricane, and a meteor is thought to have dealt with the dinosaurs, leading to the rise of mammals. The idea that systems can become fragile in an invisible way as a result of gradual trends in climate, pollution, land cover, or exploitation pressure may seem counterintuitive. However, intuition can be a bad guide, and this is precisely where good and transparent systems theory can become useful. Resilience can often be managed better than the occurrence of stochastic perturbations. For instance, a lake that is not loaded with nutrients is less likely to shift to a turbid state in a climatically extreme year than a lake that has a near-critical concentration of nutrients. We cannot prevent heat waves or storms, but we can manage the long-term trends in pollution and nutrient load.

Promoting Good Transitions

Finding smart ways to promote a self-propagating runaway shift from a deteriorated state to a good state is perhaps the most rewarding part of the work on alternative stable states in ecosystems. What is rewarding is that the transition can be relatively easy once you find the Achilles' heel of the system. In its most elegant form, this is the sequence: Determine how to reduce the resilience of the bad state first and then reverse it with little effort at all. Biomanipulation, as a shock therapy to make turbid lakes clear again is a classic example of this approach. First, the resilience of the turbid state is reduced (and that of the clear state enhanced) by decreasing the nutrient load to the lake. Subsequently, a brief intensive fishing effort converts the system into the clear state.

Using Natural Swings in Resilience

An innovative idea related to managing ecosystems with alternative stable states is that we can often make smart use of natural variation in resilience. Recognizing this is important in strategies for promoting wanted transitions as well as for preventing unwanted transitions. Natural swings may open windows of opportunity to induce a transition out of an unwanted state. For instance, a rainy El Niño year may be a window of opportunity for forest restoration, and a year with low water levels may make it easier to push a shallow lake to the clear state. The other side of the coin is that natural swings can lead to situations in which resilience of a wanted state becomes dangerously small. Rangeland managers in Australia are already advised to anticipate droughts that hit the continent during El Niño years by reducing livestock to prevent potentially irreversible degradation of the land that may easily result from overgrazing in such years. Although, I know of few other examples, one can imagine that management directed at the prevention of bad transitions in periods with naturally low resilience could be useful in other fields. For instance, rainy years may lead to higher phosphorus loading and elevated water levels that reduce resilience of the clear state of a shallow lake. Mowing of submerged vegetation that would normally have little effect could induce a shift to the turbid state in such a year. Similarly, natural changes in marine circulation patterns can alter temperature and food supply in such a way that stocks of commercially important fish species become less resilient to fishing. For instance, cod is restricted to relatively cold water. This may well imply that recruitment is less successful in populations that live at the edge of the species range as well as in years when the temperature of the water in the region is elevated. Ideally, fishing pressure should thus be tuned to such marine regime shifts if collapse of the population is to be prevented, and we want to harvest most in years when this would not harm the system too much. Such smart adaptive management approaches that use insight in natural swings in resilience seem quite rare so far.

FURTHER READING

Carpenter, S. R. 2002. Regime Shifts in Lake Ecosystems: Pattern and Variation. Oldendorf/Luhe, Germany: Ecology Institute. *Using lakes as an example, the author discusses various aspects of regime shifts. Particular attention is given to stochastic aspects and statistical problems of identifying regime shifts and alternative stable states.*

Folke, C., S. Carpenter, B. Walker, M. Scheffer, T. Elmqvist, L. Gunderson, and C. S. Holling. 2004. Regime shifts, resilience, and biodiversity in ecosystem management. Annual Review of Ecology Evolution and Systematics 35: 557–581. *This is an overview of current thinking about resilience of ecosystems written on the occasion of the 30th anniversary of the influential review by C. S. Holling in 1973.*

Holmgren, M., and M. Scheffer. 2001. El Niño as a window of opportunity for the restoration of degraded arid ecosystems. Ecosystems 4: 151–159. *This article presents an explanation of how reduced resilience during rainy years may be used to change barren semiarid regions back into a forested state by excluding grazers temporarily.*

Scheffer, M. 2008. Critical Transitions in Nature and Society. Princeton, NJ: Princeton University Press. *This book elaborates on the issues I cover in this chapter and also reviews case studies ranging from the climate system and ecosystems to socioeconomic dynamics.*

Scheffer, M., and S. R. Carpenter. 2003. Catastrophic regime shifts in ecosystems: Linking theory to observation. Trends in Ecology and Evolution 18: 648–656. *This article reviews how evidence for alternative stable states in ecosystems may be obtained.*

Scheffer, M., S. R. Carpenter, J. A. Foley, C. Folke, and B. Walker. 2001. Catastrophic shifts in ecosystems. Nature 413: 591–596. *This article reviews theory and examples of catastrophic shifts in ecosystems.*

III.18

Responses of Communities and Ecosystems to Global Changes

Erika Zavaleta and Nicole Heller

OUTLINE

1. What are global changes?
2. Global climate change
3. Elevated CO_2 in the atmosphere and oceans
4. Global nitrogen fertilization
5. Ozone depletion in Earth's stratosphere
6. Interacting global changes

Increases in the scale and extent of human activity in the last two centuries have brought about environmental changes that affect most of the globe. These global changes include directional shifts in climate, greenhouse gas concentrations, nitrogen fixation, and stratospheric ozone depletion. They also include biotic changes such as land cover change, biological invasions, and global loss of biodiversity. In this chapter, we focus on the responses of ecological communities and ecosystems to directional changes in climate, atmosphere, and global biogeochemistry. Our understanding of these responses comes from observations of trends in nature, experiments manipulating global change factors at small scales over years to decades, and predictive models. We consider all of these sources, with an emphasis on the empirical knowledge derived from observations and experimentation.

GLOSSARY

biodiversity. The totality of the inherited variety of all forms of life across all levels of variation, from ecosystem to species to gene (E. O. Wilson).

biogeochemistry. The cycles of matter and energy that transport the Earth's chemical components through time and space, and the chemical, physical, geological, and biological processes and reactions that govern the composition of the natural environment.

biomes. Generalized regional or global community types, such as tundra or tropical forest, characterized by dominant plant life forms and prevailing climate.

community. A group of interacting species living in a specified area. Communities are often defined by the dominant vegetation types, such as maple-oak or sagebrush. However, community composition is dynamic as species dominance and diversity shift in space and time.

ecosystem. An ecosystem is a complex system formed by the interactions of living (biotic) and nonliving (abiotic) components, which shape each other through exchange and material flows. An ecosystem can be bounded more or less arbitrarily and can range in scale from an ephemeral pond to the entire globe but most often refers to a landscape-scale system characterized by one or a specified range of community types (e.g., grassland ecosystems).

eutrophication. An increase in an ecosystem's plant production resulting from nutrient inputs, often with undesirable effects such as excessive plant decay, oxygen deprivation, and water quality declines in aquatic systems.

nitrogen fixation. The conversion of inert atmospheric dinitrogen (N_2) to nitrate and ammonia that can be taken up by organisms.

phenology. The timing of recurring biological phenomena, ranging from annual budburst and senescence in plants to the onset of animal migrations, egg laying, and metamorphosis.

1. WHAT ARE GLOBAL CHANGES?

Humans, like all other organisms, modify their environment. Human modification, unlike that of other organisms, however, is drastically altering ecosystems over the entire globe through an explosion in human numbers and the scope of their activities. As anthropogenic impacts on the environment cause changes in

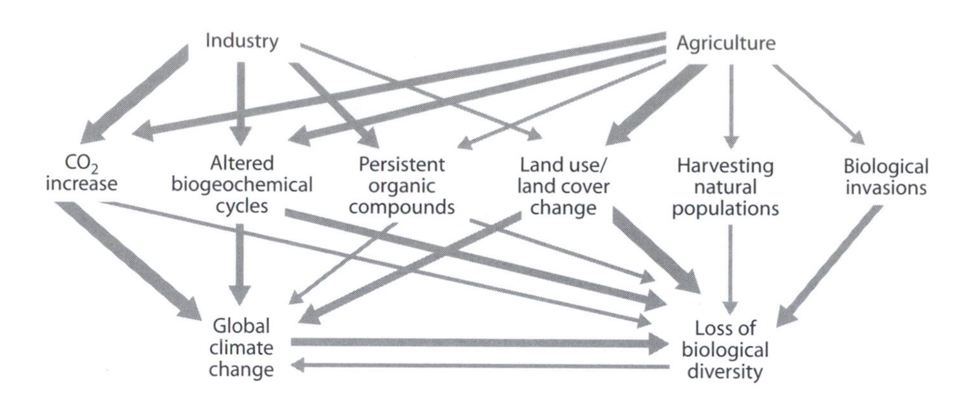

Figure 1. The interactive nature of human activities and resulting global changes. (Adapted from Vitousek et al., 1997b)

biological systems, those changes in turn cause further changes in the environment, resulting in complex feedbacks and interactions. This process is described as global change. Many, although not all, global changes originate from transported chemical effects of human activity, such as carbon dioxide (CO_2) emissions, aerosols, CFCs, and NO_x emissions. Global changes also include direct effects on biological systems, such as habitat destruction, invasive species, and biodiversity loss. Finally, global changes interact extensively—for example, tropical forest destruction releases large amounts of greenhouse gases that contribute to global climate change (figure 1). In this chapter, we focus on climate change, CO_2, nitrogen (N) deposition, and ozone depletion. Habitat loss and degradation, invasive species, and loss of biodiversity are discussed elsewhere in this volume, but we address them in the context of interactions with climate and atmospheric change.

Communities and ecosystems respond to global changes interactively (figure 2). Global changes can affect ecosystem processes, such as through altered biogeochemical cycles, in ways that then influence the resident community. Conversely, global changes can drive community shifts directly, in ways that then affect ecosystem processes, such as when a shift from tropical forest to savanna vegetation drastically reduces local rainfall. The result is a feedback loop between community and ecosystem change, one that can buffer or amplify global change effects over time and result in long-term, ongoing changes even after a particular global change driver has stabilized. For example, an abrupt increase in temperature might reduce grassland soil moisture slightly for the first several years. The same warming over decades could transform that grassland to a woody shrubland, with much more extensive effects on water relations as the dense, shallow, seasonal grass roots are replaced by deep, well-spaced, perennial taproots.

2. GLOBAL CLIMATE CHANGE

Earth's climate is strongly influenced by atmospheric concentrations of greenhouse gases and aerosols. Greenhouse gases warm the Earth's surface by trapping heat in the lower troposphere, and aerosols contribute to cloud formation and cool Earth's surface. Emissions of carbon dioxide (CO_2), the most important human-generated greenhouse gas, come primarily from fossil fuel combustion and land use change. Two other important greenhouse gases, methane and nitrous oxides (NO_x), arise mainly from agricultural production.

Human-caused increases in greenhouse gas concentrations now have accelerating effects on Earth's climate. In the twentieth century, Earth warmed by an average of 0.75°C, and precipitation increased by 2%. Sea level rose 0.2 m as a result of thermal expansion of the world's oceans. Rates of change were at least twice as fast during the later half of the century compared to the first and are continuing to increase with each year as greenhouse gas emissions increase. Changes in wind patterns, ocean salinity, and the periodicity and magnitude of extreme events have also occurred, including increased drought, tropical storms, heavy rainfall, and heat waves. If CO_2 doubles by the end of this century as predicted, global mean temperatures will likely increase in the range of 2° to 4.5°C. Local changes will be more extreme: in the last century, at high latitudes, temperature increases have been nearly double that of the global mean, and precipitation has increased or decreased by as much as 40% in some locations.

Modeled Responses

Species populations respond to changes in climate through changes in abundance, distribution, and, in some cases, rapid evolutionary change. Species differ in their capacity to tolerate, adapt, and move, causing existing communities to disassemble and re-form anew.

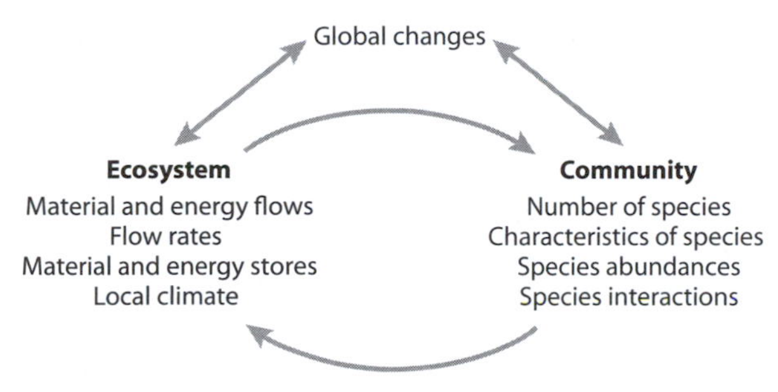

Figure 2. Interactions among global changes and ecosystem and community responses.

Many "no-analog" communities—those made up of combinations of species not currently found together—are expected to emerge. No-analog communities have existed in the past. For instance, during the last glaciation event in the United States 17,000 to 12,000 years ago, an ecosystem including spruce, sedge, oak, ash, and hophornbeam was abundant. These tree species do not occur together in the United States today. John Williams and colleagues estimate, based on climate simulation models, that 4% to 39% of the world's land area will experience novel combinations of climate variables and almost certainly will develop no-analog communities. Such novelty means it will be exceedingly difficult to forecast future conditions.

Despite this challenge, predictive modeling studies have been an important avenue of climate change research to explore community and ecosystem change. Commonly, models are used to simulate changes in the distribution of biomes or plant functional types (PFTs) rather than individual species. Two prevalent types of vegetation modeling approaches are equilibrium biogeography models and dynamic global vegetation models (DGVMs). Equilibrium models provide "snapshot" views of what the distribution of biomes might look like given a particular CO_2 level in the future. Dynamic models are gaining popularity because they help address how ecosystems might transition between states and can integrate vegetation dynamics with biogeochemical processes to explore changes in ecosystem function as well as structure.

Broadly, both equilibrium and dynamic vegetation models show major poleward shifts in cold-limited biomes as a result of climate change. For instance, tundra is predicted to decline in distribution as it will largely be "pushed off" the North American and Eurasian continents with the northward expansion of boreal forest. Boreal forest is also expected to decline as temperate forests expand northward. Dynamic models predict these vegetation changes will result in large productivity increases in northern latitudes. Biomes that are limited by water availability, rather than temperature, show more complex responses depending on the balance among precipitation change, hydrological response, and physiological adjustments by organisms.

Observed and Experimental Changes

Communities

Warming in the last century has been associated with shifts in species distributions upward in elevation and poleward. Camille Parmesan and Gary Yohe measured changes in the range boundary of 99 birds, butterflies, and alpine herbs. They found that range limits have moved on average 6.1 km per decade toward the poles or meters per decade upward in elevation. In another review, 80% of 434 species studied over time showed increased abundance in northern or high-elevation locations. For instance, J. P. Barry and colleagues compared surveys of intertidal marine invertebrates in the Monterey Bay, California, in the early 1930s to surveys in the early 1990s. These surveys show a clear pattern of increased abundance of southern species and decreased abundance of northern species.

Changing climate also affects the timing of climate-cued events in organism life cycles and, in some cases, could desynchronize important trophic or mutualistic interactions among species. For instance, a bird species that uses photoperiod to cue migration may not shift its behavior although its caterpillar food may emerge earlier in the spring because of higher temperatures. Terry Root and colleagues report that for 694 species, spring events occurred an average of 5.1 days earlier per decade since 1951. On average, birds are shifting faster than invertebrates and amphibians, and trees are shifting much more slowly. Faster rates of change for some organisms compared to others means there will be strong selection pressure on species to shift food sources, nesting behavior, or other habits. Species that are generalists will likely be favored over specialists.

In the oceans, warming has marked impacts on coral reef communities, which support an estimated 25% of marine biodiversity. Corals themselves are colonies of small animals that feed by filtering plankton out of the water and secreting calcium carbonate skeletons. Inside of corals live symbiotic algae that provide food and give corals their bright colors. In response to various stresses including thermal warming, corals "bleach" because algae are lost from hosts, exposing the white calcium carbonate skeleton underneath. Bleaching events both reduce coral abundance and change community composition because coral species vary in sensitivity to warming. Corals also have high genetic diversity, and with warming, more resistant corals and symbionts could spread, buffering the survival of populations. This evolutionary response, however, might not keep pace with warming rates in combination with other stresses including overfishing, pollution, and ocean acidification.

Ecosystems

Warming at a global scale accelerates the water cycle, with net precipitation increases. However, on land, warming also increases water losses through evapotranspiration. Net drying has occurred and is expected in many parts of the world, with strong seasonal effects. Summer drying in the boreal forest and many montane regions, as a result of both earlier snowmelt and increased water losses, has produced forest diebacks in areas such as the southwestern United States and increased wildfire frequency, extent, and severity over large regions. Warming can also affect precipitation type: in California's Sierra Nevada range, winter precipitation is shifting from snow to rain. The downstream implications are substantial, as flows shift toward winter and spring flood pulses rather than gradual snowpack melt over several months.

Warming effects on phenology influence productivity and other material and energy exchanges at local to global scales. In the northern hemisphere, the earlier onset of spring has led to an earlier and longer annual period of draw-down in atmospheric CO_2 caused by plant photosynthesis and carbon uptake. When moisture is available, warming often accelerates and extends the annual duration of decomposition, speeding the return of carbon from soil and plants to the atmosphere. This effect, although also shaped by other global changes, could provide a positive feedback to climate change by shifting carbon stores from ecosystems to the atmosphere. In field experiments simulating global changes, warming increases productivity through the winter in Mediterranean grasslands but can also accelerate the end of the growing season. The net effect of warming on productivity depends on the degree to which production is limited by temperature versus by moisture or other resources. Ultimately, it depends most on how communities change in response to warming; for example, a shift from soft-bodied plants toward woody plants in meadow or tundra ecosystems strongly affects productivity and its distribution above and below ground.

3. ELEVATED CO_2 IN THE ATMOSPHERE AND OCEANS

Atmospheric concentrations of CO_2 have increased steadily since the advent of the industrial era and are accelerating each year (figure 3). The source of these rising levels is emissions of CO_2 from human activity, mainly in the forms of fossil fuel combustion and land use change. Current atmospheric CO_2 concentrations, revealed by ice core samples, are dramatically higher than at any time in the past 650,000 years.

Besides the climate-mediated effects explored earlier, rising atmospheric CO_2 levels directly and profoundly influence communities and ecosystems because CO_2 is a necessary input for photosynthesis. Most plants on Earth acquire CO_2 for photosynthesis by opening stomatal pores in their leaves that let CO_2 in and water vapor out through simple diffusion. Higher CO_2 levels allow plants to acquire more CO_2 per unit of time their stomata are open, permitting more rapid photosynthesis (58% faster on average across 60 studies) and growth. As a result, if other necessary resources are not limiting, elevated CO_2 accelerates growth and can increase ecosystem productivity. This effect has been observed in growth chamber experiments, agricultural fields, and some natural ecosystems. In other natural ecosystems, however, other resources such as soil nitrogen limit plant growth, and increased CO_2 shifts the competitive balance among species in the plant community rather than increasing overall productivity.

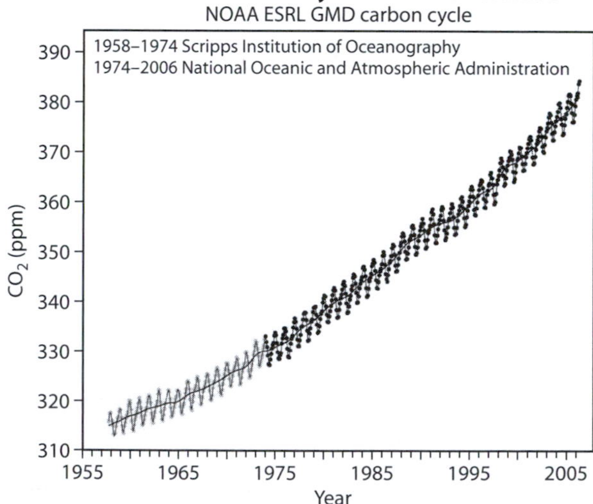

Figure 3. The longest record of Earth's atmospheric carbon dioxide concentrations was begun by Dr. Charles Keeling and is maintained by the National Oceanic and Atmospheric Administration. Annual peaks and valleys occur because most of the world's land area is in the Northern Hemisphere. CO_2 levels drop in the summer when Northern Hemisphere plants are growing and rise in the winter when plants reduce activity.

By allowing plants to open stomata less, rising atmospheric CO_2 also increases plant water use efficiency, on average, by a substantial 20%. This effect noticeably reduces ecosystem water losses to plant transpiration and increases soil moisture content in a range of herb-dominated ecosystems, including tallgrass prairie, salt marsh, crops, and several grasslands. Responses in woody ecosystems, especially conifer forests, are less clear because the behavior of woody plant stomata is less responsive to elevated CO_2. Even relatively large gains in water storage, however, could be more than offset by increased evapotranspiration driven by CO_2-induced climate warming. The net effects of rising atmospheric CO_2 on ecosystem water balance, especially in arid ecosystems with more direct evaporation from soil, could be negative.

Plant-derived nitrogen is an essential source of protein to herbivore communities. Tissue quality in plants grown in CO_2-enriched environments tends to be lower than in plants grown in ambient conditions because high CO_2 concentrations dilute tissue nitrogen (N). The feeding intensity of herbivores depends on plants' carbon-to-nitrogen ratio (C:N), leaf thickness, and concentration of defensive secondary compounds. Experiments consistently find that insects compensate for lower tissue quality by increasing consumption rates. Increased feeding may not compensate for decreased tissue quality in all cases however. Some studies show decreased growth rates of insects reared on CO_2-enriched plant tissue, depending on the plant species involved, indicating variable plant response to CO_2 enrichment. Herbivores may therefore shift feeding preferences to plants that maintain lower C:N in response to CO_2 enrichment and produce a change in top-down control on plant community composition.

A troubling change resulting from increased CO_2 emissions is acidification in marine ecosystems. Oceans currently absorb about a third of the CO_2 released through fossil fuel combustion. As dissolved CO_2 concentrations in oceans increase, so does acidity—first in the upper layers, then gradually in deeper waters. The effects of ocean acidification on marine communities are only beginning to be understood, but laboratory studies show that acidification directly threatens corals, crustaceans, and other marine organisms that use calcium carbonate to build shells and skeletons. Acidic waters dissolve calcium carbonate shells and make it increasingly difficult for organisms to precipitate calcium carbonate for shell building in the first place. Acidification rates to date vary across the globe and peak in the Atlantic; the consequences of continued acidification could include widespread losses of organisms, disrupting marine food webs and eliminating entire communities such as coral reefs.

4. GLOBAL NITROGEN FERTILIZATION

Nitrogen is abundant in the atmosphere in a biologically inert form called dinitrogen (N_2). Human activity has vastly altered the global nitrogen cycle by fixing nitrogen deliberately for fertilizer and inadvertently during fossil fuel combustion (figure 4). Until the twentieth century, nitrogen fixation was accomplished primarily by bacteria in the root nodules of legume plants. In 1913, with the advent of the Haber-Bosch process, humans began to fix atmospheric nitrogen synthetically to produce ammonia fertilizer. The amount of industrially fixed nitrogen has since increased steadily. Humans now fix more nitrogen than all other natural sources combined. Synthetic fertilization use is widespread since the green revolution in the 1970s and is anticipated to increase by 70 in the next 20 years. Fertilizer runoff from agriculture affects aquatic communities and ecosystems around the world. In addition, fossil fuel burning, especially in automobiles, emits nitrous oxides (NO_x), which are converted to nitrate and nitric acid in the atmosphere and deposited directly on vegetation and into bodies of water. Nitrogen deposition contributes to acid rain and ecosystem fertilization in large areas of the globe, with its most significant effects around and downwind from heavily industrialized and densely populated regions.

Much of the nitrogen applied to agricultural fields migrates into waterways. Nitrogen is a vital nutrient for organisms, but in excess amounts, it can restructure entire food webs and significantly decrease species diversity. In both freshwater and marine aquatic systems, excess nutrients cause algal blooms ("red tides") that block light and deplete oxygen, creating "dead" zones. In the Black Sea, for example, heavy sewage, fertilizer inputs, and industrial waste from over 15 countries

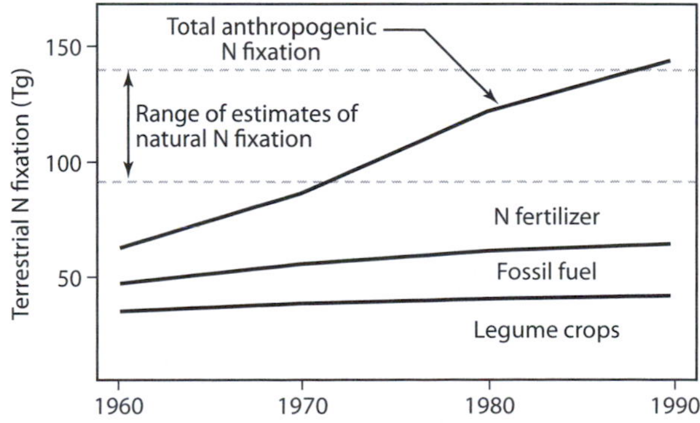

Figure 4. Human fixation of nitrogen over time through fertilizer production, fossil fuel combustion, and the growth of legume crops, compared to all natural sources of N fixation. (From Vitousek et al., 1997a)

resulted in year-round severe eutrophication, frequent red tides, and fish kills. In concert with an accidentally introduced exotic jellyfish species that benefited from the high nutrient conditions, this fertilization drove 75% of commercial fish species to extinction in the 1980s.

Nitrogen also reaches terrestrial systems from automobile and other industrial exhaust, with varying effects. In systems limited by scarce nitrogen, nitrogen deposition tends to increase productivity and carbon storage and can dramatically reduce biodiversity. In historically nutrient-poor ecosystems such as those with serpentine and calcareous soils, nitrogen deposition can dramatically increase productivity. In so doing, nitrogen deposition can reduce or eliminate endemic assemblages that are ill-adapted to compete against faster-growing invaders that are better able to exploit high soil fertility.

5. OZONE DEPLETION IN EARTH'S STRATOSPHERE

Ozone (O_3) in the Earth's stratosphere absorbs 95% of the ultraviolet radiation from the sun, which otherwise would destroy life. Chlorofluorocarbons (CFCs), used extensively for cooling in refrigerators and in aerosol products until an international ban in 1996, are mobile, long-lived, stable compounds that degrade in the stratosphere and destroy ozone. Ozone is a highly selective absorber of UV-B light, so the ratio of UV-B light to other wavelengths increases as ozone decreases. The effects of ozone depletion are unevenly distributed around the globe and pronounced at high latitudes during the winter and early spring, when the formation of stratospheric ice particles above the poles provides surfaces for rapid decomposition of CFCs and the breakdown of ozone.

Our knowledge of community and ecosystem changes that result from increased UV-B light comes largely from small-scale experiments in confined ecosystems called mesocosms. Investigators can manipulate UV-B levels directly in mescosms, increasing UV-B with radiation-emitting lamps or decreasing it with selective filters.

Organisms exhibit wide-ranging sensitivity to ultraviolet radiation, both within and across taxonomic groups such as higher plants and fungi. Species with greater pigmentation or body coverings such as hair are better protected from the harmful effects of UV-B light, whereas amphibians are poorly protected and highly sensitive. Microbes vary widely in sensitivity. Thus, increased UV-B light alters community composition toward less-sensitive species and affects species interactions. Max Bothwell and colleagues used an experimental river flume facility to test how algal and invertebrate communities exposed to different amounts of sunlight were affected. They found that the sensitivity to UV-B is higher in algal grazers than in algae, altering the balance between primary producers and consumers in shallow benthic communities.

UV-B–driven changes in trophic or community structure can alter carbon and nutrient cycling in ecosystems. In a subarctic heath system, David Johnson and colleagues showed that increased UV-B radiation alters soil microbial C:N through microbial community changes, with potentially far-reaching consequences for plant communities. Increased UV-B radiation can also directly affect ecosystem functioning. In semiarid systems, increased UV-B can accelerate decomposition and carbon cycling. Interactions between increased UV-B light and other global changes, although likely important, remain poorly understood.

6. INTERACTING GLOBAL CHANGES

Many anthropogenic global changes are occurring at once. Their interactions, rather than their individual effects, will have the biggest implications for ecological responses.

Some global changes are linked through positive feedbacks, so increases in one will cause increases in another. Habitat destruction, particularly in the world's forests, contributes to greenhouse gas emissions and warming: climate changes are increasing wildfire and the extent of forest dieback in some regions, accelerating forest loss. Other global changes are linked through negative feedbacks. For example, aerosol emissions partially counteract the warming effects of greenhouse gases by reducing the amount of solar radiation reaching Earth's surface.

In general, community and ecosystem responses to the suite of ongoing, interacting global changes cannot be predicted from studies of individual global changes. As study of global change interactions grows, it reveals a range of additive to synergistic or idiosyncratic effects. When nitrogen deposition and CO_2 increases occur together, the added nitrogen can alleviate nutrient limitation of productivity responses to CO_2. The result can be much greater productivity increases than under elevated CO_2 or nitrogen fertilization alone. Elevated CO_2 alone can lead to greater ecosystem uptake of atmospheric carbon, an effect some hope could slow the accumulation of greenhouse gases in the atmosphere. However, warming can offset this effect entirely by speeding up release of carbon into the atmosphere from decomposing matter.

Interactions among the global changes emphasized in this chapter are complicated further by their interaction with other human transformations of the environment. Many global changes—most notably nitrogen deposition, aspects of climate change such as wildfire and accelerated forest dieback, and habitat destruction—exacerbate the spread of invasive species by favoring weedy, fast-growing, disturbance-adapted species over slow-dispersing, long-lived, and specialized ones. Climate change and invasive grasses in regions from the Great Basin to Hawaii interact to transform wildfire regimes through a combination of hot, dry weather and expanding flammable fuel loads. In boreal lakes, emissions of nitrogen and sulfur (NO_x and SO_x) interact with warming to reduce dissolved organic matter (DOC). Lower DOC then allows more UV-B light (which has increased particularly at these high latitudes from stratospheric ozone depletion) to penetrate the lake surfaces, rendering large areas unable to support aquatic life. With less life in them, lake productivity declines, further reducing DOC.

A picture emerges of natural systems altered drastically by the accumulating effects of multiple global changes. Ecological science can provide some insight into the nature and magnitude of ongoing and potential future changes through experiments, historical study, and models. These insights can inform society's work to flag the undesirable effects and, through a combination of adaptive stewardship and steps to slow or reverse global change rates, avert them. In a climate of complex global change interactions and uncertainty about the precise trajectories they will take, safeguarding the capacity of ecosystems to adapt to change is a minimum step to preserve options for the future.

FURTHER READING

Lovejoy, T. E., and L. J. Hannah. 2005. Climate Change and Biodiversity. New Haven, CT: Yale University Press.

Mooney, H. A., and R. J. Hobbs. 2000. Invasive Species in a Changing World. Washington, DC: Island Press.

Vitousek, P. 1994. Beyond global warming: Ecology and global change. The Robert H. MacArthur Award Lecture, Ecology 75: 1861–1876.

Vitousek, P. M., J. D. Aber, R. W. Howarth, G. E. Likens, P. A. Matson, D. W. Schindler, W. H. Schlesinger, and D. G. Tilman. 1997a. Human alteration of the global nitrogen cycle: Sources and consequences. Ecological Applications 7: 737–751.

Vitousek, P. M., C. M. D'Antonio, L. L. Loope, M. Rejmanek, and R. Westbrooks. 1997b. Introduced species: A significant component of human-caused global change. New Zealand Journal of Ecology 21: 1–16.

Vitousek, P. M., H. A. Mooney, J. Lubchenco, and J. A. Melillo 1997c. Human domination of Earth's ecosystems. Science 277: 494–499.

III.19

Evolution of Communities and Ecosystems

Nicolas Loeuille

OUTLINE

1. How does evolution affect communities and ecosystems?
2. The ecological implications of single-species evolution
3. Pairwise coevolution and ecosystem functioning: Plant–herbivore coevolution
4. Diffuse coevolution and complex adaptive systems
5. Multiple levels of selection and community evolution
6. Impact of species evolution and coevolution on abiotic components of ecosystems

Although much of evolutionary biology focuses on explaining phenotypic trait variation and on understanding the genetic basis for this variation, evolution can also affect the structure and functioning of communities and ecosystems. Evolution by natural selection discriminates among individuals based on their relative fitness such that the process of evolution is linked to demographic parameters. Therefore, it is expected that the evolution frequently will affect the demographic dynamics of the evolving species. Effects of evolution may extend further and affect the composition of the entire community as well as energy and nutrient fluxes at the ecosystem scale. Here, I summarize the effects of evolution on community and ecosystem properties along a gradient of increasing evolutionary complexity, from the effects of single-species evolution to the effects of the coevolution of two interacting species, and finally ending with the effects of the coevolution of many species.

GLOSSARY

coevolution. A strict definition of coevolution has been given by D. H. Janzen (1980): "Coevolution may be usefully defined as an evolutionary change in a trait of the individuals in one population in response to a trait of the individuals of a second population, followed by an evolutionary response by the second population to the change in the first." As this definition requires reciprocal evolutionary feedbacks between two populations, it directly applies to what is called pairwise coevolution.

community genetics. Initially defined by J. Antonovics (1992), a more recent definition of community genetics may be found in T. G. Whitham and colleagues (2003, see Further Reading): "The role of intraspecific genetic variation in affecting community organization or ecosystem dynamics."

complex adaptive systems. Initially defined by J. Holland (1995), the definition of complex adaptive systems has been simplified by S. A. Levin (1998, see Further Reading). A system is a complex adaptive system if it fulfills three characteristics: (1) individuality and diversity of components; (2) localized interactions between those components; and (3) an autonomous process that selects a subset for replication and enhancement from among components, based on the results of local interactions.

diffuse coevolution. The extension of pairwise coevolution to multiple species. Diffuse coevolution implies the coevolution of not just two populations, as in Janzen's definition, but three or more. Under diffuse coevolution, trait variations in populations A and B have reciprocal effects not only on each other but also on any other number of other populations in the community.

niche construction. From K. N. Laland and colleagues (1999, see Further Reading): the modification of local resource distributions by organisms in a way that influences both their ecosystems and the evolution of their resource-dependent traits.

1. HOW DOES EVOLUTION AFFECT COMMUNITIES AND ECOSYSTEMS?

Phenotypic Variation as the Main Focus of Evolutionary Studies

Historically, evolutionary biologists have focused on understanding why certain phenotypes are observed in a given population. Key questions include the reasons for the dominance of certain phenotypes and the maintenance of polymorphisms (two or more distinct phenotypes). The classical example of Charles Darwin's finches perfectly illustrates this approach. Darwin observed the beak characteristics of different species of finches and how such variation might be related to the food consumed by these species. He linked the observed traits and their diversity to selective pressures. Afterward, development of genetics provided a hereditary mechanism for phenotypic transmission within lineages, but the focus of most evolutionary research has remained the same: the causes of phenotypic variation within and among populations. After this important issue, the next step is to evaluate the effects of phenotypic variation on community and ecosystem processes at larger spatial scales.

Is There Any Reason to Think That Evolution Will Influence Communities and Ecosystems?

With the phrase "survival of the fittest," Darwin coined the idea that observed traits are maintained because they provide an advantage to the individual in terms of survival or fecundity. This idea requires a definition of the fitness of an individual. J.B.S. Haldane defined individual fitness in the 1920s simply as the total number of offspring an individual begets during her lifetime.

Fitness as defined above encompasses survival and fecundity. The fact that survival and fecundity are two major components of demographic variation provides an intuitive link between phenotypic selection and demography. If different phenotypes have different fitnesses, then variation in the composition of the population in terms of these phenotypes will determine the average demographic parameters of the population, thereby affecting features such as the total density of the species or the stability of the demographic dynamics.

Evolution does not only affect the population itself but also the demography of populations of other species in the community. In the case of Darwin's finches, the various beak shapes are related to the characteristics of resources, be they seeds or insects. If the evolution of beak shape increases the efficiency of the consumption by the bird, it may decrease the plant or insect fitness, thereby modifying their population dynamics.

It is clear that, by introducing links between fitness and demographic parameters, evolution has the potential to affect communities, at least from a demographic point of view. This chapter aims to determine the extent of the influence of evolutionary forces on community structure and ecosystem functioning. The development of these aspects will follow an increasing gradient in the level of evolutionary complexity, from one evolving species to the coevolution of many.

2. THE ECOLOGICAL IMPLICATIONS OF SINGLE-SPECIES EVOLUTION

Demographic Implications of Evolution

As demonstrated earlier, species evolution is tightly linked to demography via the definition of fitness. More particularly, evolution may act:

1. On the total density of the evolving species or indirectly on the densities of other species that interact with it.
2. On the stability of these dynamics. Although many definitions of stability exist in ecology, stability is defined here as the effect of a perturbation on a system. If the system is stable, then after slight perturbations of total population size, the system is brought back to its original state by the demographic dynamics.

Regarding the first aspect of evolution, the consensus among evolutionary ecologists is that evolution does indeed modify the total population of the species and of other species of the community. By changing the densities of species that interact with the evolving species, evolution can modify the whole community.

For instance, if evolution favors high levels of defense in plants, then the amount of resources available for herbivores will decrease. The total population of herbivores is then expected to decrease. Conversely, when defenses are selected against, the herbivore population may increase. In the end, variation in herbivore and plant densities will depend on the benefits and costs (also called trade-offs) that are associated with their interaction traits, but it is unlikely that total population sizes will be left unchanged.

Both theory and experimental results suggest that the evolution of a given species can affect stability. Early theoretical work in evolutionary ecology showed the potential for a stabilizing effect of evolution. D. Pimentel (1961) created a model of a plant–herbivore interaction. Whereas the model without evolution leads to an explosion of the herbivore population, such unstable dynamics were not observed when evolution of plant defenses was accounted for. However, in other situations the model predicted the opposite result, with evolution destabilizing the demographic dynamics. Overall, there is no general prediction about how evolution will influence stability, and there are only a few experimental and empirical data making a direct link between the two. However, T. Yoshida and his colleagues (2003) showed that rapid evolution of defenses in *Chlorella* destabilized the dynamics of the rotifer–alga community by amplifying population cycles. Although theoretical studies of how evolution affects the demographic dynamics are contradictory, and empirical results are too few, it is quite clear that evolution, even of a single species in the community, may have a strong impact on population stability or community dynamics.

These direct effects of evolution on population densities and stability also can interact with other, less direct effects of single-species evolution on the community structure and ecosystem functioning. A pioneering theoretical work by Michel Loreau (1998) details how evolutionary pressures on a focal plant species differed depending on whether competition was assumed to occur inside a given nutrient cycle or among different nutrient cycles. These various evolutionary trajectories explained functional patterns of ecosystem succession, including increasing biomass and productivity, the decreasing productivity:biomass ratio, and the increasing proportion of nutrient recycled locally.

Community Genetics: A Framework to Link Single-Species Evolution with Community Structure or Ecosystem Functioning

In addition to its influence on population dynamics, there are other ways in which evolution influences communities. The emerging concept of community genetics details community aspects of the extended phenotype concept introduced by R. Dawkins (1982). Originally, Dawkins described the extended phenotype as being the effects of genes at levels that go beyond the individual. Community genetics aims to clarify how the influence of genes may extend to the structure and functioning of entire communities. Community genetics proposes mechanisms for correlations between the genetic composition of a given species and characteristics of the community in which this species is present.

It is unlikely that the genetic composition of all species will influence the structure of the community. However, if the species is abundant or if it plays an important role in the functioning of the ecosystem (e.g., a keystone species), its genetic variability likely will affect community structure. One such example is given by T. G. Whitham and his colleagues (2003). In this example, the focal species is an aspen. Different aspen genotypes possess varying levels of leaf toxins. These defenses decrease the survival of the herbivores and pathogens that normally consume the aspen. On the other hand, caterpillars may, in turn, use the toxins that are produced by the plant to protect themselves against parasitoids. This protection has a positive effect on caterpillar density. In the end, community characteristics such as total biomass of herbivores or energy transfers will depend strongly on the genotype of the aspen (figure 1).

Other similar examples of links between the genotypes of a given species and the characteristics of the community exist. Variation in salmon genotypes has been related to the structure of riparian communities (enhanced plant growth, changes in behaviors of many vertebrate species). Variation in the behavior of beavers has extended consequences for the functioning of tree communities and for the recycling of nutrient and litter composition. Genotypes of cottonwoods may constrain the composition of the aphid communities that develop on them. These changes in aphid communities can affect other arthropods such as ants and higher trophic levels.

In each case, it is possible to make a link between the genotype of the target species and a trait at the community or ecosystem level, such as total diversity, number of trophic levels, mineralization, etc. For a given community trait, part of its variance is explained by the genetic variance of the focal species. The ratio between this part and the total variance defines the heritability of the trait at the community level. This definition is analogous to the one used for individual phenotypic traits (ratio between the variance of the trait that may be explained by genetic variance and the total variance of the trait).

It is often difficult to single out the evolution of a single species without considering the community in which it is embedded. Evolution may be much faster for a given species because of its short generation time, high selective pressures, or a high potential for evolution (e.g., high genetic variation), but in many cases, the evolution of many species happens on similar time scales. The issue of how this additional evolutionary complexity influences the system as a whole then becomes a central issue.

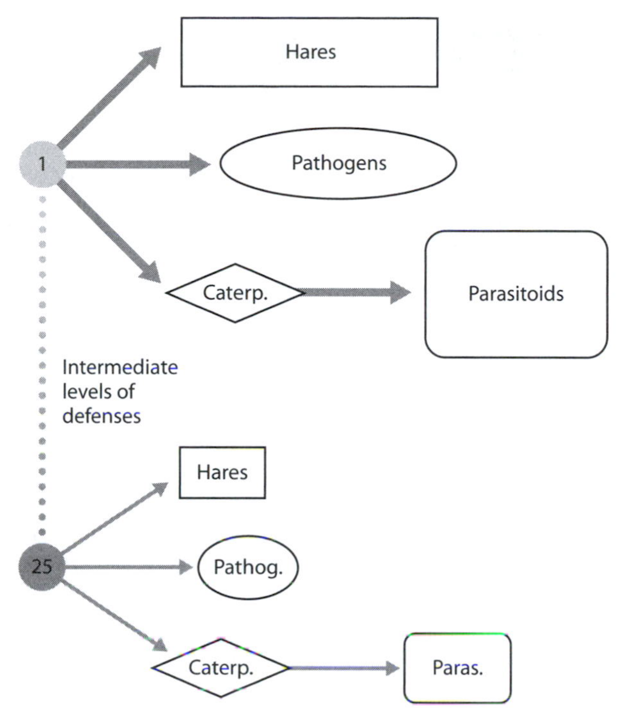

Figure 1. Community genetics. This figure illustrates the example of community genetics detailed by Whitham et al. (2003). The circles represent different genotypes of a plant species (aspen). The number inside the circle represents the amount of chemical defenses produced by the genotype relative to the smallest amount of chemical defenses observed. Interaction strength is proportional to the thickness of the arrows, and the area of each compartment is proportional to its density. The low-defense genotype (1) is heavily consumed by all of its enemies. The community that develops on the most-defended genotype (25) is very different, as the consumers of the aspen have a much lower survival rate on it. However, caterpillars that survive on this defended genotype ingest chemical compounds that protect them against their parasitoids. Many intermediate genotypes exist between 1 and 25.

3. PAIRWISE COEVOLUTION AND ECOSYSTEM FUNCTIONING: PLANT–HERBIVORE COEVOLUTION

Energy Transfer within Food Webs

Diffuse coevolution is the simultaneous evolution of several interacting species, whereas pairwise coevolution focuses on evolution of two interacting species. Pairwise coevolution may be a good approximation to the evolutionary dynamics of the community if the evolution of the two species happens sufficiently fast or occurs independently of the rest of the community. The latter could happen if both species have specialized interactions with each other (e.g., a specialized host–parasite relationship). As far as theory and modeling are concerned, understanding pairwise coevolution is also easier than when many species coevolve.

Ecosystem functioning may be affected by coevolutionary processes because it affects the interaction strength between two coevolving species. For example, nutrient acquisition and allocation to growth are supposed to determine the primary production of plants, but herbivory will transfer this production of energy upward in the food web. Evolution of plant defenses (e.g., toxins), by decreasing the attack rate of the herbivore, will decrease this transfer of energy. On the other hand, evolution may favor the production of detoxifying enzymes in herbivores. Such evolution will increase the amount of energy available for their reproduction; hence, it favors the transfer of energy upward in the food web. Consequently, plant–herbivore coevolution has the potential to impact the distribution of energy throughout the ecosystem.

Pairwise Coevolution and the Control of Biomass

The understanding of how nutrients are distributed among the species of a given ecosystem has long been debated in functional ecology. When evolution is not considered, the interplay between two effects will determine the total biomass of a given species: the top-down effect, where a population is controlled by species that consume it, and the bottom-up effect, where the total biomass is determined by the amount of resources available to the species. Understanding the distribution of biomass is important because it constrains the ecosystem's response to a disturbance. For example, figure 2A displays how an ecosystem where top-down effects are dominant would react to nutrient enrichment. The top of the food chain would benefit from such an addition of nutrient, as well as all odd-numbered levels starting from the top. Other compartments are unaffected because their biomass is controlled by the level above.

However, the evolution of plant defenses may affect this pattern by modifying the strength of top-down forces and thereby the effects of the perturbation. Without evolution occurring, the only use the plant can make of the surplus of nutrient is increasing its growth and reproduction. If plant defenses are incorporated, then part of the nutrients may be used for the production of these defenses. Then, evolution may either favor fast-growing or defended plants, as the excess of nutrient may now be invested into antiherbivore defenses. In the top-down system introduced in figure 2A, defenses are selected when nutrient enrichment occurs. This decreases the force of the top-down control in the system, and plant biomass may increase (figure 2B),

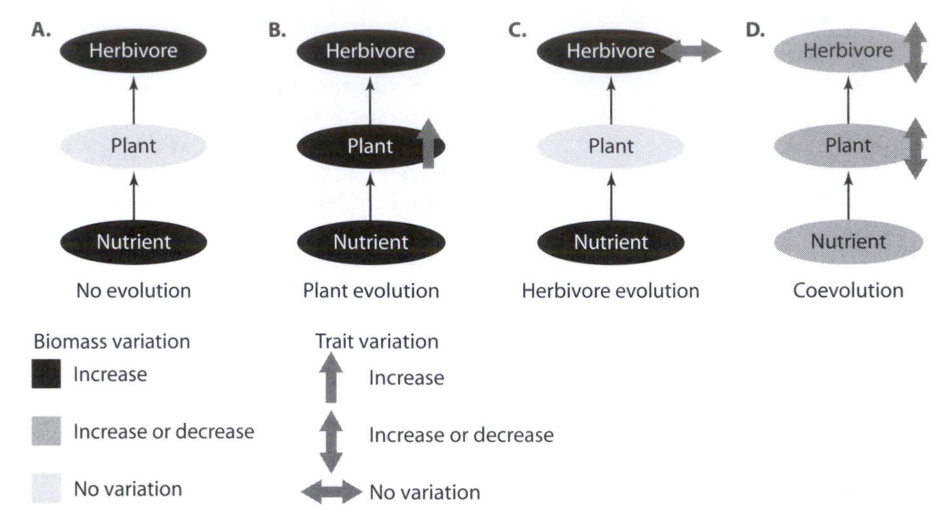

Figure 2. This figure shows how a simplified community (food chain) reacts to an enhancement of the nutrient supply for different evolutionary scenarios. In the model, the growth rate of a compartment depends linearly on the populations with which it interacts. In this type of model, without evolution, nutrient enrichment has a positive effect on the stocks of herbivores and nutrients (A). This pattern remains when the herbivore alone evolves (C). If plant defenses can evolve, nutrient enrichment produces an increase in the level of defenses and of the plant biomass (B). In the case of plant–herbivore pairwise coevolution, an increased input of nutrient affects the results of the perturbation in a way that depends on the costs associated with variation in plant and herbivore traits (D).

although it might not without evolution (figure 2A). Incorporating herbivore evolution alone does not modify the pattern obtained without evolution (figure 2C), but pairwise coevolution of plants and herbivores may change the effect of the perturbation in many different ways (figure 2D). The change will depend mainly on the physiological details of the costs of trait evolution in both plants and herbivores.

Although this example illustrates how plant–herbivore pairwise coevolution may change the forces inside a simple system, it also shows that coevolution can change the distribution of nutrients and the fluxes of energy at a community scale.

4. DIFFUSE COEVOLUTION AND COMPLEX ADAPTIVE SYSTEMS

Complex Adaptive Systems

The effect of diffuse coevolution in modifying ecosystem structure and functioning is even more intuitive. If the evolution of just one or two species influences the structure of communities, then changes in the traits of many species are even more likely to provoke large effects. However, it is difficult to assess the consequences of such diffuse coevolution. As the complexity of the system increases, theoretical predictions become very dependent on specific assumptions. Experimental knowledge becomes limited by the difficulties of maintaining many species in interaction long enough to detect demographic and evolutionary changes. Most insights about the effects of diffuse coevolution on community dynamics come from theory.

One particularly useful concept for thinking about the effects of diffuse coevolution on the structure and functioning of ecosystems is complex adaptive systems. As defined by Levin, complex adaptive systems are characterized by three characteristics:

1. The diversity and individuality of the different components.
2. Localized interactions among these different components.
3. An autonomous process that uses the outcome of these interactions to select a subset of those components for replication or enhancement.

The concept of complex adaptive systems is broad and has been applied in many different areas, from sociology to economics to natural communities. In particular, one can use complex adaptive systems to understand the link between diffuse coevolution and the structure of natural communities. By going back to the definition, it is possible to check that natural communities with diffuse coevolution are complex adaptive systems:

1. Diversity and individuality of the components: Such components may be defined in natural communities as being individuals, populations, species, or groups of species that are functionally

similar. For example, if the choice is to use species as components, then natural communities are composed of many species, so the diversity requirement is met. Because species are separated by reproductive barriers, the individuality requirement is also met.

2. Localized interactions among the different components. Natural communities may be considered to be a network of interacting species. The definition of community itself requires the definition of a locality. If species are considered to be the components of the natural community, this second characteristic of complex adaptive systems is also met.
3. Selection process. If one assumes that natural selection drives the selection process in a complex adaptive system, then natural selection acts on the individual but has consequences for the population level and demographic dynamics of the species, as illustrated above. Moreover, individual fitness is the currency of natural selection and is partly determined by the network of interactions, the structure of the community. Therefore, natural selection is a process that acts on the components defined (species) and depends on the outcome of the community structure.

Examples of Complex Adaptive System Modeling

The concept of complex adaptive systems may be linked to several models that apply diffuse coevolution to an assemblage of species to understand the structure and functioning of a community. Among natural communities, food webs have been extensively studied. Food webs are simplified compared to natural communities, as they are based on trophic interactions and largely ignore other types of interaction. The Webworld model, developed by A. J. McKane and his colleagues, is an example of a theoretical model that uses diffuse coevolution of species to understand food web structure. In these models, each species is assumed to have a large number of unidentified traits. The strength of the trophic interaction between two individuals depends on the combination of all of their traits. N. Loeuille and M. Loreau introduced a simpler model based on body size (figure 3). For any two species, the larger one will prey on the smaller one, and the strength of this

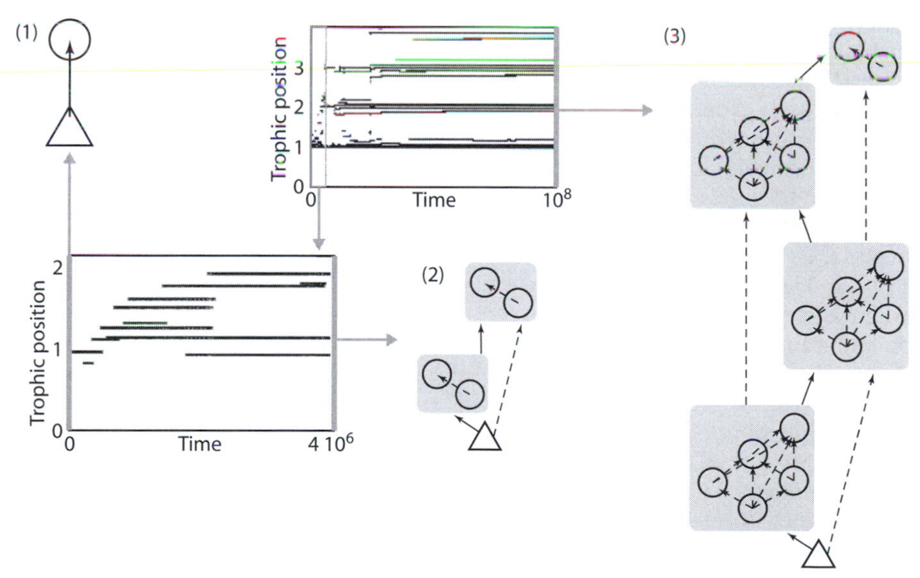

Figure 3. Complex adaptive system. Trophic and competitive interactions are based on body size (full description in Loeuille and Loreau, 2005). New morphs appear because of mutations and may disappear, immediately if they are selected against, or afterward, when the invasion of another mutant provokes their extinction. The upper rectangular panel shows the trophic position of the species of the community through time. Species having a trophic position of 1 consume the inorganic resource. Species whose trophic position is 2 consume species whose trophic position is 1, etc. The lower panel focuses on the beginning of the simulation. The emerging food web is pictured at three times: initial conditions (1), after 4% of the simulation is completed (2), and at the end of the simulation (3). The basic resource is pictured as a triangle. Coevolving species are displayed as shaded squares. In the example presented, diffuse coevolution builds a food web with four distinct trophic levels. A wide variety of other food web structures may emerge from this model.

trophic interaction is directly linked to the difference between the two species' body sizes. Species also interfere competitively if their body sizes are similar. Through diffuse coevolution of species, food webs emerge by repeated speciation events. The model includes all the characteristics of a complex adaptive system.

In the model by Loeuille and Loreau, diffuse coevolution produces a wide variety of food web structures, which can be compared to those found in empirical data sets. Such a comparison shows that the simulations indeed produce food webs whose structure is relevant to the ones observed in nature. Other complex adaptive system models also have been successful in reproducing different features of empirical food web data sets. In the Loeuille and Loreau model, it is also possible to assess how energy is transmitted throughout the food web, thereby linking coevolutionary processes to the functioning of the ecosystem.

5. MULTIPLE LEVELS OF SELECTION AND COMMUNITY EVOLUTION

So far, all the examples detailed in this chapter rely on selection at the individual level. Regarding the subject at hand, however, it is important to point out that some works suggest that evolution may act at higher levels, including the community level. Very convincing experiments show how such a high-level selection could alter the course of evolution in artificial environments. Michael J. Wade used beetles to investigate the effects of group selection in the late 1970s. The experiment was based on the selection of the groups of beetles that had the highest densities among all groups of the experiment and the observation of this group trait (density) through time. He observed that group selection was able to produce large variation in the genetic composition and traits of groups compared to the treatment with only individual selection. Because only one species was considered, insights from this experiment only apply to population-level dynamics. It is, however, possible to apply a similar experimental design to sets of communities. For instance, William Swenson and his colleagues selected soil communities on which *Arabidopsis thaliana* grew according to the biomass of the plant in a given container. For each generation of plant, the soil was inoculated with either soil communities that led to a high biomass in the previous generation or with the ones that led to low biomass. After a few generations, differences in the biomass of the plants in the two treatments became significant, so that the community trait (biomass of the plant) was considered to be heritable.

In addition to these experiments, theoretical developments detailed how such group selection was possible in nature. The first theory of group selection was written by Wynne-Edwards in the early 1960s. It is based on the idea that animal densities are not limited by the amount of resources because social interactions prevent large increases in populations, hence resource shortage. Although influential, the theory was not explicitly mechanistic and had been heavily criticized at the end of the same decade. New theoretical advances by David S. Wilson and his colleagues during the 1980s and 1990s renewed the interest of evolutionary ecologists in group-, community-, and ecosystem-level selection. The issue remains heavily debated. Most of the arguments focus on the proper definition of the units of selection and heritability of traits. Opponents to group or community selection point out that the exact boundaries of the unit of selection are clear in the case of individual selection but that it is not so for higher levels of selection. A direct consequence is that a reliable definition of heritability is difficult to determine. Finally, some opponents think that the generation length of groups and communities is too long, and genetic variability among them too low, so that individual selection generally will dominate in any system for which group selection is possible. Although debated, the framework proposed by Wilson has the advantage of including selection at any level: group or community selection but also selection at the individual level or lower. It can also tackle the partitioning of selective pressures between the different possible levels of selection.

6. IMPACT OF SPECIES EVOLUTION AND COEVOLUTION ON ABIOTIC COMPONENTS OF ECOSYSTEMS

Most studies focus on the effect of evolution on biotic structures (trophic positions, interaction strength, etc.), and the effect of such evolution on the abiotic environment is largely ignored or insufficiently detailed. Studies explicitly accounting for nutrient recycling are exceptions; other abiotic components (e.g., climate, pH) are considered even less. Two frameworks integrate the abiotic environmental conditions into an evolutionary framework.

Abiotic Components as an Extension of Natural Communities: The Gaïa Hypothesis

The Gaïa hypothesis is based on the work by J. E. Lovelock and his collaborators in the beginning of the 1980s. In this hypothesis, the abiotic environment is the product of the composition of natural communities. The initial model, called daisyworld, details how the frequencies of two types of daisies, white and black, may influence global temperature by modifying the albedo (that is to say the proportion of light reflected) of a hypothetical planet. The whole planet is then considered to be a superorganism, including both abiotic (e.g., temperature) and biotic components (the daisies). Although the hypothesis and the initial model do not contain any evolutionary processes, it is possible to include evolutionary dynamics in the Gaïa model. S. J. Lansing (1998, see Further Reading) details how this evolutionary extension can be done and discusses the robustness of the Gaïa hypothesis under these new conditions. This evolutionary extension of the initial Gaïa model is widely debated because the Gaïa hypothesis relies on a selection process that happens at very large scale and largely ignores the influence of local conditions on individual fitness, the usual currency of selection.

Niche Construction or Ecosystem Engineering

In addition to the issue of the level of selection, the Gaïa hypothesis is also criticized because the abiotic environment is considered to be completely linked to the biotic composition, as both are included in the Gaïa superorganism. Niche construction is an alternative approach that provides a more sophisticated treatment of biotic and abiotic linkages. In this framework, species can modify their environment (niche construction or, equivalently, ecosystem engineering), but part of the environment is still external to the system. The implications of the evolution for niche construction and of niche construction for the evolution of other traits were considered by Odling-Smee in the late 1990s. This model has the potential to link evolution and ecosystem functioning explicitly. However, as with the Gaïa hypothesis, the current models are usually monospecific, so that although the environment is better accounted for, the community aspect lacks ecological details or is ignored.

Understanding the effects of evolution on the structure and functioning of natural communities is very challenging. From a theoretical point of view, it involves systems with many parameters, making it difficult to check assumptions. The robustness of these theoretical studies then can become an issue. Empirical and experimental studies are confronted with timescale and complexity issues. Maintaining and observing a system that involves many species are difficult endeavors, for practical reasons.

However, it is still possible to make several statements about how evolution influences the organization of natural communities:

- It constrains the density of species and the stability of their population dynamics.
- Single-species evolution may constrain the characteristics and functioning of entire communities if the evolving species is a dominant species in the ecosystem or is a keystone species.
- Pairwise coevolution modifies the interaction strength between the two species, with possible implications for energy transfers within food webs.
- Diffuse coevolution allows the emergence of complex structures and functioning that are comparable to observed patterns.

FURTHER READING

Laland, K. N., F. J. Odling-Smee, and M. W. Feldman. 1999. Evolutionary consequences of niche construction and their implications for ecology. Proceedings of the National Academy of Sciences, U.S.A. 96: 10242–10247. *This article introduces a two-locus genetic model, where one locus controls the construction of the niche and the other locus defines how the environment affects individual fitness. This simple monospecific model allows evaluation of the interplay between evolution and abiotic components of the environment of the target species.*

Lansing, Stephen J., James N. Kremer, and Barbara B. Smuts. 1998. System dependent selection, ecological feedback and the emergence of functional structure in ecosystems. Journal of Theoretical Biology 192: 377–391. *This article uses a Gaïa-like approach but introduces an explicit selection process. After developing the evolutionary extension of the original daisy model, the authors apply their ideas to a rice-farming system.*

Levin, Simon A. 1998. Ecosystems and the biosphere as complex adaptive systems. Ecosystems 1: 431–436. *This article introduces the concept of complex adaptive systems more extensively than the present chapter. It gives several instances explaining how and why one can consider ecological communities as complex adaptive systems.*

Loeuille, Nicolas, and Michel Loreau. 2005. Evolutionary emergence of size structured food webs. Proceedings of the National Academy of Sciences, U.S.A. 102: 5761–5766. *This article shows how complex communities can emerge out of a diffuse coevolutionary process. Two features of the community emerge: the complexity itself, as the model starts with only one species and may end with hundreds, and food web structures that are diverse and comparable to those found in empirical data sets.*

Loreau, Michel. 1998. Ecosystem development explained by competition within and between material cycles. Proceedings of the Royal Society of London, Biological Sciences 265: 33–38. *This article is a useful description of how evolution may interact with nutrient flows in ecosystems. It shows how evolution favors different traits when considered within a cycle of a nutrient or between different cycles of nutrients. It links these evolutionary processes to large-scale functional properties of ecosystems such as productivity, biomass, and nutrient recycling.*

Swenson, William, David S. Wilson, and Roberta Elias. 2000. Artificial ecosystem selection. Proceedings of the National Academy of Sciences, U.S.A. 97: 9110–9114. *This article describes an experiment in which selection acts on a whole soil community. It shows large effects of community selection on the trait measured at the community level, this trait being the biomass of the plant that grows on the selected soil community. The article also details the implications of the experiment for the definition of heritability of this trait and the possible large consequences of selection at high levels.*

Whitham, Thomas G., William P. Young, Gregory D. Martinsen, Catherine A. Gehring, Jennifer A. Schweitzer, Stephen M. Shuster, Gina M. Wimp, Dylan G. Fischer, Joseph K. Bailey, Richard L. Lindroth, Woolbright Scott, and Cheryl R. Kuske. 2003. Community and ecosystem genetics: Consequence of the extended phenotype. Ecology 84: 559–573. *This article explains the concept of community genetics and demonstrates applications to several empirical situations. It also defines the conditions in which intraspecific genetic variation is likely to play a role in the structure and functioning of communities and ecosystems.*

Wilson, David S., and Eliott Sober. 1994. Reintroducing group selection to the human behavioral sciences. Behavioral and Brain Sciences 17: 585–654. *A nicely written overview of the theoretical developments of selection at group and higher levels. The article is also illustrated with many examples from the social sciences and evolutionary ecology. These examples make the article particularly easy to understand and a very good introduction to this hotly debated issue. A comment by Richard Dawkins (a proponent of selection at the gene level) follows, which makes the debate more complete.*

IV

Landscapes and the Biosphere

Brian Walker

The aim of this section of the *Princeton Guide to Ecology* is to provide an understanding of ecology at the scale of landscapes. Viewed in this way, terrestrial landscapes can be thought of as self-organizing systems of topographically determined physical/chemical factors interacting with the biological components that occupy them. The resulting patterns of biological communities are strongly influenced by human use and management activities. This becomes extreme when the landscape is fragmented by human use and consists mostly of agricultural or other nonnative cover with separated patches of native communities. Marinescapes are less subject to fragmentation effects, but their biological communities are also structured by spatial processes and ecological dynamics that are equivalent, albeit markedly different, to those that occur in terrestrial landscapes.

Starting with the terrestrial world, an understanding of ecology at landscape scales begins with the determinants of ecosystem structures and patterns across landscapes. Topography and climate are primary drivers, but the observed patterns of biological communities clearly indicate that biological processes interact strongly with the physical/chemical processes, modifying them to produce the resultant distribution of ecosystems. The first two chapters (chapter IV.1 by Tongway and Ludwig and chapter IV.2 by Fischer, Lindenmayer, and Hobbs) provide an account of how ecosystems in landscapes are structured and how they self-organize over time. The first chapter focuses on the physical–biological interactions. The second deals more with the biological processes and compares five different "models" that have been proposed to explain the structure and dynamics of landscapes.

Fewer and fewer intact landscapes remain. In all but very sparsely populated and specially protected regions, more and more of the world consists of landscapes in which native vegetation cover has been fragmented to varying degrees. The result is a disruption of the ecological processes that produced the original biological patterns, with other processes becoming dominant. Erstwhile large, single populations behave as spatially separated metapopulations in which processes such as immigration and emigration assume much greater significance than they did before. The next several chapters cover different aspects of this disruption effect of human use on the ecology of landscapes.

Chapter IV.3 (Wu) explores the dynamics of ecosystems in fragmented landscapes. As the impacts of land use intensify, landscape ecological processes become further modified, and chapter IV.4 (Moorcroft) takes this up by examining the patterns of biodiversity in managed landscapes. Chapter IV.5 (Peters, Gosz, and Collins) considers it further by focusing on disturbances in landscapes and how changes in disturbance regimes induced through different kinds of land use can result in phase shifts in ecosystem structure and composition. There is considerable overlap in chapters IV.3 through IV.5, but, with each providing its own emphasis, they collectively present an understanding of the effects of human use of ecosystems at the scale of landscapes.

Chapter IV.6 (Maurer) moves the focus up in scale to consider interactions between landscapes, over much larger spatial scales. It addresses the issues of biogeography: how species and communities change across latitudinal and other gradients, both on land and in the oceans.

Up to this point the focus has been on what determines the structure and dynamics of the ecosystems in a landscape. Chapter IV.7 (Woodward) changes the focus to interactions of the biosphere and the atmosphere. It examines how the structure and biological composition of ecosystems at landscape scales influence the physics and chemistry of the atmosphere, and vice versa. It thus provides insights into how changes in landscape cover (as dealt with in the preceding chapters) can result in changes in the chemical composition of the atmosphere and in the climate, sometimes in nonlinear and sudden ways.

The last three chapters deal with marinescapes. Chapter IV.8 (Hughes) examines the structure and dynamics of coral reefs, the marine spatial equivalent of terrestrial landscapes. It explores how coral reefs regenerate and how the connections between reefs influence the dynamics of coral reef systems, such as in the Great Barrier Reef of Australia and reefs in the Caribbean.

Chapter IV.9 (Karl and Letelier) examines seascapes, defined by the physical, chemical, and biological variables experienced by an organism during its lifetime. It is analogous to chapter IV.1 in that it emphasizes the physical/chemical dynamics in seascapes and focuses in particular on the role of microbes, a rapidly developing field of marine studies. Finally, chapter IV.10 (Pauly and Watson) considers the effects of human use on the dynamics of marine ecosystems, with an emphasis on spatial dynamics and fisheries.

IV.1

Landscape Dynamics

David J. Tongway and John A. Ludwig

OUTLINE

1. A view of landscapes as functioning systems
2. Spatial heterogeneity as an organizing principle in landscape dynamics
3. Function and dysfunction in landscape dynamics
4. The role of feedback loops
5. Assessing landscape dynamics: Thresholds and climate change
6. Concluding remarks

A terrestrial landscape can be viewed as a system of biological elements (organisms, populations, communities) forming a pattern across a topographic geomorphic unit. The dynamics of these landscape systems are driven by topography and climate and by interacting geochemical and biophysical processes. Although we recognize that important conceptual advances in landscape dynamics have been developed, such as how landscapes behave as complex adaptive and self-organizing systems, in this chapter we particularly focus on the development of the notion of landscape function, that is, how a landscape works as a geochemical–biophysical system to regulate vital resources over space and time. In highly functional landscapes, a major rainfall event will trigger runoff, but, overall, little loss of water, topsoil, and organic matter occurs from the system because these resources are dynamically captured by patterned structures within the landscape such as vegetation patches, which function as reserves or resource "banks." Vegetation patches then utilize retained resources (water, nutrients) in growth pulses to produce biomass such as seeds, most of which are cycled back into the system (soil seed banks). Biomass can also function to maintain the retentive capacity (structure) of the patch and can provide shelter and food for fauna or for consumption by livestock, which when harvested represent offtake from the system. Damaged landscapes become dysfunctional by losing their capacity to effectively regulate resources.

GLOSSARY

landscape dynamics. How a landscape, as a system of interacting components, structures, and processes, varies in space and time

landscape function. How a landscape works as a tightly coupled geochemical–biophysical system to regulate the spatial availability and dynamics of resources

landscape heterogeneity. The mix of different components, structures, and processes occurring in a landscape, such as how different organisms disperse among different vegetation patches

landscape restoration/rehabilitation. The actions and processes taken to help damaged landscapes recover toward a specified goal (landform, land use)

landscapes as self-organizing systems. How components, structures, and processes in a landscape dynamically organize to form complex, adaptive, and stable systems

landscape system threshold. A point in the dynamics of a landscape where the system changes to a different state, as, for example, a damaged landscape becomes dysfunctional to the point where available resources no longer support a species

1. A VIEW OF LANDSCAPES AS FUNCTIONING SYSTEMS

We view landscapes and marvel at the patterning of their interconnected ecosystems and wonder about what dynamic processes have caused these patterns. What have we learned about the dynamics of landscape patterns and processes in recent times? In this chapter we explore new developments in landscape dynamics by building on the work of Turner, Gardner, and O'Neill (2001) and others and by adding our Australian perspective. Disturbance-induced effects on landscapes are described in later chapters in part IV.

The importance of spatial heterogeneity and self-organization for explaining landscape dynamics has become increasingly recognized over the last 25 years as seen in the writings of Kolosa and Pickett (1991) and Rietkerk and others (2002). This recognition has led to studies on landscape function, that is, how a landscape works as a tightly coupled geochemical–biophysical system that regulates the sources and dynamics of energy, water, and nutrient resources (Tongway and Ludwig, 1997). Landscape function and its dynamics in space and time can be resolved into the availability of vital resources, which strongly affect the responses of biota, especially if stressed or disturbed.

Initially, distinctions were made between measured system complexity and functional heterogeneity, which described how ecological entities such as species perceived, related to, and responded to each other. In the last 25 years, fine-scaled ecological processes have been increasingly integrated with broad-scale geographic–geomorphologic studies to better understand overall landscape function. The underlying processes that determine how landscapes function, the need for understanding heterogeneity, and how function is affected by stress and disturbance have been studied in more detail by integrating disciplines (Lovett et al., 2005); this has benefited both the science and the management of landscapes.

A metaphor for how landscapes function as an integration of processes is a gear train (Lavelle et al., 2006). A landscape may be visualized as system of intermeshing gears, with each gear being a distinct ecosystem with its unique structures (composition) and processes (size, speed) but tightly interconnected (meshed) with other ecosystems. Those ecosystems remote from each other (distant gears) only interact through systems and processes connecting them (the meshing of gears between them), and importantly, the overall landscape is not complete as a functional system until the entire gear train is in place. In a sense, the assembling of the gear train reflects the synthesis of landscape ecology at a range of scales over the last two decades.

This gear-train metaphor is intended to highlight the interconnectivity of processes in terms of the structure and functioning of well-functioning landscapes and the importance of maintaining all of the functional links from the very finest to the coarsest scales (from the smallest, fast-spinning gear to the largest, slow-turning gear). Typically, there is a logical progression in examining landscape dynamics, commencing with detecting and characterizing the pattern at an intermediate scale, accounting for pattern in terms of geochemical–biophysical processes and then extending the analysis to both coarser and finer scales to understand the underlying interconnectivity of ecosystem processes to, for example, conserve resources.

Many disciplines have mechanistically studied ecosystems in detail at fine scales (individual cogs on gears) according to the rules of reductionist science. However, broad-scale studies looking at landscape patterns and processes have been largely descriptive and discussed retrospectively because experiments on broader scales are very difficult and expensive to conduct. Recent progress has recognized landscapes as having fundamental and crucial interactions between ecosystems (connectivity between gears) (Shugart, 1998), which are central to progressing knowledge about landscape function and dysfunction. In particular, it is critical to understand how changes in landscape dynamics are explained by the underlying processes at fine scales when landscapes are subject to stress and disturbance (see later chapters in part IV).

Understanding landscape dynamics requires studies on how pattern and process change through time in terms of how geochemical–biophysical processes link organisms to other organisms and their environment. Linking processes at all scales from the microscopic to the regional are crucial to achieving a predictive understanding of landscape dynamics. All useful models of landscape dynamics must deal with interactions across multiple scales yet make use of the basic principles most strongly expressed at each scale. Returning to the gear-train metaphor, no cog, however small, is irrelevant to understanding landscape dynamics. A cog on a gear of any size (scale) damaged by stress or disturbance can be taken as a limit to the capacity of the landscape to function efficiently. Repair may be possible, but only when our understanding of how the gear train works (how the landscape functions as a system) is adequate.

2. SPATIAL HETEROGENEITY AS AN ORGANIZING PRINCIPLE IN LANDSCAPE DYNAMICS

Interpreting changes in landscapes requires an understanding of how geochemical–biophysical processes affect the dynamic availability of vital resources such as water, nutrients, and organic matter (Ludwig and Tongway, 2000), not just changes in species populations or vegetation patterns. Switches to new alternate stability domains (basins of attraction) can be mechanistically explained in terms of resource availability in space and time (Gunderson and Holling, 2001). These include oversupply, supply cutoff, and unexpected synergisms between different resources that result in a critical threshold being crossed.

Noy-Meir, in 1981, was one of the first ecologists to emphasize the importance of the availability of vital resources in three-dimensional space over time. His dynamic models integrated structure, spatial arrangement, and persistence of life forms as being dependent on differential water availability from runoff–runon processes and recognized that rainfall itself was a poor predictor of biological outcomes. Integrating the availability of soil nutrients with water in runoff–runon redistribution processes (Tongway and Ludwig, 1997) helps to explain the landscape patterns generated by Noy-Meir's three-dimensional dynamic modeling, which also produced sigmoidal or S-shaped curves to reflect how plant production responds to water availability in arid ecosystems.

There is a continuum of scale in processes from the microscopic to regional landscapes. At a micro scale, the functional role of organisms can be characterized into four groups: microflora, micropredators, litter transformers, and ecosystem engineers (Lavelle and Spain, 2001). The activities of these groups affect and control processes such as soil gas exchange, water percolation, and soil aggregate stability, whose influence is easily recognized at larger scales. The aggregate effect of these processes and properties at landscape scale provides a mechanistic interpretation of macroscopic processes such as surface hydrology, nutrient cycling, plant nutrition, and soil erosion, which are properties readily recognized by land managers.

3. FUNCTION AND DYSFUNCTION IN LANDSCAPE DYNAMICS

The concepts outlined above are very useful in proposing explanatory frameworks for understanding landscape dynamics. A conceptual framework has been developed that depicts landscape function as a sequence of processes, commencing with rainfall as a *trigger* (figure 1), which initiates *transfer* processes such as runoff and erosion that spatially relocate resources such as water, topsoil, organic matter, and seeds across the landscape. Some of these resources may exit the landscape as *outflow*, and some may be stored in the soil *reserve*. The reserve may be considered metaphorically as a "bank" dealing in many diverse but interacting ecological "currencies": water, nutrients, seeds, and soil fauna. Some locations in the landscape, such as runon areas with patches of vegetation on deeper soils, absorb or capture more resources than other parts (interpatch runoff areas). Within vegetation patches, a *pulse* of plant growth may ensue, the magnitude of which depends on the status of soil moisture and mineralized nutrients (stored reserves). Some of the materials produced by the growth pulse may be lost from the system by fire or herbivory (*offtake*), but much is cycled back to the reserve by *biological feedback* where plant litter is reincorporated into the soil by a range of soil biota, and the seed pool is replenished. A *physical feedback* loop affects the extent to which changes in plant density at ground level regulate the amount and rate at which water and other resources are transported across the landscape. The trigger–transfer–reserve–pulse framework is depicted here as being on a fulcrum, implying that there is a crucial balance between resource losses (outflow plus offtake) and gains (feedback loops) in the landscape system; this balance dynamically fluctuates

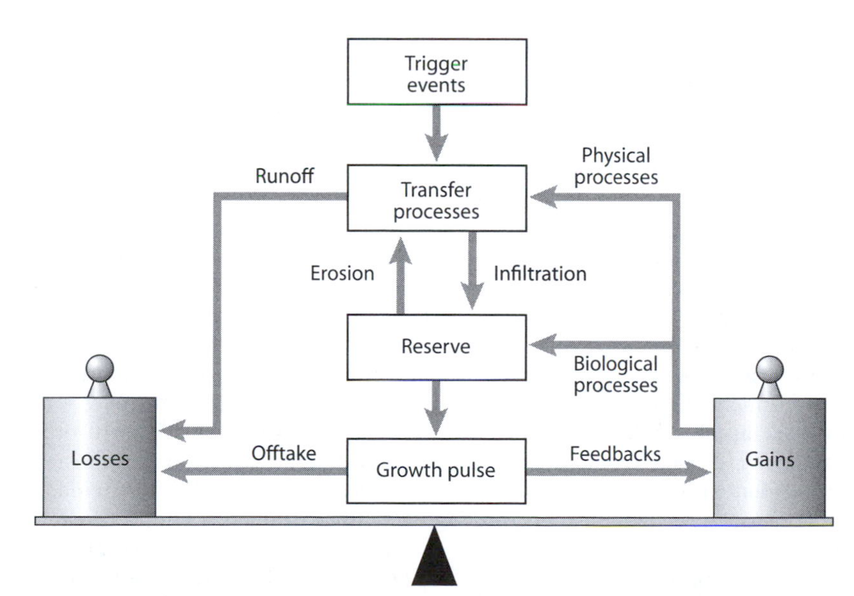

Figure 1. The trigger–transfer–reserve–pulse conceptual framework for how geochemical–biophysical processes function in landscapes to dynamically balance resource gains and losses. (After Tongway and Ludwig, 1997)

over time and can be "knocked" out of balance if disturbances damage the system.

Landscape function can be represented as a continuum from the least-damaged landscape, called highly functional, to the most-damaged, called highly dysfunctional, landscapes (Ludwig and Tongway, 2000). This continuum is evident in different landscapes such as grasslands and shrublands (plate 6). In highly functional grasslands, a high density of grass plants facilitates the capture, retention, and use of wind- and water-borne resources, whereas in dysfunctional grasslands, these grasses can be replaced by shrubs that fail to trap resources because of low ground contact cover, so that they blow and wash out of the landscape (plate 6, top). High rates of wind and water erosion and low soil nutrient concentrations characterize these dysfunctional landscapes. In contrast, highly functional shrublands effectively trap resources, but dysfunctional shrublands have bare crusted soils that allow water to run off and wind to blow materials away (plate 6, bottom).

In both of these landscapes, landscape function can be defined in terms of the dynamics of water and nutrients at scales varying from the rhizosphere to the local catchment. The grassland versus the shrubland illustrates the different scales at which these landscapes are organized (fine-grained grass tussocks versus coarse-grained shrub clumps), yet both can effectively capture and use vital resources such as water, nutrients, and organic matter. Both grasses and shrubs function above and below ground and provide goods and services (food, shelter) to other biota present.

Reductions in effective retention of vital resources by arid and semiarid landscapes have been characterized as degrees of landscape dysfunction or desertification (Tongway and Ludwig, 2002), and we noted earlier how landscape dynamics can be viewed as changes in the balance between gains and losses of resources in space and time. This concept of resource gains versus losses is very useful when working with highly disturbed sites such as lands affected by mining. If successful, the total development of rehabilitation on a mine site follows an S-shaped curve, which can be partitioned into two components: biological and physical (figure 2). Rehabilitation starts at a low level of natural landscape development (figure 2, point A). Initially this development is largely of the physical component (point B), typically caused by reshaping of landforms (e.g., overburden or spoil heaps; plate 7, photo A) into smoother surfaces that are "ripped" along contours (plate 7, photo B). The aim is to engineer a surface that will capture rainwater and minimize soil erosion. Initially the biological input is very small (figure 2, point C), typically microbial activity in any

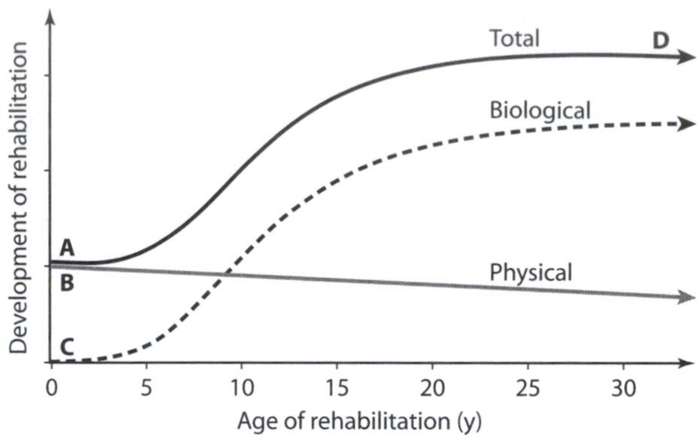

Figure 2. An example of how the rehabilitation of a landscape develops over time as an S-shaped curve as the combination of two general components: biological and physical processes.

topsoil applied to surfaces and the initial vegetation establishment (plate 7, photo C). Over time, the vegetation will markedly develop so that biological processes dominate, and the rehabilitated landscape becomes highly developed and functional (figure 2, point D; plate 7, photo D). Although the physical component remains important, it typically declines over time because, for example, riplines flatten.

As landscape function increases on rehabilitated mines, biological processes and the accumulation of resources ("natural capital"; Aronson et al., 2007) are improved, as is the capacity of the system to withstand stress and disturbance. This capacity is commonly referred to as the "buffering capacity" or "resilience" of the system (Gunderson and Holling, 2001). Typically, this capacity is described by changes in the biota of a system, but fundamentally system resilience is perhaps better explained by the robustness of interacting processes (viz., the gear-train metaphor).

4. THE ROLE OF FEEDBACK LOOPS

Mine site rehabilitation provides useful evidence of the initiation and development of feedbacks by establishing plant and soil biota communities, where, after a period, newly acquired resources can be observed to contribute to landscape stability. In particular, plant litter decomposition improves soil aggregate stability by binding soil particles with organic residues: water entry and storage improve, thus augmenting whatever properties were initially provided by engineering. Later, microclimatic conditions created by growing foliage ameliorate the immediate effects of weather and permit other biota to become established. The strengthening of these feedback mechanisms over time can be monitored, as can the "biodiversity" improvement.

As the new landscape develops, its natural capital increases (Aronson et al., 2007), as do the complexity of geochemical and biophysical processes and their interactions. A great deal of attention has been given to the "end game" of the "capital and complexity" response (Gunderson and Holling, 2001), but in heavily impacted landscapes, early information about success in establishing competent resource use is also important to understand, as excessive outflow of resources may prevent vegetation from becoming established. In the gear-train metaphor, a start needs to be made at the scale of the most readily studied scale of pattern and process, extending to both coarser- and finer-scale linked processes.

5. ASSESSING LANDSCAPE DYNAMICS: THRESHOLDS AND CLIMATE CHANGE

A useful practical concept is that of a critical threshold that marks the point where an ecosystem or landscape is self-sustaining or not. As illustrated in figure 3, as mine site rehabilitation develops over time to be highly functional, it follows an S-shaped curve. At some point along this curve, landscape functionality conceptually crosses a threshold above which the landscape system becomes increasingly self-organizing and self-sustaining because of highly functional geochemical–biophysical processes.

Process-based explanations of crossing such thresholds are more useful than biota-based explanations because of delays in some biotic responses. A critical threshold is crossed when a small change in a landscape driver results in a marked system response. In practical terms, operating a system close to the critical threshold raises the potential for a substantive change. Hence, it is important to understand the factors contributing to the system's "buffering capacity" in the vicinity of the threshold and to their mutual interactions. For example, allowing a landscape to continue to be grazed by livestock as drought conditions persist or worsen may result in surface conditions that massively erode the landscape when drought-breaking rains fall. Factors such as soil organic matter loss and exposure of unstable subsoils predispose the landscape to damage.

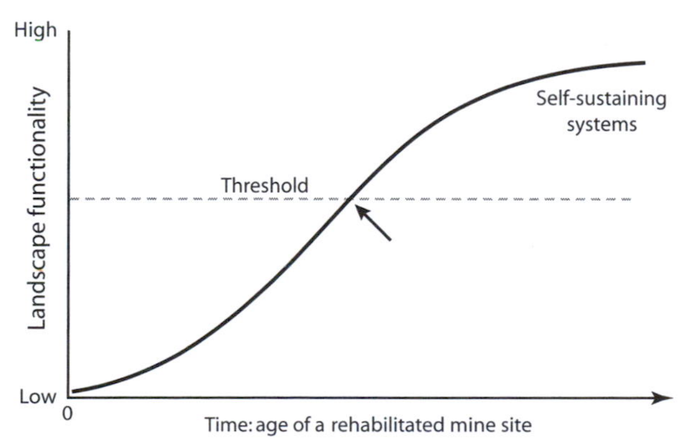

Figure 3. An S-shaped curve describing how mine site restoration typically progresses over time. At some point in time (arrow) a threshold is crossed (dashed line) when landscape functionality exceeds the level required for the site to become self-sustaining.

Because the proposed framework for landscape dynamics has climate and weather as major driver-triggering responses (see figure 1), landscape responses to modeled scenarios of climate can be addressed directly in process terms. Crucial questions proposed by Pounds and Puschendorf (2004) include: How quickly can ecosystems respond to changing climate, especially amount and seasonality of rainfall? Will there be a critical period of adjustment and uncertainty, or will there be a smooth transition to new scenarios?

6. CONCLUDING REMARKS

We have described landscape dynamics using a process-based approach in which geochemical–biophysical processes interact to regulate the economy (gains minus losses) or availability of vital resources. By understanding how landscapes function to conserve resources in terms of spatial patterns and dynamic processes, a diverse range of landscape problems and land management issues can be addressed. Here we used a mine site rehabilitation example, but the principles of landscape function can be applied to a wide range of biomes and land uses such as grasslands and savannas grazed by livestock.

FURTHER READING

Aronson, James, Suzanne J. Milton, and James N. Blignaut, eds. 2007. Restoring Natural Capital: Science, Business and Practice. Washington DC: Island Press.

Gunderson, Lance H., and C. S. Holling, eds. 2001. Panarchy: Understanding Transformations in Human and Natural Systems. Washington DC: Island Press.

Kolasa, J., and S.T.A. Pickett. 1991. Ecological Heterogeneity. New York: Springer-Verlag.

Lavelle, Patrick, T. Decaens, M. Aubert, S. Barot, M. Blouin, F. Bureau, P. Margerie, P. Mora, and J.-P. Rossi. 2006. Soil invertebrates and ecosystem services. European Journal of Soil Biology 42: S3–S15.

Lavelle, Patrick, and Alister V. Spain. 2001. Soil Ecology. Dordrecht, The Netherlands: Kluwer Academic Publishers.

Lovett, Gary M., Monica G. Turner, Clive G. Jones, and Kathleen C. Weathers, eds. 2005. Ecosystem Function in Heterogeneous Landscapes. New York: Springer-Verlag.

Ludwig, John A., and David J. Tongway. 2000. Viewing rangelands as landscape systems. In O. Arnalds and S. Archer, eds., Rangeland Desertification. Dordrecht, The Netherlands: Kluwer Academic Publishers, 39–52.

Noy-Meir, I. 1981. Spatial effects in modelling of arid ecosystems. In David W. Goodall and Ray A. Perry, eds., Arid Land Ecosystems: Structure, Functioning and Management, vol. 2. Cambridge, UK: Cambridge University Press, 411–432.

Pounds, J. A., and R. Puschendorf. 2004. Ecology: Clouded futures: Global warming is altering the distribution and abundance of plant and animal species. Nature 427: 107–109.

Rietkerk, Max, M. C. Boerlijst, F. van Langevelde, R. HilleRisLambers, J. Van de Koppel, L. Kumar, H.H.Y. Prins, and A. M. de Roos. 2002. Self organisation of vegetation in arid ecosystems. American Naturalist 160: 524–530.

Shugart, Hank H. 1998. Terrestrial Ecosystems in Changing Environments. Cambridge, UK: Cambridge University Press.

Tongway, David J., and John A. Ludwig. 1997. The Conservation of Water and Nutrients within Landscapes. In John A. Ludwig, David J. Tongway, David O. Freudenberger, James C. Noble, and Ken C. Hodgkinson, eds., Landscape Ecology, Function and Management. Melbourne, Australia: CSIRO Publishing, 13–22.

Tongway, David J., and John A. Ludwig. 2002. Reversing Desertification. In Rattan Lal, ed., Encyclopaedia of Soil Science. New York: Marcel Dekker, 343–345.

Turner, Monica G., Robert H. Gardner, and Robert V. O'Neill. 2001. Landscape Ecology in Theory and Practice: Pattern and Process. New York: Springer-Verlag.

IV.2

Landscape Pattern and Biodiversity

Joern Fischer, David B. Lindenmayer,
and Richard J. Hobbs

OUTLINE

1. Introduction
2. Conceptual landscape models
3. Key components of landscape pattern
4. Landscape pattern and biodiversity: Concluding remarks

The amount and spatial arrangement of different types of land cover are major drivers of terrestrial biodiversity. Conceptual landscape models provide the terminology needed to analyze the effects of landscape pattern on biodiversity. Conceptual landscape models vary in their degree of complexity and realism. In increasing order of complexity, conceptual landscape models include the island model, the patch–corridor–matrix model, the variegated landscape model, the hierarchical patch dynamics model, and species-specific gradient models. Less complex models are easier to communicate than complex models, but they may oversimplify the relationship between landscape pattern and biodiversity (especially in highly heterogeneous landscapes). Key components of landscape pattern that have an important effect on biodiversity include patches of native vegetation, the nature of land use outside these patches, the connectedness of patches and ecological processes (connectivity), and the variability in land cover types (heterogeneity). Landscapes with (1) large areas of native vegetation, (2) areas outside patches that are similar in structure to native vegetation, and (3) high landscape heterogeneity are likely to support a high level of biodiversity.

GLOSSARY

biodiversity. The diversity of genes, species, communities, and ecosystems, including their interactions

conceptual landscape model. A theoretical framework that provides the terminology needed to communicate and analyze how organisms are distributed through space

connectivity. The connectedness of habitat, land cover, or ecological processes from one location to another or throughout an entire landscape

landscape. A human-defined area, typically ranging in size from about 1 km^2 to about 1000 km^2

landscape heterogeneity. The variability in land cover types within a given landscape

landscape pattern. The combination of land cover types and their spatial arrangement in a landscape

matrix. The dominant and most extensive patch type in a landscape, which exerts a major influence on ecosystem processes

patch. A relatively homogeneous area within a landscape that differs markedly from its surroundings

1. INTRODUCTION

Life is not distributed uniformly across the surface of the planet. In terrestrial systems, several biophysical variables such as nutrient availability, radiation, and water availability are fundamental influences on where different organisms occur (see chapter IV.1). In addition, in a given area, the types of land cover present and where they occur have a major influence on how biodiversity is distributed through space. This chapter summarizes several conceptual models that are commonly used to help us think about landscapes. These conceptual models facilitate an understanding of the relationship between key components of "landscape pattern" and biodiversity.

2. CONCEPTUAL LANDSCAPE MODELS

The effect of landscape pattern on biodiversity can be analyzed and communicated in many different ways. Implicitly or explicitly, ecologists and nonecologists alike rely on conceptual models that summarize the most important features in a given landscape. Do we need to know where there are trees? Or what the density of buildings is? Or where the warmest areas are that are suitable basking sites for a particular reptile species? Different features of landscapes will be most important for different purposes, and therefore, different conceptual models of landscapes may be required.

A conceptual landscape model can be defined as a theoretical framework that provides the terminology needed to communicate and analyze how organisms are distributed through space. Typically, a conceptual landscape model also can be used to draw a picture of the most important features in a given landscape, that is, to visualize landscape pattern.

There are many different conceptual landscape models, and different people may perceive the same landscape in quite different ways. However, some general conceptual models and descriptions of landscapes and landscape pattern are broadly agreed on and are used by many ecologists. These conceptual models simplify the complexity of landscape patterns. This simplification is useful because it enables meaningful communication, provides a common terminology, and focuses people's attention on particular features within a landscape. A potential disadvantage of any particular simplification is that important aspects of complexity may be overlooked. In this regard, conceptual landscape models are no different from many other models that face a trade-off between useful simplicity and undesirable oversimplification.

Five conceptual landscape models are outlined in the remainder of this section: (1) the island model, (2) the patch–corridor–matrix model, (3) the hierarchical patch dynamics model, (4) the variegated landscape model, and (5) a suite of species-specific gradient or continuum models. The first two models are widely used, whereas the last three are applied less commonly, although they can be extremely useful.

The Island Model

The island model originates from the equilibrium theory of island biogeography, which was developed to explain patterns in species richness on islands surrounded by ocean (MacArthur and Wilson, 1967; see also chapter IV.3). As a conceptual model for terrestrial landscapes, the island model is based on the analogy of hospitable islands within an inhospitable ocean. Thus, the island model provides a black-and-white view of landscapes—every point in the landscape is classified to be either part of an island or part of the inhospitable surrounds (figure 1). What constitutes an island depends on the purpose for which the model is used. For example, islands might be forests, whereas all other land cover might be considered "ocean"; or islands might be areas of suitable habitat for a species of interest, in which case what constitutes an island depends on the species and its habitat requirements. For a rock fern, rocky outcrops might be suitable islands, whereas tree-cavity–dependent mammals may find suitable habitat only in islands of old-growth forest. The island model is very simple. This attractive simplicity, combined with the broad impact of island biogeography theory, means that it has been very influential. Important insights about what constitute suitable islands or patches have arisen from the

Island model
- "Islands" in an "ocean"
- Simple to map & communicate
- May oversimplify ecological complexity

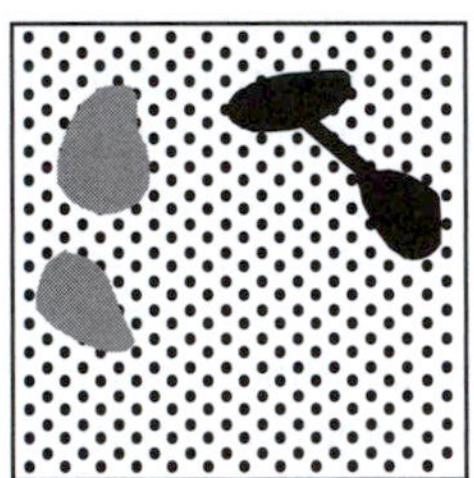

Patch–matrix–corridor model
- Different patch types in a mosaic
- May be connected by corridors
- Often based on human-defined patterns

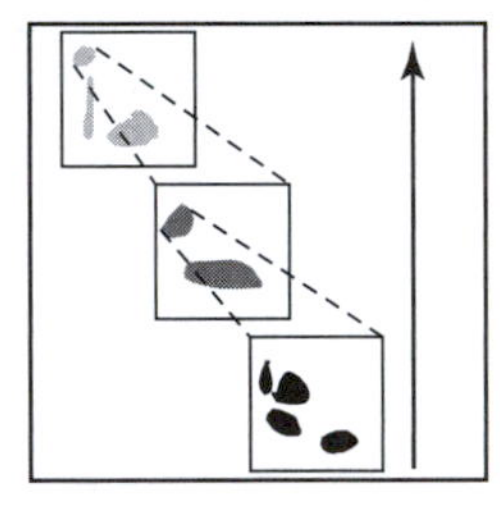

Hierarchical patch dynamics model
- Hierarchy of scales
- Sophisticated cross-scale relationships
- More difficult to communicate
- Often based on human-defined patterns

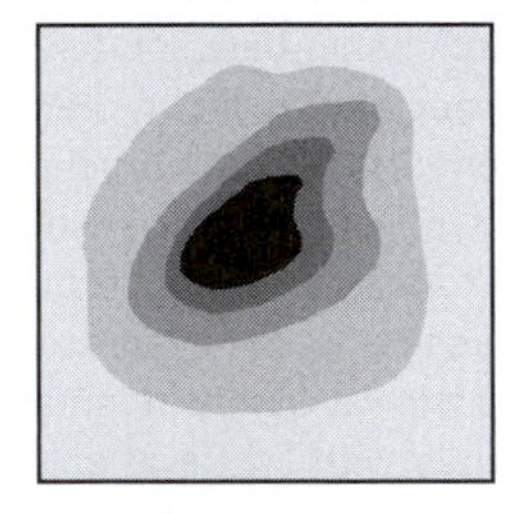

Variegation model
- Gradients in the environment, rather than patches in a matrix
- Sometimes reflects organisms' distribution
- Often based on human-defined patterns

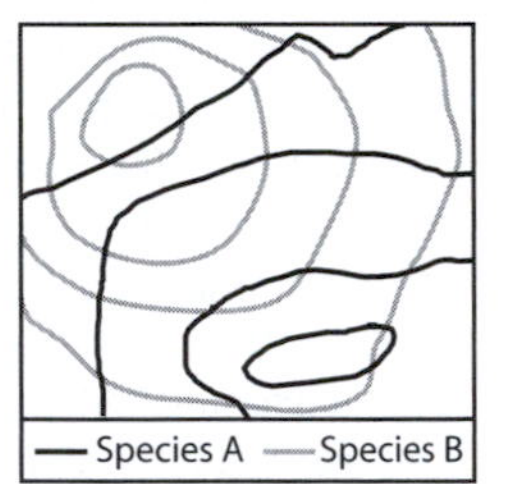

Species-specific gradient models
- Unique habitat contours for different species
- Recognizes environmental continua and differences between species
- Difficult to map & communicate

Figure 1. Schematic overview of alternative conceptual landscape models, including a short summary of their key features, strengths, and limitations.

island model (see below). However, often the island model can be too simple if considered on its own. Land cover often changes more continuously than is assumed by the island model, and what is an appropriate island for one species may be inappropriate "ocean" for another.

The Patch–Corridor–Matrix Model

The patch–corridor–matrix model is an extension of the island model. Patches in this model are broadly equivalent to islands, and the matrix is broadly equivalent to the ocean in the island model (see Glossary above). Corridors, the third landscape element, are linear features that connect the patches. The patch–corridor–matrix model is closely associated with a mosaic view of landscapes (figure 1). A mosaic view does not necessarily assume that all patches are equal but may explicitly distinguish among a number of different patch types. These patch types are typically based on human-defined attributes rather than the habitat of a particular species. For example, three patch types might be deciduous forest patches, evergreen forest patches, and mixed forest patches, and all of these might be embedded within a matrix of agricultural land. In this example, linear connections of vegetation between forest patches would be considered corridors. The patch–corridor–matrix model has several important advantages over the island model. First, it recognizes that the matrix plays several important roles in addition to isolating the patches (see below). Second, it acknowledges that corridors fulfill an important connectivity function. Third, it recognizes the role of landscape heterogeneity by considering landscape mosaics composed of various different patch types. The main limitation of the patch–corridor–matrix model is that it is usually based on a human perception of landscapes in which humans define where patches start and end and what the appropriate patch types are. This can be problematic because some organisms perceive their environment in a very different way from humans. For example, the spatial scale relevant to a beetle is much smaller than the scale at which the patch–corridor–matrix model is typically applied. Similarly, some plants and animals may respond strongly to variables that are not easily seen by humans; for example, some plants might be restricted to locations with high fertility, and some cold-blooded animals (or ectotherms) might respond strongly to subtle variations in temperature.

The Hierarchical Patch Dynamics Model

The hierarchical patch dynamics model is similar to the patch–corridor–matrix model in that it also recognizes landscape mosaics. However, it is more complex because it recognizes spatial hierarchies (Wu and Loucks, 1995). That is, it recognizes that patchiness in a given landscape does not occur at only one spatial scale, but, rather, different levels of patchiness are apparent at different scales (figure 1). At a coarse scale, for example, islands of trees might be patchily distributed through a landscape, whereas at a much finer scale, different clumps of grass might be considered as individually recognizable small patches. Hierarchy theory also provides an explicit link between different scales and how they influence one another. In particular, the ecological dynamics at coarse spatial scales is seen as a constraint or context on the ecological dynamics that occurs at finer scales. For example, individual trees in a forest might die when they reach an old age, but if the larger-scale context is continuous forest, it is highly likely that new trees are able to regenerate to replace such dead trees. Thus, the hierarchical patch dynamics model can be applied, for example, to explain the population dynamics of particular species in a given landscape (see chapter IV.3). Although the explicit recognition of spatial dynamics is a key strength of the hierarchical patch dynamics model, its level of complexity means that it is not as easily applied as the simpler island or patch–corridor–matrix models.

The Variegated Landscape Model

The landscape models discussed so far all assume that spatial discontinuities, and therefore patches, can be defined in landscapes. Although this is sometimes the case, in other cases it is unclear where patch boundaries should be drawn. The absence of obvious spatial discontinuities led to the development of the variegated landscape model. This model recognizes that land cover may change continuously through space—for example, dense tree cover may gradually blend into widely scattered trees, and these may increasingly blend into open grassland (figure 1). Superimposing islands onto a landscape with gradual changes can be a serious oversimplification. For example, many species may not be strictly dependent on predefined islands of forest but may use the continuous gradient from forest to grassland to different extents. The variegated landscape model provides a viable alternative to patch-based models in situations where it can be shown that organisms respond continuously along a gradient of landscape change. Its main contribution—to highlight that some patterns cannot easily be translated into a patchy view of the world—is important at a conceptual level because it questions one of the fundamental assumptions made by mosaic- or patch-based landscape models.

Species-Specific Gradient or Continuum Models

To different extents, the models discussed above rely on a classification of landscape pattern that can be readily perceived by people. What resonates with people, however, may not be a good way to characterize a landscape from the perspective of other species. Similarly, how one organism perceives a given landscape may be vastly different from how another perceives it. In fact, no two species are likely to perceive a given landscape in the same way. Indeed, even different humans perceive the same landscape in different ways. For this reason, a different suite of landscape models attempts to see landscapes from the perspective of a given organism of interest. Rather than predefining land cover as a mosaic or as a continuum of predefined land cover, species-specific gradient models start with a particular organism and then attempt to quantify how suitable a given point in the landscape is for that organism. This quantification can occur either empirically (using observed data) or by considering a series of key requirements of the organism, such as its needs for nutrients, shelter, space, climatic conditions, and the abundance of mutualists, competitors, and predators. Such species-specific gradient models can be visualized by drawing maps of where a given species of interest is most and least likely to occur (figure 1). Two key strengths of these models are that (1) they do not assume that the same landscape classification will necessarily work for different species, and (2) they encourage their users to think about fundamental ecological processes shaping the distribution of a species. A weakness of species-specific gradient models is that many different visualizations of a given landscape would be needed to reflect the needs of many different species. Typically, this is too complicated to be feasible, except in cases where only one or few species are targeted. In addition, such models are more difficult to communicate and are not easily depicted in Geographical Information Systems and maps. Thus, as for the variegated landscape model, the most useful contribution of species-specific gradient models may be at a conceptual level: they encourage their users to question key assumptions that are implicit in some of the other landscape models.

Overview

The different landscape models have different strengths and weaknesses and are suitable for different purposes. None of them is inherently right or wrong. To reduce the risk of oversimplifying complex ecological systems, it can be useful to think about a given scientific or management problem in more than one way, that is, to apply more than one landscape model and think about the contrasting insights obtained. Key features of the different landscape models are summarized in figure 1.

3. KEY COMPONENTS OF LANDSCAPE PATTERN

Although different landscape models will highlight different components of a landscape pattern, the important influence of some particular landscape features on biodiversity is now widely accepted (reviewed by Lindenmayer and Fischer, 2006). Four such features are discussed below: (1) patches of native vegetation, (2) modified land surrounding these patches, (3) corridors and other features that enhance connectivity, and (4) landscape heterogeneity.

Patches of Native Vegetation

The benefit of patches of native vegetation for biodiversity has long been recognized. More specifically, it is widely accepted that, other things being equal (which they may not always be), large patches of native vegetation support more species than smaller patches. Several plausible reasons that are not mutually exclusive have been proposed to explain this pattern:

1. In larger patches, the ratio of species migrating into a patch to species becoming extinct in that patch is likely to be higher (an explanation stemming from island biogeography theory; see MacArthur and Wilson, 1967).
2. Larger patches contain a larger "interior" area that is not subject to the same extent to disturbances from outside the patch. External disturbances are often termed "edge effects" and include changes in variables such as temperature, wind speed, or weed invasion.
3. Larger patches often contain a greater variety of biophysical conditions, thereby offering suitable habitat for a larger number of different species.
4. If species were randomly distributed, larger patches would contain more species simply by chance.

Similar in importance to patch size at the local scale is the total amount of native vegetation in any given landscape, which is positively related to the overall level of biodiversity at the landscape scale. Species are typically lost from a given landscape with any substantial loss of native vegetation, and the more native vegetation is lost, the more species become locally extinct. At particularly low levels of native vegetation cover, such as when native vegetation covers less than 30% of the landscape (Andrén, 1994), there is some evidence that species loss accelerates beyond the rate of loss observed as cover declines at higher levels of native vegetation.

An additional variable influencing how patches of native vegetation affect biodiversity is how these patches are distributed through space. Broadly speaking, for any given total amount of native vegetation cover, two extreme types of spatial arrangement are possible: (1) many small patches can be dispersed through the landscape, or (2) few large patches can be aggregated near one another. This has led to the "SLOSS" (single large or several small) debate (see chapter IV.3 under "habitat fragmentation"). Evidence to date on which arrangement supports more biodiversity is not clear-cut. In part, which spatial arrangement is better depends on the species of interest. A species that requires large patches would benefit from aggregated large patches, whereas dispersed small patches may be of little value to them. Other species, especially mobile ones such as some birds or bats, may be able to survive in landscapes with many dispersed small patches, partly because they can easily move between patches and thereby access different resources from different patches. The body size of organisms is also important in this context—what constitutes a small patch for a large mammal (such as an elephant) may be perceived as a very large patch by a small reptile (such as a skink).

The amount of species turnover through a landscape also influences whether many dispersed small patches support more biodiversity than few large patches. Where species composition changes substantially through space, many small patches scattered throughout a landscape may effectively sample a higher diversity of species than few large patches aggregated in only part of the landscape, provided, of course, that there are enough species that are not entirely restricted to large patches.

Modified Land Outside Patches of Native Vegetation

In the early stages of conservation ecology, interest was mostly in patches of native vegetation and their contribution to biodiversity, with little attention paid to the role of the surrounding environment. Areas of nonnative vegetation were often not considered at all, and sometimes they were explicitly considered worthless or even hostile.

Partly because many landscapes worldwide are no longer dominated by native vegetation, ecologists were forced to have a closer look at the effects of modified environments on biodiversity. Most ecologists now believe that areas outside patches of native vegetation, especially if they dominate a given landscape, have a fundamental influence on the biodiversity of that landscape.

There are at least three key ways in which areas outside patches are important. First, in areas that are largely cleared of native vegetation, other land uses typically have a dominant effect on a wide range of ecosystem processes. For example, in industrial tree plantations, wind speeds are higher at the edge of cleared stands; and in suburbia, introduced plants and animals often originate from people's gardens. Such changes in ecosystem processes, in turn, have an important effect throughout a given landscape, affecting biodiversity both within and outside patches of native vegetation.

Second, some elements of biodiversity can survive outside patches of native vegetation. For example, some bird species inhabit suburban gardens, some frog species breed in puddles adjacent to dirt roads, and some large carnivores may find food in farmland or in urban rubbish tips. The presence of native species outside patches of native vegetation does not mean that native vegetation is not important. Rather, it highlights the point that simply discounting human-modified environments as nonhabitat is overly simplistic.

Third, the nature of areas outside patches of native vegetation dictates to a large degree to what extent species can move from one patch of native vegetation to another. Roadside vegetation, for example, can assist the movement of birds between woodland patches in both rural and urban areas. Similarly, at much larger scales, whether species are able to shift their ranges in response to climate change will depend to a large degree on whether they can move through modified environments.

In summary, notwithstanding the importance of native vegetation patches for biodiversity, what happens outside these patches cannot be ignored. In many instances the patches of native vegetation and their surrounds deserve equal attention because both are fundamentally important and interact in significant ways.

Connectivity

In its broadest sense, connectivity is related to how connected biodiversity is between various locations. Although there is general agreement that connectivity is important, there is far less agreement about what its precise definition should be. Some ecologists believe that connectivity is a property of patches, whereas others think it is a property of entire landscapes. Some believe that it applies to individual species, whereas others think it should apply to a broad suite of ecosystem processes. In this section, we first define three types of connectivity that have been suggested, to overcome some of the vagueness in terminology. We then summarize how landscape pattern is related to connectivity.

Structural connectivity occurs when one part of a landscape is physically linked with another part. The most well-known example is wildlife corridors (see above) linking one patch of native vegetation with another patch. A main aim of maintaining or creating this type of structural link is to facilitate the movement of wildlife. *Functional connectivity* acknowledges that simply having a structural link may not necessarily facilitate movement for all species; that is, a given corridor may not function in the way it was intended to function if the target species does not use it. There are many reasons why some species are reluctant to move through corridors. For example, predation risk may be higher in corridors than in continuous patches of native vegetation, or certain key resources may not be available. *Ecological connectivity* is a more general term used to describe the connectedness of ecological processes in a given landscape, either abiotic (such as water flows) or biotic (the spread of weeds, or the annual migration of many bird populations).

Landscape pattern affects all three types of connectivity. Typically, having structural links makes it more likely that functional connectivity and ecological connectivity are also achieved. For this reason, corridors can be a useful strategy to enhance connectivity. Some authors warn that corridors may also have undesired consequences; for example, they may facilitate the spread of fire or weeds or introduced predators (Simberloff et al., 1992). Notwithstanding the possibility of negative consequences, in most cases, the positive consequences of corridors outweigh the risk of negative consequences (Noss, 1987).

Two other features of landscape pattern can also facilitate connectivity. The first is stepping stones. Stepping stones are small patches of vegetation scattered throughout a landscape. Some organisms can use them to move through a landscape. Second, the nature of the "matrix" itself—the dominant land cover type—has a large effect on connectivity. In general, connectivity is likely to be higher if the matrix is similar in structure to native vegetation.

Landscape Heterogeneity

Organisms differ in their habitat requirements. It follows that at a landscape scale, spatial variability in the properties of land—landscape heterogeneity—may be beneficial for biodiversity. Some agricultural landscapes can support high biodiversity because of their heterogeneity. Traditionally managed agricultural landscapes in Europe are good examples. In these landscapes, often a variety of crops is grown in relatively small fields, forest patches are scattered throughout the landscape, and field margins contain seminatural shrubland or specifically established hedgerows. Different species can use different parts of this diverse landscape mosaic, and landscape heterogeneity has been identified as a key variable enhancing biodiversity in European agricultural landscapes (Benton et al., 2003).

Landscape heterogeneity also may result from more subtle changes through space rather than abrupt changes in land cover. Gradual changes in the biophysical properties of landscapes are common in both natural and human-modified landscapes. For example, topography has a major effect on water flows and the distribution of nutrients in a landscape, and the orientation of a given slope can make it warmer and drier or cooler and wetter. Biophysical gradients have long been recognized by plant ecologists as having a major effect on species' distributions: variables such as temperature, radiation, water, and nutrient availability are particularly important for plants. Animals also may be affected by ecological gradients, although these may be related to different life history requirements such as the availability of sufficient food, space, and shelter. As with heterogeneity in land cover types, the presence of strong ecological gradients within a given landscape can be related to high levels of biodiversity at the landscape scale.

4. LANDSCAPE PATTERN AND BIODIVERSITY: CONCLUDING REMARKS

Landscapes can be conceptualized in many different ways. Some conceptual landscape models are simple (e.g., the island model), whereas others are complex (e.g., the hierarchical patch dynamics model). Similarly, some conceptual landscape models are well known, whereas others are less widely known, although they can provide useful insights (e.g., the variegated landscape model). Often, complementary insights can be gained by conceptualizing landscapes in several different ways. For example, the island model may highlight the importance of large patches of native vegetation, whereas species-specific gradient models draw attention to the value of landscape heterogeneity because what may be suitable habitat for one species may not be suitable for another species. Which landscape model is most appropriate in a given situation depends on the particular objectives of the study or management problem in question.

Table 1. Overview of important components of landscape pattern and their effects on biodiversity

Component of landscape pattern	*Effect on biodiversity*
Patches of native vegetation	• Often support many native species • Other things being equal, the larger the patch, the more species it supports
Areas outside patches of native vegetation	• In modified landscapes, dominant control of biophysical processes (such as wind speeds and water flows) • Can provide habitat for some species and link habitat patches • Can provide a buffer for patches of native vegetation • Can provide a source of introduced species and other disturbances
Connectivity	• Structural connectivity—links between patches via features such as corridors—can provide links for some species and some ecological processes (but not all species or processes benefit from structural connectivity) • Functional or habitat connectivity—the connectedness of habitat for a given species—is useful for a given species to move through the landscape • Ecological connectivity—the connectedness of particular ecological processes—can be either desirable (e.g., for natural water flows) or undesirable (e.g., for weed dispersal)
Landscape heterogeneity	• Variability in land cover types means that different species using different land cover types can occur within the same landscape • Ecological gradients in biophysical variables (temperature, nutrients, food availability) can be related to species turnover, thereby increasing landscape-scale biodiversity

Irrespective of the conceptual landscape model used, many factors determine how landscape pattern influences biodiversity (summarized in table 1). Typically, different landscape patterns will support different elements of biodiversity. In general, landscapes with a broad mix of landscape components and a large amount of native vegetation are likely to support high levels of biodiversity.

FURTHER READING

Andrén, H. 1994. Effects of habitat fragmentation on birds and mammals in landscapes with different proportions of suitable habitat—a review. Oikos 71: 355–366.

Bennett, A. F., J. Q. Radford, and A. Haslem. 2006. Properties of land mosaics: Implications for nature conservation in agricultural environments. Biological Conservation 133: 250–264.

Benton, T. G., J. A. Vickery, and J. D. Wilson. 2003. Farmland biodiversity: Is habitat heterogeneity the key? Trends in Ecology and Evolution 18: 182–188.

Fahrig, Lenore. 2003. Effects of habitat fragmentation on biodiversity. Annual Review of Ecology, Evolution and Systematics 34: 487–515.

Forman, Richard T. T. 1995. Land Mosaics: The Ecology of Landscapes and Regions. New York: Cambridge University Press.

Haila, Yrjö. 2002. A conceptual genealogy of fragmentation research: From island biogeography to landscape ecology. Ecological Applications 12: 321–334.

Lindenmayer, David B., and Joern Fischer. 2006. Habitat Fragmentation and Landscape Change. Washington, DC: Island Press.

MacArthur, Robert H., and Edward O. Wilson. 1967. The Theory of Island Biogeography. Princeton, NJ: Princeton University Press.

McIntyre, Sue, and Richard Hobbs. 1999. A framework for conceptualizing human effects on landscapes and its relevance to management and research models. Conservation Biology 13: 1282–1292.

Noss, Reed F. 1987. Corridors in real landscapes: A reply to Simberloff and Cox. Conservation Biology 1: 159–164.

Simberloff, D. A., J. A. Farr, J. Cox, and D. W. Mehlman. 1992. Movement corridors: Conservation bargains or poor investments? Conservation Biology 6: 493–504.

Wu, Jianguo, and Orie L. Loucks. 1995. From balance of nature to hierarchical patch dynamics: A paradigm shift in ecology. The Quarterly Review of Biology 70: 439–466.

IV.3

Ecological Dynamics in Fragmented Landscapes

Jianguo Wu

OUTLINE

1. Spatial heterogeneity and landscape fragmentation
2. Population and species dynamics in fragmented landscapes
3. Ecosystem dynamics in fragmented landscapes
4. Hierarchical patch dynamics

Landscapes will likely become increasingly fragmented for biological organisms and ecological processes as the human population and its demands for resources continue to escalate. Landscape fragmentation results in habitat loss and alterations in the composition and spatial arrangement of landscape elements, consequently affecting population and ecosystem processes. Thus, to protect biodiversity and ecosystem functioning and to understand how nature works in the changing world, we must understand how organisms, populations, communities, and ecosystems interact with spatially heterogeneous landscapes in which they reside—that is, ecological dynamics in fragmented landscapes. This chapter discusses the effects of landscape fragmentation on population and ecosystem processes as well as major approaches to studying these effects.

GLOSSARY

landscape connectivity. The ability of a landscape to facilitate the flows of organisms, energy, or material across the patch mosaic. Landscape connectivity is a function of both the structural connectedness of the landscape and the movement characteristics of the species or process under consideration.

landscape ecology. The science and art of studying and influencing the relationship between spatial pattern and ecological processes on multiple scales. Land use and land cover change and its ecological consequences are key research topics in landscape ecology.

landscape fragmentation. The breaking up of vegetation or other land cover types into smaller patches by anthropogenic activities, or the human introduction of barriers that impede flows of organisms, energy, and material across a landscape. *Habitat fragmentation* is a similar term to landscape fragmentation but has a more explicit focus on changes in habitat of organisms.

landscape pattern. The composition (diversity and relative abundance) and configuration (shape and spatial arrangement) of landscape elements, consisting of both patchiness and gradients.

metapopulation. The total population system that is composed of multiple local populations geographically separated but functionally connected through dispersal.

patch dynamics. A perspective that ecological systems are mosaics of patches exhibiting nonequilibrium transient dynamics and together determining the system-level structure and function.

1. SPATIAL HETEROGENEITY AND LANDSCAPE FRAGMENTATION

To study ecological dynamics in fragmented landscapes, it is necessary to characterize the spatial pattern of landscapes and understand the causes and mechanisms of the pattern. As described in chapter IV.2, landscapes are spatially heterogeneous geographic areas in which patches and gradients of different kinds, sizes, and shapes are interwoven. This spatial heterogeneity is ubiquitous in both terrestrial and aquatic systems on all scales in space and time. Several types of factors are responsible for the creation of landscape heterogeneity. First, the physical template of landscapes is usually heterogeneous in terms of geomorphological features and distribution of energy and abiotic resources. Second, disturbances, be they natural (e.g., fires, droughts, floods, and windstorms) or anthropogenic (e.g., human-induced fires, urbanization, deforestation, and highway construction), are frequently the primary cause for landscape heterogeneity. Third, biological processes (e.g., herbivory, species competition, deceases, and allelopathy) and fine-scale variability in topography and soil resources can also contribute to landscape heterogeneity. In general, abiotic conditions (e.g., climate, topography, and geomorphology) provide the context in which biological and anthropogenic processes often interact to generate landscape pattern. For example, the spatial pattern of temperature and precipitation determine the broad-scale distribution of biomes, within which the characteristics of ecosystem types are influenced by topographical features and mesoscale climatic variations. The structure and function of local ecosystems, however, are often affected significantly by biological processes.

Spatial heterogeneity gives rise to landscape pattern, of which patches—relatively homogeneous areas that differ from their surroundings—are the fundamental units. Patches can be characterized by their size, shape, content, duration, structural complexity, and boundary characteristics. Landscape pattern is usually considered to have two components: composition (the diversity and relative abundance of different kinds of patches) and configuration (the shape and spatial arrangement of patches) (see chapter IV.2 for more detail on this). Spatial heterogeneity is an important source for the biological diversity, ecosystem services, and scenic wonders of the natural world. In other words, the world is naturally and wonderfully patchy. However, landscape fragmentation—the process of breaking up contiguous landscapes or their elements by human activities—has profoundly transformed the spatial pattern of most if not all natural landscapes around the world and has become one of the greatest threats to biodiversity and ecosystem functioning. As landscapes are fragmented, extant vegetation is removed, and new land cover types are created. This process simultaneously results in both decrease in the total amount of habitat (habitat loss) and increase in the degree of isolation for remnant habitat patches (habitat fragmentation per se or habitat isolation). Also, during landscape fragmentation, the number of patches usually increases, whereas the average size of patches tends to decrease.

Quantifying landscape pattern is necessary for comparing and contrasting patterns between different landscapes, monitoring and projecting changes of a given landscape, and understanding how ecological processes are affected by, and affect, landscape pattern. Many quantitative methods have been developed to quantify landscape pattern in the field of landscape ecology. Two general types of methods can be used to quantify landscape pattern: landscape metrics and spatial statistics. Landscape metrics (Li and Wu, 2007) usually are synoptic indices designed to describe the typological, geometric, and distributional characteristics of landscapes at the levels of individual patches, patch type (or class), and the entire landscape. The underlying causes for landscape heterogeneity are spatial dependence and spatial autocorrelation (things that are closer are more similar), which are the fundamental assumptions for spatial statistical methods. As opposed to traditional statistics, spatial statistics quantifies how variables of interest are distributed and related to each other in space (Fortin and Dale, 2005). There have been numerous studies that use both approaches to characterize landscape patterns and relate them to population and ecosystem processes.

2. POPULATION AND SPECIES DYNAMICS IN FRAGMENTED LANDSCAPES

Most of the studies on ecological dynamics in fragmented landscapes have focused on the effects of landscape fragmentation on populations and species. This section discusses major findings of the effects of landscape fragmentation on population processes and species persistence and examines several main theories and approaches in this research area.

Effects of Landscape Fragmentation on Population Dynamics and Species Persistence

In reality, landscape fragmentation simultaneously leads to habitat loss and habitat isolation. These changes can certainly affect the demographic and genetic processes of populations. A great number of theoretical and empirical studies have been carried out in the past several decades to understand how habitat fragmentation affects population dynamics and species persistence. This section provides an overview of the major findings to date.

Findings on the effects of landscape fragmentation on population dynamics and species persistence have been, more often than not, incongruent because of several reasons. First, the term *landscape* (or habitat) *fragmentation* is often used to denote both habitat loss and habitat–habitat isolation (i.e., habitat fragmentation per se), and consequently, the effects of the two factors are confounded in the results of such studies. Second, various measures that reflect different aspects of landscape pattern at different scales have been used to quantify habitat fragmentation. Some measures focus on habitat loss, others are indicative of changes in habitat configuration, and still others are mixtures of both. Also, habitat fragmentation is measured at either the scale of individual patches (as in most metapopulation models) or the scale of the entire landscape (as in most landscape ecological studies). Third, different theories and models have different assumptions about what is important in fragmented landscapes in terms of population dynamics and species persistence, and these differences in assumptions often translate into discrepancies in results. Nevertheless, studies in recent decades have produced several important findings.

The relative effects of habitat loss and habitat isolation have been one of the central topics. Increasing empirical evidence indicates that habitat loss usually has much stronger effects on population dynamics and species persistence than habitat isolation. In general, the effects of habitat isolation tend to be stronger when the total amount of habitat in the landscape is small and when the species under consideration have limited dispersal abilities. The effects of habitat loss are consistently negative, whereas those of habitat isolation can be either negative or positive depending on the idiosyncrasies of the landscape pattern (e.g., the spatial configuration of habitat patches) and the species under consideration (e.g., abilities for local competition and regional dispersal). The negative effects of habitat loss are easier to understand because the removal of habitat usually results in reduction in the number of species, the abundance of populations, and the carrying capacity of the landscape.

The effects of fragmentation per se, however, are more complex because the outcome depends on how the species responds to the specific features of the fragmented habitat and altered interactions with other species in the landscape. The negative effects of habitat isolation may be caused by the disruption of dispersals, increased local extinction rates in small patches, and detrimental edge effects. The positive effects of habitat isolation may be attributable to relaxed interspecific competition, reduced predation, and disrupted spreading of disturbances. However, it is important to note that the effects of spatial patchiness occurring naturally are different from the effects of habitat fragmentation by human activities. In the latter situation, species usually do not have enough time to adapt to the newly changed environment, and thus, positive fragmentation effects are less likely, especially for nonedge species.

The size of habitat patches has significant effects on population dynamics and species persistence simply because large patches tend to have larger populations (thus with lower extinction probabilities) and more species (because of both pure area effect and higher habitat diversity). In general, patch size has strong positive effects on interior species that require a sufficiently large and relatively stable habitat. As patch size increases, the relative area of edge habitat decreases, resulting in negative effects of patch size on edge species. For generalist species that do not distinguish between edge and interior habitat, the effects of patch size usually are insignificant. The effects of patch size on population dynamics and species persistence may also vary with species that have different behavioral characteristics. For example, some studies have suggested that more mobile or dispersive species would be less strongly affected by landscape fragmentation. However, recent studies show that the opposite may be true when more mobile species suffer severe dispersal mortality in the landscape matrix (Fahrig, 2003, 2007). Other patch characteristics such as shape, orientation, and boundary conditions can also affect population processes. Their effects seem less significant than those of patch size and usually are even harder to generalize across different species and habitat types.

Landscape connectivity, which is conversely related to habitat isolation, plays a crucial role in maintaining population abundance and species persistence by affecting the movement of organisms and propagules, dispersal mortality, and gene flows. Studies from landscape ecology based on percolation theory have suggested that, as habitat area decreases to some critical value, landscape connectivity drops abruptly, indicating a possible extinction threshold for species with limited dispersal ability or high dispersal mortality (With, 2004). This finding corroborates the hypothesis that the effects of habitat isolation on population and species dynamics tend to be more important with decreasing habitat amount in the landscape. Thus, landscape connectivity exhibits threshold behavior and is species- or process-specific. Corridors, as a means of increasing habitat connectivity, can promote species persistence (by enhancing recolonization) and genetic integrity (by preventing genetic drift and bottleneck effects). However, corridors may also increase the spread of diseases and other disturbance agents across the landscape.

Theories and Approaches

Theory of Island Biogeography

The theory of island biogeography (MacArthur and Wilson, 1967) has had pervasive influences on the development of theoretical and empirical approaches to the study of ecological dynamics in fragmented landscapes. The theory asserts that the number of species on an island is determined primarily by two processes: immigration and extinction. Immigration rate decreases with distance from the continental pool of species because of variable dispersal abilities of species (distance effect), whereas extinction rate decreases with island area because of larger populations often found on larger islands (area effect). When immigration and extinction rates are equal, species diversity of the island has reached a dynamic equilibrium state. Thus, the theory relates the dynamics of species diversity directly to the size and isolation of islands. It has inspired much of the research concerning ecological dynamics in patchy environments and, in the many terrestrial applications, "islands" include individual plants, vegetation fragments, reserves or parks, and local ecosystems of all kinds.

However, both the validity of the equilibrium theory itself and its applications in terrestrial landscapes are unwarranted, although its heuristic value is still widely recognized. The theory is a typical example of the classic equilibrium paradigm, which has a number of problems when it is carefully scrutinized against reality. Because landscapes are ever-changing, and most are being increasingly fragmented, the equilibrium assumption behind the theory is hard to justify. Also, it does not consider several factors that are important to ecological dynamics in patchy environments, including habitat heterogeneity, disturbances, edge effects, multiple sources of colonizing species, and complex influences on the patch of concern from the surrounding landscape matrix.

Metapopulation Theory

The concept of metapopulation, a population of subpopulations that become extinct locally and recolonize regionally, resembles the theory of island biogeography in that both consider extinction and colonization as the two key processes. However, the former is concerned with population dynamics and species persistence, whereas the latter focuses primarily on species diversity and turnover. Also, sources for species colonization in most metapopulations are neighboring habitat patches that themselves are subject to local extinctions.

The classic (or Levins) metapopulation models are commonly known as "patch-occupancy" models in which the proportion of habitat patches occupied by a species is modeled as a function of local extinction and interpatch colonization (Fahrig, 2007). These models assume that there are an infinite number of identical habitat patches in the landscape and that within-patch population dynamics and the landscape matrix are not important to metapopulation dynamics. The classic metapopulation models are not really spatial models. More spatially sophisticated metapopulation modeling approaches have been developed in the past several decades. For example, many population models based on diffusion–reaction equations, which consider both local population processes and patch attributes (e.g., size, relative distance to other patches), are relevant to the study of metapopulation dynamics. However, these are quasispatial models that can not explicitly consider the locations and geospatial relations of habitat patches and the heterogeneity of the landscape matrix. The prevailing metapopulation modeling approach now consists of the so-called spatially realistic metapopulation models that incorporate the effects of habitat patch size and isolation on extinction and colonization rates into the classic metapopulation models. Although these models are spatially realistic, like the classic models they are concerned with only two states of habitat patches—presence and absence of a species under study—not with population processes within habitat patches. Also, the heterogeneity of the landscape matrix is usually ignored in spatially realistic metapopulation models.

Metapopulation theory has been increasingly used in conservation biology in the past three decades, replacing the prominent role of island biogeography theory. However, its use for the practice of biodiversity conservation is limited by its species-specific focus and inadequate consideration of the heterogeneity of landscape matrix and socioeconomic processes. In reality, populations neither live in habitat patches that can always be neatly delineated nor reside in a homogeneous landscape matrix. Rather, they are situated in heterogeneous and dynamically complex landscapes that are shaped by a myriad of physical, biological, and socioeconomic processes. Thus, the metapopulation approach is useful but certainly not adequate for achieving the overall goal of conserving all levels of biodiversity.

Population Viability Analysis

The question of how many individuals of a species are enough to ensure the long-term persistence of the species is important both theoretically and practically. The concept of *minimum viable population* (MVP), the smallest size of a population that can persist for a sufficiently long time with a high probability in face of demographic, environmental, and genetic stochasticities as well as natural disasters, attracted much research attention in the 1980s and the early 1990s. The MVP concept implies that there exists a threshold population size for species persistence. The process of estimating MVP or the extinction risk of species of interest has been known as *population viability analysis* (PVA) (Reed et al., 2002), and a number of conceptual procedures and computer software packages for PVA have been developed in the past few decades. Most PVA models consider multiple populations of a species in a fragmented landscape, and the general structure of PVA models is similar to that of metapopulation models. However, some recent PVA models have incorporated the effects of landscape heterogeneity on dispersal and colonization processes.

Because MVP connects the size of a population directly with its probability of extinction, its utility to species conservation is seemingly obvious. The use of MVP and PVA in conservation practices is limited by its focus on single species, demands for detailed information on the species under study, and the reductionist nature of the methodology. For many species, deriving a reliable value of MVP may not be possible simply because of data scarcity and uncertainties, and in other situations such a species-specific approach may not work simply because it is too time-consuming or costly. In addition, it is hard to imagine that the MVP or extinction risk of a given species will remain constant when the landscape in which it resides keeps changing, largely because of socioeconomic drivers. Nevertheless, PVA remains a useful tool for assessing the effectiveness of alternative conservation or management plans for protecting rare and endangered species.

Landscape Approach to Population and Species Dynamics

With the rapid development of landscape ecology, a more comprehensive approach has emerged to understand population dynamics and species persistence in fragmented landscapes. In contrast with metapopulations, "landscape populations" emphasize not only the dynamics of, and interactions between, local populations but also the effects of the heterogeneity of the landscape matrix. Landscape population models are truly spatially explicit, meaning that the size, shape, and spatial arrangement of all habitat and nonhabitat elements are represented. In metapopulation models, habitat fragmentation is usually represented in terms of patch-scale features (e.g., various measures based on the nearest neighboring patches), which are unable to capture the landscape-scale characteristics of fragmentation. In landscape population models, fragmentation is measured at the scale of the entire landscape, and thus, its effects on population processes are assessed more adequately. Also, the landscape population approach allows for explicit examination of how idiosyncratic features of habitat patches and the landscape matrix affect the dispersal of organisms or propagules. In addition, this spatially explicit approach facilitates mechanistic understanding of source–sink dynamics in which large or high-quality patches serve as sources of immigrants to small or poor-quality patches whose population growth rates are negative (sinks).

The theory of island biogeography, metapopulation theory, and most PVA models all focus on the "islands" in a homogeneous matrix, be they oceanic or habitat islands. In contrast with this island perspective, the landscape population approach explicitly considers all landscape elements and their spatial configuration in relation to population dynamics across a heterogeneous geographic area. Although landscape population models are more realistic in representation, they are structurally more complex and mathematically less tractable. They are usually implemented as computer simulation models, often linked with geographic information systems (GIS), which enable the storage, manipulation, and analysis of spatial data. In general, the metapopulation approach tends to be less detailed but more general and thus more valuable for theoretical investigations, whereas the landscape population approach is better suited to meet the practical expectations in biodiversity conservation and ecosystem management. Thus, the major type of population model used in PVA has changed from island models to metapopulation models and now is moving toward landscape population models.

3. ECOSYSTEM DYNAMICS IN FRAGMENTED LANDSCAPES

Landscape fragmentation affects not only population processes and biodiversity but also ecosystem processes such as energy flows and material cycling. These effects are caused by changes in both abiotic and biotic conditions induced by landscape fragmentation. Compared to the effects of habitat fragmentation on population and species dynamics, ecosystem effects have so far received much less attention. This section discusses the current understanding and research approaches in this area.

Effects of Landscape Fragmentation on Ecosystem Dynamics

Ecosystem ecology, the study of energy flow and material cycling within an ecosystem composed of the biotic community and its physical environment, traditionally has adopted a systems perspective that emphasizes stocks, fluxes, and interactions among components without explicit consideration of spatial heterogeneity within the system and effects of the landscape context. With the rapid development of landscape ecology since the 1980s, more and more ecosystem studies have adopted a landscape approach that explicitly deals with within-system spatial heterogeneity and between-system exchanges of energy and matter.

An increasing number of recent studies have shown that landscape fragmentation can influence ecosystem dynamics in several ways. First, the loss and creation of patches directly change the spatial distribution of pools and fluxes of energy and materials in the landscape (e.g., biomass, ecosystem productivity, nutrient cycling, decomposition, evapotranspiration). Second, the altered configuration of landscape elements, particularly introduced edges and boundaries, may affect not only the flows of organisms but also the patterns of lateral movements of materials and energy within and among ecosystems (e.g., hydrological pathways and erosion–deposition patterns). Third, landscape fragmentation can affect ecosystem processes through microclimatic modifications because of altered surface energy balance (e.g., changes in albedo, radiation fluxes, soil temperature, soil moisture, wind profile and pattern) especially near the boundaries of remnant patches (edge effects). Fourth, all the effects of landscape fragmentation on population dynamics and species persistence have bearings on ecosystem processes because both plants and animals play an important role in ecosystem processes.

Landscape Approach to Ecosystem Dynamics

A landscape approach to ecosystem dynamics is characterized by the explicit consideration of the effects of spatial heterogeneity, lateral flows, and scale on the pools and fluxes of energy and matter within an ecosystem and across a fragmented landscape (Turner and Cardille, 2007). This new approach to ecosystem studies highlights the fact that ecosystems are neither homogeneous internally nor closed externally. Such a perspective seems in sharp contrast with the traditional equilibrium view that ecosystems are self-regulatory, self-repairing, and homeostatic and is particularly appropriate when fragmented landscapes are considered. Guided by this spatial approach, several key research questions have emerged: How do the pools of energy and matter and the rates of biogeochemical processes vary in space? What factors control the spatial variability of these pools and processes? How do land use change and its legacy affect ecosystem processes? How do patch edges, boundary characteristics, within-system spatial heterogeneity, and the landscape matrix influence ecosystem dynamics and stability? How do ecosystem processes change with scale, and how can they be related across scales (i.e., scaling)? How do the responses of populations and ecosystem processes to landscape fragmentation interact? How do the composition and configuration of fragmented landscapes affect the sustainability of landscapes in terms of their capacity to provide long-term ecosystem services?

A landscape approach to ecosystem dynamics promotes the use of remote sensing and GIS in dealing with spatial heterogeneity and scaling in addition to more traditional methods of measuring pools and fluxes commonly used in ecosystem ecology (Wu, 1999). It also integrates the pattern-based horizontal methods of landscape ecology with the process-based vertical methods of ecosystem ecology and promotes the coupling between the organism-centered population perspective and the flux-centered ecosystem perspective.

4. HIERARCHICAL PATCH DYNAMICS

Understanding ecological dynamics in fragmented landscapes requires a paradigm shift away from the traditional notion of "balance of nature," which implies that ecosystems maintain a permanence of structure and function with a harmonious order if left alone. The idea of the balance of nature has profoundly influenced both the theory and practice of ecology for the past several decades. The imprints of the balance of nature are obvious in the supraorganismic concept of plant communities, the biocybernetic concept of ecosystems, and a number of similar concepts such as equilibrium, steady state, stability, and homeostasis. Such ideas have penetrated pervasively into the guiding principles and practice of biodiversity conservation and environmental protection. However, the equilibrium paradigm is of limited use in understanding real landscapes, which are heterogeneous in both space and time. Thus, since the 1980s mainstream ecological perspectives have shifted their focus from equilibrium, homogeneity, determinism, and single-scale phenomena to nonequilibrium, heterogeneity, stochasticity, and multiscale linkages of ecological systems (Wu and Loucks, 1995). The theories and approaches discussed in previous sections are examples.

This new ecological perspective has been known as "patch dynamics" and, more recently, as "hierarchical patch dynamics" (HPD) as the result of the integration between hierarchy theory and patch dynamics (Wu and Loucks, 1995; Wu, 1999; Wu and David, 2002). Although the specific meaning of a patch varies across scales and biological systems, patch dynamics has been increasingly used as a unifying concept in both marine and terrestrial systems. The major tenets of HPD include these: (1) ecological systems are spatially nested patch hierarchies, in which larger patches are made of smaller patches, (2) the dynamics of an ecological system can be studied as the composite dynamics of individual patches and their interactions at adjacent hierarchical levels, (3) pattern and process are scale dependent and interactive when operating in the same domain of scale in space and time, (4) nonequilibrium and stochastic processes are not only common but also essential for the structure and functioning of ecological systems, and (5) ecological stability frequently takes the form of metastability that is achieved through structural and functional redundancy and incorporation in space and time (Wu and Loucks, 1995).

These tenets can be illustrated by the structure and dynamics of metapopulations: metapopulations are hierarchies of patch populations; metapopulation dynamics results from the local population dynamics and between-patch interactions; metapopulation processes take place on patch and landscape scales and interact with the spatial pattern of habitat patches; local populations are subject to frequent extinctions, exhibiting nonequilibrium dynamics; and metapopulations tend to be more stable than local populations as a result of recolonization and asynchronous dynamics of individual patches across the landscape. Ecosystem processes such as primary productivity and nutrient cycling can also be perceived in similar ways. Because patches represent basic spatial units in which both population and ecosystem processes occur, the HPD paradigm provides a unifying framework for integrating population and ecosystem perspectives in fragmented landscapes.

FURTHER READING

DeFries, R., G. Asner, and R. Houghton, eds. 2004. Ecosystems and Land Use Change. Washington, DC: American Geophysical Union.

Fahrig, L. 2003. Effects of habitat fragmentation on biodiversity. Annual Review of Ecology and Systematics 34: 487–515.

Fahrig, L. 2007. Landscape heterogeneity and metapopulation dynamics. In J. Wu and R. Hobbs, eds., Key Topics in Landscape Ecology. Cambridge, UK: Cambridge University Press, 78–91.

Fortin, M.-J., and M.R.T. Dale. 2005. Spatial Analysis: A Guide for Ecologists. Cambridge, UK: Cambridge University Press.

Hanski, I., and O. E. Gaggiotti, eds. 2004. Ecology, Genetics, and Evolution of Metapopulations. Amsterdam: Elsevier Academic Press.

Levin, S. A., T. M. Powell, and J. H. Steele. 1993. Patch Dynamics. Berlin: Springer-Verlag.

Li, H., and J. Wu. 2007. Landscape pattern analysis: Key issues and challenges. In J. Wu and R. Hobbs, eds., Key Topics in Landscape Ecology. Cambridge, UK: Cambridge University Press, 39–61.

Lovett, G. M., C. G. Jones, M. G. Turner, and K. C. Weathers, eds. 2005. Ecosystem Function in Heterogeneous Landscapes. New York: Springer.

MacArthur, R. H., and E. O. Wilson. 1967. The Theory of Island Biogeography. Princeton, NJ: Princeton University Press.

Reed, J. M., L. S. Mills, J. B. Dunning, E. S. Menges, K. S. McKelvey, R. Frye, S. R. Beissinger, M.-C. Anstett, and P. Miller. 2002. Emerging issues in population viability analysis. Conservation Biology 16: 7–19.

Saunders, D. A., R. J. Hobbs, and C. R. Margules. 1991. Biological consequences of ecosystem fragmentation: A review. Conservation Biology 5: 18–32.

Turner, M. G., and J. A. Cardille. 2007. Spatial heterogeneity and ecosystem processes. In J. Wu and R. Hobbs, eds., Key Topics in Landscape Ecology. Cambridge, UK: Cambridge University Press, 62–77.

With, K. 2004. Metapopulation dynamics: Perspectives from landscape ecology. In I. Hanski and O. E. Gaggiotti, eds., Ecology, Genetics, and Evolution of Metapopulations. Amsterdam: Elsevier Academic Press, 23–44.

Wu, J. 1999. Hierarchy and scaling: Extrapolating information along a scaling ladder. Canadian Journal of Remote Sensing 25(4): 367–380.

Wu, J., and J. L. David. 2002. A spatially explicit hierarchical approach to modeling complex ecological systems: Theory and applications. Ecological Modelling 153: 7–26.

Wu, J., and O. L. Loucks. 1995. From balance-of-nature to hierarchical patch dynamics: A paradigm shift in ecology. Quarterly Review of Biology 70: 439–466.

IV.4

Biodiversity Patterns in Managed and Natural Landscapes

Paul R. Moorcroft

OUTLINE

1. Habitat loss
2. Habitat fragmentation
3. Not all species are equal
4. Invasive species and climate change

Human activities are profoundly altering the biodiversity of the earth. The principal drivers of change thus far have been the transformation of lands for human use, accompanying fragmentation of remaining natural habitats, hunting, and modification of native disturbance regimes. These forces have resulted in the extinction of numerous species and radically altered the abundances of countless others. Empirical and theoretical studies imply that many more of the world's species will experience the same fate as humanity's collective impacts on the planet further expand and intensify. Long-lived plants and animals and animal species in which individuals range widely in space appear to be particularly vulnerable because of the strong dependence of their populations on the dwindling number of regions that are free of significant human influence. In numerical terms, the impacts of humans on terrestrial biodiversity are generally larger in the tropics because of the restricted spatial distributions of many tropical species and the high species diversity of many tropical ecosystems compared to their high-latitude counterparts. Two additional, and increasingly important, modifiers of terrestrial ecosystem biodiversity are the introduction of exotic species into ecosystems and human-induced changes in climate. These more recently recognized agents of change may act independently of, or synergistically with, land cover change, habitat fragmentation, hunting, and altered disturbance regimes to yet further modify the composition of the world's ecosystems over the coming century and beyond.

GLOSSARY

early successional species. Species that appear in an ecosystem following a disturbance event, such as a fire, landslide, or logging. Early successional species typically possess *r*-selected traits, such as high dispersal ability, short generation time, and rapid growth, but at the expense of having a short lifespan and poor competitive ability. As a result, their population sizes usually increase immediately after disturbances, and then decline later as conditions become more crowded and they are competitively replaced by late successional species.

endemics. Species that have a relatively narrow geographic range, such as species found only on a particular island, or in a particular habitat or region.

fire return interval. The number of years between two successive fire events at a particular location.

invasive species. Introduced or nonindigenous species that are rapidly expanding outside of their native range.

late successional species. Species found in an ecosystem that has not experienced a disturbance for a long period of time. Late successional species typically have *K*-selected traits, such as long generation time, slow rates of growth, but long lifespan and strong competitive ability. As a result, late successional species come to dominate an ecosystem when no further disturbances occur.

species–area curve. A graph showing the number of species found in an area as a function of the area's size.

Human population growth and economic development have led to a radical transformation of the earth's land surface. As plate 8 illustrates, humanity's footprint on the planet is now pervasive, with 83% of the earth's surface being significantly affected by one or more of human land transformation, population density, power infrastructure, and transportation networks. In this chapter, I review how these and other human activities are affecting the biodiversity of terrestrial ecosystems. The focus of this chapter is on ecosystem biodiversity; however, as discussed in more detail in chapters IV.2, IV.6, and IV.8, an ecosystem's species composition has important consequences for its biophysical and biogeochemical functioning.

1. HABITAT LOSS

One of the most significant ways in which humans affect terrestrial biodiversity is the loss of species that accompanies the destruction of natural habitats. A particularly compelling study of this phenomenon is the Biological Dynamics of Forest Fragments Project (BDFFP) initiated by James Lovejoy and others in the late 1970s (Bierregaard et al., 2001), which is examining the consequences of habitat destruction in the tropical forests of the Amazon (figure 1A). Figure 1B–D shows the number of understory bird species in three forest fragments of different sizes experimentally created as part of the BDFFP. The initial species diversity of the understory bird community in the forest fragments was high, with the 3-, 11-, and 100-ha fragments, respectively, containing 90, 92, and 111 bird species (Ferray et al., 2003). However, following their creation, species diversity in the fragments declined rapidly: in a 13-year period, the number of species remaining in the 3-ha fragment decreased to approximately 30, and the diversity in the 11- and 100-ha fragments decreased to approximately 50 and 70 species, respectively (figure 1B–D). Figure 1E shows calculations of the time taken for the understory bird diversity to decrease to half its original value (t_{50}) in a series of forest fragments created as part of the BDFFP, along with similar estimates for several larger (100 to 10,000 ha) forest fragments created in a tropical forest in Kenya (see Ferraz et al., 2003). As the figure illustrates, the rate at which species diversity decreases is strongly influenced by the size of the remaining forest fragment: small 1- to 10-ha fragments lost half their species in under 10 years, whereas the diversity of the larger 100- to 10,000-ha fragments declined more slowly, taking between 10 and 100 years to lose half of their species.

Ecologists have sought to predict the magnitude of species loss when habitat is removed. One simple approach for doing so utilizes so-called species–area curves, which, as implied by the name, describe how the number of species within a region increases as a function of its area. Figure 2 shows species–area curves for reptiles on islands in the West Indies (figure 2A) and birds in the Sunda Islands (figure 2B) (MacArthur and Wilson, 1963, 1967). In both of these animal groups, the relationship between the number of species (S) and island area (A) can be reasonably described by a mathematical function of the form $S = S_0A^z$, where S_0 and z are constants. When plotted on logarithmic axes, a straight-line relationship between area and species diversity results (figure 2), where the values of S_0 and z, respectively, reflect the line's intercept and its slope.

Knowing the relationship between species number and area yields a simple prediction for the biodiversity consequences of habitat loss. For example, figure 2C shows the predicted number of species lost following 10-fold reduction in habitat area for two different groups of organisms, one with a species–area exponent of 0.2 and the other with a species–area exponent of 0.35 (most species–area exponents lie in this general range; for example, the values for reptiles and birds plotted in figure 2A,B are 0.301 and 0.303, respectively). As figure 2C shows, following a 90% reduction in habitat, the number of species predicted to be lost when the species–area exponent is 0.2 is 37%, whereas for a species–area exponent of 0.35 the predicted loss is 55%, illustrating that, for a given amount of habitat loss, the fraction of species lost is higher for groups of organisms with higher species–area exponents.

Species–area curves have been widely applied to predict the rate at which habitat destruction is causing biodiversity loss. Two important cases to which this method has been applied are losses of bird species in eastern North America and in the Atlantic coastal forests of Brazil, areas that have both experienced extensive habitat loss. Eastern North American forests were extensively cleared in the centuries following the colonization of the North American continent, with forest cover declining to approximately 50% of its original extent during the mid-1800s. Given a species–area exponent between 0.2 and 0.3, species–area curves imply that between 13% and 19% of species diversity should have been lost as a result of this land clearing. Habitat loss in the Atlantic coastal forests of Brazil has been even more dramatic: approximately 90% of the original forest has been eliminated as a result of human agriculture and development. Species–area curves imply that this should have resulted in 37 to 50% of the original Brazilian Atlantic coastal forest bird species being eliminated (see figure 2C).

The actual rates at which species have been lost from these two regions have been significantly lower than the above species–area curve estimates. Only 2% of eastern North American bird species have become extinct, and less than 1% of Brazilian Atlantic coastal forest bird species are now extinct. Several factors ac-

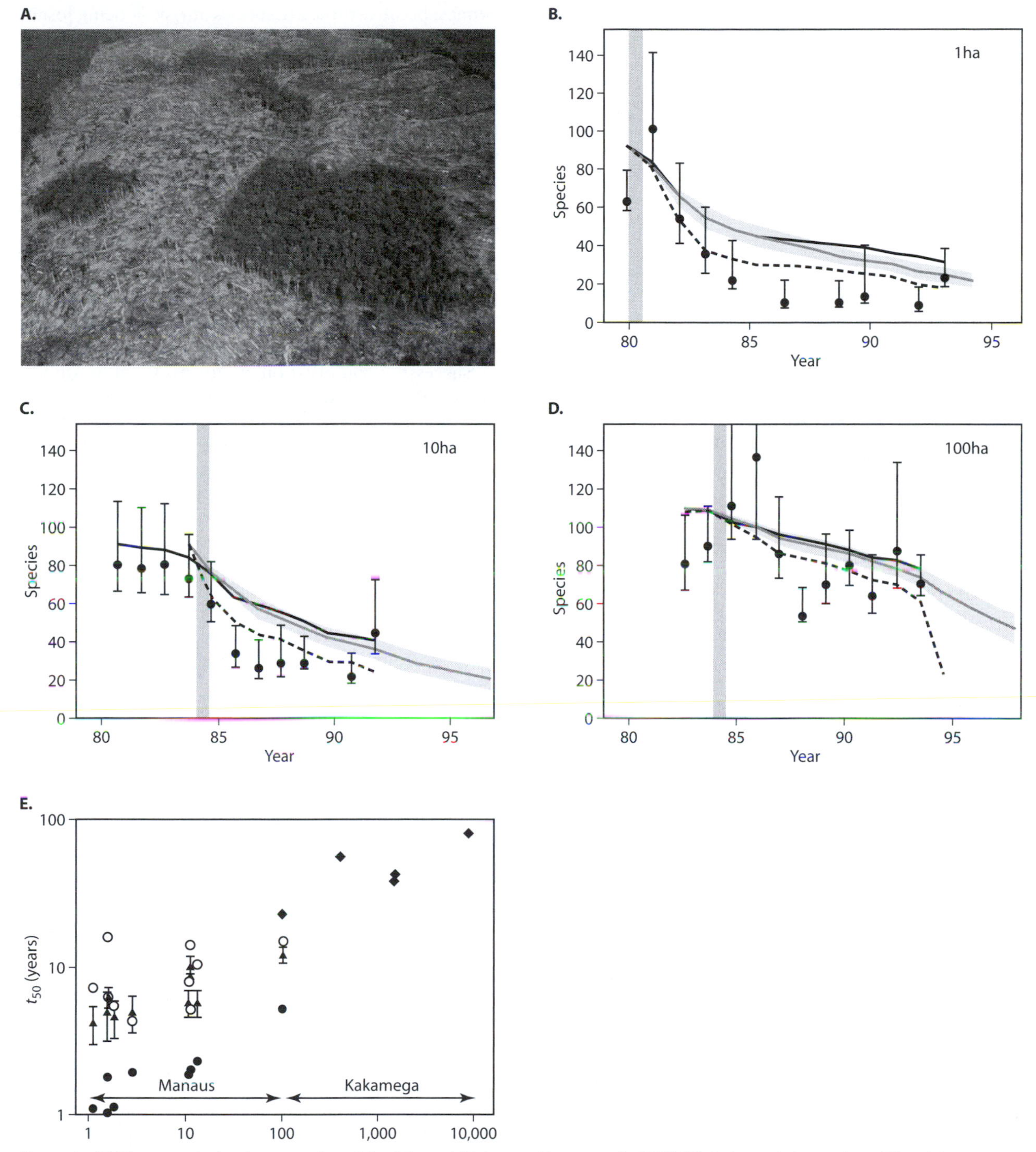

Figure 1. (A) Photograph showing examples of the 1-ha and 10-ha forest fragments created as part of the BDFFP study of Amazonian deforestation. (From Bierregaard et al., 2001) (B–D) Plots of species loss for fragments according to four different estimation methods. The gray bars indicate the timing of isolation. (From Ferraz et al., 2003) (E) Estimated time to lose 50% of the species from the BDFFP forest fragments (open circles, closed circles, and triangles) and Kakamega (diamonds). The different symbols for the BDFFP fragements are estimates obtained from three different statistical models. (From Ferraz et al., 2003)

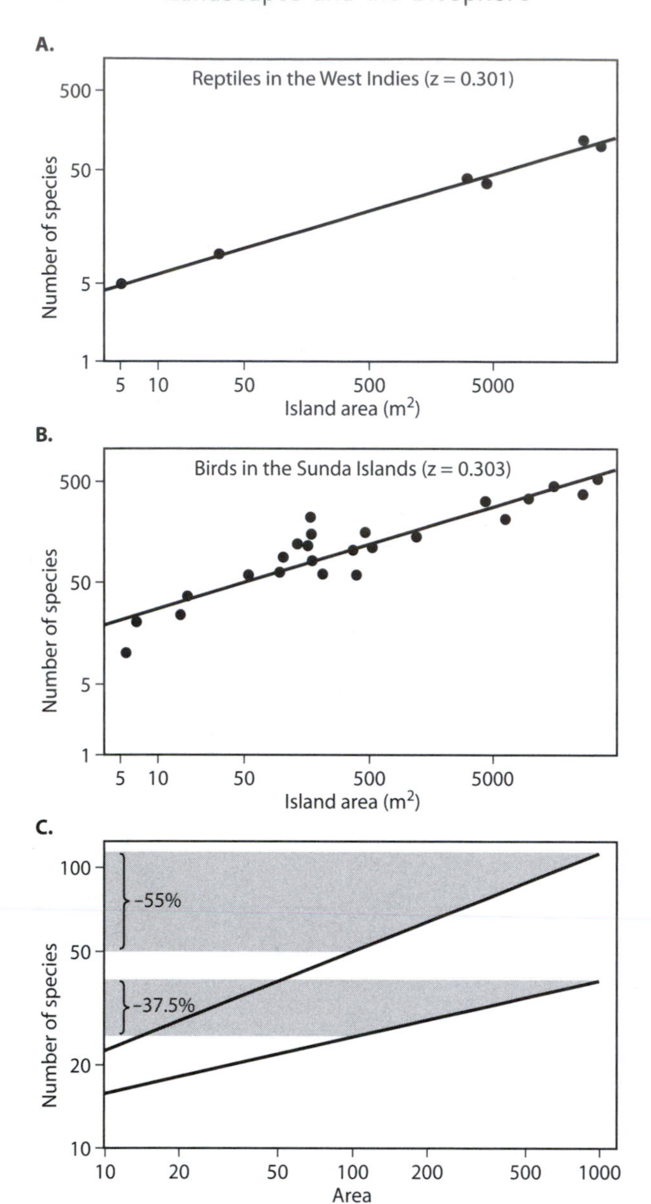

Figure 2. (A and B) Dots indicate the empirical species–area relationships for reptiles in the West Indies and birds in the Sunda Islands. Lines are fitted relations of the form $S = S_0 A^z$, where A is area, S_0 the number of species in a unit area of land, and z the species–area exponent. (C) Hypothetical species–area curves plotted for the case where $z = 0.2$ and 0.35, and S_0 is 10. The shaded areas shows how a 90% decrease in area results in predicted species loss of 37% when z is 0.2 and 55% when z is 0.35. Steeper species–area relationships (higher values of z) imply greater proportional species loss for a given amount of habitat loss.

count for this mismatch. In the case of eastern North American birds, of the 200 species found in the region, only 28 are endemic (i.e., are species found only within the region and not elsewhere). Considering only endemics, because these are species at risk of being lost as a result of habitat loss, 4 of 28 or 14% of the species were lost, a number that is consistent with the species–area curve predictions for the region (Pimm and Askins, 1995). In the case of Brazilian Atlantic coastal forest, there are 214 endemic bird species, and thus one would expect on the basis of typical species–area curve exponents that 79–108 bird species would be lost from the region (Brooks and Balmford, 1996). In reality, only one species has been lost thus far. However, because a significant amount of the coastal forest deforestation has been relatively recent, and the time scale for species extinctions is relatively long, it has been argued that it is more reasonable to consider threatened as well as extinct species because (in the absence of successful conservation intervention) it is only a matter of time before the threatened species also will become extinct. The number of endemic bird species in the Brazilian Atlantic coastal forest currently listed as threatened is 60, a number that is reasonably close to the 79–108 range predicted by species–area curve relationships.

A further implication of the above focus on endemic species is that habitat loss will have a greater impact on biodiversity in areas with large numbers of endemics than in areas with few endemics. Figure 3 shows a map of areas of high endemism—so-called "biodiversity hot spots"—around the globe. As the figure indicates, the majority of these areas are found in tropical regions, implying that the biodiversity consequences of habitat loss in tropical ecosystems will typically be considerably larger than in temperate ecosystems.

2. HABITAT FRAGMENTATION

Human activities are rarely confined to small areas and instead are dispersed across landscapes. The resulting fragmentation of remaining natural habitats that accompany human land transformation has important consequences for an ecosystem's flora and fauna over and above those caused by the reductions in the total area of natural habitat. The reason for this is various forms of "edge effects" caused by small patches of natural habitats having higher perimeter-to-area ratios compared to larger areas of natural habitat of equivalent total area (see chapter IV.5).

Species that range widely are particularly vulnerable to the deleterious effects of the edge effects caused by fragmentation. Clear evidence of this comes from a study by Woodroffe and Ginsberg (1998), who analyzed the occurrence of 10 large carnivore species as a function of reserve size (figure 4). As the figure shows, for each species there is a strong correlation between a reserve's size and its probability of being occupied.

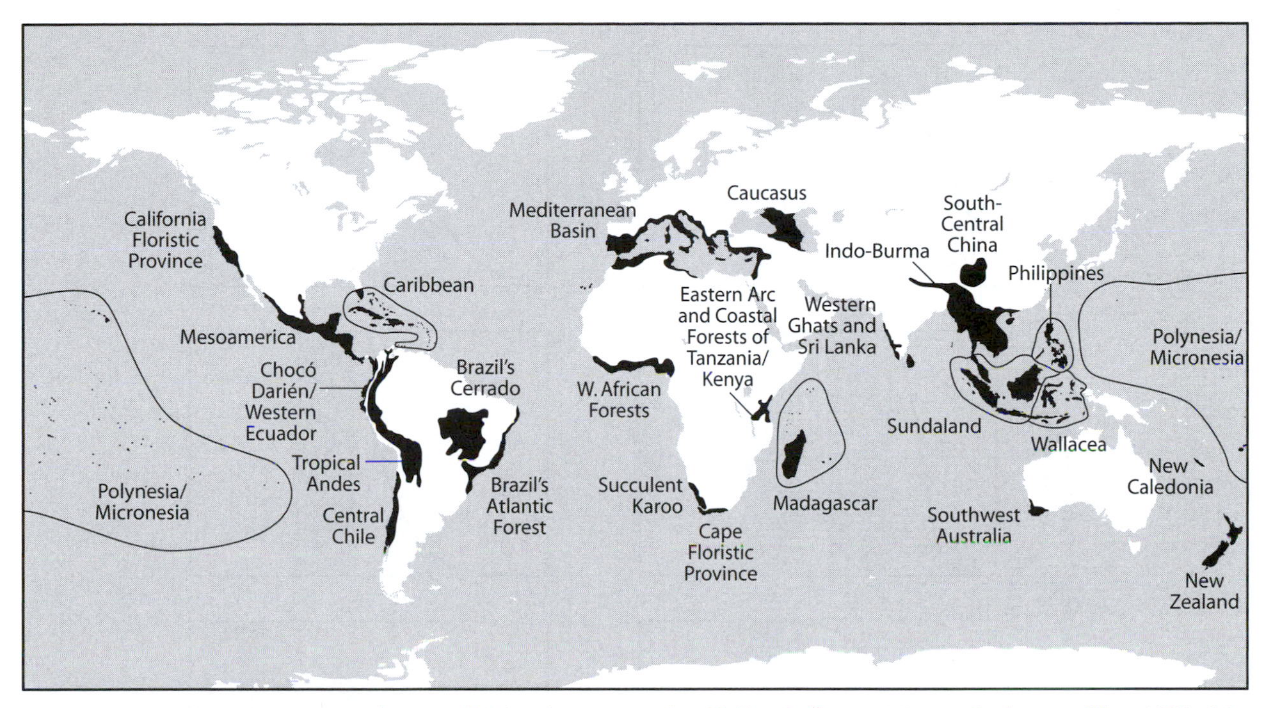

Figure 3. Twenty-five regions that contain areas of high endemism that are critical for preserving global biodiversity. Within each region, biodiversity "hot spots" comprise between 3% and 30% of the shaded area. (From Myers et al., 2000)

These edge effects arise because, in small reserves, individuals of large carnivore species are more likely to stray outside of reserve boundaries, where they are much more likely to be killed by humans. Further support for this conclusion comes from that fact that the critical reserve size for each species, the reserve size at which there is a 50% chance of the species being present (point A_{50} on each panel of the figure), is positively correlated with the average home range size of the different species.

Similar evidence for the impacts of fragmentation has come from a number of studies. A high-profile case in the United States has been the case of the northern spotted owl (*Strix occidentalis caurina*), which lives in old-growth forests of the Pacific Northwest. Studies indicate that dispersing juvenile northern spotted owls suffer greatly increased rates of predation by great horned owls and goshawks when they fly over cleared areas of forest than in intact forest, making them vulnerable to the landscape fragmentation caused by clear-cut logging operations.

The loss of biodiversity arising from habitat fragmentation can have cascading effects on ecosystem composition, structure, and function. Dramatic evidence of this phenomenon has come from a study by John Terborgh and colleagues, who studied the effects of fragmentation of tropical forests on a series of islands that were created in 1986 by the rising waters that followed the construction of the Lago Guri hydroelectric dam in Venezuela. Terborgh and colleagues found that whereas large (150 ha) islands retained nearly all their original species diversity, medium (4–12 ha) and small (0.25–0.9 ha) islands lost more than 75% of their vertebrate species. Ecological guilds that were virtually absent on the small and medium-sized islands included frugivores, which are the principal seed dispersers in tropical forests, small mammal predators (felids, mustelids, snakes, and large raptors), and, in the case of small islands, armadillos, which prey on leaf-cutter ants (Terborgh et al., 2001).

The absence of the above ecological guilds following fragmentation had ramifying impacts on the remainder of the forest ecosystem. Leaf-cutter ant densities on the small islands were 100 times greater, and rodent densities on the small and medium-sized islands 35 times greater, than those found on the large islands and on the mainland. The subsequent increase in the intensity of herbivory, in turn, affected the forest canopy: densities of tree stems less than 1 m tall on small islands were, on average, 50% lower than those found on the large islands and on the mainland. The recruitment of canopy trees appears to have been particularly

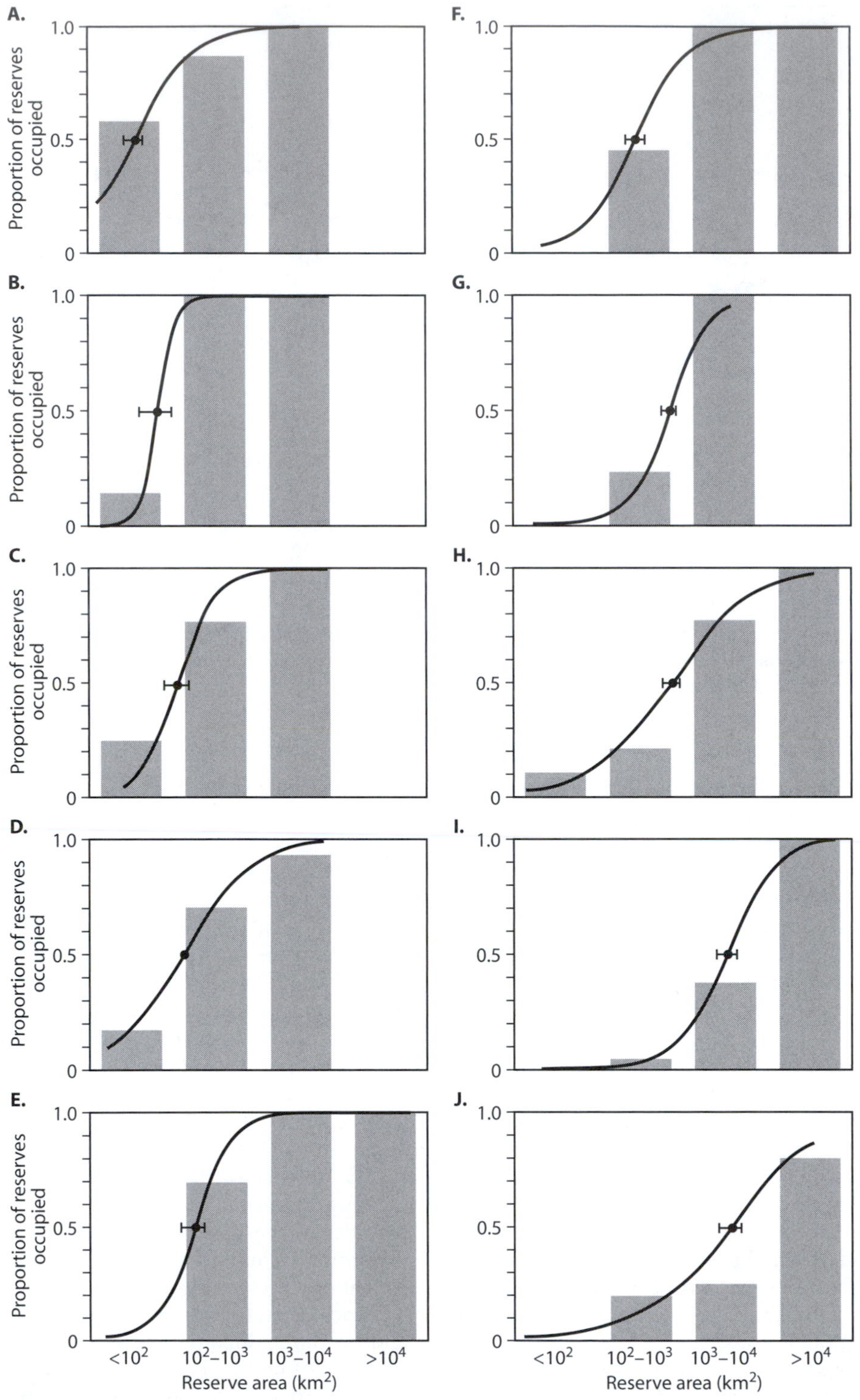

Figure 4. Proportion of reserves of various sizes in which 10 species of large carnivores have persisted. Population persistence is related to reserve area for all species. Curves show the probability of persistence predicted by a simple statistical model fitted to the binary data; filled circles show the critical reserve sizes (±SE) for which the models predict a 50% probability of population persistence (A_{50}). Species: (A) black bear; (B) jaguar; (C) snow leopard; (D) tiger; (E) spotted hyena; (F) lion; (G) dhole; (H) gray wolf; (I) African wild dog; (J) grizzly bear. (From Woodroffe and Ginsberg, 1998)

affected, with the average density stems of canopy species on the small islands being only ~20% of the average density found on the large islands and mainland. In addition to providing strong evidence for the biodiversity impacts of fragmentation on tropical forests, the Lago Guri study provides strong empirical support for the occurrence of ecological cascades within ecosystems and for the "green world hypothesis" (Hairston et al., 1960), which argues that in most ecosystems there is strong top-down regulation of herbivore abundance by predator species.

Similar, albeit less well-documented, ecological cascades arising from habitat fragmentation appear to be occurring in many of the world's tropical forests. In particular, in both the Amazon and in Southeast Asia, evidence suggests that the abundance and distribution of many species of wild pigs have been severely disrupted by habitat fragmentation. Like the frugivorous primates in the Lago Guri study, pigs are an important group of seed dispersers in many tropical forests. As a result, changes in their abundance and distribution are affecting forest canopy biodiversity in many tropical areas.

Another important edge effect occurring in fragmented tropical forest landscapes arises when surface fires started in surrounding agricultural areas spread into remaining areas of forest. Analysis of satellite imagery in Amazonia has shown that fire return intervals are reduced to less than 20 years within 500 m of a forest edge, compared to more than 100 years in the forest interior (figure 5A). Evidence from the BDFFP described earlier suggests that the increased frequency of burning in forest areas that adjoin pastures is exacerbated by changes in microclimate at

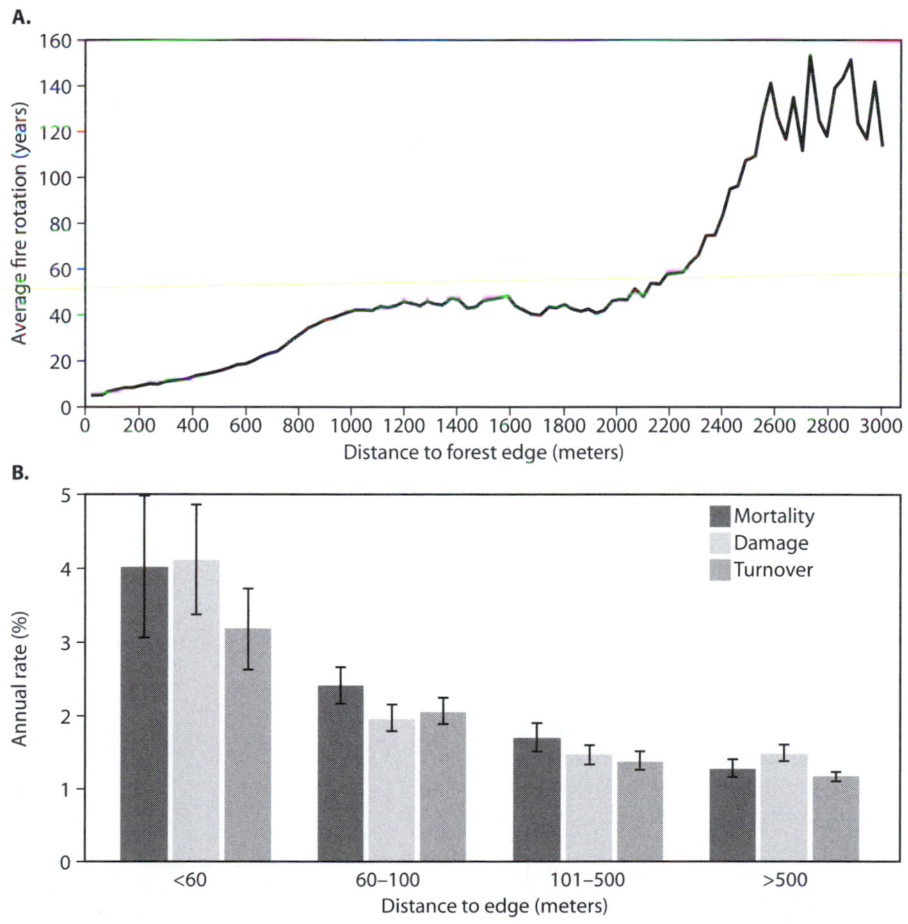

Figure 5. (A) Fire frequency as a function of proximity to forest edge in Amazonia measured by Cochrane (2003). (B) Forest dynamics measured by Laurance et al. (1998) as a function of proximity to forest edge. Panel shows annual rates of mortality, tree damage, and turnover [turnover = (mortality + recruitment)/2].

forest–pasture interfaces. These changes in microclimate extend over 100 m into surrounding forest, causing significant declines in canopy tree biomass and shifting the composition of the canopy toward early successional and fire-adapted tree species (figure 5B). Analyses of satellite measurements of cloud cover and atmospheric modeling studies also suggest that, in heavily fragmented forest landscapes, changes in the biophysical properties of the land surface arising from the conversion from forest to pasture may be altering local scale atmospheric circulation patterns and resulting spatial and temporal patterns of precipitation.

The abundant evidence regarding the detrimental effects of habitat fragmentation has led to increasing calls by a number of conservation biologists for the establishment of "megareserves"—large areas of undisturbed natural habitat that are 10,000 km^2 or larger in size. The logic for this is that only such extensive areas of natural habitats will avoid the various deleterious forms of edge effects described above and thus sustain the full complement of species within an ecosystem. Megareserves also have the additional advantage that concentrating areas of natural habitat in one area allows for easier and cheaper enforcement of boundaries than a widely scattered set of reserves of equivalent total size. For example, on a per-unit basis, the maintenance costs per unit area of Brazil's approximately 38,670 km^2 Serra do Tumucumaque National Park are about 18,000 times lower than those for the 1.1 km^2 Sauim-Castanheira Ecological Reserve.

From the perspective of biodiversity preservation, the reductions in edge effects and the cost savings associated with megareserves must, however, be tempered against another ecologically important consideration: preserving regional species diversity by maintaining reserves across the range of habitat types found within a region. In most cases, this naturally implies a network of reserves rather than one single large reserve. Consideration then has to be given to the spatial arrangement of the reserves and the desirability of providing corridors for migration and dispersal of species among the different areas. Questions about the optimal spatial configuration of habitat reserves were historically argued in qualitative terms, with ecologists debating, often heatedly, about whether a single reserve would preserve a greater diversity of species than an equivalently sized collection of smaller reserves. This so-called single large or several small (SLOSS) debate has largely been superseded in modern analyses of habitat preservation, which instead focus on developing specific recommendations for particular species and habitats. In addition to ecological concerns regarding edge effects and how species composition varies among habitat types, modern analyses often also take into account pragmatic issues such as land costs and the potential for multiple-use areas designed to meet other human land needs such as recreation.

3. NOT ALL SPECIES ARE EQUAL

Simple approaches for predicting the impacts of human activities on ecosystem biodiversity, such as the species–area curve approach described earlier, do not differentiate among species. In reality, a number of ecological factors result in certain species being more vulnerable to human activities than others. One of the strongest predictors of a species' risk is its body size. Figure 6 shows the percentage of mammalian genera that have become extinct in the past 130,000 years. As can be seen in the figure, the extinction rate has been far higher in larger mammals than in smaller ones. Although only a few of these extinctions can be categorically attributed to human activities, the timing of the losses in different regions correlates closely with patterns of human migration, and thus, it is generally thought that the vast majority of the losses were, either directly or indirectly, caused by human activities.

The increased vulnerability of larger species is a natural consequence of their accompanying life-history characteristics. Larger species tend to have longer generation times and slower rates of reproduction (so-called *K*-selected species) compared to smaller species that have shorter generation times and higher rates of reproduction (so-called *r*-selected species). As a result, larger species have slower intrinsic rates of population growth, making them more vulnerable to increases in mortality and/or declines in recruitment caused by

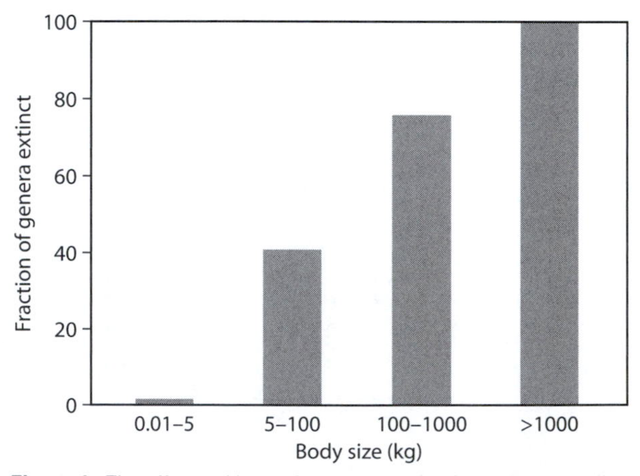

Figure 6. The effects of body size on the extinctions of mammalian herbivore genera in North America, South America, Europe, and Australia. (Figure redrawn from Owen-Smith, 1987)

human activities. Larger-sized species also tend to have larger home range sizes, making them more vulnerable to habitat fragmentation and loss (see habitat fragmentation section).

In the past, the principal mechanism of human-induced declines in species abundance and eventual species loss was almost certainly hunting, an explanation sometimes referred to as the "overkill hypothesis." Hunting continues to be a major cause of population decline in some regions. So-called bushmeat continues to be a key source of protein for human populations in a number of developing countries. For example, the bushmeat trade in Ghana is estimated conservatively to be 400,000 tons/year and is a major contributor to the declines of many of its carnivore, primate, and large herbivore species. A recent study found that from 1970 to 1988, the biomass of 41 species of mammals in nature reserves in Ghana had deceased by 76%, with many species becoming locally extinct in many of the reserves (Brashares et al., 2004).

Many modern-day species population declines are, however, the result of ecological forces other than hunting. In addition to habitat loss and habitat fragmentation discussed above, another major driver of biodiversity modification and loss is arising from human modification of disturbance regimes. Most ecosystems are subject to one or more natural forms of episodic disturbance, such as hurricanes, flooding, fire, landslides, or pathogen outbreaks. These disturbances are significant from a biodiversity perspective because they maintain so-called successional diversity within ecosystems: species that in the absence of disturbance would be excluded from an area by competitively superior species are able to persist by continually exploiting newly disturbed areas. This connection between the species composition of an ecosystem and its disturbance regime means that any shift in the intensity or frequency of disturbance will almost inevitably alter its species composition.

From a global perspective, the most significant human modification of natural disturbance regimes has been through changes in fire frequency. Figure 7A,B shows the changes in fire regimes that have occurred in South America and Southeast Asia over the past 100 years. As the figure illustrates, over the past century, tropical forests in both the New World and Old World have experienced major increases in fire frequency. A similar, though less pronounced, trend is also seen in tropical Africa. As described earlier (see habitat fragmentation section), fire profoundly alters the structure and composition of tropical forests. In particular, when the mean return time between successive fire disturbances is reduced to under 100 years, late successional old-growth tree species, such as Mahogany species in South American forests and Dipterocarp species in Southeast Asian forests, are unable to persist because of their slow growth rates and long generation times. They are replaced by early successional pioneer species, such as *Cecropia* and *Vismia* in South America and *Macaranga* species in Southeast Asia, whose fast growth rates and short generation times enable them to flourish when the time between successive disturbances is on the time scale of decades rather than centuries.

Another major factor altering the disturbance regimes of tropical forest ecosystems is timber harvesting. Figure 7E shows rates of timber harvesting around the globe, showing the marked increases in the extent of forest logging operations that have occurred over the past 300 years in tropical forests. As with body size discussed earlier, the increases in fire frequency and rates of timber harvesting in tropical forests are yet another manifestation of human activities increasingly favoring *r*-selected species over *K*-selected species.

Unlike the tropics, temperate ecosystems in the United States have experienced significant declines in fire frequency over the past 100 years (figure 7C,D). The primary cause of this has been human fire-suppression activities, which, until recently, have been a significant feature of ecosystem management in both the eastern and western United States. As would be expected from ecological theory, the decrease in fire frequency in the temperate United States that occurred over the last century has favored late successional species at the expense of early successional species. In the arid savanna and woodland ecosystems of the southwestern United States, this has led to a widespread phenomenon of "woody encroachment" in which early successional grasses and forbs are competitively replaced by woody shrubs and trees. This phenomenon has become a major management issue in the Southwestern states of Arizona, Utah, New Mexico, and Texas.

In forested regions of the United States, the declines in fire frequency arising from fire-suppression activities have been accompanied by increases in rates of timber extraction (figure 7E). In the West, logging has primarily been in the form of clear-cuts in which essentially all of the merchantable timber is removed from an area. In the East, there was large-scale clearing of the eastern forests for agriculture during the eighteenth century, followed by large-scale agricultural abandonment in the late nineteenth and early twentieth centuries. This widespread agricultural abandonment resulted in a large-scale increase in forest area compared to 150 years ago. However, there has been continual selective logging of the eastern forests, which has kept the mean time between disturbances low and thus prevented any large-scale shift back toward

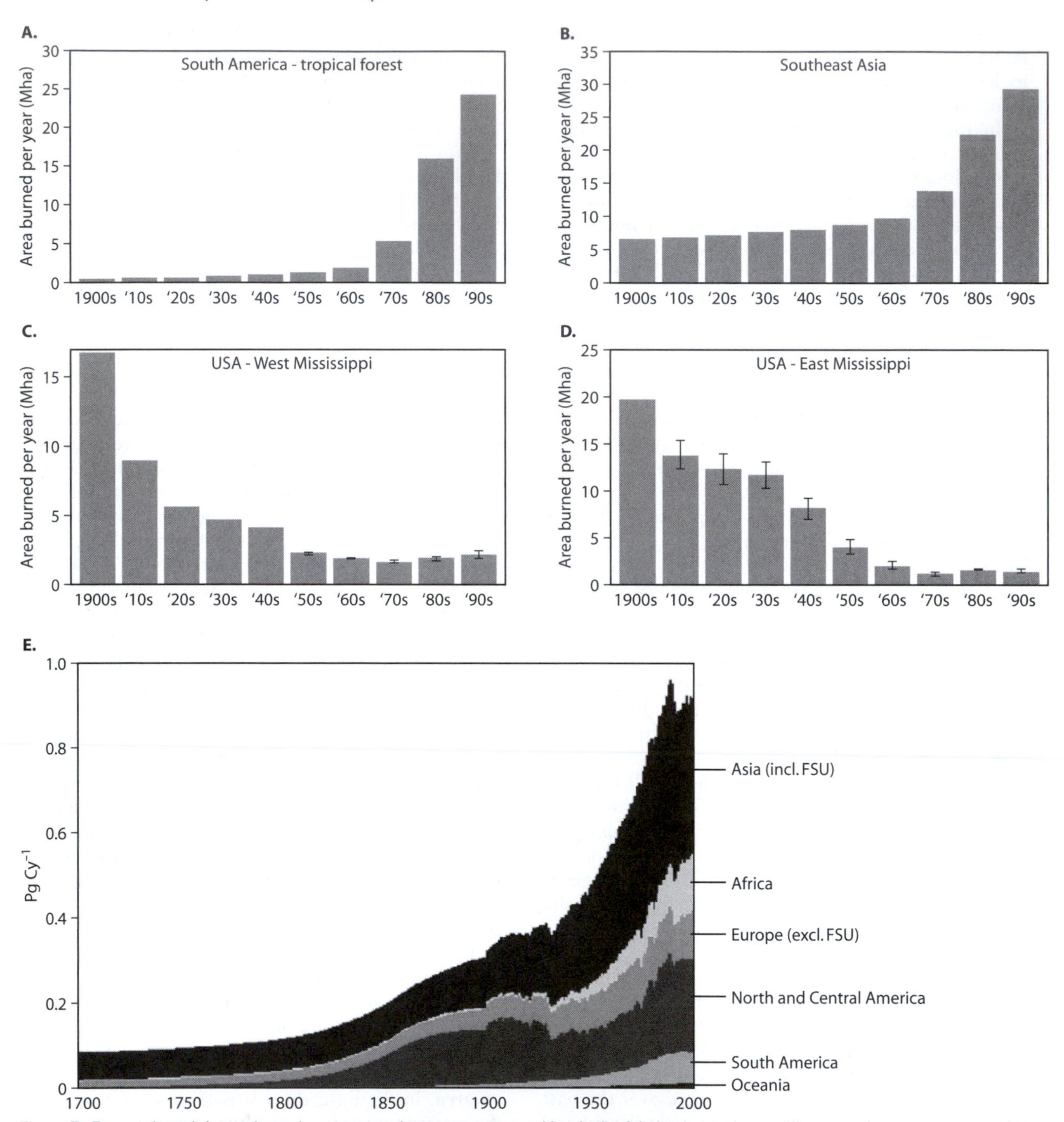

Figure 7. Temporal trends in area burned per year over the past century in (A) South America, (B) Southeast Asia, (C) the western United States, and (D) the eastern United States. (From Mouillot and Field, 2005) (E) Magnitude of timber harvesting on different continents over the past 300 years. The global integrated wood harvest for the 1700–2000 period is 86 Pg C (FSU = Former Soviet Union). (From Hurtt et al., 2006)

late successional tree species that were present before European colonization.

A complicating factor in studying the changes in disturbance regimes is the continuing debate about past human activities. For example, analyses of pollen in lake sediments suggest that in the United States Native Americans maintained elevated fire disturbance regimes in many regions that significantly impacted their composition before European colonization (Delcourt et al., 1998).

4. INVASIVE SPECIES AND CLIMATE CHANGE

Thus far, this chapter has focused on the impacts of human activities on terrestrial ecosystem biodiversity arising from habitat loss, hunting, and changing disturbance regimes. Two additional, and increasingly important, modifiers of terrestrial ecosystem biodiversity are invasive species and human-induced climate change.

The colonial era heralded a vast increase in the rate of exchange of organisms between regions, islands, and continents that were formerly isolated from each other. Continuing increases in the extent of global trade mean that this rate of biotic exchange is increasing still further. Not surprisingly, this large-scale reorganization of the global biota is having major impacts on the diversity of many terrestrial ecosystems, with certain countries, such as New Zealand, now containing as many alien species as they do native ones.

From a biodiversity perspective, the most profound consequence of an invasive species is extinction of native flora and fauna. Examples of this include the infamous introduction of the brown tree snake (*Boiga irregularis*) to Guam in the 1950s, which, in just a few decades, caused extinction of more than 10 bird species, including some endemic only to Guam (Rodda et al., 1997). Similarly, the introduction of the rosy wolf snail (*Euglandina rosea*) to the Society Islands in order to control the giant African snail (an agricultural pest) resulted in the extinction of numerous species of endemic *Partula* land snails (Strong et al., 2000; see Mooney and Cleland, 2001). To date, extinctions have arisen through the predatory and pathogenic effects of invasive species rather than as a result of competitive interactions between invasive and native species. Some ecologists have argued that it is simply a matter of time before such competition-induced extinctions occur, but others maintain that competing native and invasive species will in many cases coexist, thereby resulting in long-term increases rather than decreases in the biodiversity of many ecosystems.

Regardless of whether the long-term outcome of their competitive interactions is coexistence or eventual competitive exclusion, what is not in debate is that invasive species are having marked effects on the abundance of the native flora and fauna of many ecosystems. Sometimes these are the result of direct interactions, such as the competitive displacement of a native Californian ant species by the invasive Argentine ant (*Linepthaema humile*) or the predatory effects of the brown tree snake in Guam described earlier. Some of the most dramatic impacts of invasive species occur, however, when invasive species alter the disturbance regime of an ecosystem. A classic example of this is cheatgrass (*Bromus tectorum*), an invasive annual grass species in the United States that has resulted in a 10-fold increase in the fire frequency of over 40 million ha of land in the western states. This modified disturbance regime is, in turn, favoring yet further spread of this fire-adapted species at the expense of native vegetation. Similarly, several species of flammable grasses introduced into Hawaii for agriculture have spread into native woodlands, causing a 300-fold increase in fire frequency that is threatening to eliminate many species of woody native vegetation (D'Antonio and Vitousek, 1992).

One other biodiversity consequence of invasive species can be the production of novel species through hybridization. A classic case of this is *Spartina alterniflora*, a native coastal grass species of the eastern United States that was introduced to Great Britain around 1870. Following its introduction, *S. alterniflora* hybridized with the native *S. maritima,* producing *S. anglica*, an aggressive hybrid species that subsequently has spread widely along the British coastline (Thompson, 1991; Mooney and Cleland, 2001; Gray, 1986).

An important challenge for invasive species management has been diagnosing what traits make a species a successful invasive. An analysis of invasive and noninvasive pines in the United States found that the invasive species characteristically had *r*-selected traits, such as low seed mass, faster growth, and more frequent seeding compared to the noninvasive species. Another observation has been that a disproportionate number of thistle species in the *Cirsium* family and grass species in *Poa* and *Bromus* families are invasive. Analyses conducted for numerous other plant and animal groups have, however, failed to identify any key distinguishing traits of invasive species compared to either native species or noninvasive species. A related challenge is identifying what characteristics make an ecosystem vulnerable to invasion. A study of invasives in the United Kingdom found that plant communities could be ranked in terms of invasibility based on the proportion of bare ground and on the frequency and intensity of soil disturbance. Similar patterns have been found in other studies, indicating that invasions generally occur more readily into disturbed habitats compared to undisturbed ones.

Human-induced climate change is increasingly being recognized as another major driver of future biodiversity change. The two primary underlying causes of human-induced change are the increasing concentration of carbon dioxide in the atmosphere arising from the burning of fossil fuels and the changing biophysical properties of the land surface arising from human land-use transformation. A simple method for calculating the impacts of climate change on ecosystem diversity uses a "climate envelope" approach, in which the relationship between a

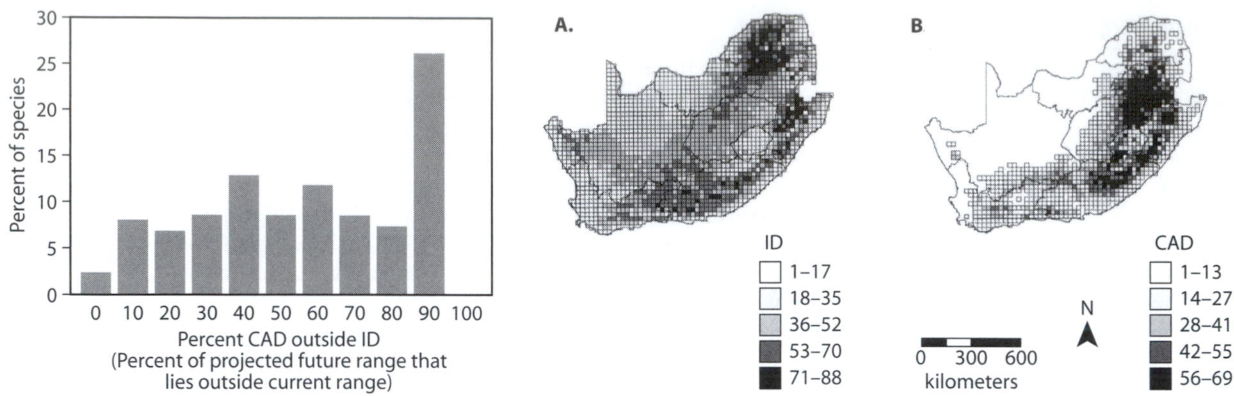

Figure 8. (left) Frequency distribution showing the predicted magnitude of range shifts for 179 South African animals following a doubling of atmospheric CO_2 concentrations calculated using climate envelope relationships. (right) Spatial pattern of species richness for the 179 South African animals under (A) current climate and (B) projected future climate. (From Erasmus et al., 2002)

species' current distribution and climatic variables, such as temperature, precipitation, and rainfall, is used in conjunction with climate projections to predict the expected change in its spatial distribution that will occur as a result of climate change. This approach makes a number of simplifying assumptions, ignoring interactions between species, and assuming that species are in equilibrium with climate, both now and in the future. However, climate envelope approach predictions are nonetheless sobering.

Figure 8A shows a histogram of the predicted magnitudes of range shifts for 179 South African animal species (including mammals, birds, reptiles, and insects) following a doubling of atmospheric carbon dioxide concentrations. As the histogram illustrates, more than 25% of the species are predicted to undergo range shifts in excess of 90%. Climate-induced range shifts of this magnitude would be potentially catastrophic for many species, especially for species that have limited dispersal abilities or that are already endangered as a result of other human activities. Figure 8B and 8C shows the resulting changes in spatial patterns of species richness per unit area, calculated by overlaying the climate envelopes of the 179 species under the current climate (figure 8B) and under the projected future climate scenario (figure 8C). The projections imply that marked losses in species diversity will occur in the western and, to a lesser extent, the southern parts of the country. A particularly sobering statistic from this analysis is that Kruger National Park, the flagship national park in South Africa, is predicted to lose up to two-thirds of its species. A recent climate envelope–based estimate for the global extent of species loss implies, controversially, that 15–37% of the world's species will be threatened by extinction by 2050 as a result of human-induced climate change.

FURTHER READING

Bierregaard, R. O., C. Gascon, T. E. Lovejoy, and R. Mesquita. eds. 2001. Lessons from Amazonia: The Ecology and Conservation of a Fragmented Forest. New Haven, CT: Yale University Press.

Brashares, J. S., P. Arcese, M. K. Sam, P. B. Coppolillo, A.R.E. Sinclair, and A. Balmford. 2004. Bushmeat hunting, wildlife declines, and fish supply in West Africa. Science 306: 1180–1183.

Brooks, T., and A. Balmford. 1996. Atlantic forest extinctions. Nature 380: 115.

Chapin, F. S., E. S. Zavaleta, V. T. Eviner, R. L. Naylor, P. M. Vitousek, H. L. Reynolds, D. U. Hooper, S. Lavorel, O. E. Sala, S. E. Hobbie, M. C. Mack, and S. Diaz. 2000. Consequences of changing biodiversity. Nature 405: 234–242.

Cochrane, M. A. 2003. Fire science for rainforests. Nature 421: 913–919.

D'Antonio, C. M., and P. M. Vitousek. 1992. Biological invasions by exotic grasses, the grass-fire cycle, and global change. Annual Review of Ecology and Systematics 23: 63–87.

Davis, M. 2003. Biotic globalization: Does competition from introduced species threaten biodiversity? Bioscience 53: 481–489.

Delcourt, P. A., H. R. Delcourt, C. R. Ison, W. E. Sharp, and K. J. Gremillion. 1998. Prehistoric human use of fire, the eastern agricultural complex, and Appalachian oak-chestnut forests: Paleoecology of Cliff Palace Pond, Kentucky. American Antiquity 63: 263–278.

Erasmus, B.F.N., A. S. Van-Jaarsveld, S. L. Chown, M. Kshatriya, and K. J. Wessel. 2002. Vulnerability of South

African animal taxa to climate change. Global Change Biology 8: 679–693.

Ferraz, G., G. J. Russell, P. C. Stouffer, R. O. Bierregaard, Jr., S. L. Pimm, and T. E. Lovejoy. 2003. Rates of species loss from Amazonian forest fragments. Proceedings of the National Academy of Sciences, U.S.A. 24: 14069–14073.

Gray, A. J., R. N. Mack, J. L. Harper, M. B. Usher, K. Joysey, and H. Kornberg. 1986. Do invading species have definable genetic characteristics? Philosophical Transactions of the Royal Society of London B 314: 655–674.

Hairston, N. G., F. E. Smith, and L. B. Slobodkin. 1960. Community structure, population control, and competition. American Naturalist 94: 421–424.

Hurtt, G. C., S. Frolking, M. G. Fearon, B. Moore III, E. Shevliakova, S. Malyshev, S. Pacala, and R. A. Houghton. 2006. The underpinnings of land-use history: Three centuries of global gridded land-use transitions, wood harvest activity, and resulting secondary lands. Global Change Biology 12: 1–22.

Laurance, W. F., L. V. Ferreira, J. M. Rankin-de-Merona, and S. G. Laurance. 1998. Rainforest fragmentation and the dynamics of Amazonian rainforest communities. Ecology 79: 2032–2040.

MacArthur, R. H., and E. O. Wilson. 1963. An equilibrium theory of insular zoogeography. Evolution 17: 373–387.

MacArthur, R. H., and E. O. Wilson. 1967. The Theory of Island Biogeography. Princeton, NJ: Princeton University Press.

Manchester, S. J., and J. M. Bullock. 2000. The impacts of non-native species on UK biodiversity and the effectiveness of control. Journal of Applied Ecology 37: 845–864.

Mooney, H. A., and E. E. Cleland. 2001. The evolutionary impact of invasive species. Proceedings of the National Academy of Sciences, U.S.A. 10: 5446–5451.

Mouillot, F., and C. B. Field. 2005. Fire history and the global carbon budget: A $1^\circ \times 1^\circ$ fire history reconstruction for the 20th century. Global Change Biology 11: 398–420.

Myers, N., R. A. Mittermeier, C. G. Mittermeier, G.A.B. da Fonseca, and J. Kent. 2000. Biodiversity hotspots for conservation priorities. Nature 403: 853–858.

Owen-Smith, N. 1987. Pleistocene extinctions: The pivotal role of megaherbivores. Paleobiology 13: 351–362.

Peres, C. A. 2005. Why we need megareserves in Amazonia. Conservation Biology 19: 728–733.

Pimm, S. L., and R. A. Askins. 1995. Forest losses predict bird extinctions in eastern North America. Proceedings of the National Academy of Sciences, U.S.A. 92: 9343–9347.

Pimm, S., P. Raven, A. Peterson, C. H. Sekercioglu, and P. R. Ehrlich, 2006. Human impacts on the rates of recent, present, and future bird extinctions. Proceedings of the National Academy of Sciences, U.S.A. 103: 10941–10946.

Rodda, G. H., T. H. Fritts, and D. Chiszar. 1997. The disappearance of Guam's wildlife. BioScience 47: 565–574.

Sanderson, E. W., M. Jaiteh, M. A. Levy, K. H. Redford, A. V. Wannebo, and G. Woolmer. 2002. The human footprint and the last of the wild. BioScience 52: 891–904.

Terborgh, J., K. Freeley, M. Silman, P. Nunez, and B. Balukjian. 2006. Vegetation dynamics of predator-free land bridge islands. Journal of Ecology 94: 253–263.

Terbough, J., L. Lopez, P. Nunez, V. M. Rao, G. Shahabuddin, G. Orihuela, M. Riveros, R. Ascanio, G. H. Adler, T. D. Lambert, and L. Balbas. 2001. Ecological meltdown in predator-free forest fragments. Science 5548: 1923–1926.

Thomas, C. D., A. Cameron, R. E. Green, M. Bakkenes, L. J. Beaumont, Y. C. Collingham, B.F.N. Erasmus, M. F. de Siqueira, A. Grainger, L. Hannah, L. Hughes, B. Huntley, A. S. van Jaarsveld, G. F. Midgley, L. Miles, M. A. Ortega-Huerta, A. T. Peterson, O. L. Phillips, and S. E. Williams. 2004. Extinction risk from climate change. Nature 427: 145–148.

Thompson, J. D. 1991. The biology of an invasive plant. BioScience 41: 393–401.

Woodroffe, R., and J. R. Ginsberg. 1998. Edge effects and the extinction of populations inside protected areas. Science 280: 2126–2128.

IV.5

Boundary Dynamics in Landscapes

Debra P. C. Peters, James R. Gosz, and Scott L. Collins

OUTLINE

Landscapes consist of a mosaic of distinct vegetation types and their intervening boundaries with distinct characteristics. Boundaries can exist along abrupt environmental gradients or along gradual changes that are reinforced by feedback mechanisms between plants and soil properties. Boundaries can be defined based on the abundance, spatial distribution, and connectivity of the underlying patches. There are three major types of boundary dynamics that differ in the direction and rate of movement of the boundary in response to climatic fluctuations: stationary, directional, and shifting. Future conditions in climate and the disturbance regime, including land use, may fundamentally alter the type of boundary as well as its location and composition through time.

GLOSSARY

boundary (ecotone). Transition area where spatial changes in vegetation structure or ecosystem process rates are more rapid than in the adjoining plant communities

corridor. Edge that promotes movement or allows unimpeded movement of organisms between local populations

directional transition. Location of a boundary between two areas that moves unidirectionally through time

edge. Well-defined area between patch types; often a barrier, constraint, or limit to the movement of animals and plants

patch. Discrete, bounded area of any spatial scale that differs from its surroundings in its biotic and abiotic structure and composition

shifting transition. Boundary location that shifts back and forth with no net change over time

state. Defined by either the dominant species or composition of species, and associated process rates

stationary transition. Boundaries that are stable with little movement through time

1. INTRODUCTION

Landscapes consist of a mosaic of distinct vegetation types and intervening boundaries (or ecotones) with different characteristics from the adjacent communities. An ecotone is a transition zone in time and space. Ecotones along spatial gradients in edaphic and climatic factors (e.g., elevation, soil texture, precipitation) have a long history in ecological studies. Gradual environmental gradients underlying vegetation transitions also exist, often related to positive feedbacks between plants and their environment. Ecotones are increasingly recognized as important elements of dynamic landscapes because of their effects on the movement of animals and materials, rates of nutrient cycling, and levels of biodiversity. Because dramatic shifts in location of vegetation types can occur at ecotones, these can also be important for management and as indicators of climate change.

2. CONCEPTUAL MODELS OF BOUNDARY DYNAMICS

Ecotones along spatial gradients, such as the treeline along an elevation gradient, have long been recognized by ecologists as important elements of landscapes. The term *zone of tension* between two plant community types dates to Clements (1904). Focused research on a broader definition of ecotones began in earnest in the late 1980s and early 1990s. Theoretical models were developed that laid the foundation for future research that considered ecotones as transitional areas between different vegetation types (Gosz, 1993). Many of these models included a hierarchy of spatial scales to depict ecotones, from plant edges to populations, patches, landscapes, and biomes. Most research initially focused on transition zones between biomes that cover large spatial extents, such as between grasslands and forests or between different grassland types. At the biome scale, ecotones were viewed as consisting of a collection of patches where the number and size of patches vary spatially across long distances (hundreds of kilometers). In general, average patch size was predicted to decrease as the distance from a biome or core population increased because of the loss of suitable habitats with increasing distance. Under conditions of increasing resource availability as distance to the biome edge decreased, patches were predicted to coalesce and shift the spatial location of an ecotone through time.

More recent conceptual models have focused on ecotones at landscape scales and have questioned this general relationship between patch size and distance from a core population. Some models focus on the properties of boundaries that influence the rate and pattern of movement of organisms, matter, and energy between adjacent areas (e.g., Wiens, 2002). In other cases, models have focused on boundary dynamics and the use of patch dynamics to provide new understanding about the structural and functional properties of ecotones. Patch dynamics theory has been integrated with hierarchy theory to relate pattern, process, and scale within the context of landscapes (chapter IV.3). A similar patch dynamics approach also allows prediction of changes in boundary location through time and across space (Peters et al., 2006).

3. PROPERTIES OF BOUNDARIES

Landscapes consist of a mosaic of different vegetation states defined by the abundance, spatial distribution, and connectivity of patches (figure 1, and see chapter IV.1). In some cases, these states represent well-defined, homogeneous plant communities that consist of highly aggregated, well-connected patches of that community (states A and B, figure 1). Patches of other states occur infrequently within that state and do not contribute significantly to its overall dynamics. Boundaries between states consist of disaggregated patches of different types with large differences in patch properties and variable connectivity (state C, figure 1). This heterogeneity in patch properties and distribution can result in nonlinear rates of ecological flows across the spatial extent of a boundary.

Patch properties are particularly important to boundary characteristics by determining both their internal function and the connections among patches that determine boundary dynamics. Patch size, type, spatial configuration, and connectivity are key properties that influence the function and dynamics of boundaries.

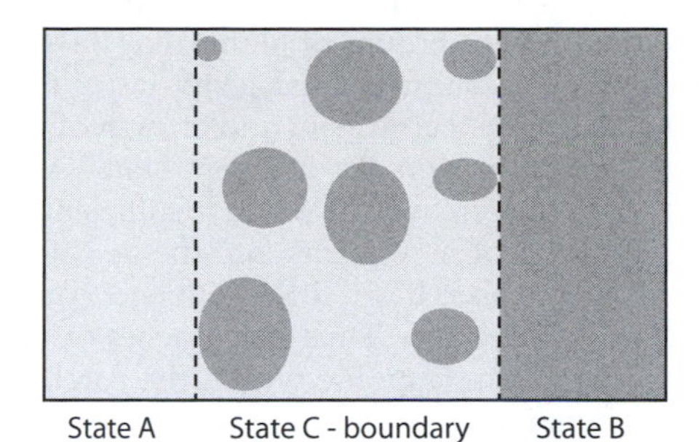

Figure 1. A biotic transition consists of two states (A, B) with a boundary state (C) between them. (Redrawn from Peters et al., 2006) The boundary consists of patches from both states that vary in size, type, spatial configuration, and degree of connectivity.

Size

The importance of patch size to function has often been recognized in landscape studies (Mazerolle and Villard, 1999). Patch size has effects on within-patch processes, such as nitrogen cycling and plant recruitment, and processes that connect patches, such as animal movement (chapter IV.1). As patch size increases, there is an exponential decrease in the perimeter-to-surface area ratio with geometric effects on patch function. Although patches have traditionally been considered internally homogeneous, edges and centers of patches can differ substantially with important consequences for ecosystem processes and animal responses. For example, small patches with large perimeter-to-surface area ratios can have high population densities of animal species that are restricted to edges, whereas large patches can contain high population densities of species associated with patch interiors.

Depending on the function concerned, the effects of patch size are effective over a particular range of sizes. For example, small animals may not forage on isolated shrubs located in large patches of grassland because there is insufficient cover, which increases the risk of predation. Once a patch consists of a larger group of shrubs, then the combined cover may be sufficient for small animals to risk moving to that patch for forage. In addition, the aggregation of plants into larger patches can have important effects on microclimate and wind and water erosion–deposition patterns that affect ecosystem processes such as seed germination, seedling establishment, and plant growth. Thus, patches consisting of groups of plants may have higher probabilities of recruitment and grow faster than expected based on the recruitment and growth of isolated individual plants within the adjacent state. As patch size continues to increase, the patch is sufficiently large that the addition of new plants and the associated increase in patch size have little influence on patch function or connectivity. Thus, for patches of this size or larger, ecosystem response within the patch is similar to the function of the state.

Type

Patch type is defined by the composition and abundance of plant species within a patch and is typically determined by the species with the highest proportion of cover that dominate patch function. Examples include grassland and forest patches that differ in structure and response to environmental drivers.

Spatial Configuration and Connectivity

Patch function and dynamics are also influenced by the spatial configuration and connections among patches. Spatial configuration refers to the distribution of patches and includes measures of richness, evenness, and dispersion (chapter IV.2). Connectivity includes the functional relationships among patches as a result of spatial properties of the landscape and movement of organisms in response to this landscape structure. Some patch types are highly connected, whereas other patch types are rarely, if ever, connected. For terrestrial systems, interactions among patches occur through the movement of water, soil particles, nutrients, or seeds. The vectors of movement are water, wind, animals, and people. Thus, both structural and functional connectivity are included in these interactions. Spatial configuration, including isolation, often interacts with patch size to influence connectivity. Patches of similar size and type located close together have a greater probability of being connected than patches of similar characteristics that are separated by large distances.

4. DYNAMICS OF BOUNDARIES IN RESPONSE TO CLIMATE AND DISTURBANCE

Boundary dynamics depends on (1) abiotic drivers (climate, disturbance) interacting with (2) properties of patches comprising the boundary and adjoining states, (3) biological processes that occur within patches (e.g., competition) and spatial processes that connect patches (e.g., seed dispersal, horizontal water movement), and (4) soil properties that determine water and nutrient availability (e.g., particle size distribution). Through time, new patches are created as plants are recruited, current patches can expand and coalesce with plant recruitment and growth, and patches can be lost as a result of plant mortality. The location and functional properties of a boundary can change through time as patches either increase or decrease in number and spatial extent.

There are three major types of dynamics that differ in the direction and rate of movement of the boundary in response to climatic fluctuations: stationary, directional, and shifting. The importance of four drivers or constraints determine which of these dynamics occurs for any given boundary: (1) abiotic drivers (climate, disturbance), (2) biotic feedback mechanisms among plants, animals, and soil biota, (3) inherent abiotic constraints (e.g., parent material, elevation), and (4) dynamic abiotic feedback mechanisms, such as organic matter and microclimate, that are influenced by the plant community. The width of the boundary and the relationship between patch size and distance from the core population vary for each type.

Stationary boundaries are stable over scales of decades with little movement of patches between states as climate fluctuates (figure 2A). However, disturbances that cause widespread plant mortality (e.g., fire, overgrazing), either along the boundary or in an adjacent state impacted by the disturbance, can shift the boundary through time and space. Although the width of the boundary is narrow and appears as an edge (figure 3A), small-scale fluctuations in boundary location can occur through time with climatic fluctuations. There is no relationship between patch size and distance from an associated state because the boundary is narrow.

Directional boundaries involve the movement by plants and patches from one state into another state such that the location of the boundary moves unidirectionally through time (figure 2B). Directional boundaries are strongly influenced by abiotic drivers, and biotic and abiotic feedback mechanisms that maintain the expansion of plants across the landscape. For example, wet summers combined with overgrazing by livestock can promote the recruitment of native woody plants and initiation of patches in neighboring grassland states located at large distances from the woody-plant-dominated state. Following patch initiation, both positive and strong abiotic feedback mechanisms act to promote the maintenance of the invading patch through time. Positive feedback mechanisms also promote patch expansion and coalescence through the increased probability of recruitment of new individuals within the patch as its size increases. Animal movement between patches leads to increased connectivity of patches as distance between patches decreases. The

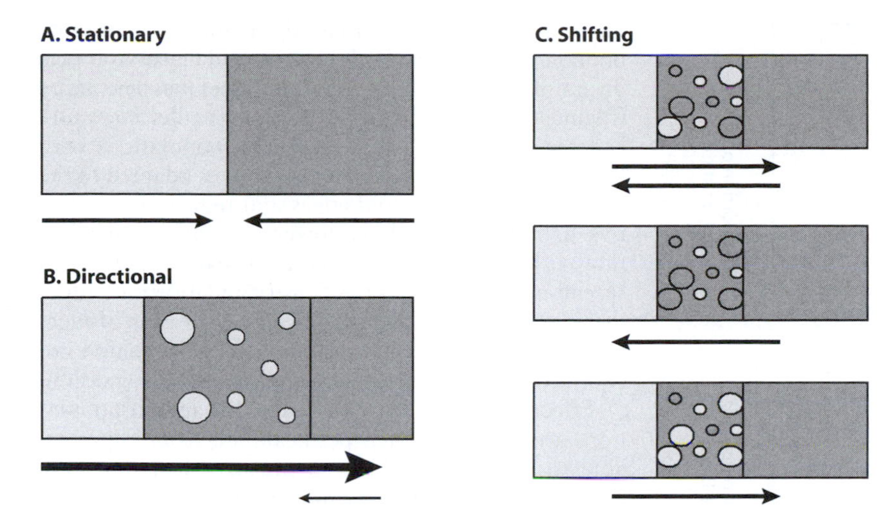

Figure 2. Three types of boundaries, (A) stationary, (B) directional, and (C) shifting, with different dynamics, patch properties, and important constraints and feedbacks.

width of the boundary is intermediate with isolated recruits located at large distances from the advancing state (figure 3B). In general, patch size is largest near its own state, where patches have had time to coalesce; patch size decreases as distance from its state increases.

Shifting boundaries occur when the location of a boundary moves back and forth through time with no net change in its location (figure 2C). These boundaries are very responsive to fluctuating climatic conditions that promote different states during different time periods. For example, a multiyear summer drought may result in high mortality of one plant species, which allows a neighboring species to expand its spatial extent across the boundary. A multiyear winter drought that favors the first species will result in a shift back to the original boundary location or farther depending on the extent of recruitment and mortality of the two species. Disturbances such as fire that reduce cover but not plant density will shift the boundary toward fire-tolerant species until the other species recovers. The boundary is typically broad (figure 3C) with no relationship between patch size and distance from the core population. Extensive periods of overgrazing by livestock and other disturbances that cause widespread plant mortality of one species over another may move these boundaries into a directional phase.

5. LANDSCAPE DYNAMICS

Landscape dynamics depends on the mosaic of boundary types that occur within the landscape. Many landscapes contain more than one type of boundary; thus, both spatial and temporal variation in dynamics are possible. For example, at the Sevilleta National Wildlife Refuge in central New Mexico, landscapes have all three types of boundaries that vary in their response to climatic fluctuations and grazing by cattle (plate 9). The stationary boundaries in this landscape occur most frequently between two grass species, *Bouteloua gracilis* (blue grama) and *B. eriopoda* (black grama), which dominate on different soils. These boundaries are controlled by soil texture constraints and soil water availability interacting with plant life-history traits. These boundaries are stable even under conditions of changes in seasonal rainfall (winter, summer) and grazing by or exclusion from cattle.

By contrast, shifting boundaries at the Sevilleta are responsive to both seasonal rainfall and cattle gazing (plate 9C,D). They occur between blue grama and black grama grasslands located on soils with intermediate sand and clay contents. Black grama expands into boundaries under summer rainfall without grazing, whereas blue grama expands with livestock grazing under a similar rainfall regime. These responses reflect different plant traits by these two species. Black grama, a characteristic species of the Chihuahuan Desert, is less grazing-tolerant than blue grama, a characteristic species of the Shortgrass Steppe, a system that evolved with heavy grazing by large herbivores, in particular bison.

Finally, directional transitions occur at the Sevilleta as a result of the invasion by the native woody plant *Larrea tridentata* (creosotebush), primarily into black grama grasslands (plate 9B). This expansion and subsequent conversion of grasslands into shrublands are promoted with grazing and either summer or winter rainfall. Shifts in dominance from grasses to shrubs are maintained by positive feedback mechanisms between shrubs and their localized soil environment.

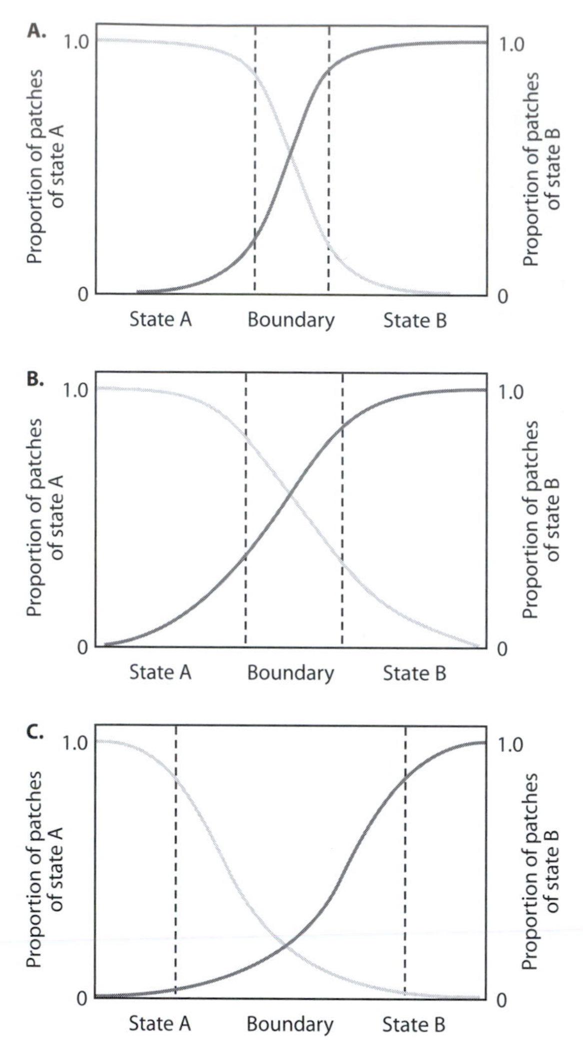

Figure 3. Boundary width for each type of boundary based on the overlap in patches from each state: (A) narrow edges of stationary boundaries, (B) intermediate widths of directional boundaries, and (C) broad transition zones of shifting boundaries.

6. CONSEQUENCES OF BOUNDARY DYNAMICS FOR ANIMAL POPULATIONS

The importance of edges as features of landscapes has a long history in animal ecology, dating to Leopold (1933). Edges between vegetation types were historically viewed as positive features with high plant productivity and biodiversity that resulted in favorable habitat for animals. However, as landscapes have become increasingly fragmented, as a result of changing patterns of land use and patchily distributed disturbances (e.g., wildfire, hurricanes, floods, drought), it is increasingly recognized that edges, and more broadly boundaries, can have both positive and negative effects on animal populations. As landscapes become more fragmented, habitat patches become smaller and more isolated, resulting in smaller animal populations with less movement between patches. Species adapted to the interior of patches avoid edges and reduce their effective habitat area or have lower growth and survival rates at habitat edges. Alternatively, exotic species can spread more rapidly as the density and extent of edges increase across a landscape. Thus, both the ratio of edge-to-interior species can change and the ratio of native-to-exotic species can change with landscape fragmentation.

Effects of boundaries and edges on animal populations are related to their permeability, from impermeable to semipermeable to permeable. Corridors are permeable edges where the movement of organisms can be promoted or unimpeded. Permeability of a corridor can vary by animal species, gender, and age, even for animals within the same landscape. For example, adult female gray-tailed voles (*Microtus canicaudis*) tend to avoid corridors as they get older and presumably more dominant socially (Lidicker and Peterson, 1999).

Animal populations can also have important feedback to boundary dynamics. Effects of animals on seed availability, from production to dispersal and storage of viable seeds, and seedling establishment can effectively shift boundary locations by favoring one plant species or another or by limiting or promoting the expansion of invasive species.

7. APPROACHES TO DETECT BOUNDARY LOCATIONS

Boundary locations can be difficult to determine depending on the scale of observation (plate 10). Both the spatial extent of the landscape and the resolution of the sampling unit need to be considered. The spatial extent needs to be sufficiently large to ensure that both end states and the boundary can be characterized (figure 1). The resolution or grain should be large enough to contain more than one individual or patch of interest yet small enough to allow edge detection. A variety of statistical methods have been developed, primarily for abrupt boundaries, although more recent advances also apply to gradual boundaries (Fagan et al., 2003). After boundaries have been identified, overlap statistics can be used to examine relationships between boundaries, for example, to compare the movement of a boundary through time. Dynamic modeling can be used to simulate the effects of boundaries on ecological processes of interest.

8. APPLYING BOUNDARY DYNAMICS TO MANAGEMENT

Management of natural systems often requires an understanding about the dynamics of boundaries in addition to information about the adjacent states. Predicting changes in the location and composition of boundaries is of particular interest to land managers. Prediction will require an understanding of the type of boundary being considered (stationary, directional, shifting) and the drivers and biological processes controlling those dynamics. The future variation in environmental drivers and consequences of that variation for the key ecological processes will also be important. If this variability in drivers is within the realm of past and current conditions, then confidence can be placed in predictions of future boundary location and composition. However, unforeseen changes in drivers or the appearance of novel disturbances, such as cultivation or housing developments, which go beyond experience to date, may result in new boundary dynamics. For example, boundaries that appear stationary under current conditions may become directional or shifting with large increases in temperatures or decreases in precipitation that either restrict plant recruitment or increase mortality.

Introduction of exotic species into a landscape containing a stationary or shifting boundary may result in a directional change as the species expands its spatial distribution. The ability of land managers to effectively manage for these spatially and temporally variable conditions will require an explicit understanding of the composition of each boundary in the landscape, including patch size, type, and spatial configuration, as it is related to changing environmental drivers.

This chapter is Sevilleta publication number 403.

FURTHER READING

Clements, F. E. 1904. The Development and Structure of Vegetation. Studies in the Vegetation of the State. III. Botanical Survey of Nebraska VII. Lincoln, NE: Woodruff-Collins Printing Co.

Curtis, J. T. 1959. The Vegetation of Wisconsin: An Ordination of Plant Communities. Madison: The University of Wisconsin Press. *A classic book on boundaries.*

Fagan, W. F., M.-J. Fortin, and C. Soykan. 2003. Integrating edge detection and dynamic modeling in quantitative analyses of ecological boundaries. BioScience 53: 730–738.

Forman, R.T.T. 1995. Land Mosaics: The Ecology of Landscapes and Regions. Cambridge, UK: Cambridge University Press. *A good general text on landscape ecology.*

Fortin, M.-J., R. J. Olson, S. Ferson, I. Iverson, C. Hunsaker, G. Edwards, D. Levine, K. Butera, and V. Klemas. 2000. Issues related to the detection of boundaries. Landscape Ecology 15: 453–466. *A discussion of methods related to boundary detection.*

Gosz, J. R. 1993. Ecotone hierarchies. Ecological Applications 3: 369–376.

Hansen, A. J., and F. di Castri, eds. 1992. Landscape Boundaries: Consequences for Biotic Diversity and Ecological Flows. New York: Springer-Verlag. *A more recent book on boundaries.*

Leopold, A. 1933. Game Management. New York: Charles Scribner's Sons.

Lidicker, W. Z., Jr. 1999. Responses of mammals to habitat edges: An overview. Landscape Ecology 14: 333–343. *An overview of animal responses to boundaries.*

Lidicker, W. Z., and J. A. Peterson. 1999. Responses of small mammals to habitat edges. In G. W. Barrett and J. D. Peles, eds., Landscape Ecology of Small Mammals. New York: Springer-Verlag, 211–227

Mazerolle, M. J., and M. Villard. 1999. Patch characteristics and landscape context as predictors of species presence and abundance: A review. Ecoscience 6: 117–124.

Peters, D.P.C., J. R. Gosz, W. T. Pockman, E. E. Small, R. R. Parmenter, S. L. Collins, and E. Muldavin. 2006. Integrating patch and boundary dynamics to understand and predict biotic transitions at multiple scales. Landscape Ecology 21: 19–23.

Risser, P. G. 1995. The status of the science examining ecotones. BioScience 45: 318–325.

Wiens, J. A. 2002. Riverine landscapes: Taking landscape ecology into the water. Freshwater Biology 47: 501–515.

Wiens, J. A., C. S. Crawford, and J. R. Gosz. 1985. Boundary dynamics: A conceptual framework for studying landscape ecosystems. Oikos 45: 421–427. *An early conceptual framework for boundaries.*

IV.6

Spatial Patterns of Species Diversity in Terrestrial Environments

Brian A. Maurer

OUTLINE

1. The physical environment
2. Dynamics of geographic populations
3. Patterns in genetic variation and adaptation
4. Synthesis: How spatial patterns in species diversity are generated and maintained
5. Empirical sampling protocols
6. Species abundance and relative abundance distributions
7. Species–area relationships
8. Abundance–distribution relationships
9. Nested subsets community pattern
10. Species diversity along environmental gradients
11. Understanding species diversity

Spatial patterns of species diversity have intrigued ecologists since European natural historians discovered that the flora and fauna of the world varied dramatically across the face of the Earth. It is only within the last few decades that a clear understanding of the processes underlying these patterns has arisen. Variation in the number of species found across space depends on several interacting sets of processes. The first set of processes affect the physical and chemical properties of the hydrosphere, atmosphere, and lithosphere. The second set of processes comprises the demographic responses of individual organisms interacting with their physical and biological environment summed up within geographic populations of different species. The final set of processes are the long-term adaptive responses of populations as natural selection shapes gene pools of different species over evolutionary time. The complex interactions of these sets of factors occur across a wide range of spatial and temporal scales, making it difficult to isolate simple explanations for data collected at single spatial and/or temporal scales.

In what follows, I provide an outline for how the three sets of processes work together to set the broad patterns of species diversity seen across geographic space. Here I focus on terrestrial patterns, although a similar argument applies to marine patterns of species diversity. After outlining the processes underlying species diversity variation, I show how different methods of sampling species diversity across geographic space produce the variety of patterns documented by ecologists and biogeographers.

GLOSSARY

adaptive syndrome. The suite of morphological, physiological, and behavioral characters that determine an organism's ability to survive and reproduce

α diversity. The species diversity of a locally sampled site

β diversity. The turnover in species diversity among different sites within a landscape, generally referring to sites that share the same metacommunity

γ diversity. Turnover in species diversity among different metacommunities

geographic population. All viable populations of a species found within the species' geographic range

geographic range. The spatial region that includes all viable populations of a species

metacommunity. For any given local community, the assemblage of all geographic populations that contribute immigrants to the community

metapopulation. A group of local populations linked together by dispersal

species diversity. The number and relative abundances of species within a specified geographic region, often divided into α, β, and γ diversity

viable population. Any population that can persist through time by a combination of local recruitment and immigration

1. THE PHYSICAL ENVIRONMENT

Spatial patterns of species diversity result from the interplay of biology with large-scale patterns in the physical properties of the Earth. For a complete description of diversity patterns, then, it is necessary to describe the dynamics of the physical system that comprises the thin layer of materials covering the Earth. Several important sources of energy drive the geophysical environment. Of these, the most important from the perspective of living systems is the sun. Energy from the sun not reflected back into space is absorbed by the surface of the Earth and the atmosphere. The absorbed energy heats air masses in the atmosphere, causing large vertical circulation patterns that distribute water in the atmosphere unevenly across the surface of the Earth. These movements result in broad patterns of climate with latitude, with wet tropical climates near the equator, bands of arid climates at approximately 30° latitude, and wetter temperate climates at 60° latitude (see chapter IV.7 for a more detailed account).

Modifications of the general patterns of heat distribution across the face of the Earth and the consequent movements of air masses across its surface incorporate a number of different processes. The rotational energy of the Earth modifies these general patterns, particularly near coastal regions. The tilt of the Earth results in seasonal patterns in climate as the Earth travels around the sun. Topographic features of the Earth's crust deflect movements of air masses upward, changing nearby patterns of precipitation. Gravitational energy from the moon causes large movements of water masses in the oceans, resulting in tidal patterns along continental margins. These factors combined result in heterogeneous patterns in the spatial distribution of water, and of its different states (i.e., gas, liquid, solid). The tremendous variability in the distribution of water determines to a large degree the distribution of life on the planet.

The crust of cooled rocks on the Earth's surface is not a static entity. The lithosphere is a dynamic system driven primarily by energy derived from the Earth's molten core. Over long spans of time (millions of years), movements of pieces of the crust redistribute continental land masses and change the configurations of the oceans. As land masses shift, so do many of the factors that determine climate. Hence, there is a continual shifting in the distribution of water across the Earth and, consequently, of living systems.

Because all living systems (with a few minor exceptions) are based on energy captured by photosynthesis, variation in the amount of solar radiation across the surface of the Earth also affects the distribution of primary production. This variation in solar energy for photosynthesis interacts with complex patterns of water distribution to form the major biomes recognized by ecologists. Additional variations in the physical conditions of the Earth's surface are imposed at a variety of scales by topography, geology, and a variety of disturbances.

2. DYNAMICS OF GEOGRAPHIC POPULATIONS

Patterns of species diversity are responses of living systems to the complex geophysical variation described in the previous section. These responses involve a wide variety of phenomena that occur across different expanses of space and time. The processes involved in generating these responses can be divided into ecological and evolutionary processes, but this division is arbitrary, for ecological and evolutionary phenomena are closely linked and constitute a single system that integrates living and nonliving components into a single, highly complex hierarchy. In this section I focus on the fundamental ecological processes that underlie species diversity patterns.

At the smallest spatial scale, individual organisms respond to environmental variation through a variety of physiological, morphological, and behavioral adaptations. By definition, these adaptations are fixed within an organism, although they may involve changes in the way an organism interacts with the environment throughout its lifetime. The fundamental importance of these adaptations with respect to species diversity patterns is that they determine the survival and reproduction of individual organisms as they interact with localized environmental conditions. Although there are many complications and variations on patterns of organismal interactions, the fundamental consequence of these interactions is that they determine the rate of change and persistence of populations of organisms belonging to the same species. Species in this sense consist of organisms that share a genetic cohesiveness that maintains the system of adaptations among all organisms belonging to the species. Many times, this means that members of the same species exchange genes through various types of reproductive activities.

Populations are often arbitrarily divided into local concentrations of individuals in space. Within a population, organisms give birth and die in response to local environmental conditions. Over time, the number of organisms in the population changes as a consequence of these organismal patterns. Most populations are not closed systems with respect to the organisms that comprise them. Organisms born elsewhere enter the population, and other organisms leave the population and move elsewhere. The rate of local population change is determined by the rates of birth, death, immigration, and emigration.

Over larger expanses of space, organisms are almost never randomly or uniformly distributed but clump into local concentrations that experience unique sets of environmental conditions. Often these localized clusters are separated sufficiently to be treated as distinct populations that are linked to one another by dispersal. Systems of local populations linked together by dispersal are called metapopulations. The boundaries of a metapopulation are rarely explicitly known, and across geographic space, there may be many metapopulations for a single species. Metapopulation dynamics has been described in two different but related ways. First, equations that describe changes in abundance within local populations that include both birth–death and immigration–emigration dynamics are used to describe spatial patterns in the dynamics of the collective metapopulation. Second, extinction–colonization processes are used to describe the spatial pattern of occupancy of local habitable "patches" within metapopulation boundaries (see chapter IV.3 for further discussion of metapopulation dynamics).

When we consider the collective population across the range of most species, we see a distinct pattern in which there are relatively few locations with large concentrations of abundance and many locations with fewer individuals. This pattern may be generated by two different patterns within local populations. First, local populations may have higher densities near abundance peaks and lower densities away from those peaks. Second, density may be relatively constant across the geographic range, but distances among local populations may be larger in many regions of the range. When averaged across space, this pattern would produce an uneven distribution of abundance across the range. Regions of high abundance often cluster close to the geographic center of the range. In some species, there may be a single region of high abundance, whereas in others there may be several regions. Differentiating between these two possibilities is difficult in practice because of the discontinuous nature of ecological conditions experienced by individuals of different species.

The abundance patterns just described are generated by spatial variation in demographic processes across the range of a species. It appears that for many species, regions of high abundance are characterized by high per capita maximum growth rates coupled with lower per capita intraspecific competition. Conversely, low-abundance regions have low maximum per capita growth rates and higher intensities of intraspecific competition. It is also possible in some cases to model the role of dispersal in maintaining these patterns. When this is done, dispersal seems to be a crucial part of local demography when there are large changes in geographic ranges, e.g., during an invasion by a species into a new geographic region. Estimation of these patterns is made more difficult by the fact that there is often a large degree of environmental variation. Additionally, many methods of estimating abundance include measurement error, which can produce poor estimates of population rates if not included in estimation procedures.

Spatial patterns in demography are linked to the conditions individuals experience in their local environments by suites of behavioral, physiological, and morphological traits, which together can be termed the "adaptive syndrome" of a species. When most of the individuals in a population possess traits that function well in a particular environment, then the fitness of individuals in that population is high, leading to a sustainable population. Conversely, when few individuals in a local population have traits that allow them to function in the local environment, the population itself will be less stable and often may be a "sink" population maintained by immigrants coming from populations with high per capita fitness. To the degree that per capita fitness is correlated with abundance, population abundance will follow spatial patterns in fitness. The fundamental insight is this: large-scale spatial patterns in the distribution of abundance are maintained by spatial variation in demographic responses of local populations to environmental conditions.

Notice that it is not necessary to invoke the concept of a "niche" in this discussion. The adaptive syndrome concept is not equivalent to what many ecologists refer to as a niche. Whereas the idea of an ecological niche invokes references to local interactions among species or some idealized hypervolume describing the ecological "needs" of a species, the necessary concepts to describe geographic distributions are not based on such vague notions. Instead, what are needed are adequate descriptions of the demographic mechanisms that result from the demographic responses of individual organisms within a population to the conditions they experience in the environment.

3. PATTERNS IN GENETIC VARIATION AND ADAPTATION

The traits that organisms employ to obtain sufficient resources for survival and reproduction play a crucial role in shaping the demographic responses of local populations to environmental conditions. These traits are the result of a long history of natural selection and other evolutionary processes shaping the current genetic makeup of each local population. The same patterns of immigration and emigration responsible for the dynamics of geographic populations spread and mix genetic variation within and among local populations. Natural selection demographically corresponds to increased mortality and/or reduced fecundity in at least some of a local population. Gene flow from populations in more productive environments can dilute genetic changes that might be more adaptive in local populations. Conversely, lower rates of gene flow to isolated populations may allow changes in genetic structure in response to selection.

The complex counterplay of migration and selection ultimately leads to suites of ecological characteristics within species (the adaptive syndrome defined previously) that link environmental conditions to demographic responses in ecological time. The adaptive syndrome of a species and the environmental context in which it is expressed together constitute what is sometimes called the "niche" of a species, although this term often has conflicting and ambiguous meanings in the ecological literature. When thinking about patterns of species diversity, it is preferable to think of the adaptive syndrome of a species as a distinct concept because it allows comparisons among species in their potential to respond to gradients of environmental conditions in space and time.

4. SYNTHESIS: HOW SPATIAL PATTERNS IN SPECIES DIVERSITY ARE GENERATED AND MAINTAINED

The fundamental concept behind understanding large-scale patterns of species diversity is the idea of a metacommunity. Defining what a metacommunity is turns out to be quite difficult. The model described in what follows incorporates most of the insights into this problem represented in the literature. A metacommunity is the collection of all species of a given trophic group that contribute populations of individuals within a specified region. For example, the metacommunity for the plants found in Yellowstone National Park, in North America, will include the geographic ranges of all species found there. Depending on the size of the geographic region involved, most species will have at least some populations outside of the region. Only when the region considered reaches the size of a continent or the degree of isolation of a remote island will most species' geographic ranges be contained within the region. In general, then, the metacommunity will cover a larger spatial region, approaching the size of a continent for some groups, than the particular community for which it serves as a source of species. The idea of a metacommunity includes what is often termed a "species pool." In some senses, a metacommunity represents the "γ diversity" of the geographic region for which it serves as the species pool.

Many spatial patterns in species diversity are based on comparisons of different geographic regions along an environmental gradient. Defining the metacommunity for a set of regions may become problematic because each region will have a different set of species that would meet the criterion proposed in the previous paragraph. Each separate region, such as regions defined by latitudinal bands, has different γ diversities, making comparisons among them complicated.

Recently, the importance of the size and shape of the geographic gradient has been shown to contribute to patterns of species diversity. The "mid-domain effect" simply states that the larger the region across which species diversity is measured, the larger the accumulation of species will be. Thus, if an environmental gradient is embedded in a geographic region that varies in size along the gradient, the effects of the gradient on species diversity will be confounded with the effects of area. This effect of the size and shape of a geographic region was first identified in studies of species diversity along latitudinal gradients. For example, species diversity of birds decreases with decreasing latitude in southeastern North America and the Florida peninsula, where the constraining effects of geographic area contribute to a reversal of the general pattern of increasing species diversity with decreasing latitude elsewhere on the continent.

Given the complications discussed above, spatial patterns of species diversity all arise from the same basic mechanism. Species that comprise the metacommunity for a particular geographic region provide the source of individuals that potentially have access to a local community. These immigrants can enter a local community under a variety of conditions. If all individuals in the metacommunity are ecologically identical regardless of species (i.e., there is ecological symmetry among species), then a stochastic process that depends on local birth–death processes coupled with immigration drives local community dynamics. When immigration is absent, the process leads to fixation of a single species in the local community. With the addition of immigration, the process will eventually equilibrate relative abundances of species in the local community with corresponding relative abundances in the metacommunity. This model of community, known as the neutral model, emphasizes the importance of dispersal processes in maintaining local species diversity.

The neutral model is often thought to be unrealistic in its assumption of ecological symmetry among species. The effect of breaking the symmetry assumption is more difficult to model dynamically, but the basic effects of ecological asymmetry on local communities are straightforward. The fundamental result is that when asymmetry exists, the environment in which a local community exists acts as a "filter" on both local demographics and immigration. Generally, this filtering effect maintains rare species in local communities by preventing extinction and encouraging immigration. The filter effect of the local environment is analogous to "natural selection" in the population genetics models. In addition, some of the filtering within a community is caused by interspecific competition. The effect of such competition is to restrict species spatially to patches in the local community where they are competitively superior. If there are insufficient "refuges" for a species, then competition may act as a strong filter, permitting only certain combinations of species.

Given the general stochastic model that couples dispersal and local environmental filtering into a community dynamic, communities can be compared among different local sites and across time. Each local community has a resistance to change based on the adaptive syndrome of each species and the nature of the environmental conditions. This "community inertia" represents a steady state that will persist until some environmental change is experienced. The response of a local community to environmental change is not instantaneous but requires a certain amount of time over which local demographics and patterns of immigration shift in response to changing conditions.

When local communities are compared across space, the resulting patterns are produced both by differences in local ecological conditions and by differences among the metacommunities that are sources of immigrants. The expectation of community inertia across space implies that there will be spatial autocorrelation among communities that declines with distance. This distance effect corresponds to the concept of β diversity. Note that β diversity, in this sense, will respond to both metacommunity properties and variation in local ecological conditions across space. Disentangling the contributions of these two factors remains a major challenge to understanding why species diversity varies across geographic space.

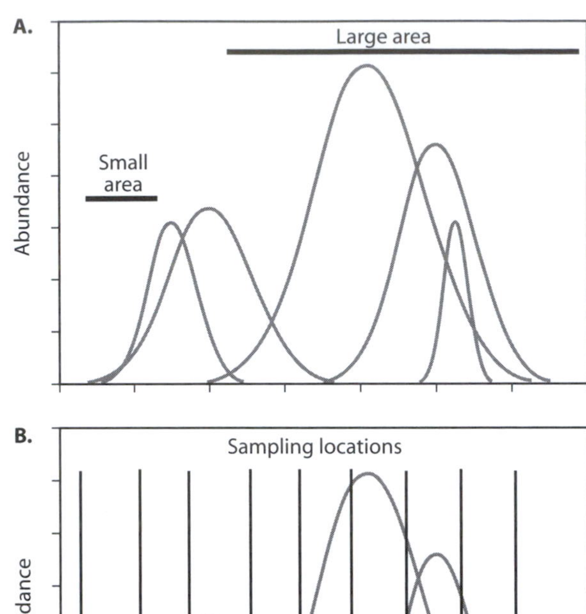

Figure 1. Schematic representation of different sampling protocols for examining spatial patterns of species diversity in space. Species distributions are shown as Gaussian distributions for simplicity. What is necessary, however, is that the geographic range be finite and abundance unevenly distributed within the boundaries. Space is represented as a single dimension, but most patterns are observed in two spatial dimensions. (A) General approach of area sampling. Samples of different sizes are located in space, and all species falling within a given area are counted (and sometimes estimates of abundance are obtained). Islands are a form of area sampling applied to ecosystems that have distinct boundaries where individuals can occur only within the boundaries. Nested-area sampling consists of nesting smaller areas within larger ones. (B) Point sampling. Point samples of the same size (often relatively small compared to the ranges of species) are located in space using some sampling protocol. Species are identified, and sometimes abundance is estimated, within the sampling plots.

5. EMPIRICAL SAMPLING PROTOCOLS

Given the general approach described above, we are now ready to examine empirical patterns that arise when researchers measure species diversity in the field. Most empirical patterns of species diversity are obtained under three general sampling regimes. The first is *area sampling*. Area sampling occurs when a geographic region is divided into areas of different sizes (figure 1A). Most often, the total number of species is counted within regions, but occasionally abundances or densities for each species within the region are available. The quintessential area-sampling regimen of this sort is counts of species number on oceanic islands of various sizes and degrees of isolation. Habitat islands are another variation on area sampling. Area sampling is useful when there are clear demarcations of areas into different ecological units. In some regions, there are no clearly established or easily identifiable ecological boundaries. In such situations, a modification of area sampling consists of nesting smaller sample areas within larger ones. This *nested-area sampling* approach (figure 1A) produces similar results to area sampling, but there is a component of spatial autocorrelation because of the accumulating total species richness from smaller, nested areas into larger areas. It is possible to reduce such autocorrelation with an appropriately designed sampling protocol. The final sampling protocol used to assess patterns of species diversity is a *point sampling*. The salient feature of point sampling is that each "point" represents a relatively small area of sampling that does not vary in size from location to location (figure 1B). Point sampling is often conducted as part of a transect placed along an environmental gradient. The size of the area actually sampled varies widely, from small plots (measured in square meters) up to regions spanning many square kilometers (i.e., 1° latitude–longitude blocks).

There are three basic assumptions about the nature of the metacommunity that are required for the sampling schemes described above to generate empirical patterns. First, the range of each species must be restricted to a part of the geographic region occupied by the metacommunity. This restriction might be a result of environmental filtering, dispersal limitation, or a combination of the two. Second, the ranges of species within the geographic region must vary in size. Although there can be overlap among ranges, both the size and region occupied by each species must vary in a nonuniform manner within the geographic region. Finally, for some patterns, it is necessary that abundances of each species be nonuniformly distributed within the boundary of their geographic range. All of these assumptions are consistent with the description provided above of the demography underlying geographic distributions of species.

6. SPECIES ABUNDANCE AND RELATIVE ABUNDANCE DISTRIBUTIONS

The most fundamental result pertaining to sampling communities within a larger metacommunity is that there are few common species and many rare species (figure 2). This pattern is nearly universal across all communities regardless of environment or taxonomic composition. Furthermore, aggregating communities in space or time does not change the pattern, although the position of individual species may change. The pattern occurs in metacommunities, geographic regions, and entire continents.

The basic assumptions described above have, as a direct consequence, this uneven distribution of abundance. Because a species cannot be found everywhere, and its abundance is unevenly distributed within its geographic range, there must inevitably be some species at a geographic location that have adaptations that more closely match environmental conditions than most of the species in the community. Other species persist in the community by finding refuges from superior competitors, using resources that are marginal

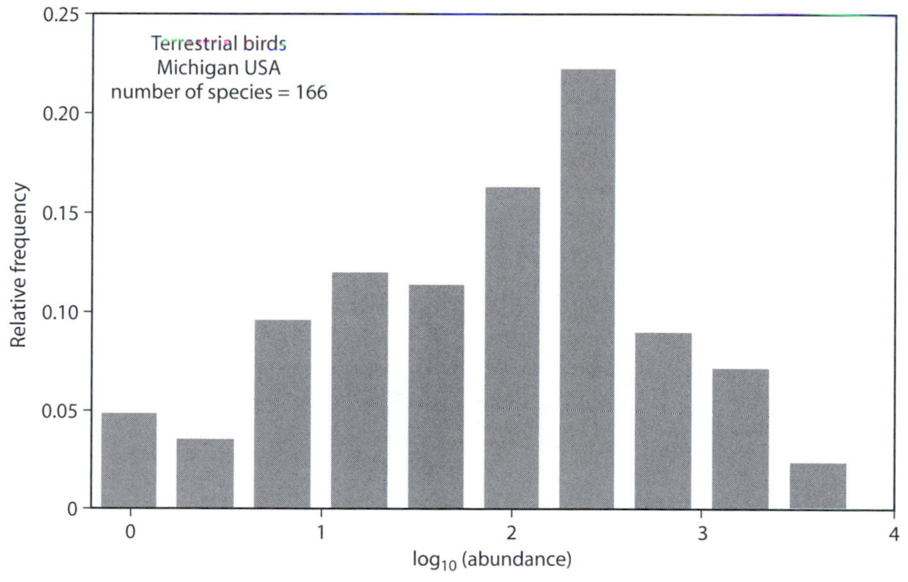

Figure 2. Distribution of abundances of 166 species of terrestrial birds found within Michigan. Data were obtained from the North American Breeding Bird Survey (BBS). Average abundances were calculated from 2000 to 2005 across all BBS routes found in Michigan.

to other species, or by persisting as a sink population maintained by immigration from a larger metapopulation.

What remains unexplained is the reason why species have different geographic range sizes. Why do the adaptive syndromes of a group of related species result in some species that are able to use a wider range of conditions than others and, hence, that have larger geographic ranges? Although we are far from understanding the answer to this question, the essence of the problem may be the necessary trade-offs that occur during microevolution. Both environments and organisms are complex and capable of change, but in order to persist in some environments, species must compromise their ability to persist in others. This leads to the potential for conflicting selection pressures in different environments. The gene pool of a species has only a limited amount of potential phenotypes that can be generated. Organisms living in environments that allow them to produce the most offspring will contribute more to the reservoir of genetic information than organisms living in other environments. This will tend to dampen out selection occurring in marginal environments via gene flow.

7. SPECIES–AREA RELATIONSHIPS

Perhaps one of the best-known of all species diversity patterns is the positive relationship between the size of an area and the number of species found in it (figure 3). This is a direct consequence of area sampling. When an area is located randomly within the ranges of a group of species, the size of that area determines how many rare species will be included in that area. Species may be rare in the region because either they are globally rare or they are at the margin of their geographic range. Smaller areas have a higher probability of not including species with restricted ranges. In other words, area alters the probabilities of species in the metapopulation appearing in local samples. In a completely mixed metacommunity, where all species are equally probable to appear in all samples, the species–area relationship disappears.

Variation in the shape of species–area relationships is related to the degree to which local communities sample the metacommunity. This, in turn, is related to the relative size of the region over which the species–area relationship is being studied. For large regions such as continents, species number increases relatively rapidly with increasing area because most species will not occur over the entire region. As the region of focus becomes smaller, such as samples taken across a biome within a region, the metacommunity being sampled does not contain rare species restricted to ecological conditions not found in the region. The increase in species with increasing area is lower. Isolation of a geographic region also affects the species–area relationship. Isolation has the effect of restricting dispersal of species within the metacommunity, effectively decreasing immigration rates. This will also lower the rate at which species richness increases with area. For example, islands typically have lower slopes for log-log plots of species richness against area than nearby mainlands.

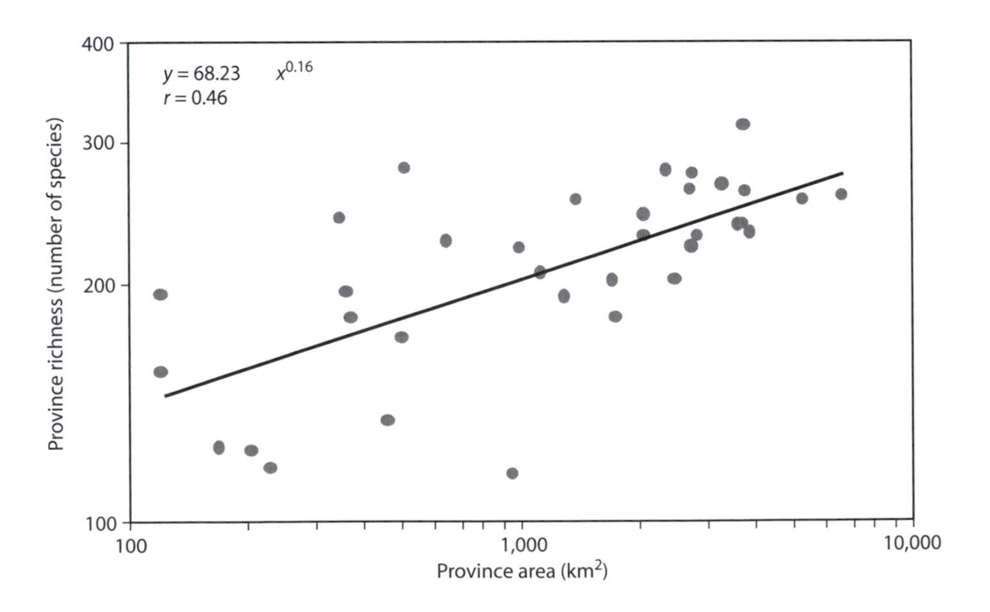

Figure 3. Relationship between number of species and area among 35 ecological provinces recognized by the U.S. National Ecological Unit Hierarchy. Only provinces found within the contiguous United States are reported.

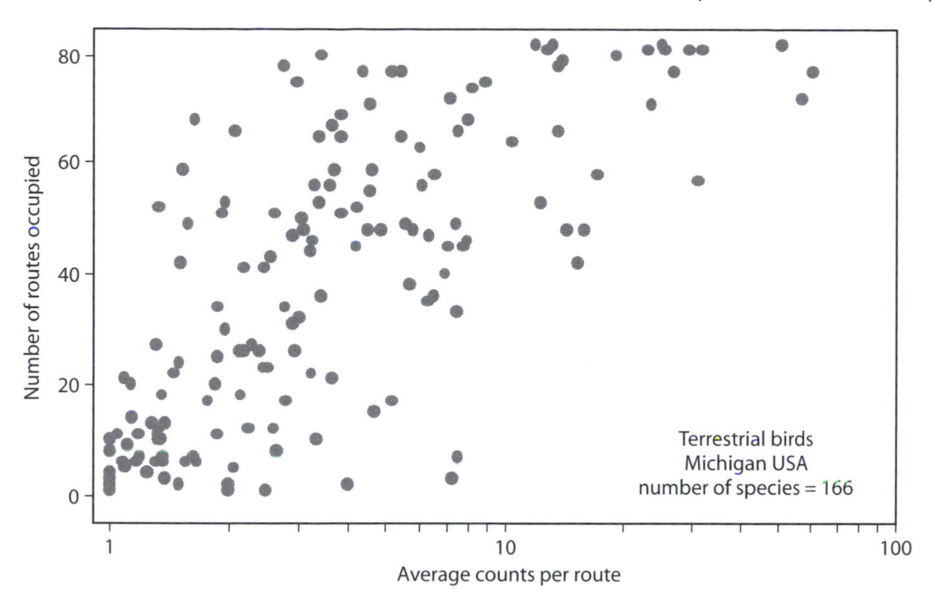

Figure 4. Relationship between distribution (number of routes occupied) and average abundance for 166 species of terrestrial birds found in Michigan.

8. ABUNDANCE–DISTRIBUTION RELATIONSHIPS

In a point sampling study, the average abundance of each species increases with the number of points the species occupies (figure 4). This is the result of the nonuniform distribution of abundance within a species' range and the fact that, if a species shows up more often across sample points, it is more likely that the metacommunity defined for that particular set of samples is located around range abundance centers for that species. There will undoubtedly be variation in the way that each species is distributed within the region, so it is not expected that the correlation between average abundance and distribution will be perfect.

Abundance–distribution relationships can be obtained at many different sampling scales. Within a single landscape, point sampling will produce a positive correlation assuming nonuniform spatial distributions of individuals of different species within the landscape. In such a situation, all samples are assumed to be drawn from the same metacommunity. As the spatial scale of the sampling increases, crossing of ecological boundaries results in some points having different metacommunities. The abundance–distribution relationship in such cases would summarize both variation among points and variation among metapopulations. The likely result of this is to increase the scatter about the abundance–distribution correlation. When the region being measured includes an entire continent, there will be a very large number of metacommunities involved in the abundance–distribution correlation, resulting in a relatively large degree of scatter among different point samples.

9. NESTED SUBSETS COMMUNITY PATTERN

Under an area-sampling scenario, larger areas are more likely to contain the rarest species in a metacommunity than smaller areas. This is because species with the smallest ranges have the lowest probability of being included in a randomly located sampling area. As the sampling area increases, the probability of including a species with a small geographic range increases. On average, then, rare species occur only in the largest sampling regions. Conversely, species with the largest geographic ranges have the highest probabilities of being included in a sample of a given size. Smaller sampling areas, on average, will tend to have only the most widespread species found in them.

The pattern described in the previous paragraph has been termed the "nested subsets" pattern because the species found in smaller sampling regions are nonrandom subsets of the species found in the largest sampling regions. The crucial assumption required for this pattern to obtain is that there must be a nonuniform distribution of geographic range sizes, with a relatively large number of ranges of small size, coupled with a smaller number of species with large ranges (figure 5).

10. SPECIES DIVERSITY ALONG ENVIRONMENTAL GRADIENTS

Because species are not capable of being found everywhere, as one moves across space, the species composition of local communities changes. The turnover in species, or β diversity, follows directly from application of the assumptions made above. There has been a long

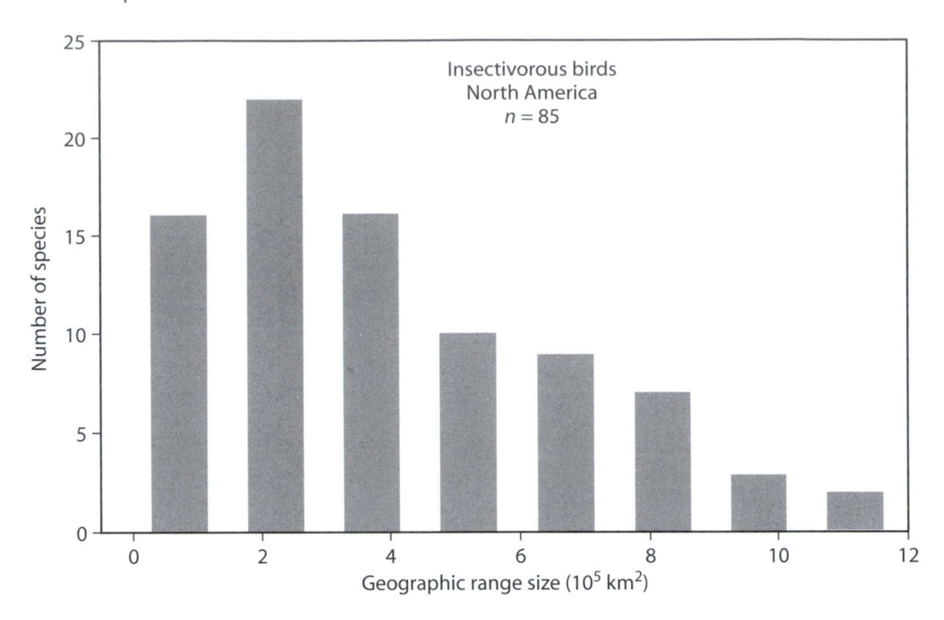

Figure 5. Distribution of geographic range sizes of 85 insectivorous birds found in North America.

tradition in ecology and biogeography that seeks to understand this turnover by observing species composition along some identifiable environmental gradient, such as a moisture, elevational, or latitudinal gradient. The patterns observed along such gradients are thought to indicate how the gradient itself affects the loss of some species and the inclusion of others in local communities. To this end, many studies of species diversity along gradients have opted to study patterns across major environmental gradients at geographic scales.

Perhaps the most controversial of these types of studies are studies of species diversity along latitudinal gradients. Many environmental conditions change with latitude, as described earlier. Generally, latitudes closer to the equator have warmer and/or wetter climates. If warm, wet climates support higher levels of primary production, it follows that species diversity should be highest in tropical latitudes. Although this is generally true, there are many qualifiers to these patterns. For example, some taxa show distinct patterns of increasing diversity with increasing latitude. Furthermore, continental areas vary strikingly with latitude, implying an overriding influence of continental area on species diversity patterns that are independent of climate. General explanations for latitudinal diversity gradients are difficult to generate and test.

Rather than collecting point samples along a latitudinal gradient, some researchers have attempted to relate species diversity directly to ecologically significant environmental factors that have complex patterns of spatial variation. Many of these studies relate species diversity directly or indirectly to some measure of energy production. For example, many studies have shown that species richness in point samples correlates positively with the rate of evapotranspiration. Evapotranspiration presumably indicates the amount of primary production. Establishing cause–effect relationships among such variables at large geographic scales, however, is difficult in the face of the complexity of the actual demographic mechanisms working within and among communities and the metacommunities from which they draw immigrants.

11. UNDERSTANDING SPECIES DIVERSITY

Patterns of species diversity across space originate from fundamental ecological mechanisms that tie demographic responses of populations of different species to complex variation in environmental conditions. The demographic mechanisms include both birth–death dynamics in local environments and dispersal dynamics that link populations together in metapopulations. Species diversity in any local community is related to both local ecological conditions and dynamics of dispersal that link local communities together into metacommunities. Longer-term evolutionary dynamics shapes the species-specific adaptive syndromes that determine how individuals within species react to particular environments. The complexities of these different processes create conceptual challenges for explanations of specific patterns observed in nature.

Observed spatial patterns in species diversity emerge from differing sampling protocols. Each of these patterns, however, implies the same underlying mechanisms. Area-sampling protocols show a distinct effect of area size on species richness. These patterns require asymmetry of geographic range sizes and locations among species. Patterns observed among point samples across geographic space are closely related to patterns derived from area sampling. By controlling for area, point sampling emphasizes the demographic aspects underlying species diversity. All patterns, regardless of sampling mode, imply the existence of metacommunity processes involving asymmetric distributions and dispersal processes among species.

FURTHER READING

Brown, J. H. 1984. On the relationship between distribution and abundance. American Naturalist 124: 255–279.

Brown, J. H. 1995. Macroecology. Chicago: University of Chicago Press.

Hengeveld, R. 1990. Dynamic Biogeography. Cambridge, UK: Cambridge University Press.

Hubbell, S. P. 2001. The Unified Neutral Theory of Biodiversity and Biogeography. Princeton, NJ: Princeton University Press.

Huston, M. A. 1994. Biological Diversity. Cambridge, UK: Cambridge University Press.

MacArthur, R. H. 1972. Geographical Ecology. New York: Harper & Row.

Maurer, B. A. 1999. Untangling Ecological Complexity. Chicago: University of Chicago Press.

Price, P. W. 2003. Macroevolutionary Theory on Macroecological Patterns. Cambridge, UK: Cambridge University Press.

Rosenzweig, M. L. 1995. Species Diversity in Space and Time. Cambridge, UK: Cambridge University Press.

Whittaker, R. H. 1975. Communities and Ecosystems, 2nd ed. New York: Macmillan.

IV.7

Biosphere–Atmosphere Interactions in Landscapes

F. I. Woodward

OUTLINE

1. Feedbacks of landscapes on the atmosphere: Theory
2. Observing feedbacks—comparing different sites
3. Time series of feedbacks
4. The nocturnal boundary layer
5. Conclusion
6. Overview

This chapter investigates some of the ways in which a vegetated landscape can influence its own climate within the planetary boundary layer. During the daytime, impacts may be exerted up to 1–2 km above the surface and are caused by changes in energy exchange, predominantly evapotranspiration, such as that resulting from deforestation or leafing out in deciduous forests. At night, it has been shown that desert vegetation can significantly warm the air above, probably by the emission of greenhouse-active hydrocarbons. A major problem exists in identifying cause and effect in the recorded changes of the planetary boundary layer, as there appears to be no certainty that the landscape is causing all of the described effects, especially where climate is subjected to chaotic dynamics of varying origins. This feature is also true for climatic change studies at the global scale, where it is difficult to identify with certainty the causes of current climatic changes.

GLOSSARY

evapotranspiration. The evaporation of water vapor from surfaces plus the evaporation of water through the plant and leaf stomata by transpiration

inversion layer. The cap of the planetary boundary layer, where there is little or no vertical mixing and where the temperature may increase or remain constant

latent energy exchange. The exchange of energy by the evaporation of water

planetary boundary layer (PBL). The lowest part of the atmosphere where the surface influences wind movements, humidity, and temperature over time periods of about 1 hour and up to 1 to 2 km above the surface

sensible heat exchange. The exchange of energy as heat

At the scale of a few meters, it is quite possible to experience feedbacks between the biosphere and the atmosphere. Under bright summer sunshine the air moving past your legs will be warmer if it has moved across dark dry sand than if it has passed through vegetation, primarily because of differences in radiation absorption and evapotranspiration. At the global scale, modifications of carbon dioxide exchange by the terrestrial and marine biospheres can change the global climate. Landscape-scale interactions are somewhere between these two extremes of scale, in the range of 10 to about 20,000 km^2.

1. FEEDBACKS OF LANDSCAPES ON THE ATMOSPHERE: THEORY

The relevant part of the atmosphere that interacts with the landscape scale defined above is the lower part of the troposphere, the planetary boundary layer (PBL). This typically extends up to 2 to 3 km above the surface during the day. Convective exchange by buoyant thermals and mechanical turbulence by wind transport lead to energy and mass being exchanged between the surface of the biosphere and the PBL. At this time the surface is warmer than the air above. The top of the PBL is defined by sudden changes in temperature and decreases in specific humidity between it and the free atmosphere above. This inversion layer acts as a cap to the PBL and may incorporate clouds. The PBL grows in depth during the day, driven by increases in solar radiation, which warm the surface and lower layers of the PBL, increasing turbulence and entraining air from the free atmosphere above. At around sunset, thermals cease to form as solar radiation reaches zero, and the PBL collapses to form the nocturnal boundary layer (NBL). At this time, the surface generally becomes colder than the air above, and a shallow inversion layer tens to hundreds of meters in depth forms above the landscape.

Plants in the landscape have the potential to influence the PBL both day and night, with active water and energy exchange during the day exerting influences on PBL development. Topography can also exert a marked and obvious effect, and valley bottoms may often be shrouded in cloud at sunrise. This results from the drainage of cold and dense air into the valley bottom; this layer underlies warmer air above, and water condensation occurs where the warmer air is cooled to dew point temperature.

The PBL, therefore, is the part of the atmosphere that is influenced in some way by the underlying landscape. In the typically heterogeneous landscapes on Earth, lateral wind movements combine with the convective upward and downward exchanges in the PBL (figure 1). The PBL consists of several layers that are affected to different extents by the surface. The canopy layer is directly influenced by the height of the vegetation and by exchanges of energy and mass with the vegetation in a way that is directly analogous to the example mentioned above of walking over dark sand or vegetation. The effects are noticeable within the canopy layer but are rapidly diminished by mixing after the air has moved over the canopy. The wake layer above is mixed by interactions between the mean wind flow and backward movements of air derived from the drag exerted by the canopy on the air flow. The surface layer above averages out the effects occurring in the canopy and wake layers, and these effects are averaged further into the mixed layer extending to the top of the PBL.

The question addressed here is how much effect does the landscape exert on the PBL above, given the extensive mixing that occurs? Observations made at meteorological stations will include these feedbacks of interest but only as a small component of the regional (synoptic) weather and climatic conditions. Two approaches can be used to identify feedbacks: one compares adjacent sites differing in some landscape characteristic, such as differences in vegetation type; the other depends on correlating time series of climate or weather with changing events, such as vegetation phenology, in the landscape.

2. OBSERVING FEEDBACKS—COMPARING DIFFERENT SITES

Forest and Pasture

Vegetation type and climate both interact in determining the daily rate of growth and the temperature and humidity of the PBL. A Brazilian tropical forest had similar heights and rates of growth of the PBL in both the wet and dry seasons (figure 2). By contrast, a nearby pasture, with shallow access to water in the soil, had very different PBL heights. In the dry season, the PBL over the pasture was higher than that over the forest because less water was available for evapotranspiration, which caused more solar radiation to be converted instead to convection. The forest has similar

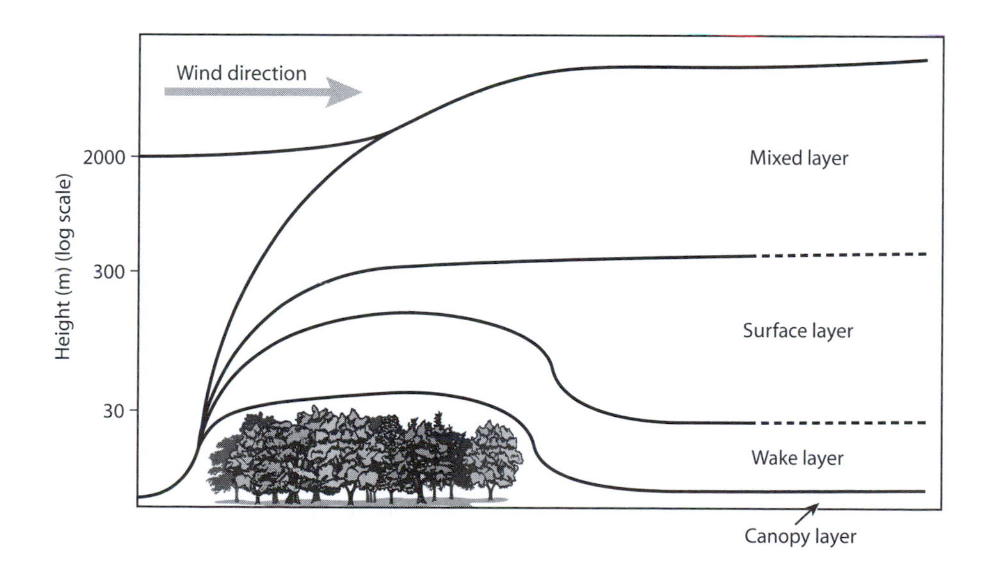

Figure 1. Diagram of the planetary boundary layer and its component layers.

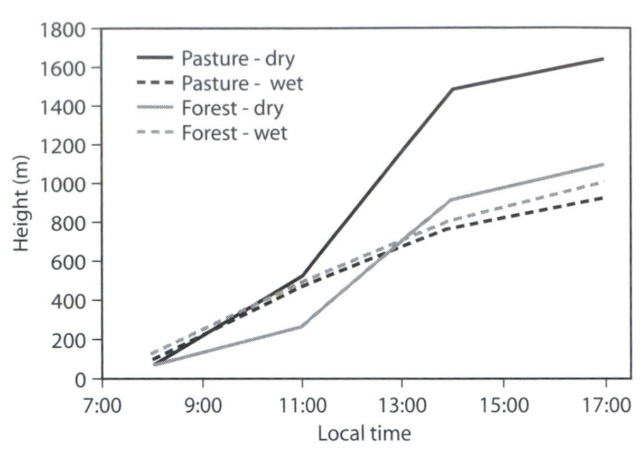

Figure 2. Change in planetary boundary layer height for tropical forest and pasture during the wet and dry seasons.

convective heat exchanges in both the wet and dry seasons, indicating that damper soils that are sheltered from the drying sun by the tree canopy account for the uniformly high convective heat exchange from the tall forest. In the wet season, the PBL was shallower over the pasture because the convective heat exchange was smaller than in the dry season, with greater heat loss by evapotranspiration. Energy exchange by evapotranspiration exerts a much smaller effect than convection on daily PBL growth. The greater dry season evapotranspiration rate from the forest exerts a significant impact on the PBL, which is cooler (−1 to −2°C) and more humid (+10%) than it is over the pasture.

The boundary between native forest and deforested pasture can itself lead to atmospheric feedbacks. In Rondonia, Brazil, the boundary develops increased cloudiness as a result of the production of humid forest breezes that interact with the warmer and drier air over the pasture. The increased convection and strong thermals drive the more humid air upward in the PBL, leading to cloud formation as cooling occurs toward the cap of the PBL.

Impacts of Irrigation

The potential for changes in landscape evapotranspiration to impact the PBL and climate has been investigated in the state of Nebraska. In this part of the Great Plains, increasing areas of native grassland have been replaced by crops over the twentieth century. In addition, much of the area of crops has also been changed from nonirrigated to irrigated, with anything up to a 40% increase in evapotranspiration, which would be expected to produce feedback on the PBL, as for the case of the pasture and forest.

Table 1. Decadal twentieth-century changes in growing season temperature (°C) over irrigated and nonirrigated areas of Nebraska

	Nonirrigated	*Irrigated*
Mean temperature	0.061	−0.035
Maximum temperature	0.045	−0.063
Minimum temperature	0.081	0.028

Source: Data from Mahmood et al. (2006).

Different trends in growing season climate were observed over the irrigated and nonirrigated areas (table 1). Mean, maximum, and minimum growing season temperatures increased where there was no irrigation. The mean minimum temperature of the growing season increased with irrigation, whereas the mean and maximum temperatures decreased. Greater rates of evapotranspiration decrease convective heat loss and cool the vegetation during the day.

Deforestation

The idea that forests control and enhance rainfall has been believed since at least the time of Christopher Columbus, who based this bold assertion on observations that the formerly wooded islands of the Azores and Canaries became desertified after deforestation by the Portuguese. More current observations indicate that anything from 10% to 70% of the water vapor in the atmosphere comes from evapotranspiration. The general connection between deforestation and precipitation is clear; however, the relationship between the extent of deforestation and the impact on precipitation is less clear.

The Atlantic forest of eastern Brazil is a biodiversity hot spot but has suffered significant loss since European settlement. In São Paulo state, forest cover was about 80% before settlement and is now as low as about 10%, with much of the remaining forest in fragments of differing size. With use of remote sensing and detailed analyses of climate records, it has proved possible to demonstrate that there exists a positive correlation between the number of rain days and the spatial extent of the remaining forest fragments (figure 3). Increasing the forest cover to an area of about 200 km^2 exerts a positive and significant impact on days of rain and precipitation. The increase then slows but is still positive as the area of tree cover increases to 1400 km^2. It is also interesting to note that the correlation is significantly less for coastal forests, and this may reflect a combination of specific maritime climatic features; in

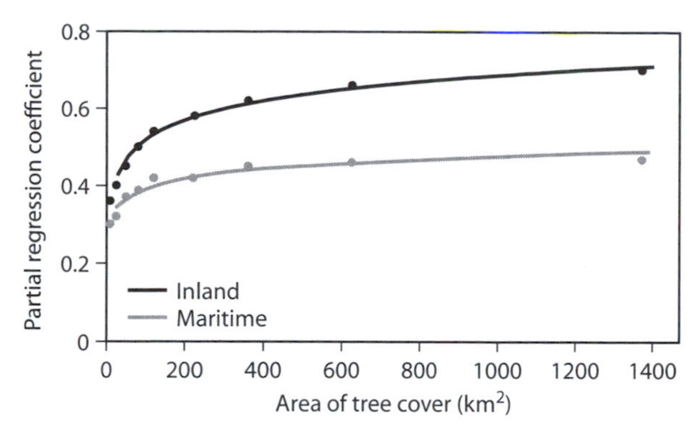

Figure 3. The correlation between area of forest fragments and rain days in Brazil for inland and maritime areas. An increasing coefficient indicates a closer correspondence between forest cover and rain days.

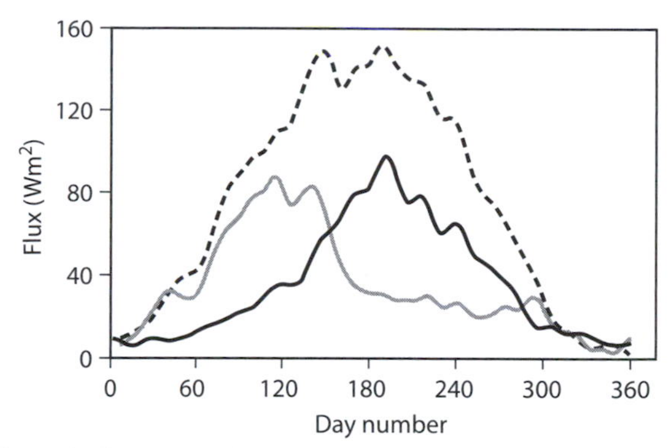

Figure 4. Energy transfer from the Harvard Forest. Dashed line, net radiation; black, latent heat (evapotranspiration); gray, sensible heat (convection).

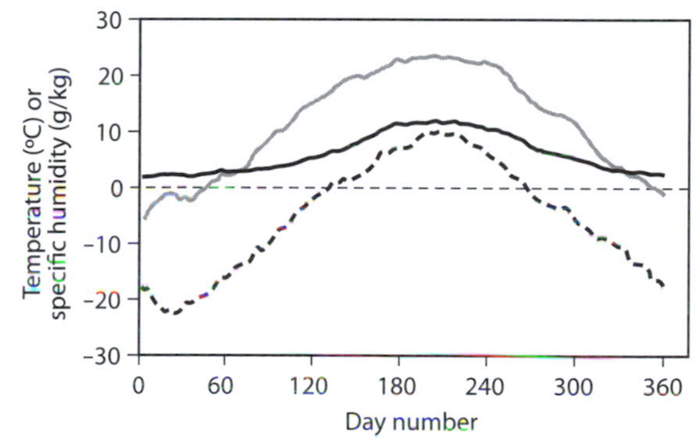

Figure 5. Climatic conditions over the Harvard Forest. Black, maximum temperature; gray, specific humidity; dashed line, minimum temperature.

particular, the sea breeze circulation and the high humidity of maritime air are relatively little influenced by landscape evapotranspiration.

3. TIME SERIES OF FEEDBACKS

Cold temperate forest landscapes have a predictable period of leafing out in the spring. Such an event provides a clear step change in the vegetation activity within the landscape, with an expectation that the onset of transpiration will lead to some humidification of the PBL. Long-term measurements of energy fluxes, climate, and vegetation phenology have been available for a range of sites globally and can provide data to test and quantify the humidification proposal. One such site is the Harvard Forest Long-Term Ecological Research Site of Harvard University (http://harvardforest.fas.harvard.edu). Flux data from the site and others are available at http://cdiac.esd.ornl.gov/programs/ameriflux/data_system/aamer.html.

Harvard Forest is located at Petersham, in the state of Massachusetts, which has about 60% cover as forest and therefore a potential to influence PBL activity and composition. Daily averages of energy fluxes for the period from 1994 to 2000 (figure 4) show a peak of net radiation (incoming less outgoing fluxes of solar and long-wave radiation) in about early to midsummer, but with different trends in sensible (convective) and latent heat (evapotranspiration) transfer. Low vegetation activity during the winter months leads to limited latent heat transfer, and convective transfer dominates. Convective transfer declines quite abruptly from day 160, whereas latent heat transfer increases up to midsummer, peaking at the same time as net radiation.

Maximum and minimum temperature and humidity (figure 5) also peak in midsummer. It is interesting to note that the completely frost-free period is from day 133 and that latent heat transfer increases markedly from just a few days before. Is there a connection, as leafing out of all the major deciduous tree species occurs between about days 115 and 152? Increasing transpiration after leaves have expanded could lead to an increase in the water vapor concentration of the PBL, which would raise the dew point temperature and reduce the impact of frosts on frost-sensitive leaves and, in addition, would be likely to increase cloudiness, another feature that reduces the occurrence of radiation frosts.

The impact of leafing out has been assessed at Harvard Forest by determining the fluxes of energy into the PBL. These are calculated by differentiating the trends of specific humidity and mean daily temperature (figure 6) and averaging over the period from 1994 to 2000. This provides a rate of change: multiplying the rate of temperature change by the specific heat of air provides the sensible heat flux into the PBL,

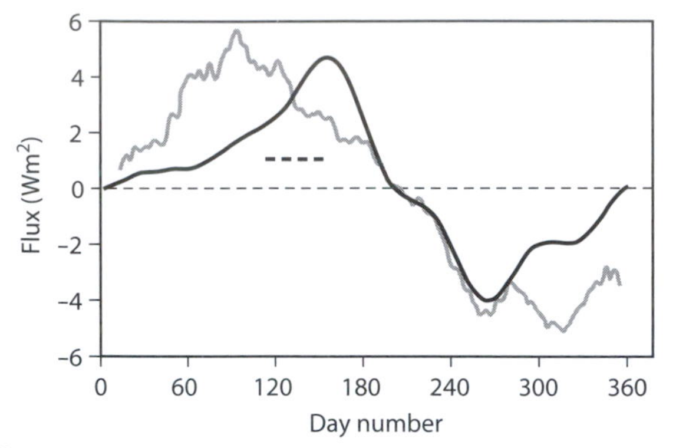

Figure 6. Energy fluxes from the landscape into the planetary boundary layer. Black, latent heat; gray, sensible heat; dashed bar, tree bud burst.

and multiplying the rate of change of specific humidity by the latent heat of vaporization provides the latent heat transfer. Averaging over years minimizes the impacts of weather fronts that in any one year may influence the PBL at a particular time. It is assumed that the PBL grows during the day to a maximum height of 1 km (figure 2).

The quite specific temporal patterns of sensible and latent heat transfer measured from the forested landscape (figure 4) can also be recognized in the PBL (figure 6). Sensible heat transfer is positive early in the year before leafing out, and this warms the PBL. Sensible heat transfer then declines from about day 95, whereas latent heat transfer increases up to day 155. Latent heat transfer increases during the period of leafing out, but there is also a noticeable increase before leafing out. It is most likely that this increase results from evapotranspiration from the ground flora and vernal species that are active before the forest trees.

The forested landscape does, therefore, influence the PBL by sensible and latent heat transfer, but this is achieved by only 5% to 6% of the landscape flux converging into the PBL, the remainder being dissipated, predominantly throughout the canopy layer (figure 1).

4. THE NOCTURNAL BOUNDARY LAYER

The nocturnal boundary develops after the convective PBL collapses when solar radiation declines toward zero. The NBL then develops, dependent on three major processes: turbulent mixing, losses of long-wave radiation, and energy exchange with the soil. Under clear, still nights, temperatures can drop significantly at plant level within the NBL, with the potential for frosts to occur. This has consequences in horticulture, for example, with frost-tender flowers on fruit trees. In such situations the flowers are sprayed with water, and external heaters are used quite regularly to minimize frost damage. Evapotranspiration from vegetation and soil increases the humidity of the NBL, which raises the dew point temperature and therefore reduces the risk of frost. It is notable, however, that in some desert areas such as the Southwest of the United States, where the air is dry and dew point temperatures are low, plants in the landscape are quite frost tender. Many desert plants emit a broad suite of hydrocarbons, both gaseous and particulate, some of which are detectable by human smell. Hydrocarbons generally absorb and emit long-wave radiation, so it is feasible that these hydrocarbons act to provide a minigreenhouse effect at the landscape scale, which is particularly important when the NBL develops on clear nights and where radiative cooling increases the potential for frosts (Hayden, 1998).

This potential effect has been investigated for the Sonoran Desert in Arizona. Two meteorological stations in the Arizona Meteorological Network (http://ag.arizona.edu/azmet/) have been selected, Roll and Marana. Maps of the desert (http://www.cast.uark.edu/pif/main/west/82table.htm) indicate that Roll has little natural vegetation, whereas Marana is vegetated with creosote bush (*Larrea tridentata*) and species of sagebrush (*Artemisia* spp.), both of which are noticeably aromatic. The evergreen nature of many desert shrubs, including *Larrea*, indicates a requirement to endure frosts during the winter.

The meteorological characteristics of the NBL for Roll and Marana were determined as averages over the period 2000 to 2006, using daily maximum, minimum, and dew point temperatures and maximum and minimum observations of relative humidity. Daily observations were extracted for January and February, when there was either no or very little precipitation (this reduces the impact of high-humidity conditions and wet soils on the NBL), when there was probably little if any irrigation for any crops, and when temperatures can potentially fall into the freezing range during the night. The analysis aimed to detect whether the vegetated landscape around Marana can exert an impact on the temperatures within the NBL, particularly at plant height, and whether this impact is more significant than at Roll, with less extensive natural vegetation.

Dew point temperature and minimum temperature at Roll (figure 7) are closely correlated (regression slope is 1.02 ± 0.02, intercept -1.26 ± 0.23), with little scatter. By contrast, there is more scatter at Marana, with many occasions when the minimum temperature is higher than the dew point temperature (slope 0.75 ± 0.03, intercept 5.6 ± 0.25). The intercept for

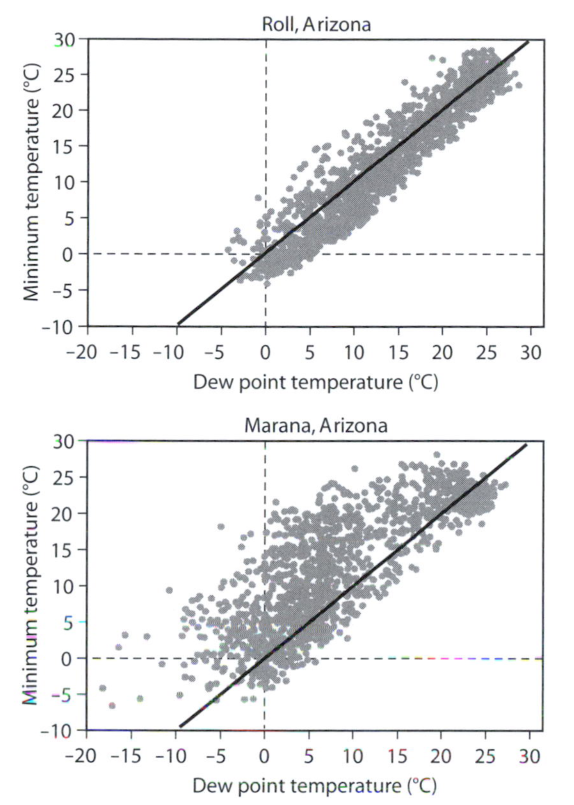

Figure 7. Relationship between daily minimum and dew point temperatures in the months of January and February for Roll and Marana, Arizona.

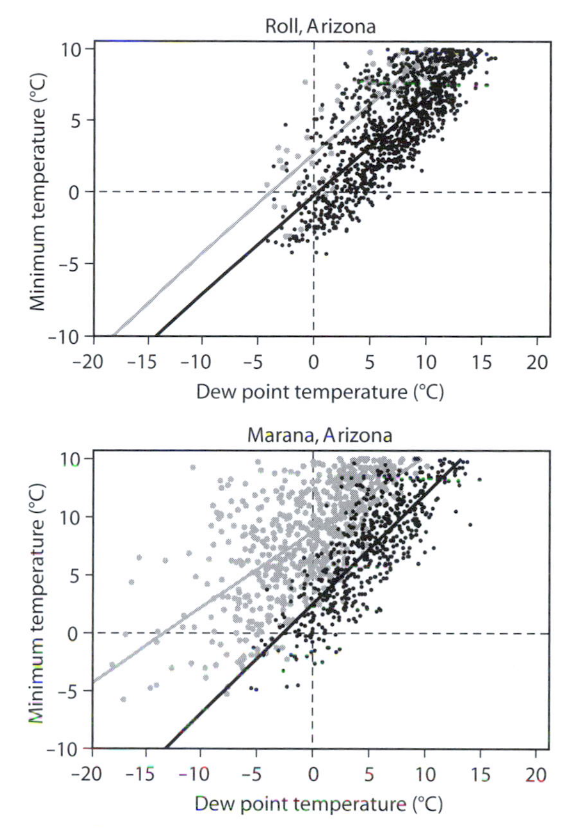

Figure 8. Relationship between daily minimum and dew point temperatures, for minimum temperatures less than 10°C in the months of January and February, for Roll and Marana, Arizona. Black circles, maximum relative humidity >70%; gray circles, maximum relative humidity <70%.

Marana indicates that, on average, the minimum temperature is 5.6°C greater than the dew point temperature, whereas at Roll, the minimum temperature is 1.3°C less than the dew point. Therefore, temperatures in the NBL at Marana appear to be elevated above the dew point temperature and more so than at Roll. The effect of this response is clearer and more ecologically and physiologically relevant when the temperatures are investigated in the chilling and freezing regions (less than 10°C) (figure 8).

Separating the responses to periods when the maximum relative humidity (RH) is greater than or less than 70% also differentiates between the two locations (figure 8). At Roll, the slopes of the relationships between dew point and minimum temperature are not significantly different, although the intercept occurs at a higher temperature when the RH is less than 70%. This reflects the greater opportunity for temperatures in the NBL to drop below the dew point when the air is dry.

At Marana, the two RH ranges differentiate two types of responses within the NBL. When the RH is greater than 70%, the response is similar to that in Roll, with close correlation between the minimum and dew point temperatures. This indicates for both sites that when the RH of the air is quite high, then nocturnal cooling will occur to about the dew point temperature. At that point, heat is released when water vapor condenses into the NBL, tending to stabilize the temperature. This effect also occurs when the RH is lower at Roll, but the stabilizing effect on temperature is less, as less condensation occurs in the drier air. At Marana, the relationship between minimum temperature and dew point at humidities less than 70% is quite different from that at higher humidities, and this difference becomes more marked when the dew point temperature drops below zero. The slope of the response is less steep, and there are many instances of high minimum temperatures. Although some of these observations may reflect the occurrence of warm weather fronts, the regularity of the occurrence suggests other activities in the NBL.

The low relative humidities at Marana are part of a crucial activity within the NBL and the PBL. When RH exceeds 70%, particulate hydrocarbons are enveloped with water, reducing any greenhouse-type effect of the hydrocarbons, and this appears generally to be the case at Roll but not Marana. During the day, this difference would be visible in the PBL with a blue haze at humidities lower than 70% and a white haze at more than 70%—visual verification of landscape impacts on the boundary layer. If the desert shrubs at Marana influence the NBL in this way, then there should be differences in long-wave energy exchange between the two sites. At Roll, increased long-wave energy loss from the landscape surface, at night, will cool the NBL and be little different between the two humidities if hydrocarbon particulates are not major components in the NBL. By contrast, particulate and other hydrocarbon emissions from the desert shrubs at Marana should exert significant impacts on long-wave energy emission in the NBL when the humidities are low.

The divergence of the minimum temperatures between the two sites has been investigated when the dew point temperature drops below zero and plants may be subjected to frost. At Roll, increasing long-wave energy loss from the landscape leads to cooling and to a decrease in the minimum temperature under all humidities with no significant differences in the slopes between the two humidity ranges (figure 9). At Marana, increasing long-wave emission leads to a decline in the minimum temperature when the RH is greater than 70%; however, there is no relationship when the RH is less than 70%. Long-wave emission is correlated with maximum temperature for both sites and all humidities, suggesting that day temperatures may play a key role in the long-wave emissions during low humidities at Marana.

The relationship between daytime maximum temperature and the daily temperature range (figure 10) indicates that at Roll there is a simple positive and linear relationship between the maximum temperature and the temperature range. This is also the case at Marana, but only when the maximum humidity is greater than 70%. At lower humidities, the relationship is significantly flatter, with a 0.29 ± 0.07°C increase in the temperature range for a 1°C increase in maximum temperature, compared with a 1.00 ± 0.16°C increase when humidities are over 70%. This suggests that high day temperatures are exerting an influence on temperatures in the NBL and not simply a warmer night following a warmer day, as seen at Roll. Hydrocarbon emissions from plants increase with temperature, and this appears the most likely explanation: the vegetated landscape is protecting itself from excessively low temperatures at night by producing a minigreenhouse effect with larger effects occurring following a warmer day.

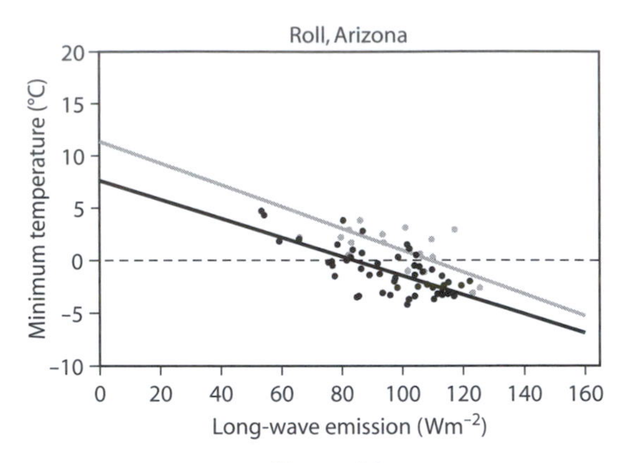

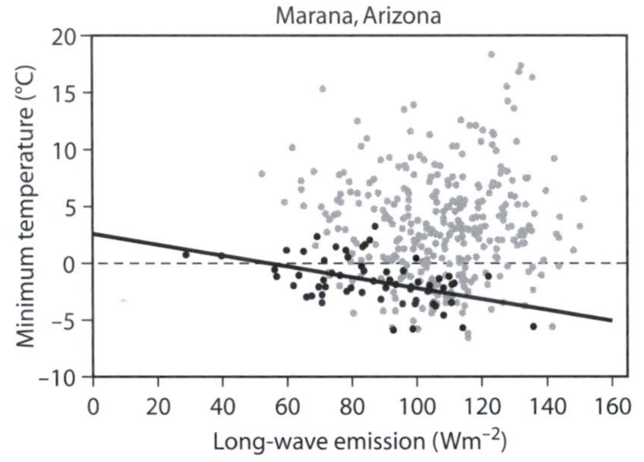

Figure 9. Relationship of minimum temperature and net long-wave emission for Roll and Marana. Black circles, maximum relative humidity >70%; gray circles, maximum relative humidity <70%.

5. CONCLUSION

The impact of the vegetated landscape on local climate is dependent on the response of the mixed PBL above the landscape. Mixing of heat and moisture from outside the area of study occurs continually, as does the inclusion of drier air from above the PBL. These continuous processes mix in the typically humid air released from the vegetated landscape. The end result of a step change in vegetation activity, such as bud burst and leafing, is an identifiable impact on the humidity and heat content of the PBL, but the landscape influence is in the order of about 5% (figure 6) for the Harvard Forest. In Nebraska, increasing the area of irrigated crops leads to a substantial 40% increase of evapotranspiration and lower temperatures (table 1). The temperature of the irrigated landscape was 1.01°C less than in 1945.

The largest impacts of the landscape appear to occur at night in the NBL. The NBL is much thinner than the

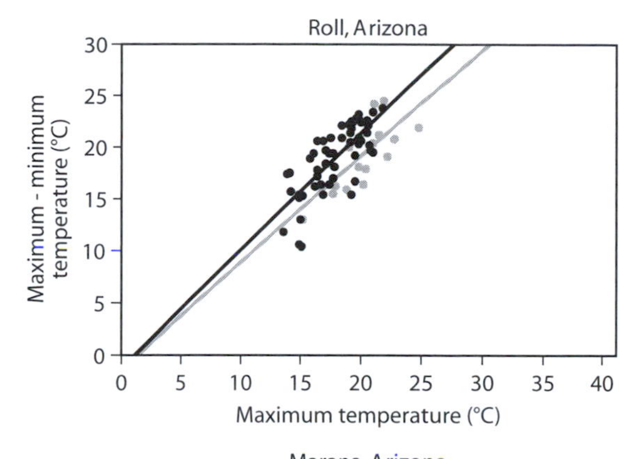

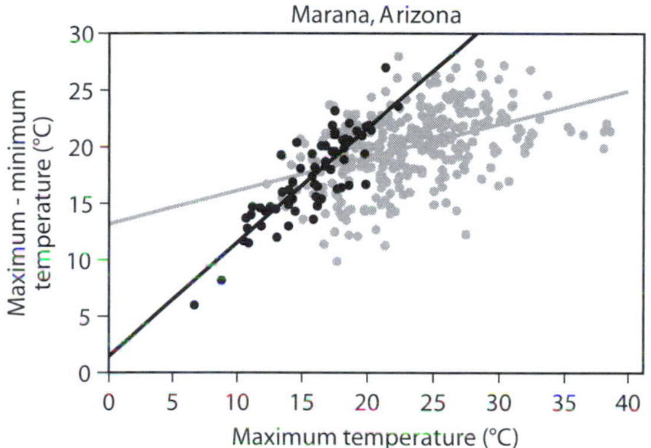

Figure 10. Relationship of maximum daily temperature range and maximum temperature for Roll and Marana. Black circles, maximum relative humidity >70%; gray circles, maximum relative humidity <70%.

PBL of the day, and with no solar radiation, long-wave radiation exchange is the key driver of the NBL climate. The impact of landscape emissions on minimum temperatures is very large, in the order of 5° to 10°C, when humidities are small and dew point temperatures are close to freezing (figures 7 to 10). Minimum temperatures at night increase with the water vapor content of the air (figure 5). As a consequence, minimum temperatures in the Sonoran desert should regularly fall below freezing, but this rarely occurs to any great extent (figure 7), which appears to be a result of significant hydrocarbon emissions from the desert plants in the landscape.

6. OVERVIEW

The work described in this chapter has been inspired by the work of Bruce Hayden, at the University of Virginia. He is an expert in climatology and ecology, and his key work (Hayden, 1998) builds on this twin expertise to describe a wide range of landscape-scale feedbacks on climate. In addition to this description, Hayden's work also identifies the problem in ascribing changes in the planetary layer to landscape activities as opposed to other independent regional climatic effects. This subject has been expanded here, and although it is possible to identify likely feedbacks, there is always the slightly nagging uncertainty associated with the absence of a certain cause-and-effect relationship between landscape and climate. Similar problems exist at the global scale when we consider human influences on climate. In this context, it is interesting to note that when global-scale models of vegetation are driven by fields of observed climate, this observed climate already includes the effects of climatic feedback.

FURTHER READING

Betts, A. K. 2004. Understanding hydrometeorology using global models. Bulletin of the American Meteorological Society 85: 1673–1688. *General description of energy transfer into the planetary and NBLs.*

Fisch, G., J. Tota, L.A.T. Machado, M.A.F. da Silva Dias, R. F. Lyra, C. A. Nobre, A. J. Dolman, and J.H.C. Gash. 2004. The convective boundary layer over pasture and forest in Amazonia. Theoretical and Applied Climatology 78: 47–59. *Describes the impacts of the wet and dry seasons on the planetary boundary layer over forested and grassland landscape; a source for figure 2.*

Fitzjarrald, D. R., O. C. Acevedo, and K. E. Moore. 2001. Climatic consequences of leaf presence in the eastern United States. Journal of Climate 14: 598–614. *A description of an approach to quantify the impact of greening up on the planetary boundary layer; a background to the calculations for figures 4 to 6.*

Geron, C., A. Guenther, J. Greenberg, T. Karl, and R. Rasmussen. 2006. Biogenic volatile organic compound emissions from desert vegetation of the southwestern US. Atmospheric Environment 40: 1645–1660. *Examples of biogenic emissions from a range of plant species from the Mojave and Sonoran deserts.*

Hayden, B. 1998. Ecosystem feedbacks on climate at the landscape scale. Philosophical Transactions of the Royal Society, Series B 353: 5–18. *A classic exposition of an array of landscape feedbacks on climate.*

Mahmood, R., S. A. Foster, T. Keeling, K. G. Hubbard, C. Carlson, and R. Leeper. 2006. Impacts of irrigation on 20th century temperature in the northern Great Plains. Global and Planetary Change 54: 1–18. *Source of data for table 1, indicating how irrigation influences temperatures at the regional scale.*

Webb, T. J., K. J. Gaston, L. Hannah, and F. I. Woodward. 2006. Coincident scales of forest feedback on climate and conservation in a diversity hot spot. Proceedings of the Royal Society, Series B 273: 757–765. *Novel analysis using climatic and remote sensing data to determine the impacts of deforestation on climate; source of data for figure 3.*

IV.8

Seascape Patterns and Dynamics of Coral Reefs

Terry P. Hughes

OUTLINE

1. Human use and abuse of coral reefs at multiple scales
2. Biogeography, hot spots, and conservation priorities
3. Population dynamics and dispersal
4. Habitat fragmentation in the sea
5. No-take areas, dispersal, and seascape dynamics

Coral reef ecosystems exhibit complex dynamics driven by multiple, interacting processes that operate across a range of scales, from local to global and from days to millions of years. Many reefs have been degraded by human action in recent decades, reducing their capacity to absorb recurrent natural and unnatural disturbances. Rebuilding and sustaining the resilience of coral reefs will depend on interventions that are based on an improved understanding of multiscale processes. The current emphasis on conservation of biodiversity hot spots and on establishing networks of no-take areas does not adequately recognize the functional role of key species groups and the critical seascape connections between protected and unprotected reefs.

GLOSSARY

biodiversity hot spots. Regions with exceptionally high species richness, often selected as priority targets for the protection of marine ecosystems.

endemics. Species with small geographic ranges.

functional group. A group of species that share a common ecological function, regardless of their taxonomic affinities. An example is the herbivores found on coral reefs, a diverse assemblage that includes many species of fish, sea urchins, and threatened species such as green turtles and dugongs.

pandemics. Species with very large geographic ranges.

planula. The free-swimming larva of corals. Planulae are released directly by brooded corals following internal fertilization. Spawning corals release both eggs and sperm, and fertilization is external.

spatial refuge. A location where a species or local population is less likely to be affected by its predators, competitors, or pathogens or other processes impacting on its survival, growth, and reproduction.

1. HUMAN USE AND ABUSE OF CORAL REEFS AT MULTIPLE SCALES

Coral reefs are iconic high-diversity ecosystems that are important for coastal human societies, primarily in developing countries. They support the livelihoods of well over 250 million people, primarily through subsistence fisheries and international tourism. Despite their intrinsic aesthetic, cultural, and social value, many coral reefs worldwide have been degraded, especially in the past 20–30 years, reducing their capacity to regenerate from natural and human disturbances. The primary causes of these declines are coastal runoff resulting from land clearing and increased urbanization, overfishing, and climate change. Through time, the scale of human impacts has grown, with even the most remote reefs being increasingly vulnerable to global warming and ocean acidification. Coral reefs are structured by spatial processes that range in scale from global to local, and their capacity to regenerate following disturbance depends on sources of resilience that operate at multiple scales. However, the scales of management of marine ecosystems are usually mismatched to the scales of important processes and to a growing array of human impacts. Interventions are often fragmented and too small in scale to be effective. An emerging approach to management highlights the importance of key multiscale processes undertaken by critical functional groups of species (including the role of humans) that sustain ecosystem resilience across temporal and spatial scales ranging from global to local.

2. BIOGEOGRAPHY, HOT SPOTS, AND CONSERVATION PRIORITIES

Many conservation groups and governments focus on the preservation of biodiversity hot spots as a priority. However, there are several new lines of evidence to suggest that regions with low species richness are more vulnerable and are of no less priority for intervention (Hughes et al., 2002). The primary coral reef biodiversity hot spot is located in the central Indo-Pacific, a large triangular region that straddles the equator, centered on the Philippines, Indonesia, Malaysia, and Papua New Guinea. The diversity of corals and other reef-associated species declines northward and southward away from the central Indo-Pacific hot spot as well to the east across the Pacific and westward across the Indian Ocean. Two secondary coral reef hot spots occur in the Red Sea and in the Caribbean. The similarity in regional-scale biodiversity patterns among major groups such as corals, reef fish, molluscs, and crustaceans reflects their shared history and a common set of mechanisms (e.g., barriers to dispersal) that exert a similar influence on many taxonomic groups.

On land, biodiversity hot spots generally contain large numbers of endemic species that are potentially vulnerable to extinction because of their restricted distribution, especially if they are also uncommon or highly specialized. However, the central Indo-Pacific hot spot is largely the result of overlapping pandemic species whose ranges include the hot spot but also extend westward across the Indian Ocean to East Africa, and/or eastward to the Central Pacific. Only 1% of Indo-Pacific corals are endemic to the central Indo-Pacific hot spot. Similarly, only 3% of reef fish have geographic ranges that lie entirely within the hot spot boundaries. Low-diversity peripheral regions (such as Hawaii, the eastern Pacific, and high-latitude subtropical reefs) have proportionately more endemics than the central Indo-Pacific hot spot. The loss of species from low-diversity locations is likely to be more important because such extinctions affect a greater proportion of an already impoverished fauna.

Low-diversity coral reefs (in the Caribbean, the Eastern Pacific, and at many high-latitude locations in the Indo-Pacific) have both fewer functional groups and lower functional redundancy within functional groups; i.e., functional groups there may be absent or represented by just a single species. For example, Caribbean reefs have about 15% of the number of coral species found on reefs throughout most of the tropical Indo-Pacific oceans. Fast-growing bushy corals with high rates of larval recruitment are diverse and abundant throughout most of the Pacific and Indian oceans and in the Red Sea, but this functional group of corals is absent entirely from the modern Caribbean fauna. An example of low functional redundancy is provided by *Acropora palmata* and *A. cervicornis*, the only two species of tall three-dimensional branching corals in the Caribbean today. These two species are now increasingly rare because of their failure to regenerate from recent mass mortalities caused by hurricanes, algal blooms, sedimentation and runoff, disease, and climate change. Their decline illustrates the vulnerability of depauperate regions that have little or no functional redundancy to compensate for the loss of one or two critically important species.

Another vulnerability of low-diversity regions is that they tend to have small populations that are mostly self-seeding and genetically isolated from elsewhere. In particular, long-distance dispersal by corals to and from geographically isolated, high-latitude reefs is very limited compared to the much higher levels of connectivity among adjacent parts of the central Indo-Pacific hot spot. Dispersal to isolated reefs or islands cannot be achieved incrementally from one generation to the next through a series of stepping-stones as it is, for example, along the 2000-km length of the Great Barrier Reef. Consequently, the depletion of isolated coral populations (e.g., because of escalating global warming) could have persistent impacts over very long periods because these distant populations cannot be rescued by larval recruitment from elsewhere once the local brood stock is lost. Furthermore, the limited genetic variation that is typically associated with isolated, inbreeding populations means that they are likely to have a reduced capacity to respond rapidly to environmental change. This triple vulnerability—a high proportion of endemics, low diversity within functional groups, and isolation by distance—makes coral reef "cold spots" much more vulnerable than hot spots.

3. POPULATION DYNAMICS AND DISPERSAL

Almost all marine species have a larval phase, and for many reef-associated species, it is the only phase of their life cycle when significant dispersal occurs. At a sufficiently local scale, most larvae come from elsewhere, and the reproductive output of a local population is dispersed. Consequently, local extinctions or depletions caused by human or natural disturbances are often quickly reversed by recruitment of larvae that come from robust populations somewhere else. Conversely, even when survivorship and fecundity are high, populations of site-attached adults such as corals and reef fish will nonetheless become locally extinct if recruitment fails to eventuate from elsewhere. Longer-lived species are buffered against fluctuations in recruitment because their populations persist through periods of low recruitment. In contrast, short-lived species are more vulnerable to recruitment failure because each new cohort of recruits represents a large proportion of the local population. Marine ecologists have traditionally assumed that the long-term supply of larvae, although often highly variable in the short term, is inexhaustible. However, there are a growing number of examples, particularly the collapse of fisheries, where widespread reductions in brood stocks have led to diminished levels of recruitment. Similarly, large-scale variation in the density of coral recruits along the Great Barrier Reef and from year to year is strongly associated with spatial and temporal changes in the fecundity of adults. Therefore, there is a two-way chicken-and-egg link between adults and recruits: more adults mean more recruits are produced; and more recruits lead (with a time lag for growth) to more adults.

The degree of connectivity between local populations varies among marine species, which has important implications for dispersal of larvae, pollutants, disease, and exotic species, for population and community dynamics, and for understanding larger, biogeographic-scale patterns of species distributions. For some species, the larval phase is very short, and local populations are largely self-seeded. More typically, larvae are dispersed varying distances among local populations, which collectively comprise a metapopulation. Two dramatic events on coral reefs have illustrated the importance of connectivity and metapopulation dynamics. One is the recurrent population explosion of the coral-eating crown-of-thorns starfish, *Acanthaster planci,* in many parts of the Indo-Pacific. On the Great Barrier Reef, for instance, there have been three cycles of outbreaks in the past 50 years, each taking several years to spread via the recruitment of starfish larvae along 10° of latitude, resulting in substantial reduction in coral cover on more than 200 reefs. The other example is the 1983–1984 population crash of the sea urchin, *Diadema antillarum*, which suffered 98–99% mortality caused by the dispersal of its pathogen from island to island throughout the Caribbean. The die-off caused persistent blooms of seaweed on many overfished reefs, where *Diadema* was the most dominant herbivore. Following the loss of most of the adult breeding population, recruitment of juvenile *Diadema* remains suppressed more than three decades later.

The answer to the question "how far do larvae go?" is complex and relates in part to the biology of larval development. Corals can be classified into two reproductive groups, broadcast spawners and brooders, which have markedly different traits that affect their dispersal. Spawners release both eggs and sperm, and fertilization occurs externally. The resulting larvae are capable of settling after 3–7 days, depending on species. In contrast, brooders release much larger, well-developed planulae that are fertilized internally. Planulae are capable of settling quickly, usually within a few hours to a day or two after release. This is somewhat surprising because planulae are much larger than the larvae of spawners and are potentially better provisioned for long-distance dispersal. However, the available evidence suggests that brooders often settle more locally than the larvae of spawners (see below). The offspring of both brooders and spawners can remain viable in the water column for weeks, but their numbers rapidly deplete with time through mortality from predation and starvation. These longer-distance larvae are few in number but are very important for maintaining gene flow, especially near biogeographic boundaries. However, there is no correlation between the breeding mode of corals (brooder versus spawner) and the size of a species' biogeographic range—the proportion of endemics and pandemics is very similar in both spawners and brooders.

The relationship between larval dispersal and the genetic composition of populations at multiple scales provides a fundamental link between ecology and evolution. Where larvae come from, how far they go, and the genetic consequences of past and present dispersal remain poorly understood for the vast majority of species. Most of the genetic studies of coral reef species have been conducted on the Great Barrier Reef, which encompasses about 2500 distinct reefs, separated from each other usually by a few tens of kilometers, stretching north–south for nearly 2000 km. Some reefs fringe parts of the mainland or inshore islands, but most occur as a broad band on the mid- and outer continental shelf, generally 40–250 km offshore. This stepping-stone physical array is markedly different from many other Pacific Ocean and Indian Ocean reefs that are isolated individually or that occur in clusters comprising remote archipelagos. Fish, echinoderms, mollusks, and other taxonomic groups with long-lived larval stages (typically 4–6 weeks) have low levels of genetic differentiation and high levels of inferred gene flow along the Great Barrier Reef. For most corals, especially brooders, larval recruitment is local, within reefs or among close neighbors. Individual reefs depend primarily on self-seeding for the maintenance of local coral populations. Long-distance dispersal of coral larvae is important over evolutionary time scales for preventing the divergence of species and the accumulation of fixed genetic differences but makes a minimal contribution demographically to the maintenance and regulation of local populations.

The species composition of larval recruits on coral reefs varies at multiple scales. The replenishment of coral populations is most commonly measured using artificial panels that provide a standardized substrate that can be experimentally deployed and retrieved, onto which larvae can attach. At a biogeographic scale, recruitment in the Caribbean is dominated by brooders, whereas spawners are predominant in the tropical Indo-Pacific, reflecting major differences in species composition of adults. Regional-scale patterns also occur along latitudinal gradients to the south and north of the central Indo-Pacific hot spot, along the length of the Great Barrier Reef, and the Ryukyu Island chain. Recruitment of larvae on reefs closest to the hot spot is dominated by spawners, whereas depauperate subtropical locations have lower rates of recruitment, principally by brooders. The species composition of recruits also varies among habitats and depths, mainly because of different larval behaviors, physiological tolerances, and postsettlement mortality, and is a major contributor to patterns of abundance and diversity at small scales.

4. HABITAT FRAGMENTATION IN THE SEA

A common perception is that seascapes are more intact and less subject to fragmentation effects than landscapes, but is that really true? Fragmentation on land continues to subdivide large populations, creating metapopulations, where immigration and emigration now play a more significant role than before. In the sea, connectivity is a natural feature of open populations. Nonetheless, there are growing signs that habitat fragmentation and the loss of reproductive adults (e.g., through overfishing, disease, or climate change) are disrupting stock recruitment relationships, leading to lower rates of larval recruitment or recruitment failure. Coastal mangroves adjoining coral reefs have been reduced to small remnants in many countries, particularly in tropical Asia, to make way for coastal settlements, tourism developments, and prawn farms. Similarly, nearshore coral reefs, seagrass beds, and associated habitats have been degraded to varying extents in different geographic regions, especially because of pollution, climate change, and disease. At least 40% of the world's coral reefs have been damaged by bleaching caused by thermal stress in the past two to three decades (Wilkinson, 2004). In most of the Caribbean, coral cover has declined by 80% or more since the 1970s (Gardener et al., 2003), with flow-on effects for many reef species that rely on the three-dimensional structure provided by branching corals. Similarly, fish stocks have been depleted by subsistence and commercial fishing almost everywhere.

In most parts of the Caribbean, the number of juvenile corals detected in reef surveys has declined sharply over the past 25 years, providing today only a very small fraction of the number of new colonies needed to maintain population sizes. Thick stands of fleshy seaweed continue to inhibit settlement of coral larvae, and new recruits are often overgrown and killed by the algae. The size of the larval pool is also likely to have decreased sharply because adult corals are fewer in number, smaller, and are often physiologically stressed. Importantly, these different mechanisms of recruitment failure offer contrasting prospects for the future. If the decline in replenishment is caused primarily by competition with seaweed, a reversal of the algal blooms (e.g., through better management of herbivorous fish) would quickly enhance coral recruitment. This would favor fast-growing species of corals that have high rates of larval recruitment, fast growth, and early reproduction. Slower-growing corals that tend to have naturally low rates of recruitment, such as the important reef frame builder *Montastrea annularis,* will take much longer, a century or more, to recover to pre-1980 levels even if their recruitment resumes. Conversely, if the recruitment failure is also caused by reduced production of larvae, there will be a much longer period of recovery and recolonization by coral recruits, even if the algal blooms were reversed. Weedy species, such as some soft corals, zooanthids, gorgonians, and sponges, are likely to rebound before most corals. For brooding corals with limited dispersal, such as *Agaricia*, recruitment rates may remain depressed until local breeding stocks can recover. Colonization by other corals that have greater long-distance dispersal may be less affected, with potentially far-reaching consequences for the long-term species composition of Caribbean reefs.

The relative susceptibility to habitat fragmentation of species with different dispersal capabilities is shown graphically in figure 1. Each patch of habitat (e.g., an island in the Caribbean or an individual reef on the Great Barrier Reef) can self-seed, receive, or export larvae. Species with long-distance dispersal should be more resistant to habitat fragmentation because the loss of nearby patches does not preclude dispersal to and from more distant locations. Therefore, habitat loss and fragmentation cause a filtering effect that impacts most on species with limited dispersal (figure 1). On the Great Barrier Reef, approximately 600 reefs out of 2500 have been significantly damaged in the past 45 years by runoff, outbreaks of crown-of-thorns starfish, and two bouts of coral bleaching in 1998 and 2002 (Bellwood et al., 2004). A much higher proportion of reefs have

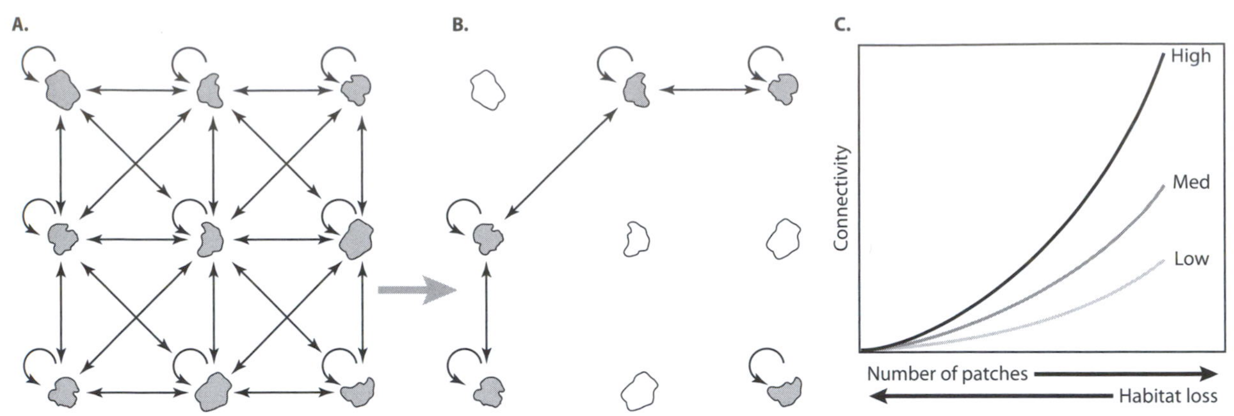

Figure 1. A graphic model showing dispersal of larvae among patches of habitat, a key process for maintaining marine populations and ecosystems. Arrows depict potential dispersal pathways among patches and self-seeding within patches (for clarity, longer-distance arrows are omitted). (A) An intact system with high connectivity. (B) A damaged ecosystem, showing reduced larval connectivity caused by habitat fragmentation and loss of brood stock. (C) The nonlinear relationship between habitat loss and the strength of larval connections for species with high, medium, and low dispersal abilities. Species with limited dispersal are more vulnerable to recruitment failure. (Modified with permission from Hughes, Terence P., David R. Bellwood, Carl Folke, Robert S. Steneck, and James Wilson. 2005. New paradigms for supporting the resilience of marine ecosystems. Trends in Ecology and Evolution 20: 380–386)

been degraded in other regions, particularly in the Caribbean, the Indian Ocean, and in densely populated parts of Southeast Asia. A major concern is that local degradation could trigger larger-scale collapses, causing the remaining "healthy" reefs to collapse once a critical threshold is reached. Importantly, because systemwide collapse is an emergent property of small-scale dynamics, even the most rigorous management of remnant areas could be too little, too late. The important lesson for management is that small-scale interventions may not be enough to prevent systemwide collapse arising from the accumulating impacts of fragmentation on metapopulation and community dynamics.

5. NO-TAKE AREAS, DISPERSAL, AND SEASCAPE DYNAMICS

The history of fisheries is dominated by the steady expansion of fishing effort to deeper and more remote locations and the elimination of spatial refuges that have helped to sustain heavily harvested areas in the past through immigration of larvae or adults. Subsistence and artisanal fisheries on coral reefs target a wide range of species, which differ hugely in their response to fishing, depending in large part on their life histories and reliance on recruitment. Slow-growing, long-lived species such as turtles and sharks can only be harvested sustainably at low intensities and take a long time to recover from overexploitation. For example, the 95% or so loss of dugongs from the southern two-thirds of the Great Barrier Reef in the past few decades will take more than 150 years to reverse, assuming the remnant population can grow at its maximum capacity of 3–4% per annum. In contrast, short-lived species can generally be harvested at high intensities and recover quickly from population crashes, so long as pulses of recruitment continue to maintain the harvested stock. Consequently, multispecies fisheries often cause a predictable change in taxonomic composition (favoring short-lived species) even where there is relatively little targeting of individual species.

No-take areas, where fishing is prohibited, are important tools for reinstating spatial refuges and rebuilding depleted stocks. When fishing is reduced, more adults of harvested species attain a larger size, and their reproductive output increases disproportionately. Some larvae may be retained within the no-take area, but most are likely to be dispersed and may help to restock the fishery outside. Apart from their utility in managing targeted species, no-take areas can also help to restore the structure of food webs and build the resilience of ecosystems. Increasingly, herbivorous fish have become a prime target of many coral reef fisheries, replacing depleted stocks of predatory fishes such as sharks and groupers that now comprise a smaller proportion of the overall catch. Herbivorous fishes, such as parrotfish, surgeonfish, and rabbitfish, play several key roles in the dynamics of tropical reefs: they graze fleshy seaweeds that compete with juvenile and adult corals for space; some erode dead coral skeletons and generate reef sediments, and their position in the food chain means they are an important energetic link between plants and predators. The removal of herbivores, especially on reefs that are also polluted, can lead to abrupt shifts from dominance by corals to persistent blooms of fleshy seaweed. Increasing concern about the combined impacts of fishing, pollution, and climate change on the Great Barrier Reef Marine Park was a major factor in recently setting aside 33% (over 100,000 km^2) as permanent no-take areas.

Most no-take areas on coral reefs are very small, often a few square kilometers or even less. Clearly, these are too small to protect highly mobile species such as dugongs, sharks, and turtles that are heavily targeted outside the no-take area. Similarly, the flow of larvae across the boundary of no-take areas is multidirectional—larvae arrive and larvae leave. Proponents of no-take areas often focus on their potential for reseeding adjoining regions. However, in many cases, the replenishment of local populations within protected areas relies on an influx of larvae from the surrounding reef matrix (including the "good" larvae of fishes and corals and the "bad" propagules of algae and diseases). Clearly, the success or failure of a network of no-take areas depends critically on areas outside that are part of the same highly connected reef system. As is the case on land, fragmented seascapes or networks of no-take areas are strongly dependent on the surrounding matrix, which typically dominates the overall dynamics. Consequently, a larger-scale approach to management is urgently required, recognizing the broader seascape as an interacting patchwork of both no-take and non-no-take areas.

FURTHER READING

Alcala, Angel C., and Garry R. Russ. 2006. No-take marine reserves and reef fisheries management in the Philippines: A new people power revolution. Ambio 35: 245–254. *This is a fascinating account of the evolution of governance and management of coastal resources, illustrating the interplay between science and local and national societies.*

Bellwood, David R., Terence P. Hughes, Carl Folke, and Magnus Nyström. 2004. Confronting the coral reef crisis. Nature 429: 827–833.

Bertness, Mark D., Steven D. Gaines, and Mark E. Hay, eds. 2001. Marine Community Ecology. Sunderland, MA: SinauerAssociates. *This volume is a comprehensive and informative textbook in three parts: (1) Processes influencing patterns in marine communities; (2) an overview of the ecology of eight community types, including coral reefs; and (3) a section on conservation and management.*

Birkeland, Charles, ed. 1996. Life and Death of Coral Reefs. London: Chapman & Hall. *This edited volume provides a very comprehensive overview of the geology and history of coral reefs, their evolution and ecology, biogeography, and human impacts on them.*

Gardner, Toby A., Isabelle M. Côté, Jennifer A. Gill, Alastair Grant, and Andrew R. Watkinson. 2003. Long-term region-wide declines in Caribbean corals. Science 301: 958–960.

Hughes, Terence P., Andrew H. Baird, David R. Bellwood, Margaret Card, Sean R. Connolly, Carl Folke, Richard Grosberg, Ove Hoegh-Guldberg, Jeremy B. C. Jackson, Joanie Kleypas, Janice M. Lough, Paul Marshall, Magnus Nyström, Steven R. Palumbi, John M. Pandolfi, Brian Rosen, and Joan Roughgarden. 2003. Climate change, human impacts, and the resilience of coral reefs. Science 301: 929–933.

Hughes, Terence P., David R. Bellwood, and Sean R. Connolly. 2002. Biodiversity hotspots, centers of endemicity, and the conservation of coral reefs. Ecology Letters 5: 775–784.

Jones, Geoff P., Maya Srinivasan, and Glenn R. Almany. 2007. Population connectivity and conservation of marine biodiversity. Oceanography 20: 43–53. *This article focuses on how knowledge of connectivity can help to improve strategies for conserving marine biodiversity.*

Karlson, Ronald. 1999. Dynamics of Coral Communities. Dordrecht: Kluwer. *This book focuses on the theory and field evidence for various processes that influence the dynamics of coral communities at multiple scales, including ecological succession, interspecific competition, predator–prey interactions, disturbance, assembly "rules," and regional enrichment of coral communities at biogeographical scales.*

Pandolfi, John M., Roger H. Bradbury, Enric Sala, Terence P. Hughes, Karen A. Bjorndal, Richard G. Cooke, Deborah McArdle, Loren McClenachan, Marah J. H. Newman, Gustavo Paredes, Robert R. Warner, and Jeremy B. C. Jackson. 2003. Global trajectories of the long-term decline of coral reef ecosystems. Science 301: 955–958.

Sobel, Jack, and Craig Dahlgren, eds. 2004. Marine Reserves: A Guide to Science, Design and Use. Washington, DC: Island Press. *This book provides a useful overview of marine reserves, focusing mainly on potential fisheries outcomes. It includes detailed case studies from California, the Bahamas, and Belize as well as a brief global overview featuring New Zealand, the Philippines, the Mediterranean, Chile, and Australia.*

Wilkinson, Clive R., ed. 2004. Status of the Coral Reefs of the World: 2004. Townsville, QLD: Global Coral Reef Monitoring Network and Australian Institute of Marine Science.

IV.9

Seascape Microbial Ecology: Habitat Structure, Biodiversity, and Ecosystem Function

David M. Karl and Ricardo M. Letelier

OUTLINE

Seascapes are marine analogs of landscapes in the terrestrial biosphere, namely the physical, chemical, and biological elements that collectively define a particular marine habitat. The field of seascape ecology, also referred to as ecological geography of the sea, seeks fundamental understanding of spatial and temporal variability in habitat structure and its relationships to ecosystem function, including solar energy capture and dissipation, trophic interactions and their effects on nutrient dynamics, and patterns and controls of biodiversity. Implicit in the study of seascape ecology is an interest in the management of global resources through the development of new theory, the establishment of long-term ecological observation programs, and the dissemination of knowledge to society at large.

GLOSSARY

euphotic zone. Upper portion of the ocean where there is sufficient light to support net photosynthesis, usually the upper 0–200 m in the clearest ocean water

genome. The complete assembly of genes present in a given organism, coded by specific nucleotide sequences of DNA, that determines its taxonomic structure, metabolic characteristics, behavior, and ecological function

microorganism. The smallest form of life (<2 μm) on our planet and the most abundant in the open sea, sometimes reaching cell densities of 1 million cells per cubic centimeter

nitrogen fixation. The process whereby relatively inert gaseous nitrogen (N_2) is reduced to ammonia (NH_3) and thus converted into a biologically available form

nutrient. One of several organic or inorganic raw materials that are required for the growth of an organism, for example, nitrogen, phosphorus, iron, and vitamins

oligotrophic. A condition of low nutrient concentration and low standing stock of living organisms, for example, the open ocean

primary production. Metabolic process during which carbon dioxide is incorporated into organic matter by bacteria and eukaryotic algae using any of a variety of energy sources, but usually solar energy

remote sensing. The indirect measurement of habitat characteristics, for example by Earth-orbiting satellites

water mass. A portion of the marine environment that has a characteristic average value of temperature and salinity that is related to its origin and global circulation pattern

1. INTRODUCTION

The global ocean covers 71% of the surface of the Earth to a mean depth of approximately 4 km. In contrast to its terrestrial counterpart, where biomes are associated with characteristic landscapes, differences in seascapes can be subtle, even when they support unique biological assemblages. Early sailors and naturalists characterized changes in oceanic habitats through differences in water temperature and clarity and the type and abundance of the fisheries they supported. However, hidden within this enormous living space is a complex mosaic of seascapes, some with well-defined horizontal and vertical limits and others with more cryptic and flexible boundaries (plate 11).

Below the well-illuminated upper layer, known as the euphotic zone, the ocean is well stratified with identifiable stable layers referred to as water masses that can be traced to specific geographic areas of formation and when they were last in contact with the atmosphere. As these water masses move through the interconnected ocean basins, their chemical and biological characteristics change as a result of coupled, integrated effects of particulate organic matter delivery and metabolism. The oldest water masses on the planet (~ 1500 years old) are in the deep North Pacific Ocean, far removed from their source in Antarctica. However, this water mass is young compared to the very old age of the North Pacific habitat itself: more than 10 million years.

The marine environment supports the growth of a diverse microbial assemblage from all three domains of life: *Bacteria*, *Archaea*, and *Eucarya*. Microbes (especially bacteria) dominate the ocean's genome, and their metabolic activities are responsible for planetary habitability and stability. However, ecologists have not traditionally used microorganisms in the development of ecological models. And although it is likely that existing theory based on the study of macroorganism species and populations is applicable to microbes, there is reason to believe that additional ecological theory may be required to explain microbial genetic and metabolic traits and their relationships to biodiversity, speciation, and evolution.

The relatively new discipline of seascape microbial ecology combines principles, theory, and models of microbiology, ecology, biogeography, genetics, and oceanography to investigate and interpret patterns in the distribution, diversity, and biogeochemistry of microbial assemblages in the sea. A revolution is under way in seascape microbial ecology, ignited in part by the application of novel molecular-based techniques. These approaches have led to the discovery of new organisms, genes, and metabolic processes that define novel marine ecosystem functions. Furthermore, major technological advances in the capability for unattended, remote ocean observation are rapidly changing our view of the structure and the four-dimensional (space and time) variability of marine ecosystems. To illustrate selected advances in seascape microbial ecology, this chapter focuses primarily on microorganisms that inhabit the sunlit portion (0–200 m) of the open sea where most of the organic matter production occurs.

2. SEASCAPE STRUCTURE, VARIABILITY, AND FUNCTION

Ecological processes in the oceanic realm have been studied for more than a century. However, the open sea is still grossly undersampled. Some of the most basic biological properties of pelagic ecosystems, e.g., rates and controls of oceanic photosynthesis, are still not well understood. Furthermore, many seascapes are currently changing as a consequence of human activities. For this reason, it is vital to establish microbial observatories in selected marine ecosystems to track, characterize, and understand changes in the health of the global ocean.

Oceanographers recognize that there are predictable abiotic and biotic properties that vary systematically with distance from shore and from the equator to the poles (see plate 11). A variety of physical forces, with spatial variability ranging from millimeters to basin scales and temporal variability of minutes to millennia (figure 1) shape seascapes, establishing unique characteristics that promote the exchange and transfer of energy and matter, including genetic information, within the global ocean. Each seascape can be defined by the physical, chemical, and biological variables experienced by an organism during its lifetime; collectively these parameters determine the success or failure of a particular strain, species, or assemblage of microbes. Hence, the size, motility, and lifespan of the organism under consideration define the spatial and temporal scale of that organism's habitat. For this reason, a given seascape is likely to be comprised of numerous microhabitats that collectively support the growth and proliferation of the microbial assemblage as a whole. And here resides one of the potential limitations when we are trying to define marine seascapes: in comparison to landscapes, our senses are unable to perceive changes at the microhabitat level.

Particulate organic matter, ranging in size from small colloids to large aggregates, constitutes the most abundant class of microhabitats in the sea; particles are ecotones or transitional boundaries within a fluid matrix. Furthermore, they are often ephemeral sites of elevated microbial biomass and accelerated metabolism, with a mean life of only a few weeks in the open sea. Life on organic-enriched particles selects for microbes with unique survival adaptations such as motility and attachment mechanisms, specialized chemoreceptors, the ability to produce extracellular enzymes

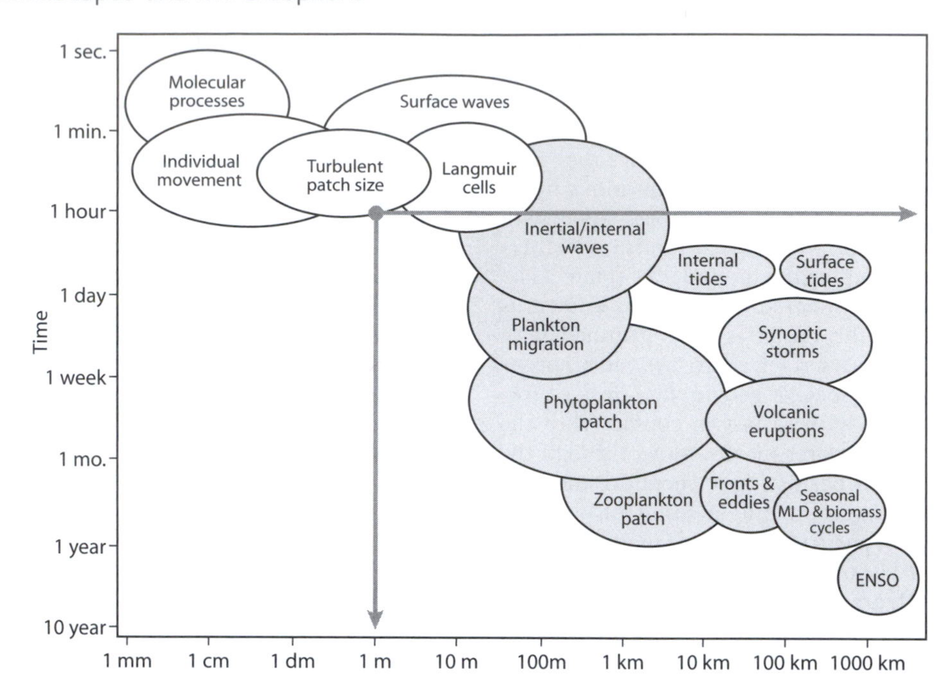

Figure 1. A schematic representation of the relevant time and space scales for key physical and biological seascape processes. The arrows define the approximate boundaries of at-sea observations using current technologies. (From Dickey, 1991)

(ectozymes), and, possibly, a starvation response for survival after the organic substrates are depleted. Once colonized by microorganisms, the environmental conditions within a given particle (e.g., pH, redox level, nutrients, and dissolved gas contents) change as a result of the metabolic activities of the associated microbial assemblage. During this organic particle aging process, structured multispecies consortia of closely interacting and cooperating microbes can lead to complex biological interactions including cell-to-cell chemical communication using one or more different "languages."

At any given moment in every seascape, there is a broad spectrum of particle types and ages with both overlapping and unique microbial populations, interacting with the background, low-nutrient populations contained within the seawater matrix. These nutrient-rich patches can provide refuges for extinction-prone, high-nutrient–requiring microbes and may help to explain the high diversity of low-number-abundance microbes that contribute significantly to the resilience of microbial assemblages in the marine environment. Spatial and temporal heterogeneity fragment large-scale biomes into a mosaic of patches and, over time, semi-isolated metapopulations and metacommunities. Therefore, particles can be viewed as ephemeral microhabitats with source and sink exchange and dynamics with surrounding particle-bound and free-living microbial assemblages.

On a much larger spatial scale, Earth-orbiting satellites equipped with sensors to measure a variety of sea surface parameters have been used to detect and map the dynamics of major biomes globally and in near–real time (plates 11 and 12). For example, the trade wind biomes, which extend from approximately 30°N to 30°S in each ocean basin, dominate our water planet. Key features of these seascapes are a stable vertical temperature/density structure and a downwelling gyrelike circulation pattern that tends to isolate the upper water column from mass exchange with bordering current systems. This stratification and insulation, broken only by an upwelling region close to the equator, lead to a permanent separation of light (above) from nutrients (below) and result in a condition of extreme oligotrophy, including low nutrient concentrations and fluxes, low standing stocks and rates of organic matter production, and, consequently, low rates of organic matter export to the deep sea.

Despite chronic nutrient limitation, oceanic subtropical gyres can support blooms of phytoplankton generally during summer months when the water column is well stratified and most depleted of essential inorganic nutrients such as nitrate and phosphate. These aperiodic and enigmatic phytoplankton blooms sequester carbon dioxide (CO_2) and recharge the upper water column with dissolved organic matter and oxygen, which support postbloom heterotrophic metabolism. More importantly, blooms contribute to the seascape mosaic that is essential for maintaining genetic diversity in these expansive habitats.

The North Pacific Subtropical Gyre (NPSG) is the largest circulation feature on Earth, but it is not yet fully characterized. Solar energy, both light intensity and spectral quality, sets the upper constraint on ecosystem productivity by determining the energy available to phototrophs. However, the most abundant oxygenic phototroph in the NPSG, the cyanobacterium *Prochlorococcus,* was discovered only two decades ago. Furthermore, recent investigations using molecular-based techniques have revealed a novel proteorhodopsin-based phototrophy as an independent pathway for solar energy capture in "nonphotosynthetic" microorganisms. It now appears that these microbes exist partly on sunlight and partly on dissolved organic matter; they are truly nature's energy hybrids.

Although ecological processes in terrestrial biomes are controlled primarily by temperature and the flow of water, marine ecological processes are controlled primarily by temperature and turbulence; the latter impacts nutrient delivery and constrains rates of primary production in marine habitats where nutrients are limiting (figure 2). Near-surface mixing also determines the mean light field of planktonic microorganisms by defining their position in the water column. Consequently, there are complex relationships among turbulence, nutrients, light, and photosynthesis that tend to select for, or against, certain traits at the individual and community levels.

In addition to a physically favorable environment, the metabolism and proliferation of microorganisms also require a renewable supply of energy, electrons for energy generation, carbon and other bioelements, and, occasionally, organic growth factors such as vitamins. Depending on how these requirements are met, all living organisms can be classified into one of several metabolic categories (table 1). Only obligate photolithoautotrophs are self-sufficient, even if they must tie their growth and survival to other microbes that are vital in sustaining nutrient availability over longer time scales. All other microbes ultimately rely on photosynthesis for a supply of energy, dissolved oxygen, or both.

Among the sea microbes there are a variety of metabolic strategies for nutrient capture, transport, and assimilation, all under genetic control. Some microbes are specialists, able to grow on only a single form of a required nutrient; others have less stringent growth requirements. For those microbes that compete for the same substrate, some specialize in the ability to capture substrate at very low ambient concentrations, whereas others have high-capacity uptake and intracellular storage capabilities, being adapted to a feast-and-famine type of existence. Maximum potential growth rates are also variable and probably under genetic control. In low-nutrient-supply habitats such as the NPSG, rapid growth and reproduction may not be the best survival strategy; in this environment the defense against protozoan grazing and viral lysis may also be of great selective advantage.

3. ASSESSMENTS OF MICROBIAL "SPECIES" DIVERSITY AND FUNCTION

Two major challenges in seascape microbial ecology are to identify the proper time and space scales to assess diversity and to define the exact nature of the diversity that should be studied. Beyond taxonomic or phylogenetic diversity, one needs to consider diversity of metabolic or physiological potential within a given species, population, or assemblage as well as niche diversity and the temporal dynamics of the seascape. The former is essential for establishing the possible flux pathways for energy and matter, and the latter is crucial for understanding interactions including competition, resource partitioning, natural selection, and speciation in marine ecosystems.

Microbes assemble in nonrandom fashion similar to patterns that are observed for macroorganisms (plate 12). If multiple environmental parameters and processes are to be compared, sample size and sampling frequency need to be matched. For example, if flow cytometric characterization of bacteria in a deep sea habitat requires a sample size of $0.1\,cm^3$, whereas microbial DNA to prepare a clone library needs $0.1\ m^3$ of seawater, diversity measured by these two procedures may be mismatched by sample size. If more than one microhabitat is combined, then the diversity may be overestimated relative to the scale that is relevant for microbial interactions to occur.

Sea microbes are ubiquitous and abundant; at typical concentrations of more than 100 million cells per liter, marine bacteria are by far the largest contributors to living organic matter in the sea. Their genetic diversity is believed to be large but at present is poorly known. Microorganisms have a very long evolutionary history (3–4 billion years) and appear to have low extinction rates, which may help to explain their enormous extant diversity. Furthermore, habitats like the open sea that are chronically nutrient stressed or energy limited can trigger a starvation–survival response in microorganisms, promoting mutations and, over

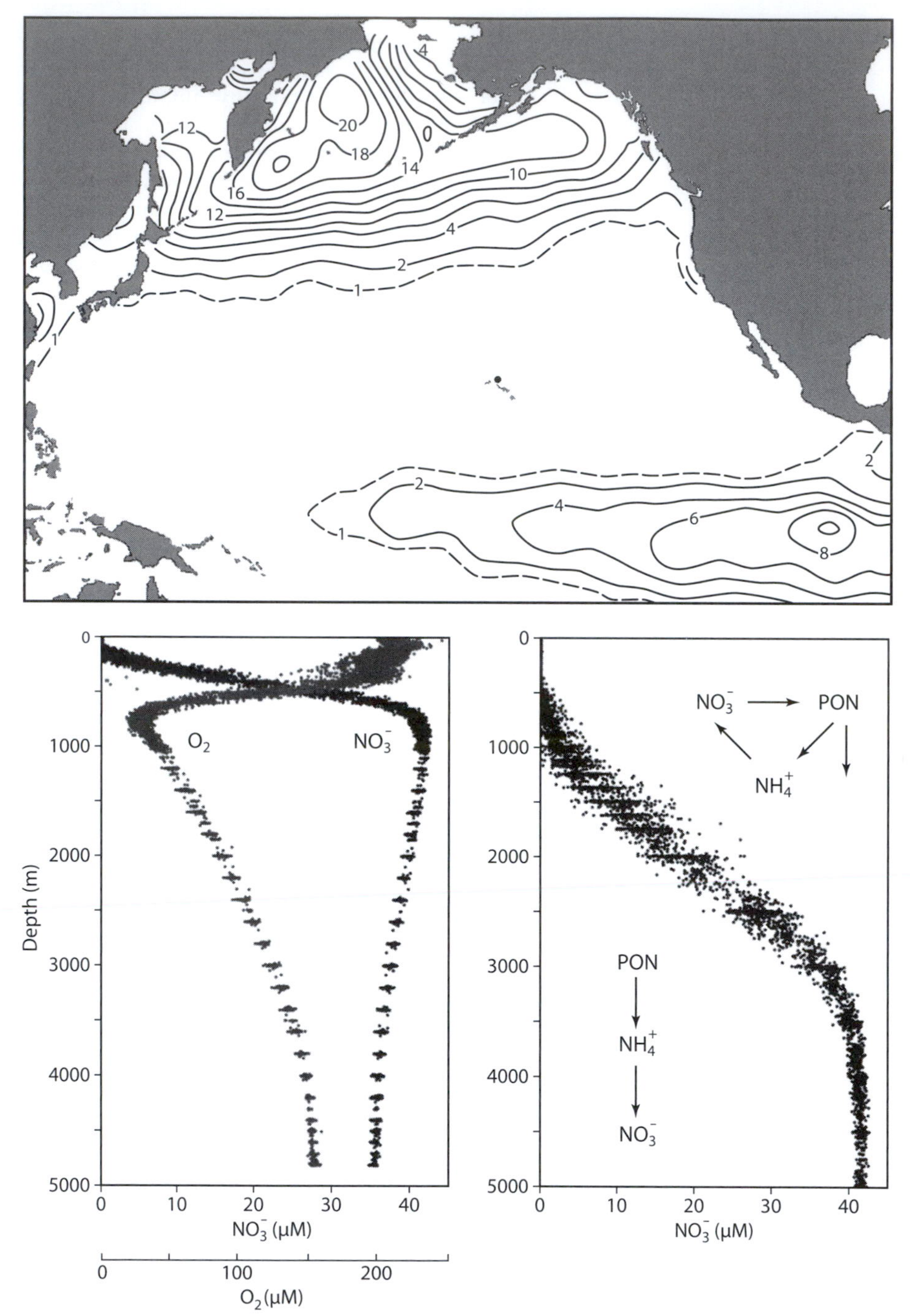

Figure 2. Map of the North Pacific Ocean basin showing the spatial variability in nitrate. Shown on top is the mean annual surface nitrate concentration (mmol NO_3^- m^{-3}) based on the World Ocean Atlas (2001) Ocean Climate Laboratory/NODC. Areas of high NO_3^- (and presumably high NO_3^- flux) correspond to high chlorophyll (>1mg m^{-3}) as a result of net plant growth. The NPSG is the central region characterized by low ambient NO_3^- concentrations (<1mmol m^{-3}) and low standing stocks of chlorophyll (<0.1mg m^{-3}). The circle in the central portion of the basin is the approximate location of Station ALOHA (22°45′N, 158°W). The data in the bottom panels, from Station ALOHA, show (left) the vertical distributions of nitrate and dissolved oxygen for the full water column (0–4800 m) and (right) for the upper 0–1000 m to emphasize the absence of nitrate in the upper euphotic zone (0–100 m) and the steep gradient in nitrate concentration versus water depth beneath the euphotic zone. Also shown are schematics of the key ecological processes of nitrate (NO_3^-) and ammonium (NH_4^+): nutrient uptake into particulate organic nitrogen (PON) and gravitational flux of PON from the upper layers coupled to remineralization back to nitrate in the deep sea.

Table 1. Variations in microbial metabolism based on sources of energy, electrons, and carbon

Source of energy[a]	*Source of electrons*	*Source of carbon*
Sunlight	Inorganic	CO_2
photo-	*-litho-*	*-autotroph*
	Organic	Organic
	-organo-	*-heterotroph*
Chemical	Inorganic	CO_2
chemo-	*-litho-*	*-autotroph*
	Organic	Organic
	-organo-	*-heterotroph*
Radioactive decay	Inorganic	CO_2
radio-	*-litho-*	*-autotroph*
	Organic	Organic
	-organo-	*-heterotroph*

Note: A "mixotroph" is an organism that uses more than one source of energy, electrons, or carbon.
[a]According to Karl (2007).

time, enhancing biodiversity. Some scientists estimate that there may be 1–10 million different "species" of microbes on Earth; others suggest the number of species may be 1 billion.

In a recent study of bacterial diversity in the North Atlantic Ocean as part of the International Census of Marine Microbes (ICoMM), tens of thousands of low-number-abundance microbes per liter were found in association with a relatively small number of very abundant microbes. This type of rank-order abundance curve, with a very long tail, is also characteristic of most macroorganism distributions. It reveals a "rare biosphere" of nearly inexhaustible genetic variability, a gene bank that can be used as necessary to maintain ecosystem stability and function if the seascape is perturbed or permanently altered.

The Linnaean paradigm of hierarchical organization of all living organisms, now 300 years old, is a basic building block of biology. However, mechanisms such as mate preference, spawning synchrony, gamete recognition, and reproductive isolation that are important for speciation in sexually reproducing populations are not applicable to most microorganisms (especially bacteria) because they reproduce by binary fission and pass genetic material vertically from parent to two identical offspring. Additionally, genes can also be passed horizontally between otherwise unrelated species, providing a vehicle for adaptation and evolution that is much more common for microorganisms than invertebrate and vertebrate taxa. Horizontal gene transfer has profound biological, ecological, and biogeochemical consequences and may promote microbial diversity in nature. Consequently, an ever-changing weblike topology rather than the traditional tree of life may be a more accurate framework to represent the evolutionary history of microbes. Indeed, we may need to consider a continuum of ecological functions within marine microbial assemblages. If so, any attempt to group or classify sea microbes may be a static representation of a dynamic system.

There is currently no widely accepted criterion for the designation of a microbial species. This has tended to isolate environmental microbiology from mainstream ecology, where species are the fundamental units of theory and models. Taxonomic assignments of microbes are usually made on genetic relatedness based on similarity of DNA or one or more marker genes (e.g., small subunit ribosomal RNA, so-called 16S rRNA). Phylogenetic surveys using 16S rRNA sequence analyses have revealed two important facts about seascapes. First, most marine diversity is microbial. Second, most of the 16S rRNA sequences recovered from natural habitats are distinct from those of the model marine microbes held captive in our laboratories. In selected seascapes, the "species" list retrieved by culture versus culture-independent approaches are distinct, suggesting that one, or both, surveys may be in error. The 16S rRNA method requires DNA amplification, cloning, and sequencing but not cell growth, whereas the pure culture isolation method requires cell growth and division. It is conceivable, even probable, that many sea microbes are not actively growing (or have growth requirements that are difficult to reproduce in the laboratory) even though their 16S rRNA genes can be isolated and identified. Furthermore, microbial diversity analysis using the 16S rRNA criterion will likely underestimate the true physiological diversity because most 16S "ribotypes" contain genetically distinct ecotypes with similar, but not identical, niches (see section 5). It is equally plausible that rare microorganisms not well represented in the gene surveys because of their low abundances in nature might be crucial for ecosystem function. Other species, although rare most of the time, might be responsible for microbial blooms following the addition of nutrients or other environmental perturbations. This disparity between phylogeny and physiology, and between contemporaneous and future potential metabolism, remains a major analytical and conceptual challenge for the discipline as a whole.

4. THE OCEAN GENOME

The genomics revolution has redirected the marine microbial research prospectus, perhaps at the expense of ecological investigation and field experimentation. However, application of these cutting-edge technologies, including whole microbial genome sequencing and marine metagenomics, has enabled major conceptual advances, helped to recruit new intellectual and funding partners, and invigorated the discipline of microbial oceanography as a whole.

A *metagenome* is the term used to describe the total inventory of genes contained within a given sample. The theoretical summation of all marine metagenomes is equal to the panoceanic genome. In relatively simple environments with limited microbial diversity, the most dominant microbial genomes can be assembled from the short fragments of environmental DNA that are isolated, cloned, and sequenced. If successful, this genome assembly provides a direct estimate of taxonomic diversity of the organisms present in that sample, regardless of whether the microbes can be cultured or not. It also provides the entire genomic parts list for the assemblage as a whole, helping to define and constrain ecosystem function. However, the marine metagenomes constructed to date have proven to be much too complex for whole genome assembly. Furthermore, novel genes and proteins (i.e., no homologs in the extant database) and even new protein families have been recovered from open-ocean samples, confirming the presence of an extremely diverse and poorly characterized microbial assemblage. Because it is impossible to know which genes are associated with which microbes from a metagenome, let alone which genes are being actively expressed and what functions they might code for, the oceanic genome provides only limited ecological information at the present time. Although metagenomics is an important first step, neither the activities of sea microbes nor their ecological function can be predicted from the parts list alone; assembly and operation manuals are also required.

5. ECOTYPE VARIABILITY AND RESOURCE COMPETITION

Competition theory predicts that, at equilibrium, the number of species cannot exceed the number of limiting resources; G. E. Hutchinson termed the observed coexistence of many species of marine phototrophic algae with similar, if not identical, growth requirements the "paradox of the plankton." Temporal and spatial heterogeneity of the habitat (contemporaneous disequilibrium), niche diversification, selective predation or viral infection (kill the winner), and allopatric (microhabitat) speciation, among other factors, have been proposed to explain this paradox.

It is now known that many closely related, even "identical" (at 16S rRNA level), microbes have fundamentally distinct genomes, physiological potentials, and ecological niches. Genetic differentiation leads to the diversification of these related strains (clones) into clusters that are referred to as ecotypes. F. Cohan's formal definition of an ecotype is: "a set of strains using the same ecological resources such that an adaptive mutant from within the ecotype leads to the extinction of all other strains of the same ecotype but does not impact the success of strains from other ecotypes." In other words, resource competition is more acute within a given ecotype than between related ecotypes, allowing these related groups to coexist within a given habitat.

Ecotypes retain many characteristics of the parent strain but are diversified genetically and ecologically because of acquisition of new genes. Much of the genetic variation that is observed between ecotypes is present as genomic islands that are inherited by virus-mediated horizontal gene transfer. These coherent blocks of functional genes confer selective advantage on the recipient strain so they are retained in the population and readily exchanged. Consequently, ecotypes should be viewed as plastic, even ephemeral, and able to respond rapidly to changing environmental conditions.

Some of the most extensive research on marine microbial ecotypes has been conducted using *Prochlorococcus* as a model system. Variation in pigmentation and photophysiology allows selected ecotypes to grow optimally under either high light intensity ($\geq$100 μmol quanta m^{-2} sec^{-1}) or low light intensity ($\leq$ 20 μmol quanta m^{-2} sec^{-1}). Additionally, these specific ecotypes have nutrient uptake mechanisms that match their habitat. For example, the high-light–adapted ecotype cannot use nitrate or nitrite as a N source and grows only on reduced forms of N including ammonium and selected dissolved organic N compounds (e.g., urea). However, because the main supply of new N from below the euphotic zone is almost exclusively as nitrate, the high-light ecotype must rely on other organisms for its supply of chemically reduced N and, possibly, other growth substrates. Field studies have shown that *Prochlorococcus* ecotypes are stratified in both horizontal and vertical space along environmental gradients in light, temperature, and nutrients. This resource partitioning enables *Prochlorococcus* to exist, indeed dominate, photosynthetic biomass over the entire euphotic zone (0–200 m) in many temperate and tropical seascapes.

Of the 12 *Prochlorococcus* genomes currently available, the number of protein-coding genes ranges from approximately 1900 to 3000 in high-light– and low-light–adapted ecotypes, respectively. A "core" genome, shared by all ecotypes, of 1250 genes has been identified, with an additional approximately 6000 unique genes (many with unknown function) collectively present in the *Prochlorococcus* "pangenome." The shapes of the accumulation curves for both the core and pangenomes suggest that the latter, but not the former, will continue to increase as the genomes of additional isolates are sequenced. The remarkable genotypic diversity, only partly charted at the present time, is undoubtedly responsible for its successful invasion into warm water marine ecosystems worldwide.

6. THE STREAMLINED GENOME OF SAR 11

The size of a microbial genome, usually expressed as total number of nucleotide base pairs (bp), is highly correlated with cell size and number of protein-coding genes. Obligate symbiotic or parasitic microorganisms have minimal genomes that reflect a reduction in the size of the genome relative to their free-living relatives; this process has been termed *genome streamlining*. Genome reduction can also occur in free-living marine microorganisms, but this evolutionary strategy usually results in an organism that is a metabolic ward of the seascape, dependent on other microbes for continued survival. Selective pressures that favor genome streamlining include growth in complete nutritional medium that contains biosynthetic organic precursors (amino acids, nucleic acid bases) and vitamins, or life in an energy-limited habitat where resource competition is keen. The most abundant microorganism in the sea, *Pelagibacter ubique* strain HTCC 1062 (SAR 11 clade), is an evolutionary product of genome streamlining. The genome of *P. ubique* is one of the smallest of any known free-living microbes, 1,308,759 bp and 1354 protein-coding genes. The genome contains no pseudogenes, introns, transposons, or extrachromosomal elements and has the shortest intergenic spacers yet reported. *P. ubique* cannot synthesize vitamins (B_6, B_{12}, thiamine, biotin), so it is dependent on the metabolic activities and biosynthetic processes of other sea microbes with more complete genomes.

The strategy of genome streamlining involves survival risk, but it appears to have been very successful for microbes in the open sea. It is possible, even likely, that the coexistence of similar but genotypically distinct microbes may be a consequence of a complex series of codependencies in such a way that no one isolate would be able to survive by itself. In this regard, community-level interactions may be much more important than population interactions.

7. STATION ALOHA: A MICROBIAL OBSERVATORY IN THE OPEN SEA

Long-term ecological studies are predicated on the assertion that certain processes, such as climate-driven changes in microbial community structure and productivity, are time dependent and must be studied as such. In October 1988, Station ALOHA (A Long-term Oligotrophic Habitat Assessment) was established in the North Pacific Ocean, approximately 100 km north of Oahu, Hawaii, for long-term observations of coupled physical, chemical, and microbiological processes as the deep water benchmark of the Hawaii Ocean Time-series (HOT) program.

Open-ocean tropical seascapes such as Station ALOHA are the aquatic analogs of terrestrial deserts because the standing stocks of living organisms and rates of photosynthesis are very low. Rather than being limited by water availability, oceanic deserts are nutrient starved, in particular by a chronic shortage of bioavailable nitrogen (N) in the surface layers where photosynthesis takes place (figure 2). This situation is a consequence of the density-induced vertical stratification between warm surface waters and the cold abyss that leads to a spatial separation of solar energy from the nutrients required to support net photosynthesis. Typically, 90–95% of the nutrients that are consumed daily in the euphotic zone at Station ALOHA are derived from local remineralization; the remaining 5–10% of the quota, termed "new" nutrients, is delivered from external sources. In the open sea, new nutrients (e.g., nitrate) are supplied mostly via the relatively slow process of vertical eddy diffusion from the deep water nutrient reservoir (figure 3); these supply rates ultimately control microbial biomass and productivity and, hence, the ecosystem's function.

Despite the chronic limitation of nitrate in the surface waters of most open ocean habitats, dissolved gaseous N_2 is present in unlimited supply (>500 μM). However, the relative stability of the triple bond of N_2 renders this form inert to all but a few specialized N_2-fixing microbes, dubbed diazotrophs. N_2 fixation in most open-ocean ecosystems is solar powered, and most diazotrophs are cyanobacteria. Diazotrophs require an ample supply of iron (Fe), which is an obligate cofactor for the enzyme nitrogenase. The Fe supply to the surface waters of most open-ocean habitats is via atmospheric dust delivery, and thus Fe flux varies considerably with geographic location and distance from dust sources (e.g., deserts). Furthermore, in order for N_2-dependent net growth to occur, diazotrophs require a suite of macro- and micronutrients, especially phosphate. If light, Fe, and phosphate are present in excess, phototrophic diazotrophs would have a competitive advantage in N-limited seascapes, and N_2 fixation may be a significant pathway for the introduction of new N into the ecosystem.

When the HOT program began, N_2 fixation was not considered to be a significant process in the NPSG.

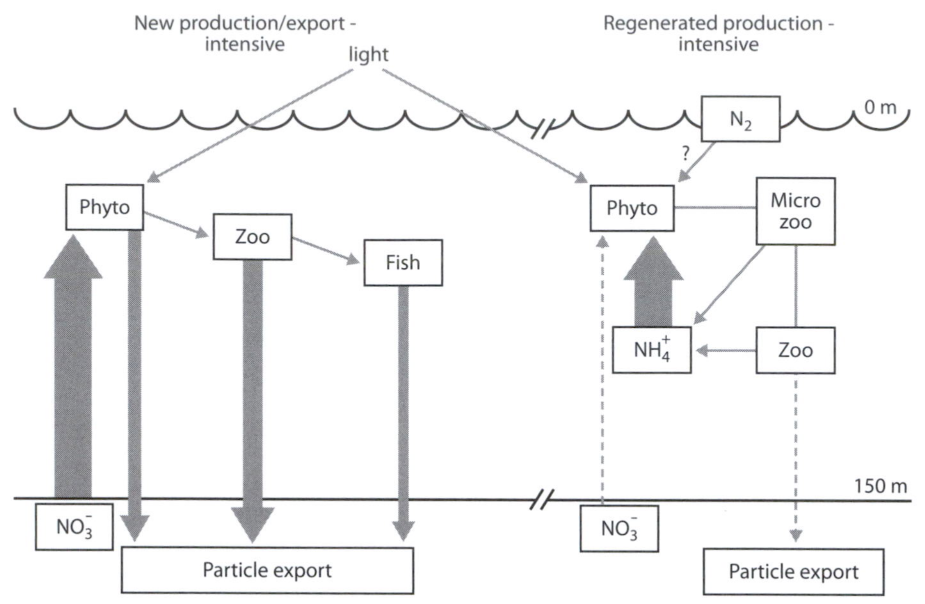

Figure 3. Conceptual view of the new versus regenerated N model for primary production in the sea. Shown are two contrasting marine ecosystems: (left) an upwelling habitat where allochthonous NO_3^--supported new production dominates total primary productivity, and (right) an open-ocean habitat where locally produced NH_4^+-supported regenerated production dominates total primary productivity. N_2 fixation represents a potential enhancement of new production in N-limited seascapes. New production-intensive biomes also support much greater export, usually in the form of sinking particulate matter, than remineralization-intensive systems such as the NPSG.

Indeed, the climax community paradigm for this seascape circa 1970 was one of a time-independent, nitrate-controlled, eukaryote-dominated, low-productivity biome. Whereas nitrate resupply from deep waters was considered the ultimate source of new N in the historical view, we now recognize N_2 fixation as an approximately equal new N flux pathway (table 2).

N_2 fixation can be viewed as a keystone ecological process in the N-stressed NPSG, supplying new nitrogen via a delivery pathway that is independent of turbulence. The impact of diazotrophs is disproportional to their relatively low abundance. With the possible exception of stochastic bloom events, diazotroph biomass rarely exceeds a few percent, at most, of phytoplankton carbon in these habitats. However, the removal of all N_2 fixers from the NPSG would likely lead to a significant decrease in phytoplankton biomass, net primary production, fish production, CO_2 sequestration, and a corresponding reduction of the export of carbon and energy to the mesopelagic and deep sea, with attendant ecological consequences—the hallmark of an ecological keystone.

We currently recognize at least three fundamentally different groups of diazotrophs at Station ALOHA (see table 3): (1) small, free-living unicellular cyanobacteria (*Crocosphaera*-like), (2) large filamentous and colonial morphologies of the cyanobacterium *Trichodesmium*, and (3) *Richelia*-like cyanobacteria living as ecto- and endosymbionts with several species of large aggregate-forming diatoms (e.g., *Rhizosolenia*, *Hemiaulus*). The N_2 fixed by each of these groups has a different impact on the ecology and biogeochemistry of the NPSG, despite the fact that all belong to the same diazotroph guild (table 3). In addition to alleviating N stress, N_2 fixation-based organic matter production enhances the sequestration of CO_2 because the import of N_2 is decoupled from the delivery of deep water nitrate that also contains a high concentration of CO_2. Gravitational settling of N_2-based particulate organic matter pumps excess carbon into the deep sea. This diazotroph-based sequestration is further enhanced because the C:P molar ratios of most diazotrophs growing under P control are higher than that of the total C:P ratio of upwelled nutrients.

It has been hypothesized that the environmental conditions necessary to promote the selection for N_2-fixing microorganisms (e.g., water column stratification, nutrient resupply rates, and N:P ratios) have systematically changed since the later 1970s, resulting in an epoch of N_2 fixation that continues today. If the biomass of N_2-fixing microbes and the rates of N_2 fixation in the NPSG are increasing over time because of climate-driven changes in the environment, then the biome is being forced into severe P limitation.

Table 2. Biological and biogeochemical indicators of N_2 fixation at Station ALOHA

Observation	*Method*
N_2-fixing microbes	Direct microscopy, *nif* H gene abundances by quantitative polymerase chain reaction (QPCR)
N_2 fixation rates	In situ measurements using acetylene reduction and ^{15}N-N_2 isotopic methods, *nif* H gene expression by reverse-transcribed QPCR
Long-term changes in the inventories of soluble and particulate phosphorus and in C:P/N:P ratios	Time-series collections and direct chemical measurements
Changes in the rate of particulate-P export and in the ^{15}N isotopic abundance and elemental ratios	Field collections using sediment traps and direct chemical and N isotopic measurements
Summertime drawdown of dissolved inorganic carbon in absence of nitrate	Direct measurements from repeat sampling of surface waters

Table 3. Diversity of form and ecological function of three major groups of diazotrophic microbes in the North Pacific Subtropical Gyre

Nanoplankton	Trichodesmium	*Diatoms*/Richelia
Small (<10 μm), high growth rate	Large (>20 μm), low growth rate	Large (>20 μm), high growth rate
"Background" population	Bloom forming	Bloom forming
Dispersed	Floaters/migrators	Sinkers/migrators
Consumed by protozoans	Not readily consumed	Consumed by zooplankton
High turnover/low export	Low turnover/low export	Variable turnover/high export

Ecological consequences might include changes in the standing stocks and turnover rates of dissolved and particulate P and altered C-N-P composition of new biomass production, which could in turn select for microorganisms that do not require as much P for growth, or for slower-growing microbes with lower P requirements because of reduced ribosomal RNA content.

Although the phosphate inventory at Station ALOHA has decreased by more than 50% over the past two decades (figure 4), most likely as a consequence of N_2 fixation, there still seems to be a surplus. Further reduction of phosphate to subnanomolar concentrations can be expected along with a selection for alternative P capture and a further shift in the cell size and activity spectra to smaller, slower-growing microorganisms. This prediction has numerous potential impacts on the trophic structure, selecting for smaller predators and thereby altering top-down grazing control of microbial populations and nutrient cycling rates. Without an adequate resupply of phosphate and other nutrients, these P-stressed ecosystems could lose biomass, biodiversity, and possibly their ability to respond to habitat fluctuations and climate change.

A key negative feedback to enhanced N_2 fixation—decoupling of N and P cycles and the export of high-N:P-ratio organic matter—is the eventual buildup of a subeuphotic zone nutrient reservoir that has an elevated nitrate:phosphate ratio relative to cellular needs (e.g., $N:P > 16:1$). As these regenerated nutrients slowly feed back into the euphotic zone, they will select against N_2 fixers because the excess phosphate is assimilated by competing nitrate-utilizing microorganisms. This would lead to another shift in community structure, ecological stoichiometry, grazing control, and organic matter export. An alternation of ecosystem states between N limitation and P (or P/Fe) limitation in the NPSG is expected to occur on an approximately 20- to 50-year cycle based on the estimated residence time of nutrients in the upper mesopelagic zone reservoir. However, the extent to which greenhouse gas–induced warming and other changes to the surface ocean will impact the dynamics of these hypothesized alternate ecosystem states is currently unknown. In global climate model simulations, higher dust deposition to the open ocean enhances global primary productivity, N_2 fixation, and CO_2 sequestration on time scales of a decade or less. Furthermore, it is almost certain that the global dimensions of subtropical gyres

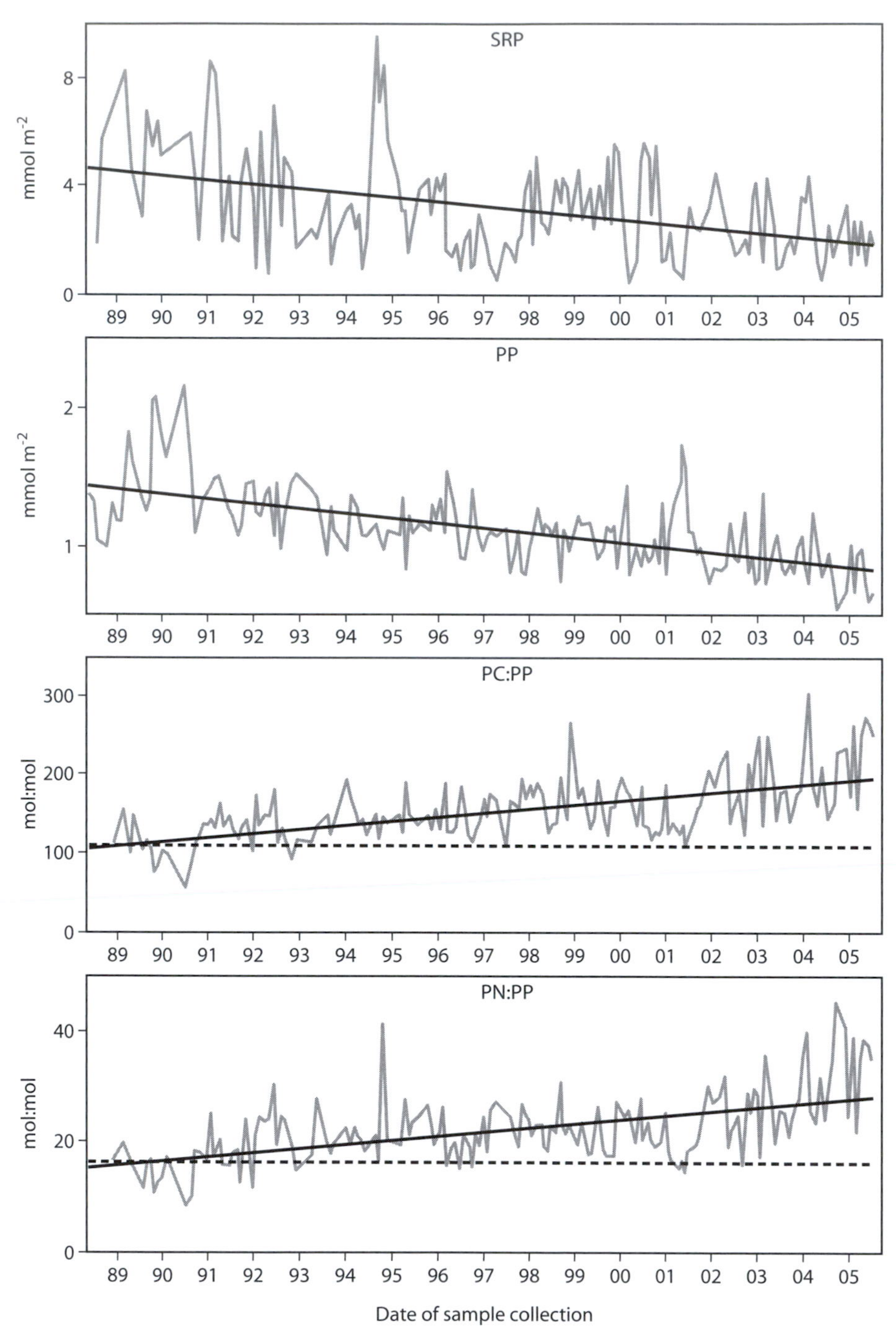

Figure 4. Station ALOHA phosphorus pool dynamics. Shown are the 15-year time-series records of soluble reactive phosphorus (e.g., phosphate, SRP) integrated over the upper 0–60m of the water column, particulate phosphorus (PP) integrated over the upper 0–75m of the water column, and the average (0–75m) carbon-to-phosphorus and nitrogen-to-phosphorus ratios of the particulate matter pools (PC:PP and PN:PP, respectively). The solid lines represent the best-fit linear regression analyses for each data set, and the dashed lines in the PC:PP and PN:PP plots represent the global average stoichiometry (106C:16N:1P) for marine particulate matter. The decreasing concentrations of SRP and PP and increasing PC:PP and PN:PP ratios are all predicted from the hypothesized enhancement of N_2 fixation at Station ALOHA over the past two decades.

will expand as the Earth warms, and the open ocean as a whole will become more stratified and nutrient depleted, setting the stage for the selection of N_2-fixing microorganisms with attendant ecological consequences.

8. CONCLUSION

Seascape ecologists are at a distinct disadvantage relative to their land-based colleagues. Whereas early terrestrial naturalists could readily observe environmental changes in space and time, open-ocean ecologists had to develop tools to observe biological changes in the vast pelagic environment; consequently, the discipline of seascape ecology is young. Furthermore, it is difficult to conduct replicated, long-term manipulation studies in the sea as has been done in lakes and on land. Finally, the ephemeral nature of many seascapes, including, but not limited to, complex, nonlinear physical–biological interactions and unpredicted climate variability, precludes the development of predictive models at present.

The field of landscape ecology has an explicit connection to humans as components that actively alter natural habitats. Likewise, seascape microbial ecology seeks fundamental information on how humans are beginning to alter some of the most remote ocean habitats. The vehicle for seascape change is global climate variability and especially the impacts of gases that affect the transmission of long-wavelength radiations through the atmosphere, such as CO_2, methane, and nitrous oxide. As these greenhouse gases accumulate in the atmosphere, the planet will get warmer. This heating, in turn, will impact the atmospheric circulation and global hydrology, thereby altering ocean circulation patterns, stratification, and dust fluxes to ocean. As the surface ocean absorbs more and more CO_2, it will become more acidic, further altering the physical/chemical characteristics and microbial biodiversity.

Seascape microbial ecology requires a sound theoretical framework for analyzing spatial and temporal patterns in the distribution, abundance, and biodiversity of microorganisms and the ecological services they perform. Continued progress in this area must be considered a priority and will require an integration of laboratory, field, and modeling efforts and new collaborations among scientists who traditionally have not interacted. The next decade should be both challenging and exciting.

FURTHER READING

Azam, F. 1998. Microbial control of oceanic carbon flux: The plot thickens. Science 280: 694–696.

Cohan, F. M. 2006. Towards a conceptual and operational union of bacterial systematics, ecology, and evolution. Philosophical Transactions of the Royal Society of London B 361: 1985–1996.

Coleman, M. L., and S. W. Chisholm. 2007. Code and context: *Prochlorococcus* as a model for cross-scale biology. Trends in Microbiology 15: 398–407.

DeLong, E. F., and D. M. Karl. 2005. Genomic perspectives in microbial oceanography. Nature 437: 336–342.

DeLong, E. F., C. M. Preston, T. Mincer, V. Rich, S. J. Hallam, N.-U. Frigaard, A. Martinez, M. B. Sullivan, R. Edwards, B. R. Brito, S. W. Chisholm, and D. M. Karl. 2006. Community genomics among stratified microbial assemblages in the ocean's interior. Science 311: 496–503.

Dickey, T. D. 1991. The emergence of concurrent high-resolution physical and biooptical measurements in the upper ocean and their applications. Reviews of Geophysics 29: 383–413.

Dykhuizen, D. E. 1998. Santa Rosalia revisited: Why are there so many species of bacteria? Antonie van Leeuwenhoek 73: 25–33.

Gevers, D., F. M. Cohan, J. G. Lawrence, B. G. Spratt, T. Coenye, E. J. Feil, E. Stackebrandt, Y. Van de Peer, P. Vandamme, F. L. Thompson, and J. Swings. 2005. Re-evaluating prokaryotic species. Nature Reviews Microbiology 3: 733–739.

Giovannoni, S. J., H. J. Tripp, S. Givan, M. Podar, K. L. Vergin, D. Baptista, L. Bibbs, J. Eads, T. H. Richardson, M. Noordewier, M. S. Rappé, J. M. Short, J. C. Carrington, and E. J. Mathur. 2005. Genome streamlining in a cosmopolitan oceanic bacterium. Science 309: 1242–1245.

Green, J., and B.J.M. Bohannan. 2006. Spatial scaling of microbial biodiversity. Trends in Ecology and Evolution 21: 501–507.

Johnson, Z. I., E. R. Zinser, A. Coe, N. P. McNulty, E.M.S. Woodward, and S. W. Chisholm. 2006. Niche partitioning among *Prochlorococcus* ecotypes along ocean-scale environmental gradients. Science 311: 1737–1740.

Karl, D. M. 1999. A sea of change: Biogeochemical variability in the North Pacific Subtropical Gyre. Ecosystems 2: 181–214.

Karl, D. M. 2007. Microbial oceanography: Paradigms, processes and promise. Nature Reviews in Microbiology 5: 759–769.

Kinzig, A. P., P. Ryan, M. Etienne, H. Allison, T. Elmqvist, and B. H. Walker. 2006. Resilience and regime shifts: Assessing cascading effects. Ecology and Society 11: Article No. 20.

Longhurst, A. 1998. Ecological Geography of the Sea. San Diego: Academic Press.

Martiny, J. B. H., B. J. M. Bohannan, J. H. Brown, R. K. Colwell, J. A. Fuhrman, J. L. Green, M. C. Horner-Devine, M. Kane, J. A. Krumins, C. R. Kuske, P. J. Morin, S. Naeem, L. Øvreås, A.-L. Reysenbach, V. H. Smith, and J. T. Staley. 2006. Microbial biogeography: Putting microorganisms on the map. Nature Reviews Microbiology 4: 102–112.

Rusch, D. B., A. L. Halpern, G. Sutton, K. B. Heidelberg, S. Williamson, S. Yooseph, D. Wu, J. A. Eisen, J. M.

Hoffman, K. Remington, K. Beeson, B. Tran, H. Smith, H. Baden-Tillson, C. Stewart, J. Thorpe, J. Freeman, C. Andrews-Pfannkoch, J. E. Venter, K. Li, S. Kravitz, J. F. Heidelberg, T. Utterback, Y.-H. Rogers, L. I. Falcon, V. Souza, G. Bonilla-Rosso, L. E. Eguiarte, D. M. Karl, S. Sathyendranath, T. Platt, E. Bermingham, V. Gallardo, G. Tamayo-Castillo, M. R. Ferrari, R. L. Strausberg, K. Nealson, R. Friedman, M. Frazier, and J. C. Venter. 2007. The Sorcerer II global ocean sampling expedition: Northwest Atlantic through Eastern Tropical Pacific. PLoS Biology 5: 0398–0431.

Sogin, M. L., H. G. Morrison, J. A. Huber, D. M. Welch, S. M. Huse, P. R. Neal, J. M. Arrieta, and G. J. Herndl. 2006. Microbial diversity in the deep sea and the underexplored "rare biosphere." Proceedings of the National Academy of Science, U.S.A. 103: 12115–12120.

Turner, M. G. 2005. Landscape ecology: What is the state of the science? Annual Reviews of Ecology, Evolution, and Systematics 36: 319–344.

Yoder, J.A.Y., and M. A. Kennelly. 2006. What have we learned about ocean variability from satellite ocean color imagers? Oceanography 19: 152–171.

IV.10

Spatial Dynamics of Marine Fisheries

Daniel Pauly and Reg Watson

OUTLINE

1. Introduction
2. Geography of fisheries' productivity
3. "Fishing down" as a major feature of contemporary fisheries
4. Mapping fisheries' interactions
5. Conclusions

Key features of the evolution of marine fisheries from their near-coastal antecedents to their present existence as industrialized, high-sea ventures are recalled along with some of the elements that led, in the early 1980s, to the emergence of the United Nations Convention on the Law of the Sea and to exclusive economic zones being granted to maritime countries. The world's marine fisheries' catches peaked in the late 1980s, as newly exploited areas ceased to compensate for the collapsing stocks of traditional fishing grounds. It is demonstrated that the expansion that until then had masked these collapses was southward (from northern temperate and boreal fishing grounds toward subtropical and tropical areas and onto the Southern Hemisphere), into deeper waters, and toward species previously not exploited and generally lower in the food webs, this last process being known as "fishing down marine food webs." We examine these trends in some detail for shelves, where most of the world catches are taken, "transition areas" (fronts, upwellings, seamounts), and oceanic waters, each characterized by a different productivity regime, determined mainly by the mechanism that lifts deep, nutrient-rich waters into the illuminated surface layers. Overcoming these trends will involve a rethinking of the resource allocation that is underlying current exploitation patterns, which presently treat the fishing industry as quasiowner of marine resources that are, in reality, public property. Other allocation issues involve the relationship between small-scale and large-scale fisheries and the wisdom of subsidizing fisheries. To fully understand the scale of these problems, however, maps of fisheries' withdrawals and other indicators of the ecosystem impacts of fisheries are essential, as they, more than any other form of presentation, can communicate complex phenomena at various scales to the public and decision makers.

GLOSSARY

bloom. A population outbreak of microscopic algae (phytoplankton) that remains within a defined part of the water column.

demersal. Organism that lives on or near the bottom of and/or feeding on benthic (bottom) organisms.

EEZ (exclusive economic zone). Area up to 200 nautical miles off the coast of maritime countries as declared under 1982 United Nations Convention of the Law of the Sea. Within their EEZs, coastal states have the right to explore and exploit, and the responsibility to conserve and manage, the living and nonliving resources.

extirpation. The process whereby an animal or plant species is rendered extinct in a particular area or country while it survives in others. When a species consists of several populations, the extirpation of the last population is equivalent to the global extinction of that species.

fishing down the marine food webs. The process wherein the fisheries within a given marine ecosystem, having depleted the large predatory fish on top of the food web, turn to increasingly smaller species, finally ending up with previously spurned small fish and invertebrates. This process is now well established in many parts of the world.

fish meal. Fish and fish-processing offal that is dried, often after cooking and pressing (for fatty fish), and ground to give a dry, easily stored product that is a valuable ingredient of animal feeding stuffs. In Peru, fish meal is made mainly of anchovies; in northern Europe, mainly capelin, sand eel, mackerel, and Norway pout are used for fish meal production. In Japan, the principal species are sauries, mackerels, and sardines, and in the United States menhaden.

gyre. Major cyclonic surface current systems in the oceans, roughly corresponding to the unproductive, highly stratified areas of the oceans that are most remote from the continents.

IUU. Illegal, unreported, and unregulated fishing.

longline. Line of considerable length, bearing numerous baited hooks, that is much used in tuna fisheries. The line is set for varying periods up to several hours at various depths or on the seafloor, depending on the target species. Longlines, which are usually supported by floats, may be 150 km long and have several thousand hooks.

neritic. Inhabiting the shallow pelagic zone over the continental shelf, i.e., waters less than 200 m deep, and deeper waters in areas of coastal submarine slopes.

pelagic. Living and feeding in the open sea, i.e., associated with the surface or middle depths of a body of water but not in association with the bottom.

purse seine. A fishing net used to encircle surface-dwelling fish. The net may be of up to 1 km length and 300 m depth and is used to encircle surface schooling fish such as mackerel or tuna. Purse seines are usually set at speed from a powered vessel, assisted by a smaller boat. During retrieval, the lower part net is closed (or pursed) by drawing a line through a series of rings to prevent the fish escaping.

seamount. An elevation rising 1000 m or more from the sea floor with limited extent across the summit.

shelf. In oceanography, continental shelves refer to the edge of continents, below the surface of the ocean, down to a depth of 200 m (approximately 600 ft). Shelves usually are the most productive parts of the ocean and sustain the bulk of the world's fisheries.

stock. Group of individuals of a species that can be regarded as an entity for management purposes; roughly corresponding to a population.

tonne. Equal to 1000 kg; different from the (short) ton of North America, which is equal to about 907 kg.

trawler. A vessel that operates a trawl net. A wide range of demersal (bottom) or pelagic (midwater) species of fish and other organisms are taken by this fishing method, which entails one or a pair of vessels towing a large bag-shaped net either along the seafloor or in midwater. It has a buoyed head rope and a weighted foot rope to keep the net mouth open. Although this is a relatively old method of fishing, bottom trawling is questioned nowadays because it destroys habitats and catches many nontarget species, which often are subsequently discarded.

trophic level. Position in the food chain, determined by the number of food-transfer steps to that level. Phytoplankton is usually given a trophic level of 1, herbivorous zooplankton 2, small fish about 3, and most large fish about 4, the variation depending on the diet of the various predators.

UNCLOS (United Nations Convention of the Law of the Sea). Concluded in 1992, it, among other things, established that maritime countries could claim 200-mile exclusive economic zones.

upwellings. Oceanographic phenomenon wherein wind induces a transport of water, usually away from a coast, with this water being replaced by water "welling up" from deeper layers. Because the upwelled water is nutrient-rich, upwellings belong to the most productive marine ecosystems.

1. INTRODUCTION

The first marine "fisheries" began in our ancestral African home and probably consisted of gathering invertebrates from their intertidal habitat and perhaps scavenging stranded marine mammals. Later, the systematic exploitation of coastal environments allowed the first migrations out of Africa and the settling of the other continents (Erlandson and Fitzpatrick, 2006).

As humans spread out of Africa, and their populations and technical sophistication grew, more species became available to a wide range of passive and later active gear, deployed farther and farther offshore by sailed vessels, in waters of increasing depth (Sarhage and Lundbeck, 1992). Many inhabited coastlines became intensely exploited in the process, and populations of large species (e.g., turtles, marine mammals) were extirpated (Jackson et al., 2001).

The development of the steam trawler in the late nineteenth century, however, set the stage for the emergence of a radically new combination. It combined the power and endurance of fossil fuel with the ability to catch everything in the path of the gear, with destruction of sea bottom habitats as an unavoidable built-in feature (Watling and Norse, 1998).

This technology package was so powerful that it led to a rapid increase of catches everywhere it was introduced (Cushing, 1998). These massive catches could not be sustained, however, and successive generations of ever-more-sophisticated trawlers responded by extending their areas of operation. Thus, British trawlers moved from the coastal waters of England and Scotland, where it all began, to the open North Sea, then further on to the North Atlantic, notably to Iceland. They were still operating there in the 1950s when, at the onset of the "Cod Wars," the Icelanders began to push them back (Kurlansky, 1997).

Similar expansions by a few of the industrialized countries of Europe and Northeast Asia led to their distant water fleets (DWFs) trawling along the coasts of the United States and Canada, Northwest Africa,

Southeast Asia, as well as in the Southern Hemisphere (Bonfil et al., 1998). In parallel, several countries, led by Japan, developed distant water fisheries for large pelagic fishes, notably tuna species, and also fished along the coast of other countries (Myers and Worm, 2003). As was the case for Iceland, these countries resented the incursions, and they gradually began to assert their own exclusive rights to these resources.

Their concerted effort, much resisted by countries with DWFs, led to the emergence in 1982 of the United Nations Convention on the Law of the Sea (UNCLOS), which, among other things, granted coastal countries a 200-mile exclusive economic zone (EEZ) and regulated conditions of access by DWFs. Many of these countries (e.g., Canada and the United States) could then force the departure of a DWF from their EEZ. However, rather than allowing the stocks that the DWF had exploited to rebuild, these countries subsidized the construction of national fleets, which substituted for the DWFs. Thus, the late 1980s became a decade in which the world's fleet capacity increased tremendously (Gelchu and Pauly, 2007).

The late 1980s was also the period when the world's fisheries' catches peaked, a feature long hidden by overreporting from China (Watson and Pauly, 2001). At their peak, global marine fisheries' landings officially reported to the Food and Agriculture Organization (FAO) were 80–90 million tonnes, and an unknown but presumably large additional catch was landed illegally or remained unreported (figure 1). The subsequent decline occurred because the catches from newly exploited areas failed to compensate for the collapsing stocks. Although the realization that overfished and collapsed stocks contribute an increasing fraction of the world catch is relatively recent (Pauly et al., 2005; Worm et al., 2006), it is based on global analyses published more than 10 years ago by Grainger and Garcia (1996).

These trends have now intensified into a full-blown crisis of fisheries. Indeed, collapses of major stocks, previously perceived as singular events, e.g., that of the California sardine (Radovich, 1982), Peruvian anchovy (Paulik, 1981), or northern cod (Walters and Maguire, 1996), have become frequent and even predicable (Cheung et al., 2007; Worm et al., 2006).

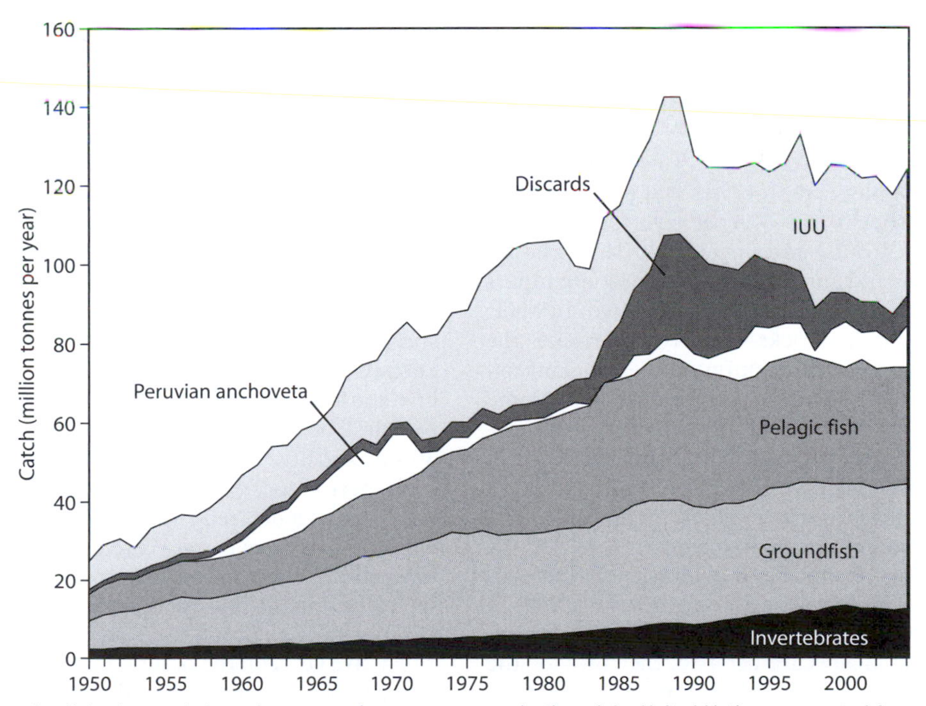

Figure 1. World marine fisheries catch, by major taxonomic groups and fishery. Shaded groups (invertebrates such as shrimp and squid, bottom fish such as cod and flounders, pelagic fish such as sardine and tuna, and the Peruvian anchoveta *Engraulis ringens*) are based on landing statistics from the Food and Agriculture Organization of the United Nations, corrected for overreporting from China (Watson and Pauly, 2001). Estimates of discarded by-catch are from Zeller and Pauly (2005), and the estimates of illegal, unreported, and undocumented (IUU) catch are from Pauly et al. (2002) and Chuenpagdee et al. (2006).

2. GEOGRAPHY OF FISHERIES' PRODUCTIVITY

Before the work of Longhurst (1998), a rigorous definition of biogeochemical "provinces" suitable for describing, in standardized fashion, the distribution of all marine organisms did not exist despite a history of oceanographic research starting with the *Challenger* Expedition (1872 to 1876). Numerous maps did exist in which this or that oceanographic parameter or the distribution of a few organisms had been used to draw provinces of some sort. However, no test had been conducted of the ability of these proposed maps to predict distributions other than those from which they were derived: circularity reigned supreme.

Reasons for this are easy to imagine, from the excessive preoccupation of various specialists with their favorite taxonomic groups to the absence, before the recent widespread availability of computers, of analytic tools that were up to the task. However, the real reason is probably that developing a truly synoptic vision of the ocean was impossible before the advent of satellite-based oceanography.

Satellites cannot see very deep into the sea, nor really can they see much—at least those satellites about which civilians know. However, what satellites do see (i.e., have sensors for) is the very stuff that generates fundamental differences between ocean provinces: sea surface temperatures and their seasonal fluctuations, and pigments (such as chlorophyll) and their fluctuations. Marine systems differ from terrestrial systems in that their productivity is essentially a function of nutrient inputs to illuminated layers. This gives a structuring role to the physical processes that enrich surface waters with nutrients from deeper layers, e.g., wind-induced mixing, fronts, and upwellings. Thus, the location, duration, and amplitude of deep nutrient inputs into different oceanic regions—as reflected in their chlorophyll standing stocks—largely determine the productivity of the phytoplankton, of the zooplankton grazing thereon, and hence the productivity of the fisheries of the regions (Bakun, 1996). This is the reason why satellite images reflect fundamental features of the ocean, whereas maps based on the distribution of various organisms, even "indicator organisms," can only reflect second-order phenomena.

As with biogeochemical "provinces," marine ecosystems can be classified according to the different mechanisms by which their productivity is maintained. When these mechanisms are detailed, they provide the basis for detailed geographies of ocean ecosystems (Watson et al., 2003, 2004, 2005a).

At the grossest level, the enrichment mechanisms yield the simple classification presented below and used here to differentiate fisheries in terms of the major ocean ecosystem types within which the species they exploit are embedded:

1. Neritic, or shelf fisheries.
2. Fisheries of transitional areas (fronts/slopes, upwellings, seamounts).
3. Oceanic fisheries.

Shelf Fisheries

Shelves consist of the often broad areas of shallow waters (by definition not deeper than 200 m) around the world's continents and major islands. They are productive because tidal or wind-driven currents frequently mix the surface waters and because nutrients washed from land often enhance coastal primary production in addition to the nutrients recycled from the deeper waters. Indeed, the shelves of the world, which cover 7.5% of the surface of the oceans (362 million km^2), have in recent years contributed about 58% of the world's marine fisheries' catch, down from above 70% in the 1950s. It is for this reason that a map of the world's fisheries, until now, essentially corresponds to a map of the world shelves (plate 13).

The shelves are strongly exploited by trawlers for their bottom fish (e.g., cod, flatfish) and invertebrates (especially shrimp). Besides contributing most of the discards of fisheries (figure 1), trawlers and similar gear strongly modify bottom habitats, both directly (Watling and Norse, 1998) and indirectly, by producing "mudtrails" often visible from space (van Houtan and Pauly, 2007), which modify the optical and chemical properties of the water column.

Shelves are also exploited for their small pelagic fish, caught with purse seiners and a variety of other gear, some of them passive (e.g., set nets and traps). Small pelagic fishes, in marine ecosystems, usually consist of species of the family Clupeidae (herrings, sardines, anchovies), Scombridae (mackerels), Osmeridae (capelins and other smelts), commonly grouped as "small pelagics" because they tend to reach sizes of only 10 to 30 cm and live in open, "pelagic" waters above the shelf (neritic species) and beyond.

Small pelagics tend to form large dense schools, which makes them easy to catch using little fuel energy, especially in comparison with ground- or demersal fish, typically caught by fuel-intensive bottom trawling (Tyedmers et al., 2005).

Passive gear, on the other hand, are deployed mainly by small-scale fishers, who, all over the world, tend to operate only in the inner shelf but still have to compete with industrial-scale operations (Pauly, 2006; Chuenpagdee et al., 2006). Small-scale fisheries are often neglected by government fisheries' management agencies, with the result that their catch, and hence contribution to food security and GDP, is often grossly underestimated (Zeller et al., 2007).

Fishing deeper and further offshore was another answer to depleted inshore, shallow-water stocks, notably targeting demersal fish (Morato et al., 2006). It is also one reason why shelves contribute less to global catches than they did before.

Fisheries of Transitional Areas

Fronts often occur as bands of high primary production parallel to the edges of the shelf and above the slope (200–500 m) that links the shelves with deeper, less productive waters. The main resources of such fronts are small pelagic fish such as mackerels and large pelagics such as tunas and billfish.

Other transitional areas that are important to fisheries are upwellings, i.e., areas where the coastal winds cause nutrient-rich water to be lifted into the surface layers and where, thanks to the extremely high primary production, immense quantities of small pelagic fish can, or could, be caught. This applies, notably, to the four large Eastern Boundary Current Systems: the California Current, which supported a huge sardine fishery (think John Steinbeck and *Cannery Row*); the Humboldt Current, which, until the early 1970s, supported the largest fishery of the world (Paulik 1981) off Peru and still produces huge catches of anchoveta (*Engraulis ringens*); the Canary Current, off Northwest Africa, still contested between DWF and the coastal countries (Kaczynski and Fluharty, 2002); and the Benguela Current, off South Africa and Namibia, where overfishing of the small pelagics has led to a system now dominated by jellyfish (Lynam et al., 2006).

In the last decade, the catch of small pelagics has been about 30–35 million tonnes per year, and most of this is used to produce fish meal and fish oil for use in both agriculture and aquaculture. The aquaculture industry, notably salmon farming, requires increasing supplies of fish meal, met in part by an increase of the fraction of global fish meal supply being diverted away from agriculture and by increased pressure on small pelagics, including species that were previously unexploited (see contributions in Alder and Pauly, 2006).

The intense pressure on small pelagics has a number of consequences, notably a depletion of the food base of marine mammals and seabirds. Indeed, this effect is so strong that it has become, in many parts of the world, the cause for massive declines of seabird and/or marine mammal populations, e.g., off Peru (Muck, 1989) or in the Mediterranean (Bearzi et al., 2003).

The search for new fishing grounds, beyond shelves, led to the discovery of large accumulations of long-lived fish, such as orange roughy (*Haplostethus atlanticus*), above seamounts, i.e., underwater mountains that rise from the seafloor, thereby representing an obstacle to currents, and locally enhance productivity (Bakun, 1996). However, the fisheries based on these resources are even less sustainable than those on shelves or slopes, seamount fish being extraordinarily long-lived, with delayed maturity, and thus unable to quickly rebuild their biomass (Morato et al., 2006; Cheung et al., 2007).

There are probably 20–30,000 large seamounts (>1000 m from the base to the top). Kitchingman and Lai (2004) give the location of about 14,000 of them. Additionally, there are probably in the order of 100,000 smaller seamounts, their number depending on the definition.

About 50% of all seamounts occur in the 40% of the ocean that is under national jurisdiction (i.e., within the EEZ of maritime countries); the other 50% are part of the High Seas, and hence the fisheries based thereon are largely unregulated, and their catch is part of the IUU catch illustrated in figure 1.

Oceanic Fisheries

The central gyre regions, which represent most of the oceanic zones, are characterized by an extremely low primary production because they are nearly permanently capped by warm, nutrient-poor water. Hence, the large fish that inhabit these waters must be swift, capable of quickly overcoming the long distances separating occasional, or seasonal, mixing events and the associated bloom of food organisms. Foremost among these fish are the tunas, which are the main target of oceanic fisheries.

These fisheries operate purse seiners, enormously long drift nets (now banned by the United Nations but still widely used), and longlines and cover all of the world ocean's tropical and subtropical regions, along with parts of the temperate regions (Fonteneau, 1998). The last two of these gear also kill a very high number of nontarget fish, turtles, seabirds, and marine mammals, another facet of a widely recognized global bycatch problem (Zeller and Pauly, 2005).

The impact on the target species is hotly contested, however. Myers and Worm (2003) suggested, based on Japanese longline catch and effort data, that tuna abundance, in all three major oceans, has declined by 90% since the 1950s, as did the abundance of most major exploited groups (see also Christensen et al., 2003). Others, although conceding that the tunas in the Atlantic and Indian oceans have been much depleted, maintain that the major stocks in the Pacific, i.e., those of yellowfin and skipjack tunas, are in good shape and that reports of their decline are inaccurate or exaggerated.

3. "FISHING DOWN" AS A MAJOR FEATURE OF CONTEMPORARY FISHERIES

Fishing down marine food webs, as defined by Pauly et al. (1998), occurs when fisheries, faced with decreasing abundance and catches of large, high-trophic-level fish (i.e., fish feeding on top of marine food chains), switch to invertebrates (shrimp, crabs, squid) and especially to small fish, notably small pelagics, i.e., to the prey of the larger fish.

Small pelagic fish usually play a crucial role in the ecosystems in which they occur, mainly because they are the group that transfers most of the biomass production by the plankton to the larger fish, seabirds, and marine mammals. This direct dependence of small pelagic fish on plankton, itself impacted by environmental fluctuations, often causes their abundance to fluctuate wildly. This has led many fisheries' scientists to conclude, erroneously, that fisheries have essentially no impact on small pelagics and that their abundance is determined overwhelmingly by environmental factors.

Another worrisome aspect of fishing down marine food webs is that it involves a reduction of the number and length of pathways linking food fishes and the primary producers, and hence a simplification of the food webs. Diversified food webs allow predators to switch between prey as their abundance fluctuates.

As the food webs are simplified by the removal of mid-trophic-level components, the remaining large predators find themselves atop short, linear food chains, incapable of buffering environmental fluctuations. This effect, combined with the drastic reduction of the number of year classes in the predators' populations, makes their overall abundance strongly dependent on annual recruitment, which contributes to increasing variability and to lack of predictability in population sizes and hence in predicted catches. The net effect is, ironically, that it will increasingly look as though environmental fluctuations drive fisheries, even where they originally did not.

Trophic reduction in landed catch caused by this phenomenon called "fishing down" can be mapped. A single trophic level, for example, represents energy flow between plants and plant eaters, or likewise between fish and their predators. Such maps reveal typical reductions of 0.05 to 0.10 trophic levels per decade, and it then becomes clear that large areas of the world's oceans have suffered a major decline in the average trophic level of reported landings (plate 14), representing large-scale and dramatic changes in the supporting marine ecosystems. Typically, most impacted are coastal areas where predatory fish species used to be the target of commercial fisheries, but their failure has refocused fishing effort on other species, often invertebrates and sometimes the prey of the target species. With the removal of the predator, populations of the prey greatly expand and support a fishery where previously one was not viable. One example is the case of the Atlantic cod fishery along the coasts of Newfoundland, where the demise of this fishery has hastened expansion of fisheries on boreal shrimp stocks.

4. MAPPING FISHERIES' INTERACTIONS

The world's fisheries cover most of the oceans' area, so it is not unexpected that there are overlaps with what is removed and the food of animals in the wild. Mapping allows studies to look at the degree of overlap and its relative significance. Studies of the distribution of marine mammals (Kaschner et al., 2006) are important, e.g., to assess the impact of the navies of the world, which, during naval exercises, deploy high-intensity sonar equipment that can harm the auditory system of whales. Another use of such distribution ranges is that, in combination with marine mammal diet composition and mapped fisheries' catches, they can be used to map interactions between marine mammals and fisheries. Such information allows assessing the veracity of claims, in various countries, that marine mammal food consumption explains the decline of fisheries (Kaschner and Pauly, 2005).

5. CONCLUSIONS

Among professional fisheries' scientists, the crisis of fisheries is still often denied. Despite frequent and fashionable references to the need for a methodological "paradigm shift," many believe that rigorous quantification of the uncertainties involved in stock assessment, and the communication of the results to fisheries' managers in the form of risk assessment, will resolve most fisheries' failures. Often what actually is lacking is a spatial focus or ability to prioritize by management jurisdiction or area. Mapping of fisheries and their interactions allows problem areas to be demonstrated. Groups funding research and conservation efforts need to channel funding to the appropriate local agencies, nongovernmental organizations, and other groups to have the maximum impact.

There are many demands for the same marine resources. They come from large multinational fishing companies, small-scale fishers, and the other animals in the marine environment. Nothing is wasted, even where fishing does not occur. The problem is one of resource allocation. Without detailed maps of resource use and spatial models to project the likely outcome of various options, there cannot be a dialogue to prevent resource and social disasters.

Thus, our key problem, really, is not "uncertainty" or even lack of knowledge by fisheries' managers. Indeed, the problem is not even one of "management" but one of public policy. This refers to the excessive role played, in allocation debates, by the users of fisheries' resources vis-à-vis the true owners of these resources: the citizens of the various countries whose fish stocks are pillaged by fleets that they subsidize for doing so.

Resolving this allocation issue requires public involvement, as occurred with, for example, the reclaiming of public waters, long perceived to "belong" to those who used such waters to cheaply dispose of toxic effluents. Indeed, reclaiming the sea from its abusers will be a key task for the twenty-first century, second only to avoiding the massive climatic change that increasing emission of greenhouse gases will give us.

Another aspect of this allocation problem is the competition between small-scale and industrial fisheries that takes place on the inner shelves of most maritime countries. As briefly mentioned above, the catches of small-scale fisheries are usually underestimated. Also, the fisheries themselves are often not an explicit part of fisheries' development plans, although their frequent use of passive gear and adjacency to the stocks they exploit make them more energy efficient than industrial fisheries and potentially more sustainable (Pauly, 2006).

Small-scale fisheries, moreover, are given only a small fraction of government subsidies to fisheries, recently reestimated at $30–34 billion per year globally, and a major driver of overfishing (Sumaila and Pauly, 2006).

One of the reasons the destruction of marine life by heavily subsidized fishing fleets went as far as it did is that the public at large retained, until recently, a romantic image of fishers and fisheries. On the other hand, the environmental nongovernmental organizations that could have helped correct this benign view of fisheries were, until recently, dependent for their analyses on fisheries' data from government laboratories, mainly assembled and pertinent to the tactical (year-to-year) management of industrial fleets, and generally useless for demonstrating the ecosystem impact of fisheries.

This is why we emphasize map-based representations of fisheries, which can communicate complex information even to lay audiences (Watson et al., 2005; Pauly, 2007; see also http://www.seaaroundus.org). Our Web site thus presents, mainly in the form of maps, and for each maritime country of the world (and also for 64 "large marine ecosystems"), what we believe is key information on the marine fisheries and ecosystems of the world. The information we provide could be far more detailed for some developed countries. However, this would leave most developing countries behind, which would seem inappropriate, given that the demonstration by Alder and Sumaila (2004) that it is fish caught along the coasts of, or exported from, developing countries that now largely supply markets in developed countries.

The close, and increasingly global, connectedness of many of the world's peoples with marine resources represents a real challenge for sustainability. Fish removed from the grasp of a small-scale fisher in Africa can end up in prestige markets of Europe within hours. What was food for a marine mammal may end up as a fish sandwich for a businessman in New York. Understanding the spatial dynamics of fish stocks, and the fisheries and fish trade dependent on them, is essential if we are to manage them for future use and equally if we are to minimize their impacts on the marine systems that support them. Protein from the sea will be increasingly important to our survival, and retaining the oceans' biodiversity and health will leave us considerably more options in an uncertain future.

FURTHER READING

Alder, J., and D. Pauly, eds. 2006. On the multiple uses of forage fish: From ecosystem to markets. Fisheries Centre Research Reports 14(3): 1–109.

Alder, J., and U. R. Sumaila. 2004. Western Africa: A fish basket of Europe past and present. Journal of Environment and Development 13: 156–178.

Bakun, A. 1996. Patterns in the Oceans: Oceans Processes and Ocean dynamics. La Jolla: California Sea Grant; and La Paz: Centro de Invstigaciones del Noroeste.

Bearzi, G., R. Randall, R. Reeves, G. Notarbartolo-di-Sciara, E. Politi, A. Canadas, A. Frantzis, and B. Mussi. 2003. Ecology, status and conservation of short-beaked common dolphins *Delphinus delphis* in the Mediterranean Sea. Mammal Review 33: 224–252.

Bonfil, R., G. Munro, U. R. Sumaila, H. Valtysson, M. Wright, T. Pitcher, D. Preikshot, N. Haggan, and D. Pauly. 1998. Impacts of distant water fleets: An ecological, economic and social assessment. Fisheries Centre Research Reports 6(6): 1–111.

Cheung, W., R. Watson, T. Morato, T. Pitcher, and D. Pauly. 2007. Intrinsic vulnerability in the global fish catch. Marine Ecology Progress Series 333: 1–12.

Christensen, V., S. Guénette, J. J. Heymans, C. J. Walters, R. Watson, D. Zeller, and D. Pauly. 2003. Hundred year decline of North Atlantic predatory fishes. Fish and Fisheries 4: 1–124

Chuenpagdee, R., L. Liguori, M. L. Palomares, and D. Pauly. 2006. Bottom-up, global estimates of small-scale marine fisheries catches. Fisheries Centre Research Report 14(8): 1–112.

Cushing, D. H. 1988. The Provident Sea. Cambridge, UK: Cambridge University Press.

Erlandson, J. M., and S. M. Fitzpatrick. 2006. Oceans, islands, and coasts: Current perspectives on the role of the sea in human prehistory. Journal of Island & Coastal Archaeology 1: 5–32.

Fonteneau, A. 1998. Atlas of tropical tuna fisheries. Paris: Edition ORSTOM.

Gelchu, A., and D. Pauly. 2007. Growth and distribution of part-based fishing effort within countries' EEZ from 1970 to 1995. Fisheries Centre Research Reports 15(4): 1–99.

Giske, J., G. Huse, and O. Fiksen. 1998. Modelling spatial dynamics of fish. Reviews in Fish Biology and Fisheries 8: 57–91.

Grainger, R.J.R., and S. M. Garcia. 1996. Chronicles of Marine Fishery Landings (1950–1994): Trend Analysis and Fisheries Potential. FAO Fisheries Technical Paper No. 359. Rome: FAO.

Jackson, J.B.C., M. X. Kirby, W. H. Berger, K. A. Bjorndal, L. W. Botsford, B. J. Bourque, R. Cooke, J. A. Estes, T. P. Hughes, S. Kidwell, C. B. Lange, H. S. Lenihan, J. M. Pandolfi, C. H. Peterson, R. S. Steneck, M. J. Tegner, and R. R. Warner. 2001. Historical overfishing and the recent collapse of coastal ecosystems. Science 293: 629–638.

Kaczynski, V. M., and D. L. Fluharty. 2002. European policies in West Africa: Who benefits from fisheries agreements? Marine Policy Journal 26: 75–93.

Kaschner, K., and D. Pauly. 2005. Competition between marine mammals and fisheries: Food for thought. In D. J. Salem and A. N. Rowan, eds. The State of Animals III. Washington, DC: Humane Society Press, 95–117.

Kaschner, K., R. Watson, A. W. Trites, and D. Pauly. 2006. Mapping world-wide distribution of marine mammal species using a relative environmental suitability (RES) model. Marine Ecology Progress Series 316: 285–310.

Kitchingman, A., and S. Lai. 2004. Inferences on potential seamount locations from mid-resolution bathymetric data. In T. Morato and D. Pauly, eds. Seamounts: Biodiversity and Fisheries. Fisheries Centre Research Reports 12(5): 7–12.

Kurlansky, M. 1997. Cod: A Biography of a Fish That Changed the World. Toronto: Knopf.

Longhurst, A. 1998. Ecological Geography of the Sea. San Diego: Academic Press.

Lynam, C. P., M. J. Gibbon, B. E. Axelsen, C.A.J. Sparks, J. Coetzee, B. G. Heywood, and A. S. Brierley. 2006. Jellyfish overtake fish in a heavily fished ecosystem. Current Biology 16: R492–R493.

Morato, T., R. Watson, T. Pitcher, and D. Pauly. 2006. Fishing down the deep. Fish and Fisheries 7: 24–34.

Muck, P. 1989. Major trends in the pelagic ecosystem off Peru and their implications for management. In D. Pauly, P. Muck, J. Mendo, and I. Tsukayama, eds. The Peruvian Upwelling Ecosystem: Dynamics and Interactions. ICLARM Conference Proceedings 18: 386–403.

Myers, R. A., and B. Worm. 2003. Rapid worldwide depletion of predatory fish communities. Nature 423: 280–283.

Paulik, G. J. 1981. Anchovies, birds and fishermen in the Peru Current. In M. H. Glantz and D. J. Thompson, eds. Resource Management and Environmental Uncertainty: Lessons from Coastal Upwelling Fisheries. New York: Wiley-Interscience, 35–79.

Pauly, D. 1999. Review of A. Longhurst's "Ecological Geography of the Sea." Trends in Ecology and Evolution 14: 118.

Pauly, D. 2006. Major trends in small-scale marine fisheries, with emphasis on developing countries, and some implications for the social sciences. Maritime Studies (MAST) 4(2): 7–22.

Pauly, D. 2007. The Sea Around Us Project: Documenting and communicating global fisheries impacts on marine ecosystems. AMBIO: A Journal of the Human Environment 34: 290–295.

Pauly, D., J. Alder, E. Bennett, V. Christensen, P. Tyedmers, and R. Watson. 2003. The future for fisheries. Science 302: 1359–1361.

Pauly, D., V. Christensen, J. Dalsgaard, R. Froese, and F. C. Torres, Jr. 1998. Fishing down marine food webs. Science 279: 860–863.

Pauly, D., V. Christensen, S. Guénette, T. Pitcher, U. R. Sumaila, C. Walters, R. Watson, and D. Zeller. 2002. Towards sustainability in world fisheries. Nature 418: 689–695.

Pauly, D., and J. Maclean. 2003. In a Perfect Ocean: Fisheries and Ecosystems in the North Atlantic. Washington, DC: Island Press.

Pauly, D., and R. Watson. 2005. Background and interpretation of the "Marine Trophic Index" as a measure of biodiversity. Philosophical Transactions of the Royal Society: Biological Sciences 360: 415–423.

Pauly, D., R. Watson, and J. Alder. 2005. Global trends in world fisheries: Impacts on marine ecosystems and food security. Philosophical Transactions of the Royal Society: Biological Sciences 360: 5–12.

Radovich, J. 1982. The collapse of the California sardine industry: What have we learned? CalCOFI Reports 23: 56–78.

Sarhage, D., and J. Lundbeck. 1992. A History of Fishing. Berlin: Springer-Verlag.

Sumaila, U. R., and D. Pauly, eds. 2006. Catching more bait: A bottom-up re-estimation of global fisheries subsidies. Fisheries Centre Research Reports 14(6): 1–114.

Tyedmers, P., R. Watson, and D. Pauly. 2005. Fueling global fishing fleets. AMBIO: A Journal of the Human Environment 34: 635–638.

van Houtan, K., and D. Pauly. 2007. Snapshot: Ghost of destruction. Nature 447: 123.

Vincent, J. R. 2007. Spatial dynamics, social norms, and the opportunity of the commons. Ecological Research 22: 3–7.

Walters, C., and J. J. Maguire. 1996. Lessons for stock assessments from the Northern cod collapse. Reviews in Fish Biology and Fisheries 6: 125–137.

Watling, L., and E. A. Norse. 1998. Disturbance of the seabed by mobile fishing gear: A comparison to forest clearcutting. Conservation Biology 12: 1180–1197.

Watson, R., J. Alder, V. Christensen, and D. Pauly. 2005a. Mapping global fisheries patterns and their consequences. In D. J. Wright and A. Scholz, eds. Place Matters—Geospatial Tools for Marine Science, Conservation and Management in the Pacific Northwest. Corvallis: Oregon State University Press, 13–33.

Watson, R., J. Alder, A. Kitchingman, and D. Pauly. 2005b. Catching some needed attention. Marine Policy 29: 281–284.

Watson, R., V. Christensen, R. Froese, A. Longhurst, T. Platt, S. Sathyendranath, K. Sherman, J. O'Reilly, P. Celone, and D. Pauly. 2003. Mapping fisheries onto marine ecosystems for regional, oceanic and global integrations. In G. Hempel and K. Sherman, eds. Large Marine Ecosystems of the World 12: Change and Sustainability. Amsterdam: Elsevier Science, 375–395.

Watson, R., A. Kitchingman, A. Gelchu, and D. Pauly. 2004. Mapping global fisheries: Sharpening our focus. Fish and Fisheries 5: 168–177.

Watson, R., and D. Pauly. 2001. Systematic distortions in world fisheries catch trends. Nature 414: 534–536.

Worm, B., E. B. Barbier, N. Beaumont, E. Duffy, C. Folke, B. S. Halpern, J.B.C. Jackson, H. K. Lotze, F. Micheli, S. R. Palumbi, E. Sala, K. A. Selkoe, J. J. Stachowicz, and R. Watson. 2006. Impact of biodiversity loss on ocean ecosystem services. Science 314: 787–790.

Zeller, D., S. Booth, G. Davis, and D. Pauly. 2007. Re-estimation of small-scale for U.S. flag-associated islands in the western Pacific: The last 50 years. Fisheries Bulletin 105: 266–277.

Zeller, D., and D. Pauly. 2005. Good news, bad news: Global fisheries discards are declining, but so are total catches. Fish and Fisheries 6: 156–159.

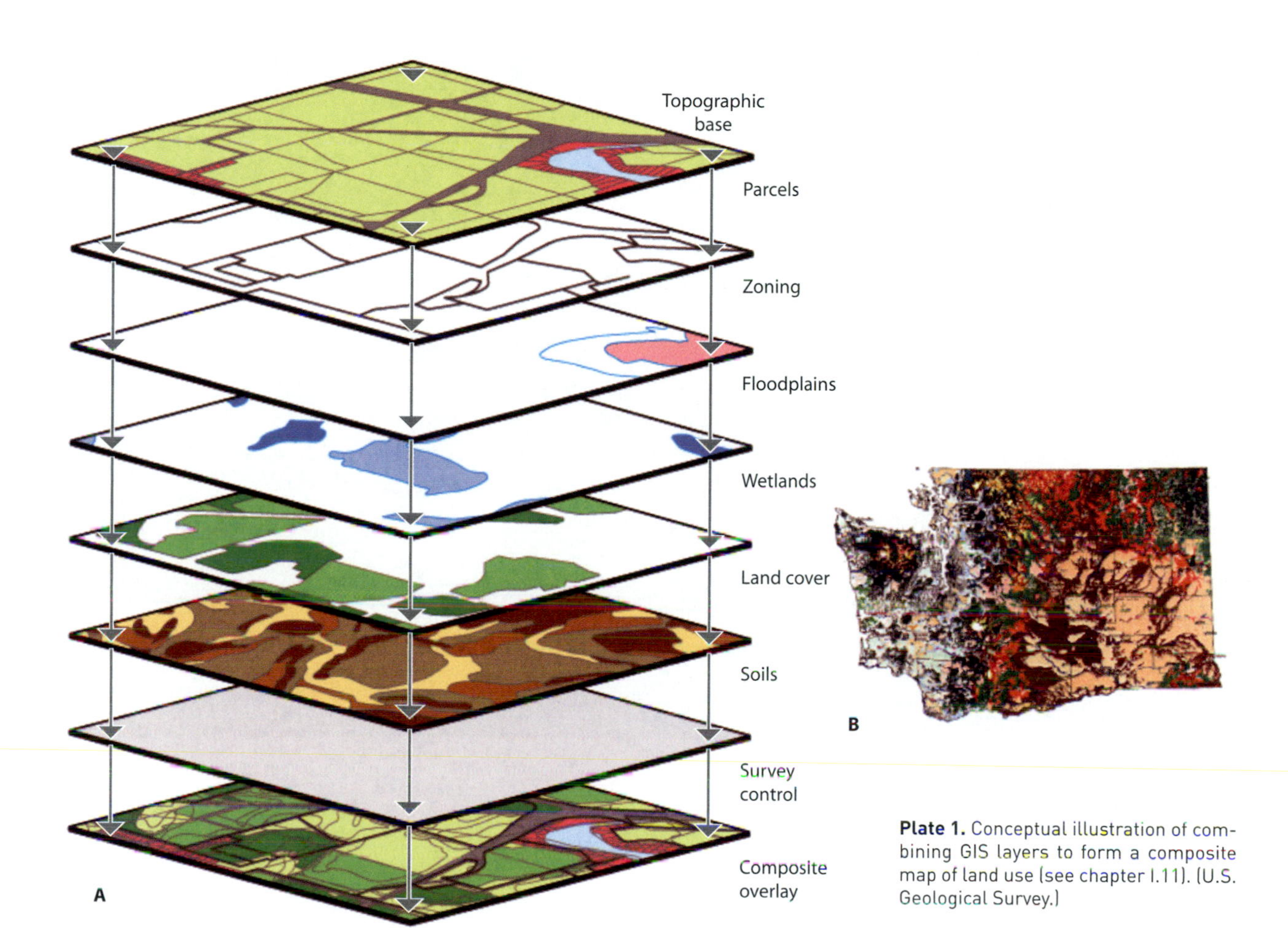

Plate 1. Conceptual illustration of combining GIS layers to form a composite map of land use (see chapter I.11). (U.S. Geological Survey.)

Plate 2. Canopy height in the Patuxent Wildlife Refuge (oblique view) as determined by a laser vegetation imaging system using a plane equipped with LiDAR (see chapter I.11).

Plate 3. (*opposite*) Hawaiian silverswords showing the diversity of forms (see chapter I.19). Clockwise from top left: *Argyroxiphium sandwicense* DC. subsp. *macrocephalum* (A. Gray) Meyrat, Haleakala silversword, on the summit crater of East Maui; *Wilkesia gymnoxiphium* A. Gray on the edge of Waimea Canyon, Kauai; *Dubautia reticulata* (Sherff) Keck from Pu'u Nianiau, East Maui; flowering plants of *Dubautia scabra* (DC) Keck subsp. *scabra* on pahoehoe lava near Pu'u Huluhulu, Hawai'i. Photos © Gerald D. Carr.

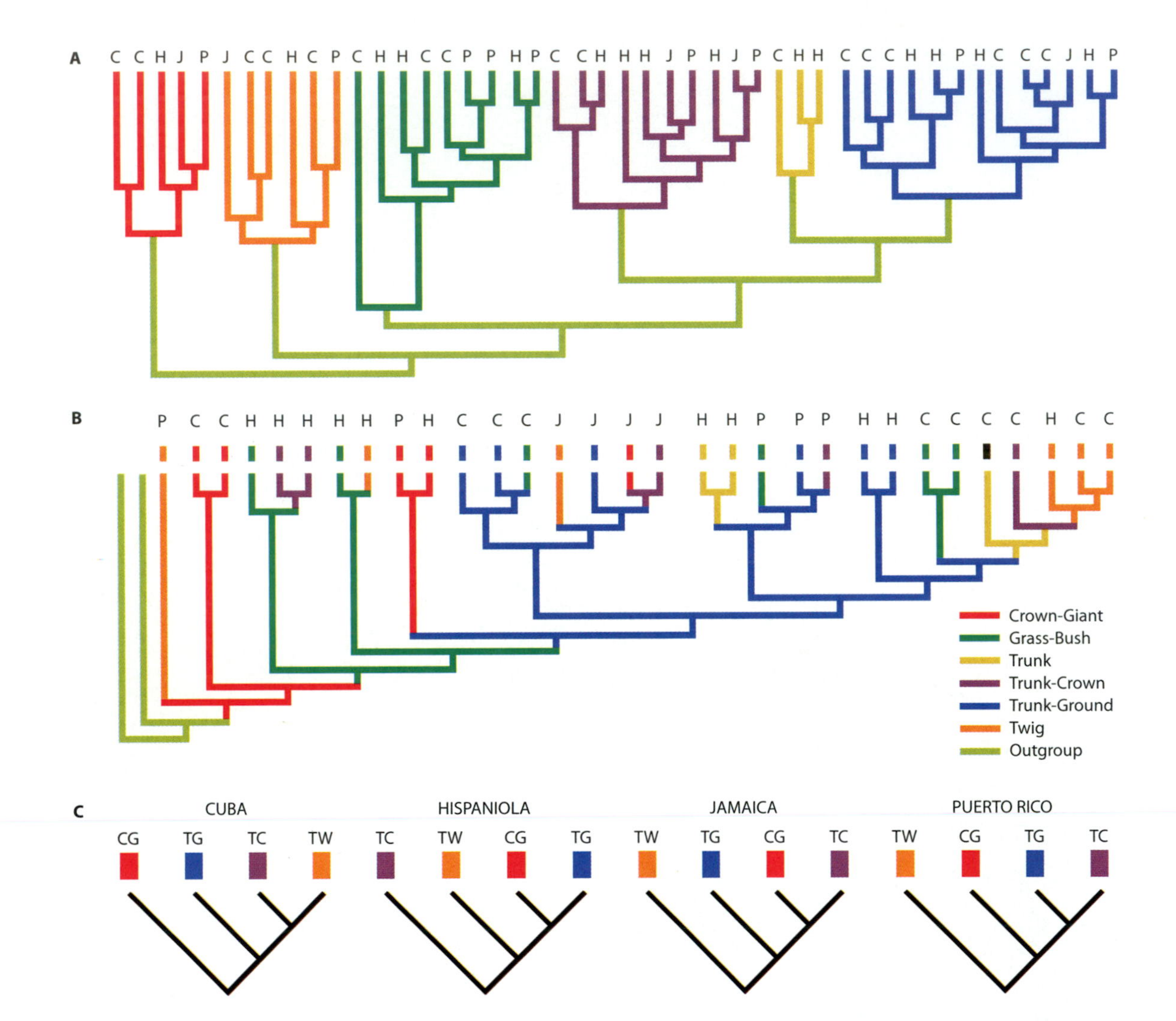

Plate 4. Diagram illustrating the pattern of adaptive radiation in Caribbean anole lizards (see chapter I.19). (A) UPGMA phenogram showing that members of the same ecomorph class cluster in morphological space regardless of geographic affinities. Branch lengths are proportional to the distance separating species or clusters in morphological space. Letters indicate the island on which a species is found (C, Cuba; H, Hispaniola; J, Jamaica; P, Puerto Rico). The shading of the branches connecting the ecomorph classes has no significance. (B) The most parsimonious tree derived from the molecular data indicates frequent transitions among ecomorph classes. The lengths of the branches have no significance. (C) Topology of the four ecomorphs common to all islands, extracted for each island separately from the most parsimonious phylogeny. (From Losos, J. B., T. R. Jackman, A. Larson, K. de Queiroz, and L. Rodríguez-Schettino. 1998. Historical contingency and determinism in replicated adaptive radiations of island lizards. Science 279: 2115–2118)

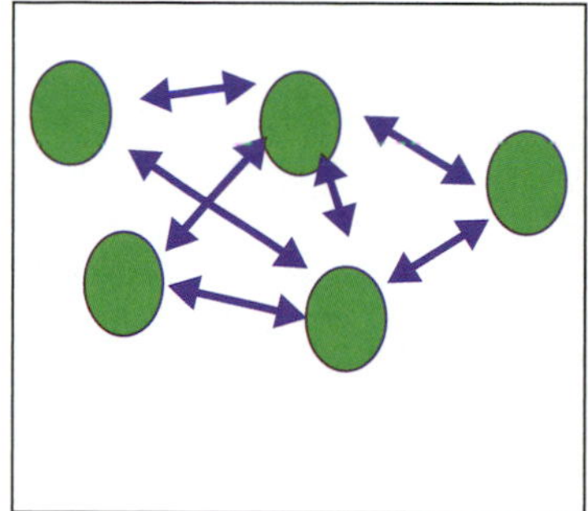

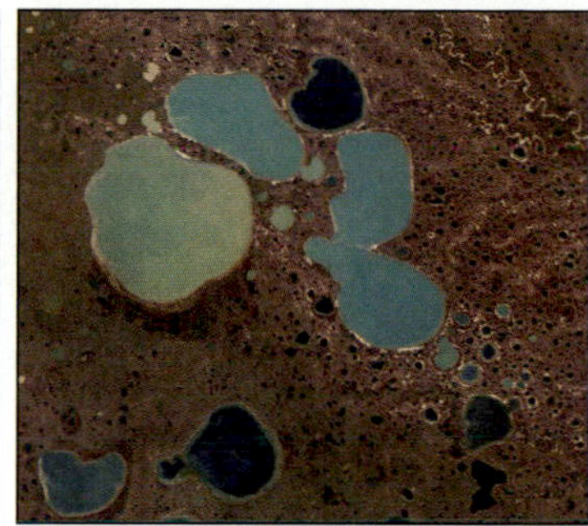

Plate 5. A schematic representation of a metacommunity and a photograph of prairie potholes in Russia (see chapter III.8). (left) The metacommunity consists of patches (circles) that are affected by each other via dispersal among them (arrows). (right) Some natural ecosystems are more likely than others to closely correspond to the metacommunity concept. Prairie pothole ponds created by glacial activity often occur in landscapes where many ponds are close enough to each other that dispersal is likely to occur among them. Nevertheless, dispersal likely varies for different organisms in these ponds, and different local environmental conditions occur in different ponds (here the ponds differ in water clarity and types of algal community). Ducks that use these ponds for their breeding grounds likely disperse very easily among them from day to day, whereas some obligate aquatic organisms with poor dispersal such as amphipods may disperse only very rarely. NASA image created by Jesse Allen, Earth Observatory, using data obtained courtesy of the University of Maryland's Global Land Cover Facility.

Highly functional — Continuum of landscape function/dysfunction — Highly dysfunctional

Semi-arid grassland

Highly functional — Continuum of landscape function/dysfunction — Highly dysfunctional

Semi-arid shrubland

Plate 6. A continuum of landscape function/dysfunction from highly functional to highly damaged or dysfunctional, with examples for semiarid grasslands (top) and shrublands (bottom) in Australia (see chapter IV.1). Photos by David Tongway.

Plate 7. A series of photos from a mine site in northern Australia illustrates the development of rehabilitation over 30 years (see chapter IV.1). Photos by David Tongway.

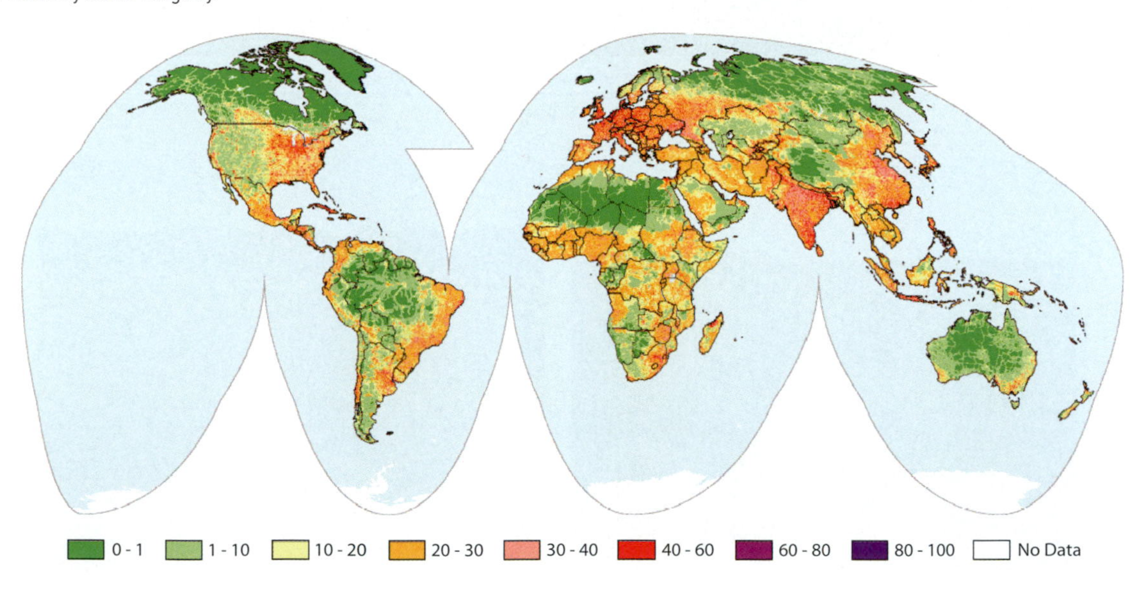

Plate 8. Human impacts on terrestrial ecosystems (see chapter IV.4). Map shows the value of an index of human influence that increases as a weighted function of population density, land transformation, human access, and power infrastructure. Eighty-three percent of the land's surface is affected by one or more of the following factors: a human population density greater than 1 person/km^2, converted to urban or agricultural land uses, lying within 15 km of a road or major river or within 2 km of a settlement or a railway, and/or producing enough light to be visible regularly to a satellite at night. (From Sanderson et al., 2002)

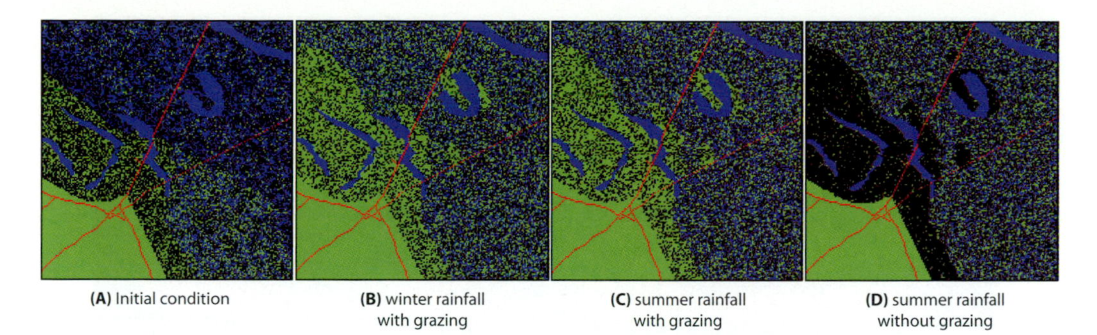

(A) Initial condition

(B) winter rainfall with grazing

(C) summer rainfall with grazing

(D) summer rainfall without grazing

	Proportion of the landscape by vegetation type		
	Creosotebush	Black Grama	Blue Grama
Initial conditions	0.27	0.41	0.32
winter rainfall + grazing	0.44	0.29	0.27
summer rainfall + grazing	0.44	0.27	0.30
summer rainfall ungrazed	0.28	0.45	0.27

Plate 9. Simulated output from a cellular automata model showing response of each boundary type to climatic fluctuations and grazing by livestock at the Sevilleta National Wildlife Refuge Long-Term Ecological Research (LTER) site in central New Mexico (see chapter IV.5). (A) At start of each simulation in 1915 and following 88 years of (B) winter rainfall and livestock grazing that promote expansion of the shrub, creosotebush (green), (C) summer rainfall and livestock grazing that promote expansion of the grazing-tolerant grass, blue grama, and (D) summer rainfall without grazing that expands the grazing-intolerant grass, black grama (black). Note that stationary boundaries (blue) do not change in any of the simulations.

Black grama grassland – Chino grama grassland ecotone

Alpine meadow – treeline

Grassland – shrubland ecotone

Plate 10. At a broad scale (large spatial extent and coarse grain), boundaries appear as lines, but the same boundaries at a fine scale are often difficult to identify. Statistical analyses can be used to detect the location of boundaries: (A) boundaries between grassland communities at Big Bend National Park, (B) alpine meadow boundaries with forests in Rocky Mountain National Park, and (C) grassland–shrubland boundaries at the Sevilleta LTER site (see chapter IV.5). (Photos courtesy of Brandon Bestelmeyer, Daniel Liptzin, and Debra Peters)

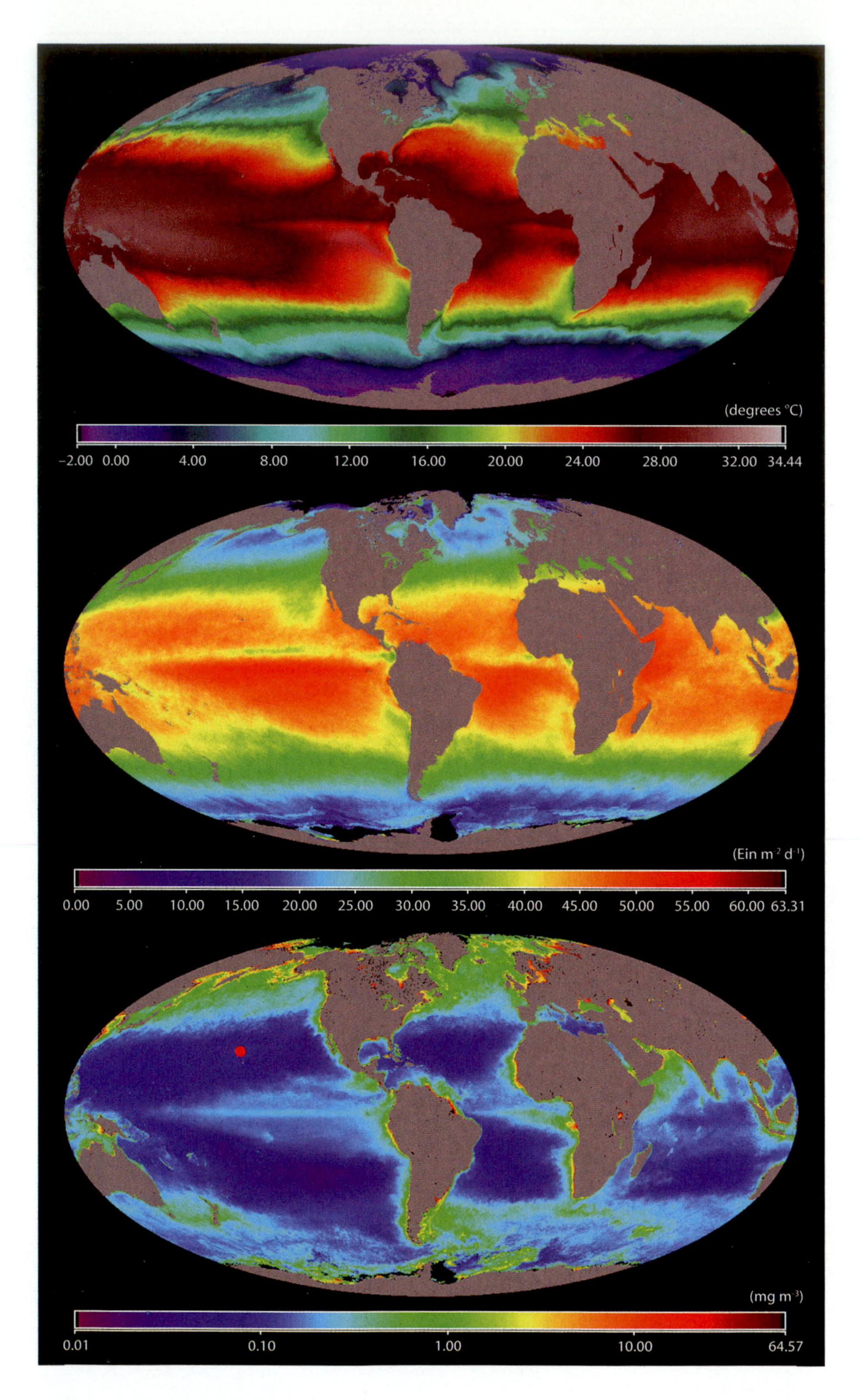

Plate 11. Satellite-derived mean annual distribution of sea surface temperature (SST; top), photosynthetically available radiation (PAR; center), and chlorophyll (chl; bottom) for the year 2006 (see chapter IV.9). SST was derived from MODIS Aqua using 4-km resolution level 3-binned data, and PAR and chl were derived from SeaWiFS using 9-km resolution level 3-binned data. All data were provided by the Ocean Biology Processing Group at NASA. The red dot in the bottom panel marks the approximate location of Station ALOHA (22°45′N, 158°W).

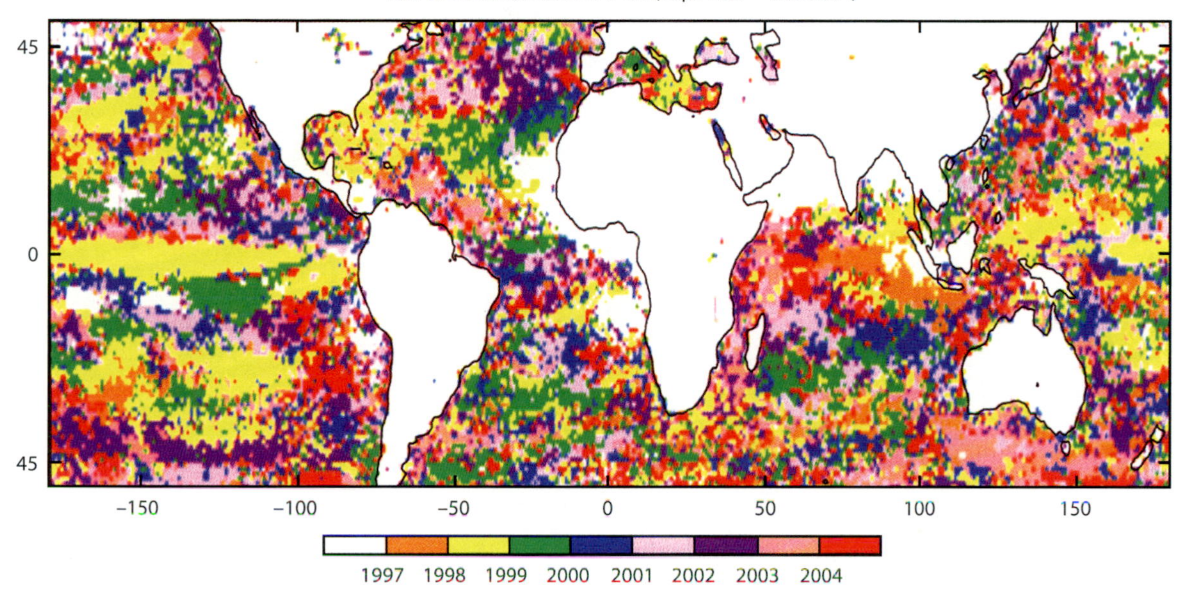

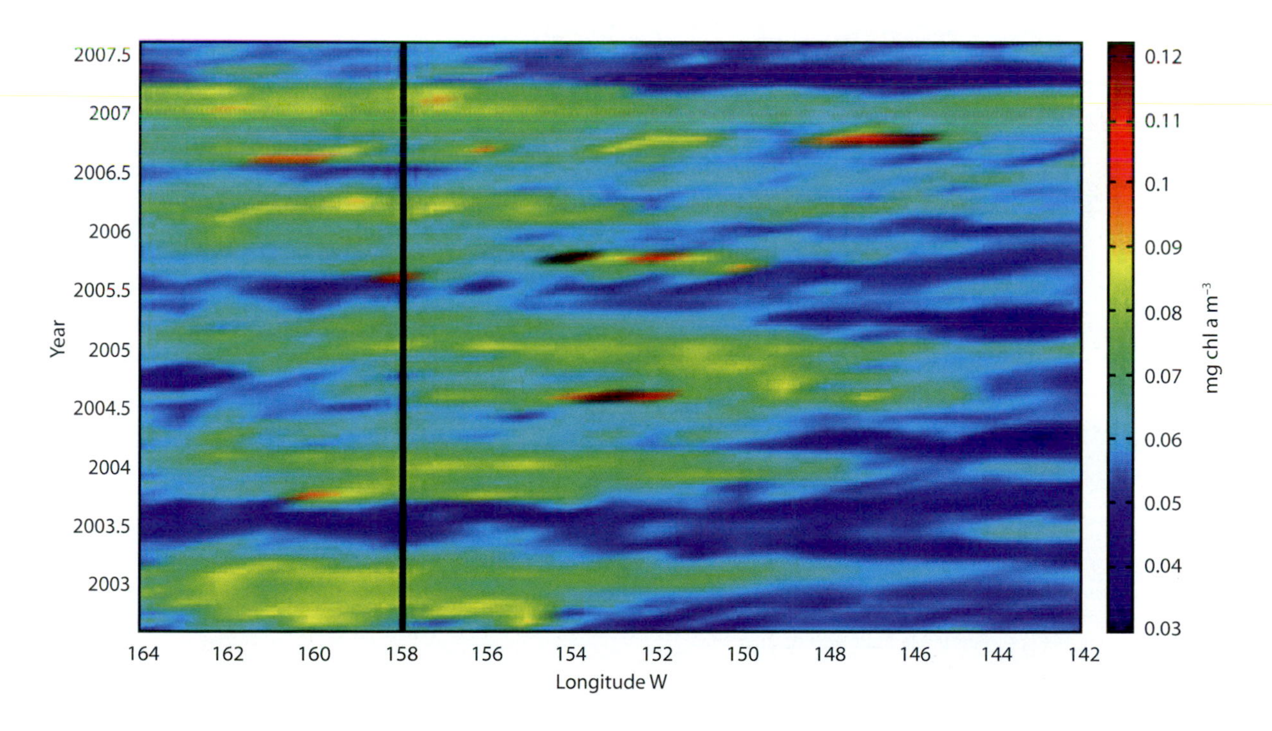

Plate 12. Time and space variability in seascape chlorophyll distributions as revealed from Earth-orbiting ocean color sensors (see chapter IV.9). (Top) SeaWiFS satellite-derived global map for period 1997–2004 showing the year in which the maximum chlorophyll concentration was achieved in each pixel (9-km resolution). (From Yoder and Kennelly, 2006) (Bottom) MODIS satellite-derived temporal and longitudinal variability in chlorophyll for the region surrounding Station ALOHA (solid vertical line at 158°W) for the period July 2002 to July 2007. (From NASA Ocean Color Time-Series Online Visualization and Analysis Center, http://reason.gsfc.nasa.gov/Giovanni/)

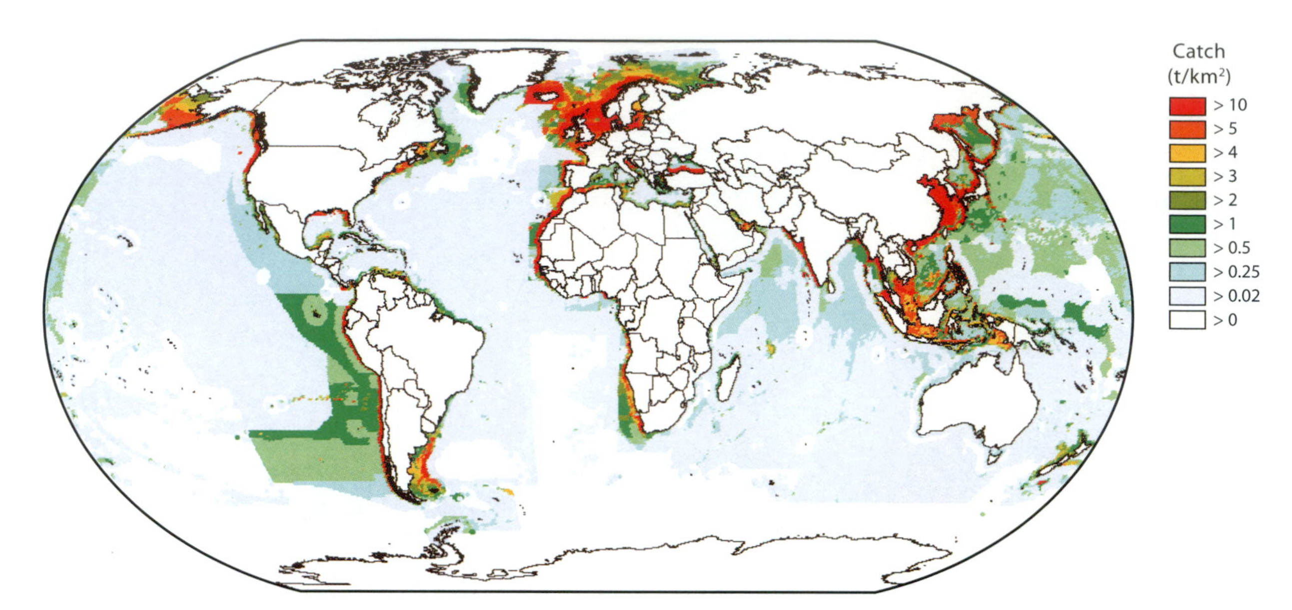

Plate 13. Map of the world marine fisheries' average annual catches since 2000, in tonnes per square kilometer per year (see chapter IV.10). Because of the high productivity of shelves (shallow areas around the continents, down to 200 m), areas of high catches (in reds and oranges) largely correspond to the distribution of shelves (see also plate 14). Catches are very low in areas nearing the poles and in midocean areas.

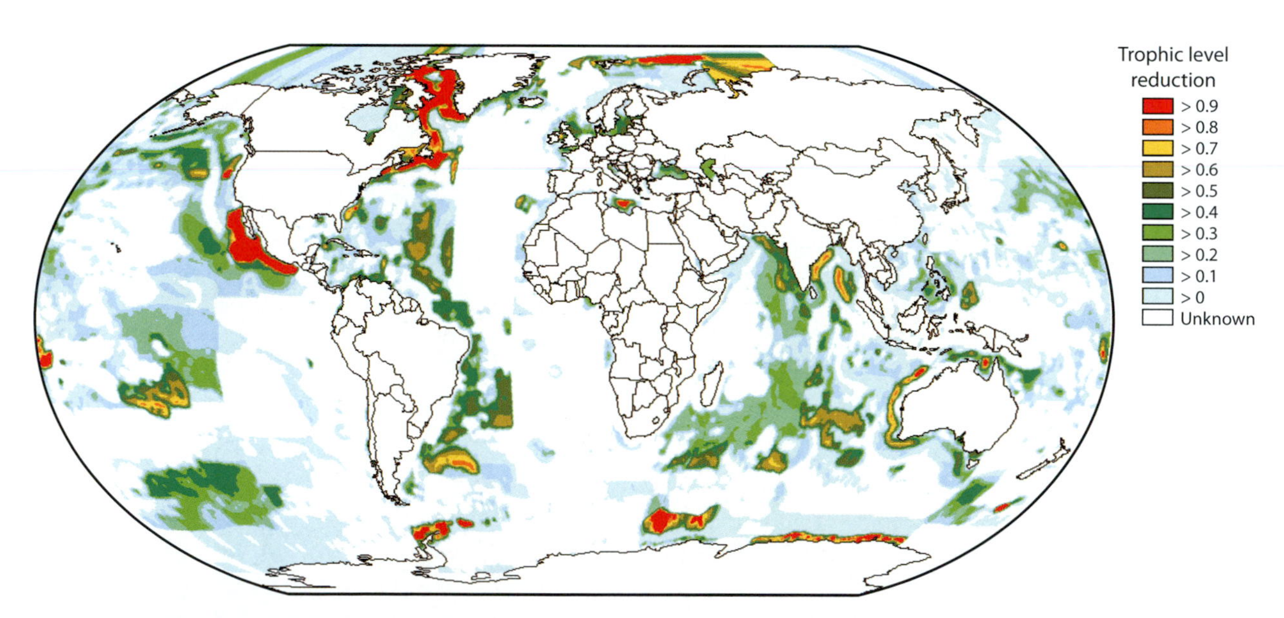

Plate 14. Global decline in trophic level of catches between the year when fisheries began in a location (defined as when >10% of the highest annual catch was first reported, but not earlier than 1950) and the most recent global catch records (2004) (see chapter IV.10). This updates figure 2 in Pauly and Watson (2005).

Plate 15. Sensitivity analysis for the loggerhead sea turtle (see chapter V.2). Inset shows the elasticity values for stage-specific survival rates. Elasticity values can be used to compare how sensitive λ is to small changes in survival rates of different stage classes. The higher the elasticity value of a stage class, the greater change in when that stage class's survivorship changes. For example, increasing the survival of small juveniles by 5% will result in change in λ of approximately 5% × 0.40 = 2%. Thus, the elasticity of a survival rate shows the percentage change that will result from a small percentage change in that survival value. Main panel shows a turtle escaping from a shrimp net equipped with a turtle excluder device. (Photo courtesy of NOAA; inset from Crowder, L. B., D. T. Crouse, S. S. Heppell, and T. H. Martin. 1994. Predicting the impact of turtle excluder devices on loggerhead sea turtle populations. Ecological Applications 4: 437–445)

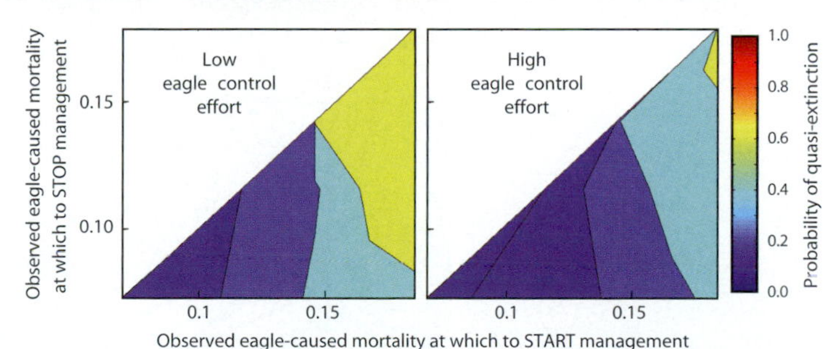

Plate 16. Several island fox populations have declined as a result of predation by golden eagles (see chapter V.2). A demographic PVA for the fox predicts population trajectories that fluctuate substantially but very rarely decline to a quasiextinction threshold of 30 animals. Adding eagle predation causes sharp increases in the probability of quasiextinction (upper right). Simulating monitoring and management for eagle control along with traditional population dynamics allowed investigation of how extinction risk is jointly affected by both management effort (how many person-hours are used to catch eagles, bottom right versus bottom left) and the monitoring triggers (fox mortality rates) used to start and stop management. The risk of quasiextinction associated with using any given combination of observed mortality rates to trigger the start and stop of eagle control, as indicated by the x and y axes, is shown by the color at that point in the figure. This use of PVA methods helps tie the different aspects of management decisionmaking directly to the future viability of a population. (Fox photograph from Stephen Francis Photography; eagle photograph from Institute for Wildlife Studies)

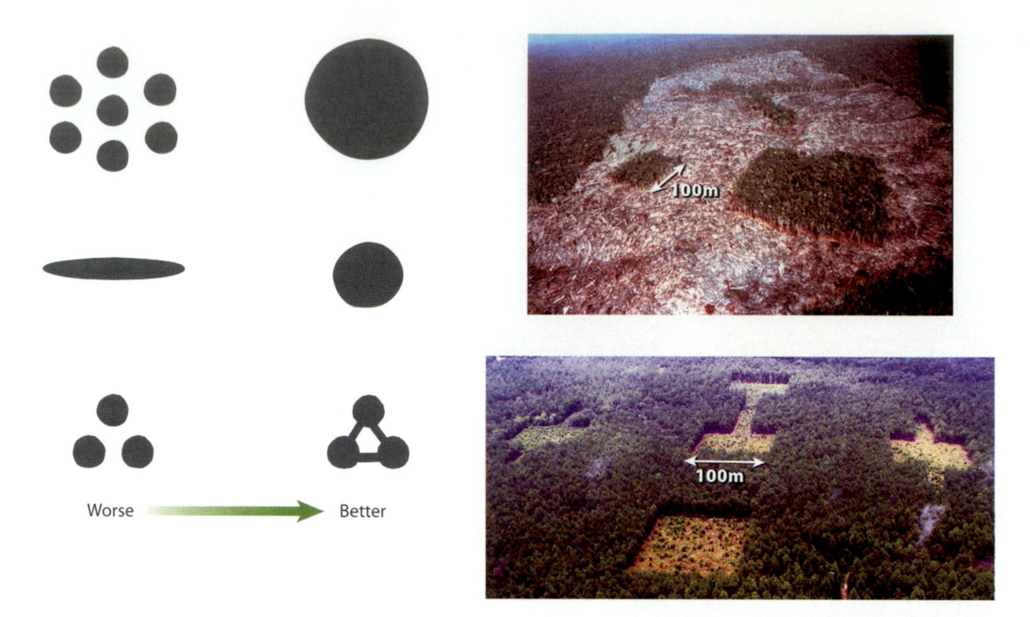

Plate 17. Some principles of reserve design that emerged from ecological theory (see chapter V.3). Photos of two large ecological experiments used to test effects of habitat fragmentation and strategies for reserve design: (top) Biological Dynamics of Forest Fragments (photo by Rob Bierregaard) (bottom) Savannah River Corridor Project, (photo by Ellen Damschen) (From Wilson and Willis, 1975)

Plate 18. Several experimental climate manipulation approaches (see chapter V.6): (upper left) suspended heat lamp warming experiment at the Rocky Mountain Biological Laboratory, Colorado (photo by S. Saleska); (lower left) open-top chamber-warming experiment at the Haibei Alpine Research Station, northeastern region of the Tibetan Plateau, China (photo by J. Klein); (right) multifactor experiment (includes elevated CO_2, warming, increased precipitation, and nitrogen deposition treatments) at the Jasper Ridge Biological Preserve, California. (Photo by J. Dukes)

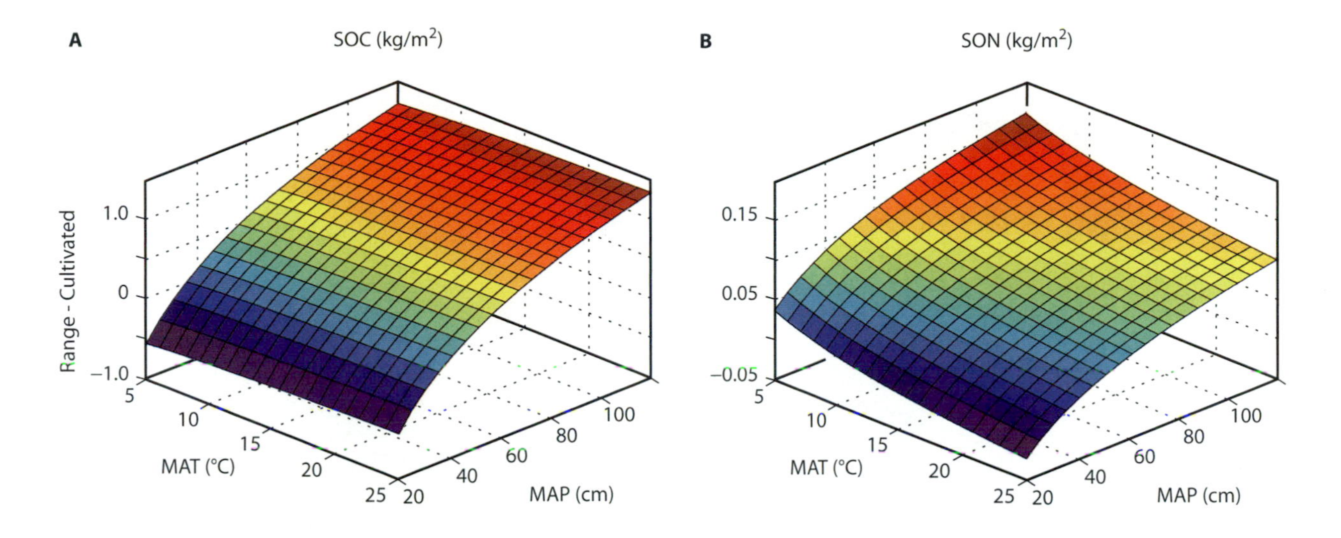

Plate 19. Multiple-regression-based comparison showing cultivation-induced changes in SOC and SON as a function of climate for the Great Plains (see chapter VI.6). Greater losses at higher precipitation and lower temperatures are closely related to greater accumulations under such conditions. From Miller, A. J., R. Amundson, I. C. Burke, and C. Yonker. 2004. The effect of climate and cultivation on soil organic C and N. Biogeochemistry 67: 57–72. Used with kind permission from Springer Science and Business Media.

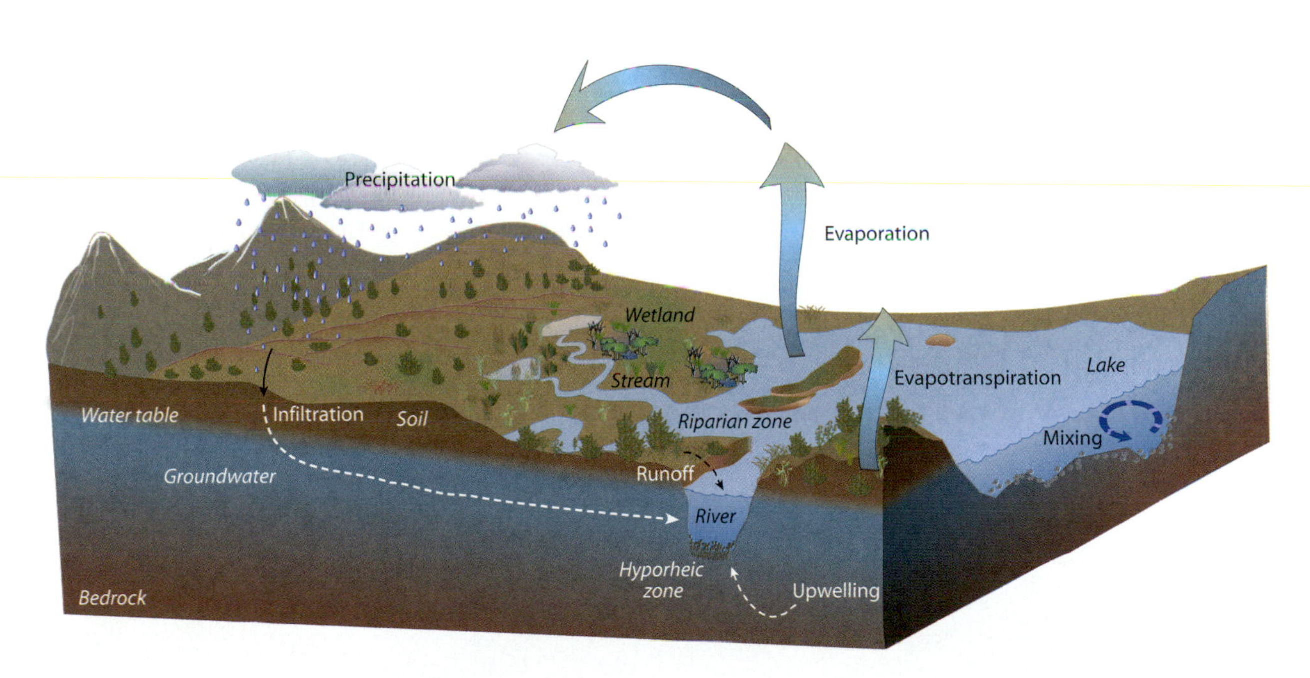

Plate 20. Schematic of the major types of freshwater ecosystems and the interaction of belowground, atmospheric, and surface water (see chapter VI.8).

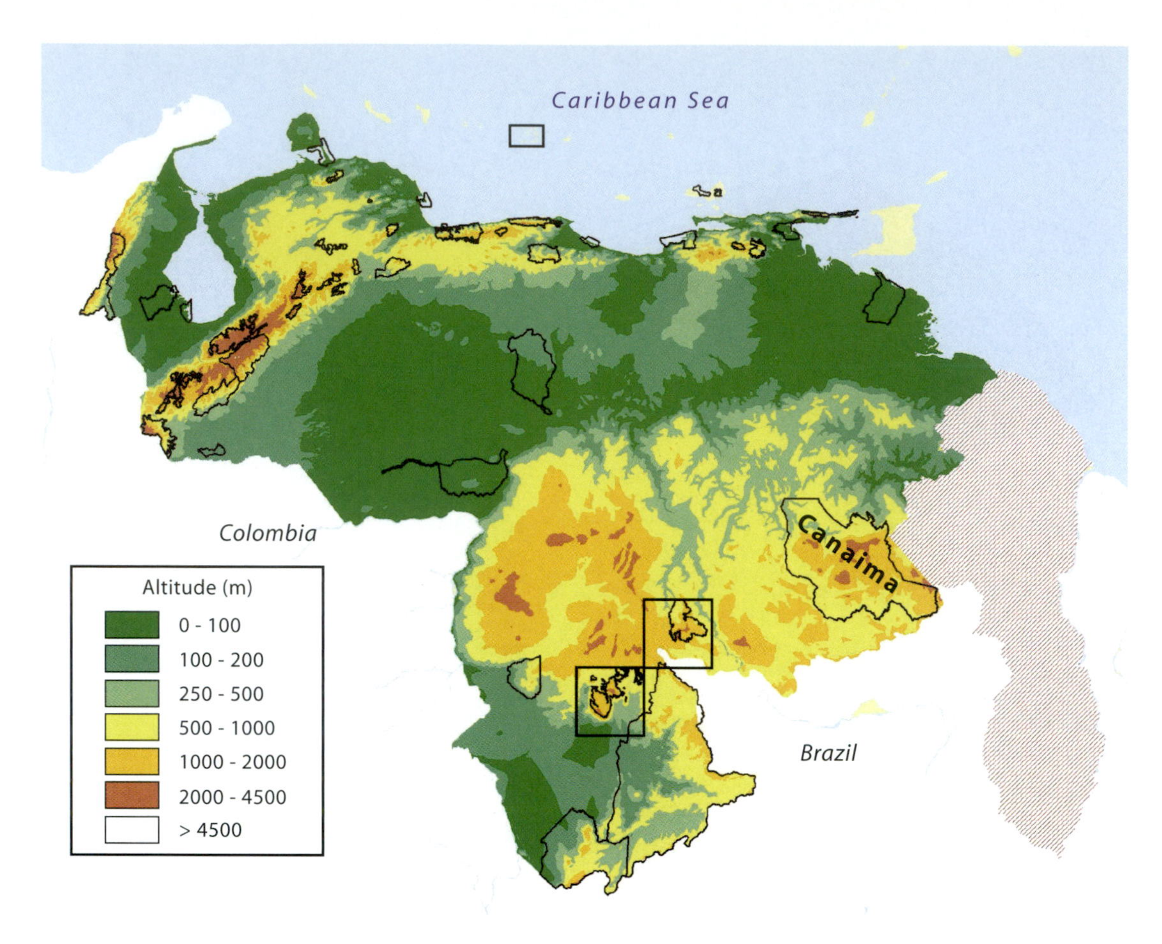

Plate 21. Venezuela's national parks (shown in black lines) are located predominantly along the country's mountain ranges (see chapter VI.13). Their establishment was primarily for the protection of the watersheds of rivers that supply water to cities and agricultural areas. Canaima, one of Venezuela's largest national parks (30,000 km^2), includes the high watershed of the Caroní River. Guri Dam, located in the lower Caroní, produces electricity equivalent to 300,000 oil barrels per day.

Plate 22. Windrows of nitrogen-fixing Cyanobacteria (bluegreen algae) blowing ashore on Lake Winnipeg, Manitoba, in late summer (see chapter VII.5). (Photo by Lori Volkart)

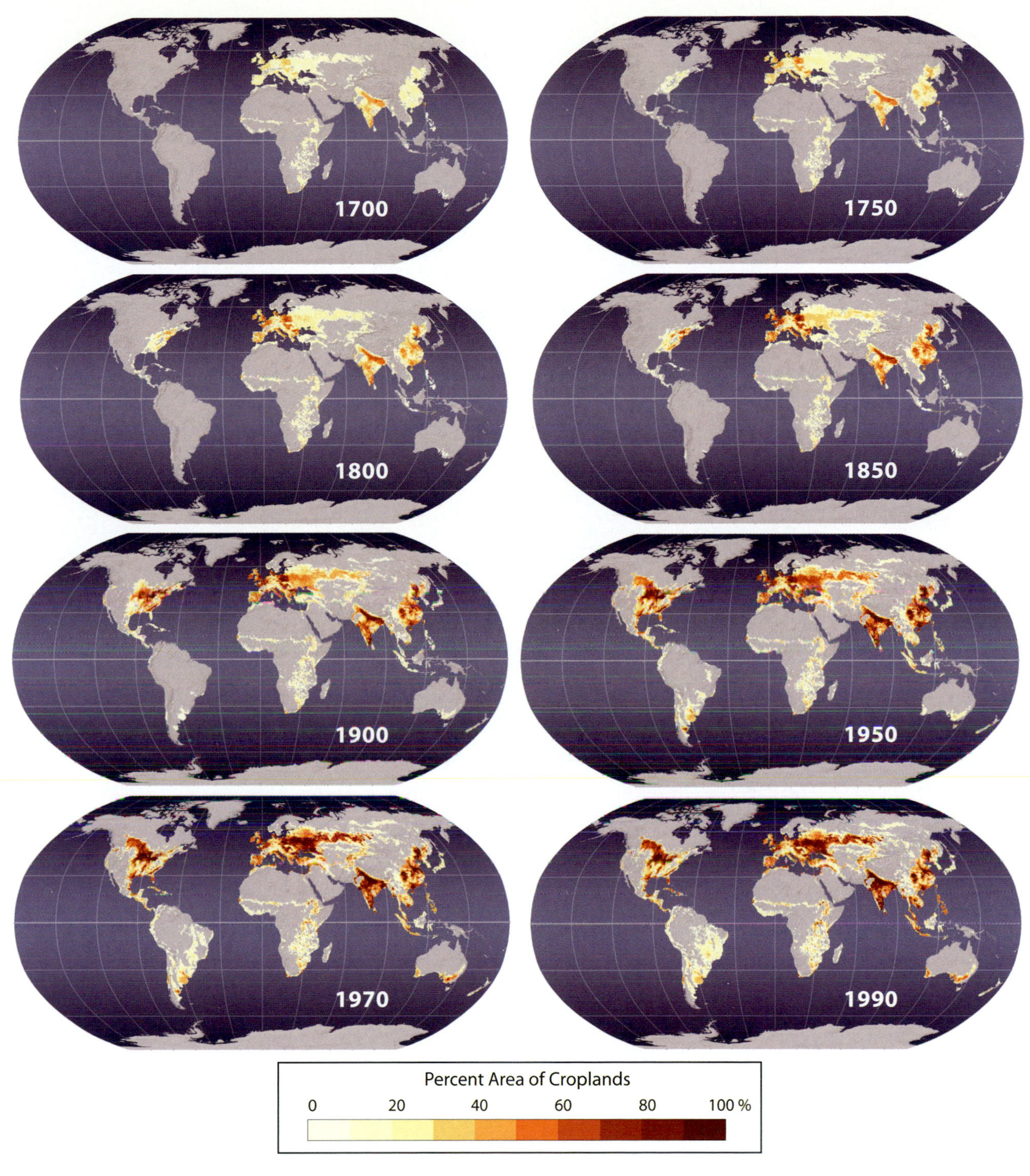

Plate 23. Historical changes in agricultural land (see chapter VII.7). Global croplands in 1700, 1750, 1800, 1850, 1900, 1950, 1970, and 1990, as estimated by Ramankutty and Foley (1999). Over the last three centuries, global cropland area increased by 12 million square kilometers (roughly 466%). In the 1700s and 1800s, croplands expanded rapidly in Europe, one of the most economically developed regions of the world at that time. After the mid-1800s, the newly developed regions of North America and the Former Soviet Union (FSU) then saw rapid cropland expansion. The rate of cropland expansion in China has been steady throughout the last three centuries. Croplands in Latin America, Africa, Australia, and South and Southeast Asia expanded very gradually between 1700 and 1850 but have experienced exponential growth rates since then. Since the 1950s, cropland areas in North America, Europe, and China have stabilized and even decreased somewhat in Europe and China. Cropland areas increased significantly in the FSU between 1950 and 1960 but have decreased since. In the last two decades, the major areas of cropland expansion were located in Southeast Asia, parts of South Asia (Bangladesh, Indus Valley, Middle East, Central Asia), in the Great Lakes region of eastern Africa, in the Amazon Basin, and in the U.S. Great Plains. The major decreases of cropland occurred in the southeastern United States, eastern China, and parts of Brazil and Argentina.

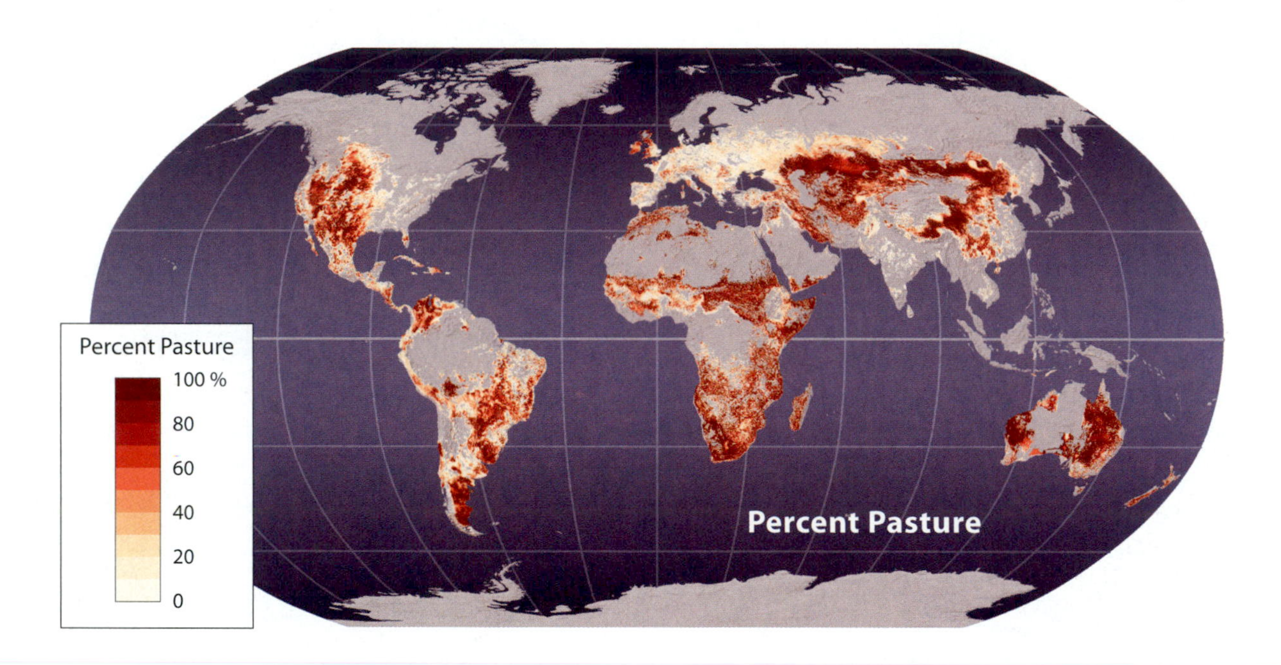

Plate 24. Global pastures, as determined by Ramankutty et al. (2008) (see chapter VII.7). With the advent of satellite remote sensing, it is now possible to monitor agricultural lands at the global scale. Using satellite images and ground-based census records, we see that croplands and pastures are now among the largest biomes on the planet, rivaling forests in geographic extent. Altogether, nearly 15 million square kilometers of land (roughly the size of South America) is used as croplands on the planet, and another ~30 million square kilometers of land (roughly the size of Africa) is used for pastures; together, croplands and pastures occupy ~35% of the ice-free land surface of the planet (Ramankutty et al., 2008).

V

Conservation Biology

David S. Wilcove

Given the rate at which humans are changing the biosphere—altering land cover and nutrient cycles, extirpating some species while spreading others around the globe, even changing the very climate of the planet—it is easy to understand why so many ecologists choose to focus their research on questions relevant to conservation. Indeed, a seemingly new discipline, conservation biology, replete with its own society and journal, arose in the late 1980s to capture the growing enthusiasm for research directed toward maintaining the earth's biodiversity. But, as many authors have noted, the roots of conservation biology go back decades, even centuries. In 1917, for example, the Ecological Society of America created a committee "charged with the listing of all preserved and preservable areas in North America in which natural conditions persist." The committee's report, published in 1926 as *Naturalist's Guide to the Americas* (Shelford, 1926) represented an early, crude "gap analysis" of protected ecosystems in southern Canada and the United States. Starting in the late 1930s, the National Audubon Society hired young biologists to study North America's rarest birds, including the ivory-billed woodpecker, California condor, and whooping crane, in an effort to prevent those species from disappearing (e.g., Koford, 1953; Tanner, 1966).

In some respects, these early assessments of declining ecosystems and imperiled species presage much of the contemporary literature in conservation ecology. The question naturally arises, what is new about conservation biology? Much of the novelty of conservation biology lies in its synthesis of many other disciplines, including evolutionary biology, ecology, economics, and sociology, for the purposes of understanding and ultimately addressing problems related to the loss of biodiversity (Groom et al., 2006). Moreover, contemporary conservation biology draws heavily from ecological and evolutionary theory with the goal of developing principles and insights that transcend particular species or ecosystems.

The chapters in this section cover some (but by no means all) of the "hot topics" in conservation biology, and they somewhat crudely trace the growth of the discipline itself. The section begins with a focus on species. Species extinction is, after all, one of the most visible and irreversible manifestations of biodiversity loss, and it remains the subject of much current research. Sodhi, Brook, and Bradshaw (chapter V.1) provide an overview of our current knowledge of human-caused extinctions. They compare the current rate of species loss (driven almost entirely by human activities) with the five great extinction events recorded in the geological record and find the destructive power of humans to be comparable to that of asteroid strikes and other abiotic events that have eliminated vast numbers of species in the past. Sodhi et al. also review the myriad ways in which human activities endanger biodiversity as well as the life-history traits that make certain species more vulnerable than others. They conclude with an alarming summary of the ways in which the loss of particular species, such as large predators and small pollinators, can trigger the extinction of many other plants and animals.

In reading the accounts of the biologists who studied ivorybills and condors more than a half-century ago, one cannot help but be impressed by their superb natural-history skills and their willingness to endure great discomfort and danger in pursuit of research. Yet the resulting work lacks predictive power. These scientists knew these animals were teetering on the brink of extinction, and in most cases they understood the reasons why. But they could not say how many individuals and populations needed to be saved in order to prevent extinction or what arrangement of habitat reserves would suffice to protect these birds for another century or two. The field of population viability analysis (PVA), discussed by Doak, Finkelstein, and Bakker in chapter V.2, strives to answer those types of questions. It entails the use of quantitative models to predict how populations of different sizes and configurations are likely to fare in the face of various natural and human-caused impacts. PVA featured prominently in the rancorous debate over the conservation of old-growth forests and northern spotted owls in the late 1980s. It has subsequently become an integral tool in conservation planning for endangered species around the world.

Given our improved knowledge of the factors influencing the persistence of populations, how do we design effective reserves for species in trouble? In chapter V.3, Haddad reviews the growing literature on reserve design. He summarizes the numerous theoretical, empirical, and experimental studies that bear on this issue, with special emphasis on habitat area, connectivity (achieved via habitat corridors and stepping-stones), and edge effects. As Haddad notes, the objective behind much of this work is to maximize the effective area of reserves—in other words, to protect blocks of habitat of the right size, shape, and distribution to preserve their target species for as long as possible. Thus, there is a clear linkage between the PVAs discussed by Doak et al. and the reserve design issues discussed by Haddad.

Although conservation biologists and practitioners continue to devote time and attention to the protection of individual species, the sheer scope of the contemporary biodiversity crisis requires them to think in terms of networks of reserves that protect multiple species. In chapter V.4, Turner and Pressey review the new and rapidly growing field of systematic conservation planning. They highlight the development of increasingly sophisticated algorithms designed to help planners deal with the painful realities of contemporary conservation: inadequate funding, incomplete data, ongoing losses of wild lands, unpredictable opportunities to acquire key parcels of land, and the differing area requirements of species. The goal of these algorithms is to identify systems or networks of conservation areas that meet explicit, quantifiable biodiversity targets with maximum efficiency and effectiveness. It is probably fair to say that, at the present time, the sophistication of these algorithms exceeds the abilities of many organizations to use them. Nonetheless, as more institutions become familiar with these tools, and as the tools themselves become more user friendly, they will surely be used far more frequently.

The final three chapters focus on topics that were, to varying degrees and for different reasons, neglected in the early years of contemporary conservation but that are now at the forefront of the field.

In chapter V.5, Jackson documents the frightening degree to which humans have degraded marine ecosystems. In particular, he highlights the pervasive impacts of overfishing and pollution, which he identifies as the main drivers of change in the world's oceans. The problem of overfishing, Jackson notes, is so vast that, across the globe, most populations of whales, sea turtles, and large predatory fish have been hunted to the point of ecological extinction; their current densities are so low that these species play no significant ecological role in the ecosystems where they persist. Readers familiar with the history of wildlife in North America may be reminded of the late nineteenth century, when a seemingly insatiable demand for pelts, flesh, and feathers, coupled with a hatred of large predators, led to the elimination or endangerment of bison, wolves, bears, mountain lions, wading birds, shorebirds, and other species across much of the continent. Conservation of the seas appears to be a century or more behind conservation of terrestrial ecosystems, a lapse made all the more dangerous by the fact that our destructive powers at the start of the twenty-first century vastly exceed those at the start of the twentieth.

A century ago, the notion that people could alter the earth's ecosystems by inadvertently altering its climate would have seemed ridiculous to most scientists. Needless to say, that is no longer the case. In chapter V.6, Debinski and Cross tackle the challenges that climate change poses for conservation. They argue that the abiotic changes stemming from climate change—for example, rising sea levels, altered precipitation patterns, and disruptions of natural disturbance regimes—will affect the distribution and abundance of species. As individual species fade in or out, other species (e.g., predators, competitors, mutualist partners) will be affected too, leading to cascading effects. Incorporating the effects of climate change into PVAs (chapter V.2) and reserve design algorithms (chapter V.4) remains a major challenge in conservation biology. The oceans, too, are hardly immune to global climate change. One such threat, highlighted by Jackson in chapter V.5, is increasing acidification of the seas; the possible consequences are almost too frightening to imagine.

Given the gaps in our network of protected areas, habitat destruction, climate change, and continuing losses of ecosystem services essential to human welfare, the ability to repair damaged ecosystems is surely one of the most important weapons in our conservation arsenal. In chapter V.7, Hobbs discusses the underappreciated but rapidly growing field of restoration ecology. Defined as the science underlying the practice of repairing damaged ecosystems, restoration ecology draws heavily from ecology and environmental engineering. Hobbs is careful to point out that ecological restoration is not simply an exercise in time travel, an attempt to restore an ecosystem to what it was like at some arbitrary point in the past. After all, many ecosystems are inherently dynamic; their composition and structure change naturally over time. Moreover, it may not be possible to restore all of the species that once occupied the ecosystem, regardless of how important they may have been. No restoration plan for the hardwood forests of eastern North America includes the passenger pigeon, and very few include the gray wolf or mountain lion. Thus, as Hobbs notes, practitioners of ecosystem restoration are increasingly of the opinion that there can be a range of outcomes for any given place. The challenge then becomes one of developing a "transparent and defensible method of setting restoration goals that clarify the desired characteristics for the system in the future, rather than in relation to what these were in the past" (chapter V.7).

The chapters presented in this section are, at best, a limited subset of the range of topics that fall under the umbrella of conservation biology. Some of the chapters in other sections (e.g., invasive species) could easily have been placed here. Conversely, many of the chapters in this section would fit equally well in other sections of the volume. This healthy ambiguity is a reflection of the fact that conservation biology has successfully borrowed from and contributed to a wide range of subjects in ecology.

FURTHER READING

Groom, M. J., G. K. Meffe, and C. R. Carroll. 2006. Principles of Conservation Biology, 3rd ed. Sunderland, MA: Sinauer Associates.

Koford, C. B. 1953. The California Condor. National Audubon Society Research Report No. 4. New York: National Audubon Society.

Shelford, V. E. 1926. Naturalist's Guide to the Americas. Baltimore: Williams & Wilkins.

Tanner, J. 1942. The Ivory-billed Woodpecker. National Audubon Society Research Report No. 1. New York: National Audubon Society.

V.1

Causes and Consequences of Species Extinctions

Navjot S. Sodhi, Barry W. Brook, and Corey J. A. Bradshaw

OUTLINE

1. Introduction
2. Extinction drivers
3. Extinction vulnerability
4. Consequences of extinctions
5. Conclusions

The five largest mass die-offs in which 50–95% of species were eliminated occurred during the Ordovician [490–443 million years ago (mya)], Devonian (417–354 mya), Permian (299–250 mya), Triassic (251–200 mya), and Cretaceous (146–64 mya) periods. Most recently, human actions especially over the past two centuries have precipitated a global extinction crisis or the "sixth great extinction wave" comparable to the previous five. Increasing human populations over the last 50,000 years or so have left measurable negative footprints on biodiversity.

GLOSSARY

Allee effects. These factors cause a reduction in the growth rate of small populations as they decline (e.g., via reduced survival or reproductive success).

coextinction. Extinction of one species triggers the loss of another species.

extinction debt. This refers to the extinction of species or populations long after habitat alteration.

extinction vortex. As populations decline, an insidious mutual reinforcement occurs among biotic and abiotic processes driving population size downward to extinction.

extirpation. This refers to extinction of a population rather than of an entire species.

invasive species. These are nonindigenous species introduced to areas outside of their natural range that have become established and have spread.

megafauna. This refers to large-bodied (>44 kg) animals, commonly (but not exclusively) used to refer to the large mammal biota of the Pleistocene.

minimum viable population. This is the number of individuals in a population required to have a specified probability of persistence over a given period of time.

1. INTRODUCTION

In the Americas, charismatic large-bodied animals (megafauna) such as saber-toothed cats (*Smilodon* spp.), mammoths (*Mammuthus* spp.), and giant ground sloths (*Megalonyx jeffersonii*) vanished following human arrival some 11,000–13,000 years ago. Similar losses occurred in Australia 45,000 years ago, and in many oceanic islands within a few hundred years of the arrival of humans. Classic examples of the loss of island endemics include the dodo (*Raphus cucullatus*) from Mauritius, moas (e.g., *Dinornis maximus*) from New Zealand, and elephantbirds (*Aepyornis maximus*) from Madagascar. Megafaunal collapse during the late Pleistocene can largely be traced to a variety of negative human impacts, such as overharvesting, biological invasions, and habitat transformation.

The rate and extent of human-mediated extinctions are debated, but there is general agreement that extinction rates have soared over the past few hundred years, largely as a result of accelerated habitat destruction following European colonialism and the subsequent global expansion of the human population during the twentieth century. Humans are implicated directly or indirectly in the 100- to 10,000-fold increase in the "natural" or "background" extinction rate that normally occurs as a consequence of gradual environmental change, newly established competitive interactions (by evolution or invasion), and occasional chance calamities such as fire, storms, or disease. The current and future extinction rates are estimated using a variety of measures such as species–area models and changes in the World Conservation Union's (IUCN) threat categories over time. Based on the global assessment of all known species, some 31, 12, and 20% of known amphibian, bird, and mammal species, respectively (by far the best-studied of all animal groups), are currently listed by the IUCN as under threat.

Just how many species are being lost each year is also hotly debated. Various estimates range from a few thousand to more than 100,000 species being extinguished every year, most without ever having been scientifically described. The large uncertainty comes mainly through the application of various species–area relationships that vary substantially among communities and habitats. Despite substantial prediction error, it is nevertheless certain that human actions are causing the structure and function of natural systems to unravel. The past five great extinctions shared some important commonalities: (1) they caused a catastrophic loss of global biodiversity; (2) they unfolded rapidly (at least in the context of evolutionary and geological time); (3) taxonomically, their impact was not random (that is, whole groups of related species were lost while other related groups remained largely unaffected); and (4) the survivors were often not previously dominant evolutionary groups. All four of these features are relevant to the current biodiversity crisis. This sixth great extinction is likely to be most catastrophic in tropical regions given the high species diversity there (more than two-thirds of all species) and the large, expanding human populations that threaten most species there as well.

The major "systematic drivers" of modern species loss are changes in land use (habitat loss degradation and fragmentation), overexploitation, invasive species, disease, climate change (global warming) connected to increasing concentration of atmospheric carbon dioxide, and increases in nitrogen deposition. Mechanisms for prehistoric (caused by humans >200 years ago) extinctions are likely to have been similar: overhunting, introduced predators and diseases, and habitat destruction when early people first arrived in virgin landscapes.

2. EXTINCTION DRIVERS

Some events can instantly eliminate all individuals of a particular species, such as an asteroid strike, a massive volcanic eruption, or even a rapid loss of large areas of unique and critical habitat because of deforestation. But ultimately, any phenomena that can cause mortality rates to exceed reproductive replacement over a sustained period can cause a species to become extinct. Such forces may act independently or synergistically, and it may be difficult to identify a single cause of a particular species extinction event. For instance, habitat loss may cause some extinctions directly by removing all individuals, but it can also be indirectly responsible for an extinction by facilitating the establishment of an invasive species or disease agent, improving access to human hunters, or altering biophysical conditions. As a result, any process that causes a population to dwindle may ultimately predispose that population to extinction.

Evidence to date suggests that deforestation is currently, and is projected to continue to be, the prime direct and indirect cause of reported extirpations. For example, it is predicted that up to 21% of Southeast Asian forest species will be lost by 2100 because of past and ongoing deforestation. Similar projections exist for biotas in other regions.

Overexploitation is also an important driver of extinctions among vertebrates and tends to operate synergistically with other drivers such as habitat loss. For example, roads and trails created to allow logging operations to penetrate into virgin forests make previously remote areas more accessible to human hunters, who can, in turn, cause the decline and eventual extirpation of forest species. It is estimated that overexploitation is a major threat to at least one-third of threatened birds and amphibians, with wildlife currently extracted from tropical forests at approximately six times the sustainable rate. In other words, the quantity, and most likely the diversity, of human prey—both fisheries and "bush" (wild) meat—are rapidly diminishing.

Megafauna—those species weighing in the tens to hundreds of kilograms—are among the most vulnerable to overexploitation. In general, a species' generation time (interval from birth to reproductive age) is a function of body mass (allometry), so larger, longer-lived, and slower-reproducing animal populations are generally unable to compensate for high rates of harvesting. Because slow-breeding large animals, such as apes, carnivores (e.g., the lion, *Panthera leo*), and African elephants (*Loxodonta africana*), are particularly vulnerable to hunting, the potential for population recovery in these animals over short time scales is low. As an example supporting this generality, there is evidence that 12 large vertebrate species have been extirpated from Vietnam, primarily because of excessive hunting, within the past 40 years. The Steller's sea cow (*Hydrodamalis gigas*), an aquatic herbivorous mammal that inhabited the Asian coast of the Bering Sea, is the quintessential example of the rapid demise of a species as a result of overexploitation. Discovered in 1741, it became extinct by 1768 because of overhunting by sailors, seal hunters, and fur traders. This species was hunted for food, its skin for making boats, and its subcutaneous fat for use in oil lamps.

The ecosystem and biological community changes precipitated by invasive species represent another leading cause of biodiversity loss. Of 170 extinct species for which causes have been identified reliably, invasive species contributed directly to the demise of 91 (54%). In particular, the rates of extinctions occurring on islands have been greatly elevated by the introduction of novel predators. Several ecological and life-history attributes of island species, such as their naturally constrained geographic range, small population sizes, and particular traits (e.g., lack of flight in birds or lack of thorns in plants) make island biotas vulnerable to predation from invading species. For example, the introduction of the brown tree snake (*Boiga irregularis*) shortly after World War II wreaked havoc on the biodiversity of the island of Guam in the South Pacific. In all likelihood, tree snakes were directly responsible for the loss of 12 of 18 native bird species, and they also reduced the populations of other vertebrates such as flying foxes (*Pteropus mariannus*), mainly because of the inability of the island's native species to recognize the novel predator as a threat. Despite an annual expenditure of US$44.6 million for the management of this problem, tree snakes on Guam are still not under control, largely because of their ability to penetrate artificial snake barriers such as fences.

The mosquito *Culex quinquefasciatus* was inadvertently introduced to Hawaii in 1826, and the disease-causing parasite (*Plasmodium relictum*) it carries arrived soon after. Since then, avian malaria (in conjunction with other threats) has been responsible for the decline and extinction of some 60 species of endemic forest birds on the Hawaiian Islands. Having evolved in the absence of the disease, Hawaiian bird species were generally unable to cope with the debilitating effects of the novel parasite. However, more than 100 years after the establishment of the disease, some native thrushes (*Myadestes* spp.) are now showing resistance to the disease. Sadly, many of the remaining species, especially forest birds in the family Drepanididae, are still vulnerable and are now restricted to altitudes where temperatures are below the thermal tolerance limits of the mosquito vector. Global warming is predicted to increase the altitudinal distribution of the mosquito, thus spelling doom for disease-susceptible birds as mosquito-free habitats disappear. The most feasible method of reducing transmission of malaria is to reduce or eliminate vector mosquito populations through chemical treatments and the elimination of larval habitats.

Perhaps one of the most infamous examples of an invasion catastrophe occurred in the world's largest freshwater lake—Lake Victoria in tropical East Africa. Celebrated for its amazing collection of over 600 endemic haplochromine (i.e., formerly of the genus *Haplochromis*) cichlid fishes (Family Cichlidae), the Lake Victoria cichlid community is perhaps one of the most rapid, extensive, and recent vertebrate radiations known. There is also a rich community of endemic noncichlid fish that inhabit the Lake. In addition to the threats posed to this unique biota by a rapid rise in fisheries exploitation, human density, deforestation, and agriculture during the past century, without doubt the most devastating effect was the introduction of the predatory Nile perch (*Lates niloticus*) in the 1950s. This voracious predator, which can grow to more than 2 m in length, was introduced from lakes Albert and Turkana (Uganda and Kenya, respectively) to compensate for depleting commercial fisheries in Lake Victoria. Although the Nile perch population remained relatively low for several decades after its introduction, an eventual population explosion in the 1980s caused the devastating direct or indirect extinction of 200–400 cichlid species endemic to the Lake as well as the extinction of several noncichlid fish species. Although many other threats likely contributed to the observed extinctions, including direct overexploitation and eutrophication from agriculture and deforestation leading to a change in the algal plankton community, there are few other contemporary examples of such a rapid and massive extinction event involving a single group of closely related species.

Human-mediated climate change represents a potentially disastrous sleeping giant in terms of future biodiversity losses. Climate warming can affect species in five principal ways: (1) alterations of species densities (including altered community composition and structure); (2) range shifts, either poleward or upward in elevation; (3) behavioral changes, such as the phenology (seasonal timing of life cycle events) of migration, breeding, and flowering; (4) changes in morphology, such as body size; and (5) reduction in genetic diversity that leads to inbreeding depression. A related threat for island and coastal biotas is the predicted loss of habitat via inundation by rising sea levels. Although large fluctuations in climate have occurred regularly throughout Earth's history, the implications of anthropogenic global warming for contemporary biodiversity are particularly pessimistic because of the rate of change and previous heavy modification of landscapes by humans. Good empirical evidence for some of these effects is rare, and speculations abound, but there are already many local or regional examples and model-based predictions that support the view that rapid climate change, acting in concert with other drivers of species loss and habitat degradation, will be one of the most pressing conservation issues global biodiversity faces over the coming centuries.

One glimpse of a possible future crisis comes from the highland forests of Monterverde (Costa Rica), where 40% (20 of 50) of frog and toad species disappeared following synchronous population crashes in 1987, with most crashes linked to a rapid progressive warming and drying of the local climate. The locally endemic golden toad (*Bufo periglenes*) was one of the high-profile casualties in this area. It has been suggested that climate warming resulted in a retreat of the clouds and a drying of the mountain habitats, making amphibians more susceptible to fungal and parasite outbreaks. Indeed, the pathogenic chytrid fungus *Batrachochytrium dendrobatidis*, which grows on amphibian skin and increases mortality rates, has been implicated in the loss of harlequin frogs (*Atelopus* spp.) in Central and South America and reductions in other amphibian populations elsewhere. It is hypothesized that warm and dry conditions may stress amphibians and make them more vulnerable to the fungal infection.

Irrespective of the reason for a population's decline from a large to small population size, unusual (and often random and detrimental) events assume prominence at low abundances. For instance, although competition among individuals is reduced at low densities and can induce a population rebound, a countervailing phenomenon known as the "Allee effect" can act to draw populations toward extinction by (for instance) disrupting behavioral patterns that depend on numbers (e.g., herd defense against predators) or by genetic threats such as inbreeding depression. Small populations, dominated by chance events and Allee effects, are often considered to have dipped below their "minimum viable population" size. Thus, once a major population decline has occurred (from habitat loss, overexploitation, or in response to many other possible stressors), an "extinction vortex" of positive feedback loops can doom species to extinction, even if the original threats have been alleviated. Further, many species may take decades to perish following habitat degradation. Although some species may withstand the initial shock of land clearing, factors such as the lack of food resources, breeding sites, and dispersers may make populations unviable, and they eventually succumb to extinction. This phenomenon evokes the concept of "living-dead" species, or those "committed to extinction." The eventual loss of such species is referred to as the "extinction debt" caused by past habitat loss. For example, even if net deforestation rates can be reduced or even halted, the extinction debt of remnant and secondary forest patches will see the extinction of countless remaining species over this interval.

3. EXTINCTION VULNERABILITY

Certain life-history, behavioral, morphological, and physiological characteristics appear to make some species more susceptible than others to the extinction drivers described above. In general, large-sized species with a restricted distribution that demonstrate habitat specialization tend to be at greater risk of extinction from human agency than others within their respective taxa (e.g., Javan rhinoceros, *Rhinoceros sondaicus*), especially to processes such as rapid habitat loss.

Because of their high habitat specificity and/or low population densities, rare species may be more prone to extinction than common species. The size of a species' range is also a major determinant of its extinction proneness. Small ranges may make species more vulnerable to stochastic perturbations, even if local abundance is high; for example, proportionally more passerines (perching birds) with relatively small geographic ranges in the Americas are at risk of extinction than their more widely distributed counterparts. Such trends are worrisome because those species with shrinking ranges as a result of adverse human activities become particularly vulnerable to other drivers such as climate change. Habitat loss also reduces the patch sizes necessary for species requiring large home ranges, making them vulnerable to extinction from a loss of subpopulation connectedness, reduced dispersal capacity, and the ensuing lower population viability.

Larger-bodied vertebrates are considered to be more extinction-prone than smaller-bodied ones when the threatening process unfolds rapidly or intensely. Indeed, threatened mammals are an order of magnitude heavier than nonthreatened ones. A common explanation for this trend is that body size is inversely correlated with population size, making large-bodied animals less abundant and more vulnerable to chronic environmental perturbations (while being buffered against short-term environmental fluctuations). The extinction proneness of large-bodied animals to human activities is further enhanced because of other correlated traits, such as their requirement of large area, greater food intake, high habitat specificity, and lower reproductive rate.

Large species can also be more vulnerable to human persecution such as hunting, whereas smaller species are generally more vulnerable to habitat loss. It is important, however, to be cautious when constructing generalized rules regarding the role of body size in the extinction process. Because they have a slower reproductive rate, larger parrots are more vulnerable to overexploitation than smaller finches, despite fewer numbers of the former being captured for the pet trade. However, some smaller species (e.g., white-eyes, *Zosterops* spp.) with small population sizes are also vulnerable to extinction because of heavy harvest rates for the pet trade, suggesting that only when the threatening processes are approximately equivalent will the larger of two species being compared demonstrate a higher risk of extinction. In addition to body size, other morphological characteristics affect extinction proneness. For instance, large investment in secondary sexual characteristics may render highly dimorphic species less adaptable in a changing environment or more attractive to specimen or pet-trade collectors.

When an environment is altered abruptly or systematically at a rate above normal background change, or beyond the capacity of adaptation via natural selection, specialist species with narrow ecological niches often bear the brunt of progressively unfavorable conditions such habitat loss and degradation. For instance, highly specialized forest-dependent taxa are acutely vulnerable to extinction following deforestation and forest fragmentation. Possible mechanisms include reductions in breeding and feeding sites, increased predation, elevated soil erosion and nutrient loss, dispersal limitation, enhanced edge effects, and other stressors. Conversely, non-forest-dependent species or those that prefer open habitats are often better able to persist in disturbed landscapes and may even be favored by having fewer competitors or expanded ranges following deforestation. It is important to be aware that in relatively stable systems, evolution engenders the speciation of taxa that occupy all available niches so both specialist and generalist species can coexist. As a result, the rapid pace of habitat and climate change renders specialization a modern "curse" in evolutionary terms.

Foraging specialization is one mechanism that can compromise a species' ability to persist in altered habitats. Many studies have shown that frugivorous and insectivorous birds are more extinction-prone than other avian feeding guilds, with the lack of year-round access to fruiting plants in fragmented forests being the culprit for the former. A number of hypotheses have been proposed to explain the disappearance of insectivorous birds from deforested or fragmented areas. First, deforestation may impoverish the insect fauna and reduce selected insectivore microhabitats (e.g., dead leaves). Second, insectivores may be poor dispersers and have near-ground nesting habits, the latter trait making them more vulnerable to nest predators penetrating smaller forest fragments. Absence of some insectivorous bird species from small fragments may not be related to food scarcity; rather, it may result from their poorer dispersal abilities. The ability to disperse in birds and insects depends on morphological characteristics such as wing loading, and physiological restrictions such as intolerance to sunlight when moving within the nonforested matrix landscape separating fragments. As a result, poor dispersal ability may make certain species vulnerable to extinction because they cannot readily supplement sink habitats (habitats in which populations cannot replace themselves), supporting otherwise unviable subpopulations, or colonize new areas. Because of poor dispersal ability, patchy distributions, and generally low population densities, the genetic diversity of species in fragmented landscapes may be difficult to maintain, with the resulting inbreeding depression further reducing population size toward extinction. However, clear and quantitative demonstrations of the role of life-history traits in the extinction process of biotas are still rare.

4. CONSEQUENCES OF EXTINCTIONS

The extinction of certain species such as large predators and pollinators may have more devastating ecological consequences than the extinction of others. Ironically, avian vulnerability to predation is often exacerbated when certain large predatory species become rarer in tropical communities. For example, although large cats such as jaguars (*Panthera onca*) do not prey on small birds directly, they exert a limiting force on smaller predators such as medium-sized and small mammals (mesopredators), which become more abundant with the former species' decline. The corollary is that abundant mesopredators inflict an above-average predation rate on the eggs and nestlings of small birds. Although this "mesopredator-release" hypothesis has been applied largely to mammals (e.g., Australian dingoes, *Canis lupus*, suppressing foxes and cats; coyotes in California controlling cat abundance), the loss of large predatory birds such as the harpy eagle (*Harpia harpyja*) may have similar ecosystem effects. Similar mesopredator release has been demonstrated for the first time in the marine environment, where the overexploitation of large pelagic sharks resulted in an increase in rays and skates that eventually suppressed commercially important scallop populations. Likewise, does the disappearance of a competitor result in the niche expansion and higher densities of subordinate species? This phenomenon has been observed between unrelated taxa—the extinction of insectivorous birds from scrub forests of West Indian islands correlated with the subsequent higher biomass of competing *Anolis* lizards.

Conservation biologists have traditionally focused on the study of the independent declines, extirpations, or extinctions of individual species while paying relatively less attention to the possible cascading effects of species coextinctions (e.g., hosts and their parasites). However, it is likely that many coextinctions between interdependent taxa have occurred, but most have gone unnoticed in these relatively understudied systems. For example, an extinct feather louse (*Columbicola extinctus*) was discovered in 1937, 23 years after likely coextinction with its host passenger pigeon (*Ectopistes migratorius*). Ecological processes disrupted by extinction or species decline may also lead to cascading and catastrophic coextinctions. Frugivorous animals and fruiting plants on which they depend have a key interaction linking plant reproduction and dispersal with animal nutrition. Thus, the two interdependent taxa are placed in jeopardy by habitat degradation. Many trees produce large, lipid-rich fruits adapted for animal dispersal, so the demise of avian frugivores may have serious consequences for forest regeneration, even if the initial drivers of habitat loss and degradation are annulled.

Essential ecosystem functions provided by forest invertebrates are also highly susceptible when species are lost after habitat loss and degradation. Acting as keystone species in Southeast Asian rainforests, figs rely on tiny (1–2 mm) species-specific wasps for their pollination. Some fig wasps may have limited dispersal ability, suggesting that forest disturbance can reduce wasp densities and, by proxy, the figs that they pollinate. Similarly, dung beetles are essential components of ecosystem function because they contribute heavily to nutrient-recycling processes, seed dispersal, and the reduction of disease risk associated with dung accumulation. In Venezuela, heavier dung beetles were more extinction-prone than lighter species on artificially created forested islands, which predicts particularly dire ecosystem functional loss given the former group's greater capacity to dispose of dung.

Almost all flowering plants in tropical rainforests are pollinated by animals, and an estimated one-third of the human diet in tropical countries is derived from insect-pollinated plants. Therefore, a decline of forest-dwelling pollinators impedes plant reproduction not only in forests but also in neighboring agricultural areas visited by these species. Lowland coffee (*Coffea canephora*) is an important tropical cash crop, and it depends on bees for cross-pollination. A study in Costa Rica found that forest bees increased coffee yield by 20% in fields within 1 km of the forest edge. Between 2000 and 2003, the pollination services provided by forest bees were worth US$60,000 to a 1100-ha farm. A forest patch as small as 20 ha located near farms can increase coffee yield and thus bring large economic benefits to the farmers. Such findings illustrate the imperative of preserving native forests near agroforestry systems to facilitate the travel by forest-dependent pollinating insects.

5. CONCLUSIONS

Although extinctions are a normal part of evolution, human modifications to the planet in the last few centuries, and perhaps even millennia, have greatly accelerated the rate at which extinctions occur. Habitat loss remains the main driver of extinctions, but it may act synergistically with other drivers such as overharvesting and pollution, and, in the future, climate change. Large-bodied species, rare species, and habitat specialists are particularly prone to extinction as a result of rapid human modifications of the planet. Extinctions can disrupt vital ecological processes such as pollination and seed dispersal, leading to cascading losses, ecosystem collapse, and a higher extinction rate overall.

FURTHER READING

Brook, Barry W., Navjot S. Sodhi, and Peter K. L. Ng. 2003. Catastrophic extinctions follow deforestation in Singapore. Nature 424: 420–423. *This is one of few papers reporting broad-scale extinctions driven by tropical deforestation.*

Clavero, Miguel, and Emili Garcia-Berthou. 2005. Invasive species are a leading cause of animal extinctions. Trends in Ecology and Evolution 20: 110. *The article highlights that invasive species represent one of the primary threats to biodiversity.*

Dirzo, Rudolfo, and Peter J. Raven. 2003. Global state of biodiversity and loss. Annual Review of Environment and Resources 28: 137–167. *The article constitutes a major review of the state of the modern global biodiversity and its associated losses.*

Fagan, William F., and E. E. Holmes. 2006. Quantifying the extinction vortex. Ecology Letters 9: 51–60. *This is the only study yet to quantify the final phases of extinction in vertebrates for which date of extinction was known.*

IUCN Red List of threatened species. Download from http://www.iucnredlist.org. *This presents an up-to-date classification of and reasons for a listed species' conservation status.*

Koh, Lian P., Robert R. Dunn, Navjot S. Sodhi, Robert K. Colwell, Heather C. Procter, and Vince S. Smith. 2004. Species co-extinctions and the biodiversity crisis. Science 305: 1632–1634. *This models how loss of a species could indirectly result in the extinction of dependent species.*

Pimm, Stuart L., and Peter Raven. 2000. Extinction by numbers. Nature 403: 843–845. *The article summarizes the likely extent of biodiversity losses as a result of human activities.*

Pounds, J. Alan, Martin R. Bustamante, Luis A. Coloma, Jamie A. Consuegra, Michael P. L. Fogden, Pru N. Foster, Enrique La Marca, Karen L. Masters, Andres Merino-Viteri, Robert Puschendorf, Santiago R. Ron, G. Arturo Sanchez-Azofeifa, Christopher J. Still, and Bruce E. Young. 2006. Widespread amphibian extinctions from epidemic disease driven by global warming. Nature 439: 161–167. *The article provides evidence on the role of climate change in recent amphibian extinctions.*

Ricketts, Taylor H., Gretchen C. Daily, Paul R. Ehrlich, and C. D. Michener. 2004. Economic value of tropical forest to coffee production. Proceedings of the National Academy of Sciences U.S.A. 34: 12579–12582. *This work shows how the loss of ecosystem services can affect pollination of commercial crops.*

Rosser, Alison M., and Sue A. Manika. 2002. Overexploitation and species extinctions. Conservation Biology 16: 584–586. *This work provides a quantitative overview of the extent of threat faced by birds and mammals from direct exploitation by people.*

Sekercioglu, Cagan H., Gretchen C. Daily, and Paul R. Ehrlich. 2004. Ecosystem consequences of bird declines. Proceedings of the National Academy of Sciences U.S.A. 101: 18042–18047. *This article provides a framework for assessing the loss of ecosystem functions caused by avian declines.*

V.2

Population Viability Analysis

Daniel F. Doak, Myra E. Finkelstein,
and Victoria J. Bakker

OUTLINE

1. Overview
2. The history of PVAs
3. Basic components and methods
4. Real-world examples
5. The future of PVAs

Population viability analysis (PVA) is the use of quantitative models to predict future population growth and extinction risks. PVA includes a variety of methods to gauge the sensitivity of population viability to natural and human-caused impacts and to estimate the efficacy of management interventions in promoting population growth and safety from extinction. PVA began as a field that borrowed tools from basic population ecology and applied them to conservation questions. From those beginnings, PVA has matured into a discipline that drives innovations in analysis methods and tries more generally to address the processes of conservation planning and priority setting. Because of their wide usage, in particular for assessing management actions, PVA approaches have been closely scrutinized, and the field continues to refine its methods to tackle key criticisms. In summary, PVA has provided specific guidance that has aided the recovery of scores of endangered species and has helped to crystallize several general principles in conservation.

GLOSSARY

demographic stochasticity. Unpredictability through time in a population's demography (how many individuals die, how many reproduce, etc.) caused by the randomness of individual fates. This type of stochasticity is usually important only at very small population sizes.

environmental stochasticity. Unpredictable changes through time in average demographic rates of a population. These changes can be caused by vacillations in weather, food, predators, or other biotic and abiotic forces influencing individuals in a population and can exert strong effects on the dynamics of populations.

genetic stochasticity. Unpredictable changes in gene frequencies as a result of processes such as random genetic drift. This type of stochasticity is usually important only at very small population sizes.

inbreeding depression. The decline in measures of individual performance (e.g., survival, growth, or reproduction) sometimes seen in offspring of parents that are closely related to one another.

lambda (λ). Annual population growth rate.

metapopulation. In general, a collection of populations that are connected by movement. More specifically, the term is usually reserved for a collection of populations each of which has reasonably high probabilities of local extinction and also of recolonization.

N_t. Population size in year *t*.

parameters. Values used to describe population dynamics in models, such as the mean or variance in fecundity or survival rate.

population viability. The probability of continued existence of a population. Viability is the converse of the risk of extinction (often defined in terms of quasi-extinction rather than complete extinction) over some time period.

quasiextinction threshold (N_{qe}). The minimum number of individuals below which a population is likely to be critically and immediately imperiled.

1. OVERVIEW

The International Union for Conservation of Nature (IUCN) currently recognizes over 15,000 species as threatened with extinction worldwide (http://www.iucnredlist.org/). However, given the uncertainty surrounding the status of numerous species or even how many species exist, the number of imperiled species on a global scale is almost certainly considerably more than those documented by the IUCN. The causes of species endangerment vary (see chapter V.1), but in all cases, conservation biologists working to avert extinctions wish to understand the degree of risk facing a particular species or population. Even more importantly, they wish to identify practical management actions that can substantially improve the viability—the long-term chances of persistence—of threatened populations. To answer these questions, the discipline of population viability analysis (PVA) has emerged over the last three decades. PVA, defined broadly, is the use of quantitative methods to predict the likely future status of populations of conservation concern and also to predict how best to manage these populations.

PVA grew into a distinct field in ecology because making predictions about population persistence is quite difficult. As Yogi Berra once said (perhaps misquoting a similar observation by Niels Bohr), "It's tough to make predictions, especially about the future," and this is particularly true for the population processes described in PVAs, which are typically known only through imperfect data and are influenced by myriad random, or stochastic, forces. PVAs are developed to generate these hard-to-make predictions in a way that is clearly reasoned and quantitative rather than based solely on expert opinion. Importantly, constructing a quantitative PVA model requires explicit articulation of what is known about a population versus what is assumed or guessed. Thus, the process of conducting a PVA hones a management team's thinking about conservation problems and data limitations while providing better and more defensible answers. Although critics of PVAs have sometimes taken aim at the accuracy of PVA predictions, there is general agreement that comparing the predictions of a PVA for one management scenario relative to another usually provides robust and useful guidance for decision makers.

PVAs have utility in a wide variety of management and basic ecological contexts. Over the years they have been used to answer such questions as: (1) What life stages should be prioritized for increased protection in order to decrease short-term extinction risks of a rare population? (2) How large do reserves need to be in order to maintain key species? (3) What parts of the life cycle should be targeted to reduce or eliminate populations of invasive species? (4) How can the harvest of populations be maximized without causing declines? (5) What features of metapopulations will allow them to persist in patches of fragmented habitat? New directions in the field include integrating PVA more directly into adaptive management decisions, evaluating the effects of sublethal threats on population persistence, and predicting the effects of multispecies interactions on extinction risks.

2. THE HISTORY OF PVAS

As of this writing, PVAs number in the hundreds to thousands and range from analyses for tiny fairy shrimp confined to vernal pools to those for whale populations spanning entire oceans. In spite of this diversity, PVAs generally share the same basic components. All PVAs are descriptions of the dynamics of a population or a collection of populations. As such, almost any population model could be considered a PVA, especially if it is used to describe imperiled or managed populations (see chapters II.1, II.2, and II.4 for basic discussions of population models). PVA first emerged as a distinct discipline within population ecology with Mark L. Shaffer's 1978 analysis of the Yellowstone grizzly population. Shaffer built a demographic model for this isolated bear population and used computer simulations to estimate the numbers of bears needed to ensure a reasonable chance (Shaffer chose 95%) of persistence over the next 100 years. By providing a mathematical description of how this population worked—the deterministic and stochastic processes that made numbers grow or shrink through time—and using it to ask about future viability under different scenarios of management, initial numbers, and other factors, Shaffer's analysis established many of the features that still characterize PVAs today.

Since Shaffer's first PVA, over 25 PVAs have been published for grizzlies, with at least 18 different models for the Yellowstone population alone. This great number of PVAs, including new and distinct PVAs for the same population, reflect both the usefulness of PVA results to management and the evolution of scientific understanding about the critical forces impacting populations. For example, during the 1980s, key concerns in many PVAs were loss of genetic variation and the resulting processes of inbreeding and inbreeding depression. However, faced with continued ignorance of how genetic inbreeding influences survival and reproduction, developers of PVAs have since shifted away from explicit consideration of genetics, focusing instead on maintaining populations at large enough numbers that loss of genetic diversity and inbreeding depression are unlikely to be serious problems. This minimum acceptable size, defined in part by genetic factors, is referred to as a quasiextinction threshold. In the 1990s, and continuing to the present, PVAs have typically concentrated on the impacts of environmental stochasticity, human-caused threats, and the spatial dynamics of metapopulations. Most recently, PVAs have begun to account for climate change and invasive species impacts because of the increased importance of these threats to population survival. These trends in the conditions and complications that PVAs emphasize reflect a key feature of this field of conservation biology: PVA methods are applied to rare species in dynamic systems for which conservation biologists generally possess incomplete knowledge of both population processes and current and future threats. Indeed, one criticism of PVAs is that they cannot account for unforeseen changing future conditions, and thus, many scientists advocate their use only for relatively short time horizons. In addition to restricting time horizons, PVA predictions can be improved by periodically refining models as methods improve, understanding of parameter values increases, or ecological systems change.

3. BASIC COMPONENTS AND METHODS

The most important rule of constructing a PVA is to keep it simple. Although it is tempting to include every possible ecological effect in modeling a population, the most robust PVA models are generally those that are less complex, based on reliable data, and tailored to fit what is known rather than simply guessed. For example, although the impacts of invasive species on a threatened plant might be expected to increase over the next 50 years, with no information on how fast or how much these impacts will change, it may be better to leave out this "realism" and instead clearly note that assuming current impact levels is optimistic. Even though the complexity of PVA models should be limited to fit the available data, some key ecological processes are almost always considered. These are outlined below along with a description of the basic model forms and useful outputs of many PVAs.

Stochasticity

PVA models can be divided into two main categories: deterministic and stochastic. Deterministic models are simple projections of population growth rate and future population sizes without consideration of the variability in model parameters from year to year. As such, deterministic models are of limited value in predicting longer-term population numbers or viability and instead are used to compare the general efficacy of different management strategies for populations with limited data. In contrast, stochastic models incorporate estimates of temporal variability in demographic rates or the overall population growth rate to better represent the "real world." Adding stochasticity to a model generally increases the estimated risk of extinction because otherwise healthy populations may experience a series of bad years by random chance. Models can include either environmental or demographic stochasticity or, frequently, both. Less frequently, models will also incorporate genetic influences, such as genetic stochasticity (i.e., random genetic drift) and inbreeding depression. As noted earlier, detailed information on the impacts of inbreeding on individual fitness of wild species is rarely available, so explicit treatment of genetic stochasticity is now uncommon in PVAs.

Environmental stochasticity is randomness in demographic rates (e.g., birth, growth, and survival rates) caused by environmental factors such as weather. Extreme forms of environmental stochasticity occur as rare years of extraordinarily good or bad conditions, which are frequently of greater importance for viability than less extreme but more frequent year-to-year variations. Demographic stochasticity is variation created by chance differences in the fates of individuals, such as the random possibility that 10 members of a population of 20 will die in a particular year, even though the true mean annual survival rate is 80%. Demographic stochasticity is considered less of a threat to population persistence than environmental stochasticity, and its influence is felt only at small population sizes (i.e., ~ 50). Nonetheless, demographic stochasticity causes populations to grow more slowly and erratically at low numbers, making predictions of complete extinction more difficult than predictions of declines to a very low population size. As with genetic stochasticity and inbreeding depression, using a quasiextinction threshold allows PVAs to account for the diminished ability to predict the fates of very small populations due to demographic stochasticity.

Adequately characterizing the effects of environmental stochasticity also requires consideration of correlations in the different effects of environmental variables. The environment often, but not always, affects many life stages similarly. For example, severe winters may reduce survival rates at all ages and also depress reproduction the following spring. Likewise, some environmental factors, such as droughts, tend to persist for several years, causing demographic rates to be similar from one year to the next, producing autocorrelation in rates through time. Finally, populations in close proximity to each other are apt to experience similar environmental conditions at the same time, or spatial autocorrelation. In general, positive correlations in demographic rates, in either time or space, tend to reduce population viability, and hence are of concern for conservation planning.

Density Dependence

To simulate competition for limited resources, many PVAs impose negative density dependence, such that population growth rates decline as populations grow, or they impose a cap on the population size at the estimated carrying capacity. PVAs can model negative density dependence in many different ways, and results are quite sensitive to the specific approach used (see chapter II.3 for further discussion of this issue in population modeling). This sensitivity, along with a scarcity of field data clearly demonstrating the presence and form of negative density dependence for many rare species, means that great care is needed in deciding whether and how to include it in a PVA. Another theoretical possibility, but one also lacking strong empirical support in most situations, is the presence of Allee effects, or positive density dependence at low densities, such that population growth rates drop when densities are sparse because of the disruption of social interactions. For example, animals that rely on group dynamics to hunt or defend themselves, such as wolves or musk oxen, are likely to have reduced survival rates once their group size falls below a certain threshold. Although Allee effects will always increase extinction risk, they are thought to operate primarily at very low densities and are often accounted for by use of a quasiextinction threshold.

Model Forms

Most PVAs are built around one of four general types of population models:

Count- or Census-Based Models

These PVAs predict future population numbers and viability using mean annual population growth rate (λ) estimated from multiple counts of total population size (or a proxy for total population size) over a specific time period and incorporate environmental stochasticity through variance in λ. At their most basic, these models take the form:

$$N_{t+1} = \lambda_t N_t, \quad (1)$$

where N_t is the number of individuals at time t, and λ_t is the population growth rate at time t. The growth rate over 1 year can be calculated from two annual counts as $\lambda_t = N_{t+1}/N_t$. The mean and variance of the natural logarithms of these λ_t values are typically used to characterize stochastic growth rates. Count-based PVAs are a direct extension of the simplest descriptions of population growth, such as exponential growth (equation 1) or slightly more complex model forms that impose a ceiling on total population size or incorporate negative density dependence in growth rates. Examples of these more complex model forms are the Ricker equation, first developed for fisheries management, or the logistic growth equation, both of which impose lower and lower growth rates as the size of a population approaches its carrying capacity (see chapters II.1 and II.2).

Demographic Models

Historically, the majority of PVAs have used life tables or population matrix models built with demographic rates (e.g., survival and reproductive rates) to describe population dynamics. These models take different forms depending on whether individuals are classified by age, life stage (e.g., larvae versus adults), or size classes. To understand the basic form of a matrix model, it is helpful to think of a simple example, such as a population that is censused in the spring just after offspring are born and consists of three age groups, or stages: newborns, 1-year-olds, and all older adults. A simple demographic model for the female part of this population (almost all demographic models track only females) uses methods from linear algebra to multiply a matrix of demographic rates (**A**) by a vector containing the number of individuals in each stage (**B**) to obtain numbers in each age class the following year (**C**) (also see chapter II.1):

$$\overset{\textbf{A}}{\begin{bmatrix} 0 & F_1 & F_2 \\ S_0 & 0 & 0 \\ 0 & S_1 & S_2 \end{bmatrix}} \overset{\textbf{B}}{\begin{bmatrix} N_{0,t} \\ N_{1,t} \\ N_{2,t} \end{bmatrix}} = \overset{\textbf{C}}{\begin{bmatrix} N_{0,t+1} \\ N_{1,t+1} \\ N_{2,t+1} \end{bmatrix}}.$$

In this equation, the $N_{i,t}$ terms indicate numbers of individuals of stage i at time t. The S_i elements in the matrix are annual stage-specific survival probabilities. The F_i elements are slightly trickier, as they delineate the stage-specific reproductive output for each individual over a 1-year period. For example, F_2 is the average number of newborns (stage 0 individuals) produced a year from now by each older adult individual (stage 2 individual) we see now. Thus, F_2 is actually composed of two parts: the probability that an adult female we see now survives for a year multiplied by the number of female offspring she will produce if she does survive. Demographic PVAs often include environmental and demographic stochasticity through computer simulation of variation in one or more demographic rates. These models offer an advantage over count- or census-based models because they can be used to directly assess the effects of threats, harvest, or management interventions targeting particular stages or demographic rates (see example 1 below).

Metapopulation and Spatially Structured Models

This category of PVA includes several types of population representations but always emphasizes a collection of distinct populations linked by movement (see chapter II.4). Spatial models may simply estimate the proportion of available sites occupied by the species of interest, based on extinction and colonization rates, or they may consist of complex matrices accounting for site-specific survival and reproductive rates as well as intersite movement rates. Their key advantage is the ability to predict overall population persistence when local groups of individuals are growing and disappearing over time as well as to distinguish the fates of different (yet linked) populations, some of which are likely to occupy better or worse habitat (see example 2 below).

Individually Based Simulation Models

Like the last type of PVA, models in this category include a range of population representations, but all rely on extensive computer simulations to track the fates and locations of individuals. They are usually used for animals and tend to emphasize the importance of individual movements and location, and thus, they are a special form of spatially structured PVA. Although these are the most data-hungry form of PVA, they also allow the tightest links with habitat and behavior, which may interact to influence population viability (see example 2 below).

Again, the type of PVA model chosen should depend on the data available to build it. For example, census-based PVAs are appropriate for species for which monitoring programs have collected count data over many years (such as many African ungulates and North American breeding bird species). At the other extreme, if detailed annual data exist for the survival, reproduction, and movements of many marked individuals, an individually based simulation may be advantageous.

Major Outputs of PVAs

Once a PVA model is constructed, it can be analyzed in various ways to yield valuable information about how threatened a population is and what management methods have the greatest chance to increase viability. The most important of these outputs are the following.

Expected Growth Rates

The most basic and often the most useful output of a PVA is the population's current annual growth rate, or λ. For deterministic models, the single estimate of mean growth rate will predict whether the modeled population is increasing ($\lambda > 1$), stable ($\lambda = 1$), or decreasing ($\lambda < 1$). In contrast, stochastic models predict not only a mean λ but also a range of possible growth rates, reflecting our uncertainty about the particular series of future environmental conditions that may occur. Critically, by ignoring variability in population performance, a predicted λ from a deterministic model will likely overestimate the long-term growth of real-world populations. Further, even if the average λ for a stochastic model is greater than 1, a population can still by chance decline or become extinct.

Future Population Size

PVAs also predict the probabilities of different future population sizes. This output is important to evaluate such things as average number of years to extinction or the probability of extinction in a specified time frame for the population of interest. Furthermore, if a recovery program is initiated, a PVA can provide managers with a prediction of how long it will take for the population to reach a target number or density.

Extinction Risk

PVAs can assess the extinction risk of a single population or compare the relative risks of two or more different populations. These risks usually evaluate the probability that a population will decline to a quasi-extinction threshold within a certain time horizon (e.g., 50 years).

Sensitivity

Sensitivity values (and the related elasticity values) determine which parameters in a model have the greatest influence on λ by observing the degree of change in λ relative to a change in an individual model parameter (e.g., adult survival). Sensitivity results can also be estimated for extinction risk or other outputs of a PVA. The results of sensitivity analysis help prioritize different conservation and management efforts, such that the most sensitive stage or age class of a population is targeted for management efforts (see example 1 below). For example, sensitivity analyses have clarified the disproportionate value of older individuals relative to young of the year for essentially all long-lived, late-maturing species, which are overrepresented in the ranks of endangered species lists worldwide.

4. REAL-WORLD EXAMPLES

Example 1: Sensitivity in a Classic PVA: Atlantic Loggerhead Sea Turtles

Loggerhead sea turtles (*Caretta caretta*) are threatened marine turtles that breed on coastal beaches and feed as juveniles and adults, in part, on pelagic and nearshore invertebrates. Two prominent threats related to these life-history characteristics are (1) loss and degradation of nesting habitat and direct harm to eggs and hatchlings, and (2) drowning of individuals in the nets of fishing boats trawling for shrimp. Despite a long-term decline in loggerheads, there was debate in the 1980s about where to focus conservation efforts, with the greatest momentum in protecting nesting habitat and individual nest sites. To understand which of these efforts was most useful in stabilizing turtle numbers, in 1987 Deborah T. Crouse and her associates constructed a demographic PVA and performed a sensitivity analysis. They demonstrated that survival of all older age classes, especially large juvenile turtles, had a much greater effect on λ than the survival of turtle eggs (plate 15, inset). This was perhaps the most influential use of sensitivity analysis in any PVA. This analysis and subsequent work led to the installation of turtle excluder devices (TEDs) on shrimp trawlers in nearshore waters of the southeastern United States by 1994. TEDs allow most large turtles caught in nets to escape unharmed (plate 15), and use of these devices has spread to other fisheries to protect other sea turtle populations and has spawned the innovation of additional measures to reduce marine by-catch.

Example 2: Spatial PVAs: Lessons for Habitat Management from Two Territorial Birds

Perhaps the best known spatial PVAs emerged during the intense scrutiny of forest management plans for the northern spotted owl (*Strix occidentalis caurina*) in the 1980s. Conservation biologists built a range of PVAs that incorporated the spatial structure of territories and the movement of juveniles between them to find vacancies within these territories. These models ranged from the initial, elegantly simple analysis of Russell Lande in 1988 that assumed general rules of movement and landscape configuration to intensive simulations of individual owls and realistic landscapes. Importantly, these different PVAs all showed that owl populations had little chance of survival without drastically altered forest management practices and, in particular, showed the need for regeneration of large blocks of habitat that were not fragmented by logging operations.

Spatial modeling is frequently hampered by a lack of data on movement, but when data are available, spatial PVAs offer very useful results. A decade after Lande published his spatial model on spotted owls, Benjamin H. Letcher and colleagues used detailed data on dispersal distances to develop an individually based, spatially explicit PVA for the threatened red-cockaded woodpecker. They concluded that protecting forest patches containing aggregated territories led to much higher population persistence than the same number of dispersed territories (figure 1), a very similar result to that of the northern spotted owl PVAs.

Example 3: Next Steps with PVA: Modeling Adaptive Management for Island Foxes

Although PVAs have dramatically increased in sophistication, most analyses do only a cursory job of analyzing the complexities of human management and monitoring activities and instead concentrate on the details of ecology. Island foxes (*Urocyon littoralis*) are endemic to six of the Channel Islands, located off the coast of southern California. Populations plummeted on four of the islands in the 1990s as a result of predation by golden eagles and a disease epidemic, propelling the island fox onto the endangered species list. For this species to persist, expensive and complicated management activities will be needed for the foreseeable future. Thus, a PVA that analyzes the details of these human actions, as well as ecological processes, is required. Two of us (D.F.D. and V.J.B.) built a demographic PVA for the fox that accounted for both data uncertainties and the strong density dependence observed in survival rates (plate 16). The model predicts rapidly increasing risk of extinction with the addition of eagle predation. Sensitivity analysis for this model also shows that adult survival is the most important life stage to manage for—but this result did not provide guidance on how best to manage threats to adult survival given available resources. To advise managers on how to use their resources to keep eagle predation, and hence risk of extinction, to acceptable levels, we simulated different levels of eagle control (capture and removal) as well as different intensities of fox mortality monitoring (the data used to decide when to start and stop eagle control). This model, along with similar models we built to assess disease risk abatement strategies, and alternative monitoring approaches, are unusual in that they simulate the managers and their actions as well as foxes and their biology, thus linking different levels of monitoring and management effort directly to estimated extinction risk (plate 16, bottom panels).

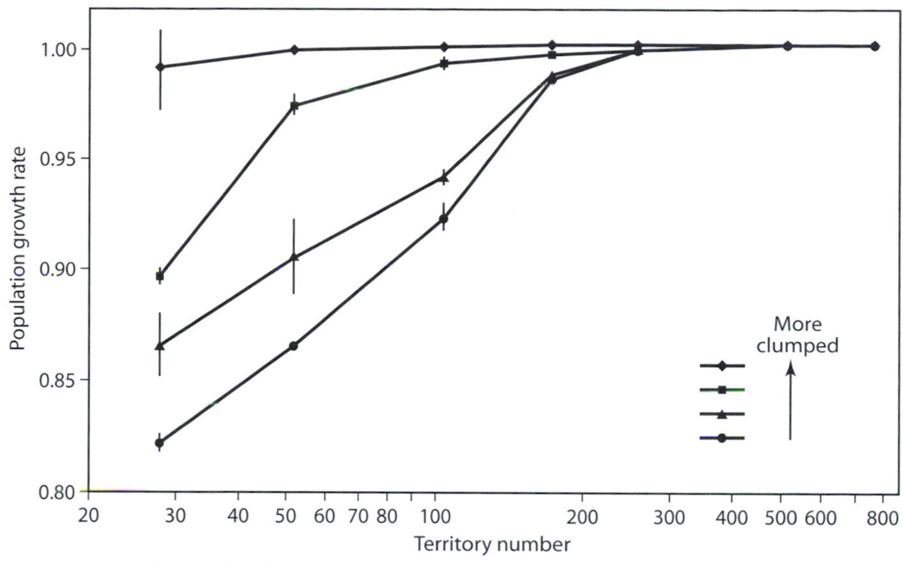

Figure 1. Population growth rate of the red-cockaded woodpecker as a function of territory number and degree of aggregation (clumping) of territories. Symbols represent growth rates in landscapes where territories have different levels of aggregation, circles being the least clumped and diamonds being the most. The same number of territories produced a higher population growth rate if the territories were aggregated than if they were randomly spaced. (Figure from Letcher et al., 1998)

5. THE FUTURE OF PVAS

Increasing Biological Realism

Scientists will continue to advance their understanding of what threats and complications are pivotal to assessments of population viability and how these should best be included in PVAs. Additional changes in the thrust of PVAs will be motivated by ever-changing global and local conditions. For example, climate change is predicted to accelerate over the coming decades, while in some (but not all) parts of the world, poaching is declining. Three processes or threats that are receiving increased attention in PVAs are (1) the health effects of toxic substances, (2) movement and spatial effects, and (3) species interactions. These factors may be fruitfully incorporated within PVAs as more and better research methods allow clearer characterization of these complex issues.

Traditionally, PVAs account for direct mortalities from sources such as hunting, habitat destruction, and predation, but not the effects of sublethal threats such as contaminant exposure and disease, which may slowly or indirectly impair survival, reproduction, and growth. For example, black-footed albatrosses are large pelagic seabirds that forage over entire ocean basins. These birds are exposed to very high concentrations of organochlorine contaminants such as PCBs and DDT, and there is mounting evidence that they suffer sublethal effects from these contaminants. Although accidental mortality from fishing practices is currently thought to be the biggest threat to black-footed albatross population viability, incorporating sublethal effects of contaminant exposure into PVAs for these and related birds will allow for a more comprehensive prioritization of conservation efforts.

Although many PVAs have incorporated spatial processes (example 2), including metapopulation and source–sink population dynamics, better information about how animals respond to habitat fragmentation and make movement decisions is broadening the range of spatial complexities that PVAs can include. In particular, the advent of GPS radiocollars for wide-ranging animals, micro-radiotags for small species, implantable sonic tags for fish, and pop-off tags for pelagic marine species heralds unprecedented gains in understanding how animals use complex habitats and how this use influences their birth, death, and growth rates. These advances in tracking technology may inject new vigor into spatial PVA modeling, and in particular, to individually based simulation models.

A final area in which PVAs are expanding is the consideration of species interactions within PVA models. Although PVAs are virtually always focused on single species, many of the forces influencing any one population are the co-occurring populations of predators, prey, or competitors. Many PVAs do include the effects of predators, prey, or humans on the focal population. Nonetheless, the way that this is done is usually static, without a full consideration of the dynamics of these interacting populations. Increasingly, there are efforts to consider how the fuller dynamics of species interactions can be incorporated into PVAs.

Adding the Human Factor: Uncertain Data and Uncertain Management

PVAs typically focus on the biology of the population being analyzed. Although not inherently unreasonable, this means that most PVAs ignore the human foibles that influence both the development of models and the carrying out of management plans. First, although the data used to build PVAs are always incomplete and imperfect, most have not tried to incorporate this uncertainty into their predictions. Thus, a major criticism of past PVAs is that they fail to account for parameter uncertainty, and in particular that they are interpreted as if their mathematical depictions of a species' biology were perfect. Solving this problem can be mathematically complex and computationally difficult, but recent PVAs have begun to take the problem of data uncertainty seriously. The resulting analyses are more robust and can also be used to analyze which data gaps most limit our ability to make powerful predictions about endangerment and management of sensitive populations. The second problem that PVAs are now tackling is the intricacy of monitoring and management programs, which themselves require careful decisions and can never be perfectly implemented. Although standard sensitivity analyses give important general answers about how best to manage, more explicit simulation of specific management actions—and ongoing data collection—allows better tailoring of PVA recommendations to real issues for endangered species managers (see example 3).

Conclusions

PVAs have become powerful tools for conservation biology that help to evaluate the risk of extinction for different populations and to guide management in ways that improve conservation efforts. One could, nonetheless, ask why conservation biologists should put so much effort into detailed analyses of single species when conservation is—or at least is argued to be—mostly concerned with multispecies communities and entire ecosystems. This is a fair question, but both biological and political realities make PVA an essential tool for today's conservation challenges. First, there is no "endangered community act," but the IUCN, the Endangered Species Act of the United States, and similar legislation in other countries do provide protection for endangered populations and species. Thus, PVA dovetails with the conservation laws we actually have. Second, although evaluating the risk of extinction can involve complex analyses, the biological reality being estimated—the risk of extinction of a population—is crystal clear. There is no correspondingly clear or biologically relevant standard of "community viability" or "community extinction." For example, communities can be highly altered without suffering any population extinctions, and conversely, communities and ecosystems may appear largely unaltered despite the extirpation of some formerly present species. Thus, even when we are concerned with the viability of a community, conducting PVAs for keystone and umbrella species may provide one of the strongest and most defensible ways to evaluate conservation risks.

Given these considerations, PVA is likely to remain a strong branch of ecology and conservation for the foreseeable future. Already we are seeing novel applications of PVA methods to reserve design, invasive species management, and other branches of conservation that are discussed in the following chapters.

FURTHER READING

Examples of PVAs

Crouse, D., L. Crowder, and H. Caswell. 1987. A stage-based population model for loggerhead sea turtles and implications for conservation. Ecology 68: 1412–1423.

Lande, R. 1988. Demographic models of the northern spotted owl (*Strix occidentalis caurina*). Oecologia 75: 601–607.

Letcher, B. H., J. A. Priddy, J. R. Walters, and L. B. Crowder. 1998. An individual-based, spatially-explicit simulation model of the population dynamics of the endangered red-cockaded woodpecker, *Picoides borealis*. Biological Conservation 86: 1–14.

Shaffer, M. L., and F. B. Sampson. 1985. Population size and extinction: A note on determining critical population size. American Naturalist 125: 144–152.

Summaries and How-to Manuals for PVA

Mills, L. S. 2006. Conservation of Wildlife Populations: Demography, Genetics, and Management. New York: Blackwell Publishing.

Morris, W. F., and D. F. Doak. 2002. Quantitative Conservation Biology: The Theory and Practice of Population Viability Analysis. Sunderland, MA: Sinauer Associates.

V.3

Principles of Reserve Design

Nick Haddad

OUTLINE

1. Overcoming the effects of habitat loss and fragmentation
2. The first principle is to preserve large habitat areas
3. Other principles reconcile ecological and economic trade-offs
4. Principles that increase the effective area of reserves
5. Identify conservation targets
6. Reduce edge effects
7. Increase connectivity
8. Revisiting the Y2Y corridor

Perhaps the greatest challenge to biodiversity conservation is overcoming the devastating effects of habitat loss. The single best approach to preserve biodiversity is to conserve or restore large habitat areas. Yet, in landscapes dominated by farming, grazing, and development that support increasing human populations, there are limits to the areas that can be conserved. Given limited areas for conservation, can reserves of fixed area be designed to increase their value for biodiversity conservation? Ecological theory suggests that strategies that increase habitat connectivity and reduce negative edge effects will have higher conservation benefits.

GLOSSARY

connectivity. The degree to which the landscape facilitates movement

corridor. Habitat that connects two or more reserves, usually the same type as found in a reserve but long and thin relative to reserve size

ecological trap. The attraction of animals to habitats where they perform more poorly, even when higher-quality habitat is available

edge effects. Changes in population sizes, species richness, or other aspects of the ecology of individuals, populations, or communities at the interface between two habitat types

habitat fragmentation. The spatial isolation of small habitat areas that compounds the effects of habitat loss on populations and biodiversity

matrix. The habitat or land use, often urban, agricultural, or degraded habitat, surrounding native habitats in reserves

The Yellowstone-to-Yukon (Y2Y) Corridor is perhaps the grandest application of ecological theory to the design of nature reserves. When complete, the reserve network would extend 1800 miles northward from Yellowstone National Park in the northwestern United States into the Yukon Territory in Canada (figure 1). It is among the most expensive applications of ecological theory in history. Conservation organizations are actively investing tens of millions of dollars in this region to protect large and connected habitats. If successful, the corridor would conserve large predators such as grizzly bears and wolves and wide-ranging ungulates such as bison and caribou by providing safe passage between the Yellowstone and Yukon regions. Ideally, conserving these large vertebrate species would create a metaphorical umbrella and also cover smaller vertebrates as well as the invertebrates, fungi, and plants that make up the bulk of biodiversity.

Conservation efforts such as Y2Y have to answer the question: Can reserves be *designed* to enhance biodiversity protection by targeting key parcels of land? To answer this question, there are a number of more basic questions we must address first. The most obvious is: How do we design reserves? Stated simply, reserve designs seek to increase the effective (if not the actual) area of reserves. The answer to this first question leads to others, such as: Which design criteria are most effective? Is reserve design simply a matter of "more is better"? And questions about ecological effectiveness must be evaluated in the context of scarce conservation resources. When one must actually invest tens of millions of dollars, which strategies do we have enough confidence in to spend money on?

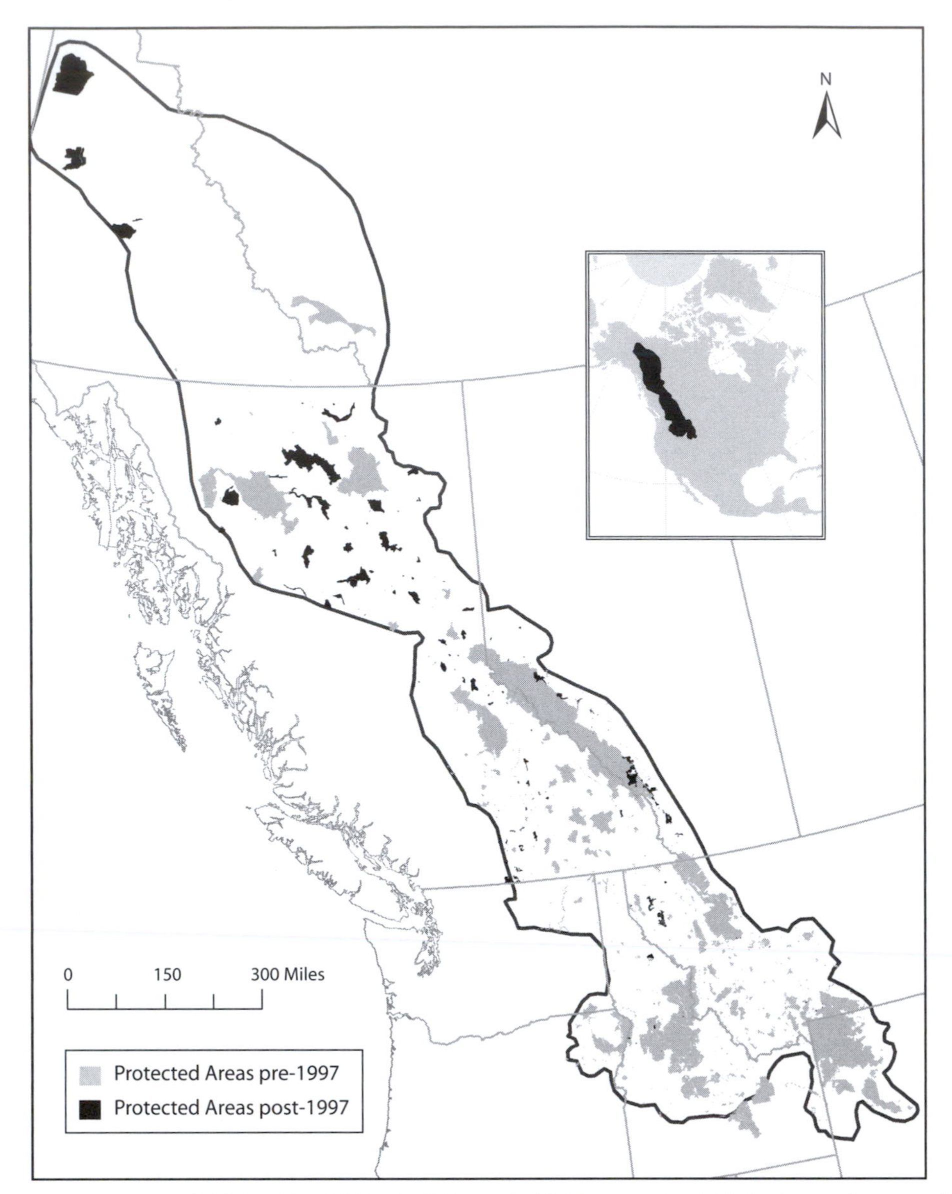

Figure 1. The Yellowstone to Yukon (Y2Y) Corridor. Within the Y2Y region, areas shaded in gray were protected before 1997, the year the Y2Y Conservation Initiative began. Areas shaded in black have received some level of protection since 1997.

Y2Y exemplifies a growing number of conservation plans that incorporate reserve design strategies and, we hope, maximize biodiversity protection. This chapter reviews the problem of habitat loss that necessitates reserve design. The easiest solution—restoring the habitat that was lost—quickly runs up against financial limitations, spurring other elements of design. The remainder of this chapter then focuses on aspects of design that seek to increase the effective area of reserves, specifically identifying conservation targets, reducing edge effects, and increasing habitat connectivity. At their core, the principles of reserve design draw on ecological theory and target conservation dollars to maximize biodiversity preservation. In this chapter, those theories—discussed in a number of other chapters—are put in their conservation context. It is here, in spending the big money, that the theoretical rubber hits the practical road.

1. OVERCOMING THE EFFECTS OF HABITAT LOSS AND FRAGMENTATION

What is our goal in designing reserves? It is an overly simple question but perhaps one worth considering. Principles of reserve design are applied to retard or reverse biodiversity loss caused by the destruction, fragmentation, and degradation of natural habitat (see chapter V.1). Over 40% of the world's land area has been transformed or degraded by humans through agriculture, forestry, and urbanization. The severity of this threat has made it the leading cause of biodiversity loss. And the threat continues to increase as the human population grows.

At its core, reserve design seeks to minimize local and global extinctions of species. The immediate threat that habitat loss poses to biodiversity is that it reduces population sizes of plants and animals. This initial consequence of habitat loss is insidious, as smaller populations are then exposed to other threats that compound extinction risk. Small populations are at greater risk to random events, such as catastrophic fires or population fluctuations that skew population structures toward one gender and limit mating potential. Smaller populations suffer more from random genetic events such as inbreeding or genetic drift that tend to reduce population fitness (see chapter V.2). All of these

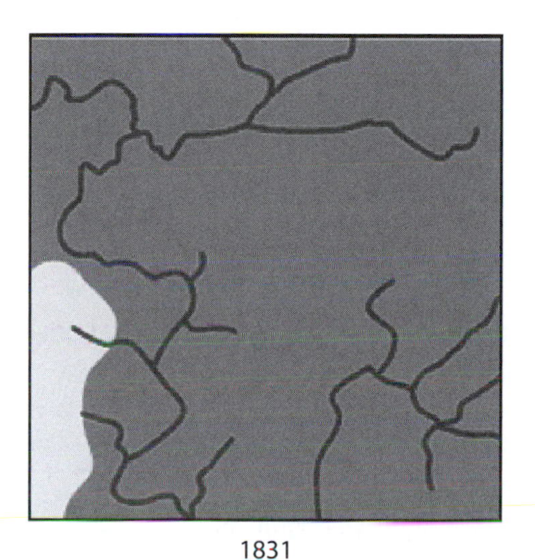

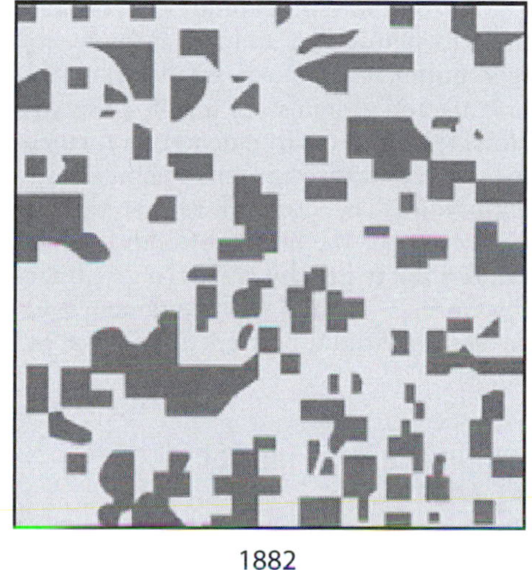

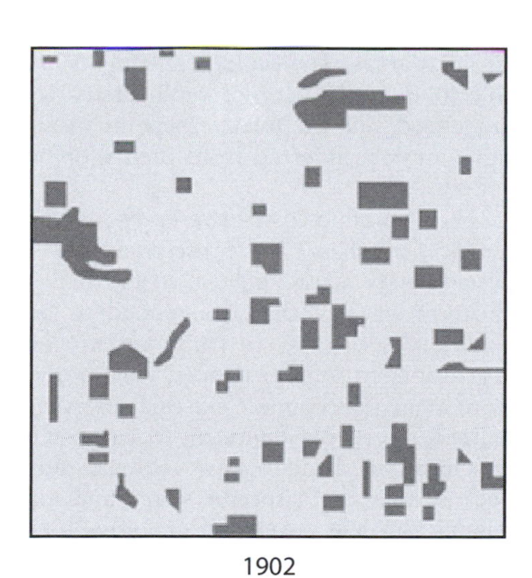

Figure 2. Habitat loss from 1831 to 1950. Cadiz Township, Wisconsin, 1831–1950. (After Curtis, J. T. 1956. The modification of mid-latitude grasslands and forests by man. In W. L. Thomas Jr., ed. Man's Role in Changing the Face of the Earth. Chicago: University of Chicago Press. Copyright Pearson Education, Inc., publishing as Benjamin Cummings)

consequences of living in small populations drive populations to smaller and smaller sizes, ultimately causing extinction. The principles of reserve design are intended to overcome these negative effects of habitat loss by increasing population size and increasing the viability of rare species.

The principles of reserve design are also intended to overcome the interrelated effects of habitat fragmentation. Figure 2 portrays a landscape in the midwestern United States where, over the course of the past 150 years, expanses of forest have been converted to agriculture. The forest left behind is not protected neatly in one reserve. Instead, the landscape is chopped into tiny fragments, woodlots that are spread among agricultural fields. Viewed in this way, populations are not just reduced by the amount of habitat lost; rather, single populations of plants and animals are now divided into larger numbers of much smaller populations. If those plants and animals are unable to move between woodlots, then habitat fragmentation further exacerbates the factors that threaten small populations and hastens extinction. The crux of reserve design principles is to account for the effects of both habitat fragmentation and habitat loss by conserving habitat where vulnerable species live and by maintaining big, interconnected reserves without barriers to dispersal of plants and animals.

2. THE FIRST PRINCIPLE IS TO PROTECT LARGE HABITAT AREAS

In planning reserves, it is worth remembering the primary cause of the biodiversity crisis: habitat loss. It follows that the most important conservation measure is to increase the amount of habitat protected as wildlands. Protecting or restoring large areas of habitat counteracts the negative effects of habitat loss and fragmentation. Large reserves harbor larger populations of target species, averting decline into the extinction vortex (see chapter V.1). They also encompass more habitat types, each of which may harbor a different set of species that are protected in larger reserves. Different habitats protected in larger reserves also provide some animals with easier access to a diverse array of resources they need to complete their life cycle. Compared to the principle of protecting large reserves, all other principles make modest improvements to biodiversity conservation.

What area must be preserved to ensure preservation of all species? There is no set rule. At the minimum, to maintain population sizes at levels high enough to be out of reach of key extinction risk factors, populations should number in the thousands. And that is a minimum estimate; to accomplish more functional goals, such as preserving the functional role each species plays within its ecosystem, the numbers would need to be higher yet, and more evenly distributed across the range. For the largest and most wide-ranging species such as northern spotted owls and grizzly bears, conservation areas must number into the millions of hectares. Areas must then cover the ranges of the Earth's millions of species. It takes only a few back-of-the-envelope calculations to show that the Earth's biodiversity requires protection of a substantial portion of the biosphere.

Inevitably, the area needed to conserve all of earth's biodiversity may outweigh the financial means of a large and growing human population. The ideal reserve network is limited, especially near cities or along coasts, where land values are high. (Conversely, vast areas are conserved on mountaintops, in deserts, and in tundra; see chapter V.4.) After attempting to conserve large areas, the other principles work within the limited areas available to conservation and with scarce conservation resources and are intended to maximize biodiversity by designing reserves in optimal configurations.

3. OTHER PRINCIPLES RECONCILE ECOLOGICAL AND ECONOMIC TRADE-OFFS

At the heart of reserve design is one key issue: funds to create reserves are limited, and land managers must make difficult choices about which areas to conserve. Should funds be invested to make reserves bigger or to restore lands that connect nearby reserves? The economic costs of reserve designs are easy to determine and involve the monetary costs of different land areas. The ecological benefits are more difficult to quantify because biodiversity is usually not monetized, and the relative benefits of alternative designs must be inferred from theory or limited observations.

The economic costs of reserves include land maintenance, purchase, and protection, and some reserve designs may carry higher costs than others. The cheapest strategy that has tended to be used most commonly is simply to conserve existing patches of vegetation. In some instances, it is more costly to restore areas that connect existing reserves than it is to expand reserves. Connecting reserves with corridors (discussed in detail below) may be more expensive because they are a specific shape and location, often passing through developed or agricultural lands that carry high values. Instead, funds could be spent to create larger reserves that are not connected. All other things being equal, managers will invest in the cheapest conservation alternative.

But the values of different reserve designs for biodiversity are likely not equal, and this is where more difficult decisions arise about ecological trade-offs. In the case of corridors versus larger reserves, ecological theories predict that connected reserves should maintain plant and animal populations and increase their diversity. An ongoing question in conservation biology is: To what extent do corridors increase population persistence and therefore preserve species? And what is that added value worth? Evaluating economic trade-offs is made more difficult still because resources available to fund different designs may not be equal. Corridors may be more expensive, but the vision of connecting reserves with corridors may excite more donors and attract more funds. Corridors may also have other uses than biodiversity protection; for example, urban greenways may serve as recreational areas. These other factors create positive feedbacks that generate more resources for conservation. Finally, sometimes a trade-off does not exist: some ecosystems are so fragmented by human activities that creating large reserves is no longer possible, leaving corridors as the best available option.

Another intense debate that highlights the complexity of ecological trade-offs in reserve designs concerned whether to conserve a fixed habitat area in a single large reserve or several small reserves (called the "single large or several small"—SLOSS—debate). The original principle had been based solely on the role of larger habitat areas to reduce extinction risk. Yet, other ecologists rightly pointed out that one large, compact reserve may not encompass the diversity of habitats required by a large-variety species. In the end, there was no one right answer to the SLOSS debate, and decisions about how to partition fixed areas into reserves depend on the conservation context. But even in light of the complexity of trade-offs, there are still some principles of reserve design about which most ecologists would agree.

4. PRINCIPLES THAT INCREASE THE EFFECTIVE AREA OF RESERVES

Many decisions about reserve design occur in landscapes that are already highly fragmented and heavily populated. In these landscapes, it is nearly impossible to create big reserves. The trade-offs that conservation biologists wrestle with are mainly those related to the biological benefits of alternative designs that are of the same area. In this context, an ecological watershed for reserve design principles came in the mid-1970s as an outgrowth of the theory of island biogeography (see chapter IV.2). Ecologists observed the parallel between islands in the ocean and habitat patches in a landscape of agriculture or development that is not used as habitat by most species. From these observations, scientists developed an initial set of principles to guide reserve design toward maximizing biodiversity conservation (plate 17). The principles were guided by two primary goals. The first was to create the largest reserves possible, reducing extinction risk. The second was to enhance dispersal, providing sources of new individuals to recolonize fragments in case of extinction and to promote genetic exchange. The initial principles were largely reinforced with the development of spatial theories in population ecology (see chapter II.4). The strong backing by theory and intuitive nature of the principles has inspired conservation planners, giving rise to the Y2Y and scores of other plans that create larger, well-connected reserves. The remainder of this chapter examines the key principles that attempt to increase the effective area of reserves.

5. IDENTIFY CONSERVATION TARGETS

In principle, reserves may be designed to preserve biodiversity in general. In practice, conservation targets are likely to be a handful of species of conservation interest—charismatic or rare species—or unique geographic or hydrologic features with which those species are associated. It is stating the obvious that areas should be protected that already contain populations that are targets of conservation. Such strategies must take into consideration that different areas may be used by species of conservation interest through the year. Some frogs and salamanders spend most of their life cycle in forests but breed in ponds; reserves must be designed to encompass both types of habitats and with pathways for individuals to move between them. One way to capture multiple habitat-specific rare species is to target geographic or hydrologic features where rare species abound (see chapter V.4). Once key areas are conserved, strategies for reserve designs vary in the degree to which degraded areas can be restored to enhance population viability.

Many reserves are designed with the assumptions that degraded habitat can be restored and that mobile animals can spread quickly to occupy restored areas. The Y2Y corridor will achieve conservation success when and if it protects expansive areas, crossing state and country borders, large enough to ensure the viability of grizzly bears, wolves, and caribou. Conservation targeted at species with large ranges may accomplish another goal of conserving biodiversity more generally. Protecting areas needed by large species may also protect the ranges of smaller ones, a strategy known as the umbrella species concept. Although applied widely, there is some debate about the utility of the umbrella species concept, especially because ranges of umbrella species do not always encompass the variety of habitats occupied by many smaller species. When conservation specifically targets smaller animals, the underpinnings of conservation remain the same as for large species: conserve an area large enough to encompass the home ranges of enough individuals to create viable populations that are protected from extinction. For mobile animals of any size, restoring corridors may promote dispersal and gene flow through the landscape.

When species have narrow habitat requirements, blocking off large and connected areas may not be the best strategy for conservation. Rare plants are often associated with very specific soil, moisture, or other environmental requirements. Where they occur, their populations can be quite dense, with thousands of individuals filling a hectare or less. These plants are less flexible in their use of nearby habitats that are conserved or restored, and distances between fragments may be too large for them to overcome via natural dispersal. It may be impossible to restore corridors with environmental characteristics needed for many plant species to spread. Thus, conservation for these kinds of plants often focuses specifically on protection of the highest-quality sites where they are found. Obviously, such species-specific plans target only the species we know well and miss all those species that have narrow requirements but are also poorly known.

6. REDUCE EDGE EFFECTS

Once reserve areas have been identified, their shape and setting may determine their effectiveness. Habitat loss and fragmentation introduce novel features into the landscape. Continuous natural habitats, once subdivided, run up against urban, agricultural, or other human-modified landscapes. These new landscape features are separated from conserved areas but can still influence protected areas from the outside. And outside influences are often strong, effectively reducing the area of a reserve for biodiversity conservation. Two interrelated landscape features are particularly important. First, and most immediate, is the boundary between conserved natural habitats and modified areas. These boundaries are often distinct and dramatic and give rise to changes in the abundances of individual plants or animals, the numbers of species, and other ecosystem attributes. Ecologists group all of these responses around habitat boundaries into the catchall phrase edge effects. The second feature depends on the type of modified habitat that is next to a conserved natural area, whether it be agriculture, buildings, or other degraded or natural habitats. Because this area is not the focus of conservation, it is referred to abstractly as the matrix. These concepts are strongly interrelated, as the type and extent of the matrix determine the types of species present at the edge and the degree to which edge effects penetrate natural areas. Later we will see how matrix habitats also influence landscape connectivity.

There has been a long history of considering positive and negative effects of edges in conservation and wildlife management. Aldo Leopold, a key figure in the early history of conservation biology, advocated the creation of edges to promote biodiversity. His views were not without merit. First, edges can provide new resources that are not available within either habitat forming them. This can happen when, for example, plants used as food by animals occur only at the edge. Second, they provide easy access to resources in the different habitats on either side of the edge; for example, adult butterflies may feed at flowers in fields, but their caterpillars may eat plants that grow in forest. For some wildlife species that are managed as game, such as white-tailed deer and bobwhite quail, high resource abundance near forest edges greatly increases their population sizes. Third, where landscapes were naturally fragmented, edges are essential in good reserve design. Many rare wetland flowering plants thrive at the boundary between bogs and upland forests. Taken together, these attributes make edges productive and diverse and help to explain why they might have some value in reserve design.

Yet we now know that early conservation biologists were mostly wrong about edges. Many imperiled species require large pristine areas of habitat. And just as there are some species that are attracted to edges, there are others that avoid them, and by large distances. Because of this avoidance, edge creation reduces the effective size of conserved habitats to sizes much smaller than the conserved area. The size can also be reduced by the presence of unwanted species that live on the edge or in the matrix. Edges create points of entry for unwanted species, including predator, invasive, or disease species, that are harmful to targets of conservation. A classic example of a harmful species that benefits from edges is the brown-headed cowbird. These birds live most of their lives in open fields but lay their eggs in the nests of other birds. Although not a forest species, brown-headed cowbirds will enter forests in search of host nests, especially when forest fragmentation increases the amount of edge and access to forest. When chicks hatch, unsuspecting mothers of forest birds feed cowbirds as their own, while brown-headed cowbird chicks work to push the mother's chicks from her nest and outcompete remaining nestmates for food.

Edges create new environments that can disrupt native ecosystems. Where forests are exposed to wind or sun at edges, new and often weedy species thrive, and trees native to continuous forest can die. Because of the effects of edges on the presence of unwanted species and changes in climate conditions, many species of conservation concern settle far from the edge. In one tropical fragmentation experiment in Brazil, called the Biological Dynamics of Forest Fragments Project (part of which is seen in plate 17), researchers have found that most edge effects, such as on soil moisture, relative humidity, abundances of understory birds, and litterfall, occur within 100m of the edge. But some edge effects, such as on wind disturbance, tree mortality, and the composition of ant communities in the leaf litter, can extend nearly a half kilometer from the edge.

The severity of edge effects differs based on the types of matrix. Whether the habitat next to the edge is a regenerating forest, an agricultural field, or a subdivision can greatly affect what happens within conserved areas. The matrix determines the types and numbers of other species that may penetrate reserves and the degree to which matrix habitats buffer harsher environmental conditions. Often, the effect of matrix habitats depends on the degree of contrast between matrix and natural habitats. Urban areas that directly abut forest reserves create dramatic edges that can lead to more severe edge effects, whereas managed or regenerating forests cause less obvious effects of edges.

Many species of concern in conservation avoid edges, but others are attracted to edges. This can be problematic for species that are tricked into settling at edges created by humans that can mimic natural edges. In large forest areas, trees fall and create small and irregular edges that support abundant wildlife. Human-created edges may seem like these natural edges and attract animals. Yet, edges created by human landscape modification are usually quite different from natural edges, and environments there can be more risky. Indigo buntings are classic edge birds, and edges may provide excellent nesting habitats and abundant food resources for their chicks. One experiment showed that when forests are cleared by humans, buntings are attracted to areas with the highest amount of edge, but so are their predators. The end result is that chicks raised in edgier environments created by humans are less likely to fledge from nests. In this regard, human-induced habitat modification may create ecological traps that exacerbate population decline caused by habitat loss.

Conservation biologists generally recognize that edges reduce the conservation value of reserves, and the degree to which edges have negative effects is usually associated with the level of human impact in the matrix. But conservation biologists also recognize that edges can have positive effects in some circumstances when they are naturally part of the landscape, when they create new environments necessary for the maintenance of wildlife populations, or when they allow access to multiple resources that occur in each habitat. As a general rule, human-created edges should be minimized, but natural or seminatural edges between reserve and matrix habitat should be retained in reserves to improve habitat for native species.

7. INCREASE CONNECTIVITY

A key goal of reserve design is to reconnect habitats that have been fragmented to allow the habitats to function as a whole rather than as a set of independent pieces. Reconnecting habitats increases genetic exchange, reducing the likelihood of inbreeding or genetic drift, and increases dispersal, aiding in the colonization of small sites where populations have become extinct. Connections also increase habitat area and may increase population sizes and biodiversity for that reason alone. However, if conservation aimed at increasing connectivity does not increase dispersal, then other strategies may add larger habitat areas for the same cost.

Corridors, or long and relatively thin strips of habitat, are the most direct way to increase landscape connectivity. By maintaining or restoring physical connections among patches, no edges or matrix must be crossed, and dispersal success should be highest. To be effective, corridors must include habitat that plants and animals would typically use during dispersal. Corridors often follow natural landscape features such as streams or mountain ridges. Connectivity can also be increased without physical connections simply by reducing the distance between reserves in the landscape. Reducing the distance between patches increases the likelihood that plants and animals can disperse between them. Lower travel distances decrease the amount of time plants or animals must spend in unsuitable, often risky matrix habitat. For species willing to leave their preferred habitat through matrix habitat, smaller reserves that form *stepping-stones* may provide a path connecting one larger patch to another.

For corridors and stepping-stones to be effective, increased connectivity must increase dispersal and gene flow. In once-continuous landscapes, it seems intuitive that reconnecting fragments should increase dispersal. This is often the case, and corridors have been shown to work for many species, including mountain lions, small mammals, birds, butterflies, and bird-dispersed plants. But physical connections will not always increase dispersal. At one extreme, increased connectivity may not benefit sedentary species if distances between reserves are too large or if corridors contain too much edge habitat. At the other, connectivity will not affect dispersal rates of highly mobile or generalist species, which fragmentation does not limit. To increase population persistence and biodiversity, connectivity must increase rare dispersal events to the point where they increase rates of colonization and reduce inbreeding and genetic drift. It is this high standard that justifies the strategy of increasing connectivity in reserve designs.

Deciding how to design landscapes that optimize connectivity can be tricky. The degree to which plants and animals will disperse through fragmented landscapes depend on characteristics such as the width, length, and degree of physical connection, which in turn depend on the quality of protected and matrix habitat. For example, some corridors (such as some urban greenways) may be so narrow that edge-avoiding species will not enter them. Alternatively, narrow corridors may enhance dispersal routes for species that move along edges. But edges may also provide habitat for predators to wait for unsuspecting dispersers. Usually, the optimal corridor width will depend on the distance at which edges affect movement behavior of animals or the distance to which negative edge effects penetrate corridors.

The optimal width of a corridor is related to its length. Animals and plants may be able to disperse through short corridors within hours or days, and easily within the lifespan of an individual. On the other hand, for gravitationally dispersed plants that disperse meters to tens of meters, it will take generations for them to pass through a kilometer-long corridor. Plants must disperse, establish, survive, and reproduce within these corridors. For these plants to succeed, habitat quality within the corridor must be very high. In landscapes where fragments are separated by long distances beyond the capacity of plants or animals to disperse through matrix, high-quality wide corridors are particularly important to increase connectivity.

The difference in quality between reserve and matrix habitat plays a key role in determining how landscape connectivity is designed. Some matrix habitats support species of conservation interest or serve as dispersal habitat. When the matrix is of moderate quality, then edge effects will be less severe, and corridors could be narrower. In such landscapes, steppingstones, rather than corridors, can be used to promote connectivity. Other matrix habitats are extremely risky for plants and animals that exit reserves. Top predators such as grizzly bears and mountain lions often exit reserves, but when they do, they are likely to be killed by humans. Roads, which are increasingly fragmenting landscapes, can cause high levels of mortality. In cases where matrix habitat increases mortality risk, corridors are especially important to increase population viability.

Maintaining landscape connectivity is particularly important in a world with a changing climate. Species distributions are determined in large part by their physiological requirements, and for many species, the limiting environmental factor is temperature. However, species are also limited by their ability to disperse to those sites where their physiologies allow them to live. Higher global temperatures will shift suitable habitats to higher latitudes and higher elevations (chapter V.6). Landscapes should be connected in anticipation of these future range shifts. To accommodate effects of a changing climate, landscapes will have to be connected at very large scales.

8. REVISITING THE Y2Y CORRIDOR

The Y2Y corridor exemplifies all of the key principles of reserve design. Its conservation targets are large predators and ungulates as well as a backbone infrastructure of established (and often prominent) reserves. It seeks to conserve the largest area possible within the region. Because of funding constraints, it targets conservation of lands that connect existing reserves, with wide corridors that reduce the effects of edges. And its scale makes it robust in the face of climate change. Y2Y has been successful in attracting investment to apply ecological theory. Time and research will tell if it is successful in conserving threatened populations and biodiversity within the ecoregion.

FURTHER READING

Beier, P., K. L. Penrod, C. Luke, W. D. Spencer, and C. Cabañero. 2006. South Coast missing linkages: Restoring connectivity to wildlands in the largest metropolitan area in the USA. In K. R. Crooks and M. A. Sanjayan, eds. Connectivity Conservation. Cambridge, UK: Cambridge University Press, 555–586.

Damschen, E. I., N. M. Haddad, J. L. Orrock, J. J. Tewksbury, and D. J. Levey. 2006. Corridors increase plant species richness at large scales. Science 313: 1284–1286.

Laurance, W. F., T. E. Lovejoy, H. L. Vaconcelos, E. M. Bruna, R. K. Didham, P. C. Stouffer, C. Gascon, R. O. Bierregaard, S. G. Laurance, and E. Sampaio. 2002. Ecosystem decay of Amazonian forest fragments: A 22-year investigation. Conservation Biology 16: 605–618.

Ricketts, T. 2001. The matrix matters: Effective isolation in fragmented landscapes. American Naturalist 158: 87–99.

Ries, L., R. J. Fletcher, Jr., J. Battin, and T. D. Sisk. 2004. Ecological responses to habitat edges: Mechanisms, models, and variability explained. Annual Review of Ecology and Systematics 35: 491–522.

Weldon, A. J., and N. M. Haddad. 2005. The effects of patch shape on indigo buntings: Evidence for an ecological trap. Ecology 86: 1422–1431.

Wilson, E. O., and E. O. Willis. 1975. Applied biogeography. In M. L. Cody and J. M. Diamond, eds. Ecology and Evolution of Communities. Cambridge, MA: The Belknap Press, 522–534.

V.4

Building and Implementing Systems of Conservation Areas

Will R. Turner and Robert L. Pressey

OUTLINE

The future of biodiversity depends critically on effective systems of conservation areas. The science underpinning the design and implementation of these systems has benefited from advances in ecology, data acquisition, and computational methods. Future success requires innovation on issues such as scale, the dynamic nature of threats and opportunities, and socioeconomic factors.

GLOSSARY

algorithm. Sequence of defined steps to achieve a result, defined by humans but often solved by computers, especially for complex conservation planning problems

conservation area. Place where action is taken to promote the persistence of biodiversity

irreplaceability. Property of a site measuring the likelihood that its protection will be required for a system of conservation areas to meet all targets or to otherwise optimize a conservation objective function

objective function. Mathematical statement of quantities to be maximized (e.g., the number of species or other biodiversity elements meeting targets) or minimized (e.g., cost)

persistence. Sustained existence of species or other elements of biodiversity both within and outside of conservation areas; as a conservation target, generally preferable to representation

representation. Sampling of biodiversity pattern, such as a number of species occurrences, within the boundaries of conservation areas; contrast with persistence

systematic conservation planning. The process of identifying and implementing systems of complementary conservation areas that together achieve explicit, quantifiable targets for the conservation of biological diversity

target. Explicit, quantifiable outcome desired for each species or other biodiversity element of interest

1. INTRODUCTION

The fraction of Earth's surface protected in conservation areas increased dramatically in the twentieth century with more than 10% of terrestrial area now under some form of protection (Chape et al., 2005). This effort could not come at a more important time: biodiversity worldwide is in jeopardy, with current species extinction rates estimated to be at least 100–1000 times higher than in prehuman times. Yet the extent of protected areas alone gives an incomplete picture. Too often these areas have been chosen on the basis of high scenic value or political expediency rather than the persistence of biological diversity. This tendency is evident in the widespread occurrence of areas ostensibly for biodiversity conservation in locations that are poorly drained, arid, remote, steep, or otherwise undesirable for homes, farms, resource extraction, and other human uses. This approach might sound like a "win–win" solution for people and other species. Yet it leaves the species most likely to become extinct—those most subject to human pressures—inadequately protected. The decline of biodiversity and the irreversible loss of conservation opportunities therefore continue even as reserve systems expand. In a seminal 2004 analysis, Ana Rodrigues and co-workers analyzed the global set of protected areas (figure 1) and found that at least 1400 terrestrial vertebrate species were not included in any protected areas, with many others underprotected. These shortfalls are likely even greater in marine and freshwater biomes, which face severe threats but have received less conservation attention in comparison. In all biomes, failure to incorporate data on biodiversity and current threats in the selection of conservation areas has limited their effectiveness.

Multiple theoretical and practical challenges must be overcome to avoid the continuing loss of biological diversity and to improve on the shortcomings of past conservation approaches. Biodiversity, threats to it, and costs of conservation are unevenly distributed at all spatial scales, from local parcels and watersheds to nations and worldwide biomes. Moreover, conservation resources are limited, so it is essential that the best decisions be made with what resources we have. Systems of conservation areas must be built and implemented in a systematic way to ensure the efficiency and success of efforts to secure biodiversity.

2. SYSTEMATIC CONSERVATION PLANNING

Conservation areas—places where action is undertaken to promote the persistence of biodiversity—are the cornerstone of conservation strategies. Successful strategies must account for the relationships among areas to create *systems*—not simple collections—of conservation areas. Conservation areas interact with one another across space through ecological processes such as animal movements, hydrological flows, and seed dispersal. Moreover, the usefulness of any one conservation area depends not on its ability to meet conservation targets on its own but on the extent to which it complements other conservation areas by improving the whole system's ability to meet targets.

How do we create effective systems of conservation areas? As with any question in science, we must first define the problem. In this case, the task is to plan systems of conservation areas to achieve a set of specified biodiversity objectives, subject to limited budgets, limited data, limited time, and constraints imposed by alternative—often conflicting—human uses for potential conservation areas.

Systematic conservation planning is the process of identifying and implementing systems of complementary conservation areas that together achieve explicit, quantifiable targets for the conservation of biological diversity. This process often applies computational methods, based on ecological and optimization principles, to extract maximum use from available data on species distributions and other key factors. Although the data, expertise, and computational and other resources used in systematic conservation planning cost

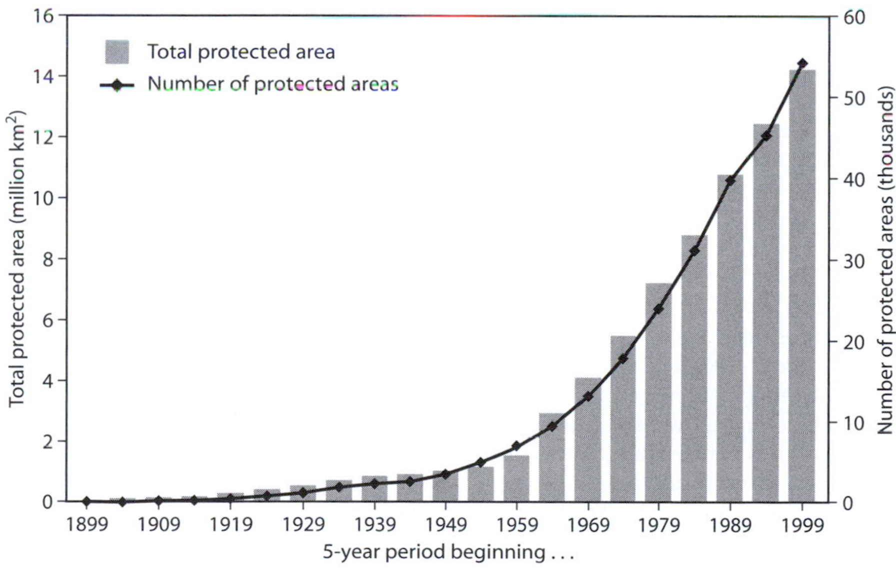

Figure 1. The number and area of protected areas (here, terrestrial protected areas) have increased dramatically worldwide. Based on data from the 2004 World Database on Protected Areas. (From Dobson, A. P., W. R. Turner, and D. S. Wilcove. 2007. Conservation biology: Unsolved problems and their policy implications. In R. M. May and A. McLean, eds. Theoretical Ecology. Oxford: Oxford University Press, 172–189)

money, nonsystematic approaches such as opportunism result in failure to meet conservation targets and lead to unnecessary loss of biodiversity—a cost even more difficult to bear.

Although conservation challenges are formidable, recent decades have seen considerable achievements in the development of more effective conservation strategies. In the remainder of this chapter, we discuss basic principles and recent advances in collecting and using data, applying systematic approaches for solving complex planning problems, dealing with the many uncertainties of real-world decisionmaking, and planning for conservation at multiple spatial scales. We conclude by examining key challenges for the future of conservation planning.

3. DATA FOR CONSERVATION PLANNING

At a minimum, conservation planning requires basic biodiversity data, some of which can be downloaded for free from the Web, quantitative conservation targets, and planning units. More sophisticated—and more effective—approaches also consider ecological and evolutionary processes, socioeconomic factors, diverse stakeholders, and the interplay of natural and human-caused dynamics to create conservation outcomes over time and space.

Biodiversity Data

All conservation planning requires biodiversity data, and the most basic forms of these data tell planners about the biodiversity pattern. Pattern refers to the distribution of biodiversity as a snapshot. Examples are maps of landscapes and vegetation types, locality records and range maps of species, predicted species distributions based on known habitat relationships, and "special elements" such as roosting and breeding sites. An important issue is that biodiversity is enormously complex, including all the species on the planet and their interactions with one another and their physical environments. Planners almost always have a very incomplete picture of this complexity and cannot wait for the whole picture to be filled in. That means that they rely on biodiversity "surrogates." Planners have to assume that protecting known species will also protect many unknown ones. They often assume also that protecting examples of vegetation types and other habitat classifications will protect many unknown species, partly by chance and partly because maps of habitats delineate environmental variation known to be important in shaping the distributions of many species. How valid are these assumptions? Tests of the effectiveness of biodiversity surrogates have produced mixed results. One general finding seems to be that rarer species are more likely to be missed by protected areas based on surrogates. This makes sense intuitively but provides little comfort because these are the species most in need of protection. Still, because data on most of biodiversity are lacking, planners have little choice but to use the best available information on known species and to use their judgment about the value of other surrogates in guiding decisions on protected areas.

Biodiversity is more than a snapshot, of course. If planners succeed only in sampling a biodiversity pattern, species will be lost from conservation areas. The reasons concern the many processes that maintain and generate biodiversity. Examples of biodiversity processes are the birth, death, and movement of individuals, the dynamics of metapopulations as species disappear from patches and recolonize them, spatiotemporal dynamics of disturbances such as fire and floods, dynamics of resources such as rainfall in deserts and upwellings of nutrient-rich water in the sea, migrations of animals, adjustment of behaviors and distributions to changing climate, and continuing evolution. If these processes stop, new species will no longer develop from existing ones, and existing species will disappear, even from conservation areas. Among the planning approaches that promote the persistence of biodiversity processes is the design or configuration of conservation areas (see chapters V.1, V.2, and V.3). Design refers to characteristics of individual conservation areas and whole conservation systems, including size, shape, directional alignment, replication, spacing, and connectivity. Many processes, such as the population dynamics of area-demanding species and regimes of disturbance (e.g., fire and postfire succession) are unlikely to persist in single conservation areas unless these areas are very large and carefully configured. More often, extensive processes can be maintained only across systems of conservation areas. Ideally, conservation planners will know enough about processes to design conservation systems accordingly. In practice, biodiversity processes, like species, are so numerous and complex that planners understand only a few of them well enough to influence conservation design. What about the others? Like unknown species, planners hope that many unknown processes will be catered to by surrogates, in this case design surrogates. Examples are rules of thumb such as these: conservation areas should be as large and well connected as possible to facilitate persistence of species and management activities; and conservation areas should be aligned along steep climatic gradients to facilitate adjustment of species distributions to climate change.

Conservation Targets

Systematic conservation planning also requires quantitative conservation objectives, most commonly expressed as targets. Targets have been mainly formulated for elements of biodiversity pattern. For example, a conservation agency might set a goal to conserve 2000 ha of vegetation type A, 10 locality records of rare species B, and 35% of the high-quality habitat of large mammal X. How do planners decide what these targets should be? Broadly speaking, targets should be the best possible interpretations of how much protection is required by a species or other biodiversity element. This means that targets will vary among species, for example, depending on how rare and how threatened they are. Targets will vary among habitat types depending on threat status, quantity, and needs of associated species, and whether some are important, for example, as critical resources during dry seasons. Targets are always limited by available data and incomplete understanding of conservation requirements. They are important, though, because they allow planners to decide how adequate existing conservation areas are, how extensive new conservation areas should be, and how important each potential new conservation area is. Ideally, conservation targets should also be based on the requirements of biodiversity processes. What would these be like? One example might be a target for the "effective area" of a species, taking into account its habitat requirements and the size, shape, and dispersion of habitat fragments, along with edge effects (see chapter V.3). Another example might concern the extent of both upland (summer) habitat and lowland (winter) habitat of a migratory species and effective connections between them. Yet another might be the minimum size of conservation areas necessary to retain recolonization sources after fires.

Planning Units

Planning units are the building blocks of an expanded system of conservation areas, and choices regarding planning units affect conservation planning in multiple ways. Cadastral units that reflect the boundaries of existing ownership or management correspond most directly to data on cost, availability, and management history and generally represent the actual units that will be involved in acquisition or other conservation actions. However, cadastral boundaries might not be available or may be too numerous (e.g., in a large planning region) to analyze conveniently, or many of them might be smaller than the resolution of some biodiversity data. Biophysical units, such as habitat patches or watersheds, align more closely with biodiversity pattern and the ecological processes (e.g., movement corridors, hydrological flows) that maintain biodiversity. Alternatively, regular grids overlaid on the study area have the advantage of being comparatively easy to create and manipulate and can be created in areas lacking cadastral or biophysical data. Both rectangular and hexagonal grids are used in conservation planning, although hexagons might be more useful when connectivity among units is a consideration, and properly constructed discrete hexagon grids are better suited to cover broad geographic areas (continental to global) without geometric distortion. The size of units matters as well. Analyses based on smaller units can generally meet targets more efficiently but often lead to less-connected sets of conservation areas if connectivity is not explicitly accounted for. Whatever size is used, targets must be specified correctly (e.g., the simplistic "protect each species at a single site" will produce wildly different results with units of different sizes), and biodiversity and other data must be available for the configuration and resolution of the planning units used. There is no one-size-fits-all choice for planning units. The best decision will vary among regions and scales, depending on such factors as the availability of cadastral data, resolution of biodiversity data, size of the region, and whether the planning exercise is likely to be interactive (requiring rapid analysis) or automated.

Beyond these minimum requirements are other kinds of data that can help planners to make effective decisions by reducing conflicts and maximizing opportunities for conservation management. Some of these data types are described below. Each type informs the planning process in a different and useful way.

Costs

Conservation actions are limited by conservation funding. Where there are options for achieving targets, selecting areas that are less expensive to protect and manage will allow scarce funds to go toward achieving additional targets elsewhere. Timing is key: it is much more efficient to consider costs at the time when systems are being designed. In the past, because of methodological and data limitations, many studies assumed that all sites have equal cost, regardless of land value or site size. However, most computational methods for conservation planning can now incorporate cost. Although the use of area or another crude surrogate can improve results somewhat where cost data are unavailable, studies have demonstrated the central importance of spatially explicit cost data to conservation planning.

Effective conservation requires more than merely buying real estate; and so the most effective conservation planning will require more than just data on land value. Additional factors, including transaction, management, and research costs, can in some cases exceed land values alone. Researchers are just beginning to confront and overcome the complexities associated with these additional costs. For example, the costs and benefits of conservation accrue differently to different people and across different spatial scales. The modeling and valuation of costs and benefits of conservation are areas ripe for future study.

Threats

Fundamentally, conservation planning is about locating management actions to separate elements of biodiversity from processes that threaten their persistence. Conservation areas can be useful in reducing threats such as those from agriculture, grazing, logging, and mining. They can also facilitate the management of invasive species and mitigation of changes to fire or other critical disturbance regimes. Information on threats can improve the planning of conservation areas in three important ways. First, information on the conservation status of an element of interest (e.g., species' global status from the IUCN Red List of Threatened Species) can aid in establishing targets, with larger targets (or larger targets in proportion to species' ranges or other measures) being assigned to those elements at greatest risk of extinction or extirpation. Second, spatially explicit data on the distribution of threats can inform the siting of conservation areas: where spatial options exist to achieve targets, areas that minimize threats are preferable. Third, estimates of the likely future distribution and magnitude of threats can guide the scheduling of conservation actions. For example, earliest conservation action can be given to those areas or species that are most imminently threatened or to those areas that have the greatest chances of retaining their biodiversity over time.

Opportunity

Areas vary in their availability for acquisition or other conservation actions. Even the most carefully crafted conservation plans can fail if the areas selected are not all available in a suitable time frame for the necessary conservation action. Data on the current or future availability of areas (the presence of willing sellers, for one) can influence the effectiveness of conservation planning in various ways. For example, ignoring areas that are less likely to become available may allow planners to focus their data collection and conservation efforts more productively elsewhere. On the other hand, avoiding difficult areas may not always be an option: focusing only on the areas most readily available often ignores those areas most useful for meeting targets. If areas of lower anticipated availability are essential for meeting targets (e.g., as the only known population of a species), an increase in the price offered for acquisition or management access might lead to higher target achievement. At the extreme end, some areas have very limited likelihood of becoming available, including those committed to residences, logging, mining leases, and the like. Sometimes these issues prevent all targets from being reached. In any case, spatially explicit opportunity data can usefully inform the planning process. Even in the absence of such data, methods that approximate site availability can produce better conservation outcomes.

Existing Investments

Rarely are systems of conservation areas built from scratch; they are more often constructed through incremental additions to existing networks. For these additions to be effective, they must account for the contributions of established reserves toward conservation targets. Thus, it is usually sensible to begin with spatial data on the boundaries of existing conservation areas and force inclusion of these areas in planning analyses. This has the added advantage of establishing nuclei of conservation management around which enlarged conservation systems can be designed. Not all conservation areas are effective, however, and areas that contribute relatively little toward targets could potentially be "unreserved" for the planning process or even removed from the actual conservation estate. The sale of these ineffective areas could in principle allow more effective conservation elsewhere, although the impact and feasibility (both economic and political) of this have yet to be explored systematically.

Data and Risk

Many kinds of data are important for conservation planning; yet all take time and money to obtain. There are risks inherent in proceeding with limited or poor data: the resulting bad decisions can waste scarce resources, compromise targets, and be difficult to undo. However in urgent conservation situations, the consequences of delaying action while waiting for better data can be just as serious. The relative consequences of inaction versus premature action must be assessed and balanced for each particular conservation situation. This is another issue that deserves urgent attention from conservation planners.

4. METHODS FOR THE SELECTION OF CONSERVATION AREAS

Although problems involving two or three possible sites and a dozen or so conservation targets can be solved with pen and paper, effective conservation in most real-world situations requires computer algorithms to identify systems of sites that will achieve the most targets subject to a limited budget. Although these analytical tools are essential for area selection, their application requires judicious consideration of the factors mentioned above and cannot substitute for carefully framing the objective, targets, and factors necessary to be considered in a given conservation situation.

Early algorithms tackled the area-selection problem with stepwise, rule-of-thumb approaches. These included a greedy, richness-based algorithm (e.g., "next, add the site that most increases the number of species represented in the system") and variants that weighted sites according to relative rarity of species and other factors. Later work framed the question as an optimization problem and used optimization methods such as mathematical programming or simulated annealing to identify optimal sets (e.g., the system of sites that maximizes target achievement). Not surprisingly, optimal formulations outperform a variety of heuristic methods (algorithms whose solutions are not provably optimal) in practice, including various stepwise approaches. These formulations require that a single mathematical objective function be specified, which is then optimized. A variety of factors (different planning units, species-specific area targets, costs, minimization of boundaries so that contiguous areas are preferred, and others) can be included in an objective function, and identification of optimal solutions with mathematical programming remains the preferred method for problems of manageable size. However, additional factors, larger numbers of areas or biodiversity elements, and more complex targets commonly prevent optimization software from attaining optimal solutions; in these cases, simulated annealing or other sophisticated heuristics are useful alternatives.

One concept that has emerged as particularly useful in conservation planning is irreplaceability. Irreplaceability values areas according to the likelihood that their protection will be required for the system to meet all targets or to otherwise maximize the conservation objective function. Values of 1.0 (completely irreplaceable) or close to 1.0 indicate that no or few spatial options exist for meeting conservation targets without the area in question. An area harboring the entire population of a species of interest, for example, is completely irreplaceable. Low values indicate that an area has many possible replacements. One of the strengths of irreplaceability is the fact that it gives planners spatial options for achieving their objectives. It thus encapsulates a measure of robustness to future area-selection decisions and is particularly suited to the dynamic, uncertain nature of real-world conservation planning.

There are now many software systems available to select conservation areas to achieve objectives and to estimate the irreplaceability of areas. Some of these have been developed specifically for conservation planning. Examples include MARXAN, C-Plan, Zonation, ResNet, and WorldMap. Other systems are less specific to conservation, including those developed, for example, to locate facilities such as fire stations, and general-purpose optimization systems that can be formulated to solve conservation problems.

5. REPRESENTATION OR PERSISTENCE? DYNAMICS AND UNCERTAINTY

Considerable research in conservation planning has focused on representation of biodiversity as its objective. Representation is the sampling of biodiversity pattern, such as the number of occurrences of each species, within the boundaries of conservation areas. But planning for representation only makes sense in a static world in which a reserve system is identified and all sites are acquired for protection simultaneously. In practice, reserve systems are often acquired over a multiyear period. During this time, unanticipated complications can wreak havoc with what had once been an optimal or otherwise perfectly sound solution. Simply recomputing new solutions based on updated data each year cannot overcome these shortcomings. To address these issues, conservation objectives must be defined in terms of the persistence of biodiversity over time, recognizing that threats change and conservation actions take place on an ongoing basis.

As with each additional factor considered in conservation planning, accounting for persistence generally requires additional data and computational complexity. But as with other factors, approaches vary in their data and computation requirements. For example, one approach combines biodiversity data (in the form of irreplaceability) with information on the relative vulnerability of prospective conservation areas to threatening processes. In this irreplaceability–vulnerability framework, those areas with high values of both vulnerability and irreplaceability are the highest priority for conservation action: they are the most likely areas to be lost, and because they have few or no replacements, their loss will most severely compromise achievement of the targets. This framework has performed well in a number of studies, but it leaves several trade-offs ambiguous. For example, which should merit higher priority, an area of moderate irreplaceability in imminent danger of destruction or a highly irreplaceable site with little evident near-term threat?

An alternative, the "minimize loss" approach, recognizes the objective function as the arbiter of all such questions. This approach embraces the persistence objective explicitly: the best choice is simply that which results in the least expected loss of biodiversity over time (whether in or out of conservation areas; note that this is different from, and generally more useful than, the maximization of biodiversity represented within conservation areas). The challenge is to develop the data and computational methods that allow this evaluation. Under this approach, a model is needed for the site-specific loss of biodiversity over time and how that loss will change if a conservation area is implemented. In principle, one could use this model and input data with a dynamic optimization technique such as stochastic dynamic programming (SDP) to select the schedule for creating conservation areas that minimizes the loss of biodiversity over some time horizon. In practice, because of the computational complexity, SDP methods are currently applicable only to trivially small problems, and thus, heuristic methods must be used to estimate optimal schedules. Heuristic or not, the key innovation of these tools is that they unify vulnerability and irreplaceability under a common framework based on the conservation objective function.

Accounting for the dynamic and uncertain nature of real-world conservation decisionmaking poses substantial and exciting challenges. Past threats, for example, are not necessarily indicative of future vulnerabilities, and more work is needed to understand biodiversity processes not only as they are changing now but also how they will change in the context of future land use and other dynamics. Further, threats are not the only dynamic processes; costs, opportunities, and other factors change over time and space as well. The importance of each of these must be weighed as individual conservation plans are developed. Moreover, not all threats can be countered by the creation of conservation areas alone. Climate change—representing extremes of both dynamics and uncertainty—poses substantial challenges for conservation decision makers. Preliminary studies suggest that siting conservation areas to protect both present and anticipated future ranges will aid species in adapting to a changing climate. However, such actions depend on the existence of spatial options—which may not exist for the many irreplaceable areas that contain rare and threatened species or other conservation elements—and more work is needed to identify alternative approaches.

Long after conservation areas have been established, additional planning challenges remain in managing them for the persistence of biodiversity. Many of the concepts and tools used to identify conservation areas can be applied to the mapping and scheduling of a variety of conservation actions within them. For example, priorities for work to remove invasive species from systems of conservation areas can be generated based on data on the irreplaceability of species (only those actually threatened by invasives), the severity of invasives, and the cost for invasive species removal for each conservation area. Yet many management actions occur not once but repeatedly over time and thus require additional data and theory for planning. For example, an optimal schedule for allocating prescribed fire among areas for species conservation might require all of the data necessary for one-time actions (e.g., species distributions among sites) plus additional information including the current successional state, habitat-specific rate of succession, species-specific ranges of acceptable fire return intervals, and human context for each area, as well as cost and effectiveness of different treatments and other factors. Optimization tools applied to a model integrating these factors could generate the optimal allocation of scarce funds for fires over both space and time. However, relatively little work has been done to address conservation planning for management. Perhaps simplified objectives—in this case, e.g., the maintenance of a particular distribution of fire-return intervals over the system—may serve as useful surrogates for management models that would otherwise be too data intensive and beyond computational capacity.

6. GLOBAL CONSERVATION PLANNING

Although the theory, data, and methods discussed so far must always be tailored to the conditions at hand, they apply to conservation planning over a broad range of circumstances, even to global scales. Is global conservation planning really necessary? After all, most conservation actions are implemented locally, and most of the US$6 billion in annual conservation spending comes from economically rich countries and is spent within their borders. Yet global biodiversity, threats to it, and the ability of countries to pay for its conservation vary in space. And each year, hundreds of millions of dollars are spent for conservation by nations, nongovernment organizations, and other funders

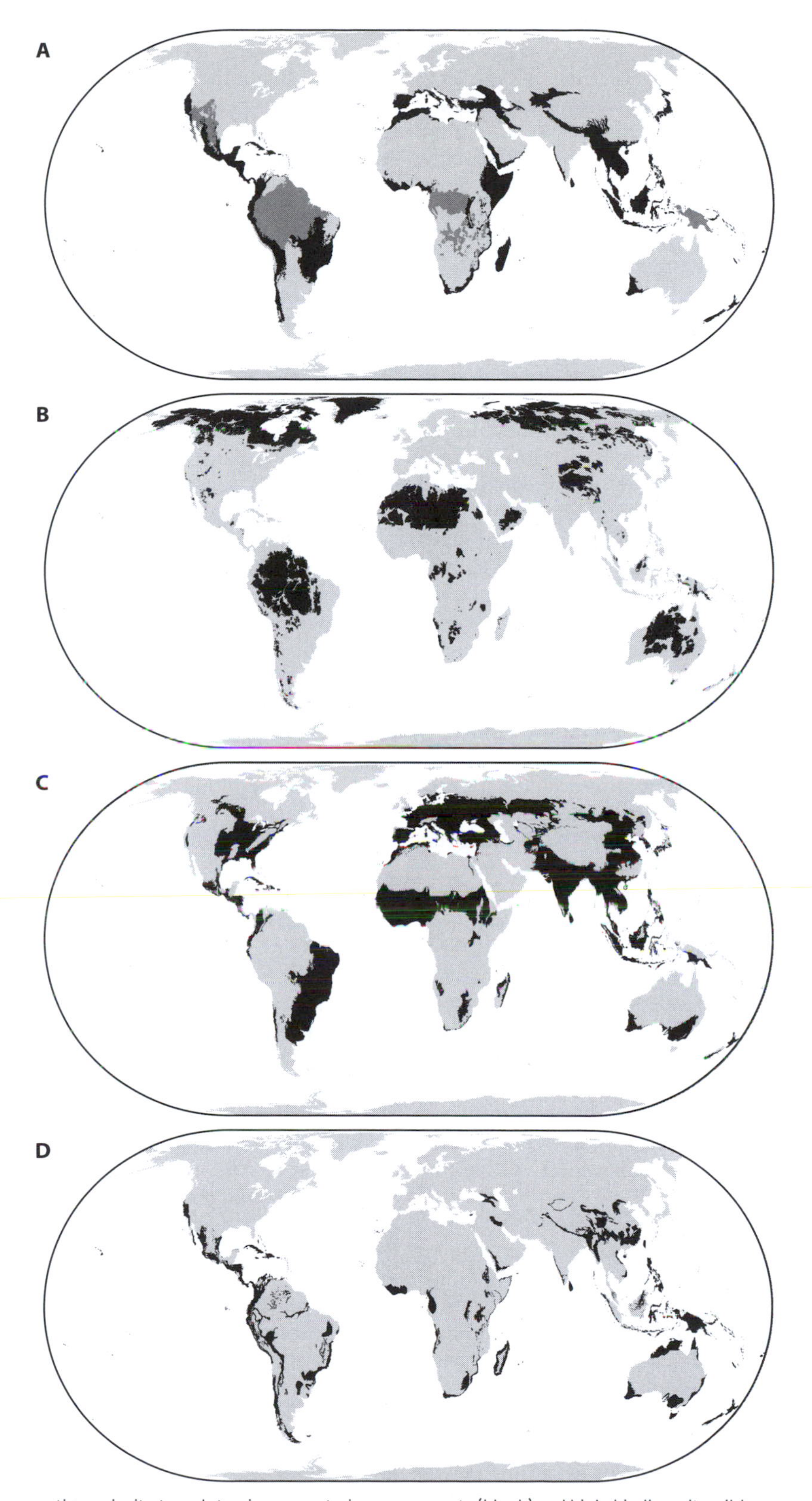

Figure 2. Global conservation priority templates incorporate irreplaceability and vulnerability in different ways. (A) Biodiversity hot spots (black) and high-biodiversity wilderness areas (dark gray). (B) Last of the wild. (C) Crisis ecoregions. (D) Endemic bird areas.

with few a priori geographic restrictions. Moreover, many conservation objectives—preventing species extinctions, for one—are inherently global in scope. Thus, the success of collective efforts to safeguard global biodiversity will depend, in part, on the context and coherence offered by global standards and global planning. Because of this, numerous institutions have developed conservation priority strategies at global scales in recent years.

To date, data necessary for conservation action at the scale of individual sites have been unavailable over a global extent. In light of this constraint, most global strategies instead determine a set of priority regions within which to focus conservation actions. At least nine sets, or templates, of global spatial priorities for terrestrial conservation have been published. These templates vary in several respects, including the metrics or taxa of interest and the approach to irreplaceability and vulnerability. The biodiversity hot spots (figure 2A, black), for example, are regions of high irreplaceability (harboring >0.5% of all plant species as endemics) and high vulnerability (>70% loss of original habitat area). High-biodiversity wilderness areas (figure 2A, dark gray), in contrast, prioritize irreplaceability roughly the same way but target low vulnerability, including only regions with <30% loss of original habitat area. Other templates account for irreplaceability but not vulnerability. The "last of the wild" (figure 2B), for example, includes the largest contiguous regions least subject to human impacts (irreplaceability neutral, low vulnerability), whereas the crisis ecoregions (figure 2C) are defined as those areas with the greatest habitat loss relative to protection (irreplaceability neutral, high vulnerability). A final category of templates, for example, endemic bird areas (figure 2D), includes areas of high irreplaceability without regard to vulnerability. Various conservation organizations use these global templates to focus their funding, capacity, and collaborative work within regions where they can, in principle, target finer-scale action to most efficiently meet their respective conservation objectives.

Each of these templates identifies priority regions, but none prescribes action at the finer scales at which most conservation actually takes place. Planning individual conservation areas requires additional data, for multiple reasons. Global patterns of endemism coincide broadly enough among taxa that plants or vertebrate classes can serve as useful (if imperfect) surrogates among regions. For example, hot spots harbor as endemics a minimum 36% of all nonfish vertebrates despite being designed around plants (of which they include >50% as endemics). But these broad-scale endemism patterns are insufficient for finer-scale conservation planning, where primary data on the conservation elements of interest are often necessary. Moving from globally significant regions to fine-scale planning presents planners with unresolved challenges. One is that aggregation of global data at the resolution of whole regions can "average out" finer resolution variations in conservation priority. This means that some low-priority regions contain high-priority patches overlooked in global assessments.

The Convention on Biological Diversity—with 189 countries as parties—has as one of its goals to "establish and strengthen national and regional systems of protected areas integrated into a global network as a contribution to globally agreed goals." This task is enormous, and no data sets exist to allow such an analysis with conventional systematic conservation planning methods. The Key Biodiversity Areas (KBAs) approach, for example, attempts to prevent global extinctions through an ongoing process identifying and protecting globally significant conservation areas. The approach first attempts to establish and test standardized, quantitative criteria for identifying sites of global conservation significance. Then, based on the premise that effective on-the-ground conservation requires local involvement, the actual application of KBA methods is done bottom-up, at regional scales (subnational-to-multinational) led by scientists within the regions. Since 2002, more than 1700 KBAs have been identified and delineated from across taxonomic groups in more than 30 countries, in addition to the more than 7000 "important bird areas" (KBAs for birds, specifically) identified worldwide in the past 25 years. An additional 595 sites, those identified so far by the Alliance for Zero Extinction as the sole known locations for endangered and critically endangered species, are among the highest priority of all KBAs.

The KBA approach potentially offers useful benefits: global standards based on irreplaceability and vulnerability, bottom-up implementation and support, reduction of data and computational needs for prioritization through prescreening of areas of global significance, and planning units designed around management or species-specific biological requirements. However, to accommodate real-world practicalities, the approach deviates from some theoretical work in conservation planning. For example, whereas traditional approaches seek spatially comprehensive input data sets, KBAs proceed in regional batches as data, capacity, and local interest are available. Regardless, as with any planning strategy, they must be continually evaluated against alternatives to identify those methods most effective in achieving conservation objectives, and revised appropriately where necessary.

7. THE FUTURE OF CONSERVATION PLANNING: RESEARCH CHALLENGES

From its early phase of rules of thumb operating on simple biodiversity data sets, systematic conservation planning has evolved to integrate various biological and social factors, larger data sets, more sophisticated computational methods, and diverse stakeholders. The field also has a growing track record of successful application to real-world problems. Yet much work remains. For example, many key factors such as cost and threat, although often investigated individually, are seldom integrated into algorithms at the same time. Essential data are often scarce, and methods for acquiring or modeling biodiversity and socioeconomic data efficiently and accurately are sorely needed. Computational methods must improve in speed and ability to handle larger data sets and real-world complexities such as the dynamic nature of planning and conservation management.

Conservation actions interact with uncertain and sometimes unknowable factors. Research must develop methods for planning conservation actions that are robust to incomplete biological knowledge, unpredictable human actions, and climate change. Conservation planning, long studied in the developed world, must be made applicable to the developing world as well. Theoretical advances must be designed and evaluated not only for their ability to obtain efficient solutions but also to encompass issues related to the socioeconomics and the needs and desires of stakeholders. Finally, systems of conservation areas are a cornerstone of effective conservation, but they must be integrated with broader landscapes and seascapes properly managed to maintain critical ecological processes, mitigate habitat fragmentation, and cope with climate change for the long-term persistence of biodiversity.

FURTHER READING

Brooks, T. M., R. A. Mittermeier, G.A.B. da Fonseca, J. Gerlach, M. Hoffmann, J. F. Lamoreux, C. G. Mittermeier, J. D. Pilgrim, and A.S.L. Rodrigues. 2006. Global biodiversity conservation priorities. Science 313: 58–61. *This article examines and compares two decades of priority strategies for conservation at the global scale.*

Margules, C. R., and R. L. Pressey. 2000. Systematic conservation planning. Nature 405: 243–253. *This comprehensive article provides an overview of the field, with a framework, since extended, for organizing the tasks and decisions involved in the conservation planning process.*

Possingham, H. P., K. A. Wilson, S. J. Andelman, and C. H. Vynne. 2006. Protected areas: Goals, limitations, and design. In M. J. Groom, G. K. Meffe, and C. R. Carroll, eds. Principles of Conservation Biology, 3rd ed. Sunderland, MA: Sinauer Associates, 507–533. *This recent overview of conservation planning provides discussion of biodiversity processes and conservation design to promote their persistence.*

Pressey, R. L., M. Cabeza, M. E. Watts, R. M. Cowling, and K. A. Wilson. 2007. Conservation planning in a changing world. Trends in Ecology and Evolution 22: 583–592. *This article summarizes what conservation planners know, and do not know, about two important scientific challenges: dealing with biodiversity processes and human-caused dynamics.*

Pressey R. L., R. M. Cowling, and M. Rouget. 2003. Formulating conservation targets for biodiversity pattern and process in the Cape Floristic Region, South Africa. Biological Conservation 112: 99–127. *This work provides an introduction to the limitations and advantages of targets, with examples of targets formulated for vegetation types and species in a global biodiversity hot spot.*

Rodrigues, A. S., J. O. Cerdeira, and K. J. Gaston. 2000. Flexibility, efficiency, and accountability: Adapting reserve selection algorithms to more complex conservation problems. Ecography 23: 565–574. *This article reviews and evaluates trade-offs between optimal methods and heuristics for a variety of conservation planning problems.*

Sarkar, S., R. L. Pressey, D. P. Faith, C. R. Margules, T. Fuller, D. M. Stoms, A. Moffett, K. A. Wilson, K. J. Williams, P. H. Williams, and S. Andelman. 2006. Biodiversity conservation planning tools: Present status and challenges for the future. Annual Review of Environment and Resources 31: 123–159. *This article provides an overview of software tools for conservation planning, summarizing their capabilities and need for further development.*

V.5

Marine Conservation

Jeremy B. C. Jackson

OUTLINE

1. Introduction
2. Causes and consequences of degradation
3. Synergistic effects
4. The future ocean
5. Coda

The synergistic effects of overfishing, pollution, and climate change pose a grave threat to all marine ecosystems. Complex food webs with abundant sharks, fishes, sea turtles, and whales are being replaced by greatly simplified ecosystems dominated by microbes, jellyfish, and disease. Runoff of excess nutrients from agricultural fields and animal wastes is causing eutrophication and the worldwide growth of anoxic coastal dead zones. The rise of carbon dioxide and other greenhouse gases is warming the ocean and making it more acidic, posing grave threats to coral reefs, polar ecosystems, and any marine organisms with calcareous skeletons. Ecosystem degradation is potentially reversible but there is very little time to act. Cessation of fishing or pollution does not always result in improved fish stocks and water quality, and there is increasing uncertainty about the potential for ecosystem recovery.

GLOSSARY

benthic. Environments or organisms on the sea floor.

bottom-up control. Regulation of ecosystem structure and function by factors such as nutrient supply and primary production at the base of the food chain, as opposed to "top-down" control by consumers.

by-catch. Nontarget species or juveniles of target species caught in a fishery that are not the intended target of the fishery. By-catch is commonly discarded dead.

dead zone. Area of the ocean with very low or no dissolved oxygen (hypoxic or anoxic) that forms in areas with low circulation and excess primary production (eutrophication).

ecological extinction. Reduction of a species distribution and abundance to the point that it no longer significantly affects the distribution and abundance of other species in the ecosystem.

El Niño–Southern Oscillation. Sustained sea surface temperature anomalies across the central tropical Pacific that are associated with the spread of warm waters from the Indian Ocean and Western Pacific to the Eastern Pacific and are a major influence on global climate, especially in the southern hemisphere. First recognized from the occurrence of warmer surface waters off the coast of Peru every 2–7 years that shuts down coastal upwelling and the anchoveta fishery.

eutrophication. Increase in chemical nutrients, most commonly nitrogen and phosphorus, and primary productivity in excess of the capacity of grazers to consume excess plant material; it is a major factor in the formation of dead zones.

exclusive economic zone (EEZ). The area bordering a nation's coast where it has special rights over the exploitation of natural resources, including fish, minerals, and petroleum. Except in areas of overlap, the EEZ extends 200 miles offshore.

fishing down the food web. The hypothesis that the observed decline in the mean trophic level of fisheries catches is caused by the selective removal and serial replacement of preferred high-trophic-level species such as swordfish, tuna, and cod by lower-level species.

fishing through the food web. The hypothesis that the observed decline in the mean trophic level of fisheries catches is caused by the serial addition of lower-trophic-level species in the presence of decreased predation by selectively removed top predatory species.

keystone species. A species that has a disproportionately large impact on ecosystem structure and function relative to its own abundance.

multiple stable state. The existence of one or more alternative ecological communities in a given habitat that persist over more than a single generation of the dominant species, contingent on the history of disturbance events that reset community composition. Most marine examples are caused by human disturbance.

pelagic. Environments or organisms of the open ocean.

plankton. Drifting organisms of the pelagic zone. Phytoplankton are photosynthetic primary producers, and zooplankton are consumers.

primary productivity. Production of reduced organic compounds from carbon dioxide and water, most commonly by photosynthesis.

resilience. The ability of an ecosystem to recover from disturbance and changes caused by the disturbance.

resistance. The ability of an ecosystem to withstand disturbance without major change in structure and function.

shifting baselines syndrome. The adoption of sliding standards for the health of ecosystems because of lack of experience and ignorance of the historical condition.

top-down control. Regulation of ecosystem structure and function by consumers rather than factors such as nutrient supply and primary production at the base of the food chain.

1. INTRODUCTION

The oceans are severely degraded by human impacts, and entire marine ecosystems are increasingly threatened with extinction or degradation almost beyond recognition. However, very few people are aware of the magnitude of the crisis because the oceans are out of sight and out of mind of our general experience. This ignorance is confounded by what Daniel Pauly called "The Shifting Baselines Syndrome," which arises because most of us accept as natural the way the world appeared to us when we were children, and unnatural as all the ways the world has changed within our lifetimes. Our children repeat the same mistake so that what seems unnatural for us is natural for them. Thus, generation by generation, our environmental standards decline to the point that we have no idea what "natural" means.

Many basic ecological features of the oceans are both qualitatively and quantitatively different from those of the land; most importantly, the properties of seawater versus air as the medium of life, the much greater connectivity of marine ecosystems over large spatial scales, and the role of the oceans as the ultimate sewer of humanity. Water is the most powerful solvent and has the greatest heat capacity of all common liquids. These are essential features of life but also allow the oceans to soak up and retain very high concentrations of pollutants that accumulate in organisms and sediments, as well as excess heat derived from the increase in greenhouse gases and climate change. Thus, pollutants may persist in the oceans for hundreds to thousands of years after pollution has been halted, so that entire coastal seas such as the Baltic, Adriatic, or Chesapeake Bay are the equivalent of "super fund" toxic clean-up sites on the land.

Transport and mixing by ocean currents distribute pollutants to all corners of the oceans as effectively, albeit much more slowly, as gases in the atmosphere; they also transport gametes, larvae, and adults over much greater distances than is typical on land. Biological connectivity in the oceans has sometimes been exaggerated, and there is increasing evidence that many organisms stay close to home throughout all stages of their life history, even in the presence of strong currents. But many larvae are routinely distributed for hundreds to thousands of kilometers, and a remarkable number of marine mammals, sea turtles, fish, and sharks actively migrate across the Pacific Ocean many times within their lifetimes.

Runoff of surface water on the land to the oceans transports ever-increasing quantities of nutrients, toxic chemicals, and sediments, altering pelagic and benthic environments and causing massive habitat destruction and loss of biodiversity. Thus, the fate of the oceans depends as much on human activities on the land as human activities in the ocean, and it is increasingly difficult to conserve ocean ecosystems based on changes in policies and practices in the oceans alone.

2. CAUSES AND CONSEQUENCES OF DEGRADATION

The major drivers of degradation are exploitation and pollution. The most important and destructive exploitation in the ocean is fishing, which severely affects ocean ecosystems worldwide. Other forms of exploitation, including mining and oil extraction, are second-order problems and have much more localized effects despite early concerns by Rachel Carson and others that these activities comprised the greatest threat to the oceans. The major causes of pollution and its consequences are much more varied. Introduced species, either accidental or for aquaculture, are the major form of biological pollution. The most important kinds of physical and chemical pollution include increased carbon dioxide and its myriad consequences from burning of fossil fuels, toxic chemicals such as mercury and PCBs, and excess nutrients from fertilizers and animal wastes.

Exploitation

The majority of fish stocks in the oceans are overfished, and the total global catch is in decline despite massive increases in effort reflected in more vessels and ever more efficient technologies. The most effective and destructive forms of industrial-scale fishing include long lining, trawling, and various forms of seining that can capture thousands to tens of thousands of fish in a single haul. Most of these methods are indiscriminate, and unwanted "by-catch" often comprises the great majority and major biomass of fish and other pelagic species killed in the process.

Pelagic long lining for tuna and billfish captures large numbers of sharks, sea turtles, and seabirds such as albatross that are in precipitous decline. The lines are many miles long and suspended by floats, deploying tens of thousands of baited hooks, and the same vessel can set many long lines simultaneously. New technologies, such as circle hooks and setting long lines at slightly greater depths, may reduce this by-catch but have yet to be proven effective on a large scale. Tuna are also caught in vast numbers using purse seines set on schools of dolphins that swim with the tuna. The dolphins are commonly located by helicopter, which greatly increases the efficiency of the fishery. Special openings in the nets were developed to allow the dolphins to escape following public outrage at the mass slaughter of dolphins. However, despite these precautions, the numbers of dolphins are still declining in the tropical Eastern Pacific, with grave implications for the dolphins as well as marketing and labeling of canned tuna as "dolphin-safe."

Trawling levels the sea floor like a bulldozer in a forest and has flattened the three-dimensional biological structure of most of the gently sloping regions on the sea floor of coastal seas and continental shelves, and even much of the deeper ocean. In the process, once complex communities of sponges, corals, bryozoans, and seaweeds have been transformed into featureless sediment plains. The resulting loss of habitat, fisheries, and biodiversity may require decades to centuries to recover. Trawling for shrimp is especially destructive, and the ratio of unwanted by-catch to shrimp is commonly greater than 1000:1.

The majority of industrial fin fisheries preferentially target large predators such as cod, tuna, and groupers high on the food chain, with the important exception of small oily fish such as sardines, anchovies, menhaden, and herring, which are commonly processed as fish meal for animal feed. Large animals are especially severely depleted by fishing. Most whales, sea turtles, sharks, and large fish of all kinds have been hunted to the point of ecological extinction, which means they play no significant ecological role in the ecosystems in which they were formerly very abundant and often functioned as "keystone species." Biological extinction may be much less common and is extremely difficult to demonstrate, but the gigantic Steller's sea cow and North Atlantic gray whale disappeared more than two centuries ago, and the Caribbean monk seal was last sighted near Jamaica in 1952.

Large cod were once so abundant they comprised the major source of protein for much of Western Europe and tropical America, but most cod fisheries have collapsed and show little or no signs of recovery. Fishers have shifted to smaller and smaller, less desirable species as preferred species are progressively overfished, a phenomenon Daniel Pauly termed "fishing down the food web." In addition, even small sardines and anchovies have been overfished, especially during natural declines related to oceanographic fluctuations such as El Niño events that suppress upwelling of deep, nutrient-rich waters that greatly reduce total primary production by phytoplankton. When this occurs, populations of anchovies decline by an order of magnitude or more, and the vast colonies of seabirds that feed on them suffer reproductive failure and mass mortality. Fishing is now prohibited during El Niño events, but before such regulations, the combination of El Niño with heavy fishing resulted in total collapse of the fisheries for a decade or more.

Overfishing is an extremely contentious subject, and fisheries biologists, managers, and conservationists still argue whether or not the losses of big fish such as tunas or billfish are 70% or 90%, or whether we are "fishing down" or "fishing through" marine food webs, whereby the net productivity of the fishery may increase as a result of release from predation with removal of top predators by fishing. But the evidence of loss in the serial economic collapse and closure of one fishery after another, as well as the total global decline in catch, is overwhelming. Responses of management, such as periodic closures of fisheries for varying intervals and establishment of marine protected areas (MPAs) are also controversial, despite the commonsense result that killing fewer fish results in their increase. Regardless of these issues and their management implications, there could be no clearer manifestation of "fishing down the food web" than the fact that the most valuable fisheries off eastern Canada, New England, and California today target invertebrates such as scallops and squid instead of fish, and similar trends are evident around the world.

Biological Pollution

Thousands of species of marine algae, shellfish, and other invertebrates are accidentally transported around the world every year attached to the hulls and in the ballast water of ships. Other species have been introduced deliberately for aquaculture, including the

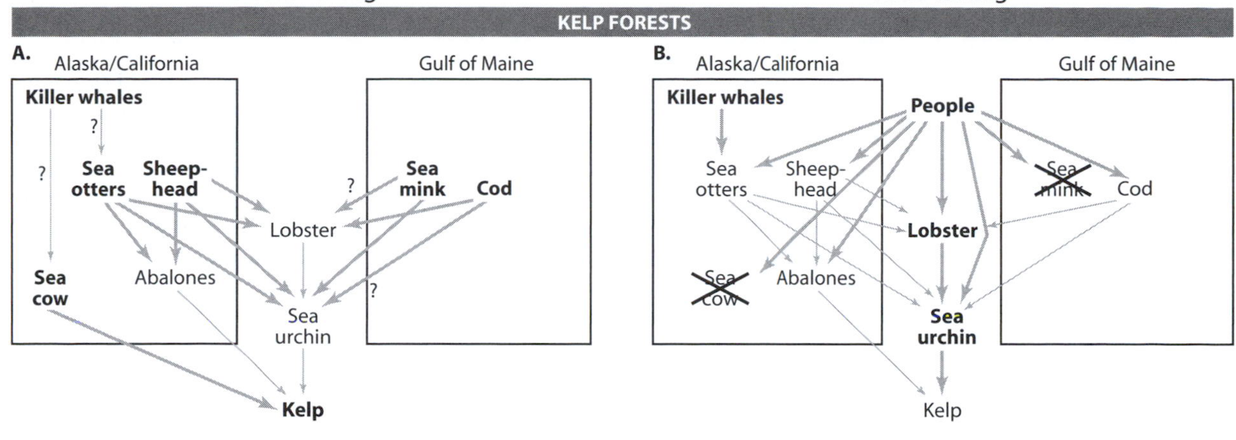

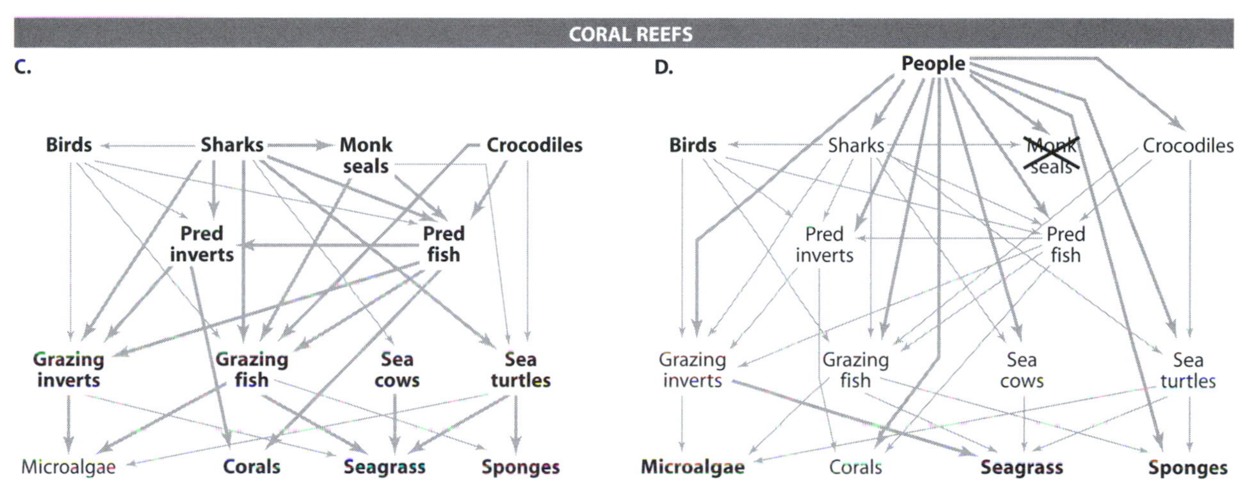

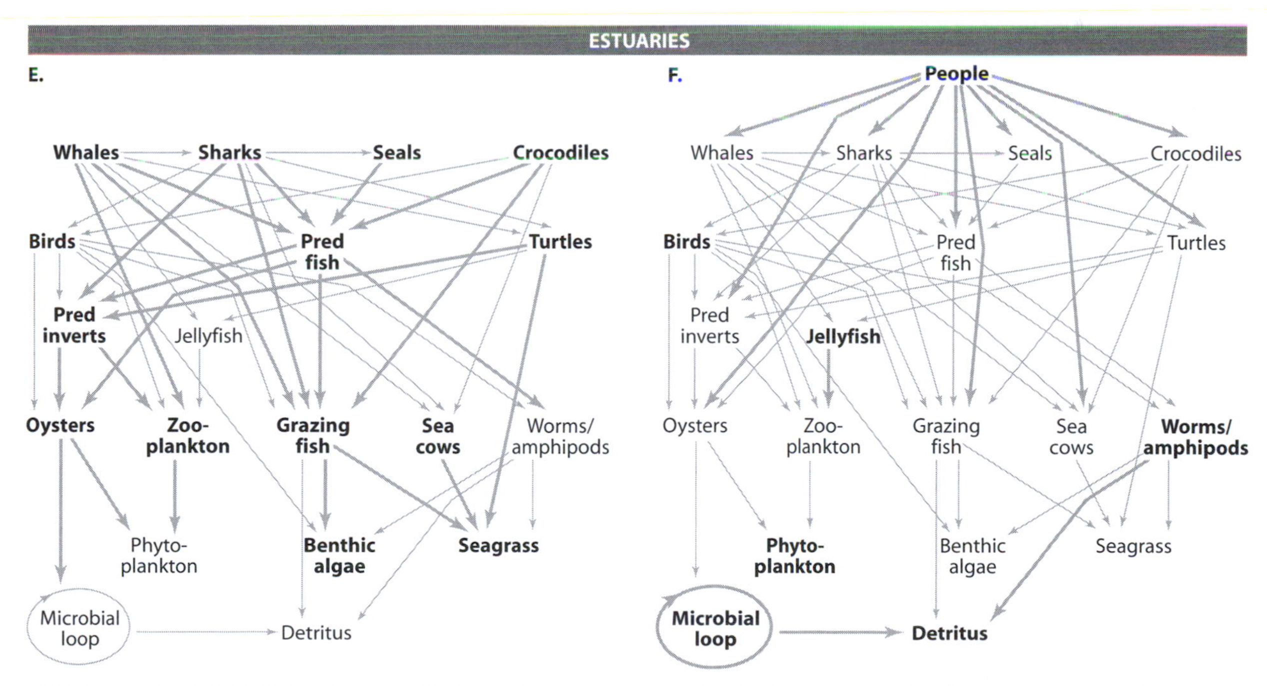

Figure 1. Simplified coastal food webs showing changes in some of the important top-down interactions due to overfishing; before (left side) and after (right side) fishing. (A and B) Kelp forests for Alaska and southern California (left box) and Gulf of Maine (right box). (C and D) Tropical coral reefs and seagrass meadows. (E and F) Temperate estuaries. The representation of food webs after fishing is necessarily more arbitrary than those before fishing because of rapidly changing recent events. For example, sea urchins are once again rare in the Gulf of Maine, as they were before the overfishing of cod, due to the recent fishing of sea urchins that has also permitted the recovery of kelp. Bold represents abundant; normal represents rare; "crossed out" represents extinct. Thick arrows represent strong interactions; thin arrows represent weak interactions.

American oyster to Western Europe and farmed Atlantic salmon that are raised throughout the Eastern Pacific with so little precaution that tens of thousands of fish escape to the wild every year. Many invasive species are extremely successful in their new environments and multiply so greatly that they displace native species. For example, the Chinese mitten crab and Asian green crab are now dominant species on North American coasts, where they compete with native fishery species. By far the most celebrated recent example is the "killer alga" *Caulerpa taxifolia* that was introduced inadvertently from the Monaco Aquarium and is smothering entire bottom communities throughout the northwestern Mediterranean, including hard grounds and seagrass beds that once supported thousands of species of invertebrates and algae. There are no known consumers or disease to help keep the alga in check. *Caulerpa* is now also threatening coral reefs in the Eastern Pacific, where it smothers reef corals and other animals and plants.

Introduced jellyfish are an equally massive problem. Jellyfish blooms are increasing dramatically in coastal seas around the world, drowning fishnets, clogging seawater intake systems for power plants, driving bathers from beaches, and severely impacting fisheries. Many of the most serious cases are a result of introduced species. Populations of the tiny Western Atlantic comb jelly *Mnemiopsis* in the Black Sea and the gigantic, half-meter Australian spotted jellyfish *Phyllorhiza* in the Gulf of Mexico have exploded in abundance to the point that they consume most of the zooplankton, including the larvae of fish and shrimp. As a result, fish catches have fallen to less than a tenth of previous levels.

Most such introductions are irreversible, and the economic costs are enormous. Microbes and viruses are also potentially harmful invasive species, but there are much less baseline data because of the need for genetic analysis for detection. The best known examples are MSX and other new forms of oyster shellfish disease caused by protists that first appeared in northeastern North America in the 1960s. Coral diseases may be another example, but it is difficult and perhaps impossible to distinguish between introductions and population explosions of formerly rare species caused by the breakdown of food webs by exploitation and pollution.

Warming, Stratification, and Acidification

The rise of carbon dioxide in the atmosphere because of the burning of fossil fuels is measurably warming the oceans, lowering ocean pH, and raising sea level. As a result, the extent of permanent summer sea ice in the Arctic has decreased by half and thinned to little more than 1 m in the past few decades and will likely disappear for the first time in the last 1 million years before the end of the century. Entire polar ecosystems are disappearing, and the future prospects of animals such as polar bears are grave. Species distributions throughout the oceans are shifting rapidly toward the poles, and tropical species such as reef corals are suffering mass mortality from the breakdown in the coral–dinoflagellate symbiosis referred to as "coral bleaching."

Symbiotic dinoflagellates ("zooxanthellae") occur within the coral tissue and leak sugar to their coral host that is essential for coral growth, calcification, and survival. In return, the corals provide essential nutrients to the zooxanthellae and protection from predators. However, when temperatures rise by about 1°C above the normal summer maximum, photosynthesis is impaired, and the dinoflagellates can no longer pass sugar to their coral hosts; whereupon the corals evict the zooxanthellae, thereby assuming a pale, ghostly color. A single bleaching event in 1998 in the Indian Ocean killed 20% of all the corals there, which is as if 20% of all the trees in North America died because of a single environmental perturbation over a few days. Episodes like this are becoming increasingly frequent and severe. However, there is growing evidence that genetically distinct strains of zooxanthellae display varying degrees of thermal tolerance, and there may already be a shift occurring in the proportions of different strains where frequent or extreme bleaching has occurred. Thus, the symbiosis may be able to adapt to global warming, although it could still take centuries for coral reefs to become reestablished where most coral have died.

Surface warming is also increasing the stratification of the oceans because warmer and lighter seawater at the surface inhibits upwelling of cooler and denser nutrient-rich waters from below. Increased sea-surface temperatures and stratification in the Northeast Pacific in the mid 1970s resulted in a marked drop in productivity, altered plankton communities, and decreases in fish recruitment that have persisted to the present. Moreover, data from planktonic protists called Foraminifera demonstrate a clear shift from temperate to tropical species along the California coast, and climate models suggest that the oceans may move into a permanent El Niño condition. Uncertainties abound regarding the degree to which upwelling could be permanently suppressed, but if ocean productivity declines, there will be an inevitable further decline in fisheries.

Of even greater concern because of the seemingly inevitable effects on all calcareous marine organisms is ocean acidification caused by the increased solution of carbon dioxide, which forms carbonic acid in seawater. Measurements have already demonstrated a drop of 0.1 pH units in the oceans, and laboratory and mesocosm experiments demonstrate that calcareous planktonic coccolithophores ("coccoliths"), pteropods, and foraminifera exhibit decreased calcification and growth under even mildly acidic conditions. These organisms are among the greatest producers of biogenic sediments in the ocean, and their skeletons carpet the deep sea floor down to about 4000 m, at which depth they begin to dissolve, and sediments become dominated by skeletons of siliceous plankton such as diatoms and radiolarians.

Ocean acidification is a massive, unplanned, and uncontrolled experiment whose implications for the biogeochemical cycles of carbon, nitrogen, and sulfur are highly uncertain but staggering. Coccolithophores are the most productive calcifying plankton that are responsible for massive blooms in temperate and subpolar waters that can be easily seen from space, as well as massive deposits of chalky sediments such as those that make up the white cliffs of Dover. We can only guess what noncalcareous plankton might take their place, or whether total productivity of these regions of the oceans will decline or increase as a result. Coccoliths are also the major producers of dimethyl sulfide, whose oxidation products affect the size and abundance of cloud condensation nuclei, cloud formation, and the albedo and heat balance of the planet. Thus, the demise of coccoliths could magnify the effects of global warming. Coccoliths also play a major role in the aggregation of organic debris as "marine snow," which is the major source of organic carbon to the deep ocean that provides food for most of the organisms on the deep ocean floor.

Comparably worrisome effects of acidification have been shown for commercially important shellfish such as oysters, mussels, and lobsters and for reef corals. At suitably low pH, corals lose their skeletons entirely and resemble small colonial sea anemones. Whether or not the corals can survive without their skeletons, reef formation would be impossible.

Toxic Chemicals

Nearly half a century after Rachel Carson warned of the effects of toxic chemicals in the environment, the oceans are increasingly polluted by mercury from the burning of coal in power plants, PCBs, insecticides, and the entire panoply of industrial chemicals that are allowed to run into the ocean. These chemicals have built up to lethal or sublethal concentrations in many marine animals and are concentrated up the food chain, especially in the Arctic, toward which emissions from lower latitudes are shunted in great quantities in the upper atmosphere. Wild salmon spawning in rivers in Alaska contain high toxic loads that are released into stream waters after they spawn and die; whereupon eagles and bears consume them, and the toxins enter terrestrial food webs. Marine mammals such as seals and killer whales, and the Inuit people who consume them, carry extremely high and potentially fatal concentrations of mercury in their tissues.

Eutrophication

Concentrations of nutrients, most importantly nitrogen and phosphorus, are increasing in coastal seas worldwide because of massive increases in runoff of nitrogen, phosphorus, and organic matter, which in turn reflect inefficient land use and industrial agriculture and animal production. Production of artificial fertilizers using chemical processes dependent on cheap oil and the widespread subsidy of fertilizers have stimulated their overuse to the point that most of the nutrients run off of the land unused. Large surpluses of grain provide cheap food for the mass production of chickens and pigs, whose wastes also typically run off unabated. The increased nutrients from all these and other sources fuel increased primary production and multiplication of phytoplankton that overwhelm the filtration capacity of marine ecosystems. The problem is exacerbated by the loss of marshlands and mangroves that formerly captured much of the nutrient runoff from the land before it reached the oceans and of formerly vast populations of suspension-feeding oysters that filtered the equivalent of the entire volume of giant estuaries such as Chesapeake Bay every few days. Extreme overfishing by dredging destroyed the physical structure of the once vast oyster reefs, hindering their regeneration. Menhaden and sponges had a similarly great filtering capacity that has also been virtually destroyed.

The result of all these nutrient inputs is runaway primary production of phytoplankton and benthic algae formally termed eutrophication, but which I prefer to call "the rise of slime." Production exceeds consumption and is amplified by positive feedbacks that produce even more nutrients and inhibit recovery of the consumers that could potentially keep the producers in check. Microbial metabolism of unconsumed phytoplankton causes anoxia, mass mortality of animals that cannot swim away, and a drop in the useful productivity of the entire ecosystem. This runaway process has resulted in the proliferation of more than 400 microbially dominated, anoxic "dead zones" around the world, with boom-and-bust cycles of toxic dinoflagellate blooms, jellyfish, and disease. Increased organic carbon in the form of simple sugars in seawater also breaks down the natural balance of microbes living on benthic organisms such as reef corals, resulting in orders of magnitude increases in microbes and increased incidence of disease.

3. SYNERGISTIC EFFECTS

The general sequence of events in the degradation of coastal ecosystems is strikingly similar in different environments including mangroves, marshes, seagrass, oyster reef, coral reef, kelp forest, and level bottom communities. First, large animals were eliminated by hunting and fishing. Second, the three-dimensional structure built by large sessile organisms such as kelps, seagrasses, and corals was lost either directly by trawling or indirectly to disease, smothering by sediments, or the effects of climate change such as coral bleaching. Third, eutrophication has resulted in development of dead zones characterized by mass mortality of animals and a drop in the useful net productivity of entire ecosystems.

These transitions in ecosystem state have been intensified through the additive and synergistic effects of the different forms of exploitation and pollution reviewed above and by the degradation of neighboring ecosystems over broad geographic scales. Consequently, adult and larval stocks of many species have become severely reduced throughout their entire ranges, and excess nutrients have degraded coastal ecosystems along the entire continental coastlines of Western Europe and the United States where *all* large coastal seas are severely polluted. Loss of species diversity further decreases the resistance and resilience of marine ecosystems to human impacts, as measured by decreased community-wide reproduction, resistance to invasive species and disease, primary and secondary productivity, and fisheries yields.

Loss or severe reduction in abundance of species also changes food web structure and function, potentially causing "trophic cascades" whereby species at different trophic levels increase or decrease dramatically because of changes in relative abundance of predators and prey. The most famous example involves the near elimination of sea otters by the hunting for their pelts all along the northwestern coast of North America. Sea otters consumed great numbers of sea urchins, which exploded in abundance following removal of the otters, whereupon the sea urchins eliminated formerly vast tracks of kelp forests. Restoration of sea otters by protection from hunting reversed the process by reducing abundance of sea urchins and enhancing the recovery of kelp. Reduction of such "top-down" control of ecosystems may reach all the way down from major predators and herbivores to microbes. For example, loss of parrotfish, surgeonfish, and sea urchins on coral reefs stimulates population explosions of macroalgae that leak profuse quantities of organic matter to the surrounding seawater, thereby increasing microbial growth and the incidence of coral disease. "Bottom-up" increases in nutrients may also lead to increased macroalgal abundance and disease, and it is sometimes extremely difficult to separate top-down from bottom-up effects, especially when ecosystem degradation has proceeded far enough that we have little or no idea of the pristine condition.

The effects of human impacts on ecosystems are often unpredictable because of the nonlinear dynamics of interactions among species and their environments and threshold effects that result in sudden changes in ecosystem composition. For example, Caribbean reefs were severely overfished by the early to mid-twentieth century, but corals remained abundant, and macroalgae were kept in check by the voracious grazing of the ubiquitous sea urchin *Diadema antillarum*. Then the sea urchin was reduced by more than 95% throughout the Caribbean in 1983 by an unidentified pathogen.

Figure 2. The slippery slope of coral reef decline through time.

Almost immediately, macroalgae overgrew entire reefs, smothering and killing the corals. Corals have failed to recover because of the continued absence of herbivores and increased coral disease and coral bleaching.

The leakage of organic matter from macroalgae may be a positive feedback mechanism that helps to maintain an alternative "stable" state of macroalgal dominance over corals. Other examples of hypothesized alternative states include lakes and estuaries with clear water, low abundance of phytoplankton, abundant submerged vegetation, and productive fisheries, versus eutrophic dead zones such as Lake Erie before it was cleaned up and the Baltic Sea and the Chesapeake and San Francisco bays. The existence of alternative stable states, rather than a continuum of ecosystem conditions, is controversial. Significantly, most of the best examples involve strong human impacts, whereas alternative states appear to be rarer in communities relatively unaffected by people. The question is fundamental because of the implications for ecosystem restoration and "recovery." Indeed, efforts to reduce nutrient pollution have seldom resulted in reduced phytoplankton or recovery of submerged vegetation, suspension feeders, and fisheries; and recovery of reef fish populations within MPAs has rarely led to recovery of corals. In general, just like Humpty Dumpty, it is far easier to break an ecosystem than to put it back together.

4. THE FUTURE OCEAN

"Business as usual" will have catastrophic consequences. Wild fisheries will be virtually eliminated. Dead zones will extend to ring the continents and move increasingly seaward. Surface warming will reduce circulation between the surface and deep ocean, reducing coastal upwelling and fisheries and the supply of oxygen to deep waters. Toxic blooms fueled by coastal runoff will become chronic with increasingly severe consequences for aquaculture and human health. Halting and possibly reversing this inexorable decline will require fundamental changes in fishing, agriculture, and energy production that are still widely perceived as unrealistic and naïve. But as the consequences of "business as usual" become more and more apparent, such changes are perhaps inevitable, barring some magical technological solution. Three main conservation actions are required for the future health of the oceans:

1. Stop all overfishing and develop responsible aquaculture on a massive scale. In the face of 6.5 billion people, increasing global equity, and continued human population increase, sustainable wild fisheries will be impossible except for weedy species such as sardines and anchovies and increasingly expensive luxury fish in the developed world. Aquaculture of species low on the food chain, such as algae, shellfish, and bottom-feeding fish such as catfish, mullet, and tilapia is the only ecologically responsible alternative, although even the most responsible aquaculture (like any agriculture on the land) has harmful ecosystem consequences in terms of habitat alteration and production of wastes. In contrast, aquaculture of such top predators as salmon and tuna only makes the problems worse because of the huge amounts of wild-caught fishmeal that are required to feed them. Of course many people prefer salmon and tuna, and it may be possible to raise them on substitute feed derived from soybeans. The pros and cons of aquaculture are highly contentious, and many environmentalists have opposed aquaculture outright because of irresponsible and unregulated practices and the predominance of salmon and tuna farming for markets in the West. But such arguments are self-defeating, and there is urgent need for agreement on standard and responsible aquaculture practices to ease the unsustainable burden of wild fisheries.

In addition to its obvious value for stabilizing the world supply of fish and shellfish, scaling back increasingly competitive and technologically intensive fishing would also contribute a modest reduction in energy consumption and emissions of greenhouse gases.

2. Reduce eutrophication by changes in agricultural practices. The rise of slime is a major threat to biodiversity and the development of aquaculture and human health in the coastal zone because of increasing frequency and severity of toxic blooms and disease. Dead zones such as Chesapeake Bay and the Gulf of Mexico could be restored by removal of subsidies for fertilizers and pesticides and taxation of their use as well as a cessation of the unregulated production and dumping of animal wastes coupled with improved sewage treatment. Reduced fertilizer and pesticide production would also significantly reduce energy consumption.

3. Cap and greatly reduce greenhouse gas emissions. Increased ocean warming, stratification, and acidification have been documented for more than 20 years, and the rates of change are increasingly nonlinear. The link to burning of fossil fuels is established, and the adverse biological consequences are clearly demonstrated by field observations and experiments. Failure to cap and reduce emissions now will almost certainly result in the loss of coral reefs and most other calcifying organisms, including major groups of primary producers and seafood species, and reduce ocean productivity.

5. CODA

Major uncertainties exist about the extent to which actions to reduce fishing, eutrophication, and carbon emissions can be effective in the absence of actions on all of them at the same time. Another important question concerns the extent to which reduction of local stress from overfishing and pollution can reduce or ameliorate the consequences of global change. The laws of open access and the "tragedy of the commons" are commonly invoked as excuses for inaction, but great progress could be made to rationalize seafood production and halt eutrophication on a case-by-case basis by nations and communities acting alone within their 200-mile EEZs. This is especially true for the wealthy nations of North America, Europe, Australia, and Japan, whose EEZs comprise nearly half the ocean, and for which the only constraint to responsible behavior is greed. In contrast, the causes and consequences of increased carbon dioxide are obviously global and will require fundamental technological as well as cultural change to bring under control. The choices are ours to make.

FURTHER READING

Feely, Richard A., Christopher L. Sabine, Kitack Lee, Will Berelson, Joanie Kleypas, Victoria J. Fabry, and Frank J. Millero. 2004. Impact of anthropogenic CO_2 on the $CaCO_3$ system in the oceans. Science 305: 362–366. *This is a report of causes and projected biological consequences of ocean acidification resulting from rising carbon dioxide from the burning of fossil fuels.*

Field, David B., Timothy R. Baumgartner, Christopher D. Charles, Vicente Ferreira-Bartrina, and Mark D. Ohman. 2006. Planktonic Foraminifera of the California Current reflect 20th century warming. Science 311: 63–86. *Data from sediment cores demonstrate major shifts in planktonic communities over the past century from a predominantly temperate to tropical assemblage of species.*

Hughes, Terence P., A. H. Baird, D. R. Bellwood, M. Card, S. R. Connolly, C. Folke, R. Grosberg, O. Hoegh-Guldberg, J.B.C. Jackson, J. Kleyapas, J. Lough, P. Marshall, M. Nystrom, S. R. Palumbi, J. M. Pandolfi, B. Rosen, and J. Roughgarden. 2003. Climate change, human impacts and the resilience of coral reefs. Science 301: 929–933. *This article discusses coral bleaching and the projected impacts of rising temperatures on the distribution and survival of reef corals.*

Jackson, Jeremy B. C., Michael X. Kirby, Wolfgang H. Berger, Karen A. Bjorndal, Louis W. Botsford, Bruce J. Bourque, Roger H. Bradbury, Richard Cooke, Jon Erlandson, James E. Estes, Terence P. Hughes, Susan Kidwell, Carina B. Lange, Hunter S. Lenihan, John N. Pandolfi, Charles H. Peterson, Robert S. Steneck, Mia J. Tegner, and Robert R. Warner. 2001. Historical overfishing and the recent collapse of coastal ecosystems. Science 293: 629–637. *This is a report of long-term, cascading consequences of fishing on marine ecosystems.*

Knowlton, Nancy. 2004. Multiple "stable" states and the conservation of marine ecosystems. Progress in Oceanography 60: 387–396. *Changes in marine ecosystems caused by human impacts such as fishing and pollution may persist long after the causes of decline are ameliorated because of threshold effects and nonlinear dynamics.*

Meinesz, Alexandre (trans. Daniel Simberloff). 1999. Killer Algae: The True Story of a Biological Invasion. Chicago: The University of Chicago Press. *This book is a chronicle of the disastrous introduction of a beautiful alga used to decorate aquaria into the northwest Mediterranean and the ecological, economic, and political consequences.*

Myers, Ransom A., and Boris Worm. 2003. Rapid worldwide depletion of predatory fish communities. Nature 423: 280–283. *This shocking and still controversial paper asserts that 90% of all the large predatory fish in the ocean have been lost as a result of overfishing.*

Norse, Elliott A., and Larry B. Crowder, eds. 2005. Marine Conservation Biology: The Science of Maintaining the Sea's Biodiversity. Washington, DC: Island Press. *This is the most complete single-volume review of marine conservation.*

Pandolfi, John M., Jeremy B. C. Jackson, Nancy Baron, Roger H. Bradbury, Hector M. Guzman, Terence P. Hughes, Christine V. Kappel, Fiorenza Micheli, John C. Ogden, Hugh P. Possingham, and Enric Sala. 2005. Are U.S. coral reefs on the slippery slope to slime? Science 307: 1725–1727. *Synergistic causes of coral reef decline and local and global strategies for coral reef recovery are described.*

Pauly, Daniel. 1995. Anecdotes and the shifting baseline syndrome. Trends in Ecology and Evolution 10: 430. *This is the fundamental paper defining the cultural amnesia underlying the collapse of fisheries and human impacts on the environment.*

Pauly, Daniel, Villy Christensen, Johanne Dalsgaard, Rainer Froese, and Francisco Torres Jr. 1998. Fishing down marine food webs. Science 279: 800–863. *Fishers shift to smaller fish lower down the food chain as stocks of preferred larger, higher-trophic-level fish are progressively exhausted.*

Rabalais, Nancy N. 2002. Gulf of Mexico hypoxia, a.k.a. "The Dead Zone." Annual Reviews of Ecology and Systematics 33: 235–263. *The causes and consequences of the most famous and thoroughly studied dead zone are reported by the heroic woman who discovered what was going on and would not let go.*

Worm, Boris, Edward B. Barbier, Nicola Beaumont, J. Emmett Duffy, Carl Folke, Benjamin S. Halpern, Jeremy B. C. Jackson, Heike K Lotze, Fiorenza Micheli, Stephen R. Palumbi, Enric Sala, Kimberley A. Selkoe, John J. Stachowicz, and Reg Watson. 2007. Impacts of biodiversity loss on ocean ecosystem services. Science 314: 787–790. *The number of species in an ecosystem affects its resistance and resilience to a wide variety of human impacts.*

V.6

Conservation and Global Climate Change

Diane M. Debinski and Molly S. Cross

OUTLINE

One of the most challenging issues for conservation during the coming decades will be preserving biodiversity in the face of climate change. It has become increasingly apparent that the climate is changing because of human activities—the chemical composition of the atmosphere has been modified, record-breaking temperatures are becoming more common on an annual basis, and polar ice caps are melting. Ecosystems will respond to these changes in a variety of ways; some may be deemed beneficial and others detrimental. The question for ecologists and conservationists then becomes how do we conserve ecosystems, ecological processes, and species under conditions of a changing climate?

GLOSSARY

assisted migration. Directed dispersal or translocation of organisms across the landscape

bioclimatic envelope models. Models that use statistical methods to correlate species occurrences with environmental predictor variables to define a species' environmental niche and predict the species' occurrence across a broader landscape

greenhouse gases (GHGs). Gases such as carbon dioxide, methane, nitrous oxide, tropospheric ozone, or chloroflorocarbons that absorb solar radiation and reflect it back down to earth, creating a "greenhouse effect" that warms the earth's surface

interannual. Between years

lake turnover. The mixing of deep anoxic (oxygen-poor) and shallow oxygen-rich water in lakes that occurs in fall and spring when water hits the threshold temperature of 4°C

oceanic conveyor belt. Ocean circulation pattern driven by temperature and salinity gradients across the globe that moves warm and cold water around the globe, moderating temperatures and salinity patterns

phenological changes. Timing of life cycle events that are related to seasonality of the organism such as hibernation, bud burst, flowering, egg laying, etc.

Quaternary period. The geologic time period beginning roughly 1.8 million years before present

stepping stones. Small, unconnected portions of suitable habitat that an organism uses to move from one place to another

trophic cascades. Changes at one level of the food chain that percolate through many other levels of the food chain, causing both direct and indirect effects on species composition

vagility. An organism's ability to move through the landscape

1. INTRODUCTION

In this chapter, we describe how climate is changing, including both paleoclimatic and anthropogenic changes. We then discuss how the Earth is responding, both from an abiotic perspective (including atmospheric changes, temperature fluctuations, and ocean circulation patterns) and from the perspective of biotic communities. We describe some of the research approaches that have been used to examine and anticipate the types of responses of ecological communities to climate change and how scientists might prioritize and manage areas for conservation under conditions of a changing climate. Finally, we end with a discussion of the missing links—the need for research that will allow us to better predict responses and to manage for change in the coming decades.

2. HOW CLIMATE IS CHANGING

Paleoclimatic Changes

Changes in the Earth's climate over the past thousands of years can be reconstructed using a combination of direct measurements from land and ocean weather stations and indirect proxy methods, which include tree rings and pollen and plankton from lakes and ocean sediment cores. These records indicate that the Earth's climate has cycled through many warming and cooling periods over geologic times. Ancient samples of atmospheric gases from ice cores reveal that the concentration of greenhouse gases (GHGs) in the atmosphere has also fluctuated in the past, with high GHG levels being correlated with warmer global temperatures. Cycles in temperature and GHG concentrations over geologic time scales have been caused by natural fluctuations in incoming solar radiation and the chemical composition of the atmosphere.

Anthropogenic Climate Changes

The 2007 Fourth Assessment Report of the Intergovernmental Panel on Climate Change (IPCC) details the unequivocal warming of the Earth's climate over the last 50 years, most of which is very likely a result of increases in anthropogenic GHG emissions. There has been a rapid 35% rise in atmospheric GHG concentrations since preindustrial times, and in 2005, atmospheric GHG concentrations were higher than any levels recorded or estimated for the previous 650,000 years. Although the magnitude of warming has varied across the Earth's surface, globally averaged temperature has risen ~0.6°C over the last 50 years. It is likely that average Northern Hemisphere temperatures during the last 50 years were warmer than during any other 50-year period over the previous 1300 years. Precipitation trends have been more variable, with some regions of the world showing significant increases and others significant decreases. Using a range of projected GHG emissions scenarios, the IPCC's best estimates of global anthropogenic warming range from 1.8 to 4.0°C over the next century, with an even greater magnitude of warming expected at high northern latitudes. Annual precipitation is very likely to increase in high latitudes, and the subtropics are likely to experience decreased precipitation over the next century.

In addition to changes in temperature and precipitation means, greenhouse warming will also lead to increased climate variability and the occurrence of climatic extremes. For example, many areas across the globe are predicted to experience more severe droughts, floods, large-scale erosion and soil wasting, landslides, and extreme heat events. Changes in these extreme events will have significant impacts on disturbance processes that are important to many ecosystems. Wildfires are influenced by the availability of fuel loads and occurrence of ignition triggers, but climate conditions are also critical in determining the severity and extent of wildfires. Extremely hot and dry conditions can lead to more ignition-prone fuel as well as to fires that burn hotter and therefore can have more extreme effects on vegetation. Fire trends from 1970 to the present in the western United States have increased in frequency and duration, and fire seasons have become longer since the mid-1980s (Westerling et al., 2006).

Reducing anthropogenic GHG emissions is necessary to avoid the most drastic climatic changes. However, even if the input of GHGs to the atmosphere is greatly reduced, scientists claim that the Earth is already "committed" to a certain level of warming and sea level rise because of the effect of GHGs that we have already emitted and oceanic thermal inertia (Wigley, 2005). Therefore, the conservation community has become focused on developing strategies that allow species to adapt and ecosystem processes to function in light of inevitable climate change.

3. ENVIRONMENTAL REPONSES TO CLIMATE CHANGE

Abiotic Responses

Abiotic changes drive biotic responses in ecosystems. Thus, in considering climate change, we must start with the abiotic responses that change the physical environment that organisms inhabit. Issues related to climate change include sea level rise, melting sea ice, melting glaciers and permafrost, altered precipitation patterns, and changes in global circulation patterns and disturbance regimes (e.g., wildfire, drought, floods, and coastal erosion).

The 2007 IPCC report summarized the state of our current knowledge. The assessment of global data shows that it is "likely that anthropogenic warming has had a discernable influence on many physical and biological systems." For example, there is high confidence that lakes and rivers will become warmer and that hydrological systems will be affected by increased runoff and spring discharge. These changes will increase ground instability, erosion, and rock avalanches. Disturbances such as flooding, wildfire, and ocean acidification will likely increase in frequency, and drought-affected areas will increase in extent. However, all of these changes are superimposed on background climatic cycles and circulation patterns of both air and water. Background climatic cycles that affect temperature and rainfall over decadal time scales include the Pacific Decadal Oscillation, El Niño Southern Oscillation, and the North Atlantic Oscillation. The oceanic "conveyor belt," which is driven by temperature and salinity gradients, moves warm and cold water around the globe, moderating temperatures and salinity patterns. We focus our discussion here on how the organisms that inhabit these ecosystems respond to abiotic changes, but it is important to remember that these abiotic cycles and circulation patterns create the backdrop within which ecological communities exist.

Biotic Responses

Paleoclimate Changes

Changes in climate are known to affect the distribution of species across landscapes. Paleoecologists document changes in vegetation over many centuries and millennia using sediment cores sampled from lake bottoms. Within the sediments are pollen granules that provide detailed chronologies of the local plant life. Over thousands of years, the same site may have been inhabited by very different vegetation communities. Davis and Shaw (2001) documented such changes in the forests of eastern North America from the Quaternary period to the present. As the climate warmed over that time period, tree species established themselves at higher latitudes and elevations. These types of large-scale changes usually occur relatively slowly, and organisms have a long time to adapt to changing conditions. The types of changes that have been documented over the past 200 years and the conditions that we expect to see in the future indicate that the rates of future change may be faster than in previous evolutionary time.

Responses to Recent Climate Changes

Scientists have already measured significant changes in local conditions and habitat suitability, shifts in phenology (timing of life-cycle events), and altered species distribution patterns in response to human-induced climate changes over the last ~30 years. Several analyses have shown that the timing of spring events (e.g., breeding, nesting, egg laying, flowering, budburst, and arrival of migrants) are occurring significantly earlier in response to recent warming for many species and all across the globe (IPCC, 2007; Parmesan, 2006; Root et al., 2005). Hibernation patterns are also being modified. Landscape responses to climate change can be monitored at continental scales using remote sensing technology. Satellite imagery allows scientists to monitor the timing of seasonal "green-up" of vegetation across a continent. Scientists have observed a trend in many regions toward earlier green-up since the early 1980s. In addition to phenological changes, many plant and animal species are showing signs of shifting range boundaries, both poleward in latitude and upward in elevation (IPCC, 2007).

Aquatic systems are also responding to recent climate changes. Changes in water temperature, dissolved oxygen, stream flow, and salinity are affecting the distributions of both freshwater and saltwater organisms from plankton and algae to large predatory fish. Glaciers are melting earlier and at an increased pace, with the potential to increase sizes of lakes and create flooding events until the glaciers are reduced in size. Sea level rise, in combination with human development, has contributed to increased erosion of coastal wetlands and mangroves. Coral reef habitats are threatened by many human activities, with warmer ocean temperatures blamed for increases in the global extent of coral bleaching. Coral bleaching is a stress response that involves the whitening of coral colonies because of the loss of symbiotic algae from the tissues of polyps. Bleaching can either slow or stop growth and reproduction of the colony. Changes in ocean acidity could have significant effects on shell-building organisms.

In some instances, the observed impact of climate change on individual species has had significant landscape-scale consequences. One example in the terrestrial world relates to changes in the mountain pine beetle life cycle that could alter forest wildfire regimes. A warmer climate allows insect pests that normally have a 2-year life cycle to complete their life cycle within 1 year. Being able to complete their life cycle within 1 year allows insect pests to survive in areas where their survival is currently limited by cold winter temperatures. For the pine beetle, this life-cycle change could portend more severe infestations with a warming climate. In fact, scientists have been monitoring the mountain pine beetle populations in Idaho since the late 1980s, and beginning in 1995, what had previously been small spot infestations in thermally favorable habitats grew into full-blown outbreak events. Extensive beetle damage to pine forests then plays a role in wildfire ecology by altering fuel loads. This example shows how minor changes in microclimate can have significant effects on species distribution patterns and disturbance regimes at large geographic scales.

The ultimate concern from a conservation perspective is how climate change might affect species survival. Research by Pounds and Crump (1994) since the late 1980s suggests that the golden toad and Monteverde harlequin frog in Costa Rica are among the first documented species to have become extinct in response to recent climate changes. Their disappearance is linked to an increase in a fungal disease outbreak during warmer and drier conditions. Extinction rates are already considered elevated over natural background levels because of habitat modification and loss. Climate change is likely to exacerbate the challenges of preservation for many species already at risk.

Projecting Responses to Future Climate Change

As climate continues to change in response to human greenhouse gas emissions, we expect species distributions to generally shift poleward and up in elevation as the climate warms (figure 1). Relationships may not be quite that simple, however. Changes in precipitation patterns across the landscape may cause species to shift along other types of gradients that may not be north–south oriented. There may be decreases in suitable habitat or limitations on dispersal. Some habitats may disappear, and other new habitats may evolve (figure 2). Of particular interest is the possibility for the creation of nonanalogous climates or combinations of climate conditions that do not currently exist anywhere in the world. For example, what happens to a tropical rainforest when it gets hotter?

Complex interactions within communities also make predicting responses to future climate change difficult. As one portion of the community changes in response to climate change, there is the potential for a cascade of events to occur. Each individual species will not necessarily shift its behavior and geographic distribution in isolation, and different species will have different abilities to respond to change in a given time period based on the breadth of their niche as well as their dispersal abilities. The addition or subtraction of key species can have significant effects. This is why it becomes so difficult to make predictions at the community level. Classic examples of what are termed "trophic cascades" have shown that changes at one level of the food chain can percolate through many other levels, causing both direct and indirect effects on species composition. Understanding the importance of complex interactions among species will be important in predicting the potential consequences of climate change on ecological communities.

It is also important to remember that extinction has many causes. In many portions of the globe, multiple threats (e.g., habitat destruction or modification) will work in concert with climate change to affect future species distribution patterns. Jetz and colleagues forecast the numbers of birds likely to be endangered by climate change versus direct habitat destruction over the next 50–100 years. They foresaw a marked latitudinal difference: in the tropics, deforestation is likely to be the main driver of species extinction, whereas in the higher latitudes, it will be climate change (Jetz et al., 2007).

The main approaches scientists use to examine potential future impacts of climate change include experimental manipulations, analyses of historical observational records, and modeling. Experiments in the field and laboratory allow us to monitor ecosystem responses to directly manipulated climate conditions (box 1; Shaw et al., 2002). These experiments are usually conducted at a scale appropriate for the structure of the plant community being examined (i.e., tens of square meters for grasslands versus hundreds of square meters for forests), but they are often small in scale from the perspective of vagile organisms that may inhabit these locations. Manipulating ecosystems at the landscape rather than the patch scale becomes much more challenging in terms of both cost and logistics. Thus, scientists can examine some phenomena experimentally, but others are primarily studied via observational methods and models.

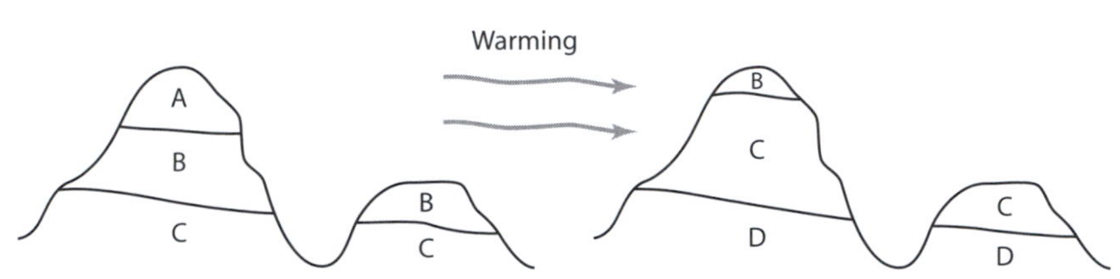

Figure 1. As climates warm, we may expect species to move up in elevation. For example, species A, B, and C exist on mountain ranges. Species A is restricted to the coldest habitats. As warming occurs, species A is left without suitable habitat, and species B and C move up the mountain. A new species, D, which previously occurred at lower elevations, is now found at the base of the mountain.

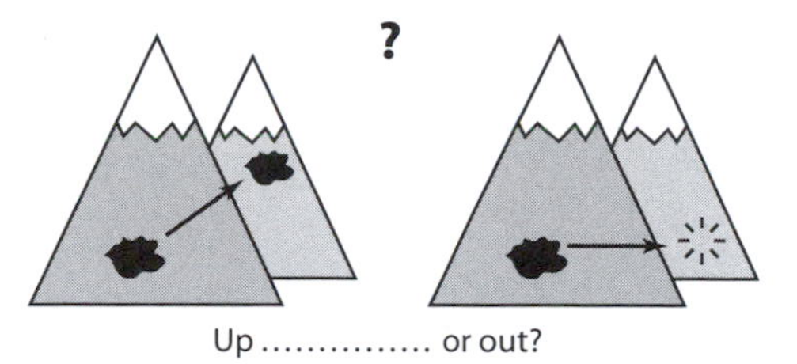

Figure 2. Some species that are associated with specific types of habitats may find that their bioclimatic envelope does not move up in elevation or closer to the poles. Rather, some habitats may disappear. For example, a mesic meadow habitat in a montane system that is associated with drainages may simply disappear if climate becomes warmer and drier.

BOX 1. CLIMATE CHANGE EXPERIMENTS

There are many research methods that scientists use to manipulate global change factors. The most common warming manipulation techniques include active warming via suspended infrared heat lamps or electric coils buried in the soil and passive warming via greenhouse-like structures that raise ground level temperatures by trapping the heat from the sun (plate 18, upper left). Warming has been shown to modify important patterns and processes such as nutrient cycling rates, plant productivity, and plant community composition and diversity.

To examine the impacts of elevated CO_2 on plant communities in the field, a new technology was developed that pipes CO_2 into experimental plots (plate 18, lower left). Elevated CO_2 experiments in a variety of forest and grassland ecosystems suggest that increasing CO_2 levels will increase plant productivity, but that availability of other nutrients (e.g., nitrogen) could limit productivity increases.

Several field experiments have manipulated precipitation levels, using rain shelters to reduce the amount of precipitation entering plots or manual water applications to augment incoming precipitation amounts. For many of these studies, the timing and variability in precipitation extremes are as important as changes in the total amount of moisture entering experimental plots.

Multifactor experiments manipulate several of these environmental factors (e.g., temperature, precipitation, and CO_2) simultaneously (plate 18, right). These experiments provide insight into the ways in which factors interact to influence ecosystems and often show that the addition of multiple change factors does not result in additive effects on ecosystem response variables. For example, the CO_2, temperature, precipitation, and nitrogen deposition field experiment at Jasper Ridge, CA, found that each of those factors separately increased plant productivity; but when CO_2 was increased in conjunction with the other treatments, it served to suppress the positive response in plant growth (Shaw et al., 2002).

Long-term observational records allow researchers to examine correlations between climate conditions and species' phenologies, behaviors, or distributions. There are a growing number of observational studies that can detect the signal of climate change over the last several decades (Parmesan, 2006; Root et al., 2005). These studies can consider climate change impacts on more mobile species and larger landscapes, but they are dependent on the existence of sufficiently long observational records to distinguish clear trends from what is often substantial interannual noise. Observational studies are also limited in their ability to establish direct causal relationships between changing climate and a species' response. Therefore, assumptions must be made about whether patterns that have been observed over the last few decades are likely to continue in a similar fashion as climate continues to change in the future. In areas where long-term observation records are not available, some studies have employed an approach called "space for time substitution" whereby changes in climate across space are used to infer the effects of changes in climate through time. These studies examine naturally occurring gradients in climate across areas that are otherwise very similar (e.g., in terms of vegetation, soils, land-use history, etc.) to determine the role of climate in determining ecosystem processes and species interactions.

Models provide an important third avenue for projecting future impacts of climate change by allowing scientists to consider larger spatial scales and different environmental conditions than currently exist. A range of modeling approaches exists. Dynamic vegetation models and species distribution models are two frequently used approaches for examining predicted community changes (Cramer et al., 2001). These models use different methods for predicting future responses to climate change, and each model has trade-offs in terms of resolution and scale, the dynamic nature of the projections, and the ability to examine species-level responses. Bioclimatic envelope models use statistical methods to correlate species occurrences with environmental predictor variables to define a species' environmental niche and predict the species' occurrence across a broader landscape (box 2; Pearson and Dawson, 2003). Although bioclimatic envelope models are able to examine species-level responses for a large number of species simultaneously, they do not capture dynamic interactions among species or between the organisms and their environment. Dynamic vegetation models, on the other hand, rely on mechanistic representations of physiological, biophysical, and biogeochemical processes to simulate changes in vegetation structure and water, energy, and carbon fluxes in response to environmental change. Although these models include dynamic interactions, they are not able to capture species-level responses and can only predict changes in vegetation structure through the use of broad vegetation classes (e.g., alpine/subalpine forest, evergreen conifer forest, mixed evergreen forest, grassland, or shrubland).

BOX 2. BIOCLIMATIC ENVELOPE MODELING

Bioclimatic envelope modeling is based on Hutchinsonian niche theory and understanding the environmental conditions necessary for a species to survive and grow. The model-predicted species distributions are therefore related to the species' climatic tolerances, and as climatic conditions change, so too will the location of a species' bioclimatic envelope. Bioclimatic models assume that species track environmental changes by moving to keep up with the changing location of their bioclimatic envelope. Pearson and Dawson (2003) reviewed the use of bioclimate models and argue that bioclimate envelopes might provide a first approximation regarding the changes that could be expected in species distribution patterns with climate change. Some problems with using bioclimatic envelope models to project impacts of climate change include the following: (1) limits to dispersal, which could make model predictions overly optimistic because some species may become extinct locally before they have a chance to move; (2) uncertainties related to model choice and future climate projections, which are often not considered; and (3) the difficulty of validating a model's ability to predict future distributions.

4. CONSEQUENCES OF CLIMATE CHANGE FOR CONSERVATION

The ecological effects of climate change will have significant consequences for biodiversity conservation. Species and ecosystems that are currently protected in reserves may shift outside of fixed reserve boundaries, possibly rendering the original conservation intent of such protected areas obsolete. The loss of some species and additions of others may lead to novel species assemblages and to possible problems with exotic invaders. Although some species may theoretically be able to move in response to changing climate conditions, habitat fragmentation may impede the ability of other species to migrate and disperse.

The challenge for conservationists is determining how to manage and conserve the landscape in order to facilitate the adaptation of species and ecosystem processes to a changing climate. This is primarily being addressed through two approaches: changes in the way we manage areas that are currently under some form of conservation, and the prioritization of land areas for future conservation given the threat of changing climate.

Managing for a Changing Climate

The science of managing for climate change is currently in its infancy, and the language of this field is still developing. Strategies include whether to manage for resistance options (e.g., those that delay the effects of climate change), resilience options (e.g., those that increase the ability of the ecosystem to return to previous conditions following a disturbance), or response options (e.g., those that facilitate ecosystem changes brought about by a changing climate) (Millar et al., 2007). Monitoring to establish baseline conditions and quantify change is a first step in providing scientists with the tools to understand how ecosystems are responding to a changing environment. Adaptive management—modifying management approaches over time as the manager obtains a better understanding of the system—will be an important approach to dealing with climate change. For example, if wildfire frequency increases with warmer temperatures, a manager might want to modify the way that wood is harvested to maximize the placement of fire breaks or minimize the amount of standing dead trees that could provide fuel for a fire. However, even if a manager knows the current status of the system, there are several challenges inherent in dealing with climate change: (1) developing a baseline for comparison; (2) understanding time lags; and (3) consideration of entirely new management approaches.

Establishing target conditions for an ecosystem is contingent on (1) knowing the preexisting conditions of an ecosystem at particular times in the past and (2) deciding what time period is appropriate to use as a baseline. With respect to carbon dioxide and other greenhouse gases, we can compare the concentrations of gases in the atmosphere to those of preindustrial times. However, it may be more difficult to decide on baselines with respect to the distribution and abundance of species or the frequency of disturbances such as fire or flooding. We know that humans have changed the frequency and the distribution patterns of both biotic and abiotic components of the ecosystem. However, determining what is a "natural" level of variation and what baseline is appropriate as a goal can provide fodder for a plethora of scientific discussions.

We also know that in studying climate change, there are ecosystem memories, time lags, and threshold responses that may blur our ability to discern direct effects of climate change. As an example of an ecosystem "memory," decades after a modification such as plowing, the attributes of the soil reflect that previous management. Time lags may be apparent in the ways that plants and animals respond to drought, one of the conditions predicted to become more severe for some portions of the globe. For example, plants with deeper root systems may be able to obtain water from deep within soil horizons, whereas plants with shallow roots may be more immediately affected. Multiple years of drought may be needed before woody plants such as trees and shrubs exhibit any significant changes (a threshold effect), whereas grasses and flowering plants may show changes much more quickly. Aquatic ecosystems may also have lag times in responding to climate change. Ice-out and freeze-up dates for lakes are already changing, modifying the length of the growing season. Lakes in temperate areas are driven by fall and spring turnover events, which occur when lake water hits a threshold temperature. If temperatures change dramatically, the frequency and/or timing of lake turnover could change, affecting the distribution of dissolved oxygen and nutrients within the water column. Such changes could have significant ramifications for all portions of the aquatic community. So, in analyzing responses to climate change, it will be essential to remember that time lags and threshold effects may blur our ability to detect change and that the patterns we discern may not always be linear.

Managers dealing with climate change will need to think outside of the box and contemplate new management approaches for conservation. For example, in some cases managers may consider assisting the movement of those species that cannot move fast enough to keep up with the rapid pace of climate change. This "assisted migration" approach to management will undoubtedly be costly, and managing so intensely for one species may preclude the options for another species. A variety of other management approaches that have not previously been used may need to be considered. There will undoubtedly be trade-offs in prioritizing such actions. The bottom line is that systems will be operating outside of the bounds of what was previously considered normal, and creative solutions may be in order.

Prioritizing Landscapes for Conservation under a Changing Climate

Our current network of fixed-boundary protected areas will be insufficient in the face of climate change. Therefore, we need to think about the conservation of additional areas and the linkages between them. Over the last few decades, one focus of conservation biology research has been on how to systematically select a portfolio of conservation sites in order to maximize their ability to protect biodiversity. A particular challenge is how to design reserves that protect the "persistence" of biodiversity through time, not just the presence of a species or a type of landcover at one point in time. Several studies have shown that efficient reserve networks that represent the optimal number of species given the cost do not necessarily protect that entire set of species over time because of fluctuating metapopulation dynamics. The challenge presented by the issue of biodiversity persistence is similar to that presented by future climate change, where we are interested in preserving the persistence of biodiversity on the landscape under changing climate conditions. A particular concern is how protected areas with fixed boundaries will be able to protect species that move across the landscape in response to changing climate.

Some initial ideas on how to design reserves in the face of climate change were based on simplified assumptions about how species distributions might respond to changing climate, including connecting reserves at different latitudes or creating reserves that stretch up and down mountains. To more rigorously examine how climate change–induced species range shifts might affect conservation planning efforts, researchers are combining bioclimatic envelope models (box 2) with systematic reserve design selection tools (see chapter V.3). In one study, Araujo et al. (2004) found that reserve designs that were optimized for species protection under current climate conditions in Europe could lose ~6–11% of plant species by 2050 because of climate change–induced shifts in species distributions. Using similar modeling approaches in Europe, Mexico, and South Africa, Hannah and colleagues demonstrated that taking into account the effect of future climate change on species ranges when planning conservation areas today was less costly (in terms of area) than delaying action to a later date (Hannah et al., 2007). Moreover, the habitat areas these species will need in a warmer world may not be around if we wait too long to acquire or protect them. These results argue for increased consideration of climate change in systematic reserve design approaches being applied today.

Across the globe, centers of species distribution may move toward poles or up in elevation. In some cases, species may be stranded and literally have no place to go. Species that live in alpine habitats may simply lose their habitat if there is no additional space further up the mountain (figure 1). Alternatively, species that exist nowhere near another suitable habitat patch may not have many options for dispersal. In many small native prairies and wetlands in the Midwestern United States, species are barely hanging on and are surrounded by a sea of inhospitable landscape dominated by intensive agriculture. Movement across such areas could effectively be "suicidal dispersal." There have been some efforts, especially at the local level to create roadside prairies and riparian buffers between cropfields and waterways. These areas may provide "stepping stones" for some species (e.g., butterflies, birds, small mammals), but it is unrealistic to envision a future where large organisms (e.g., bison) travel between prairies via roadsides.

Synergies with other environmental changes, such as habitat loss and fragmentation, will potentially have large effects on species' abilities to adapt to changing climate. Habitat loss and fragmentation caused by human modification of the landscape are some of the most significant threats to species worldwide. In many areas, the landscape is sufficiently fragmented to inhibit the movement of species as they attempt to track changing climate conditions. The concept of using corridors to connect fragmented habitats and create a more permeable landscape is a popular idea, and scientists are beginning to document the effects of corridors. However, creating corridors that facilitate movement of species from one patch to the other is easier said than done. A corridor that provides connectivity for large mammals may create a predation trap for smaller mammals, reptiles, or amphibians. In addition, the construction of large corridors can be extremely expensive, as observed by the cost involved in creating highway overpasses to connect large blocks of habitat fragmented by the Trans-Canada Highway in Alberta's Banff National Park. Furthermore, we cannot simply focus on reserves and corridors but must consider the matrix of landscapes and land uses that occur between them.

5. THE MISSING LINKS

Present models of climate change are primarily global models. Scientists are working to "scale down" these models to create regional models. However, they are still a long way from providing specific information about how each portion of the globe is expected to change. Regional predictions will be essential to planning conservation efforts for most species of concern.

Even after more specific regional models are developed, scientists will need to better understand how species move through landscapes and what types of habitats are acceptable in order to make predictions regarding the effects of climate change at a species-specific level. Reserves that were previously acceptable habitat may become unacceptable. Vagility will be a key predictor of response. Those species that are able to disperse sufficiently to keep up with the pace of changing climate may be able to maintain viable populations, whereas less mobile species may become stranded in unsuitable habitats. Behavioral and demographic studies will be needed to quantify habitat use and to classify sites as sources or sinks. Finally, many traditional conservation efforts have been species specific and focused on a few charismatic megafauna. The pervasive nature of climate change demands that landscapes be managed from the perspective of preserving biodiversity and the larger ecological communities. This does not reduce the importance of species-specific conservation measures and research, but it means that conservation must be approached from multiple perspectives.

From the community perspective, there has been some progress in combining biogeographic models with reserve design algorithms to consider the impacts of climate change on reserve design. However, there have been relatively limited on-the-ground efforts at putting that theory into practice. Experimental manipulations as well as long-term monitoring efforts will be needed to test the adequacy of existing reserve systems over time.

Finally, in planning for climate change, one of the most important things to remember is that scientists and managers will need to examine issues from a global perspective. Planning at the regional and continental scale may be adequate for some issues, but weather patterns, ocean circulation, and many climatic phenomena occur at a global scale. Narrow corridors connecting small patches of landscape may not be adequate to deal with this scale of change. Neither the forces of climate change nor the organisms of concern will pay heed to political, state, or national boundaries. Management will need to be envisioned at large geographic scales. A global approach will require that managers, politicians, and scientists work together in ways that they have not done so before. Just as our economy has become a global economy, our solutions to climate change will undoubtedly require international cooperation in terms of resource use, species conservation, and landscape management.

FURTHER READING

Araújo, Miguel B., Mar Cabeza, Wilfried Thuiller, Lee Hannah, and Paul H. Williams. 2004. Would climate change drive species out of reserves? An assessment of existing reserve-selection methods. Global Change Biology 10: 1618–1626.

Cramer, Wolfgang, Alberte Bondeau, F. Ian Woodward, I. Colin Prentice, Richard A. Betts, Victor Brovkin, Peter M. Cox, Veronica Fisher, Jonathan A. Foley, Andrew D. Friend, Chris Kucharik, Mark R. Lomas, Navin

Ramankutty, Stephen Sitch, Benjamin Smith, Andrew White, and Christine Young-Molling. 2001. Global response of terrestrial ecosystem structure and function to CO_2 and climate change: Results from six dynamic global vegetation models. Global Change Biology 7: 357–373.

Davis, Margaret B., and Ruth G. Shaw. 2001. Range shifts and adaptive responses to Quaternary climate change. Journal of Biogeography 18: 653–668.

Hannah, Lee, Guy Midgley, Sandy Andelman, Miguel Araujo, Greg Hughes, Enrique Martinez-Meyer, Richard Pearson, and Paul Williams. 2007. Protected area needs in a changing climate. Frontiers in Ecology and the Environment 5: 131–138.

Intergovernmental Panel on Climate Change. 2007. Climate Change 2007: The Physical Science Basis. Summary for Policy Makers. Contribution of Working Group I to the Fourth Assessment Report of the Intergovernmental Panel on Climate Change. Geneva: IPCC Secretariat.

Jetz, Walter, David Wilcove, and Andrew Dobson. 2007. Projected impacts of climate and land-use change on the global diversity of birds. PLoS Biology 5: 1211–1219.

Millar, Constance, Nathan Stephenson, and Scott Stephens. 2007. Climate change and forests of the future: Managing in the face of uncertainty. Ecological Applications 17: 2145–2151.

Parmesan, Camille. 2006. Ecological and evolutionary responses to recent climate change. Annual Review of Ecology, Evolution, and Systematics 37: 637–669.

Pearson, Richard G., and Terence P. Dawson. 2003. Predicting the impacts of climate change on the distribution of species: Are bioclimate envelope models useful? Global Ecology and Biogeography 12: 361–371.

Pounds, J. Allen., and Martha L. Crump. 1994. Amphibian declines and climate disturbance: The case of the golden toad and the harlequin frog. Conservation Biology 8: 72–85.

Root, Terry L., Dena P. MacMynowski, Michael D. Mastrandrea, and Stephen H. Schneider. 2005. Human-modified temperatures induce species changes: Joint attribution. Proceedings of the National Academy of Sciences U.S.A. 102: 7465–7469.

Shaw, M. Rebecca, Erika S. Zavaleta, Nona R. Chiariello, Elsa E. Cleland, Harold A. Mooney, and Christopher B. Field. 2002. Grassland responses to global environmental changes suppressed by elevated CO_2. Science 298: 1987–1990.

Westerling, Anthony L., Hugo G. Hidalgo, Daniel R. Cayan, and Thomas W. Swetnam. 2006. Warming and earlier spring increase western U.S. forest wildfire activity. Science 313: 940–943.

Wigley, Tom M. L. 2005. The climate change commitment. Science 307: 1766–1769.

V.7

Restoration Ecology

Richard J. Hobbs

OUTLINE

1. What is restoration ecology?
2. Concepts in restoration ecology
3. Key steps in ecological restoration
4. Repairing damaged ecosystem processes
5. Directing vegetation change: succession and assembly rules
6. Fauna and restoration
7. Landscape-scale restoration
8. Prevention versus restoration

Restoration ecology is the science underpinning the practice of repairing damaged ecosystems. Restoration ecology has developed rapidly over the latter part of the twentieth century, drawing its concepts and approaches from an array of sources, including ecology, conservation biology, and environmental engineering. We are faced with an increasing legacy of ecosystems that have been damaged by past and present activities, and it is increasingly recognized that, in many situations, successful conservation management will need to include some restoration. This may take many different forms, such as the reintroduction of particular species, removal of problem species such as weeds or feral animals, or the reinstatement of particular disturbance regimes (including fire and flood regimes).

GLOSSARY

alternative stable state. A relatively stable ecosystem structure or composition that is different from what was present before disturbance

disturbance. Episodic destruction or removal of ecosystem components

resilience. The ability of an ecosystem to recover following disturbance

restoration. The process of assisting the recovery of an ecosystem that has been degraded, damaged, or destroyed

succession. The process of vegetation development following disturbance, often characterized by relatively predictable sequences of species replacement over time

threshold. A situation where there has been a nonlinear (i.e., sudden or stepped) change in the ecosystem in response to a stress or disturbance, which is often difficult to reverse

1. WHAT IS RESTORATION ECOLOGY?

Restoration ecology is the science behind the term *ecological restoration*, which covers a range of activities involved with the repair of damaged or degraded ecosystems and is usually carried out for one of the following reasons:

1. To restore highly disturbed, but localized sites, such as mine sites.
2. To improve productive capability in degraded production lands.
3. To enhance nature conservation values in protected landscapes.
4. To restore ecological processes over broad landscape-scale or regional areas.

Ecological restoration occurs along a continuum, from the rebuilding of totally devastated sites to the limited management of relatively unmodified sites, and hence merges with conservation biology. Restoration aims to return the degraded system to a less degraded state that is valuable for conservation or other use and that is sustainable in the long term.

An array of terms has been used to describe these activities, including *restoration*, *rehabilitation*, *reclamation*, *reconstruction*, and *reallocation*. Generally, restoration has been used to describe the complete reassembly of a degraded system to its undegraded state complete with all the species previously present, whereas rehabilitation describes efforts to develop some sort of functional or productive system on a degraded site. In addition, some authors use the term *reallocation* to describe the transfer of a site from one land use to a more productive or otherwise beneficial use. However, the term restoration is often used to refer broadly to activities that aim to repair damaged systems.

Ecosystem characteristics to be restored can include the following:

1. Composition: species present and their relative abundances.
2. Structure: vertical arrangement of vegetation and soil components (living and dead).
3. Pattern: horizontal arrangement of system components.
4. Heterogeneity: a complex variable made up of components 1–3.
5. Function: performance of basic ecological processes (i.e., energy, water, nutrient transfers).
6. Species interactions: pollination, seed dispersal, etc.
7. Dynamics and resilience: succession and state-transition processes, recovery from disturbance.

Restoration has often been viewed as returning an ecosystem or community back to a previous state, i.e., the ecosystem that existed at the site before human disturbance or alteration. However, ecosystems are naturally dynamic entities, and hence, the setting of restoration goals in terms of static compositional or structural attributes is problematic. Often, past system composition or structure is unknown or partially known, and past data provide only static snapshots of system parameters. Current undegraded reference systems can therefore act as potential reference systems against which the success of restoration efforts in degraded systems can be measured. An alternative approach is to explicitly recognize the dynamic nature of ecosystems and to accept that there is a range of potential short- and long-term outcomes of restoration projects. Increasingly, the focus is on having a transparent and defensible method of setting restoration goals that clarify the desired characteristics for the system in the future rather than in relation to what these were in the past. Using past characteristics to guide restoration is still useful, but there is increasing recognition that continuing environmental change, including climate change, means that returning ecosystems to past states may not always be possible.

Where it is impossible or extremely expensive to restore composition and structure, alternative goals may be appropriate. These may aim to repair damage to ecological function or ecosystem services or to create a novel system using species not native to the region or suited to changed environmental conditions. Often, partial restoration or the development of alternative ecosystems with some desirable elements of structure, composition, or function can have positive conservation outcomes. For instance, plantations of timber trees may not develop all the characteristics of a native forest but still may be used by some fauna species—especially if plantation management is modified slightly to improve their value as habitat. Clearly, however, a risk analysis is needed to ensure that the changes do not lead to further problems in the future—for instance, using nonnative species in restoration may lead to these species becoming problematic in the future.

2. CONCEPTS IN RESTORATION ECOLOGY

Disturbance and Resilience

Disturbance, or episodic destruction or removal of ecosystem components, is an integral part of the functioning of many ecosystems. Disturbance often initiates massive ecosystem change and triggers a period of regeneration or recovery. Disturbance is a natural feature of many ecosystems, and disturbances range in extent and severity from localized events such as animal diggings or individual tree falls to large events such as catastrophic fires, large storms (hurricanes, cyclones), and floods. Ecosystems have a degree of resistance to disturbance, termed inertia. In other words, ecosystems can absorb a certain amount of disturbance or stress and remain more or less unchanged. For instance, a low-intensity fire in a forest may only burn surface litter and leave the major components of the ecosystem intact. Or a river system may be able to tolerate a certain level of pollution without undergoing large changes in its biota. When the disturbance is large enough, however, this inertia is overcome, and the system changes. The disturbance could be a discrete event such as a wildfire or windstorm, or it could be an accumulated chronic impact such as increasing pollution. The ability of the system to recover from that change is termed resilience. A resilient system will be able to recover quickly after a disturbance or when a degrading factor is removed and will return to more or less the same structure and composition as was present previously. Ecological restoration is required only where the system's resilience has been diminished in some way or where the normal recovery processes are too slow to achieve management goals within a desirable time frame. As an example, an arid system that has been overgrazed loses its capacity to regulate water flows because the grazing removes the mosaic structure of plants and debris that intercepts surface flows. As a result, erosion occurs, water does not enter the soil, and conditions are not suitable for plant establishment. Simply removing grazing will not initiate recovery—instead, some active intervention to reinstate some surface heterogeneity is required to "kick-start" the process of recovery.

Human activities frequently either modify the original disturbance regimes (e.g., by changing fire or grazing regimes) or add a further set of disturbances that the ecosystem had not previously experienced. In some cases, this human disturbance pushes ecosystems beyond the limits of their resilience. It is in these cases that active restoration is required.

Ecosystems become degraded when human use or alteration modifies ecosystem characteristics such that ecosystem structure and/or function is changed beyond acceptable limits. For instance, vegetation structure may be altered so that it no longer provides adequate habitat for a range of animal species, or the ecosystem may no longer provide ecosystem services, such as provision of clean water or production of food or fiber. Degradation may result in changes to the biological component of ecosystems or more fundamental changes to the system's physical or chemical characteristics.

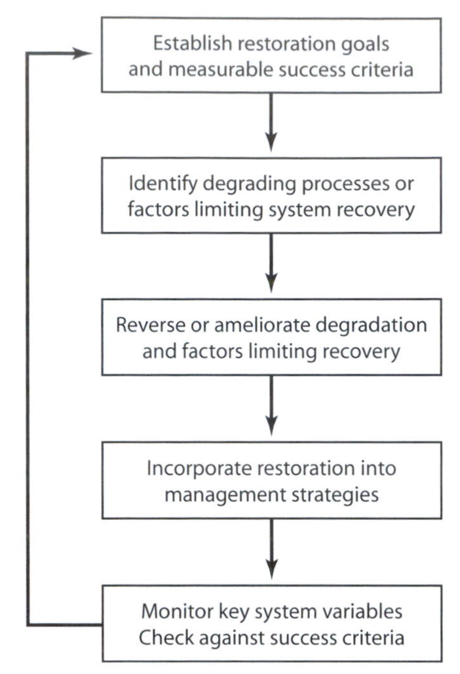

Figure 1. Processes involved in ecological restoration.

Ecological Succession

Following disturbance, ecosystems undergo a process of recovery known as succession. Succession describes the sequence of species and groups of species that are present at various times since a disturbance. Pioneer species appear early in the recovery sequence and are able to tolerate open, often harsh conditions. Later successional species are either slower growing or appear only after conditions are modified by pioneer species. The successional trajectory describes the direction and rate of change. Relay floristics describes the process whereby species appear in a recognizable sequence, whereas initial floristic composition relates to the situation in which all species that will take part in the successional sequence appear shortly after a disturbance but grow and/or assume importance at different rates. Species may either facilitate, inhibit, or tolerate other species.

Thresholds and Alternative Stable States

In some ecosystems, a relatively predictable post-disturbance recovery sequence can be expected, but in others, there is the possibility that recovery will follow different trajectories and result in different ecosystem compositions. The trajectory followed may depend on the arrival of particular species in the system, the method of management imposed, or the sequence of climatic and disturbance events during the recovery process. The ecosystem may reach a state from which little further change occurs. This state is known as an alternative stable state, which means that the ecosystem has developed a relatively stable structure or composition that is different from what was present before the disturbance. The presence of such a state often indicates the operation of system thresholds that need to be crossed before further system change can occur. A threshold usually indicates a situation where there has been a nonlinear change in the ecosystem in response to a stress or disturbance: often it may be relatively easy to degrade an ecosystem past such a threshold but much more difficult to restore the ecosystem to a less-degraded state (figure 1). For example, in some ecosystems, progressive addition of nitrogen via air pollution can tip the balance between native plant species and invasive grass species. Once invasive grasses are dominant, they prevent the reestablishment of native species, and the ecosystem becomes stuck in an altered state. Reversing this situation involves not just removing the invasive species but also dealing with the elevated nutrient status of the site. Hence, dealing with such thresholds may involve quite intensive management. The identification of system thresholds is an important element of assessing the appropriate restoration measures needed in any given situation.

Biotic and Abiotic Limitations to System Change

Ecosystem degradation can result in changes to either the biological component of an ecosystem or its abiotic (physical and/or chemical) characteristics. Biotic changes can include loss of particular species or changes in vegetation structure and composition, whereas abiotic changes can include changes in substrate physical or chemical characteristics or alteration of the hydrological regime. Damage to primary ecosystem processes may result in more fundamental system changes than simple biotic changes. Correct assessment of which ecosystem characteristics have been altered during degradation is essential for effective restoration.

System Recovery: Hysteresis and Dynamic Systems

Even with correct assessment and treatment of the problems leading to degradation, it may be impossible to return a system completely to its predisturbance state. System recovery may follow a different path from that taken during system decline (hysteresis), and the resulting system may thus differ from the original. Natural ecosystems are also naturally dynamic and hence constantly changing. This also makes it unlikely that the recovering system will return to exactly the same composition and structure as the predisturbance system. Thus, it is important to set realistic restoration goals that take into account the dynamic nature of the ecosystems we are trying to restore.

3. KEY STEPS IN ECOLOGICAL RESTORATION

There are a number of key steps in any restoration program that need to be undertaken to ensure that useful outcomes are achieved (figure 2). These include setting clear goals with associated success criteria, correctly identifying the factors limiting system recovery or leading to further degradation, and instigating restoration activities that reverse or ameliorate these factors. These activities have to be placed in the context of broader management objectives and monitored to ensure that progress is being made toward the agreed goals.

Setting Goals

Setting clear and achievable goals is an important element that is often overlooked but that greatly facilitates the process of deciding on restoration options and monitoring progress. Almost anything is possible if enough money and resources are available, but generally, goals have to be selected on the basis of cost-effective measures to overcome limiting factors to allow reasonable goals to be achieved. Goals broadly relate to the restoration of ecosystem function and/or ecosystem structure or composition. The choice of restoration options needs to be guided by both the ecological constraints operating in any given situation and the range of individual and societal goals.

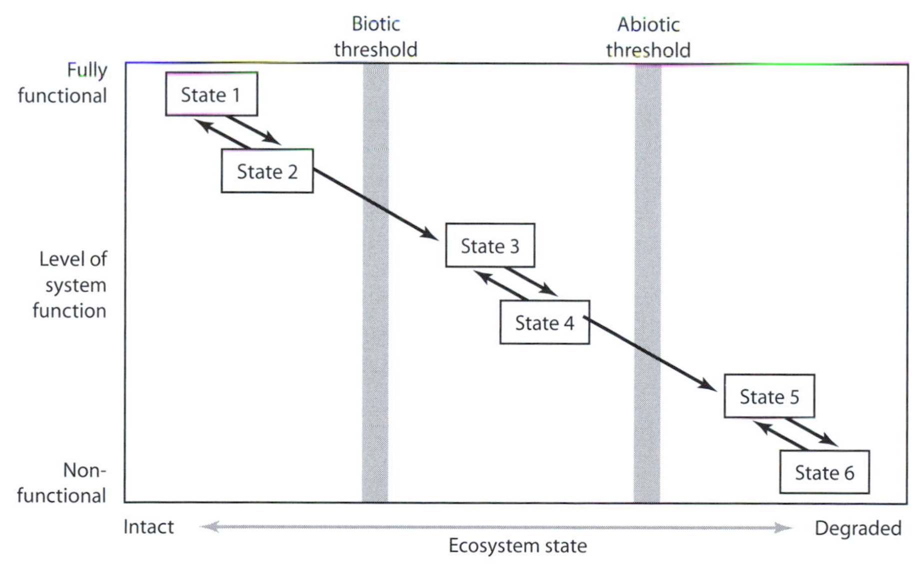

Figure 2. Summary of ecosystem degradation and restoration in terms of alternative states and transitions between them. States are indicated in boxes, and possible transitions between states are shown by arrows. Hypothesized thresholds that prevent transition from a more degraded state to a less degraded state are indicated by the vertical gray bars; such transitions can be either biotic (for instance, competition or grazing) or abiotic (for instance, changed physical or chemical conditions).

Often goals for restoration are set in relation to a particular ecosystem or species composition—for instance, a goal for postmining restoration in a forest ecosystem may be the return of the complete forest ecosystem with all the species that were there previously, or it may be simply the stabilization of the mined surface and the return of the major tree species. An alternative goal for a postmining ecosystem may be the creation of a grass pasture that can be used for livestock production.

Identifying Limiting or Degrading Factors

Although the cause of degradation may be obvious in some cases (e.g., in mine sites), the factors that are important in influencing system recovery may not be so obvious. A clear identification of the environmental factors that are either causing ongoing degradation or preventing system recovery is essential. Failure to do this properly can result in costly mistakes that do not fix the problem. These factors can affect either primary abiotic processes such as nutrient and water retention or biotic processes such as plant recolonization and survival. As an example, in a mine-site restoration project, there is no point in replanting the area if there are problems with the stability or chemical composition of the substrate. Similarly, in an area where the hydrological regime has changed, restoration of an area of riparian habitat may be ineffectual if the broader hydrological condition of the watershed is not also considered.

Incorporating Restoration into the Management Strategy

Once goals have been defined and methods identified for overcoming the degrading or limiting factors, the restoration project has to be actually implemented. This involves the practical considerations surrounding logistics, budgets, timing of operations, etc. Implementation includes not only the initial activities early in restoration but also ongoing management of the restored site to ensure that the restored system continues to develop along an appropriate trajectory that will result in the restoration goals being achieved.

Monitoring Restoration Progress

Monitoring progress is essential to the success of a project but is often not carried out effectively. The choice of variables to be monitored must relate to the goals set for the restoration. Variables must also be relatively inexpensive and simple to monitor. Restoration projects often have success criteria set that have to be met to satisfy contractual or legal obligations. Hence, tracking progress toward success is important. Monitoring can also allow adaptive management, which can identify situations where the management treatment is not having the desired result and hence can be changed.

4. REPAIRING DAMAGED ECOSYSTEM PROCESSES

Ecosystem processes include the cycling and retention of nutrients, carbon, and water. These processes are essential to ecosystem function. Degraded systems frequently have less control over ecosystem flows, and the reinstatement of structures and processes that regulate these flows is often a first step in restoration. This is also true in areas such as mine sites, where ecosystems have to be established de novo.

Removal of vegetation cover and degradation of soil structure can influence the ability of an ecosystem to mediate system flows. This can be manifested as increased erosion and loss of nutrients, altered hydrology leading to increased runoff, rising water tables, flooding, and salinization. Physical manipulation of the substrate (e.g., modifying soil structure or surface microtopography), introducing physical barriers (e.g., brushing), or reestablishing vegetative cover may be necessary to reinstate local control of water and nutrient flows. Remediation of chemical composition of soil or water may also be necessary to facilitate the reestablishment of vegetation.

An ecosystem consists of both biotic and abiotic components, which are interlinked via transfers and flows of nutrients, energy, and water. Restoration of primary processes requires attention to both the abiotic and biotic components. Attempts to reestablish vegetation on areas where primary processes have not been repaired are likely to fail. On the other hand, attention to the biotic component can also speed up repair of primary processes. For instance, nutrient capture may be enhanced by the reestablishment of mycorrhizae or the inclusion of plant species with an array of different root architectures.

5. DIRECTING VEGETATION CHANGE: SUCCESSION AND ASSEMBLY RULES

Processes in Vegetation Redevelopment

As discussed above, successional processes result in the redevelopment of vegetation on an area following a disturbance. In order for this to occur, plant propagules have to be available at a site (via dispersal or from seed stored in the soil or held on the canopy of adult trees), the site has to be suitable for establishment, and the species have to be able to grow and reproduce. In small disturbed areas surrounded by native vegetation, it may be possible to let species disperse in unaided, but often assisted reestablishment is needed. Hence, for instance, in mine-site restoration, seeds may be returned to the area in topsoil taken from adjacent areas to be mined and in seed mixes containing an appropriate mix of species. In addition, seedlings or parts of plants may be planted into the area.

Directing Vegetation Change

Restoration frequently aims to speed up vegetation development or to direct its course to a predetermined goal. In order to do this, site characteristics, plant colonization, and subsequent survival can all be manipulated. Attention to the repair of primary processes such as water retention and nutrient cycling is essential, and factors such as soil structure and chemistry and nutrient availability need to be considered. The colonization of a site by species can be effected by ensuring that seed is available on site, either in the soil or canopy seed store or by seed dispersal. Where seeds are mobile or dispersed by birds and other animals, they may be effectively dispersed without further intervention. However, this process may be too slow, and seed may need to be introduced. In extreme cases, seed germination may be too unreliable, and planting of seedlings or other plant material may be necessary. Once species are established at a site, continued vegetation development depends on their survival and how they interact with other species. Survival can be increased by ensuring that site characteristics are favorable (e.g., through water or nutrient retention) and that damage via herbivory and disease is minimized. The development of a functioning biotic community depends on achieving a good mix of species with different life forms (trees, shrubs, grasses, etc., depending on which sort of community is being restored) and the development of species interactions such as mycorrhizal associations, pollination, and seed dispersal.

In addition, large numbers of introduced plant species cause significant alteration to ecosystems around the world. Frequently, restoration has to involve the removal of these undesirable plant species, which either prevent system development or cause the successional trajectory to divert from the desired goal.

Assembly Rules

How communities are built is a central question in restoration ecology, and recently there have been attempts to consider what factors affect the characteristics of the developing community. Such factors include the timing and extent of arrival of different species on the site and a series of abiotic and biotic filters that influence the success of establishment and survival of each species based on its physiological tolerances, competitive abilities, and interactions with other species. These factors may be considered to be a set of "rules" for what species persist at any given site. Restoration efforts can be guided by these "rules" and aim to modify the abiotic and biotic filters to allow colonization and persistence of species that will allow restoration goals to be achieved.

Autogenic versus Assisted Recovery

As indicted earlier, in some instances, restoration can take advantage of autogenic processes by which species recolonize degraded areas unaided, and a plant and animal community reassembles. This is likely to occur where substrate conditions are favorable and where species are effective colonizers; it is the cheapest form of restoration. In some situations, the types of species that colonize may not be desirable (e.g., invasive weed species), or they may prevent the further development of the vegetation to a desired state. In others, there may be little or no colonization because of the continued adversity of the site or because the native species have low dispersal capabilities. In these cases, assisted recovery is necessary and involves ensuring that the desired mix of species is available on site and can persist.

6. FAUNA AND RESTORATION

Animals are important elements of ecosystems, and yet restoration projects generally focus on the reestablishment of vegetation, on the assumption that they will provide habitat for fauna to return. However, it is not always clear that this is the case, and often the precise habitat requirements for individual animal species or suites of species are poorly known. Hence, what to put back and which processes are essential to reinstate are not always clear.

Particular faunal elements, known as "ecosystem engineers," can play important roles in structuring ecosystems and modifying ecosystem processes. The removal or introduction of such species can have dramatic ecosystem effects. Particular examples include beavers in North America and Europe, large herbivores in Africa, and digging marsupials in Australia. Such animals actively modify the environment; for instance, beavers build dams on streams, which create ponds and alter the structure and flow of the water course; large herbivores such as elephants in Africa structure the vegetation by knocking over and feeding on particular types of tree; and digging marsupials in Australia locally change soil characteristics, which affects water infiltration and creates suitable sites for plant establishment. These are particularly important considerations for restoration because the reintroduction of these species can also be construed as restoration programs that are restoring key ecosystem processes.

The reintroduction of particular species of fauna is often conducted in isolation from other restoration activities. However, the reintroduced fauna may have an impact in modifying ecosystem dynamics, and this may be either to the benefit or to the detriment of the system as a whole, depending on the overall goals of management. Similarly, restoration may involve the removal of particular problematic species, including introduced predators. Many restoration projects in such places as Australia and New Zealand and many island ecosystems involve the removal of introduced herbivores and predators that have had severe impacts either on the overall habitat or on native fauna species.

7. LANDSCAPE-SCALE RESTORATION

Landscapes are heterogeneous areas of land, usually square kilometers in extent, composed of interacting ecosystems or patches. The spatial distribution and interrelationships among landscape patches determine functions such as biotic movement and fluxes of water, energy, and nutrients.

Ecosystem modification and use often lead to dramatic changes in landscape structure and function. The most obvious manifestation of this is landscape fragmentation, which occurs when the original vegetation of an area is cleared and the land is transformed for agriculture or other use. Fragmentation results in the native vegetation being left as remnants of varying sizes and degrees of isolation. Reduction in size and connectivity of habitat reduces the probability of persistence of many species and results in declining species abundances (see chapter V.1). In addition, landscape processes such as water flows can be dramatically altered, leading, for instance, to changes in wetland habitats or salinization.

As with site-based restoration, management interventions need to be based on a sound assessment of the causes of degradation. In the case of landscape fragmentation, the causes of species loss and decline are often the loss of habitat and connectivity. Thus, replacing habitat and increasing connectivity are often goals for landscape restoration. This is best done by using the existing native vegetation as a skeleton on which to build restoration efforts. Hence, additional habitat or buffer strips can be established next to existing remnants, and corridors or other connecting vegetation can be developed. Restoration actions are best planned in relation to the needs of particular species so that specific recommendations on dimensions of habitat and corridors required can be determined. In addition to habitat re-creation, broader-scale revegetation and other activities may be needed to reverse or slow down hydrological changes and influence other landscape-level processes.

8. PREVENTION VERSUS RESTORATION

Restoration activities are often relatively costly because they involve management intervention, often on a large scale. In general, it is much more cost effective to prevent damage in the first place rather than repair it once it has happened. Thus, restoration forms part of a spectrum of management options that include prevention and conservation. For biodiversity conservation, the top priority must be to retain areas that remain in good condition. The next priority will be to repair damaged areas of native vegetation. Finally, and as a last resort, restoration of areas that have been transformed (e.g., by agriculture) may be needed to increase areas of habitat or landscape connectivity. It is much more costly to re-create a natural habitat than it is to protect or repair an existing one.

FURTHER READING

Falk, Don, Margaret Palmer, and Joy B. Zedler, eds. 2006. Foundations of Restoration Ecology. Washington, DC: Island Press.

Hobbs, Richard J., and David A. Norton. 1996. Towards a conceptual framework for restoration ecology. Restoration Ecology 4: 93–110.

Society for Ecological Restoration International Science & Policy Working Group. 2004. The SER International Primer on Ecological Restoration. Downloadable at http://www.ser.org/content/ecological_restoration_primer.asp.

Temperton, Vicky M., Richard J. Hobbs, Tim J. Nuttle, and Stefan Halle, eds. 2004. Assembly Rules and Restoration Ecology—Bridging the Gap between Theory and Practice. Washington, DC: Island Press.

van Andel, Jelte, and James Aronson, eds. 2006. Restoration Ecology: The New Frontier. Oxford: Blackwell.

Walker, Lawrence R., Joe Walker, and Richard J. Hobbs, eds. 2007. Linking Restoration and Ecological Succession. New York: Springer.

Whisenant, Steve G. 1999. Repairing Damaged Wildlands: A Process-orientated, Landscape-scale Approach. Cambridge, UK: Cambridge University Press.

Young, Truman P. 2000. Restoration ecology and conservation biology. Biological Conservation 92: 73–83.

VI

Ecosystem Services

Ann P. Kinzig

OUTLINE

1. Introduction
2. History
3. Ecosystem services, trade-offs, and biodiversity
4. Scale
5. Substituting for ecosystem services
6. Ecosystem services and conservation
7. Closing thoughts

1. INTRODUCTION

Ecosystem services are defined as "the multiple benefits provided by ecosystems to humans" (The Millennium Ecosystem Assessment, 2005). In other words, ecosystem services are only services to the extent that they support human well-being and are thus an inherently anthropocentric construct. Analysts cannot understand how the delivery of ecosystem services has changed over time solely from a purely natural science analysis of ecological patterns, processes, or functions. They must also understand what people value and how much they value it. Ecological dynamics could remain constant, but the services people derive from ecosystems could still change as people's values or circumstances change. Ecosystems could degrade, from a purely ecological perspective, but that degradation of ecological systems could still support an enhanced flow of services from humanity's perspective. Any adequate assessment of the flow of ecosystem services, or any assessment of how best to manage ecological systems to maximize the benefits people receive from them, must join ecological and social analyses.

However, this is a guide to ecology, and therefore, this section primarily contains contributions by ecologists. Ecologists obviously have much to contribute to an understanding of ecosystem services. In this section, various authors cover issues of scale (Scholes, chapter VI.1), biodiversity–ecosystem functioning relationships (Naeem, chapter VI.2), and critical aspects of ecological organization (Norberg, chapter VI.3), among other things. Authors examine ecosystem services in agroecosystems (Power et al., chapter VI.4), forests (Solórzano and Páez-Acosta, chapter VI.5), grasslands (Downs and Sala, chapter VI.6), and marine ecosystems (Baskett and Halpern, chapter VI.7). Other authors analyze different types of services, from tangible and consumable goods such as fresh water (Palmer and Richardson, chapter VI.8) to the more intangible and esoteric cultural or spiritual services provided by the world's ecological systems, and other services in between (Daszak and Kilpatrick on regulating services in chapter VI.9; Pergams and Kareiva on genetic diversity in chapter VI.10). But ultimately, the ecological analyses are just the starting point—knowing how ecological functions are changing, although relevant to adequate comprehension and appropriate management of ecosystem services, is not enough. The social value placed on those services—the human desires that translate a mere ecosystem function into a beneficial service—are half the equation. I have thus chosen to close this section with two nonecological chapters (which obviously can not do full justice to the second half of the equation): one focuses on the economics of ecosystem services (Perrings, chapter VI.11) and the other on how technological innovation has altered the need for, and therefore value placed on, various services (Goklany, chapter VI.12). This section closes with an exploration of how a focus on ecosystem services, rather than species richness per se, might alter conservation practices and outcomes (Rodríguez, chapter VI.13). I return to many of these issues—on the interplay between the social and ecological components of ecosystem services—in the topics I highlight below.

2. HISTORY

The recognition that ecological systems benefit humans must, in some sense, be as old as humanity itself—every culture or social group that I know of has developed rituals aimed at influencing Nature and her bounty. Even the more scientific assessment of ecological services has a deep history; Plato, for instance, recognized the connection among deforestation, erosion, and the drying of springs in his native Greece. Similarly, the connection between Nature's services and the value humans place on those services was made, perhaps not first but certainly eloquently, by David Ricardo in 1817, when he wrote:

> The labour of nature is paid, not because she does much, but because she does little. In proportion, as she becomes niggardly in her gifts, she extracts a greater price for her work. Where she is munificently beneficent, she always works gratis. (Ricardo, D., and F. W. Kolthammer. 1817. The Principles of Political Economy and Taxation. Mineola, NY: Courier Dover Publications, reprinted 2004)

In other words, humans will value most those services that are most scarce.

More frequent scientific and economic assessments of the benefits of ecological systems, however, only began in earnest in the latter part of the twentieth century. Mooney and Ehrlich (1997) provide a more thorough history than what follows, a history that starts with Marsh in 1864 and runs through to the present day; they themselves contributed a seminal paper in 1983 (Ehrlich and Mooney, 1983). The Beijer Institute of Ecological Economics launched a biodiversity program that ran from 1991 to 1993 with the specific purpose of valuing the full range of the benefits of ecosystems as opposed to individual environmental stocks, which had been the dominant approach in the preceding 25 years. The Global Biodiversity Assessment (Heywood, 1995) addressed questions of the value of goods and services delivered by biodiversity to society, among other things. That study, combined with Gretchen Daily's volume on *Nature's Services*, released in 1997, firmly established the need to make an assessment of ecosystem services a prominent part of the scientific agenda.

More recently, The Millennium Ecosystem Assessment (MA), a globally comprehensive assessment and synthesis of the consequences of ecosystem change for human well-being, involving more than 1300 experts worldwide and focusing on both global and more regional trends, was released (MA, 2005). The MA established a consensus framework for categorizing ecosystem services (see figure 1), which included support services, provisioning services, regulating services, and cultural services. Provisioning services are those tangible and consumable items humans derive from ecosystems—food, fiber, fuel, and fresh water, to name a few. Regulating services encompass the ecological patterns and processes that contain Nature's dynamics within certain bounds; reducing, for instance, the probability of massive landslides, pandemic disease outbreaks, or catastrophic climate excursions. Cultural services are the largely intangible and "unconsumed" services that ecosystems provide (unconsumed in the sense that enjoyment by one does not preclude enjoyment by another)—these are services such as recreation, aesthetic appeal, or a spiritual communion with nature. Supporting services are critical to the maintenance of all other ecosystem services and include such things as nutrient cycling, primary production, and soil formation. These supporting services are largely not used or valued directly by people; their value is indirect because they are an essential prerequisite to the provisioning of all other services. As a result, they are generally treated differently than the other three categories of services (as they are in this volume), being more akin to ecosystem functions or processes than to ecosystem services.

Given the comprehensive treatment of ecosystem services provided by the MA, we made no such effort here. Instead, this section of *The Princeton Guide to Ecology* is intended to highlight how one might assess and enhance ecosystem services in particular ecosystem types (chapters VI.8–VI.11), or understand some of the difficulties inherent in managing particular services (chapters VI.5–VI.7). These and other chapters also emphasize some of the future scientific challenges to an improved understanding of ecosystem services.

In the remaining sections of this chapter, I emphasize what I think are the biggest challenges for scientists trying to demonstrate the benefits that ecosystems provide to humans and the biggest challenges to effective management of ecosystem services. These include (1) our inability, as of yet, to adequately anchor the flow of services to particular ecological configurations, including levels of biodiversity; (2) the challenges introduced by the multiple scales over which ecosystem services are produced and consumed; (3) the possibility of substitutions for some ecosystem services with human-made technologies and capital; and (4) what all of this might mean for formulating a more positive agenda for conservation. This is certainly not the first or last word on these issues; they find voice in many of the other chapters in this section and in the numerous articles now appearing concerning ecosystem services. Nonetheless, if we are to make the framework introduced by the MA—connecting ecosystem services to human well-being—operational, they are issues the academy must attack, and soon.

3. ECOSYSTEM SERVICES, TRADE-OFFS, AND BIODIVERSITY

Not all ecosystem services can be simultaneously maximized. There are some simple examples of this—

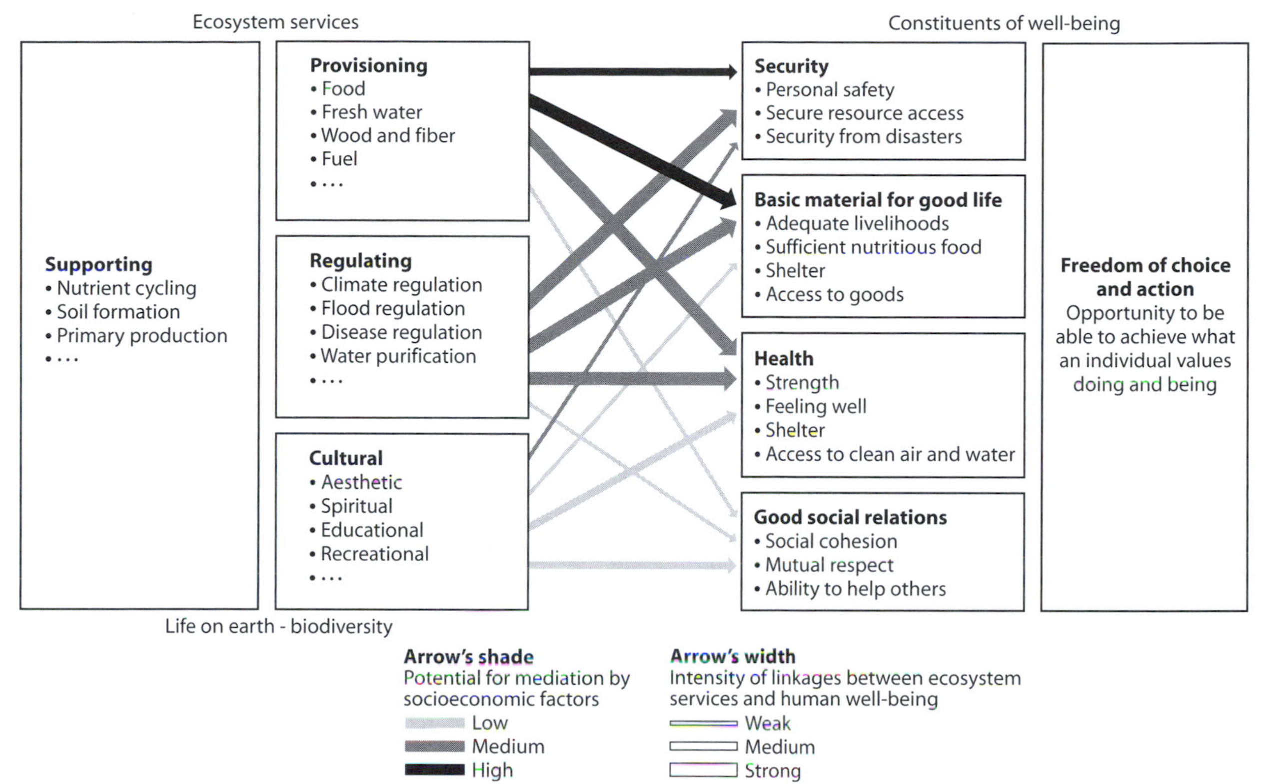

Figure 1. A conceptual framework showing the division of ecosystem services into supporting, provisioning, regulating, and cultural services and showing the connections between these service categories and human well-being. (Adapted from the Millennium Ecosystem Assessment, http://www.millenniumassessment.org/en/GraphicResources.aspx.)

high-yielding agricultural systems, for instance, are generally extremely simplified systems, often monocultures, and the dramatic increase in food provisioning attained by these agricultural systems comes at the cost of other services, such as maintenance of genetic diversity, freshwater provisioning, or recreational opportunities. (See Power et al., chapter VI.4, however, for some management strategies that at least enhance the delivery of nonprovisioning services in agricultural systems.) Similarly, wilderness areas maintained for recreational, spiritual, or aesthetic reasons may enhance some services, such as water regulation, while detracting from others, such as provisioning of fodder. Often maintaining the resilience of a system—its capacity to maintain ecosystem services at a particular level despite changing conditions—means reducing the overall average provisioning of services within that system. These issues of trade-off are covered in many of the chapters dealing with specific ecosystems or services, but they are a particular focus of chapter VI.1 by Scholes.

A corollary to this is that each service requires different ecological attributes or configurations. If the same ecological attribute were implicated each time—if the provisioning of all ecosystem services, for instance, increased with increasing species richness—then no such trade-offs would occur. All ecosystem services could be maximized by maximizing species richness. Some environmentalists assert something very close to this—that saving biological diversity (by which they often mean species richness or habitat diversity) enhances the delivery of ecosystem services. But some ecosystem services, such as food provisioning, require reductions in biodiversity if they are to be delivered effectively. Others, such as carbon sequestration, will depend on species richness only to an extent—a handful of different tree species may suffice to maximize carbon sequestration. Still others may depend on attributes of the system that are relatively unrelated to species number or identity. Erosion control, for instance, depends most directly on vegetative cover; although cover may be enhanced by having a mix of species present, species richness itself may have only a second- or third-order effect on erosion control.

Ecologists have studied the relationship between biodiversity (most often species richness or functional diversity) and ecosystem functioning. In many, but not all, cases there is a positive relationship between the two. The functions most often studied—net primary production, nutrient cycling or retention—map most directly to the supporting services described by the MA. Other functions studied—such as invasion resistance—might map more directly to regulating services. But ecologists have not, as yet, as thoroughly studied the connection between biodiversity and the broader suite of ecosystem services described in the MA. These issues of which species or ecological attributes contribute to ecosystem–service provisioning are covered in chapters VI.2 by Naeem and VI.3 by Norberg.

Of course, if one is managing for a whole suite of ecosystem services, either on a patch or across the landscape, then many different species or patches with different traits and attributes will have to be present. But managing for, say, species richness will largely not ensure maximum delivery of other services. Societal preferences and values should ultimately determine which ecosystem services are enhanced and which are allowed to degrade (informed by the constraints of possible ecological outcomes), and it will be these preferences and values that ultimately dictate our treatment of the species with whom we share the planet. Of course, the moral ("we think other species have the right to exist"), religious ("God will judge us based on how we treat the birds of the field"), and aesthetic ("there is something wondrous about an Andean Condor in flight") values come into play here, and although they may be of great importance in many decisions, they are still only part of what is driving decisions about converting, protecting, or managing the world's biodiversity. Other ecosystem services may be equally or even more valuable, particularly in the places where we live (e.g., urban areas) or in places of great poverty. The ecosystem services framework, then, delivers some particular challenges to conservation activities; these are discussed in chapter VI.10 by Pergams and Kareiva and in chapter VI.13 by Rodríguez.

4. SCALE

Of critical importance in the understanding and management of ecosystem services—indeed any ecological process—is the matter of scale (see, e.g., Levin, 1999). For some ecosystem services, the ecological configurations underpinning those services must be replicated in many sites in order to ensure adequate delivery to the world's inhabitants. Examples would include freshwater provisioning (given a limited ability to transport freshwater supplies around the globe); storm regulation (because it does local residents little good to have storm regulation a half-world or even kilometers away); or, for those residents with limited mobility, aesthetically pleasing or recreationally popular sites. Other ecosystem services may be locally located but deliver global benefits. Examples in this category would include unique and confined species that provide spiritual or medicinal benefit; or areas of high carbon sequestration, which help regulate global climate regimes regardless of where they are located; or, in these days of global trade in goods, areas of food or fiber production. There may be only one location in the world providing this global benefit (as is the case with a valuable endemic species), or several locations may be required (as is the case with food provisioning or carbon sequestration)—nonetheless, the localized ecological systems provide a global service.

Scale thus comes into play in many ways in assessing and managing ecosystem services. Over what scale is the ecosystem service actually delivered? Over what scale do people benefit from this service? Over what scale do the ecological processes underpinning this service operate? Over what scale is the ecosystem service in question managed? Rarely do these scales coincide for a particular ecosystem service, much less across the full suite of ecosystem services. Residents in a particular location hoping to benefit from a bundle of ecosystem services must rely on management institutions that operate across many scales, sometimes in very distant locations. Managers at a particular location devoted to delivering a broader public good should be cognizant of the implications of their decisions on those outside the immediate constituency. These issues of scale and the management implications are covered in many of the chapters in this section but are the particular focus of chapters VI.1 by Scholes and VI.11 by Perrings.

5. SUBSTITUTING FOR ECOSYSTEM SERVICES

Whenever humans choose to pursue some goal that erodes ecosystem services, they are making a judgment about the relative value of those services. When we convert habitat to schools, libraries, or hospitals; when we accept at least some impact on waterways and airways in exchange for mobility; when we acquire exotic goods or materials from distant places in spite of the risk this brings of invasion, we are accepting that sometimes human well-being is enhanced by degradation of ecosystem services. In the words of economists, we are substituting some other service or benefit for ecosystem services in our pursuit of happiness or utility.

There are, of course, limits to the degree to which we can substitute other types of benefits for ecosystem services. We must eat, and therefore some baseline capacity for provisioning services must be maintained.

Human quality of life is greatly eroded in the face of grave uncertainties about the future; if we want to maintain dynamic ecological processes within certain bounds—avoid the crossing of thresholds with attendant catastrophic shifts in ecosystem functioning, for instance—we should focus on regulating services. But it is simply not true that society must maintain all ecosystem services at extant levels or restore them to earlier levels. Decisions should be made, whenever possible, with a full awareness of the consequences, but sometimes human well-being can be enhanced by trading off ecosystem services for some other gain.

The sheer magnitude of technological innovation that characterized the last century makes the future of these trade-offs even more intriguing. Some ecosystem services may ultimately be able to be replaced with human-made substitutes. We have already reduced the demand for natural fibers from what it might have been through the invention of synthetic materials (many of which are better providers of the services of warmth, protection, or glamour than their natural counterparts). The advent of air-conditioning has reduced the demand for, and value placed on, localized climate regulation in many parts of the world. I have long wondered if the virtual reality machines of the future will alter the flow of cultural services humans receive from ecological systems. If my mind believes I have touched, tasted, smelled, seen, and heard the Amazon rainforest while ensconced in a virtual-reality suit, is that fundamentally different from actually being there? It may be for those in my generation, or earlier generations, with our uneasy relationships to technology, but will it be true for those who, from the earliest age, interact with technological gadgets that operate almost as extensions of the individual? And even if the technological experience differs from the "real" one, does it differ enough to justify the CO_2 emissions required to get there or the impact on the forest from tourists wanting to be there? Technology is likely to fundamentally alter the flow of services we receive from the Earth's ecosystems; it will certainly alter the strategies we employ in managing them. Goklany covers the technological aspects of ecosystem services more fully (see chapter VI.12).

6. ECOSYSTEM SERVICES AND CONSERVATION

For well over a decade now, there has been an emerging dialogue concerning the need to connect conservation activities to human well-being. For some, this connection is a normative one—people have a moral obligation both toward their fellow humans and toward the other species with whom we share the planet. Conservationists must therefore recognize and balance the trade-offs between human and nonhuman well-being where such trade-offs exist. For others, the connection is a practical one. The primary motivation is still saving species (or other aspects of the biosphere), but it is recognized that the extent and efficacy of the conservation endeavor will require support from people who may not place as much value on nonhuman species as do conservationists. Alternative motivations for conservation must be provided.

There are many challenges to integrating conservation and human well-being. It may well be, for instance, that the primary benefits of conservation do not derive to people alive now and living close to conservation areas but to more geographically distant, or even future, people. Although conservation can deliver some benefits to local people—ecotourism opportunities or provisioning of fresh water, for instance—the costs of conservation, including the opportunity costs of foreclosed, nonconservation options, may be enormous, particularly for marginalized or poverty-stricken people. When conservation activities are highly valued by people living far away, the mechanisms and institutions that allow them to secure conservation—paying local stewards so conservation is beneficial to them as well, for instance—are often lacking. The problem of scale rears its head once again.

The ecosystem services framework developed by the MA seems to be a reasonable platform on which to build a conservation agenda that joins ecological integrity with human well-being. But if conservationists were to seriously embrace an ecosystem services perspective, conservation priorities would almost certainly change. A recent article by Chan et al. (2006), for instance, examined conservation strategies under different end goals—some focused on traditional conservation goals of saving species, and others focused on ecosystem services such as forage production, recreation, and pollination, among others. There was only a partial overlap to the conservation strategies that emerged under different goals—saving species was not always the best way to enhance delivery of ecosystem services. Nonetheless, many conservation organizations are embracing the concept of ecosystem services. But it remains to be seen whether such a concept does turn out to be a constructive framework for conducting a complicated social dialogue over options, trade-offs, and priorities. This subject is the focus of Rodríguez's chapter (chapter VI.13).

7. CLOSING THOUGHTS

A focus on ecosystem services necessitates a multidisciplinary perspective. In spite of this being a guide to ecology, we need to understand that the ecological contributions to understanding ecosystem services are, of necessity, limited. *What* we should conserve and *how* we should conserve it must inevitably depend on what people value and whether there are other ways to get what they value. Scientists have their own values in this—often emphasizing, for instance, native species richness above all else, in spite of evidence that it is not always high native species richness that most effectively delivers the full suite of ecosystem services society cares about. To provide the right answer for society, ecologists must partner with economists, other social scientists, and technologists. Scientists must also try to remove their own values from scientific assessments—articulating as scientists, for instance, what the various possibilities are for the future of ecosystem services without injecting their own bias as to which of these is necessarily better. As citizens, they can hold and articulate those views, but as scientists, they must continually question the extent to which their own values influence their conclusions. Ecosystem services are the province of all of humanity, after all. If nothing else, a multidisciplinary conversation can help reveal those biases and values.

FURTHER READING

Chan, K.M.A., M. R. Shaw, D. R. Cameron, E. C. Underwood, and G. C. Daily. 2006. Conservation planning for ecosystem services. PLoS Biology 4: 2138–2152.

Daily, G. C., ed. 1997. Nature's Services: Societal Dependence on Natural Ecosystems. Washington, DC: Island Press.

Ehrlich, P. R., and H. A. Mooney. 1983. Extinction, substitution, and the ecosystem services. BioScience 33: 248–254.

Heywood, V., ed. 1995. Global Biodiversity Assessment. Cambridge, UK: Cambridge University Press.

Levin, S. A. 1999. Fragile Dominions: Complexity and the Commons. New York: Basic Books.

The Millennium Ecosystem Assessment (MA). 2005. Ecosystems and Human Well-Being: Multivolume Set. Washington, DC: Island Press. Also available on line at http://www.millenniumassessment.org/en/index.aspx.

Mooney, H. A., and P. R. Ehrlich. 1997. Ecosystem services: A fragmentary history. In G. Daily, ed. Nature's Services. Washington, DC: Island Press, 11–19.

VI.1

Ecosystem Services: Issues of Scale and Trade-Offs

R. J. Scholes

OUTLINE

1. Local, regional, and global services
2. Matching the scales of process, analysis, and management
3. Cross-scale interactions
4. Trade-offs among ecosystem services

The quantity of each individual service that a particular ecosystem delivers varies over time and place, to some degree independently of other services. It is therefore essential to specify the period and the included area when quantifying or valuing a service. It is important to match, as far as possible, the time and space scales at which the ecosystem and its services are assessed and managed to the scales at which the underlying ecological processes that deliver the services operate. It follows that ecosystem service assessments must also pay thoughtful attention to the period that the assessment covers and the location of the boundaries of the assessed area. Because each service differs somewhat in the time and space distribution of its important ecological and social processes, some compromises are necessary in practice. Very frequently the factors that control ecosystem services (the drivers) operate at scales that may only partially overlap those at which the service is used. It is also commonly found that the governance systems that determine who may use what services and in what amount operate at one or more political or economic scales, often disconnected from either the scales of the ecosystem process or management activities. These cross-scale interactions have the consequence that there is seldom a single, perfect scale for studying or managing ecosystem services: multiscale assessments and management institutions are needed. Use of one ecosystem service typically has consequences for the quantity of other services that can be used. This is known as a trade-off or, if the interaction leads to a net increase in one or more services, a synergy. Determining the appropriate mix of services to be used from a given ecosystem is a complex process for which there is no simple or perfect solution; furthermore, the appropriate mix and accompanying solutions themselves change over time. Reaching an equitable and sustainable set of trade-offs is especially difficult if the people who benefit from the services are different from the people on whose actions the continued supply of the service depends. This situation commonly arises as a result of spatial separation (e.g., between highland farmers and lowland water users), temporal separation (use of a service now versus leaving it for later generations), or differences in the jurisdiction or power of various social groups or institutions.

GLOSSARY

domain. The range of characteristic scales in time and space at which a particular process (such as the delivery of an ecosystem service) operates

resolution. The spatial or temporal interval between observations

scale. The physical dimensions, in either time or space, of a phenomenon or observation

synergy. A special case of trade-off (also known as a "positive trade-off") where the use of one service enhances the production of another

trade-off. The relationship between the quantity of one ecosystem service that is used and the quantity of one or more other ecosystem services that can be used

1. LOCAL, REGIONAL, AND GLOBAL SERVICES

Some ecosystem services are available only in a particular area or at a particular time. An example is the fruit of a wild tree species—it has a season and location of availability. Other services are effectively delivered all over the world, continuously. An example is the regulation of the global climate: because ocean currents and wind systems transport matter and energy between the equator and the poles within a few years, effects on the climate system at one location (e.g., uptake of carbon dioxide by a forest) have climate benefits throughout the world. Ecosystem services can be delivered over the full range of scales between these two extreme examples. The general observation is that ecosystem services are *patchy* (i.e., inhomogeneous) in both space and time. Throughout this chapter, the broad issues that apply to space often also apply to time. For the sake of brevity, usually only the space dimension is explicitly mentioned, but the time dimension is implied as well.

The notions of scale and resolution must not be confused. Scale is the total dimension (distance or area or years) over which a phenomenon occurs or is studied; resolution is the dimension at which individual observations are made, and specifically the interval between adjacent observations: it is the smallest detail that can be detected. Ideally, the resolution should match the fundamental "lumpiness" or "grain" of the phenomenon. For example, an ecosystem service assessment may apply to the decade 1991 to 2000 because that is a reasonable period over which change might be detected and managed, but the fundamental input data may be monthly—the scale is a decade, and the resolution is 1 month. A study with monthly resolution would be able to say something about seasonal variation, whereas one with a resolution of a year would not, even if both studies had a scale of a decade.

The term *local* does not necessarily mean a particular range of physical dimensions (e.g., 10 km, or within the influence distance of a particular organism). It means an area where a similar pattern of factors is operating, including the human factors. Thus, a reasonable basis exists for extrapolation within a local patch. *Global* is self-explanatory, but note the cautions in the next paragraph. *Regional* describes a range of scales between the local and the global. A region contains many patches with somewhat different attributes. Most countries are regions by this definition, but regions can also be defined at scales larger than the nation. It helps if there is some ecological or social rationale in the definition of the region. For example, southern Africa is a region that shares certain broad ecological and cultural features that differ, for instance, from those in West Africa.

It is important to distinguish among three different meanings of the phrase *global services*: those that are underpinned by global-scale processes; those that are generated by local processes but whose benefits are consumed globally; and those that are locally produced and consumed, but in many places around the globe. An example of the first is the regulation of the global climate system; of the second is a locally endemic species that has pharmaceutical properties of interest to distant people; and of the third might be aquifer recharge. In the last case, if the capacity to deliver the service were to be locally impaired, the consequence would be felt mainly locally. The global impact would simply be proportional to the fraction of the total supply of the service that is delivered by the affected locality. In the second case, the global impact is the same as the local impact. In the first case (underlying large-scale process), the impact of a local impairment would also be experienced beyond the local scale but could be dissipated or amplified in the process. An example is the carbon storage service provided by the frozen organic soils of the tundra. If they melted, there would be local consequences (e.g., a change in habitat suitability for certain species) but also global consequences for the concentration of carbon dioxide and methane in the atmosphere. If the release of these greenhouse gases were large enough, it would lead to further warming and further changes in the greenhouse gas budgets from this or other localities—an example of an amplifying effect. Note that positive feedback loops (amplifiers) do not automatically lead to "runaway" change. It depends on the degree of amplification and the possibility of an eventual attenuation.

The practical consequences of losing a service that is strictly local but *endemic* (i.e., it is available only in a restricted part of the world) are different from losing in one place a local service that is widespread. For instance, it is possible to grow most staple crops in many places in the world, using only local inputs. If one locality becomes degraded, others can substitute. But the services dependent on the ecosystem that was replaced to make way for the crop may be available only in that one, specific location. For example, a restricted-range species with unique pharmacological properties may have become extinct, or a landscape with deep cultural significance may have been irreparably altered.

Human systems, for example governance systems, also inhabit characteristic domains, i.e., scales of time and space at which they operate. These are often expressed as levels of organization (e.g., county, state, and federal government). However, note that level of organization is not synonymous with scale, although each level is typically associated with a particular scale domain. Levels of organization are often explicitly hierarchically nested: states are made up of counties, and federations of states, but the powers at each level may or may not be subservient to those at the higher levels. Furthermore, there are typically many institutions operating in parallel whose hierarchies may not match.

2. MATCHING THE SCALES OF PROCESS, ANALYSIS, AND MANAGEMENT

In the ideal situation, the scale at which an ecosystem service is assessed and managed would broadly match the characteristic scale of the underlying processes that deliver the service. If the scale of assessment is much larger than the scale of the process, then the real-world variability of the service in space and time becomes homogenized in the analysis. It is then possible, for instance, to conclude that no problem exists at the scale of assessment, although at the scale of delivery, ecosystem service failures may be widespread. This is a key drawback of global or regional assessments, which seldom contain adequate regional or local detail in their respective summary statements. In some cases it *is* possible to conduct global or regional assessments with a resolution that is similar to, or better than, the scale of the process—for instance if the base data are collected by moderate-resolution global coverage satellite-based remote sensing.

However, communicating the findings in such fine detail remains a problem. One solution is to illustrate the results using maps containing detail at the appropriate spatial resolution. In the time dimension, the detailed temporal variation can be displayed in a graph. Another solution is to store the fundamental data electronically at the finest resolution available and provide a mechanism for users to interrogate it at multiple scales. For effective communication, summaries usually must be made. In order to preserve a sense of the underlying variation, summaries must include, in addition to the average (or median) value, information relating to the distribution of values such as the range, standard deviation, or percentiles. For example, the average per capita income is a useful summary statistic but is greatly enhanced if you also know how that wealth is distributed—for instance, what fraction of the total does the wealthiest 10% of the population command?

At the other extreme, assessments conducted at too small a scale are likely to either overestimate or underestimate the true status of the service, depending on whether the study happened to coincide with an area of abundance or scarcity. Essentially, this is a problem of bias resulting from inadequate sampling. It can be reduced by taking a well-chosen and adequately sized sample without necessarily having to measure everything comprehensively over the full domain of the phenomenon in question. The resolution and sample size should be explicitly stated, along with measures of certainty (such as a confidence interval).

The scale mismatches described above result in misleading information. Inappropriate scale is one of the main causes of failure in ecosystem management. Local problems typically call for locally adapted solutions and global problems for global solutions, i.e., solutions implemented by institutions with global reach. In practice, even a single, apparently simple ecosystem service may rest on several ecosystem processes, each with several drivers, operating at different scales, some of which may be poorly known. Furthermore, the scale of management is often predetermined by the realities of institutional jurisdictions. Finding a workable match among process, analysis, and management scales is therefore a more-or-less messy compromise, and multiscale assessment and management are often the best solution.

3. CROSS-SCALE INTERACTIONS

Processes or drivers that operate at larger scales often constrain the range of values that processes at smaller scales can exhibit. For example, global trade patterns can have a strong influence on local prices: a commodity may be locally abundant, and therefore expected to be cheap, but because it is being sold into global markets where it has a greater scarcity value, its local price is much higher. The reverse can also happen: global abundance can suppress the price a local producer realizes. Conversely, large-scale processes are a complex aggregate of the behavior of component processes operating at smaller scales. Thus, the global trade patterns mentioned above are ultimately the product of the individual behavior of millions of individual farmers and consumers. These are examples of *cross-scale interactions*. Cross-scale interactions are another reason why multiscale assessments are usually preferable to single-scale assessments, even if the single scale has been carefully chosen. The same logic applies to multiscale institutional governance of ecosystem service use. It is very hard for an institution constrained to a particular scale simultaneously to have access to the local detail that ecosystem management needs and to the regional or global connectedness to be aware of, and able to influence, drivers at that scale.

Teleconnections are processes whose cause and effect are widely separated. For instance, drought in southern Africa (and many other parts of the world) is highly correlated with sea-surface temperature in the tropical Pacific Ocean, half a world away. Teleconnections may or may not be cross scale. They may, for instance, be local to local; for instance, they may exist between two very specific wetlands that serve as endpoints of a migratory water-bird system. They could also be local to regional or regional to local. Point-source pollution is an example of the former. Distant market demand for a particular natural resource is an example of the latter.

The presence of teleconnections makes the management of ecosystem services much more difficult because the beneficiaries are not the same as the people who influence the service delivery (and in some cases bear the direct or indirect cost of ensuring a continued supply). For example, the flow in rivers is disproportionately generated in the upper catchment, where it is affected by the land management practices of the people who live there. But most of the water users are in the lowlands. The highland dwellers have no use for the surplus water, but maximizing its quantity or quality usually requires that they forgo some action, at a cost to themselves. The notion of transfer payments for ecosystem services has arisen to address this type of situation: the lowlands pay the highlanders a fee to manage their land in a way that maximizes the flow of water. Benefit transfer payments can even apply when no physical product is exchanged and where the consumers and producers are widely separated—for instance, in the case of payments to local people to manage their land in a way that protects the biodiversity it contains.

Teleconnections can also occur in time because of the delayed or persistent consequences of many ecosystem use and management actions. Because it is impossible for future generations to pay present generations to preserve a resource for later use, the notions of intergenerational equity and the appropriate time-discounting (i.e., how do we value use in the future relative to use in the present?) are used to help manage this situation in a sustainable way.

4. TRADE-OFFS AMONG ECOSYSTEM SERVICES

The amount of one ecosystem service that is available for human use is often affected (usually negatively, but not always so) by the use of other services. This is known as a trade-off. The mechanism underlying the trade-off may be direct or indirect. It is often found that a high level of production of provisioning services (such as crops from agricultural systems, wood from forests, or livestock from rangelands) has a negative effect on regulating, supporting, and cultural services as well as on biodiversity, but this relationship is not necessarily proportional. Below a threshold level of production intensity of the provisioning service, the other services may be only slightly affected, but above this level they may be severely affected. For example, water quality typically deteriorates when crop agriculture is practiced in the watershed, but the effect may be tolerably small if set-asides (i.e., land not used for cropping) are strategically placed adjacent to the watercourses and if the level of fertilizer use is limited to that which is readily taken up by the growing crop. The crop production could be marginally increased by ploughing all the land or saturating the crop with nutrients, but the negative impacts on water quality are disproportionately high.

An ecosystem (or equivalently, a land use, or landscape, or seascape) provides not just a single service but a basket of services, even though in practice it is often managed as if only one service mattered. Estimating the costs and benefits to society of the various possible combinations of services and then deciding which is the preferred mix involve explicit or implicit consideration of the trade-offs between services and between interest groups in society.

Where all the ecosystem services under consideration can be converted to common units (for instance, by giving them a monetary value), working out a technically optimum mix is a relatively straightforward task. However, this is seldom the case. It is much easier to obtain a reliable estimate of the monetary value of a service that has an established and well-functioning market than for a service that does not. In this more usual case, other approaches to quantifying the trade-off are needed.

If the shape of the relationships between value (measured in any units, e.g., an index of preference or reduced risk of a disease) and quantity of the service can be independently established for each service, and the trade-off relationship can be established between the quantities of each of the services, it is possible to identify which ecosystem management strategies should be avoided, even if the exact optimum is elusive. For example, in the crop fertilization case described above, the net profit per unit of added fertilizer is relatively easily calculated, but the costs to the riparian ecosystem are not. But it is known that the returns to increased fertilizer use diminish above a certain point, and the leakage of nutrients to the river system increases steeply above this point, so defining a level of fertilization somewhere just below full satisfaction of plant demand as the likely social optimum seems reasonable, even if it is not the farmer optimum.

In situations where optimization (in either the quantitative or qualitative senses described above) is taking place across several different objectives, possibly held by different stakeholders, the techniques of multicriteria decisionmaking may be useful. It remains technically impossible to find an unambiguous simultaneous optimum for more than one objective, where the objectives cannot be expressed in the same units, but these techniques provide a structured framework for making the trade-offs explicit. The disadvantage is that if the technique itself is opaque, it can lend a veneer of analytical precision to what is essentially a negotiation process.

Because assigning worth to nonmarket ecosystem services and selecting a preferred mix of services from a given system both have a large subjective, value-based component that is probably inescapable, participatory approaches that involve all the significantly affected parties and that promote a reasonable degree of symmetry of power and information between them are highly desirable in managing trade-offs between ecosystem services. Given the evolution of people's needs and circumstances, the resource itself, and the long-term nature of some of the processes involved, the trade-off reached cannot be regarded as fixed for all time. It will need to be revisited periodically.

FURTHER READING

Berkes, F. 2002. Cross-scale institutional linkages: Perspectives from the bottom up. In E. Orstrom, T. Dietz, N. Dolšak, P. C. Stern, S. Stonich, and E. U. Weber, eds. The Drama of the Commons. Washington, DC: National Academy Press, 293–322.

Blöschl, G., and M. Sivapalan. 1995. Scale issues in hydrological modelling: A review. Hydrological Processes 9: 251–290. *An explanation of some fundamental ideas to do with scale.*

Chan, K.M.A., M. R. Shaw, D. R. Cameron, E. C. Underwood, and G. C. Daily. 2006. Conservation planning for ecosystem services. PLoS Biology 4: e379. doi:10.1371/journal.pbio.0040379.

Lovell, C., A. Madondo, and P. Moriarty. 2002. The question of scale in integrated natural resource management. Conservation Ecology 5: 25.

Millennium Ecosystem Assessment. 2003. Dealing with Scale. In Ecosystems and Human Well-being. Washington, DC: Island Press, 107–126. *Covers the scale issues in this article in somewhat greater depth.*

O'Neill, R. V., and A. W. King. 1998. Homage to St Michal: Or why are there so many books on scale? In D. L. Peterson and V. T. Parker, eds. Ecological Scale: Theory and Applications. New York: Columbia University Press. *An introduction, based on hierarchy theory, of some of the complexities of scale.*

Rodríguez, J. P., T. D. Beard, Jr., E. M. Bennett, G. S. Cumming, S. J. Cork, J. Agard, A. P. Dobson. and G. D. Peterson. 2006. Trade-offs across space, time and ecosystem services. *Ecology and Society* 11: 28.

VI.2

Biodiversity, Ecosystem Functioning, and Ecosystem Services

Shahid Naeem

OUTLINE

1. Biodiversity and the determinants of ecological fate
2. Biodiversity and ecosystem function
3. The implications for ecosystem services
4. Conclusions

The biological activities of plants, animals, and microorganisms influence the chemical and physical processes of their surroundings, and if one were to modify the distribution and abundance of these organisms, ecosystem functioning, or biogeochemical activity, would change. For example, trees in a forest sequester atmospheric carbon dioxide and locally enhance evaporation; invertebrates in a marine ecosystem mix and aerate sediments; and microorganisms in an aquatic ecosystem decompose organic matter. Reduce the number or mass of these organisms, and ecosystem functions, such as primary production in the forest, the rates of sediment aeration in the marine ecosystem, and rates of decomposition in the aquatic ecosystem, are likely to be altered. If ecosystem functions are altered, then it stands to reason that *ecosystem services*, which are ecosystem functions that benefit humans, are also likely to be altered. Suppose, however, rather than reducing the number or mass of organisms, we reduced only their diversity—would ecosystem functioning and services be affected? The answer is "yes." To see why, this chapter considers the fundamental relationship between biological processes and ecosystem functioning, the evidence for and mechanisms by which biodiversity influences this relationship, and how ecosystem services are likely to be affected by changes in biodiversity.

GLOSSARY

biodiversity. The genetic, taxonomic, and functional diversity of life on Earth including temporal and spatial variability.

biogeochemistry. Geochemical processes influenced by biological processes.

complementarity. Two or more species using the same resources in different ways.

Earth system. Global-scale biogeochemical processes.

ecosystem function, functioning, or process. Biogeochemical activities of ecosystems. The most common metric of ecosystem functioning is primary production, but other metrics include decomposition, nutrient mineralization, community or ecosystem respiration, or other measures of energy flow and nutrient cycling. Note that "function" refers to activity, not purpose. Compare with ecosystem property or ecosystem service.

ecosystem property. A measure of the state (e.g., species richness or standing biomass) or dynamic properties (e.g., resilience, resistance, robustness, reliability, predictability, or susceptibility to invasion) of an ecosystem. Compare with ecosystem function.

sampling effect. See selection effects.

selection effects. When the relationship between biodiversity and ecosystem functioning is significantly above or below zero, there are several possible reasons for such an effect. The simplest is that increasingly diverse ecosystems have greater probabilities of including species that have disproportionately positive or negative effects on ecosystem functioning. The former causes a positive relationship between biodiversity and ecosystem functioning and is known as a *positive selection* or sampling effect. The latter causes a negative relationship between biodiversity and ecosystem functioning and is known as a *negative* selection effect. A positive relationship can also be attributable to increasing amounts of complementarity or facilitation in a community as diversity increases. Increasing complementarity and facilitation can increase the efficiency of resource use in an ecosystem and therefore increase the amount of ecosystem functioning that occurs for a given unit of resource (e.g., light, water, or space). Loureau and Hector (2001) have developed analytical means for separating these co-occuring effects in studies of the relationship between biodiversity and ecosystem function.

1. BIODIVERSITY AND THE DETERMINANTS OF ECOLOGICAL FATE

Ecosystem in Microcosm

If one takes a bottle, fills it halfway with water, adds a rich variety of inorganic nutrients and trace metals, then seals it, sterilizes it, places it in sunlight, and then watches closely . . . for an eternity, almost nothing would happen. If, however, one were to take an identical bottle and add a single photosynthetic, nitrogen-fixing cyanobacterium, like a single cell of the heterocyst blue-green algae one finds growing in rice paddies, the bottle's environment will be utterly transformed in just a few days. With the addition of just one species, the previously clear water would become a cloudy organic soup, the bottle would warm as the dark liquid absorbed light, and the air in the bottle, or its headspace, would be completely altered in its chemical composition.

There are three possible fates for this ecosystem in microcosm. First, it could attain a life-sustaining equilibrium, and the cyanobacteria would persist indefinitely. Second, it could oscillate (predictably or chaotically) between harsh and equitable conditions. For example, the cyanobaceria could go through population swings between high and low densities, and the chemical states of the bottle could fluctuate between low and high levels of acidity. Third, the microcosm could collapse to sterility. For example, if acidity at high densities crossed a threshold of lethality and all the cyanobacteria perished, the microcosm would collapse to a sterile state.

Whatever the physical, chemical, and biological fates, or more simply the *ecological fate* of this ecosystem in microcosm, it is vastly more complex, dynamic, and interesting than its inanimate reference bottle. Even more important, our microcosm thought experiment highlights the significant influences a biota can have on ecosystem properties and ecosystem functioning, where functioning refers to biogeochemical activity.

The biosphere is similar to our microcosm in being essentially a chemically sealed sunlit system rich in nutrients and, for 2.5 of its 3.5-billion-year history, consisted largely of microorganisms. Also, like our microcosm thought experiment, Earth with a biota is vastly more complex and interesting than our inanimate neighboring planets that serve as sterile references. The biosphere, is, of course, more massive and complex than our microcosm. It is made up of roughly one thousand billion tons of biomass comprised of 10–100 million species that facilitate, feed on, or compete with one another, forming a complex web of dynamic intra- and interspecific interactions. In spite of its enormous mass and complexity, the biosphere nevertheless faces the same three kinds of possible ecological fates, only two of which have occurred in its long history. There have been long periods of life-sustaining equilibria, and there have been long periods of oscillations between harsh and equitable conditions. Interestingly, collapse to sterility has yet to occur.

Humanity's Ecological Fate

Fortunately for us, over the last 10,000 years of our 6-million-year tenure on Earth, humanity has enjoyed the relatively equitable conditions of Earth in a life-sustaining, semiequilibrium state. If population size is a measure of success, humans have been spectacularly successful. We have grown from an estimated global population of 4 or 5 million to our current 6.7 billion people. To be sure, humanity has seen periods of extreme glaciation (e.g., 30% of Earth covered by ice), and wild regional shifts of up to 10°C in 10 years (Alley et al., 2003). By and large, however, we have been spared the extremes, such as the hothouse (relatively ice-free) or snowball (glaciated from poles to the equator) states Earth experienced in the Neoproterozoic (1000–540 million years ago) (Hoffman et al., 1998).

Alhough humanity has flourished in terms of sheer numbers, environmental problems have been growing. The Millennium Assessment, a United Nations–sponsored assessment of the state of the world conducted by an international group of over 1300 individuals for 5 years, summarized its findings as follows:

> At the heart of this assessment is a stark warning. Human activity is putting such strain on the natural functions of Earth that the ability of the planet's ecosystems to sustain future generations can no longer be taken for granted. The provision of food, fresh water, energy, and materials to a growing population has come at considerable cost to the complex systems of plants, animals, and biological processes that make the planet habitable. (Millennium Assessment, 2005, p. 5)

This conclusion is founded on the Millennium Assessment's conceptual framework (Millennium Assessment, 2003), which can be summarized by a simple formula:

biodiversity → ecosystem functioning →
ecosystem services → human well-being

where the arrows describe causal linkages.

The foundation of the Millennium Assessment's stark warning is the widespread dramatic decline in biodiversity. Species-poor pastures and croplands have replaced biodiverse landscapes on over 40% of terrestrial surfaces on Earth (Foley et al., 2005); 29% of currently fished species have collapsed (Worm et al., 2006); the current background rate of species extinction is 50–500 times what it has been over the last 65 million years; several million populations and 3000–30,000 species become extinct annually; at least 250,000 species have become extinct in the last century; and 10–20 times that many will disappear this century (Woodruff, 2001). Given that biodiversity is the foundation of the Millennium Assessment's framework, it is not surprising that such staggering losses in biodiversity provoke so stark a warning.

2. BIODIVERSITY AND ECOSYSTEM FUNCTION: BIODIVERSITY ⟶ ECOSYSTEM FUNCTIONING

The study of how and why biodiversity influences ecosystem functioning, or the first linkage in the Millennium Assessment's conceptual framework (see above), has been an intense area of ecological research over the last 15 years and has provided much insight into the ecosystem consequences of changes in biodiversity. This section first considers the fundamental relationships between biological processes and ecosystem functioning and then reviews the findings of biodiversity and ecosystem functioning research. We then turn to the second linkage, ecosystem functioning → ecosystem services, in the next section.

Fundamentals

Ever since Joseph Priestley's experiments in the 1770s that showed that mice (and candles) expired in sealed jars unless plants recharged the air, we have understood that life, metabolism, and environment are linked. The microcosm thought experiment above, and indeed much of ecological research, builds on this basic idea and provides for us three fundamental principles. The first is that the collective metabolic activities of species in an ecosystem's community influence its ecosystem functioning or, more specifically, its biogeochemistry.

The second principle is that the number, quantity, and kinds of species and, although less well studied, the diversity of habitats or interactions, or more simply, any element of biodiversity, influence the magnitude and dynamics of ecosystem function. These influences vary from negligible to major, depending on the nature of the change in biodiversity.

The third principle is that extrinsic or abiotic conditions set the boundary conditions for ecosystem function. In our microcosm thought experiment, the nature and quantity of the nutrients stocked in the bottle, the temperature and pressure in the system, and the amount of sunlight hitting the bottle set the stage for what will happen in the bottle. Its sterile counterpart provides the reference for how much life can shift the bottle from its inanimate equilibrium. For the biosphere, Earth's orbital properties, size, axial tilt, distance from the sun, the nature of our sun, and the size and orbital properties of our moon, among many other factors, define the boundary conditions for Earth.

These principles, explicitly or implicitly, provide the basis for managing ecosystems such as farms, plantations, aquaculture pens, or urban systems. The intent of management is generally to improve human well-being, but short time horizons, imperfect markets, and an emphasis on favoring the production of provisioning ecosystem services tend to favor reducing native diversity to desirable or domesticated species and enhancing inputs of water and nutrients, often at the cost of other nonmarket ecosystem services (Daily et al., 1997).

Biomass versus Biodiversity

Biological processes, by definition, are integral to ecosystem functioning, but the question is whether a generic biomass, one that performs essential biogeochemical processes, is all that is needed to secure the necessary levels of ecosystem functioning that derive the bulk of ecosystem services needed to ensure human well-being. For example, does a sustainably managed maize field provide as much or more ecosystem functioning and services as a prairie grassland (with 200 species of plants) of equivalent biomass? If food provisioning is the only service of interest, the answer is "yes." Domestic hybrid maize, in fertile soils with adequate water, is more productive than most prairie grassland species, and a substantial portion of its biomass is harvestable (50–55%) for food and biofuels. However, the scientific consensus is that species-poor generic biomass rarely provides the maximum functioning possible for an ecosystem—the more diversity, the better (Hooper et al., 2005). This is especially true when one considers multiple functions. (Hector and Bagchi, 2007). A maize field, for example, lacks regulating services (e.g., atmospheric carbon sequestration, stabilization of soil against erosion), is unstable (it is readily ravaged by pests and invaded by exotics), steadily loses nutrients (in the absence of legumes), lacks many cultural services prairies provide (e.g., aesthetic and inspirational value), and comes up short on most other ecosystem functions and services even though it gets high marks for food production.

Hooper et al. (2005) summarize the evidence that supports the claim that biodiversity, not just generic biomass, is an important element to ecosystem functioning as follows:

1. Although most communities are dominated by a few species, which might suggest that large numbers of species are not needed for ecosystem functioning, in fact, many low-abundance species play critical roles. For example, keystone species, ecosystem engineers, pollinators, and biocontrol species often represent little biomass in ecosystems yet play pivotal roles in ecosystem functioning.
2. When one or a small number of invasive species homogenize a landscape, ecosystem function and ecosystem services often decline. Native communities typically represent coevolved assemblages of species that are well adapted to a wide range of local conditions. Dominance by invasive species narrows the range of environmental conditions a community can tolerate, which leads to long-term loss of ecosystem function.
3. The more species present in a community, the more likely there will be one species that can compensate for the loss of another, a process known as *biological insurance*.
4. The ubiquity of spatial and temporal heterogeneity in ecosystems means that greater numbers of differently adapted species will provide greater overall coverage and usage of natural resources than would small numbers of species.
5. Many species are complementary in their function, which improves efficiency of ecosystem functioning.

Several reviews and meta-analyses of both terrestrial and marine studies continue to confirm a variety of identifiable roles of biodiversity in shaping ecosystem function. The scope of this work, however, remains limited, and there is considerable need for further research, but the central finding that biodiversity influences the magnitude and stability of ecosystem functioning is fairly robust.

Complementarity and Redundancy

There are several reasons why biodiversity influences ecosystem functioning with functional complementarity, its converse, functional redundancy, and selection effects figuring most prominently among them. Functional complementarity arises when species are functionally different, meaning simply that they do different things. For a given space or volume, for a given amount of nutrients, communities that contain functionally complementary species are more likely to make more efficient use of available resources. This is most readily explained by way of an example of rooting depth in plants. A deep-rooting plant makes use of nutrients far beneath the surface. If all other species root to the same depth, then our deep-rooting species is functionally redundant. If, on the other hand, it is the only plant that roots as deeply as it does, then it is functionally singular. If many plants root at different depths, there is functional complementarity.

Both complementarity and redundancy are important, but they trade off against each other such that too much of either becomes a bad thing. Some degree of substitutability (redundancy) is needed to ensure ecosystem function, but some degree of complementarity is needed to improve local utilization of resources and achieve higher levels of ecosystem function and improve resistance to perturbation. The more biodiversity is reduced, the more likely an ecosystem approaches one extreme or the other.

Selection (or sampling) effects arise when one or more species in the species pool has disproportionately negative or positive impacts on ecosystem functioning in comparison with other species in the community. In such cases, higher-diversity ecosystems are simply more likely to have those species present. It is difficult to tell which mechanism accounts for the greater ecosystem functioning observed in higher-diversity replicates in experimental studies of biodiversity and ecosystem functioning, although Loreau and Hector (2001) have provided a statistical method for disentangling the two effects in experimental studies. Selection effects are primarily important in experimental research when researchers typically compare ecosystem functioning across replicate ecosystems in which each replicate's species represents a selected subset from the regional pool. It is highly likely, in such cases, that higher-diversity communities will exhibit increases in ecosystem function because of positive selection effects, although complementarity effects will often outweigh selection effects (Cardinale et al., 2007).

3. THE IMPLICATIONS FOR ECOSYSTEM SERVICES: BIODIVERSITY ⟶ ECOSYSTEM FUNCTIONING ⟶ ECOSYSTEM SERVICES

The ecosystem function → ecosystem service link in the Millennium Assessment's framework, however, is less well studied. In cases where both function and service are the same, we can obviously map our findings from the biodiversity–functioning relationship to the biodiversity–service linkage. For example, Solan et al. (2004) examined biogenic mixing depth as an ecosystem function in relation to benthic faunal diversity, but because this ecosystem function correlates directly with estuarine ecosystem services of fish production (provisioning), the degradation of organic pollutants (regulating), and the reduction of anoxic sediment, which enhances nutrient turnover (sustaining), their results apply to the relationship between marine benthic faunal biodiversity and estuarine ecosystem services. But if the ecosystem function is net primary production of largely inedible grasslands that cover less than 0.1% of their original extent, how does one map that onto ecosystem services? Below, we briefly consider the major classes of ecosystem services in light of biodiversity–functioning research.

Ecosystem Services

Sustaining

The current consensus on the importance of biodiversity to ecosystem functioning is most directly relevant to sustaining services because this research has primarily focused on key biogeochemical functions such as primary production, decomposition, nitrogen in soil and marine systems, nutrient cycling, leaf shredding in streams, community respiration, carbon dioxide drawdown, carbon storage, and other organic–inorganic matter transformation processes. These effects range from small to large, positive to idiosyncratic, and there are studies that have found no effects, but the consensus and meta-analyses demonstrate that biodiversity and biogeochemical functions are related.

Sustaining ecosystem services, however, are generally not consumed and are invisible in markets and perhaps the least familiar and understood by people.

Provisioning

The findings of biodiversity–functioning research suggest, but do not directly demonstrate, that agricultural or forest products, fresh water, bush meat, seafood, or other provisioning services may be related to biodiversity. Provisioning services often involve consumable biomass, and many biodiversity–functioning studies have shown that higher plant diversity can yield higher plant biomass. Some studies have shown that heterotrophic microbial diversity can yield higher heterotrophic microbial biomass, but these increases in biomass are not the kind of biomass people consume. There have been, however, no studies that have truly examined provisioning services (with the possible exception of a study of grassland biodiversity as a source of biofuel (Tilman et al., 2006). This lack of explicit examination of provisioning services is unfortunate because ecosystems have been primarily managed to maximize provisioning services. It would be valuable to have comparative studies in biodiversity–functioning research across a range of provisioning services.

Regulating

Perhaps the only place where biodiversity–functioning research has been directly related to ecosystem services has been with regulating ecosystem services. We use biodiversity–functioning loosely here because most of the work focuses on ecosystem properties rather than processes. Studies of the relationship between biodiversity and invasion (e.g., Fridley et al., 2007), pollination (e.g., Balvanera et al., 2005), resistance to the spread of disease (e.g., Mitchell et al., 2002,), or biocontrol (e.g., Philpott and Armbrecht, 2006) do not concern ecosystem processes but ecosystem properties. Nevertheless, these studies do link diversity with the regulation of invasive species, pests, and fruit production and are founded on similar ideas of complementarity and selection effects.

Less clear is whether empirical studies of true biodiversity–function studies (e.g., studies of ecosystem or biogeochemical functions) speak to regulating ecosystem services. Empirical biodiversity–stability studies of ecosystem function or the many theoretical studies of the same topic suggest that biodiversity can lower the variability or improve insurance of ecosystem services in the face of environmental variability. Empirical studies, however, have the aforementioned characteristics of being too small in scale, focused on too few functions, or being too short in duration, and theoretical studies have made simplifying assumptions that do not lend themselves well to extrapolation of real-world ecosystem dynamics (Cottingham et al., 2001). The idea that biodiversity stabilizes communities and ecosystems is a venerable one that dates to the 1950s studies by Eugene Odum, Robert MacArthur, and their colleagues: but its strongest support is theoretical; its next strongest support is that of highly controlled microcosm research using microorganisms; its next strongest a limited number of field experimental studies; and its weakest, because of lack of an ability to eliminate confounding factors, are observational studies such as that by Bai et al. (2004), which demonstrate greater constancy of grassland production in the face of variation in precipitation in Inner Mongolian grassland plots.

Although biodiversity–functioning research clearly supports roles for diversity in both biotic regulation (e.g., resisting invasion, the spread of disease, pollination, and biocontrol) and the regulation of ecosystem function (e.g., greenhouse gas regulation or flood control), we have a long way to go before we can make definitive statements about biodiversity and regulating ecosystem services.

Cultural

Little insight into cultural services can be gained from biodiversity–functioning research because this research has focused on functional diversity and ecosystem functions. Functional diversity is an alien concept to most nonscientists, and ecosystem functions are both largely unseen and largely unknown to people. Cultural ecosystem services rarely, if ever, concern the beneficial roles species play in ecosystem functioning in comparison to the aesthetic, inspirational, recreational, or other values. Few people visit boreal forests or rainforests to see them produce oxygen and contribute to climate regulation—they visit to see their magnificent trees.

Although theoretical and empirical biodiversity–functioning research is difficult to relate to cultural values, some of the methods used may be applicable to future research. For example, the techniques used by Solan et al. (2004), McIntyre et al. (2007), and Bunker et al. (2005) can be applied to the study of cultural services if one can derive cultural values that are related to the ecosystem functions of these systems. Perhaps polluted estuaries have lower value for recreational fishing, swimming, or boating, and if one could relate biogenic mixing depth (the focus of the Solan et al. study) to these values, then the study could be interpreted in light of cultural services. Likewise, if recreational fishing were associated with a diversity of fish or the oligotrophic status of lakes, and swimming, boating, or tourism correlated with water quality of lakes, then the study by McIntyre et al. (2007), which explores fish diversity and nutrient cycling in Lake Tanganyika, Africa, might be used to explore cultural values.

For ecosystem services in general, the cultural services are difficult to study quantitatively, and this limits our ability to synthesize diversity–functioning research and the study of cultural services.

4. CONCLUSIONS

This chapter began with a thought experiment of ecosystems in microcosm that provided a simple way to distill from the complexities of larger, more complex systems the essential elements and processes that govern system function. There is no question that life transforms an inanimate world and that the nature of that transformation is determined, at least in part, by the diversity of life, not just its mass. More than a decade of research on the relationship between biodiversity and ecosystem functioning has yielded a body of evidence that supports a strong relationship between biodiversity and ecosystem functioning. When it comes to ecosystem services, however, with the possible exception of pollination and a few other services, there is simply insufficient evidence to draw robust conclusions.

This cautionary tone, however, should not in any way suggest that the biodiversity → ecosystem function → ecosystem service → human well-being construct should not be a guiding principle for both research and application. We do know that ecosystem functioning is influenced by biodiversity because of complementarity, redundancy, and, in experimental systems, selection effects, and we do understand that ecosystem services are derived from ecosystem function. Ergo, we know for certain that changes in biodiversity affect ecosystem services.

Given that nearly half of all humans live in poverty and that these are the people for whom finding ways to improve human well-being is essential, and given that poor and vulnerable people are often the most closely reliant on natural resources and biodiversity, the utilitarian values of biodiversity as the source of ecosystem services should be widely embraced. The idea that biological conservation is a science of crisis (Soulé, 1991), however, tends to promote strategies of identifying biodiversity hot spots and taking immediate action to protect and preserve them, approaches that can adversely impact the poor and vulnerable who often shoulder the costs of such programs (Adams et al., 2004). Ecosystem services, as described in this section, provide an alternative perspective, one that takes an integrative and adaptive approach to managing human well-being by managing nature's services.

Finally, the Millennium Assessment's stark warning stems from the erosion of Earth's life support system as a result of the loss of biodiversity. The degradation of 60% of the 24 ecosystem services the Millennium Assessment examined clearly means that the ecosystem functions behind these services have changed in ways that have harmed, rather than benefited, humans. The Assessment's warning echoes Levin's *Fragile Dominion*, which begins, "Mother Earth is in trouble, at least as a habitat for humanity." The loss of biodiversity and the degradation of ecosystem functions and services do not mean that the biosphere is heading to that third ecological fate it has somehow avoided for 3.5 billion years—a collapse to sterility. It does mean, however, that humanity may suffer immensely if it continues along its current path. The good news, at least with respect to managing biodiversity, ecosystem functioning, and ecosystem services, is that our ecological fate is in our hands.

FURTHER READING

Adams, W. M., R. Aveling, D. Brockington, B. Dickson, J. Elliott, J. Hutton, D. Roe, B. Vira, and W. Wolmer. 2004. Biodiversity conservation and the eradication of poverty. Science 306: 1146–1149.

Alley, R. B., J. Marotzke, W. D. Nordhaus, J. T. Overpeck, D. M. Peteet, R. A. Pielke, Jr., R. T. Pierrehumbert, P. B. Rhines, T. F. Stocker, L. D. Talley, and J. M. Wallace. 2003. Abrupt climate change. Science 299: 2005–2010.

Balvanera, P., C. Kremen, and M. Martinez-Ramos. 2005. Applying community structure analysis to ecosystem function: Examples from pollination and carbon storage. Ecological Applications 15: 360–375.

Bunker, D. E., F. DeClerck, J. C. Bradford, R. K. Colwell, I. Perfecto, O. L. Phillips, M. Sankaran, and S. Naeem. 2005. Species loss and aboveground carbon storage in a tropical forest. Science 310: 1029–1031.

Cardinale, B. J., J. P. Wright, M. W. Cadotte, I. T. Carroll, A. Hector, D. S. Srivastava, M. Loreau, and J. J. Weis. 2007. Impacts of plant diversity on biomass production increase through time because of species complementarity. Proceedings of the National Academy of Sciences, U.S.A. 104: 18123–18128.

Cottingham, K. L., B. L. Brown, and J. T. Lennon. 2001. Biodiversity may regulate the temporal variability of ecological systems. Ecology Letters 4: 72–85.

Daily, G. C., S. Alexander, P. R. Ehrlich, L. Gouler, J. Lubchenco, P. A. Matson, H. A. Mooney, S. Postel, S. H. Schneider, D. Tilman, and G. M. Woodwell. 1997. Ecosystem services: Benefits supplied to human societies by natural ecosystems. Issues in Ecology 2: 1–18.

Foley, J. A., R. DeFries, G. P. Asner, C. Barford, G. Bonan, S. R. Carpenter, F. S. Chapin, M. T. Coe, G. C. Daily, H. K. Gibbs, J. H. Helkowski, T. Holloway, E. A. Howard, C. J. Kucharik, C. Monfreda, J. A. Patz, I. C. Prentice, N. Ramankutty, and P. K. Snyder. 2005. Global consequences of land use. Science 309: 570–574.

Fridley, J. D., J. J. Stachowicz, S. Naeem, D. F. Sax, E. W. Seabloom, M. D. Smith, T. J. Stohlgren, D. Tilman, and B. Von Holle. 2007. The invasion paradox: Reconciling pattern and process in species invasions. Ecology 88: 3–17.

Hector, A., and R. Bagchi. 2007. Biodiversity and ecosystem multifunctionality. Nature 448: 188–190.

Hoffman, P. F., A. J. Kaufman, G. P. Halverson, and D. P. Schrag. 1998. A neoproterozoic snowball Earth. Science 281: 1342–1346.

Hooper, D. U., F. S. Chapin III, J. J. Ewel, A. Hector, P. Inchausti, S. Lavorel, J. H. Lawton, D. M. Lodge, M. Loreau, S. Naeem, B. Schmid, H. Setälä, A. J. Symstad, J. Vandermeer, and D. A. Wardle. 2005. Effects of biodiversity on ecosystem functioning: A consensus of current knowledge and needs for future research. Ecological Monographs 75: 3–35.

Jonsson, M., and B. Malmqvist. 2000. Ecosystem process rate increases with animal species richness: Evidence from leaf-eating, aquatic insects. Oikos 89: 519–523.

Levin, S. 1999. Fragile Dominion. Cambridge, MA: Perseus Publishing.

Loreau, M., and A. Hector. 2001. Partitioning selection and complementarity in biodiversity experiments. Nature 412: 72–76.

McIntyre, P. B., L. E. Jones, A. S. Flecker, and M. J. Vanni. 2007. Fish extinctions alter nutrient recycling in tropical freshwaters. Proceedings of the National Academy of Sciences, U.S.A. 104: 4461–4466.

Millennium Assessment. 2003. Ecosystems and Human Well-Being: A Framework for Assessment. Washington, DC: Island Press.

Millennium Assessment. 2005. Living Beyond Our Means: Natural Assets and Human Well-being: Statement from the Board, Millennium Assessment. Washington, DC: Island Press.

Mitchell, C. E., D. Tilman, and J. V. Groth. 2002. Effects of grassland plant species diversity, abundance, and composition on foliar fungal disease. Ecology 83: 1713–1726.

Philpott, S. M., and I. Armbrecht. 2006. Biodiversity in tropical agroforests and the ecological role of ants and ant diversity in predatory function. Ecological Entomology 31: 369–377.

Solan, M., B. J. Cardinale, A. L. Downing, K.A.M. Engelhardt, J. L. Ruesink, and D. S. Srivastava. 2004. Extinction and ecosystem function in the marine benthos. Science 306: 1177–1180.

Soulé, M. E. 1991. Conservation: Tactics for a constant crisis. Science 253: 744–750.

Tilman, D., J. Hill, and C. Lehman. 2006. Carbon-negative biofuels from low-input high-diversity grassland biomass. Science 314: 1598–1600.

Woodruff, D. S. 2001. Declines of biomes and biotas and the future of evolution. Proceedings of the National Academy of Sciences, U.S.A. 98: 5471–5476.

Worm, B., E. B. Barbier, N. Beaumont, J. E. Duffy, C. Folke, B. S. Halpern, J.B.C. Jackson, H. K. Lotze, F. Micheli, S. R. Palumbi, E. Sala, K. A. Selkoe, J. J. Stachowicz, and R. Watson. 2006. Impacts of biodiversity loss on ocean ecosystem services. Science 314: 787–790.

VI.3

Beyond Biodiversity: Other Aspects of Ecological Organization

Jon Norberg

OUTLINE

Ecosystem services are provided by biological processes and structures as well as by the geophysical environment. Biodiversity is a measure of the variation in life forms and is the result of many biological and geological processes and constraints. To understand *what* species do in ecosystems, particular attention must be paid to the traits of these species, such as their optimal temperature for growth, their ability to avoid predators, or their nutrient uptake capacity. It is the distribution of traits relevant for particular ecosystem services, so-called trait spectra, that determines the performance of the biological community as a whole. The variation in particular traits within the community, such as the range of temperature tolerances, is a measure of the response capacity, i.e., the overall ability of the community, species, and individuals to respond to changes in environmental factors, such as temperature changes. Greater response capacity reduces variability in ecosystem services under environmental variability. This is particularly important for systems that are susceptible to critical transitions. Landscape patterns and processes provide a regional source of trait variability for local communities and thus maintain response capacity.

GLOSSARY

abiotic environment. The chemical, geological, and physical part of the ecosystem.

critical transitions. A change of the dominating feedback processes in an ecosystem, with implications for ecosystem structure and functioning. Systems undergoing a critical transition may be profoundly different before and after the transition.

response capacity. The ability of a local community to respond to changes in environmental drivers.

trait spectra. The abundance-weighted distribution of particular traits in the community.

Ecosystem services are sustained by an interaction between abiotic and biological processes. Thus, biological processes such as primary production account for only a part of ecosystem services. Water provisioning depends on biological processes as well as physical ones, such as those that drive the climate system. Similarly, it may be the presence of a particular striking feature, such as a blue whale or the Grand Canyon, that delivers a service, largely independent of any current biological processes (although historic biological processes were required to produce the blue whale and the Grand Canyon to begin with). The abiotic environment, such as the geomorphology of the landscape or the climate and the hydrosphere, largely sets the constraints within which biological communities develop. The abiotic environment is, however, by no means unaffected by the biological system. Water flow, for example, is determined by the climate and hydrology, but trees may play a large role in channeling water back to the atmosphere, thereby potentially affecting groundwater levels and even large-scale climate patterns (figure 1).

The role of particular species in the ecosystem processes is determined by functional and morphological characteristics, i.e., traits. Functionally similar species may be involved in sustaining particular ecosystem processes either directly (e.g., provisioning services) or indirectly (e.g., support or regulating services). The attributes (traits) of particular species and the effects of environmental variables on the species as

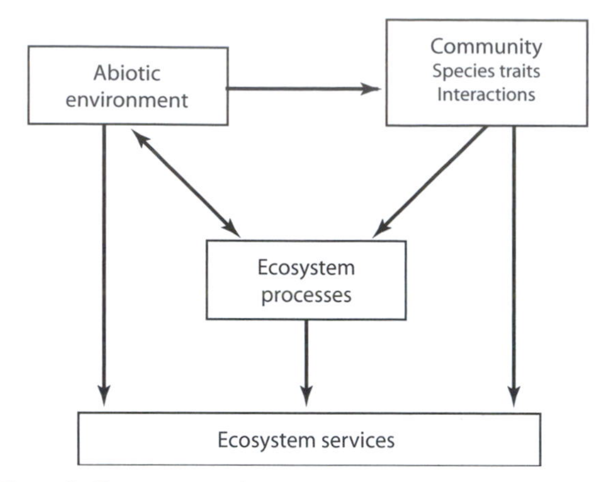

Figure 1. Ecosystem services are provided by the abiotic environment, the biological community, and the ecosystem processes. Although the abiotic environment largely constrains the biological communities, the ecological processes can also affect the abiotic environment through such processes as withering and water vapor flow.

well as on physical processes fundamentally determine which ecosystem services are provided and at what level. To understand the role of species traits and the role of environmental variables in ecosystem services, we first need to elaborate on what species do in ecosystems.

1. WHAT DO SPECIES DO IN ECOSYSTEMS?

The year 2007 marked the 300th birthday of Carl Linnaeus, the father of taxonomy. His Systema Naturae was used to give name, and thus identity, to most species known to humans. Thus, one might argue that he also was the father of the concept of species richness because, in order to name species, one needs to be able to identify them. However, a name does not reveal much about the role of a species in ecological processes. In practice, species classifications have also been based on morphology or, more recently, on their genetic code rather than on their function or what they do. Biodiversity should be a full measure of variation in life forms, and thus, when it comes to understanding how communities function, one should add some measures of traits to species' identities. In most studies, abundance has been used as a proxy for ecosystem function because it is an easy variable to measure, and one can argue that abundance is often at least proportional to some process sustained by particular organisms. However, abundance is an imperfect measure at best because (1) the ability of a set of organisms to sustain a particular function may "saturate" and thus be unrelated to abundance; and (2) the species most abundant in the system may be functionally similar or dissimilar, with real consequences for the range of services that can thus be supported. Traits represented in the community can be related to functioning and structure in a much more direct way.

Lawton (1994) posed the question "What do species do in ecosystems?" which helped inspire a research field devoted to understanding the relationship between biodiversity and ecosystem functioning. Several reviews that summarize this research field (e.g., Hooper et al., 2005) conclude that some measure of biodiversity (usually focused on species rather than, say, variation in communities or landscapes) provides a statistical estimate of the traits of the species in a community that in turn may tell us something about the functioning of a community. The more species, the more traits, for instance. Why then does one not focus more directly on trait distributions in communities in addition to species' lists and relative abundances? Relevant traits are not always easy to measure. Morphological traits are generally easier to measure and are sometimes called *soft traits*, whereas more process-relevant traits such as nutrient uptake rate, temperature response, or predator-avoidance traits are harder to quantify. Research to elucidate how to link hard and soft traits is a growing field that holds promise in helping us understand the role of the interaction among species and the environment for ecosystem services.

Species richness is a measure of the number of taxonomic units, generally species, within a biological community or ecosystem. Sometimes measures of species diversity include aspects of relative abundances of species, such as the Shannon index—communities with a more even distribution of species abundances are considered more diverse than those with a similar number of species but dominated by one or two species. In research focused on the role of species diversity for ecosystem functioning (often implying ecosystem services, i.e., processes or structures that are of human value), the contrasting hypothesis has often been species identity; i.e., does species diversity per se or species identity control ecosystem functioning on average? The notion "on average" is meant to imply that one tries to compare over many different sets of species in order to find effects that are statistically either determined by the number of species or by the particular set of species. If the analysis favors number of species over particular species, it does not imply that particular species do not have great importance for the performance and maintenance of ecosystem processes but, rather, that there is some even more important impact of the "average effect" of biodiversity. But this effect can be very hard to parse. For instance, scientists may manipulate conditions in order to create communities with different species numbers, but in manipulating conditions, they may also be influencing species identity. If a scientist uses existing patches to examine the relationship between diversity and functioning, then abiotic differences in the patches may be altering species identity as well as number. In other words, changes in species identity and species number may be correlated. When we disentangle this issue with further analysis, we will find that the crucial aspects really are to understand processes that either sustain or decrease biodiversity rather than biodiversity per se.

2. RESPONSE CAPACITY OF THE BIOTA

In a changing world, a fundamental support service is the ability of biological communities to respond to change without major loss in structures or processes that provide important ecosystem services. This response capacity is a multilevel phenomenon occurring from the level of the organisms to biomes as well as from daily to decadal time scales. I illustrate this with the response of biological communities to temperature changes, although many other changes and adaptations are important. At the most fundamental level, any protein has a temperature dependence that determines the rate at which it catalyzes chemical reactions. Proteins are also sensitive to high temperatures, which disrupt the delicate folding structures that give them their catalytic properties. In the balance between these rate-enhancing and rate-disrupting temperature dependencies, there is an optimum temperature that gives the highest rate of functioning for a particular protein. But there is also a quite broad range of temperature at which a particular protein will work well, albeit not at its optimal rate. This is the physiological response capacity. Many species have several varieties of each protein type encoded, which have slightly different temperature dependencies. With temperature-dependent cues, these proteins can be turned on or off in the genome and thus provide a phenotypic plasticity that broadens the temperature tolerance of an organism. Within a population, there are many individuals, and because of genetic recombination and mutations during reproduction, each individual may have different combinations of these proteins. Thus, within a population, the response capacity may be even broader because of additional variation in the traits attributed to protein temperature sensitivity. On evolutionary time scales, the within-species variation of these traits is also directly proportional to the rate at which natural selection can change the mean trait in the population in response to selective forces, e.g., an increase in temperature.

In regard to ecosystem services, there are additional layers of response capacity. Different species can contribute to the same process or structure that sustains an ecosystem service. This means that variation in temperature dependence between these functionally similar species also increases response capacity of the whole community, as species with better-adapted responses can replace less well-adapted species by means of competition and community succession. Again, we find that the capacity of a group of functionally similar species to change the mean trait of the group in response to changing environmental factors, e.g., temperature, is directly proportional to the variation among species and individuals within this group. In contrast to the evolutionary process, this occurs in ecological time, which often is a faster response than evolutionary responses.

Other sources of trait variation in a community are dispersal and migration processes. When the variation in traits increases with the investigated area, and dispersal/migration processes are present, the area can act as a source pool for local trait variation and thus enhance local response capacity (see section 4). Thus, distribution of traits, or so-called trait spectra, in a community is more informative than species richness per se in understanding how a community will function. Trait spectra are measured as relative abundances over a particular trait, such as the rate of nitrogen fixing (0 for nonfixers) or temperature optima for temperature response–related traits. The trait spectra for a trait such as the optimal temperature for growth are likely to be very narrow in areas with little variation in temperature such as the tropics and much wider in temperate areas that can sustain cold- and warm-adapted species that grow during different seasons. Thus, even though the tropical areas will have many more species, the variation in a particular trait, here temperature optima for growth, is likely to show much more variation in the relatively species-poor temperate areas. This calls into question the notion that species richness may be a good surrogate for functional trait diversity; different processes may be driving species richness and trait variation within communities. Whether or not species richness correlates to trait variation depends on the trait in question and on the historical forces shaping the community.

It should be noted that maintenance of response diversity can also carry a cost. Maintenance of high variability under constant conditions leads to lower functioning than if all individuals had an optimum at or near the mean condition. It is only when environmental factors change that wider trait spectra, and thus response capacities, increase overall productivity in the long term. The effect of trait variation on the capacity of a system to respond must be measured with respect to expected variation in the environment. That is, one cannot say that one trait variation is better than another, as it depends on how much environmental variation each trait variation has to buffer in the given location. "Too much" trait variation, while still maintaining response capacity, may reduce the overall functioning of the system with respect to expected conditions. There is a trade-off between optimal functioning and responsive capacity.

Trait variation, whether it is between or within species, reduces the variation in the performance of the whole community and hence reduces the variability in the ecosystem services they sustain. This can be particularly important if there are thresholds that may be crossed when particular processes fall below a certain level, leading to large changes in the ecosystem. A consequence of this is that there may be a trade-off between the provisioning of an ecosystem service at some "optimal" level and the resilience of the system.

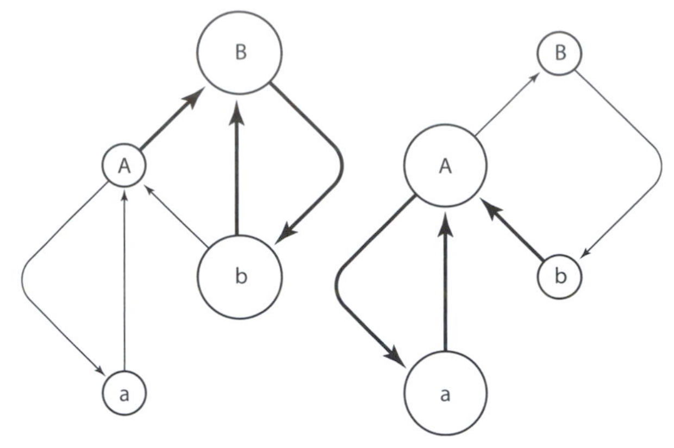

Figure 2. An example of critical transition in dominant feedback processes in food webs. Species B is slightly larger than species A and can eat the adults of A (capital letter). However, adult individuals of species A can predate on juvenile individuals of species B (lower case). The system can tend to be in either of these states because the feedback loop reinforces itself. For example, it is possible, e.g., by overfishing, to reduce adults of species B until the system flips into a state dominated by species A predation on juveniles of species B.

3. SPECIES DOMINANCE AND FEEDBACK LOOPS

Ecosystems are made up of complex interaction networks. Food webs, for example, consist of many weak interactions and a few strong ones that have a disproportionate effect on the whole structure of the food web. In addition, some interactions can cause feedback loops that either dampen or self-reinforce different configurations of the system. An example of such a self-reinforcing interaction loop is shown in figure 2. Two species, each with two size classes, for example fish, interact. Species A is generally smaller than species B. Adult individuals of species B can prey on adults of species A. However, species A's adults are larger than the juvenile class of species B, on which they can prey. This leads to two different feedback loops in this very simple food web. First, if there is a high abundance of species A and low abundance of species B (right-hand side of figure 2), the adults of species A can prevent population growth of B by preying on their young. A second scenario is when adults of species A are in low abundance, and species B is abundant. Then species B can control the population size of species A by preying on their adults.

These feedbacks are very common in ecological systems, and the example above may be illustrated with the Baltic food web, where this mechanism is present in the interaction between cod and sprat. Ecosystem services can be strongly dependent on species that are part of such an interaction feedback loop, the most well-known examples being kelp forests and coral reef systems. Furthermore, abiotic components of ecosystems, such as nutrient release from sediments, can also contribute to creating multiple self-reinforcing states through their nonlinear response to environmental variables. For instance, release of nutrients from sediments can, through a series of interactions, cause oxygen depletion in lakes, which can lead to further release of nutrients from sediments, creating a strong positive feedback loop. The sensitivity of particular ecosystem services to these feedback interactions depends on the relative strength of the interaction and its position in the whole network. The relative effect of a sudden shift from one feedback loop to another thus depends on the diversity of the system components and the relative distribution of interactions. Systems that have fewer interactions are more likely to have a few strongly dominating processes or feedbacks, whereas systems with many interactions may have more and relatively weaker interactions. This may suggest that systems with fewer interactions experience fewer but more dramatic shifts in feedback processes, whereas systems with many interactions may experience more frequent but less severe ones. However, what actually happens in real ecosystems depends very much on the particular species interactions, and little can be generalized from species richness alone. Thus, one could say that sensitivity to critical transition phenomena in ecosystems depends, first, on the traits of the individual interactions and, second, on average link density and number of nodes, which may, at least on average, be reflected in the aggregate measure of species richness.

4. THE LANDSCAPE

Communities are dynamic, and thus, so are trait spectra and species richness. Even if species richness remains constant, there can be a large turnover of species because of species responses to environmental drivers and of dispersal or movement of species. This can result in rapid changes in community characteristics such as the trait spectra or interaction networks and thus also affect major ecosystem processes. The influx of new species into the species pool is largely determined by migration and dispersal (assuming that the arriving species can thrive in the location) and, on longer time scales, by evolutionary processes.

Dispersal and migration processes are essential support services that sustain a biological community's ability to change in response to a change in environmental drivers, such as current global climate changes. For spatial variation in biodiversity to have a positive impact on communities' response capacity, there need to be (1) a positive relation between area and species richness (or rather trait variation) such that immigrating species (or individuals) are different from local species (or individuals) and can contribute to better performance, resource use, or structure that may support ecosystem services, and (2) dispersal/migration corridors or vectors that allow movement of individuals and dispersal vectors. Dispersal is the spatial process of redistribution of organisms or abiotic entities. Migration means an active movement by organisms. Dispersal vectors can be organisms such as pollinators or seed carriers, e.g., flying foxes. Sometimes dispersal vectors are important for transportation of abiotic components such as nutrients, e.g., herds of ungulates that fertilize areas in their migration track.

The effective area that is available as a regional pool of species for a local site depends on how this site is connected to other sites and the dispersal ability of the organisms involved in a particular ecosystem process. Figure 3 shows an example from Madagascar where small patches of forest, mainly protected by religious taboos, host species that act as seed disperser (lemurs) and pollinators (bees). The effective area for the potential species pool varies greatly depending on the mean travel distance of organisms (shown in the different panels of figure 3), which determines how many local sites join a cluster of connected sites. The resulting effective regional species pool and the variability in traits held by species in this pool may have a positive effect on ecosystem processes and, if these are linked to a particular ecosystem

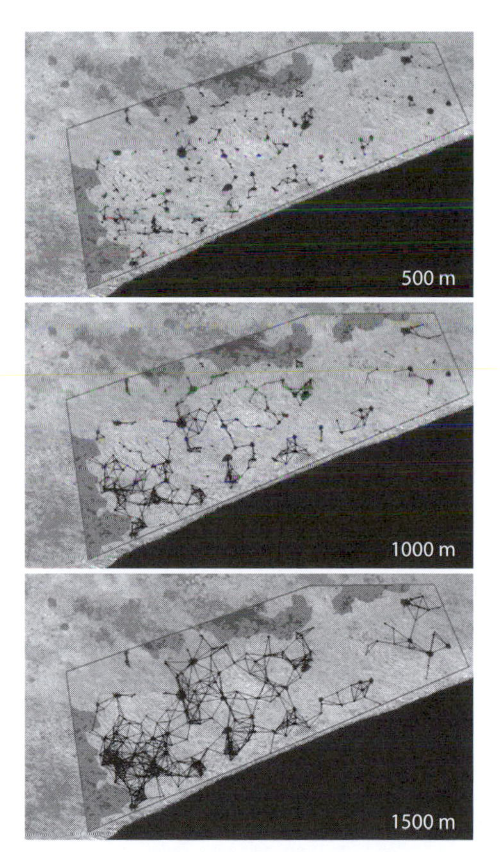

Figure 3. The relation between landscape processes and ecosystem services. Patches of habitable land, here exemplified by small areas of forests in Madagascar protected by religious taboos, are connected by the distance that species providing certain ecosystem services can travel: (top) 500 m, (middle) 1000 m, and (bottom) 1500 m. This creates different-sized components in the landscape that can often hold higher biodiversity. (Adapted from Bodin, Ö., and J. Norberg. 2007. A network approach for analyzing spatially structured populations in fragmented landscape. Landscape Ecology 22: 31–44)

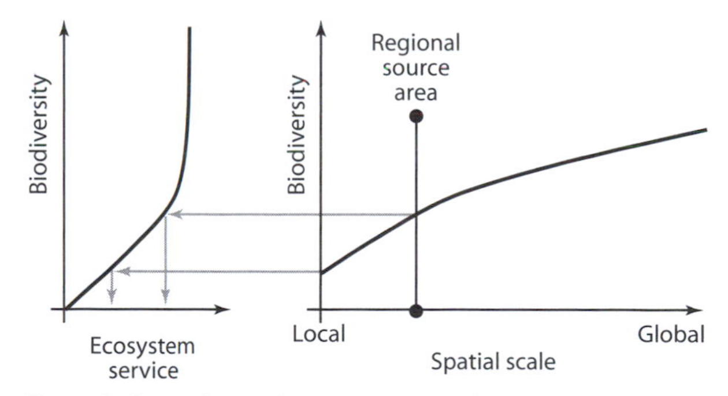

Figure 4. Depending on the average range of movement of species, the landscape is more or less connected, which determines the area that can act as an effective regional source of species for local habitats. If species biodiversity is positively related to area, then a larger regional pool may be able to sustain higher local biodiversity. When this spatial phenomenon is combined with the general positive effect of biodiversity for ecosystem processes, we can see how landscape patterns and dispersal processes could potentially increase biodiversity and thus community response capacity, leading to less variability and better provisioning of ecosystem services.

service, have a positive effect on ecosystem service provisioning as is illustrated in figure 4.

5. CONCLUSIONS

Ecologists have spent the last decade demonstrating the connection between various measures of species identity and diversity and the functioning of ecological systems. As suggested here, and by many other scientists, it may be the suite of traits present in a system, rather than species diversity itself, that is the most important determinant of ecosystem functioning and hence the provisioning of ecosystem services. Maintaining or enhancing trait diversity in natural or managed ecosystems can buffer those systems against change—enhancing their resilience—although this can also come at a cost to optimal functioning or provisioning of services for a particular set of conditions. The emerging scientific knowledge about trait spectra and how they correlate to other aspects of the ecosystem should aid in effectively balancing efficiency and resilience as the world increasingly impacts and manages the landscapes that deliver the suite of ecosystem services upon which humanity depends.

FURTHER READING

Elmqvist, T., C. Folke, M. Nyström, G. Peterson, J. Bengtsson, B. Walker, and J. Norberg. 2003. Response diversity, ecosystem change, and resilience. Frontiers in Ecology and the Environment 1: 488–494.

Hooper, D. U., F. S. Chapin III, J. J. Ewel, A. Hector, P. Inchausti, S. Lavorel, J. H. Lawton, D. M. Lodge, M. Loreau, S. Naeem, B. Schmid, H. Setätä, A. J. Symstad, J. Vandermeer, and D. A. Wardle. 2005. Effects of biodiversity on ecosystem functioning: A consensus of current knowledge. Ecological Monographs 75: 3–35.

Scheffer, M. Critical Transitions in Nature and Society. Princeton, NJ: Princeton University Press.

Tilman, D. 2001. An evolutionary approach to ecosystem functioning. Proceedings of the National Academy of Sciences U.S.A. 98: 10979–10980.

VI.4

Human-Dominated Systems: Agroecosystems

Alison G. Power, Megan O'Rourke, and Laurie E. Drinkwater

OUTLINE

1. Introduction
2. Supporting ecosystem services
3. Regulating ecosystem services
4. Supporting biodiversity
5. Managing agricultural systems for ecosystem services

Agricultural ecosystems around the globe differ radically. These systems, designed by diverse cultures under diverse socioeconomic conditions in diverse climatic regions, range from temperate zone monocultural corn production systems to species-rich tropical agroforestry systems to arid-land pastoral systems. This diversity of agricultural systems produces a variety of ecosystem services. Just as the provisioning services and products that derive from these agroecosystems vary, the support services, regulating services, and cultural services also vary. In general, agricultural activities are likely to modify or reduce the ecological services provided by unmanaged terrestrial ecosystems (except for provisioning services), but appropriate management of key processes may improve the ability of agroecosystems to provide a range of ecosystem services.

GLOSSARY

agroecosystem. An ecosystem designed and managed by humans to produce agricultural goods

agroforestry. An agricultural system in which woody perennials are deliberately integrated with crops and/or animals on the same unit of land

biological nitrogen fixation. A process carried out by specific microbes that have the ability to convert atmospheric N_2 gas into forms that can be used by plants

decomposition. The breakdown of organic residues carried out by bacteria and fungi resulting in the release of energy, nutrients, and CO_2

mineralization. The release of nutrients occurring during decomposition; nutrients such as N and P are converted from organic forms to soluble inorganic ions that can be taken up by plants

natural enemy. A predator, parasite, parasitoid, or pathogen of another organism; often describes beneficial organisms that attack pests in agricultural systems

polyculture. An agricultural system in which multiple crops are grown on the same unit of land at the same time

1. INTRODUCTION

Agricultural ecosystems cover approximately 40% of the terrestrial surface of the Earth. These highly managed ecosystems are designed by humans to provide food (both plant and animal), forage, fiber, biofuels, and plant chemicals. The primary ecosystem services provided by agriculture are these provisioning services. Influenced by human management, ecosystem processes within agricultural systems provide other services that support the provisioning services, including pollination, pest control, genetic diversity for future agricultural use, regulation of soil fertility and nutrient cycling, and water provisioning.

In addition to these provisioning services, however, agroecosystems can also provide a wide range of regulating and cultural services to human populations. Regulating services from agriculture may include flood control, water flow and quality, carbon storage and climate regulation through greenhouse gas emissions, disease regulation, and waste treatment (e.g., nutrients, pesticides). Cultural services include scenic beauty, education, recreation, and tourism, as well as traditional use. Traditional use may comprise the incorporation of agricultural places or products in traditional rituals and customs that bond human communities. One additional ecosystem service that might be classified as a cultural service is the support of biodiversity. To the extent that appreciation for nature is an explicit human value, the ability of a particular agroecosystem to maintain and enhance biodiversity may be included under cultural services. Biodiversity may, in return, provide a variety of supporting services to agricultural and surrounding systems.

In the discussion below, major ecosystem services from agriculture are described in the context of some alternative management systems. In some cases, agricultural modifications to ecosystems will undoubtedly lead to a decline in the quantity or quality of ecosystem services. Here we identify management practices that prevent or ameliorate potential loss or degradation of services where possible. Clearly, there are instances where there is a trade-off between increasing yields and supporting a broader array of ecosystem services, but some agricultural practices may both enhance yields and support ecosystem services. Not all ecosystem services are addressed in detail; in particular, cultural services are not treated extensively here.

2. SUPPORTING ECOSYSTEM SERVICES

Regulation of Agricultural Pests

Agricultural crops are inevitably attacked by insect pests and pathogens that reduce the quantity and quality of the products that humans derive from agroecosystems. Management systems that emphasize crop diversity through the use of polycultures, cover crops, crop rotations, and agroforestry can reduce the abundance of insect pests that specialize on a particular crop while providing refuge and alternative prey for natural enemies. A variety of organisms, including insect predators and parasitoids, insectivorous birds and bats, and microbial pathogens, can act as natural enemies to agricultural pests and provide biological control services in agroecosystems. These biological control services can reduce populations of pest insects and weeds in agriculture, reducing the need for pesticides.

Conservation biological control, where agricultural habitat and management practices are manipulated to conserve and enhance populations of beneficial organisms already present in a system, can be effective in reducing pest populations and pesticide usage (figure 1). The goal of conservation biological control is to sustain natural enemy populations even when pests are

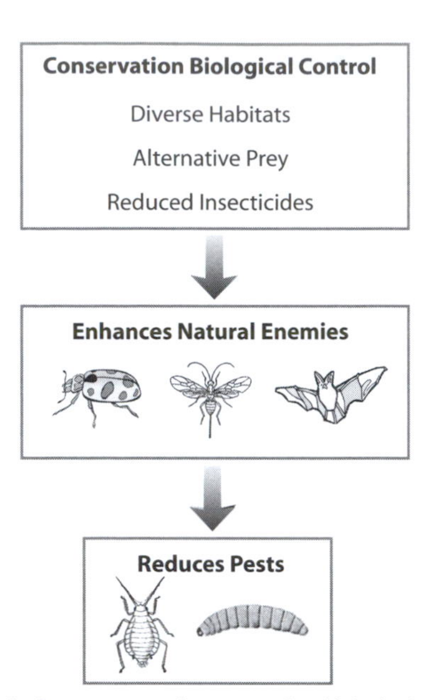

Figure 1. Components of conservation biological control.

scarce. This allows natural enemies to be abundant at the beginning of pest outbreaks, when they have the greatest chance of controlling pest populations below economically damaging levels. Conservation biological control techniques include planting polycultures where different crop species or genotypes (varieties of the same crop) are interplanted; conserving uncultivated areas throughout a farm; planting refuge strips within fields that are not tilled or sprayed with pesticides; reducing tillage; and reducing applications of broad-spectrum insecticides (i.e., those that target many insects, even beneficial ones).

Many conservation biological control techniques are complementary in providing natural enemies with refuge and alternative prey. The microclimate in field crops can often be too dry and hot for natural enemy species during the summer. Conserved areas with trees and shrubs and intentionally planted refuge areas with mixed grasses and flowers can provide cool, moist refuges for natural enemies. Many natural enemies also need an uncultivated space to overwinter, which can also be provided by natural and planted refuge areas. Uncultivated areas within farms also provide refuge for natural enemies from pesticide applications that can be directly toxic. Reducing tillage within fields, which increases plant debris, has been shown to provide refuge for natural enemies. Provisioning of alternative prey when pests are scarce can be provided by polycultures, planted refuges, and conserved natural areas, all of which can enhance the diversity of insect prey and species of nectar-bearing plants.

Even agroecosystems based on monocultures can conserve the diversity of natural enemies if pesticides are not used. For example, paddy rice monocultures managed without pesticides can have a surprisingly high diversity of herbivorous insects, predators, and parasitoids compared with similar monocultures in which pesticides are used. By refraining from applying pesticides, tropical rice farmers enable natural enemy communities of hundreds of species per hectare, providing good economic pest regulation. Pest-management programs in Southeast Asian paddy rice have taken advantage of this diversity of natural enemies and have drastically reduced pesticide inputs without sacrificing yields. In traditionally managed rice fields, predators are likely to include fish and amphibians, which contribute to pest regulation and also provide additional nutritional resources for farm families.

Pollination

Approximately 65% of plant species require pollination by animals, and 75% of crop species of global significance rely on animal pollination, primarily by insects. Although much of agriculture relies on the pollination services of domesticated honeybees (*Apis mellifera*), native bees can enhance pollination rates, fruit size, and seed set of some crops.

There is much concern about reported declines in the abundance and diversity of wild pollinators and about increasing disease problems in domesticated bees. These declines are, in part, a result of the intensification of agricultural systems. Broad-spectrum and systemic insecticides can be directly toxic to pollinators. Broadleaf herbicides decrease the abundance and diversity of flowering weeds that provide food resources for pollinators in agricultural landscapes. Expanding agricultural acreage decreases the amount of natural areas available to pollinators, areas that provide nesting sites and contain plants with a diversity of flowering times and food resources.

Specific agricultural practices can be adopted to benefit wild pollinators. These include a reduction of pesticide usage. No-till soil management has also been shown to increase the abundance of ground-nesting bees. Conserving natural habitats can increase the amount of nesting areas and food resources available to pollinators. Seminatural areas that contain mixtures of different types of flowering plants can be planted throughout a farm to increase the diversity of pollinators. Crop rotations with mass-flowering crops such as rape, clover, alfalfa, and sunflower can provide important food resources and support higher densities of native pollinators.

Nutrient Cycling

Agriculture has profound effects on cycling of nutrients at local, regional, and global scales. Nitrogen and phosphorus are the two most important nutrients limiting biological production in ecosystems, and they are the most extensively applied nutrients in managed terrestrial systems. Use of fertilizers and increased biological nitrogen fixation in agricultural ecosystems accounts for 60% of new biologically active N from anthropogenic sources. The amount of available phosphorus in the biosphere has also increased tremendously in the last 50 years, largely as a result of phosphorus applications to agricultural lands. Phosphorus flux to coastal oceans has nearly tripled. Nutrient enrichment of the environment with N and/or P has a series of complex, often detrimental, consequences for natural ecosystems.

Intensive annual crop production systems are among the most important food production systems and are also the most problematic in terms of their contributions to greenhouse gases, nutrient enrichment, and soil degradation. Annual inputs of nitrogen and phosphorus to agricultural fields consistently exceed the amounts taken out of the system by harvest. This excess, which can be anywhere from 40 to nearly 100% of what was applied, is lost to the environment.

Conventional nutrient management in agriculture is based on developing optimum delivery systems for soluble inorganic fertilizers and managing the crop to create a strong sink for fertilizer by removing all other growth-limiting factors. The problem with this strategy is that soluble inorganic forms of N and P are fast cycling and are subject to multiple pathways of loss (figure 2A). When the pool of soluble inorganic N or P is greatly increased, losses of these added nutrients from the ecosystem also increase, leading to environmental degradation. Although conventional nutrient management has resulted in greater yields, it has also resulted in poor nutrient use efficiency and major losses of fertilizers to the environment. Soil degradation is also a secondary consequence of these intensive, fertilizer-driven cropping systems, mainly because of the use of intensive tillage combined with reduced inputs of crop residues and bare fallows.

Nutrient-management strategies that target a broader range of internal cycling processes and that integrate N, P, and C cycling are more effective at supporting ecosystem services beyond yield. For example, practices such as cover cropping or polyculture enhance plant and microbial uptake of N, promote nitrogen retention in the soil organic matter, and reduce standing pools of nitrate, the form of N that is most susceptible to loss (figure 2B). Other examples of effective management practices

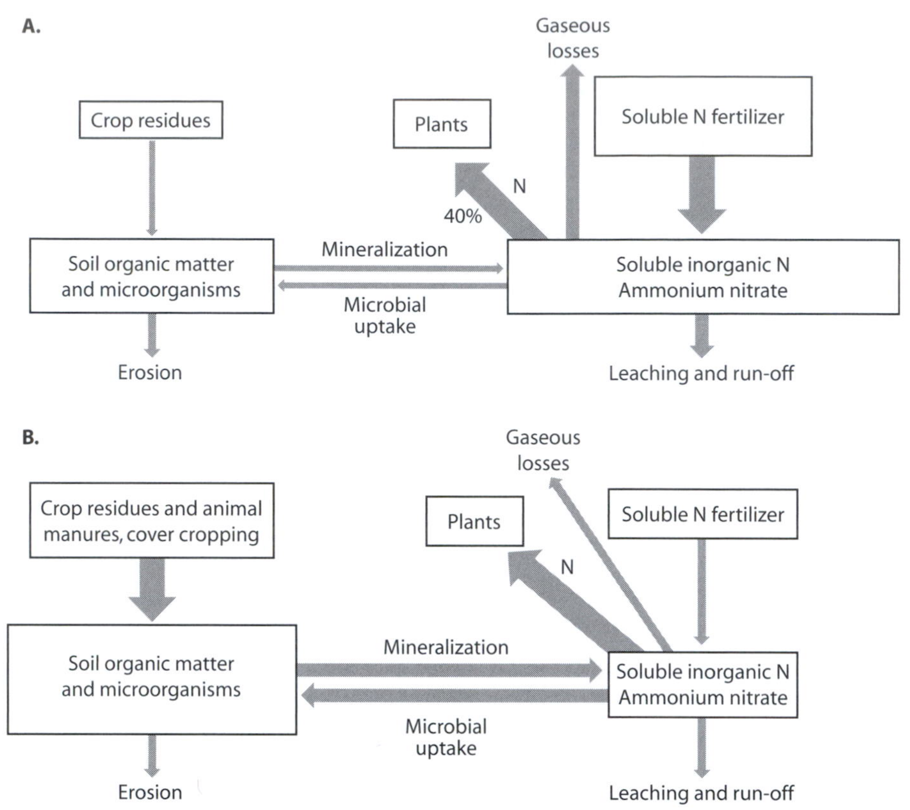

Figure 2. Contrasting nutrient management practices and their impacts on nitrogen cycling and loss pathways. (A) Conventional management with inorganic fertilizers as the main input and bare soil periods ranging from 2 to 6 months. (B) Ecologically based management relying on diversified nutrient sources and practices such as cover cropping. The size of the box or arrow indicates the size of the pool or flux. (Modified after Drinkwater and Snapp, 2007)

include using organic residues or animal manures as nutrient sources, legume intensification for biological N fixation and P-solubilizing properties, replacing soluble P fertilizers with sparingly soluble forms, diversifying rotations using plant species that promote rhizosphere (near root) processes such as soil aggregate formation, and integrating crop and animal production systems.

To maintain the full array of ecosystem services, nutrient pools such as soil organic pools, microbial biomass, and sparingly soluble P that can be accessed through plant- and microbially mediated processes, must be intentionally managed. Management practices that build these pools while minimizing processes leading to nutrient losses increase the capacity of the internal cycling processes to supply crops with nutrients. Such practices might also include the strategic use of buffer zones and hedgerows so that nutrients leaving fields are trapped before they reach other, sensitive systems (including rivers and lakes). Agroecosystems that integrate crop and animal production at the scale of either the farm or regional landscapes by, for instance, feeding crop wastes to animals and applying animal wastes as fertilizer also recouple N, P, and C cycling and can support multiple ecosystem services.

3. REGULATING ECOSYSTEM SERVICES

Water Quantity and Quality

Maintenance of water quantity and continuance of water quality are significant ecological services provided by terrestrial ecosystems. Water flow and water storage in the soil are regulated by plant cover, soil organic matter, and the soil biotic community. The plant community plays a central role in regulating water flow by retaining soil, modifying soil structure, and producing litter. Pore structure, soil aggregation, and decomposition of organic matter are also influenced by the activities of bacteria, fungi, and macrofauna such as earthworms, termites, and other invertebrates. Trapping of sediments and erosion are controlled by the architecture of plants at or below the soil surface and the amount and decomposition rate of surface litter. Macrofauna that move between the soil and litter layer influence water movement within soil as well as the relative amounts of infiltration and runoff.

Agriculture modifies the species identity and root structure of the plant community, the production of litter, the extent and timing of plant cover, and the composition of the soil biotic community, all of which influence water infiltration and retention. The intensity of agricultural production and management practices will affect both the quantity and quality of water in an agricultural landscape. Practices that maximize plant cover, such as minimum tillage, polycultures, or agroforestry systems, are likely to decrease runoff and increase infiltration. Irrigation practices may influence runoff, sedimentation, and groundwater levels in the landscape. Agricultural production systems that involve the application of significant levels of industrial nitrogen fertilizer can increase nitrate leaching and nitrate levels in drinking water, which can cause human health problems, particularly for infants. Applications of pesticides can result in pesticide residues in surface and groundwater. Hence, agricultural systems that rely heavily on agrochemicals can degrade the water-provisioning services provided by agroecosystems.

Regulation of Greenhouse Gases

Globally, agriculture is estimated to be responsible for about 14% of greenhouse gas emissions. Land use change is the second largest global cause of CO_2 emissions after fossil fuel combustion, and some of this change is driven by conversion to agriculture, largely in developing countries. In developed countries, forest conversion to cropland, pasture, and rangeland were common through the middle of the twentieth century, but current conversions are primarily for suburban development. Approximately half of global annual emissions of methane (CH_4) and a third of global annual emissions of nitrous oxide (N_2O), both greenhouse gases, are attributed to agriculture.

Agricultural activities contribute to emissions in several ways. N_2O emissions occur naturally as a part of the soil nitrogen cycle, but the application of nitrogen to crops can significantly increase the rate of emissions, particularly when more nitrogen is applied than can be taken up by the plants. Nitrogen is added to soils through the use of inorganic fertilizers, application of animal manure, cultivation of nitrogen-fixing plants (e.g., legumes), and retention of crop residues. In addition to direct N_2O emissions from fertilizer application, the production of synthetic nitrogen fertilizers is a very energy-intensive process that produces additional greenhouse gases. Flooded rice cultivation also contributes to greenhouse gas emissions through anaerobic decomposition of soil organic matter by CH_4-emitting soil microbes. The practice of burning crop residues also contributes to CH_4 and N_2O production.

Livestock also produce CH_4 and N_2O. Ruminant livestock such as cattle, sheep, goats, and buffalo emit CH_4 as a by-product of their digestive processes (enteric fermentation). Livestock waste can release both CH_4, through the biological breakdown of organic compounds, and N_2O, through microbial metabolism of nitrogen contained in manure. The magnitude of emissions depends strongly on manure-management practices (e.g., the use of lagoons, field spreading) and to some degree on the type of livestock feed.

An array of agricultural practices can reduce or offset the agricultural greenhouse gas emissions described above. Effective manure management can significantly reduce emissions from animal waste. Increasing the use of biological nitrogen fixation in place of synthetic nitrogen fertilizers can reduce CO_2 emissions from agricultural production by half. The restructuring of agroecosystems that accompanies legume intensification also modifies internal cycling processes and increases N use efficiency within agroecosystems via the recoupling mechanisms discussed above. Chronic surplus additions of inorganic N, which are currently commonplace, can be reduced under these scenarios, leading to reductions in NO_x and N_2O emissions.

Agriculture can offset greenhouse gas emissions by increasing the capacity for carbon uptake and storage in soils, i.e., carbon sequestration. The net flux of CO_2 between the land and the atmosphere is a balance between carbon losses from land use conversion and land-management practices and carbon gains from plant growth and sequestration of decomposed plant residues in soils. In particular, conservation tillage and no-till cultivation can conserve soil carbon, and planting of cover crops can reduce the degradation of subsurface carbon. Under most conditions, the increased use of legumes in rotation is also expected to increase soil C storage. Many farmers have already adopted these practices to achieve higher production and lower costs.

Finally, agricultural land can also be used to grow crops for biofuel production. Biofuels have the potential to replace a portion of fossil fuels and may lead to lower greenhouse gas emissions. Although burning fossil fuels adds carbon to the atmosphere, biofuels, if managed correctly, avoid this by recycling carbon. Although carbon is released to the atmosphere when biofuels are burned, carbon is recaptured during plant growth. The replacement of fossil fuel–generated energy with solar energy captured by photosynthesis has the potential to reduce CO_2, N_2O, and NO_x emissions. However, management practices used to grow crops and forages for biofuel production will influence net emissions. Development of appropriate biofuel systems based on perennial plant species that do not require intensive inputs such as tillage, fertilizers, and other agrochemicals have the potential to help offset fossil fuel use in agriculture and possibly in other sectors of the economy; biofuel systems that rely on annual plants such as corn may not be as beneficial.

Disease Regulation

Agricultural systems may play a role in regulating some infectious diseases of humans, both tropical and temperate. In the tropics, large-scale irrigation systems based on dams, reservoirs, and large canals can increase appropriate habitat for the snails that serve as intermediate hosts for the parasites that cause schistosomiasis, a debilitating disease that affects millions of people in the tropics. Small-scale systems, however, are less likely to lead to large increases in snail populations. In agricultural systems where soil erosion is not well managed, sedimentation and runoff can slow stream flow and decrease water depth, thereby creating excellent mosquito habitats of warm, shallow water with little or no flow. Irrigated rice paddies, in particular, can serve as excellent breeding grounds for the mosquitoes that transmit malaria and other human pathogens. Effective management of water flow and sedimentation, however, can disrupt vector development and reduce disease transmission. Moreover, traditional rice paddies in Asia that are managed without pesticides or fertilizers often harbor significant fish populations that effectively limit mosquito populations.

Changes in temperate agriculture within a suburbanizing landscape may also influence the prevalence of infectious diseases. Forest fragmentation in the northeastern United States, historically driven by agriculture but now by suburban development, has led to increased densities of the white-footed mouse, the principal natural reservoir host of Lyme disease, in the remaining forest patches. Because white-footed mice are the most competent hosts of the spirochete that causes Lyme disease, the presence of alternative hosts serves to dilute the prevalence of disease. Lyme disease risk to humans is thus correlated with the diversity of mammalian hosts, and mammal diversity is correlated with size of forest patches. Larger forest patches contain a higher diversity of mammals that, although hosts of Lyme disease, are less effective at transmitting it to humans than are white-footed mice. Small patches have relatively higher densities of white-footed mice, leading to higher disease risk. The abandonment of small areas of land from agriculture, leading to small patches of secondary forest, may also contribute to higher mouse densities and increased disease risk.

4. SUPPORTING BIODIVERSITY

It is well documented that biodiversity provides many ecological services that aid human endeavors, including agriculture. Pest regulation by naturally occurring populations of natural enemies, as described above, is one example. Biodiversity is also a cultural value embraced by most human societies. Despite the value of biodiversity to humans, it is estimated that extinction rates during the last 100 years are 100 to 1000 times higher than the average rates of extinction that preceded large-scale human modification of landscapes. Given the extensive nature of agricultural activities in terrestrial ecosystems, many of the world's species are affected by agricultural production. Arguably, agricultural production systems are a main driver of increased extinction rates through conversion of natural habitats to agriculture and increasingly intensive management. However, management options do exist to help conserve biodiversity in conjunction with agricultural production. Restructuring the agricultural system by increasing crop and livestock diversity is one approach to enhancing associated biodiversity in agroecosystems.

The spatial and temporal arrangement of domesticated plants and animals that farmers purposely include in the system may include several dimensions of diversity, including genetic diversity, species diversity, structural diversity, and functional diversity. This planned diversity may also include beneficial organisms that are deliberately added to the agroecosystem, such as biological control agents or plant-associated nitrogen-fixing bacteria. Unplanned diversity includes all the other associated organisms that persist in the system after it has been converted to agriculture or that colonize it from the surrounding landscape. As planned diversity increases along any of its dimensions, unplanned diversity also tends to increase.

The unplanned diversity that accompanies planned diversity in agricultural systems can provide many ecological services to agriculture. Uncultivated species, including wild relatives of crops that occur in and around the agroecosystem, are an important source of germplasm for developing new crops and cultivars and can provide habitat for beneficial organisms. Increasing planned crop diversity can augment the resources available to plant pollinators and to natural enemies and result in higher populations of these beneficial organisms. Increasing planned diversity may also foster beneficial soil organisms and the conservation of functional processes such as decomposition and nutrient cycling.

In addition to crop and livestock diversification, other management options to support biodiversity include reduced chemical inputs and the deliberate provision of resources and refuge habitats to wild plants and animals. Pesticides can have both direct and indirect effects on biodiversity in agroecosystems. Broad-spectrum insecticides directly reduce biodiversity by killing many nontarget insects. Loss of arthropod abundance and diversity affects other species that feed on insects including many birds and bats. Herbicides directly reduce the diversity and abundance of herbaceous plants within crop fields. This loss of plant diversity, in turn, also affects species at higher trophic levels such as pollinators and natural enemies. Fertilizer runoff from agricultural fields is linked to aquatic eutrophication and fish kills. Fertilizers may also indirectly affect plant diversity bordering fields by favoring annual plants adapted to high nutrient availability. Management practices that reduce the need for chemical inputs, such as the conservation of natural enemies of pests, crop rotations to help control pests and improve soil fertility, and the use of organic fertilizers, can all help to preserve biodiversity in and around farms.

Agricultural practices that provide a diversity of habitats can also be beneficial to wildlife. Homogenization of the agricultural landscape reduces available niches and the number of species supported. Mixed cropping systems, where a variety of crops are planted together, directly increase the diversity of plants in fields and the species supported by those plants. Agroforestry systems enhance structural diversity, which also leads to a greater diversity of habitats and resources for associated fauna and flora. On a larger spatial scale, diverse crop rotations increase the diversity of plants present in the landscape at the same time. Conserving natural field edges, planting permanent grassy areas, integrating farm operations to include arable crops and pasture and forestry land, and reducing land drainage, can all help to preserve habitat diversity, which can support biodiversity in agricultural landscapes. Often, although not always, this can be done without significant loss to yields because the productivity of the remaining agricultural area is enhanced by the management of ecosystem services.

Habitat diversity supports the diversity of plants and animals through a variety of mechanisms. Noncropped and low-intensity management areas provide a physical refuge from farm operations. They also provide nesting and overwintering sites. They modify the microclimate to provide refuge that is cool and moist compared to crop fields. Woody field borders can provide cover from predators for foraging mammals. Compared with large, intensively managed monocultures, diverse habitats provide greater temporal stability in the range of food resources for wildlife. Unmanaged field edges can also act as dispersal corridors among larger habitat patches for many species. Dispersal allows recolonization of disturbed habitats and the population mixing that prevents inbreeding, which can compromise the vigor of beneficial organisms.

Agroecological practices to promote biodiversity are not expected to conserve all species equally; effects will be highly dependent on species' life histories. Effects of management practices may also depend on the surrounding landscape. For example, changes in chemical inputs within a field or farm have been shown to have high impacts on plants with viable seed banks, whereas vertebrates with large home ranges may be influenced more by landscape composition. Furthermore, transition to organic management has been shown to have greater benefits for wildlife when the surrounding landscape is dominated by conventional agriculture than when the surrounding habitats are already diverse.

A time delay may be expected between changes in agricultural management and the effects on biodiversity. For example, a lag between rates of agricultural intensification in Great Britain and the decline of farmland bird populations has been detected. This time lag may be attributable to spatial thresholds of intensification. Below certain thresholds, species may be able to compensate for deteriorating local conditions. Conversely, a shift toward conservation-oriented management may also have delayed effects. For example, dispersal-limited species may take years to colonize newly restored habitats if those habitats are not near source populations. Shifts in crop rotations and organic inputs are likely to have long-term, accumulating effects on soil structure and chemistry. These changes influence soil biota and may have cascading food-web effects on other species.

5. MANAGING AGRICULTURAL SYSTEMS FOR ECOSYSTEM SERVICES

The particular suite of agricultural practices that will optimize ecosystem services from agroecosystems is site specific and reflects the biological (pests and pathogens, natural enemies, microbial symbionts), physical (climate, soils), and socioeconomic (government regulations, agricultural policies, market structure)

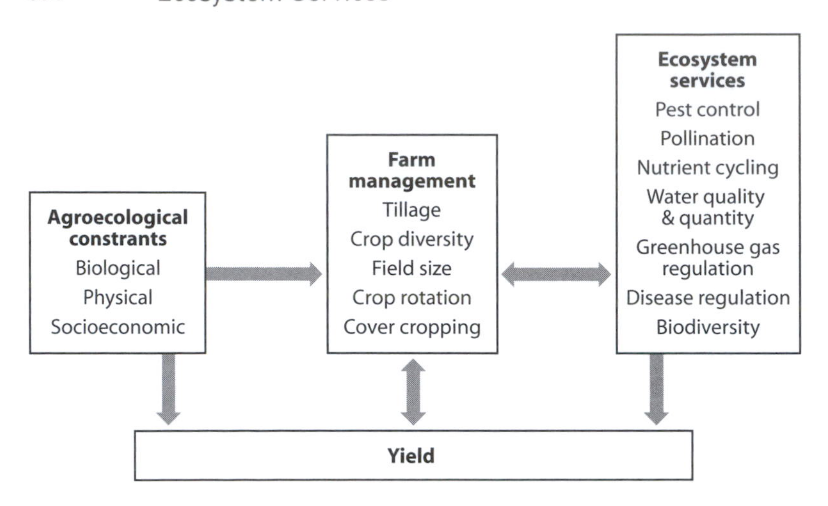

Figure 3. Interplay of constraints affecting farm management decisions, agroecosystem services, and crop yield.

environment of the agroecosystem, the crops that are being grown, resources available to the farmer, and the livelihood goals of the household (figure 3). One traditional dilemma is weighing the importance of provisioning services against the value of supporting, regulating, and cultural ecosystem services that are provided by agroecosystems. However, recent analyses have suggested that this may be a false dilemma.

Recent studies have synthesized data on the yield performance from agroecosystems around the world and found that, on average, agroecosystems using the ecological management approaches described here (e.g., conservation tillage, crop diversification, legume intensification, conservation biological control, etc.) perform as well as intensive, high-input systems (Badgley et al., 2007). That is, the provisioning services provided by the agroecosystem are not jeopardized by modifying the system to improve its ability to provide other ecological services. Moreover, the introduction of these types of practices into resource-poor agroecosystems in 57 developing countries resulted in a mean relative yield increase of 79% (Pretty et al., 2006). These synthetic analyses suggest that it may be possible to manage agroecosystems to support a full suite of ecosystem services while still maintaining the provisioning services that agroecosystems were designed to produce.

FURTHER READING

Badgley, C., J. Moghtader, E. Quintero, E. Zakem, M. Chappell, K. Aviles-Vazquez, A. Samulon, and I. Perfecto. 2007. Organic agriculture and the global food supply. Renewable Agriculture and Food Systems 22: 86–108.

Bianchi, F.J.J.A., C.J.H. Booij, and T. Tscharntke. 2006. Sustainable pest regulation in agricultural landscapes: A review on landscape composition, biodiversity and natural pest control. Proceedings of the Royal Society B 273: 1715–1727. *A meta-analysis of 28 studies of the effects of landscape structure on natural enemies and pests in agriculture.*

Drinkwater, L. E., and S. Snapp. 2007. Nutrients in agriculture: Rethinking the management paradigm. Advances in Agronomy. 92: 163–186. *This article reviews agricultural nutrient management strategies and their environmental consequences, with an emphasis on how plants and their associated microbes could be managed to improve internal nutrient cycling and reduce environmental impacts of intensive agriculture.*

Drinkwater, L. E., and S. Snapp. 2007. The rhizosphere in agricultural ecosystems. In Z. G. Cardon and J. L. Whitbeck, eds. The Rhizosphere—An Ecological Perspective. San Diego, CA: Academic Press. *This chapter describes how ecological processes occurring in the rhizosphere could restore agroecosystem functions beyond those directly related to maximizing crop growth and yields.*

Jackson, L. E., U. Pascual, and T. Hodgkin. 2007. Utilizing and conserving agrobiodiversity in agricultural landscapes. Agriculture, Ecosystems and Environment 121: 196–210. *This article addresses the contributions of agrobiodiversity to ecosystem goods and services in the context of environmental and socioeconomic risk.*

Klein, A.-M., B. E. Vaissiere, J. H. Cane, I. Steffan-Dewenter, S. A. Cunningham, C. Kremen, and T. Tscharntke. 2007. Importance of pollinators in changing landscapes for world crops. Proceedings of the Royal Society B 274: 303–313. *A review of pollination services for world food crops. This article also includes recommendations for agricultural management that may increase pollinator services.*

Pretty, J. N., A. D. Noble, D. Bossio, J. Dixon, R. E. Hine, F. de Vries, and J.I.L. Morison. 2006. Resource-conserving agriculture increases yields in developing countries. Environmental Science and Technology 40: 1114–1119. *This article provides an analysis of the effects on agricultural productivity of 286 interventions that introduced sustainable practices to 37 million hectares in 57 developing countries.*

Swinton, S. M., F. Lupi, G. P. Robertson, and S. K. Hamilton. 2007. Ecosystem services and agriculture: Cultivating agricultural ecosystems for diverse benefits. Ecological Economics 64: 245–252.

Tscharntke, T., A. M. Klein, A. Kruess, I. Steffan-Dewenter, and C. Thies. 2005. Landscape perspectives on agricultural intensification and biodiversity—Ecosystem service management. Ecology Letters 8: 857–874.

VI.5

Forests

Luis A. Solórzano and Guayana I. Páez-Acosta

OUTLINE

1. Forests and people: A long history in brief
2. Forest ecosystem services: Types and scales of delivery
3. Provisioning services: Harvest of forest products
4. Regulating services: Benefits from forests' functioning
5. Cultural services: Benefits from forests' subtle values
6. Reduction of world's forests: Prospective supply

Forest ecosystem services by definition are dependent on the use and value assigned to them by people's needs and perceptions. Humans have historically interacted with forested biomes around the globe and changed their ecological structure as well as their flow of services; consequently, forest biological states, human uses, and anthropocentrically assigned values have changed throughout human history. Although global demand for forest products and services has continuously increased, the impoverishment of the world's forests continues, and their future capacity to support human needs is at risk.

GLOSSARY

anthropocentrism. A human-centered perception and explanation of any given system, e.g., assessing a tropical forest in terms of timber value is an environmental anthropocentric perspective.

biotic impoverishment. The generalized series of transitions that occur in the structure and function of ecosystems under chronic elevated disturbance.

critical habitat. The ecosystems on which any target species—e.g., endangered and threatened pollinators—depend.

environmental uncertainty. Unpredictable sources of density-independent changes in population level parameters.

forest ecosystem management. An approach to maintaining or restoring the composition, structure, and function of natural and modified forests, based on a collaborative vision that integrates ecological, socioeconomic, and institutional perspectives, applied within naturally defined ecological boundaries.

forest fragmentation. Disruption of extensive forest habitats into isolated, smaller patches.

resilience. The capacity of an ecosystem to tolerate disturbance without collapsing into a qualitatively different state that is controlled by a different set of processes. Resilience has three defining characteristics: the amount of change the system can undergo and still retain the same controls on function and structure; the degree to which the system is capable of self-organization; and the ability to build and increase the capacity for learning and adaptation.

scale. The magnitude of a region or process, involving both spatial size and temporal rates.

1. FORESTS AND PEOPLE: A LONG HISTORY IN BRIEF

Even before the development of agriculture, human hunter-gatherers made their way onto all continents, except Antarctica, and selectively consumed and settled in forested regions. Historical evidence confirms the growth and later collapse of ancient civilizations as their forests were used, impoverished, and ultimately degraded. At the onset of Western civilization, massive and destructive forest-use patterns were repeated in Syria, Persia, Greece, and North Africa and later in Rome. The same seems to have occurred in Central America with the Mayan civilization. There are examples in contemporary nations as well, where overpopulation and deforestation have degraded landscapes to uninhabitable stages; this in turn contributed to social crises and made it difficult to create or maintain stable economic and political systems. Human use of forests over the last 8000 to 10,000 years has led to a world where today 25 countries are completely deforested and another 29 have lost more than 90% of their forest cover. At the beginning of the twenty-first century, the human population surpassed 6 billion, and dependency on forest products and services not only is still vital but continuously grows in magnitude and type, while forest use and deforestation have dramatically intensified since the second half of the last century. Today, virtually all major watersheds globally suffer some degree of disruption from forest clearing. Seeking economic growth and development, many countries are repeating forest-clearing patterns experienced by developed countries at much earlier times. Direct deforestation or climate-related forest droughts, fires, impoverishment, and degradation processes occur across extensive regions in China, India, Pakistan, Russia, Southeast Asia, the Philippines, Java, Central America, and South America—mainly in the Amazon.

2. FOREST ECOSYSTEM SERVICES: TYPES AND SCALES OF DELIVERY

Forests and their functioning depend on processes that take place over a range of spatial and temporal scales; consequently, their ecosystem services are generated at several ecological scales as well. Essential ecological, biogeochemical, hydrologic, and climate functions naturally performed by forest ecosystems have historically provided services to humanity at scales varying from short-term, site levels (e.g., food) and medium-term, regional levels (e.g., landscape-level hydrologic and climatic stability) to long-term, global levels (e.g., carbon sequestration).

In terms of human use, the types of services provided by forest can be categorized in three groups: provisioning, regulating, and cultural ecosystem services. Forest provisioning services include food, water, fuel, timber, fibers and other raw materials, biochemical and medicinal resources and genetic resources, and soil formation. Regulating services include carbon sequestration, regulation of climate, water quality and control of hydrology, erosion and sedimentation, source of critical habitat, regulation of wild plant and animal species reproduction, breakdown of pollution, source of pollinators, regulation of diseases, pests, and pathogens, biological fixation of nitrogen, other nutrient cycling, and primary productivity. Cultural services include spiritual and religious, aesthetic, and recreational services. The 2005 Millennium Ecosystem Assessment (MA) included a category of supporting ecosystem services that takes into account ecological processes underpinning the functioning of ecosystems. Here those services are incorporated into the other three categories because ecosystems services are described from a functional perspective.

3. PROVISIONING SERVICES: HARVEST OF FOREST PRODUCTS

Most forests in developing countries are being used by local people for subsistence. Forests are daily providers of the most essential goods such as food, wood for fuel and heating, timber, fibers, and plants for medicinal use; those services are so critical for survival that millions of humans depend exclusively on them. Forests supply important and diverse provisioning services to the global economy and developed nations as well.

Timber and Fibers

Wood as timber and fiber has noncommercial and commercial market value from local to global levels. Local demands for food and fuel vary from region to region depending on population density, climate, soils, and differences in culture and governance structures. Wood has value as fuel for cooking and heating; as construction material for dwellings, furniture, and tools; and as raw material for many other ingenious and artistic uses.

The sale of wood, charcoal, and timber is a reliable source of income for people and governments around the world. Global supply of wood production reached a peak of 3.4 million cubic meters in 2004, with about 59% of that harvested in developing nations. Fuel wood and charcoal accounted for about 52% of the total harvest, and 48% was used for industrial purposes, mainly as lumber, panels for construction, and pulp for paper. Developing countries accounted for about 90% of the wood cut for fuel. It is estimated that about 8% of the global forests' wood harvest for industrial uses is from illegally harvested operations across important supply regions in Brazil, Eastern Russia, Indonesia, and West Africa.

Between 2000 and 2005, global deforestation caused a total forest loss of 65 million hectares—offset by forest regrowth and the expansion of planted forests, the net loss was 36.6 million hectares. The larger losses occurred in Africa and South America, with 3.2% and 2.5% of total forest lost, respectively. Most of the losses in South America were in the Brazilian Amazon, which represents the world's largest continuous region of tropical forests left, and which lost 3.2% of its total forested area. European countries had either no change in forested area or a slight increase in forested area.

Increases in wood consumption are mainly driven by increase in population and economic growth. Analyses of future demand on industrial wood—e.g., using econometric models that take into account population growth, economic growth, land-use patterns, technological change, and other factors—estimate that, by 2010, there will be a demand of 2.5 billion cubic meters of industrial wood per year compared to 3402 million cubic meters today. Future wood production will be driven by regional developments further affecting forests in China, other Asian and Pacific region countries, Latin America, South Africa, Russia, and Eastern Europe. Even outside commercial markets, fuel from forests is a critical resource in Africa and in densely populated regions of the world including the Himalayas, the Indus and Ganges plains, the lowlands and islands of Southeast Asia, and Latin America.

The potential of forest plantations—those that are seeded with fast-growing species that produce regular harvests for commercial use—has not yet been fully realized. At the end of the twentieth century, human-created forests were planted at annual rates of about 2.6 million hectares in tropical regions and 10 million hectares in temperate zones. In 2005, the Food and Agriculture Organization of the United Nations (FAO) estimated that planted forest accounts for about 3% of the world's forests. Although industrial planted forests fulfill some of the functions and services of natural forests, planted forests are usually less diverse in habitat for other species and substantially less complex in structure, function, and their ecological capacity to absorb disturbance, and hence, their resilience is lower.

Biochemical and Genetic Resources

Substances obtained from forests' biodiversity are essential supplies in industry, medicine, and agriculture. Modern advances in molecular biology and biochemistry have allowed greater and more innovative use of chemicals and genes from the world's ecosystems, and increasing demand has turned bioprospecting attention to the abundant and as yet untapped supply of those resources provided by the biodiversity of forest ecosystems. Genetic engineering has opened new opportunities for forest genetic services; for instance, the discovery of the PCR enzyme and several other drug discoveries based on forest plants has led to a surge of confidence about the untapped economic value of forest biodiversity, with the use of biochemical and genetic forest resources increasing.

Genetic diversity has been a key raw ingredient in agricultural research, accounting for roughly half of the gains in U.S. agricultural productivity from 1930 to 1980. The human use of genes across plant species is now common industrial practice; for example, a gene responsible for a sulfur-rich protein found in the Brazil nut was isolated, cloned, and transferred into tomatoes and yeast. At the beginning of the twenty-first century, genetically modified species of food and fiber crops are commercially grown and harvested around the world.

Similarly, the study of medicines used by traditional communities and the pharmaceutical use of natural chemical compounds has intensified over the last decades. Industries not only commercialize the direct use of natural medicinal products but also use them for the design and chemical synthesis of new drugs; it is estimated that after screening, about one in 10,000 natural chemicals yields a valuable product. In the United States, 25% of prescriptions are filled with drugs whose active ingredients are extracted or derived from natural ecosystems, often forest ecosystems.

Other Nonwood Services

Nontimber products are important to local populations as well and include the provisioning of vines used as ropes; rattan; resins such as latex; cork; wild game for hunting and fishing; and other food and medicine sources such as mushrooms, tropical fruits, wild seeds, roots, flowers, fruits, stems, and leaves. This use of forest resources for medicine is particularly important because about 4 billion people have no or little access to Western medicine and in times of sickness depend on plant extracts for treatment. In the tropics, some nonwood products, including Brazil nuts, wild cacao, and açai, can produce economic yields between $80 and $100 per hectare per year.

Forest Soils: A Vital Service

Forests provide a vital service to humans in the form of healthy and intact soils. For most of the 10,000-year history of agriculture, the impact on forests was local, and only in the last century or so has the demand for forest soil services—such as soil contributions to food and timber production—become global. Although demand for agricultural land has decreased in temperate forested regions, today about one-fourth of the Earth's terrestrial surface has been transformed into cultivated systems. FAO estimates that by 2030 there will be a need for 120 million hectares of new agricultural land in developing countries, and most of these hectares will most likely come from transforming currently standing forests.

Supporting ecosystem services provided by forest soils include physical support to plant and animal communities, retention and cycling of organic matter and wastes, regulation of nutrients and major element cycles, and buffering control of hydrologic cycles—

further discussed in regulation services. The natural physical, chemical, and biological processes that produce soil's structure and productivity can take up to hundreds of thousands of years to occur; whereas their cumulative properties and services can be destroyed in scales of decades or less. Soil productivity is a critical determinant of future economic development of nations—particularly for poorer ones—and the total value of the soils' services is extremely high, as they do literally represent the physical base for survival of human societies and of millions of other species.

4. REGULATING SERVICES: BENEFITS FROM FORESTS' FUNCTIONING

In general, large-scale, long-term phenomena set physical constraints on smaller-scale, shorter-time ones, but many large-scale processes are also driven by the combined impact of small-scale ones. For example, local changes in forest growth rates can add up and influence carbon sequestration and climate at the regional and global scales. Similarly, forest soil microbes operating at scales of micrometers and minutes can control the biological fixation of nitrogen, consequently enhancing soil fertility and primary productivity at ecological scales from individual plants to the whole forest ecosystem. At larger scales, such biological regulation affects the global nutrient cycling. Large-scale processes can also constrain smaller-scale processes; for instance, global change in ocean surface temperatures and currents such as the El Niño Southern Oscillation drives changes in precipitation, e.g., drought events, that in turn impact the productivity of ecosystems and life-history cycles of plant and animal species At the landscape level, forests and their biodiversity in large patches and corridors also serve to mediate smaller-scale population dynamics of pests and diseases, regulating their spread.

Hydrology

Water flows and water quality link forests to other global ecosystems and to essential human interests. Forests act as buffers regulating the volume, quality, and timing of water flows from soils, rivers, and groundwater, which in turn divide and over time define landscape physiography and drainage basins around the globe. Water from soils is released back to the atmosphere as vapor and by percolation to streams flows, cleansed of pollution and excess nutrients, flowing at regular seasonal rates across landscapes. Highly populated continental drainage basins—including the Ganges, Danube, Mississippi, Congo, Mekong, and several others—were all covered by forests, and today their hydrologic functions are disrupted as a result of massive deforestation.

The impact of tropical rainstorms has caused more soil erosion in deforested areas than anywhere else on Earth. Deforestation destabilizes soils, causing various levels of erosion depending on the amount and frequency of precipitation, geology, topography (mainly slope), and local soil structure. Deforested systems do not have the robust vegetation required to absorb, retain, and evapotranspire water into the atmosphere. The soil's permeability and capacity to absorb and retain water are reduced after vegetation has been removed or degraded. In healthy forests, the multilayered structure of vegetation reduces the impact of drops from rain and storms when raindrops are stopped by foliage, with water then dripping down leaves, branches, and tree trunks to reach and percolate into the ground and flow into streams more gradually. Rain episodes in deforested landscapes increase surface runoff and soil erosion and diminish the recharge of the groundwater. River channels are blocked by silt from erosion, sometimes causing water to flood across landscapes, killing humans and livestock and destroying crops, living spaces, and other infrastructure of high economic value. Siltation as a consequence of deforestation is a major problem in many watersheds of the humid tropics including Indonesia, Africa, India, Asia, and America.

Climate

Forest ecosystems have a direct effect on climate by influencing the energy budget of the atmosphere and moderating local and regional temperature and rainfall regimes. Changes from forested to deforested landscapes involve alterations in albedo, heat, water pressure deficit, and leaf area index, directly impacting landscape-level evapotranspiration. It is estimated that about half of the warming that occurred in northern latitudes during the Holocene was caused by shifts in albedo from tundra to forest vegetation.

Current large-scale deforestation in the Amazon forests and elsewhere is impacting regional climates. Research has shown that Amazon trees draw water from soil layers 15–20 m deep. Once trees are removed, the landscape becomes more arid, and during the dry season, moist forests become susceptible to fire events that further impoverish the ecosystem. Over the past three decades, during El Niño–caused droughts, forest fires have been observed not only across large regions of the Amazon basin but also in southern Borneo and Mexico. Sweeping across forested landscapes, fires triggered massive tree mortality, biomass and biodiversity losses, and carbon emissions to the atmosphere, further exacerbating global climatic change. In the Amazon, forest changes at those scales have affected local and regional climate and reduced cloud formation and the intraregional precipitation cycle, further increasing the fire–drought–forest impoverishment cycle. With global warming, boreal forests are also becoming more flammable and vulnerable to natural and human-caused fires. In Canada, for instance, in the last two decades of the twentieth century, the burned area increased sixfold compared with the century trend. This trend is worrisome given that Boreal forests occupy 9.2 million square kilometers, and about half of the global forest carbon is stored in them.

Carbon Cycle and Climate Change

Through respiration and photosynthesis, forests annually release carbon dioxide equivalent to 12–14% of the atmospheric content; if there is little deforestation, this contribution to the atmosphere is offset by forest carbon uptake. Either way, forests have a significant functional role in the global carbon balance. Forests hold in their trees and soils more carbon per unit area than any other ecosystem and account for 65% of the global net plant primary productivity on land. It is estimated that forests currently hold more than 1200 billion tonnes of carbon in their vegetation and soils, a significant magnitude compared with the estimated 750 billion tonnes in the atmosphere at present. About half of the forests' carbon is stored in temperate forests, which have probably more carbon than the earth's estimated fossil fuel reserves. By sinking carbon in several biomass and soil compartments, forests take out of the atmosphere at least 10^{15} g of carbon each year, an equivalent of 14% of the total emitted by human activities. Because of their capacity to serve as carbon sinks and to control the global energy balance, effectively managed forest ecosystems will be essential in regulating and mitigating current global climatic changes.

About 25% of the increase in atmospheric CO_2 concentrations over the last 150 years came from changes in land use, mainly from clearing forests and the cultivation of their soils for food production. It has been calculated that global deforestation between 1990 and 2005 caused the carbon storage capacity of the world's forests to decline by 5%. Given current climate change conditions, carbon sequestration is a globally important forest service, potentially as valuable as any other. In 2007, carbon markets were estimated at a US$64 billion value. There is expectation that in a future carbon market, forest carbon values alone could surpass their value in timber and other products by an order of magnitude.

Pollinators and Their Regulation Services

Pollinators are regulators of plant dispersal and community structure and are significant agents of evolution. Important groups of pollinators include beetles, bees, wasps, flies, birds, and bats. The earliest seed-bearing plants were pollinated passively when large amounts of pollen blown by the wind reached their ovules. The evolution of many angiosperms is linked to their evolving ability to attract insects and other animals with their flowers and directing the behavior of pollinators so that cross-pollination occurs with higher frequency. The more attractive plants were to insects, the more frequently they were visited, and the more seed they could produce, gaining a selective advantage. Specialized groups of flower-visiting insects, such as bees and butterflies, evolved with plants for 50 million years and, in the early Tertiary—between 40 and 60 million years ago—became even more abundant and diverse. The increase and diversification of these groups of insects were directly related to the increase in diversity of angiosperms. Consequently, there is a long and profound evolutionary influence and mutual dependence between angiosperms and their pollinators. In evolutionary time, pollinators continue to allow the adaptive radiation of angiosperms into current biomes, affect the composition of floras, and influence the spatial and temporal patterns in plant communities and therefore regional and global patterns of primary productivity.

Research has shown that reproduction in many plant populations stranded in highly fragmented and degraded habitats might be pollinator limited. Scientific results have led the International Union for the Conservation of Nature to warn of the diminishing trend of pollinator diversity available to both wild and domesticated plants. Community- and ecosystem-level impacts of declines in pollinators on natural vegetation are extremely difficult to predict. Nevertheless, it is clear that the evolutionary and ecological functions of pollinators regulate the functioning and resilience of ecosystems and, therefore, provide a critical support service.

There are more analyses on the economic value of pollinators as they interact with agricultural systems than for natural forest. More than 70% of at least 1300 crop species require pollen movement by some vector, and less than 2% depend exclusively on wind. Clearly, the importance of animal vectors for agricultural crops has essential present and future economic value.

5. CULTURAL SERVICES: BENEFITS FROM FORESTS' SUBTLE VALUES

In the last half of the twentieth century, the concurrence of three processes—the emergence of so-called third-generation human rights during the 1970s and 1980s (e.g., environmental rights, international heritage patrimonies, amongst others); a broader recognition of minority groups' rights (notably, those of indigenous peoples); and the commitment and compliance made by nation-states to safeguard them both through policies and adequate structures of governance—have brought a general acknowledgment of the symbolic and cultural interactions between forests and populations. This concurrence has contributed to a new understanding of the existence of forest cultural services and their relevance for local peoples and for humanity in general.

Cultural services provided by forests are difficult to measure, and therefore, data on their use and value are still scarce. Moreover, in terms of cultural services, as stated by the MA, there is considerable uncertainty regarding the importance that people in different cultures place on them, how "importance" changes over time, and how values influence decisions that lead to trade-offs with net benefits and costs. The challenges of defining, measuring, and valuing subtle cultural services critically limit the ability to effectively conserve forests and implement the best management approaches to maintain their provisioning and regulating services in the long term.

Cultural services depend on human interpretation of forest ecosystems and their specific characteristics; in essence, they are culturally conceived, and their value is derived from the socially constructed meanings conferred to a particular forest and the services it supplies to social groups. Consideration of forests' cultural services demands assessment of the number of people benefiting from forests and the type of interaction they have with them. In general, the group of stakeholders that value forest cultural services has been growing in spatial extent; the spread of information and ease of travel have extended cultural services beyond local users to a global community. For instance, although for native communities forests have often been a constituent element of their cosmogonies and spiritual lives, such a cultural role is now broadly recognized and valued not only by each local forest's inhabitants but by distant peoples not directly affected by or involved with the service. Change in cultural services influences human well-being, affecting the sense of security, social relations, and both physical and emotional states, particularly in cultures that have retained strong connections to their local environments. The MA established three main categories of cultural services.

Spiritual and Religious Values

Loss of particular ecosystem attributes (sacred species or sacred forests), combined with social and economic changes, can sometimes weaken the spiritual benefits people obtain from ecosystems. According to MA, the tendency has been a decline in the numbers of sacred and protected areas. On the other hand, under some circumstances (e.g., where ecosystem attributes are causing significant threats to people), the loss of some attributes has enhanced spiritual appreciation for what remains.

Aesthetic Values

Following an increase in urbanization, the demand for aesthetically natural landscapes has increased. A reduction in the availability of and access to natural areas for urban residents may have important detrimental effects on public health and economies. Studies show that the quantity and quality of areas that provide this type of service have been declining, and as the remaining places continue to become scarce, the value placed on them will likely increase.

Recreation and Ecotourism

Demand for recreational use of forested landscapes is rising; therefore, more areas are managed to provide this use, reflecting a cultural change in values and perceptions. Effective management of forest for recreation and ecotourism is an ongoing learning process more advanced in temperate forests, but global standards are still being developed. The MA estimates that although more natural areas are accessible, many are undergoing degradation; to date, comprehensive data on impacts on forest caused by recreation and ecotourism are lacking. The increasing ease of travel is enabling this value to be shared spatially, and although recreation may degrade some systems, it also allows others to be preserved because the value of the forest can actually be marketed; therefore, there may be a net gain for conserving forests.

Globalization and the Cultural Valuation of Forest

Broad and democratized knowledge about forests has certainly influenced the matrix of values and cultural services that people recognize in them. From this perspective, the appropriate management of forest cultural services is highly relevant, as it will pave the way for informing decisionmaking processes and improving the management of forest resources more generally. Increased understanding and incorporation of forests' cultural services into their management plans may allow similar advances in managing the delivery of provisioning and cultural values and can ultimately affect the way societies manage forests.

Globalization has created ample recognition of the crucial services provided by forests for human well-being. Worldwide cultural changes valuing forests have been reflected in the setting of international agreements, governance structures, and mechanisms to enable the changes necessary to improve global forests management and secure their services. Some significant initiatives that focus on forests and their services include the World Commission on Forest and Sustainable Development; the Intergovernmental Panel on Forests; the UN Food and Agriculture Organization Forestry Department; the Center for International Forestry Research; The World Conservation Union Forests Program; and the UN Conference on Trade and Development and Earth Council Institute joint Carbon Market Program.

Finally, international finance agencies are also responding to cultural changes in forest valuation. For instance, the World Bank Prototype Carbon Fund, the Community Development Carbon Fund, and the BioCarbon Fund, all deal with global forests to catalyze private-sector investment to address climate change. In the field of ecotourism, the Inter-American Development Bank undertakes work to identify new avenues for investment in the rational use of forest and biodiversity conservation.

6. REDUCTION OF WORLD'S FORESTS: PROSPECTIVE SUPPLY

After the last Pleistocene glaciations, tropical, temperate, and conifer forest ecosystems occupied more than 44% of the Earth's land area. Over the past 50 years, humans have changed global forest ecosystems more rapidly and extensively than in any other period in history. The world has lost more than 46% of its forests, and most of the losses occurred during the last three decades of the twentieth century. At current deforestation rates, by 2030 there will be less than 10% of intact forests remaining with another 10% in a degraded condition, causing irreversible loss of millions of species whose genetic, chemical, and functional value cannot yet be determined.

There is enough evidence indicating that human impacts are increasing the likelihood of nonlinear dynamics causing abrupt and irreversible changes of unpredictable consequences to forest ecosystems and their services. The combined human activities of burning fossil fuels and transforming forests have changed the composition of the atmosphere, leading to a warming of the Earth at global average rates of 0.2–0.3 degree per decade. Scientific evidence shows that the rate of temperature increase is both high and fast enough to—at least temporarily—substantially change forest function.

The links between deforestation and reduction in regional rainfall patterns are clear, but there are uncertainties concerning the threshold levels at which feedbacks between different forest ecosystems and climate will trigger nonlinear abrupt changes that can negatively impact Earth's biomes and global climate. Data show geographic differences where some forests seem to be invigorated by carbon fertilization whereas others are showing ecological impoverishment and lower biomass accumulation. There are changes in species composition caused by alterations in their growth, survival, dispersion, and reproductive rates. Higher temperatures also accelerate the biogeochemical processes in forests' soils, increasing emissions of radiative gases (carbon monoxide, methane, and nitrous oxide) that further exacerbate global warming.

All these responses have clear influence on forest services primarily through direct changes in access to material well-being, health, and global security. The challenge is for all levels of society to make well-informed, responsibly agreed decisions and actions to govern the use of the world's forests. By securing forest services in the long term, humanity has better chances to continue flourishing spiritually, culturally, ecologically, and economically.

FURTHER READING

Lovejoy, T. E., and L. Hannah, eds. 2005. Climate Change and Biodiversity. Ann Arbor, MI: Sheridian. *Provides the most authoritative overview and evidence of climate change effects on biodiversity and vice versa.*

Millennium Ecosystem Assessment. 2005. Ecosystems and Human Well-being: Synthesis. Washington, DC: Island Press. *State-of-the-art scientific appraisal of the condition and trends of the world's ecosystems and their services.*

Millennium Ecosystem Assessment. 2005. Living Beyond Our Means: Natural Assets and Human Well-being: Statement from The Board. Washington, DC: World Resources Institute. *Interpretation of the key communications from the Assessment, identifying 10 key messages and conclusions.*

Myers, N. 1997. The world's forests and their ecosystem services. In G. C. Daily, ed. Nature's Services: Societal Dependence on Natural Ecosystems. Washington, DC: Island Press, 215–235. *This chapter in Daily's classic book on ecosystem services draws on well-known studies presenting an integrated picture of forests' structure and function and their overall environmental values.*

United Nations Food and Agricultural Organization (FAO). 2005. Change in extent of forests and other wooded land 1990–2005. In FAO, Global Forest Resources Assessment. Rome: FAO. *Source of the most up-to-date quantitative data on the state of the world's forests.*

Woodwell, G. M. 2001. Forests in a Full World. New Haven, CT: Yale University Press. *Comprehensive and accessible, offers an original, sound—and still up-to-date—vision of the world's forests' role and the necessary new approaches to confront the challenges of governance and effective management.*

VI.6

Grasslands

Martha Downs and Osvaldo E. Sala

OUTLINE

1. Scope of grasslands
2. Provisioning ecosystem services
3. Regulating services
4. Cultural services
5. Supporting services
6. The significance of grasslands

This section focuses on the ecosystem services provided by natural grasslands. These regions of the world are mostly limited by water availability, and they exclude anthropogenic grasslands, which derived from forests that were logged and converted into pastures, often to support cattle grazing. Grasslands account for 41% of Earth's land surface, and 38% of Earth's 6.8 billion people live in natural grasslands. Grasslands support a diversity of uses, but until recently they have been primarily used for grazing and wood gathering for fuel, with conversion to agriculture at the wet end of their climatic envelope. Alternative uses of these regions—e.g., recreation, conservation, and carbon sequestration—are gaining in societal value, particularly in developed countries. This chapter used the definition of ecosystem services presented above and the categorization developed by the Millennium Ecosystem Assessment with four types of ecosystem services: provisioning services, regulating, cultural, and supporting services.

GLOSSARY

albedo. Energy reflected from the land or water surface. Generally, white or light-colored surfaces have high albedo, and dark-colored or rough surfaces have low albedo.

carbon sequestration. The process of removing carbon dioxide from the atmospheric pool and making it less accessible or inaccessible to carbon-cycling processes.

grasslands. Short-stature vegetation dominated by grasses, characteristic of locations with a strong water limitation for at least part of the year.

petagram. One trillion million (10^{18}) grams.

soil texture. Soil texture is described by the proportions of sand (large particles), silt (intermediate-sized particles), and clay (smallest particles). Sandy, loose-textured soils allow rapid water infiltration and fast leaching of nutrients. Denser, clayey soils have poor drainage and poor soil aeration.

transpiration. The evaporation of water from the leaves, stems, and flowers of plants. Transpiration occurs through small pores, or stomata, on leaf and stem surfaces, which must remain open to take up carbon dioxide.

1. SCOPE OF GRASSLANDS

Grasslands occur where there is not enough water to support forests, although temperature also plays a role: cool locations can support forests at precipitation levels that can only support grasslands in warmer climates. For instance, rainfall in the United States generally increases from west to east, and temperature increases from north to south. The grassland–forest boundary thus runs in a diagonal fashion from southeast to northwest, reflecting the issue that lower temperatures characteristic of the north allow forest to grow at lower precipitation than in the south. Grasslands are dominated in general by herbaceous vegetation, mainly grasses and forbs, although shrubs account for an important fraction of grassland biomass in some regions, and grasslands can also support occasional trees. The proportion of shrubs and grasses depends on the texture of the soil and the seasonality of precipitation. Grasslands encompass different vegetation types with different shrub abundance from prairies to steppes.

2. PROVISIONING ECOSYSTEM SERVICES

Grasslands, through their support of grazing, produce meat, milk, and blood for many people who depend on animals for their daily protein intake. Grasslands in developed countries are primarily managed by cow-calf producers where calves are sold to fattening operations that feed grain to calves until they become mature and ready for slaughter. In contrast, in developing nations most of the meat production occurs in grasslands themselves. The type of animal used varies enormously depending on cultural and climatic conditions and ranges from goats to camels.

Especially in developing countries, residents often harvest shrubs and trees from grasslands for fuel, and the resource can provide a large percentage of household energy use in dryland regions. Overuse of this service can increase soil erosion, decrease recruitment of new plants, and degrade the ecosystem.

Grasslands support a wide variety of grazing animals, including sheep, goats, llamas, alpacas, vicuñas, and even muskoxen, which can sustainably produce fiber for household use and sale. The trade-offs between grazing density and grassland sustainability vary with the productivity of the system, which depends largely on soil type and precipitation. Most grassland plant species evolved under some grazing pressure, and a moderate population of grazers can help replicate past conditions and sustain or even increase plant species richness. Increasing grazing intensity to higher levels reduces plant cover, exposing bare ground and increasing soil erosion.

The majority of the human diet derives originally from grassland species. Annual grasses and legumes are most abundant in grasslands, and the wheat, barley, and other staple grains on which most northern countries depend today were selectively bred from wild grasses. Similarly, most of our domestic animals—including cattle, goats, and sheep—originated in grassland regions. The value of this genetic library lies not only in the sourcing of today's diet but as insurance against future pests and diseases. In grassland ecosystems, wild relatives of domesticated plants and animals continue to face evolving pests and pathogens. The defenses they develop may one day help protect domestic species from unforeseen threats.

3. REGULATING SERVICES

Grasslands play a role in regulating both local weather and global climate by influencing albedo, dust movement, evapotranspiration, and carbon storage. As grazing intensity increases, surface roughness decreases initially, increasing albedo, which defines the amount of energy reflected from the land surface. All else being equal, the increase in albedo would lead to lower surface temperatures, but it is offset by decreasing transpiration. Like sweat evaporating off the body, this evaporation of water through plants acts to reduce the local temperature and cycle water through the atmosphere. When transpiration is reduced, temperatures increase, and rainfall decreases, nudging the system toward desertification. Comparison of temperature data north and south of the United States–Mexico border, for instance, suggested that temperatures were higher south of the border, where grazing is more intense and plant cover is lower.

Beyond a certain threshold of use, grazers remove most of the herbaceous plants, promoting the establishment of woody shrubs. This change in land surface cover reduces albedo, increases wind and water erosion, and decreases infiltration of rainfall and snowfall, further accelerating the ecosystem degradation and reinforcing the local climate feedbacks.

Pollination and seed dispersal are important regulating services in grasslands, but there are few studies of their scale, distribution, or trends. In patchy landscapes, where crops are interspersed with natural grasslands, the grasslands support a diverse and abundant community of pollinators that maintain crop productivity. Reduction of the proportion of native grasslands and the population of pollinators has reduced crop yield in some cases. In these cases, farmers have resorted to bringing colonies of pollinators from a distance with the resulting cost of transportation and artificial maintenance of the pollinator populations.

Intact grasslands sequester a large amount of carbon in living biomass (both above and belowground) and in soils. Organic plus inorganic carbon in grasslands has been estimated at 770–880 Pg of carbon globally, which is equivalent to 20–25% of all the carbon stored in terrestrial systems. The amount of carbon stored varies spatially and depends mainly on long-term climatic conditions at the site. Increasing precipitation tends to increase biomass production and carbon storage in grasslands, and warmer temperatures increase decomposition, reducing soil carbon stocks (plate 19).

Established grassland systems accumulate carbon gradually but can release large quantities quickly when converted to cultivated agricultural systems. Tilling grassland soils breaks up aggregates, giving decomposers access to previously unavailable carbon compounds. Many grassland systems lose nearly 50% of carbon stocks in the first year of cultivation. However, when land is retired from cultivation and returned to natural grassland or rangeland, it can take 50 to 100 years to regain the lost carbon. This sets up a trade-off between natural grasslands, which sequester carbon, and agricultural systems, which produce food and, more recently, biofuels. Many of the world's most productive agricultural areas have already been converted from natural grasslands, and expanding agricultural conversion reduces the potential for carbon storage in terrestrial systems.

As economic markets begin to place realistic values on ecosystem services from grasslands, it may become possible to compile a package of services provided by intact grassland that approaches the economic value of agriculture conversion. Such a package could include cultural services such as recreation, ecotourism, and scenic vistas (which contribute directly to real estate value) as well as carbon sequestration services.

The Intergovernmental Panel on Climate Change (IPCC) uses benchmarks of $20, $50, and $100 per metric ton of CO_2-equivalent to estimate the future economic value of carbon mitigation strategies. The value rises as the amount of CO_2 in the atmosphere is expected to increase because damage from climate change is expected to be higher as atmospheric concentrations rise. Although the United States has no mandatory carbon limits or carbon trading markets, the Chicago Climate Exchange is beginning to place a value on the (voluntary) sequestration of carbon through offset purchases. The value of 1 metric ton of CO_2-equivalent on the Chicago Climate Exchange currently hovers around $5, which stands in stark contrast to the IPCC estimates.

4. CULTURAL SERVICES

The open spaces, arid climate, and biodiversity of many grassland regions make them attractive destinations for both tourism and recreation. Tourists pack African safaris to view large mammal species, and the trans-Sahara bird migration draws enthusiasts from across the globe. Cultural sites are especially well preserved in arid climates, from Egyptian pyramids to the Native American ruins of the American Southwest. Middle Eastern religious sites attract both local and international visitors.

The same qualities that attract tourists to these destinations also make it difficult to accommodate large numbers of visitors. Water is scarce, temperatures are high, and locations are often remote. Travelers must come long distances, and infrastructure for energy, accommodation, and water supply is often modest. Recent increased awareness of the environmental impact of tourism may open the door to more environmentally gentle development, including the use of on-site solar and wind energy, reuse of local wastewater, and structure design that incorporates passive heating and cooling.

5. SUPPORTING SERVICES

Supporting services maintain ecosystem functioning and are essential to the provision of all other services, but their effects may be indirect or observable only over long time scales. The most important supporting services in grassland systems, as in most other terrestrial systems, are primary production, soil formation, and nutrient cycling, which are closely interrelated. Plant production contributes above- and belowground biomass to the system, supplying organic matter and nutrients. Roots from growing plants hold soil in place, reducing wind and water erosion and facilitating soil development. Nutrient cycling processes, such as decomposition and microbial mineralization and immobilization, retain essential nutrients within the system and release them gradually from organic matter, maintaining a steady supply of plant nutrients and limiting leaching losses.

Natural grasslands exist primarily in areas of marginal precipitation and therefore support lower rates of primary productivity than forests or cultivated systems. Among semiarid grasslands, productivity generally increases with increasing precipitation and decreases with increasing temperature because of the effect of temperature on water availability. This pattern is illustrated nicely in the Great Plains of the United States, where mean annual temperature increases from north to south and mean annual precipitation increases from west to east. This produces a large-scale pattern of increasing net primary production from southwest to northeast. At the high end of the productivity gradient, grasslands transition to forests or agricultural systems.

Precipitation also exhibits a strong temporal influence on productivity, although the relation is not as tight as the spatial correlation, probably because of such factors as time lags in response to variable precipitation. Greater precipitation late in the season permits a longer growing season and greater production.

Soil texture clearly affects productivity, but its influence is smaller than that of climate and varies depending on water availability. In wetter areas, water loss through deep percolation limits production, and fine-textured soils promote higher yields. In drier areas, evaporative loss is a greater concern, and sandy soils may allow quicker infiltration and higher production.

It can be difficult to distinguish degraded grassland systems from those with low productivity because of low precipitation. The ratio of net primary productivity to rainfall, known as rain use efficiency, helps to separate water-limited grasslands from those that may be limited by nutrient losses, desertification, or declining organic matter.

The strong dependence of grassland production on precipitation highlights the potential for global climate change to have large impacts in these ecosystems. Reports of the IPCC predict significant changes in total precipitation and precipitation variability in the coming century. Although precipitation changes caused by climate change vary among regions, the trend for temperature increase is common across all grasslands. Increased temperature, independent of changes in precipitation, will decrease water availability because of increasing evaporative demand. Therefore, increased temperature is expected to decrease grassland production.

Although precipitation may determine how much water is delivered to grasslands, interception and retention by plants and soils determine how much water stays in the system and how much is lost as surface runoff. Soil is an almost-magical amalgam of mineral and organic particles formed over centuries by the physical-chemical breakdown of rocks and the biological recycling of organic material. Together, soil texture and organic matter content are the major determinants of soil quality, a general description for the features that make a soil hospitable to plant, animal, and microbial life. These include water infiltration, water retention, nutrient cycling and retention, and a rooting environment that is neither too acidic, too basic, nor too saline.

A strong positive feedback loop exists between plant biomass and soil formation, especially in grasslands, where belowground biomass forms a large proportion of total plant biomass. When there is sufficient water and limited grazing, grasses form strong, dense networks of roots that resist wind and water erosion and quickly intercept water and nutrients, which can otherwise be lost from the system. Moist (but not waterlogged) soils support active microbial communities, which break down plant litter and quickly incorporate it into soil organic matter, further improving soil quality. A large standing biomass of grasses also helps intercept surface runoff, retains plant litter, and contributes a larger amount of organic material each season.

Conversely, when biomass declines—either from insufficient precipitation or through overgrazing or overuse—a group of feedbacks is set in motion that leads to greater water losses and slower soil formation or even soil loss. Water, when it comes, runs quickly over bare ground, taking plant litter and surface soil with it. When it does infiltrate soils, it may not be intercepted by roots and can percolate below the rooting depth of most plants. The resulting drier soils form a less hospitable environment for decomposition and recycling of plant nutrients, making soils more vulnerable to wind erosion and accelerating the feedback loop. Soil formation, and the balance between formation and erosion, is a key supporting service for grassland systems.

Another phenomenon of arid grasslands should also be mentioned here. Crusts—composed of cyanobacteria, mosses, and lichens—form a fragile barrier that fixes atmospheric nitrogen and channels rainwater to intermittent clumps of vegetation. These create islands of active growth, soil formation, and decomposition, preserving some ecosystem services. Trampling and air pollution, however, can quickly destroy these living crusts and expose the soils beneath.

The nature of nutrient cycling, in which dead plant biomass is broken down to release the nutrients and organic matter it contains, varies with water availability. In very arid grasslands, ultraviolet radiation plays an important role in physically decomposing plant litter. In addition, termites, beetles, and other invertebrates, which can survive extremely arid conditions, prepare plant material by breaking it down into smaller particles and digesting it, which releases a large proportion of the nutrients directly into the soil matrix. Burrowing invertebrates also increase water infiltration by providing physical channels into the soil. Mesic grasslands, in contrast, provide enough water to support soil microbes and fungi, which act directly on plant material, breaking it down in place and slowly releasing nutrients and organic matter.

Mammalian grazers add another dimension to the nutrient cycling picture. Their high rates of metabolism burn off a large proportion of the carbon in the biomass they consume, reducing the amount available for recycling within the system. The meat, milk, and hair they produce is often removed from the system (a provisioning service) but is then unavailable for nutrient recycling. Large grazers also play an important role by redistributing nutrients in the landscape. Usually, these animals graze and harvest nutrients from large areas but concentrate their feces around water holes, where large nutrient losses occur.

6. THE SIGNIFICANCE OF GRASSLANDS

Grasslands cover almost 41% of the earth's surface and span the gamut from extremely arid near-deserts to highly productive systems supporting grazing animals. Agricultural systems are not included in this chapter because they are covered elsewhere in this volume, but there is much overlap between the regions that support grasslands and some of the planet's most productive agricultural areas. Because of this large extent, even services that proceed at moderate rates, such as carbon sequestration, are of global significance.

Grasslands also are home to societies experiencing some of the greatest development challenges, with little infrastructure or money, and located far from the centers of decisionmaking. To solve these development challenges, it is essential to understand the nature and value of services provided on both a local and a global scale. People living in arid and semiarid regions are among the most vulnerable to environmental degradation because they harvest goods and services directly from the natural ecosystems. In addition, grassland ecosystems are far more variable than more mesic ecosystems such as forests because the interannual variability of precipitation increases with decreasing average precipitation.

FURTHER READING

Bradford, J. B., W. K. Lauenroth, I. C. Burke, and J. M. Paruelo. 2006. The influence of climate, soils, weather, and land use on primary production and biomass seasonality in the U.S. Great Plains. Ecosystems 9: 934–950.

Brown, J. R., J. Angerer, J. W. Stuth, and B. Blaisdell. 2006. Soil carbon dynamics in the Southwest: Effects of changes in land use and management. Phase I final report of the Southwest Regional Partnership on Carbon Sequestration. http://southwestcarbonpartnership.org/.

Havstad, K. M., D.P.C. Peters, R. Skaggs, J. Brown, B. T. Bestelmeyer, E. Fedrickson, J. E. Herrick, and J. Wright. 2007. Ecological services to and from rangelands of the United States. Ecological Economics 64: 261–268.

Millennium Ecosystem Assessment. 2003. Ecosystems and Human Well-being: A Framework for Assessment. Washington, DC: Island Press.

Miller, A. J., R. Amundson, I. C. Burke, and C. Yonker. 2004. The effect of climate and cultivation on soil organic C and N. Biogeochemistry 67: 57–72.

Reynolds, J. F., M. Stafford Smith, E. F. Lambin, B. L. Turner II, M. Mortimore, S.P.J. Batterbury, T. E. Downing, H. Dowlatabadi, R. J. Fernández, J. E. Herrick, E. Huber-Sannwald, H. Jiang, R. Leemans, T. Lynam, F. T. Maestre, M. Ayarza, and B. Walker. Global desertification: Building a science for dryland development. Science 316: 847–851.

Safriel, U., and Z. Adeel. 2005. Dryland systems. In R. Hassan and R. Scholes, eds. Millennium Ecosystem Assessment: Current State and Trends Assessment. Washington, DC: Island Press, 625–656.

Sala, O. E., W. K. Lauenroth, and R. A. Golluscio. 1993. Arid and semiarid plant functional types. In T. M. Smith, H. H. Shugart, and F. I. Woodward, eds. Plant Functional Types. Cambridge, UK: Cambridge University Press, 217–233.

Sala, O. E., and J. M. Paruelo. 1997. Ecosystem services in grasslands. In G. C. Daily, ed. Nature's Services: Societal Dependence on Natural Ecosystems. Washington, DC: Island Press, 237–252.

VI.7

Marine Ecosystem Services

Marissa L. Baskett and Benjamin S. Halpern

OUTLINE

Marine ecosystems provide a variety of services: provisioning services such as fisheries, regulating services such as storm protection in coastal regions, supporting services such as primary production that can cross the land–sea boundary, and cultural services such as tourism. Understanding marine ecosystem services requires consideration of the appropriate spatial, temporal, and organizational scales for each service, and both empirical and theoretical investigations provide insight into these services and the relevant management practices. For example, theoretical population ecology has historically informed fisheries management, and disciplines such as community and spatial ecology can inform recent efforts to implement an ecosystem-based approach to managing the use of multiple marine ecosystem services. The impacts from a combination of human activities, from fishing to pollution, have led to declines in many marine ecosystem services. Future research on both the ecological and the socioeconomic aspects of marine ecosystems can guide sustainable management of the use of these services.

GLOSSARY

adaptive management. Dynamic resource management that incorporates new information gathered from scientific monitoring to systematically improve management practices

ecologically sustainable fishery. Fishery regulated to avoid any shift in the ecosystem that leads to an undesirable state, such as collapsed populations of a harvested species

ecosystem-based management (EBM). A holistic approach to resource management aimed at the sustainable delivery of multiple ecosystem services by accounting for the ecological, environmental, and socioeconomic context and explicitly addressing cumulative impacts and trade-offs among the different sectors being managed

maximum sustainable yield (MSY). The largest yield that a fishery can theoretically sustain indefinitely

stock–recruitment relationship. A mathematical description of the number of new recruits to a fishery as a function of the spawning stock size

1. WHAT AND WHERE ARE MARINE ECOSYSTEM SERVICES?

There is an old adage that the oceans are inexhaustible in their ability to provide for humans and absorb our wastes—a proverbial, and simultaneous, supermarket and waste bin. Currently, there is little doubt that both of these beliefs are wrong, but the adage captures a reality about the world's oceans: they provide a vast amount of ecosystem services to humanity. An increasingly large portion of our food comes from the oceans. Globally, about 5% of the protein in people's diet comes from seafood, but this portion is dramatically higher in places such as China, Japan, and Iceland. An estimated 15% of the world's species diversity is in the oceans, and 16 of the world's 35 animal Phyla are found only in the oceans, whereas only one is found exclusively on land. The many services this biodiversity provides include providing biochemical and medical substances for human uses, higher and more stable fisheries catch, and spiritual (aesthetic) and cultural resources. The oceans also act as a massive carbon sink by converting CO_2 into carbonic acid, which in turn slows global climate change (but increases ocean acidification). Specific ecosystems within the oceans, such as coral reefs, mangroves, and salt marshes, play important roles in buffering coastal areas from wave and storm damage. The 2004 Asian tsunami and 2005 Gulf of Mexico hurricanes dramatically illustrated the value of this service in protecting both human lives and property.

In addition to these examples of provisioning and regulating services, the oceans provide critical supporting and cultural services. Salt marshes and shellfish reefs help filter and clean the polluted waters that run off the land; in temperate coastlines, upwelling zones, where nutrient-rich water is pushed from the deep into surface waters, act as major primary productivity pumps that fuel provisioning and regulating services. Such supporting services can even extend to terrestrial ecosystems. For example, salmon mortality after spawning migrations leads to substantial deposition of originally marine nutrients into riparian ecosystems, and seabird nesting sites provide a major source of nitrogen to coastal landscapes through the guano from these fish consumers. Finally, coastal areas are a major source of recreational and aesthetic value to many people. A day spent at the beach, bird-watching in coastal salt marshes, whale-watching tours, and scuba diving in coral reefs are all services that have great intrinsic value and support tourism industries.

Patterns and Scale of Marine Ecosystem Services

Understanding the capacity of the oceans to provide ecosystem services requires an appreciation of the temporal and spatial patterns in marine ecosystems and the human uses of those ecosystems. In particular, not all habitats or locations, or times at those locations, provide equal services. Much of the provisioning and supporting services come from the relatively narrow band of shallow, nearshore ecosystems that are the most productive and diverse regions of the oceans. In contrast, the open ocean is the source of most regulating services (except storm regulation), primarily because of its vast size. Within these broad categories, there is further spatial separation of the delivery of ecosystem services. Upwelling zones, which are primarily in the coastal regions of New Zealand, Chile/Peru, South Africa, and the western United States, are key drivers of ocean productivity, which in turn fuels many fisheries. Mangroves, salt marshes, coral reefs, and kelp forests are patchily distributed; these ecosystems directly supply a variety of critical services and often serve as nursery habitats for commercially important species that are harvested in other locations. Furthermore, the delivery of marine ecosystem services can vary greatly over time. For example, El Niño events, which affect regional climate and ocean productivity, occur on a 4- to 7-year cycle, and nursery habitats often act as such for only a few months of each year, depending on the reproductive cycles of key species. In addition to this ecological variation, human interaction with and use of marine systems can vary greatly over space and time, driven by factors such as human population density and infrastructure availability.

In many cases these temporal and spatial patterns emerge from a variety of processes that act at different scales among ecosystem services and within marine systems. Regulating services, particularly climate regulation, tend to act at large and long scales, such that changes to one small patch of ocean or a single disturbance event are not likely to have a noticeable impact on the delivery of those services. Provisioning services tend to act at smaller scales such as the regional, interannual scale for coastal fishery production and the local, monthly to annual scales for wood production from mangroves. Finally, supporting and cultural services such as the aesthetic or spiritual value of a location act at a wide range of scales, from the local scale and daily basis of ecotourism to the global scale and long-term value of biodiversity and species existence.

Nonlinearities in the spatial and temporal delivery of ecosystem function further complicate the ability of different locations to deliver ecosystem services. Cumulative, and potentially synergistic, interactions among multiple natural and/or anthropogenic disturbances may cause some marine ecosystems to cross a threshold and shift to a different state that does not provide the same services. For example, overharvest of oysters and excessive fertilizer runoff in Chesapeake Bay shifted that system into a highly eutrophic estuary in a way that may have been avoided had only one of those impacts occurred. In addition, the size of a particular patch of ecosystem may affect its resistance and/or resilience to disturbance and thus its ability to sustainably provide a service. For example, a single square meter of salt marsh will have no ability to dampen storm surge, whereas a long stretch of coast covered in square kilometers of salt marsh could effectively stop large waves. These issues of scale and nonlinearities pose significant challenges for developing predictive models to forecast system behavior, models that in turn can be used to guide effective and efficient management and conservation strategies. Fortunately there is a long history of theoretical work focused on particular services from which we can draw lessons learned, inform conservation and management, and identify key information gaps and profitable directions for further research.

2. THEORY AND MANAGEMENT OF ECOSYSTEM SERVICES: EXAMPLE OF MARINE PROVISIONING SERVICES

Theoretical studies provide a powerful tool for exploring potential outcomes from management decisions and potential causes for ecological patterns at temporal and spatial scales that are often difficult to investigate empirically. Theoretical population ecology can be particularly useful in informing the management of provisioning services such as fisheries, and theoretical ecology in general has the potential to provide insight into managing the delivery of marine ecosystem services. In this section, we provide a brief overview of marine resource management theory to illustrate the contributions that ecological theory can make to the study of ecosystem services.

The Origins of Quantitative Fisheries Management

One of the original, fundamental contributions of mathematical modeling to fisheries management is the stock–recruitment relationship, which predicts the number of recruits (R) to a fishery based on its current spawning stock size (S). First developed in the 1950s with the Ricker ($R = aSe^{-bS}$ given constants a and b) and the Beverton-Holt [$R = aS/(b+S)$] stock–recruitment relationships, models such as these can predict the theoretical dynamics of a fished population under different management scenarios and therefore can inform management decisions. For example, one can calculate the harvest rate and stock size that produce the maximum sustainable yield (MSY) for a fishery, or the highest yield that can be continually caught.

Originally viewed as a target, the theoretical construct of MSY now often serves as an upper limit for fisheries management goals. This shift in management's use of theory stems from several theoretical and empirical uncertainties in determining the MSY, such as the large amount of natural variability in fisheries stock–recruitment relationships caused by additional biological complexity unaccounted for in the model, the difficulty in empirically determining the MSY without exceeding it, and additional economic properties such as discounting unaccounted for in the model. When such uncertainties lead to an overestimate of the MSY, and a stock declines or collapses, recovery may be difficult, especially when the system is nonlinear and/or has alternative stable states (e.g., in harvested species that have critical spawning masses required for successful mating). Therefore, with respect to MSY and in general, a precautionary approach that accounts for both theoretical and empirical uncertainties is necessary when applying theoretical predictions to management decisions.

Ecosystem-Based Approaches to Quantitative Marine Resource Management

The dynamics of marine provisioning services are inevitably connected to regulatory, supporting, and cultural services. Therefore, in addition to the shift noted above, marine resource management is moving toward a more holistic approach termed ecosystem-based management (EBM). EBM moves beyond single-species and single-impact management approaches in three key ways. First, EBM focuses on the sustainable delivery of multiple ecosystem services as the primary management goal. Second, recognizing that human activities do not act in isolation but, rather, may interact with each other, EBM accounts for the cumulative impacts of multiple activities. Finally, it explicitly incorporates humans into managing ecosystems and accounts for the interactions both within and between ecological and socioeconomic components. With respect to marine provisioning services, EBM places fisheries management decisions in an ecological, environmental, and socioeconomic context. Although EBM is a relatively new paradigm for ocean resource management, many of the components of EBM have a substantial history in terrestrial and marine conservation that draws from many disciplines, including a rich literature in theoretical ecology. This theory can inform the implementation of marine EBM.

For example, a key component of EBM is moving beyond a single-species approach (such as the original stock–recruitment models) to a multispecies approach. This approach can be as basic as recognizing the potential for by-catch (the accidental catch of unwanted species or individuals) or as complex as accounting for direct and indirect species interactions that influence population sizes throughout an entire food web. A wide range of theoretical approaches to multispecies fisheries exists, from simple multispecies fisheries models, such as a whale–krill model developed by May and others in 1979, to complex ecosystem-scale simulations, such as EcoSim simulations developed by Walters and others (1997) that follow biomass transfer through many community and ecosystem components. Such models have indicated the extent to which negative species interactions (e.g., predation, competition) may decrease sustainable yield and harvest rates. Applying these multispecies models to management practices can promote ecologically sustainable fisheries, or fisheries managed to protect an ecosystem state that includes healthy populations of the target species as well as any directly or indirectly interacting species (e.g., predators such as marine mammals that draw ecotourism, terrestrial species that benefit from the fertilization via anadromous organisms such as salmon). Therefore, this theory can help inform the management practices necessary to protect functional ecosystems that sustainably provide both supporting and provisioning services.

Furthermore, EBM often includes spatially explicit management such as ocean zoning, and ecological theory can readily inform such efforts. From metapopulation to advection–diffusion models, theoretical spatial ecology helps provide the understanding and predictions of movement dynamics that are necessary to make management decisions in a spatial context. For example, the theoretical critical patch size necessary for population persistence given dispersal, originally developed by Skellam (1951) and Kierstead and Slobodkin (1953), can inform the critical size of a marine reserve, or no-take zone, necessary to protect self-sustaining populations. More recently, models specific to marine reserves explore the design of reserve networks, connected by dispersal, necessary to achieve management goals. With respect to provisioning services such as fisheries, these models generally indicate that reserves are most likely to enhance fisheries' yields when they protect intensively harvested species or those with low adult movement rates. Theoretical approaches also show that reserves can help protect against uncertainty in scientific assessment, management enforcement, and naturally variable populations that may lead to overfishing. Although such implications for fisheries management have been the primary focus of marine reserve theory, these models can also help indicate the reserve design most likely to protect ecosystem health and function and, therefore, regulating, supporting, and cultural as well as provisional services within reserves, with the potential for reserve benefits to spill over to unprotected areas.

The multispecies and spatial models described above are just two examples of many cases where theory can inform management strategies, such as EBM, aimed at the sustainable use of marine ecosystem services. For example, a key advance in fisheries science is the development of more detailed models that include habitat considerations, climatic and oceanographic variability, uncertainty, and/or more realistic biological dynamics such as age structure; such models provide sophisticated quantitative tools with which to compare the outcomes of different management decisions. In addition, models built on evolutionary theory help predict the potential consequences of fisheries-based selection (e.g., the preferential removal of larger, more fecund individuals, which alters population dynamics and natural selection) and can inform evolutionarily sustainable fisheries management decisions. Overall, these theoretical contributions can help with understanding and implementing an ecosystem-based approach to management that protects the sustainability of the broad range of marine ecosystem services.

3. CURRENT AND FUTURE TRENDS

Past and Current Changes to the Delivery of Marine Ecosystem Services

Historically humans could not access most of the ocean, and marine ecosystems were capable of providing seemingly inexhaustible services relative to their terrestrial counterparts. Industrialization and rapid modernization in the last century led to the development and deployment of increasingly sophisticated extractive technology, such as large fishing boats that stay at sea for months and fish the deepest parts of the ocean, for garnering the provisioning services of the oceans. Furthermore, a burgeoning human population has been dumping increasing amounts of waste into the oceans, both directly and through runoff. The effects of these multiple human activities have significantly impacted the provisioning, regulating, and supporting services of the oceans. For example, currently nearly a third of all monitored global fish stocks have collapsed from overfishing, which is in turn having ecosystem-wide impacts. Also, humans are degrading and converting nearshore habitats at alarming rates—for instance, 35% of all mangroves have been lost globally, and coastal wetlands have been destroyed at a rate of 1.5% per year for the past several decades. Furthermore, persistent dead zones that cannot provide any services now exist at the mouths of some large river systems as a result of pollution and the subsequent extreme eutrophication. Finally, humans are altering the ocean's ability to regulate climate as anthropogenic degradation decreases its ability to absorb CO_2, which could affect the rate of global climate change. None of these impacts acts in isolation, creating cumulative impacts that have to date been largely ignored or overlooked, and there are likely to be synergistic effects of overlapping human activities that make the whole greater than the sum of the parts.

In summary, human activities are affecting every part of the oceans, and the recent Millennium Ecosystem Assessment suggests that many services have already been markedly impacted by these activities. Without long-term data, it is difficult to fully assess the extent of human effects on marine systems, a situation that is exacerbated by the "shifting baseline syndrome," where each generation has an increasingly degraded perception of what constitutes a natural and healthy ecosystem. For example, as documented by Jackson and others in 2001, in northeastern Pacific kelp forests, before modern commercial fisheries reduced fish and invertebrate populations, anthropogenic activities ranging from subsistence fishing to sea otter harvest for fur trade greatly altered ecosystem structure. Therefore, a historical perspective is necessary to determine appropriate goals for the sustainable delivery of marine ecosystem services.

Toward Sustainable Management of Marine Ecosystem Services

Given the significant anthropogenic impacts on marine ecosystem services, scientifically based management is necessary to ensure the sustainable delivery of marine ecosystem services. The appropriate management tools for the sustainable use of an ecosystem service depend on the anthropogenic impact as well as the spatial, temporal, and ecological scale of the service. For example, in addition to traditional local and state-determined limits on fisheries effort, marine reserves (often implemented at the local or state level) are an increasingly employed management tool on scales relevant to coastal fisheries and the movement patterns of their target species. However, movement scales for wide-ranging species that cross national boundaries (e.g., tuna) may be too great for marine reserves to be an effective management tool; therefore, management of such species relies on tools such as multinational agreements and extraction limits. Similarly, state and national regulation of local point and regional non-point pollution sources may reduce the negative impacts of eutrophication on marine ecosystem services, and multinational agreements on climate change are relevant to the sustainability of regulating services.

Along with combining approaches that control both where (e.g., ocean zoning) and how much (e.g., extractive or input limits) anthropogenic activity occurs at the relevant spatial and temporal scales, any management scheme can draw from both regulatory limits and market incentives, depending on the socioeconomic factors that influence the decisions by the relevant stakeholders. For heavily impacted services, going beyond regulation and incentives with restoration may be a critical component of sustainable management. Therefore, the appropriate balance of approaches depends on both the ecology of the system and the socioeconomic context. Recognition of the interactions between socioeconomic and ecological systems across multiple scales is part of an emerging science that can inform the sustainable management and use of marine ecosystem services.

Future Research Directions

Although existing research has provided key insights into management decisions relevant to marine ecosystem services, many topics remain relatively unexplored and provide promising subjects for current and future research. For example, theoretical models that describe and predict marine ecosystem services other than fisheries, such as nutrient cycling, primary production, climate control, and storm disturbance buffering, are not nearly as well developed. Empirical studies that help understand the ecology and quantify the full value of these services are also greatly needed. Because these different services occur at different spatial, temporal, and organizational scales, novel and innovative research approaches are necessary to increase scientific understanding and inform management.

One key direction for future research is the synthesis of multiple components of EBM into a quantitative framework. Along with models that cross subdisciplinary boundaries within fisheries and ecological theory, interdisciplinary approaches that model coupled social–ecological systems have great potential for contributing to management efforts focused on the sustainable delivery of marine services. Beyond the addition of more sophisticated economic analysis, accounting for socioeconomic dynamics in ecological models can improve our understanding and predictive power; such efforts are a fast-growing field of research in the discipline of natural resource economics. Furthermore, both modeling and empirically testing the trade-offs among multiple ecosystem services that may occur under different management decisions can provide critical insights into marine resource management.

Although empirical research on valuation and trade-offs can inform management goals, the complexity of and predictive power required for managing coupled socioeconomic–ecological systems necessitates a theoretical approach to help understand how to achieve such goals. Improved computational tools allow the incorporation of an increasing amount of biological and socioeconomic realism in modeling efforts, whereas simple, more tractable models remain vital to understanding how key dynamics drive model outcomes. The example of theoretical contributions to fisheries management indicates how theory can provide useful management tools, but any implementation of theory in management practice must recognize critical uncertainties. This inherent uncertainty exists because models are, by definition, simplified representations of a complex biological reality, and increases in model complexity (such as coupling social and ecological dynamics) can lead to increased model uncertainty as a result of the greater number of processes and parameters. To account for these uncertainties and any consequences of management error that may be difficult to reverse, sustainable management requires a precautionary approach that includes dynamic, adaptive management, where new information from scientific monitoring leads to the systematic improvement of management practices. Therefore, the processes of (1) using empirical and theoretical research to inform each other and enhance understanding, (2) building from simple models to large-scale, biologically detailed simulations, and (3) recognizing the importance of uncertainty through precautionary and adaptive management all provide insight into how different dynamics influence marine ecosystem services and how to manage the use of these services in a complex, uncertain world.

FURTHER READING

Botsford, L. W., J. C. Castilla, and C. H. Peterson. 1997. The management of fisheries and marine ecosystems. Science 277: 509–515. *Trends in human impacts on marine ecosystems and recommendations for ecosystem-based management.*

Jackson, Jeremy B. C., Michael X. Kirby, Wolfgang H. Berger, Karen A. Bjorndal, Louis W. Botsford, Bruce J. Bourque, Roger H. Bradbury, Richard Cooke, Jon Erlandson, James A. Estes, Terence P. Hughes, Susan Kidwell, Carina B. Lange, Hunter S. Lenihan, John M. Pandolfi, Charles H. Peterson, Robert S. Steneck, Mia J. Tegner, and Robert R. Warner. 2001. Historical overfishing and the recent collapse of coastal ecosystems. Science 293: 629–638. *Importance of historical perspective in determining the extent of human impacts on multiple marine ecosystems.*

Peterson, Charles H., and Jane Lubchenco. 1997. Marine ecosystem services. In G. C. Daily, ed. Nature's Services: Societal Dependence on Natural Ecosystems. Washington, DC: Island Press, 177–194. *Detailed discussion of anthropogenic threats to various marine ecosystem services.*

Pew Oceans Commission. Leon E. Panetta, Chair. 2003. America's living oceans: Charting a course for sea change. A report to the nation: Recommendations for a new ocean policy. Arlington, VA: Pew Oceans Commission. Available at: http://www.pewoceans.org/oceans/downloads/oceans_report.pdf. *Comprehensive recommendations for U.S. marine management policy.*

Rose, Kenneth A., and James H. Cowan, Jr. 2003. Data, models, and decisions in U.S. marine fisheries management: Lessons for ecologists. Annual Review of Ecology, Evolution, and Systematics 34: 127–151. *Overview of the current use of ecological theory and data in fisheries management.*

Worm, B., E. B. Barbier, N. Beaumont, J. E. Duffy, C. Folke, B. S. Halpern, J.B.C. Jackson, H. K. Lotze, F. Micheli, S. R. Palumbi, E. Sala, K. A. Selkoe, J. J. Stachowicz, and R.Watson. 2006. Impacts of biodiversity loss on ocean ecosystem services. Science 314: 787–790. *Global synthesis of the effect of marine biodiversity on the delivery of a variety of ecosystem services.*

Zabel, Richard W., Chirs J. Harvey, Stephen L. Katz, Thomas P. Good, and Phillip S. Levin. 2003. Ecologically sustainable yield. American Scientist 91: 150–157. *Overview of ecosystem-based management.*

See also chapters II.4, III.13, IV.10, V.3, V.5, VI.1, VI.11, VI.13, VII.2, and VII.9.

VI.8

Provisioning Services: A Focus on Fresh Water

Margaret A. Palmer and David C. Richardson

OUTLINE

1. Introduction
2. Freshwater ecosystem services and the processes that support them
3. Status of freshwater ecosystem services
4. The future of freshwater services

Healthy freshwater ecosystems play crucial roles in the global environment by controlling fluxes of minerals, nutrients, and energy, and, by providing goods and services critical to humans including water for drinking or irrigation and fish for consumption. Freshwater ecosystems also provide regulating services such as carbon sequestration, flood control, and cultural services such as recreational fishing, swimming, or aesthetic enjoyment of the open water. These goods and services are all supported by underlying ecological processes (also called ecosystem functions) such as primary production, decomposition, and nutrient processing. The well-known and dramatic decline in freshwater biodiversity that has occurred in the last several decades has been accompanied by local and regional losses of freshwater ecosystem services. These losses are being driven largely by human activities. Ecosystem services cannot be restored once lost without a focus on the underlying ecological processes that support them, and, thus, a great deal of research is ongoing to understand and quantify the linkage between services and the rates of key ecological processes.

GLOSSARY

anaerobic. Absence of oxygen, also called anoxic (e.g., anaerobic sediments).

chemotroph. An organism that makes its own food but, instead of using energy from the sun as photosynthetic organisms do, uses inorganic chemicals as an energy source; includes wetland bacteria called methanogens that produce methane (a greenhouse gas) by decomposing organic matter in anaerobic environments.

denitrification. The microbial process that converts nitrate (NO_3^-, nutrient readily available to plants) to nitrite to free nitrogen gas (N_2, generally unavailable to plants); requires a carbon source and an anaerobic environment.

geomorphology. The study of the formation, alteration, and configuration of landforms and their relationship with underlying structures.

hydrology. The study of the properties, distribution, and effects of water on the earth's surface.

hyporheic zone. The subsurface region under and lateral to a stream in which groundwater and surface water mix; considered metabolically important in streams and rivers.

organic matter, particulate and dissolved. Derived from the degradation of dead organisms, plant or animal; particulate organic matter would include leaf pieces, wood, animal body parts, etc.; dissolved organic matter refers to organic molecules that are typically less than 0.7 μm; also called dissolved organic carbon.

point- / non-point-source pollution. Point-source pollution comes from clearly identifiable local sources, includes outlet pipes from wastewater treatment plants or other industrial sources. Non-point-source pollution comes from many diffuse sources and is carried by rainfall or snowmelt as it moves over or through the ground to fresh water. These pollutants include excess fertilizers, herbicides from agricultural or residential areas, oils or other toxic chemicals from urban runoff, salt from roads or irrigation practices, bacteria or nutrients from livestock, pet waste, or pollutants from atmospheric deposition.

primary and secondary production. The production of new living material through photosynthesis by autotrophs (e.g., plants, algae) is primary production; tissue produced by heterotrophs (e.g., macrofauna, fish) is referred to as secondary production because these organisms rely on the consumption of living or dead organic material.

recharge/discharge. Movement from surface water belowground into an aquifer "recharges" the aquifer, whereas movement from the groundwater back to surface water represents discharge from an aquifer.

1. INTRODUCTION

Only about 3% of the world's water is fresh water, and most of that is bound up in glaciers, underground aquifers, or ice pack. Yet the entire human population depends on fresh water for drinking and on the goods and services provided by freshwater ecosystems. In fact, since antiquity, humans have chosen to live and work near water bodies, and entire civilizations have developed along waterways. Today, most people rely on rivers and streams for their domestic water needs as well as for irrigation, energy, and recreation. They rely on wetlands and riparian buffers to purify water, mitigate the impacts of flooding, and support diverse assemblages of plants and wildlife. Healthy freshwater ecosystems also play crucial ecological roles globally by controlling fluxes of minerals, nutrients, and energy. Indeed, all ecosystems worldwide depend to some extent on freshwater ecosystems and the complex connections that exist among terrestrial flora and fauna, groundwater, surface waters, and water vapor (plate 20). Biodiversity and ecosystem processes in terrestrial, polar, and coastal ecosystems are all influenced by inputs of fresh water and fluxes of organic matter and other materials from rivers and streams.

As outlined in earlier chapters in this section (chapters VI.1 and VI.2), ecosystem services can be categorized as provisioning, regulating, or cultural. Provisioning services are those "products" obtained from ecosystems such as fish for consumption or water for drinking or irrigation. Regulating services include nonmaterial benefits that humans receive from ecosystems such as water purification, carbon sequestration, or flood control. Cultural services represent nonmaterial benefits as well, such as recreational fishing, swimming, or aesthetic enjoyment of the open water, but we choose to focus on provisioning and regulating services in this chapter. As we describe below, each provisioning and regulating service is supported by underlying ecological processes (figure 1). In some cases, just a few processes may support a service, and in other cases, a whole suite of complex processes interact to provide the basis for a service. For example, water purification may rely on the ecological processes of denitrification, decomposition of organic matter, and algal photosynthesis, whereas riverine flood control may depend almost exclusively on the presence of healthy (intact) floodplains. Species also rely on ecosystem services to provide them with food, optimal conditions for reproduction, and dispersal routes, to name a few.

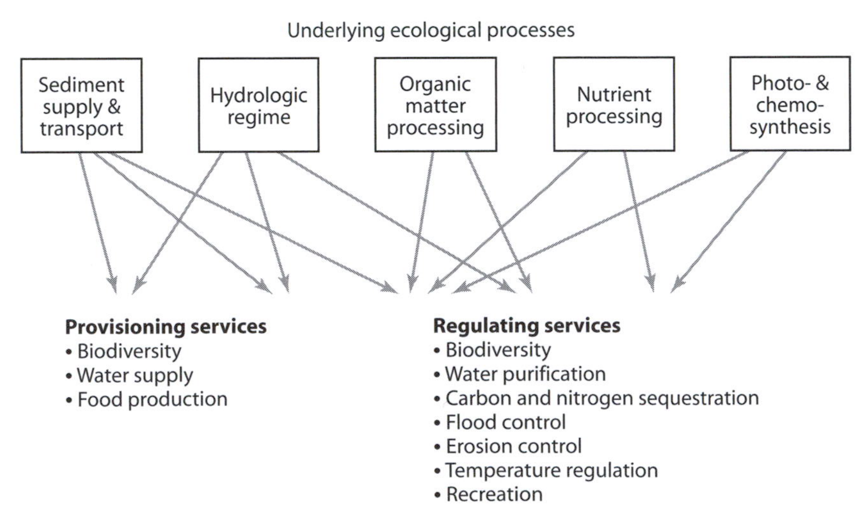

Figure 1. Examples of basic ecological processes that support the services provided by freshwater ecosystems. Subcategories of processes (e.g., denitrification as one component of nutrient processing) are described in table 1. Provisioning services are products obtained from ecosystems, whereas regulating services include nonmaterial benefits.

Because freshwater ecosystems are extremely diverse, as are the services they support, we begin with a brief description of the major types of ecosystems and then move into a detailed discussion of the services they provide.

- *Wetland ecosystems:* Any ecosystem that is regularly, but not necessarily continuously, saturated by precipitation, surface water, or groundwater and is occupied by vegetation that is adapted to saturated conditions; includes bogs, swamps, fens, and tidal or nontidal freshwater marshes.
- *Running-water ecosystems:* Any ecosystem with flowing water that is a low point in the landscape where water drains, especially after rain; perennial streams flow most of the year through well-defined channels; intermittent streams flow only part of the year; and ephemeral streams flow only after major rain events; however, hyporheic flow, which is not readily visible, may persist. Running-water ecosystems include rivers, streams, creeks, and brooks.
- *Lake, pond, and reservoir ecosystems:* Any body of water filling a depression in the landscape that may have been formed in a variety of ways including glacial retreat, tectonic events, river overflows or meanders, and natural damming of running waters (e.g., beaver ponds); their outflow may feed streams and rivers.
- *Glaciers and ice pack ecosystems:* Any body of ice on a mountaintop or in polar regions where snow accumulation exceeds or equals melting; these are slowly moving or have moved at one point in time; once believed to be barren but now known to harbor diverse flora and fauna.

2. FRESHWATER ECOSYSTEM SERVICES AND THE PROCESSES THAT SUPPORT THEM

Healthy fresh waters are living, functional systems. Valuable ecosystem services that they support include water storage and supply, the purification of water through the removal of excessive nutrients, contaminants, and sediments, carbon and nitrogen sequestration, food production in the form of invertebrates and fish, support of biodiversity in aquatic and nearby terrestrial habitats, flood control, and recreation (table 1). Each service is supported by one or more ecological processes (figure 1) that, if lost, could lead to degradation of that service. Societal preferences can lead to choices that enhance one service at the loss or decline of another, and this is often accomplished by influencing the underlying process. For example, if people want abundant trout fisheries in a region, they may stock rivers with hatchery-reared juveniles (effectively enhancing the "fish reproduction/recruitment process"), which may in turn lead to a decline in the productivity of other fish species because of competition for food. Even the act of meeting basic human needs such as providing warmth for people in cold winter months can enhance the provisioning of one service (heat production from burning wood) at the expense of another (carbon sequestration in living trees that provide wood resource).

There is a great deal of active research now to identify ways to measure ecosystem services by quantifying their linkage to the rates of the underlying ecological processes. The motivation to measure ecosystem services comes in part from recognizing that in any coupled social–ecological system, there will always be trade-offs that need to be balanced, with potential needs for compensation if something valuable is lost. Take, for example, extractive mining that can reduce water quality. Communities may find they need to engage in extractive mining as a source of income; they may partially compensate for the degradation in water quality at that site by enhancing water purification at other points in the watershed. A second example might be reforestation around housing developments; the forests lost when houses were built are reestablished to provide opportunities for recreation. No single patch can provide the full suite of services residents might want, but landscape-level management can provide a bundle of services that contribute to human well-being.

The motivation to link the rates of ecological processes to ecosystem services follows directly from the need or desire to enhance or restore a service that has been lost or degraded. Services cannot be restored without a focus on the underlying ecological processes that support them. For example, denitrification, an ecological process that contributes to the ecosystem service of water purification, occurs in many freshwater ecosystems (table 1), but knowledge of the rate of nitrogen removal is necessary to estimate whether denitrification improves water quality. Many surrogate metrics for system performance have been used in assessments of freshwater ecosystems (e.g., plant biomass, sediment carbon content); however, to date, there is no published work in which such surrogates have been directly linked to the *rate* of a specific ecological process and, therefore, to the magnitude of impact on ecosystem service. Thus, an extremely active area of research is now concerned with how to quantify these services, the spatial and temporal variability of these services across and within different biomes, and the potential use of valid surrogates for measuring the underlying process rates that are less

Table 1. Rivers and streams provide a number of goods and services that are critical to their health and provide benefits to society; the major services are outlined along with the ecological processes that support the function, how it is measured, and why it is important.

Ecosystem service	*Consequences of losing the service*	*Supporting ecological process*	*Measurements required*	*Ecosystem/ habitat*
Water purification				
Nutrient processing	Excess nutrients (eutrophication) can build up in the water, making it unsuitable for drinking or supporting life; in particular, algal blooms resulting from excess nutrients can lead to anoxic conditions and death of biota	Retention, storage, and transformation of excess nitrogen and phosphorus; decomposition of organic matter	Direct measures of rates of transformation of nutrients; e.g., denitrification (production of N_2 gas, conversion of NO_3 to more usable N forms); decomposition measured as rate of organic matter loss over time	Riparian zone, river and streambeds, wetlands, lake littoral zones
Processing of contaminants	Toxic contaminants kill biota; excess sediments smother invertebrates, foul the gills of fish, etc.; water not potable	Biological removal by plants and microbes of materials such as excess sediments, heavy metals, contaminants, etc.	Direct measures of contaminant uptake or changes in contaminant flux	Riparian zone and wetland soils and plants; bottom sediments of rivers, lakes, and wetlands
Water supply	Loss of clean water supply for residential, commercial, and urban use, irrigation supply for agriculture	Transport of clean water throughout watersheds	Measures of water movement, flow patterns, pollution load	Lakes, rivers, streams
Flood control	Without the benefits of floodplains, healthy stream corridor, and watershed vegetation, increased flood frequency and flood magnitude	Slowing of water flow from land to freshwater body so flood frequency and magnitude reduced; intact floodplains and riparian vegetation buffer increases in discharge	Measure stream and river discharge responses to rain events	Floodplains, wetlands, riparian zones
Infiltration	Lost groundwater storage for private and public use; vegetation and soil biota suffer; increased flooding in streams	Intact floodplain, riparian, wetland vegetation increase infiltration of rainwater and increase aquifer recharge	Infiltration of water in soils, water table levels in deep and shallow wells	Wetlands, streams, floodplains

Carbon sequestration				
Primary production	Water and atmospheric levels of CO_2 build up, contributing to global warming	Aquatic plants and algae remove CO_2 from the water or atmosphere, convert this into biomass, thereby storing carbon	Measure rate of photosynthesis typically based on loss of CO_2 or production of O_2	Freshwater ecosystems with sunlight but particularly shallow water habitats such as wetlands or midorder streams
Secondary production	Water and atmospheric levels of CO_2 build up, contributing to global warming	Production of biomass by microbes and metazoans stores carbon until their death	Biomass accumulation over time or measures of consumption, decomposition, or chemotrophy	All freshwater ecosystems but particularly the bottom sediments for microbes
Nitrogen sequestration				
Primary production and secondary production	Secondary production supports fish and wildlife	Creation of plant or animal tissue over time	For primary production, measure the rate of photosynthesis in the stream; for secondary, measure growth rate of organisms	All freshwater ecosystems and habitats
Food production				
Primary production	Reduction in food and food products derived from aquatic plants such as algae, rice, watercress, etc. Decreased production (secondary) by those consumers who rely on primary production as a food source	Production of new plant tissue	Measure rate of photosynthesis typically based on loss of CO_2 or production of O_2	All freshwater ecosystems and habitats with sunlight but particularly shallow water habitats such as wetlands
Secondary production	Reduction in fisheries including finfish, crustaceans, shellfish, and other invertebrates	Production of new animal tissue or microbial biomass	Measure biomass changes over time or use measures of fisheries harvest	All freshwater ecosystems and habitats but particularly the water column and surficial sediments
Biodiversity	Loss of aesthetic features, impacts aquarium trade, potential destabilization of food web, loss of keystone species can impact water quality	Diverse freshwater habitats, watersheds in native vegetation, complex ecological communities support multiple trophic levels	Identification of species, functional groups or food webs	All ecosystem and habitat types but particularly wetlands for plants and rivers for fish

(*Continued*)

Table 1. *(cont.)*

Ecosystem service	*Consequences of losing the service*	*Supporting ecological process*	*Measurements required*	*Ecosystem/ habitat*
Temperature regulation	If infiltration or shading is reduced (as a result of clearing of vegetation along stream), stream water heats up beyond what biota are capable of tolerating	Water temperature is "buffered" if there is sufficient soil infiltration in the watershed; shading vegetation keeps the water cool; water has a high heat capacity which stores excess heat	Measure the rate of change in water temperature as air temperature changes or as increases in discharge occur	Shallow water habitats, especially wetlands
Erosion/sediment control	Aquatic habitat burial impacts fisheries, decreases biodiversity, increases contaminant transport; reduction in downstream lake or reservoir storage volume	Intact riparian vegetation and minimization of overland flow	Flood-related sediment movement	Wetlands, streams, and rivers
Recreation/tourism/ cultural, religious, or inspirational values	Lost opportunities for people to relax, spend time with family; economic losses to various industries, particularly tourist-oriented ones	Clean water, particularly water bodies with pleasant natural surroundings such as forests, natural wildlife refuges, or natural wonders	Time spent using resource recreationally; revenue generated from boats, tourist attractions, hotels	Lakes, rivers, streams

time consuming than direct measurements of the process rates themselves.

Most of the ecological processes that support freshwater ecosystem services are themselves influenced by underlying hydrologic and geomorphic processes such as the flux of water and the supply (and transport) of sediment. There is now abundant scientific evidence that hydrologic interactions including groundwater/surface water interactions, the timing of low and high flows (or water cover), and the magnitude of flows in running-water systems or the period of inundation drive many ecological and biogeochemical processes.

3. STATUS OF FRESHWATER ECOSYSTEM SERVICES

Freshwater ecosystems are among the most impacted ecosystems on earth. In most industrialized countries in the world, extensive loss of wetlands and riparian ecosystems has already occurred, and the remaining habitats continue to be under threat. Degradation of lakes and streams is also extremely common worldwide. The U.S. Environmental Protection Agency reported in 2000 that more than one-third of fresh water in the United States is officially listed as impaired or polluted. Worldwide, over half of all wetlands have been altered, and over 1.3 billion people lack access to an adequate supply of safe water.

Major pollutants in freshwater ecosystems include excessive sediment, fertilizers, herbicides, pesticides, harmful pathogens, and industrial by-products such as heavy metals or PCBs. Agriculture is the source of 60% of all pollution in U.S. lakes and rivers, and in Europe, municipal and industrial sources contribute pollutant loads to lakes and rivers. In developing countries, industrialization and population growth will yield increasing pollution loads from both agricultural and industrial sources, but there are few data and little monitoring from those freshwater ecosystems. Excessive nutrients (e.g., nitrogen and phosphorus needed for plant growth) are the leading pollution problem for lakes and the third most important pollution source for rivers in the United States. The nutrient load generates algal blooms, which can spoil drinking water, make recreational areas unpleasant, and contain harmful toxins or pathogens. The eventual decomposition of the algal bloom creates anaerobic conditions in the water that can kill fish and other aquatic organisms. Heavy metals and other industrial by-products can enter the aquatic food web and accumulate in organisms, including those harvested for human consumption. According to the U.S. Environmental Protection Agency, statewide advisories for freshwater fish warn of concentrations of pollutants, namely mercury and PCBs, at levels of human health concern.

Climate change is bringing major challenges in some parts of the world as the freshwater ecosystem services are being impacted by increasing water temperatures, more precipitation in some areas, more droughts in other areas, and in many regions more intense storms. In arid regions, the extraction of surface water and groundwater is so severe that some major rivers no longer flow to the sea year round, and local communities regularly experience water shortages. Today, the Colorado River is often a dry stream at its mouth in Mexico because of damming, irrigation, and use for drinking water in the southwestern United States. The Yellow River, in China, now ends hundreds of miles from its historical mouth because of the diversion of water for irrigation and drinking. People have been forced to move from the watersheds because of periods of drought and flash floods. In wet regions, rivers and wetlands have a natural ability to absorb disturbances such as those associated with floods, but this buffering has been lost in many areas because of the diversion of surface water and development in the watershed. In urban watersheds, removing vegetation and soil and replacing them with impervious surfaces lead to higher peak discharge and greater volume and frequency of floods than in rural or forested streams.

Freshwater ecosystems near developed (urban, suburban, or residential) or agricultural areas have lost many ecosystem services because of runoff from the land, storm drains, sewers, and municipal point sources. Urban areas occupy only a small fraction of the U.S. land base, but the intensity of their impacts on local rivers can match that of agriculture. The impervious parking areas and rooftops that are associated with urban centers, and poorly managed or tiled agricultural fields, have reduced infiltration capacity and lowered aquifer recharge. This, along with channelization of stream networks, has led to lower water tables and lake levels and to more intense flooding in streams. Wetlands and all associated services are often lost entirely because wetlands are being drained and converted to agriculture, residential, or urban lands, lost as a result of road or highway construction, or stripped to harvest construction materials.

Dudgeon and colleagues grouped the major stressors on freshwater ecosystems into five categories (figure 2), and it is now well accepted that the combination of these stressors has led to very high extinction rates for freshwater species. In fact, extinction rates of freshwater fauna are estimated to be at least five times higher than those of terrestrial or avian species. Around the world, we are losing fish, amphibian, and macroinvertebrate species at alarming rates. For example, overfishing, dam construction, water diversion, and pollution have led to the severe decline of commercial and recreational fish

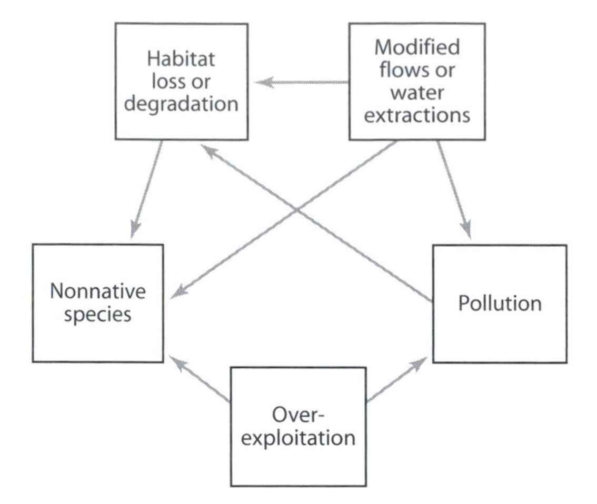

Figure 2. Five major categories of impacts that threaten freshwater biodiversity and ultimately other ecosystem services. (Modified from Dudgeon et al., 2006)

species, such as the American shad in the Hudson and Delaware rivers or Chinese paddlefish in the Yangtze River.

4. THE FUTURE OF FRESHWATER SERVICES

Among the greatest challenges that face many nations worldwide are how to sustain, restore, or judiciously manage the services that freshwater ecosystems provide and how to meet human demand for fresh water. Practices that ensure enough water for drinking and irrigation have received the most attention. In some rapidly developing countries, such as India and China, industrial water demands are beginning to increase, and agricultural and domestic needs from growing populations continue to stress water resources. These problems will be exacerbated in areas that are expected to have less water with global climate change, such as southern Africa, central and northern South America, the Middle East, and parts of China.

Urbanization and climate change, both increasing with population growth, can have negative synergistic effects; freshwater ecosystems, particularly in urban watersheds that receive more precipitation or have more intense storms, will be hard pressed to provide ecosystem services. In areas with snowpack, earlier snowmelt caused by warmer temperatures may lead to massive flooding if the melting coincides with late winter or early spring rains. In these regions, erosion of stream banks and loss of entire wetlands may occur. Lakes may experience greater stress from sediment inputs, and, if they are in urban watersheds, contaminant inputs may also increase following rains. Summer flows to rivers, wetlands, and lakes that are typically sustained by an extended period of snowmelt will be reduced.

The management responses that are needed to protect freshwater ecosystems vary depending on water stress (use-to-availability ratio), land use, how extensively runoff events change relative to current patterns, and needs of people within the watershed. In general, society will have to continue to stress many freshwater ecosystems in order to provide water for irrigation and urban populations, among other things. In fact, some ecosystem services will continue to be lost because a positive benefit is derived from the activities introducing the stress, such as water diversion for human use and crops, infrastructure in the watershed that provides housing or industry, or energy from hydroelectric dams. As indicated earlier, other freshwater systems may need to be preserved or restored to provide water purification, habitat for important aquatic species, or opportunities for recreation. A balance has to be achieved between those services humans receive from converting or stressing freshwater systems and the benefits they receive when those systems are left relatively intact.

Management decisions need to take into account the full range of costs and benefits of provisioning, regulating, and cultural services. Provisioning services, such as fisheries harvest, have a market value and identifiable profits and tend to drive decisions about converting freshwater systems. However, regulating and cultural services are often overlooked because their values are harder to define with no explicit market and diffuse societal benefits. Although a return to pristine, precolonial freshwater ecosystems is clearly unrealistic in the world today, the best management decisions will be made when there is an adequate understanding of all ecosystem services and estimations of their material and nonmaterial values.

When the goal is to restore or protect a freshwater ecosystem, a sensible approach is to take proactive measures to enhance or restore resilience and resistance because this approach may also lead to environmental benefits such as increased water quality and restored fish and plant populations. Palmer and colleagues have outlined examples of such measures including stormwater management in developed basins or, even better, land acquisition around freshwater bodies and riparian corridors to eliminate infrastructure in wetlands and floodplains and allow regrowth of vegetation. If such actions are not taken, we will be left reacting to damages and loss of ecosystem services. Some services may be replaceable by technology (e.g., water purification facilities, levees), but such measures could be extremely expensive and only tenable on fairly small scales.

Protection of freshwater ecosystem services remains a major challenge because it requires preservation, conservation, and management of freshwater ecosystems and their surrounding watersheds. Future research is needed to understand rates and variability of ecosystem processes and how those processes influence ecosystem services. These measurements will allow scientists to make better forecasts about how freshwater ecosystem services will change with global climate change, modifications in land use, and other future stresses. Finally, the scientific research community; public stakeholders; and management at local, state, and national levels must work together to ensure the sustainability of freshwater ecosystem services.

FURTHER READING

Allan, J. D., and M. M. Castillo. 2007. Stream Ecology, 2nd ed. Berlin: Springer.

Baron, J. S., N. L. Poff, P. L. Angermeier, C. N. Dahm, P. H. Gleick, N.G. Hairston Jr., R. B. Jackson, C. A. Johnston, B. D. Richter, and A. D. Steinman. 2002. Meeting ecological and societal needs for fresh water. Ecological Applications 12: 1247–1260

Dudgeon, D., A. Arthington, M. O. Gessner, Z. Kawabata, D. J. Knowler, C. Lévêque, R. J. Naiman, A. Prieur-Richard, D. Soto, M.L.J. Stiassny, and C. A. Sullivan. 2006. Freshwater biodiversity: Importance, threats, status and conservation challenges. Biological Reviews 81: 163–182.

EPA. 2000. National water quality inventory. EPA Publication 841-R-02-001, Washington, DC: EPA.

Mitsch, W. 2007. Wetlands. 4th ed. New York: John Wiley & Sons.

Naiman, R., H. Decamps, and M. E. McClain. 2005. Riparia. San Diego, CA: Elsevier Academic Press.

Wetzel, R. G. 2001. Lake and River Ecosystems. San Diego: Academic Press.

VI.9

Regulating Services: A Focus on Disease Regulation

Peter Daszak and A. Marm Kilpatrick

OUTLINE

1. Infectious diseases as disrupters of ecosystem services
2. Diseases as providers of regulating ecosystem services
3. Ecosystem regulation of infectious diseases
4. Valuing the economic impact of pathogens and their ecosystem services

Over the past few decades there has been an explosion of interest in the ecology of infectious diseases and their roles in ecosystem function. Many studies have focused on the dynamics of pathogens within human, other animal, and plant populations and their role in causing mass mortality events and population declines. Other researchers have focused on diseases that are increasing in incidence or geographic or host range—"emerging" infectious diseases—of humans, wildlife, and plants.

However, relatively few researchers have approached disease ecology in the context of ecosystem services, where diseases, parasites, or pathogens perform functions potentially useful to humans. In this chapter, we review the literature on three aspects of parasites and pathogens in the field of ecosystem services. The first is probably the most well known: their role in morbidity and mortality to their hosts, through which they disrupt the host's ability to provide an ecosystem service, i.e., disrupting the survival or life-history success of ecosystem service providers. The second is the role of pathogens as regulating service providers by suppressing populations of pest species, resisting invasion, and acting as biocontrol agents. The third aspect is poorly understood but a subject of growing interest: rather than parasites performing the service per se, it is the role of species, communities, and biodiversity in regulating the risk of infectious diseases to people, i.e., performing a regulating service, that reduces disease risk. The thrust of our chapter is to review the state of the field, and we have paid particular attention to highlighting those areas where future research is likely to be most fruitful and to identifying strategies to take the field forward. We have therefore added a fourth section that discusses efforts and strategies to estimate the value of pathogens and the cost of their impact on natural capital and ecosystem services.

GLOSSARY

density dependent. A density-dependent process varies with the population density of the species concerned. For instance, below a certain host population size, parasitic infections may not occur (there are not enough hosts for the parasite to be transmitted between them), whereas above a certain host population size, parasitic infections may become prevalent. The probability of any individual host getting infected depends on the density of surrounding hosts.

emerging infectious disease. A disease that has recently and significantly increased in impact, in the number of cases it causes, or in its geographic range; a disease that is caused by a newly evolved pathogen or has recently been transmitted from one species to another to result in an outbreak in the new host species.

parasite. An organism that resides within or on, and is nutritionally dependent on, another organism. In this article, we include all forms of infectious microbes, including viruses, prokaryotes (e.g., bacteria), and eukaryotic parasites (e.g., roundworms).

pathogen. An infectious agent or parasite that causes illness in its host, usually defined as clinical illness, i.e., causing significant pathology or damaging physiological change.

1. INFECTIOUS DISEASES AS DISRUPTERS OF ECOSYSTEM SERVICES

Infectious diseases have been reported to be the cause of morbidity and mortality in a range of key ecosystem service providers (ESPs) (table 1). In these cases, pathogens act as "mediators" of the loss of ecosystem services and effectively perform an "ecosystem disservice." The impact of pathogens is greatest when they cause population declines of keystone species or ecosystem engineers. For example, the death of one-third of the Serengeti lion population caused by canine distemper (a disease introduced via domestic dogs) had a disproportionate impact on the ecosystem. Pathogens sometimes spread rapidly through highly susceptible host populations, which include host populations that have not evolved in the presence of the pathogen. Introduced pathogens may also impact abundant species at lower trophic levels and have similarly dramatic effects on ecosystem services.

Human attempts to control or manage diseases can also have unanticipated ancillary impacts (either positive or negative) on ecosystem services. The introduction of rinderpest into Somalia in 1889 with imported domestic cattle led to a pan-African outbreak and widespread loss of livelihood as it caused the death of millions of domestic and wild ungulates and ecosystem collapse over large areas. But it also provides an example of positive ancillary impacts: the widespread removal of cattle in many regions also removed a major host for tsetse flies (*Glossina* spp., vectors for trypanosomiases including African sleeping sickness). This opened areas for productive human activities that would otherwise have been endemic zones for disease. Examples of human response with negative ancillary impacts are often found when the presence of dangerous pathogens in wildlife reservoirs leads to calls for culling, reducing, or eliminating wildlife and, thus, the positive services the wildlife might provide. For example, the presence of rabies virus in vampire bats (which feed on people as well as cattle and other livestock in Latin America) has led to indiscriminate culling of wild bats and to population declines of bat species that control agricultural pests and pollinate fruiting trees. Similarly, the controversial culling of badgers in the United Kingdom to reduce the risk of transmission of bovine TB (which they carry) to cattle reduces the population of a keystone species, which is also a subject of much cultural and ethical value. Ironically, culling in this case may increase disease incidence in surrounding areas as a result of increased movement by badgers following culling. Thus, both the direct effects of disease on wildlife populations and the impacts of human attempts to control the spread of the disease within the wildlife population or to new hosts can create ecosystem services or disservices. The complexity of disease and host interactions is a running theme of this chapter and makes the ecosystem service role of pathogens difficult to assess and anthropogenic impacts difficult to measure. For example, the introduction of West Nile virus into the United States in 1999 has led to increased use of insecticides, larvicides, and other control activities with unknown, and likely complex, impacts on ecosystem services.

2. DISEASES AS PROVIDERS OF REGULATING ECOSYSTEM SERVICES

A number of studies have demonstrated the ability of some communities to resist invasion—a regulating service that can be a function of diversity or species composition. Can this be extrapolated to infectious diseases within a host? Here, the presence of a pathogen or community of pathogen species could act to resist invasion of an introduced related pathogen. There is a growing literature on competitive or facilitative interactions in parasite coinfections, and evidence indicates that pathogen and parasite interactions play significant roles in host–parasite ecology, prevalence of infections, and impacts on hosts. The role of parasites in invasion resistance has been hypothesized for the poultry bacterium *Salmonella gallinarum*, which has been largely lost from domestic chickens following the routine prophylactic use of antibiotics to combat other ubiquitous poultry pathogens. The most common *Salmonella* sp. in domestic chickens in developed countries is now *Salmonella enteritidis*, a mouse microbe that appears to have jumped host in the absence of the chicken endemic *S. gallinarum*. Both pathogens share the same epithelial cell receptors, and it is hypothesized that the presence of the former prevented the latter from emerging—a population-scale ecosystem service. At a regional scale, it has been hypothesized that the presence of endemic flaviviruses in Central and South America may act to dampen the impact of West Nile virus (another flavivirus) in the region through cross-immunity or evolutionarily acquired resistance.

Pathogens also provide an ecosystem service by naturally suppressing pest species and through their use in the development of biotechnological tools to deal with pests or other pathogens. The most important example of the latter may be seen when pathogens have been used in or proposed as biological control agents. This has been reviewed widely in the literature, and notable examples include the use of *Myxoma* virus and rabbit calicivirus disease to control introduced rabbit

Table 1. Diseases as disrupters of ecosystem services

Disease	*Pathogen*	*Host*	*Ecosystem service*	*Impact on ESP*
Colony collapse disorder	Israeli acute paralysis virus (IAPV) in consort with other pathogens or factors	Honeybees	Pollination	Large-scale loss of populations across United States
Tasmanian devil facial tumor disease	Infectious clonal tumor	Tasmanian devil	Pest control, scavenging	Large-scale population declines
West Nile virus encephalitis	West Nile virus	Vertebrates	Seed dispersal, insect pest control, scavenging	Large-scale population declines
Amphibian chytridiomycosis	Fungus *Batrachochytrium dendrobatidis*	Amphibia	Pest control, water quality control (larvae feeding on algae), genetic resources of biomedical potential (e.g., gastric brooding frogs, dendrobatid frogs)	Large-scale declines globally, species extinctions (up to a quarter of the genus *Atelopus*)
Chestnut blight	Fungus *Cryphonectria parasitica*	American chestnut	Primary productivity, disturbance regulation	Effective extinction of codominant species in United States
Pilchard herpesvirus disease	Pilchard herpesvirus	Pilchard	Secondary productivity	Large-scale mortality events
Ebola virus disease	Ebola virus	Nonhuman primates	Ecotourism, harvesting for food	Large-scale mortality of gorillas and other nonhuman primates in Africa
Avian influenza	H5N1 avian influenza	Waterbirds, poultry	Food	High mortality rates in outbreaks in Asia, Africa, and Europe
Rabies	Rabies virus	Range of wild carnivores, bats, people, other vertebrates	Pest removal, pollination and others	Wild animals culled in response to disease risk, some species inadvertently removed when targeting reservoirs

populations, *Bacillus thuringiensis* genes to control insect pests of agricultural crops, and *Bacillus* spp. toxins to kill mosquito larvae. The challenge for this, and other efforts in biocontrol, is to achieve the desired goal (often suppression of populations of one species or group of species) without substantial impacts on non-target species.

Pathogens also perform this function naturally, through density-dependent reduction of host population growth. Where anthropogenic activities increase the population density of pest species, pathogens that are transmitted in a density-dependent manner tend to infect more hosts, cause more morbidity and mortality, and ultimately reduce the population growth of the host.

When pathogens emerge from wildlife into people, we respond as a species with high-tech strategies such as vaccines and drugs. Many of these vaccines and drugs are based on lab tests developed from other recently isolated pathogens; e.g., the serological tests for Nipah virus were originally based on antibodies to a related virus, Hendra virus; likewise, tests for H5N1 avian influenza originally relied to a large extent on genetic information and research on other strains of influenza. Thus, once a pathogen has emerged, other related pathogens perform an ecosystem service in supplying genes and reagents to be used in biotechnology. The rapid growth of biotechnological tools over the past few decades suggests that this is likely to be an important aspect of pathogen ecosystem services that is currently underestimated.

Pathogens as Providers of a Supporting Service: Maintenance of Biodiversity

Pathogens may act as ecosystem service providers by natural suppression of their hosts, which may be community dominants, pest species, or introduced species. The role of pathogens in maintaining the diversity of communities is becoming increasingly evident and indicates that pathogens are providers of a supporting ecosystem service: maintenance of biodiversity. Pathogens may suppress community dominants and thereby provide frequency-dependent selection between species, especially when pathogens are density dependent. More recently, it has been suggested that the absence of pathogens from their native ranges has facilitated the invasion of introduced species of both plants and animals. Similarly if pathogens are fundamentally involved in driving the evolution of sex, then they have played (and continue to play) a vast role in driving increased rates of evolution of biodiversity.

3. ECOSYSTEM REGULATION OF INFECTIOUS DISEASES

We can examine the role of parasites in ecosystem services from an alternative perspective, that is, not as the organisms performing the service per se but from ecosystem services that species, communities, and biodiversity perform in regulating the risk of infectious diseases to people. There is increasingly compelling evidence for this, particularly where anthropogenic disturbance has been linked to increases in either disease or disease risk, including Lyme disease, Leishmaniasis, malaria vector biting rates, and monkeypox. Several mechanisms have been proposed to explain the increased disease risk in disturbed areas, including better habitat for disease vectors, increased human incursion into natural areas, and changes in the host community, all of which lead to increased contact between reservoir hosts and humans.

It has been hypothesized that biodiverse and intact ecosystems have lower prevalence of disease-causing pathogens than do anthropogenically disturbed, species depauperate ecosystems because the more diverse communities contain, on average, less competent hosts for pathogens. This idea builds on an old disease control strategy known as zooprophylaxis in which an incompetent host (e.g., cattle for malaria) was used to deflect mosquito blood meals from competent hosts (in this case humans) that would otherwise have infected vectors feeding on them. In a natural ecosystem context, it has recently been expanded for Lyme disease, where it was hypothesized that the prevalence of *Borrelia burgdorferi* (the causative agent of Lyme) would be lower in intact forest than fragmented forest patches. The mechanism proposed for the observed pattern was a higher diversity of mammals in larger forest patches that would be, on average, less competent hosts than the few species (e.g., white-footed mice, *Peromyscus leucopus*, and eastern chipmunks, *Tamias striatus*) remaining in small forest patches. Substantial modeling work has supported this mechanism in principle, although so far, no published multisite field study has provided evidence that a mechanistic link exists between the biodiversity at a site and Lyme disease risk. A key issue is whether diversity, per se, or species composition plays the more important role in regulating pathogen prevalence.

Thus, substantial work remains to determine the mechanisms by which biodiversity and intact ecosystems regulate disease risk. Both positive and negative impacts of biodiversity on pathogen prevalence are to be expected. For example, some pathogens have complex life cycles and multiple hosts or require specific microclimates that are only present in intact communities.

Thus, loss of biodiversity may reduce the risk of these diseases affecting humans, as one or the other of their hosts disappears. In contrast, for other pathogens, the vectors or amplification hosts are human commensals (i.e., species that live near humans, like house mice), and anthropogenic disturbance thus facilitates pathogen transmission. Finally, regions of high biodiversity may be the most important source of new pathogens of humans and livestock. Many potentially zoonotic pathogens exist in biodiverse ecosystems, and the intrusion of humans into natural habitats and contact with wildlife (e.g., by hunting for bush meat) may be the most important mechanism facilitating disease emergence. One could view conservation programs that preserve high biodiversity in a region (and therefore high biodiversity of pathogens) as providing a risk of future disease emergence. However, a more accurate view is that encroachment (e.g., through road building and deforestation) into these areas provides the risk of emergence, and therefore, preserving areas of high biodiversity against development performs an ecosystem service by reducing the likelihood of human contact. This is discussed further in section 4, below.

Ecosystem Management of Disease Risk

Our review demonstrates that the successful management of disease through an ecosystem approach requires a detailed understanding of the ecology of pathogen transmission, the diversity of pathogens in the host community, and the impact of anthropogenic disturbance on host and vector communities. Successful management may entail setting aside habitat where pathogens are present and minimizing human and domestic animal intrusion. Alternatively, management may require suppression of particular host or vector species through habitat alterations. However, challenges exist with the latter strategy because benefits obtained through habitat conversion (e.g., draining wetlands for mosquito control) may be less than any cost of lost ecosystem services (water filtration, flood control). Management may involve promoting land use practices that preserve large fragments of intact forest rather than reducing forest patches to sizes that may promote the dominance of disease reservoirs. This may be a valid strategy for Lyme disease, which may be more prevalent in small forest patches because of increased density of one of its key rodent reservoirs, the white-footed mouse. Similarly, maintenance of intact ecosystems may reduce the invasion of introduced mosquitoes and the emergence of a range of diseases. However, benefits of increased disease control would have to be compared to other advantages of converting habitats or reducing patch sizes of managed habitats.

Other strategies to reduce disease risk might include social behavioral modifications. These include domestication of hunted animals to reduce high-risk contact with disease reservoirs, the introduction of domesticated food animal production to a region to reduce bush meat hunting, or anthropogenic removal of disease reservoirs. However, these can also have significant negative impacts. The first may result in the spillover of unknown pathogens into human populations [e.g., Severe Acute Respiratory Syndrome (SARS) coronavirus, which was harbored by domesticated civets and other species in wildlife markets in China]. The second approach may result in the spillover of wildlife pathogens into high-density livestock populations and ultimately into people (e.g., Nipah virus in Malaysia, which was transmitted from bats to pigs to people). The third can reduce populations of keystone species dramatically if not properly regulated (e.g., wolves in the United States) and may have the counterintuitive effect of increasing pathogen prevalence and disease risk in some situations.

4. VALUING THE ECONOMIC IMPACT OF PATHOGENS AND THEIR ECOSYSTEM SERVICES

Early publications on ecosystem services tended to focus on valuing biodiversity and intact ecosystems through the services they provide. There are very few studies that estimate the value of the positive benefits of pathogens, parasites, and diseases. The value of pathogens has largely been estimated through their negative impact on human health (i.e., morbidity and mortality) and on economic activities (e.g., trade and travel), and that is what we concentrate on in this section. The costs of disease outbreaks may be dramatically heightened by the high cost of technologically advanced health care (e.g., the cost of intensive care for acute infections of pathogenic viruses) and by the cost of quarantine and vaccination programs to prevent further spread (e.g., the high cost of the SARS outbreak was largely caused by measures taken to avoid further spread). The annual cost of *E. coli* 0157 in the United States has been estimated at US$405 million (in 2003 dollars), and the economic burden of Lyme disease treatment in the United States may reach $500 million per annum. Costs are incurred even in the absence of medical cases when the public perceive a high likelihood of infection or outbreak. For example, a single rabid kitten found in a pet store in New Hampshire in 1994 led to treatment with expensive postexposure prophylaxis (currently ~$1000/patient) for 665 people who had visited the store, even those who had not made contact with the kitten. These medical costs may be dwarfed by the costs of pathogens to livestock production and agriculture, both of which are intrinsically linked to ecosystem services. Even the costs of proactive (prophylactic) efforts to deal with pathogens can be substantial because they underlie much of the global effort in plant and animal breeding, the continual development of efficient housing and feeding regimes for livestock, and chemical treatment of crops and food products. For example, in the largest ever program to eradicate a plant disease, the U.S. government spent $200 million in citrus canker eradication in the mid-1990s, clear-cutting 1.8 million infected trees. Preharvest pest and disease damage combined to the eight most important crops accounts for 42% of attainable production, or US$300 billion globally, although it is not known how much of these are disease related versus insect pest related. Simple analyses of the cost of livestock diseases that do not account for the complexities mentioned above still produce extremely high values. For example, the 2001 food-and-mouth disease outbreak cost between $8 and $18 billion to the U.K. economy through lost production, lost travel, and lost trade.

Perhaps the greatest economic impact of infectious diseases occurs where they directly affect trade and commerce on a global scale. For example, the emergence of SARS in 2003 in China, and its subsequent spread throughout Southeast Asia, then globally, likely resulted in the loss of between $30 and $100 billion dollars to the global economy. This impact was largely because of the reduction in travel to the region, subsequent loss of trade, and ripple effects as the economies of other countries that traded with those most severely hit were themselves affected. The economic impact of H5N1 avian influenza, should it become a self-sustaining human pandemic, is estimated to be between $100 and $800 billion globally, and between $71.3 and $166.5 billion in the United States alone. Projections of future trends in globalized travel and trade suggest a steady increase in the percentage of global GDP per capita spent on these activities and a concomitant increase in the risk of future pandemic emergence. This suggests that the future cost of pathogens is likely to rise through this impact.

It is likely that the economic cost of pathogens that affect wildlife, wild plants, or other components of ecosystems will be even higher than the global economic burden of diseases to human health. However, few studies have analyzed this. One recent study suggested that diseases introduced into the United States cost around US$41 billion per annum by affecting humans, livestock, and crop plants. However, these estimates were made before the introduction of West Nile virus into the United States (which has likely increased the cost significantly), and they have not been extrapolated globally. They also do not value the complexities of ecosystem services, wherein diseases may remove a flagship forest species (e.g., loss of dogwood trees caused by dogwood anthracnose) or involve a direct risk to human health (e.g., presence of Lyme disease or West Nile virus in a region) and reduce the desirability of a region for hiking or as a place to live. Any of the recent efforts to include diseases in valuation of biological capital or ecosystem services are therefore likely to be gross underestimates. Given the rapidly increasing knowledge on the wide array of wildlife pathogens that continue to be discovered and the global nature of the problems associated with disease emergence, the true cost of diseases to ecosystem services is likely to be orders of magnitude higher than the recent estimates. Clearly, this is an area where future research will be extremely illuminating.

Valuation of the cost of pathogens in wildlife or wild plant communities can be relatively straightforward, especially where these communities are cropped or harvested by people for direct economic gain. From the ecosystem service perspective, diseases directly reduce natural capital (raw materials, food production, etc.) but also reduce the economic margins of industries that use renewable natural resources (e.g., forest harvesting, fishing). Studies that have quantified the impacts of diseases in this respect usually consider single outbreaks or single pathogens. For example, a pilchard herpesvirus emerged in Australia during the 1990s, producing repeated outbreaks, and was thought to have been introduced with South American pilchards used to fatten tuna in fish farms. This disease is estimated to have cost AU$12 million over 3 years in 1997 dollars. Chronic wasting disease of wild deer, elk, and other species in the United States and Canada is a prion disease similar to "mad cow" disease, bovine spongiform encephalopathy, which emerged in the United Kingdom in the 1980s when changes in the rendering process meant that less intensively processed cattle protein was fed back to cattle. Chronic wasting disease cost around $10 million to the state of Wisconsin and $19 million to the state of Colorado in 2002, largely because of the loss of hunting license revenue and increased surveillance and control activities. The introduction of African horse sickness into Spain in the 1990s resulted in the slaughter of 146 horses and other control measures costing an estimated $20 million at the time.

The difficulties in assessing the economic cost of pathogens are increased when the complexity of human responses to their diseases is included. For example, the control of reservoir hosts (e.g., bats for rabies, badgers for TB) has an uncalculated and likely diverse impact on the value of ecosystem services (see above). Likewise, control programs that target the (usually arthropod) vectors of human diseases can result in removal of related vectors along with other species and a cascade of ecosystem impact. For example, DDT, although effective in controlling disease vectors (mosquitoes and other insects), was bioaccumulated in the food chain and led to dramatic declines of top bird predators in the United States, Europe, and other regions. Other vector-control programs, such as the draining of wetlands, may be equally expensive if their true (or full) cost is assessed.

The accurate valuation of the cost of wildlife diseases that affect hosts without direct marketable value is also difficult and has not been attempted for most cases. If we consider the growing recognition of emerging infectious diseases as a cause of wildlife biodiversity loss, efforts to assess the cost of some of the most significant of these (e.g., amphibian declines caused by disease, and the loss of potential medical drugs) would be useful. We can hypothesize that the global spatial distribution of the risk of disease emergence to ecosystem services is likely to markedly change previous assessments. Recent work on trends in disease emergence in humans shows that the risk of emergence is greatest where the pathogen diversity is highest (i.e., the tropics), where wildlife host biodiversity (and therefore the overall number of pathogen species able to emerge) is greatest. The tropics are also an area of high ecosystem service value. Thus, the cost of anthropogenic activities in these regions that facilitate disease emergence is heightened when diseases are taken into consideration. Testing this hypothesis may provide important insights in economics, public health, ecology, and wildlife health.

These preliminary thoughts on valuation of pathogen impacts on natural capital and ecosystem services provide some interesting conclusions for balancing the cost-effectiveness of human activities. First, they support previous calls for wildlife disease emergence impact statements; second, they suggest that activities with a high risk of disease spread or emergence (e.g., global trade in animal products, bush meat hunting) have a higher economic cost than previously proposed; third, they highlight the complexity of pathogens and parasites within ecosystems such that any single human activity can have multiple outcomes when diseases are incorporated into the analysis. These all suggest that accurate analyses will be difficult but ultimately extremely worthwhile.

The examples given above all highlight the large number of studies that value the cost of diseases on humans, livestock, and (albeit less well understood) on ecosystems. There is a dearth of information on the value of the benefits of parasites to ecosystems. One potentially important "value" of an ecosystem service related to parasites is in the finding that higher biodiversity of wildlife tends to produce a higher risk of emerging diseases because of the higher diversity of pathogens that these wildlife harbor. It might be argued, therefore, that intact ecosystems provide a regulating service by preventing the emergence of these diseases. The emergence of almost all emerging infectious diseases (EIDs) is driven by a series of anthropogenic factors, including demographic changes (e.g., increases in human population density leading to the emergence of dengue hemorrhagic fever); socioeconomic changes (e.g., increased injection drug use leading to HIV/AIDS spread or global trade leading to the pandemic emergence of West Nile virus and SARS); or anthropogenic environmental changes (e.g., changes in forest cover leading to the emergence of Lyme disease). Where land use changes involving degradation of intact habitat cause disease emergence, the intact ecosystem could be considered to hold latent ecosystem service value in preventing the emergence of these diseases. For example, road building and deforestation in tropical forests have been linked to the emergence of HIV/AIDS (through increased human activity in African forests and increased contact with the wildlife reservoir of HIV-1's nearest relative, the chimpanzee), and mining activities in tropical forests have led to the emergence of Ebola and Marburg viruses. The value of these ecosystems in preventing disease regulation is in not modifying them and preventing contact that would allow disease emergence (i.e., reducing socioeconomic pressure on a region with high biodiversity). This is supported by recent analyses of the drivers of EIDs. The emergence of zoonotic diseases from wildlife, which are among the most common and highest-impact EIDs, is significantly correlated with wildlife biodiversity and socioeconomic factors such as human population density and growth. Because human population density, deforestation, road building, and globalized travel and trade are all predicted to increase in the near future, the rate at which new zoonotic diseases emerge from these biodiverse "EID hot spots" is also likely to increase.

FURTHER READING

Anderson, P. K., A. A. Cunningham, N. G. Patel, F. J. Morales, P. R. Epstein, and P. Daszak. 2004. Emerging infectious diseases of plants: Pathogen pollution, climate change and agrotechnology drivers. Trends in Ecology and Evolution 19: 535–544.

Daszak, P., A. A. Cunningham, and A. D. Hyatt. 2000. Emerging infectious diseases of wildlife: Threats to biodiversity and human health. Science 287: 443–449.

Hudson, P. J., A. P. Dobson, and K. D. Lafferty. 2006. Is a healthy ecosystem one that is rich in parasites? Trends in Ecology and Evolution 21: 381–385.

Hudson, P. J., A. Rizzoli, B. T. Grenfell, H. Heesterbeek, and A. P. Dobson. 2002. The Ecology of Wildlife Diseases. Oxford: Oxford University Press.

Jones, K. E., Nikkita G. Patel, Marc A. Levy, Adam Storeygard, Deborah Balk, John L. Gittleman, and Peter Daszak. 2008. Global trends in emerging infectious diseases. Nature 451: 990–993.

Kilpatrick, A. M., A. A. Chmura, D. W. Gibbons, R. C. Fleischer, P. P. Marra, and P. Daszak. 2006. Predicting the global spread of H5N1 avian influenza. Proceedings of the National Academy of Sciences U.S.A. 103: 19368–19373.

Kremen, C., and R. S. Ostfeld. 2005. A call to ecologists: Measuring, analyzing, and managing ecosystem services. Frontiers in Ecology and the Environment 3: 540–548.

LaDeau, S. L., A. M. Kilpatrick, and P. P. Marra. 2007. West Nile virus emergence and large-scale declines of North American bird populations. Nature 447: 710–713.

Morse, S. S. 1993. Emerging Viruses. New York: Oxford University Press.

Ostfeld, R. S., and K. LoGiudice. 2003. Community disassembly, biodiversity loss, and the erosion of an ecosystem service. Ecology 84: 1421–1427.

Patz, J. A., P. Daszak, G. M. Tabor, A. A. Aguirre, M. Pearl, J. Epstein, N. D. Wolfe, A. M. Kilpatrick, J. Foufopoulos, D. Molyneux, D. J. Bradley, and The Working Group on Land Use Change and Disease Emergence. 2004. Unhealthy landscapes: Policy recommendations on land use change and infectious disease emergence. Environmental Health Perspectives 112: 1092–1098.

Pimentel, D., L. Lach, R. Zuniga, and D. Morrison. 2000. Environmental and economic costs of nonindigenous species in the United States. BioScience 50: 53–65.

VI.10

Support Services: A Focus on Genetic Diversity

Oliver R. W. Pergams and Peter Kareiva

OUTLINE

Empirically and theoretically we know that genetic diversity is essential for rapid evolution. In the face of a rapidly changing world driven by unprecedented human impacts, the ability to evolve rapidly may be one of nature's most precious commodities. Examples of the economic value of genetic diversity are numerous and compelling, but methods for formal economic valuation of this ecosystem service are not well formulated. Even without good methods for dollarizing genetic diversity as an ecosystem service, there are ways of quantifying its value that help inform sustainable and judicious resource management strategies.

GLOSSARY

ecosystem services. Goods (food, fuel, building materials) and services (flood control, disease regulation, etc.) that benefit humans and are provided by natural ecosystems

heterozygosity. The proportion of individuals in a population that have two different alleles for a particular gene

microevolution. The occurrence of small-scale changes in allele frequencies in a population over a few generations

resilience. The ability of a system to resist or recover from disturbances and perturbation so that the key components and processes of the system remain the same

1. GENETIC DIVERSITY IS THE MOST FUNDAMENTAL OF ALL ECOSYSTEM SERVICES

The importance of genetic diversity is well known to agronomists, who for nearly a century have spoken of genetic variety as a resource to enhance crop vigor and productivity. Testimony to this value is the fact that half the yield gains in major U.S. cereal crops since the 1930s are attributed to genetic improvements (Rubenstein et al., 2005). We are able to breed and select for crops that meet different environmental challenges only because of the genetic variety in those crops, and the goal of plant breeders is typically to maintain as much genetic diversity as possible in case it is needed at some future date. More generally, a central theorem of evolution is that the rate of evolution is proportional to the amount of genetic variation. The quantitative connection between the rate of evolution and the amount of genetic variation provides the foundation for genetic diversity as perhaps the most fundamental of all supporting ecosystem services. It is clear that if there were zero genetic diversity within each species, even modest environmental change or human disturbance would imperil the species and the ecological services that species provide. In order for humans to get a return from nature (in the form of fisheries, timber, soil fertility, and so on) in a varying environment, species must harbor genetic diversity—how much we cannot say, but some for sure.

A second related appreciation for genetic diversity can be traced to the origins of conservation biology, which sought to identify minimum viable population size on the basis of genetic principles, resulting in computer models of extinction probability. The importance of minimum viable population size applies to many of the world's species, which have only small populations remaining. For example, 17% of the world's bird species are confined to small populations on islands, and of these, 23% are classified as threatened (Johnson

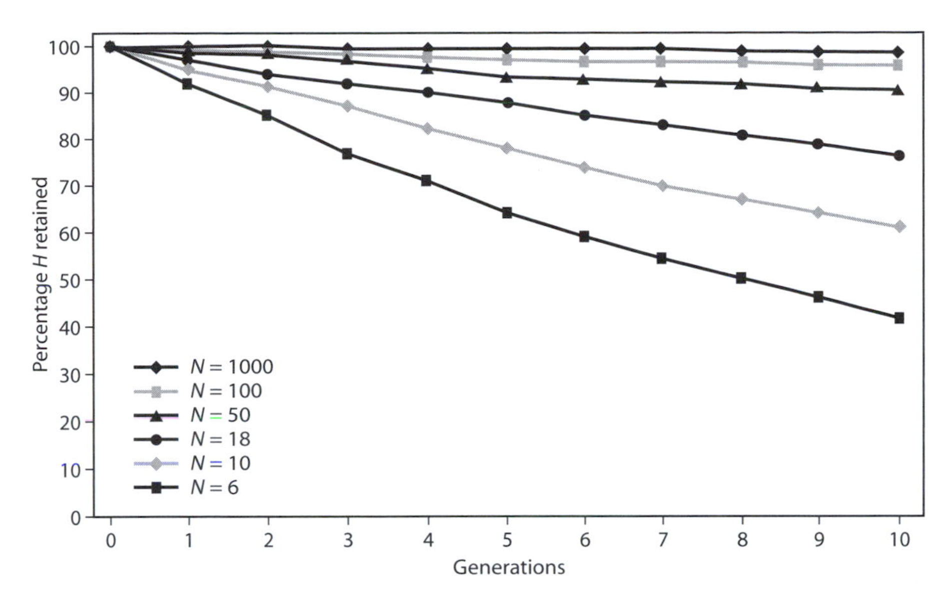

Figure 1. Predicted loss in heterozygosity (H) over time. Here H starts at 0.25 but varies by initial population size (N) from 6 to 1000, and the drop in H is rapid in low population sizes.

and Stattersfield, 1990). The International Union for Conservation of Nature (IUCN) estimates that 3032 of the 40,295 (7.5%) plant and animal species evaluated are critically endangered (category CR: IUCN, 2007), which means they are down to very small populations. It is probably fair to treat this percentage as a lower bound: some species with only small populations are considered stable enough not to be CR. It is well known theoretically that the smaller a population is, the more rapidly genetic variation is lost (figure 1), and the more likely the population is to lose rare alleles. These theoretical relationships are a matter of practical concern because we know from scores of empirical examples that lower heterozygosity (which will occur as rare alleles are lost) leads to lower rates of growth or reproduction. There is a substantial body of work on the relationship between heterozygosity and growth in annual and perennial crop plants, forest trees, birds, fish, molluscs, reptiles, and vertebrates including humans.

It is more difficult to find real-life conservation applications that focus on loss of rare alleles, but the concept is clear. When there are new evolutionary challenges, those organisms able to adapt to these challenges will be the ones to reproduce and survive. This ability to adapt may well be captured not by common genes but rather by alleles that are rare in the population. This effect is probably especially important in small populations (e.g., Sjögren and Wyöni, 1994). There has been controversy in the captive breeding community about the importance of rare alleles, and how to best manage them, with some researchers suggesting that all captive breeding be directed toward conserving rare alleles and others arguing that such a focus will sometimes reduce heterozygosity and hence introduce other vulnerabilities. The issue remains open.

In short, the applied sciences of plant and animal breeding and conservation biology provide solid empirical and theoretical reasons for valuing genetic diversity as an essential resource. Indeed, the level of provisioning or resilience of every other ecosystem service is influenced by genetic diversity, so whatever value we place on those other ecosystem services, some portion of that value should ultimately derive from genetic diversity. Ironically, however, real-world conservation and environmental projects such as those conducted by the World Bank or by major conservation nongovernmental organizations almost never pay attention to, or even mention, genetic diversity.

We argue here that genetic diversity should be given more attention in ecosystem service and conservation projects because of the much ballyhooed pace of rapid climate change as well as other landscape and environmental changes that are sweeping the world. The standard environmental message that is delivered in discussing drastic and rapid environmental change is the catastrophic extinction of species that is likely to occur. For instance, a much-cited paper in *Nature* recently concluded that if greenhouse gas emissions continued under a business-as-usual scenario, then 1 million species would be on their way to extinction by 2050 (Thomas et al., 2004). This dire prediction assumes, of course, that none of those 1 million species could evolve in a way that allowed them to adjust to a changing climate. The fate of the world is indeed gloomy beyond belief if we assume there will be no evolution. But there will be evolution, and the rate of that evolution is determined by genetic diversity. Also, we must remember that although extinctions are important, genetic diversity can prevent serious erosion of other ecosystem services even before extinction. In this chapter, we review many examples of very rapid evolution in the face of anthropogenic change, sketch how one might place a value on the genetic diversity that fuels rapid evolution, and conclude with some practical ideas of what resource and land managers might do differently if they were to fully factor in the value of genetic diversity as an ecosystem service.

2. RAPID ANTHROPOGENIC CHANGE AND THE ROLE OF EVOLUTION

Humans are changing the global environment at unprecedented rates. The major anthropogenic forces acting on species include *introductions* (of invasive species and populations), *habitat loss, harvesting, climate change*, and *pollution*. Plants and animals react to these massive environmental changes in one of two ways: either they become extinct, or they adapt. Much has been written on human-induced extinction, but relatively little has been written on human-induced microevolution, defined as rapid adaptation that takes place in species and populations within the approximate lifespan of a human being (Dobzhansky, 1941, p. 12) or <100 years (Hendry and Kinnison, 1999). Sometimes this change actually occurs over much shorter time spans, within years or even months (Hendry and Kinnison, 1999). We focus on microevolution because it is the primary natural buffer against anthropogenic extinction or the degradation of ecosystem services in this rapidly changing world and because microevolution depends on genetic diversity.

Rapid evolution is much more common than is usually realized, and the proximate selective force that causes that rapid evolution is predominantly anthropogenic change. In a review of documented cases of contemporary microevolution, 18 (86%) of the 21 cases listed (Hendry and Kinnison, 1999) involved anthropogenic causes. We surveyed the literature and present an updated list in table 1. Our criteria for inclusion are (1) that the plant or animal taxon had undergone morphological and/or genetic change over a period of <100 years and (2) that the cause of microevolution was likely anthropogenic. We further assigned each example to the likely category of anthropogenic force involved, whether introduction, habitat loss, harvesting, climate change, or pollution. Our review is by no means exhaustive, but we feel we found most well-documented examples. It is likely that many evolutionary responses to anthropogenic stresses go unnoticed and occur in species that are not as convenient to study as butterflies.

In our survey, most examples of anthropogenic microevolution were caused by pollution. Pollution was responsible for changes in 82 of 127 species listed in table 1, or about 65% of species. The bulk of these are the famous examples of industrial melanism in >70 moth species. Most of the rest (>10) entail evolution of heavy metal tolerance in plants. It is likely that there are many more, undocumented examples of response to various pollutants in other organisms.

Introductions were the other anthropogenic event most often associated with microevolution in our review. Nonnative or introduced species were linked to ~20% (25/127) of the cases of rapid evolution, usually with the introduced species itself evolving to meet the challenges of a new environment. Some of the earliest microevolution reported was caused by anthropogenic introduction of rodents, especially on islands (e.g., Berry, 1964), but very recently Chicago-area white-footed mice also showed dramatic changes in morphology and mtDNA haplotype frequencies when invading urban environments (Pergams et al., 2003; Pergams and Lacy, 2007). After introduction from Europe to North America, house sparrows changed dramatically in plumage and size (Johnston and Selander, 1964). Losos et al. (2001) showed that *Anoles* lizards introduced onto a series of Caribbean islands underwent major morphological changes in <14 years. Introductions have also resulted in behavioral evolution, including migratory behavior and escape ability. Mosquitoes rapidly evolved mating behavior, reproductive phenology, and host preference when separated into surface-dwelling populations and populations dwelling in the London Underground railway system (Bryne and Nichols, 1999).

The introduction of fish into new habitats has frequently been shown to alter selection on a great variety of traits. This is not surprising when we consider that most of the myriad examples of fish stocking of ponds, lakes, rivers, and streams are in effect the intentional (although unwitting) introduction of invasive species and populations, and most of these cases probably do result in rapid evolution of some sort. Of the myriad examples, sockeye salmon showed rapid divergence of male and female body size and male body depth in accordance with stream size (Hendry and Quinn, 1997), hatching time (Hendry et al., 1998), and reproductive isolation (Hendry et al., 2000). Chinook salmon diverged rapidly in morphology, gonadal investment, reproductive timing, and growth rate 90 years after introduction to rivers in New Zealand (see,

Table 1. Some examples of anthropogenically caused microevolution (rapid evolution occurring over <100 years)

Species	*Citation*	*Trait(s) responding*	*Selective category*
Plants			
Plantago major (plantain)	Davison and Reiling, 1995	Growth rate, resistance to ozone pollution	Climate change
>10 species	Reviewed in Bone and Farres, 2001	Heavy metal tolerance (zinc, copper, lead)	Pollution
Arabidopsis thaliana (Thale cress)	Ward et al., 2000	Seed production	Climate change
Arabidopsis halleri (penny cress)	Bert et al., 2000	Heavy metal tolerance (zinc, cadmium)	Pollution
Helianthus annuus (common sunflower), *H. debilis* (cucumberleaf sunflower)	Whitney et al., 2006	Hybridization	Introduction
Powell amaranth (Powell amaranth)	Brainard et al., 2007	Greater dormancy and longevity of seeds	Introduction
Eschscholzia californica (California poppy)	Leger and Rice, 2007	Plant and seed size, fecundity	Introduction
Brassica rapa (turnip)	Franks et al., 2007	Earlier onset of flowering	Climate change
Invertebrates			
>70 species	Kettlewell, 1973	Industrial melanism	Pollution
Jadera haematoloma (soapberry bug)	Carroll and Boyd, 1992; Carroll et al., 1997	Beak length	Introduction
Culex pipiens (house mosquito)	Bryne and Nichols, 1999	Mating behavior, reproductive phenology, preferred host, etc.	Introduction
Eurytermora affinis (marine copepod)	Lee, 1999	Physiological traits	Introduction
Prodoxus quinquepunctellus (yucca moth)	Groman and Pellmyr, 2000	Emergence time and ovipositor morphology	Introduction
Drosophila subobscura (sp. of Old World fruitfly)	Huey et al., 2000; Gilchrist et al., 2001	Latitudinal wing size clines	Introduction
Erynnis properties (Propertius duskywing), *Papilio zelicaon* (anise swallowtail)	Zakharov and Hellman, in press	Larval growth rate	Climate change
Vertebrates			
Mus musculus (house mouse)	Berry, 1964, 1986; Berry et al., 1978	Cranial and skeletal traits	Introduction
Passer domesticus (house sparrow)	Johnston and Selander, 1964	Color and size	Introduction
Rattus rattus (black rat)	Patton et al., 1975; Pergams and Ashley, 2001	Cranial and external traits	Introduction
Vestiaria coccinia (Hawaiian honeycreeper)	Smith et al., 1995	Bill size	Habitat loss
Oncorhynchus nerka (sockeye salmon)	Hendry and Quinn, 1997; Hendry et al., 1998, 2000	Sex-biased body size dimorphism, adaptive divergence in hatching time, reproductive isolation, adaptive divergence in hatching time	Introduction
Oncorhynchus tshawytscha (Chinook salmon)	Kinnison et al., 1998; Quinn et al., 2000	Morphology, gonadal investment, reproductive timing and growth rate	Introduction
Carpodacus mexicanus (house finch)	Able and Belthoff, 1998	Migratory behavior	Introduction

(Continued)

Table 1. *(cont.)*

Species	*Citation*	*Trait(s) responding*	*Selective category*
Cyprinodon tularosa (White Sands pupfish)	Stockwell and Mulvey, 1998	Allozyme allele frequencies	Introduction
5 spp. Pacific salmon	Reviewed in Hendry and Kinnison, 1999	Size selection by gill netting	Harvesting
Anolis carolinensis, A. sagrei (Anolis lizard)	Losos et al., 2001	Body shape, hind limb length	Introduction
Thymallus thymallus (grayling)	Haugen and Vøllestad, 2001	Size selection by gill netting	Harvesting
Peromyscus leucopus (white-footed mouse)	Pergams et al., 2003; Pergams and Lacy, 2007	External and cranial traits, mtDNA haplotypes	Introduction
Rana arvalis (wood frog)	Rasanen et al., 2003	Acid stress tolerance	Pollution
Ovis canadensis (bighorn sheep)	Wilson et al., 2005	Birth weight, birth date, litter size	Harvesting
Alectoris rufa (red-legged partridge), *A. chukar* (chukar partridge)	Barbanera et al., 2005	Hybridization	Introduction
Cygnus olor (mute swan)	Charmantier et al., 2006	Clutch size	Habitat loss
Salmo salar (Atlantic salmon), *Gadus morhua* (Atlantic cod)	Fraser et al., 2007	Life history traits	Harvesting
Canis lupus (wolf), *Felis silvestris* (wildcat), *Sus scrofa* (wild boar), *Coturnix coturnix* (quail), *Alectoris graeca* (rock partridge)	Randi, 2007	Hybridization w/domestic forms	Introduction
Andropadus virens (Little Greenbul)	Smith et al., 2007	Wing, tarsus, and bill size; plumage color; song	Habitat loss
Parus major (great tit)	Garant et al., 2007	Quantitative genetics of laying date, clutch size, and egg mass	Climate change
Parus major (great tit), *Ficedula hypoleuca* (pied flycatcher)	Visser, 2007	Temperature sensitivity of phenology, mistimed reproduction	Climate change

e.g., Kinnison et al., 1998). Clearly, if introduced species can evolve to a new environment they are instantly thrust into, then native species can also evolve to an environment that changes rapidly while they persist in the same location.

Microevolution can be seen not only in introduced species but also in native species interacting with alien species. Examples include the soapberry bug (different beak lengths in response to different introduced host plants; Carroll and Boyd, 1992) and yucca moths (shifts in emergence time and ovipositor morphology following colonization of a new host species; Groman and Pellmyr, 2000). In agriculture, alfalfa crop rotations selected for greater dormancy and longevity of amaranth weed seeds (Brainard et al., 2007).

Harvesting is particularly influential as an evolutionary force in fisheries. Gill netting exerts intense selection on age and size at sexual maturity by differentially removing larger fish from harvested populations (see review in Hendry and Kinnison, 1999). Harvesting selectively during early or late portions of salmon runs has been shown to alter run timing, breeding time, and sex ratio (e.g., Nelson and Soulé, 1987). Additionally, the severe reduction in spawning density and high mortality of large fish from harvesting undoubtedly shift selection pressures on social systems, including patterns of sexual selection (e.g., Foote, 1988) and competition among females for nest sites. Although most examples are related to fisheries, harvesting may cause microevolution in any organism. Bighorn sheep, for example, showed rapid changes in birth weight, birth date, and litter size as a result of harvesting (Wilson et al., 2005). We should note that although most examples of rapid change almost certainly have a genetic basis, some may be the result of environmental plasticity or behavioral changes.

Climate change has only recently begun to be implicated as a cause of microevolution, but cases are now being documented with greater frequency. Microevolution for resistance to ozone pollution was documented in plantain by Davison and Reiling (1995), and rapid selection for increased growth rate in *Arabidopsis* in response to elevated atmospheric CO_2 was shown by Ward et al. (2000). Franks et al. (2007) show rapid selection on flowering time of turnips in response to variation in length of growing season. Such microevolution in plants resulting from climate change may have profound agricultural implications in the future. Climate change has also caused rapid evolution in butterflies and birds.

Habitat loss can cause extinction, and extinctions in turn can result in rapid adaptation on the part of surviving species. For example, extinction of their food sources resulted in changes in mandible length in Hawaiian honeycreepers over the last century (Smith et al., 1995). In mute swans, recent relaxation of food constraints and predation risks caused by habitat loss have resulted in rapid selection for larger clutch size (Charmantier et al., 2006). Also, as an apparent result of habitat loss, the little Greenbul of Africa has shown microevolution of wing, tarsus, and bill size; plumage color; and song (Smith et al., 2007).

In summary, rapid evolution in response to human-caused change is common—it might well even be the norm. Unfortunately, rarely do biologists publish studies in which they find no evidence of evolutionary change. In the absence of studies regularly reporting no evolution, it is difficult to estimate the true frequency with which evolution plays a major role in species survival in the face of changing environments. It is plausible to hypothesize that much of the biodiversity that is thriving in the face of huge human impacts represents species with better than average genetic diversity as a natural asset on which to draw.

3. CAN ONE QUANTIFY THE VALUE OF GENETIC DIVERSITY?

We should first note that there are two basic philosophies of valuing biodiversity. The first philosophy is that biodiversity has value and should be protected regardless of its economic value to humans. Possibly first known in Western thought from Plato (360 BC), in modern times this view was given strong voice by Leopold (1949) and was later elaborated on by others. An alternative philosophy asserts that biodiversity should be protected because of its benefits to humans. Biodiversity performs a number of ecological services for humans that have economic, moral, aesthetic, and/or recreational value (ecosystem services), and such ecosystem services can and should be valued (e.g., Daily, 1997). This philosophy of self-interest is especially accessible to policy makers, politicians, and the general public.

We would like to emphasize a broader point here: the main reasons to do any sort of economic valuation of biodiversity are (1) to have a place at the table where ministers of commerce, transportation, defense, etc. sit, and (2) to alter behavior or policy. Until conservation is translated into value, it is viewed by most people as a curiosity. As a result, conservationists need to remind everyone that looking only at immediate commercial payoffs (say from fisheries) will consistently undervalue biodiversity, given the range of tangible and intangible services it provides. The challenge is to deduce the value that people do or should place on species ("should" because species are often providing indirect benefits not immediately appreciated by observers), including the intangible aspects of that value.

De Groot et al. (2002) provide a useful synthesis of the recent state of ecosystem services valuation, in which genetic diversity is both a habitat and a production service. Within habitat service, genetic diversity is further defined as a refugium service to "provide refuge and reproduction habitat to wild plants and animals and thereby contribute to the (in situ) conservation of biological and genetic diversity and evolutionary processes" (De Groot et al., 2002, p. 395). This categorization deals with the necessity of genetic diversity to provide material to meet evolutionary challenges, although only as they relate to the benefit of humans. As a production service, genetic diversity is defined as one of "many ecosystem goods for human consumption" (De Groot et al., 2002, p. 395). This refers to its role as source for new manufactured goods such as pharmaceuticals as well as providing continued genetic material for cultivated crops and domesticated animals.

Rephrased, genetic diversity may be said to have both a general benefit and specific benefits to humans. The specific benefits are to provide breeding material for the organisms that humans consume directly. In the cases of crops and livestock, genetic diversity can easily be valued in terms of the ability to select for increased yields as a result of the available genetic options. This calculation is not made because plant and animal breeders are so convinced of the high value of genetic diversity that they typically do everything they can to maintain this diversity at its highest possible levels, without needing any valuation to convince them. Moving away from domesticated species, the benefit of genetic diversity is that it allows the world's ecosystems to continue to function in the face of environmental and human-caused change. This more generic genetic diversity represents almost the opposite case of domesticated species—society and resource managers tend to place very little value on it. The one exception is the adoption of "evolutionarily significant units" as worthy of special protection under the U.S. Endangered Species Act (Waples, 1991). For the vast majority of species, however, there is no policy or legal framework that recognizes the link between genetic diversity and functioning ecosystems.

The job of economic valuation of genetic diversity is to reveal to people how much genetic diversity benefits them and, in most cases, to reveal this in monetary terms. Most especially, it means economic valuation should reveal to people how much the genetic diversity that they do not directly and immediately use is worth to them. People might be willing to pay generously to make sure some genetic diversity was maintained if they were led through an abstract argument that proceeded as follows:

1. The existence of humans depends on functioning ecosystems, and if
2. Functioning ecosystems require genetic diversity to meet future evolutionary challenges, then
3. Genetic diversity is required for the continued existence of humans, and so of course its total value approaches infinity.

Even if only some small portions of genetic diversity are necessary to meet evolutionary challenges over time (as is probably the case), we have no idea in advance what these challenges will be, or which portions of genetic diversity will be required to meet them. All human activities—e.g., land clearing for schools, agricultural fields, and homes; the purchase of computers or stereos that produced pollution in the making; plane or train rides to visit our relatives—result in potential declines in genetic diversity. We have to weigh the benefits of such activities against the costs of losing genetic diversity. Because there is significant uncertainty about the future value of genetic diversity, and any loss is irreversible, precaution is warranted. The precautionary principle states that if an action or policy might cause severe or irreversible harm to the public, in the absence of a scientific consensus that harm would not ensue, the burden of proof falls on those who would advocate taking the action. The most meaningful applications of the precautionary principle have been in relation to biodiversity (e.g., Cooney and Dickson, 2006).

Value means importance or desirability. It does not make sense to talk about the value of genetic diversity as though the only choice is between having genetic diversity and not having it. Instead, we should discuss value in terms of well-defined changes to genetic diversity. Economists value things in comparative terms so that the valuation of genetic diversity as an ecosystem service should define the trade-off between two or more specific situations: given the full suite of "goods" and "bads" represented in A and B, is A better than B? There is no single, objective answer to that question—it depends on the values and preferences of the parties involved in the decision. The challenge for conservation biologists and others who care about biodiversity is in getting decision makers to recognize the "goods" associated with preservation of genetic diversity and the "bads" associated with its loss when those goods and bads are often more distant and less tangible then other, more immediate, material gains that can result from particular decisions. More quantitative valuations of genetic diversity typically begin with an assumption that each species has some potential value that depends on its genetic uniqueness (Weitzman, 1998). Weitzman defines a species' distinctness as its genetic distance to its nearest neighbor. This utilization of genetic distance is quite similar to other attempts to value biodiversity using a metric of genetic distinctness, through pairwise distances between species, commonly measured with DNA–DNA hybridization. Montgomery et al. (1999) compute diversity values from a taxonomic (phylogenetic) tree and simply assign dollar weights to each diversity point along a $0 to $200 million scale. Unfortunately, this approach is unsatisfying because a species' immediate benefit to human beings hinges on its ecological relationships, which in turn may be wholly unrelated to its genetic distinctness. Also, none of these approaches gives value to those ecosystem services or intangibles that would be degraded but not necessarily eliminated by a reduction in genetic diversity.

The most sophisticated valuation approach for genetic diversity seeks to unify genetic, ecological, and economic criteria (Brock and Xepapadeas, 2003). This approach starts with some measurable economic benefit being provided by species and then generates a valuation of genetic diversity in terms of resilience and productivity. This valuation embraces the role of genetic diversity as insurance, assuming that genetically rich ecosystems are less prone to productivity loss or collapse and are more resilient as environmental conditions change.

We hypothesize that the most practical foundation for thinking about the value of genetic diversity is as insurance against catastrophe or as an investment strategy ("diverse portfolio") for maintaining ecosystem productivity in the face of the vicissitudes of nature. If major land use changes and large-scale public works projects had to pay for insurance against undesirable catastrophic ecosystem failure, then it is likely that genetic diversity would come to be valued for its ability to protect against partial or total failure (and hence provide for lower premiums). We could get closer to a "true" valuation if we could add other valuation frameworks such as direct material (food, etc.) and intangible (aesthetic, etc.) to this foundation.

4. PRACTICAL OUTCOMES FROM THE VALUATION OF GENETIC DIVERSITY

Even if we cannot calculate a dollar value for genetic diversity, we can develop public policy and management approaches that reflect the ecosystems services provided by genetic resources. A good example of resource management that explicitly values genetic diversity can be found with salmon in the Pacific Northwest. Species of pacific salmon are divided into many different populations, with each population associated with specific spawning streams, rivers, or lakes. The seasonal timing of outmigrations from fresh water to the ocean, as well as the number of years spent in fresh water and in the ocean, are under genetic control. These salmon have tremendous economic value for both recreational and commercial fisheries.

The reason for this broad array of life-history traits is genetic diversity: the salmon's stock complex is composed of several hundred discrete spawning populations structured within freshwater systems, and the fish in this system are many of the same examples we cited as displaying microevolution. This broad array of traits resulting from genetic diversity, local adaptation, and microevolution has allowed the stock complex as a whole to maintain economic productivity in spite of climate change. A recent quantitative analysis of Bristol Bay sockeye salmon in Canada indicates that the fact that different lakes are occupied by different genetic variants means that some stock do well in certain years, whereas different stocks do well in other years (under different environmental conditions). Although the environment fluctuates enormously, the Bristol Bay salmon maintain a consistent and high level of productivity because, in the aggregate, there are always some populations that are doing well.

Management and conservation of salmon are not aimed at total population size or the species but at preserving as many different life-history variants or genetically differentiated populations as possible (Ruckelshaus et al., 2001). Indeed, recognizing that many anthropogenic actions may cause microevolution beyond acceptable fisheries management limits, Jørgensen et al. (2007) propose evolutionary impact assessment as a management tool. The first step of such an assessment would describe how human actions result in biological changes, while the second step would describe how these trait changes affect the stock's utility to human society. Although salmon managers and conservationists recognize and act on the importance of genetically varied populations, the same is not true of conservation in general. Conservation that focuses on simply tallying up the number of species that occur in protected areas risks neglecting the core importance of populations, which are the level at which most genetic diversity is maintained.

Although small population size is the most obvious thing to manage against, it is worth noting that it is really small effective population size that precipitates a loss of genetic diversity. For any given population number, effective population size can be enhanced by maximizing number of breeders per generation, equalizing family sizes, equalizing the sex ratio of breeders, and reducing fluctuations in population size (Foose, 1983). These might seem impractical and out of the reach of everyday habitat management. However, by maintaining habitats of relatively equal quality (as opposed to being content with most of a population's productivity coming from a few individuals in one special place), one can maximize the number of breeders and tend to equalize family size. Even situations like fluctuations in population size are subject to management intervention—either by resource supplementation during low years or relaxation of harvest pressure during low years.

5. SUMMARY

Rapid environmental change is now placing an unprecedented premium on genetic diversity as a resource that will support ecological resilience in the face of environmental shocks. Fortunately, evolution has proven to be a remarkably potent process, even on time scales commensurate with today's environmental challenges. A prerequisite for rapid evolution is a generous supply of genetic diversity. In the world of ecosystem services, genetic diversity is rarely mentioned and almost never valued in any formal sense. The irony of this neglect of genetic diversity is the fact that it is the ultimate supporting service—without genetic diversity, ecosystem collapse is a certainty. Even though any estimates of the economic value of genetic diversity are sure to have great uncertainty, it is straightforward to recommend some basic management principles that will help maintain this ecosystem service. First, instead of species being the be-all and end-all of conservation, recognizing the value of genetic diversity implies that greater attention should be given to the conservation of populations. Second, management must work to make sure populations never get so small that genetic diversity is rapidly lost and individual fitness and population viability are dissipated. Last, if populations do begin to approach dangerously low numbers, then anything that promotes equal family sizes, a large number of breeders, and reduced population fluctuations could be critical. Genetic diversity is a proven strategy for maintaining species and ecosystem productivity and is the world's most important insurance policy. It is time to translate these values into dollars and policies that ensure the maintenance of genetic diversity for future generations.

FURTHER READING

Berry, R. J. 1964. The evolution of an island population of the house mouse. Evolution 18: 468–483.

Brainard, D. C., A. DiTommaso, and C. L. Mohler. 2007. Intraspecific variation in seed characteristics of Powell amaranth (*Amaranthus powellii*) from habitats with contrasting crop rotation histories. Weed Science 55: 218–226.

Brock, W., and A. Xepapadeas. 2003. Valuing biodiversity from an economic perspective: A unified economic, ecological and genetic approach. The American Economic Review 93: 1597–1614.

Byrne, K., and R. A. Nichols. 1999. *Culex pipiens* in London Underground tunnels: Differentiation between surface and subterranean populations. Heredity 82: 7–15.

Carroll, S. P., and C. Boyd. 1992. Host race radiation in the soapberry bug natural history, with the history. Evolution 46: 1052–1069.

Charmantier, A., R. H. McCleery, C. Perrins, and B. C. Sheldon. 2006. Quantitative genetics of age at reproduction in the mute swan: Support for antagonistic pleiotropy models of senescence. Proceedings of the National Academy of Sciences, U.S.A. 103: 6587–6592.

Cooney, R., and B. Dickson, eds. 2006. Biodiversity and the Precautionary Principle: Risk and Uncertainty in Conservation and Sustainable Use. London: Earthscan Publications.

Daily, G. C., ed. 1997. Nature's Services: Societal Dependence on Natural Ecosystems. Washington, DC: Island Press.

Davison, A. W., and K. Reiling. 1995. A rapid change in ozone resistance of *Plantago major* after summers with high ozone concentrations. New Phytology 131: 337–344.

De Groot, R. S., M. A. Wilson, and R.M.J. Boumans. 2002. A typology for the classification, description, and valuation of ecosystem functions, goods, and services. Ecological Economics 41: 393–408.

Dobzhansky, T. 1941. Genetics and the Origin of Species, 2nd ed. New York: Columbia University Press.

Foose, T. J. 1983. The relevance of captive populations to the conservation of biotic diversity. In C. M. Schonewald-Cox, S. M. Chambers, B. MacBryde, and W. L. Thomas, eds. Genetics and Conservation: A Reference for Managing Wild Animal and Plant Populations. Menlo Park, CA: Benjamin/Cummings Publishing, 374–401.

Foote, C. J. 1988. Male mate choice dependent on male size in salmon. Behaviour 106: 63–80.

Franks, S. J., S. Sim, and A. E. Weis. 2007. Rapid evolution of flowering time by an annual plant in response to a climatic fluctuation. Proceedings of the National Academy of Sciences, U.S.A. 104: 1278–1282.

Groman, J. D., and O. Pellmyr. 2000. Rapid evolution and specialization following host colonization in a yucca moth. Journal of Evolutionary Biology 13: 223–236.

Hendry, A. P., J. E. Hensleigh, and R. R. Reisenbichler. 1998. Incubation temperature, developmental biology and the divergence of sockeye salmon within Lake Washington. Canadian Journal of Fisheries and Aquatic Sciences 55: 1387–1394.

Hendry, A. P., and M. T. Kinnison. 1999. Perspective: The pace of modern life: Measuring rates of contemporary microevolution. Evolution 53: 1637–1653.

Hendry, A. P., and T. P. Quinn. 1997. Variation in adult life history and morphology among Lake Washington sockeye salmon (*Oncorhynchus nerka*) populations in relation to habitat features and ancestral affinities. Canadian Journal of Fisheries and Aquatic Sciences 54: 75–84.

Hendry, A. P., J. K. Wenburg, P. Bentzen, E. C. Volk, and T. P. Quinn. 2000. Rapid evolution of reproductive isolation in the wild: Evidence from introduced salmon. Science 290: 516–518.

Hughes, A. L. 1991. MHC polymorphism and the design of captive breeding programs. Conservation Biology 5: 249–251.

Johnson, T. H., and A. J. Stattersfield. 1990. A global review of island endemic birds. Ibis 132: 167–180.

Johnston, R. F., and R. K. Selander. 1964. House sparrows: Rapid evolution of races in North America. Science 144: 548–550.

Jørgensen, C., K. Enberg, E. S. Dunlop, R. Arlinghaus, D. S. Boukal, K. Brander, B. Ernande, A. Gardmark, F. Johnston, S. Matsumura, H. Pardoe, K. Raab, A. Silva, A. Vainikka, U. Dieckmann, M. Heino, and A. D. Rijnsdorp. 2007. Managing evolving fish stocks. Science 318: 1247–1248.

Kinnison, M. T., M. J. Unwin, N. C. Boustead, and T. P. Quinn. 1998. Population specific variation in body dimensions of adult chinook salmon (*Oncorhynchus tshawytscha*) from New Zealand and their source population, 90 years after introduction. Canadian Journal of Fisheries and Aquatic Sciences 55: 554–563.

Leopold, A. 1949. A Sand County Almanac. New York: Ballantine Books.

Losos, J. B., T. W. Schoener, K. I. Warheit, and D. Creer. 2001. Experimental studies of adaptive differentiation in Bahamian Anolis lizards. Genetica 112–113: 399–415.

Montgomery, D. R., E. M. Beamer, G. R. Pess, and T. P. Quinn. 1999. Channel type and salmonid spawning distribution and abundance. Canadian Journal of Fisheries and Aquatic Sciences 56: 377–387.

Nelson, K., and M. Soulé. 1987. Genetical conservation of exploited fishes. In N. Ryman and F. Utter, eds., Population Genetics and Fishery Management. Seattle: University of Washington Press, 345–368.

Pergams, O.R.W., W. M. Barnes, and D. Nyberg. 2003. Rapid change of mouse mitochondrial DNA. Nature 423: 397.

Pergams, O.R.W. and R. C. Lacy. 2007. Rapid morphological and genetic change in Chicago-area *Peromyscus*. Molecular Ecology (Online Early Articles). doi:10.1111/j.1365-294X.2007.03517.

Plato. 360 BC. Timaeus. Online version accessed July 23, 2007, at http://www.ac-nice.fr/philo/textes/Plato-Works/25-Timaeus.htm.

Rubenstein, K. D., P. Heisey, R. Shoemaker, J. Sullivan, and G. Frisvold. 2005. Crop Genetic Resources—An Economic Appraisal. USDA Economic Information Bulletin No. 2 accessed August 28, 2007, at http://www.ers.usda.gov/Publications/EIB2/.

Ruckelshaus, M., K. Currens, R. Fuerstenberg, W. Graeber, K. Rawson, N. Sands, J. Scott, and J. Doyle. 2001. Independent Populations of Chinook Salmon in Puget Sound. April 2001 Memo from Puget Sound Technical Recovery Team. National Marine Fisheries Service, Northwest Fisheries Science Center.

Sjögren, P., and P.-I. Wyöni. 1994. Conservation genetics and detection of rare alleles in finite populations. Conservation Biology 8: 267–270.

Smith, T. B., L. A. Freed, J. K. Lepson, and J. H. Carothers. 1995. Evolutionary consequences of extinctions in populations of a Hawaiian honeycreeper. Conservation Biology 9: 107–113.

Smith, T. B., G. Grether, I. Sepil, H. Slabbekoorn, W. Buermann, S. Saatchi, B. Milá, and J. Ollinger. 2007. Microevolutionary consequences of human disturbance in a rainforest species from Central Africa. Presentation given Feb. 8, 2007, at Evolutionary Change in Human-Altered Environments: An International Summit. Los Angeles: UCLA.

Thomas, C. D., A. Cameron, R. E. Green, M. Bakkenes, L. J. Beaumont, Y. C. Collingham, B.F.N. Erasmus, M. Ferreira de Siqueira, A. Grainger, L. Hannah, L. Hughes, B. Huntley, S. van Jaarsveld, G. F. Midgley, L. Miles, M. Ortega-Huerta, A. T. Peterson, O. L. Philips, and S. E. Williams. 2004. Extinction risk from climate change. Nature 427: 145–148.

Waples, R. S. 1991. Pacific salmon, *Oncorhynchus* spp., and the definition of "species" under the Endangered Species Act. U.S. National Marine Fisheries Service, Marine Fisheries Review 53: 11–22.

Ward, D. E., S.W.M. Kengen, J. van der Oost, and W. M. de Vos. 2000. Purification and characterization of the alanine aminotransferase from hyperthermophilic archaeon *Pyrococcus furiosus* and its role in alanine production. Journal of Bacteriology 182: 2559–2566.

Weitzman, M. L. 1998. The Noah's ark problem. Econometrica 66: 1279–1298.

Wilson A. J., D. W. Coltman, J. M. Pemberton, A.D.J. Overall, K. A. Byrne, and L.E.B. Kruuk. 2005. Maternal genetic effects set the potential for evolution in a free-living vertebrate population. Journal of Evolutionary Biology 18: 405–414.

VI.11

The Economics of Ecosystem Services

Charles Perrings

OUTLINE

Why does ecosystem change matter? Why should nonecologists care about trends that alarm most ecologists? The answers to questions like these lie in the economics of the ecosystem change. For many ecologists, however, such a statement is itself part of the problem because they understand the "economics of ecosystem change" to mean "the money that people make from ecosystem change." Of course that is part of the story. The money people make from ecosystem change does in part drive that change. But economics is not just about money. Economics is about the decisions that people make, the factors that drive those decisions, and their consequences for human well-being. The economics of the environment has much to say not just about the reasons why the pursuit of self-interest leads to undesirable ecosystem change but about why this matters to people and what can be done about it.

GLOSSARY

The following brief definitions cover some of the terms commonly used by economists that may not be familiar to readers of this book.

complementarity. Two goods are said to be complements if an increase in the price of one induces demand for the other to fall—formally, when the cross elasticity of demand is negative.

existence value. This is intended to capture peoples' willingness to pay for the mere existence of something. It is often used loosely, though, to refer to spiritual or aesthetic values, and sometimes as a substitute for intrinsic value (see below).

externality. External economies or diseconomies of production and consumption are called externalities. An externality is a third-party effect of a transaction that is not taken into account by the parties to the transaction. External effects may be positive or negative and drive a wedge between the private and social net benefits of a transaction.

intrinsic value. This is not a term in general use by economists. Noneconomists use the term to refer to the value that other species have independent of their value to people. As with existence value, the term is often used very loosely and in practice may refer to anthropocentric spiritual or aesthetic values.

joint product. When a production function (see below) generates multiple outputs, they are said to be joint products.

nonuse value. This is the value of an allocation that benefits someone other than the user. It derives from the fact that the user cares for the beneficiary. Note that the beneficiary may be some other species or a member of a future generation.

option value. The value of the option to use a resource in the future.

private optimum. The allocation that optimizes a private decision maker's objective function. If there are externalities, this will be different from the social optimum (see below).

production function. A function relating the inputs to and outputs of a production process. It embeds both the technological and the biogeochemical aspects of the production process.

renewable resources. Resources are said to be renewable when they regenerate themselves within a timeframe that is relevant to the decision process.

shadow price. This is the social opportunity cost of a resource—its true value to society. If there are externalities, implying that markets are incomplete, the shadow value will be different from the market price. Formally, it is the value of the co-state-variable along the optimal path in the solution to a state-space optimization problem.

social optimum. The allocation that optimizes the social welfare function or index of social well-being.

substitutability. Two goods are said to be substitutes if an increase in the price of one induces demand for the other to rise—formally, when the cross elasticity of demand is positive.

use value. This is the value of resources when used by the valuer. The value of resources that are used by someone other than the valuer are said to be nonuse values (see above).

1. INTRODUCTION

Ecosystem services were defined by the Millennium Ecosystem Assessment (MA) as the benefits people obtain from ecosystems (MA, 2005). It recognized that people care about the environment because of the services it offers. These services include not just the provision of foods, fuels, and fibers but a range of nonconsumptive benefits such as recreation, amenity, and spiritual renewal. Where the MA went much further than earlier studies was in identifying the indirect supporting services provided by ecosystems, including the regulation of environmental stresses and shocks, such as emergent zoonotic diseases and the role of ecosystem processes in supporting all other services.

From an economic perspective, the provisioning and cultural services together describe the environmentally derived goods and services that enter final demand—that people consume in some sense. Provisioning services comprise what have traditionally been called "renewable resources": foods, fibers, fuels, water, biochemical compounds, medicines, pharmaceuticals, and genetic material. Many of these products are directly consumed and are subject to reasonably well-defined property rights. They are priced in the market, and even though there may be important externalities in their production or consumption, those prices bear some relation to the scarcity and value of resources. Cultural services, on the other hand, define many of the nonconsumptive uses of the environment such as recreation, tourism, education, science, and learning. They include, for example, the spiritual, religious, aesthetic, and inspirational well-being that people derive from the "natural" world; their sense of place and the cultural importance of particular landscapes and species, and the traditional and scientific information, awareness, and understanding offered by functioning ecosystems. One modern expression of cultural services—ecotourism—involves well-developed markets. Most others do not. Although intellectual property rights in biochemical and genetic material drawn from ecosystems are increasingly well defined, many cultural services are still regulated by custom and usage or by traditional taboos, rights, and obligations. Nevertheless, because they are directly used by people, they are also amenable to valuation by methods designed to reveal people's preferences.

The third major category of ecosystem services identified by the MA, the regulating services, is in many ways the most interesting. For the MA, the category includes the following:

- Air quality regulation involves chemicals contributed to and extracted from the atmosphere, influencing many aspects of air quality.
- Climate regulation stems from the fact that ecosystems influence climate both locally and globally. So, for example, changes in land cover can affect both temperature and precipitation at a local scale, and changes in carbon sequestration or greenhouse gas emissions have significant effects at a global scale.
- Water regulation affects runoff, flooding, and aquifer recharge through changes in land cover and depends on the mix of plant species and soil microorganisms.
- Erosion regulation depends on vegetative cover and plays an important role in soil retention and the prevention of landslides.
- Water purification and waste treatment services are both positive and negative and include both water pollution and filtration in inland waters and coastal ecosystems. They also include the capacity to assimilate and detoxify soil and subsoil compounds.
- Disease regulation services are also both positive and negative and include change in the abundance of human pathogens, such as cholera, or disease vectors such as mosquitoes.
- Pest regulation involves the role of ecosystems in determining the prevalence of crop and livestock pests and diseases.
- Pollination services depend on the distribution, abundance, and effectiveness of pollinators.
- Natural hazard regulation covers a wide range of buffering functions, particularly in coastal ecosystems, where mangroves and coral reefs can reduce the damage caused by hurricanes and storm surges.

What these all have in common is that they affect the impact of stresses and shocks on the system. More particularly, the regulating services moderate the impact of perturbations on the provisioning or cultural services. By changing the potential cost associated with a given shock, they influence the environmental risks people face. It follows that the regulating services will be more or less valuable depending on which of the provisioning and cultural services people value, the regime of shocks to those services, and peoples' aversion to risk.

Finally, the category of support services captures the main ecosystem processes that underpin all other services. Examples offered by the MA include soil formation; photosynthesis; primary production; and nutrient, carbon, and water cycling. These services typically play out at different spatial and temporal scales. For example, nutrient cycling involves the maintenance of the roughly 20 nutrients essential for life in different concentrations in different parts of the system. It is often localized and is therefore at least partially captured by the price of the land on which it takes place. Carbon cycling, on the other hand, operates at a global scale and is very poorly captured in any set of prices. Because the supporting services are embedded in the other services, however, they are captured in the value of those services, whether or not that value is expressed in market prices.

This chapter reviews the economics of ecosystem services in the light of the MA. The MA (2005) noted that although the supply of a number of the provisioning services has been increasing over the last 50 years, the supply of many of the cultural, regulating, and supporting services has been declining. It also noted that this reflects the failure of markets to allocate resources efficiently and drew attention to the implications this has for human well-being, particularly in poorer countries. There are three dimensions to the economics of ecosystem services, each of which is explored in the following sections. The first is the measurement of their impact on production, consumption, and human well-being. The second is the identification of the gap between the socially optimal level of services relative to the level of services actually provided—i.e., the measurement of the extent to which markets fail to allocate ecological resources either efficiently or equitably. The third is the development of policies (instruments and institutions) that will close the gap.

2. THE VALUATION OF ECOSYSTEM SERVICES

A number of studies before the MA drew attention to the changes in ecosystem services and the importance of quantifying the value of these changes to human societies in terrestrial, marine, and agroecosystems. There were also attempts to identify and value ecosystem services. However, most of these failed simply because of our limited understanding of the role of ecosystem services in the production of things that people recognize and value. In part this is because ecosystem services are themselves what economists would call the "joint products" of ecosystems. Daily et al. (1997) had emphasized that most ecosystem services were the result of a complex interaction between natural cycles operating over a wide range of space and time scales. Waste disposal, for example, depends on both highly localized life cycles of bacteria and the global cycles of carbon and nitrogen. The same cycles are implicated in the provision of a range of other services. By ignoring multiple services and the interdependence among services, many early valuation studies underestimated the importance of the ecosystem stocks to the economy (Turner et al., 2003).

Another problem with many valuation studies stems from the fact that they elicit peoples' preferences for the asset being valued. When the object of valuation is familiar—is directly consumed or experienced by the person whose preferences are being elicited—this can lead to reasonably reliable estimates. But many early studies of the value of ecosystem services elicited preferences for environmental stocks from people who had little conception of the role and importance of those stocks. The problem here is that ecosystems and the services they provide are, for the most part, intermediate inputs into goods and services that are produced or consumed by economic agents. As with other intermediate inputs, their value derives from the value of those goods and services but may not be transparent to the end users (Heal et al., 2005).

In this case, the use of derived demand ("production function") methods are appropriate, and there are a growing number of studies of value of ecosystem services that use such an approach (e.g., Barbier, 2007). When output of the goods and services that enter final demand is measurable and either has a market price or one can be imputed, and when the connections among ecological functioning, ecosystem services, and human production processes are well understood, then determining the marginal value of the resource is relatively straightforward.

To illustrate, consider the following simplified description of an archetypal decision problem. The decision maker chooses a time path for the level of effort made in exploiting a number of resources, given by the vector $\mathbf{h}(t)$. The objective is to maximize some index of well-being, captured by the function $u(.)$. This is done over a time horizon, T, that could be infinite.

$$Max_{\mathbf{h}(t)} \int_{t=0}^{T} u\{\mathbf{q}(t)\{\mathbf{x}(t)[\mathbf{s}(t)]\}, \mathbf{h}(t)\}e^{-\delta t}dt.$$

The flow of net benefits depends on a vector of produced goods, $\mathbf{q}(t)$, which in turn depends on vectors of marketed inputs, $\mathbf{x}(t)$, the state of the environment, $\mathbf{s}(t)$, and effort, $\mathbf{h}(t)$. This flow is discounted at the rate δ.

The index of well-being is maximized subject to the capacity of the resources of the natural environment to grow or to regenerate, which is summarized in the equations of motion:

$$\frac{ds_i}{dt} = f_i[\mathbf{s}(t)] - h_i(t), i = 1, \ldots, n.$$

Now the value of the n ecosystem stocks in this problem is their social opportunity cost, measured by the "shadow price" obtained from the solution to the optimization problem. Specifically, if the shadow prices in the solution to the problem are denoted λ_i, then they will evolve as follows:

$$\frac{d\lambda_i}{dt} = \lambda_i(\delta - f_i') - \sum_j \frac{du}{dq_i}\frac{dq_i}{d\mathbf{x}}\frac{d\mathbf{x}}{ds_i}, i = 1, \ldots, n,$$

and in the steady state, take the value:

$$\lambda_i = \frac{\sum_j \frac{du}{dq_i}\frac{dq_i}{d\mathbf{x}}\frac{d\mathbf{x}}{ds_i}}{\delta - f_i'}, i = 1, \ldots, n.$$

So the value of the ith ecosystem stock depends (1) on its regeneration rate relative to the yield on produced capital, indicated by the discount rate, and (2) on its marginal impact on the production of the set of marketed outputs, $\mathbf{q}(t)$, through the effect it has both on other ecosystem stocks, $\mathbf{s}(t)$, and on marketed inputs, $\mathbf{x}(t)$.

If output cannot be measured directly but there is a marketed substitute for it, or the complementarity or substitutability between ecosystem services and one or more marketed inputs is understood, the same general approach can be used. If output can be measured, but there is no market for it, then stated preference nonmarket valuation techniques can be used in combination with a production function approach to derive the value of ecosystem services. Allen and Loomis (2006) use such an approach to derive the value of species at lower trophic levels from the results of surveys of willingness to pay for the conservation of species at higher trophic levels. Specifically, they derive the implicit willingness to pay for the conservation of prey species from direct estimates of willingness to pay for top predators. They make the point that it is not necessary for consumers to understand the trophic structure of an ecosystem because their willingness to pay for top predators effectively captures their willingness to pay for the whole system.

The attributes of valued ecosystems are reflected in the constituents of what some economists have defined as "total economic value" (Turner et al., 2003). In this literature, "use value" refers to benefits deriving from consumptive or nonconsumptive use by the individual, whereas "nonuse value" comprises benefits from consumptive or nonconsumptive use by others. The important point here is that an ecosystem service may have value to people even if it is not part of their consumption bundle, providing that the consumer cares about the people or species who do consume it. Moreover, even if an ecosystem service is not currently used, it may still have what is referred to as an "option value." So, for example, the option value of existing species may lie in their role in combating a currently unknown disease or pest or in ensuring ecosystem functions in currently unknown environmental conditions or in providing opportunities for future generations. Where ecosystem services have option value, and where there is a high level of uncertainty about future potential uses of biodiversity, economists have shown that it is frequently optimal to postpone irreversible decisions (such as those that might lead to extinction) in order to learn more (Heal et al., 2005).

Many economists have argued that some ecosystem attributes, including some species, have what they term "existence value." This was originally defined as people's willingness to pay to ensure the continued existence of biodiversity irrespective of any actual or potential use by present or future generations of humans. In practice, it is a form of ethically or religiously motivated altruism toward other species. In this respect, it is identical to another category of value favored by noneconomists: "intrinsic value." Intrinsic value, like existence value or many of the so-called nonuse values, reflects human preferences for the rights and well-being of other species. It is not—and cannot be—independent of peoples' preferences, including their perceptions of their own roles and responsibilities.

To return to our central argument, however, all these values refer to end uses. They can therefore be used to derive the value of the ecosystem services that support them. This applies to all of the MA services. So, for example, the regulating services such as storm protection or flood mitigation may be valued through the expected damage (losses of goods and services that enter final demand) avoided or more generally through the impact on well-being of a change in the distribution of the provisioning and cultural services.

3. ECOSYSTEM SERVICE EXTERNALITIES

Part of the problem identified by the MA is that the true value of many ecosystem services—their social opportunity cost—is ignored by decision makers. Economists typically refer to the values ignored in normal market transactions as externalities. Externalities of the kind described in the MA (2005) are often referred to as "ecosystem externalities," by which is meant the unintended effects of market transactions on human well-being through changes in biodiversity, ecological functioning, or ecosystem services. That is, an ecosystem externality involves a change in ecosystem services that impacts the well-being of others but that is ignored by the parties to a market transaction. It follows that the measurement of ecosystem externalities depends on an understanding of the biogeochemical processes involved as well as the sources of market failure. To model ecosystem externalities, economists accordingly combine human behavioral models with models of biodiversity–ecosystem functioning–ecosystem services relationships (Watzold et al., 2006).

For example, nitrogen compounds emitted from coal-fired power plants and mobile sources directly impact human health but also significantly change ecosystems. In aquatic systems, nitrogen leads to algae blooms that consume oxygen, lower dissolved oxygen levels, and work their way up food chains to reduce fish and shellfish densities. In terrestrial systems, nitrogen decreases species diversity and changes community composition. In both systems, the altered ecosystem functions generate fewer ecosystem services in the form of reduced fishing and grazing opportunities and reduced cultural values. Maximizing social well-being requires understanding of the trade-offs between the net benefits from consumptive and nonconsumptive use of ecosystem services and the costs that consumptive and nonconsumptive uses have in terms of future ecosystem services.

Economists approach this problem by distinguishing between the choices of private decision makers in existing conditions and the socially optimal outcome when all interactions are taken into account. By comparing the two decision problems, it is possible to show how and when the decisions of private agents deviate from the social optimum and hence to develop mechanisms to close the gap.

There are numerous examples of ecosystem externalities at all spatial scales. Many are a product of the disposal of wastes generated in the process of production and consumption, i.e., a consequence of emissions to land, air, and water. Others are a product of the use made of land, air, and water, i.e., result from the conversion of ecosystems for the production of foods, fuels, fibers, domestic dwellings, transport, water or power infrastructure, recreation, or amenity. In all cases, externalities are evidence of incomplete markets. Because they lie outside the set of property rights defined by society, they are not taken into account in the transactions between people. There are several reasons why markets fail to evolve to deal with scarce environmental resources, including (1) the public good nature of many environmental assets (because public goods are nonexclusive by definition, it is not possible for individuals to assert rights to them), and (2) the lack of an institutional framework within which to identify and enforce property rights, as is the case with environmental assets in areas beyond national jurisdiction. However, even where markets do exist and are reasonably complete, there are many reasons why they do not function effectively. The most serious of these are the effect of distortionary macroeconomic policies and the pervasive use of subsidies (Barbier, 2007).

The existence and persistence of international ecosystem externalities depend on the way that international markets and the rules of international trade are structured. They also depend on the incentive effects of different property right regimes. It has, for example, been argued that firms exploit the international advantages offered by relaxed labor and environmental laws and that countries will use the lack of environmental protection to induce inward investment. By this argument, ecosystem externalities are not just an incidental product of market failures. They are the outcome of strategic decisions by governments and firms seeking a competitive advantage. The claim is that where the General Agreement on Tariffs and Trade (GATT) and other trade agreements make it impossible either to induce inward investment or to protect domestic agriculture or industry through trade policy, countries use environmental policies to the same effect. Specifically, they either allow ecological dumping by relaxing environmental protection measures or use environmental regulation as trade protection measures.

The evidence for this is ambiguous, however. The relocation of polluting industries from high-income to low-income countries is certainly a part of the explanation for changes in environmental indicators observed in the literature. But studies of the incentive effects of environmental regulation in the 1990s concluded that environmental compliance costs were not generally important enough to drive location decisions. On the other hand, the environmental impacts of trade are one of the few acceptable justifications for imposing trade restrictions under GATT. The exceptions allowable under Article XX of GATT, along with the Sanitary and Phytosanitary (SPS) Agreement, authorize countries to impose restrictions on trade in order to protect human, animal, and plant life.

Nevertheless, it is the case that the liberalization of international trade through successive renegotiations of GATT has done little to address many existing international environmental market failures and has created many more. The solution to the problem has been the development of multilateral environmental agreements (MEAs) to address specific environmental problems; the most important of which are the Convention on Biological Diversity and the Convention on International Trade in Endangered Species, which deals specifically with international markets for biological resources. Beyond this there are a range of agreements dealing with particular regional issues or with particular species. In general, bilateral agreements are more effective than multilateral agreements. Although some MEAs—such as the Montreal Protocol—have been credited with making a significant difference to environmental quality, most economic research on the problem suggests that agreements with many signatories are unable to address the most important global ecosystem externalities (Barrett, 2003).

4. ECONOMIC INSTRUMENTS FOR ECOSYSTEM EXTERNALITIES

A third important aspect of the economics of ecosystem services is the development of policies and instruments to internalize ecosystem externalities and to eliminate market distortions. The most important point here is that if markets can be created, and if they include all relevant effects, they will signal the social opportunity cost of local resource uses and so will ensure that the full effects of local decisions are taken into account. They will also make it possible for those who are willing to pay for the conservation of ecosystems to do so. In other words, if properly structured, such instruments can close the gap between private and social interests.

The development of markets for ecosystem services other than the provisioning services and a few cultural services (ecotourism) is still in its infancy, but economists have evaluated a number of options. One example is transferable development rights. These are similar to cap-and-trade schemes to limit pollution emissions but involve rights to develop land in one location in exchange for conservation in another location. In Brazil, for instance, agricultural landowners not currently complying with the National Forest Code are offered the opportunity to meet conservation targets by acquiring forest reserves in other areas. A second example would be auction contracts for conservation (ACCs). These are helpful when there is an information asymmetry between farmers and conservation agencies regarding, respectively, the financial costs and ecological benefits of conservation. Landholders submit bids to win conservation contracts from the government, thereby revealing their willingness to accept compensation for taking land out of production. A pilot auctioning system for biodiversity conservation contracts in Victoria, Australia, known as *BushTender*, provides 75% more conservation than comparable fixed-price payment schemes (Stoneham et al., 2007).

Two other examples are payments for the provision of ecosystem services (PES) and direct compensation payments (DCPs). Like ACCs, PES offer compensation, in cash or kind, for ecosystem services. Services to which this has already been applied include provision of water, soil conservation, and carbon sequestration by upland farmers who manage forest lands in upper watersheds. For example, Costa Rica's 1996 Forestry Law instituted payments for four ecosystem services: mitigation of greenhouse gas emissions, watershed protection, biodiversity conservation, and scenic beauty. The National Forestry Financial Fund enters into contracts with landowners agreeing to forest preservation. DCPs are a variant of PES. They offer direct compensation to landholders for putting private land into conservation. It should be noted that in all cases, standards (establishment of levels of protection, or for the amount of money to be expended in auction) are needed in addition to the markets.

Where the creation of markets is not an option, but there exists a sovereign authority, alternative measures to internalize ecosystem externalities include both regulation and pricelike mechanisms. Regulations, including emissions standards, harvest restrictions, proscriptions, and so on, are still the mechanism of choice in many countries. However, they are not generally as efficient as the alternative, and economists typically favor either mixed regulation/market instruments, such as cap-and-trade mechanisms, or pure market instruments such as taxes, subsidies, grants, compensation payments, user fees, access fees, and charges. The best-known examples are carbon taxes, but there are a wide range of instruments of this kind currently applied. In all cases, the principle is the same. The user of a resource is confronted with the social opportunity cost—the marginal external cost—of his or her (in the case of a corporation, its) actions. This induces the socially optimal response, and that response is independent of whether the user is aware of the environmental implications of those actions.

At the international level, the lack of any sovereign authority precludes many of these instruments, and the only options for addressing ecosystem externalities are bilateral or multilateral agreements and defensive tariffs. Some of the difficulties with MEAs have already been alluded to. There are similar difficulties with defensive tariffs. Although tariffs are justified when trade carries with it some risk that is not reflected in the price of traded goods, GATT makes it difficult to implement tariffs for this purpose. Indeed, despite Article XX and the SPS agreement, many economists argue that GATT is simply too blunt an instrument to deal with the environmental effects of trade (Barrett, 2000).

5. CONCLUDING REMARKS

There are ultimately two problems to be addressed in the optimal provision of ecosystem services. The first is the problem of local market failure. The solution to this lies in the development of local or national policy responses on the provision of environmental public goods and the internalization of ecosystem externalities. The second is the problem of international market failure. It includes both the undersupply of global environmental public goods and the externalities of international trade. Both require the development of (1) incentives to decision makers to take the full costs of their actions into consideration, (2) institutions for the regulation of access to ecosystem services, and (3) an appropriate financial mechanism. The incentive problem requires both the generation of the correct incentives for biodiversity conservation and the discouragement of perverse incentives that work against conservation.

At the international level, the incentive problem requires institutions that will enable host countries to "capture" the global values associated with the provision of ecosystem services that offer global benefits. Existing institutions include both the MEAs and a financial mechanism, the Global Environment Facility. Other emerging institutions include joint implementation, bioprospecting contracts, global overlays, environmental funds, and debt-for-nature swaps. Although the Convention on Biological Diversity and the International Plant Protection Convention are critical to the development of new biodiversity institutions and mechanisms, there is limited scope for solving the problem of declining ecosystem services by negotiating cooperative outcomes in multilateral agreements of this type (Barrett, 2003).

Nonetheless, the development of appropriate incentives remains the best hope for arresting and reversing the decline in the supply of important ecosystem services identified by the MA. In this, economics has a critical role to play—both in the identification and measurement of ecosystem externalities and in the development of mechanisms to internalize those externalities. Far from being part of the problem, the economics of ecosystem services, along with the ecology of biodiversity–ecosystem functioning–ecosystem services, lies at the heart of the solution to the global crisis described by the MA.

FURTHER READING

Allen, B. P., and J. B. Loomis. 2006. Deriving values for the ecological support function of wildlife: An indirect valuation approach. Ecological Economics 56: 49–57.

Barbier, E. 2007. Valuing ecosystem services as productive inputs. Economic Policy, January: 177–229.

Barrett, S. 2003. Environment and Statecraft. Oxford: Oxford University Press.

Daily, G. C., S. Alexander, P. R. Ehrlich, L. Goulder, J. Lubchenco, P. A. Matson, H. A. Mooney, S. Postel, S. H. Schneider, D. Tilman, and G. M. Woodwell. 1997. Ecosystem services: Benefits supplied to human societies by natural ecosystems. Issues in Ecology 1(2): 1–18.

Heal, G. M., E. B. Barbier, K. J. Boyle, A. P. Covich, S. P. Gloss, C. H. Hershner, J. P. Hoehn, C. M. Pringle, S. Polasky, K. Segerson, and K. Shrader-Frechette. 2005. Valuing Ecosystem Services: Toward Better Environmental Decision Making. Washington, DC: National Academies Press.

Millennium Ecosystem Assessment (MA). 2005. Ecosystems and Human Well-Being: Synthesis. Washington, DC: Island Press.

Stoneham, G., V. Chaudhri, L. Strappazzon, and A. Ha. 2007. Auctioning biodiversity conservation contracts. In A. Kontoleon, U. Pascual, and T. Swanson, eds. Biodiversity Economics. Cambridge, UK: Cambridge University Press.

Turner, R. K., J. Paavola, P. Cooper, S. Farber, V. Jessamy, and S. Georgiou. 2003. Valuing nature: Lessons learned and future research directions. Ecological Economics 46: 493–510.

Wätzold, F., M. Drechsler, C. W. Armstrong, S. Baumgärtner, V. Grimm, A. Huth, C. Perrings, H. P. Possingham, J. F. Shogren, A. Skonhoft, J. Verboom-Vasiljev, and C. Wissel. 2006. Ecological–economic modeling for biodiversity management: Potential, pitfalls, and prospects. Conservation Biology 20: 1034–1041.

VI.12

Technological Substitution and Augmentation of Ecosystem Services

Indur M. Goklany

OUTLINE

1. Augmenting nature's productivity as technological substitution
2. Substitution possibilities for ecosystem services
3. Implementing technologies to replace or extend nature's services

This chapter briefly identifies some technologies that would augment or replace ecosystem services in order to reduce the direct human demand on nature. This identification is meant to be illustrative rather than comprehensive. This chapter does not, however, evaluate the net efficacy or desirability of listed technologies based on their costs, benefits, and impacts on nature. Those issues are outside this chapter's scope.

GLOSSARY

ecosystem services. The benefits that ecosystems provide human beings. They include critical provisioning services such as food, timber, fiber, fuel and energy, and fresh water; regulating services that affect or modify, for instance, air and water quality, climate, erosion, diseases, pests, and natural hazards; cultural services such as fulfilling spiritual, religious, and aesthetic needs; and supporting services such as soil formation, photosynthesis, and nutrient cycling. This chapter does not explicitly address supporting services; they are implicit in the ability of ecosystems to deliver the other services.

substitute (or replacement) technologies. Technologies that wholly substitute for some facet or portion of goods and services that ecosystems provide for humanity.

technological augmentation of ecosystem services. The increase, through technological intervention, in the production of goods and services that nature provides. By helping fulfill humanity's needs while limiting its direct demand on nature, such augmentation substitutes for natural inputs from ecosystems.

technology. Both tangible human-crafted objects or "hardware" (such as tools and machines) and human-devised intangibles or "software" (such as ideas, knowledge, programs, spreadsheets, operating rules, management systems, institutional arrangements, trade, and culture).

1. AUGMENTING NATURE'S PRODUCTIVITY AS TECHNOLOGICAL SUBSTITUTION

Nature once produced virtually every service, good, or material that humanity used. It supplied all food, fiber, skins, water, and much of the fuel, medicines, and building materials. Over time, human beings developed technologies to coax more of these services from nature, often at the expense of other species. Agriculture and forestry increased the production of food, fiber, and timber. Human beings also developed animal husbandry, commandeering other species to serve their needs for a steadier protein diet and for fiber and skins for bodily warmth and protection; to do work on and off the farm; and to transport goods and people. Gradually at first but faster in the past century, technological substitutes were developed that reduced human demand met directly by nature's services. Thus, synthetic fiber today limits human demand on nature to provide for clothes, skins, and leather; vinyl, plastics, and metals reduce reliance on timber for materials;

fossil fuels—themselves products of nature—and nuclear power reduce pressures on forests and other vegetation to provide humanity's energy needs; synthetic drugs reduce harvesting of flora and fauna for life-saving medicines; and fossil fuel–powered machines and telecommuting increasingly substitute for animal and human power. Nevertheless, population and economic growth continue to increase aggregate demand for most ecosystem services, and the adverse impacts of substitutions may compromise many ecosystems' abilities to provide other services.

The term *technology* as used here includes tangible human-crafted objects or "hardware" (e.g., tools and machines) and human-devised intangibles or "software" (e.g., knowledge, programs, spreadsheets, operating rules, management systems, institutional arrangements, trade, and culture) (Ausubel, 1991; Goklany, 2007). There is substantial skepticism, reinforced by the Biosphere 2 project's costly failure, about technology's ability to adequately substitute for ecosystem services (Daily et al., 1997). Nevertheless, the Millennium Ecosystem Assessment acknowledges technology's role in helping to meet human demand, particularly for provisioning services such as food, while recognizing that adverse impacts accompany these technologies (MEA, 2005a, 2005b). Recognizing this, Palmer et al. (2004) suggest the use of "designer ecosystems" to reduce humanity's load on nature. Noting that designed ecosystems are imperfect ecological solutions and may not pass muster with many conservationists and ecologists, they recommend their use as part of a future sustainable world to mitigate unfavorable conditions through a "blend of technological innovations, coupled with novel mixtures of native species, that favor specific ecosystem functions" rather than as full substitutes for natural systems (Palmer et al., 2004).

Indeed, although technology may occasionally wholly substitute for nature's goods and services, it will more frequently enhance their production. Because augmentation of nature's productivity reduces humanity's direct demand on nature, it is appropriately viewed as substituting for natural inputs from ecosystems. That is the view adopted in this chapter.

Consider food production. Had global agricultural productivity been frozen at its 1961 level, then the world would have needed over 3435 million hectares (Mha) of cropland rather than 1541 Mha actually used to produce as much food as it did in 2002 (Goklany, 2007: 161–163). Thus, technological innovation effectively substituted for over 1894 Mha of habitat, rivaling the total land reserved worldwide for conservation. Thus, arguably, in situ conservation has been enabled largely through augmentation of nature's services by agricultural technology.

Enhanced productivity was based substantially on increased pesticides, fertilizers, water, and fossil fuel inputs. However, such practices can have significant environmental costs. Preference should be given to practices that balance higher yields with lower inputs of land, water, and chemicals so that they "save" more of the environment than they destroy. And so it should be with other technologies for substituting or augmenting nature's services. However, although technology can reduce humanity's demands on nature, it cannot replace and/or substitute for nature down to the last detail. Arguably, given nature's complexity, it could not replicate itself in every detail if the clock were to be rolled back and restarted.

2. SUBSTITUTION POSSIBILITIES FOR ECOSYSTEM SERVICES

Table 1 contains a summary of various technologies that could enhance or substitute for nature's ecosystem services. The ecosystem services identified in this table are adapted from the Millennium Ecosystem Assessment (MEA, 2005a: table 1). The following provides details of some of the technological possibilities.

Food

Crops

Most food that humanity consumes today comes from technological augmentation of nature's services through agriculture. The earth's carrying capacity before agriculture has been estimated at 10 million people (Livi-Bacci, 1992: 29). However, the ecological footprint of its 6.3 billion people in 2003 was estimated to exceed carrying capacity by 23% (GFN, 2006). Therefore, assuming these estimates are accurate, present-day agriculture has boosted carrying capacity by over two orders of magnitude.

The world's population is likely to expand and become wealthier by midcentury, increasing food demand. Ideally, future agricultural practices will deliver higher yields but with lower natural and synthetic inputs (i.e., land, water, pesticides, fertilizers, and fossil fuels). Options include more intensive agriculture using conventional breeding techniques, genetically modified (GM) crops, and precision agriculture. These three approaches can coexist.

GM crops, in particular, have high potential for low-input high-yield agriculture that could produce more food per unit of land and water diverted to agriculture. Several GM crops are in various stages of development ranging from research to commercialization. (This discussion on biotechnology and GM

Table 1. Substitution possibilities for provisioning, regulating, and cultural ecosystem services

Service	*Product*	*Technologies or technological systems*
Provisioning services		
Food	Crops	High-yield agriculture, precision agriculture, GM crops
	Livestock	Cloning, breeding, artificial insemination, GM animals, fortified feeds, high-lysine feed
	Capture fisheries	Aquaculture, fish hatcheries, genetically modified fish, crop-based feeds
	Wild foods	Agriculture
Timber	Wood	High-yield tree crops, GM trees, aluminum, steel, plastics
Natural fiber	Cotton, silk, jute, flax, coir, hemp	Synthetic fibers, plastics
	Furs, skins	Synthetic fibers
Fuel	Wood, hydropower, wind	Fossil fuels, photovoltaics, higher-efficiency wind and solar, geothermal, nuclear, high-yield biofuel crops, cellulosic ethanol
Transportation and work	Beasts of burden	Bicycles, mechanized transport (i.e., trucks and cars), airplanes, tractors
Genetic resources		Polymerase chain reaction, gene banks, zoos, botanical gardens
Biochemicals, medicines, pharmaceuticals		Synthetic drugs and pharmaceuticals, GM "pharms," biofactories
Fresh water		Water purification and treatment, recycling and reuse technologies, desalination, water pricing and marketing, property rights for water
Regulating services		
Air quality regulation	Traditional air pollutants	Scrubbers, fabric filters and electrostatic precipitators for traditional air pollutants; emissions trading
Climate regulation at local, regional, and global scales		Carbon sequestration on land, oceans, geologic formations; conservation tillage; geoengineering; modification of land cover and albedo
Water regulation		Water purification and treatment, recycling and reuse technologies, desalination, water pricing and marketing, property rights for water
Erosion regulation		No- or low-till agriculture, hydroponic cultivation, cover crops
Water purification and waste treatment		Chlorination, waste water treatment, filtration, reduction in oxygen demand
Disease regulation		Chlorination, drugs and pharmaceuticals, insecticides
Pest regulation		Insecticides, integrated pest management, GM crops
Pollination		Managed pollination via nonnative/cultured pollinators (e.g., European honeybee in the United States), hand/mechanical pollination, electrostatic enhancement
Natural hazard regulation		Artificial or restored wetlands and mangroves, dams, sea walls, levees, dikes, concrete and steel houses
Cultural services		
Spiritual/religious values		Photographs, movies, videos, HD and holographic television, virtual reality
Aesthetic values, recreation, and ecotourism		Constructed or augmented landscapes and ecosystems, artificial reefs, zoos, arboretums, photographs, movies, videos, HD and holographic television, virtual tourism

Note: Most substitutes are imperfect, and some more than others; however, they provide products that would otherwise come from nature.

technologies draws liberally on Goklany [2007: chapter 9].) For example, soil and climatic conditions are frequently less than optimal for specific agricultural crops. Accordingly, bioengineered grains are being developed to tolerate such suboptimal conditions (i.e., drought, water logging, salinity, iron-deficient soil, or soils that are too acidic, too alkaline, or have excess aluminum). Similarly, staples—rice, maize, wheat, cassava, sorghum—are being bioengineered to resist biotic stresses such as insects, nematodes, bacteria, viruses, fungi, weeds, and other pests. Such crops ought to reduce pesticide usage. Spoilage-prone fruits (e.g., melons, papaya, and tomatoes) are being bioengineered to delay ripening, increase shelf life, and reduce postharvest losses. And the list goes on.

By increasing food produced per unit of land, water, and chemical inputs, GM crops would maintain or increase yields while reducing environmental impacts associated with agricultural activities. Higher yields would also reduce habitat loss, landscape fragmentation, pressures on freshwater biodiversity, pesticide and fertilizer usage, and soil erosion, which then improves water quality and conserves carbon sinks and stores. Thus, cultivation of GM crops may also displace use of more toxic pesticides with less toxic and/or less persistent ones.

Notably, GM crops have been cultivated commercially since 1996 without any detectable effects on human health. Experience worldwide indicates that they have reduced pesticide usage and increased yields and farmers' profits (see Pest Regulation, below).

Food production per unit of land, water, and chemical inputs can also be extended through precision agriculture, which uses combinations of high- and low-tech monitors, global positioning systems, computers, and process controllers to optimize the amount and timing of delivering the various inputs, based on the cultivar, soil, and climatic conditions specific to the farm (Goklany, 2007: 393).

Livestock

Both conventional and bioengineering techniques can also be applied to increase livestock productivity by improving feed crops or the livestock so they can utilize feed more efficiently and reduce nutrients excreted in their wastes. For example, lysine is an amino acid that improves protein utilization in animals. Therefore, lysine supplements or high-lysine corn and soybeans improve livestock feed and reduce overall demand for land needed to produce animal protein, perhaps by as much as three-quarters. Similarly, improving utilization of phosphorus in feed reduces phosphorus in livestock excrement and, consequently, nutrient loadings in the environment. This could be facilitated by using bioengineered corn and soybeans that are low in phytic acid and/or contain phytase, an enzyme that improves phosphorus utilization. Scientists have developed a transgenic pig that contains phytase in its saliva and excretes 75% less phosphorus. Finally, corn with high oil and energy content and forage crops with lower lignin content would also increase feed utilization by livestock, thereby also reducing demand for land and water to sustain livestock.

Capture Fisheries

The annual worldwide catch of marine and freshwater capture fisheries was approximately constant between 1995 and 2004 (FAO, 2006). But production from aquaculture—appropriately viewed as a substitute for capture fisheries—increased rapidly from 4% of total fisheries production in 1970 to 32% in 2004.

In 2005, half the marine capture fisheries stock groups monitored by FAO were fully exploited; one-quarter were underexploited or moderately exploited; the rest were overexploited, depleted, or recovering from previous overfishing. Inland capture fisheries were also generally overexploited. Thus, the potential for maintaining production from capture fisheries is currently low (Worm et al., 2006; but see Beddington et al., 2007). Accordingly, although aquaculture is not a panacea (Naylor et al., 2000), it will probably expand to meet demand for fish and other seafood (Goklany, 2007: 363–367). This can be aided by increasing production at fish hatcheries, developing GM strains that would use feed more efficiently or utilize plant-based feed, and developing methods to improve health of cultured species to reduce preconsumption losses.

Wild Food

Although the provisioning aspect of this service could be met through cultivation, that may not entirely fulfill deep-seated cultural, aesthetic, and psychic needs associated with the rituals of hunting, gathering, and consuming wild food (see below).

Timber

There are several technologies that would substitute for or augment the production of timber. These include tried-and-true approaches such as increased utilization of harvested product to reduce wastage (e.g., through the manufacture of plywood and other engineered woods or computer-controlled manufacture of veneers), using high-yield tree crops developed through conventional techniques, and meeting demand via vinyl, plastics, and other petroleum-derived materials (e.g., fiberglass for insulation, or synthetics for flooring) or inorganic materials (e.g., aluminum and steel for construction). It could also include resorting to bioengineered trees. For example, lignin in wood must be chemically separated from cellulose to make pulp used in paper production. Researchers at Michigan Technological University have bioengineered aspen trees with half the normal lignin:cellulose ratio, which, moreover, could increase pulp production by 15% from the same amount of wood.

Also, future generations, conceivably more comfortable with computer screens, may abandon hard paper copy as reading and information storage media in favor of inorganic electronic media.

Natural Fibers, Furs, and Skins

Sixty percent of the global demand for fiber is now met through synthetic fibers (e.g., polyester, nylon, vinyl, acrylic) (Kuffner, 2004). Moreover, cotton, wool, silk, flax, jute, hemp, and coir, although nominally classified as natural fibers, are produced largely through agricultural technology. In addition, cotton, the most abundant natural fiber, is increasingly produced from GM varieties, which currently occupy 40% of the world's cotton acreage (ISAAA, 2006).

Synthetic fibers also substitute for natural furs and skins, reducing pressures to either harvest wild animals or maintain livestock for those purposes, thereby diminishing demand for land, water, and chemical inputs that would otherwise be required to maintain that livestock.

Fuel and Energy

Traditionally, humanity's fuel and energy services were mostly obtained from wood, dung, solar, wind, and hydropower, occasionally supplemented by geothermal power. Since the Industrial Revolution, the fuel mix has shifted toward fossil fuels (themselves products of nature) and, to a lesser extent, nuclear. Given the present state of energy technologies, current energy demand cannot be met with nature's traditional energy services. Fossil fuels can thus be viewed as imperfect and overused substitutes for nature's services that initially conserved habitat.

Because of climate change, efforts are now under way to reduce fossil fuel usage. These include greater emphasis on new renewable technologies (e.g., photovoltaics, advanced wind and solar power devices, crop-based biofuels); nuclear; broad improvements in energy efficiency; and more exotic solutions (e.g., hydrogen fuel cells, fusion). Land-intensive energy solutions (e.g., biofuels and solar energy) could, however, have unintended adverse consequences for ecosystems and species (Ausubel, 2007). A case in point is forest conversion in Malaysia and Indonesia to produce palm oil to meet Europe's subsidized biodiesel demand, which threatens endangered orangutans and other species.

Cultivation of energy crops (e.g., corn, soybean, and oil palm) threaten to reverse last century's reductions in cropland per capita that have helped almost stabilize total habitat lost to cropland (Goklany, 2007). Also, because these crops feed both humans and livestock and, moreover, are used in numerous products, prices for milk, meat, and other food products have escalated, jeopardizing post–World War II advances against global hunger. Food costs have increased by 50% in the past 5 years in some places (Blas and Wiggins, 2007).

Just as for food, timber, and fiber, biotechnology can make crop-based fuel production more efficient. Hybrid approaches combining biology and chemistry could further increase these efficiencies.

Transportation and Work

For millennia, human beings have relied on beasts of burden to transport themselves and their goods, till the soil, and do other heavy work. These ecosystem services, although overlooked by the MEA, are still used in developing countries. However, on farms and in cities, machines, mainly fueled by fossil fuel–driven internal combustion engines, are displacing oxen, mules, and horses; today trucks and trains carry far more goods on the Silk Road than camel caravans. Although this has increased fossil fuel consumption, it has reduced habitat lost to cropland that would otherwise be required to maintain animals providing these services. In the early decades of the twentieth century, when the U.S. population was a third of what it is today, 35 Mha (or 25% of U.S. cropland) was devoted to producing feed for the millions of workhorses and mules used on and off the farm. Therefore, technology, by rendering this ecosystem service largely obsolete in rich countries, has enormously reduced their land (and water) diverted to agriculture (Goklany and Sprague, 1992).

Genetic Resources

One of nature's critical services is providing access to its vast library of genetic resources, much of which, unfortunately, is not catalogued and may be in danger of being lost. Ex situ technologies that can be used to preserve this information include gene banks, zoos, and botanical gardens. Copies of this information can be created, and access facilitated, through the use of polymerase chain reactions. Biotechnology can also aid conservation by helping to propagate threatened, endangered, and, perhaps—à la *Jurassic Park*—even extinct species.

Biochemicals, Medicines, Pharmaceuticals

Any process, substance, or quality that exists or is produced in or by a living organism can, in theory, be bioengineered into synthetic crops. Armed with such traits, bioengineered crops can be used to produce medicines and vaccines in so-called GM pharms, manufacture bioplastics, biodiesel and other biofuels, colored or other forms of processed cotton, and eliminate toxic and hazardous pollutants from soils and waters. Once a better understanding is gained about how precisely genes help to manufacture various proteins and control various processes in nature, bioengineering may help to develop products and confer traits with no natural analogs. Today's chemical, pharmaceutical, and manufacturing factories may also be supplanted by bioengineered crops, bioreactors, and biofactories, essentially substituting older risks with newer but, we hope, lesser risks (Goklany, 2007: 392).

Synthetic manufacturing techniques and processes can substitute for medicines that would otherwise have to be produced directly from natural products. For example, aspirin, perhaps the most used drug in the world, is a synthetic form of a chemical found in the leaves and bark of willow trees. Similarly, the cancer drug paclitaxel, a semisynthetic substitute for Taxol, eliminated the need to harvest the Pacific yew tree to produce the cancer drug. According to one estimate, it would take six 100-year-old Pacific yew trees to treat one patient (Edwards, 1996). The semisynthetic process, which initially used material from the more abundant European yew, has now been refined to use plant cell cultures rather than plant parts.

Freshwater Quality and Quantity

Despite some skepticism regarding cost-effectiveness of freshwater substitutes, numerous technologies are available and are used routinely worldwide to clean and purify water to enable its safe use and reuse. Unfortunately, such technologies are underused largely because institutional and cultural factors frequently preclude pricing water and charging consumers the water's replacement price. Consequently, surface waters are oversubscribed, and groundwater is overdrawn (MEA, 2005a: 39).

In addition to desalination, which can be economically and environmentally expensive, several other technologies treat, purify, recycle, and reuse water. They include chlorination, ultraviolet radiation, filtration, and chemical and biological treatment (including sewage treatment) to reduce or remove pathogens, nutrients, metals, and other chemicals to make water safe for human, agricultural, industrial, and other uses. Treatment facilities come in sizes ranging from those designed for individual households to those suitable for towns. Thus, the number of people with access to safe water and sanitation has never been higher. Nevertheless 1.1 billion people still lack access to safe water, and 2.6 billion lack adequate sanitation, adding greatly to the global burden of disease (WHO/UNICEF, 2004).

Other technologies can, in effect, also reduce human demand on fresh water. Agriculture accounts for 85% of human freshwater consumption globally. Increasing the efficiency of agricultural water use by 1% would, on average, increase water for other human and environmental uses by 5.7%.

Moreover, there is significant scope for reducing water withdrawals for municipal and industrial purposes. Municipal (and household) consumption can be reduced through restricted-flow appliances (e.g., toilets, showerheads, washing machines), and industrial water use can be limited through the use of process changes or closed-loop water systems.

Finally, it may be possible to design ecological systems to freshen water for human consumption while also providing the rest of nature access to that water. Palmer et al. (2004) note that "'designing' ecosystems goes beyond restoring a system to a past state, which may or may not be possible. It suggests creating a well-functioning community of organisms that optimizes the ecological services available from coupled natural-human ecosystems."

Air Quality Regulation

One of nature's services is to cleanse various air pollutants from the atmosphere. This can be aided by technologies such as chemical scrubbers for sulfur dioxide and nitrogen oxides, electrical and mechanical devices such as electrostatic precipitators and fabric filters to reduce particulate matter, combustion devices to oxidize chemicals such as carbon monoxide and organic compounds, or process changes such as switching to low-sulfur or no-lead gasolines. These technologies, which rich nations used successfully to reduce emissions for traditional air pollutants (that exclude greenhouse gases), are now being transferred to developing countries through knowledge transfers or trade in equipment. Consequently, developing countries are addressing environmental concerns earlier in their development cycle (Goklany, 2007). For example, the United States started replacing leaded gasoline in 1975, when its GDP per capita was $20,000 (in 2000 International dollars, adjusted for purchasing power), whereas India and China began addressing this before their GDP per capita reached $3,500 (World Bank, 2007, based on Goklany, 2007).

Climate Regulation

One of nature's services is absorption and desorption of carbon dioxide, which helps regulate climate. However, increased fossil fuel usage and changes in land use and land cover have increased atmospheric CO_2 concentrations by over 25% since industrialization started.

Technologies to reduce atmospheric greenhouse gas concentrations include biological sequestration on land through plant growth, carbon capture from CO_2 sources with subsequent sequestration in oceans or in geologic formations, and biological sequestration in oceans by stimulating the growth of plankton and other organisms through iron fertilization. Only the first of these is currently economically feasible. Specific technologies include faster-growing trees and vegetation, reduced nitrogen usage, and conservation tillage (see subsection on crops in Food and Timber, above, and Erosion Control, below). Carbon capture is technically, but not economically, feasible, and oceanic and geologic sequestration are still in the research and development phases. Their long-term environmental consequences need further evaluation.

Proposals have been floated for other exotic geoengineering options (e.g., orbiting solar power stations or climate modification through injection of sulfates into the stratosphere or covering large areas with reflective films). Their economic and technical feasibility and environmental impacts also need further analysis.

At local and, possibly, regional scales, some climate regulation can be achieved by modifying land cover and albedo (through planting trees and other vegetation, reducing paved surfaces, or painting rooftops). At much smaller scales, air conditioning and heating serve as energy-intensive substitutes.

Erosion Control

Physical disturbance of the land's surface caused by tilling, construction, or removal of vegetation can contribute to erosion. This reduces soil productivity and increases losses of carbon into the atmosphere. Because it increases sediment and any soil-associated pollutants, it also reduces water quality. Erosion can be reduced through agricultural practices that would enable low- or no-till cultivation (i.e., "conservation tillage"), maintaining ground cover through cover crops or crop residue and avoiding or postponing cultivation or disturbance of erodible soils.

Conservation tillage can be facilitated through the use of herbicide tolerant (HT) crops. These crops—developed through either conventional breeding or bioengineering—are designed to tolerate various herbicides, so that herbicide application rather than mechanical or hand weeding reduces the competition between weeds and economically valuable crops.

U.S. experience with genetically modified HT crops has generally been positive. In the United States, soybean competes with over 30 kinds of weeds that, if left unchecked, could reduce yields by 50–90%. However, a HT soybean engineered to be tolerant to a broad-spectrum herbicide, glyphosate, helps farmers get rid of weeds more effectively using smaller amounts of less-toxic and less-persistent pesticides (see Pest Regulation, below). Other popular HT varieties include those developed for corn, canola, cotton, and alfalfa.

Disease Regulation

Nature regulates disease through various mechanisms. It provides habitat both for vectors that convey pathogens that might affect human beings, their livestock, and wildlife and for the pathogens themselves. It also harbors organisms that prey on the vectors, such as "mosquitofish" that eat the larvae of mosquitoes that spread West Nile virus (see Pest Regulation, below). Substitute technologies include treatment or removal of habitat harboring the vectors (e.g., by draining swamps, ponds, or containers that could hold standing water) and segregating vectors from human hosts or targets (e.g., by using insecticide-treated bed nets, screens on doors and windows, or insecticides to repel or kill vectors) (Grieco et al., 2007). Appropriate water treatment (e.g., chlorination) would also reduce water-related diseases such as dysentery and diarrhea. Finally, any incidences of disease could be treated with medicines.

Pest Regulation

Substitutes for nature's pest regulation include the use of pesticides and control of habitat that pests need. Also, nature itself can be harnessed in integrated pest management.

A relatively new technology for controlling pests is to bioengineer crops to contain their own pesticides. GM crops that are resistant to viruses, weeds, insects, and other pests have been developed. This should reduce pesticide usage, and their residues in the environment. Real-world experience so far bears out this theory.

The most widely used crops containing their own pesticides use genes from *Bacillus thuringiensis* (Bt), a soil bacterium, which has been used as a spray insecticide in conventional agriculture for decades. Bt varieties exist for corn, cotton, potato, and rice. In the United States, such crops are sometimes used as part of integrated pest management systems in which the Bt crop farmers plant refugia with non-Bt crops to retard the development of resistance in pests targeted by the Bt crops. They also should monitor the situation, which enables adaptive management. Other strategies include crop rotation, developing crops with multiple toxin genes with each toxin targeting different sites within the target species, and inserting the bioengineered gene into the chloroplast to express Bt toxin at higher levels.

Field studies from Arizona, Mississippi, Australia, and China indicate that these strategies have effectively retarded evolution of resistant pests. In 2004, Bt cotton—planted on 7.1 million acres (or 51% of U.S. cotton area)—reduced pesticide use by 1.76 million pounds, increased yields by 82 pounds per acre, and netted farmers $42 per acre (Sankula et al., 2005: 4–5). Insecticide runoff in a watershed before and after introduction of Bt cotton showed that pesticides that are most toxic to humans, birds, and fish decreased between one-third and two-thirds (EPA, 2001: IIE36). After Bt cotton adoption, bird counts increased by10% for Texas to 37% for Mississippi (relative to the pre-adoption situation; EPA 2001: IIE38–40). Elsewhere, Bt cotton helped China reduce pesticide use in 2001 by 25% below mid-1990s levels (Pray et al., 2002), and, during the 1999/2000 season, South African farmers who adopted Bt cotton had 60% higher yields and 38% lower pesticide consumption than nonadopters (Ismael et al., 2001).

Similarly, the use of GM HT soybean, canola, corn, and cotton in 2004 reduced U.S. pesticide usage in that year by an estimated 55 million pounds (in terms of active ingredients) while it increased farmers' net income by $1.8 billion (Sankula et al., 2005).

Pollination

Pollination is an important service that improves the quantity and quality of many agricultural crops such as apples, almonds, melons, blueberries, strawberries, and alfalfa. In nature, some pollination occurs through the action of abiotic processes such as wind and water, but most is accomplished via insects (e.g., bees, butterflies, and wasps), birds, and bats. For some crops, e.g., apples, almonds, and blueberries, pollination has long been a managed activity with bee colonies being transported from location to location to coincide with the flowering season. In other words, managed pollination is itself the product of technology. In the United States, managed pollination is generally accomplished using European honeybees, a nonnative species. Cultured insects, e.g., bumblebees for greenhouse tomatoes or alfalfa leafcutter bees for alfalfa, may also be employed. Other technological substitutes include mechanical or hand pollination for small-scale applications such as greenhouses and small garden plots. The need for pollination management has increased, possibly because some monoculture crops are insufficiently attractive to native pollinators, because of declining abundance of native pollinators, the increasing size of monoculture plots, and the fact that some crops, being nonnative, lack native pollinators.

Natural Hazard Management

Although nature is responsible for many hazards such as floods, hurricanes, tornados, drought, other extreme weather and climatic events, tsunamis, earthquakes, volcanic eruptions, and other geologic hazards, it also helps to buffer some of their effects. Soils store large quantities of water, mediate transfer of surface water to groundwater, and prevent or reduce flooding while barrier beaches, coastal wetlands, mangroves, and coral reefs help absorb storm surges from hurricanes and other wave action (MEA, 2005a: 118). Although some of these services have been compromised because natural buffers have frequently been modified, if not eliminated, and human beings continue to place themselves and their property increasingly in harm's way, global mortality and mortality rates from extreme weather events have declined by 95% or more since the 1920s. The largest declines were for droughts and floods, which were responsible for 95% of all twentieth-century deaths caused by extreme events. For the United States, current mortality and mortality rates from extreme temperatures, tornados, lightning, floods, and hurricanes also peaked a few decades ago (Goklany, 2006).

These empirical trends suggest that, notwithstanding any increase that may have occurred in frequencies and intensities of extreme events, technological substitutes have more than offset any losses in nature's protective services, at least with respect to protecting human lives. Declines in mortality are probably the result of increases in societies' collective adaptive capacities from a variety of interrelated factors—increases in wealth, technological options, and human and social capital.

Technological options range from early warning systems and more accurate meteorological forecasts to artificial or restored wetlands and mangroves to defensive structures (e.g., dams, sea walls, levees, dikes) to better and smarter construction (e.g., stronger building codes, concrete and steel houses, houses built on stilts, floating structures), to improved communications and transportation systems that enable transport of people and materiel (including food, medical, and other essential supplies) in and out of disaster zones, and to the 24/7 media coverage when extreme events seem imminent.

Experience with the 2003 European heat wave and Hurricane Katrina indicates that human and social capital are as important as technological options and greater wealth. Moreover, society's greater adaptive capacity to cope with extreme events and their aftermath must be deployed more rapidly and fully. The consequences of failure of natural barriers may be less than that of poorly deployed technology-based adaptations.

Spiritual and Religious Values

Nature also helps many individuals, communities, and cultures fulfill spiritual and religious needs. However, such services are cultural constructs, inseparable from human beings. Much of this probably reflects a time when nature directly provided virtually all of humanity's provisioning services and was, therefore, endowed by humans with—for lack of a better word—"supernatural" powers. Historically, objects such as paintings, sculptures, and relics helped satisfy, to some extent, religious needs that may otherwise have had to be met by undertaking arduous and dangerous journeys to distant places. But it is almost unimaginable that such objects or their modern-day counterparts—photographs, videos, movies, DVDs, holography—can be other than weak substitutes. Nevertheless, nature may conceivably be viewed with less reverence in the future, if technology further increases its role in displacing or augmenting nature's provisioning services, thereby diminishing demand for its spiritual and religious services.

Aesthetic Values, Recreation, and Ecotourism

Artificial or human-modified landscapes and ecosystems may also partially fulfill aesthetic, recreation, and ecotourism needs provided by nature. A query for manmade sites in the Ramsar Sites database returned 514 hits (out of a total of 1675 sites) (RSIS, 2007). A particularly successful example of a constructed ecosystem is India's Keoladeo National Park, also known as the Bharatpur Bird Sanctuary. This wetland—both a Ramsar Site and a World Heritage Site—protects the village of Bharatpur from frequent floods and provides grazing for cattle while also serving as habitat for 366 bird species, 379 floral species, 50 species of fish, 13 species of snakes, 5 species of lizards, 7 amphibian species, 7 turtle species, and a variety of other invertebrates (WWF-India, undated). Other artificial systems include lakes and reservoirs created behind dams and other water projects such as the Anaivilundawa Sanctuary in Sri Lanka, Lake Mead in the United States; and Lake Kariba in Africa. Clearly, although such water projects fulfill one set of demands for ecosystem services (e.g., water for drinking, agriculture, recreation, and tourism), by diverting water, they also undermine the ecosystem's ability to meet other demands.

Human-augmented ecosystems include the Ranthambore National Park (and Tiger Preserve) in India and artificial watering holes in Chobe National Park in Botswana. Other human-made or human-modified landscapes that partially substitute for nature range from the suburban gardener's backyard to Frederick Law Olmstead's Central Park in New York to England's rural landscape from farms to hedgerows. They may also include zoos and botanical gardens, as well as artificial reefs.

Just as an Ansel Adams photograph, an Albert Bierstadt painting, or a National Geographic DVD may, for some people, compensate for the real experience of visiting Yosemite, the Rocky Mountains, or Everest, so might the aesthetic services that nature provides be substituted through other paintings, photographs, high-definition or, possibly, holographic television, videos, and movies. It should be possible, for instance, to take high-definition virtual IMAX tours of coral reefs or nature preserves in the Caribbean, the Galápagos, or in Gombe. Similarly, one may, in the future, indulge in a virtual white-water rafting trip down the Colorado, a hot air balloon ride over Victoria Falls, or a bicycling tour through the Swiss Alps.

3. IMPLEMENTING TECHNOLOGIES TO REPLACE OR EXTEND NATURE'S SERVICES

There are numerous technological options for replacing or extending the goods and services that ecosystems provide humanity. Generic options particularly applicable to provisioning services—food, timber, natural fiber, energy—include technologies that would increase harvested yield per unit of land and water, increase utilization of harvested products through reductions in postharvest and end-use losses, and enhance recycling.

Such technologies would reduce humanity's burden on nature, a burden that will otherwise increase as the global population increases and becomes wealthier. But these technologies often have severe environmental consequences. Accordingly, the trade-offs and synergies involved in meeting human needs and conserving the biosphere should be evaluated before these options are implemented. Such evaluations, which necessarily should be done on a case-by-case basis, should also consider the effects of forgoing these technological options because in a complex and imperfect world there may be no perfect solutions (Goklany, 2007: chapter 9).

FURTHER READING

Ausubel, Jesse H. 1991. Does climate still matter? Nature 350: 649–652. *This provocative question is posed in light of the fact that technologies have made human civilization more adaptable and less vulnerable to climate change.*

Ausubel, Jesse H. 2007. Renewable and nuclear heresies. International Journal of Nuclear Governance, Economy and Ecology 1: 229–243. *Based on estimates of energy produced per unit area used, the author laments reliance on land-intensive energy solutions.*

Beddington, J. R., D. J. Agnew, and C. W. Clark. 2007. Current problems in the management of marine fisheries. Science 316: 1713–1716. *These authors argue that claims of inevitable decline in fisheries are overblown and that many existing management tools have yet to be implemented widely.*

Blas, Javier, and Jenny Wiggins. 2007. UN warns it cannot afford to feed the world. Financial Times, 15 July. *A news report warns on how diverting crops for biofuel production hinders the fight against global hunger.*

Daily, Gretchen C., Susan Alexander, Paul R. Ehrlich, Larry Goulder, Jane Lubchenco, Pamela A. Matson, Harold A. Mooney, Sandra Postel, Stephen H. Schneider, David Tilman, and George M. Woodwell. 1997. Ecosystem services: Benefits supplied to human societies by natural ecosystems. Issues in Ecology 2 (Spring). Available at http://www.esa.org/science_resources/issues/FileEnglish/issue2.pdf. *This is a good summary of many of the benefits natural systems provide for human beings.*

Edwards, Neil. 1996. Taxol. Downloadable from http://www.bris.ac.uk/Depts/Chemistry/MOTM/taxol/taxol.htm. *This provides a brief history of taxol and some efforts to synthesize this cancer-fighting drug.*

FAO. 2006. State of the World's Fisheries and Aquaculture 2006. Downloadable from http://www.fao.org/docrep/009/A0699e/A0699E04.htm.

Global Footprint Network (GFN). 2006. Humanity's Footprint 1961–2003. Available at http://www.ecofoot.net/.

Goklany, Indur M. 2006. Death and death rates due to extreme weather events: Global and U.S. trends, 1900–2004. In P. Höppe and R. A. Pielke Jr., eds. Workshop on Climate Change and Disaster Losses: Understanding and Attributing Trends and Projections, Final Workshop Report. Hohenkammer, Germany.

Goklany. Indur M. 2007. The Improving State of the World: Why We're Living Longer, Healthier, More Comfortable Lives on a Cleaner Planet. Washington, DC: Cato Institute. *This presents long-term U.S. and global trends in indicators of human and environmental well-being, including trends in land and water use. Chapter 9, devoted to bioengineered crops, discusses the environmental and public health trade-offs associated with their use and nonuse. The last two chapters address sustainable development and limits to growth.*

Goklany, Indur M., and Merritt W. Sprague. 1992. Sustaining Development and Biodiversity: Productivity, Efficiency and Conservation. Policy Analysis No. 175. Washington, DC: Cato Institute. *Probably the first study to note that higher land productivity conserves habitat and biodiversity, and to provide estimates of land saved for nature by higher agricultural productivity.*

Grieco, John P., Nicole L. Achee, Theeraphap Chareonviriyaphap, Wannapa Suwonkerd, Kamal Chauhan, Michael R. Sardelis, and Donald R. Roberts. 2007. A New Classification System for the Actions of IRS: Chemicals Traditionally Used for Malaria Control. Public Library of Science One, 2(8): e716. doi:10.1371/journal.pone.0000716.

International Service for the Acquisition of Agri-Biotech Applications (ISAAA). 2006. ISAAA Brief 35-2006: Executive Summary. Downloadable from http://www.isaaa.org/resources/publications/briefs/35/executivesummary/default.html. *The work provides an authoritative annual survey of the global penetration of bioengineered crops into agriculture.*

Kuffner, Henrik. 2004. Synthetic fibres: Quantity and price from Asia, specialities from USA and Europe. Downloadable from http://www.awta.com.au/Publications/IWTO/Conf_Papers/Report_Synthetics_NL6.pdf. *A market report from the Australian Wool Testing Authority.*

Livi-Bacci, Massimo. 1992. A Concise History of World Population. English edition translated by C. Ipsen. Cambridge, MA: Blackwell. *An erudite population history of the world is presented.*

Millennium Ecosystem Assessment. 2005a. Ecosystems and Human Well-being: Synthesis. Washington, DC: Island Press.

Millennium Ecosystem Assessment. 2005b. Ecosystems and Human Well-being: Biodiversity Synthesis. Washington, DC: World Resources Institute. *These reports describe the status and trends in the world's ecosystems, the services they provide for human beings, and the options to restore, conserve, or sustainably enhance these ecosystem services.*

Naylor, Rosamond L., Rebecca J. Goldburg, Jurgenne H. Primavera, Nils Kautsky, Malcolm C. M. Beveridge, Jason Clay, Carl Folke, Jane Lubchenco, Harold Mooney, and Max Troell. 2000. Effect of aquaculture on world fish supplies. Nature 405: 1017–1024. *This article notes that some forms of aquaculture may not relieve pressure on wild fisheries because they require large inputs of wild fish for feed and have various other ecological impacts.*

Palmer, Margaret, Emily Bernhardt, Elizabeth Chornesky, Scott Collins, Andrew Dobson, Clifford Duke, Barry Gold, Robert Jacobson, Sharon Kingsland, Rhonda Kranz, Michael Mappin, M. Luisa Martinez, Fiorenza Micheli, Jennifer Morse, Michael Pace, Mercedes Pascual, Stephen Palumbi, O. J. Reichman, Ashley Simons, Alan Townsend, and Monica Turner. 2004. Ecology for a crowded planet. Science 304: 1251–1252. *The authors describe some changes that would help balance the needs and aspirations of humans with the health of ecosystems.*

Ramsar Sites Information Service (RSIS). 2007. Ramsar Sites Information Service. Database. Downloadable at http://www.wetlands.org/rsis/.

Sankula, Sujatha, Gregory Marmon, and Edward Blumenthal. 2005. Biotechnology Derived Crops Planted in 2004—Impacts on US agriculture: Executive Summary. Washington, DC: National Center for Food and Agricultural Policy. *This article provides information on acreage, costs, and benefits of GM crops planted in the United States in 2004, including farmers' net profits, and changes in pesticide use.*

WHO/UNICEF Joint Monitoring Programme for Water Supply and Sanitation. 2004. Meeting the MDG Drinking Water and Sanitation Standard: A Mid-Term Assessment of Progress. Geneva: WHO. *This is an interim report on progress toward the Millennium Development Goals for safe water and sanitation.*

World Wildlife Fund-India (WWF-India). Undated. Interpretation Programme for Keoladeo National Park. Downloadable from http://www.wwfindia.org/about_wwf/what_we_do/freshwater_wetlands/our_work/keoladeo_np/index.cfm.

Worm, Boris, Edward B. Barbier, Nicola Beaumont, J. Emmett Duffy, Carl Folke, Benjamin S. Halpern, Jeremy B. C. Jackson, Heike K. Lotze, Fiorenza Micheli, Stephen R. Palumbi, Enric Sala, Kimberley A. Selkoe, John J. Stachowicz, and Reg Watson. 2006. Impacts of biodiversity loss on ocean ecosystem services. Science 314: 787–790. *This report suggests that present trends in losses of biodiversity, unless reversed, may cause collapse of all commercial fish and seafood species by 2048.*

VI.13

Conservation of Ecosystem Services

Jon Paul Rodríguez

OUTLINE

1. Introduction
2. Conservation and utilization of ecosystem services
3. Conservation of provisioning ecosystem services
4. Conservation of regulating ecosystem services
5. Conservation of cultural ecosystem services
6. The future of ecosystem services in conservation planning

Ecosystem services have not been traditional targets of biodiversity conservation efforts. Researchers, practitioners, and policy makers have focused their attention on genes, populations, species, or ecosystems. As societal interest in ecosystem services grows, however, these may provide an improved platform for communicating and quantifying their value to humans and thus improving our understanding of the dependence of our well-being on nature and ecosystem services. Once this link is firmly established, conservation of ecosystem services should be a more natural societal choice.

GLOSSARY

biological diversity. The variety and variability of all forms of life on Earth, encompassing the interactions among them and the processes that maintain them.

ecosystem services. The benefits people obtain from ecosystems. They can be of four primary types: provisioning, regulating, cultural, and supporting ecosystem services.

ecosystem service trade-off. Reduction of the provision of one ecosystem service as a consequence of increased use of another ecosystem service. They arise from management choices made by humans, which can change the type, magnitude, and relative mix of services provided by ecosystems.

Millennium Ecosystem Assessment. A global assessment carried out between 2001 and 2005 that involved more than 1360 experts worldwide and had the objectives of assessing the consequences of ecosystem change for human well-being and establishing the scientific basis for conservation and sustainable use of ecosystem services.

systematic conservation planning of ecosystem services. A scientific process for integrating social and biological information, to support decisionmaking about the location, configuration, and management of areas designated for the conservation and sustainable use of ecosystem services.

1. INTRODUCTION

The visibility of the term *ecosystem services* recently exploded in the scientific literature: the total number of references accumulated by the late 1990s is smaller than the figure for 2005 or 2006 alone (figure 1). Attempts to preserve, restore, or enhance ecosystem services, however, clearly predate this. The first national park of the world, Yellowstone National Park, located in the northwestern United States, was established in 1872. A remarkable combination of unique geological features, striking landscapes, and abundant wildlife were preserved "for the benefit and enjoyment of the people"—in other words, for the cultural ecosystem services (ES) provided to humans. As an even earlier example, the entire global herd of Père David's deer (*Elaphurus davidianus*) descends from a few animals kept for recreation in the Imperial Hunting Park south of Beijing (a cultural ES)—the deer is believed to have become extinct in the wild during the Ming Dynasty (1368–1644). The Inca empire arose in Peru in the thirteenth century and thrived in part because of its effective management of provisioning ES, such as maize (*Zea mays*) cultivation and herding of llamas (*Lama glama*). Soils, a supporting ES, were conserved by terracing the mountainside. Between 300 BC and AD 200, Rome built 11 major aqueducts, developed by the Roman Empire to service its roughly 1 million inhabitants. Major engineering achievements allowed the

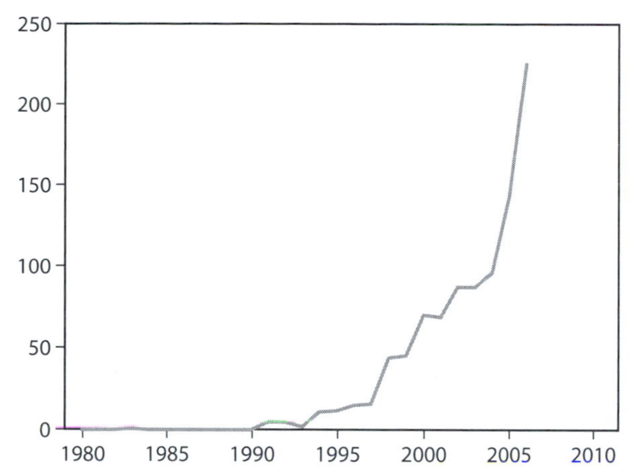

Figure 1. Number of times that the term *ecosystem services* appears in the article database of the ISI Web of Knowledge (http://www.isiwebofknowledge.com/) between 1980 and 2006. Search was conducted by entering "ecosystem services" in the topic field of the search engine.

emperor, rich citizens, and the general public to access a network of water fountains within the city, never located more than 100 m apart. Water, a provisioning ES, was actively managed and conserved by Romans two millennia ago.

Given how recently "ecosystem services" entered the conservation jargon, however, there are not many examples where conservation activities have been explicitly linked to ecosystem services. However, there are many examples where the motivation of a conservation action has been the implicit conservation of an ES. The IV Worlds Park Conference, held by the World Conservation Union (IUCN) in 1992, concluded that "protected areas are about meeting people's needs," and that they should be "part of every country's strategy for sustainable management and the wise use of its natural resources." In a follow-up guidelines document, published in 1994, IUCN identified six types of protected areas:

I. Strict protection:
   1. Strict Nature Reserve: Area of land and/or sea possessing some outstanding or representative ecosystems, geological or physiological features and/or species, available primarily for scientific research and/or environmental monitoring.
   2. Wilderness Area: Large area of unmodified or slightly modified land and/or sea retaining its natural character and influence, without permanent or significant habitation, which is protected and managed so as to preserve its natural condition.

II. Ecosystem conservation and recreation: Natural area of land and/or sea, designated to (1) protect the ecological integrity of one or more ecosystems for present and future generations, (2) exclude exploitation or occupation inimical to the purposes of designation of the area, and (3) provide a foundation for spiritual, scientific, educational, recreational, and visitor opportunities, all of which must be environmentally and culturally compatible.

III. Conservation of natural features: area containing one, or more, specific natural or natural/cultural features that are of outstanding or unique value because of their inherent rarity, representative or aesthetic qualities, or cultural significance.

IV. Conservation through active management: area of land and/or sea subject to active intervention for management purposes so as to ensure the maintenance of habitats and/or to meet the requirements of specific species.

V. Landscape/seascape conservation and recreation: area of land, with coast and sea as appropriate, where the interaction of people and nature over time has produced an area of distinct character with significant aesthetic, ecological, and/or cultural value, often with high biological diversity. Safeguarding the integrity of this traditional interaction is vital to the protection, maintenance, and evolution of such an area.

VI. Sustainable use of natural ecosystems: area containing predominantly unmodified natural systems, managed to ensure long-term protection and maintenance of biological diversity while providing at the same time a sustainable flow of natural products and services to meet community needs.

In the context of ES, types I–III focus on all but provisioning ES, whereas types IV–VI are primarily concerned with provisioning and cultural ES. In all cases, however, the goods and services provided by nature are clearly identified. Even type I protected areas, which are set aside for strict protection, still are able to provide ES, such as "genetic resources," "ecological processes," "education," and "spiritual well-being."

The recognition that human interests must be taken into consideration when areas of land are set aside for conservation purposes goes one step further in the Man and the Biosphere Programme (MAB) of the United Nations Educational, Scientific and Cultural Organization (UNESCO). The cornerstones of MAB are biosphere reserves, or "areas of terrestrial and coastal ecosystems promoting solutions to reconcile the conservation of biodiversity with its sustainable use." With over 500 sites in more than 100 countries, biosphere reserves are a global network that serves as testing grounds for "integrated management of land, water and biodiversity" (http://www.unesco.org/mab/BRs.shtml). A typical biosphere reserve fulfills conservation, development, and logistic objectives and is organized in a core area, a buffer zone, and a transition area (figure 2). Human settlements are an integral part of biosphere reserves, and research on how to best enhance human well-being while assuring environmental sustainability is key.

Although ES are never mentioned by IUCN or UNESCO, they are implicitly at the core of both initiatives. In fact, ES offer an excellent platform on which to build a general framework for all efforts to conserve biological diversity. The classical approach of focusing on genes, species, or ecosystems can be redefined in terms of the ES that they provide. For example, conservation efforts can be thought of as attempts to conserve different ES: (1) managers that enhance wild populations of birds, mammals, or fish for commercial or recreational users are conserving provisioning and cultural ES; (2) planners who seek to design an optimal reserve network for maximizing species diversity are targeting all of biological diversity and the various services that they provide; (3) ecological restoration of native vegetation around agricultural areas to increase native pollinator populations can be seen as enhancing their regulating ES; and (4) the protection of sacred burial grounds is equivalent to the conservation of a cultural ES.

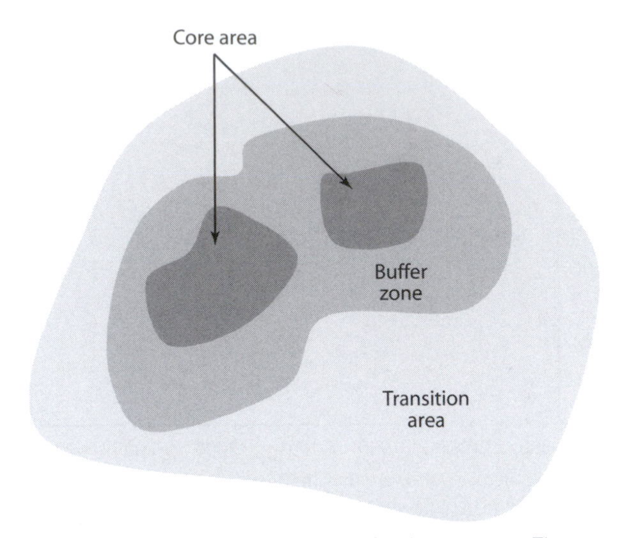

Figure 2. Schematic of the zones of a biosphere reserve. The core area is the only one that requires legal protection (e.g., national park, nature reserve) and where human activities tend to be limited to research and monitoring. Human settlements are located in the buffer zone or the transition area. Low-intensity economic activities are carried out in the buffer zone, and allowed uses in the transition area tend to be more diverse. (http://www.unesco.org/mab/BRs.shtml)

2. CONSERVATION AND UTILIZATION OF ECOSYSTEM SERVICES

The challenge of focusing biodiversity conservation on ES is that these services are not independent of each other. A decision to conserve one ES may influence the delivery of another, either positively or negatively. For example, protecting a watershed for the provision of water may enhance other provisioning ES, such as wild foods, genetic resources, or biochemicals; help improve regulating ES, such as air quality, erosion control, and water quality; and strengthen cultural ES, such as aesthetics, recreation, and tourism. But setting aside land for conservation also has opportunity costs, and some ES are forgone: timber cannot be extracted, crops cannot be planted; livestock cannot graze; and access to wild foods may be limited. How such a policy impacts human well-being will depend on the relative importance of the ES that are enhanced versus those that are forgone or degraded.

When the provision of one ES is reduced as a consequence of increased use of another, a *trade-off* is said to have taken place. Trade-offs are an integral component of decisions related to the management of ES, as they are likely to be inevitable under many circumstances. In some cases, a trade-off may be the consequence of an explicit choice, but in others, trade-offs arise without premeditation or even awareness that they are taking place. These unintentional trade-offs happen when we are ignorant of the interactions among ecosystem services or when we are familiar with the interactions but our knowledge about how they work is incorrect or incomplete.

As human societies expand across the wilderness areas of the world, the emphasis on different ES shifts as well (figure 3). Initially, people occupy wildlands, land is cleared for small-scale agriculture, and population begins to grow. The primary focus is agricultural production, and other ES are traded off against the provision of food, fiber, fuel, etc. Often subsistence lifestyles are gradually replaced by large-scale agricultural operations and urban areas, and attention shifts to regulating ES: as the human domination of the landscape increases, processes such as water regulation

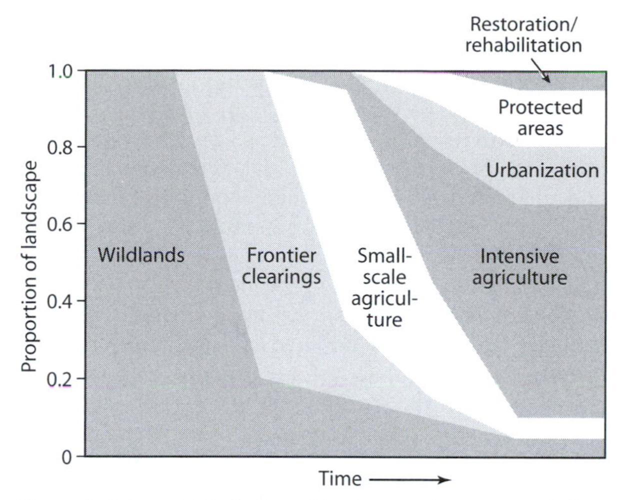

Figure 3. Schematic trajectory of land-cover changes from before human settlement to the human domination of the landscape. (From Rodríguez et al., 2006)

and purification must be enhanced. In the final stages, restoration or protection that focuses on recreation may come to dominate. The main point is that the value placed on different ES will change as societies develop and also change in ways that depend on culture and history.

In the following three sections, I present a series of examples of recent explicit or implicit efforts to conserve ES, organized around the Millennium Assessment (MA) categories for ES—provisioning, regulating, and cultural—and focusing on ES identified by the MA as degraded. The list of cases is illustrative, not exhaustive; for each type of ES, I selected two examples. Supporting services, the fourth category of ES identified by the MA, are underlying services necessary for the production of all other ecosystem services. But because supporting ES are not used directly by people, the focus of the following sections is on the other three types, which are subject to being degraded or enhanced by humans.

3. CONSERVATION OF PROVISIONING ECOSYSTEM SERVICES

Provisioning services are products obtained from ecosystems. The MA examined the condition of 11 provisioning ES, six of which were considered degraded: capture fisheries, wild food products, wood fuel, genetic resources, biochemicals, and fresh water. Timber, fiber, crops, livestock, and aquaculture either showed mixed results or appeared to have been enhanced during the period examined by the MA. Examples regarding fisheries and fresh water are briefly presented below.

Capture Fisheries

Roughly one-fourth of marine fisheries are overexploited or considerably depleted. Global marine fish harvests peaked in the late 1980s and have declined ever since. Over a billion people depend on fish as their primary or only source of protein, especially in developing countries.

Fishery exclusion areas, or marine reserves characterized by being closed to all forms of fishing, are emerging as a successful tool for managing fisheries in a wide diversity of habitats, such as coral reefs, kelp forests, seagrass beds, mangrove swamps, and the deep sea. Likewise, they work for a variety of fishing styles, including various recreational, artisanal, and industrial fisheries. Exclosures allow an increased reproductive output of fish stocks, increasing the harvest outside of the reserves (Gell and Roberts, 2003, present a comprehensive review of the topic).

The Soufrière Marine Management Area is located in the Caribbean island of Saint Lucia. It was created in 1995, spanning 11 km of coast. The reserve included several no-take areas within the island's fringing coral reefs as well as areas open to fishing located mostly along beaches. Within the following 6 years, commercial fish biomass increased fourfold within the no-take areas and threefold in the fishing beaches. Total catch and catch per unit effort of local fishers also significantly increased.

Success of this initiative was linked to the spatial design of the network, which interspersed no-take areas among other uses. Fishers complied with the reserve but took advantage of the distribution of no-take areas by placing their fishing gear near the boundaries of reserve and harvesting the fish that "spilled over" into the areas open to fishing. In this way, there was a "balance" between enhancing provisioning services and preserving biodiversity—more species, or a greater abundance of existing species, could potentially be saved by eliminating fishing altogether, but this solution contains a recognition that fishing provides an important provisioning ES.

Fresh Water

Limited access to fresh water is a major global problem, affecting 1–2 billion people worldwide, with consequences over the production of food, human health, and economic development. Roughly 1.7 million deaths occur every year because of poor water quality or inadequate sanitation. Global freshwater use is estimated to expand by 10% between 2000 and 2010. Forests and montane ecosystems remain the primary source of water for two-thirds of the global population, together accounting for 85% of the total runoff.

The establishment of protected areas is often opportunistic, without any explicit plan. Whatever is easily available is set aside first, particularly if lands are of low economic value or have no other competing uses (e.g., Yellowstone). The history of protected areas in Venezuela, however, followed another track. The country's first national park, Henri Pittier National Park, was established in 1937, protecting the headwaters of several rivers that drain into the city of Maracay and several major cacao (*Theobroma cacao*) plantations and towns along the Caribbean coast. At present, Venezuela has 43 national parks, covering over 13 million hectares. The majority of national parks are found along the country's northern mountain ranges (plate 21), in the vicinity of large cities and agricultural areas. Canaima National Park, one of Venezuela's largest national parks (30,000 km^2), includes the high watershed of the Caroní River. Guri Dam, located in the lower Caroní, produces electricity equivalent to 300,000 oil barrels per day. Water conservation has been the leading justification for the establishment of national parks. Effective and equitable distribution of this water still remains a problem to be solved.

4. CONSERVATION OF REGULATING ECOSYSTEM SERVICES

Regulating services are the benefits obtained from the regulation of ecosystem processes. The MA assessed the status and trends of 10 regulating ES and concluded that seven are degraded: air quality regulation, regional and local climate regulation, erosion regulation, water purification, pest regulation, pollination, and natural hazard regulation. The three remaining regulating ES—water regulation, disease regulation, and global climate regulation—were classified as mixed or enhanced. Below, I present examples of water purification and pollination.

Water Purification

Mangroves are woody plants that develop naturally along tropical coastlines and in subtropical areas bathed by warm currents. They form wetlands that supply numerous ecosystem services, such as coastal erosion protection, improvement of water quality (by absorbing pollutants), and organic matter accumulation. Artificial mangrove wetlands have been used in Colombia to remediate lands degraded by the oil industry and to remove pollutants from contaminated waters.

In 1994, the Colombian Petroleum Institute initiated a project for the bioremediation of an area located more than 600 km from the coast and at 75 m above sea level, inundated by wastewaters of the petroleum industry, and characterized by their high contents of chloride (>30 parts per thousand), iron (up to 13%), and barium (65 ppm). Stands of red mangrove (*Rhizophora mangle*), black mangrove (*Avicennia germinans*), and white mangrove (*Laguncularia racemosa*) were created within an artificial lagoon in the polluted area. Mangroves demonstrated a high capacity for extracting and isolating pollutants from the surrounding water. The salinity of water before treatment was 42,000 ppm, whereas at the exit of the artificial lagoon, it had declined to 3300 ppm. By 2000, 6 years after initiation of the project, plant cover had been fully restored; before treatment, no woody plants grew there.

Another project involving an artificial mangrove wetland was carried out by a shrimp-farming company in San Antero, Córdoba Department, with the objective of treating the wastewaters of their shrimp farm. The wetland was built 0.5 m above the maximum tide level, taking advantage of the topography so that the area selected would easily form an artificial lake. Red mangrove trees were planted from seedlings and embryos. Shrimp farms used water from a nearby wetland, the Ciénaga de Soledad. This "natural" water source fills the shrimp ponds, and flows into a wastewater channel that feeds the artificial wetland, or the "collector." One year of monitoring revealed that the artificial wetland was highly efficient in significantly reducing total suspended solids and biological oxygen demand, from 145 to 94 mg/liter and 14 to 9 mg/liter, respectively. In fact, water quality after exiting the facility was superior to that at the Ciénaga de Soledad itself. Six years after it was built, the wetland had become a true refuge for local biodiversity, and commercially valuable fish had naturally established themselves. In 2003, a study analyzed the catch of 24 fishing trips by artisanal fishers; each trip consisted of 5 hr in one small nonmotorized boat and three fishers. After a combined effort of 120 hr of fishing, they produced 200 kg of fish (1.7 kg/hr). Each fisher caught 2.9 kg/day, representing about US$12/day. This is equivalent to a monthly income of US$240 (assuming 20 working days/month), which is 40% higher than Colombia's 2006 minimum salary of approximately US$170/month.

Pollination

Pollinators are involved in the sexual reproduction of roughly 80% of the 300,000 species of flowering plants of the world. Their regulating ES enhances provisioning ecosystem services, such as the production of food crops, as well as any other ES linked to flowering plants (primarily provisioning and cultural ES). Over 200,000 pollinators are known, and roughly 10% of them are bees. Animal pollinators are predominantly insects (Hymenoptera: bees, wasps, and ants; Coleoptera: beetles; Diptera: flies; Lepidoptera: butterflies and moths; Thysanoptera: thrips), mammals (including bats, marsupials, monkeys, and procyonids), and birds (primarily, but not exclusively, hummingbirds).

The most important managed pollinator in the United States and Europe, the honeybee (*Apis mellifera*), recently has shown a clear decrease in population size. Although the data supporting wild pollinator trends are less reliable, there is evidence of the decline in several other bee species (especially bumblebees) as well as some butterflies, bats, nonflying mammals, and hummingbirds. In the United States, pollination by insects produces US$40 billion annually, and the value of crop pollination by honeybees alone is estimated at US$6 billion per year in Europe. The global figure for the value of pollinators has been estimated to be on the order of US$120–200 billion per year (Díaz et al., 2005).

Pollinator conservation is a very active discipline. The African Pollinator Initiative summarized a series of recommendations, which include conserving and restoring natural habitat, growing flowering plants preferred by pollinators, promoting mixed farm systems, establishing nectar corridors for migratory species, and providing nesting and feeding habitats alongside croplands. A mosaic of materials such as dry wood, bare ground, mud, resin, sand, carrion, host plants, and caves are needed to maintain pollinator diversity at any particular site.

In Mexico, several species of columnar cacti, plants in the genus *Agave*, and trees in the Family Bombacaceae rely on bats for their sexual reproduction. Founded in 1994, the Program for the Conservation of Migratory Bats (PCMB) focuses on research and environmental education for the protection of bats by conserving habitat along migratory corridors. The Brazilian free-tailed bat (*Tadarida brasiliensis*), for example, overwinters in South and Central Mexico, and migrates north each spring, forming very large breeding colonies in northern Mexico and the southwestern United states that may be as large as 20 million individuals. They perform a natural pest-control service, feeding on vast quantities of insects during the migration, but especially while at their breeding grounds; in South-Central Texas, the value of the service provided by breeding free-tailed bats feeding on cotton bollworm (*Helicoverpa zea*) is estimated at about US$1 million per year. By highlighting the economic value of bats to the Mexican government, PCMB succeeded in promoting an amendment of the national Wildlife Law to include all caves and crevices as protected areas, thus conserving key habitat for bats and enhancing their ecosystem services.

5. CONSERVATION OF CULTURAL ECOSYSTEM SERVICES

Cultural ES are nontangible benefits that people obtain from ecosystems through spiritual enrichment, cognitive development, reflection, recreation, and aesthetics. The MA assessed the status of three cultural services: spiritual and religious services, aesthetic values, and recreation and tourism. The first two were classified as degraded, and the last was considered mixed. Below, I focus on aesthetic values and, although the evidence is mixed, on recreation and ecotourism.

Aesthetic Values

Studies carried out primarily in industrialized countries have shown that people tend to prefer nonurban over built environments, and there is a range of preferences between wild and cultivated landscapes. In other words, the appreciation of aesthetic ES is a major feature of human behavior. Research shows that people rate the scenic beauty of natural scenes to be higher that urban images; natural settings that are healthy, lush, and green are perceived as more attractive, especially when contrasted with arid habitats; parklike settings, which are safe and likely to provide primary needs such as food and water, are also favored; and patients recover from surgical interventions faster and with less medical attention in rooms with a natural view than when they are looking at a blank wall.

This is not a recent phenomenon, and it appears to be prevalent over different cultures and times. The Hanging Gardens of Babylon, one of the original seven wonders of the world, were built to showcase the beauty of trees and other plants. “Imperial gardens” were carefully designed, aesthetically pleasing landscapes integrated with the palaces of Chinese dynasties. In modern times, people continue to exercise these preferences when making choices about where to live: a study of 3000 real estate transactions in the Netherlands showed that house prices were higher when the property had a garden facing large water bodies or open space. Billions of dollars are spent every year worldwide on lawn and garden maintenance in homes; in the United States, for example, the annual figure is approximately US$40 billion.

Recreation and Ecotourism

Tourism generates approximately 11% of global GDP and employs over 200 million people. Approximately 30% of these revenues are related to cultural and nature-based tourism, and nature travel is increasing between 10% and 30% per year. The tourism industry was identified by Agenda 21 as one of the few industries with the potential of simultaneously improving the economic condition of nations and the general state of the environment.

Approximately 50 million people in the United States, or 20% of the population, are bird watchers. These are people who travel away from their homes with the primary purpose of observing birds. For these trips, birders annually spend around US$20 billion on equipment and US$7 billion on travel. The overall economic output of this process is estimated at US$85 billion, generating US$13 billion in state and income taxes and creating more than 850,000 jobs. This means that each of the 1000 bird species in the United States has, on average, a value for bird watchers that is equivalent to US$85 million in overall economic output, US$13 million in tax revenue, and 850 jobs.

The Ecotourism Society defines ecotourism as "travel to natural areas that conserves the environment and sustains the well-being of local people." For developing countries, ecotourism is often seen as one of the primary economic alternatives to support the conservation of threatened ecosystems. In Kenya and Costa Rica, for example, tourism brings in several hundred million dollars per year, much of it from tourists interested in nature and wildlife. The challenges are to assure that the presence of tourists does not unacceptably degrade ecosystems, the natural capital of ecotourism, and that revenues indeed reach local communities. Evidence to date suggests that a healthy dose of skepticism and caution are warranted. The Galápagos Islands attract more than 62,000 tourists every year, drawn by the promise of spectacular seascapes and a unique array of animals and plants. The majority of the islands' inhabitants (80%) receive a share of the income generated by this economic activity, which has been designed such that human impact is kept to the minimum possible. At a first glance, one might consider this a win-win situation, but trade-offs are clearly taking place. The economic success of Galápagos has attracted people from the mainland, who migrate to the islands in search of jobs. The growth in human population has led to increased pressure on the limited infrastructure and on local provisioning ES such as fish. But the biggest problem is that local communities only receive about 15% of the income generated by tourism; most of the funds are captured by large companies that provide luxury transportation and lodging.

6. THE FUTURE OF ECOSYSTEM SERVICES IN CONSERVATION PLANNING

Systematic conservation planning seeks the adequate representation of species, or other biological or physical attributes of landscapes, in a network of nature reserves. Numerous algorithms have been developed for the identification of priority sites for inclusion in existing reserve networks. In recent years, efforts have centered on optimizing the investment of limited funds and examining the availability of human resources for implementing conservation priorities. Ecosystem services, however, are not typically considered in systematic conservation planning. In part, this is probably because ES have not been adequately quantified at the scale of large geographic areas, such as nations, continents, or the globe. In other words, data on how ES supply varies spatially or temporally are simply unavailable. But two recent studies, one carried out in the Atlantic Forests of Paraguay (Naidoo and Rickets, 2006) and another in the Californian Central Coast Region of the United States (Chan et al., 2006), have began to lay the framework for explicitly integrating ES into systematic conservation planning.

In the Paraguayan Atlantic Forests example, economic costs and benefits of the conservation of this highly threatened landscape were assessed in terms of five ecosystem services: wild food (bush meat), timber, biochemicals, carbon sequestration, and existence values. As one might expect, spatial variability of costs and benefits were large, although carbon sequestration dominated among ES values. Their contrast of three potential corridor designs to improve connectivity to the core area of the landscape allowed researchers to identify one that had net benefits that were three times higher than the other two when all five ES were considered.

The second study also carried out a spatially explicit analysis, but in this case, researchers compared priority areas based on biodiversity conservation, with those generated from the provision of water, forage production, water regulation (flood control), crop pollination, carbon sequestration, and recreation. They found relatively low correlation between the spatial distribution of ES supplies and little overlap between priority sites identified from each variable individually. Sites selected in terms of their importance to biodiversity did not capture the benefits provided by ES.

Network designs would thus have to consider the relative value placed on each ES by society in order to maximize the flow of benefits.

The question that remains is whether ES offer an appealing platform for a general framework for the conservation of biological diversity. The MA showed that humans have negatively impacted ES during the last 50 years and that natural capital has been eroded: of 24 ES examined by the MA, 15 are currently degraded globally. Tens of thousands of Web pages now refer to the MA (which generated its final published products in 2005), and nearly a million include the term *ecosystem services*. In contrast, *biodiversity*, a term first used in 1988, roughly when "ecosystem services" began to expand (figure 1), draws over 22 million hits. The "popularization" of ES, although apparently on the right track, still has a long way to go.

A global agenda for the conservation of ecosystem services would also need to overcome a series of major obstacles (Irwin et al., 2007):

1. Societies tend not to connect how their well-being is contingent on the availability of ES or biodiversity in general.
2. Local communities often have little control over the land and resources that they use or cannot influence decisions regarding them.
3. Decision makers lack coordination and frequently act without consulting their peers in governmental agencies, nongovernmental organizations, and the private sector.
4. Accountability regarding ES management is not yet well developed, leading to lack of transparency and to corruption.
5. Valuation of ES is incomplete, and incentives for responsible ecosystem stewardship are mostly nonexistent.

These obstacles will sound familiar to anyone with practical experience in the conservation of biological diversity as well as the actions required to counteract them: increased availability and application of information on ES, stronger rights in the use and management of ES, management of ES at multiple temporal and spatial scales, improvement of accountability regarding the use of ES, and development of incentives to encourage sustainable use of ES. Ecosystem services, however, may still provide one of the best conceptual frameworks for posing conservation priorities to the general public. Because ES are the benefits people obtain from ecosystems, their value should be relatively easy to understand and to communicate.

FURTHER READING

Chan, Kai M. A., M. Rebecca Shaw, David R. Cameron, Emma C. Underwood, and Gretchen C. Daily. 2006. Conservation planning for ecosystem services. PLoS Biology 4: e379. Downloadable from http://www.plosjournals.org/.

Díaz, Sandra, David Tilman, Joseph Fargione, F. Stuart Chapin III, Rodolfo Dirzo, Thomas Kitzberger, Barbara Gemmill, Martin Zobel, Montserrat Vilà, Charles Mitchell, Andrew Wilby, Gretchen C. Daily, Mauro Galetti, William F. Laurance, Jules Pretty, Rosamond Naylor, Alison Power, and Drew Harvell. 2005. Biodiversity regulation of ecosystem services. In Rashid Hassan, Robert Scholes, and Nevilae Ash, eds. Ecosystems and Human Well-being: Current State and Trends, Volume 1. Findings of the Condition and Trends Working Group. Washington, DC: Island Press, 297—329. Downloadable from http://www.maweb.org/.

Eardley, Connal, Dana Roth, Julie Clarke, Stephen Buchman, and Barbara Gemmill, eds. 2006. Pollinators and Pollination: A Resource Book for Policy and Practice. Pretoria, South Africa: African Pollinator Initiative (API). Downloadable from http://pollinator.org/Resources/Pollination%20Handbook.pdf.

Gell, Fiona R., and Callum M. Roberts. 2003. The Fishery Effects of Marine Reserves and Fishery Closures. Washington, DC: World Wildlife Fund. Downloadable from http://www.worldwildlife.org/oceans/pubs.cfm.

Hassan, Rashid, Robert Scholes, and Neville Ash, eds. 2005. Ecosystems and Human Well-being: Current State and Trends, Volume 1. Findings of the Condition and Trends Working Group. Washington, DC: Island Press. Downloadable from http://www.maweb.org/.

Irwin, Frances, Janet Ranganathan, Mark Bateman, Albert Cho, Hernan Darío Correa, Robert Goodland, Anthony Janetos, David Jhirad, Karin Krchnak, Antonio La Viña, Lailai Li, Nicolás Lucas, Mohan Munasinghe, Richard Norgaard, Sudhir Chella Rajan, Iokiñe Rodríguez, Guido Schmidt-Traub, and Frances Seymour. 2007. Restoring Nature's Capital: An Action Agenda to Sustain Ecosystem Services. Washington, DC: World Resources Institute. Downloadable from http://pdf.wri.org/restoring_natures_capital.pdf.

IUCN. 1994. Guidelines for Protected Area Management Categories. Gland, Switzerland and Cambridge, UK: Commission on National Parks and Protected Areas (CNPPA) with the assistance of World Conservation Monitoring Centre (WCMC), and The World Conservation Union (IUCN). Downloadable from http://www.unep-wcmc.org/protected_areas/categories/eng/index.html.

Millennium Ecosystem Assessment. 2003. Ecosystems and Human Well-being: A Framework for Assessment. Washington, DC: Island Press. Downloadable from http://www.maweb.org/.

Millennium Ecosystem Assessment. 2005. Ecosystems and Human Well-being: Synthesis. Washington, DC: Island Press. Downloadable from http://www.maweb.org/.

Naidoo, Robin, and Taylor H. Ricketts. 2006. Mapping the economic costs and benefits of conservation. PLoS Biology 4: e360. DOI: 310.1371/journal.pbio.0040360. Downloadable from http://www.plosjournals.org/.

Rodríguez, Jon Paul, T. Douglas Beard Jr., Elena M. Bennett, Graeme S. Cumming, Steven J. Cork, John Agard, Andrew P. Dobson, and Garry D. Peterson. 2006. Trade-offs across space, time, and ecosystem services. Ecology and Society 11: 28. Downloadable from http://www.ecologyandsociety.org/vol11/iss1/art28/.

VII

Managing the Biosphere

Stephen R. Carpenter

Human attempts to manage nature are at least as old as agriculture and possibly much older. By the time ecology was formalized as a science, applications and basic ecology were both on the agenda. Stephen Alfred Forbes (1844–1930), an influential American ecologist whose career spanned the origins and consolidation of ecology as a science, demonstrated the mix of practical and curiosity-driven science characteristic of the discipline. Forbes began his career as an economic entomologist, solving problems of pollination and pests that presage work on biological control reviewed by Murdoch in this volume (chapter VII.1). Then Forbes undertook studies of massive fish mortality in Lake Mendota, Wisconsin. He showed the connection of algae blooms and lake physics to fish kills and embarked on a remarkable research program into the ecology of lakes and rivers. His most famous paper, "the lake as a microcosm" (1887), foreshadowed the ecosystem concept as well as modern ideas of behavioral ecology and food web dynamics. As president of the Ecological Society of America and a member of the National Academy of Sciences, Forbes championed the practical uses of basic ecological science for the betterment of humankind. Many other pioneering ecologists pursued research on practical problems of society along with fundamental science questions. Basic and applied ecology have always been intertwined.

From these beginnings, ecology for management has evolved in several dimensions. There has been a progression from single-species problems, such as management of a single resource population or a single pest, to problems of managing ecosystems and social–ecological systems. This does not mean that the single-species problems are no longer relevant—in some cases elegant solutions were found, and in other cases severe problems remain. Instead, the expansion of scope is a natural response to the discovery that larger contexts—ecological or social or both—must be addressed to make progress on pressing environmental problems.

Ecological management problems often begin with suppression of a single pest species or harvest of a single wild species of fish or game and progress to consideration of the ecosystem in which the focal species is embedded. Murdoch (chapter VII.1) shows how biological control of insect pests has expanded in scope to consider dynamics of interacting resource and consumer populations. Hilborn (chapter VII.2) describes how fisheries management coalesced around the idea of maximum sustainable yield (MSY) of a single-species stock. Obvious shortcomings of MSY have led to an ecosystem management perspective considering dynamics of multiple interacting factors and the physical–chemical environment, posing a much more difficult management problem. Boyce, Merrill, and Sinclair (Chapter VII.3) point to a similar expansion of scope in wildlife management. In particular, they note the challenges posed by multiple states of wildlife populations and the ecosystem feedbacks that control them. Management of disease is a relatively new and expanding area of ecosystem management that links wildlife, ecosystem dynamics, and human health. Patz and Olson (Chapter VII.6) explain how changes in biodiversity and land use cascade through ecosystems to affect transmission, resurgence, or emergence of diseases that affect ecosystems and people.

Although management of living resources often began with a single-species perspective and then expanded in scope to consider the ecosystem, management problems of water or nutrients start with an ecosystem perspective. The scientific roots of these management perspectives lie in the geosciences, especially hydrology and geochemistry, rather than in ecology. Nonetheless, research on water and nutrients has long been a part of ecology, and the boundaries between ecology and the geosciences are not as sharp as they once were. Fresh water is frequently a limiting and nonsubstitutable resource for ecosystems and people, so ecosystem management at local, continental, or global scales often centers on water. Alcamo (chapter VII.4) summarizes the challenges of managing freshwater supply. Problems of water quality often go hand in hand with those of water quantity.

Eutrophication, the excessive fertilization of fresh waters associated with toxicity, odor, deoxygenation, and fish kills, is a pervasive global problem. Management of eutrophication is addressed by Schindler (chapter VII.5). Increasing human populations and consumption affect the entire biosphere, and the complexities of managing the expanding human footprint are especially evident for Earth's terrestrial ecosystems. Changing land use and land cover are critical factors in managing living resources, water, carbon, and associated ecosystem services. Challenges of managing land use are addressed by Foley et al. (chapter VII.7). Invasive species, reviewed by Chisholm in chapter VII.8, are a pervasive environmental problem that crosses traditional divisions of ecological science, from landscapes to ecosystems to communities and populations. Although globalized trade is a driver of species invasions, climate change, land use change, and nutrient mobilization all affect the vulnerability of ecosystems to invasion and the success of novel species once they are introduced.

The expanding scope of ecological management inevitably embraced humans as well as other components of ecosystems. All of the chapters of this section address the management of social–ecological systems. Three chapters focus in particular on the human dimensions of ecosystem management. One of the most important institutions for understanding social–ecological systems is the Beijer Institute of Ecological Economics (http://www.beijer.kva.se/), a branch of the Royal Swedish Academy of Sciences. Economics and ecology share more than etymological roots (deriving from the Greek root "oikos" or "house"). They are also quantitative sciences that address dynamics of resources. Thus, ecological economics is a natural starting point for collaboration of ecology with the social sciences. Xepapadeas (chapter VII.9) reviews ecological economics, which has now become a key science for designing incentives and regulations for guiding sustainable behavior. Ostrom (chapter VII.10) synthesizes insights about institutions for governing the commons. This field is important because of the central role of institutions in creating the context in which ecosystem management occurs. A new process of interdisciplinary synthesis, the assessment, has emerged as an important policy tool for sustainability. Miller (chapter VII.11) reviews the status and trends of global environmental assessments.

Adaptive management is addressed by several chapters. Adaptive management originated in the 1960s when researchers associated with C. S. Holling explored a process of "learning by doing" to address problems of regional development. Adaptive management acknowledges the complexity and uncertainty of ecosystem management by starting from an assumption that management actions are not answers but hypotheses to be tested. In successful projects, creative interactions of stakeholders and scientists diversify the tools and approaches available for managing ecosystems. Since the 1990s the term *adaptive comanagement,* reflecting the importance of partnerships among stakeholder groups and experts from various disciplines, has come into vogue. More recently, the idea of "adaptive governance" has expanded the scope further to consider the institutional and social contexts that make adaptive management possible. These ideas are explored by Ostrom in her chapter on governance (chapter VII.10).

The concept of resilience is central to current thinking about adaptive management. Resilience was introduced by Holling to address the coexistence of persistence and change in living systems, including social–ecological systems. Resilience is a key idea in adaptive comanagement because of the crucial importance of open discussions about change. Which aspects of the past can or should be preserved? Which future conditions are attainable? Which are most desirable? And what pathways lead from the current conditions to the preferred future conditions? Questions such as these are fundamental for managing change in social–ecological systems, and resilience concepts have been elaborated to help guide such questions.

Studies of many social–ecological management projects showed that certain characteristic phases could be recognized. In a general way, these phases occurred in each project, even though the projects were conducted in different regions of the world, in different ecosystems, and in different cultures. In each of the different phases, particular approaches and tools are needed to evoke constructive change in a social–ecological system. Many failures of ecosystem management are explainable by use of the wrong approach for a given phase of the program. A model called the adaptive cycle links the phases. For any given social–ecological system, several adaptive cycles may be in play at different scales. A system of interacting adaptive cycles is called a *panarchy*. It seems possible to understand many changes in social–ecological systems using panarchy as a framework, and this idea is now being explored by many practitioners and scientists associated with the Resilience Alliance (http://www.resalliance.org) and Stockholm Resilience Centre (http://stockholmresilience.su.se).

Four of the suggestions for Further Reading provide more information about resilience and related ideas. The article by Holling explains the adaptive cycle, its connection to resilience, and its origins as an organizing framework for adaptive management projects.

Folke, Hahn, Olsson, and Norberg in their article provide a review of concepts of adaptive governance and the facilitation of adaptive ecosystem management. Walker and Salt, in *Resilience Thinking*, offer a straightforward account of the use of resilience, adaptive cycle, and panarchy concepts in pragmatic ecosystem management, using several case studies. *Getting to Maybe*, by Westley, Zimmerman, and Patton, presents a practitioner's perspective on steering a project through the adaptive cycle.

The analysis of social–ecological systems, motivated by the desire to understand and manage the life support system of the planet, is a vast and rapidly changing field. Although this enterprise uses insights and tools from many older disciplines, many practitioners and scientists believe that it has transcended the original disciplines leading to emergence of a completely new and unique discipline called *sustainability science*. The diverse roots of sustainability science are traced in four books cited in Further Reading. *Our Common Journey* describes the elements of sustainability science that had emerged by the 1990s. *The Drama of the Commons* synthesizes insights from the behavioral sciences about management of shared resources. *Global Change and the Earth System* is a concise summary of the major global changes that confront ecosystem managers. The Millennium Ecosystem Assessment or MA is the first global assessment of ecosystem services. The synthesis volume cited here is a brief overview of the MA; the four major reports downloadable from the Web site (http://www.MAweb.org) cover status and trends of ecosystem services, scenarios for future ecosystem services, policy response options, and examples of multiscale assessments of ecosystem services. In the present volume, Miller's chapter (chapter VII.11) includes a discussion of the process for conducting the MA and other assessments.

From roots in ecology, the geosciences, and the social sciences, approaches for managing the biosphere have evolved into a vibrant collaboration of practitioners and researchers. The chapters in this volume present a sampler of some of the most important topics, emphasizing aspects most relevant to our overarching focus on ecology.

FURTHER READING

Folke, C., T. Hahn, P. Olsson, and J. Norberg. 2005. Adaptive governance of social–ecological systems. Annual Review of Environment and Resources 30: 441–473.

Holling, C. S., 1986. The resilience of terrestrial ecosystems: Local surprise and global change. In W. C. Clark and R. E. Munn, eds. Sustainable Development of the Biosphere. London: Cambridge University Press, 292–317.

Millennium Ecosystem Assessment. 2005. Ecosystems and Human Well-Being: Synthesis. Washington, DC: Island Press. Available on the internet at http://www.MAweb.org.

National Research Council. 1999. Our Common Journey. Washington, DC: National Academy Press.

National Research Council. 2002. The Drama of the Commons. Washington, DC: National Academy Press.

Steffen, W., A. Anderson, P. D. Tyson, J. Jager, P. A. Matson, B. Moore III, F. Oldfield, K. Richardson, H. J. Schellnhuber, B. L. Turner II, and R. J. Wasson. 2004. Global Change and the Earth System. Berlin: Springer-Verlag.

Walker, B., and D. Salt. 2006. Resilience Thinking: Creating Space in a Shrinking World. Washington, DC: Island Press.

Westley, F., B. Zimmerman, and M. Q. Patten. 2006. Getting to Maybe. Toronto: Random House Canada.

VII.1

Biological Control: Theory and Practice

William Murdoch

OUTLINE

1. Settings and protagonists
2. Locally persistent pest–enemy interactions
3. Locally nonpersistent systems
4. Ecological theory and biological control

Biological control—defined here as the suppression of insect pests by other insects that attack them—has been pursued by entomologists for more than a century, in part because it is typically cheap and can yield very large economic returns on investment. Over this period, the empirical record is one of often spectacular successes mixed with rather more failures. It is also a history of trial and error. In an extreme example, entomologists introduced about 50 species of enemy insects before achieving great success in controlling California red scale on citrus, a case discussed below. Such applied population dynamics has naturally attracted the interest of ecologists who, together with many of the entomologists themselves, have worked to develop theory that would explain the essential features underlying success. This theory and its connection to real biological control comprise the subjects of this chapter. The theory's domain, however, is much broader—it is the dynamics of interacting resource populations and the consumer populations that attack them.

GLOSSARY

consumer–resource interactions. These include interactions between populations of predators and prey, parasitoids and hosts, indeed any interaction in which one species depends on another for sustenance. These terms are used interchangeably here.

density-dependent processes. These cause the per-head rate of increase of the population to decrease when the population's density increases.

equilibrium. Density at which the population will remain, once it is reached, if the population is not perturbed.

parasitoid. An insect that parasitizes another insect (the host species) by laying egg(s) in or on the host, which is eaten by the immature parasitoid.

scale insects. Plant-sucking bugs that stay attached to the plant for almost their entire life history.

stable equilibrium. An equilibrium is stable when the population tends to return to it after the population is perturbed.

unstable equilibrium. The population moves away from the equilibrium following a perturbation. The result may be cycles in abundance, extinction, or chaos, in which the densities are always bounded, but there are no repeated sequences of abundance.

1. SETTINGS AND PROTAGONISTS

It is useful to distinguish two settings in which biological control occurs. First, most successful biological control has occurred in relatively long-lived agricultural systems such as orchards and, less successfully, forests. It is known for at least some cases that the pest and enemy populations persist together at a small spatial scale such as a single orchard. This is the kind of dynamics for which the bulk of ecological consumer–resource theory has been developed.

In the majority of successes in this class, the pest is an alien species as, almost always, is its successful enemy. Typically, the entomologist faced with an introduced pest travels to its place of origin, searches for enemy species in that environment and, after some preliminary laboratory work, releases the enemy, which usually confines its attacks to the pest species—it is effectively a specialist in the introduced agricultural ecosystem. Again, most consumer–resource theory is developed for specialist consumers.

The second class of control occurs in temporary crops, an environment where success is more elusive, at least in part because enemy species (and the pest) usually do not persist locally. Pest species in these situations may be native or introduced. Less is known about control here; it is likely to involve a number of enemy species, some of which may be introduced and some of whom are likely to be generalist feeders. There is also less relevant theory available.

N. J. Mills has shown that the greatest number (and proportion) of successful cases of biological control are against plant-sucking bugs (Homoptera), for example, scale insects and whiteflies. Caterpillars of moths and butterflies (Lepidoptera) constitute the second largest group, although the number of successes and especially the success rate are lower. The majority of successful control agents are parasitoids, although predatory insects have also been successful.

2. LOCALLY PERSISTENT PEST–ENEMY INTERACTIONS

The central assumption in classical biological control, in which an introduced pest is suppressed by an introduced specialist enemy species, is that the natural enemy controls the pest by maintaining it locally at a low stable equilibrium density. This is because most theory for population dynamics is equilibrium centered, and such a framework provides the most straightforward way to account for long-term persistence of the interaction. Extirpation of the pest is generally thought to be logistically unfeasible, although it has sometimes been achieved, for example in isolated situations on islands.

Successful biological control in this context exemplifies a fortiori a famous problem in predator–prey dynamics—the so-called *paradox of enrichment*. Basic mathematical predator–prey theory, regardless of its formulation, predicts that the more a predator suppresses its prey below the capacity of the environment to support the prey, the more the interaction will be unstable. Predicted instability takes the form of large-amplitude cycles in population abundance. Successful enemies often suppress the pest well below 1% of the pest's carrying capacity, so we should see large-amplitude oscillations in density or even population extinction. The dilemma is sharpened by the fact that mechanisms added to models to stabilize the interaction, with few exceptions, cause prey density to increase. Obviously, in successful, persistent, locally stable biological control situations, we see severe pest suppression without instability. This is a general mismatch between theory and observation and is also observed in natural ecosystems.

A number of well-documented cases establish that successful biological control can indeed operate through the creation of a low stable equilibrium, although there have been few efforts to confirm that this is the general case. The best evidence that this is generally true is that later visits to successful instances of biological control frequently find both pest and enemy rare but present. Scale insects (which are plant-sucking bugs that stay attached to the plant for almost their entire life history) have experienced many successful biological control efforts; a large fraction of efforts in this group have been successful, and they show the best evidence for control via establishment of a stable equilibrium. One thoroughly studied case, described below, shows a nice match with theory.

Continuous Breeders with Overlapping Generations

Insects may reproduce only once, and have only one generation per year (so generations do not overlap), or they breed more than once and have multiple, often overlapping, generations per year. The illustrative case discussed below is in this category. It is a Homopteran (the group showing greatest success), and within that is an armored scale, a group that alone accounts for two-thirds of the biological control cases considered to be complete successes. The enemy is a specialist parasitoid, a feature that greatly increases its chances of success. This example thus represents a large fraction of, and many aspects that typify, successful biological control.

California red scale (*Aonidiella aurantii*), a worldwide pest of citrus, was accidentally introduced into California about 100 years ago from China via Australia. When sufficiently abundant, it can kill trees and, on several occasions in the last century, almost destroyed the California citrus industry. It has spread to most citrus-growing areas of the world.

Among the many natural enemies introduced for control, three were in the genus *Aphytis*. These are tiny (less than 1 mm long) parasitic wasps (parasitoids). The female (males do not kill hosts) lays eggs in most of the juvenile scale stages, feeds on the smaller stages to get nutrients for maintenance and future egg production, and lays male eggs on smaller and female eggs on larger immature scale hosts. Adult scale are not attacked. The scale passes through two or three generations per year, and *Aphytis* has about three times as many generations as does the scale. The pest and parasitoid generations are overlapping, so that different life stages coexist at the same time. As a result, vulnerable stages of the pest are present when adult parasitoids are present and searching. This parasitoid genus has been used successfully throughout the world on many different pests of citrus and other crops.

In California, one *Aphytis* species was present for most of the twentieth century and apparently had little effect on scale densities. *Aphytis lingnanensis* was introduced in the late 1940s and reduced scale densities but not quite enough for economic control. Finally, *Aphytis melinus* was introduced around 1959. It was spread throughout much of southern California, immediately both reduced scale densities below the economic threshold and displaced *lingnanensis*, and has remained a spectacular success for almost 50 years except when sprayed, for example by DDT to which the scale were resistant. All aspects of this history are satisfactorily explained by ecological theory, and indeed, this example provides one of the strongest demonstrations that consumer–resource theory can explain population dynamics in the field.

Over about 20 years, W. Murdoch and colleagues showed that none of the existing notions that had been proposed to explain stability and control in general applied in this system. Field observations and experiments showed stability was not caused by parasitoids concentrating their attacks in a small fraction of hosts or by the existence of a spatial refuge from the parasitoid. Ecologists have come to suspect that spatial processes are frequently the key to the stability of field populations, but these workers also established that spatial processes, i.e., movement of pests or parasitoids among trees, played no role in control and stability. Thus, the mechanisms of control and stability operated on a local spatial scale.

Finally, these workers developed mathematical "stage-structured" models to explore and test the possibility that control and stability are explained by life-history features of the pest and how the parasitoid responds to them. Such models are designed precisely for organisms such as insects that have distinct life stages (e.g., egg, larvae of different sizes, adult) and that reproduce more or less continuously so that generations overlap. They are written in the form of delay-differential equations and keep track of the numbers entering and leaving each pest and parasitoid stage, of losses in each stage to "background" deaths and those imposed on the pest by the parasitoid, and of reproduction of pests surviving to be adults. Equilibrial densities can be calculated analytically, as can the local stability behavior of these equilibria. Dynamics following large perturbations of density far from equilibrium are investigated by computer simulation guided by results from local stability analysis.

A detailed stage-structured model of the red scale–*Aphytis* interaction was developed in which the main stabilizing mechanisms are the invulnerable adult stage of scale, a potentially powerful stabilizing force, and the fact that the parasitoid goes through three generations in the time it takes the pest to pass through one generation. This model was independently parameterized and was then able to predict with astonishing accuracy the control by *Aphytis* of experimental scale outbreaks on individual trees in a lemon orchard. Control by the resident *Aphytis* population in each tree was very rapid—the scale population was effectively brought under control in approximately a single scale generation.

This case speaks to the larger issue in ecological theory raised above, namely the almost universal tradeoff in models between stability and prey suppression: almost all mechanisms that stabilize the predator–prey equilibrium also increase the prey equilibrium. The model shows that stability in this case hinges on two main mechanisms: the invulnerable adult stage of the pest and the much shorter development time of the parasitoid relative to that of the pest. An invulnerable class of prey involves the stability–suppression tradeoff. But when the invulnerable stage is a highly fecund adult stage, it requires relatively few adults to maintain the interaction, so overall suppression is consistent with stability. The second mechanism, unlike virtually all other stabilizing processes, does not involve a tradeoff: stability is more likely, and pest suppression is stronger the shorter the parasitoid's development (or generation) time is in comparison with that of the pest. We return to these properties below.

The *Aphytis*–red scale model is also able to explain the subtle life-history features that account for the rapid displacement of *Aphytis lingnanensis* by *melinus*, itself a classic example of "competitive displacement." *Aphytis melinus* is able to produce female offspring from smaller immature hosts than is *lingnanensis*, and the model shows that this leads to increased suppression of scale and rapid displacement of *lingnanensis*. There is thus quite strong evidence that this model contains the essential features of the pest–parasitoid interaction.

The features of *Aphytis*'s control of red scale appear to be shared by many other successful cases of biological control. As noted above, of the two large groups of pest insects, the greatest number of successes and the greatest rate of success per unit effort are in the Homoptera (plant-sucking bugs); efforts on Lepidoptera have been much less successful. Mills reviewed many cases of biological control and showed that the Homoptera typically reproduce continuously, have several overlapping generations a year, and their parasitoids have markedly shorter generation times than the pest species. Lepidoptera, by contrast, typically have fewer, discrete, nonoverlapping generations per year, and the generations of their parasitoids are also discrete and synchronized with those of the pest, so they have equal development times. Mills showed, furthermore, that, among Homopteran control efforts, short parasitoid generation times were more prevalent among successes than among failures. An invulnerable adult stage is the norm in parasitoid–host relationships.

One or Multiple Enemy Species?

As noted, the displacement of *lingnanensis* by *melinus* illustrates ecology's classical competition theory. Where there is a resource population maintained by a consumer at a stable equilibrium, theory states that that consumer species will win that suppresses the resource to such a low density that it cannot support less efficient, competing, consumer species. This is how *melinus* displaced *lingnanensis* in both theory and reality. Mills notes that there have now been numerous documented examples of competitive displacement, leading in all cases to improved biological control. (As an aside, exotic ladybugs and other enemies introduced for pest control have also been shown in some cases to competitively displace native species in natural habitats.)

Theory also says, however, that competing consumers can coexist if they interfere with their own increase more than they do with that of their competitors (here there is another, analogous trade-off between pest suppression and coexistence of enemy species). Such coexistence would not seem, a priori, to be a recipe for successful biological control, and arguments have raged for decades over releasing the most efficient single enemy species (if such can be found) versus releasing multiple species. Of course, if the "best" species wins, the issue is moot. But it is possible in theory to release an enemy that wins by, for example, directly attacking the otherwise most efficient enemy and thereby inducing a higher pest density. Whether this is in reality an important issue is not clear.

In fact, although Mills found, in biological control systems apparently near equilibrium over long periods, that a single, successful, enemy species was a frequent occurrence, it is also common to observe several species of natural enemies coexisting. Obviously there are differences among such species, such as exploiting the pest on different parts of the plant, that may explain their coexistence. Typically, however, we do not know if the inevitable differences among species actually explain coexistence, nor do we typically know the roles played by the different species, if any, in control. The presence of "extra" generalist predators (which attack species other than the pest) does not need explanation because they can coexist by being supported by other resources. More vexing is the coexistence of effectively specialist parasitoids attacking the same pest.

In the red scale example, in addition to *Aphytis*, there is usually both one rare generalist predator and one rare specialist parasitoid. The above experiment and model established that neither was important in control of red scale, and there is evidence that a single species also dominates in other similar examples of control of scale. Thus, the mere presence of multiple enemy species does not imply they are all necessary or even useful for control. By contrast, long-running control of olive scale in northern California was known to require two specialist parasitoids, each of which was effective in a different season. As in ecology in general, we do not yet have a complete explanation for, or explication of the roles of, competitors coexisting at equilibrium.

Mills (2006a) examined the record of hundreds of efforts to introduce natural enemies. The overall message is that as more enemy species are introduced, a smaller proportion is established—and by far the most common outcome is that only one enemy species is established. On the other hand, the fraction of introductions that led to successful control was independent of the number of enemy species established.

Discrete Breeders

Much equilibrium-centered theory for biological control, in contrast to the systems discussed above, is written for a species whose adults are present and breed in a single episode at one time of year—they are discrete breeders, and the theory is formulated in discrete time (difference equations). Insects with such a life history are typically found at higher latitudes. Unfortunately, there is no deeply studied example that explores experimentally the mechanisms of control and applicability of the theory. Recent work by D. N. Kimberling suggests that success rates are lower in such systems.

Perhaps the best example is the winter moth in North America (mainly Nova Scotia and British Columbia), where it was a severe pest of hardwood forests and urban shade trees. This Lepidopteran was introduced from England, where it was studied in some detail. Of several parasitoid species introduced to control the moth, a fly (*Cyzenis*) and a wasp (*Agrypon*) parasitoid became established. The system has been most thoroughly studied in British Columbia, where J. Roland and D. G. Embree have shown, interestingly, that although initial suppression of the pest appears to have been induced by the parasitoid *Cyzenis*, maintenance of persistent low populations appears to rely on high and density-dependent mortality induced by native predators, mainly beetles, attacking pupae in the soil. Both the moth and fly populations in British Columbia have persisted at low densities for around 20 years except for small, local, and short-lived increases in the moth.

If control of these winter moth populations is indeed induced by the generalist beetle predators, there is likely to be no trade-off between prey suppression and stability. Theory does suggest, however, that unless the pest is strongly preferred by the generalist predators, there is an omnipresent danger that favorable conditions for the pest will allow it to escape their control.

An interesting situation, intermediate between those above and the locally nonpersistent system discussed next, is one in which the pest may be well regulated at a spatial scale larger than the local level, at which there is instability or extinction. This is the ecological notion of metapopulation dynamics. Apparent examples include the famous cottony-cushion scale controlled by the predatory beetle *Rodolia*. Both species were introduced to California citrus more than a century ago, and successful control continues to this day. *Rodolia* is seen to drive the pest density to zero on a spatial scale of at least individual trees, but movements among different trees or small groups of trees have allowed both species to persist. A similar pattern has been seen in some mite populations in apple orchards, and some greenhouse pests.

3. LOCALLY NONPERSISTENT SYSTEMS

Many pest–enemy interactions in agriculture appear not to persist at the local spatial scale of interest—a tract of forest, an orchard, a field, or a rice paddy. This is true especially in seasonal field crops, where the pest and its enemies do not persist even in a local collection of crop units. Instead, regional persistence requires that the pest and/or the natural enemy population invade from some other habitat each growing season, and the pest may or may not be kept below the density at which it causes economic damage before harvesting, tilling, etc. drive it locally extinct or to very low numbers.

We would not expect the insights of classical consumer–resource theory from the previous section—in particular the efficacy of a dominant specialist enemy species—to be applicable where local dynamics does not fit the equilibrium paradigm and the pest is known to arrive at the crop from "elsewhere." Because an enemy cannot rely on the pest as a year-round resource, we expect generalist predators to be important. Because the pest is mobile, we expect enemy population movements to be a central feature of successful control. Because local dynamics does not go to a persistent equilibrium state, multiple species of enemies are likely to persist, and local competitive exclusion will not occur. Because local extinction of the pest is consistent with global control and persistence of generalist enemies, an invulnerable pest stage is not a requirement for persistence and may interfere with control. These expectations are borne out in the rather few cases where we know biological control is successful and where mechanisms, or at least dynamics, have been investigated.

Aphid pests exemplify mobile and often nonpersistent pests and, probably as a result, provide few examples of successful biological control. There is, however, one well-studied case in a temporary crop, and it is consistent with the above expectations. W. E. Snyder and A. R. Ives have studied pea aphid (*Acyrthrosiphon pisum*), an introduced pest in alfalfa fields in Wisconsin, which is under control by a range of natural enemies. The enemies include an introduced specialist parasitoid, *Aphidius ervi*, and a wide range of mainly native generalist predators, including *Nabis* and *Orius* (bug species), ladybirds, and carabid beetles. As expected, the specialist parasitoid is relatively ineffective and is unable to control the aphid on its own. Generalist predators are essential for control, although in some instances they probably decrease the effectiveness of the specialist parasitoid.

The generalists' adult densities are determined mainly in other habitats, where they feed on other prey. The predators exert control by moving into alfalfa fields and both feeding there and producing predatory larvae, even though the latter may not even complete their development before harvest. Most of the generalists probably eat all stages of aphids—there is no invulnerable stage. In alfalfa fields, however, the predator–prey interactions are not self-maintaining—the enemies exist in the crop only because of processes at a much larger scale. Short generation time is no longer relevant, and indeed, the generalists have much longer development times than the prey. L. E. Ehler has uncovered a remarkably similar story, also in alfalfa fields, but this time in the control of an introduced moth (the caterpillar is the beet armyworm) in California.

These examples appear to illustrate a general situation. Thus, it has long been known that multiple enemies coexist in a number of successful cases of pest control in temporary crops. Indeed, historically, entomologists working in temporary crops have often maintained populations of generalists through various cultural practices, such as interplanting and judiciously timed cutting of noncrop vegetation as a replacement for spatially dispersed other habitats.

There is extensive ecological theory that incorporates spatial processes into models of consumer-resource dynamics. There is also some theory for generalist predators. There is, however, little theory that combines these two features of temporary crop systems, and, so far, biological control in such crop systems has not provided the same rich opportunity for testing general ecological theory. As the work on aphids described above illustrates, however, there is reason to be sanguine about future progress in this area.

4. ECOLOGICAL THEORY AND BIOLOGICAL CONTROL

Successful biological control of pest insects, especially in long-lived crops, has been a fertile field for ecological theorists. It has been a substantial inspiration for much, ever-more-sophisticated consumer–resource theory that has been shown to be powerful in explaining the dynamics of many natural and nonagricultural systems. Great unexploited opportunities remain for exploring and testing dynamic theory because successful control occurs in simple and species-sparse systems—features that facilitate experimental manipulation. Although these are nonnatural systems, some of them still exhibit, for example, dynamic stability, which is a central ecological phenomenon to be explained.

Benefit has flowed less conspicuously in the opposite direction: biological control has remained a largely empirical, trial-and-error process. This is to some extent inevitable—whether an enemy species will succeed or fail must depend to some extent on local and particular contingencies. Nonetheless, as we saw above, potentially useful guidelines have emerged, and the recent analyses by Mills and others of large sets of historical cases have established that these guidelines apply in a wide range of real pest control cases. Further probing along these lines is likely to be rewarding.

FURTHER READING

Ehler, L. E. 2007. Impact of native predators and parasites on *Spodoptera exigua*, an introduced pest of alfalfa hay in northern California. Biological Control 52: 323–338. *Establishes that control of this pest is by native generalist predators.*

Kimberling, D. N. 2004. Lessons from history: Predicting successes and risks of intentional introductions for arthropod biological control. Biological Invasions 6: 301–318. *A statistical analysis of the factors affecting success and failure of biological control using information from historical examples.*

Mills, N. J. 2006a. Interspecific competition among natural enemies and single versus multiple introductions in biological control. In J. Broeder and G. Boivin, eds. Trophic and Guild Interactions in Biological Control. Berlin: Springer, 191–220. *An analysis of many historical cases of biological control.*

Mills, N. J. 2006b. Accounting for differential success in the biological control of homopteran and lepidopteran pests. New Zealand Journal of Ecology 30: 61–72. *An analysis of the life-history features that influence success in biological control, using numerous case studies.*

Murdoch, W. W., C. J. Briggs, and R. M. Nisbet. 2003. Consumer–Resource Dynamics. Princeton, NJ: Princeton University Press.

Murdoch, W. W., C. J. Briggs, and S. Swarbrick. 2005. Host suppression and stability in a parasitoid–host system: Experimental demonstration. Science 309: 610–613. *Presents the key experiment and model establishing the mechanisms causing control and stability in the red scale example.*

Snyder, W. E., and A. R. Ives. 2003. Interactions between specialist and generalist natural enemies: Parasitoids, predators, and pea aphid biocontrol. Ecology 84: 91–107.

VII.2

Fisheries Management

Ray Hilborn

OUTLINE

1. The history of fisheries and ecological thought
2. The nature of a fishery
3. The biological basis of sustainable harvesting
4. Ecosystem effects and interactions
5. The human ecology of fisheries
6. How fisheries are managed
7. The current state and future of the world's fisheries

Fisheries constitute the most significant example of exploitation of natural ecosystems to produce protein for human consumption. Fisheries are an important part of ecology, both because of their importance for humans and because of the impact fisheries have on almost all oceans, lakes, and rivers. A fishery consists, at a minimum, of an ecosystem and a collection of humans who exploit it. In most cases there is now a third component, a management system. This chapter explores the ecology of the exploited ecosystems and how they interact with the human ecology of exploiters and the managers.

GLOSSARY

bionomic equilibrium. The balance between fish stock abundance and the fishing fleet that a fishery will evolve to in the absence of regulation

by-catch. The unintended catch of a nontarget species in a fishery

fish stock. A population of a single species that is geographically distinct enough to be managed separately from other populations of the same species

fishery. The interaction between humans and an exploited fish stock

maximum sustained yield. The highest long-term average yield that can be obtained from a fish stock on a sustainable basis

trophic interaction. Interaction between species in an ecosystem as a result of predation or the consequences of predation

unfished or virgin biomass. The average stock size in an unexploited condition

1. THE HISTORY OF FISHERIES AND ECOLOGICAL THOUGHT

Regulations to promote the conservation of fisheries originated in Europe in the fourteenth century over concern about Atlantic salmon. Similarly, in North America, where the salmon runs seemed inexhaustible, the need for restrictions on catch was recognized soon after the salmon fishery began to develop. However, the susceptibility of marine fish to overexploitation was recognized much more slowly, and Thomas Huxley (known as Darwin's bulldog for his advocacy of the theory of evolution) championed the school of thought that the fecundity of fishes was so large that fishing could not have an impact on the abundance of fish in the sea, and "that the cod fishery, the herring fishery, the pilchard fishery, the mackerel fishery, and probably all the great sea-fisheries, are inexhaustible; that is to say that nothing we do seriously affects the number of fish. And any attempt to regulate these fisheries seems consequently . . . to be useless."

Numerous European and North American scientists led the way in providing evidence that fishing not only could affect the abundance of marine fish but could reduce the abundance of certain fish stock so much that their potential yield was reduced. These early scientists also developed the theoretical basis for understanding fisheries production, as elaborated in a later section of this article, and from their work emerged the concept of maximum sustained yield (MSY), the maximum long-term average catch that could be removed from a population on a sustainable basis.

By the 1950s the concept of MSY was firmly entrenched in fisheries thinking as an objective of management and as an intrinsic property of most natural populations, and there had developed an elaborate set of methods for calculating it. The simple view of MSY of the 1950s has gradually faded as we have come to recognize the complexity of society's objectives, the difficulty in estimating the productive potential of natural populations, the natural variability of ecosystems, the interaction between species in ecosystems, and the problems inherent in regulating the exploiters of the resources.

At the same time, economists were studying the nature of fisheries and determined that in the absence of regulation, fisheries would evolve toward the "bionomic equilibrium," in which fishing boats and fleets could just meet their costs but not produce any true profit. So long as a fishery was profitable, more fishermen would enter the fishery and reduce the profit of everyone participating.

As a general rule, fisheries in Europe and North America were barely managed until the stocks had been depleted enough that fishermen were economically struggling, and some of the users recognized the need for regulation to reduce fishing effort. The consequence of this was that the natural trajectory of most stocks was to be exploited much harder than MSY and reduced to lower levels, and there was the now familiar situation of "too many boats chasing too few fish." For the more aggressive fishermen, there was always another new fishery to explore and develop, and individuals and fleets pushed into deeper and farther waters in search of a fishery that had not been overexploited.

There were early attempts to regulate catch and keep stocks from becoming overexploited; the International Pacific Halibut Commission, founded in 1923, is one of the most successful national or international agencies. It has maintained the stock of halibut in a healthy condition for 85 years through strict catch limits. The major change in fisheries management came in the late 1970s with the signing of the United Nations Law of the Sea, which allowed states to extend their territorial waters out to 200 miles, thus bringing most of the world's major fisheries under the control of coastal states. The Law of the Sea provided these states with the legal authority to exclude other countries and manage the fisheries for their own benefit. In the United States this coincided with the passing of the Magnuson Fisheries and Conservation act of 1976, which established regional Fisheries Management Councils that provide the ongoing basis for the attempt to produce MSY from U.S. fisheries.

2. THE NATURE OF A FISHERY

A fishery consists of two essential elements: an exploited aquatic ecosystem and a human system of exploiters. In managed fisheries there is a third element, the management system.

Aquatic ecosystems are almost always highly complex and include everything from bacteria to whales, yet the species targeted by humans are usually a very small subset of the ecosystem. Most commonly, the fisheries initially target the species that are valuable in the marketplace, often top predatory fishes and high-value invertebrates. But as fisheries and markets have developed, fish lower down the food chain are increasingly being targeted, and it is now common for a very high proportion of the fish within an ecosystem to be exploited.

The second essential ingredient in ecosystem management is people and how they change aquatic ecosystems. We concentrate on fishing fleets and their impacts, but people also have major impacts through the introduction of exotic species, both intentionally and accidentally, habitat change, pollution, and climate change. It is important to remember that almost all forms of aquatic resource management involve changing the actions of people. Fisheries management is widely recognized to be people management. This is just as true when we consider the broader perspectives on aquatic ecosystems: you rarely manage the ecosystem; you almost always manage the people modifying the ecosystem.

Therefore, to understand why management regimes are relatively successful or unsuccessful, you need to look primarily at the people and how they behave and interact with the management regime. Changes in the ecosystem may be what you are interested in, but understanding the people is the way to understand why the ecosystem is changing.

The third key ingredient in understanding aquatic ecosystem management is the management system itself, the legal, political, and social constructs that modify human impacts. Some systems have no management, but this is increasingly rare, and most aquatic ecosystems are embedded within a framework of laws, institutions, and customs. What we see as we look deeper into a range of ecosystems is that it is these laws, institutions, and customs that determine the outcomes.

Within U.S. federal waters, a range of key legislation establishes the governance system. Most important is the Magnuson-Stevens Fisheries Management and Conservation Act, first enacted in 1976 and reauthorized by Congress in 2007. This act laid down the U.S. claim to the 200-mile exclusive economic zone and established the regional Fisheries Management Councils that effectively manage U.S. fisheries in federal waters. In one sense the Magnuson-Stevens act is primarily about regulation of harvest, but successive reauthorizations have included more and more frameworks for dealing with habitat and fishing impacts on nontarget species.

3. THE BIOLOGICAL BASIS OF SUSTAINABLE HARVESTING

The biological basis for all sustainable harvesting is reproductive surplus. All natural populations are capable of net population growth under favorable conditions. A single pair of fish in a good habitat can often produce dozens of offspring that survive to breed again. The simple theory of single-species dynamics suggests that at low densities resources are abundant: there is plenty of food for each individual; the best hiding places from predators are not taken; and individuals have a high probability of survival and reproduction. As densities increase, food per individual becomes scarcer, the best protection from predators is taken, and the net population growth decreases until at some point there is no net population growth.

Competition for resources occurs in most animal populations, but in the study of exploited fish populations, the relationship between population size and reproductive surplus has been the subject of much controversy. It is now widely accepted that fishing does have a major impact on fish population abundance, but the debate over the relationship between spawning stock abundance and reproductive surplus continues. By the 1990s, however, it became increasingly accepted that heavy fishing pressure also reduces recruitment to the fishery.

The large yields available during the early stages of a fishery's development often lead to expectations of larger yields than are sustainable. The exploiting industries almost inevitably develop infrastructure to harvest and process the nonsustainable yield, and once that yield is gone, they usually create great economic and political pressure to delay the needed reduction in fishing effort and catch. Any such delay is then likely to drive the stock below its most productive level, resulting in even more severe reductions in catch when the inevitable decline does come.

4. ECOSYSTEM EFFECTS AND INTERACTIONS

The theory of exploitation was initially developed for single-species population dynamics, and most of the world's fisheries management agencies regulate fisheries on a species-by-species basis. Nevertheless, it is widely recognized that there are strong interactions among species, and fisheries frequently impact multiple species. One form of interaction is the catching of nontarget species, called "by-catch." By-catch first became a major concern for charismatic species such as marine mammals, birds, and turtles that were incidentally caught in fishing gear. This has led many fisheries agencies to restrict fishing locations or gear or to require the use of "by-catch" avoidance devices to reduce or eliminate the by-catch of these species.

Another form of by-catch is the capture of nontarget fish that are often discarded because of low economic value. In the 1980s about 80 million tons of fish were landed worldwide, and 27 million tons were discarded. Globally, discards have declined dramatically since then because many of the fish that were formerly discarded are now retained and often used as feed in aquaculture operations.

A second form of ecosystem impact of fishing is trophic interaction between and among species. Such trophic interactions may take the form of fishing down predatory species, thereby allowing their prey to increase in abundance, or fishing down prey species and causing their predators to decline. As top predators such as cod have been fished down, many of their invertebrate prey items, such as lobsters, scallops, and prawns, have increased. In some cases these species may now provide more valuable catch than their predators did before they were depleted. Fisheries agencies are only now beginning to recognize that they cannot have maximum yield of both predators and prey.

Some fishing gear also impact the physical environment. In habitats with considerable vertical structure on the bottom, fishing gear such as trawls and dredges will eliminate much of this structure. For instance, coral communities are highly vulnerable to bottom-contact fishing gear. On soft-bottom habitats, the impacts of fishing gear are much less dramatic but still cause changes in the communities that live in the sand and mud habitats.

5. THE HUMAN ECOLOGY OF FISHERIES

Fisheries are not static systems that can be manipulated and reshaped at will by management. Rather, the human element in fisheries has its own dynamics and consists of individuals or firms seeking to maximize their own well-being. Fisheries commonly begin with a period of discovery and spread of information about the existence of a potentially valuable stock. A few individuals discover that money is to be made exploiting a certain stock in a certain way. Then there follows a period of rapid growth of effort as others are attracted by the success of the initial fishermen. In this phase, profitability is high, individuals often increase their boat size or purchase additional boats, and new individuals are drawn into the fishery. Governments often subsidize boat construction during this development stage.

Next, the fishery reaches full development, where yields are near or perhaps a little above a long-term sustainable level. The rapid development results in declining rates of fishing success as the stock is reduced and more fishermen compete for the remaining fish. At this stage profits are reduced and possibly negative, and the incentives to build bigger boats or enter the fishery are gone. There is often a lot of political pressure both for increased subsidies and for permission to maintain allowable catches despite scientific advice that catches need to be reduced. The fishery often then enters an overexploitation stage, which may be followed by a collapse. If the collapse is not too catastrophic, there is often a period of declining fishing pressure as the less successful fishermen find it no longer worthwhile to pursue the stock. The stock may or may not recover somewhat on its own during this period. Where possible, fishermen seek new fishing opportunities and move on to less-exploited stocks.

At each stage in this process, individuals and firms are doing what is best for them given the incentives and information available. Some individuals will be able to catch more fish or have lower costs and still be profitable, whereas less efficient fishermen will, as the stock declines, no longer be making any profit and may not even be able to meet the costs of fishing. Where there are other opportunities available, these inefficient fishermen will leave the fishery, but often there are no alternatives available.

The most common consequences of unregulated fisheries are excess fleet size, depleted stocks, and impoverished fishing communities. To the extent that governments have subsidized fleet development, the level of overexploitation will be worse. Because of changes in human and natural environments, fishery systems may never really be at a true bionomic equilibrium, but it serves as a construct that lets us understand the human and biological dynamics.

The process described earlier is an idealized description in which each individual or firm acts independently. Garret Hardin outlined this process as typical of human society and "the tragedy of the commons." Elinor Ostrom and her colleagues have shown how many societies manage commons to avoid this tragedy (see chapter VII.10). In looking at fisheries around the world, we often find that societies do have mechanisms to avoid the apparently inevitable overexploitation and collapse.

6. HOW FISHERIES ARE MANAGED

Fisheries management in Western countries has evolved through a series of stages. The first stage is before there is any form of active management and is commonly referred to as unregulated open access. Individuals or companies are generally free to fish what they want, when they want, and where they want, with the only constraints being the economics of the marketplace and the technology currently available. The history of industrial whaling is a classic demonstration of unregulated open access, from its beginning with Basque whalers almost 1000 years ago until the advent of active regulation by the International Whaling Commission. As easily accessible stocks close to ports were depleted, the fisheries moved farther and farther offshore and often suffered economic declines when stocks declined or products became less valuable.

However, at some point almost all fisheries reach a point where there is an obvious need for management. Initial management usually consists of restrictions on fishing gear, perhaps banning highly efficient methods or restricting boats to a maximum size. Closed seasons or areas are also common elements of initial management. Such limitations usually fail to stop the decline in fish abundance, and Western countries generally then proceed either to restrict the number of fishing vessels (called limited entry) or the total catch by shortening fishing seasons. The management of Pacific halibut in Alaska, for example, is commonly cited as one of the great success stories of sustainable management. Catches were restricted by an increasingly short fishing season (declining to less than 2 days per year by the late 1980s) but did not employ a limitation on the number of vessels until the early 1990s.

At present, most Western fisheries are managed by a complex combination of gear restrictions, time and area closures, limited entry, and commonly restrictions on the total catch. When combined with effective enforcement and good science, these systems can be effective at maintaining stocks in a healthy, sustainable condition. However, the normal outcome is also one of little economic profitability and often severe economic hardship. The key cause of the poor economic profitability is excess fleet capacity. When limited-entry programs are put into place, there are usually too many boats already—it is the too many boats that has led to the need for additional regulation and limited entry.

Increasingly, Western countries are adopting some form of what is now known as "dedicated access" in which incentives are established to encourage the fishing fleets to match their harvesting capacity to the biological productivity of the resource. The most common form of dedicated access is Individual Transferable Quotas, in which individual license holders are allocated a share of the total catch, and these shares can be traded. The result is almost always that some individuals choose to remain in the fishery and buy catch share from other individuals who chose to sell out and leave the fishery. The consequence is a smaller fleet that is profitable, but with a significant drop in total employment.

Another form of dedicated access includes formation of cooperatives among harvesters who agree on how to share the catch and then operate only enough vessels needed to catch the available harvestable surplus. Again, the consequence is usually a more profitable smaller fleet.

A third form of dedicated access employs territorial fishing rights, in which communities or individuals are granted exclusive access to some portion of the ecosystem. The common theme of all of these forms of dedicated access is providing incentives to match the harvesting capacity to the productive capacity of the resource.

All of these Western management systems require a strong central government agency that collects data, evaluates stock size, and determines and enforces regulations.

Much of the world has evolved different approaches. In the coastal waters of Japan, regional cooperatives have been granted management authority: they determine how their members share in the harvest and often engage in very intensive enhancement activities by releasing juvenile fish or outplanting valuable invertebrates. The cooperatives are self-governing and largely autonomous.

The tradition in the Pacific Islands was for village control of local inshore resources. The villages employed a wide range of techniques including closed areas and times, limitations on fishing gear, and, above all, restrictive access.

> The most important form of marine conservation used in Palau and many other Pacific Islands was reef and lagoon tenure. The method is so simple that its virtues went virtually unnoticed by Westerners. Yet it is probably the most valuable fisheries management measure ever devised. Quite simply, the right to fish in an area is controlled, and no outsiders are allowed to fish without permission. (Johannes, 1988)

In essence, the recognition of the potential for territorial fishing rights in Western fisheries is a belated recognition of the effectiveness of the management system in the Pacific Islands.

The management of Chilean artisanal fisheries has undergone an interesting transformation. The Chilean government went through the conventional Western sequence of attempting to regulate fisheries by total allowable catch and found that this system failed. In the last 15 years, they have switched to a system of territorial fishing rights with local community fishing cooperatives given exclusive access to sections of the coastline.

7. THE CURRENT STATE AND FUTURE OF THE WORLD'S FISHERIES

Popular media have publicized a number of alarmist articles in the last few years with such striking conclusions as "all the large fish in the oceans were gone by 1980" and "all the world's fish stocks will be collapsed by 2048." More careful analysis of the data reveals a more complex and generally less pessimistic view of world fisheries. Worldwide, fish landings reached a peak in the late 1980s and have been slightly declining on average since that time. Fisheries continue to employ millions of people around the world and provide the economic basis for thousands of fishing communities. The picture is very different region by region, with some areas such as the North Atlantic showing considerable declines in landings, while other areas continue to increase landings.

Few places conduct a systematic analysis of the status of their fished stocks. The Food and Agriculture Organization of the United Nations estimates that between 20 and 30% of the economically most important fish stocks in the world have been overexploited, and this number has fluctuated in that range since the late 1980s—no increasing trend in overexploitation is seen. In the United States the percentage of fish stocks that are overexploited has been declining and, as of 2006, was 26%, with a resultant loss of potential yield of perhaps 15%. In other places the situation is not nearly so optimistic. All cod stocks are overexploited to some extent in the North Atlantic; some, such as the stock off Newfoundland, are almost completely gone, but others in Iceland and Norway are quite abundant and technically only slightly overfished.

Within industrial nations, there is a broad range of fisheries health. The United States stands out as perhaps the most intensely managed for conservation, with restrictive catch regulations now the norm and more stocks increasing rather than decreasing. Iceland, Australia, New Zealand, and Canada also stand out as countries that have adopted quite conservative harvesting regimes. The European Union presents a more complex picture, where scientific advice for lower catches is frequently ignored in the political process of consensus required within the EU fisheries policy, and as a result, many European fish stocks remain heavily overexploited.

We know much less about the status of fish stocks in most of Africa and Asia, where the fisheries most commonly consist of a mixture of small-scale village artisanal fisheries and industrial fisheries based in the major ports. There is no infrastructure of scientific study to determine stock status, and the countries rarely have the institutional ability to enforce the regulations they may have on the books. It is therefore difficult to be at all optimistic about the future of most of these fisheries, where intense exploitation and potential collapse would seem to be a likely future.

The high seas present a quite different picture. The major fisheries on the high seas are for tuna, and with the exception of the highly valued bluefin tuna, most of the high-seas tuna stocks of the world are not yet overexploited and remain healthy and productive. However, the international organizations that regulate these fisheries generally require consensus in order to adopt catch regulation. It seems highly likely that, should the economics of these fisheries prove profitable enough to drive the stocks into overfished states, the management agencies will be unable to prevent the decline.

FURTHER READING

Cushing, D. 1982. Climate and Fisheries. London: Academic Press.

Hall, S. J., and B. M. Mainprize. 2005. Managing by-catch and discards: How much progress are we making and how can we do better? Fish and Fisheries 6: 134–155.

Hilborn, R., T. A. Branch, B. Ernst, A. Magnusson, C. V. Minte-Vera, M. D. Scheuerell, and J. L. Valero. 2003. State of the world's fisheries. Annual Review of Environment and Resources 28: 359–399.

Ostrom, E. 1990. Governing the Commons: The Evolution of Institutions for Collective Action. Cambridge, UK: Cambridge University Press.

Smith, T. D. 1994. Scaling Fisheries: The Science of Measuring the Effects of Fishing, 1855–1955. Cambridge, UK: Cambridge University Press.

VII.3

Wildlife Management

Mark S. Boyce, Evelyn H. Merrill, and Anthony R. E. Sinclair

OUTLINE

Wildlife populations occur within both protected areas and human-dominated ecosystems. In both cases, populations are monitored to ensure they coexist with other species in their habitats in a stable way or are harvested as a resource in a sustainable fashion. Management may be limited to monitoring or it may involve active change in the ecosystem. Conditions that require active management include altering competition and predation effects, adjusting habitats, and counteracting effects of exotic species. Ecosystems exhibit complex behavior such as the rapid switch from one set of species to another when the environment changes gradually, a phenomenon called multiple states. Such rapid changes may require active management to ensure the persistence of valued species.

GLOSSARY

density dependence. The relationship of mortality or births to the size of a population, with proportional mortality increasing or births decreasing as numbers in the population increase.

ecosystem. The interaction of the biotic community, made up of all the species present, and the physical and chemical environment.

multiple states. Alternative composition of species that occurs when a threshold of environmental change is reached.

trophic cascade. The alternating changes in populations of each trophic level when a top level is perturbed.

trophic level. The position of a species in the chain of energy or nutrients. In a three-level chain, the top level is taken by predators, the next level below by herbivores, and the bottom level by plants.

wildlife. Typically refers to vertebrates such as mammals, birds, reptiles, and amphibians.

1. WHAT IS WILDLIFE MANAGEMENT?

Wildlife, which generally refers to the higher land vertebrates such as mammals, birds, reptiles, and amphibians, is valued by society for social, economic, and esthetic reasons.

People enjoy observing wildlife and spend considerable income on feeding wildlife in their backyards or traveling to their natural habitats to view wildlife. In 2001 in the United States, more than 50 million Americans spent over $3.7 million feeding wild birds, and over 66 million people made trips primarily to view wildlife, spending $38.4 billion. Wildlife also is harvested for meat and other products and for sport hunting. Reindeer (*Rangifer tarandus*) husbandry provides livelihood for Lapps in Scandinavia, safari hunting accounts for the bulk of revenue earned in communal areas in Zimbabwe, and in the European Union, hunting generates about 100,000 jobs. Iguana meat and eggs are the traditional Easter substitute for red meat in Central America, and felt made from beaver fur is used to make wide-brimmed Stetson hats so often seen at rodeos in western North America.

Humans and wildlife do not always live in harmony. In North America and Europe, humans commonly lose crops to geese and rodents, plantation trees to bears (*Ursus* spp.) and pigs (*Sus scrofa*), and livestock to lynx (*Lynx lynx*) and wolves (*Canis lupus*). The European badger (*Meles meles*) has been implicated in the maintenance and transmission of tuberculosis (*Mycobacterium bovis*) to cattle in England. In the extreme, humans are killed by bears (*Ursus* spp.) in North America, lions (*Panthera leo*) in Africa, and tigers (*P. tigris*) in India. Management of wildlife is thereby motivated by a desire to (1) enjoy wildlife nonconsumptively, (2) obtain a sustainable harvest, and (3) minimize conflicts with humans.

The implementation of wildlife management falls into two categories: either manipulation or protection of wildlife populations. Manipulative management means that management does something by direct means or indirectly by influencing food supply, habitats, predators, or disease. Manipulative management occurs when a population is harvested, when it declines to an unacceptably low density, or when density expands to an unacceptably high level, coming in conflict with human interests. In contrast, preventative or protective management aims to minimize external forces on the population and its habitat. This might involve allowing populations to fluctuate according to the natural ecological processes within the system. Such management is usually appropriate in national parks, as we discuss below.

No matter what the motivation or approach, wildlife management is conducted in the context of the ecosystem. Aldo Leopold, an early proponent of wildlife management, proposed that "To keep every cog and wheel is the first precaution of intelligent tinkering." Theodore Roosevelt, an avid hunter and 26th president of the United States, who created many national parks and wildlife and bird refuges and established the national forests, recognized that it was necessary to protect entire ecosystems to preserve species such as the brown pelican (*Pelecanus occidentalis*) in Florida. Indeed, ecosystem characteristics determine the distribution and abundance of wildlife, and these environmental conditions vary over space and time. Conversely, wildlife can have profound influences on ecological processes. Further, most animals are mobile, sometimes migrating vast distances across many ecosystems. How wildlife influences ecosystems and how the ecosystem shapes wildlife interactions can be complex. Scientific approaches to wildlife management in an ecosystem context have been developed during the past century that can assist those who are responsible for managing wildlife resources.

2. SPECIES INTERACTIONS AND WILDLIFE MANAGEMENT

Wildlife populations are embedded in ecosystems and are inherently linked to other species. Ecologically, we define species by (1) their trophic level, e.g., predators feed on herbivores, and in turn, herbivores feed on plants, (2) their influence on ecosystem functioning, e.g., keystone species are those that have disproportionate influence beyond what would be expected by their biomass, and (3) their ability to indicate a range of species, i.e., umbrella or indicator species. Wildlife managers use these groups as a framework for understanding the role of species in an ecosystem context. Where management involves either population reductions or restoration of predators to a system, "cascading" effects are likely to result in changes at lower trophic levels. In Banff National Park, Canada, wolves recovered from extirpation, but because they avoided humans, they were excluded from portions of the park. In areas with abundant wolves, elk (*Cervus elaphus*) abundance was reduced. Where there was less elk herbivory, there were increases in aspen (*Populus tremuloides*) recruitment and willow (*Salix* spp.) growth, beaver (*Castor canadensis*) colony density was higher, and consequently, there was greater riparian songbird diversity and abundance. In contrast, in ecosystems without top predators across Europe and North America, increasing deer populations have been associated with undesirable effects on plant communities. Heavy browsing by deer in forests has killed tree seedlings or reduced height growth, slowed stand rotations, and increased risk of fungal infections. Conifers, which become a major food for deer during the scarcity of winter, may be intolerant of browsing because they do not relocate nutrients to stems and roots as much as broadleaf species but invest heavily in leaves and conserve nutrients by retaining leaves. In the upper Midwest of the United States, hemlock (*Tsuga canadensis*) seedlings and saplings have become rare across much of their range, apparently in response to deer browsing. Mature hemlock trees produce abundant seeds and new seedlings, and saplings might establish if deer abundance were reduced, but this might require as long as 70 years for this slow-growing understory species.

The effects of deer herbivory also are evident in the understory vegetation of many forests, and these changes can influence forest dynamics. Forests of Pennsylvania often are invaded by the thorny shrub *Rubus allegheniensis*, which promotes the establishment of tree seedlings. Where deer reduce its abundance, a competitor, the hay-scented fern, *Dennstaedia punctilobula*, which deer avoid, becomes abundant. Where the hay-scented fern becomes dominant, the establishment of tree seedlings is restricted, and smaller-stature herbs are excluded. Cessation of browsing, once hay-scented fern has become established, rarely results in a recovery by *Rubus* spp. or other species. Thus, browsing by deer alters the forest understory to a state resistant to invasion by the original dominant species. Large and often unacceptable reductions of deer can be necessary to release plants from the effects of heavy browsing.

Although managers often attribute changes in plant communities to herbivory, the effects of herbivores are sometimes contingent on other environmental forces. In portions of the Serengeti, elephants (*Loxodonta africana*) were considered to be the culprits behind declines of mature woodland savanna during the 1960s because they were seen pushing over trees. However, it became evident that fires limited recruitment of the Serengeti woodlands. The extent of burning during the dry season had to be less than 30% if recruitment were to replace adult tree mortality, but burning was typically 80% of the woodlands during the mid 1960s. Most tree mortality was from old age, not elephants. Even though burning was reduced in the 1980s and was maintained at relatively low levels because of heavy grazing by wildebeest (*Connochaetes taurinus*), elephants in portions of the Serengeti were able to limit tree recruitment. Thus, ecosystems can occur naturally in more than one combination of interactions among species.

Anticipating the repercussions of changing species abundances is often not easy. For example, at the mouth of the Stikine River in SE Alaska, a colony of harbor seals (*Phoca vitulina*) plagued salmon fishermen, feeding on salmon caught in gill nets. During the 1940s, the U.S. Fish and Wildlife Service responded to demands from fishermen and killed the seals using depth charges. The seals also fed on flounders (*Platichthys stellatus*) at the mouth of the Stikine River, and rather than enhancing the salmon fishery, what happened was an increase in flounders that fed on salmon smolt. The ultimate outcome was a decline in the salmon fishery for many years until the harbor seal population recovered.

Similarly, management actions in multipredator and multiprey systems can pose difficult challenges. In northern Alberta, woodland caribou (*Rangifer tarandus tarandus*) have faced recent declines attributed to the expansion of a road network associated with timber harvest and oil/gas development. These land disturbances have improved forage in regenerating clearcuts and along roadsides for moose (*Alces alces*) and white-tailed deer (*Odocoileus virginanus*). Increases in these alternate prey species have supported an increase in the wolf population, facilitated by enhanced movement of wolves along the roads and other linear clearings made by seismic exploration. As a result, wolf predation has been identified as the major source of mortality for threatened woodland caribou populations.

Management sometimes involves manipulating the competitive interactions between species. Thus, in Kruger Park of South Africa in the 1960s the abundant wildebeest were reduced in number to facilitate the survival of the rare and threatened roan antelope (*Hippotragus equinus*). Although this treatment had the desired effect, it also had unintended consequences because reduced grazing allowed taller grass, which both facilitated predation by lions and changed the habitat. Thus, wildebeest were reduced to much lower levels than intended.

Nowhere has altering competitive interactions been more controversial as when wildlife managers are contending with the legacy of exotic introductions. In 1881 ring-necked pheasants (*Phasianus colchicus*) were introduced to Oregon, and within 30 years the species had spread across extensive areas of the North American continent. The species was welcomed as a game bird in many areas, but the consequences on native birds have been poorly documented. In particular, pheasants are aggressive toward lesser prairie chickens (*Tympanuchus cupido*), although habitat changes may have favored pheasants over prairie chickens as well. Competition between the two species is poorly documented even though prairie chicken populations declined rapidly coincident with the expansion of pheasants.

3. HABITAT MANAGEMENT

One of the most effective tools for wildlife management involves manipulations of habitats. Burning grasslands to improve forage quality, implementing marsh water-level drawdowns to establish emergent vegetation, harvesting aspen (*Populus tremuloides*) to improve brood-rearing cover for ruffed grouse (*Bonasus umbellus*), and constructing water guzzlers in the desert are examples of common habitat-management tools used to benefit wildlife. Clearly, habitat manipulations, as well as protection, have ramifications for the entire ecosystem.

The longleaf pine (*Pinus palustris*) ecosystem originally covered approximately 35 million hectares in several states of southeastern United States but remains naturally only as scattered remnants today. Under natural fire regimes, fires burned in this ecosystem every 2 to 4 years, reducing debris from the forest floor and returning nutrients to the soil. Burning maintains open areas, allowing birds such as the Eastern wild turkey (*Meleagris gallopavo*) and northern bobwhite quail (*Colinus virginianus*) to feed on grass and plant seeds exposed by the burning. Frequent fires also retard woody-stemmed growth and promote herbaceous growth, which provides components for insects to thrive. These insect-infested areas are used by young birds, especially during late spring and early summer. Although manipulations of habitats can provide immediate benefits, they may have unintended long-term consequences. Again in Kruger Park habitats were manipulated for several decades by a variety of measures. Water holes were installed using windmills to provide year-round water, and burning of the grassland was set out in a strict rotation. The outcome of these manipulations was a radical change in movement patterns of the ungulates because they no longer needed the dry-season migrations to find water in the rivers. This in turn changed the grazing patterns, distribution of predators, and predation rates. Rare ungulates, which could originally survive when predators departed with the migrants, now suffered much higher predation and in some cases extirpation.

Where regulations exist requiring the maintenance of viable populations for native species, such as on lands subject to the National Forest Management Act in the United States, broad-scale landscape planning uses a variety of habitat-management approaches and monitors a range of species to see if habitat manipulations produce the desired effects. In the Chequamegon National Forest in northern Wisconsin, a 5-year Forest Plan recognized the need for integrating uncut forest reserves along with rotational cutting to maintain a diversity of stand ages. In the planning process, a suite of management indicator wildlife species were identified to help monitor whether overall biodiversity was being retained.

In other situations, managers use threatened or flagship species to gain the social and political clout to manage for important habitats or linkages for a range of species. Thus, the kiwi (*Apteryx* sp.) is an important icon for protecting New Zealand forests, as is the tiger in Asia, the Indian one-horned rhinoceros (*Rhinoceros unicornis*) in Chitwan National Park of Nepal, and the elephant in many savanna parks of Africa, and primates fill that role in several forest reserves. Protection of old-growth forest habitats for the northern spotted owl (*Strix occidentalis caurina*) in the Pacific Northwest of the United States ensured persistence for a diversity of old growth–dependent species such as the marbled murrelet (*Brachyramphus marmoratum*) and sharp-tailed snake (*Contia tenuis*). In fact, the claims of "jobs versus the spotted owl" soon gave way to "jobs for loggers versus jobs for commercial fishermen" as the value of old-growth habitats in protecting salmon-spawning streams was realized.

Grizzly bears (*Ursus arctos*) require large areas to sustain viable populations. A population in the Selkirk Mountains of northern Idaho and Washington and southern British Columbia, Canada has become genetically isolated by barriers to movements created by agriculture-dominated valleys and roads. Managers have been able to locate movement corridors to maintain the connectivity of landscapes for grizzly bears using habitat models in geographic information systems (GIS) based on data from GPS radio-collared animals. Management for grizzly bears, a high-profile threatened species, draws coordinated support from conservation groups across political boundaries in international efforts such as the Yellowstone-to-Yukon Initiative.

Hunting interests also have played a key role in habitat conservation. Most of the national parks and wildlife areas of India were former hunting reserves established by maharajas during British rule. In Canada, since 1985, $32 million in revenues associated with the Canadian Wildlife Habitat Conservation Stamp has funded hundreds of habitat conservation projects using partnerships with landowners, communities, industry, government, and nongovernmental organizations. These lands also provide habitats for shorebirds, grassland birds, and other wildlife. Game ranching for big game hunting in South Africa has focused on conserving native species and their habitats, requiring formal management plans approved by the government that identify suitable habitats for targeted species. Habitat management including prescribed burning is encouraged to restore native communities. Although top predators can be excluded because they compete with hunting demands for game animals, most of the native fauna and flora are maintained on 13% of South Africa's landbase, and some species such as the black wildebeest (*Connochaetes gnou*) and bontebok (*Damaliscus dorcas*) persist only on these properties.

As habitat changes over time, these interactions among wildlife species are likely to change, and wildlife managers must anticipate the consequences. For example, studies in the Serengeti have shown that the hunting success of lions is determined by the availability of dense cover for ambush rather than by the density of available prey. These studies predict that the increase in dense cover as a result of increases in young trees has benefited lions, and the greater hunting by lions appears to be responsible for declines of many medium- and small-sized antelope in woodland areas. Habitat changes occur more rapidly where anthropogenic changes occur. For example, in Wood Buffalo National Park in northern Alberta, Canada, wet meadows are key wintering habitats for the bison that originally motivated the creation of one the world's largest national parks. Despite cessation of slaughter for meat and the introduction of predator control, the bison population fell from an all-time high of 12,500 after reintroduction to less than 4000 by the late 1980s. Although still the subject of debate, managers have suspected that changes in the local hydrologic regime associated with the W.A.C. Bennett dam may have altered the pattern of meadows and forests, concentrating bison into areas more susceptible to wolf predation. In fact, where habitat manipulations are being planned, managers must anticipate the dynamics of such interactions. For example, although burning scattered stands of aspen in high-elevation areas in Banff National Park was thought to benefit the declining migratory segment of the Ya Ha Tinda elk population, it may in fact expose them to higher predation by concentrating them predictably in areas.

4. ECOLOGICAL-PROCESS MANAGEMENT

In national parks and other protected areas, the objective is to protect natural ecological processes and to minimize human influence. Such an approach to wildlife management provides baselines by which we can evaluate the consequences of human alterations to ecosystem outside of these protected zones. As long as all natural components of the ecosystem are intact, density-dependent food limitation or predator–prey interactions will limit herbivore populations, and human intervention should not be necessary. In some instances, human alterations to the ecosystem require that intervention occurs to replace missing ecological processes. For example, fencing of a park or natural area can limit the ability of animals, such as elephants, to disperse from high-density areas as has occurred in Kruger National Park. Similarly, in Botswana, fencing on cattle ranches has stopped the migration of wildebeest there. In Hwange Park, Zimbabwe, dry-season water sources for a large area of woodland were excluded in the 1930s by fences around farms that expropriated the river valleys. To maintain the area for wildlife, boreholes were installed in the woodland that later became Hwange Park. The consequent increase in elephant browsing around these artificial water holes during the subsequent 70 years has required manipulation of the elephant populations. If access to the original water sources had been available, such elephant controls might have been much reduced. Another example involves winter feeding of elk, which is conducted at the National Elk Refuge to replace winter range now occupied by the town of Jackson, Wyoming. During spring, elk migrate into Yellowstone National Park and Grand Teton National Park where they are protected. Because of the winter feeding program at the National Elk Refuge, enabling legislation for Grand Teton National Park allows for an annual cull of elk inside the park to limit herd size.

When an ecological process is disrupted, the best solution might be to restore the missing component or ecological process. In a celebrated example, wolves were restored to Yellowstone National Park in 1995 where they had been eliminated about 70 years earlier. Wolves have influenced herbivore distribution and abundance as well as the scavenger community and other predators. Although not fully understood, wolf recovery has precipitated a "trophic cascade" in some localities where woody vegetation has been released from herbivory because elk avoid areas where there is a high risk of wolf predation. This in turn has restored habitats for beavers, songbirds, and other animals.

5. MULTIPLE STATES

Ecosystems can occur naturally in more than one combination of species populations; that is, there are multiple ecosystem states. Multiple states usually exhibit the sudden switch from one community to another because of thresholds to disturbance. Management needs to take into account such characteristics of a system because one can sometimes change a system through interference. Thus, in central Australia there are large areas of dense scrub, called "mulga" composed of *Acacia* spp. This scrub has developed over the past century with progressively intense human interference. Originally, the habitat was composed of grassland with a low density of shrubs kept down by heavy browsing from the indigenous small marsupial, the burrowing bettong (*Bettongia lesueur*), supplemented by burning of the grass. As farming spread, the exotic red fox (*Vulpes vulpes*) eliminated the bettongs, cattle grazing reduced fire frequency, and the shrubs grew to their present density. Now grazing is no longer possible, and fire cannot penetrate the dense stands. In essence the system has become locked into another state. Similar switches in state have been observed when fire has changed Serengeti *Acacia* savannas to grassland and then elephants have taken over and maintained the grassland by preventing regeneration of the trees. In this case a woodland state and a grassland state both occur with elephants. A disturbance such as fire is required to change the state from woodland to grassland, but once grassland occurs, elephants can maintain it in that state. When an ecosystem crosses into another state, it often takes considerable resources to revert, if reversion is possible at all.

Ultimately, it is the ecosystem that has to be conserved for the protection of individual species. Management of ecosystems has to accommodate the effects of major natural disturbances such as fires, floods, and storms and the consequences of natural changes in state. Equally, management would benefit from knowing where thresholds to human disturbance occur to avoid such thresholds and allow recovery from unwanted disturbance.

6. MANAGING WILDLIFE WITHIN ECOSYSTEM COMPLEXITY

Understanding the ecosystem consequences of wildlife management requires a continuing iteration among ecological modeling, manipulation, and monitoring as we gradually discover how management actions influence ecological systems. The lesson for wildlife professionals managing in an ecosystem context is that the complexity of ecological interactions can yield unexpected consequences. How should wildlife management proceed given this uncertainty? Avoiding introductions of exotic species seems simple, but restoration of native species requires an understanding not only of the original limiting factors but of how the species is likely to fare in current circumstances. Because altering the availability and pattern of habitats influences not only wildlife distribution but also their interactions with other species, managers must evaluate land management and design habitat improvements with this in mind. Small-scale perturbations that promote high, predictable use, such as food plots and water sources, might not achieve the desired benefits if prey become predictably available to predators.

In many instances management actions may need to target several species simultaneously to accommodate the interactions among species. Conducting pilot studies on a small scale can help mangers anticipate consequences but provides no assurances as to the final results when full-scale management actions are taken. Clearly, careful monitoring of a spectrum of species needs to be in place to document consequences to management actions. An adaptive management framework where alternative responses to the anticipated results can be evaluated might be key to making progress.

FURTHER READING

Boyce, Mark S., and Alan Haney, eds. 1997. Ecosystem Management: Applications for Sustainable Forest and Wildlife Resources. New Haven, CT: Yale University Press. *This book provides the rationale for sustainable management of wildlife, habitats, and resources.*

Côté, Steeve D., Thomas P. Rooney, Jean-Pierre Tremblay, Christian Dussault, and Donald M. Waller. 2004. Ecological impacts of deer overabundance. Annual Review of Ecology, Evolution, and Systematics 35: 113–147. *This article reviews complexity of deer herbivory and management in an ecosystem context.*

Lessard, Robert, S. Martell, Carl Walters, T. Essington, and J. Ketchell. 2005. Should ecosystem management involve active control of species abundances? Ecology and Society 10(2): 1.

Sinclair, Anthony R. E., and Andrea Byrom. 2006. Understanding ecosystems for the conservation of biota. Journal of Animal Ecology 75: 64–79. *This article reviews the information on the management of wildlife in ecosystems.*

Sinclair, Anthony R. E., Simon A. R. Mduma, J. Grant C. Hopcraft, John M. Fryxell, Ray Hilborn, and Simon Thirgood. 2007. Long-term ecosystem dynamics in the Serengeti: Lessons for conservation. Conservation Biology 21: 580–590. *This article provides an example of a long-term data set used for the management of wildlife within protected areas.*

Smith, Douglas W., Rolf O. Peterson, and Douglas B. Houston. 2003. Yellowstone after wolves. BioScience 53: 330–340. *This article discusses an example of a reintroduction of a native species into its original range and the changes that occurred in the ecosystem.*

VII.4

Managing the Global Water System

Joseph Alcamo

OUTLINE

1. Uncovering the worldwide connectivity of water
2. Intervening in the global water system

An important new insight is that water in its various forms operates as a system on scales much larger than a single lake, river basin, aquifer, or municipality. Although the global cycling of water through the earth's physical system (ocean, atmosphere, terrestrial freshwater bodies) has long been recognized, researchers are only now uncovering a much wider net of connectivities that binds together the flow of water on a global scale. The connectivities are physical (e.g., upstream storages of water cause large-scale changes in the residence time of surface water), economic (e.g., water is embedded in food and other products and traded internationally), and even institutional (e.g., decisions about trade of water technology have a global impact). This new awareness of connectivities has spawned the concept of the "global water system." Recent research has also made it clear that the global water system is undergoing large-scale, unparalleled, and poorly understood changes that pose major risks to ecosystems and society. The policy community needs to respond immediately to these risks, and this response should take place at all levels, from local to global. At the global level, there are three main tasks to take on. First, we need to expand our knowledge base about the global water system by extending the scope of earth observations, by conducting new large-scale field experiments, and by developing new tools for the simulation of the global water system. Second, we should expand global governance of the water system through various means (as a complement to governance at the local and other levels). Options include invoking an international convention on environmental flows, instituting water labeling of products at the international level, and enforcing water efficiency standards of internationally traded products. Finally, we should challenge current assumptions about water use in the world by stimulating a public debate on the definition of "essential water needs" and by broadening the viewpoint of water professionals to include the global perspective.

GLOSSARY

teleconnection. A cause-and-effect chain that operates through several intermediate steps and leads to a linkage between two parts of a system that (to researchers at least) is unexpected or surprising.

virtual water. The volume of water that circulates in an economic system as an embedded ingredient of food and other traded products. This concept originated from the idea that arid countries compensate for water deficits by importing water-intensive commodities rather than domestically producing these commodities.

1. UNCOVERING THE WORLDWIDE CONNECTIVITY OF WATER

Although the Earth is known as the "water planet," most water researchers and managers focus on scales much smaller than the planetary. Indeed, most freshwater studies concentrate on lakes, streams, or perhaps watershed-scale hydrologic processes, and nearly all water managers concern themselves with planning the water supply in their community or perhaps river basin. Water science and management were basically local activities until "watershed thinking" revolutionized these endeavors in the 1960s and 1970s. Afterward, it became more common for water researchers and engineers to incorporate watershed-wide relationships among climate, runoff, and water use into their work.

Now we are again called on to broaden the perspectives of water science and management. The motivation comes from recent research showing that water is interconnected on a planetary level more tightly and in more ways than previously appreciated.

Table 1. Agents of change in the global water system and their impacts

Agents of change	Environmental changes	Global issues						
		A	*B*	*C*	*D*	*E*	*F*	*G*
1. Climate change	Change in flow regime (runoff volume and timing)		•	•		•		•
	Indirect effects caused by vegetation change		•	•		•		•
	Development of nonperennial rivers		•	•	•	•	•	•
	Segmentation of river networks					•	•	
	Alteration of extreme flow events		•			•	•	•
	Changes in wetland distribution	•	•	•	•		•	•
	Changes in chemical weathering				•			•
	Changes in erosion and sedimentation				•	•		
	Saltwater intrusion in coastal groundwater	•						
	Accelerated salinization through evaporation	•	•				•	
2. Water management (including dams, diversions, and channelization)	Nutrient and carbon retention				•			•
	Retention of particulates				•	•		•
	Change in flow regime (runoff volume and timing)		•	•		•		•
	Streamflow variability and extremes		•					
	Loss of longitudinal and lateral connectivity						•	
	Creation of new wetlands	•		•	•		•	
3. Land use change	Wetland filling or draining		•	•	•		•	
	Change in sediment transport				•	•		•
	Change in vegetation cover		•					
	Alteration of first-order streams					•	•	
	Nitrate and phosphate increase	•		•	•			•
	Pesticide increase	•		•				•
4. Irrigation and water transfer	Change in flow regime (runoff volume and timing)		•	•		•		•
	Salinization through evaporation		•	•				
5. Release of industrial and mining wastes	Heavy-metal increase	•		•				
	Acidification of surface waters			•			•	
	Salinization	•		•			•	
6. Release of urban and domestic wastes	Eutrophication	•		•	•		•	•
	Development of water-borne diseases	•						
	Organic pollution	•		•			•	
	Persistent organic pollutants	•		•				•

Source: Global Water System Project: Science framework and implementation activities. 2005. http://www.gwsp.org.
A: human health, B: water cycle, C: water quality, D: carbon balance, E: fluvial morphology, F: aquatic biodiversity, G: coastal zone impacts. Only the major links between issues and impacts are listed here.

Although the existence of a global hydrologic cycle has been recognized for decades, science is now uncovering a vastly wider web of biological, biogeochemical, and even socioeconomic connectivities that bind water globally. Furthermore, we are only beginning to understand the nature of these interconnections and their implications for society and the rest of nature.

This new awareness of connectivities has spawned the concept of "global water system" (Framing Committee of the Global Water System Project, 2005). Water is considered to be a global *system* in the conventional sense of being an entity made up of components linked together and working as a unit. What are its functions? Although researchers are a long way from uncovering these and other attributes of the global water system, a first hypothesis might be that it redistributes moisture in the most thermodynamically efficient way, transports energy and materials around the world through climatologic and geologic processes, makes moisture available where it is needed by organisms, and, overall, contributes to the sustenance of life on earth. Humans are part of the system and also have their own particular goals in appropriating water

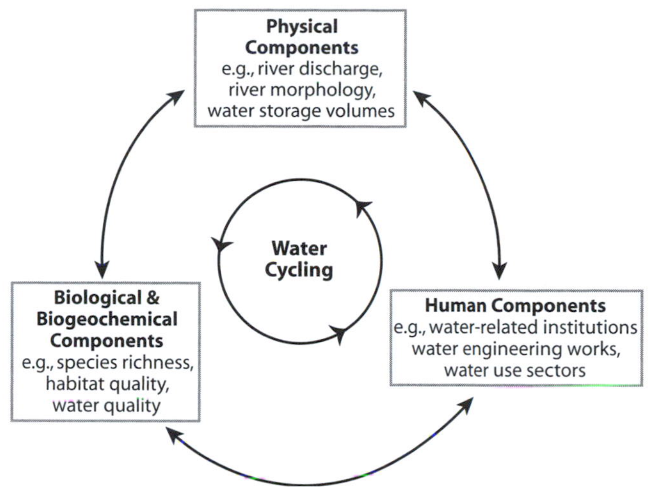

Figure 1. Components of the global water system. (From Global Water System Project of the Earth System Science Partnership. Framing Committee of the Global Water System Project. 2005. The Global Water System Project: Science Framework and Implementation Activities, Earth System Science Partnership, Global Water System Project Office, Bonn, Germany. Downloadable from http://www.gwsp.org)

from the system. It is also clear that in pursuing their goals, humans have caused a drastic transformation of the system, as described in table 1.

In this chapter, we elaborate the concept of the global water system, especially its freshwater part, and describe the components of the system as well as some key connectivities that bind it together. We then discuss the widespread transformations taking place in the system and the many uncertainties that remain about these changes. Finally, we discuss threats to water security as a type of failure of fulfilling the functions of the system and discuss the kinds of interventions that may help us to cope with these threats.

Components and Connectivities

The global water system can be understood as a structure made up of three types of components—physical, biological/biogeochemical, and human—that are linked internally and with each other through a network of connectivities or teleconnections with spatial scales of hundreds to thousands of kilometers (Framing Committee of the Global Water System Project, 2005) (figure 1). What are the major features of these components and their connectivities?

Physical Components

Decades of research in climatology and hydrology have firmly established the physical connectivities of water in a worldwide system of stocks and flows. The cycling of water in a physical sense is the most obvious part of the global water system. By far the largest stocks are the world's oceans, storing about 1.35 million km^3 of water and providing around 86% of the total continuous source of water for the atmosphere (Framing Committee of the Global Water System Project, 2005). Ice caps take a distant second place as a repository of moisture. Each year evaporation from the world's oceans combined with evaporation/transpiration from the land provide a flow of nearly 500,000 km^3 of water to the atmosphere. This volume is returned to the earth's surface as precipitation. The cycle is closed in that about 40,000 km^3 of this precipitation finds its way back to the ocean each year through rivers and subsurface watercourses.

Other physical connectivities arise from the interplay of land, atmosphere, and hydrology, which affects energy and moisture fluxes and can influence precipitation patterns over large areas. As an example, scientists hypothesize that moisture feedbacks among vegetation, soil, and the atmosphere play a key role in the persistence of both dry and wet conditions over the Sahel in Africa (Nicholson, 2000). Another well-known example is the link between deforestation of the Loess Plateau of China and changes in sediment and flow characteristics of the Yellow River for hundreds of kilometers downstream of the plateau. Uncovered soils wash off to the river, and this substantially increases the river's sediment load. Sediment settles out in the delta of the Yellow River, raises the riverbed, and thus contributes to more frequent downstream flooding.

Scientists have documented many other examples worldwide of engineering works and land use changes that have caused major changes in sedimentation and flow characteristics of rivers for very long distances downstream.

Biological and Biogeochemical Components

Water is essential for maintaining the integrity and biodiversity of both terrestrial and aquatic ecosystems. Hence, the living parts of the world's freshwater ecosystems, both aquatic and riparian organisms, are part of the global water system. Biogeochemical processes are also covered here as well as the sum of processes determining water quality in freshwater systems. Various biogeochemical connectivities occur in the global water system because water is an important medium for transporting carbon, nitrogen, phosphorus, and other elements through the earth system and serves as an important repository for these elements (Cole et al., 2007). Hence, the global biogeochemical cycles are intertwined with the global water system. Through its linkages with the carbon and other biogeochemical cycles, water helps regulate the release and sequestration of CO_2 and other radiatively important trace gases. On a global basis, the hydrologic cycle is one of the principal vehicles controlling the mobilization and transport of chemicals and other constituents from the continents to the oceans.

Human Components

The many manifestations of society's manipulation of water resources make up an essential part of the global water system. These include water engineering structures (reservoirs, canals), water-related organizations (financers of water infrastructure, water planning agencies, water companies), and water use sectors (municipal water utilities, thermal power plants). Society is not only a component of the global water system but also a major agent of change within the system (table 1).

We are only now beginning to realize the extensive connectivities that bind the socioeconomic part of the global water system together. Some linkages are formed by economic relationships, as in the case of the international flow of "virtual water" embodied in cross-boundary food trade. The basic idea of virtual water is that arid countries compensate for their water deficits by importing "virtual water" in the form of food products rather than using their own scarce water resources for growing food themselves. Because large volumes of water are needed to grow crops (e.g., cultivating 1 kg of grain requires approximately 1000 to 1500 liters, depending on location and type), it follows that the enormous international trade in foodstuffs involves a similarly huge trade in virtual water. The annual global volume of virtual water imported in food is around 1250 km^3 (Oki and Kanae, 2004), a substantial volume as compared with the global total of 2400 km^3 of water withdrawn for irrigation (Alcamo et al., 2007). On one hand, the virtual water concept can be thought of as a new variation on the old principle of competitive advantage. On the other hand, it provides new insight into how humanity mobilizes and controls a significant part of the hydrologic cycle.

Another form of socioeconomic connectivity arises between centralized organizations and worldwide development of water infrastructure. Only now are researchers beginning to uncover the sweeping influence of centralized development agencies, banks, private water companies, and other organizations on the worldwide water system. Decisions taken in a few world capitals about the structure of water pricing or the sale of water engineering are having wide-scale impacts on water use and supply (Pahl-Wostl, 2002). As the world economic system becomes ever more integrated, it should be expected that more water-related connectivities will emerge.

The System Transformed

Far from being static, the global water system is undergoing changes that are widespread, worldwide, and concurrent. In the following paragraphs, we review some of these important changes.

The *physical characteristics* of freshwater systems are undergoing a major transformation, which includes persistent changes in precipitation and hydrologic patterns, changes in runoff and the retention time of fresh water on the continents, modification of the sedimentation characteristics of rivers, and alteration of the moisture fluxes between the atmosphere and terrestrial environment (Vörösmarty et al., 2003; Meybeck, 2003). One sign of widespread changes is that the flow and storage characteristics of 172 of 292 of the largest river systems in the world have been significantly altered by impoundments (Nilsson et al., 2005).

Climate change will have an increasingly noticeable impact on freshwater systems throughout the world over the coming decades. Some semiarid and arid regions (e.g., northeastern Brazil, the western United States, southern Africa) are expected to have significantly declining average river discharge and groundwater recharge (Kundzewicz et al., 2007). As recently reported by the Intergovernmental Panel on Climate Change, more than one-sixth of humanity lives in river basins fed by snow and glacier melt, and it is expected that winter flows will temporarily increase here (relative to annual river discharge) because of warmer temperatures (Kundzewicz et al., 2007).

Although the local mechanisms of physical changes are fairly well understood, many questions remain about the global manifestation of these changes as well as their intensity. For example, what will be the combined impact of climate change as compared with continued flow diversions and impoundments on freshwater inflow to the world's estuaries?

The *biological and chemical characteristics* of freshwater systems are undergoing widespread modifications, including major alterations in dissolved oxygen levels and other important water quality parameters (Meybeck, 2003) as well as long-term changes in the flux of sediments and nutrients delivered by freshwater systems to oceans. Chemical and physical modifications of freshwater systems have constricted the habitat of aquatic organisms and severely impacted aquatic ecosystems (Naiman et al., 1995; Polunin, 2005). An example of this is the Rhine River, which has experienced more than a century of channelization and riparian development, leaving it isolated from 90% of its original floodplain. Some rivers, such as the Colorado and the Yellow rivers, often do not reach the ocean. More than 20% of freshwater fish species have become threatened, endangered, or extinct within the past few decades (Dudgeon et al., 2005).

The continent of Africa could endure particularly wide-ranging transformations of its water chemistry and biology. In the absence of specific action, recent scenarios point to a four- to eightfold increase in wastewater loadings over most of Africa within the next four decades, suggesting a likely worsening of freshwater quality (Alcamo et al., 2005). What will be the implications of these increased loadings on water chemistry and biology? What will be the spinoff effects on aquatic ecosystems and the freshwater fishery, which is an important protein source for inland African countries?

Widespread changes are also occurring in the anthropogenic use of water, with declining trends in water withdrawals in some industrialized countries and rapid increases over most river basins in the developing world (Alcamo et al., 2003, 2007). The structure of the water economy is also rapidly changing in the developing world as water use in the domestic and manufacturing sectors claims a larger and larger fraction of total water withdrawals (Alcamo et al., 2007). Furthermore, only now have scientists begun to study the underlying causes of global changes in water use, and an open question is which factor—demographic change, economic growth, technological change, consumption patterns, or other—will be most important and where (Alcamo et al., 2003, 2007). Because water abstraction is rapidly expanding, we should expect sharper competition among households, irrigated farmers, electrical utilities, and other water users. What impact will this competition have on achieving the Millennium Development goal of halving the world's population without access to sustainable water supply by 2015?

Although changes are taking place throughout the global water system, not all parts of the world will be affected to the same degree. Key questions are, where will the most important changes occur, and how much

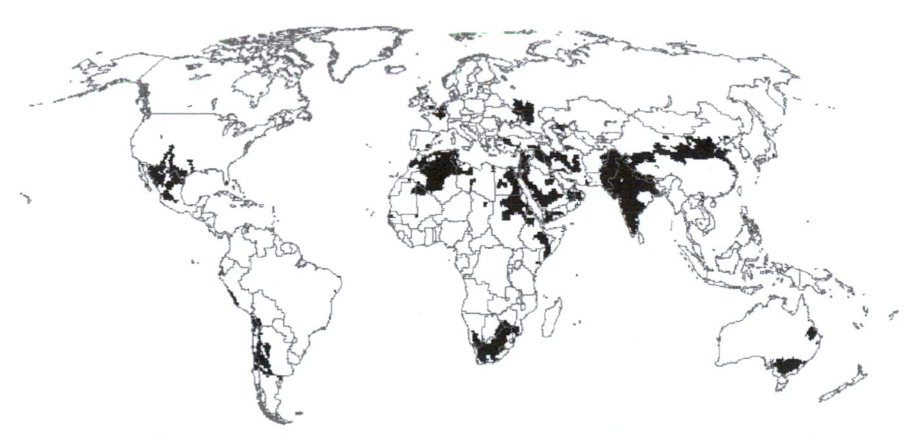

Figure 2. An example scenario of "hot spots" of increasing water stress. This map depicts (in black) the river basins where the ratio of water withdrawals increases between 1995 and 2030 because of a combination of increasing water withdrawals and decreasing water availability related to climate change. Watersheds depicted in black were already in the "severe water stress" category in 1995. This simulation is driven by the climate and socioeconomic assumptions of the "Market First" scenario of the Third Global Environmental Outlook of UNEP. (From Alcamo, J., and T. Henrichs. 2002. Critical regions: A model-based estimation of world water resources sensitive to global changes. Aquatic Sciences 64: 352–362)

of the world will be affected? According to one estimate, future "hot spot" areas with increasing abstraction and/or decreasing availability related to climate change will cover 7–13% of the world's river basin areas (Alcamo and Henrichs, 2002; figure 2).

2. INTERVENING IN THE GLOBAL WATER SYSTEM

Why Intervene at the Global Level?

We have seen that the global water system is undergoing important changes in its physical composition, its chemical and biological characteristics, and in its human dimensions. What implications do these have for water security? From the viewpoint of the global water system, water security can be seen as a long-term but temporary balance between water users and water availability or services. On one hand, water availability and quality are determined by the spatial and temporal patterns of precipitation and other meteorological phenomena, the structure of watercourses, the characteristics of the Earth's surface, and wastewater loadings. On the other hand, human water users (households, municipalities, industries, water sports enthusiasts) or nonhuman water users (aquatic and riparian organisms and their ecosystems) adjust over time to the spatial and temporal pattern of water availability or services. The system is shaken out of this temporary balance by rapid or abrupt changes in the global water system, some of which are described above. It is also made unstable by continuing pressure on the water system from water users aspiring to a higher level of water use to support a higher standard of living.

From the systems viewpoint, a breakdown in water security represents a *systems failure* in that the global water system cannot fulfill the goals of water users. The failure is most often manifested as too little water or, more precisely, a gap in many parts of the world between the water available to water users and their current aspirations or requirements. Depending on local and regional circumstances, this gap can disrupt aquatic ecosystems, cause temporary or persistent water shortages, displace current water users with those having a competitive advantage, and cause a decline in living standards or hinder their improvement.

Too much water, as in severe flooding, is also a systems failure in that a temporary but extreme state of the system poses a threat to one of its own components, namely people. (Although droughts and floods may pose a security threat to humans, they may be beneficial in some ways to aquatic ecosystems. For example, studies of river hydroecology are beginning to show that very low and very high river discharges sometimes play a key role in the life cycle of aquatic organisms.)

How can we as human agents in the global water system cope with global threats to water security? How should we intervene to address failures of the system? In reality, society acts every day through conventional water management to ensure or enhance water security at the local and river basin level. The wide palette of response options is shown in table 2.

But when is it appropriate to intervene at the global level as compared to the local or watershed level? We suppose that global action would be worthwhile in a few different cases: (1) when the driving forces of change are global in scale, as in the case of climate change impacts on water resources; (2) when changes are driven by worldwide institutions such as multinational water companies or international funding agencies; (3) when connectivities are global or large-scale in nature, as in the case of the strong feedbacks among land, atmosphere, and hydrology in the Sahel region or the large volume of virtual water that links nations together through international food trade; (4) when a threat arises to a globally important part of the system, such as an impending extinction of an aquatic or riparian species or the deterioration of vital ecosystem services; and (5) when an important change occurs concurrently throughout much of the world, as in the case of rapidly increasing water withdrawals and wastewater discharges in developing countries. Given these justifications for intervening globally, what form should these interventions take? The following paragraphs describe three clusters of action.

Intervention 1: Extending Our Knowledge Base of the System

A prerequisite for selecting the right way to intervene in the global water system is to have enough knowledge at hand to act wisely. But the reality is that the knowledge base is quite weak, and special effort should be expended in improving this base. Below, we consider three approaches: global monitoring, large field experiments, and new modeling and assessment tools. These activities should all work toward enhancing our understanding of the intensity, location, and causes of change in the water system. A particularly important task is to identify "hot spot" areas of the world of rapid change or particular sensitivity to change (see figure 2 for an example of hot spot analysis).

Expand the Scope of Remote Earth Observations of the Global Water System

The past decades have seen enormous progress in the use of satellites and aircraft in collecting data about the global environment. One effort particularly relevant to

Table 2. Response options for fresh water and related services from inland water ecosystems

Legal and *regulatory* interventions include:

- Ownership and use rights at different administrative levels
- Regulation of pollution
- Regulation of environmental flows and artificial flood releases
- Legal agreements for river basin management
- Regulations related to ecosystem and species conservation and preservation

Economic interventions include:

- Markets and trading systems for flow restoration and water quality improvements
- Payments for ecosystem rehabilitation
- Point source pollution standards and fines/fees, taxes, incentives
- Demand management through water pricing
- Payments for watershed services

Governance interventions include:

- Participatory mechanisms (e.g., watershed/catchment councils and farmer-based irrigation management systems)
- River basin organizations (international or regional scale)
- Integrated water resource management and basin planning
- Private sector participation
- Institutional capacity building (e.g., for regulatory agencies)

Technological interventions include:

- Water infrastructure projects (such as dams, dikes, water treatment and sanitation plants, desalinization)
- Soil and water conservation technologies (such as physical and vegetative measures for soil and water conservation)
- End-use and transmission efficiency options (such as drip irrigation and canal lining/ piping)
- Demand management/technologies for higher end-use efficiency (such as low-flow showerheads, energy conservation programs/ incentives)
- Research into water-saving technologies and breeding crops for drought tolerance

Social, cultural, and *educational* interventions include:

- Environmental education and awareness
- Making explicit the value of nonprovisioning water ecosystem services
- Research into land–water interactions in a watershed context

Source: Aylward, B., J. Bandyopadhyay, and J.-C. Belausteguigotia. 2005. Ecosystems and Human Well-Being: Volume 3. Policy Responses. Washington, DC: Millennium Ecosystem Assessment, Island Press, 213–255.

the global water system is the Soil Moisture and Ocean Salinity (SMOS) Mission of the European Space Agency. Beginning in 2008, SMOS will collect planetwide data on soil moisture, a key parameter of the earth's water cycle (Berger et al., 2002). Although they are very useful, the SMOS and other space-based missions tend to concentrate on the physical side of the global water system despite the urgent need for planetwide data on ecological, biogeochemical, and anthropogenic variables (e.g., spatial variation of water quality, state of aquatic ecosystems, and locations of human appropriation of water resources). Collecting these types of data will certainly pose technical challenges, but the scientific community has already shown that satellite sensors can meet the challenge. For example, satellite sensors have been used to measure changes in the spectral signature of radiance of freshwater systems, and these measurements have been used to derive various water quality parameters including temperature, turbidity, salinity, and chlorophyll concentrations (IGOS, 2004).

Conduct New Large-Scale Field Experiments and Surveys

Although remote earth observations are ideal for providing a global picture of changes in the water system, intensive field experiments are useful for providing better understanding about the details of processes and feedbacks in the system. Programs such as the "African Monsoon Multidisciplinary Analyses" (AMMA) collect vast amounts of hydrologic and climatological data by concentrating the capacity of scientists in an efficient way over a short period of time. From the perspective of global water research, AMMA-type experiments bring us further along in understanding large-scale teleconnections among land use, climate, and the hydrologic cycle. Another useful type of field campaign consists of flow manipulation experiments (as conducted on the Colorado, Snowy, and other rivers) in which experimental flows are released from dams in order to study their downstream ecological effects (Arthington et al., 2003). Data from these campaigns provide valuable new insight into the flow requirements of aquatic and riparian ecosystems. But field campaigns are needed not only by natural scientists but also by the social scientists. It is urgent to conduct planetary-scale social science surveys covering a wide range of social groups and countries. Data from these surveys are needed to improve our knowledge about human vulnerability to changes in the water system, in particular about its spatial variability and variety. This knowledge will allow researchers to better identify locations of both rapid change and vulnerable populations.

Develop and Use New Tools for Simulating the Global Water System

Collecting new data is important, but these data must also be analyzed by new types of analytical tools. A new generation of global- and continental-scale water models is required for comprehending and anticipating future changes in the global water system. To be useful for addressing policy-relevant questions, these models must be able to integrate a very wide range of global-scale information about the socioeconomic system, land use, climate, hydrology, and aquatic ecosystems. Model builders will have to team up with groups collecting data to identify the current and future "hot spots" of changes in the global water system. Because of the importance of simulating the global water system, the new generation of water models must be hooked into worldwide Internet-based "user support systems" that store key model outputs and make them widely available to researchers, policy analysts, and interest groups. New integrated assessment procedures are needed for systematically tracking the state of the global water system and computing scenarios of future changes and policy responses and especially for communicating this information to society. Perhaps the water community can gain from the experience of the Intergovernmental Panel on Climate Change, which periodically assesses and interprets the state of understanding of climate change issues in a way relevant for policymakers and other stakeholders.

Intervention 2: Expanding Governance of the System

As compared to global monitoring, field experiments, and the like, a more direct intervention would be to expand the global governance of water. "Water governance" is defined by the United Nations Development Programme as "the political, economic and social processes and institutions by which governments, civil society and the private sector make decisions about how best to use, develop and manage water resources" (UNDP, 2004). The first steps to govern water globally were already taken in 1921 with the adoption of the "Convention and Statute on the Regime of Navigable Waterways of International Concern," which prohibits states from impeding the navigation of important international waterways passing through their territory. Two years later, a convention concerning "Hydraulic Power" established guidelines for states to negotiate about hydropower projects affecting international waters. The much more recent "Convention on the Law of the Non-Navigational Uses of International Watercourses" (1997) also intervenes in international waters by urging the prevention, reduction, and control of pollution; by hindering the further introduction of alien species; and by fostering cooperation between and among states in the management of water resources.

The aim of the preceding three conventions was to influence the development and management of international watercourses. But what about the rest of the global water system? The Ramsar Convention (1971) ("Convention on Wetlands of International Importance Especially as Waterfowl Habitat") established the principle that the international community can intervene even if a particular issue does not involve international waters. Ramsar promulgated international guidelines for protecting wetlands *within the borders of countries* because these wetlands are internationally significant to "ecology, botany, zoology, limnology or hydrology." As an example of "international significance," the Convention argues that wetlands are vital to migrating waterfowl whose habitat can include many other countries outside of the wetlands locations.

As noted in the above definition, water governance can also be carried out by nonpolitical or quasipolitical institutions. Examples at the global level include the Global Water Partnership, the World Water Council, the World Water Forums, the World Conservation Union, and the World Bank. These institutions, and especially the political conventions noted above, have established the basic legitimacy of governing water on a global basis. How should we now build on this experience? Some ideas are described in the following paragraphs.

International Convention on Environmental Flows

A timely follow-up to the Ramsar Convention would be an international convention establishing universal compliance with environmental flows. Such a convention would set up international guidelines for the natural flow regimes needed for protecting or restoring aquatic ecosystems and would require that these flows be protected in undeveloped river basins and reestablished where possible in developed basins. These guidelines would have to be quite general and flexible because of the large differences between flow requirements of different ecosystems. The convention would cover both international and noninternational rivers following the precedent established by the Ramsar Convention. Rivers within country borders would be covered by the agreement because of their "international importance" in providing vital ecosystem services such as regulation of the global nutrient cycle and provision of food. Of course, such a convention could not provide full protection for aquatic biota because it would not address the physical modification of aquatic and riparian habitats or the degradation of water quality or other factors endangering ecosystems. Nevertheless, universally protecting natural flow regimes where they still exist, and restoring some semblance of these patterns where they do not, would be important steps in protecting the biological side of the global water system.

International Water Labeling

Product labeling falls somewhere between consumer protection and public education; it is used to inform consumers about the performance of a product with the aim of reducing the use of dangerous or environmentally harmful products. A prominent example regarding water is Australia's national water labeling program in which notices are placed on dishwashers and other appliances. These notices indicate the water use intensity of the appliance and whether it conforms to minimum water efficiency standards. The promoters of labeling programs believe that well-informed consumers will voluntarily seek out water-saving products. Hence, Hoekstra and others have proposed that water labeling be tried on an international basis to stimulate global water conservation (Hoekstra, 2006). Although water labeling does not exist internationally, the forest industry has established a valuable precedent that could be built on. Internationally traded wood products carry a certification label of the Forest Stewardship Council if they comply with "responsible forest management" criteria. The water community could adopt a similar approach and carry out its own "certification" of the water performance of internationally traded appliances. Alternatively, labeling could be introduced by governments through a convention of the type described above. Either way, a labeling program would require information on water use efficiency to be placed on all internationally traded products that use significant amounts of water, and this could ultimately become a powerful tool to stimulate worldwide water conservation.

International Regulation of Water Use Standards

The international community could go beyond labeling and pass a law setting a maximum permitted water use on internationally traded products. Agreement would be needed on reasonable water consumptions for different technologies, and the law would have to be updated periodically to keep pace with technological improvements in water use efficiency. Such a statute could require that major technology exports such as power plant turbines be water efficient, and this would encourage not only industrialized countries but also developing countries to use the most up-to-date water-saving equipment. Because these technologies may be costlier than their less water-efficient counterparts, measures have to be taken to avoid burdening developing countries with unfairly high compliance costs.

The Human Right to Water

Many groups and organizations are advocating an explicit international declaration of the human right to adequate water supply and sanitation. With around 1.1 billion people lacking access to safe water and 2.4 billion to basic sanitation (UNDP, 2004), it is thought that such a declaration would pressure governments to comply with Millennium Development Goal 7: "To halve, by 2015, the proportion of people without sustainable access to safe drinking water." But the international acceptance of this human right is unclear. To date, the strongest official statement is "General Comment No. 15" published in 2002 by the Committee on Economic, Social and Cultural Rights of the United Nations. This statement decrees that "the human right to water entitles everyone to sufficient, safe, acceptable, physically accessible and affordable water, for personal and domestic uses" and makes important statements about the obligations of governments to deliver clean water and adequate sanitation to their citizens (see Committee on Economic, Social and Cultural Rights: http://www.unhchr.ch/html/menu2/6/cescr.html). Although it represents "decisive progress" on this question, General Comment No. 15 is a recommendation rather than a legal document. An unfinished task of the international community is to make an unequivocal legally binding statement about the human right to water. It is also time to think about how such a right will be enforced. A first step could be for UNESCO and UNICEF to expand their current surveys of compliance with international goals for water and sanitation to include a broader examination of government compliance with basic rights to water (WHO and UNICEF, 2004).

Intervention 3: Challenging the Goals of the System

The educator and systems theorist Donella Meadows conjectured that systems have particular "leverage points" where humans can intervene most effectively to change the system's behavior (Meadows, 1999). Furthermore, she claimed that the most sensitive of these points was challenging the goals of the system and their underlying assumptions. How does this idea apply to the global water system? What would it mean to challenge its goals? From the human standpoint, these goals are to provide humanity with adequate water for its perceived aspirations or requirements. To challenge these goals would be to ask: Do we really need the volume of water we now use and aspire to use? In the following paragraphs, we review two ways to address this basic question.

Stimulate a Public and Institutional Debate on Water Needs

Just as the many impacts of conventional energy use have stimulated a worldwide debate about how much energy we really need, so too the consequences of human abstraction of water justify a serious public and institutional debate about the volume of water sufficient for the needs of humanity and nature. The bottom line is our physical requirement for water. The United Nations High Commission on Refugees recommends a minimum allocation of 15 liters per day for each person in a refugee camp, but regards 7 liters per day as the "minimum survival allocation." Going beyond survival, the United Nations estimates 20 liters per day as a guideline for "reasonable access to water." In the "World Water Vision Scenarios," the Secretariat of the World Water Commission assumed that 40 liters per day per person was the minimum needed for basic personal and household use (Rijsberman, 2001). Another widely quoted figure is 100 liters per day per person, which Falkenmark and Lindh (1993) call a "fair level of domestic supply." Hence, the range of minimum personal needs outside of crisis situations is estimated to be around 20–100 liters per day per person. In the actual situation, the average daily water use of a sub-Saharan African is about 25 liters, which is not much above minimum personal needs. (These and other estimates of water use following in this paragraph are taken from World Resources Institute, Freshwater Resources for 2000: http://earthtrends.wri.org/pdf_library/data_tables/wat2_2005.pdf. These are water withdrawal data and are therefore somewhat larger than water requirements. To convert water withdrawals to water requirements it is necessary to subtract losses between the point of withdrawal and the point of use.) At the other extreme, the current European lifestyle requires 233 liters per person per day, and North Americans use 638 liters per person per day, a considerably higher figure than estimates of minimum personal requirements. In the face of these data, we need to seriously examine the questions, "What is an equitable level of water use, and how can this be universally achieved and complied with?" In the same way, the assumptions behind the water needs of industry and agriculture also have to be critically examined.

Reform the Education and Training of Water Researchers, Engineers, and Managers

If our aim is to make a lasting change in society's attitudes about water, we will eventually have to train a new generation of researchers, engineers, and managers to think in a new way about water. The reality is that conventional education and training tend to reinforce current assumptions about water resource development. Students learn how to design water infrastructure and to develop water management plans but much less about competition between water sectors or the global factors discussed in this chapter. An exception to this rule, called "integrated water resources management," is slowly finding its way into university curricula. This is a management approach that "integrates" many different aspects of river basin development by promoting a long-term perspective to planning, by encouraging the participation of diverse interest groups in the planning process, by reconciling the water needs of many different human users together with needs of aquatic ecosystems, and by advocating a strengthening of water use efficiency as an alternative to expansion of water supply.

Adopting integrated water resources management in the routine training of water researchers and professionals would be a major step in encouraging new thinking about water. But a further step is needed. It is just as urgent to expand university curricula in ecology, economics, hydrology, water and wastewater management, and other water-related disciplines to encompass the global perspective. The new generation of water researchers and professionals must understand that water can no longer be considered just a local or river basin issue. On the contrary, research has uncovered widespread and large-scale connectivities showing that water is also a global system. Moreover, pervasive changes going on in this system pose risks to humanity and the rest of nature that require global attention. In summation, a major task for the new generation of water specialists is to enlarge the scope of water research and management from the local, watershed, and regional levels to include the global scale.

FURTHER READING

Alcamo, J., M. Floerke, and M. Maerker. 2007. Future long-term changes in global water resources driven by socio-economic and climatic changes. Hydrological Sciences 52: 247–275.

Alcamo, J., D. van Vuuren, C. Ringler, W. Cramer, T. Masui, J. Alder, and K. Schulze. 2005. Changes in nature's balance sheet: Model-based estimates of future worldwide ecosystem services. Ecology and Society 10: 19. Downloadable from http://www.ecologyandsociety.org/vol10/iss2/art19/.

Framing Committee of the Global Water System Project. 2005. The Global Water System Project: Science Framework and Implementation Activities. Earth System Science Partnership. Bonn, Germany: Global Water System Project Office. Downloadable from http://www.gwsp.org.

Hoekstra, A. Y. 2006. The global dimension of water governance: Nine reasons for global arrangements in order to cope with local water problems. Value of Water Research Report Series No. 20. UNESCO-IHE Institute for Water Education, Delft, The Netherlands. Downloadable from http://www.waterfootprint.org.

Kundzewicz, Z. W., L. J. Mata, N. W. Arnell, P. Döll, P. Kabat, B. Jiménez, K. A. Miller, T. Oki, Z. Sen, and I. A. Shiklomanov. 2007. Freshwater resources and their management. In M. L. Parry, O. F. Canziani, J. P. Palutikof, P. J. van der Linden, and C. E. Hanson, eds. Climate Change 2007: Impacts, Adaptation and Vulnerability. Contribution of Working Group II to the Fourth Assessment Report of the Intergovernmental Panel on Climate Change. Cambridge, UK: Cambridge University Press, 173–210.

Naiman, R. J., J. J. Magnuson, D. M. McKnight, and J. A. Stanford, eds. 1995. The Freshwater Imperative: A Research Agenda. Washington, DC: Island Press.

Polunin, N.V.C., ed. 2005. State of the World's Waters. Cambridge, UK: Cambridge University Press.

Vörösmarty, C., M. Meybeck, B. Fekete, K. Sharma, P. Green, and J. Syvitski. 2003. Anthropogenic sediment retention: Major global-scale impact from the population of registered impoundments. Global and Planetary Change 39: 169–190.

VII.5

Managing Nutrient Mobilization and Eutrophication

D. W. Schindler

OUTLINE

1. History of the term
2. The role of thermal stratification
3. Natural and cultural eutrophication
4. Ratios and sources of key nutrients
5. Whole-lake experiments and their role in eutrophication control policy
6. Nonpoint sources of nutrients
7. The trophic cascade
8. Internal recycling of phosphorus
9. Eutrophication in flowing waters
10. Eutrophication and the quality of drinking water
11. Eutrophication of estuaries

Increasing the inputs of the nutrients phosphorus and nitrogen to freshwater bodies and estuaries causes increased growth of nuisance algae, termed eutrophication. In lakes, eutrophication can be prevented by controlling inputs of phosphorus. In estuaries, there is still controversy over whether nitrogen, phosphorus, or both must be controlled.

GLOSSARY

epilimnion. The uniformly warm upper layer of a lake when it is thermally stratified in summer.

eutrophic. Eutrophic lakes are richly supplied with plant nutrients and support heavy plant growths.

eutrophication. The complex sequence of changes initiated by the enrichment of natural waters with plant nutrients.

hypolimnion. The uniformly cool and deep layer of a lake when it is thermally stratified in summer.

mesotrophic. Mesotrophic lakes are intermediate in characteristics between oligotrophic and eutrophic lakes. They are moderately well supplied with plant nutrients and support moderate plant growth.

oligotrophic. Oligotrophic lakes are poorly supplied with plant nutrients and support little plant growth.

thermocline. Thermal or temperature gradient in a thermally stratified lake in summer. Occupies the zone between the epilimnion and hypolimnion.

Eutrophication is the word used by scientists to describe the result of overfertilization of lakes with nutrients. The first symptom noticeable to casual observers is that the fertilized lakes turn green with plant growth. Paradoxically, we value the increased growth of plants that follows fertilization on land but abhor similar effects in our waters.

1. HISTORY OF THE TERM

Eutrophication is derived from the German word *eutrophe,* which means "nutrient-rich." The two nutrients that are responsible for increasing growth of algae and other aquatic plants are nitrogen and phosphorus. Eutrophic lakes typically have dense algal blooms. They can also have dense beds of rooted aquatic plants if the lakes have shallow areas with mud or sand bottoms.

The term *eutrophication* was coined by the German wetland scientist C. A. Weber in 1907 to refer to the rich wetlands in areas of Europe that received nutrient runoff from surrounding lands. The term was first applied to lakes by Einar Naumann roughly a decade later. The term *oligotrophic* (nutrient poor) was applied to nutrient-poor lakes, which generally have clear water and deep waters that contain high concentrations of oxygen. Lakes that are between these two extremes are generally termed mesotrophic. All three categories of lakes can undergo eutrophication if nutrient concentrations are increased. Recently, extremely eutrophic lakes have been termed hypereutrophic.

Table 1. General trophic classification of lakes and reservoirs in relation to phosphorus and nitrogen

Parameter (annual mean values)	*Oligotrophic*	*Mesotrophic*	*Eutrophic*	*Hypereutrophic*
Total phosphorus (mg m^{-3})				
Mean	8.0	26.7	84.4	—
Range	3.0–17.7	10.9–95.6	16–386	750–1200
Total nitrogen (mg m^{-3})				
Mean	661	753	1875	—
Range	307–1630	361–1387	393–6100	—
Chlorophyll *a* (mg m^{-3}) of phytoplankton				
Mean	1.7	4.7	14.3	—
Range	0.3–4.5	3–11	3–78	100–150
Secchi transparency depth (m)				
Mean	9.9	4.2	2.45	—
Range	5.4–28.3	1.5–8.1	0.8–7.0	0.4–0.5

Source: Modified from Wetzel, R. G. 2001. *Limnology*, 3rd ed. Amsterdam: Elsevier.

The early use of the terms was to refer to a lake's appearance. Measurable indices of productivity, such as algal abundance, chlorophyll *a*, photosynthesis, or nutrient concentration were developed later and are usually used now to define trophic conditions in lakes (e.g., see table 1).

The term *eutrophication* became widely used by limnologists (scientists who study lakes and other fresh waters) to describe the complex sequence of changes in aquatic ecosystems caused by an increased rate of supply of plant nutrients to water.

The immediate response of an aquatic ecosystem to increased nutrients is an increase in photosynthesis and abundance of plants. This can give rise to increased productivity at all levels of the food chain, up to and including fish. But, as described in greater detail below, changes can also occur in the kinds of organisms inhabiting aquatic ecosystems during eutrophication to disrupt this transfer of energy up the aquatic food chain.

2. THE ROLE OF THERMAL STRATIFICATION

In order to understand all of the changes caused by eutrophication, a working knowledge of a lake's thermal characteristics is necessary. Some eutrophic lakes are deep enough to have a thermocline, a sharp boundary separating the warm upper waters of a lake (known as the epilimnion) from cold deep layers (termed the hypolimnion). This occurs because cooler waters are more dense than warmer waters. Most swimmers have experienced the thermocline as they pass suddenly from warm water to cold during a deep dive. The depth of a thermocline is determined by the wind velocity and the size of a lake. It can be as shallow as a few meters in small lakes to 15 meters or more in large lakes. Typically, a thermocline in a north-temperate lake will form when a lake warms in May, and last until cooling of the overlying air in September or October causes convective mixing, slowly deepening the thermocline until eventually there is no difference in the density difference between upper and deeper layers of water, and the lake mixes totally.

The production of algae and other plants occurs chiefly in the epilimnion of a lake because that is where light for photosynthesis is greatest and where most nutrients enter the lake. But as the algae and other organisms in the epilimnion die, they sink slowly through the thermocline to decompose in the hypolimnion. This decomposition consumes oxygen. If the rain of organic matter increases as it does when nutrient supplies are increased, oxygen in the hypolimnion can be depleted to very low concentrations. If oxygen concentrations become very low, it becomes impossible for air-breathing organisms to survive in the deeper layers of a lake. As a result, as a lake becomes more eutrophic, the species of fish and bottom-living invertebrates change from those that require high concentrations of oxygen to those that can tolerate low oxygen.

3. NATURAL AND CULTURAL EUTROPHICATION

Eutrophic lakes can occur naturally in terrain with naturally rich soils and geologic sources of nutrients. But lakes can also become eutrophic very rapidly as the result of human influences. *Cultural eutrophication* is the term used to describe lakes that have rapidly increasing concentrations of nutrients and algal blooms as the result of human activity. Typical nutrient sources are sewage, manure, agricultural fertilizer, and in some countries, phosphorus-based detergents. Cultural eutrophication was first noticed in European lakes at about the turn of the twentieth century, as land was cleared and populations of humans and livestock increased. Similar observations were made soon after in North American waters. Also, by the early twentieth century, water was used as a vector to transport human wastes from populous areas to prevent diseases that were prevalent in earlier times. Typically, wastes were piped to the nearest lake or river where they were discharged. The eutrophying effect of nutrients was unknown at the time. More recently, studies of algal remains in lake muds have been used to deduce that eutrophication occurred in much earlier times. For example, G. E. Hutchinson, one of the pioneers in the study of eutrophication, deduced that Lago di Monterosi, a small lake on the road from Rome to Siena, Italy, underwent cultural eutrophication after the Romans built the Via Appia in 171 BC, which brought many more people to the lake. The modern road follows the same route.

Fish kills resulting from low oxygen in the hypolimnion are frequently observed during cultural eutrophication. Several species of fish, including whitefish, cisco, and lake trout in North America, will suffocate rather than leave the cold hypolimnions that are their normal summer habitat. Some invertebrates, such as the opossum shrimp *Mysis*, disappear for similar reasons. The sight of hundreds or thousands of dead or dying fish on the surface of a eutrophic lake in midsummer has led those unfamiliar with the eutrophication problem to conclude that eutrophic lakes are dying. To the contrary, they are teeming with life, though not necessarily of the type that humans value.

4. RATIOS AND SOURCES OF KEY NUTRIENTS

Most plants require nutrients in rather set proportions in order to grow. Typically, algae contain roughly 40 g of carbon to every 7 g of nitrogen to every 1 g of phosphorus. This ratio is known as the *Redfield ratio* after the oceanographer Alfred Redfield who first discovered it. If any one of these three key elements is in short supply with respect to plant growth, it can limit plant production.

Phosphorus is the element that is usually the primary culprit in the eutrophication of lakes. In most lakes, phosphorus is very scarce with respect to the ratio in plants, compared to nitrogen or carbon. In many cases, precipitation falling on a lake's surface or in its catchment is the only source of the element. But most of the phosphorus falling with precipitation on a lake's catchment is typically taken up by terrestrial vegetation, so that only a few percent of what falls reaches the lake. Only in the case of lakes with catchments set in phosphorus-rich geologic substrates are lakes naturally eutrophic.

Nitrogen, too, enters largely with precipitation, but in most areas it is not as scarce as phosphorus, with respect to the nutrient demands of plants. It is usually in the form of nitrate (NO_3) or ammonium (NH_4), both of which are highly available to plants. But the atmosphere contains high concentrations of gaseous nitrogen. Some algae, most notably certain species of *Cyanobacteria*, are capable of fixing atmospheric nitrogen. (Cyanobacteria are commonly called bluegreen algae because of the color of certain diagnostic pigments. Although they contain chlorophyll, they are technically not algae because the pigments are not contained in a chloroplast as they are in true plants.)

Carbon is even more plentiful with respect to phosphorus. Most lakes have abundant supplies of carbon dioxide as a result of exchange with the atmosphere and the decomposition of organic material. In addition, the weathering of rocks and soils in a lake's catchment supply bicarbonate (HCO_3), which can be used by algae as well. In contrast to nitrogen, all species of algae can utilize CO_2. Thus, of the three primary nutrients, phosphorus is the only one without a major source in the form of an atmospheric gas.

In a typical eutrophication scenario, human waste or animal manure is washed into a lake. These wastes contain high phosphorus with respect to nitrogen and carbon. Algae typically respond by increasing rapidly in abundance. Often, the new supply of phosphorus will allow them to outstrip the supply of nitrogen. This situation favors the nitrogen-fixing Cyanobacteria mentioned above. Many species of nitrogen fixers float on the surface of the water, where they form unsightly "blooms." Most are too large to be eaten effectively by the typical invertebrates that occur in lakes, so they tend to accumulate, eventually falling to the bottom where their decay consumes oxygen, or washing ashore in unsightly windrows that rot and cause terrible odors (plate 22).

Early attempts to control eutrophication did not focus on nutrient control. Instead, attempts were made to poison the algal blooms by applying algal toxins such as copper sulfate or synthetic herbicides. These allowed only short-term control, and the algae were usually as abundant as ever in a few weeks. Little thought was given to the control of nutrient sources.

In the mid-twentieth century, a new source of nutrients caused eutrophication to accelerate: phosphate detergents. Before the advent of detergents, soaps were used to wash clothing. These did not work well in many waters, especially those that were high in calcium, magnesium, and bicarbonate, where scummy residues were often left on clothes. Detergents were developed by industrial chemists as more effective cleaning agents, and they quickly replaced soaps. Unfortunately, early detergents were very resistant to biodegradation, and they accumulated in lakes. As a result, it was not uncommon in cities such as Chicago or Milwaukee to be met by huge clouds of foam blowing down the streets when winds were onshore. Manufacturers responded by making detergents biodegradable, so that aquatic microorganisms could degrade them quickly. Unfortunately, many of these contained 20% phosphorus by weight. In the lower St. Lawrence and Great Lakes in the 1960s, roughly half of the phosphorus supplied by humans was as detergents, with most of the rest from human sewage. The industrial phosphorus effectively doubled the rate of cultural eutrophication.

Concern for the Great Lakes and many large European lakes led limnologists of the 1960s to search for ways to solve the eutrophication problem. There were literally thousands of studies of the nutrient requirements of algae, but until that time, researchers were not really focused on solving the problem in lakes. There were, however, two particularly important studies that helped to convince scientists and regulators that the key to controlling eutrophication was to control phosphorus.

The first of these was a long-term study of Lake Washington done by Tom Edmondson. He and his students had documented the increasing eutrophication of the lake during the early twentieth century. Edmondson convinced local regulators to divert sewage effluent from the lake. His analysis, published in *Science,* showed that the concentration of algae decreased in proportion to the decrease in the concentration of phosphorus.

The second influential study was a review of eutrophication done by Richard Vollenweider for the Organization for Economic Cooperation and Development. Vollenweider analyzed an extensive literature, deducing that phosphorus was the key to controlling eutrophication. He published the first models that related phosphorus input or "loading" to the state of eutrophication in lakes. He later revised these models to include the effects of flow through the lakes.

These two studies, and small lake experiments described later, were among those used by Jack Vallentyne, then cochair of the Scientific Advisory Board on the Great Lakes for the International Joint Commission (IJC), to convince the IJC to recommend to the Canadian and American governments that regulating the phosphate content of detergents and removing phosphorus from sewage were essential first steps to solving the eutrophication problem in the Great Lakes.

5. WHOLE-LAKE EXPERIMENTS AND THEIR ROLE IN EUTROPHICATION CONTROL POLICY

Controlling phosphorus was opposed by manufacturers of phosphate detergent and their allies. They mounted a Madison Avenue–style campaign to discount the evidence for phosphorus control, first arguing that carbon was far more likely the cause of eutrophication in the Great Lakes. This theory was tested at the Experimental Lakes Area (ELA) in a whole-lake fertilization experiment. Lake 227 had extremely low concentations of available carbon, much lower than in the Great Lakes. When fertilized with phosphorus and nitrogen, huge algal blooms were formed despite the scarcity of carbon. Although the algae showed symptoms of extreme carbon shortage, this was slowly made up by exchange from atmospheric supplies. The detergent people quickly changed their arguments to nitrogen as the cause, arguing that phosphorus recycled too quickly to be controlled effectively so that nitrogen control would be more effective. Of course, restricting nitrogen had no implications for the detergent industry. Its removal at the sewage treatment plant was also much more costly than removing phosphorus.

A second experiment at ELA proved the nitrogen control theory to be wrong as well. A double-basin lake, Lake 226, was separated into two basins by a watertight plastic curtain. Nitrogen and carbon were added to both basins but phosphorus only to one basin. The basin receiving phosphorus became very eutrophic; the basin receiving only nitrogen and carbon remained in a natural state. These simple experiments proved convincing to reluctant policy makers.

The experiments in these two lakes allowed another dimension of eutrophication to be investigated. As described above, floating blooms of nitrogen-fixing Cyanobacteria are often one of the most visible and objectionable effects of eutrophication. Lake 227 in the early years of fertilization did not have this group of algae. Although fertilization caused algae to increase, the species of algae did not change. We had added N:P at a ratio of 14:1 by weight in order to ensure that we were not confounding our investigation of carbon limitation by causing nitrogen to be limiting.

In contrast, the N:P ratio used in fertilizing Lake 226 was only 4:1 because it was designed to mimic the ratio in sewage. Nitrogen-fixing Cyanobacteria were the predominant algae to respond to fertilization, and the natural species of algae remained rare. To test whether this shift was a coincidence or whether the low N:P favored nitrogen-fixing Cyanobacteria, the N:P ratio in Lake 227 was decreased to 5:1 in the seventh year of fertilization. Within a few weeks, nitrogen-fixing Cyanobacteria became as dominant in Lake 227 as they were in Lake 226. Measurements of nitrogen fixation confirmed that they were using gaseous N_2 to overcome shortages of ionic forms. Obviously, although the amount of algal increase was caused by phosphorus addition, the species that became dominant were affected by the N:P ratio.

Although the arguments made by detergent manufacturers and their allies delayed legislation, science prevailed, and in 1973 phosphorus-control legislation was passed in Canada. In the United States, progress was slower, because phosphorus control was decided by individual states. But eventually, all states in the Great Lakes catchment regulated phosphorus. The resulting improvement in the state of the lower Great Lakes is a success story that should make limnologists proud.

In Europe, the debate over carbon did not occur, and there was minimal debate over the need to control nitrogen in fresh water. Instead, there was a rather fierce debate over whether to regulate the phosphorus content of detergents or to simply remove the phosphorus at the sewage treatment plant. Some countries decided to regulate detergents, others not, but all western European countries eventually regulated phosphorus loading to lakes.

6. NONPOINT SOURCES OF NUTRIENTS

In North America, the political focus on eutrophication diminished after phosphorus was controlled in detergents and sewage, even though clear evidence was emerging that *nonpoint sources* of phosphorus, such as fertilized land, feedlots, storm runoff from urban areas (which contains lawn fertilizer, pet excrement, and other high-nutrient materials), and leaky septic tanks were important. Fortunately, a few scientists kept studying these problems. Many lakes have become eutrophic as a result of excessive cottage development, land clearing, fertilizer applications, and urbanization in their catchments. These problems are much more difficult to control, and many of the necessary conditions are in the hands of local or municipal regulators who are not aware of the consequences of increased nutrient inputs for lakes.

7. THE TROPHIC CASCADE

Another cause of eutrophication was found to be the decline in predatory fish species. Steve Carpenter and Jim Kitchell and their colleagues at the University of Wisconsin deduced that declining piscivorous predators, as has happened in most lakes as the result of angling pressure, allowed increases in minnows and other small fish species because of reduced predation. The high populations of these small fish in turn depleted the populations of grazing zooplankton that under normal circumstances helped to control the abundance of algae through their grazing. The resulting *trophic cascade* caused lakes that were in a low algal state when all four levels of the food chain were intact to assume a high algal state when the top level of the food chain was removed. Carpenter and Kitchell proved their theories in a series of whole-lake experiments, showing that both the trophic cascade and phosphorus loading were important to controlling eutrophication.

8. INTERNAL RECYCLING OF PHOSPHORUS

Although control of nutrients and the integrity of biotic communities have allowed many lakes to recover, these efforts have not been universally successful. In some lakes, internal recycling of phosphorus from lake sediments keeps lakes from recovering. This phosphorus originated outside the lake, but in some lakes, anoxic conditions in deep water allow phosphorus to be remobilized into the lake during summer stratification or under winter ice. This recycling can go on for many years after external sources of phosphorus have decreased. The lakes with the greatest internal recycling appear to be those that have low concentrations of iron. The lack of iron is thought to lessen the precipitation of ferric phosphate, or coprecipitation of phosphates with ferric hydroxides, allowing phosphorus that would otherwise be immobilized in sediments to be released from sediments. Many mechanisms have been proposed to reduce this recycling, including bubbling of the hypolimnion with air or oxygen to prevent low oxygen, mixing of upper and lower strata in the lakes, or addition of iron, alum, or lime to attempt to keep the phosphorus in sediments from being released. None of these techniques has been totally successful in all systems, for reasons that are not well understood.

9. EUTROPHICATION IN FLOWING WATERS

Streams and rivers can also suffer from cultural eutrophication. Slow-flowing streams show many of the same symptoms as lakes, with algal scums and low oxygen becoming major problems. In shallow clear-water rivers with rocky bottoms, huge mats of attached algae rather than plankton blooms are typical symptoms. As streams undergo eutrophication, typical groups of benthic invertebrates such as Trichoptera, Ephemeroptera, and Plecoptera typically decline and are replaced by chironomid (insect) and oligochaete species that are more tolerant of low oxygen.

Few states or provinces have guidelines for controlling the eutrophication of flowing waters.

10. EUTROPHICATION AND THE QUALITY OF DRINKING WATER

In recent years, considerable attention has been paid to the role of eutrophication in degrading drinking water sources. Obviously, where sewage or manure is the source of nutrients, it is also the source of bacteria such as *Escherichia coli* and protozoans such as *Cryptosporidium* and *Giardia* that cause gastrointestinal disease. But nutrients have indirect effects via their effect on aquatic organisms. Some species of bloom-forming Cyanobacteria also produce potent liver toxins, which occasionally cause the deaths of livestock or pets that drink water containing them. Several species of algae can also cause problems with taste and odor. As a result of these problems, high-quality drinking water becomes difficult and costly. In many cases, conventional filtration and chlorination of water will not suffice. In the worst cases, reverse osmosis must be used to eliminate chemicals of concern.

11. EUTROPHICATION OF ESTUARIES

Most recently, much attention has been paid to the eutrophication of the coastal oceans, particularly in bays and estuaries. Although these are not generally used as drinking water (except in a few areas where shortage of fresh water has required desalinization of seawater), they support important fisheries.

The debate over what nutrient to control is alive and well in estuaries. Physiological studies show that algae are usually nitrogen limited. As was the case in fresh water, many scientists have interpreted this as a sign that nitrogen control is necessary to reduce eutrophication.

Several studies suggest that this interpretation is incorrect, as it was in freshwater lakes. In the 1970s, phosphorus inputs were reduced to the Stockholm Archipelago, a part of the Baltic Sea just off the most populous part of Sweden. At the time, the main algal species in the Archipelago were nitrogen-fixing Cyanobacteria, and they showed signs of extreme nitrogen limitation. But a decline in phosphorus concentration was followed by a huge decline in algae. Unfortunately, the recovery was not complete because of phosphorus return from the sediments of the Archipelago, as described above for eutrophic lakes. As a result, whether or not to control nitrogen inputs in an attempt to cause further declines in algal blooms is still hotly debated.

A few studies suggest that if humans are determined to control nitrogen, phosphorus control is necessary as well. If the ratio of N:P decreases below the Redfield ratio, Cyanobacteria tend to become dominant, at least in estuaries where salinity is far below that of seawater, as in the Baltic Sea.

FURTHER READING

Carpenter, Stephen, James Kitchell, and J. Hodgson. 1985. Cascading trophic interactions and lake productivity. BioScience 35: 634–639.

Edmondson, W. Thomas. 1991. The Uses of Ecology: Lake Washington and Beyond. Seattle, WA: University of Washington Press.

Hutchinson, G. Evelyn. 1973. Eutrophication. American Scientist 61: 269–279.

Schindler, David W. 1974. Eutrophication and recovery in experimental lakes: Implications for lake management. Science 184: 897–899.

Schindler, David W., and John R. Vallentyne. 2008. The Algal Bowl: Overfertilization of the World's Freshwaters and Estuaries. Edmonton, AB: University of Alberta Press.

VII.6

Managing Infectious Diseases

Jonathan A. Patz and Sarah H. Olson

OUTLINE

Changes in biodiversity and habitat change affect the transmission or emergence of a range of infectious diseases. These environmental factors include agricultural encroachment, deforestation, road construction, dam building, irrigation, wetland modification, mining, the concentration or expansion of urban environments, coastal zone degradation, and other activities. As a result, a cascade of factors can exacerbate infectious disease resurgence, such as forest fragmentation, disease introduction, pollution, poverty, and human migration. Subsequent biological mechanisms of disease emergence that are affected include altered vector breeding sites or reservoir host distribution, niche invasions or interspecies host transfers, changes in biodiversity (including loss of predator species and changes in host population density), human-induced genetic changes of disease vectors or pathogens (such as mosquito resistance to pesticides or the emergence of antibiotic-resistant bacteria), and environmental contamination of infectious disease agents.

GLOSSARY

emerging disease. As defined by the Centers for Disease Control and Prevention, emerging infectious diseases are diseases of infectious origin whose incidence in humans has increased within the past two decades or threatens to increase in the near future. In general, an emerging disease can be a completely new disease or an old disease occurring in new places or new populations or that is newly resistant to available treatments.

gonotrophic cycle. The complete cycle of time between a mosquito's blood feeding and subsequent laying of eggs.

reservoir host. A reservoir host can harbor human pathogenic organisms without acquiring the disease, and so serves as a source from which the infectious disease may spread.

vector-borne disease. Infectious diseases spread indirectly via an insect or rodent. Often, part of the pathogen's life cycle occurs within the insect vector. Examples include malaria, dengue fever, West Nile virus, Lyme disease, plague, and Hantavirus.

zoonotic disease. Any disease that is spread from animals to people. These are also called "zoonoses" (as opposed to "anthroponoses," which are diseases transmitted directly from person to person). Examples of zoonotic diseases include rabies, Lyme disease, and bat-borne Nipah and Hendra viruses.

1. INTRODUCTION

Widespread deforestation and habitat destruction not only threaten biodiversity worldwide, but land use change influences a range of infectious diseases. Anthropogenic (human-created) drivers that especially affect infectious disease risk include destruction or encroachment into wildlife habitat, particularly through logging and road building; changes in the distribution and availability of surface waters, such as through dam construction, irrigation, or stream diversion; agricultural land use changes, including proliferation of both livestock and crops; deposition of chemical pollutants, including nutrients, fertilizers, and pesticides; uncontrolled urbanization or urban sprawl; climate variability and change; migration and international travel and trade; and either accidental or intentional human introduction of pathogens (table 1 and figure 1).

Table 1. Mechanisms and processes of land use change that affect health

Agricultural development
Urbanization
Deforestation
Population movement
Increasing population
Introduction of novel species/pathogens
Water and air pollution
Biodiversity loss
Habit fragmentation
Road building
Macro- and microclimate changes
Hydrological alteration
Decline in public health infrastructure
Animal-intensive systems
Eutrophication
Military conflict
Monocropping
Erosion

Source: Patz et al., 2004.
Note: Ranked from highest to lowest public health impact by meeting participants. Criteria for ranking were based on estimated impacts on both the number of infectious diseases and the prevalence of those diseases.

Ecological changes can affect specific biological mechanisms of disease transmission. Several biological mechanisms have been identified and are reported in the Millennium Ecosystem Assessment. These include niche invasions or interspecies host transfers; changes in biodiversity (including loss of predator species and changes in host population density); altered vector breeding sites or reservoir host distribution; human-induced genetic changes of disease vectors or pathogens (such as mosquito resistance to pesticides or the emergence of antibiotic-resistant bacteria); and environmental contamination of infectious disease agents.

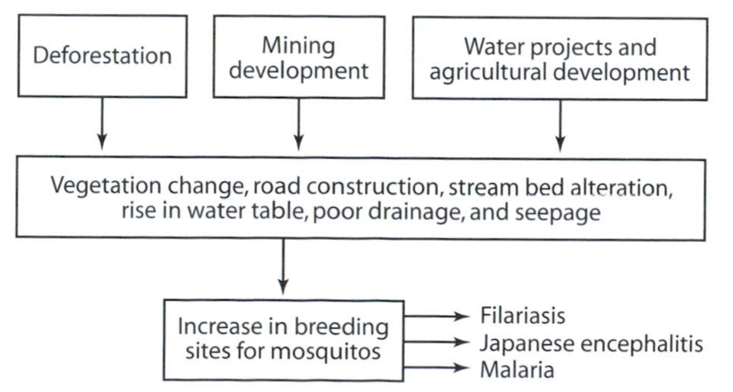

Figure 1. Examples of habitat change and vector-borne disease. Rising water table, irrigation, mining, and deforestation can increase the risk of mosquito-borne diseases. (From Patz et al., 2005)

2. NICHE INVASION AND CROSS-SPECIES TRANSFER OF PATHOGENS

Many widespread diseases of today originally stemmed from domestication of livestock. For example, measles, smallpox, and tuberculosis resulted from the domestication of wild cattle. Pathogens that are currently passed from person to person (anthroponotic), including some influenza viruses and human immunodeficiency virus (HIV), were formerly zoonotic but have diverged genetically from their ancestors that occurred in animal hosts.

Rapid population growth and population movements have quickened the pace and extensiveness of ecological change over the past two centuries. New diseases have emerged even as some pathogens that have been around for a long time have been eradicated or rendered insignificant, such as smallpox. Ecological change, pollutants, the widespread loss of top predators, persistent economic and social crises, and international travel, which drive a great movement of potential hosts, have progressively altered disease ecology, affecting pathogens across a wide taxonomic range of animals and plants.

According to estimates, nearly 75% of human diseases are zoonotic and stem from wildlife or domestic animals. The emergence of many diseases has been linked to the interface between tropical forest communities, with their high levels of biodiversity, and agricultural communities, with their relatively homogeneous genetic makeup but high population densities of humans, domestic animals, and crops. For instance, expanding ecotourism and forest encroachment have increased opportunities for interactions between wild nonhuman primates and humans in tropical forest habitats, leading to pathogen exchange through various routes of transmission.

When severe acute respiratory syndrome (SARS) virus erupted and almost became pandemic, the cause was linked to an animal–human interaction occurring in customary wet markets in China—a known source of influenza viruses since the 1970s. (Live-animal markets, termed "wet markets," are common in most Asian societies and specialize in many varieties of live small mammals, poultry, fish, and reptiles.) The majority of the earliest reported cases of SARS were of people who worked with the sale and handling of wild animals. The species originally at the center of the SARS epidemic was the palm civet cat, but further research implicated the Chinese horseshoe bat as the definitive reservoir host.

Bush-meat hunting in the deep jungle has afforded easier exchange of pathogens between humans and nonhuman primates. In Central Africa, 1–3.4 million tons of bush meat is harvested annually. In West Africa, a large share of protein in the diet comes from bush meat. The bush-meat harvest in West Africa includes a large numbers of primates, so the opportunity for interspecies disease transfer between humans and nonhuman primates is significant, providing ample opportunity for cross-species transmission and the emergence of novel microbes into the human population.

The "Taxonomic Transmission Rule" states that the probability of successful cross-species infection increases the closer hosts are genetically related (chimpanzees are closer genetically to humans, for example, than birds or fish are) because related hosts are more likely to share susceptibility to the same range of potential pathogens. Bush-meat consumption has been implicated in the early emergence of HIV (and workers collecting and preparing chimpanzee meat have become infected with Ebola).

Unfortunately for wildlife, transmission across species can also go from humans to wildlife. For example, the parasitic disease Giardia was introduced to the Ugandan mountain gorilla by humans through ecotourism activities. Gorillas in Uganda also have been found with human strains of *Cryptosporidium* parasites, presumably from ecotourists. Human tuberculosis has also jumped species, infecting the banded mongoose. Such transfer and emergence events not only affect ecosystem function but could possibly result in a more virulent form of a human pathogen circling back into the human population from a wildlife host.

3. WHY CAN HIGH BIODIVERSITY PREVENT DISEASE EMERGENCE?

Habitat fragmentation generally reduces species biodiversity. Infectious diseases, especially those involving intermediate reservoir host species in their life cycle, can thereby be affected. Organisms at higher trophic levels usually exist at a lower population density (per classic food webs) and are often quite sensitive to changes in food availability. Smaller forest patches left after fragmentation, for example, may not have sufficient prey for top predators, resulting in local extinction of predator species and a subsequent increase in the density of their prey species.

Lyme disease in particular is influenced by the level of species diversity in the biome. In eastern U.S. oak forests, studies on the interactions among acorns, white-footed mice (*Peromyscus leucopus*), moths, deer, and ticks have linked defoliation by gypsy moths with the risk of Lyme disease. Most tick vectors feed on a variety of host species that differ dramatically in their function as a reservoir for the bacterium that causes Lyme disease—that is, the probability of a tick picking up the bacterium from different reservoir hosts varies substantially. Increasing species richness has been found to reduce disease risk, and the involvement of a diverse collection of vertebrates in this case may dilute the impact of the main reservoir, the white-footed mouse.

Also, small woodlots tend not to have the range required of predator mammalian species, and probable competitors occur at lower densities in these areas than in more continuous habitats. Therefore, habitat fragmentation causes a reduction in biodiversity within the host communities, increasing disease risk though the increase in both the absolute and relative density of a primary reservoir, the white-footed mouse. Other diseases are known to have resurged following land use change, including cutaneous leishmaniasis, Chagas disease, human granulocytic ehrlichiosis, babesiosis, plague, louping ill, tularemia, relapsing fever, Crimean Congo hemorrhagic fever, and LaCrosse virus.

4. HARMING HABITATS CAN HARM HUMAN HEALTH: TROPICAL RAINFOREST DESTRUCTION AND THE RISE OF MALARIA

The global rate of tropical deforestation continues at staggering levels with nearly 2–3% of global forests lost each year. Recent evidence from Africa, the Amazon, and parts of Asia now identify deforestation as one of the causes for the increase in malaria across the tropics and for habitat fragmentation (e.g., "fishbone" pattern in the Amazon from road building, creating breeding sites and "edge effects").

The World Health Organization (WHO) estimates that malaria is responsible for 13% of global disability and mortality from infectious and parasitic diseases. It is the world's most widespread fatal or debilitating vector-borne disease, killing nearly 2 million (mostly children) annually. Southeast Asia, Africa, and the Amazon have experienced increased malaria risk accompanying both human population growth and environmental change.

Habitat disturbance has been changing mosquito distributions for centuries. For example, the draining of swamps surrounding Rome reduced mosquito populations and malaria in the city. When agricultural practices spread across Europe, the resultant social and land transformations contributed to the eradication of malaria. However, adverse effects resulted from the removal of forests from within the ancient Indus valley and are proposed to have shifted the habitat preferences of the dangerous *A. stephansi* mosquito from the forest to urban areas and waterways and thereby to have contributed to the civilization's collapse circa 2000 BC. To this day, *A. stephansi* remains the most

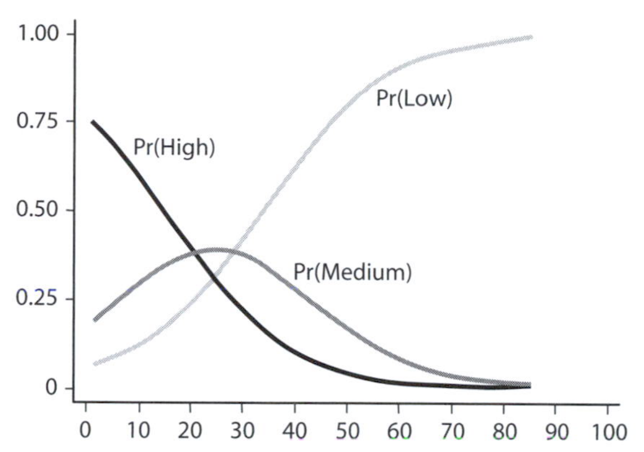

Figure 2. Biting rates of *An. darlingi* mosquitoes (*y*-axis) versus percentage forest contained in a 2 km × 2 km grid (*x*-axis). The black line is the high-probability curve and shows that mosquito biting rates (or abundance because each mosquito is captured when it comes to bite) decline rapidly as the amount of mature forest increases within the grid area. These curves account for potential confounding by human population density. (From Vittor et al., 2006)

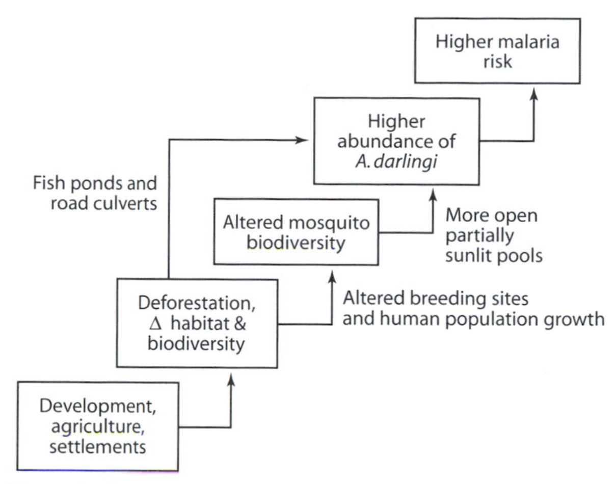

Figure 3. Schematic summary of likely steps explaining how deforestation may increase malaria risk in the Amazon region. Note that habitat change can affect mosquito biodiversity (by altering niches) and, subsequently, relative mosquito species abundance.

prolific vector for urban malaria transmission across the Middle East.

Soon after the discovery that malaria was transmitted by mosquitoes, attention shifted away from the influence of Land Use and Cover Change (LUCC) on malaria risk. Priority of malaria programs became vector eradication, via insecticides and clinical health interventions. If LUCC was recognized as a risk factor, it was used to focus the application of DDT to mosquito breeding sites.

Malaria entomology studies have examined sites for adult or larval abundance in different habitats and or gradients of one LUCC type. For example, a project in western Venezuela found greater mosquito species richness in tall forests in comparison to short forests and open areas. Larval species richness in open areas was higher for swamps and flooded pastures than ground pools, lagoons, or rivers. In Africa, breeding pools for *Anopheles* larvae had significantly less canopy cover than breeding pools for other mosquito species, and the larval habitat size was significantly different across several different LUCC categories. These and other observations continue to show that LUCC can alter the distribution of a disease-causing vector.

Throughout Amazonia, *Anopheles darlingi* is the most efficient and principal interior vector of falciparum malaria. In a recent field study from the Peruvian Amazon, sites surrounded by deforestation had *An. darlingi* biting rates nearly 300 times larger than forested sites, even after controlling for human population density. Sites with greater than 80% deforestation had a mean biting rate of 8.33, whereas sites with less than 30% deforestation had a mean biting rate of 0.03 (figure 2). These findings suggest that environmental risk factors for malaria are related to the LUCC changes associated with human expansion into forested areas. Additionally, larvae of *An. darlingi* were more often found in breeding sites with more sunlight (less forest canopy), with emergence grasses, and with algae. Figure 3 illustrates likely steps between deforestation and malaria risk. Current indicators of LUCC related to malaria risk from localized studies remain to be tested across a broader region, and investigations of ecology have not been performed.

Changing landscapes can significantly affect local weather more acutely than long-term climate change. Land cover change can influence microclimatic conditions including temperature, evapotranspiration, and surface runoff, all key to determining mosquito abundance and survivorship. In Kenya, open treeless habitats average warmer midday temperatures than forested habitats and also affect indoor hut temperatures. As a result, the gonotrophic cycle of female *An. gambiae* was found to be shortened by 2.6 days (52%) and 2.9 days (21%) during the dry and rainy seasons, respectively, compared to forested sites. Similar findings have been documented in Uganda, where higher temperatures have been measured in communities bordering cultivated fields compared to those adjacent to natural wetlands, and the number of *An. gambiae s.l.* per house increased along with minimum temperatures after adjustment for potential confounding variables. Also, survivorship of *An. gambiae* larvae in sunlit open areas is much greater than that in forested areas. In short, deforestation and cultivation of natural swamps in the African highlands create conditions favorable for the survival of *An. gambiae* larvae, making analysis of land use change on local climate, habitat, and biodiversity key to malaria risk assessments.

5. AGRICULTURAL DEVELOPMENT, CROP IRRIGATION, AND BREEDING SITES

Land converted for agriculture represents the Earth's largest land surface change since human existence began. Migrant farmers appear to be the primary *direct* agents of tropical deforestation around the world, although the initial causes are often roads built by logging, mining, or petroleum interests. Agricultural development in many parts of the world has resulted in an increased requirement for crop irrigation, which reduces water availability for other uses and increases breeding sites for disease vectors. An increase in soil moisture associated with irrigation development in the southern Nile Delta following the construction of the Aswan High Dam has caused a rapid rise in the mosquito *Culex pipiens* and a consequent increase in the arthropod-borne disease Bancroftian filariasis (or elephantiasis). Onchocerciasis (river blindness) and trypanosomiasis (sleeping sickness) are further examples of vector-borne parasitic diseases that may be triggered by changing land-use and water management patterns.

One example of a very specific agricultural development that created a new ecological niche that caused a malaria epidemic occurred during the 1930s in Trinidad with the production of cacao. Cacao trees establish a lighter and drier environment than the surrounding natural forests. Thus, large shade trees (called *immortelles*) were planted to provide shade to the cacao trees. But immortelles support epiphytes, mostly bromeliads. In Trinidad, these immortelles supported a bromeliad tank species that naturally collects a small amount of water that is an ideal breeding site for *A. bellator*. *A. bellator* is a malaria vector that also prefers drier areas and the subcanopy elevation of the forests. Public health officials noted that splenomegaly among schoolchildren correlated with the areas cultivating cacao and that the vector *A. bellator* was not found outside of the cacao farm area. Removal of the bromeliads, by hand or with herbicides, reduced *A. bellator* populations and returned malaria rates to prior endemic levels.

Another example of agricultural land use practices and disease emergence arises from Malaysia, where Nipah virus first occurred in pig farmers in the late 1990s. Throughout the peninsula, pig farms were expanding and were often colocated with fruit orchards. The leading theory is that contaminated bat saliva on fruit dropped into pig pens and infected the pigs when they ingested the fruit. Porcine Nipah virus causes severe coughing in pigs, and it is believed that farmers contracted the disease by inhaling the aerosolized form of the virus.

6. CONCLUSIONS

The notion of sustainable health—mirroring the concept of sustainable development—means to provide health for today's generation without compromising that same opportunity for future generations. This chapter has illustrated how managing our natural resources is tightly linked to sustaining human population health. Along the path toward global economic development, health risks must be considered at various levels. These levels include (1) specific health risk factors, (2) landscape or habitat change, and (3) institutional (economic and behavioral) levels. It is essential that societies shift away from dealing primarily with specific risk factors and look "upstream" at land-use and biodiversity changes for causative factors of effects on infectious disease. As such understanding increases, it will become more feasible to plan approaches to prevent new infectious disease emergence.

Inherent trade-offs become evident when land-use change and health are correlated. These involve ethical values, environmental versus health choices, and disparities in knowledge and economic class. Trade-offs are between short-term benefit and long-term damage. For example, draining swamps may reduce vector-borne disease hazards but also destroy the wetland ecosystem and its inherent services (such as water storage, water filtration, biological productivity, and habitats for fish and wildlife). Research can help decisionmaking by identifying and assessing trade-offs in different land-use-change scenarios. Balancing the diverse needs of people, livestock, wildlife, and the ecosystem will always be a prominent feature.

As illustrated, biodiversity loss, habitat destruction, and agricultural practices can lead to infectious disease emergence or resurgence; the public, therefore, needs to be attentive to entire ecosystems rather than simply their local environs. Although we may not live within a certain environment, its health may indirectly affect our own. For example, intact forests support complex ecosystems and provide essential habitats for species that are specialized to that flora and, in turn, may end up as relevant to our health. The challenge then is to identify and promote optimal situations whereby the maximum number of people—including future generations (as well as a number of species)—can benefit from policies geared toward sustaining both health and the environment.

7. RESOURCE SITES

http://www.ecohealth.net. The International Association for Ecology and Health (and its flagship journal *EcoHealth*). *The association and journal now provide a gathering place for research and reviews that integrate the diverse knowledge of ecology, health, and sustainability, whether scientific, medical, local, or traditional.*

http://sage.wisc.edu/pages/health.html. *Information on global climate and land-use change impacts on ecosystems and human health at the University of Wisconsin.*

FURTHER READING

Aron, J. L., and J. A. Patz, eds. 2001. Ecosystem Change and Public Health: A Global Perspective. Baltimore, MD: Johns Hopkins University Press.

Foley, J. A., R. DeFries, G. P. Asner, C. Barford, G. Bonan, S. R. Carpenter, F. S. Chapin, M. T. Coe, G. C. Daily, H. K. Gibbs, J. H. Helkowski, T. Holloway, E. A. Howard, C. J. Kucharik, C. Monfreda, J. A. Patz, I. C. Prentice, N. Ramankutty, and P. K. Snyder. 2005. Global consequences of land use. Science 309: 570–574.

Millennium Ecosystem Assessment. 2005. Ecosystems and Human Well-Being: Current State and Trends. Findings of the Condition and Trends Working Group. Millennium Ecosystem Assessment Series. Washington, DC: Island Press. Downloadable from http://millenniumassessment.org.

Ostfeld, S. R., and F. Keesing. 2000. Biodiversity and disease risk: The case of Lyme disease. Conservation Biology 14: 722–728.

Patz, J. A., P. Daszak, G. M. Tabor, A. A. Aguirre, M. Pearl, J. Epstein, N. D. Wolfe, A. M. Kilpatrick, J. Foufopoulos, D. Molyneux, D. J. Bradley, and Members of the Working Group on Land Use Change and Disease Emergence. 2004. Unhealthy landscapes: Policy recommendations on land use change and infectious disease emergence. Environmental Health Perspectives 101: 1092–1098.

Patz, J. A., U.E.C. Confalonieri (Convening Lead Authors), F. Amerasinghe, K. B. Chua, P. Daszak, A. D. Hyatt, D. Molyneux, M. Thomson, L. Yameogo, M. Malecela-Lazaro, P. Vasconcelos, and Y. Rubio-Palis. 2005. Health health: Ecosystem regulation of infectious diseases. In Millennium Ecosystem Assessment, Ecosystems and Human Well-Being: Current State and Trends. Findings of the Condition and Trends Working Group Millennium Ecosystem Assessment Series. Washington, DC: Island Press.

Patz, J. A., and S. H. Olson. 2006. Malaria risk and temperature: Influences from global climate change and local land use practices. Proceedings of the National Academy of Sciences U.S.A. 103: 5635–5636.

Taylor, L. H., S. M. Latham, and M.E.J. Woolhouse. 2001. Risk factors for human disease emergence. Philosophical Transactions of the Society of London B 356: 983–989.

Vittor, A. Y., R. H. Gilman, J. Tielsch, G. Glass, T. Shields, W. Sánchez Lozano, V. Pinedo-Cancino, and J. A. Patz. 2006. The effect of deforestation on the human-biting rate of *An. darlingi*, the primary vector of falciparum malaria in the Peruvian Amazon. American Journal of Tropical Medicine and Hygiene 74: 3–11.

VII.7

Agriculture, Land Use, and the Transformation of Planet Earth

Jonathan A. Foley, Chad Monfreda, Jonathan A. Patz, and Navin Ramankutty

OUTLINE

1. What are we farming? Geographic patterns of major crop types
2. How are we farming? Changing agricultural management
3. Agriculture as a force of global environmental change
4. Summary and conclusions

It is fair to say that our planet's most precious resource is *land.* Land is the source of the vast majority of our food and fresh water, nearly all of our fiber and raw materials, and many other important goods and services. It is also our home. But our relationship to the land has been dramatically changing over the history of our species, mainly through the invention and evolution of agriculture. Today, with the emergence of modern agricultural practices, coupled with the population growth and technological developments of recent centuries, we have transformed a staggering amount of the Earth's surface into highly managed landscapes. Even more startling: the widespread use of irrigation and chemical fertilizers has fundamentally altered the flows of water and nutrients across large regions of the globe. These modifications to the land have driven fundamental changes to the ecology of our planet. Even the effects of future climate change may not have such a major, transformative effect on the environment and on human society as agriculture. However, despite the importance of agriculture in the global environment, we still know relatively little about how it affects ecological systems across local, regional, and global scales.

GLOSSARY

cropland. Land used for growing crops. The UN Food and Agriculture Organization defines this as the sum of arable lands and permanent crops. Arable land is defined by FAO as including "land under temporary crops (double-cropped areas are counted only once), temporary meadows for mowing or pasture, land under market and kitchen gardens, and land temporarily fallow (less than 5 years). The abandoned land resulting from shifting cultivation is not included in this category. Data for arable land are not meant to indicate the amount of land that is potentially cultivable." Permanent crops are defined as "land cultivated with crops that occupy the land for long periods and need not be replanted after each harvest, such as cocoa, coffee, and rubber; this category includes land under flowering shrubs, fruit trees, nut trees and vines, but excludes land under trees grown for wood or timber."

extensification. The practice of increasing the amount of agricultural land that is under cultivation.

human appropriation of net primary production (HANPP). How much of the biological productivity of a given location is used, consumed, or co-opted by human activities.

intensification. The practice of stimulating more agricultural production per unit area, mainly through increasing use of agricultural chemicals, irrigation water, high-yielding plant varieties, and machinery.

land cover. Describing the physical state of the land surface, such as "rainforest," "cropland," or "desert."

land use. The practices employed on a particular piece of land, such as rotating grazing or intensive maize cultivation.

net primary production (NPP). The biological productivity of the landscape, that is, the rate of conversion of physical energy (sunlight) into biological energy (through photosynthesis) in a given location.

pasture. Agricultural land used for animal grazing. The UN FAO defines this category as "land used permanently (5 years or more) for herbaceous forage crops, either cultivated or growing wild (wild prairie or grazing land). The dividing line between this category and the category 'Forests and woodland' is rather indefinite, especially in the case of shrubs, savannah, etc., which may have been reported under either of these two categories."

Since the dawn of agriculture, some 9000 years ago, humans have progressively transformed the landscapes of our planet.

Over time, agricultural land use steadily spread across the globe, reaching into nearly every region, setting the stage for an explosion of agricultural activity after the rise of the Industrial Revolution. Equipped with new technologies, and rapidly increasing population and income levels, agriculture quickly expanded to meet increased food demand over the last 300 years (see plate 23). But this global expansion of farmland was not uniform. Instead, it has traced a path determined largely by the history of European economic and political control. In particular, the direct impact of European settlement was seen in the rapid expansion of agricultural land through North America, Argentina, South Africa, and Australia/New Zealand. The rest of the world also experienced significant cropland expansion as regions became connected to European markets and spheres of influence.

Understanding the agricultural expansion over the last three centuries is especially critical because of the tremendous growth in global cropland area (an increase of ~12 million square kilometers, or ~466%) that happened during this time. During the 1700s and 1800s, croplands expanded most rapidly in Europe, one of the most economically developed regions of the world at that time. After the mid-1800s, the newly developing regions of North America and what would become the Soviet Union witnessed rapid cropland expansion. Cultivation in tropical developing nations expanded only gradually between 1700 and 1850 but has experienced exponential growth rates since that time. Since the 1950s, cropland areas in North America, Europe, and China have stabilized and even decreased somewhat in Europe and China.

Today, the most productive landscapes of the planet —those with the best climate and soil conditions—are already used for cropland agriculture. "Breadbaskets" of cultivated land are found largely in the temperate and subtropical zones, especially in regions of rich soils and adequate rainfall, such as the midwestern United States, the Prairie Provinces of Canada, the Argentinian Pampa, through Europe and the former Soviet Union, the major river basins of India and China, and Australia. New frontiers of cultivation are found in Southeast Asia and southeastern Brazil. Cultivation is also prevalent in West Africa, the Ethiopian highlands, the Rift Valley, and southern Africa, and at lower intensities surrounding the major growing areas. Pastures, or "Meat Baskets," are predominantly found in the grasslands and savannas of the world, in drier regions compared to croplands. The greatest extent of pastures are seen in the western United States and Canada, Patagonia, southeastern Brazil, Sahelian and southern Africa, through much of central Asia, and Australia (plate 24).

Altogether, nearly 15 million square kilometers of land (roughly the size of South America) is now used as croplands on the planet, and another ~30 million square kilometers of land (roughly the size of Africa) is now used for pastures; together, croplands and pastures occupy ~35% of the ice-free land surface of the planet (Ramankutty et al., 2008).

Because most of the fertile lands of the world are already under cultivation, there are only relatively few opportunities for further agricultural expansion. Today, only the rainforest regions of Latin America, Africa, and Indonesia offer any significant remaining cultivable lands, especially for crops such as soybeans and oil palm. However, the expansion of agricultural lands would come at the expense of ecologically valuable rainforests, posing a serious dilemma of balancing the needs of economic development and ecological sustainability.

1. WHAT ARE WE FARMING? GEOGRAPHIC PATTERNS OF MAJOR CROP TYPES

Although agriculture has come to shape landscapes across the entire planet, it assumes many different forms. Mediterranean olive groves, short-season North American maize, and the perennial plantain rhizomes of tropical Africa all fall under the rubric of agriculture, yet they hold radically different implications for people and the environment. Fortunately, new data sets illustrating the geographic patterns of individual crop types (Monfreda et al., 2008) have begun to fill in the contours of global cropland maps to answer a question crucial to understanding how agriculture has transformed the planet: Which crops grow where? And why?

History, culture, climate, and economics have shaped the complex patterns of crop production seen today. Globally, three major crop groups make up the majority of the 13.4 million square kilometers of crops harvested every year: cereals (6.6 million square kilometers), oil crops (1.8 million square kilometers), and fodder crops (1.4 million square kilometers). Cereal crops are the only crop group to occur in every growing region of the world. Three cereal crops—wheat, rice, and maize—are by far the most widespread of all crops and together make up 38% of the global crop area. Of these, wheat occupies the greatest area, extending across the fertile soils of the Gangetic Plain, the Canadian Prairie Provinces, the Former Soviet Union, northern China, and parts of Australia and Argentina. Maize is spread across the greatest array of climates, whereas rice is overwhelmingly concentrated in the densely populated lowlands stretching from eastern India to the lower reaches of the Yellow River in China. Soybeans, rapeseed, groundnuts, and sunflower are the most widely grown oil crops, covering about 11% of all crop area. Although oil crops do not cover as wide an area as cereals, they follow a similar geographic distribution and often grow in rotation with cereals. By contrast, fodder crops, which are distinct from pastureland and include alfalfa and other hay crops, are for the most part confined to the wealthy, high-latitude countries that have a largely animal-based diet.

The three major crop groups—cereals, oil crops, and fodder crops—often appear in combination with minor crops, which may be very important locally but occupy a lesser extent worldwide. These minor groups include pulses (0.67 million square kilometers), roots and tubers (0.50 million square kilometers), fruits (0.48 million square kilometers), vegetables (0.44 million square kilometers), fiber crops (0.35 million square kilometers), and sugar crops (0.26 million square kilometers).

The major and minor crop groups occur in various combinations to form a patchwork of farming systems across the planet. Just one or two crops dominate the least agriculturally diverse crop belts, which occur in areas that are dry, major grain exporters, or both. These low-diversity crop belts include the maize-soybean rotations of the U.S. Midwest and the enormous soybean monocultures expanding into the Brazilian Amazon. Extensive low-diversity croplands in dry regions include the wheat-barley fields of southern Australia, the wheat fields of the western United States, and the maize-millet-sorghum belt of the Sahel. By contrast to those regions that grow just one or two crops, the regions of the greatest crop diversity cultivate a high proportion of noncereal crops. These areas include the intensive lowland rice-vegetable-oil crop systems of East Asia and maize-rice-potato belt of the Peruvian Andes. Exceptional agricultural diversity also occurs in the Mediterranean region, which grows substantial amounts of grapes, olives, sunflowers, and other noncereal crops in near equal proportions with maize, barley, and wheat.

Developing a good picture of these farming systems is key to understanding the effect of agriculture on the global environment for at least two reasons. First, different crops are more or less suited to the climate and soils of different regions. Understanding how well crops grow across different regions indicates where and how much agricultural expansion may need to occur to meet growing demand for food, fiber, and biofuel. Second, different crops require different kinds of farming practices, including the use of fertilizer or irrigation. The geographic distribution of farming practices has major implications for the environment, including freshwater resources, the carbon cycle, biodiversity, and human health. Getting a better handle on the diversity of crops and farming practices across the planet is therefore an important part of grasping the key questions facing agriculture and the global environment.

2. HOW ARE WE FARMING? CHANGING AGRICULTURAL MANAGEMENT

Although changes in the geographic extent of our agricultural lands have been considerable, they do not provide the entire story.

Although significant agricultural *expansion* has occurred in the past few decades, the *intensification* of agricultural practices—under the aegis of the "Green Revolution"—has dramatically increased, completely changing the relationship among humans, agriculture, and environmental systems across the world. Specifically, since the 1950s, there has been a major shift toward agricultural intensification, in lieu of expansion, enabled by the widespread development of irrigation systems, the invention of inorganic fertilizers in the early 1900s, and the development of crops that better exploit water and nutrients and are more resistant to pests and diseases.

Simply put, the world's existing agricultural lands are being used much more intensively as opportunities for agricultural expansion are being exhausted elsewhere.

In the last 40 years, global agricultural production has more than doubled—although global cropland area has increased by only ~12%—mainly through the use of high-yielding varieties of grain, increased reliance on irrigation, massive increases in chemical fertilization, and increased mechanization (Foley et al., 2005). As a result, the growing demand for food in the past few decades has been increasingly met through higher yields on existing croplands rather than through agricultural expansion.

Indeed, in the past 40 years, there has been an ~700% increase in global fertilizer use and an ~70% increase in irrigated cropland area (Foley et al., 2005). These represent significant changes to the planet's hydrologic and biogeochemical cycles: for example, we now apply more nitrogen fertilizer than is naturally fixed in the biosphere. And the diversion of freshwater flows for irrigation alone exceeds the changes in water availability expected from future climate change. So although these modern agricultural practices have successfully increased food production, they have caused extensive environmental damage across many portions of the planet (Foley et al., 2005).

3. AGRICULTURE AS A FORCE OF GLOBAL ENVIRONMENTAL CHANGE

The expansion and intensification of agriculture have become major forces in shaping the human impact on the global environment. Whether it is through clearing tropical rainforests, practicing subsistence agriculture on marginal lands, or intensifying industrialized farmland production in temperate croplands, modern agricultural practices are changing the world's landscapes in many ways. Although the precise character of agricultural land use varies greatly across the world, the ultimate outcome is generally the same: the production of new agricultural goods for human needs, often at the expense of degrading environmental conditions.

Agricultural practices can have dramatic effects on many environmental processes, ranging across local, regional, and global scales. At these different scales, it may be useful to define "first-order" responses of the environment to agricultural land-use change as well as "second-order" environmental effects.

"First-order" environmental effects of agricultural land use are those that directly, and immediately, affect the environment. For example, the expansion of agricultural land immediately and directly diminishes the extent and increases the fragmentation of natural ecosystems, often degrading critical habitats and diminishing biological diversity. Furthermore, agricultural land use can directly impact the ecological functioning of the landscape, especially in terms of water, carbon, and nutrient cycling. These changes in ecosystem processes can directly affect the amount and availability of freshwater flows, the fixation and sequestration of carbon, and the flow of nutrients in an ecosystem.

But the impacts of our land-use practices go far beyond their immediate surroundings, generating "second-order" environmental effects. For example, clearing forests for croplands or pastures turns the carbon in living trees into carbon dioxide—a greenhouse gas in the atmosphere that is contributing to global warming. Changes in land cover also have profound impacts on climate through altering the flows of energy and water from the surface to the atmosphere, thereby changing the atmospheric circulation and the climate system. Furthermore, changes in the hydrologic and nutrient balance of watersheds (a first-order effect) resulting from intensive agricultural practices can lead to downstream problems of water quality and degradation in streams, lakes, wetlands, and coastal areas. Other examples include human health consequences brought about by changes in the ecology of disease organisms and their vectors that may accompany land-cover change.

Below, we briefly describe some of the many effects agricultural land-use practices have on the global environment.

Effects of Agriculture on Terrestrial Ecosystems

Our civilization's growing demand for land comes at the expense of natural ecosystems. Croplands and pastures are established in lands that used to be covered with forests, savannas, and grasslands—so there is a clear, direct relationship between agricultural expansion and ecological degradation.

In terms of biodiversity, it has been shown that agricultural land-use practices have caused significant losses of species, mainly through habitat loss, modification, and fragmentation (Sala et al., 1995). Agricultural land use can also degrade biodiversity through changes in the quality and supply of freshwater resources, the degradation of soil, or the introduction of nonnative species (Sala et al., 1995).

Beyond biodiversity, agricultural land-use practices can strongly influence the structure and functioning of terrestrial ecosystems. One key aspect of how human actions are altering ecosystems is through changes in *productivity*. In a groundbreaking study, Peter Vitousek and colleagues (Vitousek et al., 1986) asked how human practices, including agriculture, affect the terrestrial biosphere by estimating the "human appropriation of net primary productivity" (HANPP). HANPP is defined as the share of the world's biological productivity that is used, managed, or co-opted by human actions. From their analysis, Vitousek et al. estimated that roughly 30% of the planet's terrestrial net primary productivity is appropriated by human actions, largely through agriculture and forestry. This surprising fact—that something like 30% of terrestrial biological production ends up in human hands—is one of the most quoted facts in modern ecological science.

One of the key problems in understanding the human impact on global productivity is first determining how productivity has changed from land-use practices—including productivity *decreases* from landscape degradation (e.g., soil erosion, deforestation) or possible productivity *increases* from agricultural technology (e.g., plant breeding, inputs of fertilizer and irrigation water). On the basis of an analysis of global cropland productivity and patterns of natural ecosystem productivity (Foley et al., 2007), we see major changes in global productivity in many parts of the world. Although much of the terrestrial biosphere experiences a significant decrease in NPP from land-use practices, several regions see a substantial increase in productivity—especially regions that are heavily influenced by irrigation, fertilizer inputs, and tree crop plantations, such as the western United States, the upper midwestern United States, western Europe, northwestern India, northeastern China, and large parts of Indonesia and Malaysia.

It is also interesting to consider the fate of this human-appropriated production—in other words, how are we using ecosystem production, where are the products going, and who is consuming them? Using global trade and agricultural statistics, we can examine the how global crop production is allocated to different human uses, such as food and nonfood uses. Furthermore, we can document the different economic roles of cropland products, including the percentage of crop production used for domestic consumption versus international exports.

Effects of Agriculture on Freshwater Resources

Agricultural land use has also caused significant changes to the quantity and quality of freshwater resources around the world through their impacts on hydrology and nutrient cycles.

To begin, agricultural land-cover change (e.g., converting a landscape from natural vegetation to an agricultural system) can have a major impact on hydrology by altering the amount (and seasonal timing) of key hydrologic processes, including evapotranspiration from the surface, soil moisture storage, water yield into surface and groundwater flows, and the discharge of streams and rivers. Numerous studies have shown how agricultural land use can significantly affect the water balance and freshwater flows of large watersheds across the world.

In addition to these changes in the water balance, many watersheds and aquifers have been heavily affected to withdraw water for irrigation and other agricultural uses. As a result, many large rivers, especially those in semiarid regions, have greatly reduced flows, and some routinely dry up before reaching the ocean. Altogether, agriculture accounts for $\sim$85% of global consumptive water use; and it has been estimated that nearly 50% of the available renewable freshwater supply is currently withdrawn by human activity (Postel et al., 1996).

Furthermore, agricultural land-use practices—especially agricultural chemical use—can dramatically affect freshwater quality over large regions. In particular, nutrient inputs from agriculture—mainly from chemical fertilizers and livestock wastes—now exceed the natural sources to the biosphere and have widespread effects on water quality and coastal and freshwater ecosystems (Matson et al., 1997; Bennet et al., 2001). The resulting degradation of inland waters and coastal ecosystems causes oxygen depletion, fish kills, increased blooms of toxic cyanobacteria, and increased episodes of water-borne disease.

Effects of Agriculture on Atmospheric Composition and Climate

Agricultural land-use practices have also played a critical role in changing the greenhouse-gas composition of the atmosphere and, therefore, the global climate system. In particular, it is estimated that roughly 35% of the world's cumulative CO_2 emissions since 1850 resulted directly from land use, and $\sim$15% of current-day anthropogenic CH_4 emissions come from flooded rice fields (Prentice et al., 2001). In addition, a large portion of global N_2O (another important greenhouse gas) emissions comes from heavily fertilized agricultural fields. The fact that land-use practices constitute a significant portion of our greenhouse gas emissions is often unrecognized. In the future, changing agricultural land-use practices will likely play a significant role in the mitigation of greenhouse gas emissions.

Agricultural land use can also affect the physical climate system directly through changes in surface energy and water balance: changes in land cover strongly affect the physical properties of the land surface and how it interacts with the atmosphere. Replacing forest cover with pasture, for example, reduces the amount of water evaporated back into the atmosphere, leaving less water and energy available to fuel weather systems, large-scale convection, and atmospheric circulation.

Numerous computer modeling studies have shown that changes in land cover can produce large changes in climate over several areas of the world—sometimes larger than the changes in climate expected from greenhouse gas emissions, at least on a regional scale (Foley et al., 2003). It is therefore critical to consider the effects of agricultural land-use change on climate when considering the future behavior of our climate system.

Effects of Agriculture on Human Health

During the course of recent history, agricultural land-use practices have had many positive impacts on human health, largely by enhancing access to nutrition and medicinal products. Nevertheless, land-use practices have also led to many serious, unintended health consequences.

In particular, habitat modification, changing hydrologic conditions, and increased proximity of people and livestock can all modify the transmission of infectious agents and can lead to serious disease outbreaks (Patz et al., 2004). For example, several recent studies have shown that increasing tropical deforestation coincides with an upsurge of malaria and/or its vectors in Africa, Asia, and Latin America (Vittor et al., 2006). Disturbing wildlife habitat for agricultural land use is of particular concern because ~75% of human diseases have links to wildlife or domestic animals. And irrigation in tropical areas often increases the habitat and breeding sites for disease vectors and infectious agents, including schistosomiasis, Japanese encephalitis, and malaria (Patz et al., 2005).

The loss of biological diversity from agricultural land use can also increase the risk of infectious disease. For instance, many agricultural practices promote rodent populations—important reservoirs and vectors of many diseases—by decimating their natural predators and supplementing their food supply. In addition, forest fragmentation in the eastern United States alters biodiversity and predator abundance in ways that promote deer populations, with a subsequent rise in the density of ticks that can carry Lyme disease. Moreover, biodiversity loss favors expansion of mouse populations; because the white-footed mouse is the most competent reservoir host for the bacterium that causes Lyme disease, such land-use change can ultimately increase the risk of contracting Lyme disease (Ostfeld and Keesing, 2000).

Furthermore, the combined effects of agricultural land-use practices and extreme weather events can also have serious impacts, both on direct health outcomes (e.g., injuries and fatalities from storms) and on infectious diseases. For example, Hurricane Mitch, a devastating storm that hit Central America in 1998, demonstrates the combined effects of land use and extreme weather: thousands of people perished, widespread illness from water- and vector-borne diseases ensued, and an estimated 1 million people were left homeless. It has been widely reported that areas with extensive deforestation and poor agricultural practices experienced far greater morbidity and mortality. The southern part of Lempira Province, Honduras, however, escaped Hurricane Mitch with only minor damage and no loss of life, even though it endured some of the most intense rainfall and winds of the hurricane. The practice of sparing large shade trees, and planting crops interspersed underneath, was reported as the key protective factor compared with other regions that experienced disastrous mud slides (Patz et al., 2004).

4. SUMMARY AND CONCLUSIONS

Throughout history, agriculture has played a crucial role in sustaining the health, nourishment, and economy of the world's population. This is especially true today, as human population and consumption continue to grow, increasing our reliance on secure food, fiber, and biofuel supplies. At the same time, many agricultural practices can disturb the environment in ways that affect the quality of ecosystem services and natural resources—including ecosystems, soils, waterways, climate, and even the air we breathe.

Agricultural land-use practices have been, and will continue to be, a major driver of global environmental change, especially in terms of changing key ecosystem services, natural resources, and human health. In fact, it is possible that the environmental effects of land-use practices could rival, and even exceed, the effects of global warming. (Naturally, this does *not* diminish the importance of global warming as a serious scientific and policy issue. Rather, we argue that land-use practices and global warming are *both* crucial global environmental issues and that a balanced science and policy framework will consider the combined, synergistic effects of land use, greenhouse gas emissions, and other major drivers of global environmental change. Fortunately, such frameworks are already emerging from the International Geosphere-Biosphere Programme, the Millennium Ecosystem Assessment, and from many national research programs.)

Tackling the widespread environmental challenges of agricultural land use requires decisionmaking and policy actions that reduce the negative impacts of land-use practices while maintaining the positive societal and economic benefits. Fortunately, there are numerous opportunities for simultaneously improving the environmental and economic benefits of agricultural land use. Examples include precision agricultural techniques that increase production per unit land area, per unit fertilizer applied, and per unit water consumed; land-use practices designed to maintain water flows and water quality, such as the use of buffer strips near sensitive streams and rivers; and agroforestry practices that provide food and fiber but also maintain critical habitats for threatened species. Land-use policies should also aim to enhance the resilience of these critical systems, making them more robust to outside disturbances (such as a new disease, invasive species, or pest) and environmental "surprises" (including sudden changes in climate or water availability).

Developing new land-use strategies that balance immediate human needs with those required for long-term environmental sustainability will be a critical challenge to ecological science in the coming decades. Ultimately, it will require a much tighter collaboration between scientists, policy makers, corporations, and real-world practitioners than we often see today. However, such collaborative ventures offer tremendous potential for better managing our landscapes—and the air, water, and biological diversity on which all life depends—sustainably into the future.

FURTHER READING

Bennett, E. M., S. R. Carpenter, and N. Caraco. 2001. Human impact on erodable phosphorus and eutrophication: A global perspective. BioScience 51: 227–234.

DeFries, R., G. P. Asner, and J. A. Foley. 2006. A glimpse out the window: Landscapes, livelihoods and the environment. Environment 48: 22–36.

DeFries, R. S., J. A. Foley, and G. P. Asner. 2004. Land-use choices: Balancing human needs and ecosystem function. Frontiers in Ecology and the Environment 2: 249–257.

Foley, J. A., M. H. Costa, C. Delire, N. Ramankutty, and P. Snyder. 2003. Green surprise? How terrestrial ecosystems could affect earth's climate. Frontiers in Ecology and the Environment 1: 38–44.

Foley, J. A., R. DeFries, G. P. Asner, C. Barford, G. Bonan, S. R. Carpenter, F. S. Chapin, M. T. Coe, G. C. Daily, H. K. Gibbs, J. H. Helkowski, T. Holloway, E. A. Howard, C. J. Kucharik, C. Monfreda, J. A. Patz, I. C. Prentice, N. Ramankutty, and P. K. Snyder. 2005. Global consequences of land use. Science 309: 570–574.

Foley, J. A., C. Monfreda, N. Ramankutty, and D. Zaks. 2007. Our share of the planetary pie. Proceedings of the National Academy of Sciences U.S.A. 104: 12585–12586.

Matson, P. A., W. J. Parton, A. G. Power, and M. J. Swift. 1997. Agricultural intensification and ecosystem properties. Science 277: 504–509.

Monfreda, C., N. Ramankutty, and J. A. Foley. 2008. Farming the planet. Part 2: The geographic distribution of crop areas, yields, physiological types, and NPP in the year 2000. Global Biogeochemical Cycles 22: GB1022, doi:10.1029/2007GB002947.

Ostfeld, S. R., and F. Keesing. 2000. Biodiversity and disease risk: The case of Lyme disease. Conservation Biology 14: 722–728.

Patz, J. A., and U.E.C. Confalonieri (convening lead authors), F. Amerasinghe, K. B. Chua, P. Daszak, A. D. Hyatt, D. Molyneux, M. Thomson, L. Yameogo, M. Malecela-Lazaro, P. Vasconcelos, and Y. Rubio-Palis. 2005. Human health: Ecosystem regulation of infectious diseases. In Millennium Ecosystem Assessment. Ecosystems and Human Well-Being: Current State and Trends. Findings of the Condition and Trends Working Group, Millennium Ecosystem Assessment Series. Washington, DC: Island Press.

Patz, J. A., P. Daszak, G. M. Tabor, A. A. Aguirre, M. Pearl, J. Epstein, N. D. Wolfe, A. M. Kilpatrick, J. Foufopoulos, D. Molyneux, D. J. Bradley, and members of the Working Group on Land Use Change and Disease Emergence. 2004. Unhealthy landscapes: Policy recommendations on land use change and infectious disease emergence. Environmental Health Perspectives 112: 1092–1098.

Postel, S. L., G. C. Daily, and P. R. Ehrlich. 1996: Human appropriation of renewable fresh water. Science 271: 785–788.

Prentice, I. C., G. Farquhar, M. Fashm, M. Goulden, M. Heimann, V. Jaramillo, H. Kheshgi, C. L. Quéré, and R. J. Scholes. 2001. The carbon cycle and atmospheric carbon dioxide. In J. T. Houghton, Y. Ding, D. J. Griggs, M. Noguer, P. J. van der Linden, X. Dai, K. Maskell, and C. A. Johnson, eds. Climate Change 2001: The Scientific Basis. Contribution of Working Group I to the Third Assessment Report of the Intergovernmental Panel on Climate Change. Cambridge, UK: Cambridge University Press, 183–237.

Ramankutty, N., A. T. Evan, C. Monfreda, and J. A. Foley. 2008. Farming the planet. Part 1: The geographic distribution of global agricultural lands in the year 2000. Global Biogeochemical Cycles 22: GB1003, doi:10.1029/2007GB002952.

Ramankutty, N., and J. A. Foley. 1999. Estimating historical changes in global land cover: Croplands from 1700 to 1992. Global Biogeochemical Cycles 13: 997–1027.

Ramankutty, N., J. A. Foley, J. Norman, and K. McSweeney. 2002. The global distribution of cultivable lands: Current patterns and sensitivity to possible climate change. Global Ecology and Biogeography 11: 377–392.

Ramankutty, N., J. A. Foley, and N. J. Olejniczak. 2002. People on the land: Changes in global population and croplands during the 20th century. Ambio 31: 251–257.

Sala, O. E., F. Stuart Chapin III, Juan J. Armesto, Eric Berlow, Janine Bloomfield, Rodolfo Dirzo, Elisabeth Huber-Sanwald, Laura F. Huenneke, Robert B. Jackson, Ann Kinzig, Rik Leemans, David M. Lodge, Harold A. Mooney, Martín Oesterheld, N. LeRoy Poff, Martin T. Sykes, Brian H. Walker, Marilyn Walker, and Diana H. Wall. 2000. Global biodiversity scenarios for the year 2100. Science 287: 1770–1774.

Vitousek, P. M., P. R. Ehrlich, A. H. Ehrlich, and P. A. Matson. 1986. Human appropriation of the products of photosynthesis. BioScience 36: 368–373.

Vittor, A. Y., R. H. Gilman, J. Tielsch, G. E. Glass, T. M. Shields, W. Sanchez-Lozano, V. V. Pinedo, and J. A. Patz. 2006. The effects of deforestation on the human-biting rate of *Anopheles darlingi*, the primary vector of falciparum malaria in the Peruvian Amazon. American Journal of Tropical Medicine and Hygiene 74: 3–11.

VII.8

The Ecology, Economics, and Management of Alien Invasive Species

Ryan Chisholm

OUTLINE

Biological invasions have been a feature of global ecology since the origin of life: plants and animals invaded the land from the sea, and chance dispersal events have occasionally allowed species to invade new continents, islands, or bodies of water. The current wave of biological invasions is qualitatively different from these prehistoric invasions because it is mediated by human activities. It is also quantitatively different because the frequency of invasions is orders of magnitude higher than background levels.

GLOSSARY

alien invasive species. An alien species that becomes established in an ecosystem and threatens native biological diversity or has other negative ecological and economic impacts.

alien (equivalently: nonnative, nonindigenous, foreign, exotic) species. A species, subspecies, or lower taxon occurring outside its natural range (past or present) and dispersal potential (i.e., outside the range it occupies naturally or could occupy without direct or indirect introduction or care by humans); includes any part, gamete, or propagule of such species that might survive and subsequently reproduce.

ecosystem services. The conditions and processes through which ecosystems, and the species that make them up, sustain and fulfill human life.

introduction. The movement, by human agency, of a species, subspecies, or lower taxon (including any part, gamete, or propagule that might survive and subsequently reproduce) outside its natural range (past or present). This movement can be either within a country or between countries.

1. INTRODUCTION

Before mass migrations of humans across the globe, natural dispersal of plants and animals was restricted by geographic barriers such as oceans, mountain ranges, and deserts. These barriers to migration have been lowered by human activity. The current wave of biological invasions began at the end of the Quaternary glacial period as humans began to disperse across the globe. The process began in earnest with the age of exploration in the fifteenth century and subsequent European colonization of the New World. Biological invasions accelerated rapidly in the twentieth century with the advent of international shipping and aviation, the construction of highways, and the destruction of large swathes of natural habitat. Today, virtually every location on the planet, from mountain peaks to remote oceanic islands, has recently been invaded by species that originated elsewhere. Invasive species occur in all taxonomic groups, from mammals to fungi to viruses.

Charles Darwin, on his *Beagle* voyage, was perhaps the first scientist to observe and note the process of biological invasion. Many of these observations led to insights that were incorporated into *The Origin of Species*. Charles Elton's seminal book, *The Ecology of Invasions by Animals and Plants*, was published in 1958 and is considered the classic text of invasion biology. The field of invasion biology burgeoned in the latter decades of the twentieth century and is now the subject of numerous books and several specialized journals.

In this review, I first examine which species invade which habitats and address how and why these invasions occur. I next investigate models of how invasive species spread and then discuss the ecological and economic impacts of invasive alien species across the globe. Finally, I discuss management and policy options for preventing unwanted species introductions and for mitigating invasions when they do occur.

2. WHICH SPECIES INVADE WHICH HABITATS AND WHY?

Perhaps the most infamous examples of biological invasions have resulted from deliberate introductions. In the western United States, saltcedars (*Tamarix* spp.), which choke out native vegetation, elevate soil salinity, and reduce river flow, were originally introduced as garden ornamentals. In eastern North America, the gypsy moth (*Lymantria dispar*), which causes massive defoliation of trees during periodic outbreaks, was deliberately introduced to investigate its potential for silk production in the nineteenth century. In Australia, cane toads (*Bufo marinus*) were originally imported to control pests of sugar cane in the 1930s but rapidly spread across the continent, poisoning pets and native wildlife that attempted to eat the toads and consuming smaller native species.

Accidental introductions are also the source of many biological invasions. Examples are zebra mussels (*Dreissena polymorpha*), which were introduced to the North American Great Lakes via ballast water in the late 1980s, and black rats (*Rattus rattus*), which have invaded numerous oceanic islands as ship stowaways. Historically, most plant and vertebrate animal invasions have resulted from deliberate introductions, whereas invertebrate animal and microbe invasions have resulted from accidental introductions. Exceptions include bumblebees (*Bombus* spp.) in New Zealand, which were deliberately introduced for crop pollination, and the sea lamprey (*Petromyzon marinus*), which was accidentally introduced into the North American Great Lakes where it parasitizes native fish species.

Although thousands of species have successfully invaded different habitats across the world, this represents only a small fraction of the species that are introduced or have the opportunity to invade. According to the "tens rule" of invasion biology, approximately 1 in 10 species that are imported will appear in the wild (become introduced), 1 in 10 of these introduced species will become established, and only 1 in 10 of these established species will actually become invasive. The tens rule encapsulates crudely what is observed in many statistical patterns of species invasions. A deeper treatment of the problem of invasions considers the factors that make particular alien species more successful invaders and particular habitats more invasible.

An overwhelming factor in the success or failure of biological invasions is propagule pressure. This conforms to intuition: species that are introduced in larger numbers are more likely to become successful invaders; and habitats that are subjected to a greater seed rain or immigration rain from potential invaders are more likely to become invaded. Studies of species invasions ought always to be set in this context because of the potential for propagule pressure to confound other factors. For example, the contrast between the high prevalence of invasive species near roads, seaports, and airports and the low prevalence of invasive species in nature reserves is at least partly explained by differences in propagule pressure. Similarly, large-flowered plants are often overrepresented among invasive taxa, not necessarily because they are inherently more invasive but because they are disproportionately selected for import by horticulturalists and therefore exert greater propagule pressure in their introduced range than small-flowered species.

The observation that many species have become successful invaders in multiple parts of the world following independent introduction events suggests that there is more to invasions than propagule pressure and random chance. Indeed, several studies have found the best predictor of invasiveness among a given group of species to be simply a previous history of invasions or widespread distribution elsewhere. This motivates the search for ecological traits associated with invasiveness. A central goal is the development of tools that predict which species will become invasive, although any such tools are unlikely to be perfect simply because highly invasive species often have physiologically similar noninvasive congeners.

Traits associated with invasive species include fast growth, generalist resource use, high reproductive output, high physiological tolerance, asexual reproduction, and long-distance dispersal capability. Invasive species are also often taxonomically distant from species in the invaded habitat. Examples of successful generalist invaders include feral cats (*Felis catus*) and brown tree snakes (*Boiga irregularis*), both of which are predators that have invaded many habitats globally that were naive to these taxonomic groups. Further insights come from taxonomic groups containing many species that have been introduced outside their natural range. For example, an analysis of the different *Pinus* species introduced across the globe revealed that dispersal ability, competitive ability, and adaptability to varying disturbance regimes explain almost all of the variation in invasiveness among species.

The most straightforward hypothesis to explain the invasiveness of particular alien species is the enemy release hypothesis. In their native ranges, populations of plants and animals are kept in check by a suite of predators, herbivores, and pathogens. When a species is introduced outside its native range, it will often escape the biotic constraints imposed by these enemies. This gives the invader a fitness advantage because it can reallocate resources from enemy defense to growth and reproduction. For example, the Australian brushtail possum (*Trichosurus vulpecula*) is a major pest in New Zealand, where it occurs at 10 times the densities seen in Australia, has fewer competitors and predators than in its home range, and has no microparasites and only 20% the number of macroparasites. The spectacular success of certain biological control programs (see section 4, below) also provides strong evidence for the enemy release hypothesis.

Although the enemy release hypothesis is certainly appealing as an explanation of alien invasions, empirical evidence for it is mixed. An observation that an alien species exhibits increased vigor and fecundity and lacks enemies from its home range is not sufficient to establish a causal relationship. Moreover, many alien species rapidly acquire enemies in their new range, especially as they spread and encounter a wider range of native species. The enemy release hypothesis may be especially applicable to specialist alien plants that have evolved chemical defenses against specialist herbivores in their home ranges—such species are less likely to acquire new enemies in the invaded range.

The complement to the question of which species are likely to invade is the question of which habitats are likely to be invaded. Characteristics of invasible habitats are geographic isolation, the occurrence of anthropogenic or natural disturbance, and high availability and quality of food, light, and other resources. That isolated habitats are more vulnerable to invasion is evidenced by the proliferation of alien species on oceanic islands such as Hawaii and New Zealand and in freshwater systems such as North America's Great Lakes. Today, alien plant richness on islands is often equal to native plant richness, whereas on continents alien plant richness is typically about 20% of native plant richness. Disturbances that facilitate invasions include changes in fire and hydrological regimes, changes in nutrient levels, and changes in grazing regimes. Extensive plant invasions across grasslands in Australia and the Americas have been linked to changes in the fire regime.

An unresolved question in invasion biology is whether high native species richness facilitates or inhibits invasion by alien species. The "invasion paradox" is that independent lines of research support both negative and positive relationships between native species richness and invasibility. Resolution of the invasion paradox requires the synthesis of data from different types of study across a broad range of spatial scales: small-scale observational and experimental studies usually report negative relationships between native and exotic species richness, whereas most larger-scale studies are observational and report positive relationships. An explanation for a positive relationship at large scales is that large areas exhibit greater spatial heterogeneity and hence greater habitat diversity, and that such regions tend to support both more native and more alien species. Notably, theoretical models have demonstrated that the different native/alien diversity patterns at different scales would be expected as statistical artifacts even in the absence of any species differences.

3. SPREAD OF ALIEN INVASIVE SPECIES

Once invasions are under way, an understanding of the spatial and temporal propagation of invasive species is essential for predicting their impacts and designing effective management programs. The spread of an invasion is governed by many variables, including initial population size, the age structure and breeding system of the invading species, and characteristics of the invaded environment. Despite these complications, mathematical models of the spread of invasive species have been one of ecological modeling's great success stories.

In 1951, J. G. Skellam developed the now-classic reaction–diffusion model of invasion biology, which describes the dynamics of a population that is both growing and spreading. Skellam's model predicts that the front of an invasion should move at a constant velocity and has been successfully applied to case studies such as muskrats (*Ondatra zibethicus*) invading central Europe and sea otters (*Enhydra lutris*) recovering from near extinction along the coast of California.

Invasions that do not fit the Skellam model have motivated the development of extended models such as stratified diffusion models (which include occasional long-distance dispersal events), advection–reaction–diffusion models (where advection accounts for species that tend to drift in a particular direction, perhaps because of wind or water flow) and models that allow for environmental heterogeneity (such as rivers and mountain ranges). An important insight from the stratified diffusion models is that occasional long-distance dispersal events can dominate a species' spread rate. The stratified diffusion models predict that the front of an invasion should move at an accelerating speed, and these models generally provide better fits than the basic

Skellam model to data of invasive plant spread. The dynamics of systems in which the invading species is a predator or competitor of a native species have been successfully modeled using variations of the classic Lotka-Volterra predator–prey and competition models.

A common feature of invasions is an initial lag phase during which the species persists at low densities. This precedes a phase of rapid growth during which the species is recognized as a problem. Although some invasive species, such as zebra mussels, have exhibited only a brief lag phase or none at all, others may persist at low densities for decades before becoming abundant. An understanding of this feature of the spread of invasions is essential for identifying potential invaders early on and developing effective management strategies. There are several possible explanations for this lag phase. The first explanation, consistent with the basic Skellam model, is that an invading species may remain below detection thresholds even though its population is growing. A second explanation, consistent with the stratified diffusion model, is that invasions often begin from only a single introduced population, whereas species spread more rapidly from multiple foci. A third explanation is that invading species may require time to adapt genetically or behaviorally to environmental conditions in the new habitat. A fourth explanation is that an invading species may require a "window of opportunity" associated with favorable environmental conditions that allows it to grow above a critical threshold beyond which more rapid growth can occur.

4. IMPACTS OF ALIEN INVASIVE SPECIES

By definition, alien invasive species are associated with negative ecological and economic impacts, but any discussion of these negative impacts must be set in the context of benefits provided by alien species. Agricultural, pastoral, horticultural, and forestry industries across the world depend heavily or entirely on alien species. Alien species provide more than 98% of the food grown in the United States. These economic benefits have been the primary motivation for past deliberate introductions of alien species, only a small proportion of which have actually become invasive.

Ecological and Evolutionary Impacts

The ecological impacts of alien invasive species can be grouped into three categories: drivers of extinction, modifiers of ecosystem processes, and modifiers of evolution. Perhaps the most widely recognized ecological impact of alien invasive species is the extinction and decline of native species following the introduction of new predators, competitors, and pathogens. In Guam, the introduced brown tree snake has been blamed for the extinction of over 10 native bird species. In East Africa's Lake Victoria, the introduction of the predatory Nile perch (*Lates niloticus*) contributed to the extinction of about 200 species of native cichlid fish. In Hawaii, the introduction of avian malaria continues to be a major factor behind the collapse of the local avifauna. The American chestnut (*Castanea dentata*) was once a dominant tree in the forests of eastern North America but was virtually exterminated in the early twentieth century by chestnut blight, a disease caused by an introduced fungus. In Australia, introduced red foxes (*Vulpes vulpes*) have been implicated in the extinctions of 10 to 15 native mammal species. Troublingly, it is likely that many parts of the world are currently in "extinction debt" because of invasive species, meaning that invasive species have driven some native species below a minimum long-term viable population size, and further extinctions can be expected even in the absence of further invasions.

At a global scale, alien invasions are undoubtedly contributing to a decline in species diversity and the homogenization of the Earth's biota: a few successful invasive species, such as black rats and rock doves (*Columba livia*), have proliferated across the world at the expense of many localized endemic species. At local and regional scales, however, invasions have often led to increases in diversity: in states of the United States and Australia, the average plant species diversity has increased by about 20%; on oceanic islands average plant species diversity has increased by about 100%. Similar or greater increases in diversity are also observed at higher taxonomic levels.

Although there are many examples of extinctions caused by alien predators and pathogens, there are far fewer examples of extinction caused by competition from alien species. In Britain, the decline of the native red squirrel (*Sciurus vulgaris*) is often blamed on competition with the introduced eastern gray squirrel (*Sciurus carolinensis*) but may be driven more by a disease carried by the introduced squirrel. Other cases of competition from alien species causing extinctions have occurred over long time scales, suggesting that the competitive effects are relatively weak. For instance, in Australia, competition with the dingo (*Canis lupus dingo*) contributed to the extinction of the thylacine (*Thylacinus cynocephalus*) before European settlement, but the process apparently took several hundred years.

An important observation, which has implications for management (see section 5, below), is that the relationship between alien invasions and extinctions is often correlative rather than causative. Causal inference is confounded by other threatening processes. For instance, feral pigs and alien plants are both blamed for the decline of native plants in Hawaii, but the proliferation of alien plants may be merely a secondary outcome of disturbance by pigs. In Florida's Everglades, extensive alien tree invasions are at least partly to blame for declines in native biodiversity, but both of these processes are outcomes of changes to the region's hydrology that resulted from a water-diversion project.

Thorny conservation issues arise when alien invasive species are themselves endangered in their home range. In such cases, the goal of conservation at a global level may conflict with goals of eradication and control at a local level. For example, the world's largest population of wild banteng (*Bos javanicus*) occurs in Australia, a country to which it was introduced and where it is considered by many to be a pest. In some cases win-win situations are possible: animals from a population of introduced tammar wallabies (*Macropus eugenii*) in New Zealand were repatriated to Australia after they were found to be descendants of an extinct Australian population.

Less widely acknowledged, but perhaps of greater concern, are the impacts of alien invasive species on ecosystem-level processes such as nutrient cycling, fire regimes, siltation rates, and hydrology. Particularly large alterations may be caused by the introduction of species with novel physiological traits, such as nitrogen fixers. In Hawaii, the invasion of nitrogen-limited ecosystems by the nitrogen-fixing alien tree *Myrica faya* has altered ecosystem development by increasing nitrogen inputs to ecosystems by about a factor of four. In South Africa, invasion by alien trees (*Acacia*, *Hakea*, and *Pinus* spp.) has converted native Fynbos shrublands into woodlands, accelerated nitrogen input and nitrogen cycling, reduced stream flows, and modified the fire regime to one that is characterized by less frequent, more severe fires.

By altering environmental conditions, some invaders facilitate further invasion or prevent the reestablishment of native communities. In the western United States, invasive cheatgrass (*Bromus tectorum*) completely alters ecosystems by increasing fire frequency to the point where native shrub-steppe communities cannot recover. In Hawaiian wet tropical forests, leaf-litter decay rates of alien understory species are, on average, several times greater than those of native understory species, suggesting that alien invasion accelerates nutrient cycling and further facilitates the invasion of alien plants adapted to nutrient-rich soils. Such circumstances pose difficulties for management because the invaders have, in effect, moved the ecosystem into an alternate stable state. Perhaps most worrying is the potential for widespread invasions to transform ecosystems and create feedbacks that influence regional or even global processes such as climate.

Ultimately, the most profound impacts of biological invasions may be their effects on evolution, which occur through three mechanisms: evolutionary diversification of alien species in new environments; evolutionary adaptation of native species to altered ecological conditions; and hybridization between previously allopatric taxa.

Alien species tend to undergo genetic drift in new environments because their populations are isolated from the source population and contain only a fraction of the genetic material (the founder effect). This, combined with directional selection imposed by new ecological challenges and freedom from previous ecological constraints, provides opportunities for evolutionary innovation among alien species. The codling moth (*Cydia pomonella*), a pest of fruit crops, invaded North America in about 1750 and has since evolved into genetically distinct races that specialize on apple, plum, and walnut. Studies of the fruitfly *Drosophila subobscura* suggest that it has diversified along environmental gradients in its introduced range in the New World and that the genetic basis of this diversification is different from that along similar environmental gradients in its native Old World range.

The altered environmental conditions imposed by alien invasive species may also promote directional selection in native species. For example, native black snakes (*Pseudechis porphyriacus*) in Australia have evolved resistance to the toxins of invasive cane toads. Alien invasion can even promote diversification of native species. The most striking examples of this are herbivorous insects that have evolved distinct ecotypes to feed on alien plants. Diversification of native species can also occur via allopatric speciation in cases where only some populations of native species are invaded.

Hybridization occurs more rapidly than directional selection and diversification and provides the most compelling examples of evolutionary change associated with alien invasions. The most commonly observed impact of hybridization on biodiversity is negative: the loss of distinct endemic species. Both the New Zealand Gray Duck (*Anas superciliosa*) and the Hawaiian Duck (*A. wyvilliana*) are at risk of extinction through hybridization with the introduced North American Mallard (*A. platyrhynchos*). But hybridization can also have positive effects on biodiversity. A new, reproductively isolated species of cordgrass (*Spartina anglica*) evolved in England in the nineteenth century from a hybrid of one native and one exotic cordgrass. A recent case study of hybridization in fruit flies (*Rhagoletis* spp.) demonstrates that this phenomenon is also possible in animals.

Although this review has focused mostly on the negative impacts of alien invasions, a longer-term view reveals that the speciation processes facilitated by invasions will at least partly offset the biodiversity losses that we are currently observing. Another positive aspect of alien invasions is that they provide model systems for addressing basic research questions in ecology and evolutionary biology. Alien invasions can be viewed as experiments, albeit uncontrolled and imperfectly replicated, that would be unfeasible or unethical across the large range of spatial and temporal scales over which they occur. Examples of insights that have stemmed from studies of alien invasions are that species are not optimally adapted for their environment, that communities are usually not saturated with species, and that reproductive isolation can take millions of years.

Economic Impacts

The economic impacts of alien invasive species can be separated into damages and the costs of control. Alien invasive species are associated with economic damages to infrastructure, crops, pastures, livestock, forestry, fisheries, and human health. To the extent that environmental values can be quantified monetarily, economic damages are also associated with damages to ecosystem services, which include water supply, pollination, and the provision of recreational opportunities.

The annual economic costs of alien invasive species have been estimated at US$120 billion in the United States alone. To provide a context for these costs, alien species contribute about US$800 billion to the U.S. food system annually. About 80% of the costs in the U.S. study are attributable to a few groups of alien invaders: pests and pathogens of crop plants; crop weeds; introduced rats, which consume grain and other food intended for human consumption; feral cats, which eat native birds that have an associated recreational value; and introduced diseases of livestock and humans. Two other notable invaders of the United States, each of which is associated with roughly US$1 billion in annual economic costs, are the zebra mussel, which clogs water pipes and other equipment, and the red imported fire ant (*Solenopsis invicta*), which impacts wildlife, livestock, and public health.

In many cases, the high economic costs of alien invasive species can justify spending on alien species control programs. The U.S. experience with red imported fire ants prompted Australian government agencies to launch a fire ant eradication campaign that has spent over US$100 million since fire ants were detected in Queensland in 2001. In South Africa, cost-benefit analyses of mountain catchment areas invaded by alien trees demonstrated that the costs of controlling alien trees were less than the projected benefits of increased water flows from catchments—across the country, approximately 3 billion cubic meters of water is lost annually to alien trees. These analyses motivated the South African government's Working for Water Program, which has spent more than US$400 million controlling alien trees since its inception in 1996.

Such cost-benefit analyses, although useful, are plagued by uncertainty in the underlying economic and biological data. Moreover, they can only provide a lower-bound estimate of the costs associated with alien invasive species, because it is difficult or even impossible to attach a dollar value to certain impacts, such as the loss of biodiversity or the depletion of the aesthetic values of natural areas. Thus, failure to demonstrate cost efficacy of alien species control does not mean that a proposed program would not be beneficial from the standpoint of overall societal welfare.

5. MANAGEMENT OF ALIEN INVASIVE SPECIES

Prevention of unwanted introductions is the most cost-effective method of dealing with the invasive species problems. For a relatively small investment in quarantine and screening procedures, government agencies can reduce accidental and unwanted deliberate introductions. As discussed earlier, predicting invasiveness based on species traits is a difficult task, and this limits the effectiveness of quarantine procedures. Several countries presently use an "innocent until proven guilty" approach, whereby only species on a list of known offenders are prohibited entry. A "guilty until proven innocent" approach, placing the burden of proof on the importer, would obviously be more effective at excluding invasive species. However, because only a small proportion of introduced species actually become invasive, such screening systems are inevitably plagued by false positives. The issue of screening policies incites conflict among environmental groups, agricultural agencies, free-trade advocates, and commodity importers such as horticulturalists, fish and game agencies, and the pet industry. A step toward resolving this problem would be the introduction of legislation holding the importer of an alien species responsible for any damages caused if the species becomes invasive. Similar considerations are also pertinent to the debate over the release of genetically modified organisms into the environment.

Once a species has been introduced and become invasive, the goal becomes control, containment, or eradication. Eradications are easiest and most cost effective early on in the process of an invasion, before an alien species has become widespread or abundant. The problem with this is that the vast majority of alien species do not become serious problems, and it is unclear which alien species should be the focus of early eradication programs. There are few examples of alien species that have been successfully eradicated once an invasion is well established. Exceptions are the successful 50-year campaign to eradicate the American nutria (*Myocastor coypus*) from Britain, the successful eradication of introduced mammals from several offshore islands in New Zealand, and the apparent success of the Australian government's expensive fire ant eradication program. Factors contributing to successful eradication programs are the use of specific knowledge of the target organism's biology and the application of sustained funding and control efforts even after the immediate ecological and economic threats have been alleviated. When eradication is not feasible, control (maintaining the alien species at acceptably low densities) or containment (restricting the geographic distribution of the alien species) may still deliver economic and ecological benefits.

Strategies for the management of alien plants include chemical control, mechanical control, and biopesticides. Chemical control is widely used for pest control in agriculture and forestry but may have adverse environmental and public health impacts. The cost of such campaigns can escalate when repeated application is required or when the target species evolves resistance to the chemical. Moreover, campaigns involving chemicals often arouse considerable public resistance. Mechanical control arouses less public resistance and can often be effective, especially for large woody plants, but is difficult or impossible for widespread species.

Strategies for the management of alien animals include poison baiting, trapping, and shooting. Poison baiting has been particularly effective on introduced mammals. The Western Shield operation in Western Australia targets introduced foxes with baits containing a poison to which many native mammals are resistant. This program has led to the resurgence of native mammal species and even the removal of three species from the endangered list. However, as with chemical control of plants, poison baiting can impact nontarget species and engender public opposition. Trapping and shooting can be effective for alien animal species with restricted distributions. Again, public opposition is a factor here, as evidenced by outcry over feral horse control in the United States, Australia, and New Zealand. Recreational hunting may help maintain feral animal populations at acceptable densities, but it is generally ineffective as a means of eradication.

Classical biological control, the deliberate introduction of predators or pathogens can be particularly effective for regulating populations of invaders. Although only about 30–40% of biological controls on weeds and 10–15% of biological controls on arthropods are successful, the net benefit of biological control programs is positive because some are so spectacularly successful. On St. Helena, the gumwood tree (*Commidendrum robstum*) was threatened with extinction by an alien herbivorous insect (*Orthezia insignis*) but saved by the deliberate introduction of a predatory insect (*Hyperaspis pantherina*). In Australia, dense infestations of prickly pear cactus (*Opuntia* spp.) covered an area the size of the British Isles until the introduction of the cactoblastis moth (*Cactoblastis cactorum*) in the 1920s. In South Africa, the introduction of herbivores and pathogens of alien trees has had some success: an introduced gall-rust fungus of *Acacia saligna* has reduced stem densities by up to 98%.

Other forms of biological control include the release of sterile individuals into a population of an alien species or the introduction of novel genetic material. In the United States, the screwworm fly (*Cochliomyia hominivorax*) was successfully eradicated by introducing sterile males into the population. In Australia, current research efforts seek to control the European carp (*Cyprinus carpio*) by releasing transgenic individuals carrying a "daughterless gene."

There is rising concern about the potential impacts of biological control on nontarget species. The small Indian mongoose, released onto numerous islands across the world to control introduced rats, has contributed to the decline of numerous native bird species. In Hawaii, the introduction of a predatory snail (*Euglandina rosea*) to control the giant African snail (*Achatina fulica*) led to the extinction of numerous native snail species. The myxomatosis virus from South America was used successfully to control introduced European rabbits (*Oryctolagus cuniculus*) in Australia in the 1950s; however, the virus subsequently spread to Europe, where it devastated native rabbit populations and thereby endangered species that prey on rabbits, such as the Spanish lynx. Even biological control agents that have been subject to stringent host-specificity testing have attacked nontarget species: a Eurasian weevil, *Rhinocyllus conicus*, introduced for the control of an invasive thistle in North America, has begun to attack native thistles. Furthermore, host-specificity testing cannot guard against indirect impacts of alien species on, for example, food chains or virus reservoirs. Although the likelihood of biological control agents having unintended consequences may be small, the magnitude of these consequences may be large, and so this possibility needs to be incorporated into cost-benefit analyses of biological control programs.

Habitat or ecosystem management is often prescribed as a holistic approach to alien invasive species problems. Such approaches target the overall condition of the ecosystem rather than individual alien species by focusing on processes such as native vegetation restoration, fire regimes, grazing regimes, and nutrient inputs. This can be seen as a shift from the symptoms of environmental problems to the fundamental causes. Numerous studies have shown that fertilization and disturbance promote alien plant invasion in grassland ecosystems. Success of native grassland restoration projects depends on the reduction of nutrient loads, the introduction of appropriate fire regimes, the existence of sufficient propagules of native species, and the control of alien species. In South Africa, alien plant management programs use a multispecies approach that incorporates numerous control methods and acknowledges the need for native vegetation restoration and appropriate fire regimes.

Management of alien invasive species becomes problematic when invasion is attributable to global change processes. Research on the impacts of global climate change has shown that some invasive species such as cheatgrass and kudzu (*Pueraria lobata*) respond positively to elevated carbon dioxide and that other species respond positively to elevated temperature and rainfall. Although habitat management and other control methods can still be used to manage these species, it is unclear what the desirable end state should be if external conditions, such as climate and nitrogen deposition rates, have changed.

6. CONCLUSIONS

Biological invasions have occurred throughout Earth's history, but the current wave of anthropogenic invasions is occurring on an unprecedented scale. The small proportion of introduced alien species that become invasive cause massive ecological and economic damage. Ecological theories of invasions give us some guidance as to which species are likely to invade, how they will spread, and how they can be controlled or eradicated. The most effective method of managing the global biological invasion would be to strengthen international quarantine regulations. Once a species has already invaded, eradication is sometimes possible but difficult; more commonly, the invasion will necessitate ongoing control costs. Cost-benefit analyses can help determine which invasive species are worth controlling and where conservation funds can be directed. In conducting such cost-benefit analyses, it should be acknowledged that part of the problem with invasive species, as with other environmental issues, is that ecological values cannot easily be translated into dollar values.

The news on alien species is, however, not all bad. Large sectors of our economies are based on products derived from introduced alien species, most of which are not invasive. In many cases, alien invasions are symptoms of other problems such as climate change, changes to the nutrient cycle, changes to hydrology, and habitat clearance. Historically, alien invasions have increased local biodiversity at the expense of global biodiversity and heterogeneity, but over longer time scales, evolutionary diversification promoted by alien invasions may at least partly compensate for this lost biodiversity. Furthermore, alien invasions provide unparalleled opportunities for understanding the forces that shape ecological communities.

Invasion by alien species is now, along with climate change, habitat clearance, and changes to the nitrogen cycle, a major global change process. The biological communities of the future will likely be assembled from collections of species that originated in different corners of the globe and are able to adapt to the new environmental conditions. Understanding how alien invasions will interact with other global change processes to shape these communities is a fundamental challenge for ecologists.

A final point is that there is a strong geographic bias in research on biological invasions: most studies are of invasions in North America, Western Europe, South Africa, Australia, and New Zealand. These areas have probably historically been more prone to invasion because of their economic history and trading patterns, but the science of biological invasions would benefit from more research into invasions in the tropics and other understudied regions.

FURTHER READING

Cox, G. W. 2004. Alien Species and Evolution: The Evolutionary Ecology of Exotic Plants, Animals, Microbes, and Interacting Native Species. Washington, DC: Island Press. *A comprehensive review of the interactions between alien invasions and evolution.*

Elton, C. S. 1958. The Ecology of Invasions by Animals and Plants. Chicago: University of Chicago Press. *Charles Elton's classic book.*

Fridley J. D., J. J. Stachowicz, S. Naeem, D. F. Sax, E. W. Seabloom, M. D. Smith, T. J. Stohlgren, D. Tilman, and B. Von Holle. 2007. The invasion paradox: Reconciling pattern and process in species invasions. Ecology 88: 3–17. *A review of the observational, experimental, and theoretical evidence describing the invasion paradox.*

Mack, R. N., D. Simberloff, W. M. Lonsdale, H. Evans, M. Clout, and F. A. Bazzaz. 2000. Biotic invasions: Causes, epidemiology, global consequences, and control. Ecological Applications 10: 689–710. *A review of invasive species with a global perspective.*

Mooney, H. A., R. N. Mack, J. A. McNeely, L. E. Neville, P. J. Schei, and J. K. Waage, eds. 2005. Invasive Alien Species: A New Synthesis. Washington, DC: Island Press. *An edited volume reviewing the ecology, economics, and management of biological invasions.*

Pimentel, D., L. Lach, R. Zuniga, and D. Morrison. 2000. Environmental and economic costs of nonindigenous species in the United States. Bioscience 50: 53–65. *An analysis of the costs of alien invasive species in the United States.*

Sax, D. F., J. J. Stachowicz, and S. D. Gaines, eds. 2005. Species Invasions: Insights into Ecology, Evolution and Biogeography. Sunderland, MA: Sinauer Associates. *An edited volume examining the scientific insights provided by alien invasions.*

Shigesada, N., and K. Kawasaki. 1997. Biological Invasions: Theory and Practice. New York: Oxford University Press. *A clear and accessible introduction to mathematical models of biological invasions.*

Trends in Ecology and Evolution. 2005. Volume 20, issue 5. *A special issue on invasions including articles on the management and risk assessment of invasions, and the role of propagule pressure in explaining invasions.*

Trends in Ecology and Evolution. 2007. Volume 22, issue 9. *An issue containing several articles on biological invasions, covering biological control, the evolutionary impacts of alien species, and the interaction of habitat modification and species invasions.*

Internet Sources

IUCN. Invasive species specialist group's global invasive species database. Download from http://www.issg.org/database/. *A database that facilitates information sharing on invasive species internationally.*

IUCN. Guidelines for the prevention of biodiversity loss caused by alien invasive species. Download from http://www.issg.org/infpaper_invasive.pdf. *Information intended to assist governments and management agencies in preventing the introduction of alien species, and controlling or eradicating those that become invasive.*

VII.9

Ecological Economics: Principles of Economic Policy Design for Ecosystem Management

Anastasios Xepapadeas

OUTLINE

1. Introduction
2. Ecological modeling and resource dynamics
3. Economic modeling for ecosystem management
4. Instruments of economic policy and policy design

Ecological economics studies the interactions and coevolution in time and space between ecosystems and human economies. The rate at which humans exploit or harvest ecosystems services exceeds what might be regarded as a desirable level from society's point of view. The consequences of this overexploitation are well known (e.g., climate change, biodiversity loss and extinction of species, collapse of fisheries, overexploitation of water resources). The objective of designing economic policy is to develop a system of regulatory instruments so that the state of the regulated ecosystems will converge toward the socially desirable outcome. The purpose of this chapter is to present an approach describing how economic policies might be designed to achieve this objective.

GLOSSARY

control variable. A variable whose values can be chosen by a decision maker in order to affect the path of the state variables.

ecological economics. The study of the interactions and coevolution in time and space between ecosystems and human economies.

economic policy. The intervention by a regulator through policy instruments in private markets so that a desired market outcome is attained.

externality. An externality is present when the well-being (utility) of an individual or the production possibilities of a firm are directly affected by the actions of another agent in the economy.

internalization of an externality. A situation in which the agent who generates the externality bears the cost that the externality imposes on other agents.

market failure. A market failure exists when competitive markets fail to attain Pareto optimum.

Pareto optimum. A situation in which it is not possible to make someone better off without making someone else worse off.

production function. A real-valued function that shows the maximum amount of output that can be produced for any given combination of inputs.

public good. A commodity for which use of one unit of the good by one agent does not preclude its use by other agents.

state variable. A variable that characterizes the state of a system at any point in time and space.

utility function. A real-valued function that shows that if a consumer prefers the bundle of goods x to the bundle of goods y, then the utility of x is greater than the utility of y.

1. INTRODUCTION

Ecological economics studies the interactions and co-evolution in time and space between ecosystems and human economies. Human economies in the process of their operation and development use the flows of services generated by ecosystems. In using these services, humans make decisions about the size and the time profile of the harvested flows of ecosystems services as well as about the growth rates of different types of natural capital that are embedded in the ecosystems and that generate the flows of desirable services. Long series of empirical observations have established that, given the institutional structure of the economies (e.g., markets, allocation of property rights, regulatory authorities, international agreements), the rate at which economic agents exploit (or harvest) ecosystems services exceeds what might be regarded as a desirable level from society's point of view. The consequences of this overexploitation are well known and include serious interrelated environmental problems such as climate change, biodiversity loss and extinction of species, collapse of fisheries, and overexploitation of water resources. To put this point differently, the market outcome, or the outcome stemming from individual actions, regarding the harvesting of ecosystem services and the time paths of the stocks of natural capital (or natural resources) is different from an outcome (or a state) that is socially desirable.

The challenge of designing economic policy in this context is to develop a system of regulatory instruments or incentive schemes that will affect the behavior of economic agents (individuals, firms, nations) regarding the harvesting of ecosystem services in such a way that harvesting rates and time paths of the stock of natural capital under the economic policy will converge toward the socially desirable outcome. The purpose of this chapter is to present an approach describing how these economic policies might be designed.

2. ECOLOGICAL MODELING AND RESOURCE DYNAMICS

The building of meaningful ecological–economic models capable of helping in the design of policies for ecosystem management requires the development of two interacting modules: an ecological module describing the evolution of the state of the ecosystem and the ways that the interventions of the economic agents influence this evolution; and an economic module describing, in broad terms, the net benefits accruing to economic agents from the use of the ecosystem's flow of services.

The traditional resource models presented, for example, by Clark (1990) or Dasgupta and Heal (1979) describe the evolution of the population (or biomass or stock) of a biological, a renewable, or an exhaustible resource when exploitation (harvesting) by economic agents takes place. Let $x(t)$ denote the stock of a resource at time t, which generates a flow of valuable services to economic agents. Following the Millennium Ecosystems Assessment (2005) classification, these services may include provisioning services (e.g., food, water, fiber, fuel), regulating services (e.g., climate regulation, disease), cultural services (e.g., spiritual, aesthetic, education), or supporting services (e.g., primary production, soil formation). It should be noted that some of the above services, mainly the provisioning, can be used after harvesting the resources stock (e.g., fishing, water pumping), whereas others, mainly regulating and cultural services, are associated with the existing stock of the resource (e.g., aesthetic services and preservation of a forest). Let $F(x(t))$ be a function describing the net growth of the resource stock. This growth function embodies factors such as birth, death, migration in case of biological resources (e.g., fisheries), natural inflows, and seepage in case of renewable resources (water resources or accumulation of pollutants), whereas in the case of exhaustible resources with no discoveries, $F(x(t)) \equiv 0$. If we denote by $h(t)$ the harvesting of the resource, so that provisioning services are used, then the evolution of the resource can be described by an ordinary differential equation (ODE), which can be written, for some initial stock, x_0, as:

$$\frac{dx(t)}{dt} \equiv \dot{x}(t) = F(x(t)) - h(t), \; x(0) = x_0 > 0. \quad (1)$$

The most common specification of the growth function $F(x(t))$ is the logistic function, $F(x) = rx(1 - x/K)$, where r is a positive constant called *intrinsic growth rate,* and K is the *carrying capacity* of the environment, which depends on factors such as resource availability or environmental pollution. If $h(t) = F(x(t))$, then the population remains constant because harvesting is the same as the population's net growth. This harvesting rate corresponds to *sustainable yield.* Harvesting rate is usually modeled as population dependent or $h = qEx$ where q is a positive constant, referred to as a *catchability coefficient* in fishery models, and E is *harvesting effort.* The activities of economic agents can affect the resource stock, in addition to harvesting, by affecting parameters such as the intrinsic growth rates or the carrying capacity. Assume, for example, that the intrinsic rate of growth and the carrying capacity of the environment for the resource described by equation 1 are affected by the stock of environmental pollution that accumulates on the ecosystem (e.g., a lake). Let $S(t) = \sum_{i=1}^{n} s_i(t)$ denote the sum of emissions generated by $i = 1, \ldots, n$ sources at time t, and let $P(t)$ be the stock of the pollutant accumulated in the ecosystem (e.g., phosphorus accumulation from agricultural leaching). Then the evolution of the pollutant stock can also be described by an ODE:

$$\dot{P}(t) = S(t) - bP(t), \ P(0) = P_0 > 0 \qquad (2)$$

where $b > 0$ is a constant reflecting the environment's self-cleaning capacity. The negative impact of the pollutant's stock on the intrinsic growth rate and the environment's carrying capacity can be captured by functions $r(P)$, $r'(P) < 0$ and $K(P)$, $K'(P) < 0$. Then the evolution of the resource is described by:

$$\dot{x}(t) = r(P(t))x(t)\left[1 - \frac{x(t)}{K(P(t))}\right] - h(t). \qquad (3)$$

The ODE system (equations 2 and 3) is an example of a simple ecosystem model in which economic agents affect the resource stock in two ways, through harvesting and through emissions generated by their economic activities. The agents that harvest the resource and the agents that generate emissions are, in the majority of the cases, not the same, and it is hard to coordinate their decisions. Furthermore, the pollutant can generate additional environmental damages to individuals, which can be summarized in a damage function.

The simple model of resource dynamics described by equation 1 can be generalized in many ways (see, e.g., Murray, 2003). Generalizations may include age-structured populations, multispecies populations and Lotka-Volterra predator–prey models, mechanistic resource-based models of species competition, models with spatial variation including metapopulation models, and models with resource diffusion over space.

A general multispecies model with J prey populations denoted by $x_j(t)$ and J predator populations denoted by $y_j(t)$ can be written, for $j = 1, \ldots, J$, as:

$$\dot{x}_j(t) = x_j(t)\left[a_j - \sum_{k=1}^{J} \beta_{jk} y_k(t)\right], \ x_j(0) = x_{j0},$$
$$\dot{y}_j(t) = y_j(t)\left[\sum_{k=1}^{J} \gamma_{jk} x_k(t) - \delta_j\right], \ y_j(0) = y_{j0}, \qquad (4)$$

where all parameters are positive constants.

In the mechanistic resource-based models of species competition emerging from the work of Tilman (e.g., Tillman, 1982), species compete for limiting resources. In these models, the growth of a species depends on the limiting resource, and interactions among species take place through the species' effects on the limiting resource. Let $\mathbf{x}(t) = (x_1(t), \ldots, x_J(t))$ be the vector of species biomasses, and $R(t)$ the amount of the available limiting resource. Then a mechanistic resource-based model with a single limiting factor in a given area can be described by the following equations:

$$\frac{\dot{x}_j(t)}{x_j(t)} = g_j(R(t)) - d_j, \ x_j(0) = x_{j0}, \ j = 1, \ldots, J,$$
$$\dot{R}(t) = S(t) - aR(t) - \sum_{j=1}^{J} w_j x_j(t) g_j(R(t)),$$
$$R(0) = R_0 \qquad (5)$$

where $g_j(R)$ is resource-related growth for species j, d_j is the species' natural death rate, $S(t)$ is the amount of resource supplied, a is the natural resource removal rate (leaching rate), and w_j is the specific resource consumption by species j.

Another important characteristic of ecosystems, in addition to the temporal variation captured by the models described above, is that of spatial variation. Biological resources tend to disperse in space under forces promoting "spreading" or "concentrating." These processes, along with intra- and interspecies interactions, induce the formation of spatial patterns for species in a given spatial domain. A central concept in modeling the dispersal of biological resources is that of *diffusion*. Biological diffusion when coupled with population growth equations leads to general reaction–diffusion systems (e.g., Okubo and Levin, 2001; Murray, 2003). When only one species is examined, the coupling of classical diffusion with a logistic growth function leads to the so-called Fisher-Kolmogorov equation, which can be written as

$$\frac{\partial x(z,t)}{\partial t} = F(x(z,t)) + D_x \frac{\partial^2 x(z,t)}{\partial z^2}, \qquad (6)$$

where $x(z, t)$ denotes the concentration of the biomass at spatial point z at time t. The biomass grows according to a standard growth function $F(x)$ but also disperses in space with a constant diffusion coefficient D_x. In general, a diffusion process in an ecosystem tends to produce a uniform population density, that is, spatial homogeneity. However, under certain conditions reaction–diffusion systems can generate spatially heterogeneous patterns. This is the so-called Turing mechanism for generating diffusion instability.

Spatial variations in ecological systems can also be analyzed in terms of metapopulation models. A metapopulation is a set of local populations occupying isolated patches, which are connected by migrating individuals. Metapopulation dynamics can be developed for single or many species. For the single species occupying a spatial domain consisting of $s = 1, \ldots, S$ patches, the dynamics becomes

$$\frac{\dot{x}_s(t)}{x_s(t)} = F(x_s(t)) + \sum_{k=1}^{S} d_{sk} x_k(t), \; s = 1, \ldots, S, \qquad (7)$$

where $x_s(t)$ is the species population in patch s, and d_{sk} is the rate of movement from patch k to patch s, $(s \neq k)$. Thus, dynamics is local with the exception of movements from one patch to the other.

If harvesting is introduced into the ecological models of equations 4–7, and growth functions depend on pollutants generated by economic activities, then the ecological model is extended to include economic variables whose time paths are chosen by economic agents.

3. ECONOMIC MODELING FOR ECOSYSTEM MANAGEMENT

Choosing time paths for harvesting or other variables that might affect the state of the ecosystem, which are called *control variables*, implies management of the ecosystem. In economics, the most common type of management is the *optimal management*, which means that the control variables are chosen so that an *objective function* is optimized (maximized or minimized). Of course, other types of management rules can be applied such as adaptive rules or imitation rules, but the focus of the present article is on optimal rules. To provide a meaningful presentation of the optimal rules, some fundamental economic concepts are useful.

Some Fundamental Economic Concepts

Preferences, Utility, Profits

Individuals have preferences summarized by the preference relationship , which means "at least as good as." Let n goods be indexed by $i = 1, \ldots, n$, and the combinations of different quantities from these goods $\mathbf{x} = (x_1, \ldots, x_n)$, $\mathbf{y} = (y_1, \ldots, y_n)$. Then a consumer's preferences regarding the two combinations or bundles of goods could be described as:

- $\mathbf{x} \succsim \mathbf{y}$ means that combination $\mathbf{x}$ is at least as good as combination $\mathbf{y}$.
- $\mathbf{x} \succ \mathbf{y}$ means that combination $\mathbf{x}$ is better than (is preferred to) combination $\mathbf{y}$.
- $\mathbf{x} \sim \mathbf{y}$ means that the consumer is indifferent between $\mathbf{x}$ and $\mathbf{y}$.

A utility function is a real-valued function of the combinations of goods, such as:

$$\text{if} \begin{Bmatrix} \mathbf{x} \succsim \mathbf{y} \\ \mathbf{x} \succ \mathbf{y} \\ \mathbf{x} \sim \mathbf{y} \end{Bmatrix} \text{ then } \begin{Bmatrix} U(\mathbf{x}) \geq U(\mathbf{y}) \\ U(\mathbf{x}) > U(\mathbf{y}) \\ U(\mathbf{x}) = U(\mathbf{y}) \end{Bmatrix}.$$

A central paradigm of modern economic theory (e.g., Mass Colell et al., 1995) is that, for exogenously determined prices and income, consumers choose the combinations of goods they consume by maximizing their utility function subject to a budget constraint, whereas competitive firms, for exogenously determined prices of inputs and outputs, maximize profits subject to the constraints imposed by technology, which are usually summarized by a production function.

Pareto Efficiency

Economic Allocation

Consider an economy consisting of $i = 1, \ldots, I$ consumers, $j = 1, \ldots, J$ firms, and $l = 1, \ldots, L$ goods. The consumption for individual i is given by the vector $\mathbf{x}_i = (x_{1i}, \ldots, x_{Li})$, and production by firm j is given by the vector $\mathbf{y}_i = (y_{1j}, \ldots, x_{Lj})$. Consumers maximize profits subject to their budget constraint, whereas firms maximize profits subject to technology.

An *economic allocation* $(\mathbf{x}_1, \ldots, \mathbf{x}_I, \; \mathbf{y}_1, \ldots, \mathbf{y}_J)$ is a specification of a consumption vector for each consumer and a production vector for each firm. The allocation is *feasible* if

$$\sum_{i=1}^{I} x_{il} \leq \sum_{j=1}^{J} y_{jl}, \; l = 1, \ldots, L.$$

Pareto Optimality

A feasible allocation $(\mathbf{x}_1, \ldots, \mathbf{x}_I, \mathbf{y}_1, \ldots, \mathbf{y}_J)$ is Pareto optimal or Pareto efficient, if there is no other allocation $(\mathbf{x}'_1, \ldots, \mathbf{x}'_I, \mathbf{y}'_1, \ldots, \mathbf{y}'_J)$ such that $u(\mathbf{x}'_i) \geq u(\mathbf{x}_i)$, $\forall i = 1, \ldots, I$ and $u(\mathbf{x}'_i) > u(\mathbf{x}_i)$ for some i. To put it differently, a feasible allocation $(\mathbf{x}_1, \ldots, \mathbf{x}_I, \; \mathbf{y}_1, \ldots, \mathbf{y}_J)$ is Pareto optimal or Pareto efficient if society's resources and technological possibilities have been used in such a way that there is *no* alternative way to organize production and distribution that makes some consumers better off without making someone worse off.

Competitive Equilibrium

An allocation $(\mathbf{x}_1{}^*, \ldots, \mathbf{x}_I{}^*, \mathbf{y}_1{}^*, \ldots, \mathbf{y}_J{}^*)$ and a price vector $\mathbf{p}^* = (p_1{}^*, \ldots, p_L{}^*)$ comprise a *competitive* or *Walrasian equilibrium* if

- Firms maximize profits by taking equilibrium prices $\mathbf{p}^*$ as given.
- Consumers maximize utility subject to their budget constraint determined by their given income w, taking equilibrium prices $\mathbf{p}^*$ as given.
- Markets clear, or demand equals supply, at the equilibrium prices $\mathbf{p}^*$, or $\sum_{i=1}^{I} x_{il} = \sum_{j=1}^{J} y_{jl}$, $l = 1, \ldots, L$.

First Welfare Theorem

If the price vector $\mathbf{p}^*$ and the allocation $(\mathbf{x}_1{}^*, \ldots, \mathbf{x}_I{}^*, \mathbf{y}_1{}^*, \ldots, \mathbf{y}_J^*)$ constitute a competitive equilibrium, then this allocation is Pareto optimal.

Second Welfare Theorem

Suppose that $(\mathbf{x}_1{}^*, \ldots, \mathbf{x}_I{}^*, \mathbf{y}_1{}^*, \ldots, \mathbf{y}_J{}^*)$ is a Pareto efficient allocation, then there is a price vector $\mathbf{p}^*$, such that $(\mathbf{x}_1{}^*, \ldots, \mathbf{x}_I{}^*, \mathbf{y}_1{}^*, \ldots, \mathbf{y}_J{}^*)$ and $\mathbf{p}^*$ constitute a competitive equilibrium.

Welfare Efficiency

A Pareto efficient allocation maximizes a linear social welfare function of the form $W = \sum_{i=1}^{I} \alpha_i u_i(\mathbf{x}_i)$.

Externalities

Environmental and resource economics have long been associated with the concepts of externalities and market failure. An *externality* is present when the well-being (utility) of an individual or the production possibilities of a firm are directly affected by the actions of another agent in the economy. When externalities are present, the competitive equilibrium is not Pareto optimal.

Public Goods (Bads)

A public good is a commodity for which use of one unit of the good by one agent does not preclude its use by other agents. Public goods (bads) are not *depletable*. Environmental externalities (air pollution, water pollution) are nondepletable public bads and are mainly associated with missing markets or missing property rights. Competitive markets have the following characteristics in the presence of externalities:

- Competitive markets cannot obtain the Pareto optimal levels of public goods or public bads.
- Competitive markets cannot obtain the Pareto optimal levels of harvesting for open-access or common-pool natural resources.

Economic Policy

The above results imply that competitive markets fail to produce a Pareto optimal outcome or a socially optimal outcome in the presence of environmental externalities and open-access resources. When competitive markets *fail* to produce the Pareto optimal allocation, there is a need for market *intervention* and *economic policy* to achieve the Pareto optimal allocation. Because environmental externalities and open-access characteristics are predominant in ecosystems, competitive (and of course imperfectly competitive) markets *fail* to attain the socially optimal ecosystem state. Thus, there is a need for economic policy for ecosystem management.

Optimal Ecosystem Management

The economic concepts defined above can help formulate optimal ecosystem management and methods for designing economic policy to achieve a socially optimal state for ecosystems. The approach is to define an objective function for the economic agent(s), that will be optimized subject to the constraints imposed by the ecological model of the ecosystem, which will be along the lines of models described in section 2. In principle the objective function will include profits associated with harvesting or utility associated with the ecosystem services. The way in which the objective function is set up, the ecological constraints that are taken into account, determine the solution of the ecological–economic model. By solution we mean the paths for the control variables and the stock of ecosystems resources, which are the *state variables*, and the equilibrium state of the ecosystem under a specific management rule. Two types of solution are distinguished in general, a *socially optimal solution* and a *privately optimal solution.*

The Socially Optimal Solution

The socially optimal solution corresponds to a solution in which social welfare is maximized. This means that the objective function includes utility accruing from harvesting (which is sometimes called *consumptive utility* and is mainly associated with provisioning ecosystem services) and utility associated with the other services such as regulation, cultural or supporting services, existence values, or benefits associated with productivity or insurance gains (which is sometimes called *nonconsumptive utility*). The objective function for the social welfare maximization problem also includes damages from environmental degradation, which are environmental externalities (nondepletable public bads), as well as stock effects that negatively affect production functions in the case of management of open access resources. The socially optimal solution is sometimes referred to as the so-called problem of the *social planner*, where a fictitious social planner maximizes social welfare by taking into account all the externalities not accounted for by competitive markets.

Let $U^c(\mathbf{h}(t))$ denote consumptive utility at time t associated with harvesting $\mathbf{h} = (h_1, \ldots, h_n)$ species, and $U^{nc}(\mathbf{x}(t))$ denote nonconsumptive utility associated with ecosystem services generated by species biomasses existing in the ecosystem and not removed by harvesting. The total flow of utility at time t can be written as $U^c(\mathbf{h}(t)) + U^{nc}(\mathbf{x}(t))$. Because the objective in the dynamic context is, in the majority of cases, to maximize the present value of the utility flow over an infinite time horizon, the objective can then be written as:

$$\max_{\{\mathbf{h}(t)\}} \int_0^\infty e^{-\rho t}[U^c(\mathbf{h}(t)) + U^{nc}(\mathbf{x}(t))]dt, \qquad (8)$$

where $\rho \geq 0$ is a utility discount rate, subject to the constraints imposed by the structure of the ecosystem. A solution to this problem will produce the socially optimal paths for the controls and the states ($\mathbf{h}^*(t)$, $\mathbf{x}^*(t)$) and a long-run equilibrium state ($\mathbf{h}^*, \mathbf{x}^*$) as $t \to \infty$, provided that the solution satisfies appropriate stability properties. It should be noted that, in principle, benefits associated with consumptive utilities can be approximated using market data from concepts such as consumer and producer surplus, whereas benefits associated with nonconsumptive utility and environmental externalities are hard to estimate because markets for the larger part of the spectrum of ecosystem services and environmental pollution are missing.

The Privately Optimal Solution

The privately optimal solution is distinguished from the socially optimal one by the fact that only consumptive utilities or profits enter the objective function. In particular, when the market outcome regarding the ecosystem's state is analyzed, the basic assumption is that management is carried out by a "small" profit-maximizing private agent that in general ignores "stock effects," the general nonconsumptive flows of ecosystem services, or other externalities generated by the agent's activities. Thus, the private agents do not take into account, or do not *internalize*, externalities associated with their management. There are some very well-known examples.

In the case of an open-access commercial fishery, usually each harvester takes the landing price as fixed but ignores the fact that his/her own harvesting reduces the stock of fish and thus increases costs. Because the resource has open-access characteristics, each harvester enters in competition to catch the fish first before someone else does. As a result, in the open-access or bionomic equilibrium, the stock of fish is smaller relative to the social optimum, which internalizes "stock effects." Overfishing and stock collapse can be attributed to this type of externality, also known as *the tragedy of the commons*. In the case of pollution control, emissions are generated by a group of agents (e.g., an industry), but the damages affect another group of agents (e.g., inhabitants of a certain area). Because the cost of emissions is not internalized by the emitters, in the absence of regulation, emissions exceed the socially desirable level, which is determined by internalizing environmental damages. In other cases harvesters do not take into account nonconsumptive utility associated with the stock of the harvested resource (e.g., existence values), which increases even more the deviation between the social and the private optimum that results from open access. There are also situations in which the harvester does not take into account the fact that harvesting the specific resource might harm the stock of other resources (e.g., by-catch in fishing), which is an additional externality. Another type of externality can be associated with strategic behavior in resource harvesting if more than one economic agent harvests the resource. If many small harvesters are present, then the privately optimal solution can be obtained as an *open loop* or *feedback* Nash equilibrium, which also deviates from the social optimum.

The fact that general "stock effects" are not taken into account at the private optimum implies that $U^{nc}(\mathbf{x}(t)) = 0$ in equation 8. As a result, the privately optimal solution will deviate from the socially optimal solution. Furthermore, because *all* the ecological constraints are operating in the real ecosystem, there will be discrepancies between the perceived evolution of ecosystems under management that ignores certain constraints and the actual evolution of the ecosystem. These discrepancies might be a cause for *surprises* in ecosystem management.

4. INSTRUMENTS OF ECONOMIC POLICY AND POLICY DESIGN

The inability of privately optimal solutions realized in the context of unregulated market economies to attain the socially optimal outcome regarding the state of an ecosystem calls for environmental policy (detailed analysis can be found in Baumol and Oates, 1988; Xepapadeas, 1997), which is assumed to be designed and implemented by a regulator. The classic instruments of environmental policy can be divided into two broad categories.

Economic Incentives or Market-Based Instruments

- *Environmental taxes ("ecotaxes" or "green taxes")*. These are taxes imposed on emissions, harvesting, and polluting inputs or outputs. In particular environmental taxes include:
  - Emissions taxes (tax payments related to measured or estimated emissions).
  - Landing fees (tax related to the amount of harvested resource from an ecosystem).
  - Product charges (consumption taxes, input taxes, or production taxes, which are substitutes for emission taxes when emissions are not directly measurable or estimable).
  - Tax differentiation (variation of existing indirect taxes in favor of clean products or activities that are environmentally and ecologically friendly).
  - User charges (payments related to environmental service delivered).
  - Tax reliefs (tax provisions to encourage environmentally or ecologically friendly behavior).
- *Subsidies for reduction of harvesting or emissions*. These include subsidies for *land-set-aside programs*, or *buffer zone programs* in agriculture, which aim at reducing overproduction, protect and expand ecosystems (e.g., wetlands), or reduce agricultural runoff, as well as subsidies for introducing environmentally friendly or resource-saving technologies.
- *Tradable quotas or tradable emission permits*. Under these systems resource users or emitters operate under an aggregate limit on resource use or emissions and trading is allowed on permits or quotas adding up to a specific limit. For example, a cap-and-trade system in fisheries management includes a total allowable commercial catch limit and assigns individual transferable quotas (ITQ). ITQs are rights to harvest fish from a particular area and are distributed to each commercial fishing permit holder based on some rule (e.g., holder's historic catch levels). This instrument essentially creates a market for the environmental good, which was missing because of absence of well-defined property rights. Similar markets can be created for biodiversity preservation by assigning rights for bioprospecting.
- *Voluntary agreements (VAs)*. Voluntary approaches to environmental regulation have more recently been regarded as alternative instruments of environmental policy. They are expected to increase economic and environmental effectiveness as well as social welfare, relative to traditional policy instruments, because they allow economic agents greater flexibility in their pollution or harvesting control strategies and also have the potential to reduce transaction and compliance costs. VAs can be classified into three basic categories, based mainly on the degree of public intervention:
  - Negotiated agreements imply a bargaining process between the regulatory body and an economic agent to jointly set the environmental goal and the means of achieving it.
  - Unilateral agreements are environmental improvement programs prepared and voluntarily adopted by economic agents themselves.
  - Public voluntary agreements are environmental programs developed by a regulatory body, and economic agents can only agree to adopt them or not.

In general, participation in a VA program exempts the economic agent from stricter regulation.

Direct Regulation or Command and Control

This type of regulation includes the use of limits on inputs, outputs, or technology at the firm level. When the objective of direct regulation is the firm's harvesting (e.g., harvesting rates, harvesting periods, "no-take" reserve areas) or emissions, then the type of regulation is called a *performance standard*. When the regulator requires the use of a specific technology, then the regulation is called a *design standard*. Performance standards can be associated with a maximum allowed amount of emissions or harvesting, whereas design standards can be associated with the use of *best available technologies*.

The above classification is by no means exhaustive, and it should be noted that instruments can be used in combinations and that they can be characterized by spatial and temporal variation.

Optimal environmental policy can be designed by using the following approach:

- Obtain the socially optimal solution as the solution of the social planner's problem that internalizes all externalities. The paths for the control and the state variable are determined.
- Obtain the privately optimal solution as the solution of a representative profit-maximizing agent or as market equilibrium without internalization of externalities. The paths for the control and the state variables are determined, and the deviations from the corresponding socially optimal paths are established.

- Introduce an instrument or a menu of economic policy instruments and derive the regulated privately optimal solution as a function of the policy instruments.
- Choose the policy instruments so that the privately optimal solution converges to the socially optimal solution.

The following example from the literature of determining optimal emission taxation can help clarify this approach. We choose the emission problem instead of an ecosystem management problem because the latter requires the use of more complicated optimal control techniques.

We start by considering a market of $i = 1, \ldots, n$ firms that behave competitively. The firms produce a homogeneous output q_i and, during production, generate emissions e_i. A derived profit or derived benefit function can be defined as:

$$B_i(e_i) = \max_{q_i \geq 0} \pi_i = \max_{q_i \geq 0} [pq_i - c_i(q_i, e_i)], \quad B_i''(e_i) < 0, \tag{9}$$

where p is the exogenously determined output price, and $c_i(q_i, e_i)$ is a convex cost function decreasing in e_i. A reduction in emissions will increase costs because this involves the use of resources for pollution abatement.

Social welfare is defined as total benefits from production less social damages from emissions. Using the derived profit function (equation 9) and a social damage function $D(E)$ which is an increasing and convex function reflecting environmental damages caused by emissions, the social planner solves the problem:

$$\max_{(e_1, \ldots, e_n) \geq 0} \sum_{i=1}^{n} B_i(e_i) - D(E), \quad E = \sum_{i=1}^{n} e_i. \tag{10}$$

The necessary and sufficient first-order conditions for the socially optimal emissions $e_i{}^*$ generated by the ith firm are:

$$B_i'(e_i{}^*) - D'(E^*) \leq 0, \text{with equality if } e_i{}^* > 0. \tag{11}$$

Thus, when positive emissions are generated, marginal benefits equal marginal social damages. The polluting firms fully internalize external social damages if they are confronted with an emission tax per unit of waste released in the ambient environment equal to marginal social damages. This price incentive for emission control is the well-known "Pigouvian tax" or emission tax.

Let the emission tax τ be defined as $\tau = D'\left(\sum_{i=1}^{n} e_i{}^*\right)$. The firm solves the problem

$$\max_{e_i \geq 0} B_i(e_i) - \tau e_i$$

with necessary and sufficient first-order conditions:

$$B_i'(e_i^0) - \tau \leq 0, \text{with equality if } e_i^0 > 0. \tag{12}$$

Because $\tau = D'\left(\sum_i e_i{}^*\right)$, it can be seen by comparing equation 11 to equation 12 that the emission tax leads to the socially optimal emissions for firm i, for all i.

In many cases the design and/or the implementation of optimal policy might not be possible because of informational constraints, cost of implementation, and so on. In this case, another approach is for the regulator to set a given standard, such as ambient pollution standards, maximum harvesting rates, minimum safety margins for species populations, and then choose the instrument or the menu of instruments from those described above to achieve the standard at a minimum cost. The policy instruments can be revised or updated as the state of the ecosystem changes or as more information is acquired about the responses of the economic agents and the ecosystem to economic policy. This type of policy is not optimal, but if it is combined with the general insights obtained by having determined the structure of the optimal policy, it might be a useful approach to policy design and implementation.

FURTHER READING

Baumol, W. J., and W. E. Oates. 1988. The Theory of Environmental Policy, 2nd ed. Cambridge, UK: Cambridge University Press.

Clark, C. 1990. Mathematical Bioeconomics: The Optimal Management of Renewable Resources, 2nd ed. New York: Wiley.

Dasgupta, P. S., and G. M. Heal. 1979. Economic Theory and Exhaustible Resources. Oxford: James Nisbet and Co. and Cambridge University Press.

Mas-Colell, A., M. D. Whinston, and G. R. Green. 1995. Microeconomic Theory. New York: Oxford University Press.

Millennium Ecosystem Assessment. 2005. Ecosystems and Human Well-Being: Synthesis. Washington, DC: Island Press.

Murray, J. D. 2003. Mathematical Biology. Berlin: Springer-Verlag.

Okubo, A., and S. Levin, eds. 2001. Diffusion and Ecological Problems: Modern Perspectives, 2nd ed. Berlin: Springer.

Tilman, D. 1982. Resource Competition and Community Structure. Princeton, NJ: Princeton University Press.

Xepapadeas, A. 1997. Advanced Principles in Environmental Policy. Aldershot: Edward Elgar.

VII.10

Governance and Institutions

Elinor Ostrom

OUTLINE

1. The diversity of social–ecological systems
2. Common-pool resources
3. The conventional theory of common-pool resources
4. Self-organized resource governance systems in the field
5. Attributes of a resource and resource users that increase the likelihood of self-organization
6. Types of ownership used in self-organized field settings
7. The importance of larger governance regimes
8. The advantages of polycentric resource governance systems

Governance is a multilevel process established by humans to craft institutions—rules—that affect who can do what in relation to specific aspects of a linked social–ecological system (SES), who will monitor conformance to these rules, and how these rules may be modified over time in light of feedback from the SES itself and from those involved in its use, management, and conservation. Governance processes may be undertaken by governments (which are one type of organization) as well as by organizations of all types.

GLOSSARY

common-pool resource. A resource system in which it is costly to exclude potential beneficiaries, but one person's use subtracts resource units from those available to others

governance. The process of crafting institutional rules to fit diverse settings

institutional rules. Rules defining rights and responsibilities of participants in a repeated setting

polycentric systems. A governance system in which citizens are able to organize multiple governing authorities at differing scales

social–ecological system. An ecological system and a linked social system of resource users and their governance arrangements (if present)

1. THE DIVERSITY OF SOCIAL–ECOLOGICAL SYSTEMS

Readers of this *Princeton Guide to Ecology* will be well informed about the immense diversity of ecological systems. Ecological systems vary in regard to their geographic range, density of specific plant and animal populations, patterns of species diversity, nutrient cycling, landscape dynamics, and disturbance patterns—to name just a few of the subjects included in the sections of this Guide. Ecological systems are complex systems with interactions occurring at multiple spatial and temporal scales.

In addition to the diversity of ecological systems considered independent of human interactions, the variety of linked social–ecological systems (SESs) that exist in the world is even larger. The "social" side varies in regard to the size and socioeconomic attributes of users, the history of their use, the location of their residences and their work places, the types of leadership and entrepreneurship experienced, the cultural norms they share, the level of human and social capital they have, their knowledge about the ecological system, their dependence on the system for diverse purposes, and the technologies available to them, to name just a few of the most important general characteristics.

Relevant organizations include families, private for-profit and not-for-profit firms, neighborhood groups, and communities living in or near to an ecological system. The rules crafted in a governance process regulate one or more of the following:

- Who is authorized to harvest specific types of resource units from a particular SES and for what mix of purposes?
- The timing, quantity, location, and technology of harvesting.

- Who is obligated to contribute funds, labor, and other resources to provide infrastructure for or to maintain key attributes of an ecological system?
- How harvesting and contribution activities are to be monitored and enforced.
- How conflicts over harvesting and contributions are to be resolved.
- How the rules affecting the above may be changed over time with changes in the performance of the resource system and the strategies of participants.

Although some policy advocates recommend using one type of institution—such as the creation of private property or the establishment of ownership by a national government—for all ecological systems, considerable evidence exists that all types of institutions fail under some circumstances and succeed under others. The challenge facing those who are involved in the governance of SESs or study governance processes is matching institutional arrangements to the structure of a focal SES and other linked SESs at larger or smaller scales. Because the structure of an SES changes over time, it is also important to enable institutional rules to adapt over time.

2. COMMON-POOL RESOURCES

Most ecological systems used by multiple individuals can be classified as common-pool resources. Common-pool resources generate finite quantities of resource units. One person's harvesting of resource units from a common-pool resource subtracts from the quantity of resource units available to others. Examples of common-pool resources include both natural and human-made systems such as groundwater basins, forests, grazing lands, fisheries, and irrigation systems. Examples of the resource units derived from common-pool resources include water, timber, fodder, and fish. Most common-pool resources are sufficiently large that multiple actors can simultaneously use the resource system, and efforts to exclude potential beneficiaries are costly.

When resource units are highly valued and many actors benefit from harvesting them for consumption, exchange, or as a factor in a production process, the harvests made by one individual are likely to create negative externalities for others. Nonrenewable resources, such as oil, may be withdrawn in an uncoordinated race that reduces the quantity of the resource units that can be withdrawn and greatly increases the cost of harvesting and use. Renewable resources, such as fisheries, may be congested in a particular time period but may also be so overharvested that the stock generating a flow of resource units is destroyed. An *open-access* common-pool resource (meaning one for which there are *no* rules related to the use of the resource) that generates highly valued resource units is likely to be overharvested and may even be destroyed.

3. THE CONVENTIONAL THEORY OF COMMON-POOL RESOURCES

Many textbooks in resource economics and in law and economics present a conventional view of a simple open-access common-pool resource as the only theory needed for understanding how to design better governance systems. In this theory, the users are presented as being trapped in a "tragedy of the commons" and unable to extract themselves from the processes of overuse and potential destruction of the system. The users face a "social dilemma" in that they would all be better off if they all found a way of cooperating together, but no one acting alone has an incentive to bear the costs of such cooperation.

Empirical examples exist where the absence of any property rights and the independence of actors capture the essence of the problem facing harvesters. For many scholars, the collapse of many ocean fisheries confirms the worst predictions to be derived from this theory. Since users are viewed as being trapped in these dilemmas, repeated recommendations have been made that external authorities *impose* institutions on such settings. Some recommend private property as the most efficient form of ownership, but others recommend government ownership and control. Implicitly, theorists assume that government officials will act in the public interest and understand how ecological systems work and how to change institutions so as to induce socially optimal behavior.

The possibility that the users themselves would find ways to organize themselves is not seriously considered in some of the public policy literature. Organizing to create rules that specify rights and duties of participants creates a public good for those involved. Everyone included in the community of users would benefit from this public good, whether they contribute or not. Thus, getting "out of the trap" is itself a second-level dilemma. Further, investing in monitoring and sanctioning activities so as to increase the likelihood that participants follow the agreements they have made also generates a public good. Thus, investing in monitoring and sanctioning is a third-level dilemma.

Much of the initial problem is thought to exist because the individuals are stuck in a social dilemma. It is not consistent with the conventional theory that the "helpless" participants solve a second- and third-level dilemma in order to address the first-level dilemma under analysis. Growing evidence from many studies of common-pool resources in the field, however, has called for a serious rethinking of the theoretical foundations for the analysis of common-pool resources. Empirical studies do not challenge the empirical validity of the conventional theory *where it is relevant*, but rather its generalizability to *all* common-pool resources.

4. SELF-ORGANIZED RESOURCE GOVERNANCE SYSTEMS IN THE FIELD

Most common-pool resources are more complex than the base theory of homogeneous users taking one type of resource unit from a resource system that generates a predictable flow of units. A rich case-study literature illustrates the wide diversity of settings in which users dependent on common-pool resources have organized themselves to achieve much higher outcomes than is predicted by the conventional theory.

Evidence from field research challenges the universal generalizability of the conventional theory. Although the conventional theory is generally successful in predicting outcomes in settings where large numbers of resource users have no links to one another and cannot communicate effectively, it does not provide an explanation for settings where users are able to create and sustain agreements to avoid serious problems of overappropriation. Nor does it predict well when government ownership will perform appropriately or how privatization will improve outcomes.

5. ATTRIBUTES OF A RESOURCE AND RESOURCE USERS THAT INCREASE THE LIKELIHOOD OF SELF-ORGANIZATION

Scholars familiar with the results of field research substantially agree on a set of variables that enhance the likelihood of users organizing themselves to avoid the social losses associated with open-access common-pool resources. Considerable consensus exists that the following attributes of resources and of resource users increase the likelihood that self-governing organizations will form and try to increase the probability of a sustainable common-pool resource.

Attributes of the Resource System

1. Feasible improvement: Resource conditions are not at a point of deterioration such that it is useless to organize or so underutilized that little advantage results from organizing.
2. Indicators: Reliable and valid indicators of the condition of the resource system are frequently available at a relatively low cost.
3. Predictability: The flow of resource units is relatively predictable.
4. Spatial extent: The resource system is sufficiently small, given the transportation and communication technology in use, that those users can develop accurate knowledge of external boundaries and internal microenvironments.

Attributes of the Resource Users

1. Salience: Resource users are dependent on the resource system for a major portion of their livelihood.
2. Common understanding: Resource users have developed over time a shared image of how the resource system operates (Resource System attributes 1, 2, 3, and 4 above) and how their actions affect each other and the resource system.
3. Low discount rate: Resource users do not heavily discount benefits to be achieved from the resource in the future time periods as contrasted to the present.
4. Trust and reciprocity: Resource users trust one another to keep promises and relate to one another with reciprocity.
5. Autonomy: Resource users are able to determine access and harvesting rules without external authorities countermanding them.
6. Prior organizational experience and local leadership: Resource users have learned at least minimal skills of organization and leadership through participation in other local associations or learning about ways that neighboring groups have organized.

When a group of resource users shares these attributes about the resource system and about themselves, they are more likely to agree that all would be better off if they could develop and generally abide by a set of institutional rules for governing their common-pool resource.

6. TYPES OF OWNERSHIP USED IN SELF-ORGANIZED FIELD SETTINGS

The rules adopted by the users of a common-pool resource may approximate those of private property in some settings. When resource users develop private property rights (utilizing some legal mechanisms available to them via a court system or administrative law), the flows of resource units can usually be measured accurately. This is needed in order to achieve a record of the volume of resource flows (acre-feet of water, board-feet of timber, or tons of fish) that the private right conveys to the owner. Other self-organized institutional rules may define a group as the common owner of a resource and develop specific rules for when and by whom harvesting may be undertaken. Limits may exist on the harvesting technology to be used or the purpose of harvesting (for family consumption and/or commercial purposes) and the responsibilities that co-owners have for maintenance or monitoring. When governments declare ownership of common-pool resources, rules related to who can use and for what purpose are defined and monitored by an administrative agency of the governmental owner.

There are many well-documented examples of private property, community property, and government property systems that work effectively over time to keep the common-pool resource sustainable. Unfortunately, there are also a multitude of empirical examples where private, community, or government ownership is faltering or has collapsed. There are other examples where resource users have not succeeded in overcoming common-pool dilemmas—usually when the resource system is very large.

7. THE IMPORTANCE OF LARGER GOVERNANCE REGIMES

Many of the variables listed above—particularly those related to the resource users—are strongly affected by the larger political regime in which users are embedded. Larger regimes can facilitate local self-organization by providing accurate information about natural resource systems, providing arenas in which participants can engage in discovery and conflict-resolution processes, and providing mechanisms to back up local monitoring and sanctioning efforts. Perceived benefits of organizing are greater when users have accurate information about the resource itself, about the users of it, and about the threats facing a resource.

The costs of monitoring and sanctioning those who do not conform to rules devised by users are very high if the authority to make and enforce these rules is not recognized by higher governmental authority. Thus, the probability of users adapting more effective rules when they live in a governmental regime that facilitates their efforts over time is higher than when they live in regimes that ignore resource problems entirely or, at the other extreme, presume that all decisions about governance and management need to be made by central authorities. If local authorities are not recognized by larger regimes, it is difficult for users to establish an enforceable set of rules. On the other hand, if rules are imposed by outsiders without consulting local participants in their design, local users may not consider such rules to be legitimate and may try to evade them.

The search for rules that improve the outcomes obtained in commons dilemmas is an incredibly complex task whether undertaken by users or by government officials. It involves a potentially infinite combination of specific rules that could be adopted in any effort to match the rules to the attributes of the resource system. Instead of assuming that designing rules that improve performance of common-pool resources is a relatively simple analytical task that can be undertaken by distant, objective analysts, we need to understand the institutional design process as involving an effort to tinker with a large number of component parts. Those who tinker with any tools—including rules—try to find combinations that work together more effectively than other combinations. Policy changes are experiments based on more or less informed expectations about potential outcomes and the distribution of these outcomes for participants across time and space. Whenever individuals agree to add a rule, change a rule, or copy another successful system's rule set, they are conducting a policy experiment. Further, the complexity of the ever-changing biophysical world combined with the complexity of rule systems means that any proposed rule change faces a nontrivial probability of error.

Policy makers working in a *single* authority for a large region have to experiment simultaneously with *all* of the common-pool resources within their jurisdiction. And, once a change has been made and implemented, further changes will not be made rapidly. The process of experimentation will usually be slow, and information about results may be contradictory and difficult to interpret. Thus, an experiment that is based on erroneous data about one key structural variable or one false assumption about how actors will react can lead to a very large disaster. In any design process where there is substantial probability of error, having redundant teams of designers has repeatedly been shown to have considerable advantage.

8. THE ADVANTAGES OF POLYCENTRIC RESOURCE GOVERNANCE SYSTEMS

Thus, we need to address why a series of nested but relatively autonomous, self-organized, resource governance systems can do a better job in policy experimentation than a single central authority. A polycentric system is one in which citizens are able to organize not just one but multiple governing authorities at differing scales. Thus, a polycentric system would have some units at a smaller scale corresponding to the size of the basic common-pool resources in the system.

Among the advantages of authorizing the users of smaller-scale common-pool resources to adopt policies regulating the use of common-pool resources are:

- Local knowledge. Resource users who have lived near and harvested from a resource system over a long period of time develop relatively accurate mental models of how the biophysical system itself operates because the very success of their efforts depends on such knowledge. They also know others living in the area and what norms of behavior are considered appropriate in what circumstances.
- Inclusion of trustworthy participants. Local resource users can devise rules that increase the probability that others will be trustworthy and use reciprocity. This lowers the cost of relying entirely on formal sanctions and paying for extensive guarding.
- Reliance on disaggregated knowledge. Feedback about how the resource system responds to changes in actions of resource users is provided in a disaggregated way. Fishers are aware, for example, if the size and species distribution of their own catch are changing over time and tend to discuss the size of their catch with other fishers. Irrigators learn whether a particular rotation system allows most farmers to grow the crops they most prefer by examining the resulting productivity of specific fields or talking with others about yields at a weekly market.
- Better-adapted rules. Given the above, resource users are more likely to craft rules that are better adapted to each of the local common-pool resources than any general system of rules.
- Lower enforcement costs. Because local resource users have to bear the cost of monitoring, they are likely to craft rules that make infractions obvious to other users so that monitoring costs are lower. Further, if rules are seen as legitimate, rule conformity will tend to be higher.
- Redundancy. The probability of failure throughout a large region is greatly reduced by the establishment of parallel systems of rule making, interpretation, and enforcement.

There are, of course, limits to all ways of organizing the governance of common-pool resources. Among the limits of a highly decentralized system are these:

- Some resource users will not organize. Although the evidence from the field is that many local resource users do invest considerable time and energy in their own regulatory efforts, other groups do not do so. Many reasons exist for why some groups do not organize, including the presence of low-cost alternative sources of income and thus a reduced dependency on the resource, considerable conflict among resource users along multiple dimensions, lack of leadership, and fear of having their efforts overturned by outside authorities.
- Some self-organized efforts fail. Given the complexity of the task involved in designing rules, some groups will select combinations of rules that generate failure instead of success. They may be unable to adapt rapidly enough to avoid the collapse of a resource system.
- Local tyrannies. Not all self-organized resource governance systems will be organized democratically or rely on the input of most resource users. Some will be dominated by a local leader or a power elite who only make rule changes that they think will advantage them still further. This problem is accentuated in locations where the cost of exit is particularly high and reduced where resource users can leave when local decision makers are not responsible to a wide set of interests.
- Stagnation. Where local common-pool resources are characterized by considerable variance, experimentation can produce severe and unexpected results leading resource users to cling to systems that have worked relatively well in the past and stop innovating long before they have developed rules likely to lead to better outcomes.
- Limited access to scientific information. Although time and place information may be extensively developed and used, local groups may not have access to scientific knowledge concerning the type of resource system involved.
- Conflict among resource users. Without access to an external set of conflict-resolution mechanisms, conflict within and across common-pool resource systems can escalate and provoke physical violence. Two or more groups may claim the same territory and may continue to make raids on one another over a very long period of time.
- Inappropriate discrimination. Determining who has a right to use a resource based on ascribed characteristics can be the basis for excluding some individuals from access to sources of productive endeavor that has nothing to do with their trustworthiness.
- Inability to cope with larger-scale common-pool resources. Without access to some larger-scale jurisdiction, local users may have substantial difficulties regulating only a part of a larger-scale common-pool resource. They may not be able to

> exclude others who refused to abide by the rules that a local group would prefer to use. Given this, local users have no incentives to restrict their own use and watch others take away all of the valued resource units that they have not harvested.

Many of the capabilities of a parallel adaptive system exist in a polycentric governance system. Each unit may exercise considerable independence to make and enforce rules within a circumscribed scope of authority for a specified geographic area. In a polycentric system, some units are general-purpose governments, whereas others may be highly specialized. Self-organized resource governance systems, in such a system, may be special districts, private associations, or parts of a local government. These can be nested in several levels of general-purpose governments that also provide civil, equity, as well as criminal courts.

In a polycentric system, the users of each common-pool resource would have authority to make at least some of the rules related to how that particular resource will be utilized and thus would achieve most of the advantages of utilizing local knowledge and the redundancy and rapidity of a trial-and-error learning process. On the other hand, problems associated with local tyrannies and inappropriate discrimination can be addressed in larger, general-purpose governmental units that are responsible for protecting the rights of all citizens and for the oversight of appropriate exercises of authority within smaller units of government. It is also possible to make a more effective blend of scientific information with local knowledge where major universities and research stations are located in larger units but have a responsibility to relate recent scientific findings to multiple smaller units within their region. Because polycentric systems have overlapping units, information about what has worked well in one setting can be transmitted to others who may try it out in their settings. Associations of local resource governance units can be encouraged to speed up the exchange of information about relevant local conditions and about policy experiments that have proved particularly successful. And, when small systems fail, larger systems can be called on, and vice versa.

Polycentric systems are themselves complex, adaptive systems without one central authority always dominating all of the others. No guarantee exists that such systems will find an effective combination of rules at diverse levels that are sustainable in any particular environment. In fact, scholars and policymakers should not expect that any governance system will operate over time at optimal levels given the immense difficulty of fine-tuning complex, multitiered systems. Experimentation, feedback, and adaptation are continuing processes required of any governance system for a social–ecological system to be sustainable over time.

FURTHER READING

Berkes, Fikret, Johan Colding, and Carl Folke, eds. 2003. Navigating Social–Ecological Systems. Cambridge, UK: Cambridge University Press.

Dietz, Thomas, Elinor Ostrom, and Paul Stern. 2003. The struggle to govern the commons. Science 302: 1907–1912.

Gunderson, Lance H., and C. S. Holling. 2002. Panarchy. Washington, DC: Island Press.

Hardin, Garrett. 1968. The tragedy of the commons. Science 162: 1243–1248.

Janssen, Marco. 2002. Complexity and Ecosystem Management. Cheltenham, UK: Edward Elgar.

Levin, Simon. 1999. Fragile Dominion. Reading, MA: Perseus Press.

National Research Council. Committee on the Human Dimensions of Global Change. 2002. The Drama of the Commons. Elinor Ostrom, Thomas Dietz, Nives Dolšak, Paul Stern, Susan Stonich, and Elke Weber, eds. Washington, DC: National Academy Press.

North, Douglass. 2005. Understanding Institutional Change. Princeton, NJ: Princeton University Press.

Ostrom, Elinor. 1990. Governing the Commons: The Evolution of Institutions for Collective Action. New York: Cambridge University Press.

Ostrom, Elinor. 2005. Understanding Institutional Diversity. Princeton, NJ: Princeton University Press.

VII.11

Assessments: Linking Ecology to Policy

Clark A. Miller

OUTLINE

1. Assessment purpose and function
2. Assessment design and organization
3. An illustrative example: The Millennium Ecosystem Assessment
4. Conclusions: The politics of assessment

Finding ways to deliberate and communicate knowledge and ideas among ecologists, policy officials, and the public has not always proven easy or straightforward. In response to these challenges, governments have sought ways to systematize and rationalize the flow of ecological and other scientific knowledge into policy processes. The tool that they invented to do so is called an assessment. This chapter examines the purpose and functions of assessment, identifies central questions that confront assessment organizers, and argues for careful attention among ecologists to assessment design and management choices. The chapter also explores the need to think about assessments not only in scientific terms but also as an important location for fostering the societal deliberation of ecological knowledge and ideas.

GLOSSARY

assessment. Assessment is a tool for accomplishing three tasks: first, identifying, synthesizing, and evaluating a wide range of claims to knowledge; second, certifying a particular set of knowledge claims as relevant to policy decisionmaking; and third, fostering necessary communication among scientists of many disciplines, others with relevant knowledge, and policy and public audiences.

Millennium Ecosystem Assessment. The Millennium Ecosystem Assessment was a 5-year effort to assess global trends in ecosystem services and to transform the resultant knowledge into political action to reduce ecological threats worldwide.

Ecology is a highly policy-relevant science. As a discipline, ecology, perhaps as much as any other field of science, creates knowledge and ideas that are essential to the proper design and conduct of environmental policy. Yet, finding ways to communicate and deliberate knowledge and ideas among ecologists, policy officials, and the public has not always proven easy or straightforward. In a democracy, politics is inevitably contested, and ecological ideas, like all policy-relevant scientific ideas, are often uncertain, subject to tacit assumptions and models and open to interpretation from divergent scientific and political perspectives. Hence, beginning in the 1970s, government agencies sought ways to systematize and rationalize the flow of ecological and other scientific knowledge into policy processes. The tool that they invented to do so is called an assessment.

Today, assessments are central to the practice of ecology. Each year, ecologists contribute to thousands of assessments. High-profile assessments such as the Millennium Ecosystem Assessment (MA) and the Intergovernmental Panel on Climate Change provide ecologists with the opportunity to convey their ideas on a global stage. Far more mundane, but no less important, ecologists contribute every day to drafting the multitude of assessments required by the environmental laws of the United States and other countries: risk assessments, biological assessments, ecological assessments, wetlands assessments, endangered species assessments, biodiversity assessments, ecosystem assessments, etc. The collective impact of this work has enormous consequences for shaping the knowledge and ideas that are ultimately brought to bear on the making of public policy and, therefore, for what aspects of the environment are, and are not, preserved and protected for future generations. Hence, it is vital that ecologists give serious attention to the conceptual and practical foundations of assessment.

My objective in this chapter is to introduce and describe assessment, with special attention to questions of assessment purpose and function, assessment design and organization, and the broader politics of ecological and environmental assessment. In these discussions, I draw heavily on the experience of the MA, given its prominence in the recent history of ecology. Of necessity, these discussions are brief. I hope, however, that they are sufficient to express the need for the ecological community to invest significant time and energy in the innovative design and practice of ecological assessments. Assessments, as processes that link science and policy, are essential to the design of well-reasoned environmental policies. They are also important elements in the infrastructure of democratic decisionmaking. For both reasons, they deserve careful consideration.

1. ASSESSMENT PURPOSE AND FUNCTION

Assessments have emerged since the 1970s as arguably one of the most important means for connecting science to public policy choices. In this task, they are joined by two other prominent mechanisms: (1) the informal work of scientists who, whether in their day-to-day work or as an exceptional event, seek to communicate the results of their own work and the work of their colleagues to policy officials through targeted publication, informal conversation, or public hearings; and (2) the formal work of scientific advisory processes that are convened either through the work of organizations such as the National Academy of Science or as permanent bodies, such as the U.S. President's Science Advisor or U.S. Environmental Protection Agency's Science Advisory Board.

Of these three approaches to science advice, assessments generally respond to the needs of policy agencies and officials for synthetic surveys of relevant scientific material that has undergone some form of formalized review and can thus be taken as an official statement certifying what is known regarding a particular subject. In this fashion, assessments perform several key functions. First, they identify, synthesize, evaluate, and assess scientific facts, data, studies, and theories as well as other knowledge and ideas for their relevance and significance with respect to the policy question at hand. This is crucial, as science, at least as currently organized, tends to produce a vast array of publications containing data, analyses, theories, and conjectures. Individually, often, these publications are extremely narrow, contain relatively greater or lesser amounts of uncertainty, and provide little or no information regarding their relevance to a range of policy questions. Nor do they include all that scientists know regarding their subjects. There is a considerable need, therefore, for assessments to sift and winnow through the available publications and combine the resulting information with the tacit knowledge of participating scientists to produce a synthetic statement of what is known about the policy question of relevance.

Second, assessments serve as mechanisms for the political certification of science and other relevant knowledge and information. This is an underappreciated but essential function of assessments. Especially since the 1970s, science has become increasingly important to environmental policy, including, in the United States, for example, legal requirements for regulatory agencies to defend their decisions in explicitly scientific terms. As a consequence, disputes about what science says about a particular policy question have become commonplace, and more and more organizations have become adept at deploying science to serve their own, narrow interests. Both industry groups and environmental activist organizations have turned to scientists and science to support their positions in policy debates, and both increasingly fund extensive scientific research portfolios. Assessments, in turn, have emerged as one approach that policy agencies have developed in the attempt to politically certify or warrant a certain view of science as the officially recognized version for purposes of a particular policy question, thus removing or reducing scientific debate. As a result, assessments are not simply scientific activities; rather, they are hybrid processes that combine scientific and political elements—a fact that has important consequences for their design and organization.

The third key function of assessments is communication. Often it is assumed that the purpose of assessments is to communicate knowledge and ideas from scientists to policy officials. This is a significant oversimplification of the communication function performed by assessments, however. First, it is important to recognize that assessment communication among scientists, policy officials, stakeholders, and other publics or audiences is often bidirectional. It is as essential for scientists to be aware of policy questions and concerns while conducting assessments as it is for policy audiences to hear scientific conclusions. Second, assessments also serve to foster communication among scientists, especially across the boundaries of discipline, specialization, or nation. Policy concerns are almost always highly interdisciplinary and, in some cases, transnational or even global. In these cases, scientific communities and other knowledge holders that rarely, if ever, interact systematically with one another may need to work closely together to fruitfully integrate the specialized knowledge, ideas, data, models, and theories from each. This requires effective strategies for cross-disciplinary communication and learning.

2. ASSESSMENT DESIGN AND ORGANIZATION

Creating an assessment that effectively serves the functions of synthesizing and vetting knowledge and ideas, certifying science as policy relevant, and facilitating effective communication among a range of participants and audiences are not simple matters. Questions of design and organization are thus essential to successful assessments. Key choices focus on the framing of the assessment; the organization of processes for identifying, articulating, integrating, reviewing, and certifying knowledge and ideas; participation in these processes; the form the final products of the assessments will take; and strategies for communication among those who commission and conduct assessments and to assessment audiences. These choices can impact the accuracy and comprehensiveness of an assessment, its credibility and legitimacy, and its ultimate uptake and effectiveness.

Assessors will not always have complete freedom to choose assessment design, however. In almost all cases, a policy body sets the terms and often also key aspects of the design and organization of the assessment. In many cases, e.g., environmental impact assessments or wetlands assessments, the form and conduct of assessments are tightly constrained by requirements set by appropriate environmental laws and regulations. In other cases, assessments are commissioned by a policy agency that desires scientific input to a policy decision or by a stakeholder who desires to influence a decision being made by a policy agency. In these cases, the organization that commissions the assessment will typically set its terms and at least exert some influence over its design. In still other cases, such as the Millennium Ecosystem Assessment and Intergovernmental Panel on Climate Change, a governing board, comprised of politically or socially sanctioned representatives, is responsible for key design choices. In almost all cases, however, scientists will be involved to at least some extent in key aspects of designing and organizing assessments.

Framing

One set of choices associated with assessment design where scientists typically play a significant role is the choice of its conceptual framework, orientation, and focus. Framing can be defined as the systematic lenses, interpretative frameworks, or narrative storylines through which assessments make sense of and give meaning to facts, evidence, theories, uncertainties, etc. Framing is crucial because it provides not only the filters to be used in selecting the knowledge and ideas relevant to the assessment but also the integrative connections that tie together the many parts of a given assessment. Framing also provides the communicative resonance that gives assessments the ability to shape ideas for audiences and for use in processes of policy decisionmaking.

Standards of Evidence

A second important set of questions involves how assessors interpret and weigh evidence and uncertainty during the process of synthesis and integration. How much uncertainty is tolerated in assessment findings and how is that uncertainty expressed? How are competing knowledge claims weighed and adjudicated, e.g., when multiple scientific data and analyses arrive at different conclusions, scientists from different disciplines disagree, or scientific findings are in conflict with knowledge held by stakeholders or other assessment participants? What standards must be met for knowledge to be included in an assessment? Must it be quantitative or peer reviewed? Or will the assessment include more qualitative evaluations or perhaps even narratives and stories from relevant participants? Will dissent be allowed in the assessment, and, if so, in what form?

Participation

Another important set of choices that assessment designers face includes who will participate in the assessment, in what capacity, and at what points in time. Participation determines who will have a voice in shaping the knowledge and ideas that emerge from the assessment and, thus, who will have a voice in shaping the knowledge and ideas that shape policy decisions. Designers must decide which scientists will participate and how (e.g., as authors or reviewers), what disciplines they will represent, whether nonscientists will be allowed to participate and in what ways (either as knowledge holders, e.g., indigenous communities who hold traditional ecological knowledge, or as reviewers of assessment findings), how duties will be divided among different participants (e.g., between the governing board and scientific leaders), and many more complex questions concerning who will be able to take part and therefore help to shape assessment outcomes.

Processes

A final important set of choices relates to the processes by which the assessment operates. These choices determine the rules and procedures by which various aspects of the assessment will be conducted. If the assessment includes teams of authors writing different sections of a final report, how will these teams be selected, and how will their duties be divided up? Will assessment products be reviewed before they are published? If so, how often, at what point in the process, and how will reviewers be selected? Process is also essential for fostering effective communication. What mechanisms will be used to facilitate communication between and among different groups of scientists, knowledge holders, policy officials, stakeholders, and potential audiences? How will the commissioning organization communicate its requirements to assessors? How will assessors communicate among themselves? How will the final assessment products be prepared and communicated? Still other process questions involve the transparency and openness of the assessment to nonparticipants. How much information is made available about the assessment, when, and to whom, during the course of its activities?

3. AN ILLUSTRATIVE EXAMPLE: THE MILLENNIUM ECOSYSTEM ASSESSMENT

For ecologists, perhaps the most prominent and important assessment of the past decade has been the MA. Designed to synthesize and call attention to ecological knowledge showing widespread degradation of ecosystems around the globe—and the potential implications of ecosystem decline for human welfare and well-being—the MA is an illustrative and illuminating example of the value of innovation in assessment design and practice. Inspired by its predecessor, the Intergovernmental Panel on Climate Change (IPCC), the MA sought to influence global policy on the environment by assessing and communicating ecological knowledge to policy officials across the globe. As it evolved, however, the MA departed significantly from key design features of IPCC, offering a new model for assessments in the twenty-first century. It is well worth a careful look, therefore.

MA Origins

The origins of the MA lie in three intersecting strands of history. One strand took place in parallel developments in the politics of climate change and biodiversity loss. By the late 1990s, many observers were concerned that global climate policies were advancing more rapidly than corresponding efforts to protect global biodiversity. Among scientists, blame was frequently laid for this lag on the absence of an equivalent to IPCC for biodiversity. Especially after the negotiation of the Kyoto Protocol, a number of high-profile scientists argued strongly for the creation of an IPCC for biodiversity to give a scientific boost to global biodiversity negotiations. During the same time period, many ecologists were becoming concerned about the degradation of ecosystems worldwide—some related to concerns about biodiversity but even more often not. In many parts of the world, fisheries were collapsing, agricultural yields were in decline, and key ecological groups, such as amphibians, were being decimated by disease. These concerns raised the specter of widespread ecological crises that demanded policy action.

The third strand occurred in the emerging field of ecological economics, where growing interest in measuring the benefits of environmental protection gave rise during the 1990s to a new concept of ecosystem services (or sometimes ecosystem goods and services). Unlike prior concepts of nature and biodiversity, the idea of ecosystem services focused not on the intrinsic value of wilderness or nature but, rather, on the value of ecosystems to human welfare and well-being. What services, in other words, were ecosystems providing to humans that would need to be replaced if the underlying ecological processes degraded or collapsed?

These strands came together in the late 1990s when scientists at the World Resources Institute (WRI) suggested the idea of conducting an assessment of global ecosystem services that would measure ecological decline around the planet and highlight its impacts on human welfare and well-being. To showcase its ideas, WRI funded and organized a pilot assessment and brought its ideas to the attention of UN Secretary General Kofi Annan. In turn, Annan, as part of his efforts to launch a major effort to promote human welfare and well-being around the globe known as the Millennium Development Goals, called in 2000 for the conduct of an MA to explore connections between global ecological degradation and the needs of human communities.

MA Design

The governance of the new MA was ultimately a hybrid that built on but did not entirely mimic the IPCC. WRI agreed to withdraw from the position of principal organizer in favor of the creation of an independent Board of Directors. WRI's influence, nevertheless, was crucial in shaping the design and organization of the MA as it subsequently evolved. Walt Reid, who led the initiative at WRI, was appointed the MA Director and cochair of the Board of Directors. The other cochair was Robert Watson, a key figure in the organization and leadership of the IPCC.

The MA Board of Directors played a similar role to that of the IPCC Plenary. The IPCC Plenary is comprised of government representatives who establish the rules and procedures of the IPCC, set its work plan, and formally approve its final products. The MA Board of Directors served essentially identical purposes but was organized very differently. Instead of government representatives, Board members were drawn from a range of public and private-sector organizations. The argument behind this design choice was that ecological degradation was a problem not just for governments but also for business groups, not-for-profit organizations, and indigenous communities, and, hence, all of these groups should have a stake in the MA governance and its outcomes. Ideally, therefore, individuals in all of these sectors would come to see the value of the MA and help persuade others to take up and use its ideas and findings. Recognizing the importance of keeping governments involved, as well, the MA requested and received approval to conduct the assessment from the governing organizations of several key international environmental treaties addressing biodiversity, climate change, wetlands, migratory species, and others. Prominent efforts were made by MA leaders to engage with the UN Convention on Biological Diversity, the Ramsar Convention, the Convention to Combat Desertification, and other treaty organizations, to formally approve of the MA, and to identify questions that would be important in guiding the assessment. Whether this indirect approach to securing government participation and approval was sufficient remains an open question.

A second design choice—to frame the MA in terms of ecosystem services—also reflected the preliminary influence of the WRI and had important implications for other design decisions. For example, participation in the assessment was significantly broadened beyond ecologists and biologists to include substantial participants from the social sciences and also from nontechnical communities, including indigenous groups. In this way, the MA departed significantly from the IPCC model, which had focused almost exclusively on peer-reviewed science and included almost exclusively natural scientists. To some extent, social scientists had been included in marginal aspects of the IPCC, such as the assessment of climate change impacts, but nonscientists were never involved in the core conceptual formation of IPCC assessments (as they were in the MA), nor did the IPCC allow knowledge to be included that had not been subjected to careful peer review and publication in a recognized scientific journal.

In other aspects, the MA followed the IPCC model much more closely. The basic output of the MA was a multivolume report that contained three primary sections: the *State and Trends* of global ecosystems, *Scenarios* of future ecosystem change, and *Policy Responses* for limiting future ecological degradation. Each volume was written in chapters by teams of authors. Each team had lead authors, who assembled groups of specialists to draft the chapter in question. Subsequently, chapter drafts were distributed for review to government experts as well as to experts working in private sector, non-for-profit, and university settings. Chapter authors then modified chapter drafts to take reviewer comments into account, overseen by an editorial board that worked to ensure that authors had not inappropriately dismissed or failed to take into account reviewer comments. Finally, the entire document was submitted to the MA Board of Directors for approval. Here, however, the process once again differed from the IPCC. Unlike the IPCC Plenary, the Board did not undertake a line-by-line review of the final assessment summary. Instead, the Board wrote its own summary of the assessment conclusions. In addition, assessment participants wrote a series of short synthesis pieces that excerpted key findings from the overall assessment relevant to different policy sectors.

Innovative Communication

Beyond its innovative approach to governance, the MA also sought novel approaches to communicating its ideas and findings to policy and public audiences. One element of this strategy involved the compilation of elaborate scenarios of global ecological futures. The idea of scenarios as tools for aiding policy decisionmaking has acquired considerable attention in the past 20 years, especially through use of this model by business leaders at Royal Dutch/Shell. After the work of the Club of Rome and its *Limits to Growth* report in the early 1970s, however, scenarios have played only a limited role in international environmental assessment and governance. For example, the IPCC used a stripped-down version of scenarios but only to develop standardized technical inputs for climate models.

By contrast, the MA adopted a much more narrative-centered approach to building scenarios. Seeking a tool that could help communicate the meaning and implications of ecological change for human societies, the MA assigned authors for one of its volumes the task of developing a series of distinct scenarios that would highlight divergent possible policy responses to contemporary ecological crises. In this fashion, MA leaders hoped, policy officials would be able to envision the broad outlines of what might happen should they adopt a range of possible strategies. Scenarios, in this instance, were not meant to be predictive but merely plausible, scientifically grounded visions of what might happen in the absence of clear policy action to arrest ecological decline.

Another innovative approach to communication involved the decision to focus considerable attention on what the MA termed "subglobal" or regional assessments. MA leaders believed that although a global assessment was needed to draw attention to worldwide ecological degradation, such an assessment would be unlikely to be of much value to policy officials at a range of subglobal scales (local, national, or regional) whose decisions would be most influential in determining the future trends of individual ecosystems. A major goal of the MA became, therefore, to pioneer the concept of multiscale assessments that would link from the local to the global.

This decision was important because it took the MA into novel territory regarding the organization of international scientific assessments. First, it demanded the creation of a new and different kind of knowledge and synthesis that did not solely focus on global trends in ecological degradation. Instead, detailed knowledge of specific ecosystems and their intersections with human welfare and well-being became necessary. Local experts became essential participants, edging the MA away from the elite international scientific leadership that dominated the IPCC and other parts of the MA. Although subglobal assessments were required to focus on ecosystem services, their methodological approaches were not standardized, leading to a proliferation of alternative methods and practices, each adapted to local sources of knowledge and the needs of local policy agencies and officials. The results were surprisingly diverse and led MA leaders to recognize the vast differences among potential audiences for their work. Consequently, the MA sought to develop novel approaches to disseminating their ideas and conclusions to local communities and audiences, including, for example, the use of theater and dance to communicate with indigenous communities. Much of this work was highly experimental but illustrated the potential value that could be achieved in going beyond the traditional limits of international scientific assessments.

4. CONCLUSIONS: THE POLITICS OF ASSESSMENT

Organizers inevitably face complex trade-offs in making decisions about the design and organization of assessments. These choices significantly impact the shape that the final knowledge developed by the assessment takes as well as the assessment's ability to foster effective communication among diverse audiences. It is crucial, therefore, that assessors give these choices careful consideration.

Assessors must also give careful consideration for another reason: namely, the potential political dimensions of their choices. There is a tendency, at times, for people to assume that assessments are essentially tools of science, but they are just as much tools of politics. Assessments are instruments not only for aggregating and synthesizing scientific knowledge but also for integrating scientific and other forms of knowledge and certifying the results as policy relevant and significant. As a consequence, assessments are hybrid entities: fully scientific and political instruments for determining what knowledge and ideas should and should not shape policy decisions.

Consider the MA choice to frame their assessment in terms of ecosystem services. MA designers made this choice largely on the basis of what they believed would be persuasive to policymakers, but it also had the implication of downplaying the significance of knowledge and ideas regarding the intrinsic value of natural ecosystems irrespective of their value to humans. This was a serious political choice that was in fact disputed among assessment authors and criticized by other players in the policy process.

Assessment designers and practitioners should recognize that, in designing assessments, they are creating important elements of policy and political processes that will have significant impacts on what (and also whose) knowledge and ideas come to be seen as politically influential. They are, in other words, helping to shape a critical aspect of democratic decisionmaking. Democratic theorists remind us that the deliberation of knowledge and ideas is perhaps the single most important element in the creation of a robust and healthy democracy. It is thus crucial for assessments to go beyond a narrow focus on science alone. Equally important for assessments is the challenge of enhancing the deliberation of scientific and other ideas underpinning policy choices on ecology and the environment.

FURTHER READING

Farrell, Alexander E., and Jill Jager. 2006. Assessments of Regional and Global Environmental Risks: Designing Processes for the Effective Use of Science in Decisionmaking. Washington, DC: Resources for the Future.

Jasanoff, Sheila, and Marybeth Long Martello. 2004. Earthly Politics: Local and Global in Environmental Governance. Cambridge, MA: MIT Press.

Miller, Clark A., and Paul N. Edwards. 2001. Changing the Atmosphere: Expert Knowledge and Environmental Governance. Cambridge, MA: MIT Press.

Mitchell, Ronald B., William C. Clark, David W. Cash, and Nancy M. Dickson. 2006. Global Environmental Assessments: Information and Influence. Cambridge, MA: MIT Press.

Reid, Walter V., Fikret Berkes, Tom Wilbanks, and Doris Capistrano, eds. 2006. Bridging Scales and Knowledge Systems: Concepts and Applications in Ecosystem Assessments. Washington, DC: Island Press.

Milestones in Ecology

Compiled by Christopher Morris

This timeline presents a view of the successive advancement of the field of ecology, first through earlier developments that provided a foundation for the field and then through developments within the formal discipline of ecology itself, from the late 1800s onward. The timeline concludes with a cutoff date in the late twentieth century, based on the principle that contemporary developments need a certain interval of time before their significance can be properly evaluated.

500,000 BC. Proposed date for the earliest use of fire in a controlled manner, the first major alteration of the natural environment by human activity. By about 3000 BC, many forest regions of the Middle East will be stripped of trees for the fuel demands of the Bronze Age.

8000 BC. Estimated time for the beginnings of agriculture, crop irrigation, and village formation in various areas of the world, especially in parts of the so-called Fertile Crescent such as Mesopotamia and the Nile Valley. Evidence also indicates that plants such as gourds were being cultivated at about the same time in the Oaxaca Valley of Mexico.

ca. 2000 BC. The Indus Valley civilization, one of the three great sites of early civilization along with Mesopotamia and Egypt, declines and eventually collapses. Cited as a leading cause of this is the large-scale removal of forests, which is thought to have shifted the habitat preferences of the mosquito *Anopholes stephansi,* a dangerous malaria vector, from forest to urban areas.

900s BC (?). The Bible states in Genesis 1:26: "And God said, Let us make man in our image, after our likeness: and let them have dominion over the fish of the sea, and over the fowl of the air, and over the cattle, and over all the earth, and over every creeping thing that creepeth upon the earth." This passage has been interpreted in contrasting ways in the modern era, on the one hand as the Christian basis for the concept of environmental stewardship, and conversely, as a God-given right to exploit the natural world for human benefit.

500s BC. Ancient Chinese writers describe feeding patterns in animal communities with aphorisms such as "The large fish eat the small fish"; "Large birds cannot eat small grain"; and "Each hill can shelter only a single tiger." In the twentieth century, animal ecologist Charles Elton will cite these sayings to show ancient awareness of the principle of food pyramids.

400s BC. The Greek philosopher Empedocles postulates that animals had originally been formed at random from individual parts, with those in which the parts formed a natural body shape surviving and reproducing over time, whereas those with mismatching parts died out. This concept of survival of certain body types and extinction of others roughly anticipates the nineteenth-century theory of natural selection.

400s BC. Herodotus, known as the father of history, reports on the wildlife he observes in areas of the Mediterranean. He identifies an example of mutualism, involving the Nile crocodile (*Crocodylus niloticus*) and a bird (the Egyptian plover, *Pluvianus aegyptius*) that removes and eats parasitic leeches in the crocodile's mouth. Herodotus also describes a balance-of-nature concept by noting that prey animals such as the rabbit have greater reproductive capacity than the predators that feed on them.

ca. 380 BC. A striking example of resource depletion is described by Plato, who decries loss of forest cover, and subsequent erosion, in the mountains of his native region of Attica. He portrays the area as "A mere relic of the original country. . . . What remains is like the skeleton of a body wasted by disease. All the rich soil has melted away, leaving a country of skin and bone."

300s BC. Aristotle establishes a classification system for animals, placing those with red blood in a different category from those without blood. This in effect corresponds to the contemporary distinction between vertebrates and invertebrates. He further divides the blooded animals into five groups similar to the modern system of mammals, birds, reptiles, amphibians, and fish.

300s BC. Chinese philosophers of the Taoist (Daoist) tradition develop a concept of living as one with nature, based on the idea that humanity is only a single component of the wholeness of the natural world, rather than the master of it. Restraint is urged in the use of resources to maintain the harmony and balance of nature. Taoism has thus been described as a model for the modern environmental philosophy of deep ecology.

ca. 300 BC. Aristotle's pupil Theophrastus produces an exhaustive study of plants, describing them according to such criteria as method of reproduction, size, habitat, method of cultivation, practical uses, and appearance, smell, and taste. This is considered the first significant step in the systematic classification of plant life.

ca. 220 BC. The Qin Dynasty of ancient China enacts the world's earliest known environmental protection laws. According to documents recently discovered, bans or restrictions were placed on the cutting of trees, the burning of grass or picking of germinating plants, the killing of baby animals and birds, and the use of poison, traps, or nets to catch fish and game.

100s BC. Ancient Rome provides contrasting examples of the management of ecosystem services. On one hand, the Romans successfully manage and conserve water through an elaborate system of aqueducts and water fountains. On the other hand, they fail to properly manage forest resources, and evidence suggests this deforestation is one of the causes of the eventual collapse of the Roman Empire.

ca. AD 65. Greek physician Dioscorides travels widely in the area of the Mediterranean and Asia Minor and writes *De Materia Medica,* a five-volume compendium on "the preparation, properties, and testing of drugs." In this work he describes more than 600 different plants that have pharmacological effects and establishes the practice of organizing botanical information on a species-by-species basis.

AD 800s. The Arab scholar Al-Jahiz provides the first explicit description of a food chain, stating that "All animals, in short, can not exist without food, neither can the hunting animal escape being hunted in his turn." He also is among the first to describe the effect of environmental factors on animal life.

1100s. Ibn al-'Awwam, an Arab agriculturist active in Moorish Spain, writes a comprehensive encyclopedia of botany in which he discusses hundreds of different kinds of crop plants, including 50 types of fruit trees. This work includes valuable information on soil science, fertilization, grafting, and plant pathology.

1100s. The Cistercians, a Roman Catholic monastic order, develop methods of sustainable agriculture that allow their monasteries to derive all necessary sustenance and income from their own farm production. In particular, they are noted for the practice of an early form of restoration ecology, converting desolate or abandoned terrain into fertile agricultural land.

early 1200s. St. Francis of Assisi becomes widely known for his love of nature. According to legend, he is able to gather flocks of birds around him and preach to them. He comes to represent the concept of living in harmony with the environment rather than trying to dominate it.

1240s. Frederick II, emperor of the Holy Roman Empire, publishes a six-volume treatise entitled *De Arte Venandi cum Avibus* [*The Art of Hunting with Birds*]. It is lavishly illustrated and contains many valuable observations on avian biology, dealing with topics such as comparative anatomy, feeding habits, characteristic habitats, and diurnal/nocturnal patterns.

1273. King Edward I of England enacts what is considered the first air pollution control law. The law bans the use of coal for fuel within the city of London. It applies to "sea coal," a low-grade form of soft coal that exudes excessive smoke.

ca. 1300. Forests of England and France have become depleted to such an extent that wood has to be imported from other countries. Demand for wood comes not only for heating and building but also for industrial uses such as ironworking, brewing, dyeing, and glassmaking.

before 1492. Native Americans engage in the cultivation strategy of the "three sisters" (corn, beans, and squash) over large areas of the Americas. They also practice land-use management through the controlled burning of forests and grasslands, which promotes the success of fire-resistant (or fire-dependent) plant species.

1493. Christopher Columbus returns to Spain from his momentous voyage to the West Indies, carrying specimens of various plants, birds, and animals previously unknown to the Old World. These include important food plants such as maize (corn), sweet potatoes, peppers, bananas, and pineapples. Columbus is also said to be one of the first to note that forests influence and enhance rainfall, through his observations of the landscape of the Azores and Canaries.

1503. The German artist Albrecht Dürer creates a watercolor known as "The Large Piece of Turf," which is described as the first work of art with an ecological theme. Before this, artists had included plants as background for scenes with human subjects, but Dürer's painting realistically portrays a plant community separate from human society.

1543. Andreas Vesalius, a Flemish physician, publishes *De Fabrica Corporis Humani* [*Concerning the Structure of the Human Body*], a book that founds the scientific discipline of anatomy. The influence of Vesalius extends beyond anatomy as his work inspires various similar studies in zoology, as by Pierre Belon on birds (1555) and Conrad Gesner (see next).

1556. Swiss naturalist Conrad Gesner completes the fourth volume of his *Historiae Animalium,* a far-ranging study of animal life. The books are profusely illustrated and deal with, respectively, mammals, reptiles and amphibians, birds, and fish and other aquatic animals. Gessner's work is regarded as the beginning of the field of vertebrate zoology.

1583. Italian scientist Andrea Cesalpino publishes *De Plantis Libri XVI,* which is considered the first textbook of botany in that he studies plants for their own sake rather than in terms of their applications in medicine, agriculture, and horticulture. In this book, he employs a system of binomial nomenclature that anticipates the work of Linnaeus by almost two centuries.

1603. The Lincean Academy is founded by Federico Cesi in Rome. Galileo Galilei will become a member in 1611. This is one of the first scientific communities, and it carries out important research in fields such as entomology and paleobotany, especially with the use of the newly invented microscope.

1623. In his book *Pinax Theatri Botanici* [*An Illustrated Exposition of Plants*], Swiss botanist Gaspard Bauhin refines the earlier work of Cesalpino in taxonomy, by classifying thousands of plants according to their genus and species names.

1661. English diarist John Evelyn publishes one of the first known books on air pollution, entitled *Fumifugium, or the Inconveniencie of the Aer and Smoake of London Dissipated.* To improve the poor air quality of London, he suggests alternative energy (burning aromatic woods instead of sea coal) and an early version of the greenbelt concept (moving energy-intensive industries away from the city center and replacing them with gardens and orchards).

1662. John Graunt founds the scientific field of demography. He makes a rigorous study of the population of London, compiling statistics for birth and death rates, age distribution, and sex ratio. He then extends this methodology to the animal kingdom by using similar criteria to study fish populations.

1665. Robert Hooke publishes *Micrographia*, a book that describes the microbiological environment for the first time. Hooke studies sections of the cork plant and identifies patterns of tiny cylindrical structures therein, which he likens to small enclosed rooms and thus calls *cells*.

1668. Francesco Redi of Italy makes the first step in disproving the theory of spontaneous generation by conducting an experiment to show that maggots cannot appear spontaneously on rotting meat. He is influenced by the thinking of the physiologist William Harvey, who had stated *"Ex ovo omnia"* ("Everything comes from the egg").

1669. Dutch microscopist Jan Swammerdam conducts extensive research on insects and produces accurate descriptions of the anatomy and life history of numerous species, accompanied by meticulously detailed drawings. He determines that the successive forms of egg, larva, pupa, and adult are not different organisms but actually different life stages of one organism.

1669. The Danish scientist Nicholas Steno recognizes that "tongue stones," unknown solid objects found in rock, look very much like the teeth of living sharks. He concludes that these tongue stones are not mineral matter but the remains of animal and plant organisms, thus establishing the organic origin of fossils.

1674–1683. Using a microscope of his own construction, Anton van Leeuwenhoek is the first person to detect and describe "animalcules," known today as bacteria or protozoa, which he views in such media as rainwater, lake and well water, and the human mouth. His discoveries awaken humankind to a previously unknown world of microorganisms and establish the science of microbiology.

ca. 1681. The flightless dodo bird of the Raphidae family becomes extinct on its native island of Mauritius in the Indian Ocean. This is attributed to a combination of hunting by European settlers, deforestation of habitat by these settlers, and predation by exotic species introduced by the settlers, such as pigs and monkeys. This is the first notable extinction of a species in historic times, and it heightens awareness of the effect of human activity on the natural world.

1686. English naturalist John Ray publishes the first of three volumes describing nearly 20,000 plant species. He classifies plants according to overall morphology and focuses on the species as the fundamental level at which organisms should be distinguished from one another. He also notes that the fossil record seems inconsistent with the Biblical account of the Great Flood.

1700s. European explorers and naturalists in the Americas encounter many species of animals and plants not found in the Old World, though some are mistakenly linked to similar known species (e.g., the American bison, the wild turkey). In particular, they note the greater general abundance of wildlife in the New World than in Europe.

1730s. French entomologist René A. F. de Réaumur studies the reproductive rate of aphids and calculates that each individual is capable of producing about 6 billion offspring over successive generations, within a period of just 6 weeks. Because such a population does not exist, he speculates that there must be natural limiting factors to regulate population.

1735. Carolus Linnaeus publishes the first edition of *Systema naturae* [*Systems of Nature*]. In this work, he is the first to consistently use a system of binomial nomenclature based on observable physical characteristics, and thus he is regarded as the father of the modern field of taxonomy.

1741. German naturalist and explorer Georg Wilhelm Steller describes an aquatic herbivorous mammal that becomes known as Steller's sea cow (*Hydrodamalis gigas*), which he observes near the Asian coast of the Bering Sea. By 1768 Steller's sea cow will be hunted to extinction, in a classic example of the rapid disappearance of a species through overexploitation.

1749. George-Louis Leclerc, Comte de Buffon, publishes the first of 44 volumes of his *Histoire naturelle* [*Natural History*]. He defines the concept of a species on the basis of reproductive exclusivity, and he would later be cited by Darwin as a major influence on his concept of evolution. Buffon also considers issues such as the role of geography in biodiversity, the relationship of humans to other primates, and the formation of the earth from molten matter.

1751. Pierre-Louis Moreau de Maupertuis publishes *Système de la Nature,* in which he records his views on heredity, in particular the occurrence of mutant traits. He also will be recognized as a forerunner to Darwin for his observation that stronger animals in a population produce more offspring.

1760. The prominent Swiss mathematician Leonhard Euler develops an equation to describe and forecast the dynamics of age-structured populations. This model for population growth rate will be refined by Alfred Lotka in the twentieth century (and thus become known as the Euler-Lotka equation).

1764. Linnaeus reexamines his earlier belief that species are fixed and immutable; he especially notes obvious examples of hybridized plants. He theorizes that God must have created only a limited number of species, which hybridized over time to form the great diversity of species existing in the present day.

1773. Georg Foster, a botanist accompanying Captain James Cook to New Zealand, provides one of the earliest descriptions of a species invasion unintentionally caused by human activity. He notes that canary grass, a plant native to the Mediterranean, has become established in several sites, possibly through windborne seeds from Cook's vessel in an earlier voyage.

1774. Joseph Priestley publishes his description of oxygen, which he and several others had independently discovered 2 years earlier. In his experiments, Priestley establishes that plants convert the carbon dioxide breathed out by animals back into oxygen, thus providing the basis for an understanding of photosynthesis. He also learns that mice will die in a sealed environment if plants are not present to reoxygenate the air, indicating that the collective metabolic activities of species in an ecosystem will influence its biogeochemistry.

1775. Linnaeus states a balance-of-nature concept, declaring that "In order to perpetuate the established course of nature in a continued series, the divine wisdom has thought fit, that all living creatures should constantly be employed in producing individuals; that all natural things should contribute and lend a helping hand to preserve every species; and lastly, that the death and destruction of one thing should always be subservient to the restitution of another."

1779. Dutch physiologist Jan Ingenhousz refines the findings of Priestley concerning the ability of plants to reoxygenate the air by establishing that this occurs only in the presence of light. He also learns that photosynthesis takes place through the actions of the green plant pigment chlorophyll, rather than the entire plant.

1785. Scottish geologist James Hutton rejects the prevailing idea of the time that the Earth has an age of about 6000 years. He states that the time required for the gradual formation of the natural structures of the planet has to be much longer than this, so long in fact that the age of the Earth is inconceivably great.

1789. Clergyman Gilbert White publishes *The Natural History and Antiquities of Selborne,* a comprehensive and precise description of the natural environment in and around his hometown in southern England. It becomes the most widely read book of natural history ever published in the English language, and it has remained in print continuously up to the present day.

1794. Erasmus Darwin develops one of the first formal statements of the theory of evolution. In the form of a poem, he presents the idea that microorganisms in the ocean evolved over successive generations into the plant and animal life of the present day. However, he does not conceive of natural selection as the mechanism driving this process, something that will be left to his grandson Charles.

1798. French scientist Georges Cuvier publishes a set of drawings showing that there is a significant difference between the jaw structure of an Indian elephant and that of a mammoth, indicating that the mammoth is an extinct animal and not a different type of elephant. This definitely establishes the fact of extinction; before Cuvier the general view was that species that had apparently vanished must still be living somewhere else on Earth.

1799. Baron Alexander von Humboldt, a Prussian naturalist, begins his 5-year exploration of the colonial empire of Spain in the New World. He carries out many activities of zoological, botanical, and ethnographic research, including the collection and documentation of more than 60,000 tropical plants. Humboldt strives to describe his findings within the context of a single integrative science that unites all forms of natural phenomena. This provides the foundation for the current holistic approach to describing the complexity of the Earth's environmental system.

1800s. The nineteenth century experiences a significant rise in the level of carbon dioxide gas in the atmosphere, according to later measurements of ancient ice. This will be attributed to the significant increase in fossil fuel use associated with industrialization, population growth, and improved living standards.

1801. In *Systeme des animaux sans vertebres* [*System for Animals Without Vertebras*], Jean-Baptiste Lamarck becomes the first to provide a rigorous classification system for *invertebrates* (a term he coins), comparable to that of Linnaeus for vertebrates.

1809. Lamarck presents what has been described as the first truly modern and comprehensive theory of evolution. He states that more complex life forms such as mammals evolved from simpler forms and that the behavior of organisms will cause unused body parts to degenerate or new parts to develop as needed for survival. His thinking differs from later evolutionary biologists in that he believes morphological changes in an animal during its lifetime can be passed on to its descendants.

1815. Appreciation for the natural world is expressed by the Romantic Movement, especially in the poetry of William Wordsworth and the journals of his sister Dorothy. The urbanization and extensive land clearance associated with the Industrial Revolution have made Europeans more conscious of the loss of natural habitat and brought greater intellectual awareness of Nature as an entity distinct from human civilization.

1817. Cuvier publishes the first edition of a monumental work on the animal kingdom, noted especially for his classification of animals into four large groups (vertebrates, mollusks, articulates, and radiates), each with a discrete type of anatomical organization. This departs from the earlier idea that animal life is arranged in a hierarchy of complexity from the simplest organisms up to humans.

1824. Through his studies of steam engines, French engineer Nicolas Léonard (Sadi) Carnot establishes that heat moves from a system of higher temperature to one of lower temperature and that through this process work is done. This provides the basis for the second law of thermodynamics (which will be formally stated by Clausius in 1850) and eventually leads to an understanding of energy flows in ecosystems.

1826. Thomas Robert Malthus publishes the final edition of *An Essay on the Principle of Population,* in which he contends that "The power of population is indefinitely greater than the power in the earth to produce subsistence for man." He argues that food supply cannot keep pace with population growth, and this "Malthusian" view inspires many subsequent analyses of the relationship between population and resources, in particular Darwin's ideas about natural selection.

1827–1828. John James Audubon's *The Birds of America* is published and becomes an immediate best seller. It includes superbly illustrated descriptions of nearly 500 bird species of North America and is generally regarded as the greatest illustrated book ever published. In the early twentieth century, a pioneering conservation group will honor the painter by taking the name The National Audubon Society.

1830. The first volume of *The Principles of Geology,* a landmark work by Scottish scientist Charles Lyell, is published in London. In it he promotes the principle of uniformitarianism, the idea that the visible features of the Earth were formed over vast time spans by physical processes that are the same as those taking place at the present time.

1831–1836. Charles Darwin, on his *Beagle* voyage, observes and describes the effects of biological invasion, in particular the presence in the Galápagos Islands of numerous European animal and plant species. Many of these observations lead to insights that will be incorporated into *On the Origin of Species.*

1836. Louis Agassiz studies the glaciers of his native Switzerland and observes that the effects of glaciation can be seen in other places where glaciers no longer are present. He then advances the idea that a great Ice Age occurred at some time in the past, during which huge glaciers covered much of the Earth's surface. Agassiz thus can be considered one of the first to recognize the phenomenon of climate change.

1838. Belgian mathematician Pierre F. Verhulst publishes what is known as the logistic equation to describe the maximum number of individuals an environment can support. He indicates that the factors that tend to limit population growth will increase in proportion to the ratio of the excess population to the overall population. In the 1920s this equation will be utilized by Raymond Pearl and Lowell Reed for their studies of historic population growth in the United States.

1838. German botanist August Grisebach examines plant communities in terms of their geographic distribution and their relationship to climate, an early example of the study of ecosystem structure and function. He also describes the forces that can affect plant distribution.

1840. Justus von Liebig publicizes the Law of the Minimum, which states that if all the essential nutrients but one are available in the quantities required for the growth of a plant, the deficiency of that one nutrient will prevent growth. Later scholars will apply this concept of the limiting factor to larger contexts such as ecosystems.

1842. Sir Richard Owen coins the term *Dinosauria,* meaning "terrible lizard," to describe a category of large extinct terrestrial vertebrates. Through history dinosaur bones had been discovered, but they were thought to be from mythical creatures or from giant forms of existing reptiles. It was not until the discoveries of Gideon Mantell and others in the early nineteenth century that dinosaurs were recognized as a distinct taxonomic group.

1847. Hermann von Helmholtz publishes *Uber die Erhaltung der Kraft* [*On the Conservation of Energy*], establishing that the amount of energy in the universe is constant and that energy can neither be created nor destroyed, only converted from one form to another. This becomes known as the first law of thermodynamics, and, like the second law (which actually was identified earlier), it will become fundamental to ecologists' understanding of bioenergetics.

1854. Henry David Thoreau publishes *Walden Pond, or Life in the Woods*, his account of a self-sustaining lifestyle apart from industrial society. Thoreau, with his mentor Emerson, will initiate an intellectual movement in North America comparable to that of Rousseau and the Romantic poets in Europe. This is based on the realization that the pastoral landscape, so long taken for granted, is threatened by human activity.

1858. Alfred Russel Wallace completes the manuscript of an essay presenting his conclusions concerning the origin and diversity of species. He sends the manuscript for review to a more prominent scientist, Charles Darwin, who notes the similarity with his own ideas on evolution, as yet unpublished.

1859. Darwin publishes *On the Origin of Species,* a book describing his theory of evolution by natural selection. It is greeted with great interest and controversy, and it becomes the foundation of a new form of biology. Its influence will eventually surpass that of any other book in the history of science, with the possible exception of Newton's *Philosophiae Naturalis Principia Mathematica.*

1859. Thomas Austin introduces two dozen European rabbits onto his property in the state of Victoria, Australia, for the purpose of sport hunting. In a classic example of unchecked population growth in an invasive species, within a decade the descendants of these rabbits will reach a population that numbers in the millions and spreads over much of the continent.

1862. English naturalist Henry Walter Bates recognizes and illustrates what will become known as Batesian Mimicry. This is the ability of a prey species to evolve a defense against a common predator through similarity of coloration and patterning with other species that are poisonous or otherwise less palatable.

1862. The theory of spontaneous generation, widely believed since the time of Aristotle and first challenged by Redi in the seventeenth century, is finally put to rest by Louis Pasteur. He shows that a sterilized broth will remain sterile as long as bacteria are prevented from entering it in some way, even if it is exposed to air.

1865. German physicist Rudolf Clausius introduces the concept of entropy (disorder or randomness) to measure the amount of energy available to do work. He establishes that entropy cannot decrease in a physical process and can only remain constant in a reversible process. This becomes formalized as the second law of thermodynamics, and it will come to have great influence on the thinking of ecologists.

1865. Gregor Mendel reads a paper summarizing his experiments with pea plants, dating back to 1854. During his lifetime his work is noted but not frequently cited, and its implications for the theory of evolution are not recognized. However, in the early twentieth century, his findings will be reexamined, and he will ultimately be described as the founder of the science of genetics.

1871. Darwin publishes *The Descent of Man,* in which he directly applies his principles of evolution to the human race. He states that "man is descended from a hairy, tailed quadruped, probably arboreal in its habits, and an inhabitant of the Old World." He describes this animal as a *Quadrumana*, an older term for the great apes.

1872. British scientist Robert Angus Smith publishes *Air and Rain: The Beginnings of a Chemical Climatology,* in which he summarizes his extensive investigations of precipitation throughout the British Isles. He reports high levels of acidity in the rainwater of large manufacturing cities such as Manchester and Glasgow. Smith employs the phrase "acid rain" to describe this.

1872. President Ulysses S. Grant signs into law an act establishing Yellowstone National Park, the first national park of the world. Located in the northwestern United States, it has a remarkable combination of unique geological features, striking landscapes, and abundant wildlife.

1873. London experiences the first in a series of "killer fogs" that are responsible for thousands of deaths. These occur sporadically until the mid-twentieth century, when stricter controls on emissions are finally enacted.

1873. The term *ecology* enters the English language, having been coined shortly before this in German as *Ökologie* by the biologist Ernst Haeckel. He combines two Greek terms meaning "household" or "dwelling" and "science" or "study." The concept is that ecology is "the study of the house of nature." The spelling *ecology* (as opposed to *oecology*) will be officially adopted as the correct English version in 1893.

1874. *Man and Nature, or Physical Geography as Modified by Human Action* is published by the American scholar George Perkins Marsh. As the title indicates, Marsh examines the role of human activity as an agent of environmental change, noting especially the desertification of once-fertile areas of the Mediterranean. He thus departs from the conventional view of the time that the physical landscape of the Earth is essentially the product of natural phenomena.

1875. Geologist Eduard Suess introduces the term *biosphere,* which he describes as "the place on earth's surface where life dwells." An alternate term, *ecosphere,* will be introduced in 1953. Today both words are also used to describe an enclosed, self-contained ecosystem.

1876. Alfred Russel Wallace produces a definitive work of biogeography, *The Geographical Distribution of Animals,* which provides support for the theory of evolution. In the companion work *Island Life* (1880), he blazes the trail for the study of island biogeography.

1877. Zoologist Karl Möbius studies natural oyster banks in the North Sea, employing what is thought to be the first use of the term *community* in the modern sense. He describes the complexity and interrelationships of an oyster bed community and recognizes that the nature of the community would be transformed if the population of any given species were to decrease or increase through human intervention.

1880s. The folk belief that "Rain follows the plow" encourages agricultural development in the Great Plains region of the United States. It was thought that plowing under native vegetation to plant crops such as wheat and corn will lead to increased rainfall, an idea that will be proved catastrophically wrong during the Dust Bowl era of the 1930s.

1881. German naturalist Karl Semper publishes *Animal Life as Affected by the Natural Conditions of Existence.* He provides an accurate description of a generalized food chain, outlining the pyramid of mass from plant material to herbivores to carnivores. This is regarded as one of the earliest studies of energy flows in nature.

1883. The German biologist August Friedrich Leopold Weismann emphasizes that inherited characteristics are transmitted only via the sex cells and not the somatic cells, thus discounting the Lamarckian idea of inheritance of acquired characteristics. He describes a special hereditary substance possessed by all organisms, which he calls germ plasm; this roughly corresponds to the modern understanding of DNA.

1883. Thomas Henry Huxley, known as "Darwin's Bulldog" for his fierce advocacy of the theory of evolution, declares that the abundance of fish is so great that the fishing industry cannot seriously impact general fish populations. He says that "The cod fishery, the herring fishery, the pilchard fishery, the mackerel fishery, and probably all the great sea-fisheries, are inexhaustible; that is to say that nothing we do seriously affects the number of fish. And any attempt to regulate these fisheries seems consequently . . . to be useless."

1887. American zoologist Stephen A. Forbes publishes "The Lake as a Microcosm," in which he writes, "One finds in a single body of water a far more complete and independent equilibrium of organic life and activity than on any equal body of land. . . . It forms a little world within itself." His description of the interrelationship of the organisms at a definable site leads to the development of the ecosystem concept as well as to modern ideas of food web interactions and of the analogy between wildlife and human communities.

1889. The importing of domestic cattle into Somalia provides an example of contrasting effects of an introduced species. On one hand, this leads to an epidemic of the viral disease rinderpest, causing the death of millions of domestic and wild ungulates. On the other hand, it has a positive impact in that the decimation of cattle in many regions removes a major host for Tsetse flies (*Glossina* spp.), the vector of destructive diseases including African sleeping sickness.

1889. William Temple Hornady publishes *The Extermination of the American Bison,* predicting the imminent demise of a species (*Bison bison*) that just half a century earlier was estimated to consist of at least 60 million individuals. This creates public support for saving the species, and in the twentieth century its numbers will rebound to several hundred thousand.

1892. America's first environmental advocacy group, the Sierra Club, is founded in San Francisco with 182 charter members. Famed naturalist John Muir is the club's first president. He pledges "to do something to make the mountains glad."

1893. The earliest book in English with the explicit term *ecology* in its title is published: Louis H. Pammel's *Flower Ecology.* Pammel will go on to publish a textbook simply called *Ecology* in 1903, and he will also become the mentor of the famous African-American botanist George Washington Carver.

1895. Johannes Eugenius Warming of Denmark publishes *Plantesamfund: Grundtræk af den økologiske Plantegeografi,* the first in-depth textbook on plant ecology. Based on a series of lectures at the University of Copenhagen, it describes plant communities throughout the world in the context of the environmental factors affecting them.

1896. Polish botanist Jozef Paczoski establishes the field of phytosociology, the study of the organization and distribution of plant communities. He later publishes a textbook dealing with the modification of the environment by plants through the creation of microenvironments.

1898. Andreas Schimper of Germany is credited with originating the term *tropical rain forest.* Building on the pioneering work of von Humboldt, he and others recognize that similar vegetation types arise under similar climatic conditions in different parts of the world, and they define basic principles of plant form and function to explain these global patterns.

1899. Henry Chandler Cowles of the University of Chicago studies in detail the plant life of the Indiana Dunes bordering Lake Michigan. He notes the natural changes that occur in this vegetation as the sand dunes recede or advance, and from this, he develops a formal concept of ecological succession.

1901. Swiss scientist François-Alphonse Forel publishes *Handbuch der Seenkunde,* based on his decades-long studies of the waters of Lake Geneva. This is the first textbook in the field of limnology and establishes it as a scientific discipline.

1904. Christen Raunkiaer, a student of Eugenius Warming, prepares a life-forms classification system for plants using quantitative methods. In this system plants are categorized according to the physical position and degree of protection of their perennating organs (buds) during adverse growing conditions. He observes that most plant species in a given community fall within the polar frequency categories of very common or very rare (Raunkiaer's Law).

1905. Conservation of the natural environment becomes government policy in the administration of Theodore Roosevelt. He establishes the U.S. Forest Service, headed by Gifford Pinchot, and creates many national parks and wildlife and bird refuges. Roosevelt recognizes that it is necessary to protect entire ecosystems in order to preserve certain endangered species.

1905. Frederic Edward Clements publishes *Research Methods in Ecology,* considered the earliest work to set out a systematic approach to ecological research. It emphasizes experimental evidence and advocates the use of mathematical and graphical presentations. At about this time he begins to use the term *ecotone,* or "zone of tension," to describe boundary areas between adjacent communities.

1905. The concept of a limiting factor in photosynthesis is shown by the British botanist Frederick Frost Blackman. The limitations on this process are the supply of carbon dioxide, the relative temperature, and the amount of light, and Blackman demonstrates that the rate of photosynthesis is controlled by whichever of these factors is least available.

1907. The term *eutrophication* is coined by the German scientist C. A. Weber to refer to the rich wetlands in Europe that receive nutrient runoff from surrounding areas. The term is then applied to lakes by Swedish limnologist Einar Naumann, roughly a decade later.

1908. What becomes known as the Hardy-Weinberg Principle is proposed independently by Godfrey Hardy of Britain and Wilhelm Weinberg of Germany. One of the basic concepts of population genetics, it states that the frequency of genotypes in a large random-mating population is simply the product of their relative frequencies. This indicates that gene pool frequencies are inherently stable unless evolutionary mechanisms cause them to change.

1912. Geneticist Sewall Wright publishes his first scholarly paper, on the anatomy of the trematode; his last paper will appear 76 years later. In the 1920s, Wright will introduce the concept of genetic drift, and he, along with J.B.S. Haldane and Ronald A. Fisher, will found the field of population genetics.

1913. Charles C. Adams publishes *Guide to the Study of Animal Ecology,* the first major work devoted specifically to this topic. He draws on research carried out in the Lake Superior area of northern Michigan (1905) and at Isle Royale (1909). Adams argues that the term *ecology* should become standard usage for animal studies as well as plants, saying "To use a different name for the same subject or process in botany and zoology is as undesirable as to use a different term for heredity in plants and animals."

1913. The *Journal of Ecology* is founded by the British Ecological Society as the first international, peer-reviewed ecological journal in the world. It will be joined in 1920 by *Ecology,* the flagship journal of the Ecological Society of America, which is the successor to an earlier publication *The Plant World* (founded 1897).

1913. Victor Ernest Shelford develops the Law of Tolerance, an extension of Liebig's Law of the Minimum. It states that for an organism to succeed in a given environment, conditions must remain within a maximum and minimum range of tolerance for that organism. In 1915 Shelford will help to organize the Ecological Society of America and will serve as its first president.

1914. The last known passenger pigeon (*Ectopistes migratorius*) dies in captivity in a Cincinnati zoo. This marks the end of a species that in the mid-nineteenth century was so numerous that a flock passing overhead could darken the sky for an entire day, in the manner of a solar eclipse. According to tradition, the last credible sighting of a passenger pigeon in the wild was by Theodore Roosevelt in 1907.

1916. Clements presents his view of ecological succession in *Plant Succession: An Analysis of the Development of Vegetation.* He conceives of a community as in effect a superorganism, whose component species are tightly bound together both in the present and in their common evolutionary history. This advances the idea of a community functioning as an integrated unit, with interactions among its plants, animals, microorganisms, and so on.

1917. Ornithologist Joseph Grinnell of the University of California establishes the concept of the ecological niche in his paper "The Niche Relationships of the California Thrasher." The word *niche* is from the Middle French *nicher,* meaning "to nest," and the term is used to indicate that the ecological niche of a species is in effect its "nook" or "cubbyhole," i.e., its specific, limited, and accustomed place within a larger community.

1921. Olof Arrhenius, son of the famous physical chemist Svante Arrhenius, describes the species–area relationship. He counts the number of species occurring in different-sized units of vegetation for certain Swedish island systems and then presents the results as a mathematical formula. He observes that "the number of species increases continuously as the area increases."

1924. J.B.S. Haldane publishes the first of a series of papers under the title *A Mathematical Theory of Natural and Artificial Selection,* in which he offers a mathematical description of fitness. He defines individual fitness simply by reproductive success, i.e., the total number of offspring that the individual produces in a lifetime.

1925–1926. The mathematicians Alfred Lotka of the United States and Vito Volterra of Italy independently develop standard models to describe the interactions of predator and prey species. The so-called Lotka-Volterra equations demonstrate the inherent tendency of predator–prey populations to oscillate; e.g., a large predator population will reduce available prey to the point where predators decline from lack of food, but this will result in a population increase for the prey species and a subsequent resurgence of the predator.

1926. August Thienemann, a German limnologist, publishes a unique food web of lakes, in which he develops the basic concepts of nutrient cycling in water and food cycle relationships among producers, consumers, and decomposers. Thienemann had earlier observed that biodiversity is greater with habitat diversity and is reduced with habitat disturbance.

1926. Botanist Henry Gleason presents an alternative to the prevailing view of Clements that ecological communities are the result of tight species associations. He argues that a community is not an organic entity but rather "is merely the resultant of two factors, the fluctuating and fortuitous immigration of plants and an equally fluctuating and variable environment . . . not an organism, scarcely even a vegetational unit, but merely a coincidence."

1926. Russian scientist Vladimir Vernadsky develops the modern concept of the term *biosphere* and notes the tendency of human activity to influence this entity. He states that all organisms on earth "are inseparably and continuously connected—first and foremost by feeding and breathing—with their material-energetic environment." Vernadsky also recognizes that the oxygen, nitrogen, and carbon dioxide in the Earth's atmosphere result from life processes.

1926. The Ecological Society of America creates a committee "charged with the listing of all preserved and preservable areas in North America in which natural conditions persist." The committee's report is published as *Naturalist's Guide to the Americas* (edited by Shelford). This is an early example of a "gap analysis" of protected ecosystems in southern Canada and the United States.

1926. Warder Clyde Allee begins a series of papers with the heading "Animal Aggregations." He concludes that there is a natural tendency for organisms of the same species to assemble in social groups and that negative effects can arise not only from overcrowding in a habitat but also undercrowding. He also proposes that animals will unconsciously cooperate in the interest of group survival, which he terms *proto-cooperation.*

1927. Charles Sutherland Elton publishes *Animal Ecology,* in which he stresses the importance of feeding (energy) relationships among organisms as the basis for understanding nature. Elton describes the niche of a species as "its place in the biotic environment, its relations to food and enemies," and he is credited with saying, "When an ecologist says 'There goes a badger,' he should include in his thoughts some definite idea of the animal's place in the community to which it belongs, just as if he had said, 'There goes the vicar.'"

1930. British scientist Roy Clapham is reportedly the first to use the term *ecosystem* to describe the fundamental unit of nature. German entomologist Karl Friedrich had introduced the word *holocoen* for this concept in 1927, and a comparable term *biogeocoenosis* will be employed by the Soviet forest scientist Vladimir N. Sukachev in 1944. However, neither of these terms is as widely used as *ecosystem,* which becomes the standard word.

1930. The *Genetical Theory of Natural Selection* is published by Ronald Fisher, the first definitive effort to explain Darwin's evolutionary theories in a genetic context. This becomes a classic text of modern evolutionary biology and has been described as the single most important work in the field after *On the Origin of Species* itself.

1931. American economist Harold Hotelling develops a model for the efficient use of nonrenewable resources over time. According to Hotelling, it can be economically rational to degrade an ecosystem or deplete a species even when market prices reflect the true value of these resources. His approach to resource theory has become the dominant one in contemporary ecological economics.

1932. Josias Braun-Blanquet of Switzerland develops a standard method of sampling for vegetation classification, based on the visual abundance of a species within a defined area. He uses this *relevé* method to classify much of the vegetation of Europe.

1934. Russian biologist Georgyi Frantsevitch Gause presents pioneering research on competition for resources. His experiments on microbial communities show that two similar species will grow adequately when cultured separately, but in mixture, one will drive the other to extinction. This indicates that when two populations with ecologically similar requirements compete for limited resources in a stable environment, eventually one will persist and the other disappear.

1935. A. J. Nicholson and V. A. Bailey of Australia publish "The Balance of Animal Populations," presenting a model that can be used to describe the population dynamics of a coupled host–parasite (or predator–prey) system. A central characteristic of the Nicholson-Bailey model is that both populations undergo oscillations with increasing amplitude until first the host dies out and then the parasitoid population as well.

1935. British scientist Arthur G. Tansley refines the term *ecosystem*, which had been coined 5 years earlier at his request by his colleague Roy Clapham. Tansley promotes the use of *ecosystem* as the most useful term for an interactive system consisting of all the organisms functioning in a given area and all the physical (nonliving) factors affecting the area.

1935–1937. Clements publishes two papers in which he describes how human activity has disrupted the climax community of the Great Plains, leading to the Dust Bowl conditions that exacerbated the Great Depression in the United States. He and others recommend an organismic approach to land use in which the natural grasslands will be saved or restored.

1937. Theodosius Dobzhansky, a Ukranian-American geneticist, publishes *Genetics and the Origin of the Species*; this effectively provides the link between the Darwinian theory of evolution through natural selection and the Mendelian theory of mutation in genetics. Dobzhansky will later become known to the general public through his 1973 paper "Nothing in Biology Makes Sense Except in the Light of Evolution," which is often cited as a dismissal of the antievolution worldview known as creationism.

1938. British engineer Guy S. Callendar studies historic records of temperature and determines that a warming trend is taking place. He compares this with historic measures of CO_2 concentrations and concludes that an increase in CO_2 levels correlates with this warming. Callendar then publishes "The Artificial Production of Carbon Dioxide and Its Influence on Climate," in which he directly implicates fossil fuel combustion as an agent of climate change.

1939. Frederic Clements and Victor Shelford collaborate on *Bio-Ecology*, which is intended to correlate plant ecology and animal ecology with the interest of advancing the science of ecology in general. They argue for the importance of plant–animal interactions within what Clements refers to as the *biome*, which is defined as "an organic unit comprising all the species of plants and animals at home in a particular habitat."

1939. Paul Hermann Müller of Switzerland develops the powerful organic insecticide DDT, which proves very effective in controlling insect populations that are the vectors for diseases such as malaria and typhus. In 1948 Müller will receive the Nobel Prize for this achievement, but the use of DDT will later become controversial because of its perceived detrimental effects on wildlife.

1940. Chancey Juday, an American aquatic scientist, publishes a study of the energy budget of lakes, based mainly on extensive studies carried out by himself, Edward A. Birge, and others at Lake Mendota, Wisconsin. Birge and Juday develop the concept of primary production, i.e., the rate at which food energy is generated, or fixed, by photosynthesis.

1942. Ernst Mayr presents the biological species concept, stating that "Species are groups of interbreeding natural populations that are reproductively isolated from other such groups." Although various other valid descriptions of a species have been proposed, this concept of reproductive isolation remains the most widely recognized approach to the issue.

1942. Working on his doctoral thesis in zoology, graduate student Raymond Lindeman studies Cedar Creek Bog, a senescent lake on the Anoka Sand Plain of east central Minnesota. He describes the area in a paper entitled "The Trophic-Dynamic Aspect of Ecology" (published after his death at age 27). Lindeman classifies the organisms into trophic (feeding) levels and then determines the energy flows between these levels. His trophic-dynamic model establishes the "bottom-up" perspective as the dominant paradigm for a generation of ecologists.

1943. Agronomist Norman Borlaug joins a program funded by the Rockefeller Foundation to help farmers in Mexico increase their wheat production. This program will develop high-yield, disease-resistant crops and eventually lead to the improved agricultural practices of the Green Revolution of the 1960s and 1970s.

1944. Based on experiments with bacteria, Oswald T. Avery and colleagues establish that the nucleic acid DNA is the essential material that carries genetic properties in virtually all living organisms. This is considered the foundation of modern DNA research, although it would remain for Watson and Crick to describe the structure of the DNA molecule.

1945. David Lack reports that beak size for Darwin's finches in the Galápagos islands depends on whether or not a given species co-occurs with other finch species. In 1959 G. Evelyn Hutchinson concludes that such differences in size evolve to allow different species to partition the food web structure even though their ecological requirements are the same. This issue of body size ratios in co-occurring species becomes one of the classic tests of competition theory

1945. P. H. Leslie develops the Leslie matrix model, which combines the mortality and fertility functions of a population in a single expression. This becomes one of the most widely used tools to determine the increase or decrease of a population over time, as well as the age distribution within the population.

ca. 1947. In one of the most noted examples of an invasive species, the predatory tree snake *Boiga irregularis* is accidentally introduced on the island of Guam. Having neither competitors nor predators on the island, it soon becomes the top predator there. It will be responsible for a severe decline in Guam's native bird population and the presumed extinction of at least 10 bird species, as well as having major impacts on lizard and bat populations.

1949. *A Sand County Almanac*, a collection of essays by naturalist Aldo Leopold, is published a year after the author's death. It brings about widespread public appreciation of the value of biodiversity for its own sake and is considered a founding work of environmental ethics. The book arguably ranks second only to Rachel Carson's *Silent Spring* in terms of its influence on the popular environmental movement.

1949. Wolves appear on Isle Royal in Lake Superior, where the moose population had lived free of predation since first reaching the island about 1900. This establishes a predator–prey relationship that will be extensively studied by ecologists over the ensuing decades, as they monitor the size of the wolf and moose populations, their interactions with each other, and the effects of relative population size on the island's vegetation.

1951. Studies of plant communities by Robert Whittaker and John Curtis lead to a challenge to the concept of rigid plant associations advocated by Frederic Clements. Whittaker presents the theory and technique of gradient analysis to describe continuous distributions of species along environmental gradients, thus supporting Henry Gleason's contrasting concept of individualistic communities.

1951. To account for the fact that, just by chance, a species may be absent from a site where it would have thrived, John G. Skellam develops the reaction–diffusion model of invasion biology. Skellam's mode describes the dynamics of a population that is both growing and spreading, and it predicts that the front of an invasion will move at a constant velocity.

1952. Ernst Mayr describes what is termed the "founder effect," an instance in which a new population is established by a relatively small number of individuals with limited genetic variation. Such effects are particularly important on islands where organisms arrive from other sites, as by wind or wave action.

1953. Eugene P. Odum publishes the comprehensive work *Fundamentals of Ecology,* along with his brother Howard. This is recognized as the first textbook to examine ecology from a holistic, macroscopic perspective. It uses a "top-down" approach starting at the ecosystem level and describes community stability as being based on the sharing of energy throughout the food web.

1953. James Watson of the United States and Francis Crick of Britain present an accurate model of the double-helix structure of the DNA molecule. This leads to a revolution in the science of biology, in particular the advancement of the field of molecular biology.

1955. The concept of Maximum Sustained Yield (MSY) is adopted as the goal of international fisheries management at a conference sponsored by the United Nations. MSY is the maximum long-term average catch that can be removed from a fish population on a sustainable basis.

1956. Lotka's 1925 work *Elements of Physical Biology* gains a new audience when it is reissued as *Elements of Mathematical Biology*. This book contains the theoretical basis for much of modern ecology, as it covers issues such as evolutionary change, biogeochemical cycles, growth and reproduction, interspecies equilibrium, energy balance, the operations of the senses, and the function of consciousness.

1957. Howard T. (Tom) Odum measures primary production in a number of freshwater spring communities of Florida. He characterizes the transfer of carbon between trophic levels with the goal of understanding how energy moves through an ecosystem.

1958. Charles Elton publishes *The Ecology of Invasions by Animals and Plants,* which becomes the classic text of invasion biology. He concludes that simplified food webs such as monocultures seem to be more vulnerable to invaders than complex ones, probably because they offer greater opportunity to establish new niches. This implies that greater biodiversity tends to produce greater stability,

1958. The oceanographer Charles Keeling begins his decades-long record of Earth's atmospheric carbon dioxide concentrations, as measured at Mauna Loa, Hawaii, and other locations. He becomes the first to confirm definitively the rise of atmospheric carbon dioxide by means of a data set now known as the "Keeling Curve."

1958–1959. G. Evelyn Hutchinson refines and popularizes the concept of the niche, the multidimensional space of resources (light, nutrients, structure, etc.) that is available to and specifically used by a species. He formulates the ecological niche as a quantitative description of the range of environmental conditions that allow a population to persist in a given location, i.e., to have a positive or at least zero (break-even) growth rate.

1959. Crawford Stanley Holling of Canada introduces the concept of functional response, which describes the relationship between prey density in a certain area and the amount of prey consumed by each predator in that area. This will become an important principle of modern population ecology. Holling will also make important contributions to chaos theory and to the development of the field of ecological economics.

1960. Nelson G. Hairston, Frederick E. Smith, and Lawrence B. Slobodkin argue that predators protect the "green world" from herbivores such as rabbits by restricting their densities to levels that allow plant life to flourish. This "HSS hypothesis" provides a top-down alternative to the bottom-up paradigm that fertility is the key to understanding plant biomass. Robert T. Paine later uses the term *trophic cascade* to describe this indirect effect of predation on vegetation.

1960s. F. Herbert Bormann, Eugene E. Likens, Robert S. Pierce, and Noye Johnson serve as the core of the Hubbard Brook Ecosystem Study group for ecological and biogeochemical research in the White Mountain National Forest of New Hampshire. The group makes important contributions in areas such as nutrient cycling, biomass measurement, forest management, and anthropomorphic environmental disruptions.

1960s. The modern field of ethology (innate animal behavior) is established, mainly through the work of Konrad Lorenz, Niko Tinbergen, and Karl von Frisch. A fundamental principle of the field is that behavior has an evolutionary basis. Thus, there is interest in behaviors that seem disadvantageous to the individual performing them; e.g., warning behavior.

1962. Frank W. Preston publishes the last of his three major papers on the mathematical characteristics of ecological commonness and rarity. He points out a consistent relationship between the commonness and rarity of individuals and species in many forms of life, which in his view can be best represented by a lognormal curve.

1962. The naturalist author Rachel Carson publishes *Silent Spring,* an account of the harmful effect on wildlife of pesticides such as the insecticide DDT. The book heightens public awareness of the extent to which the natural environment is vulnerable to human intervention. It becomes a best seller and is generally credited with launching the modern U.S. environmental movement at the popular level. In 1972 the use of DDT will be generally banned in the United States.

1963. Harold Barnett and Chandler Morse provide the first empirical analysis of the historic supply of natural resources in the United States. It shows that from 1890 to 1957 the cost in capital and work hours of resources such as timber, fish, minerals, and fuels declined significantly, which is attributed to increased efficiency. This is used by some to argue that the only real limitation on resource exploitation is human ingenuity, not physical supply.

1963. Stanley C. Wecker and Peter H. Klopfer study the question of how a species chooses the place it inhabits. They conclude that this choice is partly genetic, having evolved as a means for detecting the most favorable environment for survival, but also partly physiological/experiential, e.g., affected by the natal habitat or the nature of parenting. Later workers will develop the ideal free distribution model (IFD) to predict the area that an animal will select.

1963–1964. William D. Hamilton proposes a method of accounting for the evolution of apparently altruistic behavior (actions beneficial to the recipient but detrimental to the actor). According to Hamilton's Rule, a behavior that lowers the chance for survival of the individual may occur when it increases the chance for survival of close relatives with similar genetic makeup.

1964. Paul R. Ehrlich and Peter H. Raven collaborate on "Butterflies and Plants: A Study in Coevolution." This popularizes the use of the term *coevolution* to describe a pattern in which reciprocal evolutionary change occurs in two interacting species.

1966. George C. Williams challenges the idea of group selection, the concept that behaviors detrimental to an individual can evolve because they benefit the group. He argues that most seemingly group-selected traits really are advantageous to the individuals performing them. The intuitive basis for this is that if an individual feature were indeed good only for the group, it could not evolve because the individual performers would all die out.

1967. In his studies of the blue-gray gnatcatcher (*Polioptila caerulea*), Richard B. Root employs the term *guild*. He defines this as "a group of species that exploit the same class of environmental resources in a similar way" (in this case, foliage-gleaning insectivorous birds). The concept derives from the medieval practice of the artisans in a particular industry forming a guild to promote their mutual interests.

1967. Lynn Townsend White Jr. recommends that St. Francis of Assisi be made the patron saint of ecology, stating "He proposed what he thought was an alternative Christian view of nature and man's relation to it; he tried to substitute the idea of the equality of all creatures, including man, for the idea of man's limitless rule of creation."

1967. Robert MacArthur and Edward O. Wilson present the theory of island biogeography, an equilibrium theory designed to predict the number of species that will exist on a given island. They propose that the number of species on any island reflects a balance between the rate at which new species immigrate to colonize it and the rate at which established species become locally extinct. An "island" in this context is not only a body of land surrounded by water but can be any insular area, such as a formerly continuous natural habitat now fragmented by encroaching civilization.

1968. Garrett Hardin publishes an essay entitled "The Tragedy of the Commons." He argues that "Freedom in a commons brings ruin to all." That is, if a society allows unregulated use of public resources, the inevitable result will be depletion of the resources, because each individual actor will behave in his own best interest and exploit the commons.

1968. Motoo Kimura of Japan introduces the neutral theory of molecular evolution, which assumes that genetic variation results from a combination of mutation-generating variation and genetic drift eliminating it. The theory is called neutral because allele and genotype differences at a gene are selectively neutral (or nearly so) with respect to each other.

1969. Richard Levins employs the term *metapopulation* to describe an assemblage of local populations living in a network of habitat patches. The Levins model presents the essence of the metapopulation concept, i.e., that a species may persist in a balance between stochastic local extinctions and recolonization of currently unoccupied patches. Metapopulation theory will be further developed by Ilkka Hanski from the 1980s onward.

1969. Robert Paine describes the role that a Pacific Northwest starfish and a large snail from Australia's Great Barrier Reef play in their respective ecosystems. He notes that removal of these two predator species would cause a population explosion of certain of their prey, with severe consequences for the rest of the ecosystem. He borrows a term from architecture to designate such an organism as the *keystone species* for its ecosystem.

1972. Daniel B. Botkin and colleagues present a so-called gap model for forest growth, based on the growth of the individual trees that make up the forest stand. The concept of a "gap" is based on the idea that the death of a large tree creates a space in the forest canopy in which certain species then become established. In the 1980s Herman H. Shugart and others will provide additional gap models.

1972. In "Acid Rain," Gene Likens and colleagues describe the long-term effects of acid precipitation, based on observations at Hubbard Brook, New Hampshire. The report demonstrates an explicit link between the use of fossil fuels in North America and increased acidification of rain and snow. This is consistent with the first description of acid rain, made exactly 100 years earlier by Robert Angus Smith.

1972. John Maynard Smith introduces the concept of an evolutionarily stable strategy (ESS), defined as a strategy so effective that it cannot be displaced by a rare new (mutant) strategy if it has become fixed in a population. He bases this on game theory, reasoning that a population will resist the development of new traits or behaviors because this may decrease the likelihood of successful reproduction.

1972. Robert MacArthur summarizes his views in *Geographical Ecology: Patterns in the Distribution of Species.* The book's theme is that "the structure of the environment, the morphology of the species, the economics of species behavior, and the dynamics of population changes are the four essential ingredients of all interesting biogeographic patterns." Although he acknowledges the role of history in shaping species assemblages, MacArthur regards useful patterns of species diversity as the result of repeatable phenomena, not chance events.

1973. Leigh Van Valen proposes the Red Queen Hypothesis, referencing an episode in *Alice in Wonderland* in which the Red Queen tells Alice that "It takes all the running you can do to keep in the same place." The analogy is that a species may have to evolve continuously in order to maintain fitness relative to the evolving species with which it interacts.

1973. With "Resilience and Stability of Ecological Systems," Holling initiates interest in the phenomenon of resilience. He describes resilient systems as those that tend to maintain, or restore, their integrity when subject to disturbance or rapid change. The concept of resilience thus becomes influential not only in population and community ecology but also in disparate fields such as systems theory, social science, and economics.

1974. Simon A. Levin publishes two articles (one with Paine) providing the theoretical foundations for the fields of spatial ecology and patch dynamics. Two decades later, in a paper that becomes the most-cited work in ecology in the 1990s, Levin will argue that "the problem of pattern and scale is the central problem in ecology." He perceives the biosphere as a complex adaptive system with patterns of regularity emerging from self-organizing processes; this contrasts with the "Gaiaesque" view of the Earth as a unified superorganism.

1976. Hal Caswell develops the neutral theory of biodiversity, based on the idea that the importance of biotic factors such as competition and predation can be assessed by comparing their empirical patterns with the results of a stochastic model that does not assume their existence. This approach will later be expanded by Stephen Hubbell.

1976. Robert May publishes an influential review for *Nature* entitled "Simple Mathematical Models with Very Complicated Dynamics." May applies chaos theory to mathematical ecology, stating that simple nonlinear equations describing the growth of biological populations can exhibit a wide spectrum of dynamic behavior.

1977. Harold Mooney provides evidence for the theory of convergent evolution, which maintains that different species in widely separated ecosystems, but with comparable climates, will develop similar ways of adapting to their environment. He also takes an economic approach to plant evolution, showing how plants strive to obtain the greatest effect from resources with the lowest expenditure of energy.

1978. Mark L. Shaffer analyzes the Yellowstone grizzly bear population and uses computer simulations to estimate the numbers of bears needed to ensure a reasonable chance of persistence over the next century. This marks the emergence of population viability analysis (PVA), a process of identifying the threats faced by a species and evaluating the likelihood that the species will persist for a given time into the future.

1980. The U.S. National Science Foundation establishes the Long-Term Ecological Research Program, with an initial set of six research sites. LTER's mission is to provide knowledge to protect and manage ecosystems, their biodiversity, and the services they provide. Other large-scale research programs are subsequently created with similar aims of developing ecological knowledge and disseminating it to policy makers and the general public. These include the International Geosphere–Biosphere Programme (1987), DIVERSITAS (1991), and the Sustainable Biosphere Initiative (1992).

1982. David Tilman begins an ongoing study of a series of grassland plots at Cedar Creek, Minnesota. Over time this will yield valuable insights into such issues as plant competition and the relationship of biodiversity to ecosystem function, and Tilman's papers will be among the most cited of any contemporary researcher in ecology/environment. Other highly cited authors of this era include John Lawton (population dynamics), Peter Vitousek (forest ecosystems), and Kevin Jones (environmental contaminants).

late 1980s. The world's marine fisheries catch reaches a historic peak, as newly exploited areas are no longer able to compensate for the decline of traditional fishing grounds. Northern cod, once so abundant they constituted the major source of protein for much of Western Europe, are now severely depleted through overfishing.

1988. E. O. Wilson edits the volume *Biodiversity,* which calls attention to the rapidly accelerating loss of plant and animal species as a result of increasing human population pressure and the demands of economic development. This concept of "biological diversity" had been developing since the nineteenth century, but it is this publication that establishes *biodiversity* as the correct term for the variety of organismal life in a given system of reference at all levels of organization.

1988. Norman Myers uses the term *hot spots* to describe areas of the world that have the combined qualities of high levels of species endemism and high rates of depletion of vegetative cover. He identifies certain tropical forests that meet these criteria and argues that conservation efforts should be prioritized to focus on these areas because the risk of extinction is greatest there and the potential payoff from conservation is the highest.

1988. The Intergovernmental Panel on Climate Change (IPCC) is formed; it will become the leading scientific authority on this issue. In a series of reports (1995, 2001, 2007), the IPCC will state in progressively stronger language that a warming trend has been occurring since the mid-twentieth century and that this trend is directly tied to human activity, specifically increases in greenhouse gas emissions.

1989. James H. Brown and Brian A. Maurer formally introduce the research program known as macroecology, the study of how species divide resources (energy) and space at large spatial and temporal scales. Reportedly the first use of the word *macroecology* was in a 1971 monograph by the Venezuelan researchers Guillermo Sarmiento and Maximina Monasterio.

1992. General concern about the impact of anthropogenic biodiversity loss is voiced at the Rio de Janeiro Earth Summit. The Convention on Biological Diversity (CBD) is then founded to promote sustainable development and the protection of biodiversity. This leads to an intensified effort by ecologists to understand the effects that changes in biodiversity can have on ecosystem functioning and the likely significance of such changes for humankind.

1992. Roy Anderson and Robert May publish *Infectious Diseases of Humans,* which will become the key reference in the field of ecological epidemiology. This book summarizes the authors' work of the 1970s and 1980s in pioneering the use of mathematical models for studying the movement of infectious diseases through populations and the effect of programs of immunization and control to combat them.

Glossary

abiotic. Having to do with the chemical, geological, and physical aspects of an entity; i.e., the nonliving components.

absolute decomposition. The amount of detritus consumed by microbial decomposers (e.g., bacteria, fungi) and detritivores (e.g., earthworms).

acid–base reactions. A class of chemical reactions involving the transfer of protons without electrons.

acquisition. Any of various processes of acquiring resources from the environment, such as photosynthesis in leaves and nutrient uptake by roots.

adaptation. The evolution of a population by a process of natural selection in which hereditary variants most favorable to organismal survival and reproduction are accumulated, and less advantageous forms are discarded.

adaptive management. Dynamic resource management that incorporates new information gathered from scientific monitoring to systematically improve management practices.

adaptive radiation. The rapid diversification of an ancestral species into several ecologically different species, associated with adaptive morphological, physiological, and/or behavioral divergence.

adaptive syndrome. The suite of morphological, physiological, and behavioral characters that determine an organism's ability to survive and reproduce.

aerobe. An organism that requires the presence of atmospheric oxygen to live. Distinguished from ANAEROBE.

aerobic. Relating to or occurring in the presence of oxygen. Distinguished from ANAEROBIC.

age structure. The distribution of various chronological ages in a given population.

agroecosystem. An ecosystem designed and managed by humans to produce agricultural goods.

agroforestry. An agricultural system in which woody perennials are deliberately integrated with crops and/or animals on the same unit of land.

airshed. A region sharing a common flow of air.

albedo. Energy reflected from the land or water surface. Generally, white or light-colored surfaces have high albedo, and dark-colored or rough surfaces have low albedo.

alien. Describing a species that occurs outside its natural range and dispersal potential, especially one that becomes established in an ecosystem and threatens native biological diversity or has other negative ecological and economic impacts.

Allee effect. An inverse relationship between population density and per capita population growth rate. Allee effects can accelerate the decline of a shrinking population. (First described by Warder Clyde Allee.)

allele. One of two or more alternative forms of a gene occupying the same chromosomal locus.

allocation. The partitioning of resources among alternative structures or functions within a plant. The principle of allocation states that resources used for one purpose will be unavailable for other purposes, creating trade-offs that strongly influence plant growth and life cycles.

alternative stable state. A relatively stable ecosystem structure or composition that is different from the stable state which was present before a disturbance.

altruism. Behavior that is detrimental to the individual actor performing it but beneficial to one or more other individuals; costs and benefits are measured in terms of effects on fitness, which can be quantified by lifetime reproductive success. Thus, **altruistic.**

anaerobe. An organism that can live in an environment in which atmospheric oxygen is absent. Distinguished from AEROBE.

anaerobic. Describing or occurring in the absence of oxygen. Distinguished from AEROBIC.

anoxia. The absence of oxygen. Thus, **anoxic.**

anthropocentrism. 1. Human-centeredness; the perspective that humans are the central entity of the universe. 2. Specifically, the fact of viewing the natural environment primarily in terms of its direct benefit to humans. Thus, **anthropocentric.**

anthropogenic. Human-caused; describing a phenomenon or condition of the natural world that results from, or is significantly influenced by, human activity.

anthroponosis. A disease transmitted directly from person to person. Compare ZOONOSIS.

apparent competition. An indirect interaction between prey species in which a given prey species experiences more intense predation because of the presence of the alternative prey, as a result of changes in either predator abundance or predator behavior.

assisted migration. A directed dispersal or translocation of organisms across the landscape.

attenuation. A decline in the number of species represented on islands with distance from a source of colonists.

autotroph. A self-feeder; i.e., an organism that can convert inorganic carbon to organic materials and thus does not need to ingest or absorb other living things. Green plants use light energy to make this conversion. Thus, **autotrophic.**

balance of nature. A popular term for the concept that different species in an ecosystem will tend to interact with each other in a manner that produces a stable state, with populations remaining relatively constant over an extended time.

basic reproductive number. For microparasites, the average number of new infections that would arise from a single infectious host introduced into a population of susceptible hosts. For macroparasites, the average number of established, reproductively mature offspring produced by a mature parasite throughout its life in a population of uninfected hosts. Usually denoted R_0.

batch culture. A method of cell culture in which strains are grown for a fixed period (e.g., a few days) before being transferred to a fresh medium. Compare CONTINUOUS CULTURE.

Batesian mimicry. See MIMICRY.

benthic. Referring to environments or organisms on the sea floor (the **benthos**).

bioclimatology. The scientific study of the effects of climatic conditions on living organisms.

biocontrol. See BIOLOGICAL CONTROL.

biodiversity. The genetic, taxonomic, and functional variety of all forms of life on Earth, encompassing the interactions among them and the processes that maintain them.

biodiversity hot spot. See HOT SPOT.

bioenergetics. The processes by which energy flows take place in living systems, or the study of such processes.

biogenic. Having to do with life; produced by or involving living things.

biogeochemistry. The scientific study of the physical, chemical, geological, and biological processes and reactions that govern the cycles of matter and energy in the natural environment.

biogeography. The geography of life; i.e., the scientific study of the way in which living organisms are distributed over the Earth, in terms of space and time.

biological control. A nonchemical pest control strategy involving the purposeful release of natural enemies of a pest (often from the pest's area of origin), with the goal that the enemy will both suppress the density of the pest species and also persist to suppress future outbreaks of the pest.

biological diversity. See BIODIVERSITY.

biological nitrogen fixation. See NITROGEN FIXATION.

biomass. 1. The total mass of living biological material present in a given ecosystem at a certain time. 2. The totality of organic material that can be employed for use as fuel or for other industrial purposes.

biome. Any of various generalized regional or global community types, such as tundra or tropical forest, that are characterized by dominant plant life forms and prevailing climate.

biomechanics. The application of mathematical and biophysical theory to understand animal movement.

bionomic equilibrium. A term for the balance between fish stock abundance and the fishing fleet that a fishery will evolve to in the absence of regulation.

biosphere. The living world; the total area of the Earth that is able to support life.

biotic. Having to do with or involving living organisms.

biotic impoverishment. The generalized series of transitions that occur in the structure and function of ecosystems under chronic elevated disturbance.

bloom. A population outbreak of microscopic algae (phytoplankton) that remains within a defined part of the water column.

bottom-up. Describing strategies and efforts for conservation and restoration that rely on individual, localized initiatives rather than on large-scale government mandates.

bottom-up control. The regulation of ecosystem structure and function by factors such as nutrient supply and primary production at the base of the food chain, as opposed to "top-down" control by consumers.

boundary. Another term for an ECOTONE.

by-catch. The unintentional catch in a fishery of a species not targeted for capture, or of immature members of a targeted species.

carapace. The hard outer shell surrounding the bodies of small animals such as waterfleas and larger animals such as turtles.

carbon sequestration. The process of removing carbon dioxide from the atmospheric pool and making it less accessible or inaccessible to carbon-cycling processes.

carbon sink. 1. A site or reservoir in the environment that takes up released carbon from some other part of the atmospheric carbon cycle; e.g., oceans or forests. 2. More generally, any process or mechanism that removes carbon dioxide from the atmosphere.

cavitation. Another term for EMBOLISM.

chaos. The property of an attractor in a dynamic system that can be roughly characterized as aperiodic and sensitively dependent on initial conditions, and that can be detected by the presence of a positive Lyapunov exponent. In popular use *chaos* describes random, unpredictable, and disorderly conditions, but the phenomena given the technical name *chaos* have an intrinsic feature of determinism and some characteristics of order.

character adaptation. A character that evolved gradually by natural selection for a particular biological role, through which organisms possessing the character have a higher average rate of survival and reproduction than do organisms having contrasting conditions that have occurred in a population's evolutionary history.

character displacement. The evolution of enhanced differences between two species in the same geographic location, as a result of selection against members of one or both species that use the same resources as members of the other species (i.e., ecological character displacement) or against individuals that tend to hybridize with members of the other species (i.e., reproductive character displacement).

chemoautotrophy. A mode of nutrition by which an organism can reduce inorganic carbon to organic matter in the absence of light, using preformed bond energy contained in other molecules.

chemotroph. An organism that makes its own food but, instead of using energy from the sun as photosynthetic organisms do, uses inorganic chemicals as an energy source.

classical biological control. See BIOLOGICAL CONTROL.

climax community. A stable community of organisms in equilibrium with existing environmental conditions; this represents the final stage of an ecological succession.

cline. A geographic gradient in the frequency or mean value of a phenotype or genotype.

coalescence. The point at which common ancestry for two alleles at a gene occurs in the past.

coancestry. The probability that two alleles sampled from two different individuals are identical by descent.

coefficient of relatedness. The probability that one animal shares an allele carried by another as a result of descent from a common ancestor.

coevolution. A process of reciprocal evolutionary change in two interacting species, driven by natural selection.

coevolutionary cold spot. A geographic region in which one of a set of interacting species does not occur or in which the interaction, although occurring, does not result in reciprocal evolutionary change.

coevolutionary hot spot. A geographic region in which the interaction between two or more interacting species does result in reciprocal evolutionary change.

coexistence. The indefinite persistence of two or more species within the same community; this involves species that will continue to persist in the face of perturbations in their abundances. Species that co-occur may or may not be stably coexisting, because one or more of them may be on the way to local extinction at a time scale that is too slow to be immediately apparent.

coextinction. A process in which the extinction of one species triggers the loss of another species.

colonization. The successful occupation of a new habitat by a species not previously found in this locale.

cometabolism. The simultaneous metabolism of two substrates in such a manner that the metabolism of one substrate occurs only in the presence of the second substrate.

common-pool resource. A resource system in which it is costly to exclude potential beneficiaries, but in which one person's use subtracts resource units from those available to others.

community. An assemblage of species found together in a specific habitat at a certain time, interacting with each other in this area.

community genetics. The study of the role played by intraspecific genetic variation in community organization or ecosystem dynamics.

community module. A small number of species involved in a clearly defined pattern of interactions, such as two consumers competing for a shared resource, or two prey species interacting indirectly through their impacts on a shared predator.

community organization. A general term for the number of species found in a community, their relative abundances, and their pattern of interconnections by means of competition, exploitation, and mutualism.

competition. Ecological interaction in which two or more species negatively affect one another by consuming common resources or by other harmful means.

complementarity. The fact of two or more species using the same resources in different ways.

complementarity effect. The influence that combinations of species have on ecosystem functioning as a consequence of their interactions (e.g., resource partitioning, facilitation, reduced natural enemy impacts in diverse communities).

complex adaptive system. A system characterized by individuality and diversity of components, localized interactions between those components, and an autonomous process that selects a subset for replication and enhancement from among components, based on the results of local interactions.

conceptual landscape model. A theoretical framework that provides the terminology needed to communicate and analyze how organisms are distributed through space.

connectivity. The linkage of habitat, land cover, or ecological processes from one location to another or throughout an entire landscape.

conservation. An action taken to promote the persistence of biodiversity.

conspecific. An organism of the same species as another or others.

conspecific attraction. The attraction of individuals to members of the same species during the process of habitat selection.

constraint. Any of various factors that can absolutely limit certain actions of an organism.

consumer. An organism that consumes food from other organisms; a HETEROTROPH.

consumer–resource interaction. Any interaction in which one species depends on another for sustenance, including interactions between populations of predators and prey, or parasitoids and hosts.

context dependency. Spatial and temporal variation in the strength and/or outcome of mutualism that can be attributed to the local environmental context.

continental shelf. See SHELF.

continuous culture. A method of cell culture in which there is a continuous input of nutrients and output of spent medium, resulting in constant environmental conditions. Compare BATCH CULTURE.

continuum. A distribution of many species along a gradient, in which each species appears to be distributed randomly with respect to other species.

convergence. The development of increasing similarity over time, usually applied to species somewhat unrelated evolutionarily. Also called **convergent evolution.**

cooperation. 1. Mutually beneficial interactions among individuals of the same species, often involving social interactions such as foraging or parental care. 2. More generally, any behavior that benefits two or more interacting individuals.

corridor. A relatively linear area connecting two or more habitats, facilitating the movement of organisms between local populations.

critical habitat. The ecosystem on which any target species depends, such as endangered or threatened pollinators.

critical population size. The population size of susceptible hosts for which $R_0 = 1$, where R_0 is the BASIC REPRODUCTIVE NUMBER, and which must therefore be exceeded if an infection is to spread in a population.

critical transition. A change in the dominating feedback processes in an ecosystem, with implications for ecosystem structure and functioning. Systems undergoing a critical transition may be profoundly different than before the transition.

cropland. An official term for all land used for growing crops, including the sum of arable lands (land under temporary crops or temporarily fallow) and permanent crops (long-growing crops such as cocoa, coffee, rubber, fruit trees, nut trees, and vines).

culturability. The ability to grow strains in a laboratory in pure culture. Thus, **culturable.**

cultural services. The nonmaterial benefits people obtain from ecosystems through such factors as spiritual enrichment, cognitive development, reflection, recreation, and aesthetic experiences.

dead zone. A portion of the ocean or another body of water with very low levels of dissolved oxygen, forming in areas with low circulation and excess primary production (EUTROPHICATION).

decomposition. The breakdown of organic residues carried out by bacteria and fungi, resulting in the release of energy, nutrients, and carbon dioxide.

deforestation. A large-scale process of clearing land of trees or forest, as for logging, agriculture, housing, or the like.

demersal. Describing an organism that lives on or near the bottom of the ocean or a deep lake, especially one that feeds on benthic (bottom) organisms.

demographic stochasticity. Unpredictability through time in a population's demography, caused by the randomness of individual fates. This type of stochasticity is usually important only at very small population sizes.

demography. The statistical study of trends in population, including how many individuals die, how many reproduce, how they are distributed by age, their geographic location, and so on.

denitrification. The microbial process that converts nitrate (NO_3^-, a nutrient readily available to plants) to nitrite to free nitrogen gas (N_2, generally unavailable to plants); this requires a carbon source and an anaerobic environment.

density. The relative number of individuals of a given species that are found in a certain area.

density dependence. The fact of the growth rate of a population varying in accordance with the abundance or density of the organism in question, as when the per capita rate of increase of the population decreases as the population's density increases.

density-dependent. Describing a process that varies in accordance with the population density of the species concerned. For instance, below a certain host population size, parasitic infections may not occur (there are not enough hosts for the parasite to be transmitted between them), whereas above a certain host population size, parasitic infections may become prevalent.

density-dependent transmission. Parasite transmission in which the rate of contact between susceptible hosts and the source of new infections increases with host density; the probability of any individual host getting infected thus depends on the density of surrounding hosts.

desertification. The development of desert conditions in an area that was previously not an arid environment, as a result of climatic changes and/or human activity.

detrital. Having to do with or consisting of detritus (dead primary producer material).

detrital production. The amount of net primary production not consumed by herbivores, which senesces and enters the detrital compartment.

detritivore. An organism that feeds on DETRITUS. Thus, **detritivorous.**

detritus. Dead or decomposing primary producer material, which normally becomes detached from the primary producer after senescence.

developmental constraint. A bias in the morphological forms that a population can express, caused by the mechanisms and limitations of organismal growth and morphogenesis.

diapause. A physiological condition in which an organism can remain dormant to survive long periods of challenging conditions such as low temperatures or drought.

diauxie. Literally "double growth"; a description of the manner in which bacterial populations feed on mixtures of substrates (usually sugars). **Diauxic growth** is characterized by an initial growth phase, followed by a lag as the strain switches from the first to the second substrate, which in turn is followed by a second growth phase as the second substrate is utilized.

diet choice. The decisions made by foragers regarding which encountered food items to consume and which to reject. The abundances of different food types, their ease of being found and manipulated, and their value to the forager generally influence the decision to eat or not to eat.

diffuse coevolution. The extension of the coevolution of two populations to multiple other populations in the community.

dilution rate. In cell culture, the rate at which nutrients are input into (and output from) the microcosm.

direct effect. The immediate impact of one species on another's chance of survival and reproduction, through a physical interaction such as predation or interference.

directional transition. The location of a boundary between two areas that moves in one direction through time.

disaptation. A character that decreases its possessor's average rate of survival and reproduction relative to contrasting conditions evident in a population's evolutionary history. A **primary disaptation** is disadvantageous when it first appears; a **secondary disaptation** acquires a selective liability not present at its origin as a consequence of environmental change or an altered genetic context.

discharge. Movement of the water in an aquifer back to surface water.

dispersal. The movement of individuals among local populations in a larger population.

disruptive selection. A process of selection that favors opposite extremes of a trait within a single population.

dissolved matter. A term for organic matter derived from the degradation of dead organisms, consisting of molecules that are typically less than 0.7 μm in size. Compare PARTICULATE MATTER.

disturbance. An episodic event that results in a sustained disruption of an ecosystem's structure and function, generally with effects that last for an extended time. This may be a physical disturbance (e.g., fire, flood, drought, volcanic eruption), a biogenic disturbance (e.g., colonization by herbivorous insects or mammals), or an anthropogenic disturbance (e.g., deforestation, drainage of wetlands, chemical pollution, alien species introduction).

divergent (natural) selection. Selection arising from environmental forces acting differentially on phenotypic traits (morphology, physiology, or behavior) resulting in divergent phenotypes.

diversity. The fact of being varied or different. See BIODIVERSITY.

α, β, γ diversity. The species diversity (or richness) of a local community or habitat (α diversity); the difference in diversity associated with differences in habitat or spatial scale (β diversity); the total diversity of a region or other spatial unit (γ diversity).

diversity index. A mathematical expression that combines species richness and evenness as a measure of diversity.

domain. The range of characteristic scales in time and space at which a particular process operates, such as the delivery of an ecosystem service.

dynamics. The changes through time in the size of a population, or in a related measure such as density.

early successional species. Species that appear in an ecosystem following a disturbance event, such as a fire, landslide, or logging. Early successional species typically possess *r*-selected traits, such as high dispersal ability, short generation time, and rapid growth, but at the expense of having a short lifespan and poor competitive ability. As a result, their population sizes usually increase immediately after disturbances, and then decline later as conditions become more crowded and they are competitively replaced by LATE SUCCESSIONAL SPECIES.

ecological. 1. Having to do with the natural environment or the science of ecology. 2. Having to do with the protection or sustainable use of the natural environment.

ecological character displacement. Divergence in ecological traits caused by competition for shared resources; this may lead to reproductive isolation as a by-product.

ecological economics. The study of the interactions and coevolution in time and space between ecosystems and human economies.

ecological epidemiology. The study of infectious diseases in the context of the interactions between parasites and hosts.

ecological extinction. A reduction of the distribution and abundance of a species, to the point that it no longer significantly affects the distribution and abundance of other species in the ecosystem.

ecologically based divergent selection. Selection arising from environmental differences and/or ecological interactions (e.g., competition) that acts in contrasting directions on two populations (e.g., large body size confers high survival in one environment and low survival in the other) or favors opposite extremes of a trait within a single population (i.e., disruptive selection).

ecologically sustainable. A descriptive term for a fishery regulated so as to avoid any shift in the ecosystem that leads to an undesirable state, such as collapsed populations of a harvested species.

ecological network. A set of species that are connected to one another via some form of interaction, either trophic (as in a food web) or nontrophic (e.g., pollination, seed dispersal).

ecological niche. See NICHE.

ecological release. The expansion of habitat or use of resources by populations into areas of lower species diversity with reduced interspecific competition.

ecological speciation. A process by which barriers to gene flow evolve between populations as a result of ecologically based divergent natural selection.

ecological stoichiometry. The balance of multiple chemical substances in ecological interactions and processes, or the scientific study of this balance.

ecological succession. See SUCCESSION.

ecological trap. The attraction of animals to habitats where they perform more poorly, even when higher-quality habitat is available.

ecology. The branch of science concerned with the interrelationships of organisms with each other and with their environment.

ecomorph. An organism, population, or species whose physical appearance is determined by its environment.

ecosphere. Another term for BIOSPHERE.

ecosystem. A natural unit consisting of all the plants, animals, and microorganisms (biotic) factors in a given area, interacting with all of the nonliving physical and chemical (abiotic) factors of this environment. An ecosystem can range in scale from an ephemeral pond to the entire globe, but the term most often refers to a landscape-scale system characterized by one or a specified range of community types (e.g., a grassland ecosystem).

ecosystem-based management (EBM). A holistic approach to resource management aimed at the sustainable delivery of multiple ecosystem services by accounting for the ecological, environmental, and socioeconomic context and explicitly addressing cumulative impacts and trade-offs among the different sectors being managed.

ecosystem processes. The biogeochemical flows of energy and matter within and between ecosystems, e.g., primary production and nutrient cycling. Also, **ecosystem function(ing).**

ecosystem property. A measure of the status (e.g., species richness or standing biomass) or dynamic properties (e.g., resilience, resistance, susceptibility to invasion) of an ecosystem.

ecosystem services. The conditions and processes by which ecosystems and the species that comprise them sustain and fulfill human life. These include critical provisioning services such as food, timber, fiber, fuel and energy, and fresh water; regulating services that affect or modify factors such as air and water quality, climate, erosion, diseases, pests, and natural hazards; cultural services such as fulfilling spiritual, religious, and aesthetic needs; and supporting services such as soil formation, photosynthesis, and nutrient cycling.

ecosystem service trade-off. The reduction of the provision of one ecosystem service as a consequence of increased use of another service, due to management choices made by humans.

ecotone. A transition area where spatial changes in vegetation structure or ecosystem process rates are more rapid than in the adjoining plant communities.

ectotherm. An organism that uses external sources of heat for metabolism and whose rate of metabolism is closely linked to ambient temperatures; e.g., invertebrates, fish, amphibians, or reptiles. Thus, **ectothermic.**

edge. A well-defined area between patch types; this is often a barrier, constraint, or limit to the movement of animals and plants.

edge effects. Changes in population sizes, species richness, or other aspects of the ecology of individuals, populations, or communities at the interface between two habitat types.

effective population size. An ideal population that incorporates such factors as variation in the sex ratio of breeding individuals, the number of offspring per individual, and numbers of breeding individuals in different generations.

El Niño–Southern Oscillation (ENSO). Sustained sea surface temperature anomalies across the central tropical Pacific that are associated with the spread of warm waters from the Indian Ocean and Western Pacific to the Eastern Pacific and that are a major influence on global climate, especially in the Southern Hemisphere. (From the Spanish term *El Niño,* meaning "the Christ child," because the phenomenon was first observed at around Christmas time.)

embolism. In plants, the blockage of water transport by air bubbles in the xylem (water-transporting cells), causing reduced water transport and, potentially, plant death.

emerging (infectious) disease. A disease of infectious origin whose incidence in humans has increased in the recent past or threatens to increase in the near future. It can be a completely new disease or an old disease occurring in new places or new populations, or one that is newly resistant to available treatments.

endangered. Describing a species that is predicted to be in danger of extinction throughout all or a significant portion of its range.

endemic. 1. A species that has a relatively narrow geographic range, such as one that is found only on a particular island or in a particular habitat or region. 2. Having to do with or describing such a species. Thus, **endemism.**

energetic equivalence. A concept that denotes the equivalence of species in terms of the amount of energy that their populations use within natural communities.

energy budget. A calculation or description of the relative energy flows into and out of a living system, such as a lake or an individual organism.

entropy. A measure of the randomness or disorder of a system, which in a living system will increase over time as its energy content degrades to usable heat. Thus animals must compensate for ever-increasing entropy by constantly acquiring energy via food.

environmental stochasticity. Unpredictable changes through time in the average demographic rates of a population. These changes can be caused by vacillations in weather, food, predation, or other biotic and abiotic forces influencing individuals in a population and can exert strong effects on the dynamics of populations.

environmental uncertainty. Unpredictable sources of density-independent changes in population level parameters.

epilimnion. The uniformly warm upper layer of a lake when it is thermally stratified in summer.

equilibrium. The level of density at which a population will remain, once it has reached this level, if the population is not perturbed.

euphotic zone. The upper portion of the ocean where there is sufficient light to support net photosynthesis, usually the upper 0–200 m in the clearest ocean water.

eutrophic. Describing a lake or other body of water that is richly supplied with plant nutrients and supportive of heavy plant growths.

eutrophication. The overenrichment of an ecosystem resulting from excessive additions of chemical nutrients; eutrophication may create anaerobic conditions ("dead zones") in aquatic ecosystems.

evapotranspiration. The evaporation of water vapor from surfaces as well as the evaporation of water through the plant and leaf stomata by transpiration.

exaptation. The appropriation of a character through natural selection for a biological role other than the one by which the character was constructed through natural selection.

exclusion. A condition in which a species is driven to local extinction as a consequence of a competitive interaction.

exclusive economic zone (EEZ). The area bordering a nation's coast where it has special rights over the exploitation and management of natural resources, including fish, minerals, and petroleum. Except in areas of overlap, the EEZ extends 200 nautical miles offshore.

existence value. The worth that humans place on the mere fact of knowing that certain species or ecosystems exist, even though they may not experience or make use of them personally.

exotic. Describing any nonnative species deliberately or accidentally introduced into a new habitat.

exploitation. The use of a natural resource by humans. See also OVEREXPLOITATION.

exploitative competition. A process in which individuals have indirect negative effects on other individuals by acquiring a resource and thus depriving those others of access to the resource.

extensification. The practice of increasing the amount of agricultural land that is under cultivation.

extent of occurrence. The area that falls within the outermost geographic limits of the occurrence of a species.

externality. In ecological economics, a third-party effect of a transaction that is not taken into account by the parties to the transaction. External effects may be positive or negative and can drive a wedge between the private and social net benefits of a transaction.

extinction. 1. The complete disappearance of a species, involving the death of all its members; more precisely termed **global extinction.** 2. The disappearance of a species from a given habitat or ecosystem, though it still exists elsewhere; more precisely termed **local extinction.**

extinction debt. A future loss of species that are subject to habitat destruction or fragmentation, but that have not yet reached an EXTINCTION THRESHOLD; these species are predicted to become extinct over time though this has not yet occurred.

extinction threshold. The point at which a species' metapopulation begins a critical decline, as from habitat loss and fragmentation, so that recolonizations will not be sufficient to compensate for local extinctions, and the entire metapopulation will become extinct even if some habitat patches exist in the landscape.

extinction vortex. As populations decline, an insidious mutual reinforcement occurring among biotic and abiotic processes to drive population size downward to extinction.

extirpation. The process by which a species is rendered extinct in a particular area or country while it survives in others. When a species consists of several populations, the extirpation of the last population is equivalent to the global extinction of that species.

facilitation. The positive effect of one species on another.

facultative mutualism. A type of mutualism that increases an organism's success but that is not absolutely required for its survival and/or reproduction.

fecundity. The innate capacity to produce offspring.

fertility. 1. The fact of producing offspring. 2. The ability of soil to support plant life, especially agricultural plants. Thus, **fertile.**

fertility rate. The number of offspring produced by a female over a lifetime or during a specific age interval.

fire regime. The historic pattern of fires in a given ecosystem, in terms of frequency, severity, and extent.

fire return interval. The number of years between two successive fire events at a particular location.

fire suppression. An intentional effort to prevent wildfires in a forest ecosystem.

fishery. 1. A site for harvesting or catching fish or other aquatic life. 2. The sum of activities involved in the obtaining of fish or other aquatic life.

fishing down (through) the food web. The process by which the fisheries within a given marine ecosystem, having depleted the preferred large predatory fish at the top of the food web, turn to increasingly smaller species, finally ending up with previously spurned small fish and invertebrates.

fish stock. A population of a single species that is geographically distinct enough to be managed separately from other populations of the same species.

fitness. The extent to which an individual contributes its genes to future generations relative to other individuals in the same population; i.e., the individual's relative reproductive success.

food chain. A description of an ecological system in terms of the feeding linkages and energy and materials flows among major groups of species therein (plants, herbivores, decomposers, carnivores).

food web. A network of feeding relationships among organisms in a local community.

food web compartment or **channel.** A subweb; i.e., a frequently and strongly connected set of species that connect with much lower frequency and strength to other species in the larger web.

food web connectance. The number of actual links or interactions in a food web divided by the maximum possible links.

food web pathway. A directed set of interactions from any one species to another (e.g., a resource to a consumer to a predator of the consumer).

foraging games. The behavioral challenges facing both predator and prey when the prey can perceive and respond to the hunting tactics of the predator and the predator can perceive and respond to the antipredator tactics of its prey.

forest fragmentation. The disruption of extensive forest habitats into isolated, smaller patches.

forest management. An approach to maintaining or restoring the composition, structure, and function of natural and modified forests, based on a collaborative vision that integrates ecological, socioeconomic, and institutional perspectives, applied within naturally defined ecological boundaries.

founder control. A condition in which the dominant species in a competitive interaction is the species that is initially most abundant.

founder event. The establishment of a new population with few individuals that contain a small, and hence unrepresentative, portion of the genetic diversity relative to the original population, potentially leading to speciation.

frequency-dependent transmission. Parasite transmission in which the rate of contact between susceptible hosts and the source of new infections is independent of host density.

F_{ST}. A measure of genetic differentiation among populations, expressing the proportion of variance within a set of local populations that results from the differentiation among them.

function. The use, action, or mechanical role of phenotypic features.

functional diversity. The variety and number of species that fulfill different functional roles in a community or ecosystem.

functional group. A group of species that share a common ecological function, regardless of their taxonomic affinities; e.g., the diverse assemblage of herbivores found on coral reefs.

functional response. The relationship between per capita resource consumption and resource abundance.

Gaia hypothesis. A description of the entire Earth as a unified superorganism. (Named for the ancient Greek goddess of the Earth.)

gap dynamics. A form of natural disturbance in certain forest types, in which an opening (gap) in the overstory alters the competitive environment, thus favoring the establishment or regeneration of certain species in the understory.

gene flow. Movement between groups that results in genetic exchange.

genetic bottleneck. A period during which only a few individuals survive and become the only ancestors of the future generations of the population.

genetic drift. Chance changes in allele frequencies that result from small population size.

genetic load. A decrease in average population fitness (relative to the fittest genotype) caused, for example, by immigration of locally less-adapted immigrants (**migration load**), mating among relatives (**inbreeding load**), fixation of deleterious alleles (**drift load**), or any other population process.

genetic stochasticity. Unpredictable changes in gene frequencies as a result of processes such as random genetic drift. This is usually important only at very small population sizes.

genome. The complete assembly of genes present in a given organism, coded by specific nucleotide sequences of DNA, that determines its taxonomic structure, metabolic characteristics, behavior, and ecological function.

geographic population. All the viable populations of a species found within the species' geographic range.

geographic range. The spatial region that includes all the viable populations of a species.

geomorphology. The study of the formation, alteration, and configuration of landforms and their relationship with underlying structures.

geophagy. The eating of dirt. This behavior may be used to balance mineral intake for animals living in environments of low food quality.

geospatial. Having to do with the distribution of information in a geographic sense in such a way that entities can be located by some coordinate of a reference system (e.g., latitude and longitude), which places these entities at some point on the globe.

global carbon balance. The long-term net flux of carbon between terrestrial and marine ecosystems and the atmosphere.

global extinction. See EXTINCTION.

global positioning system (GPS). A set of 24 satellites that orbit the Earth and communicate their position to a ground receiving device in order to determine the geographic location of that receiver.

gonotrophic cycle. The complete cycle of time between a mosquito's blood feeding and subsequent laying of eggs.

gradient. A transitional geographic area in which environmental conditions vary.

gradualism. The accumulation of individually small quantitative changes in a population leading to qualitative change. Compare SALTATION.

grasslands. Short-stature vegetation dominated by grasses; characteristic of locations with a strong water limitation for at least part of the year.

greenhouse gases (GHGs). Gases such as carbon dioxide, methane, nitrous oxide, tropospheric ozone, or chlorofluorocarbons that absorb solar radiation and reflect it back down to Earth, creating a "greenhouse effect" that warms the Earth's surface.

gross primary production. The amount or rate of organic matter (sugars) produced from carbon dioxide by green plants through photosynthesis.

group selection. The concept that behaviors detrimental to an individual can evolve because they benefit the larger group of which it is a member.

growth rate hypothesis. The principle that differences in organismal C:N:P (carbon–nitrogen–phosphorus) ratios are caused by differential allocations to RNA necessary to meet the protein synthesis demands of rapid biomass growth and development.

guild. A group of species that exploit the same class of environmental resources in a similar way.

gyre. A major cyclonic surface current system in the ocean, roughly corresponding to the unproductive, highly stratified areas of the oceans that are most remote from the continents.

habitat. The place where an organism or population lives.

habitat fragmentation. The spatial isolation of small habitat areas that compounds the effects of habitat loss on populations and biodiversity.

habitat selection. The process by which individuals choose areas in which they will conduct specific activities.

HANPP. A measure of how much of the biological productivity of a given location is used, consumed, or co-opted by human activities. (An acronym for human appropriation of net primary production.)

Hardy-Weinberg principle. The concept that after one generation of random mating, single-locus genotype frequencies can be represented as a binomial function of the allele frequencies. (Named for G. H. Hardy and Wilhelm Weinberg.)

herbivore. An animal that feeds solely on plant tissue. Thus, **herbivorous.**

herbivory. The consumption of living plant material.

herd immunity. A condition in which a population contains too few susceptible hosts (either because of natural infection or immunization) for infection to be able to establish and spread within the population.

heritability. 1. The fraction of the total phenotypic variation in a population that can be attributed to genetic differences among individuals. 2. More specifically, that fraction of the total phenotypic variation that results from the additive effects of genes.

heterosis. An increase in fitness resulting from matings among individuals from different populations (e.g., as a result of superdominance or drift-load effects).

heterospecific attraction. The attraction of individuals to other potentially competing species during the process of habitat selection.

heterotroph. An organism that must consume organic compounds as food for growth (e.g., animals, most bacteria, and fungi). Thus, **heterotrophic.**

heterotrophic respiration. The metabolic process by which consumers (heterotrophs) convert sugars to carbon dioxide, releasing energy.

heterozygosity. The proportion of individuals in a population that have two different alleles for a particular gene.

holistic community. The concept that species within a community are highly interdependent, forming organism-like units.

homeostasis. The fact of an organism, or a system, maintaining constant internal conditions in the face of externally imposed variation.

hormone. Any of various substances acting as chemical messengers to carry information from one part of an organism (e.g., the brain) to another (e.g., the gonads), often via the blood transport system. Hormones bind to receptors on target cells and thus regulate the function of their targets. Various factors influence the effects of a hormone, including its pattern of secretion, transport processes, the response of the receiving tissue, and the speed with which the hormone is degraded.

host. The organism from which a parasite obtains its nutrition or shelter.

host plant. The plant on which an insect herbivore feeds.

hot spots. Regions with exceptionally high species richness, termed "hot" because they are often selected as priority targets for efforts to protect and conserve ecosystems.

human appropriation of net primary production. See HANPP.

Hutchinsonian ratio. The body size ratio of larger species over smaller species in a pair of species; niche theory predicts that co-occurring species should have larger body size ratio than expected by chance. (Developed by G. Evelyn Hutchinson.)

hydraulic lift. The process by which some plant species passively move water from deep in the soil profile, where water potentials are high, to more shallow regions where water potentials are low.

hydrology. The study of the properties, distribution, and effects of water on the Earth's surface.

hypolimnion. The uniformly cool and deep layer of a lake when it is thermally stratified in summer.

hyporheic zone. The subsurface region under and lateral to a stream in which groundwater and surface water mix; considered metabolically important in streams and rivers.

hypoxia. A low level of atmospheric oxygen. Thus, **hypoxic.**

hysteresis. The phenomenon that the forward shift and the backward shift between alternative attractors happen at different values of an external control variable.

inbreeding depression. The decline in measures of individual performance (e.g., survival, growth, or reproduction) sometimes observed in offspring of parents that are closely related to one another.

indirect cues. Stimuli produced by factors that are correlated with other factors with direct effects on intrinsic habitat quality.

indirect effect. The impact of one species on another's chance of survival and reproduction mediated through direct interactions with a mutual third-party species.

indirect interaction. An interaction between two species that is modified by a third species.

individualistic communities. The principle that communities are in essence groups of populations that occur together mainly because they share adaptations to the same abiotic environment; i.e., communities do not have organism-like qualities.

instability. The fact of being unstable. See STABILITY.

insular biogeography. See ISLAND BIOGEOGRAPHY.

intensification. A term for the practice of stimulating more agricultural production per unit area, mainly through increasing use of agricultural chemicals, irrigation water, high-yielding plant varieties, and machinery.

interaction strength (IS). The dynamic influence of one species on another, either direct or indirect.

interference competition. Competition in which individuals have direct negative effects on other individuals by preventing access to a resource through aggressive behaviors such as territoriality, larval competition, overgrowth, or undercutting.

interspecific. Having to do with or involving individuals of different species.

intraguild predation. A predation event in which one member of the feeding guild preys on another member of the same guild (predators consuming predators).

intraspecific. Having to do with or involving individuals of the same species.

intrinsic habitat quality. The expected fitness of an individual when it lives in or uses a given habitat, after controlling for any effects of conspecifics on fitness.

intrinsic rate of increase. The maximum per capita population growth rate for a population with a stable age structure (i.e., the proportions of the population in different age groups remain the same). The intrinsic rate of increase is often achieved when the population is at low density.

intrinsic value. A term for the value that other species have independent of their value to humans.

introduced species. A species established by human action in an area outside its natural range.

introduction. The movement, by human agency, of a species, subspecies, or lower taxon (including any part, gamete, or propagule that might survive and subsequently reproduce) into an area outside its natural range (past or present).

invasion. The fact of a nonnative species becoming established in a novel location, often with negative ecological consequences.

invasive species. A nonnative species that has become established in a new area outside its natural range. Also, **invader.**

inversion layer. The cap of the planetary boundary layer, where there is little or no vertical mixing and where the temperature may increase or remain constant.

irreplaceability. The status of a given site whose protection will be required for a system of conservation areas to meet all targets or to otherwise optimize a conservation objective function.

island biogeography. A branch of biogeography that studies the factors that affect the number of different species on a given island; "island" in this context can refer not only to the literal sense of a body of land surrounded by water, but also by extension to any isolated ecosystem surrounded by unlike ecosystems. The **theory of island biogeography** (Robert H. MacArthur and Edward O. Wilson) proposes that the number of species in a given island results from the dynamic equilibrium of the opposite processes of immigration from a source and local extinctions.

isotopic record of carbon. The changes in the ratio of carbon-13 to carbon-12 over geological time in marine carbonates or in organic matter in sediments or sedimentary rocks.

iteroparous. Describing a reproductive pattern in which individuals reproduce more than once in their lives. Thus, **iteroparity.**

IUU. An acronym for illegal, unreported, and unregulated fishing.

juvenile. A preadult stage in the development of an organism, resembling the adult but not yet reproductively mature.

keystone predator. A predator that strongly interacts with its prey and facilitates their coexistence; its removal from an ecosystem would significantly impact community organization.

keystone species. A species that has a disproportionately large impact on ecosystem structure and function relative to its own abundance.

kinematics. 1. Animal movement; the angles, velocities, and rates at which different body parts move throughout space. 2. The scientific study of such processes.

kin selection. Selection resulting from the effects of an organism on the fitness of its relatives, as well as through the organism's own reproduction.

kinetics. 1. The forces produced by organisms during dynamic movements. 2. The scientific study of such movements. Thus, **kinetic.**

lake turnover. The mixing of deep anoxic (oxygen-poor) and shallow oxygen-rich water in lakes, occurring in fall and spring when water reaches the threshold temperature of 4°C.

land cover. The physical state of a land area in terms of its surface features, such as "rainforest," "cropland," or "desert."

landscape. A human-defined area of the natural terrestrial world, typically ranging in size from about 1 km^2 to about 1000 km^2.

landscape connectivity. The ability of a landscape to facilitate the flows of organisms, energy, or material across a patch mosaic; this is a function of both the structural connectedness of the landscape and the movement characteristics of the species or process under consideration.

landscape dynamics. The manner in which a landscape, as a system of interacting components, structures, and processes, varies in space and time.

landscape ecology. The science and art of studying and influencing the relationship between spatial pattern and ecological processes on multiple scales. Land use and land cover change and its ecological consequences are key research topics in this discipline.

landscape fragmentation. The breaking up of vegetation or other land cover types into smaller patches by human activities, or the human introduction of barriers that impede flows of organisms, energy, and material across a landscape.

landscape function. The manner in which a landscape works as a tightly coupled geochemical–biophysical system to regulate the spatial availability and dynamics of resources.

landscape heterogeneity. The mix of different components, structures, and processes occurring in a given landscape, such as how different organisms disperse among different vegetation patches.

landscape pattern. The combination of land cover types and their spatial arrangement in a landscape.

landscape restoration. The actions and processes taken to help damaged landscapes recover toward a specified goal (landform, land use).

landscape system threshold. A point in the dynamics of a landscape where the system changes to a different state, as, for example, a damaged landscape becomes dysfunctional to the point where available resources no longer support a species.

land use. The practices employed on a particular piece of land, such as rotating grazing or intensive maize cultivation.

latent energy exchange. The exchange of energy by the evaporation of water.

late successional species. The species found in an ecosystem that has not experienced a disturbance for a long period of time. Late successional species typically have *K*-selected traits, such as long generation time, slow rates of growth, long lifespan, and strong competitive ability. As a result, late successional species come to dominate an ecosystem when no further disturbances occur. Compare EARLY SUCCESSIONAL SPECIES.

leaf energy balance. The balance of energy inputs and outputs that influence leaf temperature. Solar radiation is the most important input, and transpirational cooling and convective heat loss are the most important outputs.

life table. A table summarizing age-specific survivorship and fertility, used to calculate the net reproductive rate.

limiting factor. Of the various components of an ecosystem that can potentially restrict the ability of a certain organism or species to grow, survive, and reproduce in that area, the one that is least available at a given time. Limiting factors may be abiotic (e.g., temperature) or biotic (e.g., predators). Also, **limiting resource.**

limnology. The scientific study of the ecological and physical characteristics of lakes and other inland bodies of water.

lineage. A single line of ancestor–descendent relationship, connecting nodes within a phylogeny.

linkage disequilibrium. A statistical association between alleles at one locus and alleles at a different locus, the consequence of which is that selection on one locus (e.g., a locus affecting an ecological trait such as color pattern) causes a correlated evolutionary response at the other locus (e.g., a locus affecting mating preference).

load. See GENETIC LOAD.

local adaptation. The adaptation of populations to the immediate physical environment or to the resident populations of other species with which they interact.

local competition. Competition among relatives for limiting resources (including mates).

local extinction. See EXTINCTION.

local population. An assemblage of individuals sharing a common environment, competing for the same resources, and reproducing with each other.

longline. A line of considerable length, bearing numerous baited hooks, that is often used in commercial fishing; e.g., for tuna or swordfish. The line is set for varying periods up to several hours at various depths or on the seafloor, depending on the target species. Longlines, which are usually supported by floats, may be 150 km long and have several thousand hooks.

macroecology. The study of how species divide resources (energy) and area at large scales of space and time.

macroparasite. A parasite that grows but does not multiply in its host, producing infective stages that are released to infect new hosts; the macroparasites of animals mostly live on the body or in the body cavities (e.g., the gut); in plants, they are generally intercellular.

Malthusian. Having to do with or reflecting the views of the English economist and demographer Thomas Robert Malthus, especially the proposition that population is naturally limited by the available food supply.

market failure. 1. A situation in which competitive markets do not attain Pareto optimality; i.e., it is not possible to benefit one individual without harming someone else. 2. More generally, any situation in which competitive markets are not able to bring about the efficient and equitable distribution of resources.

mass effects. The quantitative effects of dispersal on local population dynamics. Emigration from a population may have negative effects on its demography, while immigration may have positive (rescue) effects.

matrix. The dominant and most extensive patch type in a landscape, which exerts a major influence on ecosystem processes.

maximum sustained (sustainable) yield (MSY). The highest long-term average yield that can be obtained from a fish stock on a sustainable basis.

megafauna. 1. A term for large-bodied (>44 kg) animals. 2. Specifically, the large mammal biota of the Pleistocene.

mesopredator. A predator that is fed on by another predator, usually a top carnivore.

mesotrophic. Describing lakes that are intermediate in characteristics between oligotrophic and eutrophic lakes. They are moderately well supplied with plant nutrients and support moderate plant growth.

metabolic. Having to do with or involved in METABOLISM.

metabolic rate. The energy expenditure of an organism per unit time. Metabolic rate is normally expressed in terms of rate of heat production (kilojoules per time).

metabolism. A network of chemical reactions that take place in living entities, by which energy and materials are taken up from the environment, transformed into the component of the network that sustains it, and allocated to perform specific functions.

metacommunity. A set of local communities that are linked by the dispersal of its components and potentially interacting species.

metapopulation. 1. A total population system that is composed of multiple local populations geographically separated but functionally connected through dispersal. 2. More specifically, a collection of populations each of which has reasonably high probabilities of local extinction and also of recolonization.

metapopulation capacity. A measure of the size of a habitat patch network, taking into account the total amount of habitat as well as the influence of fragmentation on metapopulation viability.

methanogens. Wetland bacteria that produce methane (CH_4, a greenhouse gas) by decomposing organic matter in anaerobic environments.

microbe. Another term for a MICROORGANISM, especially a disease-causing bacterium.

microbiology. The scientific study of microorganisms.

microevolution. The occurrence of small-scale changes in allele frequencies in a population over a few generations.

microhabitat. A small, localized habitat within a larger ecosystem, used for a specific type of activity (e.g., foraging, oviposition, nesting).

micronutrient. A chemical element necessary in relatively small quantities for organism growth.

microorganism. An organism that is generally too small to be observed without the aid of a microscope, either single-celled or a microscopic cell cluster; e.g., bacteria, cyanobacteria, unicellular algae, fungi, and viruses.

microparasite. A small, often intracellular parasite that multiplies directly within its host.

Millennium Ecosystem Assessment (MA). A five-year effort (2001–5) to assess global trends in ecosystem services and to transform the resulting knowledge into political action to reduce ecological threats worldwide.

mimicry. 1. Evolution by natural selection in which a character is favored because it closely resembles one found in a different species. 2. The fact of different species sharing similarly perceived characteristics in this manner. In **Batesian mimicry** (named for Henry Walter Bates), the mimic species is a desired prey item that tricks its potential predator by adopting the warning coloration of a similar distasteful or dangerous species. In **Müllerian mimicry** (named for Hans Müller), two species that are both distasteful or dangerous share common warning systems.

mineralization. The microbially mediated conversion of organically bound nutrients such as nitrogen and phosphorus to soluble inorganic forms that can be taken up by plants.

minimum viable population. The smallest number of individuals in a population required for the population to have a specified probability of persistence over a given period of time.

model. A term for the species whose character is copied by a mimic. See MIMICRY.

modularity. The evolution of developmental constraints by which one of two or more alternative, qualitatively different suites of characters can be activated by particular genetic or environmental cues.

monoculture. 1. An agricultural setting or system in which only one crop is cultivated; e.g., corn. 2. Any area in which a single plant species dominates the landscape. Thus, **monocultural.**

Monod equation. An equation describing the relationship between substrate concentration and the growth rate of a microbial population. (Developed by Jacques Monod.)

monophyletic. Describing a group of species that are more closely related to each other than any of them are to other species outside the group. Thus, **monophyly.**

morphology. 1. The observable form and structure of a given organism or taxon, considered as a whole. 2. The study of the form and structure of organisms, especially their external form. Thus, **morphological.**

MRCA. Most recent common ancestor; the most recent node that is shared by any two taxa in a phylogenetic tree.

MSY. See MAXIMUM SUSTAINED YIELD.

Müllerian mimicry. See MIMICRY.

multiple (stable) state. The existence of one or more alternative ecological communities in a given habitat, persisting over more than a single generation of the dominant species; this is contingent on the history of disturbance events that reset community composition.

mutualism. A two-species interaction that confers survival and/or reproductive benefits to both partners.

mycorrhizae. A relationship of symbiosis between the roots of most higher plants and several groups of fungi, in which the fungal partner typically derives energy from the plant and the plant receives nutrients from the fungus.

natal habitat. The area in which an animal lives immediately after birth.

natural enemy. A species that utilizes another species as a resource and harms that other species in so doing. Natural enemies include true predators, parasites, parasitoids, pathogens, and herbivores.

natural resource. Any feature of the environment that is utilized by humans in its natural state, through activities such as forestry, fishing, and mining.

natural selection. A difference, on average, between the survival or fecundity of individuals with certain phenotypes compared with individuals of the same species with other phenotypes.

neritic. Having to do with or inhabiting the shallow pelagic zone over the continental shelf, i.e., waters less than 200 m deep, and deeper waters in areas of coastal submarine slopes.

net ecosystem production. The amount or rate of organic material produced by green plants after both autotrophic and heterotrophic respiration.

net primary production (NPP). The biological productivity of the landscape, that is, the rate of conversion in a given location of physical energy (sunlight) into biological energy (through photosynthesis), in the form of organic carbon that becomes available for other trophic levels in the ecosystem.

net reproductive rate. The average number of offspring to which a newborn female gives birth over her entire life.

neutral dynamic. A variation in community composition determined by stochastic effects of dispersal and demography among species with equivalent niches.

neutrality. A term for the assumption of equivalence in individuals' prospects of reproduction or of death, irrespective of the species they belong to.

neutral theory. The principle that genetic change is primarily the result of mutation and genetic drift, and that different molecular genotypes are neutral with respect to each other.

niche. The specific role and requirements of a particular population or species within a larger community.

niche complementarity. A condition in which different niches result in variation in the utilization of resources or space.

niche construction. The modification of local resource distributions by organisms so as to influence both the ecosystem and the evolution of resource-dependent traits.

niche dimension. An environmental variable along which a species' niche is characterized, such as food size; typically represented as the axis of a graph.

nitrification. The biologically mediated oxidation of ammonium (NH_4) to nitrate (NO_3^-); specialized microorganisms derive their energy from this transformation.

nitrogen fixation. A process by which inert atmospheric dinitrogen (N_2) is converted into chemical forms (e.g., nitrate, ammonia) that can be used by organisms. Nitrogen fixation is carried out by certain microorganisms.

nonadaptive radiation. An elevated rate of speciation in the absence of noticeable ecological shifts.

nonaptation. A term for a character that cannot be selectively distinguished from contrasting conditions present in the evolutionary history of a population.

non-point-source pollution. Pollution that comes from many diffuse sources and that is carried by rainfall or snowmelt as it moves over or through the ground to fresh water. These pollutants include excess fertilizers, herbicides from agricultural or residential areas, oils or other toxic chemicals from urban runoff, or salt from roads or irrigation practices. Compare POINT-SOURCE POLLUTION.

nontrophic interaction. A direct interaction that changes the behavior, morphology, or chemical composition of a species in response to the threat of being consumed.

nonuse value. The value of an allocation that benefits someone other than the user, deriving from the fact that the user cares for the beneficiary. The beneficiary may be some other species or a member of a future generation.

nullcline. A set of points in an ecological model where the rate of change of one species is zero (at equilibrium). In community models, intersections of nullclines indicate points where more than one species is at equilibrium.

numerical response. The relationship between the number of predators in an area and prey density.

numerical stability. A steady-state equilibrium in population size; i.e., the numbers of individuals to which a system will return if it is perturbed; stability in predator–prey systems refers to the numerical stability of both predator and prey that allows them to coexist indefinitely.

nutrient. One of the organic or inorganic raw materials required for the growth and survival of an organism; e.g., nitrogen, phosphorus, iron, or vitamins.

nutrient concentration. The percentage of a given element such as nitrogen and phosphorus within producer biomass or detritus on a dry weight basis.

nutrient content. The quantity of an element in an organism's biomass. This may be measured as moles or grams per organism, as the percentage of mass made up by a given element, or as the X:C ratio where X is a nutrient such as N or P.

nutrient foraging. The noncognitive foraging behaviors of plants to influence the uptake of water, light, nitrogen, and other nutrients, as by adjusting allocations to roots and shoots or altering uptake kinetics.

nutrient limitation. A condition that occurs when the rate of a biological process such as productivity or decomposition is constrained by a low supply of one or more biologically essential elements.

objective function. A mathematical statement of quantities to be maximized (as in the case of the number of species or other biodiversity elements meeting targets) or minimized (e.g., cost).

obligate mutualism. A type of mutualism without which an organism will fail to survive and/or reproduce.

oceanic conveyor belt. An oceanic circulation pattern driven by temperature and salinity gradients to move warm and cold water around the globe, thus moderating temperatures and salinity patterns.

oligotrophic. Describing a condition of low nutrient concentration and low standing stock of living organisms. Oligotrophic lakes are poorly supplied with plant nutrients and support little plant growth. Thus, **oligotrophy.**

omnivorous. Describing an animal that feeds on both plants and other animals as a primary food source. Such an animal is an **omnivore.**

omnivory. The fact of feeding at more than one trophic level, such as occurs when a predator consumes herbivores as well as other predators.

outbreeding depression. A decline in fitness resulting from mating among distantly related individuals (as from the disruption of coadapted gene complexes).

overexploitation. The excessive, unsustainable use of a natural resource by humans, to the extent that the resource becomes depleted, or, in the case of wildlife, suffers extinction or loss of genetic diversity.

overfishing. Fishing activities that deplete a fishery to a point beyond the capacity of species to reproduce and maintain their population; e.g., the overfishing of cod in the Baltic Sea.

oviposition. The act of laying an egg on or in a host.

ovipositor. The specialized structure in many adult female parasitoids that allows them to lay an egg on or in a host.

pandemic. 1. An outbreak of a disease that spreads globally or throughout a very large region. 2. A species with a very large geographic range.

parasite. An organism that resides within or on another organism and is nutritionally dependent on that organism.

parasitoid. An insect in which the adult female lays one or more eggs on, in, or near the body of another insect (the host), and the resulting parasitoid offspring use the host for food as they develop, killing the host in the process.

Pareto optimality or **optimum.** An equilibrium situation in which economic resources and output are allocated so efficiently that no individual can be made better off without at least one other individual becoming worse off. (Described by economist Vilfredo Pareto.)

particulate matter. A term for organic matter derived from the degradation of dead organisms, consisting of materials such as leaf pieces, wood, animal body parts, and so on.

pasture. Agricultural land used for animal grazing, officially defined as land used permanently for herbaceous forage crops, either cultivated or growing wild.

patch. A relatively homogeneous area within a landscape that differs markedly from its surroundings in its biotic and abiotic structure and composition.

patch dynamics. The perspective that ecological systems are mosaics of patches exhibiting nonequilibrium transient dynamics and together determining system-level structure and function.

patch network. A series of discrete patches in a fragmented landscape, each of which may be occupied by a local population, and which together make up a system that may be occupied by a metapopulation.

pathogen. An infectious agent or parasite that causes illness in its host, usually defined as clinical illness, i.e., significant pathology or damaging physiological change. Thus, **pathogenic.**

pelagic. Having to do with environments or organisms of the open ocean, specifically the surface or middle depths of the oceans rather than the bottom.

per capita growth rate. The rate at which a population changes per individual in the population, as a result of reproduction, mortality, emigration, and immigration.

performance. A quantitative measure of the capability of an organism to conduct an ecologically relevant task such as sprinting, jumping, or biting.

persistence. The sustained existence of species or other elements of biodiversity both within and outside of conservation areas.

phenology. The timing of recurring biological phenomena, ranging from annual budburst and senescence in plants to the onset of animal migrations, egg laying, and metamorphosis. Thus, **phenological.**

phenotype. The outward characteristics of organisms, such as their form, physiology, and behavior.

phenotypic plasticity. The ability of an individual to express different features (i.e., alter its phenotype) in response to different environmental conditions.

phoresis. A mechanism of dispersal involving the attachment of the organism or a part to another actively dispersing organism. Thus, **phoretic.**

photoautotroph. An organism that converts inorganic carbon to organic materials and therefore does not need to ingest or absorb other living things. Green plants (as well as certain algae and cyanobacteria) are photoautotrophs because they use light energy to make this conversion.

photoautotrophy. A mode of nutrition by which an organism can reduce inorganic carbon to organic matter using light energy. Thus, **photoautotrophic.**

photosynthesis. The fundamental chemical process in which green plants (and blue-green algae) utilize the energy of sunlight and other light to convert carbon dioxide and water into carbohydrates, with chlorophyll acting as the energy converter. This releases oxygen and is the chief source of atmospheric oxygen. Photosynthesis provides green plants with their complete energy requirement and allows other organisms to obtain their own nutrients from these plants, either directly or indirectly.

photosynthetic. Relating to or involved in a process of photosynthesis.

photosynthetic pathways. Alternative photosynthetic pathways (C3, C4, and CAM) that differ in underlying biochemical and physiological mechanisms, resulting in contrasting performance depending on temperature and the availability of light, water, and nutrients.

phylogenetic clustering/overdispersion. The tendency of species to be on average more (or less) evolutionarily related in a sample than in the larger species pool.

phylogenetic distance. In a phylogenetic tree, the sum of branch lengths from one tip (or internal node) down to the MRCA (most recent common ancestor) and back up to another tip (or node).

phylogenetics. The scientific study of the evolutionary relationships within and between groups.

phylogenetic tree. A branching diagram showing the hierarchy of evolutionary relationships among a group of taxa (extant and/or extinct). Terminal taxa or tips are connected by branches to internal nodes that indicate a hypothesized ancestor. A clade includes all of the taxa (extant and extinct) that descend from a node.

phylogeny. 1. The evolutionary history of a species or other taxonomic group. 2. See PHYLOGENETIC TREE.

phytoplankton. Microscopic, mostly single-celled photosynthetic organisms living in the ocean or another body of water where they drift with the currents.

phytosociology. The study of the organization and distribution of plant communities.

planetary boundary layer (PBL). The lowest part of the Earth's atmosphere where the surface influences wind movements, humidity, and temperature over time periods of about 1 hour and up to 1–2 km above the surface.

plankton. A collective term for various drifting organisms of the pelagic zone. Phytoplankton are photosynthetic primary producers, and zooplankton are consumers.

planula. The free-swimming larva of corals. Planulae are released directly by brooded corals following internal fertilization.

pleiotropy. The multiple phenotypic effects of a gene (e.g., a gene affecting color pattern that also affects mating preferences). Thus, **pleiotropic.**

point-source pollution. Pollution that comes from clearly identifiable local sources, such as outlet pipes from waste water treatment plants or other industrial sources. Compare NON-POINT-SOURCE POLLUTION.

policing. A term for actions by group members that suppress or punish selfish behavior by other group members.

polyculture. An agricultural system in which multiple crops are grown on the same unit of land at the same time.

polygyny. A mating system in which a few males monopolize many females.

polymorphism. The existence of two or more forms in the same population that differ in morphology or some other way.

population. A group of individuals of the same species occupying a certain geographic area over a specified period of time.

population cycles. Changes in the numbers of individuals in a population that repeatedly oscillate between periods of high and low density.

population dynamics. The variation in time and space in the size and density of a population.

population genetics. The study of the genetic composition of biological populations, the factors that lead to changes in this genetic composition over time, and the ways in which these changes affect evolution and speciation.

population growth rate. The per capita rate at which a population changes size over time, typically computed as the birthrate minus the death rate.

population regulation. The tendency of a population to persist within bounds.

population viability. See VIABILITY.

population viability analysis (PVA). A formal process of identifying the threats faced by a species and evaluating the likelihood that the species will persist for a given time into the future.

postmating isolation. Barriers to gene flow that act after mating; e.g., intermediate trait values of hybrids that make them poor competitors for resources, reducing their fitness.

predator. A natural enemy that kills its victim in order to utilize resources contained in that victim.

predator–prey relationship. The ongoing ratio in a given habitat of the population size of a predator and that of its target prey.

premating isolation. Barriers to gene flow that act before mating; e.g., divergent mate preferences that prevent copulation between individuals from different populations.

primary producer. An organism capable of converting atmospheric carbon dioxide into organic matter; an AUTOTROPH.

primary production or **productivity.** The production of new living material by autotrophs (e.g., plants, algae), most commonly through photosynthesis. Compare SECONDARY PRODUCTION.

private good. A commodity whose consumption by one agent reduces the amount of the good available for use by others; e.g., a loaf of bread. Compare PUBLIC GOOD.

private optimum. The allocation that optimizes a private decision maker's objective function. If there are externalities, this will be different from the SOCIAL OPTIMUM.

producer–scrounger games. A term for the contrasting behavior patterns in socially foraging animals of either searching for their own food (producer) or searching for opportunities to join the food discoveries of others (scrounger).

production function. A real-valued function that shows the maximum amount of output that can be produced for any given combination of inputs.

propagule. Any part of an organism used for the purpose of dispersal and propagation.

provisioning services. Tangible and consumable items humans derive from ecosystems, such as food, fiber, fuel, and fresh water.

pseudointerference. A form of temporal density dependence in which the parasitoid efficiency decreases at high parasitoid densities because an increasing fraction of parasitoid attacks are wasted on already parasitized hosts.

public good. A commodity whose consumption by one agent does not preclude the availability of the same amount for use by others; e.g., clean air. Compare PRIVATE GOOD.

purse seine. A large fishing net used to encircle surface-schooling fish such as mackerel or tuna. The net may be of a size up to 1 km length and 300 m depth. Purse seines are so called because during retrieval, the lower part of the net is closed (or pursed) by drawing a line through a series of rings to prevent the fish from escaping.

PVA. See POPULATION VIABILITY ANALYSIS.

quantitative trait. A trait that shows continuous rather than discrete variation; such traits are determined by the combined influence of many different genes and the environment.

quasiextinction. The fact of a population collapsing to the point where extinction is likely to occur in the foreseeable future if existing conditions and trends persist.

quasiextinction threshold (N_{qe}). The minimum number of individuals below which a population is likely to be critically and immediately imperiled.

Quaternary period. The geologic time period beginning roughly 1.8 million years before present.

random walk. In population genetics, a change in allele frequencies from their initial values as a result of repeated episodes of genetic drift.

range edge or **limit.** The outermost geographic occurrences of a species, usually excluding vagrant individuals.

rarefaction curve. The statistical expectation of the number of species in a survey or collection as a function of the accumulated number of individuals or samples, based on resampling from an observed sample set.

recharge. The movement of surface water (e.g., rainwater) below ground into an aquifer.

redox reaction. One of a class of chemical reactions that involve the transfer of electrons with or without protons (i.e., hydrogen atoms); a contraction of the terms *reduction* (addition of electrons or hydrogen atoms to a molecule) and *oxidation* (removal of electrons or hydrogen atoms from a molecule).

reductionism. An analytical approach by which understanding of complex systems can be obtained by reducing them to the interactions among their constituent parts. Thus, **reductionist.**

regime. The specific, relatively stable state of a given ecosystem.

regime shift. A relatively rapid transition from one persistent dynamic regime to another; e.g., from a grassy savanna with low shrub biomass to a woody savanna with low grass biomass.

regulating services. Benefits obtained from the regulation of ecosystem processes, including disease regulation, climate regulation, erosion regulation, and pollination.

rehabilitation. See RESTORATION.

relative abundance. The quantitative pattern of rarity and commonness among species in a sample or a community.

remote sensing. The indirect measurement of habitat characteristics, for example by Earth-orbiting satellites.

renewable. Describing resources that are able to regenerate themselves within a relatively short time frame through natural processes. For example, wind energy and hydropower are renewable resources; a natural gas deposit or a coal seam is nonrenewable; a forest or a fish stock is potentially renewable but may become nonrenewable if overexploited.

replacement technology. See SUBSTITUTE TECHNOLOGY.

representation. A term for a sampling of biodiversity pattern, such as a number of species occurrences, within the boundaries of conservation areas.

reproductive isolation. A reduction or lack of genetic exchange (gene flow) between taxa.

reproductive success. The number of an animal's offspring that survive to maturity, relative to the number produced by others in the same population.

reservoir host. A host that can harbor human pathogenic organisms without acquiring the disease and thus serve as a source from which an infectious disease may spread.

resilience. The ability of an ecosystem to recover from or resist disturbances and perturbation, so that the key components and processes of the system remain the same.

resilient. Describing ecosystems that are able to maintain, or restore, essentially the same state when subject to disturbance or rapid change.

resistance. 1. The ability of an ecosystem to withstand disturbance without major change in structure and function. 2. The ability of an individual organism to limit or suppress the effects of an infectious disease.

resource. Any aspect of the environment that may be consumed by one individual such that it is no longer available to another organism; e.g., resources for plants include nutrients such as nitrogen, phosphorus, and potassium, along with light, water, and carbon dioxide.

response capacity. The ability of a local community to respond to changes in environmental drivers.

restoration. The process of assisting the recovery of an ecosystem that has been degraded, damaged, or destroyed.

restoration ecology. The study of the ways in which active human intervention can aid in the recovery of disturbed ecosystems.

richness. See SPECIES RICHNESS.

richness estimator. A statistical estimate of the true species richness of a community or larger sampling universe, including unobserved species, based on sample data.

saltation. The evolution of a large, qualitative change in phenotype in a single mutational step. Compare GRADUALISM.

scale. The physical dimensions, in either time or space, of a phenomenon or observation.

scale insect. Any of various plant-sucking insects that stay attached to a plant for almost their entire life history.

scrounger. See PRODUCER–SCROUNGER GAMES.

seamount. An elevation rising 1000 m or more from the sea floor with limited extent across the summit.

secondary production. The production of new living material through tissue produced by heterotrophs (e.g., fish); so called because these organisms rely on the consumption of living or dead organic material of other organisms. Compare PRIMARY PRODUCTION.

selection. See NATURAL SELECTION.

selection effects. The influence that species have on ecosystem functioning simply through their species-specific traits and their relative abundance in a community. Positive selection effects occur when species with higher-than-average monoculture performance dominate communities. Negative selection effects involve the dominance of species that do not contribute significantly to ecosystem functioning.

selection gradient. A measure of the strength of selection acting on quantitative traits.

selective sweep. Favorable directional selection that results in a region of low genetic variation closely linked to the selected region.

selfish genetic elements. Genes that spread at a cost to the organism; stretches of DNA that act narrowly to advance their own proliferation or expression and typically cause negative effects on nonlinked genes in the same organism.

selfishness. Behavior that benefits the individual performing it at a cost to one or more other individuals.

self-organization. In social species, the spontaneous development of group organization, without central control, because of the actions and interactions of multiple individuals.

semelparous. Describing a reproductive pattern in which individuals reproduce only once in their lives and often die shortly after reproduction. Thus, **semelparity.**

sensible heat exchange. The exchange of energy as heat.

sexual selection. A difference among members of the same sex between the average mating success of individuals with a particular phenotype and that of individuals with other phenotypes; can be based on factors such as the ability to dominate rivals of the same sex, or attributes that are more attractive to the opposite sex.

shadow price. The social opportunity cost of an ecological resource; i.e., its true value to society. If there are externalities, implying that markets are incomplete, the shadow value will be different from the market price.

shelf. An area at the edge of a continent, below the surface of the ocean, down to a depth of 200 m (approximately 600 ft). Shelves usually are the most productive parts of the ocean and sustain the bulk of the world's fisheries.

shifting baselines syndrome. The adoption of sliding standards for the health of ecosystems because of lack of experience and ignorance of the historical condition.

shifting transition. A boundary location that shifts back and forth with no net change over time.

sink. 1. A term for any population that consistently receives more immigrants than it sends out emigrants. Contrasted with SOURCE. 2. See CARBON SINK.

sink population. A local population that has negative expected growth rate, and that therefore would go extinct without immigration. Its habitat is termed a **sink habitat.** Contrasted with SOURCE POPULATION.

social-ecological system. An ecological system and a linked social system of resource users and their governance arrangements (if present).

social foraging. A process of collective feeding by groups of the same or different species. Social foraging may allow for information sharing, PRODUCER–SCROUNGER GAMES, group hunting, task specialization, and, very often, safety in numbers.

social optimum. The allocation that will optimize the social welfare function or index of social well-being.

soil texture. See TEXTURE.

source. Any population that consistently sends out more emigrants than it receives immigrants. Contrasted with SINK.

source population. A local population that has sufficiently high growth rate when small to persist even without immigration. Its habitat is termed a **source habitat.** Contrasted with SINK POPULATION.

spatial. Having to do with the space or area in which an organism or population is found, or in which an ecological process takes place.

spatial ecology. A discipline that studies the fundamental effects of space on the dynamics of individual species and on the structure, dynamics, and diversity of communities.

spatial refuge. A location where a species or local population is less likely to be affected by its predators, competitors, or pathogens, or by other processes impacting on its survival, growth, and reproduction.

speciation. The process by which new species develop through evolutionary forces.

species. A fundamental category for the classification and description of organisms, defined in various ways but typically on the basis of reproductive capacity; i.e., the members of a species can reproduce with each other to produce fertile offspring but cannot do so with individuals outside the species.

species abundance distribution (SAD). The relative abundance of different species in a given community.

species accumulation curve. The observed number of species in a survey or collection as a function of the accumulated number of individuals or samples.

species–area curve. A graph showing the number of species found in an area as a function of the area's size.

species–area relation(ship). A relationship that describes how the number of species increases with the area sampled or with the size of the system under analysis; e.g., a lake, habitat fragment, or island.

species diversity. See DIVERSITY.

species richness. 1. The number of species in a community, in a landscape or marinescape, or in a region. 2. The fact of having a relatively large diversity of species in a given ecosystem.

species sorting. Variation in community composition determined by the optimization of fitness among species across patches.

stability. The fact of being stable; the ability of an ecological entity to maintain an equilibrium state, or to return to some previous equilibrium state following a perturbation; e.g., a population whose variability is small relative to the level of environmental variability in its habitat.

stable coexistence. The status of competing species that are able to maintain positive abundances in the long term and are able to recover from perturbations that cause them to deviate from their long-term or steady-state abundances.

stable point. A level of population to which, if the density is initially near this point, the actual population will move generally closer through time.

state. The prevailing conditions of an ecosystem at a given point in time and space, especially as defined by either the dominant species or composition of species, and associated process rates.

stepping stones. A term for small, unconnected portions of suitable habitat that an organism uses to move from one place to another.

stewardship. The ethical concept that the proper role for humans with respect to the natural environment is to act as a steward, or watchful caretaker.

stochasticity. Random (unpredictable) variability that is described by a probability distribution giving the mean, variance, and other properties of the random process. Thus, **stochastic.**

stock. A group of individuals of a species that can be regarded as an entity for management purposes; roughly corresponding to a population. See also FISH STOCK.

stock–recruitment relationship. A mathematical description of the number of new recruits to a fishery as a function of the spawning stock size.

substitutability. A condition in which an increase in the price of a certain good or service will induce greater demand for another good or service (the substitute).

substitute technology. A form of technology that can wholly take the place of some aspect of ecosystem services.

succession. The process of vegetation development following a disturbance, often characterized by relatively predictable sequences of species replacement over time.

support(ing) services. Processes that are critical to the support of all other ecosystem services, such as nutrient cycling, primary production, and soil formation.

survivorship. The probability that a newborn survives to or beyond a specified age.

sustainable. Able to be maintained over an extended period of time based on current conditions and practices; e.g., an ecosystem or a renewable resource. Thus, **sustainability.**

switching. A behavioral response by predators to relative prey abundance, such that common prey are disproportionately attacked.

symbiosis. An interaction (positive, negative, or neutral) in which two species exist in intimate physical association for most or all of their lifetimes and are physiologically dependent on each other. Thus, **symbiotic.**

sympatric speciation. A geographic mode of speciation in which a single population splits into two species in the absence of any geographic separation, often via disruptive selection. Also known as **sympatry.**

synergy. A situation in which two agents act together to enhance each other, producing a greater positive effect than could be obtained from their separate individual efforts. Such an interaction is **synergistic.**

syntrophy. A mutualistic interaction allowing two strains to utilize a substrate that neither could utilize if the other were absent.

systematic conservation planning. A scientific process for integrating social and biological information to support decisionmaking about the location, configuration, and management of target areas designated for the conservation of biological diversity.

target. 1. The explicit, quantifiable outcome desired for each species or other ecological element of interest. 2. The particular prey species that is the focus of a given predator's efforts at predation.

taxon, *plural* **taxa.** Any named group (e.g., Vertebrata, Mammalia, *Homo sapiens*) at any taxonomic rank (e.g., Kingdom, Class, Species); higher taxa are more inclusive.

taxon cycle. A temporal sequence of the geographic distribution of species, from colonizing to differentiating to fragmenting to specializing.

taxonomy. The scientific discipline that is concerned with the naming and classification of organisms. Thus, **taxonomic.**

technological augmentation. The increase, through technological intervention, in the production of goods and services that nature provides.

technology. A broad term for both tangible human-made objects, such as tools and machines, and human-devised intangibles for the use of such objects, such as processes, programs, and services.

teleconnection. A cause-and-effect chain that operates through several intermediate steps and that leads to a linkage between two parts of a system which is unexpected or novel.

texture. A description of soil in terms of the proportions of sand (large particles), silt (intermediate-sized particles), and clay (smallest particles). Sandy, loose-textured soils allow rapid water infiltration and fast leaching of nutrients. Denser, clayey soils have poor drainage and poor soil aeration.

thermocline. A thermal or temperature gradient in a thermally stratified lake in summer, occupying the zone between the epilimnion and hypolimnion.

threshold. A situation in which there has been a nonlinear (i.e., sudden or stepped) change in an ecosystem in response to a stress or disturbance; this is often difficult to reverse.

threshold element ratio (TER). The nutrient ratio of an organism's food when that organism shifts from limitation by one element to limitation by another. For example, in the case of C:P, when food is above the TER, that organism will be limited by P, and when food is below the TER, that organism will be limited by C.

top carnivore. See TOP PREDATOR.

top-down. Describing strategies and efforts for conservation and restoration that rely on large-scale government mandates rather than on individual, localized initiatives.

top-down control. Regulation of ecosystem structure and function by consumers rather than factors such as nutrient supply and primary production at the base of the food chain.

top predator. A predator at the top of the food chain feeding on organisms at lower trophic levels (e.g., mesopredators and herbivores).

trade-off. 1. The relationship between the quantity of one ecosystem service that is used and the quantity of one or more other ecosystem services that can be used. 2. More generally, the loss of one quality or aspect of something in return for gaining another quality or aspect.

trait spectra. The abundance-weighted distribution of particular traits in the community.

transmission threshold. The condition $R_0 = 1$, where R_0 is the BASIC REPRODUCTIVE NUMBER, which must be crossed if an infection is to spread in a population.

transpiration. The evaporation of water from the leaves, stems, and flowers of plants. Transpiration occurs through small pores, or stomata, on leaf and stem surfaces, which must remain open to take up carbon dioxide.

trawling. A fishing method in which one or a pair of vessels (trawlers) tow a large bag-shaped net (trawl net) either along the sea floor or in midwater. Although this is a relatively old method of fishing for many species of fish, bottom trawling is currently questioned because it destroys habitats and catches many nontarget species, which often are subsequently discarded.

trophic. Having to do with food or feeding, especially the feeding of one species on another.

trophic cascade. Reciprocal predator–prey effects that alter the abundance, biomass, or productivity of a community across more than one trophic link in the food web; e.g., removing predators enhances herbivore density, which in turn diminishes plant biomass.

trophic interaction. A direct interaction between species in an ecosystem involving the consumption of a resource species by a consumer species.

trophic level. The position of a given species in the chain of energy or nutrients. In a three-level chain, the top level is taken by predators, the second level by herbivores, and the bottom level by plants.

trophic link. A feeding relationship between two species in an ecosystem.

turnover event. A process of extinction and recolonization in a local population.

unfished biomass. See VIRGIN BIOMASS.

unstable equilibrium. The condition of a population that has moved away from equilibrium following a perturbation. The result may be cycles in abundance, extinction, or chaos, in which the densities are always bounded, but there are no repeated sequences of abundance.

upwelling. An oceanographic phenomenon in which wind induces a transport of water, usually away from a coast, with this water being replaced by water "welling up" from deeper layers. Because the upwelled water is nutrient-rich, upwellings belong to the most productive marine ecosystems.

use value. The value of resources when used by the valuer, as opposed to the value of resources that are used by someone other than the valuer (nonuse values).

utility function. A real-valued function showing that if a consumer prefers the bundle of goods x to the bundle of goods y, then the utility of x is greater than the utility of y.

vagility. An organism's ability to move through the landscape.

vector. An organism carrying parasites from one host individual to another, within which there may or may not be parasite multiplication.

vector-borne disease. Infectious diseases spread indirectly via an insect or rodent; e.g., malaria, dengue fever, West Nile virus, plague, Hantavirus. Often, part of a pathogen's life cycle occurs within the insect vector.

viability. The probability of continued existence of a population. Viability is the converse of the risk of extinction (often defined in terms of QUASIEXTINCTION rather than complete extinction) over some time period.

viable. Describing any population that can persist through time by a combination of local recruitment and immigration.

virgin biomass. 1. The average fish stock size in an unexploited condition. 2. Living vegetation potentially available for use as an energy source.

virtual water. The volume of water that circulates in an economic system as an embedded ingredient of food and other traded products; a concept based on the idea that arid countries compensate for water deficits by importing water-intensive commodities rather than domestically producing these commodities.

water mass. A portion of the marine environment having a characteristic average value of temperature and salinity that is related to its origin and global circulation pattern.

water/nutrient efficiency. The efficiency of photosynthesis relative to investment of water or nutrients.

watershed. All of the land area from which water that drains from it or is under it flows to a certain lower point such as a river or the ocean.

weathering. The breakdown of rocks and minerals, at least partially into soluble and biologically available components.

wildlife. A broad term for all uncultivated plants and undomesticated animals living freely in nature, especially vertebrates such as mammals, birds, reptiles, and amphibians.

within-system cycle. Transfers of nutrients among plants, animals, microorganisms, and soil and/or solution, within the boundaries of an ecosystem.

yield. The number of microbial cells produced per unit of substrate.

zoonosis. An infection that occurs naturally and that can persist in a wildlife species, and that also can infect and cause disease in humans.

zoonotic disease. Describing a disease that can be spread from animals to people, such as rabies or Lyme disease.

Index

References in *italics* refer to figures.